AF614989

INTERNATIONAL SERIES OF MONOGRAPHS ON PHYSICS

INTERNATIONAL SERIES OF MONOGRAPHS ON PHYSICS

157. K.H. Bennemann, J.B. Ketterson: *Novel Superfluids, Volume 2*
156. K.H. Bennemann, J.B. Ketterson: *Novel Superfluids, Volume 1*
155. C. Kiefer: *Quantum gravity, Third edition*
154. L. Mestel: *Stellar magnetism, Second edition*
153. R. A. Klemm: *Layered Superconductors, Volume 1*
152. E.L. Wolf: *Principles of electron tunneling spectroscopy, Second edition*
151. R. Blinc: *Advanced ferroelectricity*
150. L. Berthier, G. Biroli, J.-P. Bouchaud, W. van Saarloos, L. Cipelletti: *Dynamical heterogeneities in glasses, colloids, and granular media*
149. J. Wesson: *Tokamaks, Fourth edition*
148. H. Asada, T. Futamase, P. Hogan: *Equations of motion in general relativity*
147. A. Yaouanc, P. Dalmas de Réotier: *Muon spin rotation, relaxation, and resonance*
146. B. McCoy: *Advanced statistical mechanics*
145. M. Bordag, G.L. Klimchitskaya, U. Mohideen, V.M. Mostepanenko: *Advances in the Casimir effect*
144. T.R. Field: *Electromagnetic scattering from random media*
143. W. Götze: *Complex dynamics of glass-forming liquids - a mode-coupling theory*
142. V.M. Agranovich: *Excitations in organic solids*
141. W.T. Grandy: *Entropy and the time evolution of macroscopic systems*
140. M. Alcubierre: *Introduction to 3+1 numerical relativity*
139. A. L. Ivanov, S. G. Tikhodeev: *Problems of condensed matter physics - quantum coherence phenomena in electron-hole and coupled matter-light systems*
138. I. M. Vardavas, F. W. Taylor: *Radiation and climate*
137. A. F. Borghesani: *Ions and electrons in liquid helium*
136. C. Kiefer: *Quantum gravity, Second edition*
135. V. Fortov, I. Iakubov, A. Khrapak: *Physics of strongly coupled plasma*
134. G. Fredrickson: *The equilibrium theory of inhomogeneous polymers*
133. H. Suhl: *Relaxation processes in micromagnetics*
132. J. Terning: *Modern supersymmetry*
131. M. Mariño: *Chern-Simons theory, matrix models, and topological strings*
130. V. Gantmakher: *Electrons and disorder in solids*
129. W. Barford: *Electronic and optical properties of conjugated polymers*
128. R. E. Raab, O. L. de Lange: *Multipole theory in electromagnetism*
127. A. Larkin, A. Varlamov: *Theory of fluctuations in superconductors*
126. P. Goldbart, N. Goldenfeld, D. Sherrington: *Stealing the gold*
125. S. Atzeni, J. Meyer-ter-Vehn: *The physics of inertial fusion*
123. T. Fujimoto: *Plasma spectroscopy*
122. K. Fujikawa, H. Suzuki: *Path integrals and quantum anomalies*
121. T. Giamarchi: *Quantum physics in one dimension*
120. M. Warner, E. Terentjev: *Liquid crystal elastomers*
119. L. Jacak, P. Sitko, K. Wieczorek, A. Wojs: *Quantum Hall systems*
118. J. Wesson: *Tokamaks, Third edition*
117. G. Volovik: *The Universe in a helium droplet*
116. L. Pitaevskii, S. Stringari: *Bose-Einstein condensation*
115. G. Dissertori, I.G. Knowles, M. Schmelling: *Quantum chromodynamics*
114. B. DeWitt: *The global approach to quantum field theory*
113. J. Zinn-Justin: *Quantum field theory and critical phenomena, Fourth edition*
112. R.M. Mazo: *Brownian motion - fluctuations, dynamics, and applications*
111. H. Nishimori: *Statistical physics of spin glasses and information processing - an introduction*
110. N.B. Kopnin: *Theory of nonequilibrium superconductivity*
109. A. Aharoni: *Introduction to the theory of ferromagnetism, Second edition*
108. R. Dobbs: *Helium three*
107. R. Wigmans: *Calorimetry*
106. J. Kübler: *Theory of itinerant electron magnetism*
105. Y. Kuramoto, Y. Kitaoka: *Dynamics of heavy electrons*
104. D. Bardin, G. Passarino: *The Standard Model in the making*
103. G. C. Branco, L. Lavoura, J.P. Silva: *CP Violation*
102. T. C. Choy: *Effective medium theory*
101. H. Araki: *Mathematical theory of quantum fields*
100. L. M. Pismen: *Vortices in nonlinear fields*
99. L. Mestel: *Stellar magnetism*
98. K. H. Bennemann: *Nonlinear optics in metals*
94. S. Chikazumi: *Physics of ferromagnetism*
91. R. A. Bertlmann: *Anomalies in quantum field theory*
90. P. K. Gosh: *Ion traps*
87. P. S. Joshi: *Global aspects in gravitation and cosmology*
86. E. R. Pike, S. Sarkar: *The quantum theory of radiation*
83. P. G. de Gennes, J. Prost: *The physics of liquid crystals*
73. M. Doi, S. F. Edwards: *The theory of polymer dynamics*
69. S. Chandrasekhar: *The mathematical theory of black holes*
51. C. Møller: *The theory of relativity*
46. H. E. Stanley: *Introduction to phase transitions and critical phenomena*

Novel Superfluids

Volume 1

Edited by

K. H. Bennemann

Department of Physics, Freie Universität, Berlin

J. B. Ketterson

Department of Physics, Northwestern University, Illinois

Great Clarendon Street, Oxford, OX2 6DP,
United Kingdom

Oxford University Press is a department of the University of Oxford. It furthers the University's objective of excellence in research, scholarship, and education by publishing worldwide. Oxford is a registered trade mark of Oxford University Press in the UK and in certain other countries

Impression: 1

British Library Cataloguing in Publication Data
Data available

ISBN 978–0–19–958591–5

Printed and bound by
CPI Group (UK) Ltd, Croydon, CR0 4YY

Acknowledgements

We thank Oxford University Press, in particular Sönke Adlung for kind patience and general help and furthermore C. Bennemann for technical assistance.

Contents

Volume 1

Volume 2

Contributors

Alford, Mark G. Washington University, St Louis, MO 63130, USA

Bennemann, K. H. FU-Berlin, Arnimallee 14 Str. 16–18, 14195 Berlin, Germany

Bunkov, Y. M. Institut Néel, CNRS, 25 rue des Martyrs BP 166, 38042 Grenoble cedex 9, France

Chevy, Frédéric Laboratoire Kastler Brossel, École Normale Supérieure, 24 rue Lhomond, 75005 Paris, France

Dalibard, Jean Laboratoire Kastler Brossel, École Normale Supérieure, 24 rue Lhomond, 75005 Paris, France

Haley, Richard P. Lancaster University, Lancaster LA1 4YB, UK

Kasamatsu, Kenichi Kinki University, 3-4-1 Kowakae, Higashiosaka City, Osaka 577-8502, Japan

Ketterson, J. B. Northwestern University, Department of Physics, 2145 Sheridan Road, Evanston, Illinois, 60208, USA

Klaers, Jan Institut für Angewandte Physik, Universität Bonn, Wegelerstr. 8, 53115 Bonn, Germany

Kobayashi, Michikazu University of Tokyo, Komaba 3-8-1, Meguro-ku, Tokyo 153-8902, Japan

Kuwata-Gonokami, M. University of Tokyo, Hongo 7-3-1, Bunkyo-ku, Tokyo 113-0033, Japan

Lee, Yoonseok University of Florida, Gainesville, FL 32611, USA

Rajagopal, Krishna Center for Theoretical Physics, Massachusetts Institute of Technology, Cambridge, MA 02139, USA

Schäfer, Thomas North Carolina State University, Raleigh, NC 27695, USA

Schmitt, Andreas Institute for Theoretical Physics, Vienna University of Technology, 1040 Vienna, Austria

Tsubota, Makoto Osaka City University, 3-3-138 Sugimoto, Sumiyoshi-ku, Osaka 558-8585, Japan

Volovik, G. E. Low Temperature Laboratory, Aalto University, Finland, and L.D. Landau Institute for Theoretical Physics, Moscow, Russia

Weitz, Martin Institut für Angewandte Physik, Universität Bonn, Wegelerstr. 8, 53115 Bonn, Germany

Yamamoto, Y. Stanford University, 348 Via Pueblo Mall, Stanford, CA 94305-4088, USA

List of Symbols

$\alpha_{0(1)}$	zero (first) sound attenuation
α_c	sound attenuation at transition temperature
β	tipping angle
β_i	β-parameters in G-L free energy expansion
β_{ijk}	$\beta_i + \beta_j + \beta_k$
γ	gyromagnetic ratio of ^{3}He nuclei
δ_a	diameter of an aerogel strand
δ_v	viscous penetration depth
ζ_a	aerogel correlation length
η	viscosity
θ_{cw}	Curie–Weiss temperature
θ_L	Leggett angle
κ	thermal conductivity
λ_D	dimensionless dipole coupling strength
λ_F	Fermi wavelength
μ_o	permeability of vacuum
ξ_o	temperature-independent coherence length
$\xi(T)$	temperature-dependent coherence length
ξ_a	coherence length in aerogel
$\xi_D^{A(B)}$	dipole healing length in the A(B) phase
$\xi_S^{A(B)}$	surface healing length in the A(B) phase
$\xi_H^{A(B)}$	magnetic healing length in the A(B) phase
ρ_x	mass density of material x
ρ_a	mass density of aerogel
ρ_n	normal fluid density
ρ_s	superfluid density
$\rho_{s(n)}^a$	superfluid (normal fluid) density in aerogel
ρ_s^b	bare superfluid density
σ	strain
τ	relaxation time
τ_f	frictional relaxation time
τ_t	transport relaxation time
$\tau_{e(i)}$	elastic (inelastic) scattering time
$\chi_{l(g)}$	magnetic susceptibility of a Fermi liquid (Fermi gas)
ω	angular frequency
ω_L	Larmor frequency
Δ_o	zero temperature superfluid gap
$\Delta(T)$	temperature-dependent superfluid gap

$\Delta_a(T)$	temperature-dependent superfluid gap in aerogel
$\Delta_{\parallel(\perp)}$	parallel (perpendicular) component of order parameter
Γ	period
$\Omega_{A(B)}$	Leggett frequency in the A(B) phase
$\Omega^a_{A(B)}$	Leggett frequency in the aerogel A(B)-like phase
$\tilde{a}$	global anisotropy parameter
$c_{0,1,2,4}$	speed of zero, first, second, and fourth sound, respectively
c_a	speed of longitudinal sound in aerogel only
$c_{f(s)}$	speed of fast (slow) sound mode
$c_{\ell(t)}$	speed of longitudinal (transverse) sound
$\hat{d}$	$\hat{d}$-vector, preferred direction in spin space of the A phase
f	frequency
$f_{A(B)}$	G–L free energy density in the A(B) phase
$f_D^{A(B)}$	dipole energy density in the A(B) phase
$f_H^{A(B)}$	magnetic energy density in the A(B) phase
g_D	dipole coupling strength in superfluid
$\hbar$	Dirac constant ($h/2\pi$)
k	wavenumber
k_B	Boltzmann constant
k_F	Fermi wavenumber
ℓ	(total) mean free path
ℓ_a	geometric mean free path in aerogel
ℓ_e	elastic mean free path
ℓ_i	inelastic mean free path
$\hat{l}$	$\hat{l}$-vector, orbital angular momentum vector
$n(0)$	normalized density of states at Fermi energy
$n_{\ell(s)}$	spin density in liquid (solid)
$\hat{n}$	$\hat{n}$-vector, direction of spin-orbit rotation in the B phase
p	pressure
q	momentum transfer
v_F	Fermi velocity
x	pair-breaking parameter
$A_{\mu j}$	order parameter (μ: spin index, j: orbital index)
C_o	Curie constant per spin
C_p	isobaric heat capacity
$C_{l(g)}$	heat capacity of normal Fermi liquid (Fermi gas)
$C_{n(s)}$	normal fluid (superfluid) heat capacity
$D_{f(s)}$	volume (surface) fractal dimension
D_M	spin diffusion coefficient
E_F	Fermi energy
$F_l^{s(a)}$	symmetric (asymmetric) Fermi liquid parameter of angular momentum l
$\vec{H}$	magnetic field
M	total magnetization
$M_{l(s)}$	magnetization of normal Fermi liquid (solid)
$N_o, D(\varepsilon_F)$	density of states at Fermi energy for both spins
P	porosity of aerogel in percentage

Q	quality factor
$R_{\mu j}$	spin-orbit rotation matrix
S	entropy
T_F	Fermi temperature
T_F^{**}	magnetic Fermi temperature
T_t	turnaround or trajectory temperature
T_c	superfluid transition temperature in bulk
T_{ca}	superfluid transition temperature in aerogel
T_{AB}	A–B transition temperature
T_{ABa}	A–B transition temperature in aerogel
X_a, X^a, X	physical quantities in aerogel
$Y(T)$	Yoshita function
Z_a	acoustic impedance

Preface

A central goal of this book is to present a compilation of recent developments in the general area of superfluidity as it arises from Bose–Einstein condensation (BEC) and Bardeen–Cooper–Schrieffer (BCS) fermion pairing (and generalizations thereof). Much of the discussion focuses on novel non-metallic systems, but we also include a discussion of recent advances in metallic superconductivity associated with new or rapidly evolving classes of compounds. The contributions are all authored by active workers in the various subfields. It is hoped that these discussions will shed new light on the fascinating phenomenon of superfluidity and on related fundamental problems in many-body physics.

Superfluidity and superconductivity are perhaps the most dramatic of all the condensed matter phenomena.[1] Each is associated with a loss of viscosity of an underlying liquid (e.g. superfluid atoms in liquid helium or superconducting electrons in metals). In the 1970s and 1980s we assembled the two-volume book: *The Physics of Liquid and Solid Helium* (John Wiley). These books focused primarily on superfluidity in the helium liquids. More recently we assembled the two-volume book: *Superconductivity: Conventional and Unconventional Superconductors* (Springer). The general idea behind both of these efforts was for experts to summarize the present status of the various subfields of these two disciplines so as to provide a coherent, authoritative overall account.

[1] As a measure of the importance of superfluidity one can use the number of Nobel prizes that have been awarded which *directly* involve the phenomenon: Abrikosov, Ginzburg, Leggett (2003); Cornell, Ketterle, Wieman (2001); Lee, Osheroff, Richardson (1996); Bednorz, Müller (1987); Kapitsa (1978); Giaever, Josephson (1973); Bardeen, Cooper, Schrieffer (1972); Landau (1962), a total of eight (eighteen individuals). Closely related work would include: Chu, Cohen-Tannoudji, Phillips (1997); Onnes (1913).

However the superfluid helium liquids and the metallic superconductors do not exhaust the general appearance of superfluidity in both the laboratory and nature. So to complete the effort (which now spans more than thirty years) we initiated the project leading to the present book: *Novel Superfluids*. Most of these superfluids have been experimentally confirmed but others are only conjectured.

Einstein's prediction of a new phase transition in a gas, now called Bose–Einstein condensation (BEC), was not directly realized in the laboratory for approximately 70 years. It had to wait for the development of laser trapping and cooling techniques (at NIST, Stanford, and Collège de France) and their application to alkali gas atoms before it was realized in experiments at Colorado and MIT. The properties of this superfluid phase will be described in various chapters.

In addition to BEC, superfluidity involving trapped bosonic atoms, Bardeen–Cooper–Schrieffer (BCS) superfluidity (the cause of superconductivity in metals) has been realized in gases of fermionic atoms. In fact trapped fermion gases can be continuously (and reversibly) tuned so as to pass from one classification to the other, via the so-called Feshbach resonance in which the atom–atom interaction strength (typically set by the s-wave scattering length) is continuously varied via an external magnetic field. On the BEC side one has molecules consisting of pairs of fermions, although typically in a high vibrational state, and on the BCS side one has Cooper pairs. Three length scales enter here: (i) the interatomic spacing; (ii) the range of the interatomic potential; and (iii) the size of the bound molecule (on the BEC side) or the coherence length (on the BCS side). This "state tunability" has reawakened a long-standing debate in fermion superfluidity as to what is the relation between BCS superfluidity and BEC of paired-fermion molecules. The ability to open a p-wave or higher angular momentum channel would allow trapped atom analogs of ^{3}He, heavy fermion, and proton/neutron BSC-like states.

In addition to the study of uniform superfluids, cold gas systems can now be studied in the presence of a background optical lattice which, in addition to adding another length scale (the optical lattice constant), allows the creation of solid state analogs. So along with the earlier ability to prepare clouds of varying number density, the amplitude of this lattice potential, which determines the Bloch band width, W, can be tuned. These capabilities can be combined with the Feshbach tunable interaction between the particles, characterized by the Hubbard parameter, U. As an example one can now pass from Mott localized phase to a superfluid. Such controls are a theorist's delight since they allow direct application of various tools (models) in their tool kit, including Hubbard Hamiltonians and time-dependent Ginzburg–Landau theories. Some researchers hold the hope that cold gas systems may allow progress with understanding high temperature superconductivity; when appropriately normalized by T_F, achievable transition temperatures are the highest known. And there is evidence that the electrons in some high T_c materials form molecules which later become globally coherent in the superconducting phase, behavior which might emerge in the right cold gas system.

Also, in view of the many recently discovered superfluids and novel superconductors, the goal emerges as to what are the most general conditions for the occurrence of superconductivity and superfluidity, to find generalizations of BCS, Ginzburg–Landau theory, etc., and to understand various mechanisms that can lead to a superfluid state.

Perhaps the most exotic superfluid of all is the universe itself. Many theorists regard the vacuum in which our world is embedded as a kind of superfluid and the presence of this superfluid (or the phase transition underlying it) gives the mass that most of the elementary particles have. Finding the elusive Higgs particle, a major goal of the recently commissioned Large Hadron Collider (LHC), can be taken as proof of the underlying theory on which much hangs. The final chapter of Volume 1 of this book is devoted to examining the analogy between superfluidity and properties of our universe.

At the other end of the spectrum is the center of a neutron star where, within its core, an exotic phase of matter may form in which quarks leave the confines of the nucleons in which they are normally bound and form a kind of quark superfluid, one form of which involves a *three-particle* bound state of the u, d, and s quarks; it is in some ways analogous to the electron superfluid, but with more bizarre properties, which are discussed in Volume 1.

Surrounding the possible superfluid quark matter core, there almost certainly exists a neutron (and a minority proton) superfluid whose properties (discussed in Volume 2) affect the behavior of the star itself. A phenomenon thought to be related to neutron superfluidity is the sudden increases in the rotation rate (spin-ups). New information on neutron star interiors is emerging from their cooling behavior; on a longer time scale structural information may emerge from their binary merger with a black hole, via the gravitational signals the LIGO antennas are hoping to see with their planned upgrades.

Electrons and holes in semiconductors can bind to form bosonic entities called excitons that are expected to undergo BEC under the right conditions. The search for this phenomenon in bulk materials has gone on for 30 or so years and evidence has now appeared of its occurrence in locally strained Cu_2O. Conditions are more favorable in quantum-well structures engineered to confine photons and several groups have reported BEC in such structures. Experiments involving both bulk and quantum-well confined excitons are discussed in these volumes.

Last but not least, and in two separate experiments, a room temperature BEC has been observed for a gas of photons (in a cavity in which dye molecules equilibrate the transverse modes) and magnons (in the ferrimagnet yttrium iron garnet). These experiments and the associated theory are discussed in Volume 1 and 2 respectively.

Although many recent advances in superconductivity were reviewed in our earlier Springer books, some new and surprising classes of high temperature superconductors have appeared, and some chapters in Volume 2 are devoted to these discoveries. Similarly some important advances involving the helium liquids have appeared including: (i) vortices in both ^{3}He and ^{4}He, (ii) spin superfluidity in ^{3}He, and (iii) the effect of confinement on superfluidity ^{3}He, all three of which are covered in Volume 1.

To set the stage for this book, the first chapter gives a brief review of some of the basic elements of superfluidity and superconductivity including: BEC in non-interacting and interacting gases; the physics underlying superfluid ^{4}He, and ^{3}He; and the Ginzburg–Landau and BCS theories of superconductivity. The following chapter gives brief introductions to some of the novel superfluids described more thoroughly in later chapters, with particular emphasis on the cold gases.

K. H. Bennemann
J. B. Ketterson

January 2013

An introduction to superfluidity and superconductivity

K. H. Bennemann and J. B. Ketterson

Superfluidity is a vast area of research which cannot begin to be covered in a single chapter, especially if we are to do more than simply list various subfields of study. However to give some grounding for what follows we will briefly cover some of the basics.[1] The study of superfluids began with the observation of a vanishing electrical resistance of mercury below about 4.23 K by Heike Kamerlingh Onnes in 1911 [1], a phenomenon now called superconductivity [2, 3].[2] It is said that Lord Kelvin expected that electrons flowing through a conductor would become attached to the host lattice ions at very low temperatures, i.e. the resistivity of a metal would approach infinity, perhaps an all time record for the disagreement of theory and experiment. A second key property, the near complete expulsion of magnetic flux from pure superconducting metals below their transition temperature [4], was the observed by Meissner and Ochsenfeld in 1933. That this property occurs independent of the magnetic history of a sample established that superconductivity was associated with a *phase transition*, and not some peculiar electrical transport property. The Meissner–Ochsenfeld effect, along with superconductivity itself, was accounted for, phenomenologically, by the London brothers in 1935, and this constituted the first major theoretical insight into superconductivity [5–7].[3] The vanishing of the electrical resistance can also be termed electron superfluidity, the idea being that at least some of the electrons flow as a non-viscous fluid in a superconductor, those electrons then behaving as a superfluid.

In parallel, but evolving at a slower pace, evidence for a second kind of superfluidity was accumulating for liquid ^{4}He. After Onnes showed that helium (^{4}He) could be liquefied, boiling at 4.2 K at atmospheric pressure (and with a critical temperature of 5.2 K), it was discovered that it too had unusual properties, not the least of which was that it remains a liquid to absolute zero (for pressures below about 25 atmospheres), a phenomenon ascribed to quantum zero-point motion of the atoms in the presence of a weak inter-particle force. Remarkably, when lowering the temperature of the liquid by pumping on its vapor, it was observed that the boiling suddenly ceased at about 2.19 K [8]. Further measurements on a host of phenomena, established that this behavior

[1] References in this chapter will largely be limited to books, review articles, and a few papers which initiated major lines of research.

[2] For a fascinating discussion of the discovery of superconductivity involving a recently discovered note book of Kammerlingh-Onnes see [2].

[3] A still valuable discussion of superconductivity is given in [6].

resulted from a transition to another phase, called He II, the high temperature form then being He I [9–11].[4,5] In particular the specific heat shows a singularity, the shape of which resembles the Greek lower case letter lambda, the transition then being called the lambda point [12–14].[6,7] That He II is a superfluid, an example being that that some part of the liquid can flow through a small capillary under a vanishingly small pressure differential, was discovered much later (in 1938), independently, by J. F. Allen [15] and P. Kapitza [16]. In a remarkable insight, F. London proposed that the lambda transition and superfluidity were associated with a Bose–Einstein condensation (BEC) [17]; the latter arises from a peculiar behavior of an ideal gas obeying Bose statistics [18] noted by Einstein [19]: when the particle number is held constant, particles start collecting into a zero-momentum state below some temperature.

[4]A still valuable discussion of liquid helium is given in [9].

[5]A standard reference is [10]. More recent developments in the field can be found in [11].

[6]See also [13].

[7]For a detailed discussion of the behavior of the heat capacity near the lambda point see [14].

All aspects of superfluidity, which is the focus of these volumes, can be argued to arise from BEC in one way or another, so we will begin our discussion with the simplest case: BEC in an ideal gas.

1.1 Bose–Einstein condensation in an ideal gas

For a gas consisting of particles with integral spin s, resulting in a spin degeneracy $g = 2s + 1$, the total number of particles, N, in a volume, L^3, is given by

$$N = \frac{gL^3}{(2\pi)^3} \int \frac{d^3k}{e^{[\varepsilon(\mathbf{k})-\mu]/k_BT} - 1}; \tag{1.1}$$

here $\varepsilon(\mathbf{k}) = \hbar^2k^2/2m$ where m is the particle mass. This equation fixes the chemical potential μ for a given temperature T. Writing (1.1) in terms of ε alone we have

$$N = \frac{gL^3m^{3/2}}{2^{1/2}\pi^2\hbar^3} \int \frac{\varepsilon^{1/2}d\varepsilon}{e^{[\varepsilon-\mu]/k_BT} - 1}. \tag{1.2}$$

The corresponding expression for the total energy, E, is

$$E = \frac{gL^3}{(2\pi)^3} \int \frac{\varepsilon(\mathbf{k})d^3k}{e^{[\varepsilon(\mathbf{k})-\mu]/k_BT} - 1} \tag{1.3}$$

or in terms of the ε alone

$$E = \frac{gL^3m^{3/2}}{2^{1/2}\pi^2\hbar^3} \int \frac{\varepsilon^{3/2}d\varepsilon}{e^{[\varepsilon-\mu]/k_BT} - 1}. \tag{1.4}$$

Introducing $z = \dfrac{\varepsilon}{k_BT}$ we may rewrite Eq. (1.2) as

$$\frac{N}{L^3} = \frac{g(mk_BT)^{3/2}}{2^{1/2}\pi^2\hbar^3} \int_0^\infty \frac{z^{1/2}dz}{e^{z-(\mu/k_BT)} - 1} \tag{1.5}$$

and Eq. (1.4) as

$$\frac{E}{L^3} = \frac{g(mk_BT)^{5/2}}{2^{1/2}\pi^2\hbar^3} \int_0^\infty \frac{z^{3/2}dz}{e^{z-(\mu/k_BT)} - 1}. \tag{1.6}$$

To avoid a divergence, the chemical potential in the Bose distribution must be *greater than or equal to zero*;[8] were $\mu > 0$ there would be a value of $\varepsilon(\mathbf{k})$ for which $\varepsilon(\mathbf{k}) - \mu = 0$ or $e^{[\varepsilon(\mathbf{k})-\mu]/k_BT} = 1$ resulting in a vanishing denominator and hence a diverging integral. The case $\mu = 0$ is special (we then have a zero in the numerator and denominator of (1.2) the ratio of which, by itself, is indeterminate); as we will see shortly, this situation corresponds to the presence of a finite occupation of the $\varepsilon = 0$, or equivalently $k = 0$, state. As noted above, when the total number of particles is fixed, the chemical potential has that value which satisfies (1.1) for some temperature T. As the temperature falls, the chemical potential approaches zero from below. However at some critical temperature, $T = T_c$, μ reaches 0 and below that temperature Eq. (1.1) can no longer be satisfied unless a macroscopic number, N_0, of the particles have zero energy; these particles are then said to be in a *condensate*. This is the phenomenon of Bose–Einstein condensation, which occurs in momentum space rather than the usual condensation which occurs in real space.

[8]For the case of a Fermi gas the denominator in Eq. (1.2) is replaced by $e^{[\varepsilon(\mathbf{k})-\mu]/k_BT} + 1$ which cannot vanish and hence the chemical potential may have either sign. At $T = 0$, μ corresponds to the Fermi energy, ε_F.

As the temperature continues to fall the number of particles in the condensate grows. At $T = 0$ we have $N = N_0$; i.e. all the particles are in the condensate. The value of N_0 below T_c follows from the solution of the equation[9]

$$N = \frac{gL^3m^{3/2}}{2^{1/2}\pi^2\hbar^3}\int_0^\infty \frac{\varepsilon^{1/2}d\varepsilon}{e^{\varepsilon(\mathbf{k})/k_BT} - 1} + N_0. \tag{1.7}$$

[9]Equation (1.7) can also be written as $N = \int d\varepsilon \left[N_0\delta(\varepsilon) + \frac{gL^3m^{3/2}}{2^{1/2}\pi^2\hbar^3}\frac{\varepsilon^{1/2}}{e^{\varepsilon/k_BT}-1}\right]$.

The critical temperature T_c is the solution of the equation

$$N = \frac{gL^3m^{3/2}}{2^{1/2}\pi^2\hbar^3}\int_0^\infty \frac{\varepsilon^{1/2}d\varepsilon}{e^{\varepsilon/k_BT_c} - 1}, \tag{1.8}$$

which corresponds to a temperature where $\mu = 0$, but where the number of particles in the condensate is zero. Eq. (1.8) can be solved numerically with the result

$$T_c = \frac{3.31}{k_Bg^{2/3}}\frac{\hbar^2}{m}\left(\frac{N}{L^3}\right)^{2/3}. \tag{1.9}$$

Note that, in order of magnitude, the onset of a condensate occurs when the thermal de Broglie wavelength, $\lambda\!\!\!^{-} = \hbar/\left(2\pi mk_BT\right)^{1/2}$, is of the order of the mean separation of the particles, $\left(L^3/N\right)^{1/3}$.[10] From (1.7) and (1.8) it follows that

[10]This criterion also applies to the onset of degeneracy in a Fermi system.

$$N_0 = N\left[1 - \left(\frac{T}{T_c}\right)^{3/2}\right]. \tag{1.10}$$

Further analysis shows that there is a maximum of the heat capacity and a discontinuity in its *derivative*, rather than a discontinuity in the heat capacity itself, as at a second-order phase transition; we might refer to this as a *third-order* phase transition.

1.2 The weakly interacting Bose gas; the Bogoliubov transformation

The above treatment neglected particle-particle interactions. This problem was solved in the limit of weak interactions by N. N. Bogoliubov in 1947 [20]. The theory has recently been applied to BEC in trapped atoms and to excitonic systems, as will be discussed in the chapters by Yamamoto (Chapter 9) and by Chevy and Dalibard (Chapter 7). The method Bogoliubov developed had far reaching implications, and was later adapted to superconductors, as will be discussed in Section 1.6.2.

1.2.1 The Hamiltonian

To treat the effects of interactions we use the following second quantized model Hamiltonian for a translationally invariant system

$$\hat{H} = \sum_{ik} H_{ik}^{(1)} \hat{a}_i^\dagger \hat{a}_k + \frac{1}{2} \sum_{iklm} U_{iklm}^{(2)} \hat{a}_i^\dagger \hat{a}_k^\dagger \hat{a}_m \hat{a}_l + \cdots \tag{1.11}$$

where the two-body matrix element associated with the inter-particle potential $U^{(2)}(\mathbf{r}_1 - \mathbf{r}_2)$ is

$$\begin{aligned} U_{iklm}^{(2)} &= \langle ik | U^{(2)} | lm \rangle \\ &= \iint d^3r_1 d^3r_2 \psi_i^*(\mathbf{r}_1) \psi_k^*(\mathbf{r}_2) U^{(2)}(\mathbf{r}_1 - \mathbf{r}_2)\, \psi_l(\mathbf{r}_1) \psi_m(\mathbf{r}_2). \end{aligned} \tag{1.12}$$

The $\psi_i(\mathbf{r})$ denote the wave functions of the single particle states with quantum number(s) i associated with the (destruction) operators $\hat{a}_i$; we assume $g = 1$ in what follows. Since we are here dealing with a gas, plane waves of the form $\psi_{\mathbf{k}}(\mathbf{r}) = L^{-3/2} e^{i\mathbf{k}\cdot\mathbf{r}}$ are appropriate basis functions; hence we can write

$$\begin{aligned} U^{(2)}_{\mathbf{k}_1'\mathbf{k}_2'\mathbf{k}_2\mathbf{k}_1} &= \frac{1}{L^6} \iint e^{-i\mathbf{k}_1'\cdot\mathbf{r}_1} e^{-i\mathbf{k}_2'\cdot\mathbf{r}_2} U(\mathbf{r}_1 - \mathbf{r}_2) e^{i\mathbf{k}_2\cdot\mathbf{r}_2} e^{i\mathbf{k}_1\cdot\mathbf{r}_1} d^3r_1 d^3r_2 \\ &= \frac{1}{L^6} \iint e^{-i\mathbf{k}_1'\cdot(\mathbf{r}_1-\mathbf{r}_2)} e^{-i(\mathbf{k}_2'+\mathbf{k}_1')\cdot\mathbf{r}_2} U(\mathbf{r}_1 - \mathbf{r}_2) e^{i(\mathbf{k}_2+\mathbf{k}_1)\cdot\mathbf{r}_2} e^{i\mathbf{k}_1\cdot(\mathbf{r}_1-\mathbf{r}_2)} \\ &\quad d(\mathbf{r}_1 - \mathbf{r}_2) d^3r_2 \\ &= \frac{1}{L^6} \iint U(\mathbf{k}_1 - \mathbf{k}_1') e^{i(\mathbf{k}_2+\mathbf{k}_1-\mathbf{k}_2'-\mathbf{k}_1')\cdot\mathbf{r}_2} d^3r_2 \end{aligned}$$

where in the third step we have introduced the Fourier transform of the two-body potential; the remaining integral vanishes unless $\mathbf{k}_2 + \mathbf{k}_1 - \mathbf{k}_2' - \mathbf{k}_1' = 0$ (corresponding to momentum conservation in our translationally invariant system in the presence of two-body interactions only). When this condition is satisfied the integral results in a factor L^3 or

$$U^{(2)}_{\mathbf{k}_1'\mathbf{k}_2'\mathbf{k}_2\mathbf{k}_1} = L^{-3} U(\mathbf{k}_1 - \mathbf{k}_1') \delta_{\mathbf{k}_2+\mathbf{k}_1-\mathbf{k}_2'-\mathbf{k}'}.$$

Inserting this matrix element into (1.11), and writing our single particle energies as

$$H^{(1)}_{\mathbf{k}\mathbf{k}'} = \frac{\hbar^2 k^2}{2m}\delta_{\mathbf{k}\mathbf{k}'},$$

our Hamiltonian becomes

$$\hat{H} = \sum_{\mathbf{k}} \frac{\hbar^2 k^2}{2m}\hat{a}^\dagger_{\mathbf{k}'}\hat{a}_{\mathbf{k}} + \frac{1}{2L^3}\sum_{\mathbf{k}_1,\mathbf{k}_2,\mathbf{k}'_1,\mathbf{k}'_2} U(\mathbf{k}_1 - \mathbf{k}'_1)\,\delta_{\mathbf{k}_2+\mathbf{k}_1-\mathbf{k}'_2-\mathbf{k}'_1}\hat{a}^\dagger_{\mathbf{k}'_1}\hat{a}^\dagger_{\mathbf{k}'_2}\hat{a}_{\mathbf{k}_2}\hat{a}_{\mathbf{k}_1};$$

defining $\mathbf{k}'_1 = \mathbf{k}_1 - \mathbf{q}$ and $\mathbf{k}'_2 = \mathbf{k}_2 + \mathbf{q}$ this expression can then be written as

$$\hat{H} = \sum_{\mathbf{k}} \frac{\hbar^2 k^2}{2m}\hat{a}^\dagger_{\mathbf{k}}\hat{a}_{\mathbf{k}} + \frac{1}{2L^3}\sum_{\mathbf{k}_1,\mathbf{k}_2,\mathbf{q}} U(\mathbf{q})\,\hat{a}^\dagger_{\mathbf{k}_1-\mathbf{q}}\hat{a}^\dagger_{\mathbf{k}_2+\mathbf{q}}\hat{a}_{\mathbf{k}_2}\hat{a}_{\mathbf{k}_1}. \tag{1.13}$$

We have gone through the previous derivation in some detail since this model Hamiltonian is widely used; it is also valid for fermions and will be used later in our discussion of the BCS theory of superconductivity. For the case at hand we have the Bose commutation relations

$$\hat{a}_{\mathbf{k}}\hat{a}^\dagger_{\mathbf{k}'} - \hat{a}^\dagger_{\mathbf{k}'}\hat{a}_{\mathbf{k}} = \delta_{\mathbf{k}\mathbf{k}'};\, \hat{a}_{\mathbf{k}}\hat{a}_{\mathbf{k}'} - \hat{a}_{\mathbf{k}'}\hat{a}_{\mathbf{k}} = 0. \tag{1.14a,b}$$

1.2.2 The weakly interacting Bose gas at $T = 0$

We confine ourselves here to $T = 0$ and limit the discussion to weak interactions; furthermore we simplify to the case $U(\mathbf{q}) = U(\mathbf{q} = 0) \equiv U_0$.[11] With these changes our Hamiltonian becomes

[11] This is equivalent to an inter-particle potential $U^{(2)}(\mathbf{r}_1 - \mathbf{r}_2) = U_0\delta(\mathbf{r}_1 - \mathbf{r}_2)$.

$$\hat{H} = \sum_{\mathbf{k}} \frac{\hbar^2 k^2}{2m}\hat{a}^\dagger_{\mathbf{k}}\hat{a}_{\mathbf{k}} + \frac{U_0}{2L^3}\sum_{\mathbf{k}_1,\mathbf{k}_2,\mathbf{q}} \hat{a}^\dagger_{\mathbf{k}_1-\mathbf{q}}\hat{a}^\dagger_{\mathbf{k}_2+\mathbf{q}}\hat{a}_{\mathbf{k}_2}\hat{a}_{\mathbf{k}_1}. \tag{1.15}$$

Bogoliubov made the following important observation: if there is a macroscopic number of particles N_0 in the $\mathbf{k} = 0$ state, then, since the operator for the number of particles is $\hat{N}_{\mathbf{k}} = \hat{a}^\dagger_{\mathbf{k}}\hat{a}_{\mathbf{k}}$, $\hat{a}^\dagger_0\hat{a}_0 = N_0 \cong N$. Furthermore $\hat{a}^\dagger_0 \cong \hat{a}_0 = N_0^{1/2}$ and, since $N_0 >> 1$, the order of the operators in the commutation relation, $\hat{a}_{\mathbf{k}}\hat{a}^\dagger_{\mathbf{k}} - \hat{a}^\dagger_{\mathbf{k}}\hat{a}_{\mathbf{k}} = 1$, becomes *irrelevant*; i.e. we can be treat the $\mathbf{k} = 0$ occupation numbers as *macroscopic* or *thermodynamic* quantities. On the other hand, for $\mathbf{k} \neq 0$, the order of $\hat{a}_{\mathbf{k}}$ and $\hat{a}^\dagger_{\mathbf{k}}$ remains important and both are much smaller than $\hat{a}^\dagger_0 \cong \hat{a}_0 = N_0^{1/2}$. The leading contribution to the interaction term in (1.15) involves the form

$$\hat{a}^\dagger_0\hat{a}^\dagger_0\hat{a}_0\hat{a}_0 = a_0^4. \tag{1.16}$$

The next largest terms would be of order in a_0^3, an example of which is $\hat{a}^\dagger_{\mathbf{k}_1-\mathbf{q}}\hat{a}^\dagger_{\mathbf{k}_2+\mathbf{q}}\hat{a}_{\mathbf{k}_2}a_0$, however these vanish since they do not conserve momentum. The second-order terms, which is as far as we will carry the calculation, have the form

$$a_0^2\sum_{\mathbf{k}\neq 0}(\hat{a}_{\mathbf{k}}\hat{a}_{-\mathbf{k}} + \hat{a}^\dagger_{\mathbf{k}}\hat{a}^\dagger_{-\mathbf{k}} + 4\hat{a}^\dagger_{\mathbf{k}}\hat{a}_{\mathbf{k}}); \tag{1.17}$$

the terms in this expression arise from the following six combinations in Eq. (1.15): $\mathbf{k}_1 - \mathbf{q} = \mathbf{k}_2 + \mathbf{q} = 0$; $\mathbf{k}_1 = -\mathbf{k}_2 = 0$; $\mathbf{q} = \mathbf{k}_1 = 0$; $\mathbf{q} = \mathbf{k}_2 = 0$; $\mathbf{k}_1 - \mathbf{q} = \mathbf{k}_2 = 0$; and $\mathbf{k}_2 + \mathbf{q} = \mathbf{k}_1 = 0$. We now examine the relation between N, N_0, a_0^2, and a_0^4. We start by writing

$$N = \sum_{\mathbf{k}} \hat{a}_{\mathbf{k}}^\dagger \hat{a}_{\mathbf{k}} = N_0 + \sum_{\mathbf{k}\neq 0} \hat{a}_{\mathbf{k}}^\dagger \hat{a}_{\mathbf{k}} = a_0^2 + \sum_{\mathbf{k}\neq 0} \hat{a}_{\mathbf{k}}^\dagger \hat{a}_{\mathbf{k}}.$$

Therefore

$$a_0^2 = N - \sum_{\mathbf{k}\neq 0} \hat{a}_{\mathbf{k}}^\dagger \hat{a}_{\mathbf{k}} \tag{1.18a}$$

and

$$a_0^4 \cong N^2 - 2N \sum_{\mathbf{k}\neq 0} \hat{a}_{\mathbf{k}}^\dagger \hat{a}_{\mathbf{k}}. \tag{1.18b}$$

The forms (1.16) and (1.17) may then be combined as

$$N^2 + N \sum_{\mathbf{k}\neq 0} (\hat{a}_{\mathbf{k}} \hat{a}_{-\mathbf{k}} + \hat{a}_{\mathbf{k}}^\dagger \hat{a}_{-\mathbf{k}}^\dagger + 2\hat{a}_{\mathbf{k}}^\dagger \hat{a}_{\mathbf{k}}), \tag{1.19}$$

where, to the same accuracy, we have replaced N_0 by N in (1.17). Our Hamiltonian then becomes

$$\hat{H} = \frac{N^2}{2L^3} U_0 + \sum_{\mathbf{k}} \frac{\hbar^2 k^2}{2m} \hat{a}_{\mathbf{k}}^\dagger \hat{a}_{\mathbf{k}} + \frac{N}{2L^3} U_0 \sum_{\mathbf{k}\neq 0} (\hat{a}_{\mathbf{k}} \hat{a}_{-\mathbf{k}} + \hat{a}_{\mathbf{k}}^\dagger \hat{a}_{-\mathbf{k}}^\dagger + 2\hat{a}_{\mathbf{k}}^\dagger \hat{a}_{\mathbf{k}}). \tag{1.20}$$

To lowest order, the first term gives the ground state energy at $T = 0$, and its derivative with respect to N is the chemical potential:

$$E_0 = \frac{N^2}{2L^3} U_0; \quad \mu = \frac{N}{L^3} U_0. \tag{1.21a,b}$$

The Hamiltonian (1.20) is *quadratic* in the three forms $\hat{a}_{\mathbf{k}} \hat{a}_{-\mathbf{k}}$, $\hat{a}_{\mathbf{k}}^\dagger \hat{a}_{-\mathbf{k}}^\dagger$, $\hat{a}_{\mathbf{k}}^\dagger \hat{a}_{\mathbf{k}}$. By an appropriate *canonical transformation* it can be brought into *diagonal* form where a number operator appears, as will be clear shortly. The transformation that will accomplish this has the form

$$\hat{a}_{\mathbf{k}} = u_{\mathbf{k}} \hat{b}_{\mathbf{k}} + v_{\mathbf{k}} \hat{b}_{-\mathbf{k}}^\dagger \tag{1.22a}$$

and

$$\hat{a}_{\mathbf{k}}^\dagger = u_{\mathbf{k}}^* \hat{b}_{\mathbf{k}}^\dagger + v_{\mathbf{k}}^* \hat{b}_{-\mathbf{k}}. \tag{1.22b}$$

This procedure is referred to as a *Bogoliubov transformation*; a similar trick (although now adapted to Fermi operators, see Eqs. (1.221a,b)) plays an important role in the theory of superconductivity. When Eqs. (1.22a,b) are substituted in Eqs. (1.14a,b) it is found that the operators $\hat{b}_{\mathbf{k}}^\dagger$ and $\hat{b}_{\mathbf{k}}$ will satisfy the *same* commutation relations (1.14a,b)

$$\hat{b}_{\mathbf{k}} \hat{b}_{\mathbf{k}'}^\dagger - \hat{b}_{\mathbf{k}'}^\dagger \hat{b}_{\mathbf{k}} = \delta_{\mathbf{k}\mathbf{k}'} \tag{1.23a}$$

and

$$\hat{b}_{\mathbf{k}} \hat{b}_{\mathbf{k}'} - \hat{b}_{\mathbf{k}'} \hat{b}_{\mathbf{k}} = 0 \tag{1.23b}$$

provided that we require $u_{\mathbf{k}}$ and $v_{\mathbf{k}}$ to satisfy the relation[12]

$$|u_{\mathbf{k}}|^2 - |v_{\mathbf{k}}|^2 = 1. \tag{1.24}$$

The phases of $u_{\mathbf{k}}$ and $v_{\mathbf{k}}$ may be chosen to be real here.[13] We will not go through the algebra in detail but simply remark that the transformation (1.22a,b) that diagonalizes the Hamiltonian involves the forms

$$u_{\mathbf{k}} = \frac{1}{\sqrt{1 - L_{\mathbf{k}}^2}}, \; v_{\mathbf{k}} = \frac{L_{\mathbf{k}}}{\sqrt{1 - L_{\mathbf{k}}^2}} \tag{1.25a,b}$$

where[14]

$$L_{\mathbf{k}} \equiv \frac{1}{mu^2}\left[\varepsilon(\mathbf{k}) - \xi(\mathbf{k}) - mu^2\right], \tag{1.26}$$

with

$$\xi(\mathbf{k}) = \frac{\hbar^2 k^2}{2m}, \tag{1.27a}$$

$$\varepsilon(\mathbf{k}) = \sqrt{2mu^2\xi(\mathbf{k}) + (\xi(\mathbf{k}))^2}, \tag{1.27b}$$

and

$$u = \sqrt{\frac{U_0 N}{mL^3}}; \tag{1.27c}$$

this can be verified by substituting (1.25a,b) and (1.22a,b) into (1.20). The quantities $u_{\mathbf{k}}$ and $v_{\mathbf{k}}$ can also be rewritten as

$$u_{\mathbf{k}}^2 = \frac{1}{2}\left[\frac{\xi(k) + mu^2}{\varepsilon(k)}\right] \tag{1.28a}$$

and

$$v_{\mathbf{k}}^2 = \frac{1}{2}\left[\frac{\xi(\mathbf{k}) + mu^2}{\varepsilon(\mathbf{k})} - 1\right]; \tag{1.28b}$$

these forms bare some similarity to corresponding expressions that arise in the theory of superconductivity. The Hamiltonian now takes the form

$$\hat{H} = E_0 + \sum_{\mathbf{k}\neq 0} \varepsilon(\mathbf{k})\hat{b}_{\mathbf{k}}^{\dagger}\hat{b}_{\mathbf{k}} \tag{1.29}$$

where

$$E_0 = \frac{1}{2}Nmu^2 + \frac{1}{2}\sum_{\mathbf{k}\neq 0}\left[\varepsilon(\mathbf{k}) - \frac{\hbar^2 k^2}{2m} - mu^2 + \frac{m^3 u^4}{\hbar^2 k^2}\right]. \tag{1.30}$$

This Hamiltonian has the same *structure* as non-interacting particles; the only difference is that the single-particle energies, $\hbar^2 k^2/2m$, have been replaced by $\varepsilon(\mathbf{k})$ as given by (1.27) and we have a shift in the zero of energy; $\varepsilon(\mathbf{k})$ is plotted in Fig. 1.1. Qualitatively the spectrum (1.27) has the following properties. At long wavelengths the effects of the interparticle potential dominate

[12] If a transformation of coordinates is applied to the generalized momenta and coordinates of a mechanical system which preserves the form of Hamilton's equations it is said to be canonical; historically a transformation of the operators of a quantum system that preserves the commutation relations is said to be canonical.

[13] With (1.24) we may write (1.23a,b) in the alternative form

$$\begin{pmatrix} \hat{a}_{\mathbf{k}} \\ \hat{a}_{\mathbf{k}}^{\dagger} \end{pmatrix} = \begin{pmatrix} \cosh\theta_{\mathbf{k}} & \sinh\theta_{\mathbf{k}} \\ \sinh\theta_{\mathbf{k}} & \cosh\theta_{\mathbf{k}} \end{pmatrix} \begin{pmatrix} \hat{b}_{\mathbf{k}} \\ \hat{b}_{\mathbf{k}}^{\dagger} \end{pmatrix};$$

this is to be compared with the corresponding transformation for a superconductor given by Eq. (1.228).

[14] Equivalently, $L_{\mathbf{k}} = \operatorname{sech}\theta_{\mathbf{k}}$.

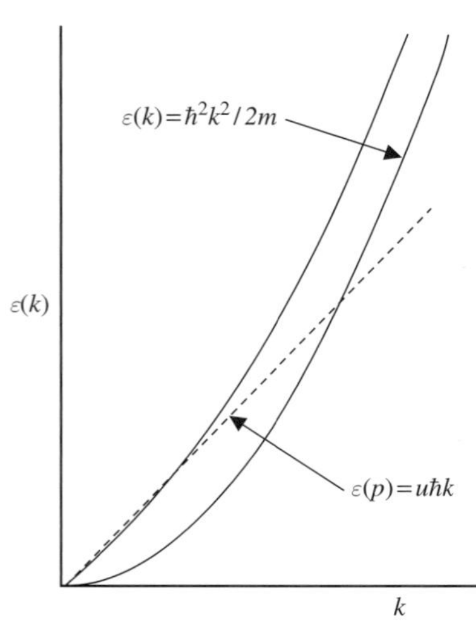

Fig. 1.1 The energy spectrum of a weakly interacting Bose gas. The spectrum for a non-interacting gas is also shown for comparison.

and the spectrum has a linear frequency versus wavevector behavior; this is characteristic of sound waves with a velocity u. On physical grounds we expect that any interacting system will possess low frequency sound waves. At the other extreme the high-energy excitations behave as free particles with energy $\hbar^2 k^2/2m$, which is also expected on physical grounds. At intermediate wavevectors there is a gradual transition between these two limiting behaviors.

1.2.3 The weakly interacting Bose gas for non-zero temperatures

At absolute zero there are no excitations relative to the interacting ground state; i.e.

$$n_{\mathbf{k}} = \left\langle \hat{b}^{\dagger}_{\mathbf{k}} \hat{b}_{\mathbf{k}} \right\rangle = 0. \qquad \text{(ground state)} \tag{1.31}$$

The operator $\hat{b}^{\dagger}_{\mathbf{k}} \hat{b}_{\mathbf{k}}$ is the number operator for what are called *elementary excitations*, also called *quasiparticle* or *dressed particle* excitations, from the ground state. Since, to the order of the approximation given by (1.29), the interactions have been removed, the mean number of excitations at finite temperature would be the same as the expression for an ideal gas

$$n_{\mathbf{k}} = \frac{1}{e^{\varepsilon(\mathbf{k})/k_B T} - 1}; \tag{1.32}$$

however the energy spectrum $\varepsilon(\mathbf{k})$ is now given by (1.27) rather than the free particle form, $\hbar^2 k^2/2m$. Note that the chemical potential is now set to zero, since there is no constraint on the number of excitations, as there was earlier in Section 1.1 on the number of atoms; the situation is entirely analogous to the case of phonons in a crystal or, more relevant to the present context, the phonon-like excitations in superfluid ^{4}He, which we take up in the next section.

We can also ask how the occupation numbers $\left\langle \hat{a}^{\dagger}_{\mathbf{k}} \hat{a}_{\mathbf{k}} \right\rangle$ of the original or bare particles are affected by temperature. From (1.22) we have

$$\hat{a}^{\dagger}_{\mathbf{k}} \hat{a}_{\mathbf{k}} = (u_{\mathbf{k}} \hat{b}^{\dagger}_{\mathbf{k}} + v_{\mathbf{k}} \hat{b}_{-\mathbf{k}})(u_{\mathbf{k}} \hat{b}_{\mathbf{k}} + v_{\mathbf{k}} \hat{b}^{\dagger}_{-\mathbf{k}}).$$

Since the operators $\hat{b}^{\dagger}_{\mathbf{k}} \hat{b}^{\dagger}_{-\mathbf{k}}$ and $\hat{b}_{-\mathbf{k}} \hat{b}_{\mathbf{k}}$ have no diagonal matrix elements we have from (1.25a,b)

$$\begin{aligned} \left\langle \hat{a}^{\dagger}_{\mathbf{k}} \hat{a}_{\mathbf{k}} \right\rangle &= u^2_{\mathbf{k}} \left\langle \hat{b}^{\dagger}_{\mathbf{k}} \hat{b}_{\mathbf{k}} \right\rangle + v^2_{\mathbf{k}} \left\langle \hat{b}_{-\mathbf{k}} \hat{b}^{\dagger}_{-\mathbf{k}} \right\rangle \\ &= [n_{\mathbf{k}} + L^2_{\mathbf{k}}(n_{\mathbf{k}} + 1)]/(1 - L^2_{\mathbf{k}}) \end{aligned}, \tag{1.33}$$

where we have used $\left\langle \hat{b}^{\dagger}_{\mathbf{k}} \hat{b}_{\mathbf{k}} \right\rangle = n_{\mathbf{k}}$, and $\left\langle \hat{b}_{-\mathbf{k}} \hat{b}^{\dagger}_{-\mathbf{k}} \right\rangle = n_{\mathbf{k}} + 1$, and the expression is of course valid only for $\mathbf{k} \neq 0$. The number of particles in the $\mathbf{k} = 0$ condensate is

$$\begin{aligned} N_0 &= N - \sum_{\mathbf{k} \neq 0} \tilde{n}_{\mathbf{k}} \\ &= N - \frac{L^3}{(2\pi)^3} \int \tilde{n}_{\mathbf{k}} d^3 k \end{aligned}. \tag{1.34}$$

For $T = 0$ where $n_{\mathbf{k}} = 0$ Eqs. (1.26) and (1.33) give

$$\tilde{n}_{\mathbf{k}} = \frac{m^2 u^4}{2\varepsilon(\mathbf{k})\left[\varepsilon(\mathbf{k}) + \hbar^2 k^2/2m + mu^2\right]}. \tag{1.35}$$

Carrying out the integration in (1.34) using (1.35) gives

$$N_0 = N\left[1 - \frac{8}{3}\sqrt{\frac{Na^3}{\pi L^3}}\right] \tag{1.36}$$

where we have introduced a scattering length, a, which is given in lowest order by $U_0 = 4\pi\hbar^2 a/m$; to second order this relation is

$$U_0 = \frac{4\pi\hbar^2 a}{m}\left(1 + \frac{4\pi a}{L^3}\sum_{\mathbf{k}\neq 0}\frac{1}{k^2}\right). \tag{1.37}$$

Note that although $N_0 < N$ it is none the less finite, from which we will surmise that it is a general feature of a Bose gas that should persist in the case of a strongly interacting system, like liquid helium II.

At non-zero temperature, $\tilde{n}_{\mathbf{k}}$ increases; at some temperature we would reach a point where N_0 in Eq. (1.34) would vanish. Although the validity of the theory would break down prior to this point due to interactions among the particles, this would qualitatively mark the disappearance of the condensate (and the superfluidity that is associated with it).

1.2.4 The weakly interacting Bose gas with a moving condensate

We now generalize the Bogoluibov model to the case where the condensate is in motion with wavevector $\mathbf{k}_0$ and

$$a_{\mathbf{k}_0} = \sqrt{N_{\mathbf{k}_0}}. \tag{1.38}$$

Starting with (1.15) and carrying out calculations similar to those leading to Eq. (1.20) we obtain the Hamiltonian to second order as

$$\begin{aligned}\hat{H} = {} & \frac{N^2}{2L^3}U_0 + N\frac{\hbar^2 k_0^2}{2m} + \sum_{\mathbf{k}\neq\mathbf{k}_0}\left(\frac{\hbar^2 k^2}{2m} + \hbar\mathbf{k}\cdot\mathbf{v}_0\right)\hat{a}_{\mathbf{k}}^\dagger\hat{a}_{\mathbf{k}} \\ & + \frac{N}{2L^3}U_0\sum_{\mathbf{k}\neq\mathbf{k}_0}\left(\hat{a}_{\mathbf{k}_0+\mathbf{k}}^\dagger\hat{a}_{\mathbf{k}_0+\mathbf{k}} + \hat{a}_{\mathbf{k}_0-\mathbf{k}}\hat{a}_{\mathbf{k}_0-\mathbf{k}}^\dagger + \hat{a}_{\mathbf{k}_0+\mathbf{k}}^\dagger\hat{a}_{\mathbf{k}_0-\mathbf{k}}^\dagger + \hat{a}_{\mathbf{k}_0+\mathbf{k}}\hat{a}_{\mathbf{k}_0-\mathbf{k}}\right)\end{aligned} \tag{1.39}$$

where $\mathbf{v}_0 = \hbar\mathbf{k}_0/m$ is the drift velocity of the condensate, and the additional terms

$$N\frac{\hbar^2 k_0^2}{2m} + \sum_{\mathbf{k}\neq\mathbf{k}_0}\hbar\mathbf{k}\cdot\mathbf{v}_0\hat{a}_{\mathbf{k}}^\dagger\hat{a}_{\mathbf{k}}$$

can be viewed as the change in energy resulting from a transformation to a reference frame moving with velocity $\mathbf{v}_0$. In place of (1.21a,b) we now have (again to lowest order)

$$E_0 = \frac{N^2}{2L^3}U_0 + N\frac{\hbar^2 k_0^2}{2m}, \mu = \frac{N}{L^3}U_0 + \frac{\hbar^2 k_0^2}{2m}. \tag{1.40a,b}$$

As with Eq. (1.20), Eq. (1.39) can be diagonalized with the transformations

$$\hat{a}_{\mathbf{k}} = u_{\mathbf{k}}\hat{b}_{\mathbf{k}} + v_{\mathbf{k}}\hat{b}^{\dagger}_{-\mathbf{k}} \tag{1.41a}$$

and

$$\hat{a}^{\dagger}_{\mathbf{k}} = u^{*}_{\mathbf{k}}\hat{b}^{\dagger}_{\mathbf{k}} + v^{*}_{\mathbf{k}}\hat{b}_{-\mathbf{k}}. \tag{1.42b}$$

The resulting Hamiltonian is

$$\hat{H} = E_0 + \sum_{\mathbf{k}\neq\mathbf{k_0}} (\varepsilon(\mathbf{k}) + \hbar\mathbf{k}\cdot\mathbf{v}_0)\hat{b}^{\dagger}_{\mathbf{k}}\hat{b}_{\mathbf{k}} \tag{1.43}$$

where

$$E_0 = \frac{1}{2}Nmu^2 + N\frac{\hbar^2 k_0^2}{2m} + \frac{1}{2}\sum_{\mathbf{k}\neq\mathbf{k}_0}\left[\varepsilon(\mathbf{k}) - \frac{\hbar^2 k^2}{2m} - mu^2 + \frac{m^3u^4}{\hbar^2 k^2}\right]. \tag{1.44}$$

In place of the Bose distribution function (1.32) we now have

$$n_{\mathbf{k}} = \frac{1}{e^{(\varepsilon(\mathbf{k})+\hbar\mathbf{k}\cdot\mathbf{v}_0)/k_BT} - 1}; \tag{1.45}$$

Note that for **k** vectors such that $\varepsilon(\mathbf{k}) + \hbar\mathbf{k}\cdot\mathbf{v}_0 \leq 0$ the distribution will be singular; this will occur for $v_0 > u$ which represents a *critical velocity above which superfluidity is destroyed.* We will find a similar phenomena in helium II, to be discussed next.

1.3 Superfluid helium II

Liquid He is the only substance which, below some temperature-dependent pressure, does not solidify on cooling to absolute zero. The fact that it does not freeze is due to the large quantum zero point energy and the weak, isotropic, interatomic force. Two isotopes of helium exist, ^{4}He and ^{3}He. Both nuclei have two protons but ^{4}He and ^{3}He have, respectively, two neutrons and one neutron, corresponding to nuclear angular momenta of $I = 0$ and $I = 1/2$. The $S = 0$ electronic ground state has no angular momentum and hence the total quantum number, F, is identical to I. The very low temperature properties of the liquid phase of these two isotopes are governed by the associated statistics: Bose–Einstein for ^{4}He and Fermi–Dirac for ^{3}He. Since quantum effects play such a dominant role the helium liquids are often called *quantum liquids*.

On cooling the gas at atmospheric pressure, liquefication of ^{4}He occurs at 4.2 K. On further cooling, a second-order liquid–liquid phase transition is observed to occur at 2.19 K. This is a unique phase transition not occurring elsewhere in nature and the properties of the new (low temperature) liquid are quite unusual. Most important is the property of superfluidity, where, depending on the experimental conditions, a flow can take place with no associated viscosity. However, under a different set of conditions the liquid may behave as an ordinary viscous liquid. In fact it turns out that, in a mathematically consistent way, the properties of the liquid may be described in terms of two inter-penetrating and non-interacting fluids: a normal viscous fluid, and a *superfluid* which has no viscosity. To rationalize this picture requires a microscopic model. The one we will use here is the phenomenological model of Landau [21, 22].

1.3.1 Landau's phenomenological model for ^{4}He

We can extend some of the reasoning used in Section 1.3 for the weakly interacting Bose gas to the case of the "Bose liquid", ^{4}He. A basic notion in low temperature physics is the idea that quantum many-body systems may be described in terms of *elementary excitations*. At absolute zero there are no excitations, and the system is said to be in its ground state. On heating the system, excitations will appear. The translational invariance of liquid helium suggests that we associate a momentum and energy with each of these excitations and, provided they are few in number, they will behave as a weakly interacting gas. Being made up of atoms obeying Bose statistics, our excitations will also be bosons. If we have the energy–momentum dispersion relation, $\varepsilon(\mathbf{p})$, for the excitations, which we expect to differ in some ways from that of the weakly interacting Bose gas, we can use the Bose distribution for an ideal gas to determine the thermal properties. The total energy will then be given by

$$E = \frac{V}{(2\pi\hbar)^3}\int d^3p\, n(\varepsilon)\,\varepsilon \tag{1.46}$$

where

$$n(\varepsilon) = \frac{1}{e^{\varepsilon(p)/k_BT} - 1}. \tag{1.47}$$

At long wavelengths (small p) the excitations must correspond to longitudinal sound waves.[15] Here, our excitations are the phonons we encountered for the weakly interacting Bose gas, which are familiar from the Debye theory of specific heats. Since the frequency–wavevector relation of a sound wave is (where we now replace u by c)

$$\omega = ck \tag{1.48}$$

where $c = 2.38 \times 10^4$ cm/sec is the sound velocity (for $T = P = 0$), our phonons obey the dispersion relation $\varepsilon = cp$. As the wavelength begins to approach inter-atomic distances (for $T = P = 0$ the average interatomic spacing is 3.8 A), significant deviations can be expected to occur.[16] Fortunately, the method of inelastic neutron scattering permits a complete determination of ε(p), the results of which are shown in Fig. 1.2 [23, 24]. For historical reasons the excitations near the minimum (at $p = p_0$) are called rotons. For temperatures above 1 K they dominate the thermal properties.[17] In the vicinity of p_0 we may write the roton energy

$$\varepsilon(p) = \Delta + \frac{(p-p_0)^2}{2\mu} \tag{1.49}$$

where $\Delta = 8.9$ K, $p_0 = 2.1 \times 10^{-19}$ g cm/sec; $\mu = 1.72 \times 10^{-24}$ g.

[15] A liquid, which will not support a shear, can have no transverse waves.

[16] Although first principles many-body quantum theory permits a reasonably accurate calculation of ε(p), the overall shape may be rationalized by considering two effects. Since the liquid possesses a structure on an interatomic distance scale one might expect a diffraction-like effect analogous to the turning down of a phonon dispersion curve as it approaches the first Brillouin zone; for larger momenta the curve should ultimately take up a free-particle behavior where $\varepsilon(p) = p^2/2m$.

[17] This is a "phase space effect"; since all statistical integrations involve a factor $d^3p = 4\pi p^2 dp$, excitations with large p can be important even though they are separated from the ground state by a large energy.

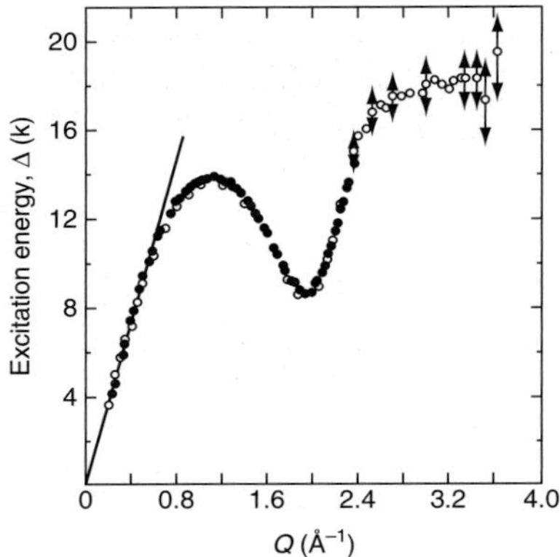

Fig. 1.2 The dispersion curve for He II for the elementary excitations at $T = 1.1$ K. From [23].

Let us first calculate the thermal properties of the phonon gas where ε is given by Eq. (1.48). Doing the angular integration and defining the dimensionless variable $x = cp/k_BT$

$$E_p = \frac{V(k_BT)^4}{2\pi^2\hbar^3c^3}\int dx\frac{x^3}{e^x-1};$$

at low temperature the upper limit of the integration, which is ordinarily cut off by the number of degrees of freedom in the Debye theory, may be extended to ∞. Using $\int dxx^3/(e^x-1) = \pi^4/15$ we obtain

$$E_p = \frac{V\pi^2(k_BT)^4}{30\hbar^3c^3}. \tag{1.50}$$

The phonon heat capacity, $c_p = \partial E/\partial T$, is given by

$$c_p = \frac{2}{15}\frac{V\pi^2k_B^4T^3}{\hbar^3c^3}. \tag{1.51}$$

We next calculate the roton thermal properties. Since $\Delta >> k_BT$ we may use the Boltzmann occupation factor. Using Eq. (1.49) for $\varepsilon(p)$, Eq. (1.46) becomes

$$E_r = \frac{V}{(2\pi\hbar)^3}\int d^3pe^{-[\Delta+\frac{1}{2\mu}(p-p_0)^2]/k_BT}\left[\Delta+\frac{1}{2\mu}(p-p_0)^2\right] \tag{1.52}$$

To sufficient accuracy we may write $p^2dp = p_0^2\,dp$ and using

$$\int_0^\infty dxe^{-ax^2} = \left(\frac{\pi}{a}\right)^{1/2} \quad \text{and} \quad \int_0^\infty dxe^{-ax^2}x^2 = \frac{\pi^{1/2}}{2a^{3/2}}$$

we obtain

$$E_r = \frac{2Vp_0^2\,(\mu k_BT)^{1/2}\,e^{-\Delta/k_BT}}{(2\pi)^{3/2}\quad\hbar^3}\left[\Delta+\frac{1}{2}k_BT\right]. \tag{1.53}$$

The roton heat capacity is given by

$$C_r = \frac{2k_BVp_0^2(\mu k_BT)^{1/2}}{(2\pi)^{3/2}\quad\hbar^3}e^{-\Delta/k_BT}\left[\frac{3}{4}+\frac{\Delta}{k_BT}+\left(\frac{\Delta}{k_BT}\right)^2\right]. \tag{1.54}$$

1.3.2 Superfluidity

Consider the situation shown in Fig. 1.3a; liquid helium in its ground state (i.e. at $T = 0$) is flowing uniformly[18] through a pipe with velocity $\mathbf{v}$. Let us now go into a frame of reference in which the liquid is at rest; in this frame (Fig. 1.3b) the pipe moves with velocity $-\mathbf{v}$. Suppose we now introduce an excitation of energy $\varepsilon(p)$ into the liquid in the rest frame. From the law of transformation of energy to a moving frame (Galilean invariance), the energy E in the lab frame is given by

[18] Such a flow is only possible if the viscosity vanishes, which we will shortly rationalize.

$$E = \varepsilon + \mathbf{p}\cdot\mathbf{v} + \frac{Mv^2}{2}, \tag{1.55}$$

where $Mv^2/2$ is just the kinetic energy of the liquid as a whole. On the other hand $\varepsilon + \mathbf{p} \cdot \mathbf{v}$ is the change in energy due to the appearance of the excitation which must be less than zero if the liquid is to decrease its energy, i.e. if drag is to occur. Thus,

$$\varepsilon + \mathbf{p} \cdot \mathbf{v} < 0; \tag{1.56}$$

this condition is most easy to satisfy if $\mathbf{p}$ and $\mathbf{v}$ are antiparallel; i.e. if

$$\varepsilon - pv < 0 \tag{1.57}$$

or

$$v > \frac{\varepsilon}{p}. \tag{1.58}$$

This inequality is first satisfied for the excitations that lie at the point of tangency of a line drawn from the origin to the curve $\varepsilon(p)$, which occurs near the roton minimum. The corresponding velocity, which we will call the critical velocity, v_{crit}, is about 5×10^3 cm/sec. For fluid velocities less than this value the energy of the fluid cannot be lowered by the introduction of excitations; the fluid will therefore flow with no viscous loss, i.e. it will be a *superfluid*.

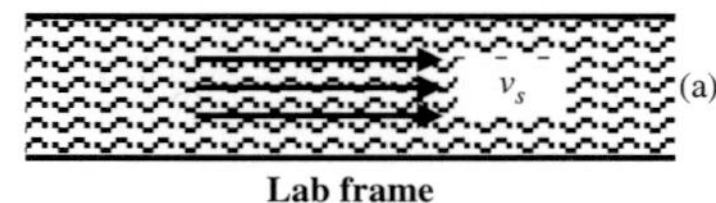

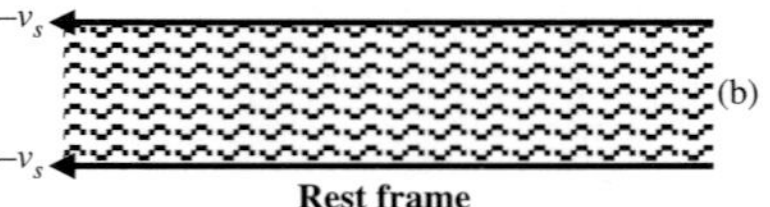

Fig. 1.3 (a) A superfluid moving uniformly through a pipe with velocity v_s in the laboratory frame; (b) The same fluid as viewed from a frame where it is at rest but the pipe moves with a velocity $-v_s$.

1.3.3 The two-fluid model

At finite temperature some distribution of excitations exists. Let us calculate the effect of moving this gas of excitations with some velocity $-\mathbf{v}$ relative to the fixed ground state (Fig. 1.3b). The total momentum associated with the moving excitation gas is

$$\mathbf{P} = \frac{V}{(2\pi\hbar)^3} \int d^3p\, \mathbf{p}\, n(\varepsilon') \tag{1.59}$$

where ε' is the energy of the excitation in a frame moving with the gas, which we take to be the equilibrium distribution. We use our Galilean transformation to evaluate the energies in terms of quantities measured in the frame where the ground state is at rest i.e. $\varepsilon' = \varepsilon - \mathbf{p} \cdot \mathbf{v}$. Thus,

$$\mathbf{P} = \frac{V}{(2\pi\hbar)^3} \int d^3p\, \mathbf{p}\, n(\varepsilon - \mathbf{p} \cdot \mathbf{v}). \tag{1.60}$$

or

$$\mathbf{P} \cong \frac{V}{(2\pi\hbar)^3} \int d^3p\, \mathbf{p} \left[n(\varepsilon) - \frac{\partial n}{\partial \varepsilon} \mathbf{p} \cdot \mathbf{v} + \cdots \right]. \tag{1.61}$$

Noting the average of $\mathbf{p}$ for an equilibrium distribution vanishes, and performing the angular integration on the remaining term we have

$$\mathbf{P} = -\mathbf{v} \frac{4\pi}{3} \frac{V}{(2\pi\hbar)^3} \int dp\, p^4 \frac{\partial n}{\partial \varepsilon}. \tag{1.62}$$

We now evaluate Eq. (1.62) for the phonons. Integrating by parts and using $\varepsilon = cp$ we obtain

$$\mathbf{P} = \mathbf{v}\frac{16\pi}{3c}\frac{V}{(2\pi\hbar)^3}\int p^3 n\, dp = \frac{4}{3c^2}E_p\mathbf{v}. \tag{1.63}$$

If we associate $|\mathbf{P}|/|\mathbf{v}|$ with a mass, then the corresponding mass per unit volume will be given by

$$\rho_{np} = \frac{4}{3c^2}\frac{E_p}{V} = \frac{2\pi^2}{45\hbar^3c^5}(k_BT)^4. \tag{1.64}$$

This quantity we identify as the phonon contribution to the *normal density* of the liquid. To evaluate the roton contributions we again employ Boltzmann statistics and note $\partial n/\partial\varepsilon = -n/k_BT$; thus using Eq. (1.62)

$$\rho_{\mathrm{nr}} = \frac{4\pi}{3k_BT(2\pi\hbar)^3}\int dp\, p^4\, n(\varepsilon). \tag{1.65}$$

We approximate p^4 by p_0^4; the resulting integral is

$$\rho_{\mathrm{nr}} = \frac{2\mu^{1/2}p_0^4}{3(2\pi)^{3/2}(k_BT)^{1/2}\hbar^3}e^{-\Delta/k_BT}. \tag{1.66}$$

The total normal density is then

$$\rho_n = \rho_{np} + \rho_{nr}. \tag{1.67}$$

For temperatures below the phase transition, $\rho_n < \rho$; the transition temperature, or λ point as it is called,[19] is then identified as the temperature at which $\rho_n = \rho$. That fraction of the liquid which is not normal we call the *superfluid*; i.e.

[19]The name arises from the shape of the heat capacity curve which looks like the Greek letter λ near the phase transition.

$$\rho_s = \rho - \rho_n \tag{1.68}$$

and, for velocities below the critical velocity, it moves with *no friction.* At $T = 0$ all of the liquid is superfluid.

At some temperature one would have $\rho_n(T) = \rho$ at which point superfluidity would be expected to disappear. As with the case of the dilute Bose gas, interactions of the elementary excitations would make such an estimate of T_c only qualitative.

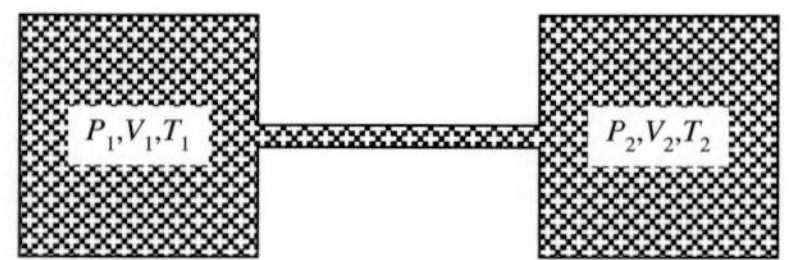

Fig. 1.4 Two vessels maintained at different temperatures connected by a superleak.

1.3.4 The thermo-mechanical effect

Consider two vessels which are connected by a superleak[20] as shown in Fig. 1.4. It is observed that if a temperature differential is established between the two vessels, an exchange of superfluid occurs resulting in a pressure differential ΔP. We will analyze this situation thermodynamically. The combined first and second laws may be written

[20]A superleak is a connection constructed so that the resistance to the flow of normal fluid is exceedingly high and hence only superfluid can be transported.

$$dE_i = T_i\, dS_i - P_i dV_i + \mu_i\, dN_i \tag{1.69}$$

where the index i distinguishes the two vessels and E_i, S_i, V_i, and N_i are the energy, entropy, volume, and particle number for each vessel. Since the superfluid cannot transport entropy, the entropies of the two vessels must remain constant, i.e. $dS_1 = dS_2 = 0$; also the volumes are assumed to be constant requiring $dV_1 = dV_2 = 0$. Since the total energy remains constant we have

$$\mu_1 dN_1 + \mu_2 dN_2 = 0. \tag{1.70}$$

But the total number of particles N_1+N_2, is constant which requires $dN_1 = -dN_2$; thus

$$\mu_1(T_1, P_1) = \mu_2(T_2, P_2). \tag{1.71}$$

Since the pressure dependence of μ is small, we can expand it obtaining

$$\begin{aligned} \mu(T_2, p_2) &= \mu(T_2, p_1) + \left(\frac{\partial \mu}{\partial P}\right)_T (P_2 - P_1) \\ &= \mu(T_1, p_1) + \int_{T_1}^{T_2} dT \left(\frac{\partial \mu}{\partial T}\right)_P + \left(\frac{\partial \mu}{\partial P}\right)_T (P_2 - P_1). \end{aligned} \tag{1.72}$$

Using the standard expression for the differential of the Gibbs free energy we have

$$\left(\frac{\partial \mu}{\partial T}\right)_p = -\frac{S}{N}; \qquad \left(\frac{\partial \mu}{\partial P}\right)_T = \frac{V}{N}. \tag{1.73}$$

Defining the mass density, $\rho = M/V$, introducing an entropy per unit mass,[21] $s = S/M$, and using Eq. (1.71), Eq. (1.72) becomes

$$P_2 - P_1 = \int_{T_1}^{T_2} dT \rho s. \tag{1.74}$$

[21] Note that in hydrodynamics it is common to introduce an *entropy per unit mass*, which in some ways is more natural than an entropy per unit volume; the former is constant under an adiabatic volume change (as when an adiabatic sound wave propagates), whereas the latter would not be constant.

The thermo-mechanical effect, or fountain effect as it is also called, is analogous to the *osmotic pressure* which occurs in solutions where the solvent (here the superfluid) can be exchanged between two vessels containing different quantities of solute (the normal fluid) through a semi-permeable membrane (the super leak).

1.3.5 Hydrodynamics of superfluids and sound propagation

Associated with the densities $\rho_n(\mathbf{r}, t)$ and $\rho_s(\mathbf{r}, t)$, characterizing the normal and superfluid components, we will in general have two velocity fields, $\mathbf{v}_n(\mathbf{r}, t)$ and $\mathbf{v}_s(\mathbf{r}, t)$ and we now give a simplified discussion of the accompanying two fluid hydrodynamics [25].[22] Since the superfluid component represents the fraction of the liquid in its ground state, we associate the transport of entropy with the normal fluid moving at $\mathbf{v}_n$. If we write the entropy per unit volume as ρs, where ρ is the mass per unit volume and s is the entropy per unit mass, and ignore dissipative (entropy generating) effects, then the rate of change of entropy within

[22] For a brief history see [25].

some volume, V, arises from a flux $\rho s\mathbf{v}_n$ of entropy through the surrounding surface, S, and is therefore given by

$$\frac{\partial}{\partial t}\int_V \rho s\, d^3r = -\int_S \rho s\mathbf{v}_n \cdot d^2\mathbf{r}; \tag{1.75}$$

recognizing that this must hold independent of the chosen volume and applying Green's theorem we obtain a law for conservation of entropy as

$$\frac{\partial}{\partial t}(\rho s) + \nabla \cdot (\rho s\mathbf{v}_n) = 0. \tag{1.76}$$

Our next task is to obtain equations of motion that describe the time evolution of our two velocity fields. In what follows we will continue to ignore dissipation and limit our discussion to the linear approximation using the intuitive approach given by London, which in turn utilized earlier ideas of Tisza [26] (rather than the more formal derivation of Landau).[23] For a general state of motion, the kinetic energy density, e_K, and momentum density, $\mathbf{g}$, are given by

[23]It seems Tisza was the first to propose a two-fluid model of He II which he based on a Bose–Einstein condensation. Although this model does not give the correct excitation spectrum, which is better captured by Landau's model, it does introduce the idea that He II possesses a condensate, which is now generally accepted as the defining characteristic of a superfluid. Curiously, Landau appears to have not-placed much emphasis on this feature. For additional discussion of these points see [26A].

$$e_K = \frac{1}{2}\rho_n v_n^2 + \frac{1}{2}\rho_s v_s^2 \tag{1.77}$$

and

$$\mathbf{g} = \rho_n\mathbf{v}_n + \rho_s\mathbf{v}_s. \tag{1.78}$$

We can rewrite Eq. (1.77) in the form

$$\begin{aligned} e_K &= \frac{1}{2\rho}(\rho_s\mathbf{v}_s + \rho_n\mathbf{v}_n)^2 + \frac{\rho_s\rho_n}{2\rho}(\mathbf{v}_s - \mathbf{v}_n)^2 \\ &= e_K^+ + e_K^-; \end{aligned} \tag{1.79}$$

comparing (1.78) and (1.79) we have $e_K^+ = \mathbf{g}^2/2\rho$. We therefore associate e_K^+ with the kinetic energy density of the liquid as a whole, whereas e_K^- can be viewed as an "internal" kinetic energy density.

He II supports two kinds of propagating collective modes. The first is the usual sound waves where the momentum and density oscillate out of phase. The other mode is called *second sound* and corresponds to a counter motion of the normal and superfluid velocities but with no change in the overall momentum density or mass density of the fluid; since this is a new feature, unique to the superfluid state, we will treat it first. A consistent treatment of this mode emerges by assuming $\mathbf{g} = 0$ from which it follows that $\mathbf{v}_s = -(\rho_n/\rho_s)\,\mathbf{v}_n$, $e_K^+ = 0$, and $e_K^- = \rho(\rho_n/2\rho_s)\mathbf{v}_n^2$. Next we recall the thermodynamic identity which for a fixed volume of fluid takes the form,

$$d\varepsilon = Td(\rho s) - \mu' d\rho \tag{1.80}$$

where ε is the energy per unit volume and μ' is the chemical potential per unit mass. Carrying out a Legendre transformation to define the Gibbs free energy per unit volume, $\rho\mu' = \varepsilon + P - T\rho s$ where P is the pressure, we obtain

$$\rho d\mu' = dP - \rho s dT. \tag{1.81}$$

Now the change in free energy density $d\left(\rho\mu'\right)$ at constant pressure and density must arise from the change in the internal kinetic energy density, $de_{\bar{K}} = (\rho\rho_n/\rho_s)\,\mathbf{v}_n \cdot d\mathbf{v}_n$. Suppose we consider a slab of area A and thickness dx. The change in free energy of this slab is then given by $Ade_{\bar{K}}dx$. This change must result from an influx of entropy density taken as moving with velocity v_n normal to the slab which is driven by a temperature differential dT that acts for a time dt yielding $Av_n\,\rho s\,dTdt$. Equating these two forms we obtain $(\rho_n/\rho_s)\,(\partial v_n/\partial t) = -s\partial T/\partial x$ or, extending to an arbitrary volume element,

$$\frac{\rho_n}{\rho_s}\frac{\partial \mathbf{v}_n}{\partial t} + s\nabla T = 0. \tag{1.82}$$

Taking the divergence of (1.82), using (1.76), and ignoring terms of second order gives

$$\frac{\partial^2 s}{\partial t^2} - \frac{\rho_s}{\rho_n}s^2\nabla^2 T = 0;$$

introducing the heat capacity per unit mass at constant pressure, $c_p = T\partial s/\partial T$, this equation becomes

$$\frac{\partial^2 T}{\partial t^2} - \frac{T\rho_s}{c_P\rho_n}s^2\nabla^2 T = 0. \tag{1.83}$$

Eq. (3.39) is a wave equation having solutions with a mode velocity

$$c_2^2 = \frac{\rho_s}{\rho_n}\frac{s^2 T}{c_P} \tag{1.84}$$

in which the temperature oscillates in space and time. Since the density and pressure associated with the propagation of this mode remain constant it cannot be excited by a conventional piezoelectric transducer. However passing an oscillating electrical current through a thin resistive film will cause its temperature to oscillate and launch a second sound wave (propagating at twice the frequency); the mode can be detected using a second film having a high differential change in resistance with temperature (such as a superconductor biased close to its transition temperature with a magnetic field). Second sound was first observed by V. Peshkov in 1944 [27].

To complete our discussion of the linearized hydrodynamics of He II we must include motions in which $\mathbf{g} \neq 0$. When $\mathbf{g} = 0$ we have $\tilde{\mathbf{v}} \equiv \mathbf{v}_n - \mathbf{v}_s = (\rho/\rho_s)\mathbf{v}_n$ and from Eq. (1.82) it then follows that

$$\begin{aligned} \dot{\tilde{\mathbf{v}}} &= \dot{\mathbf{v}}_n - \dot{\mathbf{v}}_s \\ &= -(\rho/\rho_n)s\,\nabla\, T. \end{aligned} \tag{1.85}$$

This equation, involving the relative motion of the two fluids should also hold in a frame where $\mathbf{g} \neq 0$ moving with a velocity $\mathbf{v}$ defined by $\rho\mathbf{v} = \rho_s\mathbf{v}_s + \rho_n\mathbf{v}_n$. Now the Euler equation of conventional hydrodynamics should continue to hold for the motion of the fluid as a whole; hence

$$\rho \dot{\mathbf{v}} = -\nabla P. \tag{1.86}$$

Using Eqs. (1.85) and (1.86) to separately solve for $\mathbf{v}_n$ and $\mathbf{v}_s$ we obtain

$$\dot{\mathbf{v}}_s = -\frac{1}{\rho}\nabla P + s\,\nabla T \tag{1.87a}$$

and

$$\dot{\mathbf{v}}_n = -\frac{1}{\rho}\nabla P - \frac{s\rho_s}{\rho_n}\nabla T. \tag{1.87b}$$

To complete the description, we require the equations for conservation of mass and entropy

$$\frac{\partial \rho}{\partial t} + \nabla \cdot (\rho \mathbf{v}) = 0 \tag{1.88a}$$

and

$$\frac{\partial}{\partial t}(\rho s) + \nabla \cdot (\rho s \mathbf{v}_n) = 0. \tag{1.88b}$$

By combining Eqs. (1.86) and (1.88a) and writing $c_1^2 = \partial P/\partial \rho$ we obtain the wave equation for first sound

$$\nabla^2 \rho - \frac{1}{c_1^2}\frac{\partial^2 \rho}{\partial t^2} = 0, \tag{1.89}$$

where c_1 is the velocity of ordinary density waves in a fluid, which in the present case is designated as first sound.

Analogous to our discussion of the fountain effect, we can identify a propagating mode in which the superfluid oscillates but the normal fluid is "locked"; the latter is achieved by filling the environment with a finely divided powder or by propagating through a capillary, which via the normal fluid viscosity suppresses its motion. We start by writing (1.88a) in the form $\partial\rho/\partial t + \nabla \cdot (\rho_s \mathbf{v}_s + \rho_n \mathbf{v}_n) = 0$ and setting $\mathbf{v}_n = 0$ which gives $\partial \rho/\partial t = -\rho_s \nabla \cdot \mathbf{v}_s$. Taking the divergence of (1.87a) and inserting this last form gives

$$\frac{\partial^2 \rho}{\partial t^2} = \frac{\rho_s}{\rho}\nabla^2 P - \rho_s s\,\nabla^2 T. \tag{1.90}$$

On the other hand, setting $\dot{\mathbf{v}}_n = 0$ in (1.87b) gives

$$\frac{\partial}{\partial t}(\rho s) = 0, \tag{1.91}$$

or $\delta s = -(s/\rho)\delta\rho$. Writing $P = P(\rho, s)$, expanding the first term in (1.90), and then using (1.91) we have

$$\begin{aligned}\nabla^2 P &= \left(\frac{\partial P}{\partial \rho}\right)_s \nabla^2 \rho + \left(\frac{\partial P}{\partial s}\right)_\rho \nabla^2 s \\ &= \left[\left(\frac{\partial P}{\partial \rho}\right)_s - \left(\frac{\partial P}{\partial s}\right)_\rho \frac{s}{\rho}\right]\nabla^2 \rho;\end{aligned} \tag{1.92a}$$

similarly

$$\nabla^2 T = \left[\left(\frac{\partial T}{\partial \rho} \right)_s - \left(\frac{\partial T}{\partial s} \right)_\rho \frac{s}{\rho} \right] \nabla^2 \rho. \tag{1.92b}$$

Eq. (1.90) now becomes

$$\frac{\partial^2 \rho}{\partial t^2} = \left[\frac{\rho_s}{\rho} \left(\frac{\partial P}{\partial \rho} \right)_s - \frac{\rho_s s}{\rho^2} \left(\frac{\partial P}{\partial s} \right)_\rho - \rho_s s \left(\frac{\partial T}{\partial \rho} \right)_s + \frac{\rho_s s^2}{\rho} \left(\frac{\partial T}{\partial s} \right)_\rho \right] \nabla^2 \rho. \tag{1.93}$$

Using the definitions of c_1 and c_2 we may rewrite this as

$$\omega^2 - k^2 \left\{ \frac{\rho_s}{\rho} c_1^2 + \frac{\rho_n}{\rho} c_2^2 \left[1 - \frac{1}{\rho s} \left(\frac{\partial P}{\partial T} \right)_\rho - \frac{\rho}{s} \left(\frac{\partial T}{\partial \rho} \right)_s \left(\frac{\partial s}{\partial T} \right)_\rho \right] \right\} = 0. \tag{1.94}$$

The last term may be rewritten

$$\begin{aligned} -\frac{\rho}{s} \left(\frac{\partial T}{\partial \rho} \right)_s \left(\frac{\partial s}{\partial T} \right)_\rho &= \frac{\rho}{s} \left(\frac{\partial T}{\partial \rho} \right)_s \left(\frac{\partial s}{\partial \rho} \right)_T \left(\frac{\partial \rho}{\partial T} \right)_s \\ &= \frac{\rho}{s} \left(\frac{\partial s}{\partial \rho} \right)_T = -\frac{1}{\rho s} \left(\frac{\partial p}{\partial T} \right)_\rho , \end{aligned} \tag{1.95}$$

where in the last step we have used a Maxwell relation. Defining $\beta = (\partial p / \partial T)_\rho$, we may write

$$c_4^2 = \frac{\omega^2}{k^2} = \frac{\rho_s}{\rho} c_1^2 + \frac{\rho_n}{\rho} c_2^2 \left(1 - \frac{2\beta}{\rho s} \right). \tag{1.96}$$

The second term is usually small and to a good approximation we have $c_4 = (\rho_s / \rho)\, c_1$. Fourth sound has been a powerful probe in studying the superfluid density.

1.3.6 The Landau–Ginzburg theory of ^{4}He[24]

In Landau's theory of the second-order phase transition, one assumes that the free energy may be expanded as a power series in an *order parameter*, $\psi(\mathbf{r})$, the structure of which depends on the type of phase transition involved. For ^{4}He it is a complex scalar function, ψ, the interpretation of which is discussed in Section 1.3.7. We assume that the expansion of the free energy, F, involves only even powers of $|\psi|$ and for a uniform system we write it as

$$F = F_0(T) + \int d^3 r \left[\alpha(T) \, |\psi|^2 + \beta(T) \, |\psi|^4 \right] \tag{1.97}$$

or

$$F = F_0(T) + \int d^3 r \, [F_2(T) + F_4(T)], \tag{1.98}$$

where $F_0(T)$ is the free energy of the normal phase and F_2 and F_4 are quadratic and quartic contributions to the free energy density of the superfluid phase.

[24] The analogous theory for the case of a superconductor is discussed in Section 1.5.4. and for superfluid ^{3}He in Sections 1.7.3ff.

To assure overall system stability we will assume $\beta > 0$. When $\alpha(T) > 0$, the minimum energy obviously occurs for $\psi = 0$; however for $\alpha < 0$, ψ will be finite. Thus, we can model a second-order phase transition by writing $\alpha = a(T - T_c)$ and $\beta(T) = \beta(T_c)$. The magnitude of ψ is determined by minimizing Eq. (1.97) with respect to ψ^* :

$$\alpha(T)\psi + 2\beta\psi\,|\psi|^2 = 0 \tag{1.99}$$

which has two solutions

$$\psi = 0; \qquad T > T_c \tag{1.100}$$

$$|\psi| = \left[\frac{a(T_c - T)}{2\beta}\right]^{1/2}; \quad T < T_c. \tag{1.101}$$

Thus the system orders at T_c with a strength proportional to $(T_c - T)^{1/2}$.

In the presence of spatial inhomogeneities we must include an additional term in F to account for the extra free energy that arises when the order parameter is position dependent; such a contribution involves $\nabla\psi$. From symmetry the new contribution cannot change sign under $\mathbf{r} \to -\mathbf{r}$ and to the lowest order we may then write

$$F_{g2} = \frac{1}{2}K\,|\nabla\psi|^2 \tag{1.102}$$

where now

$$F = F_0(T) + \int d^3r\,\left[F_2 + F_4 + F_{g2}\right]. \tag{1.103}$$

In their treatment of superconductivity Landau and Ginzburg treated ψ phenomenologically as a *macroscopic wave function* in the spirit of F. London's idea. The effect of a magnetic field is then incorporated by noting that in quantum mechanics one uses the prescription $\nabla \to \nabla - (ie/\hbar c)\mathbf{A}$ where $\mathbf{A}$ is the vector potential. One may derive the particle current using the following trick; suppose we assign an infinitesimal *fictitious charge e^* to each He atom.* Classically the change in free energy produced by a change in $\delta\mathbf{A}$, in the presence of an electric current $\mathbf{j}$, is given by

$$\delta F = \frac{1}{c}\int d^3r\,\mathbf{j}\cdot\delta\mathbf{A}. \tag{1.104}$$

Thus, we define the superfluid mass current, $\mathbf{g}$, by the functional derivative

$$\mathbf{g} = \frac{m_4 c}{e^*}\left(\frac{\delta F}{\delta\mathbf{A}}\right)_{e^*=0} \tag{1.105}$$

where m_4 is the mass of the ^{4}He atom. In the presence of a vector potential we have

$$F_{g2} = \frac{1}{2}K\left|\left(\nabla - \frac{ie^*}{\hbar c}\mathbf{A}\right)\psi\right|^2 \tag{1.106}$$

which gives

$$\mathbf{g} = \frac{i}{2}\left(\frac{m_4}{\hbar}\right)K\left[-\psi^*\nabla\psi + \psi\nabla\psi^*\right] \tag{1.107}$$

Writing $\psi(\mathbf{r}) = |\psi(\mathbf{r})|\, e^{i\varphi(\mathbf{r})}$ we obtain

$$\mathbf{g} = K\left(\frac{m_4}{\hbar}\right)|\psi|^2\,\nabla\Phi \tag{1.108}$$

To make contact with the Landau theory where the superfluid momentum is $\rho_s\mathbf{v}_s$, we write

$$\mathbf{v}_s = \frac{\hbar}{m_4}\nabla\Phi \tag{1.109}$$

and

$$\rho_s = K\left(\frac{m_4}{\hbar}\right)^2 |\psi|^2. \tag{1.110}$$

Note that if we literally interpret ψ as a wave function and use the quantum mechanical expressions for particle flux, we are led to the expression

$$\mathbf{g} = \frac{\hbar}{2i}\left(\psi^*\nabla\psi - \psi\nabla\psi^*\right) \tag{1.111}$$

where $\rho_s = m_4\,|\psi|^2$ and $\mathbf{v}_s = \dfrac{\hbar}{m_4}\nabla\varphi$, implying $K = \dfrac{\hbar^2}{m_4}$.

1.3.7 Quantized vorticity

A basic assumption of the Landau's hydrodynamics was that $\nabla\times\mathbf{v}_s = 0$ (so-called irrotational flow) and we now look at this assumption in more detail. We start by examining the integral

$$\int d^2\mathbf{r}\cdot\nabla\times\mathbf{v}_s = \oint d\mathbf{l}\cdot\mathbf{v}_s, \tag{1.112}$$

which is sometimes called the *circulation*. If we consider individual superfluid (condensate) atoms moving with velocity $\mathbf{v}_s$, each having a momentum $\mathbf{p} = m_4\mathbf{v}_s$, we can rewrite the integral on the right-hand side of (1.112) in the form $(1/m_4)\oint d\mathbf{l}\cdot\mathbf{p}$. We now assume our condensate atoms are executing circular "orbits" around some imaginary central line, as one has for a vortex line in conventional hydrodynamics; if we now interpret the line integral in the light of the Bohr–Sommerfeld quantization rule, $\oint d\mathbf{l}\cdot\mathbf{p} = 2\pi n\hbar$, and carry out this integral on a circle of radius r around the central line, we obtain the condition

$$\oint d\mathbf{l}\cdot\mathbf{p} = m_4\oint d\mathbf{l}\cdot\mathbf{v}_s = 2\pi r m_4 v_s = 2\pi n\hbar \tag{1.113}$$

or

$$v_s = \frac{n\hbar}{m_4}\frac{1}{r}, \tag{1.114}$$

where $h = 2\pi\hbar$ is the usual Planck constant. We have obtained the somewhat remarkable result that the flow is *quantized* for a superfluid vortex. Note that for all $r \neq 0$ the curl of (1.114) vanishes; in this sense the flow is still irrotational. But the velocity predicted by this expression diverges for small r and hence must break down; we expect this breakdown to occur at a radius r_c where the superfluid velocity v_s is of the order of the Landau critical velocity (50 m/sec):

$$r_c = \frac{nh}{M_4}\frac{1}{v_{\rm crit}}. \tag{1.115}$$

We can interpret r_c as a *core radius* for our vortex and it is of the order of an angstrom. For values of r of the order of or smaller than r_c we must have $\nabla \times \mathbf{v}_s \neq 0$ when $n \neq 0$. In practice vorticies with $n > 1$ are unstable to a break-up into vortices with $n = 1$.

We can equally well consider a vortex from the perspective of the Ginzburg–Landau theory. According to Eq. (1.109) the superfluid velocity is related to the gradient of the phase of the superfluid (condensate) wave function. Requiring that the change in phase on circumventing the vortex core be restricted to a multiple of 2π also leads to Eq. (1.114). For the case of a superconductor one has quantized flux lines as will be discussed in Section 1.5.3.

Quantized vorticity will be discussed at length in the chapter by Tsubota, Kasamatsu and Kobayashi (Chapter 3). It is also discussed in the chapters by Chevy and Dalibard Chapter 7 and by Yamamoto Chapter 9, as well as in other sections of this book.

1.4 Microscopic interpretation of the Bose order parameter

For the dilute Bose gas it follows from Eqs. (1.34)–(1.36) that the expectation value for the number operator, $n_{\mathbf{k}} = \left\langle \hat{a}^{\dagger}_{\mathbf{k}} \hat{a}_{\mathbf{k}} \right\rangle$, can be written in the form,

$$n(\mathbf{k}) = (2\pi)^3 n_0 \delta(\mathbf{k}) + n'(\mathbf{k}) \tag{1.116}$$

corresponding to a condensate number, N_0, and number density, n_0,

$$\frac{N_0}{L^3} = n_0. \tag{1.117}$$

That some fraction of the particles will have zero momentum is now accepted as the *defining property* of a Bose fluid at $T = 0$. We now examine this property from an alternative point of view.

1.4.1 Off-diagonal long-range order

The one-body density matrix can be defined as follows

$$\rho^{(1)}(\mathbf{r}_1, \mathbf{r}_2, t) = \left\langle \hat{\psi}^{\dagger}(\mathbf{r}_2, t) \hat{\psi}(\mathbf{r}_1, t) \right\rangle. \tag{1.118}$$

For a translationally invariant system (a fluid) in equilibrium this function depends only on $\mathbf{r} = \mathbf{r}_1 - \mathbf{r}_2$. The momentum distribution is taken as the Fourier transform of $\rho^{(1)}$ and hence

$$n(\mathbf{k}) = \int \rho^{(1)}(\mathbf{r})\, e^{-i\mathbf{k}\cdot\mathbf{r}} d^3 r \tag{1.119a}$$

and

$$\rho^{(1)}(\mathbf{r}) = \frac{1}{(2\pi)^3}\int n(\mathbf{k})\, e^{i\mathbf{k}\cdot\mathbf{r}/\hbar} d^3k. \tag{1.119b}$$

If $n(\mathbf{k})$ has the form (1.116), then it immediately follows that

$$\rho^{(1)}(\mathbf{r} \to \infty) = n_0$$

and *does not vanish*, a property first pointed out by O. Penrose [28];[25] one then says that the density matrix displays *off-diagonal long-range order* (a term due to C. N. Yang). Note that for a classical liquid we always have

$$\rho^{(1)}(\mathbf{r} \to \infty) = 0. \tag{1.120}$$

[25] For a detailed discussion of off-diagonal long-range order in both Bose and Fermi superfluids see [29].

In the above discussion of the G–L theory the order parameter ψ was left undefined. In Bogoliubov's theory we found that $\langle \hat{a}_0 \rangle = a_0$ does not vanish and is a c-number. In a similar manner the expectation value of a $\hat{\psi}$-operator does not vanish in the superfluid state and this quantity is generally chosen to be the order parameter in a superfluid; i.e.

$$\langle \hat{\psi}(\mathbf{r}) \rangle \equiv \psi(\mathbf{r}) \tag{1.121a}$$

and

$$\langle \hat{\psi}^\dagger(\mathbf{r}) \rangle \equiv \psi^\dagger(\mathbf{r}). \tag{1.121b}$$

Note these will in general be complex quantities.

The question then arises as to presence of and size of a condensate in superfluid ^{4}He. For small momentum transfers a neutron scattering inelastically from the liquid will excite elementary excitations, $\varepsilon(\mathbf{k})$, which can be thought of as a collective response of the liquid as a whole. This is how the excitation spectrum shown in Fig. 1.1 is obtained. On the other hand very energetic neutrons should scatter from individual atoms, some of which are expected to be in the condensate. The distribution of the recoiling neutrons would then consist of two fractions: one arising from $n'(\mathbf{k})$ and the other from n_0 (see Eq. (1.116)). In practice, the experimental data can be fitted by a single Gaussian-like distribution above the lambda point, but are best fitted by two Gaussians, rather than a Gaussian and a delta function, below the transition, with the second component increasing down to about 1.5 K. The absence of a delta function arises from two causes: (i) the usual instrumental resolution, and (ii) a so-called final state interaction in which the recoiling neutron interacts with the liquid, thereby adding momentum uncertainty; this latter feature is intrinsic and cannot be eliminated. None the less the overall picture is consistent with a condensate fraction of about 10% falling to about 5% at the solidification pressure. The results are in general agreement with quantum-many-body calculations of the condensate fraction.[26]

[26] For a general review see [30].

1.4.2 The Gross–Pitaevskii equation

The ground state wave function of an inhomogeneous weakly interacting Bose condensate was studied by Gross [31] and Pitaevskii [32] in connection with the superfluid vortex The second quantized Hamiltonian for a system of interacting bosons in an external potential, V^{ext}, is given by

$$\hat{H} = \int d^3r \hat{\psi}(\mathbf{r},t)\left[-\frac{\hbar^2}{2m_4}\nabla^2 + V^{\text{ext.}}(\mathbf{r})\right]\hat{\psi}(\mathbf{r},t) + \frac{1}{2}\int d^3r \int d^3r' \hat{\psi}^\dagger(\mathbf{r},t)\hat{\psi}^\dagger(\mathbf{r}',t)U(\mathbf{r}-\mathbf{r}')\psi'(\mathbf{r}',t)\psi'(\mathbf{r},t) \tag{1.122}$$

where $\hat{\psi}(\mathbf{r},t)$ and $\hat{\psi}^\dagger(\mathbf{r},t)$ are the boson field operators that annihilate and create a particle at the position $\mathbf{r}$. Restricting ourselves to s-wave scattering only, the interaction potential has the form

$$U(\mathbf{r}-\mathbf{r}') = U_0\delta(\mathbf{r}-\mathbf{r}') \tag{1.123}$$

where $U_0 = 4\pi\hbar^2 a/m_4$ is the interaction strength discussed earlier, and a is the scattering length.

The Heisenberg equation of motion is

$$\frac{\partial\hat{\psi}(\mathbf{r},t)}{\partial t} = \frac{i}{\hbar}[\hat{H},\hat{\psi}(\mathbf{r},t)]. \tag{1.124}$$

Plugging in the Hamiltonian and using the commutation relations for boson particles,

$$\hat{\psi}(\mathbf{r})\hat{\psi}(\mathbf{r}') - \hat{\psi}(\mathbf{r}')\hat{\psi}(\mathbf{r}) = 0 \tag{1.125a}$$

and

$$\hat{\psi}(\mathbf{r})\hat{\psi}^\dagger(\mathbf{r}') - \hat{\psi}(\mathbf{r}')\hat{\psi}(\mathbf{r}) = 0\hat{\psi}(\mathbf{r})\hat{\psi}^\dagger(\mathbf{r}') = \delta^{(3)}(\mathbf{r}-\mathbf{r}'), \tag{1.125b}$$

we obtain

$$i\hbar\frac{\partial\hat{\psi}(\mathbf{r},t)}{\partial t} = \left[-\frac{\hbar^2}{2m}\nabla^2 + V_{\text{trap}}(\mathbf{r})\right]\hat{\psi}'(\mathbf{r},t) + U_0\hat{\psi}^\dagger(\mathbf{r},t)\hat{\psi}(\mathbf{r},t)\hat{\psi}(\mathbf{r},t). \tag{1.126}$$

At this point one assumes the presence of a condensate and writes the operator $\hat{\psi}$

$$\hat{\psi}(\mathbf{r},t) = \psi(\mathbf{r},t) + \hat{\psi}'(\mathbf{r},t). \tag{1.127}$$

Here $\psi = \langle\hat{\psi}(\mathbf{r})\rangle$ is a classical or "c-number" quantity, i.e. an ordinary function, which corresponds to the condensed fraction through $|\psi|^2 = N_0/V$ where N_0 is the number of particles in the condensate, and $\hat{\psi}'(\mathbf{r})$ is an operator describing the remaining, or "fluctuating", part of the system. Neglecting the fluctuating part, we obtain the non-linear Schrodinger equation or Gross–Pitaevskii equation (GP)

$$i\hbar\frac{\partial\psi(\mathbf{r},t)}{\partial t} = \left[-\frac{\hbar^2}{2m_4}\nabla^2 + V^{\text{ext.}}(\mathbf{r}) + U_0|\psi(\mathbf{r},t)|^2\right]\psi(\mathbf{r},t). \tag{1.128}$$

This equation is solved subject to the constraint

$$\int d^3r|\psi(\mathbf{r},t)|^2 = N_0 \tag{1.129}$$

To find a stationary solution for the condensate wave function in this mean field theory, one can substitute the form

$$\psi(\mathbf{r},t) = e^{-i\mu t/\hbar}\psi(\mathbf{r}) \tag{1.130}$$

which gives the time-independent equation

$$\left[-\frac{\hbar^2}{2m_4}\nabla^2 + V^{\text{ext.}}(\mathbf{r}) + U_0\,|\psi(\mathbf{r},t)|^2\right]\psi(\mathbf{r},t) = \mu\psi(\mathbf{r},t). \tag{1.131}$$

For a slowly varying condensate we can neglect the ∇^2 term in (1.130) and identify

$$\mu = V^{\text{ext.}}(\mathbf{r}) + U_0 n(\mathbf{r}) \quad \text{(slowly varying system)} \tag{1.132}$$

as the chemical potential with $n(\mathbf{r})$ the local condensate density.

Writing the energy as $E = \langle\hat{H}\rangle = E_0 + E'$, where we assign E_0 to the condensate, and using Eqs. (1.122) and (1.127), this "Hartree" approximation yields

$$E_0 = \int d^3r\left[\frac{\hbar^2}{2m_4}|\nabla\psi(\mathbf{r})|^2 + V^{\text{ext.}}(\mathbf{r})|\psi(\mathbf{r},t)|^2 + \frac{1}{2}U_0|\psi(\mathbf{r})|^4\right] \tag{1.133a}$$

with an associated condensate energy density

$$e(\mathbf{r}) = \frac{\hbar^2}{2m_4}|\nabla\psi(\mathbf{r})|^2 + V^{\text{ext.}}(\mathbf{r})|\psi(\mathbf{r},t)|^2 + \frac{1}{2}U_0|\psi(\mathbf{r},t)|^4. \tag{1.133b}$$

In the absence of an external potential this form of the equation is structurally the same as that obtained using the Landau–Ginzburg free energy functional when $K = \hbar^2/m_4$, $\alpha = \mu$, and $\beta = N_0U_0$.

A plane wave solution would have the form

$$\psi(\mathbf{r}) = \sqrt{\frac{N_0}{V}}e^{i\mathbf{k}\cdot\mathbf{r}} \tag{1.134}$$

with a corresponding energy per particle given by

$$\varepsilon(\mathbf{k}) = \frac{\hbar^2k^2}{2m} + V^{\text{ext.}}(\mathbf{r}) + \frac{1}{2}U_0\frac{N_0}{V}, \tag{1.135}$$

where the second and third terms can be viewed as a contribution to chemical potential from which the excitation energies are measured [33].[27] However the excitations do not have the Bogoliubov linear behavior at long wavelengths; i.e. the G–P approximation misses the collective character of the excitations.

[27] According to the Hugenholtz–Pines theorem the excitation spectrum of an interacting Bose gas will not have a gap; see [33].

Eq. (1.131) and (1.128) have one-dimensional soliton solutions. For the static case one has the form,

$$\psi(x) = \psi_0 \tanh\left(\frac{x}{\sqrt{2}\xi}\right), \tag{1.136}$$

where $\xi = \hbar/\sqrt{2mU_0N_0/V}$, while for the dynamic case one has

$$\psi(x,t) = \psi_0 e^{-i\mu t/\hbar} \frac{1}{\cosh\left(\sqrt{2m\,|\mu|\,/\hbar^2}x\right)}, \tag{1.137}$$

with $\mu = U_0\,|\psi_0|^2\,/2$.

1.5 Phenomenological theories of superconductivity

1.5.1 The London equations

We begin with two alternative derivations of a pair of equations that are useful in describing many of the magnetic properties of superconductors. The older approach starts with the Drude–Lorentz equation of motion for electrons in a metal which is just Newton's law for the velocity, $\mathbf{v}$, of an electron with mass, m, and charge, e, in an electric field, $\mathbf{E}$, with a phenomenological viscous drag proportional to $\mathbf{v}/\tau$:

$$m(\dot{\mathbf{v}} + \frac{1}{\tau}\mathbf{v}) = e\mathbf{E}. \tag{1.138}$$

For a perfect conductor $\tau \to \infty$. Introducing the current density $\mathbf{j} = ne\mathbf{v}$, where n is the conduction electron density, Eq. (1.138) can be written

$$\frac{d\mathbf{j}}{dt} = \frac{ne^2}{m}\mathbf{E}, \tag{1.139}$$

which is referred to as the first London equation.

The time derivative of Maxwell's fourth equation is (in c.g.s. units)

$$\nabla \times \frac{\partial \mathbf{H}}{\partial t} = \frac{4\pi}{c}\frac{\partial \mathbf{j}}{\partial t} + \frac{\varepsilon}{c}\frac{\partial^2 \mathbf{E}}{\partial t^2}, \tag{1.140}$$

where ε is the host dielectric constant. Taking the curl of (1.140) and using (1.139) we have

$$\nabla \times \left(\nabla \times \frac{\partial \mathbf{H}}{\partial t}\right) = \left(\frac{4\pi ne^2}{mc} + \frac{\varepsilon}{c}\frac{\partial^2}{\partial t^2}\right)\nabla \times \mathbf{E}; \tag{1.141}$$

using $\nabla \times \mathbf{E} = -\frac{1}{c}\frac{\partial \mathbf{H}}{\partial t}$ yields

$$\nabla \times (\nabla \times \frac{\partial}{\partial t}\mathbf{H}) + \left(\frac{1}{\lambda_L^2} + \frac{\varepsilon}{c^2}\frac{\partial^2}{\partial t^2}\right)\frac{\partial}{\partial t}\mathbf{H} = 0 \tag{1.142}$$

where we have introduced the London depth, λ_L, defined by

$$\frac{1}{\lambda_L^2} = \frac{4\pi ne^2}{mc^2}. \tag{1.143}$$

Equation (1.142) has been obtained for a perfect conductor model. In order to conform with the experimentally observed Meissner effect, we must exclude time-independent field solutions arising from integrating (1.142) once with respect to time, and we therefore write

$$\nabla \times (\nabla \times \mathbf{H}) + \left(\frac{1}{\lambda_L^2} + \frac{\varepsilon}{c^2} \frac{\partial^2}{\partial t^2} \right) \mathbf{H} = 0; \tag{1.144}$$

this is referred to as the second London equation. In what follows we will refer to Eq. (1.144) simply as the London equation.

An alternative derivation of (1.144) is motivated by the idea that some of the moving electrons behave collectively as a superfluid, a liquid possessing no viscosity, a concept borrowed from the two fluid model of liquid ^{4}He (see Section 1.3.3). We start by assuming that the total free energy of a superfluid consists of three parts

$$F = F_N + E_{\text{kin}} + E_{\text{mag}}, \tag{1.145}$$

where F_N is the free energy associated with the normal liquid, E_{kin} is the kinetic energy of the moving superfluid, and E_{mag} is the magnetic field energy. We may write these latter two terms as

$$E_{\text{mag}} = \frac{1}{8\pi} \int H^2(\mathbf{r}) d^3r \tag{1.146}$$

and

$$E_{\text{kin}} = \frac{1}{2} \int \rho(\mathbf{r}) v^2(\mathbf{r}) d^3r, \tag{1.147}$$

where $\rho(\mathbf{r})$ is the mass density associated with the superfluid. Writing $\rho = nm$ and $\mathbf{v} = (1/ne)\mathbf{j}$ and using the fourth Maxwell equation $\nabla \times \mathbf{H} = (4\pi/c)j$, Eq. (1.147) becomes

$$E_{\text{kin}} = \frac{1}{8\pi} \int \frac{mc^2}{4\pi ne^2} (\nabla \times \mathbf{H})^2 \, d^3r; \tag{1.148}$$

n is now interpreted as the density of superconducting electrons. We will assume that the superconducting electrons adjust their motion so as to minimize the total free energy; this requires $\delta(E_{\text{mag}} + E_{\text{kin}}) = 0$ or

$$\int \left[\mathbf{H}(\mathbf{r}) \cdot \delta\mathbf{H}(\mathbf{r}) + \frac{mc^2}{4\pi ne^2} (\nabla \times \mathbf{H}(\mathbf{r})) \cdot (\nabla \times \delta\mathbf{H}(\mathbf{r})) \right] d^3r = 0, \tag{1.149}$$

where $\delta\mathbf{H}(\mathbf{r})$ is a variation of the (initially unknown) function $\mathbf{H}(\mathbf{r})$. Integrating the second term by parts (and placing the resulting surface outside the superconductor) we obtain

$$\int \left[\mathbf{H}(\mathbf{r}) + \lambda_L^2 \nabla \times (\nabla \times \mathbf{H}) \right] \delta\mathbf{H}(\mathbf{r}) d^3r = 0. \tag{1.150}$$

Since the variation $\delta\mathbf{H}(\mathbf{r})$ is arbitrary, the term in the square brackets must vanish; therefore

$$\nabla \times (\nabla \times \mathbf{H}) + \frac{1}{\lambda_L^2} \mathbf{H} = 0, \tag{1.151a}$$

which is equivalent to (1.144) (including the displacement term in Maxwell's equation yields the last term in (1.144), which is negligible for most applications).

Using the vector identity

$$\nabla \times (\nabla \times \mathbf{H}) = \nabla(\nabla \cdot \mathbf{H}) - \nabla^2 \mathbf{H},$$

we can write (1.151a) as

$$\nabla^2 \mathbf{H} - \frac{1}{\lambda_L^2} \mathbf{H} = 0. \tag{1.151b}$$

Another way to present (1.151a) is to write $\nabla \times \mathbf{H} = (4\pi/c)\mathbf{j}$ in the first term, and $\mathbf{H} = \nabla \times \mathbf{A}$ in the second term and work in a gauge where $\nabla \cdot \mathbf{A} = 0$ which yields the equation

$$\mathbf{j} = -\frac{ne^2}{mc} \mathbf{A}. \tag{1.152}$$

As a simple application of Eq. (1.151) we now discuss the behavior of a superconductor in a magnetic field near a plane boundary. Consider first the case of a field perpendicular to a superconductor surface lying in the x–y plane with no current flowing in the z-direction. From the second Maxwell equation, $\nabla \cdot \mathbf{H} = 0$, we obtain $\partial H_z/\partial z = 0$ or $H = \text{const}$. From the fourth Maxwell equation, $\nabla \times \mathbf{H} = (4\pi/c)\mathbf{j}$, the first term in (1.151a) vanishes and hence $\mathbf{H} = 0$ is the only solution. Thus a superconductor exhibiting the Meissner effect cannot have a field component perpendicular to its surface.

As the second example consider a field lying parallel to the superconductor surface, e.g. $\mathbf{H} || \hat{\mathbf{x}}$ which we may write as $\mathbf{H} = H(z)\hat{\mathbf{x}}$ (which satisfies $\nabla \cdot \mathbf{H} = 0$). Eq. (1.151b) is then

$$\frac{\partial^2 H_x}{\partial z^2} - \frac{1}{\lambda_L^2} H_x = 0 \tag{1.153}$$

or

$$\mathbf{H}_x(z) = \hat{\mathbf{x}} H_x(0) e^{-z/\lambda_L}. \tag{1.154}$$

A field parallel to the surface is therefore allowed; however it decays exponentially, with a characteristic length, λ_L, in the interior. The length $\lambda_L(T = 0)$ ranges from 500–10,000 Å, depending on the material. Accompanying this parallel field is a surface current density, which, from Maxwell's fourth equation, is

$$\mathbf{j}(z) = \frac{-\lambda_L c}{4\pi} H_x(0) e^{-\lambda_L/z} \hat{\mathbf{y}} \tag{1.155}$$

This current density shields or screens the magnetic field from the interior of the superconductor.

1.5.2 Normal-metal/superconducting interface energy: type I and type II superconductors

A London superconductor has the unusual property that the surface energy separating the normal and superconducting phases is *negative*. For a liquid vapor system this would be equivalent to a negative surface tension. Thermodynamically the latter system would then prefer to break up into many small droplets so as to maximally take advantage of the free energy gained by forming additional surfaces. An analogous process can happen in a superconductor, although the entities that form are *flux lines* rather than droplets, as we discuss below in Section 1.5.3.

A semi infinite superconducting slab occupying the half-space $x > 0$ in a uniform tangential magnetic field will expel that field up to the critical value, H_c, above which it would penetrate. The change in free energy per unit volume, F, on penetrating the superconductor would then be $H_c^2(T)/8\pi$ (neglecting any paramagnetic contribution), which must be equal to the "condensation" energy density, F_S, of the superconducting state in the absence of that field. Including the region governed by the London penetration phenomena we would write the total free energy F in the superconductor as

$$F = A\int_0^\infty dx\, F(\mathbf{r}) = A\int_0^\infty dx\left[F_S + \frac{\mathbf{H}^2(\mathbf{r})}{8\pi} + \frac{\lambda_L^2}{8\pi}(\nabla\times\mathbf{H}(\mathbf{r}))^2\right], \quad (1.156)$$

where A is the interface area, and the second and third terms are the magnetic field energy density and the superfluid electron-kinetic energy density, respectively. Actually $\mathbf{H}(\mathbf{r})$, being the response of the media to the constant external field $\mathbf{H}_c$, is more properly viewed as the magnetic induction, $\mathbf{B}(\mathbf{r})$, and when the latter is the independent variable, the equilibrium condition involves the magnetic Gibbs free energy; i.e. we must have

$$G_N = G_S. \quad (1.157)$$

By definition

$$G = F - \frac{A}{4\pi}\int_0^\infty \mathbf{H}_c\cdot\mathbf{B}(x)dx. \quad (1.158)$$

where we set $\mathbf{B}(x) = \mathbf{H}(x)$; from Eq. (1.154) we have $\mathbf{H}(x) = H_c\hat{\mathbf{z}}e^{-x/\lambda_L}$. Inserting this form in the combined Eqs. (1.158) and (1.156) we obtain

$$G = A\int_0^\infty F_S\, dx + \gamma A \quad (1.159)$$

where γ is the surface energy per unit area, i.e. the "surface tension", given by

$$\begin{aligned}\gamma &= \frac{H_c^2}{4\pi}\left(-\frac{\lambda_L}{2}e^{-2x/\lambda_L} + \lambda_L e^{-x/\lambda_L}\right)\Bigg|_0^\infty \\ &= -\lambda_L\frac{H_c^2}{8\pi},\end{aligned} \quad (1.160)$$

which as noted earlier is negative.

It turns out that there are two kinds of superconductors, those with positive surface energy, called type I, and those with negative surface energy, called type II. Physically a positive contribution to the surface energy, arises because superconductivity is destroyed over a region of order ξ, called the coherence length (see Section 1.5.7), at a normal-metal/superconductor interface; i.e. we lose the superconducting *condensation energy* over a volume of order $A\xi$. This is equivalent to a positive contribution to the surface energy of order

$$\gamma \cong \xi \frac{H_c^2}{8\pi}. \tag{1.161}$$

So in a type I material $\xi > \lambda_L$ and hence the positive contribution (1.160) outweighs the negative contribution (1.160) and the interface is stable. For a type II material the system does, in some sense, try to *maximize* the amount of internal interfacial area above some field (referred to as the lower critical field); however it is subject to a constraint imposed by quantum mechanics, as we discuss in the next section.

We will not enter a discussion of the magnetic properties of the type II materials, but simply summarize the qualitative behavior.[28] It turns out that they show a Meissner effect up to a so-called lower critical field, H_{c1}, where flux lines first enter the sample. This field can be accurately calculated from the interaction energy between vortices. As the field is further increased the density of flux lines increases up to an upper critical field, H_{c2}, where the (generally triangular) lattice spacing is of order ξ, and above which superconductivity disappears. H_{c2} follows directly from the lowest eigenvalue of the linearized G–L equation.

[28]For a nice discussion of the magnetic properties of type II superconductors see in [34].

1.5.3 Quantized flux lines in the London theory

The previous discussion of the surface energy of a normal-metal/ superconductor interface suggests that type II materials, where $\xi \lesssim \lambda$, are unstable to the formation of domain structures which in some way maximize the amount of interface area. Two possible domain geometries are: (i) an array of nested sheets (closed or open, depending on the geometry)[29] and (ii) a two-dimensional lattice of flux filaments. Calculations show the latter domain structure to be more stable.

[29]For a superconducting slab we envision an array of interfaces parallel to the surface, and for a cylindrical sample an array of coaxial cylinders. Other shapes would have more complex structures.

Since the filaments (are presumed to) admit flux into the interior of the superconductor, we envision them as having a normal core with a diameter of order ξ, outside of which super-currents flow in a diameter of order λ, which produce the internal field via Ampere's law. As a primitive model of a single flux filament we consider the extreme limit $\xi \to 0$ for which the London approach should provide a good description. We recall Eq. (1.142) associated with our first derivation of the London equation

$$\frac{\partial}{\partial t}\left[\nabla \times (\nabla \times \mathbf{H}) + \frac{4\pi n e^2}{mc^2}\mathbf{H}\right] = 0. \tag{1.162}$$

We next integrate (1.162) over an area A intersecting the filament (for convenience we choose a plane perpendicular to its axis) and use Ampere's law (Maxwell's fourth equation) in the form

$$\frac{\partial}{\partial t}\int\left[\frac{mc}{ne^2}\nabla\times\mathbf{j}+\mathbf{H}\right]\cdot d^2\mathbf{r}=0. \tag{1.163}$$

In integrating (1.163) with respect to time, we now allow the possibility of a non-zero constant of integration (since the flux filament phenomena violate the Meissner behavior); thus

$$\int\left[\frac{mc}{ne^2}\nabla\times\mathbf{j}+\mathbf{H}\right]\cdot d^2\mathbf{r}=\phi. \tag{1.164}$$

Applying Stokes law to the first term in (1.164), yields

$$\frac{mc}{ne^2}\oint\mathbf{j}\cdot dl++\int d^2\mathbf{r}\cdot\mathbf{H}=\phi. \tag{1.165}$$

If we choose the contour to enclose a large area, we may expect the first term to be exponentially small (since the currents fall off exponentially with a characteristic length λ); we can then identify the constant of integration, ϕ, as the total flux contained within the filament (most of which also falls inside a radius of order λ).

To gain further insight we substitute $\mathbf{H}=\nabla\times\mathbf{A}$ into (1.165) and again apply Stokes' law to obtain

$$\oint\left[\frac{mc}{ne^2}\mathbf{j}+\mathbf{A}\right]\cdot dl=\phi; \tag{1.166a}$$

writing $\mathbf{j}=ne\mathbf{v}=ne\frac{\mathbf{p}}{m}$, Eq. (1.166a) becomes

$$\oint\left[\mathbf{p}+\frac{e}{c}\mathbf{A}\right]\cdot dl=\frac{e}{c}\phi. \tag{1.166b}$$

We identify the integrand as the canonical momentum, sometimes denoted as **P**, associated with the motion of a charged particle in the Hamiltonian formulation of mechanics. F. London correctly concluded that superconductivity was a macroscopic quantum phenomenon and, guided by this insight, suggested that (1.166) must conform to the Bohr–Sommerfeld quantization rule for the (quasiclassical) motion of an electron; i.e. $(|e|/c)\phi=nh$ or $\phi=nhc/|e|$, where n is an integer. However this assumes the orbiting entities are single electrons; Ginzburg and Landau allowed for a more general case where $e\to e^*$; we then have

$$\varphi=n\varphi_0 \tag{1.167a}$$

where

$$\phi_0=\frac{hc}{|e^*|}. \tag{1.167b}$$

From the BCS theory it is known that $e^* = 2e$; i.e.

$$\phi_0 = \frac{hc}{2|e|}$$
$$= 2.07 \times 10^{-7} \text{ Gauss cm}^2, \tag{1.167c}$$

which is referred to as the flux quantum. Hence flux enters a type II superconductor as an array of quantizied flux filaments; the lowest energy situation corresponds to singly quantized ($n = 1$) filaments each carrying a flux quantum ϕ_0.

We next examine Eq. (1.166) for a contour of radius $\lambda >> r >> \xi$; the amount of flux contained is then vanishingly small and the first term in (1.166b) dominates, yielding the condition $2\pi pr = nh$ or $p = n\hbar/r$; BCS theory also dictates that the mass of the orbiting entity is $m^* = 2m$; hence

$$\mathbf{v}(\mathbf{r}) = \frac{\hbar}{2mr}\hat{\boldsymbol{\theta}}, \tag{1.168}$$

where $\hat{\boldsymbol{\theta}}$ is an azimuthal unit vector. This velocity profile corresponds to the large r behavior of a vortex in a fluid, although here the vorticity is quantized. One then refers to the filaments as *quantized vortex lines* or *vortex lines* for short; note the similarity to what was found in superfluid ^{4}He (see Eq. (1.114)).

Vorticity involves a non-vanishing curl of the velocity; i.e. $\nabla \times \mathbf{v} \neq 0$; but the curl of $\mathbf{v}$ in (1.168) vanishes for $r \neq 0$. However the circulation $\Gamma \equiv \oint \mathbf{v} \cdot d\boldsymbol{l} \neq 0$ and thus all the vorticity is located in an infinitesimal region near the origin. Physically it would be spread out over a coherence length, ξ. In our $\xi \to 0$ model we can obtain an approximate description by adding a singular *source term* to the London equation in the form,

$$\mathbf{H} + \lambda_L^2 \nabla \times (\nabla \times \mathbf{H}) = \phi_0 \hat{\mathbf{z}} \delta^{(2)}(\mathbf{r}), \tag{1.169}$$

where $\delta^{(2)}(\mathbf{r})$ is a two-dimensional δ function and $\hat{\mathbf{z}}$ is a unit vector (parallel to the vortex axis). Equation (1.169) may be written in cylindrical coordinates as

$$H_z - \frac{\lambda_L^2}{r}\frac{d}{dr}\left(r\frac{dH_z}{dr}\right) = \phi_0 \delta^{(2)}(\mathbf{r}). \tag{1.170}$$

The left side of (1.170) is a special case of Bessel's equation[30] and the solution having the required singular behavior near $r = 0$ is

[30] Here we adopt the definitions of Abramowitz and Stegun, (1970), Dover Publications, New York, Pg. 374 ff.

$$H_z = \frac{\phi_0}{2\pi\lambda_L^2} K_0\left(\frac{r}{\lambda_L}\right) \tag{1.171}$$

where K_0 is the zeroth order modified Bessel function of an imaginary argument. From Ampere's law

$$\mathbf{j} = \frac{c}{4\pi}\nabla \times \mathbf{H} = -\frac{c}{4\pi}\frac{dH_z}{dr}\hat{\boldsymbol{\theta}}. \tag{1.172}$$

For small x, $K_0(x) \cong \ln(1/x)$; hence

$$j_\theta = \frac{c}{4\pi} \cdot \frac{\phi_0}{2\pi\lambda_L^2 r} = \frac{neh}{4\pi mr} \tag{1.173}$$

or $v = \hbar/(2mr)$, in agreement with (1.168). For large x we use the form,

$$K_0(x) \cong \left(\frac{\pi}{2x}\right)^{1/2} e^{-x} \tag{1.174}$$

or

$$H_z \cong \frac{\phi_0}{2\pi\lambda_L^2}\left(\frac{\pi\lambda_L}{2r}\right)^{1/2} e^{-r/\lambda_L} \tag{1.175}$$

and the field drops off exponentially, as argued above.

The energy of a vortex line follows from Eqs. (1.145)–(1.148),

$$E = \frac{L}{8\pi}\int \left[\mathbf{H}^2(\mathbf{r}) + \lambda_L^2(\nabla \times \mathbf{H})^2\right] d^2r \tag{1.176}$$

where L is the length of the vortex line. We integrate (1.176) by parts to obtain

$$\frac{E}{L} = \frac{\lambda_L^2}{8\pi}\oint \mathbf{H} \times (\nabla \times \mathbf{H}) \cdot dl + \frac{1}{8\pi}\int \mathbf{H} \cdot \left[\mathbf{H} + \lambda_L^2 \nabla \times (\nabla \times \mathbf{H})\right] d^2r. \tag{1.177}$$

The quantity in square brackets in the second term of (1.177) is equal to the left side (1.173). Were we to simply replace it by $\delta^{(2)}(\mathbf{r})$ and integrate we would obtain $H(r \to 0)$, which is logarithmically divergent. This indicates that our simple model for the vortex core is not sufficiently accurate to evaluate the energy. To avoid the divergence problem we assume that H is finite everywhere, thus eliminating $\delta^{(2)}(\mathbf{r})$ from the right side of (1.173), which results in the second term in (1.177) vanishing. We must still account for the energy in the core of the vortex, however, which we do by separating the line integral in (1.176) into three parts: a circle at a very large radius (which makes a vanishingly small contribution due to the exponential fall-off of $\mathbf{H}(\mathbf{r})$) at large r), two counter traversing radial paths from the outer circle to an inner circle of a very small radius (which cancel each other), and finally a path around the inner circle (which makes the only non-vanishing contribution). For small r the first term in (1.177) is

$$\frac{E}{L} = \frac{\lambda_L^2}{8\pi}\oint \frac{\phi_0^2}{(2\pi\lambda_L^2)^2}\frac{1}{r}\ln\left(\frac{\lambda_L}{r}\right) dl \tag{1.178}$$

or

$$\frac{E}{L} = \frac{\phi_0^2}{(4\pi\lambda_L)^2}\ln\frac{\lambda_L}{\xi}, \tag{1.179}$$

here we choose a radius $r = \xi$ for the (inner) line integral; this corresponds to the physically reasonable assumption that the field divergence is removed at the coherence length scale (as a more complete theory confirms). We expect that the $1/r$ divergence of the superfluid velocity ultimately destroys superconductivity in the vortex core. By extending this treatment one can calculate the interaction energy between two vortices.

Rather than using the London theory, one can numerically integrate the Ginzburg–Landau equations to obtain $|\psi(r)|$. The qualitative behavior of $H(r)$ and $|\psi(r)|$ are shown in Fig. 1.5.

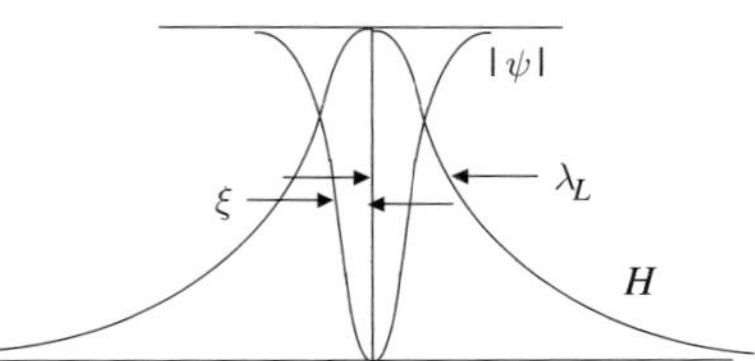

Fig. 1.5 Schematic representation of an isolated Abrikosov vortex line showing the radial dependence of the order parameter, $\psi(r)$, and the magnetic field, $H(r)$, in the limit $\xi << \lambda_L$.

1.5.4 The Ginzburg–Landau theory

Our earlier discussion has brought out the idea that superconductivity is some kind of macroscopic quantum state. Ginzburg and Landau built this idea into the Landau second-order phase transition theory by assuming the existence of a macroscopic "wave function", ψ, which they took as the order parameter associated with superconductivity. Since wave functions can be complex, only the form $\psi\psi^*$ may enter the free energy expansion; we therefore write

$$\mathrm{F} = \mathrm{F(T)} + \alpha|\psi|^2 + \frac{1}{2}\beta|\psi|^4. \tag{1.180}$$

Minimizing with respect to $|\psi|$ yields

$$|\psi| = 0 \qquad T > T_c \tag{1.181a}$$

$$|\psi| = \left[a\frac{(T_c - T)}{\beta}\right]^{1/2}. \qquad T < T_c \tag{1.181b}$$

To describe situations where the superconducting state is inhomogeneous we must generalize (1.180). Taking F in (1.180) to be the free-energy density we can write

$$\begin{aligned} F &= F_0 + \int \mathrm{F}(\mathbf{r})d^3r \\ &= F_0 + \int d^3r\left[\alpha\,|\psi(\mathbf{r})|^2 + \frac{1}{2}\beta\,|\psi(\mathbf{r})|^4\right]; \end{aligned} \tag{1.182}$$

F is now the *total* free energy. Equation (1.182) in its present form does not model the increase in energy associated with a spatial distortion of the order parameter, required to model effects associated with a coherence length, ξ. To account for such effects Ginzburg and Landau added a "gradient energy" term to (1.182) of the form

$$F_G = \int \frac{\hbar^2}{2m^*}\,|\nabla\psi(\mathbf{r})|^2\,d^3r \tag{1.183}$$

with m^* as a parameter; the choice of the coefficient $\hbar^2/2m^*$ makes (1.183) mimic the quantum mechanical kinetic energy (introduced earlier in Eq. (1.147)).

Ginzburg and Landau went further and assumed that if (1.183) was to be regarded as the kinetic energy contribution to the Hamiltonian density of the superconducting electrons, then as in quantum mechanics the interaction of the electrons with an electromagnetic field would be accomplished by the Hamiltonian prescription

$$\nabla \to \nabla - \frac{ie^*}{\hbar c}\mathbf{A}; \tag{1.184}$$

the use of e^* allows the superconducting entities to carry a different charge ($e^* = 2e$ in BCS theory). Combining the above we have

$$F_G = \int d^3r\frac{\hbar^2}{2m^*}\left|\left(\nabla - \frac{ie^*}{\hbar c}\mathbf{A}(\mathbf{r})\right)\psi(\mathbf{r})\right|^2. \tag{1.185}$$

Finally we must add the contribution of the magnetic field to the energy density

$$F_H = \frac{1}{8\pi} H^2(\mathbf{r}).$$

Combining the above we have

$$\begin{aligned} F = F_0 + \int d^3r \Big[\alpha \, |\psi(\mathbf{r})|^2 + \frac{1}{2}\beta \, |\psi(\mathbf{r})|^4 \\ + \frac{\hbar^2}{2m^*} \left| \left(\nabla - \frac{ie^*}{\hbar c} A(\mathbf{r}) \right) \psi(\mathbf{r}) \right|^2 + \frac{1}{8\pi} H^2(\mathbf{r}) \Big]. \end{aligned} \tag{1.186}$$

The minimization of (1.186) must be carried out using the methods of the calculus of variations since F is a functional involving the free energy density,

$$F \left\{ \psi(\mathbf{r}), \psi^*(\mathbf{r}), \nabla\psi(\mathbf{r}), \nabla\psi^*(\mathbf{r}), \mathbf{H}(\mathbf{r}) \right\},$$

which in turn involves the unknown functions $\psi(\mathbf{r})$, $\psi^*(\mathbf{r})$, and $\mathbf{H}(\mathbf{r})$. Minimizing (1.186) with respect to $\psi^*(\mathbf{r})$ yields[31]

$$\begin{aligned} \delta F = \int d^3r \left[-\frac{\hbar^2}{2m^*} \left(\nabla - \frac{ie^*}{\hbar c} A(\mathbf{r}) \right)^2 \psi(\mathbf{r}) + \alpha\psi(\mathbf{r}) + \beta \, |\psi(\mathbf{r})|^2 \, \psi(\mathbf{r}) \right] \delta\psi^*(\mathbf{r}) \\ + \int d^2\mathbf{r} \cdot \frac{\hbar^2}{2m^*} \left(\nabla - \frac{ie^*}{\hbar c} A(\mathbf{r}) \right) \psi(\mathbf{r}) \delta\psi^*(\mathbf{r}) \end{aligned} \tag{1.187}$$

(variation with respect to ψ, which is an independent variable, yields the complex conjugate of (1.187)). To minimize F we set the integrand of the first part of (1.187) to zero; this yields the first Ginzburg–Landau equation

$$-\frac{\hbar^2}{2m^*} \left(\nabla - \frac{ie^*}{\hbar c} A(\mathbf{r}) \right)^2 \psi(\mathbf{r}) + \alpha\psi(\mathbf{r}) + \beta |\psi(\mathbf{r})|^2 \psi(\mathbf{r}) = 0. \tag{1.188}$$

The surface term (which was generated by an integration by parts) can be used (with caution) to establish certain boundary conditions.

Variation of (1.186) with respect to $\mathbf{A}$ (with $\mathbf{H} = \nabla \times \mathbf{A}(\mathbf{r})$) yields Ampere's law

$$\nabla \times \mathbf{H}(\mathbf{r}) = \frac{4\pi}{c} \mathbf{j}(\mathbf{r}) \tag{1.189}$$

provided we identify $\mathbf{j}$ as

$$\mathbf{j}(\mathbf{r}) = \frac{e^*}{2m^*} \left[\psi^*(\mathbf{r}) \left(-i\hbar\nabla - \frac{e^*}{c} \mathbf{A}(\mathbf{r}) \right) \psi(\mathbf{r}) + \psi(\mathbf{r}) \left(+i\hbar\nabla - \frac{e^*}{c} \mathbf{A}(\mathbf{r}) \right) \psi^*(\mathbf{r}) \right], \tag{1.190a}$$

or equivalently,

$$\mathbf{j}(\mathbf{r}) = \frac{-ie^*\hbar}{2m^*} \left(\psi^*(\mathbf{r}) \nabla\psi(\mathbf{r}) - \psi(\mathbf{r}) \nabla\psi^*(\mathbf{r}) \right) - \frac{e*^2}{m^*c} |\psi(\mathbf{r})|^2 \mathbf{A}(\mathbf{r}). \tag{1.190b}$$

[31] The variation of F with respect to ψ^* involves writing

$$\delta F = \int d^3r \left[\frac{\partial F}{\partial \psi^*} \delta\psi^* + \sum_{i=1}^{3} \frac{\partial F}{\partial(\partial\psi^*/\partial x_i)} \delta \left(\frac{\partial \psi^*}{\partial x_i} \right) \right]$$

and then performing an integration by parts on the second term which in turn generates a surface term.

Equation (1.190) is the second GL equation; we note that (1.190) is the same as the expression for the current density in quantum mechanics. Note the current density satisfies the equation $\mathbf{j}(\mathbf{r}) = c\delta F/\delta \mathbf{A}(\mathbf{r})$; i.e. it is the variable conjugate to $\mathbf{A}(\mathbf{r})$.

1.5.5 Gauge invariance

The simplest solution of (1.188) is the case of a uniform superconductor, $\psi \neq \psi(\mathbf{r})$, with $\mathbf{A} = 0$. However (1.188) possesses a continuum of other solutions having the same free energy, as we now show. As with any complex function, we can write $\psi(\mathbf{r}) = |\psi(\mathbf{r})|\, e^{i\Phi(\mathbf{r})}$, where $|\psi(\mathbf{r})|$ and $\Phi(\mathbf{r})$ are the position dependent amplitude and phase respectively. Let us examine a class of solutions which satisfy the complex equation

$$\left(\nabla - \frac{ie^*}{\hbar c}\mathbf{A}(\mathbf{r})\right)\psi(\mathbf{r}) = 0 \tag{1.191}$$

which is equivalent to the two real equations

$$\nabla\,|\psi(\mathbf{r})| = 0 \tag{1.192a}$$

and

$$\left[\nabla\Phi - \frac{e^*}{\hbar c}\mathbf{A}\right] = 0. \tag{1.192b}$$

From (1.192a) we see that the only allowed solutions of (1.191) involve a constant amplitude, $|\psi(\mathbf{r})| = const.$; Eq. (1.192b), on the other hand, has infinitely many solutions involving a vector potential and a position dependent phase (which does not affect the free energy), related by

$$\mathbf{A} = \frac{\hbar c}{e^*}\nabla\Phi. \tag{1.193}$$

Any vector potential satisfying (1.193) results in a uniform free energy and (on substituting (1.191) into (1.190)) a vanishing current density (note that $\mathbf{H} = \nabla \times \mathbf{A} = 0$ for all $\mathbf{A}$ of the form (1.193)).

The above exercise shows that the symmetry broken in superconductivity is *gauge symmetry,* or equivalently, *phase symmetry*. Superconductors having different phase functions, $\Phi(\mathbf{r})$, are in a real sense physically distinct; this arbitrariness of the phase is the analog for a superconductor of the property that the magnetization may point in any direction in an isotropic (liquid) ferromagnet.

1.5.6 The Josephson effects

Recalling Eq. (1.190b) for the G–L current density, and writing ψ in the form $|\psi(\mathbf{r})|\, e^{i\Phi(\mathbf{r})}$ we obtain

$$\mathbf{j}(\mathbf{r}) = \frac{e^*\hbar}{m^*}\,|\psi(\mathbf{r})|^2\left[\nabla\Phi(\mathbf{r}) - \frac{e^*}{\hbar c}\mathbf{A}(\mathbf{r})\right]; \tag{1.194}$$

i.e. the current in a superconductor involves the gradient of the (gauge invariant) phase (recall that real G–L wave functions carry no current). If the form

$$\left[\nabla - \frac{ie^*}{\hbar c}\mathbf{A}(\mathbf{r})\right]\psi(\mathbf{r}) = 0 \tag{1.195}$$

in Eq. (1.187) vanishes, it follows from Eq. (1.190b) that the super current will also vanish.

We now consider the interesting case where two superconductors, 1 and 2, are joined by a "junction" through which superconducting electrons can tunnel (see Fig. 1.6); this typically involves a thin insulating layer. To treat such a system we must modify Eq. (1.195). Taking the surface normals, $\hat{\mathbf{n}}$, for superconductors 1 and 2 parallel to $\pm\hat{\mathbf{x}}$, we will adopt the more general boundary conditions

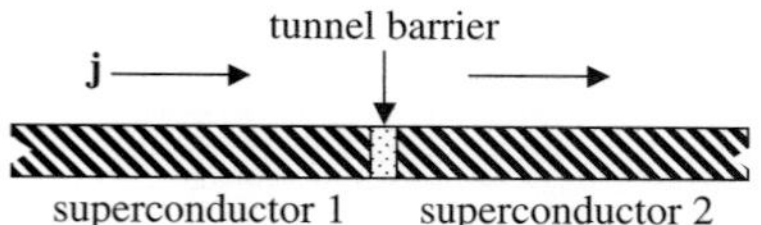

Fig. 1.6 Schematic representation of a Josephson junction.

$$\frac{\partial\psi_1}{\partial x} - \frac{ie^*}{\hbar c}A_x\psi_1 = \frac{\psi_2}{\lambda} \tag{1.196a}$$

and

$$\frac{\partial\psi_2}{\partial x} - \frac{ie^*}{\hbar c}A_x\psi_2 = -\frac{\psi_1}{\lambda}; \tag{1.196b}$$

here the parameter λ, which measures the coupling strength and has units of length, is a property of the junction itself and is taken as being the same in these two equations. Inserting Eq. (1.196a) and its complex conjugate into Eq. (1.190b) yields

$$\begin{aligned} j_x &= -\frac{ie^*\hbar}{2m^*}\left[\psi_1^*\frac{\partial\psi_1}{\partial x} - \psi_1\frac{\partial\psi_1^*}{\partial x}\right] - \frac{e^{*2}}{m^*c}|\psi_1|^2 A_x \\ &= -\frac{ie^*\hbar}{2m^*}\left[\psi_1^*\left(\frac{\psi_2}{\lambda} + \frac{ie^*}{\hbar c}A_x\psi_1\right) - \psi_1\left(\frac{\psi_2^*}{\lambda^*} - \frac{ie^*}{\hbar c}A_x\psi_1^*\right)\right] - \frac{e^{*2}}{m^*c}|\psi_1|^2 A_x \\ &= -\frac{ie^*\hbar}{2m^*}\left[\frac{\psi_1^*\psi_2}{\lambda} - \frac{\psi_1\psi_2^*}{\lambda^*}\right]. \end{aligned} \tag{1.197}$$

In the absence of magnetic atoms, the superconducting properties are invariant under time reversal, which results in $\psi \to \psi^*$, $\mathbf{j} \to -\mathbf{j}$, and $\mathbf{A} \to -\mathbf{A}$. Under these operations both sides of Eqs. (1.196a,b) turn into their complex conjugates and hence λ must be real. We then obtain from (1.197)

$$j_x = \frac{-ie^*\hbar}{2m^*\lambda}\left(\psi_1^*\psi_2 - \psi_1\psi_2^*\right). \tag{1.198}$$

Writing $\psi_i = |\psi_i|e^{i\Phi_i}$, and assuming both sides are prepared from the same kind of superconducting material, $|\psi_1| = |\psi_2|$, we have

$$j = j_m \sin\Phi_{21} \tag{1.199}$$

where

$$j_m = \frac{e^*\hbar}{m^*\lambda}|\psi|^2 \tag{1.200a}$$

and

$$\Phi_{21} = \Phi_2 - \Phi_1. \tag{1.200b}$$

Note j_m is the maximum super current density that may be carried by the junction.

In deriving Eq. (1.199) we have assumed there are no electric or magnetic fields in the junction. To generalize to the case when fields are present we will use gauge invariance arguments.[32] Under a gauge transformation the vector and scalar potentials change as follows:

[32]The arguments we will give here are supported by microscopic calculations based on BCS theory, as first given by Josephson; the predictions were somewhat controversial prior to the experimental observations.

$$\mathbf{A} \to \mathbf{A} + \nabla\chi \tag{1.201}$$

and

$$V \to V - \frac{1}{c}\frac{\partial \chi}{\partial t} \tag{1.202}$$

where $\chi(\mathbf{r}, t)$ is an arbitrary single-valued function. (Recall $\mathbf{H} = \nabla \times \mathbf{A}$ and $\mathbf{E} = -\nabla V - \partial\mathbf{A}/c\partial t$ and hence the fields themselves are not altered under these transformations.)

Recalling Eq. (1.193), $\mathbf{A}(\mathbf{r}, t) = (\hbar c/e^*)\nabla\Phi(\mathbf{r}, t)$, suppose we introduce a gauge function, $\chi(\mathbf{r}, t)$ and transform $\mathbf{A}(\mathbf{r}, t)$ according to (1.201). If (1.193) is still to be satisfied, we must simultaneously change the phase of the G–L wave function according to

$$\Phi(\mathbf{r}, t) \to \Phi(\mathbf{r}, t) + \frac{e^*}{\hbar c}\chi(\mathbf{r}, t). \tag{1.203}$$

Comparing (1.203) with (1.204) we see that gauge invariance of $V(\mathbf{r}, t)$ requires

$$\frac{\partial\Phi(\mathbf{r}, t)}{\partial t} + \frac{e^*}{\hbar}V(\mathbf{r}, t) = 0. \tag{1.204}$$

Assuming we have a constant potential difference V_{21} across the junction, and an accompanying phase difference $\Phi_{21}(t)$, then integration of (1.204) yields

$$\Phi_{21} = \Phi_{21}^{(0)} - \frac{e^*}{\hbar}V_{21}t; \tag{1.205}$$

accompanying this time-dependent phase is an *oscillating supercurrent*

$$j(t) = j_m \sin\left[\Phi_{21}^{(0)} - \frac{e^*}{\hbar}V_{21}t\right]. \tag{1.206}$$

Introducing the frequency $\omega_J = \partial\Phi_{21}/\partial t$ we see that (1.205) leads to

$$\begin{aligned} \omega_J &= \frac{|e^*|}{\hbar}V_{21} \\ &= \frac{2|e|}{\hbar}V_{21}. \end{aligned} \tag{1.207}$$

The oscillating supecurrent current $j(t)$ caused by the static voltage V_{21} will be associated with an oscillating voltage which will be superimposed on the static voltage. Note that in order to have a constant voltage V_{21} across the junction we must have an accompanying static current which exceeds j_m; this results in dissipation that accompanies the oscillating supercurrent.

We next examine the Josephson effects in the presence of a magnetic field. We will restrict ourselves to the case of a relatively weak field where a quasiclassical description is adequate; i.e. the dominant effect of a magnetic field, which is described by a vector potential **A**, is to make the phase position dependent. From the discussion surrounding Eq. (1.192a,b) we know that for a "pure" gauge field (one derived from a gauge function $\chi(\mathbf{r}, t)$ and not involving a field $\mathbf{H}(\mathbf{r})$) the *only* effect of the vector potential is to produce a position-dependent phase; this suggests that the vector potential associated with a *weak* field, $\mathbf{H}(\mathbf{r})$, may be approximately accounted for by incorporating a position-dependent phase in the wave function. Comparing Eqs. (1.201) and (1.203) we have the gauge invariant form analogous to (1.204) as

$$\nabla \Phi - \frac{e^*}{\hbar c}\mathbf{A}(\mathbf{r}) = 0. \tag{1.208}$$

Therefore, the gauge-invariant phase difference is given by

$$\Phi_{21} = \Phi_{21}^{(0)} + \frac{2\pi}{\phi_0}\int_1^2 \mathbf{A} \cdot dl. \tag{1.209}$$

where $\phi_0 \equiv (hc/2e)$ is called the flux quantum.

The fundamental equations governing the behavior of Josephson junctions are the current–phase relation (1.199), the voltage–phase relation (1.205), and the gauge-invariant phase relation (1.209). They are believed to be exact. Equation (1.207) is now used as the basis for defining the standard volt in terms of a measured frequency and the fundamental constants, e and h (Taylor, Parker, and Langenberg (1969)).

1.5.7 Boundaries

In discussion the surface tension associated with a normal-metal/superconductor boundary in Section 1.5.2. we introduced the idea of a coherence length ξ. Such a length arises naturally in the Ginzburg–Landau theory as we now demonstrate. Here we limit ourselves to the case involving an inhomogeneous order parameter that is generated by the presence of a boundary, in the absence of a magnetic field. Assume we have a superconducting half-space occupying the region $x > 0$. We further assume that the order parameter is driven to zero at this interface. Experimentally this can be accomplished by coating the surface of the superconductor with a film of ferromagnetic material.[33] We then seek a solution to the one-dimensional Ginzburg–Landau equation

$$\frac{\hbar^2}{2m^*}\frac{d^2\psi}{dx^2} + \alpha\psi + \beta\psi^3 = 0. \tag{1.210}$$

Noting α is negative in the superconducting state ($\alpha = -|\alpha|$), defining the Ginzburg–Landau coherence length as

$$\xi^2 \equiv \frac{\hbar^2}{2m^* |\alpha|}, \tag{1.211}$$

[33] Paramagnetic impurities (those bearing a spin in a host material) or interfaces with a ferromagnetic metal strongly depress superconductivity. A normal metal interface has a much smaller effect and an insulator or vacuum has a negligible effect for most purposes.

and writing $\frac{\beta}{|\alpha|}\psi^2 = f^2$ we may rewrite (1.210) as

$$-\xi^2 f'' - f + f^3 = 0. \tag{1.212}$$

Multiplying by f′ we may rewrite (1.212) as

$$\frac{d}{dx}\left[-\frac{\xi^2 f'^2}{2} - \frac{1}{2}f^2 + \frac{1}{4}f^4\right] = 0; \tag{1.213}$$

hence the quantity in square brackets must be a constant. Far from the boundary, $f' = 0$ and $f^2 = 1$ (equivalent to $\psi^2 = |\alpha|/\beta$); then (1.213) becomes

$$\xi^2 (f')^2 = \frac{1}{2}(1 - f^2)^2, \tag{1.214}$$

which has the solution $f = \tanh\left[x/(\sqrt{2}\xi)\right]$ or

$$\psi = \left(\frac{|\alpha|}{\beta}\right)^{1/2} \tanh\left(\frac{x}{\sqrt{2}\xi}\right). \tag{1.215}$$

From (1.215) we see that ξ is a measure of distance over which the order parameter responds to a perturbation. Since $\alpha = a(T - T_c)$ we have

$$\xi(T) = \left(\frac{\hbar^2}{2m^* a T_c}\right)^{1/2} \left(1 - \frac{T}{T_c}\right)^{-1/2}. \tag{1.216}$$

We see that the G–L coherence length diverges as $(1 - T/T_c)^{-1/2}$; this divergence is a general property of the coherence length at all second-order phase transitions (although the exponent differs in general from this "mean field" value of 1/2 close to T_c).

1.6 The BCS theory of superconductivity

Superconductivity remained unexplained from its discovery in 1911 up to 1957 when the first satisfactory microscopic theory was provided by Bardeen, Cooper, and Schrieffer (BCS) [35]. Part of this delay was inevitable since the theory is fully quantum mechanical and, in part, rested on progressive developments in many-body theory. As might be expected after such a long gestation period the final form the theory took was rather subtle.

1.6.1 Pairing

In beginning our discussion of superconductivity we will be motivated by Bogoliubov's earlier treatment of the interacting Bose gas, discussed in Section 1.2, and the emergence of a non vanishing macroscopic expectation value for the operator associated with the creation of particles in the $\mathbf{k} = 0$ state, the so-called condensate; i.e. $\left\langle \hat{a}^\dagger_{\mathbf{k}=0} \right\rangle \neq 0$. In Section 1.4.1 we showed that this creates an unusual kind of order, off-diagonal long-range order. Below we will identify an analogous quantity that replaces $\left\langle \hat{a}^{\dagger}_{\mathbf{k}=0} \right\rangle$ for a superconductor.

Now the electrons in a metal occupy a Fermi sea, and those electrons that can alter their behavior to cause some kind of a condensation would have an energy of order $k_B T_c$ and therefore must lie near the Fermi energy, ε_F; these electrons have very large wavevectors $|\mathbf{k}| \cong k_F$ and the latter is of order a^{-1} where a is the lattice constant. We can form entities having a net value of $k = 0$ if we combine, somehow bind, states with $+\mathbf{k}$ and $-\mathbf{k}$ to make *standing waves*, so called pairs, such that the total wavevector is zero. In so doing we have a combined "Bose-like" entity. By superimposing such standing waves formed from electrons with wavevectors lying within some range $\delta k \sim \xi^{-1}$ in the radial direction we can make states that are localized in real space over a range ξ called the coherence length; the electrons involved are depicted in Fig. 1.7. We may loosely interpret ξ as the size of the pair, as we will discuss shortly. Continuing this process all around the Fermi surface we make a collection of $k = 0$ states from which one might conceivably form some kind of condensate. In conventional superconductors the paired electrons also have opposite spins; i.e. $+\mathbf{k}\uparrow$ pairs with $-\mathbf{k}\downarrow$ to make $S = 0$ spin singlets.[34] This allows the electrons to more closely approach other and hence bind more strongly. Note that in all this we have used each electron only once, which preserves the one-particle-per-state Pauli requirement.

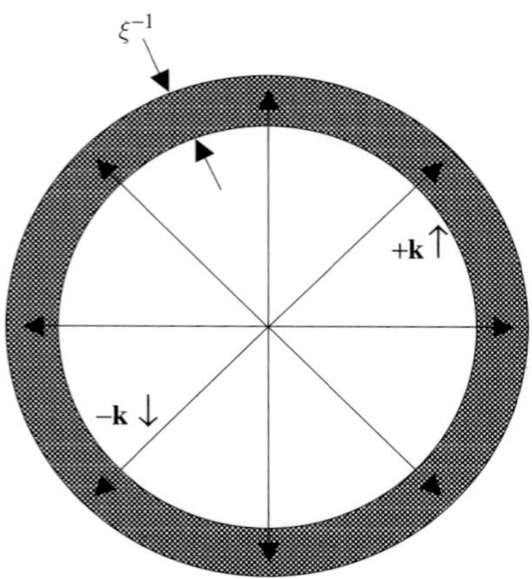

Fig. 1.7 Pairing of electrons above a filled Fermi sea acting over a range of wavevectors of the order of the reciprocal of a coherence length, ξ^{-1}.

[34]Superfluid ^{3}He is an exception where the pairs form in an $S = 1$ triplet state that will be discussed in Section 1.7.

BCS correctly identified the electron–phonon interaction as the source of the pairing in the then known superconductors.[35] The basic idea is that a rapidly moving electron polarizes the background ions as it progresses through the lattice. Since the lattice responds more slowly than the electron system this polarization persists after the electron has "left". A second electron, which arrives later in time, then encounters this polarization and is attracted to it and, indirectly, to the first electron.[36] The interaction is expected to be "cut off" for time separations greater than a response time for the ions, which is typically taken to be the reciprocal of a Debye frequency. However the details of this process are complex and will not be dealt with here [36].[37]

[35]Other pairing mechanisms have now been identified, particularly with respect to the high temperature cuprate superconductors and the so-called heavy fermion materials such as UPt_3.

[36]Since this involves a total of two processes it is second order in perturbation theory, which will make an overall negative, or attractive, contribution to the energy.

[37]For a detailed discussion of this mechanism see [36].

Evidence preceded the BCS theory that there was a gap [37],[38] now universally denoted as Δ, in the energy spectrum of the electronic excitations in a superconductor below the transition temperature, a point emphasized by Bardeen [38]. On dimensional grounds one can combine Δ with the Fermi velocity v_F to form a length that is conventionally taken as $\xi = \hbar v_F / \pi \Delta$; this is the coherence length referred to above. It would be natural to associate this length with the "size" of a bound pair. This is indeed what is found if one performs a calculation in which only a single pair above a filled Fermi sea is examined; a calculation first performed by Cooper [39]. Cooper's calculation helped set the stage for the BCS theory. And the pairs are now referred to as Cooper Pairs.

[38]One influential experiment was [37].

However one immediately encounters a problem. If we ask how many other electrons would be present in a volume of order ξ^3 one finds a number of order 10^8. So clearly all electrons must be treated on an equal footing.

1.6.2 The Bogoliubov–Valatin transformation

Analogous to the Bose commutation relations given by Eq. (1.14a,b) we have the Fermi commutation rules

$$\left[\hat{c}^{\dagger}_{\mathbf{k}\sigma}, \hat{c}_{\mathbf{k}'\sigma'}\right]_{+} = \hat{c}^{\dagger}_{\mathbf{k}\sigma}\hat{c}_{\mathbf{k}'\sigma'} + \hat{c}_{\mathbf{k}'\sigma'}\hat{c}^{\dagger}_{\mathbf{k}\sigma} = \delta_{\mathbf{k}\mathbf{k}'}\delta_{\sigma\sigma'} \tag{1.217a}$$

and

$$[\hat{c}_{\mathbf{k}\sigma}, \hat{c}_{\mathbf{k}'\sigma'}]_{+} = \hat{c}_{\mathbf{k}\sigma}\hat{c}_{\mathbf{k}'\sigma'} + \hat{c}_{\mathbf{k}'\sigma'}\hat{c}_{\mathbf{k}\sigma} = 0. \tag{1.217b}$$

Here $\hat{c}^{\dagger}_{\mathbf{k}\sigma}$ and $\hat{c}_{\mathbf{k}\sigma}$ are the Fermi creation and destruction operators for electrons with wavevector **k** and spin $\sigma = \uparrow, \downarrow$.

Recalling that we had a non-vanishing expectation value of the $k = 0$ creation operator for the interacting Bose gas, $\left\langle \hat{a}^{\dagger}_{\mathbf{k}=0} \right\rangle \neq 0$, it is natural to seek an analogous feature for the superconductor; with hindsight knowledge this turns out to be the quantity $\left\langle \hat{c}^{\dagger}_{\mathbf{k}\uparrow}\hat{c}^{\dagger}_{-\mathbf{k}\downarrow} \right\rangle \neq 0$ associated with operators that create our net $\mathbf{k} = 0$ pairs. Such expectation values will arise naturally if one introduces a canonical transformation which "mixes" $\hat{c}^{\dagger}_{\mathbf{k}\uparrow}$ with $\hat{c}_{-\mathbf{k}\downarrow}$ to make new operators according to the prescription

$$\hat{\gamma}^{\dagger}_{\mathbf{k}\uparrow} \equiv u_{\mathbf{k}}\hat{c}^{\dagger}_{\mathbf{k}\uparrow} - v_{\mathbf{k}}\hat{c}_{-\mathbf{k}\downarrow} \tag{1.218a}$$

and

$$\hat{\gamma}_{-\mathbf{k}\downarrow} \equiv u_{\mathbf{k}}\hat{c}_{-\mathbf{k}\downarrow} + v_{\mathbf{k}}\hat{c}^{\dagger}_{\mathbf{k}\uparrow}; \tag{1.218b}$$

this is called a Bogoliubov–Valatin (B–V) transformation and the signs have been chosen such that the resulting pairs will have zero net spin. To preserve the over all statistics we require that the transformation should maintain the Fermi commutation relations; i.e. we require

$$\left[\hat{\gamma}^{\dagger}_{\mathbf{k}\sigma}, \hat{\gamma}_{\mathbf{k}'\sigma'}\right]_{+} = \delta_{\mathbf{k}\mathbf{k}'}\delta_{\sigma\sigma'}, \left[\hat{\gamma}_{\mathbf{k}\sigma}\hat{\gamma}_{\mathbf{k}'\sigma'}\right]_{+} = 0; \tag{1.219a,b}$$

the transformation is then said to be canonical. Inserting (1.218a,b) into (1.217a,b) we find that (1.219a,b) will be obeyed if we choose $u_{\mathbf{k}}$ and $v_{\mathbf{k}}$ such that

$$|u_{\mathbf{k}}|^2 + |v_{\mathbf{k}}|^2 = 1 \tag{1.220}$$

which is to be compared with Eq. (1.24) for the Bose case. However, apart from this constraint (1.220) $u_{\mathbf{k}}$ and $v_{\mathbf{k}}$ are undetermined at this point. We will also need the inverse of (1.218a,b) for which a short calculation yields

$$\hat{c}^{\dagger}_{\mathbf{k}\uparrow} = u_{\mathbf{k}}\hat{\gamma}^{\dagger}_{\mathbf{k}\uparrow} + v_{\mathbf{k}}\hat{\gamma}_{-\mathbf{k}\downarrow} \tag{1.221a}$$

and

$$\hat{c}_{-\mathbf{k}\downarrow} = u_{\mathbf{k}}\hat{\gamma}_{-\mathbf{k}\downarrow} - v_{\mathbf{k}}\hat{\gamma}^{\dagger}_{\mathbf{k}\uparrow}. \tag{1.221b}$$

At this point we note that the B–V operators are useful even in the absence of superconductivity. Consider the ground state $|\psi_{\text{gnd}}\rangle$ of a free Fermi gas, which in terms of the $\hat{c}^{\dagger}_{\mathbf{k}\sigma}$ operators can be written

$$|\psi_{\text{gnd}}\rangle = \prod_{|\mathbf{k}|<k_F,\sigma} \hat{c}^{\dagger}_{\mathbf{k}\sigma} |\psi_{\text{vac}}\rangle \tag{1.222}$$

where $|\psi_{\text{vac}}\rangle$ is the empty or vacuum state. Now consider a special case of (1.218a,b) where

$$u_{\mathbf{k}} = \theta(\varepsilon_{\mathbf{k}} - \varepsilon_F), \quad v_{\mathbf{k}} = \theta(\varepsilon_F - \varepsilon_{\mathbf{k}}); \tag{1.223}$$

here $\theta(x)$ is the theta function. For this special case we have

$$\hat{\gamma}^{\dagger}_{\mathbf{k}\uparrow} = \begin{cases} \hat{c}^{\dagger}_{\mathbf{k}\uparrow}; & k > k_F \\ -\hat{c}_{-\mathbf{k}\downarrow}; & k < k_F \end{cases} \tag{1.224}$$

and

$$\hat{\gamma}^{\dagger}_{-\mathbf{k}\downarrow} = \begin{cases} \hat{c}^{\dagger}_{-\mathbf{k}\downarrow}; & k > k_F \\ \hat{c}_{\mathbf{k}\uparrow}; & k < k_F. \end{cases} \tag{1.225}$$

To create an electron excitation, $|\mathbf{k}, \uparrow\rangle$, when $k > k_F$ one would add an electron with quantum numbers, $\mathbf{k}, \uparrow$, whereas to add an excitation with quantum numbers $\mathbf{k}, \uparrow$ when $k < k_F$ we must *destroy* the *existing* electron state with quantum numbers $-\mathbf{k}, \downarrow$ thereby *creating* a *hole* state; both of these states are captured by the form $\hat{\gamma}^{\dagger}_{\mathbf{k}\uparrow} |\psi_{\text{gnd}}\rangle$.

From (1.218a,b) and the constraint (1.220) it is apparent that we can write

$$u_{\mathbf{k}} = \cos\theta_{\mathbf{k}}, \quad v_{\mathbf{k}} = \cos\theta_{\mathbf{k}}. \tag{1.226}$$

With this definition we can rewrite (1.218a,b) as

$$\begin{pmatrix} \hat{\gamma}^{\dagger}_{\mathbf{k}\uparrow} \\ \hat{\gamma}_{-\mathbf{k}\downarrow} \end{pmatrix} = \begin{pmatrix} \cos\theta_{\mathbf{k}} & -\sin\theta_{\mathbf{k}} \\ \sin\theta_{\mathbf{k}} & \cos\theta_{\mathbf{k}} \end{pmatrix} \begin{pmatrix} \hat{c}^{\dagger}_{\mathbf{k}\uparrow} \\ \hat{c}_{-\mathbf{k}\downarrow} \end{pmatrix}, \tag{1.227}$$

and from this perspective we see that the B-V transformation is equivalent to a *rotation in particle–hole space.* The inverse transform (1.221a,b) then corresponds to a rotation in the opposite direction

$$\begin{pmatrix} \hat{c}^{\dagger}_{\mathbf{k}\uparrow} \\ \hat{c}_{-\mathbf{k}\downarrow} \end{pmatrix} = \begin{pmatrix} \cos\theta_{\mathbf{k}} & \sin\theta_{\mathbf{k}} \\ -\sin\theta_{\mathbf{k}} & \cos\theta_{\mathbf{k}} \end{pmatrix} \begin{pmatrix} \hat{\gamma}^{\dagger}_{\mathbf{k}\uparrow} \\ \hat{\gamma}_{-\mathbf{k}\downarrow} \end{pmatrix}, \tag{1.228}$$

where we replace $\theta_{\mathbf{k}}$ by $-\theta_{\mathbf{k}}$.[39]

[39] Eq. (1.228) is the Fermi counterpart of the Bose relation (1.23a,b); see also footnote 12 in Section 1.2.

1.6.3 The order parameter of a superconductor

In choosing an order parameter for a superconductor, we are guided by the requirement that it must vanish in the normal state and, in addition, reflect a property characteristic of the ordered state. Based on the Bose gas where we had $\left\langle a^{\dagger}_{\mathbf{k}}\right\rangle \neq 0$, we might anticipate that a suitable order parameter will involve a *pair amplitude*, which we denote as $F^*_{\uparrow\downarrow}(\mathbf{k})$:

$$F^*_{\uparrow\downarrow}(\mathbf{k}) = \left\langle \hat{c}^{\dagger}_{\mathbf{k}\uparrow}\hat{c}^{\dagger}_{-\mathbf{k}\downarrow}\right\rangle. \tag{1.229}$$

As a first step in evaluating (1.229) we substitute (1.221a) and the Hermitian conjugate of (1.221b) to obtain

$$F^*_{\uparrow\downarrow}(\mathbf{k}) = \left\langle \left(u_{\mathbf{k}}\hat{\gamma}^{\dagger}_{\mathbf{k}\uparrow} + v_{\mathbf{k}}\hat{\gamma}_{-\mathbf{k}\downarrow}\right)\left(u^*_{\mathbf{k}}\hat{\gamma}^{\dagger}_{-\mathbf{k}\downarrow} - v^*_{\mathbf{k}}\hat{\gamma}_{\mathbf{k}\uparrow}\right)\right\rangle. \tag{1.230}$$

Under the assumption that our mean field superconducting state is *diagonal* in the B–V operators, we ignore terms that do not pair a creation operator with a destruction operator and apply the prescriptions

$$\left\langle \hat{\gamma}^{\dagger}_{\mathbf{k}\sigma}\hat{\gamma}_{\mathbf{k}\sigma}\right\rangle = f_{\mathbf{k}\sigma} \tag{1.231a}$$

and

$$\left\langle \hat{\gamma}_{\mathbf{k}\sigma}\hat{\gamma}^{\dagger}_{\mathbf{k}\sigma}\right\rangle = 1 - f_{\mathbf{k}\sigma}, \tag{1.231b}$$

where

$$f_{\mathbf{k}\sigma} = \frac{1}{e^{\varepsilon_{\mathbf{k}}/k_BT} + 1} \tag{1.232}$$

is the Fermi distribution for non-interacting particles with the excitation energies $\varepsilon_{\mathbf{k}}$ (defined in the next subsection). Equation (1.230) then takes the form

$$F^*_{\uparrow\downarrow} = u_{\mathbf{k}}v_{\mathbf{k}}\left(1 - f_{\mathbf{k}\uparrow} - f_{\mathbf{k}\downarrow}\right). \tag{1.233}$$

From the discussion surrounding the free Fermi gas where we have $u_{\mathbf{k}}v_{\mathbf{k}} = 0$ (see Eqs. (1.223)), it is clear that F^* vanishes in the normal state, as required. For the case of an unpolarized superconductor where $f_{\mathbf{k}\uparrow} = f_{\mathbf{k}\downarrow} = f_{\mathbf{k}}$ (the overwhelming majority of materials), we can rewrite (1.233) as

$$\begin{aligned} F^*(\mathbf{k}) &= u_{\mathbf{k}}v_{\mathbf{k}}\left(1 - 2f_{\mathbf{k}}\right) \\ &= u_{\mathbf{k}}v_{\mathbf{k}}\tanh\frac{\varepsilon_{\mathbf{k}}}{2k_BT}. \end{aligned} \tag{1.234a,b}$$

1.6.4 The BCS Hamiltonian

Although not the route used by BCS, one arrives more quickly and naturally at the accepted description of a superconductor by applying the B–V transformation to a suitable *model* Hamiltonian, in a manner analogous with Bogoliubov's theory of the weakly interacting Bose gas. The *true* Hamiltonian would of course include both the electron and lattice degrees of freedom along with their self and mutual interactions. BCS suppressed the individual phonon degrees of freedom but modeled their effect on the electron system through an effective electron–electron interaction which, as argued above, will be chosen to be attractive; the property that it is delayed in time will also be introduced in a model dependent manner, as we will see. The starting Hamiltonian we will utilize is analogous to that used for the Bose case (see Eq. (1.13)) but now generalized to include the spin quantum numbers

$$\hat{H} = \sum_{\mathbf{k},\sigma} \xi_{\mathbf{k}} \hat{c}^{\dagger}_{\mathbf{k}\sigma} \hat{c}_{\mathbf{k}\sigma} + \sum_{\substack{\mathbf{k},\mathbf{k}',\mathbf{q} \\ \sigma,\sigma'}} V_{\mathbf{k},\mathbf{k}',\mathbf{q};\sigma,\sigma'} \hat{c}^{\dagger}_{\mathbf{k}+\mathbf{q}\sigma} \hat{c}^{\dagger}_{\mathbf{k}'-q\sigma'} \hat{c}_{\mathbf{k}'\sigma'} \hat{c}_{\mathbf{k}\sigma} \tag{1.235}$$

where at this point all electron–electron interactions are included. Note that in writing (1.235) we have written the single particle energies as $\xi_{\mathbf{k}}$ and will measure them relative to the Fermi energy, as is common in many-body theory; i.e. $\xi_{\mathbf{k}} \equiv \hbar^2 k^2/2m - \varepsilon_F$ with $\varepsilon_F = \hbar^2 k_F^2/2m$.

Direct (Coulomb) interactions between electrons will now be suppressed; their effect can be assumed to already be incorporated into the normal state single particle excitation energies, which we might model by writing the kinetic energy as $\hbar^2/2m^*$ where m^* is some effective mass.[40] BCS theory focuses on the electron–phonon induced electron–electron coupling between the $+\mathbf{k}\uparrow$ and $-\mathbf{k}\downarrow$ states that will form the pairs; only this contribution, written as $\hat{H}_{IR}$, will be retained in the interaction term of a *reduced Hamiltonian*, $\hat{H}_R = \hat{H}_0 + \hat{H}_{IR}$, and with this approximation we have

[40] Here the idea is that the excitations of a many-body quantum system, usually called quasiparticles (a term due to Landau), already include most of the direct interaction effects and therefore interact rather weakly.

$$\hat{H}_R = \sum_{\mathbf{k},\sigma} \xi_{\mathbf{k}} \hat{c}^{\dagger}_{\mathbf{k}\sigma} \hat{c}_{\mathbf{k}\sigma} + \sum_{\mathbf{k},\mathbf{k}'} V_{\mathbf{k}\mathbf{k}'} \hat{c}^{\dagger}_{\mathbf{k}'\uparrow} \hat{c}^{\dagger}_{-\mathbf{k}'\downarrow} \hat{c}_{-\mathbf{k}\downarrow} \hat{c}_{\mathbf{k}\uparrow}. \tag{1.236}$$

1.6.5 Calculating thermodynamic properties using the Bogoliubov–Valatin method

The next step is to rewrite the reduced Hamiltonian of our system in terms of the B–V operators. To do this we substitute Eqs. (1.228a,b) into (1.236). As was the case for the pair amplitude, we generate two types of terms: (i) those containing products of an equal number of creation, $\hat{\gamma}^{\dagger}_{\mathbf{k}\sigma}$, and destruction, $\hat{\gamma}_{\mathbf{k}\sigma}$, operators; and (ii) those containing an unequal number which are assumed to vanish in a mean field theory. Carrying out this process we obtain

$$\begin{aligned} \hat{H}_0 = {} & \sum_{\mathbf{k}} \xi_{\mathbf{k}} \left[2v_{\mathbf{k}}^2 + \left(u_{\mathbf{k}}^2 - v_{\mathbf{k}}^2\right) \left(\hat{\gamma}^{\dagger}_{\mathbf{k}\uparrow} \hat{\gamma}_{\mathbf{k}\uparrow} + \hat{\gamma}^{\dagger}_{-\mathbf{k}\downarrow} \hat{\gamma}_{-\mathbf{k}\downarrow} \right) \right] \\ & + \left[\text{terms with unequal numbers of } \hat{\gamma}^{\dagger}_{\mathbf{k}\sigma} \text{ and } \hat{\gamma}_{\mathbf{k}\sigma} \right] \end{aligned} \tag{1.237}$$

and

$$\begin{aligned} \hat{H}_{IR} = \sum_{\mathbf{k},\mathbf{k}'} V_{\mathbf{k}\mathbf{k}'} \Big\{ & u_{\mathbf{k}} v_{\mathbf{k}} u_{\mathbf{k}'} v_{\mathbf{k}'} \left[1 - \hat{\gamma}^{\dagger}_{\mathbf{k}\uparrow} \hat{\gamma}_{\mathbf{k}\uparrow} - \hat{\gamma}^{\dagger}_{-\mathbf{k}\downarrow} \hat{\gamma}_{-\mathbf{k}\downarrow} \right] \left[1 - \hat{\gamma}^{\dagger}_{\mathbf{k}'\uparrow} \hat{\gamma}_{\mathbf{k}'\uparrow} - \hat{\gamma}^{\dagger}_{-\mathbf{k}'\downarrow} \hat{\gamma}_{-\mathbf{k}'\downarrow} \right] \\ & + \left[\text{terms with unequal numbers of } \hat{\gamma}^{\dagger}_{\mathbf{k}\sigma} \text{ and } \hat{\gamma}_{\mathbf{k}\sigma} \right] \Big\}. \end{aligned} \tag{1.238}$$

Using Eqs. (1.231a,b) and dropping the unpaired terms the expectation value of our reduced Hamiltonian becomes

$$\begin{aligned}\langle \hat{H}_R \rangle &= 2\sum_{\mathbf{k}} \xi_{\mathbf{k}} \left[v_{\mathbf{k}}^2 + \left(u_{\mathbf{k}}^2 - v_{\mathbf{k}}^2 \right) f_{\mathbf{k}} \right] \\ &+ \sum_{\mathbf{k},\mathbf{k}'} V_{\mathbf{k}\mathbf{k}'} u_{\mathbf{k}} v_{\mathbf{k}} u_{\mathbf{k}'} v_{\mathbf{k}'} \left[1 - 2f_{\mathbf{k}} \right] \left[1 - 2f_{\mathbf{k}'} \right].\end{aligned} \tag{1.239}$$

The Helmholtz free energy in the pairing approximation is given by

$$F = \langle H_R \rangle - TS \tag{1.240}$$

where the entropy for fermions is given by the usual expression from statistical mechanics

$$\begin{aligned}S &= -k_B \sum_{\mathbf{k},\sigma} \left[f_{\mathbf{k}} \ln f_{\mathbf{k}} + (1 - f_{\mathbf{k}}) \ln(1 - f_{\mathbf{k}}) \right] \\ &= -2k_B \sum_{\mathbf{k}} \left[f_{\mathbf{k}} \ln f_{\mathbf{k}} + (1 - f_{\mathbf{k}}) \ln(1 - f_{\mathbf{k}}) \right].\end{aligned} \tag{1.241}$$

Introducing $u_{\mathbf{k}} = \cos\theta_{\mathbf{k}}$ and $v_{\mathbf{k}} = \sin\theta_{\mathbf{k}}$ and minimizing F with respect to the parameters $\theta_{\mathbf{k}}$ yields (after some algebra)

$$u_{\mathbf{k}}^2 = \frac{1}{2}\left(1 + \frac{\xi_{\mathbf{k}}}{\varepsilon_{\mathbf{k}}} \right) \tag{1.242}$$

and

$$v_{\mathbf{k}}^2 = \frac{1}{2}\left(1 - \frac{\xi_{\mathbf{k}}}{\varepsilon_{\mathbf{k}}} \right), \tag{1.243}$$

where we have identified excitation energies as

$$\varepsilon_{\mathbf{k}} = \sqrt{\xi_{\mathbf{k}}^2 + \Delta_{\mathbf{k}}^2} \tag{1.244}$$

with $\Delta_{\mathbf{k}}$ defined by

$$\Delta_{\mathbf{k}} = -\sum_{\mathbf{k}'} V_{\mathbf{k}\mathbf{k}'} u_{\mathbf{k}'} v_{\mathbf{k}'} (1 - 2f_{\mathbf{k}'}). \tag{1.245a}$$

This equation may also be written using (1.234b) as

$$\Delta_{\mathbf{k}} = -\sum_{\mathbf{k}'} V_{\mathbf{k}\mathbf{k}'} u_{\mathbf{k}'} v_{\mathbf{k}'} \tanh\left(\frac{\varepsilon_{\mathbf{k}'}}{2k_B T} \right). \tag{1.245b}$$

(It is useful to point out that since $\varepsilon_{\mathbf{k}}$ is positive definite, $f_{\mathbf{k}} = 0$ for all $\mathbf{k}$ at $T = 0$.) If we minimize F with respect to $f_{\mathbf{k}}$ (i.e. $\partial F/\partial f_{\mathbf{k}} = 0$) we obtain (1.232); the Fermi occupation factor follows immediately from expression (1.241) for S which is based on the exclusion principle. Using the expressions for $u_{\mathbf{k}}$ and $v_{\mathbf{k}}$ we may rewrite the gap equation, (1.245) as

$$\Delta_{\mathbf{k}} = -\sum_{\mathbf{k}'} V_{\mathbf{k}\mathbf{k}'} \frac{(1 - 2f_{\mathbf{k}'})}{2\varepsilon_{\mathbf{k}'}} \Delta_{\mathbf{k}'}. \tag{1.246}$$

Let us first discuss the behavior of the gap function in the simplest (Cooper) model

$$V_{\mathbf{kk}'} = \begin{cases} -V & \text{for} \quad |\xi_{\mathbf{k}}| \text{ and } |\xi_{\mathbf{k}'}| \leq \hbar\omega_D \\ 0 & \text{for} \quad |\xi_{\mathbf{k}}| \text{ or } |\xi_{\mathbf{k}'}| > \hbar\omega_D. \end{cases} \tag{1.247}$$

Our "cutting off" of the potential at a frequency ω_D models the property that our interaction is delayed in time and cannot respond to times shorter than ω_D^{-1}, as discussed in Section 1.6.1. With this model Eq. (1.245) may be written as the condition

$$1 = N(0)V \int_{-\hbar\omega_D}^{+\hbar\omega_D} d\xi \frac{1 - 2f\left(\sqrt{\xi^2 + \Delta^2(T)}\right)}{2\sqrt{\xi^2 + \Delta^2(T)}}, \tag{1.248}$$

where $N(0)$ is the density of states at the Fermi energy. The solution of Eq. (1.248), which must be obtained numerically, gives $\Delta(T)$ which is shown graphically in Fig. 1.8. It is easy to show that for T near T_c we have $\Delta(T) \propto \sqrt{T_c - T}$, supporting its interpretation as the order parameter and therefore consistent with Eq. (1.157b) of the G–L theory.

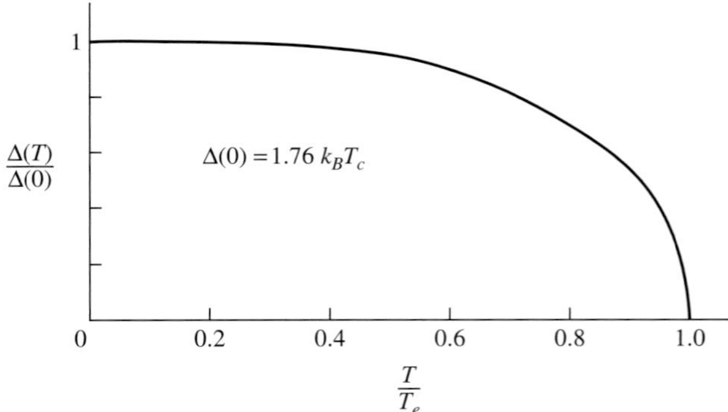

Fig. 1.8 The reduced energy gap as a function of T/T_c.

The transition temperature, T_c, corresponds to $\Delta(T_c) = 0$, which from (1.248) is obtained from

$$\begin{aligned} 1 &= N(0)V \int_{-\hbar\omega_D}^{+\hbar\omega_D} d\xi \frac{1 - 2f(\xi, T_c)}{2\xi} \\ &= N(0)V \int_0^{+\hbar\omega_D} \frac{d\xi}{\xi} \tanh \frac{\xi}{2k_B T_c}. \end{aligned} \tag{1.249}$$

For large ξ, $\tanh(\xi/2k_B T_c) \to 1$ and the integral has the asymptotic form $\ln(\hbar\omega_D/k_B T_c) + C$; a numerical calculation yields $C = \ln 1.13$. Therefore

$$1 = N(0)V \ln \frac{1.13\hbar\omega_D}{k_B T_c} \tag{1.250}$$

or

$$k_B T_c = 1.13\hbar\omega_D e^{-1/N(0)V}. \tag{1.251}$$

We now evaluate the free energy. The $-TS$ contribution, Eq. (1.241), can be rewritten as

$$-TS = \sum_{\mathbf{k}} \left[\varepsilon_{\mathbf{k}}(1 - 2f_{\mathbf{k}}) - 2k_B T \ln(2\cosh \frac{\varepsilon_{\mathbf{k}}}{2k_B T}) \right]. \tag{1.252}$$

The contribution of $\langle \hat{H}_{IR} \rangle$ to the energy, E', follows from Eqs. (1.239) and (1.245):

$$\begin{aligned}\langle \hat{H}_{IR} \rangle &= -\sum_{\mathbf{k}} \frac{\Delta_{\mathbf{k}}^2}{2\varepsilon_{\mathbf{k}}}(1 - 2f_{\mathbf{k}}), \\ &= -\frac{\Delta^2(T)}{V}.\end{aligned} \tag{1.253}$$

From Eq. (1.239) and our expressions for $u_{\mathbf{k}}$ and $v_{\mathbf{k}}$ we have

$$\langle H_0' \rangle = 2\sum_{\mathbf{k}} \left[\xi_{\mathbf{k}} v_{\mathbf{k}} + f_{\mathbf{k}} \varepsilon_{\mathbf{k}} \right]. \tag{1.254}$$

From (1.253) and (1.254) we obtain the energy E' as

$$\langle H_R' \rangle = 2\sum_{\mathbf{k}} f_{\mathbf{k}} \varepsilon_{\mathbf{k}} + \left[2\sum_{\mathbf{k}} \xi_{\mathbf{k}} v_{\mathbf{k}} - \frac{\Delta^2}{V} \right]. \tag{1.255}$$

The free energy is the sum of (1.252) and (1.255):

$$F' = \sum_{\mathbf{k}} \left[\varepsilon_{\mathbf{k}} - 2k_B T \ln \left(2\cosh \frac{\varepsilon_{\mathbf{k}}}{2k_B T} \right) \right] - \frac{\Delta^2(T)}{V} + 2\sum_{\mathbf{k}} \xi_{\mathbf{k}} v_{\mathbf{k}}. \tag{1.256}$$

Although we will not show it, the condensation free energy is related to the thermodynamic critical field by

$$F_N - F_S = \frac{1}{8\pi} H_c^2(T) \tag{1.257}$$

From thermodynamics the heat capacity is

$$\begin{aligned}C &= T\frac{dS}{dT} \\ &= -2k_B T \sum_{\mathbf{k}} \left[\ln f_{\mathbf{k}} - \ln(1 - f_{\mathbf{k}}) \right] \frac{\partial f_{\mathbf{k}}}{\partial T} \\ &= 2k_B T \sum_{\mathbf{k}} \frac{\varepsilon_{\mathbf{k}}}{k_B T} \frac{d}{dT} \left(\frac{1}{e^{\varepsilon_{\mathbf{k}}/k_B T} + 1} \right).\end{aligned}$$

Noting that $\varepsilon_{\mathbf{k}} = \varepsilon_{\mathbf{k}}(T)$ and writing $f'(x)$ in terms of $f(x)$ we obtain

$$\begin{aligned}C &= \frac{-2}{k_B T^2} \sum_{\mathbf{k}} f_{\mathbf{k}}(1 - f_{\mathbf{k}}) \left(-\varepsilon^2 + T\Delta \frac{d\Delta}{dT} \right) \\ &= \frac{2}{k_B T^2} N(0) \int_0^\infty d\xi f(\varepsilon)(1 - f(\varepsilon)) \left(\varepsilon^2 + T\Delta \frac{d\Delta}{dT} \right).\end{aligned} \tag{1.258}$$

At very low temperature, $f_{\mathbf{k}} << 1$ and $\Delta(d\Delta/dT) \cong 0$, and the heat capacity approaches the limiting form,

$$\begin{aligned} C &\cong \frac{2N(0)\Delta^2(0)e^{-\Delta(0)/k_BT}}{k_BT^2} \int_0^\infty e^{-\xi^2/2k_BT\Delta(0)} d\xi \\ &\cong 2N(0)\Delta(0)\sqrt{2\pi}k_B\left(\frac{\Delta(0)}{k_BT}\right)^{3/2} e^{-\Delta(0)/k_BT}; \end{aligned} \tag{1.259}$$

we see that the heat capacity approaches zero asymptotically as $e^{-\Delta(0)/k_BT}$ due to the presence of the gap in the energy spectrum.

1.6.6 Inhomogeneous superconductors

Apart from its relative simplicity, one of the strengths of the Ginzbug–Landau theory is its ability to treat problems in which the order parameter varies with position, which includes all cases where a magnetic field is present. As presented above, the BCS theory was limited to the case of a uniform superconductor and we now discuss how the theory can be generalized to handle inhomogeneous cases.

We start by assuming we have a second quantized Hamiltonian of the form

$$\begin{aligned} \hat{H} &= \int d^3r\hat{\psi}^*_\alpha(\mathbf{r})\left[-\frac{\hbar^2}{2m}\left(\nabla - \frac{ie}{\hbar c}\mathbf{A}(\mathbf{r})\right)^2 - U(\mathbf{r})\right]\hat{\psi}_\alpha(\mathbf{r}) \\ &+ \frac{1}{2}\int d^3r'\int d^3r\hat{\psi}^*_\alpha(\mathbf{r})\hat{\psi}^*_\beta(\mathbf{r}')V(\mathbf{r},\mathbf{r}')\hat{\psi}_\beta(\mathbf{r}')\hat{\psi}_\alpha(\mathbf{r}), \end{aligned} \tag{1.260}$$

where $U(\mathbf{r})$ and $V(\mathbf{r},\mathbf{r}')$ are one- and two-body potentials, repeated indeces are summed, and the operators $\hat{\psi}$ obey the following fermion commutation relations

$$\hat{\psi}_\alpha(\mathbf{r})\hat{\psi}_\beta(\mathbf{r}') + \hat{\psi}_\beta(\mathbf{r}')\hat{\psi}_\alpha(\mathbf{r}) = 0 \tag{1.261a}$$

and

$$\hat{\psi}^\dagger_\alpha(\mathbf{r})\hat{\psi}_\beta(\mathbf{r}') + \hat{\psi}_\beta(\mathbf{r}')\hat{\psi}^\dagger_\alpha(\mathbf{r}) = \delta_{\alpha\beta}\delta^{(3)}(\mathbf{r}-\mathbf{r}'). \tag{1.261b}$$

For most applications involving superconductivity it is safe to assume that the two-body potential occurring in the second term in (1.260) is very short-range on the scale over which the gap can vary, so we write it as

$$V(\mathbf{r},\mathbf{r}') = V\delta(\mathbf{r}-\mathbf{r}') \tag{1.262}$$

and (1.260) then becomes

$$\begin{aligned} \hat{H} &= \int d^3r\hat{\psi}^*_\alpha(\mathbf{r})\left[-\frac{\hbar^2}{2m}\left(\nabla - \frac{ie}{\hbar c}\mathbf{A}(\mathbf{r})\right)^2 + U(\mathbf{r})\right]\hat{\psi}_\alpha(\mathbf{r}) \\ &+ \frac{V}{2}\int d^3r\hat{\psi}^\dagger_\alpha(\mathbf{r})\hat{\psi}^\dagger_\beta(\mathbf{r})\hat{\psi}_\beta(\mathbf{r})\hat{\psi}_\alpha(\mathbf{r}). \end{aligned} \tag{1.263}$$

More generally the potentials could depend on spin and, in addition, V might vary with position, as when passing from one material to another.

In a (singlet) mean field theory of superconductivity we can anticipate expectation values of the following three products: $\langle\hat{\psi}_\alpha^\dagger(\mathbf{r})\hat{\psi}_\alpha(\mathbf{r})\rangle$, $\left\langle\hat{\psi}_\uparrow^\dagger(\mathbf{r})\hat{\psi}_\downarrow^\dagger(\mathbf{r})\right\rangle$, and $\langle\hat{\psi}_\downarrow(\mathbf{r})\hat{\psi}_\uparrow(\mathbf{r})\rangle$. An effective Hamiltonian that incorporates these forms and is simultaneously Hermitian would have the form[41]

[41] In addition to "singlet" pairing of the form $\langle\hat{\psi}_\downarrow(\mathbf{r})\hat{\psi}_\uparrow(\mathbf{r})\rangle$, the potential can be attractive in a "triplet channel" leading to mean values of $\langle\hat{\psi}_\uparrow(\mathbf{r})\hat{\psi}_\uparrow(\mathbf{r})\rangle$, $\langle\hat{\psi}_\downarrow(\mathbf{r})\hat{\psi}_\downarrow(\mathbf{r})\rangle$, $\langle\hat{\psi}_\downarrow(\mathbf{r})\hat{\psi}_\uparrow(\mathbf{r})\rangle$, and $\langle\hat{\psi}_\uparrow(\mathbf{r})\hat{\psi}_\downarrow(\mathbf{r})\rangle$, together with their complex conjugates; this situation occurs in superfluid ^{3}He.

$$\hat{H}^{(e)} = \int d^3r\left[\hat{\psi}_\alpha^\dagger(\mathbf{r})H_0\hat{\psi}_\alpha(\mathbf{r}) + V(\mathbf{r})\hat{\psi}_\alpha^\dagger(\mathbf{r})\hat{\psi}_\alpha(\mathbf{r})\right.$$
$$\left. + \Delta(\mathbf{r})\hat{\psi}_\uparrow^\dagger(\mathbf{r})\hat{\psi}_\downarrow^\dagger(\mathbf{r}) + \Delta^*(\mathbf{r})\hat{\psi}_\downarrow(\mathbf{r})\hat{\psi}_\uparrow(\mathbf{r})\right], \tag{1.264}$$

where

$$H_0 = -\frac{\hbar^2}{2m}\left(\nabla - \frac{ie}{\hbar c}\mathbf{A}\right)^2 + U(\mathbf{r}) - \mu \tag{1.265}$$

and we continue to measure the single particle energies from the chemical potential, μ; here $V(\mathbf{r})$ ia a self-consistent potential and $\Delta(\mathbf{r})$ is interpreted as a position dependent gap function. To form a mean field theory which is bilinear in the $\hat{\psi}$ operators from Eq. (1.263) we replace the operator product $\hat{\psi}_\alpha^\dagger(\mathbf{r})\hat{\psi}_\beta^\dagger(\mathbf{r})\hat{\psi}_\beta(\mathbf{r})\hat{\psi}_\alpha(\mathbf{r})$ by the following combinations:

$$\hat{\psi}_\uparrow^\dagger(\mathbf{r})\hat{\psi}_\downarrow^\dagger(\mathbf{r})\langle\hat{\psi}_\downarrow(\mathbf{r})\hat{\psi}_\uparrow(\mathbf{r})\rangle,$$
$$\hat{\psi}_\downarrow(\mathbf{r})\hat{\psi}_\uparrow(\mathbf{r})\left\langle\hat{\psi}_\uparrow^\dagger(\mathbf{r})\hat{\psi}_\downarrow^\dagger(\mathbf{r})\right\rangle, \quad -\hat{\psi}_\alpha^\dagger(\mathbf{r})\hat{\psi}_\alpha(\mathbf{r})\left\langle\hat{\psi}_\beta^\dagger(\mathbf{r})\hat{\psi}_\beta(\mathbf{r})\right\rangle,$$

where the signs follow from the commutation relations (1.261a,b). Comparing (1.263) with (1.264) we have

$$V(\mathbf{r}) = -V\left\langle\hat{\psi}_\uparrow^\dagger(\mathbf{r})\hat{\psi}_\uparrow(\mathbf{r})\right\rangle = -V\left\langle\hat{\psi}_\downarrow^\dagger(\mathbf{r})\hat{\psi}_\downarrow(\mathbf{r})\right\rangle \tag{1.266}$$

and

$$\Delta(\mathbf{r}) = V\langle\hat{\psi}_\uparrow(\mathbf{r})\hat{\psi}_\downarrow(\mathbf{r})\rangle = -\langle\hat{\psi}_\downarrow(\mathbf{r})\hat{\psi}_\uparrow(\mathbf{r})\rangle. \tag{1.267}$$

If we are to have a non-vanishing gap function, $\Delta(\mathbf{r})$, we must introduce a B–V-like transformation for the $\hat{\psi}(\mathbf{r})$ operators in our Hamiltonian (1.264). For a non-superconductor we would expand $\hat{\psi}(\mathbf{r})$ as

$$\hat{\psi}_\alpha(\mathbf{r}) = \sum_n w_n(\mathbf{r})\,\hat{c}_{n\alpha}, \tag{1.268}$$

where $w_n(\mathbf{r})$ is some suitable complete set of eigenfunctions for the problem at hand (e.g. plane waves for a Fermi gas) and the $\hat{c}_{n\alpha}$ destroy these states. For a superconductor Bogoliubov generalized (1.268) to read

$$\hat{\psi}_\uparrow(\mathbf{r}) = \sum_n[u_n(\mathbf{r})\hat{\gamma}_{n\uparrow} - v_n^*(\mathbf{r})\hat{\gamma}_{n\downarrow}^\dagger] \tag{1.269a}$$

and

$$\hat{\psi}_\downarrow(\mathbf{r}) = \sum_n\left[u_n(\mathbf{r})\hat{\gamma}_{n\downarrow} + v_n^*(\mathbf{r})\hat{\gamma}_{n\uparrow}^\dagger\right], \tag{1.269b}$$

where to retain Fermi commutation rules we require (see (1.219a,b))

$$\hat{\gamma}_{n\alpha}\hat{\gamma}_{n'\beta} + \hat{\gamma}_{n'\beta}\hat{\gamma}_{n\alpha} = 0 \tag{1.270a}$$

and

$$\hat{\gamma}_{n\alpha}\hat{\gamma}^{\dagger}_{n'\beta} + \hat{\gamma}^{\dagger}_{n'\beta}\hat{\gamma}_{n\alpha} = \delta_{nn'}\delta_{\alpha\beta}. \tag{1.270b}$$

Note Eqs. (1.269a,b) contain *two independent sets of eigenfunctions,* $u_n(\mathbf{r})$ and $v_n(\mathbf{r})$; our task is now to obtain the appropriate equations that these eigenfunctions satisfy.

If our generalized Bogoliubov transformation is to diagonalize the Hamiltonian it must result in the form

$$\hat{H}_e = E_{0S} + \sum_{n,\alpha} \varepsilon_n \hat{\gamma}^{\dagger}_{n\alpha}\hat{\gamma}_{n\alpha} \tag{1.271}$$

where E_{0S} is the ground state energy of our superconductor and ε_n are the excitation energies. From Eqs. (1.270a,b) and (1.271) we have

$$\left[\hat{H}_e, \hat{\gamma}_{n\alpha}\right] = -\varepsilon_n \hat{\gamma}_{n\alpha} \tag{1.272a}$$

and

$$\left[\hat{H}_e, \hat{\gamma}^{\dagger}_{n\alpha}\right] = \varepsilon_n \hat{\gamma}^{\dagger}_{n\alpha}. \tag{1.272b}$$

Our strategy will be to evaluate the commutator of $\hat{H}_e$ with $\hat{\psi}_{\uparrow}$ and $\hat{\psi}_{\downarrow}$ in two non-equivalent ways. For the first of these we use (1.269a,b) and Eq. (1.272) to obtain

$$\left[\hat{H}_e, \hat{\psi}_{\uparrow}(\mathbf{r})\right] = -\left(H_0 + U(\mathbf{r})\right)\hat{\psi}_{\uparrow}(\mathbf{r}) - \Delta(\mathbf{r})\hat{\psi}^{\dagger}_{\downarrow}(\mathbf{r}) \tag{1.273a}$$

and

$$\left[\hat{H}_e, \hat{\psi}_{\downarrow}(\mathbf{r})\right] = -\left(H_0 + U(\mathbf{r})\right)\hat{\psi}_{\downarrow}(\mathbf{r}) + \Delta(\mathbf{r})\hat{\psi}^{\dagger}_{\uparrow}(\mathbf{r}), \tag{1.273b}$$

after which we insert Eqs. (1.270a,b) into the right-hand side. The second way to evaluate the commutator is to substitute Eqs. (1.269a,b) directly, evaluating the resulting commutators of the $\hat{\gamma}$ operators using Eqs. (1.270a,b). Comparing the coefficients of the $\hat{\gamma}$ operators in the two sets of expressions so formed we obtain *Bogoliubov's equations,*

$$\varepsilon_n u_n(\mathbf{r}) = \left[H_0 + U(\mathbf{r})\right] u_n(\mathbf{r}) + \Delta(\mathbf{r}) v_n(\mathbf{r}) \tag{1.274a}$$

and

$$\varepsilon_n v_n(\mathbf{r}) = -\left[H_0^* + U(\mathbf{r})\right] v_n(\mathbf{r}) + \Delta^*(\mathbf{r}) u_n(\mathbf{r}). \tag{1.274b}$$

To solve this set of coupled partial differential equations we require separate expressions for the mean potentials $U(\mathbf{r})$ and $\Delta(\mathbf{r})$. These follow from substituting Eqs. (1.269) into (1.266) and (1.267) and evaluating expectation values with our earlier prescriptions (1.231a,b) from which we obtain

$$U(\mathbf{r}) = -V\sum_n \left[|u_n(\mathbf{r})|^2 f_n + |v_n(\mathbf{r})|^2(1 - f_n)\right] \tag{1.275}$$

and

$$\Delta(\mathbf{r}) = +V \sum_n v_n^*(\mathbf{r}) u_n(\mathbf{r})(1 - 2f_n). \tag{1.276}$$

The functions $u_n(\mathbf{r})$ and $v_n(\mathbf{r})$ obey orthogonality relations which can be obtained from (1.274a,b). We multiply Eq. (1.274a) for u_n on the left by u_m^*, multiply the complex conjugate of Eq. (1.274a) for u_m^* on the left by u_n and subtract to obtain

$$(\varepsilon_n - \varepsilon_m) \int d^3r \, u_m u_n = \int d^3r \, (u_m^* H_0 \, u_n - u_n H_0^* u_m) + \int d^3r \left[u_n^* \Delta v_n - u_n \Delta^* v_m^* \right].$$

Performing similar steps with (1.274b) yields

$$(\varepsilon_n - \varepsilon_m) \int d^3r \, v_m^* v_n = - \int d^3r \, (v_m^* H_0^* v_n - v_n H_0 \, v_m^*) + \int d^3r \left[v_n^* \Delta^* u_n - v_n \Delta u_m \right].$$

Performing integrations by parts on those terms on the right involving $\boldsymbol{\nabla}$ and adding these expressions we have

$$(\varepsilon_n - \varepsilon_m) \int d^3r \left(u_m^* u_n + v_m^* v_n \right) = 0.$$

The integral must therefore vanish for $n \neq m$. For $n = m$ the integral is undetermined and, adopting the usual normalization condition, we have

$$\int d^3r \left(u_m^*(\mathbf{r}) u_n(\mathbf{r}) + v_m^*(\mathbf{r}) v_n(\mathbf{r}) \right) = \delta_{mn}. \tag{1.277a}$$

We next multiply Eq. (1.274a) for u_n on the left by v_m and Eq. (1.274b) for v_n on the left by u_m and subtract; performing the same operations with n and m interchanged and subtracting the resulting expressions yields a second orthogonality expression

$$\int d^3r \left[v_m(\mathbf{r}) u_n(\mathbf{r}) - u_m(\mathbf{r}) v_n(\mathbf{r}) \right] = 0. \tag{1.277b}$$

A generalized set of completeness relations can be obtained by substituting (1.269a,b) into the commutation relations (1.261a,b) and then using (1.277a,b) which yields

$$\sum_n \left[u_n^*(\mathbf{r}) u_n(\mathbf{r}') + v_n(\mathbf{r}) v_n^*(\mathbf{r}') \right] = \delta^{(3)}(\mathbf{r} - \mathbf{r}') \tag{1.278a}$$

and

$$\sum_n \left[u_n(\mathbf{r}) v_n^*(\mathbf{r}') - v_n^*(\mathbf{r}) u_n(\mathbf{r}') \right] = 0. \tag{1.278b}$$

The Bogoliubov method is extensively developed in the book by de Gennes [34] where many other applications are described, in particular a derivation of the Ginzburg–Landau equations and the diamagnetic (Meissner) response;[42] it is also developed in the book by Ketterson and Song [41]. The method has been used to describe vortex lines at the microscopic (as opposed to G–L) level [42]. Although having the advantage of relative simplicity, the more formal,

[42]The first derivation of the Ginzburg–Landau equations was carried out by L. P. Gorkov using Green's function methods [40].

and powerful, method of discussing non-uniform (and non-equilibrium) superconductivity is based on Green's functions [43].

1.6.7 Off-diagonal long-range order in a superconductor

The concept of off-diagonal long-range order introduced in Sec. 1.4.1 for a Bose system can be generalized to apply to a superconductor. Condensation occurs for pairs in a superconductor, so rather than considering the one-particle density matrix, as in Eq. (1.118), we must examine the two-particle density matrix

$$\rho^{(2)}_{\alpha\beta\gamma\delta}\left(\mathbf{r}_1,\mathbf{r}'_1;\mathbf{r}_2,\mathbf{r}'_2,t\right)=\left\langle\hat{\psi}^\dagger_\gamma(\mathbf{r}_2,t)\hat{\psi}^\dagger_\delta(\mathbf{r}'_2,t)\hat{\psi}_\alpha(\mathbf{r}_1,t)\hat{\psi}_\beta(\mathbf{r}'_1,t)\right\rangle; \qquad (1.279)$$

here $\hat{\psi}^\dagger_\gamma(\mathbf{r},t)$ and $\hat{\psi}_\alpha(\mathbf{r},t)$ are the creation and destruction operators for electrons. We now examine (1.279) for the case $\mathbf{r}_1=\mathbf{r}'_1$, $\mathbf{r}_2=\mathbf{r}'_2$ where (1.279) becomes

$$\rho^{(2)}_{\alpha\beta\gamma\delta}\left(\mathbf{r}_1,\mathbf{r}_2,t\right)=\left\langle\hat{\psi}^\dagger_\gamma(\mathbf{r}_2,t)\hat{\psi}^\dagger_\delta(\mathbf{r}_2,t)\hat{\psi}_\alpha(\mathbf{r}_1,t)\hat{\psi}_\beta(\mathbf{r}_1,t)\right\rangle. \qquad (1.280)$$

As we have seen (see Eqs. (1.229) and (1.267) for a conventional superconductor), the pair creation operator has a finite expectation value, $F^*_{\gamma\delta}(\mathbf{r},t)\equiv\left\langle\hat{\psi}^\dagger_\gamma(\mathbf{r},t)\hat{\psi}^\dagger_\delta(\mathbf{r},t)\right\rangle$, and hence

$$\rho^{(2)}_{\alpha\beta\gamma\delta}\left(\mathbf{r}_1,\mathbf{r}_2,t\right)=F^*_{\gamma\delta}(\mathbf{r}_2,t)F_{\alpha\beta}(\mathbf{r}_1,t),\quad |\mathbf{r}_1-\mathbf{r}_2|\to\infty; \qquad (1.281)$$

this constitutes off-diagonal long range for the Fermi superfluid case.

For a conventional superconductor we would write the pair ampliude as $F^*_{\uparrow\downarrow}$; when writing it in the form $F^*_{\gamma\delta}$ we allow for more general kinds of pairing, the first realization of which was discovered in ^{3}He, which we take up next.

1.7 ^{3}He: the first unusual superfluid[43]

Liquid ^{3}He is a system which has some features in common with electrons in a metal. We recall that ^{3}He has no net electronic spin and its statistics are therefore governed by the $I=1/2$ nuclear spin. It may be thought of as having an even larger particle–particle interaction since the "empty volume" for the atoms to move in is small; most of the liquid volume is occupied by the essentially impenetrable "hard core" associated with the filled 1s^2 shell of the He-atoms.[44] But at temperatures of order ten times lower than that for a Fermi gas the heat capacity does approach a constant, as one would predict from the Pauli–Sommerfeld model of a Fermi gas.[45] This suggests that the excitations in ^{3}He retain some characteristics of an interacting Fermi gas, on which the BCS theory is based. This being the case it is natural to expect that at some temperature it might become a superfluid, in analogy with BCS superconductors. This indeed happens at temperatures between approximately 1×10^{-3} K and 3×10^{-3} K [46], depending on the pressure, as shown in the phase diagram

[43] For an extended discussion of the properties of superfluid ^{3}He see [44].

[44] The Lenard-Jones form of the potential acting between two ^{3}He atoms is strongly repulsive at short distances. Of course electrons are also strongly repulsive at short range, however this interaction is screened in a metal over a Thomas–Fermi screening length which is generally much smaller than an inter-electron spacing.

[45] For a discussion of the normal state properties see [45].

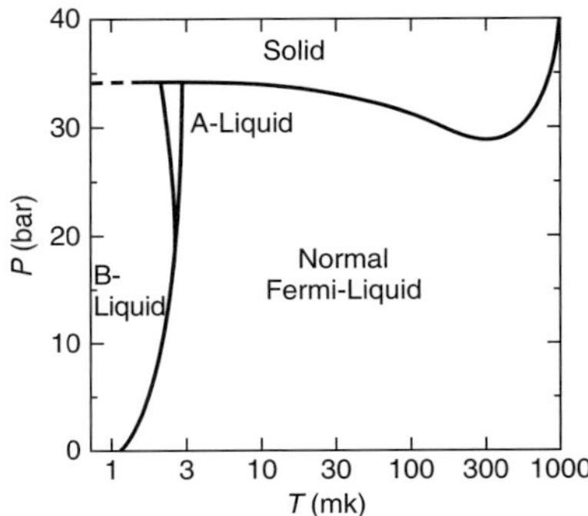

Fig. 1.9 The phase diagram of liquid and solid ^{3}He in the P–T plane. Only temperatures below the liquid–gas line are shown. In this regime we have three liquid phases: normal liquid and two superfluid phases, denoted as A and B. At pressures above about 33 bar we encounter the solid phase, which is also divided into several phases.

presented in Fig. 1.9. Note that the ratio T_c/T_F, where T_F is the ^{3}He Fermi temperature, is of the same order as for a conventional metallic superconductor and in this sense its transition temperature is not low.

1.7.1 The A and B phases; triplet pairing

The BCS pairing in ^{3}He differs from the commonly encountered electron case. At zero magnetic field there are two different superfluid phases which are referred to as the A phase and B phase (for historical reasons). The first of these occurs in a narrow wedge in P–T space at higher temperatures and near the melting curve; the latter exists over a broader region. The existence of multiple phases arises from the fact that the two fermions making up a pair are bound in a state with an orbital angular momentum of $l = 1$. There are potentially three such states associated with the magnetic quantum number, m $(0, \pm 1)$. Since such states are odd on interchanging the two ^{3}He atoms (the parity of spherical harmonics is $(-1)^l$), the spin part of the quantum-mechanical state function must be even in order that the overall state function is antisymmetric and thereby satisfies the Pauli principle; i.e. it must involve the $s = 1$ spin states, of which there are also potentially three. The most general state would then be a superposition of all of the possible nine products that can be formed from the three orbitals and three spin states with complex amplitudes (18 parameters in all). The two equilibrium states actually occurring in nature at zero field correspond to very specific combinations of subsets of these nine states, which we will discuss below. The validity of the states identified with the A and B phases has been unambiguously established as a result of numerous experiments,[46] those involving nuclear magnetic resonance [48][47] and ultrasonic propagation [50] yielding the most conclusive evidence.

[46]For a detailed discussion of the experimental properties see [47].

[47]See also [49].

Before discussing the nature of the two BCS superfluid ground states, we seek a qualitative explanation of the underlying factors leading to the pairing in a higher angular momentum state. This result has its origin in a combination of the attractive Van der Waals-like nature of the He–He interaction potential at large distances, which then becomes strongly repulsive at short separations (the so-called hard core effect). One can consider the scattering by such a potential of two He atoms with momenta $+\mathbf{k}_F$ and $-\mathbf{k}_F$ (from which a Cooper pair might form) and examine the various partial-wave phase-shifts, $\delta_l(2k_F)$, governing the scattering. The net result of such a calculation is that the s-wave phase shift, δ_0, is negative corresponding to repulsion, while for the next highest angular momentum state, δ_1, associated with $l = 1$ or p-wave scattering, it is positive, corresponding to attraction. Physically this occurs because the particles would "see" the repulsive barrier for s-wave scattering; however the centrifugal barrier keeps the $l = 1$ and higher states from seeing the repulsive hard core,[48] and thereby allows them to sample the attractive Van der Waals tail. As noted above, particles in an $l = 1$ orbital must be in a triplet spin state.

[48]At momenta significantly higher than the Fermi momentum the particles can of course approach each other more closely and thereby sample the hard core.

There is an alternative way to think about the attraction needed to form a Cooper pair. ^{3}He atoms in the liquid ground state can locally avoid each other by aligning their spins—the Pauli principal will then keep them further apart, thereby avoiding an increase in energy arising from the hard-core repulsion.

This property favors local spin alignment (on a fluctuating basis), thereby enhancing the magnetic susceptibility, but not to the point where the liquid acquires a spontaneous moment. None the less another ^{3}He atom encountering a polarized region would have its energy lowered by temporarily aligning; this idea forms the basis of the spin fluctuation model, or paramagnon model as it is sometimes called, of the attractive interaction leading to superfluidity in ^{3}He. This qualitative idea is supported by a model, but realistic, many-body calculation.

1.7.2 The order parameter of superfluid ^{3}He

As discussed in Section 1.6, a superconductor has the unusual property that the expectation value of the destruction (or creation) operators for a pair of electrons with opposite momentum and spin, the so-called pair amplitude, has a non-zero value. To account for the additional degrees of freedom associated with the $l = 1$ orbital and $s = 1$ spin degrees of freedom of superfluid ^{3}He we now generalize Eq. (1.229) to read

$$F_{\alpha\beta}(\mathbf{k}) = \langle \hat{c}_{\mathbf{k}\alpha}\hat{c}_{-\mathbf{k}\beta} \rangle \tag{1.282}$$

where α, β denote spin projections. To define a macroscopic G–L order parameter we can integrate over the magnitude of the wavevector $\mathbf{k}$ and rewrite (1.282) as

$$\psi_{\alpha\beta}(\mathbf{p}) = \overline{F_{\alpha\beta}(\mathbf{k})} \equiv \sum_{|\mathbf{k}|} \langle \hat{c}_{\mathbf{k}\alpha}\hat{c}_{-\mathbf{k}\beta} \rangle \tag{1.283a}$$

or in matrix form as

$$\underset{\sim}{\psi}(\mathbf{p}) \equiv \begin{pmatrix} \psi_{\uparrow\uparrow}(\mathbf{p}) & \psi_{\uparrow\downarrow}(\mathbf{p}) \\ \psi_{\downarrow\uparrow}(\mathbf{p}) & \psi_{\downarrow\downarrow}(\mathbf{p}) \end{pmatrix}; \tag{1.283b}$$

here $\mathbf{p} = \mathbf{p}(\theta, \varphi)$ is a unit vector parallel to $\mathbf{k}$ that accounts for a possible variation of the order parameter over the Fermi surface; note the temperature, pressure, and field dependencies of the order parameter have been suppressed here.

From the commutation relations obeyed by $\hat{c}_{\mathbf{k}\alpha}$, $\psi_{\alpha\beta}$ has the property

$$\begin{aligned} \psi_{\alpha\beta}(\mathbf{p}) &= \sum_{|\mathbf{k}|} \langle \hat{c}_{\mathbf{k}\alpha}\hat{c}_{-\mathbf{k}\beta} \rangle \\ &= -\sum_{|\mathbf{k}|} \langle \hat{c}_{-\mathbf{k}\beta}\hat{c}_{\mathbf{k}\alpha} \rangle \\ &= -\psi_{\beta\alpha}(-\mathbf{p}). \end{aligned} \tag{1.284}$$

The behavior of $\psi_{\alpha\beta}(\mathbf{p})$ under inversion of $\mathbf{k}$ involves two cases:

Singlet pairing (space symmetric, spin antisymmetric)

$$\psi_{\alpha\beta}(\mathbf{p}) = -\psi_{\beta\alpha}(\mathbf{p}) = \psi_{\alpha\beta}(-\mathbf{p}). \tag{1.285}$$

Triplet pairing (space antisymmetric, spin symmetric)

$$\psi_{\alpha\beta}(\mathbf{p}) = \psi_{\beta\alpha}(\mathbf{p}) = -\psi_{\alpha\beta}(-\mathbf{p}). \tag{1.286}$$

It is common to write the spin basis states, $\underset{\sim}{\chi}_\mu$, of $\underset{\sim}{\psi}(\mathbf{p})$ in the form $i\underset{\sim}{\sigma}_\mu\underset{\sim}{\sigma}_2$ where the subscript μ runs from zero to three with σ_0 defined as the unit matrix while $\mu = 1$ to 3 denote the three Pauli matrices. The resulting states are then

$$\begin{aligned}\underset{\sim}{\chi}_0 &= (\uparrow)(\downarrow) - (\downarrow)(\uparrow)\\ &= i\underset{\sim}{\sigma}_0\underset{\sim}{\sigma}_2 = i\begin{pmatrix}1 & 0\\ 0 & 1\end{pmatrix}\begin{pmatrix}0 & -i\\ i & 0\end{pmatrix} = \begin{pmatrix}0 & 1\\ -1 & 0\end{pmatrix}\end{aligned} \tag{1.287}$$

for $\mu = 0$ and

$$\begin{aligned}\underset{\sim}{\chi}_1 &= -(\uparrow)(\uparrow) + (\downarrow)(\downarrow)\\ &= i\underset{\sim}{\sigma}_1\underset{\sim}{\sigma}_2 = i\begin{pmatrix}0 & 1\\ 1 & 0\end{pmatrix}\begin{pmatrix}0 & -i\\ i & 0\end{pmatrix} = \begin{pmatrix}-1 & 0\\ 0 & 1\end{pmatrix}\end{aligned} \tag{1.288a}$$

$$\begin{aligned}\underset{\sim}{\chi}_2 &= i(\uparrow)(\uparrow) + i(\downarrow)(\downarrow)\\ &= i\underset{\sim}{\sigma}_2\underset{\sim}{\sigma}_2 = i\begin{pmatrix}0 & -i\\ i & 0\end{pmatrix}\begin{pmatrix}0 & -i\\ i & 0\end{pmatrix} = \begin{pmatrix}i & 0\\ 0 & i\end{pmatrix}\end{aligned} \tag{1.288b}$$

$$\begin{aligned}\underset{\sim}{\chi}_3 &= (\uparrow)(\downarrow) + (\downarrow)(\uparrow)\\ &= i\underset{\sim}{\sigma}_3\underset{\sim}{\sigma}_2 = i\begin{pmatrix}1 & 0\\ 0 & -1\end{pmatrix}\begin{pmatrix}0 & -i\\ i & 0\end{pmatrix} = \begin{pmatrix}0 & 1\\ 1 & 0\end{pmatrix}\end{aligned} \tag{1.288c}$$

for $\mu = 1$–3. The form (1.287) is antisymmetric on the interchange of α and β and thus represents the $s = 0$ singlet state; the states (1.288a–c) are symmetric and form a representation of the three $s = 1$ triplet states.

Writing the total spin operator as

$$\hat{\mathbf{S}} = \frac{1}{2}(\hat{\boldsymbol{\sigma}}^{(1)} + \hat{\boldsymbol{\sigma}}^{(2)}), \tag{1.289}$$

where (1) and (2) refer to the two ^{3}He atoms making up a pair, and operating on the first of the forms (1.288a–c) we obtain

$$\hat{S}_1\underset{\sim}{\chi}_1 = \hat{S}_2\underset{\sim}{\chi}_2 = \hat{S}_3\underset{\sim}{\chi}_3 = 0; \tag{1.290}$$

i.e. $\hat{\mathbf{S}} \cdot \underset{\sim}{\boldsymbol{\chi}} = 0$. It then follows that the states $\underset{\sim}{\boldsymbol{\chi}}$ *transform as a vector.*

In the absence of a magnetic dipole–dipole interaction (associated with the ^{3}He nuclei) or spin-orbit coupling (very small), we can write the order parameter as a product of the orbital states, containing the variation over the Fermi surface, and spin–space states taken from Eqs. (1.287) and (1.288a–c). The singlet order parameter would then be[49]

$$\underset{\sim}{\psi}^{\text{singlet}} = d_0(\mathbf{p})\underset{\sim}{\chi}_0 = \begin{pmatrix}0 & d_0(\mathbf{p})\\ -d_0(\mathbf{p}) & 0\end{pmatrix} \tag{1.291}$$

[49]For an $l = 0$ (s-wave) superfluid d_0 would be a constant. For $l = 2, 4, \ldots$ it would be angle dependent.

while from (1.288) a triplet state would have the form

$$\underset{\sim}{\psi}^{\text{triplet}} = \mathbf{d}(\mathbf{p}) \cdot \underset{\sim}{\chi} = \begin{pmatrix} -d_x(\mathbf{p}) + id_y(\mathbf{p}) & d_z(\mathbf{p}) \\ d_z(\mathbf{p}) & d_x(\mathbf{p}) + id_y(\mathbf{p}) \end{pmatrix}, \tag{1.292}$$

where $d_\mu(\mathbf{p})$ is a vector in spin space.

We also require a suitable set of orbital states. For the special case involving $l = 1$ pairing these can be chosen to coincide with the three components of the unit vector $\mathbf{p}$, and hence they also transform as a vector:

$$\begin{aligned} \mathbf{p} &= (p_1, p_2, p_3) \\ &= (\sin\theta\cos\varphi, \sin\theta\sin\varphi, \cos\theta). \end{aligned} \tag{1.293}$$

The most general order parameter would be a linear superposition of various possible products of the orbital and spin–space states, $p_i(\theta,\varphi)\underset{\sim}{\chi}_\mu$, with arbitrary complex amplitudes which we denote as $A_{i\mu}(T, P, \mathbf{H})$. As indicated, these parameters will be functions of the external thermodynamic variables (temperature, T, pressure, P, and magnetic field $\mathbf{H}$); in a non-uniform system they would also be a function of the spatial coordinate, $\mathbf{r}$. Our order parameter then has the form

$$\underset{\sim}{\psi}(\mathbf{p}) = \sum_{i=1,\mu=1}^{3} A_{i\mu}(T, P, \mathbf{H}) p_i \underset{\sim}{\chi}_\mu \tag{1.294}$$

while the associated $\mathbf{d}$ vector is

$$d_\mu = \sum_{i=1}^{3} A_{i\mu}(T, P, \mathbf{H}) p_i. \tag{1.295}$$

1.7.3 The G–L free energy of superfluid ³He

Because it was the first firm example of an unconventional order parameter we will treat the Ginzburg–Landau theory of ^{3}He in some detail. To obtain the free energy we must find all of the real quadratic and quartic forms which can be constructed from the order parameter (1.294) and its Hermitian conjugate. In the absence of magnetic dipole–dipole or spin–orbit interactions we must exclude any terms that couple spin space with orbital space; the terms must then be rotationally invariant in both spin and orbital space. To construct a quadratic invariant we average the form

$$\overline{\sum_{\mu,\nu,i,j} A_{\mu i} A^*_{\nu j} p_i p_j \underset{\sim}{\chi}_\mu \underset{\sim}{\chi}^\dagger_\nu} \propto \sum_{i\mu} A_{i\mu} A^*_{i\mu} = \mathrm{Tr}(\mathbf{A} \cdot \mathbf{A}^\dagger),$$

where we have exploited the orthogonality of our orbital and spin states (normalization constants are suppressed here as they can be absorbed into the G–L parameters which will be introduced shortly).

In the fourth order it turns out that there are *five* invariants. They are most easily found by introducing the diagrams shown in Figs. 1.10 and 1.11. Here the dashed and solid lines refer to the spin and orbital space indices

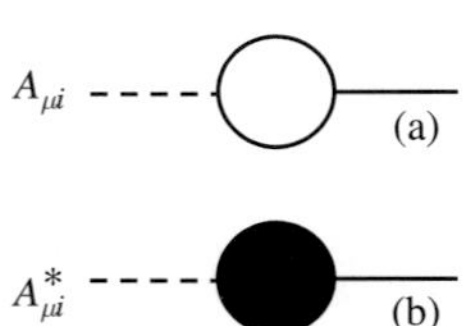

Fig. 1.10 Diagrams depicting **A** and **A***.

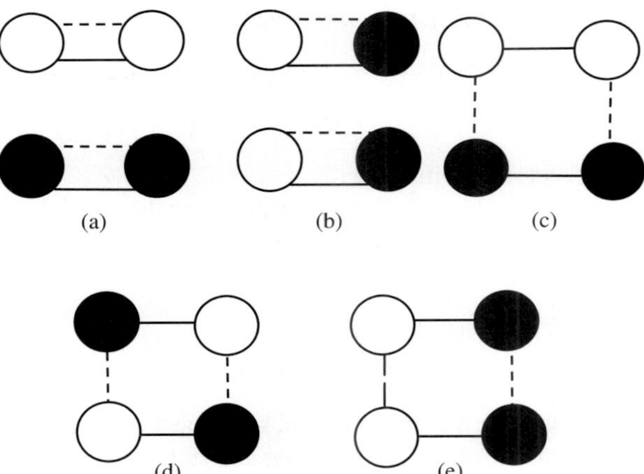

Fig. 1.11 The five quadratic invariants involving $A_{\mu i}$ and $A^*_{\mu i}$.

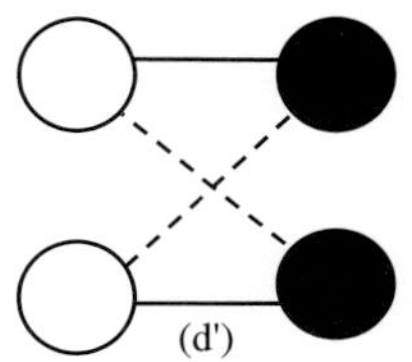

Fig. 1.12 An alternative form for diagram 1.11d.

of **A** respectively, and the filled circle corresponds to its complex conjugate. Connecting a line between two circles corresponds to summing over an index. Since we are excluding terms mixing the spin and orbital degrees of freedom we only connect a solid line to a solid line and a dashed line to a dashed line. Note we could also draw diagram (d) in Fig. 1.11 in the "twisted" form shown in Fig. 1.12. The diagrams (a) through (e) in Fig. 1.11 correspond to the following five quartic invariants.

$$\left(\sum_{i\mu} A_{i\mu} A_{i\mu}\right)\left(\sum_{i\mu} A^*_{i\mu} A^*_{i\mu}\right) = \left|\mathrm{Tr}(\mathbf{A}\cdot\tilde{\mathbf{A}})\right|^2 \tag{1.296a}$$

$$\left(\sum_{i\mu} A_{i\mu} A^*_{i\mu}\right)^2 = \left[\mathrm{Tr}(\mathbf{A}\cdot\mathbf{A}^\dagger)\right]^2 \tag{1.296b}$$

$$\sum_{ij\mu\nu} A_{i\mu} A_{i\nu} A^*_{j\nu} A^*_{j\mu} = \mathrm{Tr}\left[(\mathbf{A}\cdot\tilde{\mathbf{A}})\cdot(\mathbf{A}\cdot\tilde{\mathbf{A}})^*\right] \tag{1.296c}$$

$$\sum_{ij\mu\nu} A_{i\mu} A^*_{i\nu} A_{j\nu} A^*_{j\mu} = \mathrm{Tr}\left[(\mathbf{A}\cdot\mathbf{A}^\dagger)\cdot(\mathbf{A}\cdot\mathbf{A}^\dagger)\right] \tag{1.296d}$$

$$\sum_{ij\mu\nu} A_{i\mu} A^*_{i\nu} A^*_{j\nu} A_{j\mu} = \mathrm{Tr}\left[(\mathbf{A}\cdot\mathbf{A}^\dagger)\cdot(\mathbf{A}\cdot\mathbf{A}^\dagger)^*\right], \tag{1.296e}$$

where Tr denotes the trace, and $\tilde{\mathbf{A}}$ is its transpose of **A**. Introducing the G–L coefficients α and β_i where i runs from 1 to 5 we may now write the free energy density as

$$\begin{aligned} F(\psi) = F(0) &- \alpha\mathrm{Tr}(\mathbf{A}\cdot\mathbf{A}^\dagger) + \beta_1|\mathrm{Tr}(\mathbf{A}\cdot\tilde{\mathbf{A}}^{\mathrm{tr}})|^2 \\ &+ \beta_2[\mathrm{Tr}(\mathbf{A}\cdot\mathbf{A}^\dagger)]^2 + \beta_3\mathrm{Tr}[(\mathbf{A}\cdot\tilde{\mathbf{A}}^{\mathrm{tr}})\cdot(\mathbf{A}\cdot\tilde{\mathbf{A}}^{\mathrm{tr}})^*] \\ &+ \beta_4[\mathrm{Tr}(\mathbf{A}\cdot\mathbf{A}^\dagger)\cdot(\mathbf{A}\cdot\mathbf{A}^\dagger)] + \beta_5\mathrm{Tr}[(\mathbf{A}\cdot\mathbf{A}^\dagger)\cdot(\mathbf{A}\cdot\mathbf{A}^\dagger)^*]. \end{aligned} \tag{1.297}$$

1.7.4 Unitary states

In what follows we will restrict ourselves to so-called *unitary states* which are defined by the condition[50]

$$\underset{\sim}{\psi}\underset{\sim}{\psi}^{\dagger} \propto \underset{\sim}{1}; \tag{1.298}$$

[50] One can show that such states are not magnetically polarized, which is the case for ^{3}He in the absence of a magnetic field.

for the case of a triplet order parameter given by Eq. (1.292),

$$\begin{aligned}
\underset{\sim}{\psi}\underset{\sim}{\psi}^{\dagger} &= \left(\mathbf{d}\cdot\underset{\sim}{\chi}\right)\left(\mathbf{d}\cdot\underset{\sim}{\chi}\right)^{\dagger} \\
&= \begin{pmatrix} \mathrm{d}_z & \mathrm{d}_x + \mathrm{i}\mathrm{d}_y \\ \mathrm{d}_x - \mathrm{i}\mathrm{d}_y & -\mathrm{d}_z \end{pmatrix}\begin{pmatrix} \mathrm{d}_z^* & \mathrm{d}_x^* + \mathrm{i}\mathrm{d}_y^* \\ \mathrm{d}_x^* - \mathrm{i}\mathrm{d}_y^* & -\mathrm{d}_z^* \end{pmatrix} \\
&= |\mathbf{d}|^2\,\underset{\sim}{\sigma}_0 - \mathrm{i}(\mathbf{d}\times\mathbf{d}^*)\cdot\underset{\sim}{\sigma}.
\end{aligned} \tag{1.299}$$

Thus unitarity requires $\mathbf{d}\times\mathbf{d}^* = 0$.

The property $\mathbf{d}\times\mathbf{d}^* = 0$ will be satisfied if $\mathbf{d}$ has the form

$$\mathbf{d} = a\mathbf{n} \tag{1.300}$$

where $\mathbf{n}$ is a unit vector and a is a complex parameter. As discussed above, $\mathbf{d}$ is linear in $\mathbf{p}$ for p-wave pairing. There are then two possible cases.

$$d_\mu = a n_{i\mu} p_i, \quad \text{(case I)} \tag{1.301}$$

where $n_{i\mu}$ are components of a real matrix and a is a complex constant; in this case the vector $\mathbf{n}$ in Eq. (1.300) depends on the angles specifying the orientation of axes defining the p_i, whereas the constant a is independent of the angles. Alternatively we could have

$$d_\mu = a_i p_i n_\mu \quad \text{(case II)} \tag{1.302}$$

where a_i is a complex vector and n_μ are components of a real vector. In this case the parameter a in (1.300) depends on the angles defining p_i and n_μ is independent of the angles.

1.7.5 Minimization of the free energy of an $l = 1, s = 1$ superfluid

We now discuss the minimization of the free energy given in Eq. (1.297) for the two cases discussed above.

1.7.5.1 *Case I*

For this case $\mathbf{A}$ is real except for an irrelevant complex constant. It then follows that $\mathbf{A}\cdot\mathbf{A}^\dagger = \mathbf{A}\cdot\tilde{\mathbf{A}}$ is a *real symmetric matrix*, which can always be diagonalized by a proper rotation of the coordinate system. Since the trace is invariant under such a rotation, the free energy may be expressed in terms of the three real, non-negative, eigenvalues of $\mathbf{A} \cdot \tilde{\mathbf{A}}$: λ_1, λ_2, and λ_3. In evaluating the free energy given in Eq. (1.297) we need

$$\mathrm{Tr}\left[\mathbf{A} \cdot \tilde{\mathbf{A}}\right] = \sum_i \lambda_i \tag{1.303}$$

$$\mathrm{Tr}\left[(\mathbf{A} \cdot \tilde{\mathbf{A}}) \cdot (\mathbf{A} \cdot \tilde{\mathbf{A}})\right] = \sum_i \lambda_i^2. \tag{1.304}$$

Using Eq. (1.363) and (1.364), we may write Eq. (1.299) as $\Delta F = F - F_0$ where

$$\Delta F = -\alpha \sum_i \lambda_i + (\beta_1 + \beta_2)\left(\sum_i \lambda_i\right)^2 + (\beta_3 + \beta_4 + \beta_5)\sum_i \lambda_i^2. \tag{1.305}$$

There are three different cases depending on whether no eigenvalues vanish (three dimensional case), one eigenvalue vanishes (two-dimensional case), and two eigenvalues vanish (one-dimensional case). Three non-vanishing λ_i corresponds to a superfluid involving all three orbital states, p_i; analogously two non-vanishing λ_i corresponds to two orbital states, and a single λ_i to one state. We first minimize Eq. (1.305) for the case of no vanishing eigenvalues:

$$\frac{\partial \Delta F}{\partial \lambda_i} = -\alpha + 2(\beta_1 + \beta_2)\sum_{j=1}^{3} \lambda_j + 2(\beta_3 + \beta_4 + \beta_5)\lambda_i = 0 \tag{1.306}$$

$$\lambda_i = \frac{-\alpha + 2(\beta_1 + \beta_2)\sum_{j=1}^{3} \lambda_j}{2(\beta_3 + \beta_4 + \beta_5)}. \tag{1.307}$$

Summing Eq. (1.307) over i and solving for $\sum_{i=1}^{3} \lambda_i$ we obtain

$$\sum_{i=1}^{3} \lambda_i = \frac{3\alpha}{6(\beta_1 + \beta_2) + 2(\beta_3 + \beta_4 + \beta_5)}; \tag{1.308}$$

inserting (1.308) into Eq. (1.307) we obtain

$$\lambda_i = \frac{\alpha}{6(\beta_1 + \beta_2) + 2(\beta_3 + \beta_4 + \beta_5)}. \tag{1.309}$$

Finally, plugging (1.308) and (1.309) into (1.305) we obtain the free energy

$$\Delta F = -\frac{3}{4}\alpha^2 \frac{1}{3(\beta_1 + \beta_2) + (\beta_3 + \beta_4 + \beta_5)}. \quad \textit{(3D state)} \tag{1.310}$$

The fact that all three eigenvalues are equal implies $\mathbf{A} \propto \mathbf{R}$, a rotation matrix, since $\mathbf{R} \cdot \tilde{\mathbf{R}} = \mathbf{1}$. We may then express $\mathbf{d}$ in the form,

$$\mathbf{d} = \Delta \mathbf{R}(\boldsymbol{\theta}) \cdot \mathbf{p} \quad \textit{(3D state)} \tag{1.311}$$

where Δ is a complex constant, referred to as the gap parameter, and $\mathbf{R}(\boldsymbol{\theta})$ is a rotation matrix for an arbitrary angle about an arbitrary axis. The 3D state is also called the isotropic state.

We will only quote the results for two and one dimensional cases since they do not occur in practice. For the two-dimensional, also called planar, case where we have two non vanishing eigenvalues λ_i, the free energy is

$$\Delta F = \frac{\alpha^2}{2} \frac{1}{2(\beta_1 + \beta_2) + (\beta_3 + \beta_4 + \beta_5)}. \quad \textit{(planar state)} \tag{1.312}$$

The corresponding order parameter is

$$\mathbf{d} = \Delta \left[\boldsymbol{l}_1 (\mathbf{m}_1 \cdot \mathbf{p}) + \boldsymbol{l}_2 (\mathbf{m}_2 \cdot \mathbf{p}) \right]; \quad \textit{(planar state)} \tag{1.313}$$

here $\boldsymbol{l}_1, \boldsymbol{l}_2$ are a pair of orthogonal unit vectors that project out two orthogonal combinations of the three orbital basis functions, $\mathbf{p}$, while $\mathbf{m}_1, \mathbf{m}_2$ denote another orthogonal pair, arbitrarily oriented with respect to the first pair, that project out two orthogonal combinations of the three spin basis functions, $\boldsymbol{\chi}$.

The free energy for the one-dimensional or polar case, where there is a single eigenvalue λ_i, is given by

$$\Delta F = \frac{\alpha^2}{4(\beta_1 + \beta_2 + \beta_3 + \beta_4 + \beta_5)}. \quad \textit{(polar state)} \tag{1.314}$$

The corresponding **d** vector is

$$\mathbf{d} = \Delta \boldsymbol{l} (\mathbf{m} \cdot \mathbf{p}), \quad \textit{(polar state)} \tag{1.315}$$

where $\boldsymbol{l}$ and $\mathbf{m}$ are two arbitrarily directed unit vectors defining an axis in the spin and orbital spaces respectively.

1.7.5.2 *Case II*

In this case we had

$$\mathbf{d} = (\mathbf{a} \cdot \mathbf{p})\mathbf{n} = \mathbf{A} \cdot \mathbf{p} \tag{1.316}$$

where **A** has the form

$$\mathbf{A} = \begin{pmatrix} a_x n_x & a_y n_x & a_z n_x \\ a_x n_y & a_y n_y & a_z n_y \\ a_x n_z & a_y n_z & a_z n_z \end{pmatrix} \tag{1.317}$$

for which we calculate

$$\mathrm{Tr}[\mathbf{A} \cdot \mathbf{A}^\dagger] = \left(a_x a_x^* + a_y a_y^* + a_z a_z^*\right)\left(n_x^2 + n_y^2 + n_z^2\right) = \mathbf{a} \cdot \mathbf{a}^*; \tag{1.318}$$

similarly

$$\mathrm{Tr}\left|\mathbf{A} \cdot \tilde{\mathbf{A}}\right|^2 = |\mathbf{a} \cdot \mathbf{a}|^2 \tag{1.319a}$$

$$\mathrm{Tr}\left(\mathbf{A} \cdot \mathbf{A}^\dagger\right)^2 = \left(\mathbf{a} \cdot \mathbf{a}^*\right)^2 \tag{1.319b}$$

$$\mathrm{Tr}\left[(\mathbf{A} \cdot \tilde{\mathbf{A}}) \cdot (\mathbf{A} \cdot \tilde{\mathbf{A}})^*\right] = |\mathbf{a} \cdot \mathbf{a}|^2 \tag{1.319c}$$

$$\mathrm{Tr}\left[(\mathbf{A}\cdot\mathbf{A}^{\dagger})\cdot(\mathbf{A}\cdot\mathbf{A}^{\dagger})\right]=(\mathbf{a}\cdot\mathbf{a}^{*})^{2} \tag{1.319d}$$

$$\mathrm{Tr}\left[(\mathbf{A}\cdot\mathbf{A}^{\dagger})(\mathbf{A}\cdot\mathbf{A}^{\dagger})^{*}\right]=(\mathbf{a}\cdot\mathbf{a}^{*})^{2} \tag{1.319e}$$

Substituting Eq. (1.315) and (1.319a–e) into Eq. (1.297) we obtain

$$\Delta F=-\alpha(\mathbf{a}\cdot\mathbf{a}^{*})+(\beta_1+\beta_3)\,|\mathbf{a}\cdot\mathbf{a}|^{2}+(\beta_2+\beta_4+\beta_5)(\mathbf{a}\cdot\mathbf{a}^{*})^{2}. \tag{1.320}$$

Note that $0\le|\mathbf{a}\cdot\mathbf{a}|\le(\mathbf{a}\cdot\mathbf{a}^{*})$; in particular $|\mathbf{a}\cdot\mathbf{a}|$ can be zero. Therefore if $(\beta_1+\beta_3)>0$, the free energy is minimized by having $|\mathbf{a}\cdot\mathbf{a}|=0$. We write $\mathbf{a}=\mathbf{a}_1+i\mathbf{a}_2$; then $\mathbf{a}\cdot\mathbf{a}=a_1^2-a_2^2+2i\mathbf{a}_1\cdot\mathbf{a}_2$. Thus if $|\mathbf{a}_1|=|\mathbf{a}_2|$ and $\mathbf{a}_1\perp\mathbf{a}_2$, we have $\mathbf{a}\cdot\mathbf{a}=0$ and $\mathbf{a}\cdot\mathbf{a}^{*}=a_1^2+a_2^2=+2a_1^2$. In this case

$$\Delta F=2\alpha a_1^2+4(\beta_2+\beta_4+\beta_5)a_1^4. \tag{1.321}$$

Minimizing Eq. (1.321) with respect to a_1 we obtain

$$a_1^2=\frac{\alpha}{4(\beta_2+\beta_4+\beta_5)}, \tag{1.322}$$

from which it follows that

$$\Delta F=-\frac{\alpha^2}{4}\frac{1}{\beta_2+\beta_4+\beta_5}. \tag{1.323}$$

This state is called the axial or Anderson–Brinkman–Morel (ABM) state. The associated vector $\mathbf{d}$ has the form.

$$\mathbf{d}=\mathbf{n}\,(\mathbf{a}\cdot\mathbf{p}) \tag{1.324}$$

and

$$\mathbf{a}=\Delta(\mathbf{m}_1+i\mathbf{m}_2) \tag{1.325}$$

where $\mathbf{m}_1$ and $\mathbf{m}_2$ are two orthogonal unit vectors which project two independent combinations of orbital states; we define a third direction, $\boldsymbol{l}$, the gap axis (which microscopically is also the angular momentum axis), by

$$\boldsymbol{l}=\mathbf{m}_1\times\mathbf{m}_2. \tag{1.326}$$

For the case $\beta_1+\beta_3<0$ the minimal occurs when $|\mathbf{a}\cdot\mathbf{a}|$ is maximum, i.e. when it is equal to $\mathbf{a}\cdot\mathbf{a}^{*}$;[51] minimization results in the same free energy obtained for the one-dimensional case. It follows that, regardless of the signs of $(\beta_3+\beta_4+\beta_5)$ and $(\beta_1+\beta_2)$ that the 2D state is *never stable*. For $\beta_1+\beta_2>0$, and $\beta_3+\beta_4+\beta_5>0$ the 3D state will have a lower energy than the 1D state.

[51] For this case the vector $\mathbf{a}$ may be written $\mathbf{a}=\Delta e^{i\phi}$, where Δ is a real vector.

From the weak coupling theory of superconductivity (which we do not discuss) one can show that $\beta_1=-\beta$, $\beta_2=\beta_3=\beta_4=2\beta$, and $\beta_5=-2\beta$; here β is a parameter. Inserting these results in Eqs. (1.310), (1.314) and (1.320) we obtain

$$\left.\begin{aligned} &\text{3D}: \quad \Delta F = -\frac{3\alpha^2}{20\beta} \\ &\text{1D}: \quad \Delta F = -\frac{\alpha^2}{12\beta} \\ &\text{Axial}: \quad \Delta F = -\frac{\alpha^2}{8\beta} \end{aligned}\right\} \text{weak coupling} \tag{1.327}$$

It follows that in weak coupling theory, the 3D state has the lowest energy and ^{3}He B is identified with this state (also called the Balian–Werthamer state). However, in the strong coupling theories of Anderson and Brinkman the possibility that the axial phase may be more stable in some situations is rationalized, and indeed ^{3}He A is identified as the ABM phase.

1.7.6 Anisotropy of the energy gap in an $l = 1$ $s = 1$ superfluid; order parameter of ^{3}He

For the 3D phase $\mathbf{d} = \Delta\mathbf{R}\cdot\mathbf{p}$ by Eq. (1.311); since the axes are arbitrary we take

$$\mathbf{R} = \begin{pmatrix} 1 & 0 & 0 \\ 0 & 1 & 0 \\ 0 & 0 & 1 \end{pmatrix}, \quad \mathbf{A} = \Delta\begin{pmatrix} 1 & 0 & 0 \\ 0 & 1 & 0 \\ 0 & 0 & 1 \end{pmatrix} \tag{1.328}$$

Using $\underset{\sim}{\psi} = \mathbf{d}\cdot\underset{\sim}{\chi}$ we obtain

$$\underset{\sim}{\psi} = \Delta\begin{pmatrix} p_x + ip_y & -p_z \\ -p_z & -p_x + ip_y \end{pmatrix} \tag{1.329}$$

The square of the gap energy is proportional to Tr $[\underset{\sim}{\psi}\underset{\sim}{\psi}^\dagger]$ for which we have

$$\begin{aligned} \mathrm{Tr}\left[\underset{\sim}{\psi}\underset{\sim}{\psi}^\dagger\right] &= 2\Delta^2\hat{p}^2 \\ &= 2\Delta^2; \end{aligned} \tag{1.330}$$

thus we note the gap is *isotropic* in the 3D phase.

For the 1D case, $\mathbf{d} = \boldsymbol{l}(\mathbf{m}\cdot\mathbf{p})\Delta$ by Eq. (1.315). The axes are again arbitrary so we choose $\boldsymbol{l} \parallel \mathbf{m} \parallel \mathbf{z}$ obtaining

$$\mathbf{A} = \Delta\begin{pmatrix} 0 & 0 & 0 \\ 0 & 0 & 0 \\ 0 & 0 & 1 \end{pmatrix} \tag{1.331}$$

and

$$\underset{\sim}{\psi} = \Delta\begin{pmatrix} 0 & -p_z \\ -p_z & 0 \end{pmatrix} \tag{1.332}$$

from which it follows, using $k_z = \cos\theta$,

$$\begin{aligned}\mathrm{Tr}\left[\underset{\sim}{\psi}\underset{\sim}{\psi}^{\dagger}\right] &= 2\Delta^2 k_z^2 \\ &= 2\Delta^2\cos^2\theta;\end{aligned} \tag{1.333}$$

thus the gap has a maximum at the pole and vanishes along the equator and it is for this reason that the state is referred to as the polar state.

For the axial state $\mathbf{d} = \Delta\mathbf{n}((\mathbf{m}_1 + i\mathbf{m}_2)\cdot\mathbf{p})$ by Eqs. (1.324) and (1.325); if we take $\mathbf{n} \parallel \mathbf{y}$, $\mathbf{m}_1 \parallel \mathbf{x}$, $\mathbf{m}_2 \parallel \mathbf{y}$, then $\mathbf{A}$, $\underset{\sim}{\psi}$ and $\mathrm{Tr}[\underset{\sim}{\psi}\underset{\sim}{\psi}^{\dagger}]$ have the form

$$\mathbf{A} = i\Delta\begin{pmatrix} 0 & 0 & 0 \\ 1 & i & 0 \\ 0 & 0 & 0 \end{pmatrix}, \tag{1.334}$$

$$\underset{\sim}{\psi} = i\Delta\begin{pmatrix} p_x + ip_y & 0 \\ 0 & p_x + ip_y \end{pmatrix} \tag{1.335}$$

and

$$\begin{aligned}\mathrm{Tr}\left[\underset{\sim}{\psi}\underset{\sim}{\psi}^{\dagger}\right] &= 2(p_x^2 + p_y^2) \\ &= 2\Delta^2\sin^2\theta;\end{aligned} \tag{1.336}$$

thus in the axial state the gap vanishes on the pole and has a maximum on the equator.

1.7.7 Perturbations to the equilibrium G–L free energy

For completeness we will list, without proof, three important additional contributions to the G–L free energy. For an excellent discussion of these terms see the review by Richardson and Lee [47].

1.7.7.1 *The nuclear dipole–dipole energy*

The free energy which we minimized to determine the structure of the equilibrium order parameter of the A and B phases contained no terms that couple spin to the orbital degrees of freedom and hence nothing fixes their relative orientation. In practice their relative alignment is fixed by the dipole–dipole interaction which contributes a term to the free energy of the form

$$F_D = g_D\left[|\mathrm{Tr}\mathbf{A}|^2 + \mathrm{Tr}\left(\mathbf{A}\cdot\mathbf{A}^*\right) - \frac{2}{3}\mathrm{Tr}\left(\mathbf{A}\cdot\mathbf{A}^{\dagger}\right)\right], \tag{1.337}$$

where $g_D > 0$ is a coupling constant. Minimizing the energy arising from this term for the A phase results in $\mathbf{d} \parallel \boldsymbol{l}$ where $\boldsymbol{l}$ is defined by Eq. (1.326). For the case of the B phase only the angle between $\mathbf{d}$ and $\mathbf{k}$ is fixed, at a value $\theta = \cos^{-1}(1/4)$. The magnetic dipole–dipole energy plays an important role in nuclear magnetic resonance experiments as interpreted by Leggett, which provided the first clues as to the nature of the A and B phases.

1.7.7.2 *Effect of an external magnetic field*

An external magnetic field interacts with the superfluid ^{3}He through both a linear term and a quadratic Zeeman term according to the following forms

$$F_H = \eta \sum_{ijk\mu} \varepsilon_{ijk} H_i A_{j\mu} A_{k\mu} + g_Z \sum_{ij\mu} H_i A_{i\mu} A^*_{j\mu} H_j, \tag{1.338}$$

where η and g_z are parameters. Near the transition to the superfluid A phase, the first term in (1.338) causes a *splitting* of the A phase into the so-called A_1 phase followed by the A phase. The second term results in $\mathbf{d}\perp\mathbf{H}$ in the equilibrium A phase and an orientation independent energy in the B phase. High fields suppress the B phase and for sufficiently high fields it is absent.

1.7.7.3 *Gradient energies*

The last contribution which we list arises from spatial inhomogeneities in the order parameter and has the form

$$F_G = K_1 \sum_{i\nu\lambda} \frac{\partial}{\partial x_\nu} A_{i\nu} \frac{\partial}{\partial x_\lambda} A^*_{i\lambda} + K_2 \sum_{i\nu\lambda} \frac{\partial}{\partial x_\nu} A_{i\lambda} \frac{\partial}{\partial x_\nu} A^*_{i\lambda} + K_3 \sum_{i\nu\lambda} \frac{\partial}{\partial x_\lambda} A_{i\nu} \frac{\partial}{\partial x_\nu} A^*_{i\lambda}. \tag{1.339}$$

These energies play a crucial role in determining the structure of the liquid-crystal-like structures that form in the presence of boundaries, and when the liquid flows or is in rotation.

1.8 Superfluidity: recent developments

1.8.1 General remarks on superfluid transitions

Recently there has been much progress in the theory of superfluidity in Bose and Fermi systems in the presence of *externally controlled* one- and two-body potentials. This was stimulated, in part, by the development of new experimental techniques that allow whole new classes of experiments to be performed, some of which are discussed in Sections 2.5 and 2.6 of the next chapter. These include tuning the inter-particle coupling with a magnetic field, through the so-called Feshbach resonances, and application of a background potential, imposed by optical standing waves, the so-called optical lattices [51]. These experiments and the accompanying theories have shed new light on the nature and occurrence of superfluidity [52–54].

Two key parameters controlling composite-fermion (BEC) versus paired-fermion (BCS) superfluidity are: (i) the size of the molecular bosons or Cooper pairs (controlled by the interaction strength between the fermions) and (ii) their mutual overlap. The latter involves the inter-particle separation, d, which follows from the superfluid density n_s, scaling as $d \propto n^{-1/3}$ in three dimensions. On physical grounds, and motivated by the Heisenberg uncertainty relations ($\Delta x \Delta p \sim \hbar, \ldots$), both limits should occur: local Bose molecules and small Cooper pairs with weak overlap, as well as larger (delocalized) bosons and Cooper pairs with stronger overlap. For weakly interacting local bosons we expect a BEC transition where all particles condense into the $\mathbf{p} = 0$ state. For larger entities, Cooper pairs form consisting of fermions near ε_F with momenta $\mathbf{p}$ and $-\mathbf{p}$ and opposite spin (due to the Pauli principle), and one sees a collective transition into the BCS state involving pair sizes of order the coherence length ξ with the overlap proportional to $n^{-1/3}$.

The superfluid transition for varying coupling strength, and specifically a crossover from BCS to Bose superfluidity, has been intensively studied theoretically over the years [55–59]. In particular, boson-like behavior of the Cooper pairs for strong coupling, short coherence length, and low superfluid density has been expected for a long time. As discussed by Leggett, Nozieres and others, fundamental considerations, e.g. the Heisenberg uncertainty relationships, suggest a continuous BEC⇔ BSC transition should occur between these superfluid states upon varying the interaction. This has been confirmed by the experiments of Greiner et al. and Regal et al. [60, 61].

Furthermore, assuming repulsive interaction between bosons, one expects a transition from a superfluid phase to a Mott insulator at low temperatures and high densities, which is enhanced by the presence of an optical lattice. For a large interaction parameter, U, the bosons are essentially frozen on the lattice sites, thereby forming an ordered "insulating" state in which the kinetic energy, characterized by a hopping parameter J (see Eq. (1.348) below), is too small to overcome the inter-particle repulsion energy present when two (or more) atoms are on the same site [62–64]. The energy gap characterizing the insulating state results in a splitting from the repulsive interaction U between the bosons. Of course, the phase diagram depends on the density of bosons, specifically the number of bosons per lattice site.

The interplay of the two most important superfluid control parameters can be studied in an optical lattice where the depth of the lattice potential and superfluid density can be varied. For the fermions, which in the limit of strong coupling form bosonic molecules, one may use the Hamiltonian

$$\hat{H} = \hat{\psi}_\sigma^\dagger \left(-\frac{\hbar^2 \nabla^2}{2m} - \mu \right) \hat{\psi}_\sigma - g\, \hat{\psi}_\uparrow^\dagger \hat{\psi}_\downarrow^\dagger \hat{\psi}_\downarrow \hat{\psi}_\uparrow; \tag{1.340}$$

here μ is the chemical potential, determined by the particle density n, and g is the (local) coupling among the fermions. Although we will not go through the derivations here, one obtains from $\hat{H}$ the usual BCS-theory with an order parameter given by $\Delta \propto \langle \hat{\psi}_\uparrow \hat{\psi}_\downarrow \rangle \equiv \psi$, where the latter defines the G–L order parameter. Using Green function methods [65] and functional integral analysis (see Sa de Melo et al.) one can obtain a *time-dependent* Ginzburg–Landau equation (TDGL) equation generalizing Eq. (1.188) as

$$\left(a + b|\psi(\mathbf{r},t)|^2 - c\frac{\hbar^2}{2m}\nabla^2 \right) \psi(\mathbf{r},t) = i\hbar d \frac{\partial \psi(\mathbf{r},t)}{\partial t}, \tag{1.341}$$

along with expressions for the coefficients.[52] This equation determines the superfluid phase transition and the properties of the superfluid. Note that d is complex with the real part implying $\psi(\mathbf{r},t)$ has a propagating (non-diffusive) component. For weak coupling, $g \to 0$, one expects the BCS-state with delocalized singlet Cooper pairs of size ξ; for strong coupling, $g \to \infty$, we expect localized fermion pairs ($\xi \to 0$), i.e. bosonic molecules.

[52] Using a functional integral technique and the Hubbard–Stratonovich transformation, Sa de Melo et al. obtain the coefficients in Eq. (1.341) as

$$a = -\frac{m}{4\pi a_s} + \sum_{\mathbf{k}} \left(\frac{1}{2\varepsilon_{\mathbf{k}}} - \frac{X}{2\xi_{\mathbf{k}}} \right) \to N(0) \ln\left(\frac{T}{T_c}\right)$$

$$b = \sum_{\mathbf{k}} \left(\frac{X}{4\xi_{\mathbf{k}}^3} - \frac{\beta Y^2}{8\xi_{\mathbf{k}}^2} \right) = \frac{7\zeta(3)N(0)}{8\pi^2 (k_B T_c)^2}$$

$$c = \sum_{\mathbf{k}} \left[\frac{X}{4\xi_{\mathbf{k}}^2} - \frac{\beta Y}{8\xi_{\mathbf{k}}} + \left(\beta^2 XY + \frac{\beta Y}{\xi_{\mathbf{k}}} - \frac{2X}{\xi_{\mathbf{k}}^2} \right) \frac{(k\cos\theta)^2}{8m\xi_{\mathbf{k}}} \right] \to \frac{7\zeta(3)N(0)\varepsilon_F}{12\pi^2 (k_B T_c)^2}.$$

$$d = \sum_{\mathbf{k}} \frac{\tanh(\beta\,\xi_{\mathbf{k}}/2)}{4\xi_{\mathbf{k}}^2} + i\frac{\pi}{8} N(\varepsilon_F)\beta\sqrt{\frac{\mu}{\varepsilon_F}} \cdots \to i\frac{\pi N(0)}{8k_B T_c}\left(1 - 2i\frac{k_B T_c}{\pi\varepsilon_F}\right),$$

(for energies between Δ and $\min(k_B T_c, |\mu|)$),

where $X = \tanh(\beta\xi_{\mathbf{k}}/2)$, $Y = \operatorname{sech}^2(\beta\xi_k/2)$, and $\beta = 1/k_B T$. The arrows refer to results for the weak coupling (BCS) limit, $g \to 0$, and $1/k_F a_s \to -\infty$.

Note, that the Gross–Pitaevskii equation (see Sec. 1.4.2)

$$\left(-\mu'+U|\psi(\mathbf{r},t)|^2 - \frac{\hbar^2}{2M}\nabla^2\right)\psi(\mathbf{r},t) = i\hbar\frac{\partial\psi(\mathbf{r},t)}{\partial t}, \tag{1.342}$$

with $U = 4\pi ab/M$, $\mu' = E_b - 2|\mu|$, $M = 2m$, is a *special form* of the TDGL equation and describes a dilute gas of bosons of mass 2 m and binding energy E_b with repulsive interaction U between particles.[53] Note, the chemical potential of the bosons controls the transition towards BCS Cooper pairs via the change of sign in μ'.

[53]From the TDGL equation one gets in the strong coupling limit the Gross–Pitaevskii equation for a dilute gas of bosons with mass $M = 2\,m$ and repulsive interaction $U = 8\pi a_s/M$. Note, the chemical potential plays the role of the energy in the Schrodinger equation.

1.8.2 BCS, BEC transitions and the crossover transition

The phase transitions occurring in a system of interacting fermions for varying coupling strength and density may be analyzed using standard statistical mechanics and many-body theory.

As usual the chemical potential determines the particle density

$$n = \sum_{\mathbf{k}}\left[1 - \tanh\left(\frac{\xi_{\mathbf{k}}}{2k_BT}\right)\right] + \Delta n, \tag{1.343}$$

where $\xi_{\mathbf{k}} = \varepsilon_{\mathbf{k}} - \mu$ and Δn results from fluctuations which become more important as g increases.[54]

[54]Note, the fluctuations become important near the transition from bosonic molecules to BCS Cooper pairs. For strong coupling ($g \to \infty$) one gets (see Sa de Melo et al. [55])

$$\Delta n = \sum_{\mathbf{k}} \frac{1}{\exp(\beta\hbar\omega_q) - 1}(\hbar\omega_q - 2\mu),$$

where $\hbar\omega_q = -E_b + \hbar^2q^2/4\,m$.

The coupling, g, between the fermions is expressed in terms of the s-wave scattering length a_s through the relation

$$m/4\pi a_s = -\frac{1}{g} + \sum_{\mathbf{k}}\frac{1}{2\varepsilon_{\mathbf{k}}}, \tag{1.344}$$

where the limit $g \to 0$ corresponds to BCS-theory while $g \to \infty$ corresponds to BEC. Then

$$-\frac{m}{4\pi a_s} = \sum_{\mathbf{k}}\left[\frac{1}{2\xi_{\mathbf{k}}}\tanh\left(\frac{\xi_{\mathbf{k}}}{2k_BT_c}\right) - \frac{1}{2\varepsilon_k}\right], \tag{1.345}$$

where $\Delta(T_c) = 0$ and fluctuations are neglected. Note the generalized BCS coupling constant determines the size of the Cooper pairs or bosonic molecules.

Clearly, the self-consistent solution of the last two equations should yield weakly coupled Cooper pairs, as determined in BCS-theory; however it should also yield g for stronger coupling corresponding to a small coherence length, ξ, or localized bosonic molecules. Detailed analysis by Sa de Melo et al. in [55] supports this physical expectation and gives the following results:

(1) weak coupling, $g \to 0$, $1/k_Fa_s \to -\infty$,

$$\mu = \varepsilon_F; k_BT_c = \alpha\exp\left(-\pi/2k_F|a_s|\right); \tag{1.346}$$

with $\alpha = 8\gamma\varepsilon_F/\pi e^2$ where $\gamma = 1.781$; this is the BCS result for T_c.

(2) strong coupling, $g \to \infty$, $\mu(T_c) = -E_b/2$, where E_b is the binding energy of molecule,

$$k_BT_c \simeq 0.218\varepsilon_F. \tag{1.347}$$

Here T_c is the superfluid transition temperature for weakly interacting bosons of mass $2m$ and density $n/2$.

Thus, in an elegant way, one is able to treat both the BCS and BEC superfluid transitions as well as a continuous BCS $\Leftrightarrow$ BEC crossover; this also follows from the TDGL theory. For physical reasons this crossover transition has been expected for a long time [56–65]. For illustration of the crossover phase transition see Fig. 1.13 [55] and Fig. 1.14 [64].

1.8.3 The Mott insulator—superfluid phase transition

The Hamiltonian $\hat{H}$ can be rewritten in a Wannier or tight-binding form in terms of indices referring to (optical) lattice sites. This might be particularly useful for treating alloy systems (different bosonic molecules, etc.), lattice vacancy effects etc. and to apply coherent potential approximation (CPA) which has been successfully used in tight-binding, Wannier-based theories for electrons in crystals.

As an example, one frequently uses the Bose–Hubbard Hamiltonian to treat optical lattices

$$\hat{H} = -J\sum_{i,j}\hat{b}_i^{\dagger}\hat{b}_j - \mu\sum_i \hat{n}_i + \frac{U}{2}\sum_i \hat{n}_i(\hat{n}_i - 1). \qquad (1.348)$$

The first term, with $J > 0$, describes the hopping of bosons between lattice sites i, j. The second term is used to fix the concentration; here $\hat{n}_i = \hat{b}_i\hat{b}_i$ is the on site occupation number operator and μ is the chemical potential.[55] The last term accounts for interactions where $U > 0$ is the repulsive interaction between bosons. Using this Hamiltonian the behavior of the system can be analyzed; in particular it yields the transition from the insulating Mott state, where the bosons are frozen on the lattice sites, to the superfluid state, where the bosons can flow unimpeded between lattice sites.

[55] The on-site binding energies, $\sum_i \varepsilon_i\hat{n}_i$ have been suppressed by shifting the zero of energy and taking ε_i identical.

Assuming the temperature is sufficiently low, qualitatively we expect a Mott insulator state for $U > J$ and a superfluid state for $U < J$, where coherent hopping of the bosons between lattice sites dominates. For analysis one may use tight-binding Green's function methods (where $G = G_0 + G_0 J G$), as in the case of phase transitions for electrons on a lattice. To include fluctuations, mean field theory is extended via the Monte Carlo method. One obtains a phase diagram μ/U versus J/U with an insulating state characterized by an energy gap and a vanishing gap for the superfluid state. Simplifying the analysis (for $U > J$). For the insulating phase Nogueira finds a gap in the energy spectrum which is controlled by U; as J increases relative to U one transitions to a superfluid state.

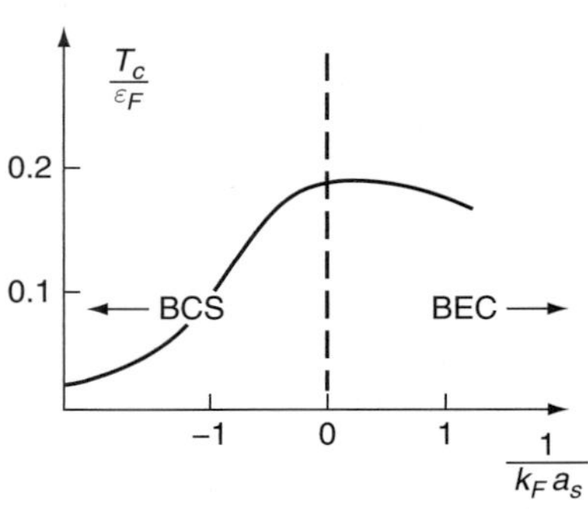

Fig. 1.13 Schematic illustration of the BCS $\Leftrightarrow$ BEC transition. (After Sa de Melo et al. [55])

Detailed analysis using standard methods, Green function theory etc. yields the following properties as a function of coupling U for the energy spectrum approximately

(1) weak coupling limit ($U \ll J$, frequent inter-site hopping)

$$\mu = -2dJ - \frac{U}{2} + n \qquad (1.349)$$

and

$$E_k = \pm\left[\left(\varepsilon_k - \mu - \frac{U}{2} + 2Un\right)^2 - U^2n^2\right]^{1/2}; \qquad (1.350)$$

hence we have a gapless spectrum, as required for a superfluid state.

(2) strong coupling ($U \gg J$), where hopping is suppressed and the energy spectrum is given approximately by

$$E_{\pm}(k) = -\mu + \frac{1}{2}\left(\varepsilon_k + (2\nu+1)U\right) \pm \frac{1}{2}\left(\varepsilon_k^2 + 2(2\nu+1)\varepsilon_k U + U^2\right)^{1/2}. \qquad (1.351)$$

Hence, for large U one has an energy gap,

$$\Delta_g = E_+ - E_- = \left((2dJ)^2 - 4d(2\nu+1)UJ + U^2\right)^{1/2} \qquad (1.352)$$

and this indicates that the system is an insulator. Note, approximately $\Delta \sim U + \ldots$ for $U \gg J$.

As the repulsive interaction U decreases a critical value, U_c, is reached where the system becomes superfluid; the condition

$$\Delta_g(U_c) = 0 \qquad (1.353)$$

determines the phase boundary between the insulating and superfluid state, with the critical value given by

$$U_c = 2dJ\left(2\nu+1 + 2(\nu(\nu+1))^{1/2}\right). \qquad (1.354)$$

Note this depends on ν, the number of particles per site. One gets a series of "Mott" lobes with decreasing maximal values J/U for μ/U at 0.5, 1.5 etc. and minima at $\mu/U = 1, 2, 3$, etc.

Using Monte Carlo methods one may treat fluctuations which can become important near the phase transition. In particular near the maximal critical values J/U such fluctuations contribute important corrections to the mean field results. As an example, where as the mean field theory gives for ($\nu = 1$) $J/U_c \approx 0.028$ the Monte Carlo calculation gives 0.034.

We refer the reader to the literature for the behavior of bosons on a lattice [66, 67] and for various interesting physical situations. Such studies may also be related to those of Cooper pairs treated by Wannier type or tight-binding type methods on a lattice [68]. It is possible that BCS → BEC transitions can also be induced by special geometries for systems in reduced dimensions.

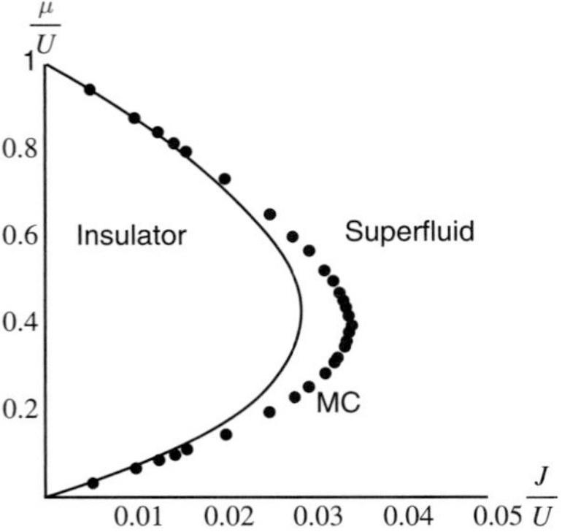

Fig. 1.14 Illustration of transition from Mott insulator to superfluid phase for a system of bosons with repulsive interaction U on a lattice ($\nu = 1$). MC refers to Monte Carlo calculation and J to hopping between lattice sites. (Nogueira, unpublished.)

1.8.4 Summary

We note that extending the Bose–Hubbard Hamiltonian $\hat{H}$ to fermion operators, thereby allowing for the decay of the bosons into two fermions and their subsequent reformation into bosons, a similar simple analysis will yield a phase diagram with BEC and BCS phases and a (continuous) transition between them. It would be interesting to relate theories of superconductivity describing Cooper pairs with Wannier lattice site indices to the problem of BEC–BCS crossover.

This discussion of superfluidity, extending the earlier one for the classical superfluid, ^{4}He, already explains many recent studies. Of course, the analysis may be extended to spin-dependent transitions, triplet pairing (as in ^{3}He), dynamics of the phase transitions (as governed by the TDGL and Gross–Pitaevskii equation) etc. Clearly this field will remain active for many years to come.

References

[1] Kamerlingh Onnes, H. The Superconductivity of Mercury. *Comm. Phys. Lab. Univ. Leiden*; Nos. 122 and 124 (1911).

[2] van Delft, D. and Kes, P. The discovery of superconductivity, *Physics Today*, September 2010.

[3] de Bruyn Ouboter, R. Heike Kamerlingh Onnes's Discovery of Superconductivity, *Sci. Am.* **276** (3): 98–103 (1997).

[4] Meissner, W. and Ochsenfeld, R. Ein neuer effekt bei eintritt der supraleitfähigkeit, *Naturwissenschaften* **21** (44): 787–788 (1933).

[5] London, F. and London, H. The electromagnetic equations of the supraconductor, *Proc. Roy. Soc.* (London), **A149**, 71 (1935).

[6] *Superfluids: The Macroscopic Theory of Superconductivity*, Vol. I, London, F. reprinted by Dover Publications, Inc., New York (1961).

[7] Keesom, W. H. *Helium*, Elsevier, Amsterdam (1942).

[8] McLennan, J. C. Smith, H. D. and Wilhelm, J. O. *Phil. Mag.* **14**, 161 (1932).

[9] *Superfluids: The Macroscopic Theory of Superfluid Helium*, Vol. II, London, F. reprinted by Dover Publications, Inc., New York (1961).

[10] J. Wilks, *The Properties of Liquid and Solid Helium*, Clarendon Press, Oxford (1967).

[11] Wilks, J. and Betts, D. S. *An Introduction to Liquid Helium*, Clarendon Press, Oxford (1987).

[12] Dana, L. I. and Kemerlingh Onnes, H. Verslag Akad. Wetenschappen Amsterdam *Proc. Acad. Sci. Amsterdam*, **29**, 1051 and 1061 (1926).

[13] *Cryogenic Science and Technology*: Contributions by Leo I. Dana, edited by Donnelly, R. J. and Francis, A. W. Union Carbide Corp. (1985).

[14] Ahlers, G. Experiments near the superfluid transition in ^{4}He and ^{3}He – ^{4}He Mixtures in *The Physics of Liquid and Solid Helium*, Bennemann, K. H. and Ketterson, J. B. Ed., Wiley Interscience. Part I, Chapter 2, Pg 85 (1976).

[15] Allen, J. F. and Misener, A. D. Flow of liquid helium II, *Nature* **141**, 75 (1938).

[16] Kapitza, P. Viscosity of Liquid Helium below the λ-Point, *Nature* **141**, 74 (1938).

[17] London, F. On the Bose–Einstein Condensation *Phys. Rev.* **54**, 947 (1938).

[18] Bose, S. N. Planck's Gesetz und Lichtquantenhypothese, *Zeitschrift für Physik* **26**:178–181 (1924).

[19] Einstein, A. Quantentheorie des einatomigen idealen Gases. *Sitzungsberichte der Preussischen Akademie der Wissenschaften* **1**, 3 (1925).

[20] Bogoliubov, N. N. On the theory of superfluidity, *J. Phys. USSR*, **11**, 23, (1947).

[21] Landau, L. The theory of superfluidity in helium II, *J. Phys. USSR*, **V**, 71 (1941).

[22] Landau, L. D. On the theory of superfluidity in helium II, *J. Phys. USSR*, **XI,** 91, (1947).

[23] Cowley, R. A. and Woods, A. D. B. *Phys. Rev. Lett,* **21** 787 (1968); *Can. J. Phys.* **49** 177 (1971).

[24] Woods, A. D. B. and Cowley, R. A. Structure and excitations of liquid helium, *Rep. Prog. Phys.* **36** 1135 (1973).

[25] Donnelly, R. J. The two-fluid theory and second sound in liquid helium, *Physics Today* October 2009, Pg. 34.

[26] Tisza, L. *Nature*, **141**, 913 (1938); *Compt. Rend.* **207**, 1035 (1938); *Compt. Rend.* **207**, 1186 (1938); *J. Phys. Radium* **1**, 164 (1940); *J. Phys. Radium* **1**, 350 (1940).

[26A] Griffin, *A. Laszlo Tisza (1907–2009): An Appreciation, J. Low. Temp. Phys.* **157**, 1 (2009).

[27] Peshkov, V. Second sound in helium II, *J. Phys. USSR*, **8**, 381 (1944).

[28] Penrose, O. *Phil. Mag.* **42**, 1473 (1951).

[29] Pitaevskii, L. P. *The Physics of Superconductors*, Ed. K. H. Bennemann and J. B. Ketterson, Springer, Berlin.

[30] Sokol, P. E. Bose–Einstein Condensation in Liquid Helium in *Bose Einstein Condensation*, Ed. A. Griffin, D. W. Snoke, and S. Stringari, Cambridge University Press, Cambridge G.B. Pg. 51 (1995).

[31] Gross, E. P. Structure of a quantized vortex in boson systems. *Il Nuovo Cimento* (Italian Physical Society) **20**, 454 (1961).

[32] Pitaevskii, L. P. Vortex lines in an imperfect Bose gas, *Soviet Physics JETP-USSR* **13**, 451 (1961).

[33] Hugenholtz, N. M. and Pines, D. Ground-state energy and excitation spectrum of a system of interacting bosons. *Phys. Rev.* **116**, 489 (1959).

[34] de Gennes, P. G. *Superconductivity in Metals and Alloys*, W. A. Benjamin and Sons, New York, 1967.

[35] Bardeen, J. Cooper, L. N. and Schrieffer, J. R. Theory of superconductivity, *Phys. Rev.* **104**, 1175 (1957).

[36] Marsiglio, F. and Carbotte, J. P. Electron–phonon superconductivity, *Superconductivity* Vol 1, Ed. Bennemann, K. H. and Ketterson, J. B. Springer-Verlag, Berlin (2008).

[37] Corak, W. S. Goodman, B. B. Satterthwaite, C. B. and Wexler, A. Exponential temperature dependence of the electronic specific heat of superconducting vanadium, *Phys. Rev.* **96**, 1442 (1954).

[38] Bardeen, J. Theory of Superconductivity: Theoretical Part, *Handbuch der Physik*, vol. 15 Springer, Berlin, pp 274–369 (1956).

[39] Cooper, L. N. *Phys. Rev.* **104,** 1189 (1956).

[40] Gorkov, L. P. *Soviet Physics JETP* **9**, 1364 (1959).

[41] Ketterson, J. B. and Song, S. N. *Superconductivity,* Cambridge University Press, (2000).

[42] Bardeen, J. and Stephen, M. J. *Phys. Rev.* **140**, A1197 (1965).

[43] Abrikosov, A.A. Gorkov, L.P. and Dzyaloskinski, I. E. *Methods of Quantum Field Theory in Statistical Physics.* Prentice-Hall, New York (1963).

[44] Vollhardt, D. and Wolfle, P. *The Superfluid Phases of Helium 3*, Taylor and Francis, London (1990).

[45] Baym, G. and Pethick, C. Landau Fermi-liquid theory and the low temperature properties of normal liquid ^3He, in *The Physics of Liquid and Solid Helium*, Part II, Edited by Bennemann, K. H. and Ketterson, J. B. John Wiley, Inc., pg. 1 (1978).

[46] Osheroff, D. D. Richardson, R. C. and Lee, D. M. *Phys. Rev. Lett.* **28**, 885 (1972).

[47] Lee, D. M. and Richardson, R. C. *Superfluid* ^3He, in *The Physics of Liquid and Solid Helium*, Part II, Edited by Bennemann, K. H. and Ketterson, J. B. John Wiley, Inc., pg. 287 (1978); Wheatley, J. C. *Rev. Mod. Phys.* **47**, 415 (1975).

[48] Anderson, P. W. and Brinkman, W. F. in Theory of anisotropic superfluidity in ^{3}He in *The Physics of Liquid and Solid Helium*, Part II, Edited by Bennemann, K. H. and Ketterson, J. B. John Wiley, Inc., pg. 177 (1978).

[49] Leggett, A. J. *Rev. Mod. Phys.* **47**, 331 (1975).

[50] Sarma, B. K. Levy, M. Adenwalla, S. and Ketterson, J. B. *Physical Acoustics*, Vol. XX, pg. 107, Edited by M. Levy, Academic Press, Inc., Boston (1992).

[51] Greiner, M. Ultracold quantum gases in 3d optical lattice potentials, Thesis, T. U. München, 2003.

[52] Annett, J. F. *Superconductivity, Superfluidity and Condensates*, Oxford University Press, Oxford U.K., (2004).

[53] Leggett, A. J. Bose–Einstein condensation in the alkali gases: Some fundamental concepts, *Rev. Mod. Phys.*, **73**, 307 (2001).

[54] Pethick, C. J. and Smith, H. *BEC in Dilute Gases*, Cambridge University Press (2002).

[55] Sa de Melo, C. A. Randeria, M. and Engelbrecht, J. R. Crossover from BCS to Bose superconductivity:transition temperature and time-dependent *Ginzburg–Landau theory*, *Phys. Rev. Lett.* **71**, 3202 (1993).

[56] Nozieres, P. and Schmitt–Rink, S. Bose condensation in an attractive fermion gas: From weak to strong coupling superconductivity, *J. Low Temp. Phys.* **59**, 195 (1985).

[57] Eagles, D. M. Possible pairing without superconductivity at low carrier concentrations in bulk and thin-film superconducting semiconductors, *Phys. Rev.* **186**, 456 (1969).

[58] Leggett, A. J. in *Modern Trends in the Theory of Condensed Matter*, Proc. Karpacz Winter School of Theor. Physics, Springer Berlin (1980).

[59] Drechsler, M. and Zwerger, W. Crossover from BCS-superconductivity to Bose-condensation, *Ann. Phys.* **1**, 15 (1992).

[60] Greiner, M. Regal, C. and Jin, D. S. Probing the excitation spectrum of a Fermi gas in the BCS–BEC crossover regime, *Phys. Rev. Lett.* **94**, 070403 (2005).

[61] Greiner, M. Regal, C. A. and Jin, D. S. Emergence of a molecular Bose–Einstein condensate from a Fermi gas, *Nature* **426**, 537 (2003).

[62] Greiner, M. Mandel, O. Esslinger, T. Hänsch, T. W. and Bloch, I. Quantum phase transition from a superfluid to a Mott insulator in a gas of ultracold atoms, *Nature* **415**, 39 (2002).

[63] Krutitsky, K. V. Pelster, A. and Graham, R. Mean-field phase diagram of disordered bosons in a lattice at nonzero temperature, *New J. Phys.* **8**, 187 (2006).

[64] F. Nogueira, private communication (F. U. Berlin, 2010).

[65] Abrikosov, A. A. Gorkov, L. P. Ye. Dzyaloshinskii, I. *Quantum Field Theoretical Methods in Statistical Physics*, Pergamon, (1965).

[66] Zwierlein, M. High temperature superfluidity in an ultracold Fermi gas, Thesis M.I.T., (2006).

[67] Kleinert, H. Schmidt, S. and Pelster, A. Reentrant phenomenon in the quantum phase transitions of a gas of bosons trapped in an optical lattice, *Phys. Rev. Lett.* **93**, 160402–1 (2004).

[68] Kerker, G. and Bennemann, K. H. Theory of Superconductivity for Transition-Metal Alloys, *Solid State Commun.* **14**, 399 (1974).

2 Survey of some novel superfluids

J. B. Ketterson and K. H. Bennemann

In recent years there has been a rapid increase in the number of systems that can be made to be superfluid. Perhaps the most dramatic of these was the demonstration of Bose–Einstein condensation (BEC) in trapped alkali metal atoms, following advancements in the technology to trap and cool atoms. This field has literally exploded since that time. But almost as dramatic has been the discovery that several other kinds of systems can be rendered superfluid under the right circumstances.

In this chapter we briefly review some of the work on trapped gases along with a discussion of some other superfluid systems including: photons, excitons, and magnons. At the end of the chapter we take up astrophysical superfluids, among which we can surely include the protons and neutrons within neutron stars but which also include the more exotic possibility of quark superfluids at their core.

2.1 BEC of photons

Although Bose proposed his statistics to apply to Planck's photons, these light quanta can appear and disappear individually via absorption and emission processes and hence their number is not constrained by any criteria other than the equilibrium energy density, which is in turn related to the temperature through Plank's formula, and hence varies. It may therefore appear surprising that photons might Bose condense under appropriate circumstances, as first reported by Klaers, Schmitt, Vewinger, and Weitz [1, 2]; this is the topic of the chapter by Klaers and Weitz in this book (Chapter 6) but will be briefly introduced here.

The first step required for photon BEC is to confine the quanta within a "box" and, in addition, to arrange for the energy spectrum to have a minimum at which the optical excitations can accumulate. This can be accomplished using a miniature Fabry–Perot cavity, as shown schematically in Fig. 2.1. The modes of this cavity have the well-known Hermit–Gaussian form,[1] the spectrum of which we can obtain using simple arguments, as we now discuss.

[1] For a discussion of these modes see [3].

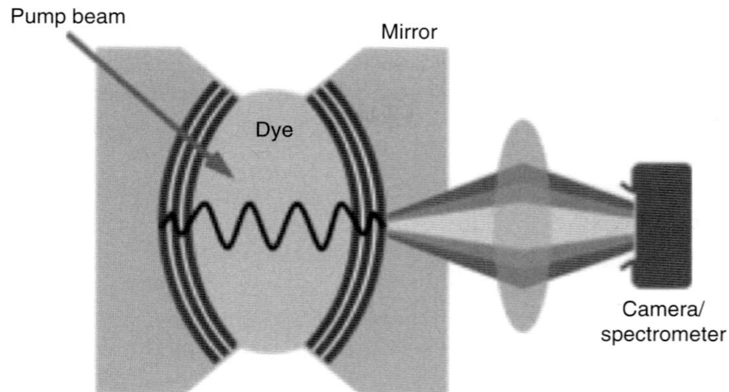

Fig. 2.1 Cross-section of a dye containing Fabry–Perot cavity defined by multi-layer spherical dielectric mirrors separated by a distance D, between which standing wave modes with varying numbers of longitudinal modes (along with transverse modes) can be confined. The dye is pumped at 532 nm and directed at 45° so as to bypass the mirrors. The spectrum of modes within the cavity is monitored with a spectrometer, from transmitted light. In addition, a CCD camera monitors the radial profile arising from the transverse modes. (After Klaers et al. (2010).)

In addition to the usual standing wave modes along the cavity axis, which will be numbered by q, there are modes in the radial direction. As we will argue below, the energies associated with these modes correspond to an isotropic 2D harmonic oscillator, where energy levels accompanying the associated potential energy, $V(r)$, are shown schematically in Fig. 2.2 for a given longitudinal mode number, q (which is 7 in the experiments to be described, corresponding to 3.5 wavelengths).

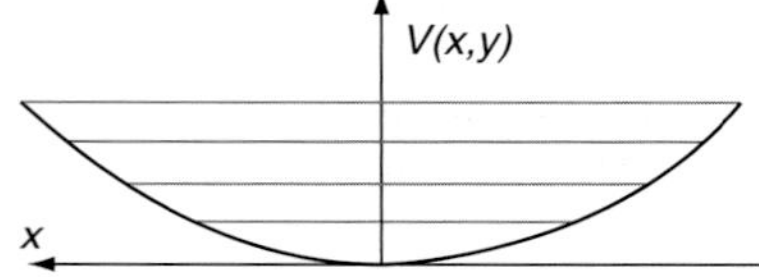

Fig. 2.2 Energy levels of photons confined in the radial direction within a Fabry–Perot cavity. (After Klaers et al. (2010).)

2.1.1 Free photons in a 1D box

Suppose we initially neglect the curvature of the mirrors. The frequency of a wave reflecting back and forth between the (now flat) mirrors, $\omega = \omega_q(k_r)$, with wavevector components k_z and k_r, is given by,

$$\omega_q(k_r) = c\sqrt{k_r^2 + k_z^2} = c\sqrt{k_r^2 + (\pi q/D)^2}, \tag{2.1}$$

where c is the velocity of light in the dye solution, $k_z = \pi q/D$, and D is the effective mirror separation (including any effects introduced by the mirrors themselves). Figure 2.3 shows these cavity frequencies schematically for some mode number q; note that at large k_r the slope approaches c. In the opposite limit of small k_r we can expand (2.1) as

$$\begin{aligned}\omega_q(k_r) &= \omega_q(0) + \frac{1}{2}\left.\frac{\partial^2}{\partial k_r^2}\right|_{k_r=0} + \cdots \\ &= \omega_q^c + \frac{\hbar k_r^2}{2m} + \cdots\end{aligned} \tag{2.2}$$

where $\omega_q^c = \pi cq/D$ is the q-dependent cavity cut-off frequency and we introduced an effective mass as

$$m^{-1} = \frac{1}{\hbar}\left.\frac{\partial^2\omega}{\partial k_r^2}\right|_{k_r=0} = \frac{cD}{\pi\hbar q} = \frac{c^2}{\hbar\omega_q^c}. \tag{2.3}$$

From this perspective one can view the transverse excitations of the cavity as being associated with free particles of mass m moving in two dimensions.

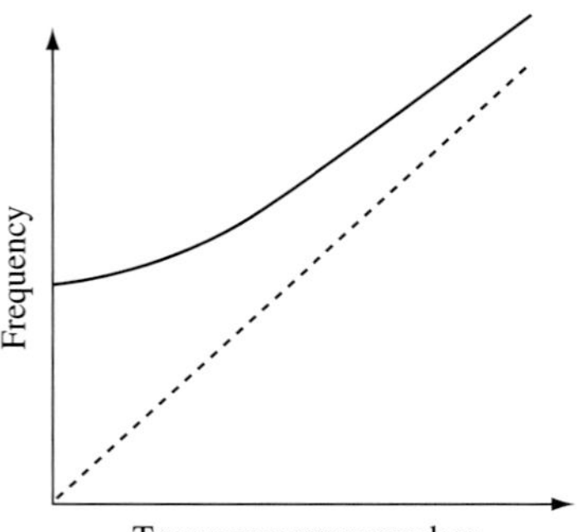

Fig. 2.3 A plot of the cavity frequencies for some longitudinal mode number q. (After Klaers et al. (2010).)

2.1.2 Bound photons in a Fabry–Perot cavity

We must now incorporate effects associated with the radius of curvature of the mirrors, R (assumed the same for both mirrors). The spacing D then becomes r-dependent and is given by

$$D(r) = D_0(r) - 2\left(R - \sqrt{R^2 - r^2}\right) = D_0 - \frac{r^2}{R} + \cdots \tag{2.4}$$

Our cut-off frequency then has the form $\omega_c = \omega_c(r)$ where

$$\omega_q^c(r) = \frac{\pi cq}{D(r)} \cong \omega_q^c + \frac{\pi cq}{D_0^2 R}\left(x^2 + y^2\right). \tag{2.5}$$

We can now associate the function $\hbar\omega_q(k_r, r)$ with a harmonic oscillator Hamiltonian,

$$\hat{H} = \hbar\omega_q^c + \frac{\hbar^2}{2m}\left(\frac{\partial^2}{\partial x^2} + \frac{\partial^2}{\partial y^2}\right) + \frac{1}{2}\kappa(x^2 + y^2), \tag{2.6}$$

where the second and third terms on the right represent the kinetic and potential energies with $\kappa = 2\hbar\pi cq/D_0^2 R$ being the spring constant. The energy levels of the 2D oscillator are then (see Fig 2.2)

$$\varepsilon_{q,n_x,n_y} = \hbar\omega_{q,n_x,n_y} = \hbar\omega_q^c + \left(n_x + n_y + 1\right)\hbar\Omega, \tag{2.7}$$

where the frequency is given by

$$\Omega = \sqrt{\kappa/m} = c\sqrt{\frac{2}{D_0 R}} \tag{2.8}$$

and n_x and n_y are the quantum numbers associated with oscillations along x and y.

2.1.3 BEC in two dimensions

Having determined the spectrum of 2D photon excitations in a Fabry–Perot cavity, the next task is to determine the condition for BEC. We start by recalling the expression for the number of excitations for the case when the excitation spectrum is continuous,

$$N(T) = \begin{cases} \displaystyle\int_0^\infty \frac{N(\varepsilon)d\varepsilon}{e^{[\varepsilon - \mu(T)/k_B T} - 1}; & T > T_c, \\ \displaystyle N_0 + \int_0^\infty \frac{N(\varepsilon)d\varepsilon}{e^{\varepsilon/k_B T} - 1}; & T < T_c, \end{cases} \tag{2.9}$$

where $N(\varepsilon)$ is the density of states for free particles of mass m in a 2D box of area L^2, and is given by $N(\varepsilon) = mL^2/2\pi\hbar^2$. The critical density $N(T_c)$ is then given by

$$N(T_c) = \frac{mL^2}{2\pi\hbar^2}\int_0^\infty \frac{d\varepsilon}{e^{\varepsilon/k_BT_c}-1} = \frac{mL^2k_BT_c}{2\pi\hbar^2}\int_0^\infty \frac{dx}{e^x-1}. \tag{2.10}$$

This integral is divergent at the lower limit and hence BEC of a non-interacting gas of free particles does not occur in 2D [4].

However, for the case of a cavity with curved mirrors, approximated by *particles in a harmonic trap,* the energy levels are discrete; the expressions analogous to (2.9) are then

$$N(T) = \begin{cases} \displaystyle\sum_{n=0}^{\infty} \frac{g_n}{e^{[(\varepsilon_n-\mu]/k_BT_c}-1}; & T > T_c, \\ \displaystyle N_0 + \sum_{n=1}^{\infty} \frac{g_n}{e^{\varepsilon_n/k_BT_c}-1}; & T < T_c, \end{cases} \tag{2.11}$$

where g_n is the level degeneracy. The expression corresponding to (2.10) then has the form,

$$N(T_c) = \sum_{n=0}^{\infty} \frac{g_n}{e^{\varepsilon_n/k_BT_c}-1} = \sum_{n=0}^{\infty} \frac{n+1}{e^{(n+1)\hbar\Omega/k_BT_c}-1}, \tag{2.12}$$

where we have written $\varepsilon_n = (n+1)\hbar\Omega$ and noted that $g_n = n+1$. Taking the limit of large quantum numbers (valid near T_c and above) we may approximate (2.12) as

$$N(T_c) \cong \left(\frac{k_BT}{\hbar\Omega}\right)^2 \int_0^\infty dx\frac{x}{e^x-1} = \frac{\pi^2}{3}\left(\frac{k_BT}{\hbar\Omega}\right)^2. \tag{2.13}$$

Hence we conclude that BEC can occur for the bound 2D photons in a Fabry–Perot cavity [5, 6].

2.1.4 Requirements for achieving BEC of cavity photons

To achieve BEC we must (i) have photon numbers $N>N(T_c)$, as calculated from Eq. (2.11) at a temperature $T<T_c$ and (ii) arrange for the photons to indirectly interact so as to thermally equilibrate the distribution function $n(\varepsilon_{q,n})$ while N is maintained constant. The first condition is achieved by pumping the dye with a laser which creates *electronic excitations* which, in turn, populate various modes $\omega_{q,n}$ of the cavity; note the purely thermal excitation of cavity modes is suppressed by a factor $\exp(\hbar\omega_q^c/k_BT) \approx \exp(-80)$. The second condition is achieved through the rapid absorption and emission that occurs if a family of cavity modes q, n lies in a region where the absorption and emission spectra of the dye overlap, the latter thus acting as a heat bath (involving the rotation/vibration levels of the dye molecules) that equilibrates the transverse degrees of freedom. Qualitatively one expects the number of electronic excitations N_e of the dye to be related to the number of cavity excitations N by $N \sim N_e w\tau_c$, where τ_c is the lifetime of a cavity excitation and w is the transition rate.

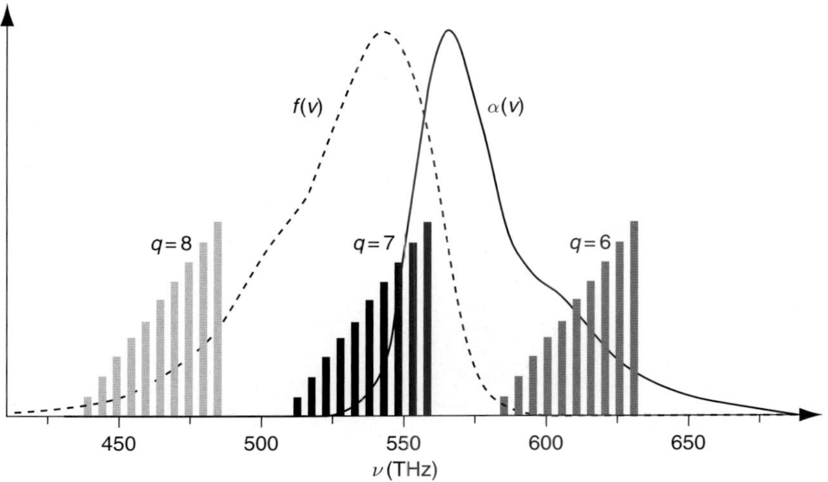

Fig. 2.4 Schematic representation of the cavity modes. The vertical bars show allowed values of the cavity frequencies, $\nu_{q,n}$, with their heights proportional to the degeneracy of the level, g_n. Shown also are the absorption strength, $\alpha(\nu)$ (dashed line), and emission efficiency, $f(\nu)$ (solid line), versus frequency, ν, of the dye rhodamine 6G dissolved in methanol. Note the strong overlap of $\alpha(\nu)$ and $f(\nu)$ in the vicinity of the $q = 7$ longitudinal mode for which $\nu_q^c = 5.1 \times 10^{14}$ Hz. (After Klaers et al. (2010).)

Figure 2.4 shows the frequency dependence of the absorption strength, $\alpha(\nu)$, and emission efficiency, $f(\nu)$, arising from rhodamine 6G dye; note there is a strong overlap for frequencies of order 550 THz. The vertical bars depict cavity frequencies $\nu_{q,n}$ $(\omega_{q,n}/2\pi)$ for this cavity where $D = 1.46\,\mu$; the height of the bars is proportional to g_n. Note also that the transverse modes, n, associated with the longitudinal mode $q = 7$ are concentrated in the overlap region.

2.1.5 Experimental observation of cavity photon BEC

We now summarize the experiments of Klaers et al. Fig. 2.5 shows the free-space wavelength dependence of the output of the spectrometer monitoring the cavity as a function of the cavity pump power, where the latter is normalized to an experimentally determined pump power, P_c, associated with the onset of BEC. Note that above $P > P_c$ there is a pile-up of spectral density at a wavelength close to that which one would calculate from the cut-off frequency, $\nu_{q=7}^c$; the width of this peak is reported to be limited by the spectrometer resolution.

The inset to Fig. 2.5 shows the predicted wavelength dependence of the intensity, $I(\nu)$, based on a non-interacting Bose–Einstein model in which the number of photons in the cavity is proportional to the pump power; i.e. $N \propto P$, and the excitations are at room temperature.

The last piece of evidence we will present is the observed *spatial narrowing* of the photon distribution function seen in Fig. 2.6. For temperatures above T_c the excitations are broadedly distributed among the radial harmonic oscillator states, as seen to the left. However, below the transition a macroscopic fraction of the excitations occupy the ground state; this behavior manifests itself as a bright dot in the center of the image to the right. This behavior is the 2D analog of a similar phenomenon observed with alkali atoms cooled in a magneto-optical trap.

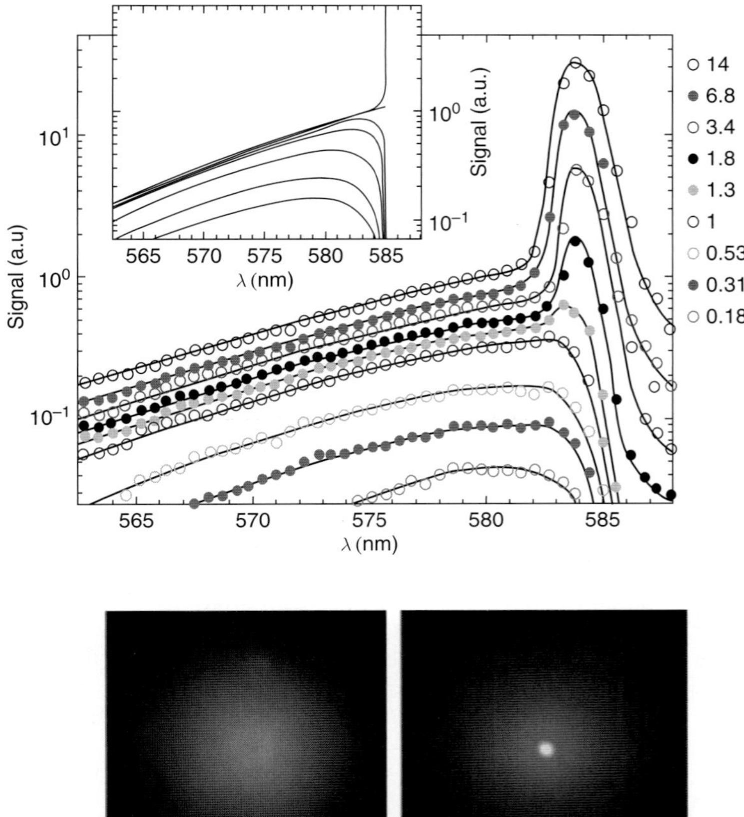

Fig. 2.5 The output of the spectrometer shown in Fig. 2.1 as a function of the free-space wavelength for various ratios of the pump power relative to an experimentally determined critical pump power, $P_c = (1.55 \pm 0.6)\,\mathrm{W}$. Note the increase in spectral density at higher pump powers at a wavelength of 584 nm, near the value expected for the cavity cut-off frequency for $\nu^c_{q=7}$. The insert shows the behavior predicted by the theory. (After Klaers et al. (2010).)

−100 µm 0 100 µm

−100 µm 0 100 µm

Fig. 2.6 Images of the lateral intensity distribution in the cavity. The image to the left is recorded at a temperature $T > T_c$, while that to the right is for $T < T_c$. Note the distinct narrowing and that excitations progressively fall into the ground state harmonic oscillator level as the temperature falls. (After Klaers et al. (2010).) (See Plate 1 for color image).

2.2 BEC of excitons

One can identify two classes of electronic excitations from the ground state of a crystal. In the first class one brings an electron from infinitely far away into the material, in which case the charge changes by $-|e|$ (this of course requires the presence of a free surface); conversely, one can remove an electron in a similar manner, carrying it off to infinity and changing the charge by $|e|$. The excitations created in this way are termed *electron-like* and *hole-like* excitations respectively. They are also called *one-particle excitations.*

Excitations in the second class involve removing an electron from the ground state and promoting it to an excited state, in which case we say we make an *electron–hole pair.* In analogy with the one-particle excitations the electron–hole pairs are classed as *two-particle excitations.* The question then arises as to what are the effects of the (attractive) Coulomb interaction between the pair so created. In a semiconductor we may regard the band gap as that energy required to excite an electron from the top of the highest valence band to the bottom of the lowest conduction band, in a manner such that the final electron and hole wave functions are uncorrelated, requiring that they be widely separated (spread out).

However, there is another kind of electron–hole excitation in which an electron and hole are created, but they do not escape their mutual Coulomb attractions and end up in a bound electron–hole state. Such states are termed *excitons.* The binding energy and characteristic size of excitons vary widely among different materials. One may construct simple models for the limiting cases where the size of the excitons is (i) much larger than, or (ii) of order of, the spacing between the atoms. These limiting cases are referred to as Wannier [7] and Frankel [8] excitons respectively and will be discussed in the next two subsections.

Consisting of two spin 1/2 particles, excitons are necessarily bosons and hence can, in principle, undergo BEC [9].[2] For this to occur their number density, n, and the effective temperature, T, must satisfy the condition $T < T_c$ where T_c is given by Eq. (1.9) of Chapter 1 in three dimensions or in two dimensional harmonic well by Eq. (2.13) of the current chapter.

[2] For a more extended discussion see [9].

Large exciton densities are generally achieved by laser pumping. A further requirement is that the excitons have long lifetimes; the latter can arise from either intrinsic properties of the material hosting the excitons themselves, or as a result of an engineered structure, as we discuss later in this section. In either case any equilibrium achieved is necessarily dynamic in character, and associating it with a temperature (necessarily larger than that of the lattice) must be done with caution.

2.2.1 Wannier excitons

For the Wannier exciton one assumes that the electron and hole interact via the classical Coulomb potential, but modified by the dielectric constant of the host crystal, ε, i.e. $V(r) = -e^2/\varepsilon r$. To keep the treatment as simple as possible we assume that the electron and hole are, respectively, associated with non-degenerate band minima and maxima located at the Γ point of a cubic crystal, in which case we can assume locally parabolic bands with isotropic effective masses m_e and m_h. Replacing e^2 by e^2/ε and introducing the reduced mass $\mu = m_e m_h/(m_e + m_h)$ in the Bohr formula for the hydrogen atom, we can immediately write the energy levels as

$$\varepsilon_n = \varepsilon_g - \frac{\mu e^4}{2\hbar^2\varepsilon^2 n^2} + \frac{\hbar^2 k^2}{2(m_e + m_h)}, \tag{2.14}$$

leading to various Rydberg-like spectral series; the corresponding Bohr radius is given by $a_0 = \hbar^2\varepsilon/e^2\mu$. The last term in (2.14) accounts for the translational kinetic energy of the exciton.

The excitons may form in either singlet ($S = 0$ or paraexciton) or triplet ($S = 1$ or orthoexciton) states. A spin splitting (usually referred to as exchange splitting) of the excitonic levels then occurs with $\Delta E \propto \langle \boldsymbol{\mu}_e \cdot \boldsymbol{\mu}_h \rangle$. In addition to the excited states of the exciton, the electron and hole can recombine leading to exciton annihilation. When the coupling to light via stimulated emission and absorption of excitons is included one obtains the so-called exciton–polariton; we will discuss exciton–polaritons shortly.

A well-known exciton spectral series is the so-called yellow series in cuprous oxide (Cu_2O) [10]. The exciton absorption spectrum is shown in Fig. 2.7,[3] and a set of parameters describing these excitations is given in Table 2.1. In Cu_2O there are other low-lying states, leading to additional exciton spectral series (the so-called green, blue, and violet series). Cuprous oxide is further distinguished by the fact that symmetry of the electron and hole levels leading to yellow exciton formation is such that decay of the lowest $S = 1$ spin triplet state proceeds only by *quadruple* emission while the $S = 0$ ground-state annihilation is electromagnetically forbidden in all orders. For this reason Cu_2O has been the center of much attention in the quest to achieve excitonic BEC.

[3]For some more recent measurements see [12].

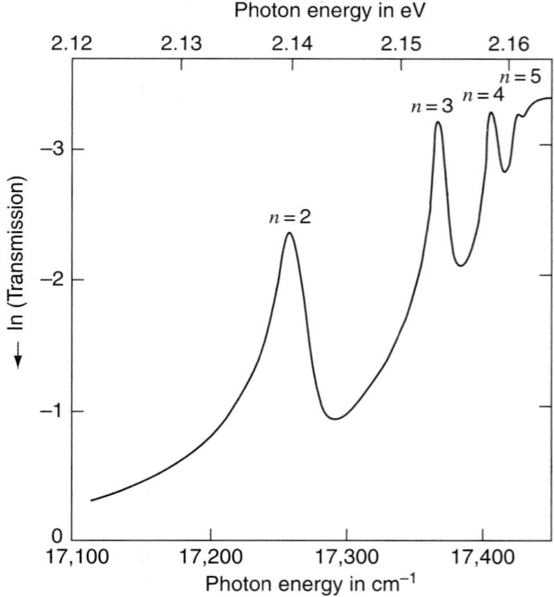

Fig. 2.7 Absorption spectrum of cuprous oxide showing the Rydberg-like series of energy levels $n = 2$ to $n = 5$ in Eq. 2.14. The $n = 1$ level is not observed due to the selection rules. (Baumeister (1961).)

Table 2.1 Parameters characterizing the yellow exciton series in Cu_2O

Quantity	Symbol	Value
dielectric constant	ε_∞	7
electron mass	m_e	1.0 m_o
hole mass	m_h	0.7 m_o
$n = 1$ exciton mass	m_{ex}	2.7 m_o
Rydberg ($n > 1$)	Ry_{ex}	97 meV
Bohr radius ($n > 1$)	a_{ex}	1.11 nm
$n = 1$ binding energy	Ry_{ex}	150 meV
$n = 1$ Bohr radius	a_{ex}	0.5 nm
band gap	E_{gap}	2.17 eV(4K)
ortho–para exchange energy	Δ	2 meV

2.2.2 Frankel excitons

One can also consider the limit in which the electron–hole separation of the exciton bound state approaches atomic or molecular dimensions. Here we might picture the electron and hole as actually residing on the same molecule or atom; i.e. we have an excited-state orbital (the electron) and the vacant orbital which it previously occupied (the hole). This picture is applicable to rare gas crystals (e.g. solid krypton) and various molecular crystals (e.g. anthracene) where the coupling between neighboring atoms/molecules is small. However, there is some overlap of orbitals on adjacent sites and hence the excitation can hop (make transitions) between neighboring atoms/molecules.

2.2.3 Exciton–polaritons

The term exciton polariton is used to describe the extitations in the region where the energies of excitons and light waves are approximately the same. Since this will be important in connection with our discussion of exciton–polariton condensation in two dimensions (see Sec. 2.2.5). We now give a simple model to describe this region, motivated by the Lorentz model of a dielectric in which one pictures electrons as being bound to atoms by springs. In place of the atoms we have the holes and the resulting equations of motion for the electron and hole are,

$$m_e\ddot{\mathbf{r}}_e + \kappa(\mathbf{r}_e - \mathbf{r}_h) = -|e|\mathbf{E}(r_e) \tag{2.15a}$$

and

$$m_h\ddot{\mathbf{r}}_h + \kappa(\mathbf{r}_h - \mathbf{r}_e) = |e|\mathbf{E}(\mathbf{r}_h), \tag{2.15b}$$

where κ is the spring constant of the bound exciton. Introducing the coordinate $\boldsymbol{\rho} = \mathbf{r}_e - \mathbf{r}_h$ we can combine these equations as,

$$\begin{aligned}\ddot{\boldsymbol{\rho}} + \omega_T^2\boldsymbol{\rho} &= -\left(\frac{|e|}{m_e}\mathbf{E}(\mathbf{r}_e) + \frac{|e|}{m_h}\mathbf{E}(\mathbf{r}_h)\right)\\ &\cong -\frac{|e|}{\mu_X}\mathbf{E}(\mathbf{r}),\end{aligned} \tag{2.16}$$

where $\omega_T^2 = \kappa/\mu_X$ and $\mu_X = m_e m_h/(m_e + m_h)$ is the reduced mass. In the second step we made the assumption that the electron and hole are tightly bound, with a separation small compared with the wavelength of light, so the electric field is essentially the same at each; we will evaluate $\mathbf{E}$ at the exciton center of mass, $\mathbf{r}$.[4]

[4] Since we do not have translational symmetry, due to the periodic electric field, only an approximate separation of variables is possible.

On the other hand the electro-magnetic wave equation in the media is given by

$$\nabla^2\mathbf{E} - \frac{1}{c^2}\frac{\partial^2\mathbf{D}}{\partial t^2} = 0, \tag{2.17}$$

where $\mathbf{D} = \mathbf{E} + 4\pi\,\mathbf{P}$, $\mathbf{P}(\mathbf{r}) = -n|e|\boldsymbol{\rho}(\mathbf{r})$ is the electric polarization of the lattice and n is the number of oscillators per unit volume.

If we assume a plane wave propagating parallel to the z-axis and polarized parallel to the x-axis, $\mathbf{E}(z, t) = \mathrm{Re}(E_0\, \hat{\mathbf{x}}\, e^{ikz-i\omega t})$, we can then write the pair of equations (2.16) and (2.17) as

$$-\frac{ne^2}{\mu_X} E_0 + \left(-\omega^2 + \omega_T^2\right) P_0 = 0 \tag{2.18a}$$

and

$$\left(-k^2 + \frac{\omega^2}{c^2}\right) E_0 + \frac{4\pi\omega^2}{c^2} P_0 = 0. \tag{2.18b}$$

Setting the determinant of the matrix of the coefficients of E_0 and P_0 to zero, we obtain the dispersion relation as

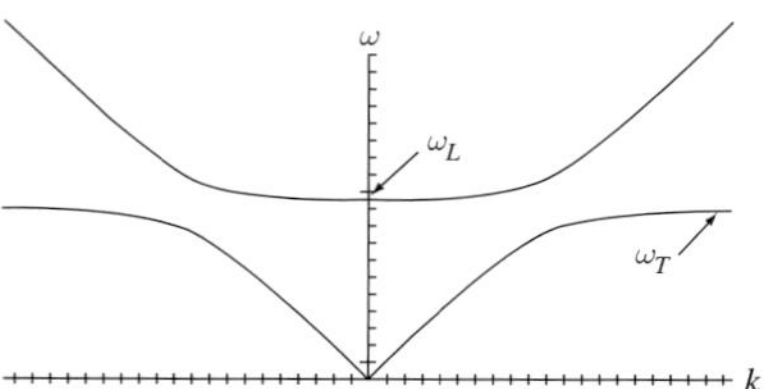

Fig. 2.8 The frequency versus wavevector for a coupled system involving an exciton and a light wave. Note the splitting between the lower and upper branches.

$$\left(\frac{\omega^2}{c^2} - k^2\right)\left(\omega_T^2 - \omega^2\right) + \frac{\omega^2\omega_X^2}{c^2} = 0; \tag{2.19}$$

where we have introduced an exciton plasma frequency as $\omega_X^2 = 4\pi ne^2/\mu_X$. The two branches associated with this equation are plotted in Fig. 2.8. At $k = 0$ the solution is

$$\omega^2 = \omega_T^2 + \omega_X^2 \equiv \omega_L^2, \tag{2.20}$$

where we have defined another frequency, ω_L; the frequencies ω_L and ω_T are referred to as the longitudinal and transverse optical frequencies, respectively, and are identified in the figure.

We may define the dielectric constant as either $(ck/\omega)^2$ or $1 + 4\pi P/E$. Allowing for a contribution ε_∞ arising from a background electronic polarization coming from atom cores (which we take to be frequency independent in the range under consideration) we then have

$$\varepsilon(\omega) = \varepsilon_\infty + \frac{\omega_X^2}{\omega_T^2 - \omega^2}. \tag{2.21}$$

If we define the dielectric constant at $\omega = 0$ as ε_0 then from (2.21) we have,

$$\varepsilon_0 = \varepsilon_\infty + \frac{\omega_X^2}{\omega_T^2}. \tag{2.22}$$

In terms of these quantities we can write $\varepsilon(\omega)$ in the form,

$$\varepsilon(\omega) = \varepsilon_\infty + (\varepsilon_0 - \varepsilon_\infty)\frac{\omega_T^2}{\omega_T^2 - \omega^2} = \varepsilon_\infty \frac{\omega_L^2 - \omega^2}{\omega_T^2 - \omega^2}; \tag{2.23}$$

note this has a zero at $\omega = \omega_L$, and a pole at $\omega = \omega_T$ (if we add a damping term of the form $\dot{\boldsymbol{\rho}}/\tau$ to the left-hand side of (2.14) the unphysical divergence in (2.23) will be removed). Waves do not propagate between these two frequencies where $\varepsilon(\omega)$ is negative, since this corresponds to an imaginary k-vector implying that the waves are exponentially damped.[5] The frequency dependence

[5] An analogous phenomenon, referred to as the *Reststrahlen* effect, occurs in alkali halides where oppositely charged ions oscillate out of phase when coupling strongly to infrared light.

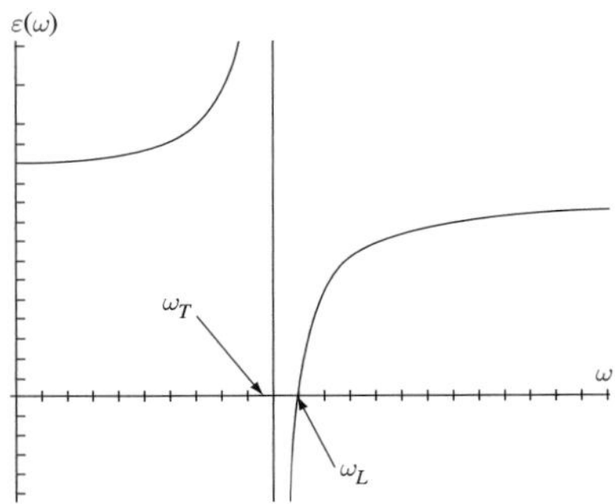

Fig. 2.9 The frequency dependence of the dielectric constant $\varepsilon(\omega)$; note that it goes through zero at ω_L and, in the absence of damping, goes through infinity at ω_T.

of the dielectric constant is plotted schematically in Fig. 2.9. Equation (2.23) gives us the following relation between ω_L and ω_T:

$$\varepsilon_\infty + (\varepsilon_0 - \varepsilon_\infty)\,\frac{\omega_T^2}{\omega_T^2 - \omega_L^2} = 0. \tag{2.24}$$

Equation (2.22) may also be written in the form,

$$\frac{\omega_L^2}{\omega_T^2} = \frac{\varepsilon_0}{\varepsilon_\infty} \tag{2.25}$$

for the case where optical phonons replace excitons this expression is referred to as the *Lyddane–Sachs–Teller relation.*

The effect of the exciton's kinetic energy can be included by replacing $\omega_T \to \omega_T + \hbar k^2/2\,(m_e + m_h)$ in the denominator of Eq. (2.23), in which case $\varepsilon(\omega) \to \varepsilon(\mathbf{k}, \omega)$. The justification for such a crude treatment lies in the fact that the quantum-mechanical result for the polarizability has the same form as the Drude model, albeit with an altered oscillator strength (involving the dipole matrix element).

Cu_2O represents a special case, since light interacts with the $n = 1$ orthoexcitons only via a quadrupole coupling and paraexcitons do not couple at all. This results in a form for the dielectic function that differs from Eq. (2.23) [13],

$$\varepsilon(\mathbf{k}, \omega) = \varepsilon_\infty + \frac{f\,c^2k^2}{\omega_0^2 + ak^2 - \omega^2 - i\omega/\tau}, \tag{2.26}$$

where, as mentioned above, we have included the contribution from the kinetic energy in the denominator (through the constant a) along with a damping term. But most important is the form of the numerator which is now proportional to k^2, reflecting the fact that the quadrupole coupling enters at a higher order in a multipole expansion (with a strength measured by the parameter f). The weaker coupling of the quadrupole orthoexcitons to light, resulting in lower attenuation (and the appearance of quantum beats), makes Cu_2O an ideal material to study exciton–polariton formation [14].

2.2.4 Probing the exciton distribution function in cuprous oxide

The most common way to excite excitons is with a pulsed laser; Fig. 2.10 shows, schematically, a typical setup when the sample is cooled to cryogenic temperatures. If the photon energy, $\hbar\omega$, exceeds the energy gap, ε_g, a high

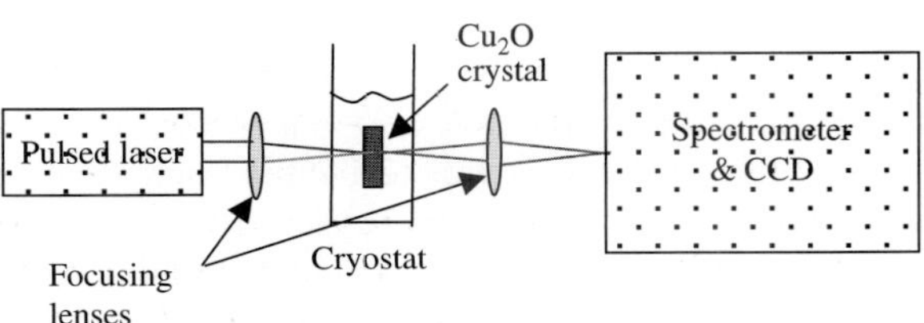

Fig. 2.10 A schematic of a typical photoluminescence experiment. The sample is illuminated by a short laser pulse with $\hbar\omega > \varepsilon_g$ creating a high density of electron–hole pairs. The resulting emissions are collected and focused onto a spectrometer equipped with a detector (preferably a CCD).

concentration of far-from-equilibrium electron–hole (e–h) pairs is generated. Various inelastic processes, most leading to annihilation, quickly set in, but some fraction of the e–h pairs combine to form excitons. Additional inelastic processes now drive the exciton distribution function toward equilibrium, those involving phonons allowing it to approach the lattice temperature (which is usually of order a Kelvin or less in experiments seeking BEC). If the time scale for this equilibration is fast compared with the decay rate of the exciton density itself, then a quasi-equilibrium is established with an accompanying chemical potential at higher temperatures. If the number density and temperature satisfy the condensation criteria, BEC should occur.

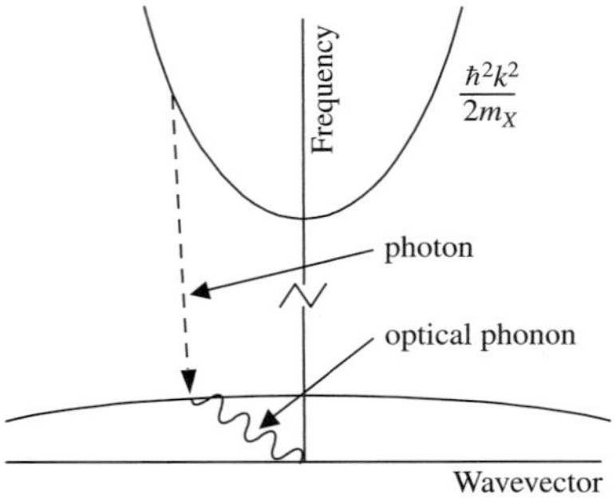

Apart from those excitons in the polariton regime, which can leave the crystal as light quanta of the same energy, direct decay of the excitons is kinematically forbidden and hence inelastic processes involving optical phonons are used to monitor the exciton distribution function; the energetics are depicted in Fig. 2.11. The so-called *phonon-assisted* decay processes involve the simultaneous emission of a photon (the dashed line) and an optical phonon (the wavy line). Note that the curvature of the optical phonon branch is greatly exaggerated in the figure; in practice it is much smaller than that of the exciton. Because of this the energies of the emitted photons are shifted by an essentially constant amount (the optical phonon frequency Ω) relative to the energy of the decaying exciton. Moreover, the intensity, I, of photons of a given wavelength (and corresponding energy) is directly proportional to the number of excitons having that energy prior to decay i.e. $I \propto N(\varepsilon)n(\varepsilon, \mu, T) \propto \varepsilon^{1/2}n(\varepsilon)$, where $N(\varepsilon)$ is the exciton density of states and $n(\varepsilon)$ is the distribution function; the wavelength dependence of the emitted intensity is then a kind of *replica* of the energy distribution of the excitons. If such spectra are collected at different times relative to the laser pulse that excites the system, then the time dependence of the distribution function can be mapped out. It has been common to assume that the excitons achieve a dynamic equilibrium that can be approximated by a Bose distribution, $n(\varepsilon, \mu, T_e)$ with some effective temperature T_e. If the observed $n(\varepsilon)$ is fitted to a Bose distribution at various times one can extract the time dependence of μ and T_e; the exciton number density is then inferred from integrating over the energy. The goal in this approach is then to see if $\mu \to 0$, followed by a pile-up of spectral density into a (resolution broadened) condensate fraction, as the exciton number density increases.

Fig. 2.11 The decay of an exciton via the emission of both a photon and an optical phonon.

2.2.5 Searches for excitonic BEC in cuprous oxide

Until recently Cu_2O has been a prime candidate for achieving excitonic BEC.[6] This choice arose from the relatively long lifetimes observed for both the ortho and para excitons, which are on the order of nano- and microseconds respectively. As discussed above, one strategy to search for BEC in Cu_2O is to infer the exciton distribution function from the time-dependent phonon-assisted luminescence following an intense laser pulse. Such pulses are typically obtained from a cavity-dumped argon (517 nm) or a pulsed, frequency-doubled, Nd:YAG laser (532 nm). The resulting spectrum is then observed, preferably in a mode where specific time intervals can be recorded.

[6]For a review of earlier work see [15].

Historically, various experiments have found that the low temperature data can be well fitted by a Bose distribution with a chemical potential $\mu \equiv 0$, but that a condensate does not emerge as the number density, n, increases. In fact as n increases it "hugs the boundary" defined by the condition $\mu \equiv 0$ but does not cross it, a phenomena that was termed "quantum saturation" [16]. A compelling explanation is that the distribution function, although mimicking an equilibrium Bose form, is actually a non-equilibrium, non-uniform form that arises from effects such as diffusion (nothing confines the excitons and the more energetic ones diffuse faster and further); if this is the case, number densities predicted by fitting the distribution function are not reliable. Later experiments addressed this possibility by performing an absolute calibration of the emitted light so as to *directly measure the exciton number* with the result that the density so determined was *two orders of magnitude smaller* than that required for BEC.

To explain the observed behavior it has been proposed that some two-body annihilation process, termed an Auger process, sets in at higher densities which strongly limits the achievable densities. Such a behavior can be modeled by a rate equation for the exciton density of the form,

$$\frac{dn}{dt} = -\frac{n}{\tau} - An^2. \tag{2.27}$$

It is found that such a model gives a satisfactory description of the exciton kinetics. The mechanism proposed is *biexciton formation*, wherein two excitons temporarily form a bound state, within which one exciton subsequently recombines to emit a photon and, in the process, transfers sufficient energy to the remaining exciton to cause it to dissociate it into an e–h pair [17–20].

The Wolfe group has pioneered a technique for strain confinement of excitons that can be used to address the problem of spatial inhomogenity arising from diffusion [21]. Here one presses a curved rod against a surface of the free crystal. This produces an internal stress profile within the crystal which can be calculated from the theory of Hertz [22].[7] The stress field couples to the excitons vie a deformation potential tensor, the parameters of which are known. The extremum of the stress field lies close to, but not at, the contact point and the parameters are such that some of the excitons can be attracted to this point. Figure 2.12 shows an image of excitons being attracted to and residing in such a trap. The strain lifts the three-fold degeneracy of the ortho excitons, and the middle- and lower-branch orthoexcitons, as well as the paraexcitons, can be trapped in the well whereas the upper orthoexciton branch cannot. Trapping the excitons requires an inelastic process so they can lose energy and fall in.

[7] For a discussion of this theory for elastically isotropic materials see [23].

An alternative to relying on such processes is to directly fill the trap by resonant optical excitation. Two-photon excitation [24], which has different selection rules [25, 26], allows penetration deep into the crystal in order to *directly* generate excitons in the trap [27, 28]. As discussed, this is a promising approach to BEC.

Another approach to creating a coherent $k = 0$ state (at least initially) involves two-photon pumping with counter propagating lasers, and a four-wave mixing approach has been proposed to probe the resulting time evolution [29, 30]. Finally, we note that if polaritons could be effectively confined in

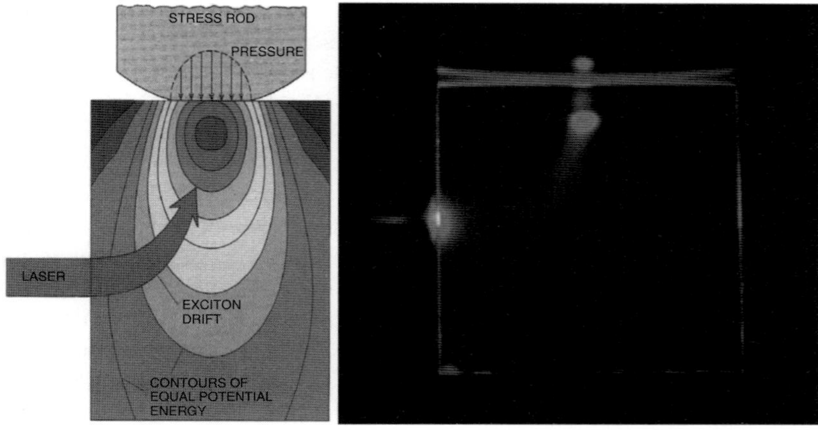

Fig. 2.12 In the schematic drawing to the left we see a depiction of the strain contours arising from a curved piston which is pressed against the upper surface of the crystal; the maximum lies slightly below this surface. Also shown is an incoming laser beam (green) which produces excitons (red) which drift normal to the strain contours and are ultimately trapped in the vicinity of the maximum strain point. To the right we see the actual luminescence profile; the two very bright regions are where the laser beam enters the crystal and where the excitons accumulate respectively. (Courtesy of J. Wolfe.) (See Plate 2 for color image).

Cu_2O, e.g. by using a con-focal cavity (similar to that used to confine and equilibrate photons described in the previous section), this could provide an alternative route to BEC. Experiments show that the half light/half matter character of polaritons suppresses Auger recombination [31].

2.2.6 Evidence for strain trapped excitonic BEC at sub-Kelvin temperatures

If exciton recombination mechanisms place intrinsic limits on achievable densities,[8] realizing excitonic BEC in bulk Cu_2O may require cooling to lower temperatures. This is the strategy employed by Yoshioka et al. [33], and discussed by Kuwata-Gonokami in Chapter 8 of this volume. These workers utilized a pumped ^{3}He cryostat together with a cigar-shaped strain trap (with trap frequencies:[9] $\omega_{x,y} = 2\pi \times 17\,\text{MHz}$; $\omega_z = 2\pi \times 25\,\text{MHz}$) to confine paraexcitons at high densities and low temperatures (down to 278 mK). Rather than exciting over the gap at a remote point and having the excitons drift to the well, as shown above from the Illinois experiments (see Fig. 2.12), the Tokyo group excited carriers directly in the trap by passing the laser beam through it; stress was applied through contact with a convex lens. A c.w. laser generated incoherent 1s ortho excitations (which lie 12 meV above the 1s paraexcitons) through a phonon-assisted absorption process. The orthoexcitons subsequently convert to 1s paraexcitons which releases their excess kinetic energy via acoustic phonons, thereby relaxing to energies near the bottom of the well, where they accumulate. Within the well, local strain relaxes the selection rule suppressing direct paraexciton radiative recombination, thereby making them visible.

The critical temperature for BEC in a harmonic trap is given by[10]

$$k_B T_c = \frac{\hbar \overline{\omega} N^{1/3}}{[\zeta(3)]^{1/3}} \cong 0.94 \hbar \overline{\omega} N^{1/3}, \tag{2.28}$$

where we have introduced an average trap frequency as $\overline{\omega} = (\omega_x \omega_y \omega_z)^{1/3} = 2\pi \times 19\,\text{MHz}$. In order for the motion in the trap to be coherent one must have $\omega\tau_{ep} > 1$, where τ_{ep} is the exciton–phonon scattering time. If exciton decay is

[8] In addition to the work of the Illinois group, the Tokyo group has studied paraexciton recombination; see [32].

[9] See Section 2.5.7 for a discussion of trap frequencies.

[10] See Eq. (2.72) in Section 2.4.1.1.

limited by a one-body process their number, N, will scale as G, the generation rate; i.e. $N \propto G$. If, on the other hand, it is limited by a two-body recombination rate, which is proportional to N^2, and which is anticipated for the exciton numbers required for BEC, one has $N \propto G^{1/2}$. Now from Eq. (2.27) the BEC transition temperature, T_c, scales as $N^{1/3}$. Combining these arguments we have $T_c \propto G^{1/6}$; i.e. the required power scales as T_c^6, suggesting that BEC should be easier to achieve at lower temperatures.

Trapped paraexciton cloud temperatures at lower concentrations were obtained by measuring the width of the luminescence profile (correcting for the strain-dependent coupling) and comparing it with the Maxwell–Boltzmann prediction for particles in a harmonic trap with known curvatures. A typical image of an exciton cloud is shown in Fig. 2.13a. A plot of the measured widths versus lattice temperature is shown in Fig. 2.13b; from this plot it follows that the temperature of the exciton gas lies above that of the lattice for temperatures below about 800 mK, implying the phonon cooling mechanism cannot keep up with heat generated within the cloud, presumably from one-photon decay in the dilute limit. Given this limitation on the achievable exciton temperatures, higher concentrations are required if BEC is to be achieved.

Yoshioka et al. performed measurements over a wide range of exciton numbers N at a lattice temperature of 354 mK; the data for $N = 2 \times 10^7$ and 2×10^9 are shown in Fig. 2.14a. Note that at the higher concentrations there is a tendency for the cloud to "explode"; i.e. it ranges over energies far removed from the well minimum. Their interpretation is that BEC onsets at higher concentrations and that the excitons experience an enhanced two-body

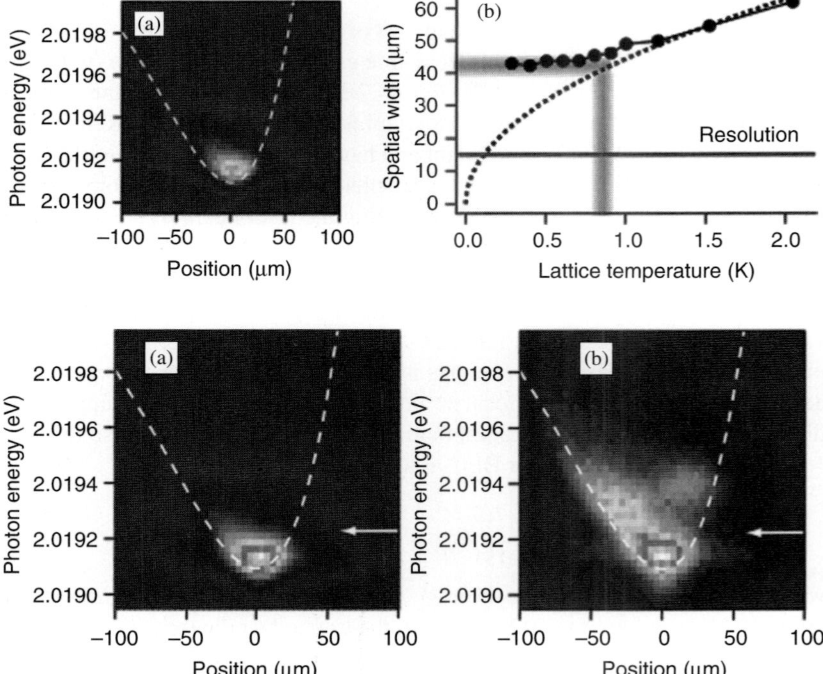

Fig. 2.13 (a) A typical cloud luminescence profile. The dashed line shows the position dependence of the potential energy of the strain trap. (b) The temperature dependence of the spatial width of the cloud luminescence profile. Note the observed width tracks the lattice temperature down to about 800 mK, below which it is larger, implying an elevated temperature. The dashed line shows the trajectory for thermal equilibrium. (After Yoshioka et al. (2011).) (See Plate 3 for color image).

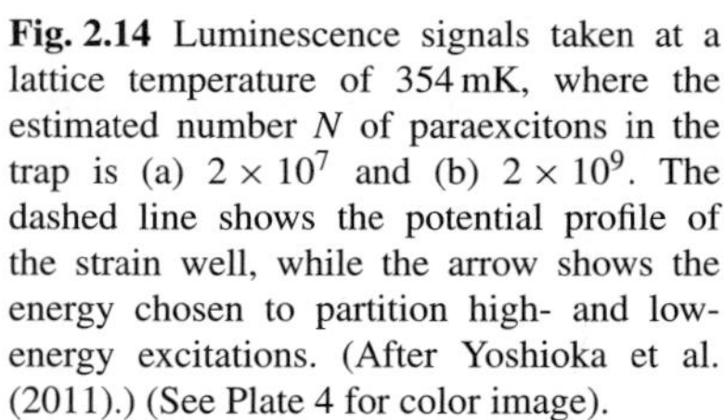
Fig. 2.14 Luminescence signals taken at a lattice temperature of 354 mK, where the estimated number N of paraexcitons in the trap is (a) 2×10^7 and (b) 2×10^9. The dashed line shows the potential profile of the strain well, while the arrow shows the energy chosen to partition high- and low-energy excitations. (After Yoshioka et al. (2011).) (See Plate 4 for color image).

decay process on entering the condensate, thereby promoting excitons to higher energies within the well. Similar data showing an explosive onset were taken as a function of temperature at fixed exciton number and are shown in Fig. 2.15b.

To characterize explosive behavior these workers split the spectrum into low and high energy components and lumped the luminesence into one or the other; the arrow in Fig. 2.14 shows the chosen energy about which to carry out this partitioning. The key feature to observe here is the rapid rise in this ratio as one either raises the exciton number (at fixed temperature) or lowers the temperature (at fixed exciton number).

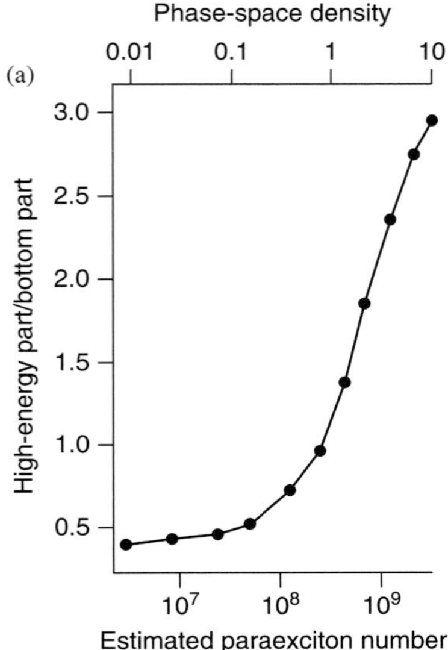

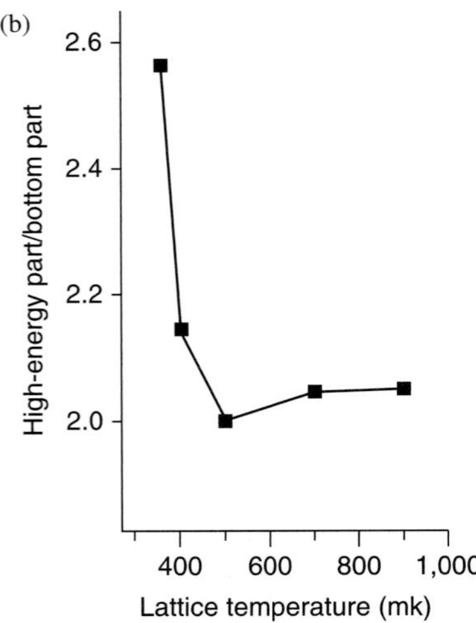

Fig. 2.15 (a) The ratio of the high energy to low energy parts of the spectrum as a function of the estimated exciton number. Note the rapid increase starting at exciton numbers above 10^8; here the temperature is fixed at a 354 mK. (b) The same quantity but now plotted as a function of the temperature; here the exciton number is fixed at $N = 2 \times 10^9$.

2.2.7 Polaritons in a 2D photonic cavity

As discussed in Section 2.2.3, excitons couple strongly with light in the region where their dispersion relations ω versus $\mathbf{k}$ would intersect in the absence of coupling, the resulting quantum particles being the exciton–polaritons (or polaritons for short). The curvatures of the resulting branches, $\partial^2\omega(\mathbf{k})/\partial k^2$, are generally large in the vicinity of the crossing (much larger than for the excitons by themselves) implying small effective masses, which would ordinarily favor BEC. However, in the absence of a local minimum in $\omega(\mathbf{k})$ at which excitations can accumulate, BEC of polaritons will not occur. An additional problem is that in most semiconductors the dipole coupling with the light results in very short exciton lifetimes. Both of these problems can be addressed by confining the excitons to two dimensions and engineering an appropriate photonic structure to keep the light confined, as we now discuss [34, 35].

Figure 2.16 shows a schematic diagram of a two-dimensional GaAs quantum well sandwiched between two multilayer dielectric Bragg mirrors. The inclusion of several closely spaced quantum wells can increase exciton lifetimes while minimizing their direct interaction for a given overall density. For GaAs quantum wells, the alternating layers of the mirrors involve AlAs and an $Al_{1-x}Ga_xAs$ alloy. Since the materials making up the mirrors have a larger band gap than the GaAs, within which the excitons lie slightly lower in energy, they will not absorb energy. The layer spacing is chosen to reflect light at the polariton frequencies themselves, and is approximately 1/4 the average wavelength of light. In the absence of coupling to excitons we can write the dispersion relation of the light confined between the mirrors as,

$$\omega(k_z, k_\perp) = c\sqrt{k_z^2 + k_\perp^2}, \tag{2.29}$$

where c is the velocity of light in GaAs and k_z and $k_\perp$ are the components of the wavevector parallel and perpendicular to the normal to the layers. The presence of the mirrors quantizes k_z in terms of the mirror separation, d, yielding the frequencies,[11]

$$\begin{aligned}\omega(n, d, k_\perp) &= c\sqrt{\frac{n^2\pi^2}{d^2} + k_\perp^2} \\ &= \omega_0 + \frac{1}{2}\frac{c^2}{\omega_0}k^2 - \cdots\end{aligned} \tag{2.30}$$

[11]Note the similarity of our discussion here and that in the beginning of Section 2.2.8.

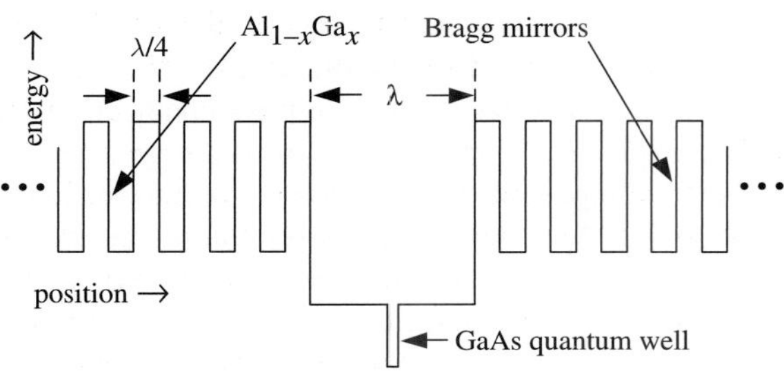

Fig. 2.16 A photonic structure consisting of a GaAs quantum well confined between $Al_{1-x}Ga_xAs/AlAs$ multilayer Bragg mirrors which suppress spontaneous exciton–polariton decay. (After Weisbuch et al. (1992).)

where we have ignored a correction accounting for the fact that the nodes will not be precisely at the leading edges of the mirrors and we have also defined $\omega_0 = cn\pi/d$.

We now focus our attention on Fig. 2.17. The flat and parabolic dashed curves show, schematically, the dispersion curves for the 2D quantum-well excitons and the mirror-confined photons respectively. Note the curvature of the excitons is negligible relative to the photons, and that the mirror spacing has been chosen so that at $k_\perp = 0$ the excitons and photons are degenerate or nearly so.

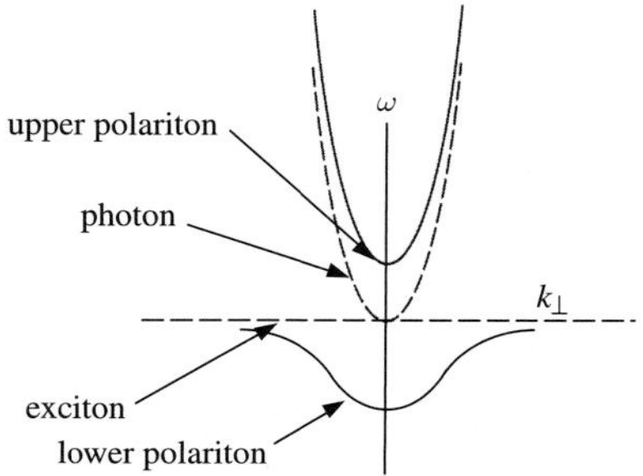

Fig. 2.17 The band structure in the vicinity of the upper branch of the gap produced by the Bragg mirrors, together with the additional splittings introduced by photon–exciton coupling to produce polaritons possessing a local extremum.

When the coupling of the light to the excitons is turned on, and taking account of the fact that k_z is quantized through Eq. (2.30), in place of Eq. (2.18b) we now have,

$$\left(-k_\perp^2 - \frac{n^2\pi^2}{d^2} + \frac{\omega^2}{c^2}\right) E_0 + \frac{4\pi\omega^2}{c^2} P_0 = 0; \tag{2.31}$$

and in place of Eq. (2.19),

$$\left(\frac{\omega^2}{c^2} - k_\perp^2 - \frac{n^2\pi^2}{d^2}\right)\left(\omega_T^2 - \omega^2\right) + \frac{\omega^2\omega_X^2}{c^2} = 0. \tag{2.32}$$

For our assumption of degeneracy at $k_\perp = 0$, i.e. $\omega_0 = \omega_T$, we would have,

$$\left(\omega_0^2 - \omega^2 + c^2k_\perp^2\right)\left(\omega_0^2 - \omega^2\right) + \omega^2\omega_X^2 = 0. \tag{2.33}$$

The mode repulsion resulting from the coupling manifests itself in the two solid curves in Fig. 2.17, now referred to as the upper and lower polariton branches. Excitons can now accumulate in the minimum of the lower polariton branch (which can be the global minimum) and BEC can occur. In addition, the mixing with the light component *greatly decreases* the effective mass of the excitations thereby allowing BEC at higher temperatures and lower densities.

2.2.8 BEC of exciton–polaritons in two dimensions

BEC of exciton–polaritons in cavities of the kind just described has been reported in GaAs-based devices by the Yamamoto group [36] and by the Dang group [37] for CdTe structures. Condensation in a spatial trap created by the application of a localized external strain has also been studied [38]; here we picture a pin pressed against the surface of a quantum-well device. As a result of these and other experiments, work in this field has greatly expanded and is the subject of the chapter by Yamamoto of this volume (Chapter 9) and an earlier review article [39].

The first step in creating a BEC is to flood the quantum well(s) with excitons followed by monitoring their momentum distribution. This initial pumping is best achieved by bringing in laser light at a large angle such that it is not reflected by the Bragg mirrors, but has an energy sufficient to create excitons in the quantum well for the corresponding value of $k_\perp$ (which is fixed by the incoming angle). These excitons will lie above the minimum energy at $k_\perp = 0$ and must therefore lose momentum and energy through inelastic processes if they are to condense; these processes involve acoustic and optical phonon interactions as well as exciton–exciton scattering and will not be discussed here. The distribution in momentum space can be monitored from the angle of the emerging recombination emission of the excitons in the well, vertical emission corresponding to $k_\perp = 0$. Although the mirrors transmit only a fraction of this light there is still sufficient emission to record the angular dependence of the emitted light which, in turn, corresponds to the in-plane momentum distribution.

Figure 2.18 shows the intensity of light emitted perpendicular to a GaAs-based device similar to that described above (however with a total of 12 wells) as a function of the pump power as observed by the Yamamoto group. We note that at a critical intensity, P_{th}, an abrupt change of slope is observed in the emission versus pump power behavior *signaling the formation of a condensate.* It is important that the densities involved in these experiments are much smaller than the so-called Mott limit where the exciton separation approaches the exciton radius, at which point the excitons would lose their identity.

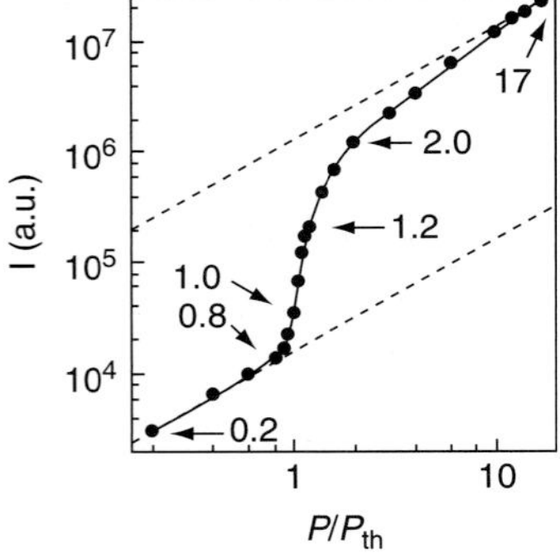

Fig. 2.18 Lower polariton emission intensity versus the normalized pump intensity, P/P_{th}, for light leaving perpendicular to the sample (corresponding to $k_\perp = 0$); note the superluminal rise at the critical pumping level. Various specific pumping levels are shown by arrows. (After Deng et al. (2002).)

The $\omega(\boldsymbol{k}_\perp)$ relation for both the upper and lower polariton branches can be measured through the angular dependence of the reflected light; when conservation of energy between the incoming light and the resulting cavity excitations can be satisfied, a dip occurs in the reflectivity. This energy dependence of the reflectivity for a selected angle is shown as the upper trace in Fig. 2.19. Both

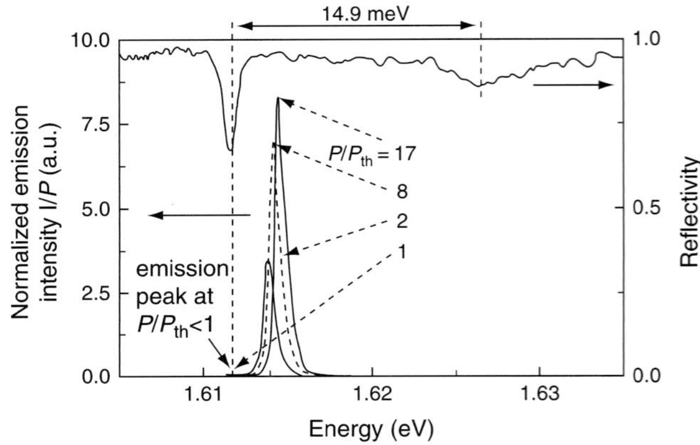

Fig. 2.19 The upper curve shows a *reflection* spectrum arising from the lower (narrow) and upper (broad) polariton bands. The lower curves show the emission spectrum for four different pumping levels. (After Deng et al. (2002).)

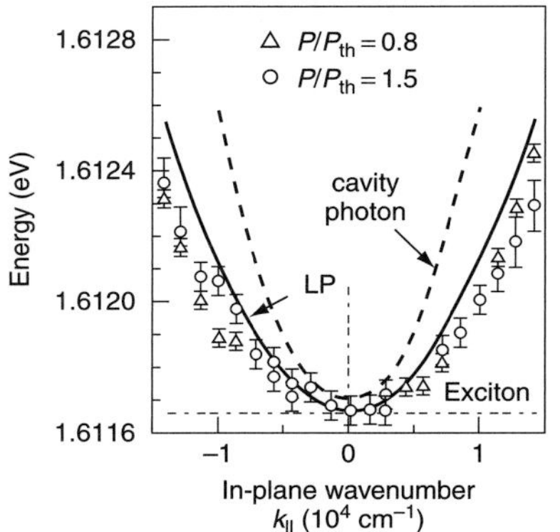

Fig. 2.20 The solid curve shows the calculated lower polariton curve in the presence of coupling together with the data above and below the threshold pump power. The dotted lines represent the uncoupled cavity and exciton dispersion curves which have been shifted for easy comparison with the coupled case. (After Deng et al. (2002).)

the lower and upper branches are seen. The spectrum of the lower branch, where the excitons reside following pumping, can be observed from the emission spectrum; this is also shown in Fig. 2.16 for four different pumping levels. Note that at higher levels it is blue shifted due to a phase-space filling effect (which we do not discuss). By assembling such data the authors construct the dispersion curve for the lower polariton branch, shown in Fig. 2.20. Shown as dashed lines are the predictions of the uncoupled photon and exciton modes, although shifted vertically for easy comparison with the predictions for the lower polariton branch, shown by a solid line. In addition to the static properties we have described, Deng et al. also obtained evidence for the formation of macroscopic quantum coherence by measuring the second-order coherence function of the light emitted from the polaritons in the time domain, which is beyond the scope of our discussion here.

Figure 2.21 shows a color-enhanced rendition of the angular distribution of the recombination radiation from a CdTe quantum-well cavity for three different pumping powers, by the Dang group. Note that at the highest pump intensity, on the far right, a narrow peak emerges which is taken as a signature of the formation of a condensate.

Fig. 2.21 Far-field emission measured at 5 K for three excitaton intensities which are increasing to the right. Shown as pseudo-3D images of the farfield emission within an angular cone of $\pm 23°$ with the emission intensity displayed on the vertical axis (in arbitrary units). With increasing excitation power, a sharp and intense peak is formed in the center of the emission distribution corresponding to emission angles $\theta_x = \theta_y = 0$, implying that they are associated with $k_\perp = 0$. (After Kasprzak et al. (2006).) (See Plate 5 for color image).

2.3 BEC of magnons

Magnons are the quantized excitations of a magnetic system; they are also bosons. However they can appear and disappear in three-particle scattering processes involving phonons, and without a constraint on their number they will not Bose condense under equilibrium conditions. Like the photons and excitons discussed in the previous two sections, dynamic condensation could occur if: (i) a dense concentration can be created by pumping the system to generate excitation densities well above those present in equilibrium, and (ii) the non-equilibrium magnons created in this way can scatter with the thermal magnons in a sufficiently number-conserving manner to yield a dynamic, quasi-equilibrium state characterized by a non-zero chemical potential.

Magnon BEC has been reported in the material YIG, yttrium iron garnet, by Demokritov et al. [40–42]. These workers estimate that non-number-conserving processes involving phonons occur on time scales of order 0.2–0.5 μs, whereas magnon–magnon scattering occurs on 10–50 ns scales;

taken together these estimates imply that a dynamic condensate can form. The experiments were performed under special conditions involving strong microwave pumping; they are the subject of the chapter by Demokritov and Slavin in Volume 2 of this book. YIG with the composition $Y_3Fe_5O_{12}$ or $Y_3Fe_2(FeO_4)_3$, is a well-known ferrimagnetic material with a Curie temperature of 550 K, and is unique in that it has unusually narrow ferromagnetic resonance (FMR) linewidths ($\Delta H \cong 0.5$ G), implying naturally long magnon lifetimes. (This property is exploited commercially to make field-tunable microwave filters and other devices.)

To understand how BEC can happen, in particular where the condensation point occurs in k-space, we will have to digress and discuss the behavior of magnons in thin films, since the associated phenomena are not treated in most books on solid state physics and magnetism. Those familiar with this topic may proceed directly to the experiments described in Section 2.3.4.

2.3.1 Damon and Eshbach theory of spin waves in a platelet [43][12]

[12] See also [44]. The topic is also discussed in the book [45].

2.3.1.1 *Basic equations*

The geometry utilized in the experiments is that of a platelet of thickness s and we will only consider the case of an in-plane external magnetic field, which in our chosen coordinate system (see Fig. 2.22) we write, $\mathbf{H}_0 = H_0\mathbf{z}$, where $\mathbf{z}$ is a unit vector along the z-axis. The modes must simultaneously satisfy the Landau–Lifshitz equation for $\mathbf{M}$ as well as Maxwell's equations. The first of these, which governs the precession dynamics of the magnetization $\mathbf{M}$, is given in the absence of dissipation by,

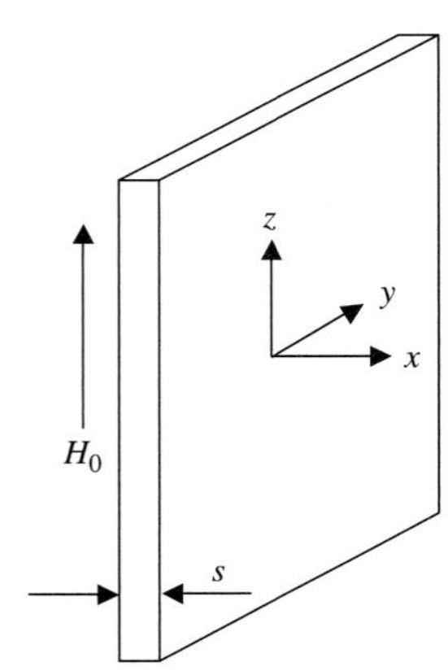

Fig. 2.22 Coordinate system for spin waves propagating in a platelet of thickness s.

$$\frac{d\mathbf{M}}{dt} = \gamma\,\mathbf{M} \times \mathbf{H}, \tag{2.34}$$

where γ is the electron gyromagnetic ratio and $\mathbf{H}$ is the *internal* magnetic field (with the geometry-dependent demagnetization effects included). Since the phase velocities of a spin wave turn out to be many orders of magnitude smaller than the velocity of light (and here we assume insulating samples where there are no ohmic currents), the Maxwell equations reduce to

$$\nabla \cdot \mathrm{B} = \nabla \cdot (\mathbf{H} + 4\pi\mathbf{M}) = 0 \tag{2.35a}$$

and

$$\nabla \times \mathbf{H} = 0; \tag{2.35b}$$

this is referred to as the *magneto-static limit.*

We will assume that $\mathbf{H}_0$ is large enough that the internal static magnetization $\mathbf{M}_0$ is uniform in our (infinite) platelet geometry. From the Maxwell boundary condition on the continuity of tangential fields, the static internal field is also $\mathbf{H}_0$, along which $\mathbf{M}_0$ aligns in equilibrium. Allowing for dynamic contributions to these two fields, which we write as $\mathbf{h}(\mathbf{r}, t)$ and $\mathbf{m}(\mathbf{r}, t)$, and restricting ourselves to a single frequency, we can then write,

$$\mathbf{H}(\mathbf{r}, t) = H_0\mathbf{z} + \mathbf{h}(\mathbf{r})e^{i\omega t} \tag{2.36a}$$

and

$$\mathbf{M}(\mathbf{r},t) = M_0\mathbf{z} + \mathbf{m}(\mathbf{r})\,e^{i\omega t}. \tag{2.36b}$$

Here we ignore any external driving field $\mathbf{h}^{ext}(\mathbf{r},t)$ and hence $\mathbf{h}(\mathbf{r},t)$ will arise solely from $\mathbf{m}(\mathbf{r},t)$; i.e. we solve for the homogeneous problem for the collective modes of the system. Eq. (2.35b) allows us to write $\mathbf{h}(\mathbf{r},t) = \nabla\psi(\mathbf{r},t)$, where $\psi(\mathbf{r},t) = \psi(\mathbf{r})e^{i\omega t}$ is a magnetic scalar potential; inserting this form for $\mathbf{h}(\mathbf{r},t)$ into Eq. (2.35), Maxwell's equations reduce to

$$\nabla^2\psi + 4\pi\nabla\cdot\mathbf{m} = 0. \tag{2.37}$$

Returning to Eq. (2.34), substituting our forms for **H** and **M** from (2.36a,b), with **h** written in terms of ψ, and retaining only terms linear in **m** and ψ yields, after some algebra,

$$4\pi m_x = \kappa\frac{\partial\psi}{\partial x} - i\nu\frac{\partial\psi}{\partial y} \tag{2.38a}$$

and

$$4\pi m_y = i\nu\frac{\partial\psi}{\partial x} + \kappa\frac{\partial\psi}{\partial y}, \tag{2.38b}$$

where we have introduced the dimensionless parameters (used in much of the literature)

$$\kappa = \frac{\Omega_H}{\Omega_H^2 - \Omega^2}; \qquad \nu = \frac{\Omega}{\Omega_H^2 - \Omega^2}; \tag{2.39a,b}$$

with

$$\Omega_H = \frac{H_0}{4\pi M_0}; \qquad \Omega = \frac{\omega}{4\pi\gamma M_0}. \tag{2.40a,b}$$

Substituting Eqs. (2.38a,b) for **m** into Eq. (2.37) we obtain the following equation for ψ:

$$(1+\kappa)\left(\frac{\partial^2\psi^i}{\partial x^2} + \frac{\partial^2\psi^i}{\partial y^2}\right) + \frac{\partial^2\psi^i}{\partial z^2} = 0; \tag{2.41a}$$

outside the sample, where $M_0 = 0$ and hence $\kappa = 0$, we have

$$\nabla^2\psi^{\mathrm{o}} = 0. \tag{2.41b}$$

2.3.1.2 *Boundary conditions, solution forms, and dispersion relations*

Matching the tangential component H_y requires (apart from an irrelevant additive constant)

$$\psi^i\big|_{x=\pm s/2} = \psi^{\mathrm{o}}\big|_{x=\pm s/2}. \tag{2.42a}$$

Matching the normal component B_x requires $h_x^i + 4\pi m_x = h_x^{\mathrm{o}}$ or, on using (2.38a),

$$\left[(1+\kappa)\frac{\partial\psi^i}{\partial x} - i\nu\frac{\partial\psi^i}{\partial y}\right]\Bigg|_{x=\pm s/2} = \frac{\partial\psi^{\mathrm{o}}}{\partial x}\Bigg|_{x=\pm s/2}. \tag{2.42b}$$

We seek a solution to the above equations of the form,

$$\psi(x, y, z) = X(x)Y(y)Z(z). \tag{2.43}$$

Given that Eqs. (2.41a,b) are linear second-order partial differential equations with constant coefficients, the solutions $X(x)$, $Y(y)$, and $Z(z)$ will either be oscillatory or exponential forms. Demanding that the boundary condition (2.42a) be satisfied for all y and z requires,

$$\psi^i(\mathbf{r}) = X^i(x)Y(y)Z(z) \tag{2.44a}$$

and

$$\psi^{\mathrm{o}}(\mathbf{r}) = X^{\mathrm{o}}(x)Y(y)Z(z), \tag{2.44b}$$

where X^i and X^{o} apply to the regions inside and outside the sample respectively. Our interest here is in modes propagating in the sample plane where we can write,

$$Y(y) \propto e^{-ik_y y} \tag{2.45a}$$

and

$$Z(z) \propto e^{-ik_z z}. \tag{2.45b}$$

Outside the sample $X(x)$ must fall off exponentially; hence,

$$X^{\mathrm{o}}(x) = Ce^{-\kappa_x x}, \quad x > s/2 \tag{2.46a}$$

and

$$X^{\mathrm{o}}(x) = De^{+\kappa_x x}, \quad x < -s/2. \tag{2.46b}$$

Inside the sample we anticipate oscillatory behavior and write,

$$X^i(x) = A \sin k_x x + B \cos k_x x. \tag{2.47}$$

Inserting the forms (2.45a,b), (2.46a,b), and (2.47) into (2.41a,b) then gives,

$$(1 + \kappa)(k_x^2 + k_y^2) + k_z^2 = 0 \tag{2.48a}$$

and

$$\kappa_x^2 - k_y^2 - k_z^2 = 0. \tag{2.48b}$$

Applying our first boundary condition (2.42a) we obtain,

$$A = \frac{C - D}{2} \frac{\exp(-\kappa_x s/2)}{\sin(k_x/2)} \tag{2.49a}$$

and

$$B = \frac{C + D}{2} \frac{\exp(-\kappa_x s/2)}{\cos(k_x/2)}. \tag{2.49b}$$

Applying our second boundary condition (2.42b) at $x = +s/2$ while also eliminating A and B using (2.49a,b) gives,

$$(1+\kappa)k_x\left[\frac{(C-D)}{2}\cot\left(\frac{k_x s}{2}\right)-\frac{(C+D)}{2}\tan\left(\frac{k_x s}{2}\right)\right]-C\nu k_y=-C\kappa_x; \tag{2.50a}$$

in the same way matching $x=-s/2$ gives,

$$(1+\kappa)k_x\left[\frac{(C-D)}{2}\cot\left(\frac{k_x s}{2}\right)+\frac{(C+D)}{2}\tan\left(\frac{k_x s}{2}\right)\right]-D\nu k_y=D\kappa_x. \tag{2.50b}$$

Setting the determinant of the coefficients of the homogeneous set of equations (2.50a,b) for C and D to zero yields,

$$\kappa_x^2+2\kappa_x k_x(1+\kappa)\cot(k_x s)-k_x^2(1+\kappa)^2-\nu^2 k_y^2=0. \tag{2.51}$$

Equation (2.51) together with Eqs. (2.48a,b) govern the dispersion of in-plane propagating magnons in an infinite platelet in terms of the wavevector components $\kappa_x, k_x\ k_y, k_z$ and the parameter $\kappa=\kappa(\omega, H_0, M_0)$. For fixed H_0 and M_0 we can in principle solve (2.51) together with (2.48a,b) for $\omega=\omega(k_x,k_y)$, or equivalently $\omega=\omega(k,\phi)$, where $k=\sin\phi k_z+\cos\phi k_y$. We start our analysis of Eq. (2.51) by discussing propagation perpendicular and parallel to the applied field.

2.3.1.3 *Propagation perpendicular to the field*

Here we have $k_z=0$ ($\phi=0$) and Eqs. (2.48a,b) lead to

$$k_x^2=-k_y^2;\quad \kappa_x^2=k_y^2 \tag{2.52a,b}$$

or

$$\kappa=-1;\quad \kappa_x^2=k_y^2. \tag{2.53a,b}$$

From the definition (2.39a), Eq. (2.53a,b) is equivalent to $\Omega^2=\Omega_H^2(\Omega_H+1)$, or on using (2.40b),

$$\omega=\gamma\sqrt{H_0(H_0+4\pi M_0)}. \tag{2.54}$$

Note that (2.53a,b) together with (2.51) require $\kappa_x^2=k_y^2=0$, and hence (2.54) corresponds to a *uniform precession* of the magnetization; This is the well-known Kittel equation obtained in most textbooks on solid state physics by more elementary methods.

For the case when (2.52a) applies, k_x must be imaginary if we are to have a propagating mode (real k_y). Together with (2.52b) we conclude that only a *surface wave* is allowed, i.e. one that decays exponentially on going into or out of the film (and in the present case is governed by the same decay length, κ_x^{-1}). For these waves (2.51) becomes

$$1+(1+\kappa)^2-\nu^2+2i(1+\kappa)\cot(ik_y s)=0. \tag{2.55}$$

In the thick film limit ($s\to\infty, \cot(ik_y s)\to -i$) we obtain $\nu=\pm(\kappa+2)$, with the lower sign corresponding to the physical root which, in turn, leads to $\Omega=\Omega_H+1/2$ or

$$\omega=\gamma\,(H+2\pi M). \tag{2.56}$$

This mode is generally referred to as the Damon–Eshbach mode; its frequency lies *higher* than the Kittel mode (or other volume modes discussed below).

2.3.1.4 *Propagation parallel to the field*

Here we have $k_y = 0$ or $\phi = \pi/2$ and Eqs. (2.48a,b) require,

$$k_x^2 = -\frac{k_z^2}{(1+\kappa)}, \quad \kappa_x^2 = k_z^2. \tag{2.57a,b}$$

The first of these says that we have propagating bulk waves (real k_x) provided that $1+\kappa < 0$ or $\kappa < -1$. From (2.39a) this requires $\Omega^2 < \Omega_H(\Omega_H + 1)$ or, equivalently, $\omega < \gamma\sqrt{H_0(H_0 + 4\pi M_0)}$; i.e. all such modes lie *below* the Kittel frequency.

Inserting (2.53b) into (2.51) gives

$$\cot(k_z s) = \frac{(1+\kappa)^2 - 1}{2(1+\kappa)}. \tag{2.58}$$

The numerical solution of this equation yields $\kappa = \kappa(k_z s, n)$ which is multi-valued with branches numbered by the index n. For a given value of $k_z s$ and branch n, we can use the associated value of κ to find $\Omega = \Omega(\Omega_H)$ for those parameters using the definition (2.39a). The uniform mode ($k_z = 0$) corresponds to $\kappa = -1$; i.e. we recover the Kittel mode. For $k_z s \to \infty, \Omega \to \Omega_H$; i.e. $\omega \to \gamma H_0$, the free spin Larmor frequency. These various solutions are referred to as *magneto-static modes*.

2.3.1.5 *The general case*

For propagation at an arbitrary angle ϕ we return to (2.48a,b) and solve for κ_x and k_x in terms of k_y and k_z and substitute these expressions in (2.51); introducing the quantity $\eta = k_z/k_y$, where $\phi = \tan^{-1}\eta$ and is measured from the z-axis, one can then eliminate k_z to obtain,

$$(1+\eta^2)^{1/2} + 2|(1+\eta^2)^{1/2}|\left(-\frac{1+\eta^2+\kappa}{1+\kappa}\right)^{1/2}(1+\kappa)\cot\left[|k_y|s\left(-\frac{1+\eta^2+\kappa}{1+\kappa}\right)^{1/2}\right]$$
$$+(1+\kappa)^2\frac{1+\eta^2+\kappa}{1+\kappa} - \nu^2 = 0, \tag{2.59}$$

which for arbitrary parameters must be solved numerically. We note in passing that the surface mode does not exist at all angles, but merges with the highest bulk mode at an angle given by $\phi_s = \tan^{-1}\left(\Omega_H^{1/2}\right)$, and beyond this point a surface wave ceases to exist. A curious property of the surface waves is that in the limit of large thicknesses they propagate in only one direction on a given surface, and in the opposite direction in the opposing surface: switching the magnetic field direction reverses the directions on the two surfaces.

2.3.2 Adding exchange

In the presence of exchange, and for a cubic material, one must add a contribution to the free energy of the form,

$$U_{\text{non}-u} = \frac{1}{2}\alpha\left[\left(\frac{\partial \mathbf{M}}{\partial x}\right)^2 + \left(\frac{\partial \mathbf{M}}{\partial y}\right)^2 + \left(\frac{\partial \mathbf{M}}{\partial z}\right)^2\right], \tag{2.60}$$

where the parameter α measures the strength of the effect. Defining an effective magnetic field as the functional derivative of this expression gives,

$$\begin{aligned} \mathbf{H}_{\text{grad}} &= -\frac{\delta U_{\text{non}-u}}{\delta \mathbf{M}} = \sum_{i=1}^{3} \frac{\partial}{\partial x_i} \frac{\partial U_{\text{non}-u}}{\partial \mathbf{M}/\partial x_i} \\ &= \alpha \nabla^2 \mathbf{M}. \end{aligned} \tag{2.61}$$

Extending the dipolar theory of Daman and Eshbach to include exchange greatly increases the algebraic complexity of the calculations [46, 48].[13] A perturbative treatment of spin waves in a thin platelet that combines magnetostatic and exchange effects has been given by Kalinikos and Slavin [49]. From a different perspective, Kreisel et al. [50] studied a microscopic model of spins on a cubic lattice coupled by a nearest neighbor Heisenberg interaction, together with the long-range dipole–dipole interaction; the system was infinite in y- and z-directions, but had a different number of layers along the x-direction to represent the sample thickness. The surface mode and the lowest-lying (uniform in x) volume mode are in good agreement with the Damon–Eshbach theory for small in-plane k-vectors. However all of their higher-order volume modes, which involve a progressively increasing numbers of nodes along x, are shifted upward in frequency. This arises because the wavevectors along x are quantized such that $k_x \cong n\pi/s$ and this generates an exchange contribution to the mode frequencies of $\gamma\,\alpha n^2\pi^2/s^2$. We find that the following form gives good overall agreement with exact calculations by Kreisel et al., *including the higher volume modes*:

[13] For a discussion of exchange effects when the field is perpendicular to the sample plane see [47].

$$\omega = \omega(n, k_y, k_z, H) + \gamma\,\alpha \frac{n^2\pi^2}{s}, \tag{2.62}$$

where $k_{\|}$ is the in-plane component of the wavevector and

$$H = H_0 + \alpha k_{\|}^2. \tag{2.63}$$

In Fig. 2.23 we plot some of the low-lying mode frequencies as a function of wavevector along the z (// to $\mathbf{H}_0$) and y ($\perp$ to $\mathbf{H}_0$), including the effects of exchange.

2.3.3 The experiments

As noted above, BEC of magnons has been reported by Demokritov et al.; the apparatus used is shown in Fig. 2.24. Here, a 5 μm-thick YIG film was positioned adjacent to a microwave strip line resonator. The sample and resonator were placed in an in-plane magnetic field of 700 G. The spectrum for these parameters is shown schematically in Fig. 2.25 and the minimum

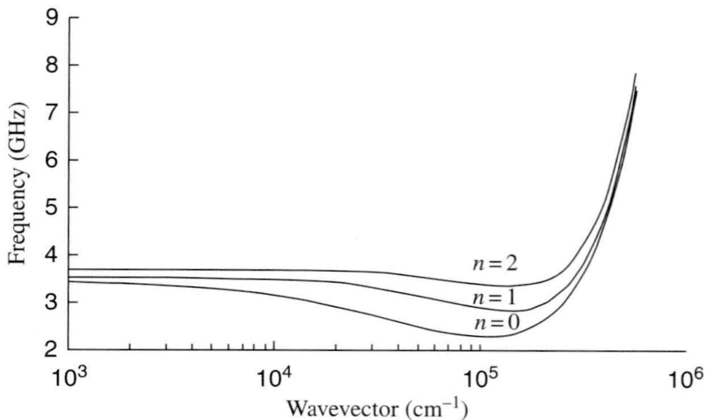

Fig. 2.23 The three lowest magnon dispersion curves in yttrium iron garnet for propagation perpendicular to the magnetic field as calculated for: $H = 700\,\text{G}$, $s = 0.495$ microns, $\mu = 2\mu_B$, $M_0 = 139.3\,\text{G}$, and $\gamma\alpha = 5.17 \times \text{Gm}^2$. (J. Sklenar, unpublished.)

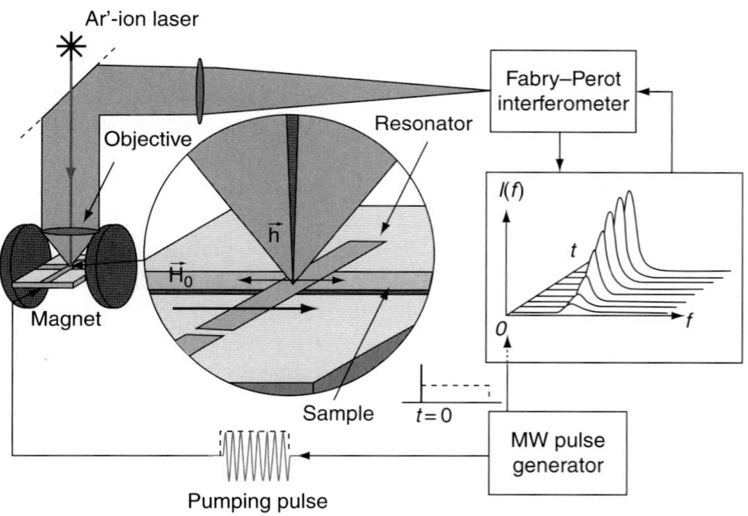

Fig. 2.24 The experimental setup used to study BEC of magnons in YIG. The excitation spectrum is obtained using Brilloun light scattering. The sample is excited by a microwave pulse applied to a strip line resonator. (Courtesy of S. O. Demokritov.)

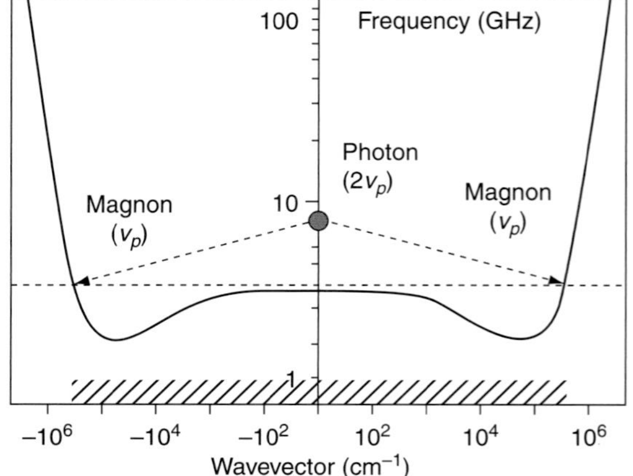

Fig. 2.25 The dispersion relation of magnons in the direction parallel to H_0. The system is pumped at 8.1 GHz and via a parametric process produces two magnons at half the pump frequency, as shown by the two arrows. Via an interaction with thermally excited magnons the pumped magnons thermalize, while their number remains essentially constant. Since the density then exceeds that which can be accommodated by the equilibrium Bose distribution, a condensate forms. (After Demokritov et al. (2006).)

referred to above occurs for $\omega_{\rm min} \cong 2.1$ GHz. The resonator was pumped with 1–100 μs pulses on a 5 percent duty cycle at 8.1 GHz. In order to generate magnons with finite wavevectors, the experiment utilizes a *parametric process* in which a microwave photon with angular frequency ω (and with $k \cong 0$ on the scale of magnons) generates two magnons with frequency $\omega/2$ and wavevectors $\pm k$ that satisfy the dispersion relation. In this way they flood the sample with a non-equilibrium population of magnons which they argue subsequently undergoes BEC. The pumped magnons then have a frequency of 4.05 GHz, which is larger than the dispersion minimum. From here the authors present arguments and estimates that magnon–magnon scattering processes thermalize the gas such that it undergoes BEC at the minimum frequency, located at $f_{\rm min} = 2.1$ Ghz, with a corresponding wavevector $k_{\rm min} = 5 \times 10^{-4}\,{\rm cm}^{-1}$; they present strong arguments that four-magnon processes (which conserve the number of magnons) will dominate over the non-number-conserving three-particle processes involving phonons. The authors monitor the time evolution of the magnon distribution via a dynamic Brillouin light scattering (BLS) technique.[14] With such instrumentation the frequencies of the scattered photons were collected and analyzed to construct an overall spectral representation of the magnon distribution function versus time within some frequency range.

[14] In a Brillouin scattering experiment one directs a laser beam at the surface of a sample. Photons are then scattered by excitations naturally present in the material due to thermal excitation; magnons in the present case. From the change in the angle and frequency of the scattered photons one can infer the density of magnons of a given wavelength and frequency. The frequency shifts are small and resolving the frequency shift requires a sensitive multi-pass Fabry–Perot optical cavity. Carrying out such measurements as a function of time requires additional instrumentation.

The data in the absence of pumping are shown in Fig. 2.26; the maximum occupation corresponds to $f_{\rm min}$ located at the wavevector $k_{\rm min}$. This data, together with the magnon theory, can be used to experimentally establish the density of states in the vicinity of the minimum. Figure 2.27 shows the BLS spectra at various times following the application of a 0.7 W microwave pulse at the frequency $f_{\rm pump} = 8.1$ GHz. Note that for short times the magnons are concentrated in the vicinity of $f_{\rm pump}/2$, but as time evolves the magnons pile up at $f_{\rm min}$. The authors find that the spectrum at later times can only be fitted with a Bose distribution containing a smoothed delta-function-like contribution and a chemical potential corresponding to $f_{\rm min}$. That a condensate is actually present is supported by later measurements from this group, showing that the emitted radiation from the condensate is actually *coherent* (as expected for a condensate) [57]. In yet another experiment the group showed that the condensate can be created by pumping with broadband noise [52] (mimicking a hot thermal source). These and other experiments constitute strong evidence for magnon BEC. Finally, we note that the preparation of two separate, but interacting, condensates has been studied [53].

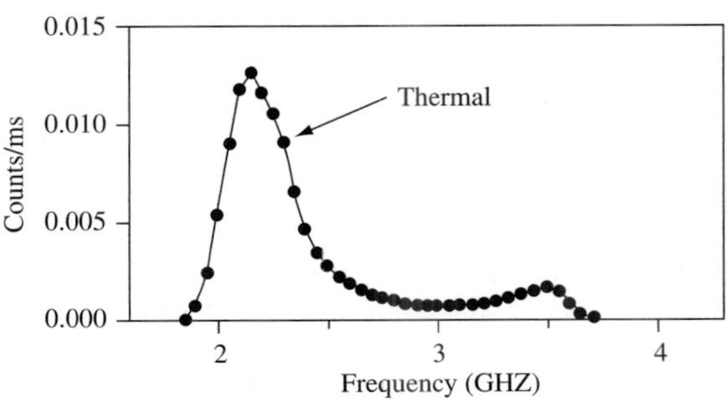

Fig. 2.26 The BLS spectrum from a YIG film in the absence of pumping showing the thermal distribution of the magnons. (After Demidov et al. (2007.)

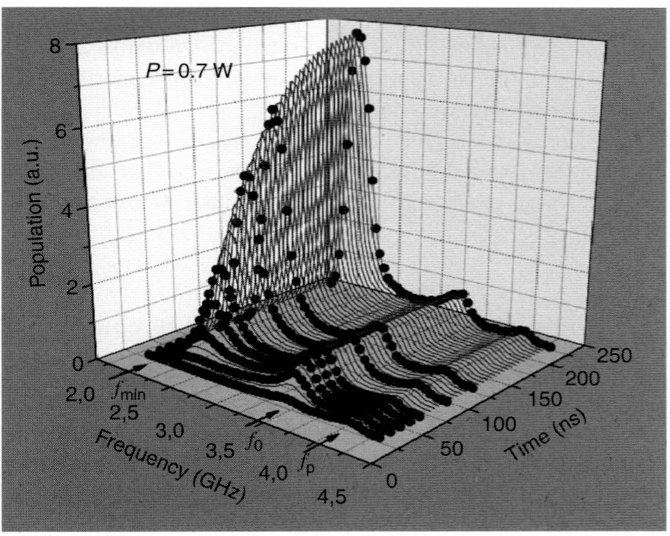

Fig. 2.27 The time dependence of the BLS spectrum of magnons following the initiation of pumping. As time evolves the strength of the feature located at $f_{\text{min}} = 2.1\,\text{GHz}$ grows at the expense $f_0 = f(\mathbf{k} = 0) = 3.5\,\text{GHz}$ due to increasing magnon occupation (note the pump frequency, f_p, has been suppressed by filtering). (Courtesy of S. O. Demokritov.)

2.4 Cold atom superfluids: theoretical background

In 1995, BEC was achieved in a gas of optically trapped and magnetically cooled Rb^{87} atoms [54], a weakly interacting system approximating Einstein's ideal Bose gas model, and some 70 years after his proposal. Shortly after, and with some improvements, this was confirmed by an independent group for Na^{23} (the only stable isotope) [55]. Rather than being an end point, these developments spawned a multitude of new lines of research, some of which will be reviewed by Chevy and Dalibard in Chapter 7; additional aspects will be reviewed by Zwierlein and by Chin and Gemelke, in Volume 2 along with a theoretic review by Trivedi and L. Loh. The achievement of BEC has built on a chain of developments by the relevant communities. Here we only briefly review a few of the techniques used, along with their application to the study of some interesting physical phenomena; a more complete treatment is available in some recent books [56–58]. Our discussion will largely be limited to alkali metal vapors.

Once vaporized (by heating a solid source within an ultra-high vacuum environment), the first stop on the road to preparing an ultra-cold atom ensemble is to use pairs of opposing laser beams (three beam pairs propagating along mutually orthogonal axes and intersecting at a common point) to slow the atoms. To understand how slowing can happen we must examine the dynamic response of our atoms to the laser beams, which we outline in Section 2.4.2. This is followed by a discussion of how atoms can be slowed (and trapped) in a so-called magneto-optical trap (MOT), which exploits the phenomena of optical molasses and Sisyphus cooling, which are treated in Section 2.4.3. Although necessary steps, neither of these techniques can achieve low enough temperatures or high enough densities to achieve BEC, so it is necessary to take the atom cloud in the MOT and first trap it, with procedures discussed in

Section 2.4.4, and then evaporatively cool it, by letting the higher-energy atoms escape from the trap, using techniques discussed in Section 2.4.5. Finally in Section 2.4.6 we will briefly mention a few of the many experiments involving superfluidity of atoms and even molecules.

Since the statistical mechanics of BEC for atoms in a confining potential differs from that for free particles, treated earlier in Section 2.1, we begin our discussion here with a treatment of the non-interacting Bose gas confined in a harmonic potential.

2.4.1 BEC of non-interacting atoms in a harmonic potential

For simplicity we will approximate the central region of typical traps involving magnetic fields or pairs of Gaussian laser beams by the quadratic form,

$$V(x, y, z) = \frac{M}{2}\left(\omega_x^2 x^2 + \omega_y^2 y^2 + \omega_z^2 z^2\right), \tag{2.64}$$

where $\omega_i^2 = \kappa_i/M$, κ_i is a spring constant and M is the mass of the atom. The corresponding energy levels are given by the usual form,

$$\varepsilon\left(n_x, n_y, n_z\right) = \hbar\left(\omega_x\left(n_x + \frac{1}{2}\right) + \omega_y\left(n_y + \frac{1}{2}\right) + \omega_z\left(n_z + \frac{1}{2}\right)\right). \tag{2.65}$$

2.4.1.1 *The density of states in a harmonic trap*

As with the free Bose gas treated in Section 2.1 we seek an expression for the density of states. Suppressing the zero-point energy, we can rewrite (2.65) as $\varepsilon = \varepsilon_x + \varepsilon_y + \varepsilon_z$, where $\varepsilon_i = \hbar\omega_i n_i$ or $n_i = \varepsilon_i/\hbar\omega_i$. Now the total number of states, n_ε, with an energy less than some value ε is

$$\begin{aligned} N_\varepsilon &= \int dn_z \int dn_y \int dn_x \\ &= \int_0^{\varepsilon} \frac{dn_z}{d\varepsilon_z} d\varepsilon_z \int_0^{\varepsilon-\varepsilon_z} \frac{dn_y}{d\varepsilon_y} d\varepsilon_y \int_0^{\varepsilon-\varepsilon_z-\varepsilon_y} \frac{dn_x}{d\varepsilon_x} d\varepsilon_x \\ &= \frac{1}{\hbar^3\omega_x\omega_y\omega_z} \int_0^{\varepsilon} d\varepsilon_z \int_0^{\varepsilon-\varepsilon_y} d\varepsilon_y \int_0^{\varepsilon-\varepsilon_y-\varepsilon_x} d\varepsilon_x. \end{aligned} \tag{2.66}$$

Carrying out the integrations gives,

$$N_\varepsilon = \frac{\varepsilon^3}{6\hbar^3\omega_x\omega_y\omega_z}, \tag{2.67}$$

and hence the density of states $N(\varepsilon) = dN_\varepsilon/d\varepsilon$ is given by,

$$N(\varepsilon) = \frac{\varepsilon^2}{2\hbar^3\omega_x\omega_y\omega_z}. \tag{2.68}$$

2.4.1.2 *Fixing the chemical potential and the BEC transition temperature*

Suppose we have a harmonic trap of the form (2.64) into which we have loaded N atoms at a temperature T. Based on the Bose distribution function $n(\varepsilon)$ we have the following constraint (see Eq. (1.1)) on the chemical potential μ:

$$N = \int \frac{N(\varepsilon)d\varepsilon}{e^{(\varepsilon-\mu)/k_BT} - 1}. \quad (2.69)$$

As the temperature is lowered $\mu \to 0$ from below, and condensation occurs when it reaches zero, which occurs at the temperature T_c governed by

$$N = \int \frac{N(\varepsilon)d\varepsilon}{e^{\varepsilon/k_BT_c} - 1}. \quad (2.70)$$

The integral is of the form,

$$\int_0^\infty dx \frac{x^{\alpha-1}}{e^x - 1} = \Gamma(\alpha)\zeta(\alpha), \quad (2.71)$$

where $\Gamma(\alpha)$ $(= (\alpha - 1)!)$ and $\zeta(\alpha)$ are the gamma and Riemann zeta functions respectively, with argument α. The final result for the transition temperature is then

$$k_BT_c = \frac{\hbar\overline{\omega}N^{1/3}}{[\zeta(3)]^{1/3}} \cong 0.94\hbar\overline{\omega}N^{1/3}, \quad (2.72)$$

where we have used $\zeta(3) = 1.202$ and we write, $\overline{\omega} = [\omega_x\omega_y\omega_z]^{1/3}$.

2.4.1.3 *Properties below T_c*

For $T < T_c$, Eq. (2.72) can be interpreted as the number of atoms excited out of the condensate. The condensate fraction, N_0/N, is then given by

$$\frac{N_0}{N} = 1 - \left(\frac{T}{T_c}\right)^3. \quad (2.73)$$

Recall that we have assumed a quasi-continuous distribution of levels in obtaining a unique (sharp) transition temperature. To be valid this requires $k_BT_c >> \hbar\overline{\omega}$ or equivalently $N^{1/3} >> 1$.

Below the transition temperature where the chemical potential vanishes we can evaluate the total energy of our Bose gas as,

$$E = \int \frac{\varepsilon N(\varepsilon)d\varepsilon}{e^{\varepsilon/k_BT_c} - 1}. \quad (2.74)$$

Combining this expression with (2.71) and (2.68) gives

$$\begin{aligned} E(T) &= \frac{\Gamma(4)\zeta(4)}{2\hbar^3\overline{\omega}^3}(k_BT)^4 \\ &= \frac{\pi^4}{30\hbar^3\overline{\omega}^3}(k_BT)^4 \qquad (T < T_c), \end{aligned} \quad (2.75)$$

where we have used $\Gamma(4) = 6$ and $\zeta(4) = \pi^4/90$. Note that this expression does *not* involve the number of particles. This is related to the fact that below T_c particles progressively go into the condensate, where they do not contribute to the energy, and T_c goes down with the number of particles (as $N^{-1/3}$). The heat capacity is given by,

$$C(T) = \frac{\partial E}{\partial T} = 4\frac{E(T)}{T} \qquad (T < T_c). \quad (2.76)$$

A difference between the uniform gas and the gas trapped in a harmonic potential is that the heat capacity of the latter undergoes a discontinuity at the

BEC transition temperature, a property characteristic of a second-order phase transition. For the uniform gas the heat capacity only has a change in slope at the transition, a property London associated with a third-order transition.

2.4.1.4 *The condensate wave function*

If interactions are neglected, the condensate wave function has the usual Gaussian form, which for our anisotropic potential well is,

$$\psi(x, y, z) = \frac{1}{\pi^{3/4}(a_x a_y a_z)^{1/2}} \exp\left(-\frac{x^2}{2a_x^2} - \frac{y^2}{2a_y^2} - \frac{z^2}{2a_z^2}\right), \tag{2.77}$$

where

$$a_i^2 = \frac{\hbar}{M\omega_i}. \tag{2.78}$$

The condensate is often too small to be imaged directly; in this case the trapping potential is removed and the condensate is allowed to expand, the rate of which is controlled by the velocity or momentum distribution within the condensate; the latter follows from the Fourier transform of $\psi(x, y, z)$ and is given by,

$$\phi(p_x, p_y, p_z) = \frac{1}{\pi^{3/4}(c_x c_y c_z)^{1/2}} \exp\left(-\frac{p_x^2}{c_x^2} - \frac{p_y^2}{c_y^2} - \frac{p_z^2}{c_z^2}\right), \tag{2.79}$$

where

$$c_i = \frac{\hbar}{a_i} = \sqrt{M\hbar\omega_i}. \tag{2.80}$$

Note the expansion front is *anisotropic*, reflecting the underlying potential. On the other hand, in the absence of a condensate ($T > T_c$) the velocity profile is governed by the Maxwell distribution, which is isotropic, leading to a much more isotropic form on expansion.

Unlike the case of the ideal gas in a box, where the condensate is spread throughout the box, condensation in a harmonic well involves progressively higher particle densities in the central region and with it enhanced interaction effects. Assuming a repulsive interaction we would expect an expanded condensate wave function, i.e. larger than the mean size of the ground-state Gaussian form given by (2.77) for the non-interacting case. This effect can be modeled by solving the Gross–Pitaevskii equation (see Section 2.4.2) in the presence of the harmonic potential.

2.4.2 BEC of interacting atoms in the presence of a potential

Having considered the non-interacting gas in a trap our next task is to consider interactions. If the gas largely consists of a condensate, that part of the problem can be treated with the Gross–Pitaevskii approach.

2.4.2.1 *The Gross–Pitaevskii equation*

Here one starts with the expression for the total energy of the condensate,

$$E = \int d^3r \left[\frac{\hbar^2}{2M} |\nabla\psi(\mathbf{r},t)|^2 + V^{\text{ext.}}(\mathbf{r})\,|\psi(\mathbf{r},t)|^2 + \frac{1}{2}U_0|\psi(\mathbf{r},t)|^4 \right], \quad (2.81)$$

where the inter-particle potential has been assumed to have the form $U(\mathbf{r}) = U_0\delta(\mathbf{r})$. On performing a variation subject to the condition

$$\int d^3r\,|\psi(\mathbf{r})| = N, \quad (2.82)$$

we obtain the Gross–Pitaevskii (G–P) equation, also called the non-linear Schrödinger equation (see Eq. (1.131) of Chapter 1), as

$$\left[-\frac{\hbar^2}{2M}\nabla^2 + V^{\text{ext.}}(\mathbf{r}) + U_0|\psi(\mathbf{r},t)|^2 \right] \psi(\mathbf{r},t) = \mu\psi(\mathbf{r},t); \quad (2.83)$$

here $\psi(\mathbf{r},t)$ is the condensate wave function and μ is the associated Lagrange multiplier, which is also the chemical potential. For the case of a uniform Bose gas the G–P equation reduces to

$$\mu = U_0|\psi(\mathbf{r},t)|^2 = U_0 n, \quad (2.84)$$

where $n = N/V$ is the particle density.

An approximate solution of the G–P equation for the case of a harmonic potential can be obtained using a variational approach and a Gaussian form for the trial wave function,

$$\psi(x,y,z) = \frac{N^{1/2}}{\pi^{3/4}(b_x b_y b_z)^{1/2}} \exp\left(-\frac{x^2}{2b_x^2} - \frac{y^2}{2b_y^2} - \frac{z^2}{2b_z^2} \right), \quad (2.85)$$

similar to the solution for the non-interacting case; here the b_i are variational parameters which, physically, we expect to be somewhat larger than the a_i for a repulsive interaction between particles. Inserting this form into (2.81) with the potential given by (2.64) and carrying out the integrations gives,

$$E(b_x,b_y,b_z) = N\sum_{i=1}^{3} \hbar\omega_i \left(\frac{b_i^2}{a_i^2} + \frac{a_i^2}{b_i^2} \right) + \frac{N^2 U_0}{2(2\pi)^{3/2} b_x b_y b_z}, \quad (2.86)$$

where $i = x,\ y,\ z$. Minimizing this expression with respect to the three parameters b_i leads after some algebra to

$$b_i = \left(\frac{2}{\pi}\right)^{1/10} \left(\frac{Na}{\bar{a}}\right)^{1/5} \frac{\overline{\omega}}{\omega_i}\,\bar{a}, \quad (2.87)$$

where we have introduced the (Fermi) scattering length, a, in place of the strength U_0 of the delta function employed in Eq. (2.81) through the expression

$$U_0 = 4\pi\hbar^2 a/M. \quad (2.88)$$

Substituting Eq. (2.87) into (2.86) we obtain the energy as

$$E = N\frac{5}{4}\left(\frac{2}{\pi}\right)^{1/5} \left(\frac{Na}{\bar{a}}\right)^{2/5} \hbar\overline{\omega}. \quad (2.89)$$

When the scattering length is negative the condensate will contract and at some particle number, N_c, becomes unstable (collapses). The critical value for some direction i is found by setting the first *and* second derivatives of E with respect to the parameter b_i equal to zero. For the isotropic case where we have a single frequency ω one finds (see Pethick and Smith [57]),

$$\frac{N_c|a|}{a_{\mathrm{osc}}} = \frac{2(2\pi)^{1/2}}{5^{5/4}} \cong 0.67, \tag{2.90}$$

where from Eq. (2.78) the characteristic length is $a_{\mathrm{osc}} = \sqrt{\hbar/M\omega}$.

2.4.2.2 *The Thomas–Fermi approximation*

Another commonly used approximation is to ignore the kinetic energy arising from the first term on the right in Eq. (2.81), which for our Bose case gives,

$$\left(V(\mathbf{r}) + U_0 \left|\psi(\mathbf{r})\right|^2\right) \psi(\mathbf{r}) = \mu\psi(\mathbf{r}), \tag{2.91}$$

or

$$n(\mathbf{r}) = \left[\mu - V(\mathbf{r})\right] / U_0; \tag{2.92}$$

this is called the Thomas–Fermi approximation. The chemical potential is fixed by the condition

$$\int d^3r\, n(\mathbf{r}) = N. \tag{2.93}$$

2.4.3 Non-interacting Fermi particles in a harmonic trap

We can easily adapt the above discussion of the statistical mechanics of non-interacting bosons in a harmonic trap to the fermion case. The number of states, N_ε, with energy less than ε, and density of states, $N(\varepsilon)$ are, apart from the degeneracy, g (2 for a spin 1/2 particle), the same for both cases. Referring to Eqs. (2.67) and (2.68) we can now generalize them as,

$$N_\varepsilon = \frac{g\,\varepsilon^3}{6\hbar^3\overline{\omega}^3}, \quad N(\varepsilon) = \frac{g\,\varepsilon^2}{2\hbar^3\overline{\omega}^3}, \tag{2.94a,b}$$

where $\overline{\omega} = (\omega_x\omega_y\omega_z)^{1/3}$. The condition fixing the chemical potential is now

$$N = \int \frac{N(\varepsilon)d\varepsilon}{e^{(\varepsilon-\mu)/k_BT} + 1}, \tag{2.95}$$

while the total energy is

$$E = \int \frac{\varepsilon N(\varepsilon)d\varepsilon}{e^{(\varepsilon-\mu)/k_BT} + 1}. \tag{2.96}$$

For an arbitrary temperature these integrals must be done numerically.

At absolute zero all states are filled so the Fermi energy corresponds to $N_\varepsilon = N$ for $\varepsilon = \mu$ leading to

$$\mu = \hbar\overline{\omega}\,(6N/g)^{1/3}; \qquad \text{(3D Fermi oscillator)} \tag{2.97}$$

this is to be compared with the free-particle case where

$$\mu = \frac{\hbar^2}{2M}\left(\frac{6\pi^2 N}{gV}\right)^{2/3}; \qquad \text{(3D Fermi gas)} \tag{2.98}$$

the two cases scale as $N^{1/3}$ and $N^{2/3}$ respectively. At $T = 0$ the total energy of all the states is

$$\begin{aligned} E(0) &= \int_0^\mu d\varepsilon\, \varepsilon N(\varepsilon) = \int_0^\mu d\varepsilon \frac{\varepsilon^3}{2\hbar^3\overline{\omega}^3} = \frac{\mu^4}{8\hbar^3\overline{\omega}^3} \\ &= \frac{3N}{4g}\mu, \qquad \text{(3D Fermi oscillator)} \end{aligned} \tag{2.99}$$

which is to be compared with

$$E(0) = \frac{3N}{5g}\mu. \qquad \text{(3D Fermi gas)} \tag{2.100}$$

As with the free Fermi gas one can carry out a Sommerfeld expansion for the total energy in the limit $T << T_F = \mu/k_B$ having the usual form,

$$E(T) = E(0) + \frac{\pi^2}{6}N(\mu)(k_B^2 T^2). \tag{2.101}$$

From this we can calculate the heat capacity at constant particle number, $C_N = (\partial E/\partial T)_N$.

In the limit of large quantum numbers the motion is semiclassical. A one-dimensional classical oscillator with spring constant κ_x, mass M, and frequency $\omega_x = \sqrt{\kappa_x/M}$, with an energy equal to the Fermi energy, μ, would have a "Fermi amplitude" at $T = 0$ given by

$$x^F(0) = \sqrt{2\mu/\kappa_x} = \sqrt{2\mu/M\omega_x^2}, \tag{2.102}$$

at which point $v = 0$, while the maximum velocity, the "Fermi velocity", would be

$$v_x^F(0) = \sqrt{2\mu/M}, \tag{2.103}$$

which occurs at $x = 0$. An image of the first would exhibit anisotropy while the second would be isotropic. For temperatures above absolute zero we have amplitudes that exceed $x^F(0)$; likewise we should have velocities that exceed $v_x^F(0)$. Unlike a free Fermi gas, where the box walls oppose the Fermi pressure, $P = \partial E/\partial V$, our Fermi gas in the harmonic well is confined by the potential itself.

A readily measured quantity on trapped atom clouds is the velocity distribution following the rapid removal of the trapping potential. Although one must be careful to properly define what is measured, such distributions can be calculated, but must be done numerically at finite temperatures.

2.4.4 Thomas–Fermi theory of interacting Fermi particles in a trap

The Thomas–Fermi theory discussed above for Bose particles can easily be extended to Fermi particles. To do this at $T = 0$ all we have to do is add to the potential energy, $V(\mathbf{r})$, in Eq. (2.91) the Fermi energy of a gas, $\varepsilon_F(n(\mathbf{r}))$, having a density $n(\mathbf{r})$ at position $\mathbf{r}$ in the potential well. Equation (2.89) then takes the form,

$$(V(\mathbf{r}) + \varepsilon_F(n(\mathbf{r})) + U_0 n(\mathbf{r})) = \mu, \tag{2.104}$$

where (for the case of a single polarized spin where $g = 1$) this is given by,

$$\varepsilon_F(n(\mathbf{r})) = \frac{\hbar^2}{2m}(6\pi^2 n(\mathbf{r}))^{2/3}. \tag{2.105}$$

The chemical potential is again fixed by the condition (2.93).

2.4.5 Feshbach resonances

A special class of cold gas interactions involves the Feshbach resonances, which have attracted great interest in recent years and led to some spectacular phenomenology. The effect has a long history in nuclear physics. In experiments on slow neutron scattering from nuclei, very narrow resonances, termed Feshbach resonances, are encountered which result from the formation of a long-lived "compound" nucleus. The long lifetimes result from the minimal overlap between two respective wave functions: that representing the compound nucleus, called the closed channel, and that corresponding to the product of the wave functions of the incoming neutron and starting nucleus, called the open channel.[15] With respect to cold trapped gases, we can imagine two incoming atoms in some particular hyperfine state, with $S = 1$, being nearly degenerate with a long-lived singlet molecular state with $S = 0$, as shown in Fig. 2.28.

[15] There is of course only one time-evolving many-particle wave function so these separations are only approximate. One then says the wave function of the separated atoms is concentrated in a different part of Hilbert space from the virtual diatomic molecule.

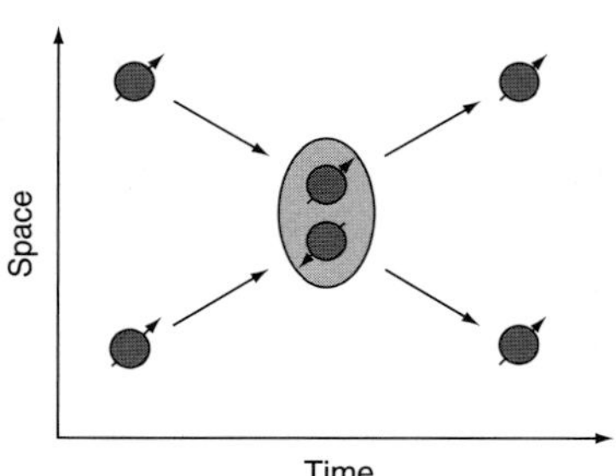

Fig. 2.28 The time evolution of two incoming atoms, each in a hyperfine state with $F \neq 0$, which form a long-lived quasi-bound molecular state, with $F = 0$, (generally in some excited state with vibrational quantum number ν) which after some time decays back into unbound atoms. (After Duine and Stoof (2004).)

2.4.5.1 *Scattering in a single channel*

We start by recalling some elements of scattering theory. At low energies, ε, the scattering can be taken as an s-wave ($l = 0$) and the scattering amplitude, $f(\theta)$, and total cross-section, σ, are then given by,

$$\begin{aligned} f_0 &= \frac{1}{2ik}(e^{2i\delta_0(k)} - 1) \\ &= \frac{1}{k\cot(\delta_0(k)) - ik} \end{aligned} \tag{2.106}$$

and

$$\sigma = (4\pi/k^2)\sin^2\delta_0(k), \tag{2.107}$$

where $k = (2\mu\varepsilon/\hbar^2)^{1/2}$ with μ the reduced mass and k and ε the center of mass wavevector and energy.

Following Duine and Stoof [59] we can consider the textbook example of a square-well potential (see Fig 2.29a which is drawn for a case where $V_0 < 0$)

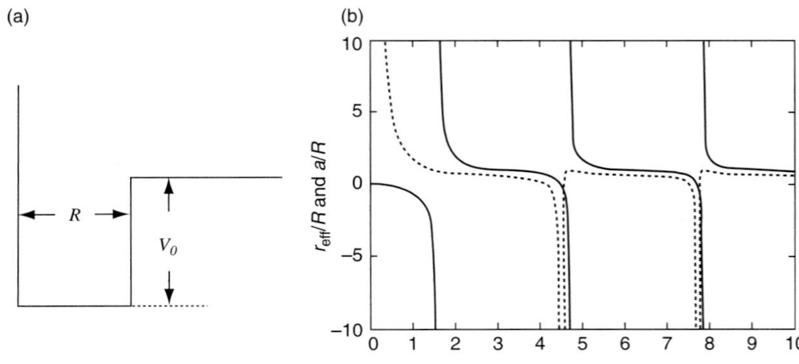

Fig. 2.29 (a) A model square-well potential. (b) A plot of the normalized scattering length, a, and the effective range, r_{eff}, versus $\gamma = R(2mV_0\hbar^2)^{1/2}$ for the square well. (After Duine and Stoof (1994).)

where, on matching the wave functions inside and outside the well in the usual way, one finds the following expression for the s-wave phase shift:

$$\delta_0(k) = -kR + \tan^{-1}\left(\frac{k}{K}\tan(KR)\right), \tag{2.108}$$

where $K = (2m(\varepsilon - V_0)/\hbar^2)^{1/2}$. As discussed in most elementary texts on nuclear physics, at low energies one can expand $\delta_0(k)$ for a wide class of potentials as

$$k\cot(\delta_0(k)) = \frac{1}{a} + \frac{1}{2}r_{eff}k^2 + \cdots, \tag{2.109}$$

where a is the *scattering length*, defined as $a = -\lim_{k\to 0}(\delta_0(k)/k)$, and r_{eff} is called the *effective range*. As $k \to 0$, $\sigma \to 4\pi a^2$. For the square-well potential a and r_{eff} turn out to be

$$a = R\left(1 - \frac{\tan\gamma}{\gamma}\right) \tag{2.110}$$

and

$$r_{eff} = R\left[1 + \frac{3\tan\gamma - \gamma(3+\gamma^2)}{3\gamma(\gamma - \tan\gamma)^2}\right], \tag{2.111}$$

where $\gamma = R(m|V_0|/\hbar^2)^{1/2}$; for the case $V_0 \to +\infty$, $R = a$. The quantities $r_0(\gamma)$ and $a(\gamma)$ are plotted in Fig. 2.29b. Note that the scattering length a can have either sign, pass through zero, and diverge; the latter occurs when $\gamma = (n + 1/2)\pi$ and is referred to as a *resonance*. With increasing V_0 this occurs each time the potential can support *a new bound state*; a large positive scattering length implies a bound state with an energy just below the continuum.

2.4.5.2 *Scattering in two channels*

We now discuss a simple two-channel model in which Feshbach resonances can occur. Alkali atoms with a lone unpaired electron in an s-state are expected to bond much more strongly in a singlet state (as in the usual pairing of

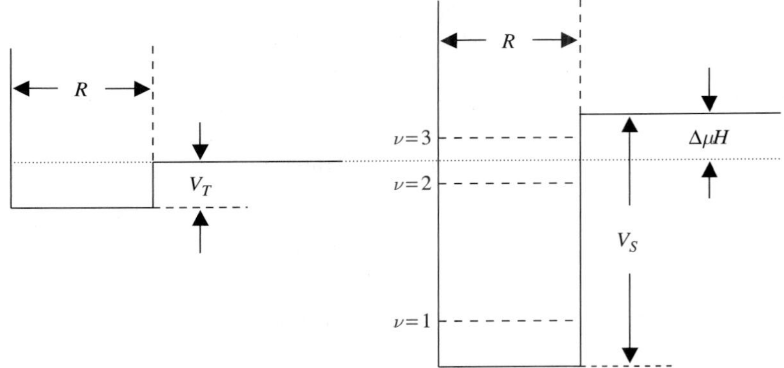

Fig. 2.30 A two-square-well model leading to Feshbach resonances. At the left we see the shallow well representing the triplet state; here we assume it has no bound states. To the right we see the singlet channel which has a deep well and several bound states numbered by a vibrational quantum number, ν. The singlet well is displaced upward, representing the fact that the triplet states have lower energy since they are trapped in the external field, H. The two wells are coupled by the hyperfine interaction. (After Duine and Stoof (1994).)

electrons in forming a chemical bond). Note that when using magnetic traps the electron spins are aligned parallel and potential pairs are then in triplet states. In a square-well model for the interaction between the two atoms we can assign different depths V_S and V_T to represent the singlet and triplet bonding (taking R to be the same for both), leading to bound state energies $E_{\nu S}$ and $E_{\nu' T}$ where ν and ν' number the bound states. But the singlet and triplet channels are coupled by the hyperfine interaction $V_{hf} = a_{hf}\mathbf{I}\cdot\mathbf{S}$ (with larger values of I leading to more channels). In the presence of an external magnetic field, $\mathbf{H}$, we also have the Zeeman energy, $E_Z = -\boldsymbol{\mu}\cdot\mathbf{H}$ arising from the interaction of the moments with that field. As a very simple model we can introduce a Hamiltonian of the form,

$$\hat{H} = \begin{pmatrix} \hat{H}_T & V_{hf} \\ V_{hf} & \hat{H}_S + \Delta\mu H \end{pmatrix}, \tag{2.112}$$

where $\hat{H}_T$ and $\hat{H}_S$ are the Hamiltonians in the absence of magnetic effects and $\Delta\mu$ is the difference between the magnetic moments in the two channels. The energies resulting from diagonalizing Eq. (2.112) are

$$\varepsilon_{\pm} = \frac{1}{2}\left[\Delta\mu H + \varepsilon_S + \varepsilon_T \mp \sqrt{(\Delta\mu H + \varepsilon_S + \varepsilon_T)^2 + 4V_{hf}^2}\right], \tag{2.113}$$

where ε_S and ε_T are the singlet and triplet energies in the absence of the Zeeman and hyperfine effects. We will assume that there are many singlet bound states, numbered by a vibrational quantum number ν, and write $\varepsilon_S = -|\varepsilon_{S\nu}|$; in practice ν is large (of order 10) in order that the bound state energies in the well be comparable to typical Zeeman energies. For simplicity one can assume no bound triplet states and take ε_T as the kinetic energy of an unbound pair (e.g. of two atoms in a BEC). When $\Delta\mu H - |\varepsilon_{S\nu}| + \varepsilon_T = 0$ the free particle triplet states are degenerate with the bound singlet molecules, and in the presence of the hyperfine coupling the singlet and triplet states will be *superpositions* of each other and split by $\pm V_{hf}$. If H is adiabatically swept through the resonance starting from a low field, a system consisting of N spin aligned atoms will be converted into $N/2$ singlet molecules; the effect reverses itself on lowering the field;[16] this behavior is shown schematically in Fig. 2.31.

[16]This phenomenon is known as adiabatic fast passage in nuclear magnetic resonance.

One can extend the single-channel square-well analysis discussed above to the case of two coupled square wells (see Duine and Stoof [59]); the scattering length then depends on both k and H; i.e. $a = a(k, H)$; the magnetic field dependence of a can be represented by the expression,

$$a(H) = a\left(1 - \frac{\Delta}{H - H_\nu}\right), \qquad (2.114)$$

where Δ and H_ν involve $V_{S,T}$, R, V_{hf}, and $\Delta\mu$. The system is resonant for $H = H_\nu$ and the width Δvanishes in the absence of coupling ($V_{hf} = 0$).

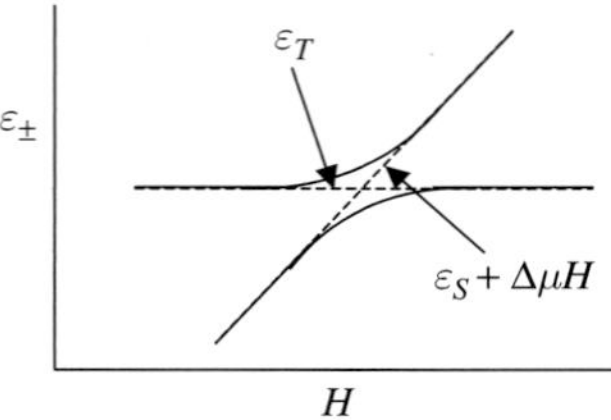

Fig. 2.31 The expected behavior near a level crossing of the singlet and triplet channels. The solid lines show how the system would evolve under a slow (adiabatic) increase of the magnetic field, where the triplet state would change continuously into the singlet state with increasing magnetic field. The dashed lines show the evolution for the case of a rapid field sweep where the system would stay in the triplet state.

2.4.6 Atom–laser interactions

2.4.6.1 *The dynamic susceptibility*

The response of the atoms to light involves their induced electric dipole moment, which is the expectation value, $\overline{\mathbf{d}}(t)$, of the dipole moment operator, $\hat{\mathbf{d}}(\mathbf{r})$. Restricting to first order in time-dependent perturbation theory, we can write this expectation value as,

$$\begin{aligned}\overline{\mathbf{d}}(t) &= \langle\Psi(\mathbf{r},t)|\,\hat{\mathbf{d}}(\mathbf{r})\,|\Psi(\mathbf{r},t)\rangle \\ &\cong \left\langle\Psi^{(0)}(\mathbf{r},t)\right|\hat{\mathbf{d}}(\mathbf{r})\left|\Psi^{(1)}(\mathbf{r},t)\right\rangle + \left\langle\Psi^{(1)}(\mathbf{r},t)\right|\hat{\mathbf{d}}(\mathbf{r})\left|\Psi^{(0)}(\mathbf{r},t)\right\rangle,\end{aligned}$$

where we recall that atoms have no dipole moment in their ground state; choosing the atom center as our origin we can write $\hat{\mathbf{d}}(\mathbf{r}) = -|e|\hat{\mathbf{r}}$.

From the classical expression for the energy of a dipole in an electric field it follows that the perturbing Hamiltonian is $\hat{H}'(\mathbf{r},t) = -\hat{\mathbf{d}}\cdot\boldsymbol{\mathcal{E}}(\mathbf{r},t)$; here $\boldsymbol{\mathcal{E}}(\mathbf{r},t)$ is the external electric field which we take to be a single plane wave and write it in the form,

$$\boldsymbol{\mathcal{E}}(\mathbf{r},t) = \frac{1}{2}\left(\boldsymbol{\mathcal{E}}_0(\omega)e^{i\mathbf{k}\cdot\mathbf{r}-i\omega t} + \boldsymbol{\mathcal{E}}_0^*(\omega)e^{-i\mathbf{k}\cdot\mathbf{r}+i\omega t}\right). \qquad (2.115)$$

We can then write our induced dipole moment in the form,

$$\overline{\mathbf{d}}(\omega) = \alpha(\omega)\boldsymbol{\mathcal{E}}_0(\omega), \qquad (2.116a)$$

where $\alpha(\omega)$ is called the *polarizibility*. Using the standard expression for $\Psi^{(1)}$ from first-order time-dependent perturbation theory and taking $\boldsymbol{\mathcal{E}}(\mathbf{r},t)$ to be essentially constant over the size, a, of an atom, one obtains,

$$\alpha(\omega) = \sum_n \left|\langle n|\,\hat{\mathbf{d}}\cdot\boldsymbol{\varepsilon}\,|g\rangle\right|^2\left(\frac{1}{E_n - E_g + \hbar\omega} + \frac{1}{E_n - E_g - \hbar\omega}\right), \qquad (2.116b)$$

where we have assumed that the atom is in its ground state with energy E_g when $\boldsymbol{\mathcal{E}}_0 = 0$ and the index n denotes the excited states; here $\boldsymbol{\varepsilon}$ is the polarization, a unit vector parallel to $\boldsymbol{\mathcal{E}}_0$ (for the case of linear polarization).

In what follows we will assume that we are working close to a resonance involving a transition to some state n = e (for excited) and ignore all other levels along with the second (non-resonant) term in Eq. (2.116);[17] to conform with commonly used conventions we also relabel the ground state E_0 as E_g.

[17]This is the so-called rotating wave approximation where the contribution from the non-resonant denominator is neglected.

As a model for the effects of spontaneous decay from the excited state we will write $E_e \to E_e - i\hbar\Gamma_e/2$, where Γ_e is the (probability) *decay rate*;[18] we then have,

[18]The time-dependent amplitude of a quantum state with energy E_e is proportional to $\exp(-iE_e t/\hbar)$; rewriting the energy of this state in the form $E_e \to E_e - i\hbar\Gamma_e/2$ causes the absolute amplitude to decay as $\exp(-\Gamma_e t/2\hbar)$ and the probability as $\exp(-\Gamma_e t/\hbar)$.

$$\alpha(\omega) = \frac{\left|\langle e|\,\hat{\mathbf{d}}\cdot\boldsymbol{\varepsilon}\,|g\rangle\right|^2}{E_e - E_g + \hbar\omega - i\hbar\Gamma_e/2} \tag{2.117a}$$

$$= \frac{(E_e - E_g + \hbar\omega)\left|\langle e|\,\hat{\mathbf{d}}\cdot\boldsymbol{\varepsilon}\,|g\rangle\right|^2}{(E_e - E_g)^2 + \hbar^2\omega^2 + \hbar^2\Gamma_e^2/4} + i\frac{\hbar}{2}\frac{\Gamma_e\left|\langle e|\,\hat{\mathbf{d}}\cdot\boldsymbol{\varepsilon}\,|g\rangle\right|^2}{(E_e - E_g)^2 + \hbar^2\omega^2 + \hbar^2\Gamma_e^2/4}. \tag{2.117b}$$

Defining a transition frequency, $\hbar\omega_{eg} = (E_e - E_g)$, and a detuning factor, $\delta = \omega - \omega_{eg}$, we can write the polarizibility as,

$$\alpha(\omega) = \frac{\hbar^{-1}\delta\left|\langle e|\,\hat{\mathbf{d}}\cdot\boldsymbol{\varepsilon}\,|g\rangle\right|^2}{\delta^2 + \Gamma_e^2/4} + \frac{i}{2}\frac{\hbar^{-1}\Gamma_e\left|\langle e|\,\hat{\mathbf{d}}\cdot\boldsymbol{\varepsilon}\,|g\rangle\right|^2}{\delta^2 + \Gamma_e^2/4} \tag{2.118a}$$

or

$$\alpha(\omega) = \alpha'(\omega) + i\alpha''(\omega). \tag{2.118b}$$

2.4.6.2 *The optical trapping potential*

The real part of α can be used to define an effective "trapping potential." Representing the added energy of the induced dipole as $V = -\overline{\mathbf{d}}\cdot\boldsymbol{\mathcal{E}}$ with $\overline{\mathbf{d}} = \alpha'\boldsymbol{\mathcal{E}}$, and introducing a quantity called the Rabi frequency as,

$$\Omega_R = \left|\langle e|\,\hat{\mathbf{d}}\cdot\boldsymbol{\mathcal{E}}_0\,|g\rangle\right|/\hbar, \tag{2.119}$$

the shift in the ground the energy of the atom in the optical field is described by

$$V_g = \frac{\hbar\Omega_R^2\delta}{\delta^2 + \Gamma_e^2/4}. \tag{2.120}$$

The imaginary part of α represents dissipation; it may be also be used to define a "loss rate" (see below), Γ_g, of atoms in the *ground state*, and is given by

$$\Gamma_g = \frac{1}{\hbar}\alpha''(\omega)\overline{\mathcal{E}^2}. \tag{2.121}$$

Taken together, Eqs. (2.120) and (2.121) can be interpreted as a *level shift* and a *level width* of the ground state;[19] i.e.

[19]The energy levels in the presence of the external driving field, which both shift and acquire a lifetime, are said to be associated with "dressed states".

$$E_g \to E_g + V - i\hbar\Gamma_g/2. \tag{2.122}$$

In order to be able to consider ensembles of atoms in space, each with different origins, we must distinguish between an *internal coordinate*, $\mathbf{r}_k^{(i)}$, of an electron within the kth atom and an *external coordinate*, $\mathbf{r}_k^{(e)}$, marking the position of that nucleus in space; i.e. $\mathbf{r}_k = \mathbf{r}_k^{(i)} + \mathbf{r}_k^{(e)}$. In obtaining Eq. (2.118) we have already integrated over the internal coordinates and, in addition, neglected the variations in $\mathcal{E}$ on the atomic scale, a. With this understanding we can now move the origin of the electric field to some global reference point, $\mathbf{r}^{(e)}$, and again denote it as $\mathbf{r}$, and continue to write the electric field at any space–time point by $\mathcal{E}(\mathbf{r}, t)$ (and its value at atom k as $\mathcal{E}(\mathbf{r}_k, t)$) .

In practice, the pairs of laser beams form standing waves with maxima and minima separated by a distance $\lambda/2$. In addition they have an overall intensity profile (typically Gaussian) which is slowly varying on the scale of the wavelength λ; i.e. $\mathcal{E}_0 = \mathcal{E}_0(\mathbf{r})$ in Eq. (2.115). Hence $\Omega_R = \Omega_R(\mathbf{r})$ and therefore $V_g \to V_g(\mathbf{r})$ in Eq. (2.120). When $\delta < 0$, ω is said to be "red shifted" relative to the atomic transition frequency. The potential is then *negative* and the atoms are attracted to regions of higher intensity with a force $\mathbf{F} = -\nabla V_g(\mathbf{r})$. When macroscopic objects are involved (micron-sized particles) this is the regime of laser tweezers [60] which has been widely used in biological experiments.[20] When $\delta > 0$, the blue shifted case, atoms are *expelled* from the beam.

[20]The utility of laser tweezers was demonstrated in [61].

2.4.7 The optical lattice

If an atom cloud has been pre-cooled to the extent that it can be spatially trapped by an optical potential, as governed by Eq. (2.120), the nature of the internal motion within the illuminated region can then become important. Let us suppose that the trap is defined by three pairs of coherent[21] incoming laser beams which are all derived from the same source having wavelength λ, with associated wavevector $K = 2\pi/\lambda$, and that enter along three orthogonal axes; for simplicity we take the intensities of all these beams to be the same. The atoms then see a standing optical wave which produces a trapping potential of the form,

[21]This requires that the path lengths associated with all the beams differ by amounts much smaller than the coherence length of the laser.

$$V(x, y, z) = V_0(\sin^2 Kx + \sin^2 Ky + \sin^2 Ky). \qquad (2.123)$$

Rather than being plane waves, the one-particle translational states of the atoms are then Bloch waves, $\psi_{\mathbf{k}}(\mathbf{r}) = u_{\mathbf{k}n}(\mathbf{r})e^{i\mathbf{k}\cdot\mathbf{r}}$, with energies that lie in bands $\varepsilon = \varepsilon_n(\mathbf{k})$, where the index n numbers the bands.[22] As V_0 increases there is an increasing tendency of the atoms to concentrate near the potential minima and when $V_0 >> \hbar^2 K^2/M$ the low-lying states will be narrow tight binding bands formed from the low-lying harmonic oscillator-like levels centered on each lattice site, with the lowest being a Gaussian of width $a_0 \cong (\hbar/M\omega_0)^{1/2}$, where $\omega_0 = (2V_0K^2/M)^{1/2}$.[23]

[22]In one dimension the wave functions are Mathieu functions.

[23]Note, this Gaussian is unrelated to the envelope wave function of the condensate as a whole.

A natural question is what happens if we start at a low temperature with the cloud in a Bose condensate and gradually increase V_0. In practice V_0 will itself have a spatial dependence arising from the over all intensity profile of the beams; i.e. $V_0 = V_0(\mathbf{r})$; this dependence is what defines the width of the condensate (see Eq. (2.78)). Qualitatively we expect the condensate wave function to be modulated in all three directions at the period of the lattice potential (corresponding approximately to the factor $u_{\mathbf{k}=0,n=1}$ associated with the lowest Bloch state). Figure 2.32 shows, qualitatively, the expected behavior of the condensate wave function in one dimension. The Fourier transform of the cloud

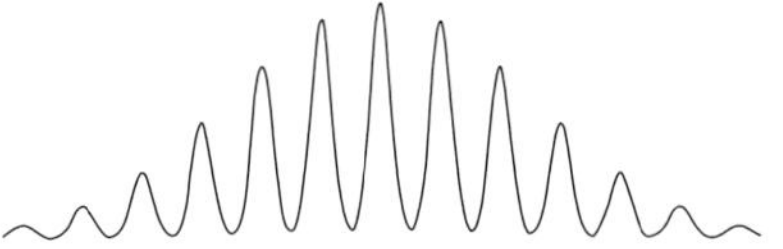

Fig. 2.32 Schematic representation of the condensate wave function, $\psi(x)$, in a one-dimensional optical lattice. Here we have a Gaussian-like envelope function centered on the minimum of the assumed harmonic optical trapping potential, $V_0(x)$, that is in turn modulated by a local Gaussian-like Bloch function, $u(r)$, associated with an increased amplitude for the atoms to reside at the standing wave potential minima making up the lattice.

intensity will then contain peaks at the reciprocal lattice vectors $\mathbf{K}$, the number of which increases with increasing V_0, corresponding to a higher degree of localization. However tunneling is present between all sites and the wave function is not localized on any particular site. Hence no qualitative change occurs on cooling; i.e. there will be no transition to a different phase. We still have single-particle physics.

Let us now ask what happens when we turn on atom–atom interactions which we will assume are repulsive. In the limit of large V_0, the atoms are highly localized; equivalently, the inter-site overlap integral is small and therefore the Bloch band width, W, is very narrow. Now in the limit of high densities, when $Kn^{-1/3} \simeq 1$, where n is the number density of the atoms, we can have more than one atom per lattice site. Let the increase in energy associated with double occupancy of a site be U. If this energy is of order the Bloch band width, W, the energy alignment between neighboring sites that facilitates inter-site tunneling will be disrupted. It turns out that when this happens the system undergoes a phase transition in which the atoms become fully localized on the lattice sites. For a Bose system long-range phase coherence is then lost and with it the condensate. This transition is called the Mott transition, after Sir Neville Mott, who was a pioneer in the study of similar transitions involving electrons in solids. The related transition in cold atom systems was predicted by Jaksch et al. [62].

Rather than using the full Hamiltonian to describe such systems, most studies have utilized a simplified site-based Bose–Hubbard model Hamiltonian [63],

$$\hat{H} = \sum_{i}^{M} \varepsilon_i \hat{n}_i - J \sum_{i}^{M} \sum_{j}^{nn} \hat{a}_i^{\dagger} \hat{a}_j + \frac{1}{2} U \sum_{i=1}^{M} \hat{n}_i(\hat{n}_i - 1), \tag{2.124}$$

where M is the number of lattice sites; i, j denote the individual sites; nn denotes nearest neighbors; $\hat{n}_i$ is the site number operator, $\hat{a}_i^{\dagger}\hat{a}_i$; ε_i are the optical on-site binding energies; and J is the inter-site coupling strength. The last term, which describes the many-body atom–atom interactions, vanishes if the ith site contains 0 or 1 atom, but contributes an added energy U when the site is doubly occupied. Note that in this site-based model the single particle states are Wannier functions as opposed to Bloch states, with J playing the role of the inter-site overlap integral, which results in a band width W. Detailed calculations[24] show that for an average of one atom per lattice site. i.e. when $N \cong M$, the phase transition in three dimensions will occur for $U/J = 5.8\,z$, where z is the number of nearest neighbors of a given lattice site (six for our simple cubic lattice).

[24]Along with the work of Fisher et al. this problem was studied in: [64–66].

2.5 Cold atom superfluids: cooling and trapping

Cold atom experiments necessarily involve two components: cooling and confinement. In this section we will discuss both of these requirements, which generally exploit appropriately configured combinations of magnetic fields and laser beams. We begin with the most widely used optical cooling technique, optical molasses. What is generally the final stage in cooling, adiabatic demagnetization, assisted by r.f. or Raman sideband evaporation of the confined cloud, will be taken up as part of Section 2.6.

2.5.1 Optical molasses

Returning to Eq. (2.121), we note that whenever a ground state atom is "lost", a photon is absorbed from the beam. Hence Γ_g can be be also interpreted as the rate at which the atom absorbs photons; these lost photons later reappear as isotropically emitted spontaneous radiation. We can then write the rate at which photons are irreversibly lost from the excitation beam as

$$\dot{N}_{ph}(\omega) = \Gamma_g = \frac{1}{2\hbar^2} \frac{\Gamma_e \left|\langle e|\, \hat{\mathbf{d}} \cdot \boldsymbol{\varepsilon}\, |g\rangle\right|^2}{(\omega - \omega_{eg})^2 + \Gamma_e^2/4} \overline{\mathcal{E}}^2, \tag{2.125a}$$

which we rewrite as

$$\dot{N}_{ph}(\omega) = \frac{\pi}{\hbar^2} A(\omega) \left|\langle e|\, \hat{\mathbf{d}} \cdot \boldsymbol{\varepsilon}\, |g\rangle\right|^2 \overline{\mathcal{E}}^2, \tag{2.125b}$$

where we have introduced the function,

$$A(\omega) = \frac{\Gamma_e/2\pi}{(\omega - \omega_{eg})^2 + \Gamma_e^2/4}, \tag{2.126}$$

which is the familiar Lorentzian line shape.

2.5.1.1 *Exploiting the Doppler effect*

Optical molasses exploits the Doppler shift, $\Delta\omega = \mathbf{k} \cdot \mathbf{v}$, seen by an atom moving with velocity $\mathbf{v}$ with respect to the wave with wavevector $\mathbf{k}$. Assume we have counter propagating beams with $\mathbf{k} = \pm k\mathbf{n}_x$, where $\mathbf{n}_x$ is a unit vector parallel to x. On average, a photon scattered from the beam moving to the right gives a kick $+\hbar\mathbf{k}$ to the atom; scattering from the beam moving to the left gives a kick $-\hbar\mathbf{k}$. Assume we are overall red shifted with respect to the atomic resonance ω_{eg}. A particle moving to the right with velocity v_x will see a (positive) frequency shift kv_x from the left propagating beam and hence an increasing $\dot{N}_{ph}$ from that beam (since we are then closer to the resonance in $A(\omega)$); at the same time it will see a frequency $-kv_x$ for the right propagating beam and a decrease in the associated $\dot{N}_{ph}$. The total momentum change per unit time from both beams is then,

$$\frac{dp_x}{dt} = -\hbar k \left[\dot{N}_{ph}(\omega + kv_x) - \dot{N}_{ph}(\omega - kv_x)\right]. \tag{2.127}$$

Note that if we reverse the sign of v_x, dp_x/dt also changes sign. We can then write the equation of motion as,

$$\frac{dp_x}{dt} = -\gamma\, v_x, \tag{2.128}$$

where

$$\gamma(v_x) = \frac{k\pi}{v_x\hbar} \left|\langle n|\, \hat{\mathbf{d}} \cdot \boldsymbol{\varepsilon}\, |0\rangle\right|^2 \overline{\mathcal{E}}^2 \left[A(\omega + kv_x) - A(\omega - kv_x)\right], \tag{2.129}$$

which to lowest order gives,

$$\gamma \cong \frac{2k^2\pi}{\hbar} \left|\langle e| \hat{\mathbf{d}} \cdot \boldsymbol{\varepsilon} \, |g\rangle\right|^2 \overline{\mathcal{E}}^2 \frac{\partial A}{\partial \omega}. \tag{2.130}$$

This “viscous drag” rapidly slows the particles, thereby *lowering their temperature* [67, 68].

2.5.1.2 *Limiting temperatures achievable with optical molasses*

Atom recoil, which initially cools the atoms from some starting temperature (typically room temperature), ultimately results in a limiting temperature governed by the random nature of the recoil processes themselves. Kicks are involved during both the absorption and emission processes and hence, on average, the rate of change of the mean square momentum is given by,[25]

[25]For the cooled atoms we may neglect the Doppler shift associated with their motion.

$$\frac{d\left\langle p_x^2\right\rangle}{dt} = 2(\hbar k)^2 \frac{dN_{ph}}{dt}. \tag{2.131}$$

Explicitly differentiating the left-hand side and using Eq. (2.128) we have,

$$2\left\langle p_x \dot{p}_x\right\rangle = -2\gamma \left\langle p_x \dot{v}_x\right\rangle = -\frac{2\gamma}{M}\left\langle p_x^2\right\rangle. \tag{2.132}$$

Substituting this result in place of the left-hand side of (2.131) gives,

$$\frac{\left\langle p_x^2\right\rangle}{2M} = -\frac{(\hbar k)^2}{2\gamma}\frac{dN_{ph}}{dt}. \tag{2.133}$$

Identifying the left-hand side of (2.133) as the average thermal energy of the x degree of freedom of a Maxwell gas,[26] $k_B T/2$, and inserting Eqs. (2.125) and (2.131) on the right-hand side we obtain,

[26]Here we explicitly assume the recoil processes result in a Maxwell distribution.

$$k_B T = \hbar \frac{A(\omega)}{dA(\omega)/d\omega}. \tag{2.134}$$

Note that in the process of obtaining this result *the laser intensity canceled out*; using higher intensities will not lead to lower temperatures. All that remains is to find the lowest achievable temperature by minimizing Eq. (2.132) with respect to ω, or equivalently the detuning parameter, δ; carrying out this operation yields $\delta = \Gamma_e/2$, which on substituting in (2.134) gives

$$k_B T_{\min} = \frac{\hbar \Gamma_e}{2}. \tag{2.135}$$

In applying the optical molasses cooling method it was found that the temperatures achieved in the experiments were *lower* than the theoretical limit given by Eq. (2.135); i.e. another cooling mechanism must be at work, which will be discussed later in this section.

2.5.2 The magneto-optical trap (MOT)

Although the optical molasses technique offers a way to cool atoms it does not trap them, a necessary step on the road to BEC. However, by combining this technique with a properly configured magnetic field, and exploiting the magnetic moment associated with the excited states of our alkali atoms, one can simultaneously cool *and* trap the atoms. For simplicity we initially assume an $S = 0$ atom with an S_0 ground state and P_1 excited state. The ground state has no moment, but the excited state can have one and this results in an energy,

$$E_Z = -\mu_B H\, l_z \tag{2.136}$$

with $\mu_B = -|\mu_B|$ and $l_z = 0, \pm 1$ for our P state. This causes an l_z - dependent frequency shift in the spectral functions $A(\omega)$ governing the absorption,

$$A(\omega) \to A\left(\omega - (|\mu_B|/\hbar)H\, l_z\right). \tag{2.137}$$

By using left and right circularly polarized beams the $l_z = +1$ and $l_z = -1$ P_1 states may be excited *separately*. Now suppose there is a magnetic field that varies linearly along z, $H(z) = H'z$.[27] If an atom is at $z = 0$, where the magnetic field vanishes, it will, on average, receive equal, but oppositely directed, kicks from the up and down laser beams (assumed to have equal amplitude) due to the resulting spontaneous radiation; hence this atom experiences no net force. However if our atom is located away from the origin, say at $z > 0$, its energy is Zeeman shifted and it will receive more kicks from that beam for which the detuning factor $|\delta| = \left|\omega - \omega_{eg} - (|\mu_B|/\hbar)l_z H'z\right|$ is smaller; if that corresponds to the beam propagating downward there will be a restoring force. The sign of the force for an atom with $z < 0$ is reversed and hence we have an overall restoring force; both cases can be assured by picking the right sign for H'. Using arguments analogous to those used to obtain (2.125) we have,

[27] We will take the intersection of our three orthogonal counter-propagating beam pairs as the origin.

$$\frac{dp_z}{dt} = -\hbar k\left[\dot{N}_{ph}\left(\omega - (|\mu_B|/\hbar)H'z\right) - \dot{N}_{ph}\left(\omega + (|\mu_B|/\hbar)H'z\right)\right], \tag{2.138}$$

where $\dot{N}_{ph}(\omega)$ is given by (2.125b). To first order in z we can write (2.138) as

$$\frac{dp_z}{dt} = -\kappa\, z \tag{2.139}$$

or

$$F_z = -\kappa\, z,$$

where κ is a spring constant given by

$$\kappa \cong \frac{2k\pi(|\mu_B|/\hbar)H'}{\hbar}\left|\langle e|\,\hat{\mathbf{d}}\cdot\boldsymbol{\varepsilon}\,|g\rangle\right|^2 \overline{\mathcal{E}}^2 \frac{\partial A}{\partial\omega}. \tag{2.140}$$

The mean square extent along the z-direction resulting from this mechanism is given by

$$\frac{1}{2}\kappa\left\langle z^2\right\rangle = \frac{1}{2}k_B T. \tag{2.141}$$

Hence the mean square size of the cloud scales as $1/H'\overline{\mathcal{E}}^2$.

In a MOT we necessarily have field variations in other directions which serve to confine the atoms in the x, y plane.

2.5.2.1 *Inclusion of hyperfine coupling*

Isolated atoms, if they are to be bosons, must have an integral total angular momentum quantum number, F, where $F = I + L + S$. Therefore alkali atoms, with the electronic configuration ns^1, must have a half integer nuclear spin, I, that combines with the $s = 1/2$ electron spin to yield integer F; these states will have the usual $2F + 1$-fold degeneracy. Such atoms will necessarily have hyperfine shifts due to the associated nuclear moments.

Consider the case of Rb with a $5s^1$ ground state electronic configuration corresponding to an $^2S_{1/2}$ state. Rb has two stable isotopes: ^{85}Rb (72.2%) with $I = 5/2$ and ^{87}Rb (27.8%) with $I = 3/2$. For ^{85}Rb there are two hyperfine states with $F = 2$ and $F = 3$, while for ^{87}Rb we have $F = 1$ and $F = 2$ states. These latter states are separated by approximately 6 GHz. The relevant excited electronic states are $^2P_{1/2}$ and $^2P_{3/2}$.

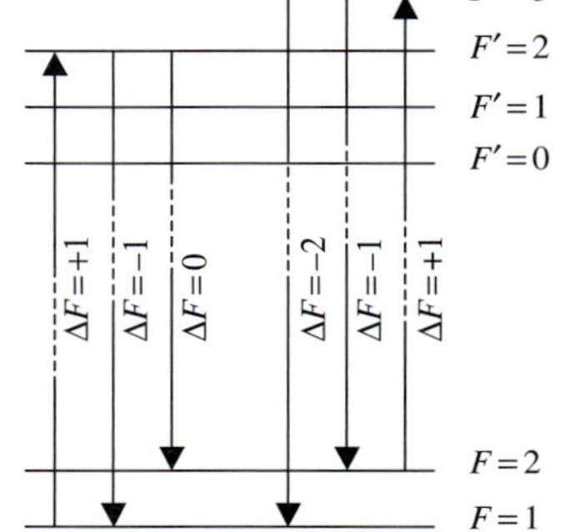

Fig. 2.33 Trapping and re-pumping cycles.

Since it involves a smaller number of hyperfine levels, we further restrict the discussion to the trapping of ^{87}Rb and excitations to the $^2P_{1/2}$ state (a $6p^1$ configuration), for which we can have $F' = 0, 1, 2,$ and 3. The dipole allowed transitions will involve $\Delta F = 0, \pm 1$. The three pairs of laser trapping beams associated with a MOT are tuned near (and to the red side of) the allowed $F = 2 \rightarrow F' = 3$ transition ($\Delta F = +1$) which then decays back to the $F = 2$ state ($\Delta F = -1$) after which the cycle repeats. Occasionally "forbidden" $F' = 3 \rightarrow F = 1$ transitions ($\Delta F = -2$) will occur, which slowly fill the $F = 1$ state. These three transitions are shown to the right in Fig. 2.33. To avoid the loss of atoms to the $F = 1$ state (referred to as a dark state) we utilize a second laser (since it must have a different wavelength) to pump $F = 1 \rightarrow F' = 2$ transitions ($\Delta F = +1$), half of which will fall to the $F = 2$ ($\Delta F = 0$) state where they rejoin that cycle. The other half that fall back to $F = 1$ will be repumped to $F' = 2$, and that cycle is repeated until most atoms end up in the $F = 2$ state. These three transitions are shown to the left in Fig. 2.33.

2.5.2.2 *The dark-spot MOT or SPOT*

Re-absorption of light scattered by the atoms in a MOT ultimately places a limit on the achievable densities. As the density increases, the scattered light, rather than escaping, can be reabsorbed by other atoms in the trap, the probability for which scales as the number of atoms already in the trap. The resulting recoil generates a kind of internal pressure that acts against the optical trapping pressure. Since both scale in the same way, increasing the laser intensity does not lead to higher densities.

One can exploit the presence of the dark states discussed in the previous subsection to increase the density of cold atoms [69]. Rather than applying a re-pump beam with the usual Gaussian cross-section one can mask its center, thereby creating a tube-like (annular) beam with a greatly reduced intensity at its center; the intersection of the rims associated with two such beams applied at right angles makes a kind of bottle in their interior. What happens then is that cold Rb atoms in the $F = 1$ hyperfine ground state, which do not absorb

$F = 2 \to F' = 3$ light from the trapping beams, can build up to much higher densities ($\sim 100\times$). Those atoms that do move to the perimeter of the beams will eventually cycle back to the $F = 1$ state; but the colder the $F = 1$ atoms are, the less they see the perimeter.

Rather than cooling the $F = 2$ ground state in a subsequent evaporation stage one can equally well cool the higher density population of $F = 1$ states (which carry the same electronic magnetic moment); this was one of the tricks initially used to achieve BEC. This kind of MOT is referred to as a SPOT.

2.5.3 Sisyphus cooling[28]

As noted above it was found that the temperatures achieved experimentally in MOTs were *lower* than the theoretical limit given by Eq. (2.133). Here we discuss another cooling mechanism that explains the discrepancy. It has been termed Sisyphus, or polarization gradient cooling. Consider a trap involving counter-propagating beams propagating along z. Let the one propagating toward increasing z, $\boldsymbol{\mathcal{E}}_{\text{up}}(\mathbf{r}, t)$, be polarized along x and that propagating toward decreasing z, $\boldsymbol{\mathcal{E}}_{\text{down}}(\mathbf{r}, t)$ be polarized along y. Choosing the amplitude $\mathcal{E}_0(\omega)$ to be real, we can then write the total field as

$$\begin{aligned} \boldsymbol{\mathcal{E}}(\mathbf{r}, t) &= \boldsymbol{\mathcal{E}}_{\text{up}}(\mathbf{r}, t) + \boldsymbol{\mathcal{E}}_{\text{down}}(\mathbf{r}, t) \\ &= \frac{\mathcal{E}_0(\omega)}{2}\left[\boldsymbol{\varepsilon}_x\left(e^{+ikz-i\omega t} + e^{-ikz+i\omega t}\right) + \boldsymbol{\varepsilon}_y\left(e^{-ikz-i\omega t} + e^{+ikz+i\omega t}\right)\right]. \end{aligned} \tag{2.142}$$

Let us examine the field seen by an atom located at positions such that $kz = 0, \pi/4, \pi/2$, and $3\pi/4$; inserting these values into (2.142) we obtain,

$$\boldsymbol{\mathcal{E}}(kz = 0, t) = \mathcal{E}_0(\boldsymbol{\varepsilon}_x + \boldsymbol{\varepsilon}_y)\sin\omega t(kz = 0, t); \qquad \text{linear, } +45^\circ;$$

$$\boldsymbol{\mathcal{E}}(kz = \pi/4, t) = \mathcal{E}_0\left[\boldsymbol{\varepsilon}_x\cos(\omega t - \pi/4) - \boldsymbol{\varepsilon}_y\sin(\omega t - \pi/4)\right]; \qquad \text{left circular;}$$

$$\boldsymbol{\mathcal{E}}(kz = \pi/2, t) = \mathcal{E}_0(\boldsymbol{\varepsilon}_x - \boldsymbol{\varepsilon}_y)\sin\omega t; \qquad \text{linear, } -45^\circ;$$

$$\boldsymbol{\mathcal{E}}(kz = 3\pi/4, t) = \mathcal{E}_0\left[-\boldsymbol{\varepsilon}_x\cos(\omega t + \pi/4) - \boldsymbol{\varepsilon}_y\sin(\omega t + \pi/4)\right]; \text{ right circular.} \tag{2.143a–d}$$

As indicated these are successively linearly polarized at $+45^\circ$, left circularly polarized, linearly polarized at -45°, and right circularly polarized. The position dependence of the polarization results in position-dependent transition rates, which has important consequences, as we now discuss.

To be specific we examine an alkali atom with ground g, and excited, e, states corresponding to $n\mathrm{S}_{1/2}$ and $(n+1)\mathrm{P}_{3/2}$. For simplicity we only consider a nucleus with $I = 0$. In terms of their orbital, $|l, m\rangle$ and spin, $|s_z\rangle$ parts, the ground and excited states with quantum numbers $|l, j, j_z\rangle$ can be written,

[28] The name *Sisyphus* is derived from the mythological Greek character who was condemned to roll a heavy stone to the top of a hill only to have it roll down again whereupon he was forced to repeat the process, endlessly.

$$\left|g, \frac{1}{2}\right\rangle = \left|0, \frac{1}{2}, \frac{1}{2}\right\rangle = |0,0\rangle \left|\frac{1}{2}\right\rangle$$
$$\left|g, -\frac{1}{2}\right\rangle = \left|0, \frac{1}{2}, -\frac{1}{2}\right\rangle = |0,0\rangle \left|-\frac{1}{2}\right\rangle \qquad (2.144a,b)$$

and

$$\left|e, \frac{3}{2}\right\rangle = \left|1, \frac{3}{2}, \frac{3}{2}\right\rangle = |1,1\rangle \left|\frac{1}{2}\right\rangle$$
$$\left|e, \frac{1}{2}\right\rangle = \left|1, \frac{3}{2}, \frac{1}{2}\right\rangle = \sqrt{\frac{2}{3}}\,|1,0\rangle \left|\frac{1}{2}\right\rangle + \sqrt{\frac{1}{3}}\,|1,1\rangle \left|-\frac{1}{2}\right\rangle$$
$$\left|e, -\frac{1}{2}\right\rangle = \left|1, \frac{3}{2}, -\frac{1}{2}\right\rangle = \sqrt{\frac{1}{3}}\,|-1,0\rangle\,|1/2\rangle + \sqrt{\frac{2}{3}}\,|1,0\rangle \left|-\frac{1}{2}\right\rangle$$
$$\left|e, -\frac{3}{2}\right\rangle = \left|1, \frac{3}{2}, -\frac{3}{2}\right\rangle = |1,-1\rangle \left|-\frac{1}{2}\right\rangle. \qquad (2.145a–d)$$

From the Clebsch–Gordan coefficients implicit in Eqs. (2.143a–d), it follows that in a region where the light polarization is pure σ_- (with the selection rule $\Delta j_z = \ -1$) we only couple $|g,-1/2\rangle$ with $|e,-3/2\rangle$, and $|e,+1/2\rangle$ with $|e,-1/2\rangle$; the associated probabilities are 1 and 1/3 respectively. Corresponding couplings apply for σ_-. These properties are summarized in Fig. 2.34.

We will assume a red detuning. From the fact that the state admixture differs for atoms having various values of kz along the beam, they will experience *position-dependent* trapping potentials; i.e. $V_g \to V_{g\pm}(z)$ (see (2.120)), since the matrix element implicit in the Rabi frequency (see (2.119)) then has the form $\Omega_R = \Omega_R(z)$. We will assume red shifted laser light is applied ($\omega < \omega_e - \ \omega_g$). As noted above, at points where the polarization is pure σ_- we couple the $|g,-1/2\rangle$ and $|e,-3/2\rangle$ states, producing a shift in $-|V_g|$ and simultaneously couple the $|g,+1/2\rangle$ and $|e,-1/2\rangle$ producing a shift $-|V_g/3|$. The net shift between the $|g,-1/2\rangle$ and the $|g,+1/2\rangle$ states is then $-|2V_g/3|$; this is the deepest point in the potential landscape for the $|g,-1/2\rangle$ state. A distance $\lambda/4$ away, where we have pure σ_+ polarization we have the deepest well for the $|g,+1/2\rangle$ state.

The following processes occur in the presence of the polarization gradient associated with our counter-propagating waves.[29] In the vicinity of the σ_- maximum, atoms in a $|g,-1/2\rangle$ state are strongly trapped by coupling to the $|e,-3/2\rangle$ state; if an atom undergoes a transition it will exclusively go back by

[29]For a detailed discussion see [70].

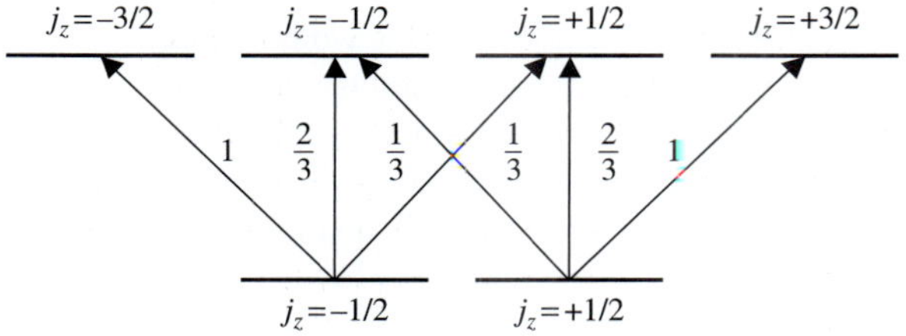

Fig. 2.34 Allowed transitions according to the selection rule $\Delta j_z = 0, \pm 1$. The numbers give the transition probabilities which follow from the squares of the amplitudes in Eq. (2.143a–d).

spontaneous emission to $|g, -1/2\rangle$. On the other hand, if σ_- induces a transition from $|g, +1/2\rangle$ to $|e, -1/2\rangle$, it will spontaneously decay with probability 2/3 by linear (π) emission to $|g, -1/2\rangle$ or by σ_+ emission with probability 1/3 to $|g, +1/2\rangle$. After many fluorescence cycles these processes tend to concentrate $|g, -1/2\rangle$ states at the σ_- maximum. Similar processes concentrate atoms in $|g, +1/2\rangle$ states at the maximum of σ_+.

We now give a qualitative discussion of the cooling process. First we must recognize that the atoms have kinetic energy and will therefore traverse the optical standing waves. Consider the trajectory of an atom (see Fig. 2.35) that starts at a σ_- maximum in a $|g, -1/2\rangle$ state. We assume the time required to optically transfer between $|g, -1/2\rangle$ and $|g, +1/2\rangle$ states is long compared to the time for an atom to drift from a potential minimum to a maximum. Since the total energy is conserved, the kinetic energy is lower in the region of the maximum. If, while there, the atom makes a $|g, -1/2\rangle$ to $|e, +1/2\rangle$ transition to an excited state followed by spontaneous π emission to the $|g, +1/2\rangle$ state (with the higher probability 2/3), the atom will be near the bottom of the potential well for that state, the excess energy being carried away by the emitted photon in a kind of anti-Stokes Raman transition. Here the atom continues to drift on this branch and can undergo transitions near the maxima of that potential. The accumulation of such branch changing transitions extracts kinetic energy, thereby cooling the atom. A detailed analysis shows that the limiting temperature of this process is given by,

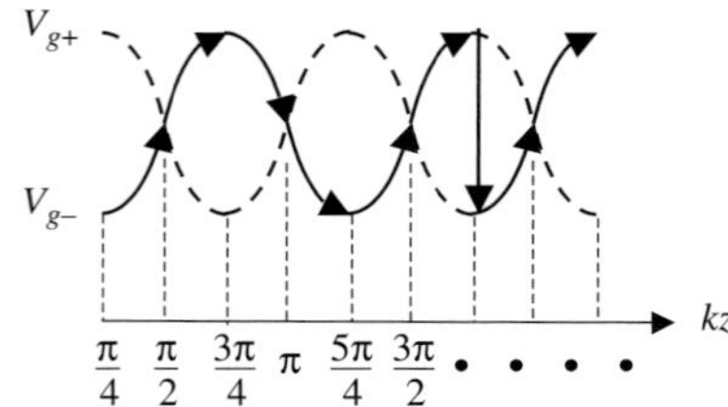

Fig. 2.35 The sinusoidal position variation of the trapping potentials $V_{g\pm}$. The solid line shows a trajectory in time of an atom starting from a minimum of the potential.

$$k_B T \approx \left\langle \frac{p^2}{2M} \right\rangle \approx |V_g|. \tag{2.146}$$

Note, $V_g \propto \mathcal{E}_0^2$, the laser intensity, as well as the detuning, δ^{-2}, both of which are confirmed experimentally [71].

Quantum mechanically, atoms moving within an optical lattice will exhibit a band structure. Atoms in states with kinetic energy, K, significantly less than the potential barrier between wells, should lie in narrow bands spaced approximately by the vibration frequency in the well where the latter is of order,

$$\Omega \approx k\sqrt{V_g/M}$$

and k is the wavevector. A condition for observation is that $\Omega\tau >> 1$, where τ is a characteristic time between optical transitions. Such levels have been seen as side bands on a probe beam.

2.5.4 The optical–dipole trap

Although the MOT allows us to cool atoms, the temperatures (or equivalently the densities) are still too high (or too low) to achieve BEC. Two more ingredients are required: (i) a means to further cool the atoms and (ii) a means to trap them, without inputting heat, while they are being cooled. Process (i) invariably involves some form of evaporative cooling, which as noted above, we take up in Section 2.6. Process (ii) is accomplished with either far off-resonant laser beams, referred to as a dipole trap, or appropriately configured magnetic fields. We start with a discussion of optical trapping.

In our discussion of atom–laser interactions we arrived at Eq. (2.120) for the ground state energy shift, or equivalently the trapping potential, of an atom in an optical field. In order to avoid heating we need to work far off the resonance: note the dissipation falls off as δ^{-2}, while the trapping potential only falls off as δ^{-1}, where δ is the detuning parameter. Far off resonance we can write (2.120) as,

$$V_g = \frac{\hbar \Omega_R^2 \delta}{\delta^2}; \tag{2.147}$$

this will be attractive for $\delta < 0$ and repulsive for $\delta > 0$.

Figure 2.36 shows, schematically, a red shifted dipole trap (left) and a blue shifted "bottle" (right). Note only two intersecting beams are required to trap an atom cloud tightly. Doped yttrium aluminium garnet (YAG) lasers are suitable for trapping alkali atoms, utilizing the fundamental (1064 nm) for a red shifted trap and the second harmonic (532 nm) for the blue shifted bottle. Typically 5 W is required. Note that by controlling the intensity profile of the laser beams, traps that are more "box-like" (resulting in a more uniform cloud), as opposed to the more typical "harmonic" traps, become possible. Later we will discuss a case where red and blue detuned beams are combined to make a partitioned trap.

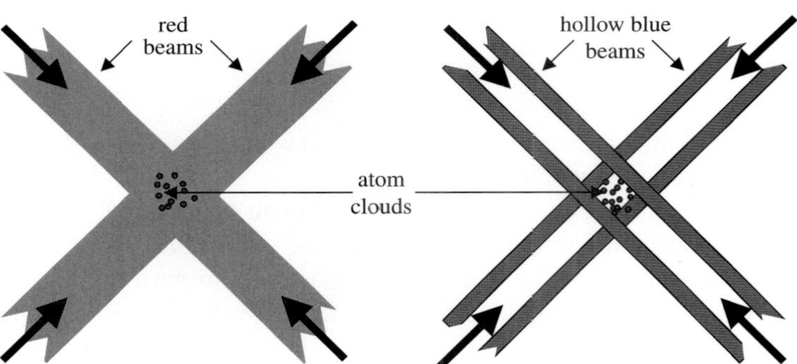

Fig. 2.36 Left: a red shifted dipole trap; right a blue shifted dipole bottle.

2.5.5 Static magnetic atom traps

We now discuss how atoms can be trapped by exploiting their Zeeman energy in a magnetic field. We start by briefly reviewing the behavior of time-independent magnetic fields in empty space.

The energy of an atom in a field **H** with total electronic angular momentum $j \neq 0$ is given by $E_H = -\mu_L H j_z$, with j_z the projection along the local direction of **H**, $\mu_L = g_L \mu_B$, and g_L the Lande g-factor,

$$g_L = 1 + \frac{j(j+1) + s(s+1) - l(l+1)}{2j(j+1)}. \tag{2.148}$$

For states with j_z such that $E_H < 0$, atoms can in principle accumulate, provided the magnetic field has a *global* extremum; further analysis shows that in a source-free region this can only be a minimum [72]. Now readily accessible fields (a few Tesla) have equivalent temperatures that are of order a few Kelvins. Hence atoms must be *pre-cooled* to sub mK temperatures if they are to be trapped magnetically. We will discuss techniques that can do this shortly; however they exploit various kinds of magnetic field profiles which we now examine.

In the static limit, and in free space where there are no current sources, the fourth Maxwell equation reduces to $\nabla \times \mathrm{H} = 0$, from which it follows that the field can be obtained from a scalar potential, Φ; i.e. $\mathbf{H} = -\nabla\Phi$. From the second Maxwell equation $\nabla \cdot \mathbf{H} = 0$ it further follows that Φ obeys Laplace's equation,

$$\nabla^2\Phi = 0. \tag{2.149}$$

The form $\Phi = -H_0 z$ leads to a constant magnetic field, H_0, which we have chosen parallel to z.

2.5.5.1 *The gradient or quadrupole magnet*

We next examine the field arising from the second-order terms in the expansion of Φ; for axial symmetry along z we can write,

$$\Phi(x, y, z) = a_{\|}z^2 + a_{\perp}(x^2 + y^2). \tag{2.150}$$

To satisfy (2.149) we require $a_{\perp} = -a_{\|}/2$, and defining $H' = -2a_{\|}$, the slope of the field along z, we have,

$$H_z = H'z; \quad H_x = -\frac{1}{2}H'x; \quad H_y = -\frac{1}{2}H'y; \tag{2.151a}$$

the corresponding magnitude is given by

$$H = H'\sqrt{z^2 + (x^2 + y^2)/4}. \tag{2.151b}$$

Such a field can be produced by a pair of identical coils, one directly above the other, with *oppositely* directed currents, a configuration which is sometimes called an anti-Helmholtz, or Maxwell-gradient, pair.[30]

[30] Since a single coil corresponds to a magnetic dipole, two coils producing oppositely directed fields corresponds to a quadrupole.

2.5.5.2 *The Ioffe–Pritchard trap*

In regions where H is very small the spin state of the atoms in the gas is very sensitive to random external field perturbations, as well as perturbations induced by the atom's own motion in the local field; either effect can induce *spin flips* that depolarize the atoms. For this reason the pure gradient field described by Eqs. (2.151a,b) is not suitable as an atom trap (although it is used in the magneto-optical trap (MOT)). It is therefore desirable to create a field configuration where the field magnitude has a minimum (see above discussion), but is locally non-zero so as to keep the atoms polarized (as well as trapped), and where the magnitude increases on moving away from that point in any direction.[31] In search of such a configuration we start by considering the axially symmetric case (corresponding to magnets that are easiest to wind) including the first- and third-order terms in the expansion of Φ. Axially

[31] Since we may trap atoms with positive or negative j_z, the field may point in any direction.

symmetric solutions to Laplace's equation have the form,

$$\Phi(r,\theta) = \sum_l a_l r^l P_l(\cos\theta), \tag{2.152}$$

where $\cos\theta = z/r$. In order to have a minimum, we seek a field that is symmetric with respect to z and we therefore retain only the first two terms in Φ that are odd in l; recalling $P_3(\cos\theta) = (1/2)(5\cos^3\theta - 3\cos\theta)$ we then have,

$$\begin{aligned}\Phi(\rho,z) &= a_1 z + (a_3/2)\left(5z^3 - 3zr^2\right)\\ &= a_1 z + 3a_3(z^3 - z\rho^2/2),\end{aligned} \tag{2.153}$$

where we have introduced the cylindrical coordinate ρ; Eq. (2.153) yields the field components

$$H_z == H_0 + 3H''\left(z^2 - \frac{1}{2}\rho^2\right); \quad H_\rho = -3H''z\rho, \tag{2.154}$$

where $H_0 = -a_1$ and $H'' = -a_3$. Such a field may be produced by a pair of identical and concentric coils (the circular coils in Fig. 2.37) with currents circulating in the *same* direction; we will call this configuration a Helmholtz pair.[32] The *absolute value* of H (which our goal is to make non-zero) is given to second order by

[32]Formally, the Helmholtz pair, is one for which the ratio of the spacing to the diameter of the coils is chosen so that the field starts in fourth order on leaving the midpoint; for the Maxwell pair a different ratio is chosen so that the field starts with fifth-order terms.

$$H = H_0 + 3H''\left(z^2 - \frac{1}{2}\rho^2\right); \tag{2.155}$$

depending on the sign of H'' this expression can be made to have a minimum along z or ρ *but not both.*

Curiously, we can address this problem by adding a second set of coils which produces a more general gradient field. To do this we alter Eq. (2.150) to allow one producing different gradients in the three orthogonal directions,

$$\Phi = a_{xx}x^2 + a_{yy}y^2 + a_{zz}z^2, \tag{2.156}$$

which can be produced by a pair of rectangular coils. We will assume the coil axis lies parallel to x, and that the long axis of the rectangle is parallel to z, as shown in Fig. 2.37. To satisfy Laplace's equation we require,

$$a_{xx} + a_{yy} + a_{zz} = 0. \tag{2.157}$$

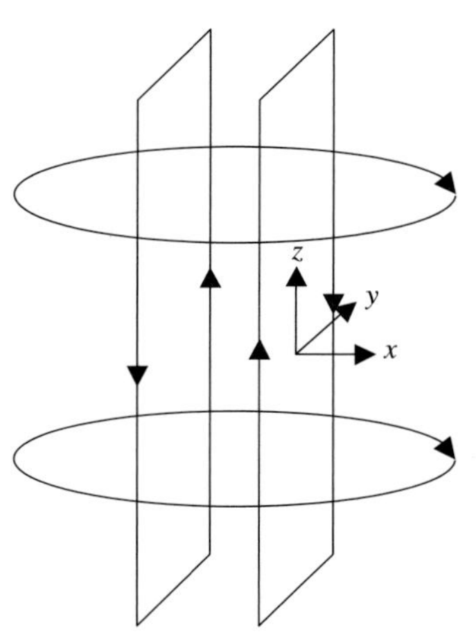

Fig. 2.37 A circular Helmholtz pair along z and a rectangular Maxwell pair along x. The arrows show the directions of the circulating currents. The combination of both coils makes an Ioffe–Pritchard trap.

When the extent of these coils in the z-direction is large, the field, (as measured from the midpoint of the coils) varies slowly in that direction and as an approximation we can take $a_{zz} \cong 0$, which gives $a_{xx} = -a_{yy} = H'_{IP}$, where the parameter H'_{IP} fixes the gradient along the x- and y-directions; the field components are then given by,

$$H_x = H'_{IP}, \quad H_y = -H'_{IP}, \quad H_z \cong 0. \tag{2.158}$$

If we now *combine* the fields from these configurations as given by (2.154) and (2.158) and calculate the absolute value of the total field, we obtain in leading order,

$$|\mathbf{H}^{\text{total}}| = H_0 + H''\left(z^2 - \frac{3}{2}\rho^2\right) + \frac{H'^2_{IP}}{2H_0}\rho^2. \tag{2.159}$$

Hence, if $3H_0H'' < H'^2_{IP}$ the field will have a global minimum at $z = \rho = 0$. A trap of this kind is called an Ioffe–Pritchard trap [73, 74].

A variant of the Ioffe–Pritchard trap producing a similar field profile is the *cloverleaf* magnet. Rather than having two coil pairs wound perpendicularly to each other, the clover leaf magnet incorporates four additional coil pairs with axes displaced with respect to the primary coil pair. We note in passing that one can combine magnetic and dipole–optical traps [75].

2.5.6 Dynamic magnetic traps: the time-averaged rotating potential (TOP) trap [76]

If we remove the constraint that the magnetic field is static, other configurations leading to trapping emerge. Suppose we start with a pure gradient field which we now write in the form,

$$H_x = H'x; \quad H_y = H'y; \quad H_z = -2H'z. \tag{2.160}$$

To this static field we add a uniform field that rotates in the x–y plane,

$$\mathbf{H}_{\perp}(t) = H_0(\cos \Omega t\, \mathbf{n}_x + \sin \Omega t\, \mathbf{n}_y), \tag{2.161}$$

where $\mathbf{n}_x$ and $\mathbf{n}_y$ are unit vectors along x and y and Ω is the angular frequency. The components of the total field are then,

$$H_x = H'x + H_0 \cos \Omega t; \quad H_y = H'y + H_0 \sin \Omega t; \quad H_z = -2H'z. \tag{2.162}$$

Calculating the magnitude of the total field resulting from (2.162), and expanding it through second order assuming $H'^2d^2 << H_0^2$, where d is some typical trap size, we obtain,

$$\begin{aligned} H(t) &\cong H_0 + H'(x \cos \Omega t + y \sin \Omega t) \\ &+ \frac{H'^2}{2H_0}\left(x^2 + y^2 + 4z^2 - (x \cos \Omega t + y \sin \Omega t)^2\right). \end{aligned} \tag{2.163}$$

If we now *average* over times t such that $\Omega t >> 1$, we are left with

$$\langle H(t) \rangle \cong H_0 + \frac{H'^2}{4H_0}\left(x^2 + y^2 + 8z^2\right). \tag{2.164}$$

Note the overall sign of (2.164) is fixed by H_0. The rotation frequency Ω must be chosen high enough that the moments do not move too far in the trap, but low enough that transitions between spin states, which are of order the spin precession (Larmor) frequency ω_L, do not occur, and provided $\Omega << \omega_L$ the spins will adiabatically follow $\mathbf{H}(t)$.

2.5.7 Trap oscillation frequencies

Actually determining the trapping potential can be a problem, especially for optical–dipole traps where the beam profile resulting from focusing inside the vacuum chamber can be difficult to characterize. However, by disturbing the cloud from equilibrium and measuring the principal oscillation frequencies one can directly measure the principal curvatures of the potential at the extremum.

For the magnetic case, the potential of a magnetic dipole with moment $\boldsymbol{\mu}$ in a magnetic field is given by $V = -\boldsymbol{\mu} \cdot \mathrm{H}$, and the sign can be arranged to favor trapping. If the atoms are displaced from the minimum of the potential they will experience a force $\mathbf{F} = -\nabla V$ which is linear in displacement from the origin. The resulting equation of motion is then $\mathbf{F} = M\ddot{\mathbf{r}}$, where M is the atom mass. Supposing that $\boldsymbol{\mu}$ lies parallel to the local field $\mathbf{H}(x, y, z)$, the associated spring constants for an axially symmetric trap field are

$$\kappa_i = -\frac{\partial^2 V}{\partial x_i^2} = \mu \frac{\partial^2 H}{\partial x_i^2}. \tag{2.165}$$

For the case of the TOP trap described by Eq (2.164) we have the following oscillation frequencies parallel and perpendicular to the z-axis:

$$\omega_z^2 = \frac{4\mu H'^2}{MH_0} \tag{2.166a}$$

and

$$\omega_\perp^2 = \frac{\mu H'^2}{2MH_0}. \tag{2.166b}$$

Here we are working in a limit where $\omega_{z,\perp} << \Omega$. The trap frequencies for an Ioffe–Pritchard trap can also be calculated using (2.165).

2.5.8 The Zeeman slower

The simplest way to introduce atoms into a magneto-optical trap is to evaporate them directly into the vacuum chamber. However this introduces a background pressure that increases the scattering rate, thereby raising the trapped atom temperatures and reducing the lifetime of a subsequently prepared condensate. A better approach is to create an atomic beam which directs the atoms into the trap. If the velocity of the atoms is not reduced prior to insertion they will just pass through the trap and be lost. Slowing the atoms is accomplished using a laser beam directed opposite to the atomic beam. To work efficiently the Doppler-shifted atomic transition frequency, in the reference frame of the atoms moving at some average velocity $v(z)$ at position z along the beam path, must be matched to the laser frequency; this is accomplished by applying a position-dependent axial magnetic field, $H(z)$, which introduces a Zeeman shift to the atomic transition frequency. By arranging for the Zeeman shift to cancel the Doppler shift a constant radiation force is applied which slows the atoms.

The initial transverse velocity of the atoms emerging from the oven is reduced by collimating the beam prior to entering the field region. However, the subsequent act of slowing the beam imparts a transverse component to the velocity profile, one of the parameters that can be optimized to obtain a desired trade-off between the beam intensity and its velocity distribution. Note the baffles used to collimate the beam allow the pressure in the oven to be higher than the rest of the apparatus

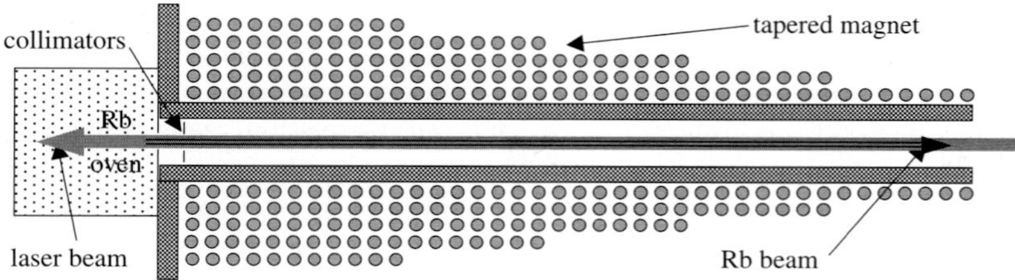

Fig. 2.38 The Zeeman slower. Atoms emerge from an oven at the mean Maxwell velocity and encounter a counter-propagating laser beam. A tapered coaxial magnet provides a local field $H(z)$ that Zeeman-shifts the Doppler-shifted atomic transition frequency to match the local atom velocity $v(z)$.

A schematic drawing of a device, called a Zeeman slower, which carries out these operations is shown in Fig. 2.38. Not shown is cryogenic pumping that extracts atoms not passing through both collimating baffles which would otherwise raise the pressure in the rest of the apparatus.

2.6 Cold atom superfluids: representative results

In this section we will briefly review some of the highlights of superfluid and related phenomena that have been achieved in recent years. This is by now a vast and rapidly evolving field and our discussion here is necessarily limited and somewhat arbitrary.

2.6.1 A sketch of the first BEC experiment; r.f. assisted evaporative cooling

The final step in achieving BEC generally involves cooling the atom cloud via evaporation. Of course this process must be carried out over time scales that are long compared to the inelastic collisional relaxation time of the atom cloud. To carry out their evaporation, Anderson et al. used a "dynamic" strategy. Since the atoms in their TOP were highly polarized, flipping their spins caused them to be expelled from the trap. Atoms in the trap experience a (rotating) Zeeman energy shift, so by applying an r.f. field with a frequency $\hbar\omega(\mathbf{r}) \cong 2\mu_B H(\mathbf{r})$, where $H(\mathbf{r})$ is the local TOP field given by Eq. (2.164), those electrons can flip their spin and be ejected from the trap. Provided this happens adiabatically, the cloud cools. By starting at a frequency corresponding to spins on the periphery of the trap, where they experience a higher field magnitude, and then slowly decreasing the frequency, progressively lower and lower energy atoms are evaporated, thereby cooling the remaining ensemble. Of course the process is not extended to the bottom of the trap, as it would then empty.

In practice the evaporation follows a series of other steps. Here we will describe the original experiments of Anderson et al. in 1995. Their setup included the following subsystems: (i) a dark spot MOT (Sec. 2.5.2.2.), (ii) two orthogonal Helmholtz pairs to generate a (7.5 kHz) rotating field in the x–y plane (which combined with the gradient magnet of Sec. 2.5.5.1 generates a TOP configuration, as described in Sec. 2.5.6); (iii) the ability to briefly apply a homogeneous magnetic field and laser pulses of appropriate wavelengths; and (iv) coils to apply a frequency-swept r.f. pulse.

We now recount the steps carried out in the actual experiments: (i) To begin, a Rb atom cloud was collected in the MOT; (ii) the cooling cycle commenced with a quick compression of the cloud by ramping up the gradient field; (iii) a small magnetic field was then applied, followed by a short pulse of circularly polarized laser light to optically pump the moments so as to lie parallel with the magnetic field (a Raman transition from the $F = 1, M_F = 1$ dark state to a $F = 2, M_F = 2$ state). (iii) The MOT trapping beams were then removed and the rotating H_0 field quickly applied to make a TOP trap; the H' component is then adiabatically increased to further compress the cloud (and shorten its relaxation time).

The final step was to apply, and ramp down in frequency, an r.f. field to progressively expel spins from the cloud; in parallel, the H_0 field was reduced to simultaneously perform an adiabatic demagnetization (in a rotating frame). After this, and following a short (2 sec) interval to allow equilibration, the cloud was expanded by further lowering the H_0 field. This was done in two steps: first slowly, in order to adiabatically expand the cloud, so it maintained its initial shape, and then suddenly to cause all atoms to move ballistically with velocities given by their values at that point in time. The resulting profile of the velocity distribution within the cloud can be compared with that predicted by Eq. (2.79), which in turn is derived from the shape of the original condensate profile given by Eq. (2.77). Note that the field-dependent spring constants of the Gaussian ground state can be determined independently by measuring the field-dependent transverse and longitudinal resonant frequencies of the TOP trap and using Eqs. (2.166a,b); this calibration can be done prior to carrying out the final cooling steps or in an independent run.

Imaging was carried out by applying a laser pulse to drive the $F = 1 \rightarrow F' = 2$ transition and recording the shadow so created with a CCD. Since the spring constants of the trap are anisotropic, the resulting images, reflecting the shape of the condensate wave function, will also be anisotropic. As noted in Sec. 2.4.1, the velocity distribution will be isotropic in the absence of a condensate, since it is only governed by the trap temperature. Figure 2.39 shows

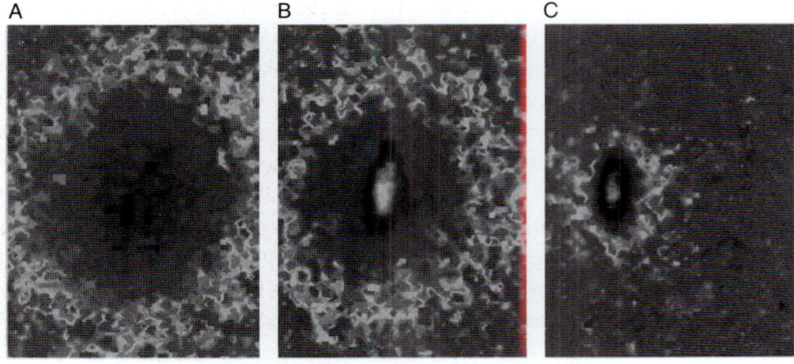

Fig. 2.39 False color images of an expanded atom cloud. (A) Just before the appearance of a condensate; note the essentially isotropic (green & yellow) profile expected for a Maxwell distribution. (B) Slightly after the formation of a condensate; the elliptical blue-white region arises from the probability density of the condensate wave function emerging from the green & yellow background associated with the excited atoms. (C) A case where the condensate occupies a much larger fraction of the cloud (note the magnification is lower here). (After Anderson et al. (1995).) (See Plate 6 for color image).

the various stages in the formation and imaging of a Rb^{87} condensate by the Colorado group. Similar results were obtained on the element Cs by the MIT group.

2.6.2 Evaporative cooling from an optical–dipole trap

The later stages in the formation of a BEC as described above utilized a magnetic trap. However further cooling of atom clouds pre-cooled by a MOT can also be achieved, more simply, using lasers alone. As an example, BEC has been achieved in Rb^{87} using such an approach [77]. Here an optical–dipole trap was formed from tightly focused ($\cong 50\,\mu m$), 12 W, CO_2 laser beams crossing at right angles; each beam passed through an acousto-optic modulator which allowed independent control of their intensities. (Heating due to spontaneous emission of the atoms turns out to be completely negligible.) With the CO_2 beams present, the initial cloud was prepared in the usual way in a MOT, after which the MOT beams and magnets were shut off; however just before shut off, the re-pump beam was shuttered so as to allow most atoms to transition into the $F = 1$ state. BEC was achieved simply by lowering the intensity of the beams, resulting in evaporative cooling from this dipole trap.

"Exotic" atoms can also be optically trapped, and BEC has been achieved in Yb via this approach [78]. But more importantly one has more flexibility in controlling atom cloud shapes with laser traps; as an example a two-dimensional condensate has been achieved using evanescent waves above the surface of a prism [79]. Patterned substrates would open a wide range of options, especially if they involved plasmon resonant structures allowing the production of strongly non-local trapping fields.

We also mention that there can be advantages to using combinations of optical and magnetic traps [80]. In the simplest case the dominant confining force can be supplied by the Gaussian beam profile of the optical–dipole trap and one would load it with a MOT as just described. However one is then free to apply various combinations of magnetic fields by differentially driving the coil pair used for the MOT itself and by additional coils. These fields can be configured so as to preferentially trap spin-up or spin-down atoms via a constant H_0 field. But one can also cause atoms to leave the trap by making it asymmetric with a gradient field H', resulting in a barrier over which atoms can escape. The latter allows cooling by adiabatic demagnetization.

2.6.3 The observation of Feshbach resonances

As an example of a Feshbach resonance we consider ^{23}Na [81]. The lowest state in a magnetic field has $M_S = -1/2; M_I = +3/2$, corresponding to the hyperfine state $F = 1$, $M_F = 1$. Nearby in the field is a molecular state with $S = 1, M_S = +1, I = 1, M_I = +1$. Figure 2.40 shows results from the Ketterle group. The system starts in a condensate. If the field is swept to a value close to H_ν and held there, the molecular component will be lost and when the field is swept back to its starting value the number of atoms remaining gives a measure

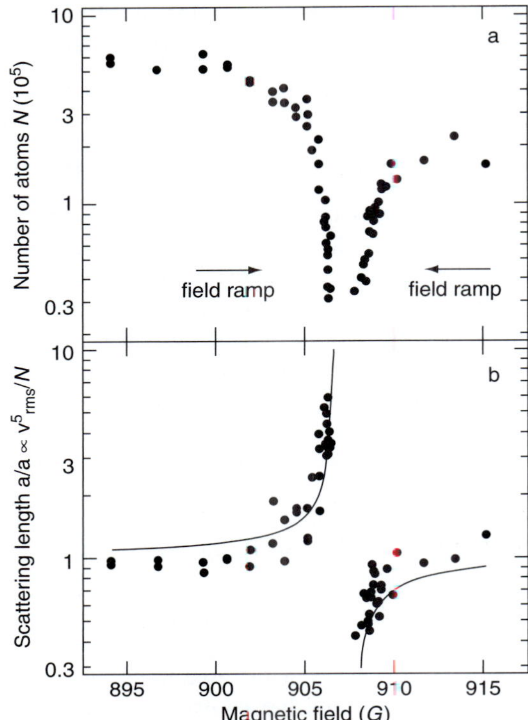

Fig. 2.40 Measurements of a field-dependent Feshbach resonance in ^{23}Na. (After Inouye et al. (1998).)

of the degree of mixing of the atomic and molecular wave functions at the hold point near H_ν. By repeating the experiment many times with different hold fields the data in Fig. 2.40a were obtained. We note that if when approaching from the high field side the condensate is swept suddenly through H_ν, it will remain there (although it is now only metastable). Experiments are then performed in which the field is swept *down* to various hold fields.

Figure 2.40b shows the measured values of the field-dependent scattering length, $a(H)$. These data are obtained by expanding the condensate; by measuring its shape and comparing with that expected for an interacting condensate with an inter-atom potential $V(H)$ derived from $a(H)$, the investigators inferred the latter.

In another set of experiments Roberts et al. [82] used the Feshbach resonance to study the elastic scattering collision rate between atoms in the $F = 2$, $M = -2$ state of ^{85}Rb. The technique used involved creating a non-isotropic energy distribution in a magnetically trapped cloud by changing the anisotropy of a magnetic trap, and observing, via absorption imaging, the time necessary for the cloud to re-equilibrate through elastic collisions; the measurements were repeated for various values of the field as the system was tuned through the Feshbach resonance. From such data they extracted information on the long-range (r^{-6}) behavior of the inter-atom van der Waals interaction along with singlet and triplet scattering lengths.

2.6.4 Condensate collapse

As noted in Section 2.4.2, a condensate can be rendered unstable in the presence of attractive interactions (see Eq. (2.90)). One of the early applications of the Feshbach resonance was to tune the scattering length to negative values and observe this collapse, and the instability of a ^{85}Rb condensate was observed by Roberts et al. [83, 84] by slowly changing the atom–atom interaction from repulsive to attractive using a Feshbach resonance. The signature of collapse was taken to be a sudden ejection of atoms from the condensate. By measuring the onset of this transition as a function of atom number N and attractive s-wave interaction strength, a, the stability condition was found to be $N_c|a|/a_{\text{osc}} = 0.459$ where $a_{\text{osc}} = \sqrt{\hbar/M\bar{\omega}}$. This value can be compared with the Gaussian variational prediction of the Gross–Pitaevskii equation of 0.67 (see Eq. (2.90)) or the value 0.574 obtained by direct numerical integration [85].

2.6.5 Condensate interference

It is common not to distinguish between the idea of a condensate and a macroscopic wave function with a locally well-defined phase. Whether a wave function exists, or in what sense it exists, for an isolated sample with a *fixed atom number* has been widely discussed. That a temporal oscillation is present between two coupled superfluids held at different chemical potentials is of course the basis of the overwhelmingly demonstrated Josephson effect.[33] A related question is what will happen if two isolated samples with fixed atom numbers are subsequently allowed to inter-penetrate; in particular, will one see a spatial interference pattern. That a spatial interference pattern indeed develops when two condensates merge has been confirmed experimentally [87]. The underlying theory is reviewed by Pethick and Smith [88].

[33] For a discussion in the context of neutral superfluids see [86].

To observe condensate interference the Ketterle group first cooled sodium atoms in a magnetic trap. Atoms in the trap were partitioned into two groups by passing a planar, blue-detuned, 517 nm argon ion laser beam through the trap center. The condensates were subsequently merged by simultaneously removing the partitioning beam and the magnetic trapping field, thereby causing the condensate to both expand and mix. The profile of the atoms in the trap(s) and the evolving pattern on removing the trap/partition were imaged using a sensitive phase contrast method.

Figure 2.41 shows a color enhanced image of sodium atoms in the double trap. To the left we see a condensate in the absence of the laser beam separating the condensates. The middle two panels show double traps with two different separations. The final panel to the right shows a single condensate obtained after sweeping one half away.

Figure 2.42 shows the imaged interference pattern generated by switching off the trap and removing the partition. Qualitatively one pictures the combined system as involving the superposition of two expanding coherent wave fronts emanating from two extended sources (the original trapped clouds). The

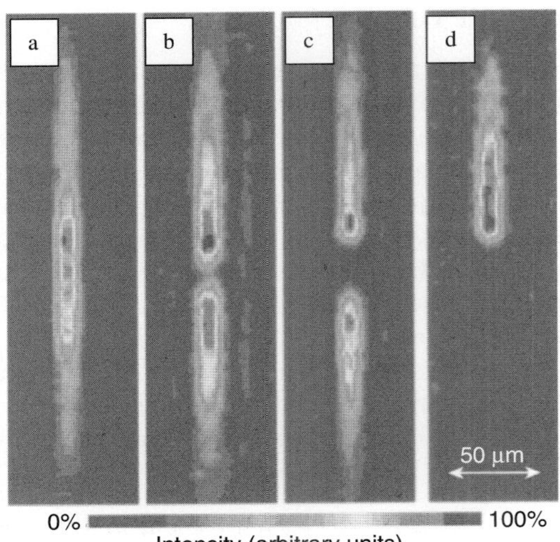

Fig. 2.41 Phase-contrast images of: (a) a single Bose condensate; (b) and (c) double Bose condensates with different separations (achieved by altering the argon laser power); and (d) an initial double condensate after sweeping one half away. (Anderson et al. (1997).) (See Plate 7 for color image).

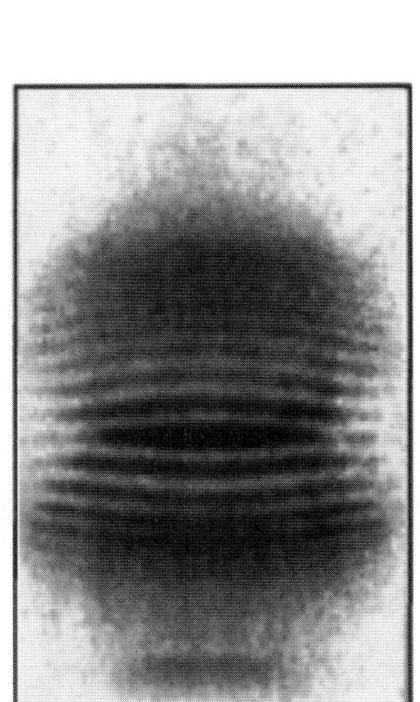

Fig. 2.42 An interference pattern obtained by removing the partition and expanding a double condensate (separated by $\approx 3\,\mu\text{m}$) taken after 40 ms time of flight.

interference of these wave fronts generates a time evolving standing wave pattern, which is what is imaged in Fig. 2.42. Being in relative motion (from the expansion) the de Broglie wavelength associated with the moving condensate fronts from the two clouds forms the standing wave; the wavelength is $\lambda = h/p$ with $p = Mv$, and taking the effective velocity to be $v = d/t$, where t is the time of flight and d is the condensate separation, we then have,

$$\lambda = \frac{ht}{Md}. \tag{2.167}$$

If the sources were point-like and the resulting waves spherical, one would expect hyperbolic nodes in interference pattern. Modeling with extended sources was reported to well represent the observed patterns.

2.6.6 Other condensate properties

Most of the trapped gas phenomena we are discussing involve the static properties. Among the most important of these is vortex states in rotating condensates; vortices can exist quite generally in condensates. Below we will discuss vortices in connection with Fermi condensates. Vortex physics in superfluids in general is reviewed in the chapter by Tsubota, Kasamatsu and Kobayashi (Chapter 3).

The dynamical properties of condensates and trapped gases are also of interest. As an example a superfluid critical velocity is expected, observation of which has been reported [89]. Another example is the excitation of sound [90]; here the pressure, the variations of which constitute the sound wave, is provided by the trap that confines the gas.

2.6.7 Trapped and cooled fermions

Cooling a Fermi system to temperatures lower than the Fermi temperature, T_F, is problematic due to suppression of scattering, since the final states, into which the atoms must scatter, progressively fill as the temperature is lowered, and further occupancy is forbidden by the Pauli principle; this leads to the familiar T^{-2} increase in the lifetimes of Fermi systems. Strategies have been developed for various atoms to cope with this problem, some of which we outline below.

2.6.7.1 *Potassium*

Cooling a trapped atom cloud of fermions below T_F was first achieved by DeMarco and Jin [91, 92]. The system studied was an enriched sample of ^{40}K (the natural abundance is 0.012% due to a half life of 0.28×10^9 years); this isotope has a rather high nuclear spin $I = 4$, leading to hyperfine ground states with $F = 7/2$ and 9/2, the latter lying lower. The cloud was pre-cooled in one MOT and transferred to a second MOT (in a better vacuum), the magnets of which could be operated in an Ioffe–Pritchard configuration which held the cloud when the MOT was shut down.

To partially address the problem of falling scattering rate, the cloud was compressed to increase this rate (by ramping up the trap field after cloud injection). But a key component of the cooling strategy was to exploit energy exchange between the two Zeeman baths (with $F = 7/2$, 9/2) through evaporative cooling using swept microwave fields to progressively eject atoms by driving them to untrapped spin states. By applying appropriate microwave frequencies along with ramping the trapping field, they evaporatively cooled the sample while maintaining a roughly equal mixture of the $M_F = 7/2$ and 9/2 states, these states being metastable against M_F-changing collisions at low temperatures. As a final step the $M_F = 7/2$ component was removed with a microwave sweep, thereby leaving a single-component spin-polarized gas with $M_F = 9/2$.

The shape of the expanded cold clouds was absorption imaged by driving the $4S_{1/2}$, $F_{9/2} \rightarrow 4P_{3/2}$, $F_{11/2}$ transition and the resulting images were recorded. From such data the number of atoms (from the overall absorption) and the velocity distribution function (from measurement of the expanded column densities) were obtained for different cloud starting conditions. These data were compared with computed distribution functions expected for given initial particle numbers, N, and temperatures, T, and the results were consistent with the presence of an equilibrium Fermi distribution prior to the expansion of atoms in a harmonic trap with parameters (the constants κ_i) that were determined independently from the trap oscillation frequencies, which can be calculated using Eq. (2.165) and the measured field characteristics. In particular, and in contrast with a classical system, a pronounced non-Gaussian velocity distribution is both expected and observed for smaller values of T/T_F. For details of the procedures used to confirm degenerate behavior the reader is referred to DeMarco's thesis [92]. The lowest temperature deduced in these experiments was $T/T_F \cong 0.5$.

2.6.7.2 *Lithium*

Truscott et al. [93] trapped a mixture of the two stable isotopes of lithium; ^{6}Li (natural abundance 7.5%) with spin 1 and ^{7}Li (92.5%) with spin 3/2. The presence of the boson ^{7}Li enables the fermion ^{6}Li to thermalize by scattering with the ^{7}Li, thereby minimizing the Pauli blocking effects. The initial cloud was injected into a MOT from a pre-cooled beam containing both isotopes that emerged from a Zeeman slower. The trapped ^{7}Li atoms were then optically pumped into the $F = 2$, $M_F = 2$ state, the desired state for magnetic trapping, while the ^{6}Li were pumped into the $F = 3/2$ state. The gradient magnet and lasers associated with the MOT were then removed; thereafter, the cloud was trapped by a clover-leaf (Ioffe–Pritchard-like) magnet. The final cooling of the cloud was by microwave induced evaporation of the ^{7}Li, while the ^{6}Li atoms cooled sympathetically through collisions with the ^{7}Li, but were not themselves ejected. These two baths were assumed to be in equilibrium.

Once cooled and equilibrated, the clouds are expanded and absorption imaged using pulses from a laser tuned near the $^2S_{1/2} \rightarrow {}^2P_{3/2}$ transition. The frequency of this transition differs between the two isotopes allowing them to be *separately imaged* (or ejected) from the trap. Figure 2.43 shows three pairs of images for the two isotopes. The anisotropy arises from a combination of the trapping potential, which as discussed above is determined independently through the trap frequencies, and for the Fermi system, the pressure associated with the Pauli exclusion. As was the case for potassium, discussed above, the number of atoms and the temperature can be obtained from the images by fitting them to the Bose or Fermi distribution functions. The Bose system shows a greater sensitivity at low temperatures which makes it the more reliable thermometer.

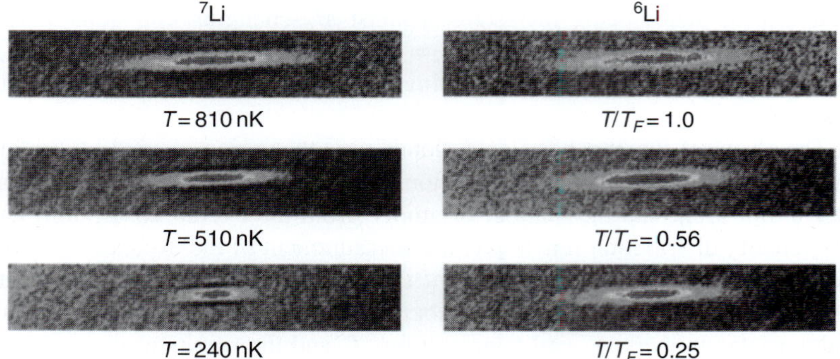

Fig. 2.43 Absorption profiles of expanded clouds on ^{7}Li and ^{6}Li for various starting temperatures. Points to note are: (i) because of the Fermi pressure, the Fermi cloud (right) undergoes much less of a contraction relative to the Bose cloud (left) as the temperature is reduced, and (ii) the boson cloud undergoes a greater change in shape with falling temperatures. (After Truscott et al. (2001).) (See Plate 8 for color image).

Sympathetic cooling can also be achieved using mixtures of Bose and Fermi gases composed of different atoms, e.g. the species ^{6}Li–^{40}K and ^{6}Li–^{23}Na.

2.6.8 Diatomic molecule formation

The first class of experiments involved the use of Feshbach resonances to control the formation of diatomic molecules [94]. Depending on the circumstances, such molecules can be in high vibrational states. However, very near a Feshbach resonance the binding energy of the molecules approaches zero; hence the molecules become extremely large and the number of bound states falls. The field at which the resonance occurs can be determined precisely from molecule dissociation measurements at low densities.

In experiments with cold ^{40}K ($T/T_F = 0.13$ to 0.33) in a magnetic trap (which retains molecules and atoms alike), Regal et al. [95] measured the number of atoms in their trap after sweeping through a Feshbach resonance separating unbound atoms and diatomic molecules. When sweeping the field from the unbound to bound regime they observed a large reduction in the number of atoms in a subsequent absorption image following expansion (which is only sensitive to unbound atoms). But if they cycled the field passing from the atomic to the molecular regime and back again, no significant loss was seen on expansion. In yet another experiment, they carried out fast field sweeps between the atomic and molecular regimes and found that the atom number was largely maintained. Taken together these measurements showed that one could *reversibly* make molecules by slowly sweeping through the Feshbach resonance.

To image the molecules a radio-frequency pulse was applied that dissociates the molecules into free atoms (in the $M_F = -5/2$ and $M_F = -9/2$ spin states) and is detuned beyond the molecule dissociation threshold so that it does not affect the residual unpaired atoms (in the $M_F = -7/2$ state). Immediately after this dissociation, a spin-selective absorption image is taken (of the previously unoccupied $M_F = -5/2$ state). Alternatively, the atoms in the expanded cloud can be imaged by tuning to the $M_F = -7/2$ spin state. Similar experiments were performed with the ^{7}Li sympathetically cooled ^{6}Li [96, 97].

2.6.9 Molecular BEC

The ability to form molecules opens the way to making a molecular BEC, which has been observed in ^{40}K using the cooling and imaging techniques discussed in the previous section [98]. If molecule formation resulting from sweeping through the Feshbach resonance is sufficiently isentropic and the temperature of the initial atom gas is low compared to the Fermi temperature (the Fermi temperature and BEC temperatures being quite similar on dimensional grounds), then the molecular sample should have a sizeable condensate fraction.

The group found that in a gas with T/T_F below 0.17, a molecular condensate was observed in time-of-flight absorption images taken immediately following the magnetic-field sweep. Note the condensate was not formed by any active cooling of the molecules, but rather by reversibly passing from the cold Fermi gas (later shown to be superfluid) into a BEC. Figure 2.44 depicts the formation of a molecular condensate from fermionic ^{40}K atoms cooled by evaporative cooling from an optical dipole trap followed by sweeping through a Feshbach resonance. Similar behavior has been observed in sympathetically cooled ^{6}Li [99, 100].

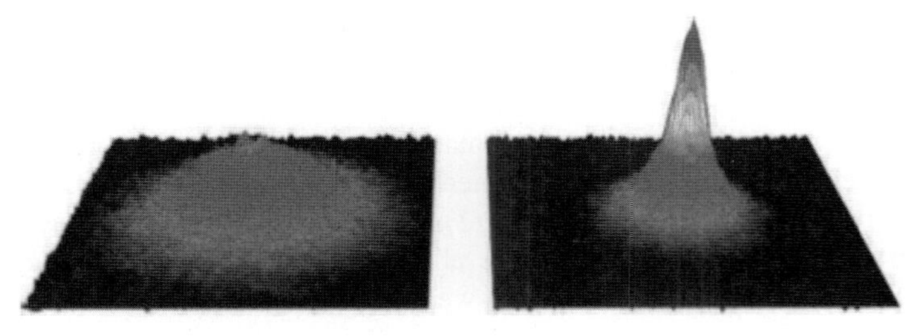

Fig. 2.44 Surface plot of the optical density for a molecular cloud created by applying a magnetic-field sweep to a Fermi gas of ^{40}K atoms. The two experiments start with fermionic atoms at different temperatures: $T = 0.19T_F$ (left) and $T = 0.06T_F$ (right). A molecular cloud is then created by passing through the Feshbach resonance. For the higher starting temperature, shown on the left, a condensate does not form, while for the lower starting temperature, shown to the right, a condensate clearly forms. (After Greiner et al. (2003).) (See Plate 9 for color image).

2.6.10 BEC–BCS crossover [101]

In a trapped Fermi system, the Feshbach resonance allows the exploration of different regimes through a continuous variation of the scattering length. As we have discussed, on the low-field (attractive) side of the resonance atoms can bind as bosonic molecules; here real two-body bound states exist, and molecules are readily formed by three-body recombination. On the high-field (repulsive) side BCS pairs can form, in analogy with neutral ^{3}He (although with s-state pairing). What happens between these limits has long been of interest [102, 103]. In particular, is there a phase transition and if so what is its nature?

The BEC–BCS crossover region has been studied in ^{6}Li [104, 105]. A crossed (as opposed to a single)-beam optical dipole trap was utilized which leads to tighter confinement (and hence higher atom densities). When combined with the large scattering lengths produced by the Feshbach resonance, a regime of strong interactions is entered where $na^3 \cong 0.3$ and the chemical potential is of the same the order as $k_B T_C$, where $T_C \cong 1.5$ K is the condensation temperature. As a consequence the thermal cloud in a time-of-flight expansion profile is *modified* due to the large condensate mean field.

In considering the crossover theoretically, relevant energy scales are the molecular bonding energy, ε_0, for the former and the BCS energy gap, Δ, and the Fermi energy for the latter; relevant length scales are the atom separation in the molecule for the former and the mean particle spacing, $n^{-1/3}$ and coherence length $\xi = \Delta/\hbar V_F$ for the latter. A BCS state is predicted [106, 107], with rather high transition temperatures, $T_c \approx 0.5T_F$; for comparison with other Fermi superfluids we have: ordinary metals, $10^{-4}T_F$; ^{3}He $10^{-2}T_F$, and high temperature superconductors, $10^{-2}T_F$. Figure 2.45 shows the predicted behavior of the chemical potential.

Perhaps the best evidence that superfluidity is present on the BCS side of the BEC–BCS crossover is the observation of vortices [108]. The experiments were performed on ^{6}Li atoms. Cloud preparation involved: (i) sympathetic cooling by ^{23}Na atoms in a magnetic trap; (ii) transfer to an optical–dipole trap, where a 50%/50% spin mixture of the two lowest hyperfine states was prepared; (iii) further evaporative cooling as the field utilized to sweep the Feshbach resonance was ramped up; and (iv) stirring the condensate with a rotating pair of blue-detuned laser beams located symmetrically on each side of the cloud center. Figure 2.46 shows the presence of vortices in a rotating

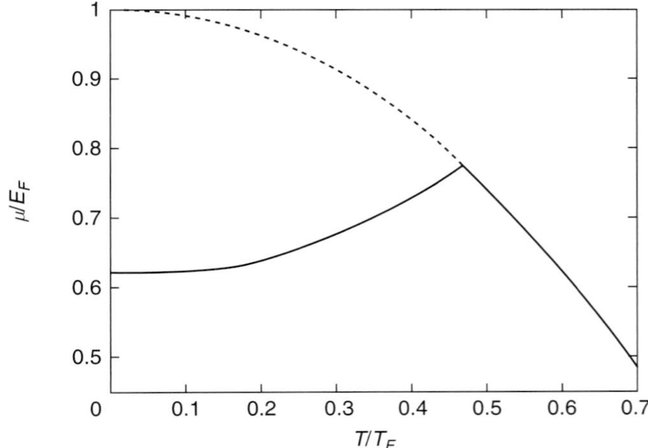

Fig. 2.45 The chemical potential for a Feshbach resonance pairing model of a trapped Fermi gas (solid line) versus T/T_F. The dashed line shows the behavior in the absence of pairing. The superfluid and normal regimes are separated by a second-order phase transition. (After Holland et al. (2001).)

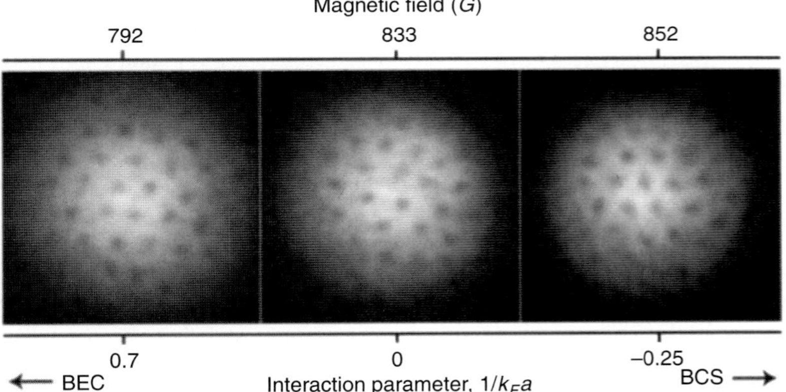

Fig. 2.46 Three images of a rotating condensate as the magnetic field is swept through the Feshbach resonance in a rotating cloud of ^{6}Li atoms. Note, vortices are present on both sides as well as in the region of the transition. (After Zwierlein et al. (2005).)

condensate at three fields associated with a Feshbach resonance. Starting on the left, the cloud is swept from the BEC side, through the crossover to the BCS side; note that a vortex lattice is present in all three regions, providing convincing evidence that superfluidity is present on the proposed BCS side. This group went on to study various properties of vortex states such as their lifetime and nucleation properties.

2.6.11 Imbalanced condensates and phase separation

With the wide level of control that various combinations of trapping parameters now afford (near-resonant laser beams; r.f./microwave state mixing; static, far-off-resonance, optical–dipole traps; magnetic field traps; ...), one can now prepare clouds involving different atomic species, different hyperfine states of the same species, or with imbalanced spin populations in the same hyperfine

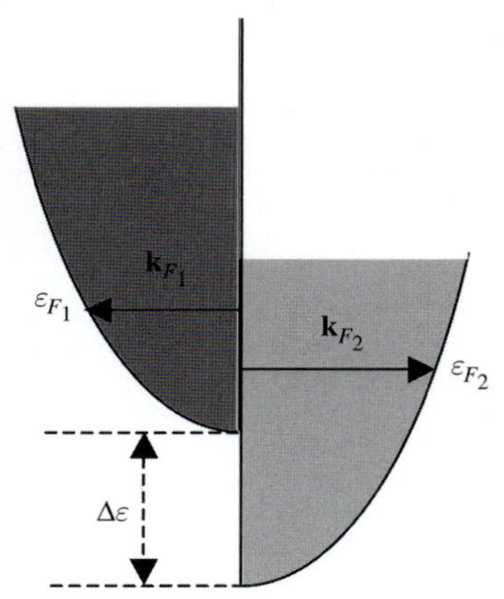

Fig. 2.47 Schematic representation of a fully imbalanced two-species Fermi system.

[34]In superconductors in a magnetic field the paired BCS state competes with the normal state in the presence of Pauli paramagnetism; see [109].

state. The number of trapped states is not limited to two. Such systems are of special interest in connection with the imbalanced color–flavor locked u–d–s superfluid quark state (see Chapter 10), thought to exist in the core of neutron stars, the imbalance there arising from the smaller well depth caused by larger strange quark mass.

Figure 2.47 shows a schematic representation of an imbalanced Fermi system with two species. In the presence of a magnetic field (and when not using it to control a Feshbach resonance) the two "bands" can be displaced according to

$$\Delta\varepsilon = (\boldsymbol{\mu}_1 + \boldsymbol{\mu}_2) \cdot \mathbf{H}, \tag{2.168}$$

where $\boldsymbol{\mu}_1$ and $\boldsymbol{\mu}_2$ are the magnetic moments. Depending on the preparation history the Fermi energies ε_{F_1} and ε_{F_2}, and hence the Fermi wavevectors $\mathbf{k}_{F_1}$ and $\mathbf{k}_{F_2}$, can differ. Clearly this system offers an enormous parameter space to explore new phenomena. If one or both subsystems is BCS paired, with gaps Δ_1 and Δ_2, these compete with the relative shift, $\varepsilon_{F_1} - \varepsilon_{F_2}$.[34]

In the presence of a spin imbalance one has a kind of cold gaseous ferromagnet. As noted, one or both spin states can be a superfluid; depending on the temperature and magnetic field, these states exist in superfluid ^{3}He. Another possibility is that the system phase separates, say into a spin paired state and a polarized state, with one in the center of the trap and the second on the periphery. In a spin imbalanced system the differing densities result in differing Fermi wavevectors and hence a BCS paired state will have a net center of mass motion, making the construction of a ground state more complicated [110, 111].

Bedaque, Caldas, and Rupak [112] considered Fermi systems composed of two-particle species with different densities, which includes imbalanced spin species. They argued that a phase separation into normal and superfluid components is the energetically favored ground state, and that the most likely ground state is a mixed phase containing bubbles of a normal state with unequal populations, immersed in a "sea" of a BCS phase with equal populations. The Hamiltonian considered was for a homogeneous system, so in the presence of a trap potential the separated phases can segregate concentrically about the trap center.

Phase separation in an imbalanced cloud of of spin-polarized ^{6}Li atoms was observed by both the Rice [113] and MIT groups [114]. We will focus on the latter, who studied additional behaviors by creating a wide range of spin mixtures. The two lowest hyperfine states, labeled 1 and 2, were created by sweeping an r.f. field at a controlled rate. As before, by positioning the magnetic field on each side of the Feshbach resonance the BEC and BCS states are obtained. Superfluidity as a function of the spin imbalance in the mixtures was probed by setting the cloud into rotation on the BEC side using two rotating, blue-detuned laser beams, as discussed above. The BCS side was probed by sweeping the magnetic field to that side of the resonance.

Figure 2.48 shows spectroscopically imaged expanded clouds of the two hyperfine states on both the BEC and BCS sides of the transition for a range of state imbalances,

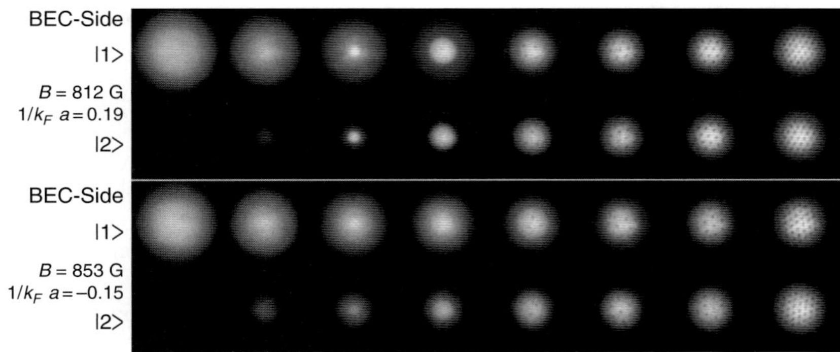

Fig. 2.48 Clouds prepared on the BEC ($1/k_F a = 0.2$) and BCS ($1/k_F a = 0.15$) sides of the transition. The upper and lower rows in each panel show the clouds for the two trapped species 1 and 2. On the BEC (upper) side the imbalances, $(N_2 - N_1)/(N_2 + N_1)$, from left to right are 100, 90, 80, 62, 28, 18, 10, and 0%; on the BCS (lower) side the imbalances are 100, 74, 58, 48, 32, 16, 7, and 0%. The dark dots distributed within the clouds arise from vortices and their presence implies superfluidity. (After Zwierlein et al. (2006).)

$$\delta = \frac{N_2 - N_1}{N_2 + N_1}, \tag{2.169}$$

ranging from 0% to 100%. The dots imbedded in the clouds arise from vortices. Note vortices are not visible in the presence of large spin state imbalances [115, 116].

On the basis of their measurements Shin et al. [117] have proposed the phase diagram shown in Fig. 2.49 for the boundaries separating normal, superfluid, and unstable phases in terms of the spin polarization and the reduced temperature, T/T_F.

Other important properties of trapped Fermi gases have been measured. As a single example we point out that the velocity of sound has been measured [118].

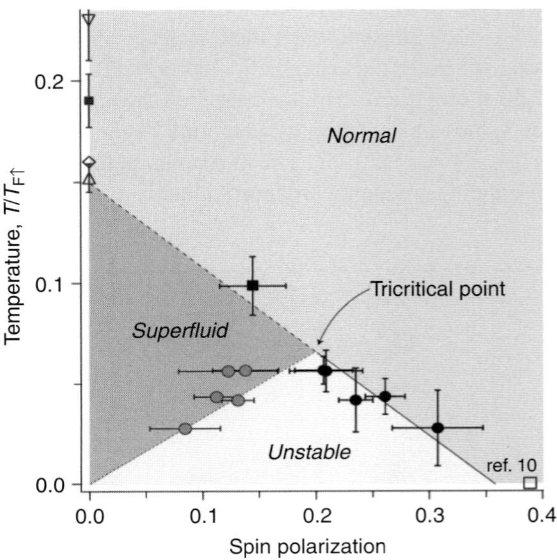

Fig. 2.49 Phase diagram for an imbalanced superfluid as a function of the spin polarization and the reduced temperature T/T_F. (After Shin et al. (2008).)

2.6.12 The Bose superfluid to Mott insulator transition

The BEC to Mott insulator transition (see Section 2.4.7) was first studied in ^{87}Rb by Greiner et al. [119]. In these experiment spin-polarized clouds of laser-cooled atoms in the $F = 2$, $M_F = 2$ state were transferred into a cigar-shaped magnetic trapping potential from which radio-frequency evaporation was carried out to produce a BEC with up to 2×10^5 atoms and no detectable thermal component. The magnetic trap profile was then made spherical; the chosen trap frequency of 24 Hz resulted in an isotropic condensate with a Thomas–Fermi diameter of 26 μ (see Section 2.4.2.2.).

The laser beams used to create the optical lattice were derived from an injection seeded tapered amplifier and a laser diode operating at a wavelength of 852 nm. The three beams used were spatially filtered and guided to the experimental chamber using optical fibers so as to be mutually perpendicular; in addition their polarizations were mutually perpendicular. Standing waves were generated by three mirrors which reflect each beam back along its incoming axis. Each beam was passed through a separate acousto-optic modulator (AOM) such that the frequencies of all three beams *differed* by amounts of order 30 MHz; if this is not done complex polarization-dependent intensity patterns will result; the frequency shifting allows the cross beam, polarization-dependent, interference effects to average out. Only in this way can a simple cubic lattice potential of the form given in Eq. (2.123) result. The AOMs also allow the intensity of the beams, and with it the strength of the lattice potential, to be continuously adjusted via the applied r.f. power, thereby driving the transition in the optical lattice.

The cloud was "transferred" to the optical lattice by ramping up the latter within the magnetic trap over a period of 80 ms; this is slow enough that the condensate stays in its many-body ground state. There were of order 65 lattice periods spanning the condensate width corresponding to 150,000 sites in all, with average occupancies of up to 2.5 atoms per site. Phase coherence between the different lattice sites (present when there is an overall condensate) can be probed by turning off both traps (magnetic and optical) thereby releasing the cloud. Figure 2.50 shows the absorption images of the resulting expansion profiles for various values of the trapping potential V_0 measured in units of the atom recoil energy, $E_R = \hbar^2 K^2/2M$. The maximum potential V_0 corresponded to an intra-well trapping frequency of approximately $\upsilon \cong 30$ kHz.

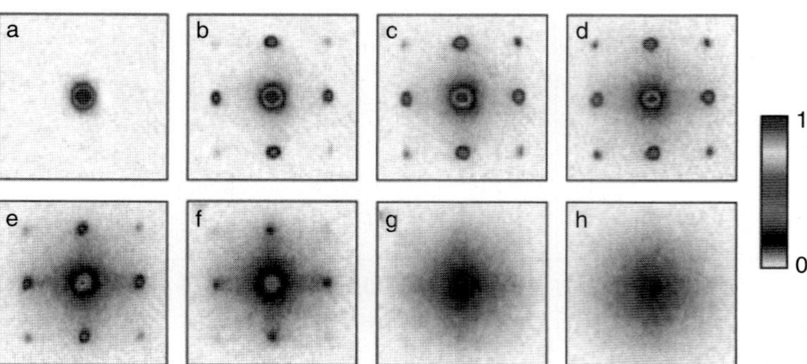

Fig. 2.50 Absorption images of expanded cloud profiles for various values of the trapping potential V_0 after a time of 15 ms. In units of V_0/E_R the various frames correspond to: (a) 0, (b) 3, (c) 7, (d) 10, (e) 13, (f) 14, (g) 16, (h) 20. (After Greiner et al. (2002).) (See Plate 10 for color image).

In frame (a) we see the profile generated by the Bose condensate in the absence of a lattice that is strongly concentrated in the center, corresponding to $k \cong 0$ as appropriate for a condensate. In frame (b) we see four spots arising from the (1,0), (0,1), $(-1{,}0)$, and $(0, -1)$ vectors of the reciprocal lattice of the first Brillouin zone; also seen weakly are the (1,1), $(-1{,}1)$, $(1, -1)$, and $(-1, -1)$ vectors from the second zone. These correspond to the constructive interference of delocalized atoms the amplitude of which is modulated by the lattice and which have an overall phase that is imposed by the condensate. As the lattice potential continues to increase, a diffuse background develops, in addition to the condensate diffraction spots, where the former presumably arises from an atom fraction that has started to localize into "patches". However by frame (g) the condensate driven spots are only barely visible and by frame (h) we have only a diffuse pattern, signifying essentially complete localization of the atoms on the lattice sites with no phase correlation among them. In the vicinity of the transition the system may be thought of as a dynamically fluctuating mixture of superfluid and localized regions. To the extent that it is at $T \cong 0$, these fluctuations must be quantum mechanical in character; i.e. the system is passing through a so-called *quantum critical point*. A similar transition occurs at the superconductor–insulator transition that is driven by alloying a superconductor and a semiconductor.

By performing the necessary calculations to map the known or measured parameters associated with the optical Bloch band structure onto the parameters ε, J, and U of the model Hamiltonian (2.79), Grainer et al. estimate that the superfluid–Mott transition in a well with $V_0/E_R = 13$ occurs at $U/J = 36$; this is to be compared with the theoretical prediction (see Section 2.4.7) of $U/J = 5.8\,z$, which for a simple cubic lattice where $z = 6$, yields 34.8.

Support for the interpretation of the exhibited behaviors as involving a phase transition, follows from the observation that overall phase coherence in the condensate is quickly restored when the trapping potential is adiabatically reduced to values on the BEC side of the quantum critical point; i.e. the behavior is reversible. Qualitatively, the timescale to restore coherence is of order tunneling time, $\tau_{\text{tunnel}} \approx \hbar/J$ between neighboring lattice sites, which the authors estimate is of order 2 ms for $V_0/E_R = 9$. The time necessary to establish phase coherence was measured quantitatively by the following proceedure: (i) bring the system to equilibrium on the Mott side; (ii) adiabatically lower V_0 to a value on the BEC side in a time t; and (iii) expand the cloud and examine the width of the condensate peak. Here it is found that coherence was largely re-established in a time of order 4 ms.

We now examine the excitation spectrum from the Mott ground state of a system having one atom per site. Deep in the Mott phase, where $J << U$ and all atoms are localized, an excitation consists of (i) removing an atom from some site, thereby lowering the energy (see Eq. (2.124) of that site by the optical binding energy, ε_0 (including the zero-point energy in the well), and (ii) adding it to some neighboring (already occupied) site in a state ε_n, raising the total energy of the latter by $\varepsilon_n + U$ when the inter-atom potential is included. The lowest excitation energy occurs when $\varepsilon_n = \varepsilon_0$ and is therefore equal to U. Since we create a vacancy or hole at the first site and add a particle to the second, such excitations are referred to as particle–hole excitations.

Suppose a potential gradient is applied to the lattice such that the energy difference ΔE between neighboring sites is equal to U; resonant inter-site tunneling can then occur via particle–hole excitations. If the potential gradient is applied for some time and then removed, a population of particle–hole excitations remains. If V_0 is then reduced, so the system passes into the superfluid phase, the effect of the Mott excitations is passed on as phase fluctuations in the superfluid, which on expansion are manifested as broadened diffraction peaks in the absorption images. A plot of the width of the diffraction peaks as a function of the potential gradient shows a sharp maximum, which was identified by Grainer et al. as arising from a particle–hole excitation of the Mott ground state. A peak is also observed at a value for ΔE of approximately twice that of the first, possibly corresponding to a two-particle–hole excitation.

The Mott transition, and optical lattices in general, have remained an active research area which will be reviewed in Volume 2 of this book by Chin and Gemelke.

2.7 Superfluidity in neutron stars

There are three possible end points in the life cycle of a star that in order of increasing mass are:[35] the white dwarf,[36] the neutron star, and the black hole. Superfluidity is not relevant to the first, and the last is governed by general relativity.

[35]For a discussion of models pertaining to white-dwarf and neutron stars see Chapter XI of *Statistical Physics Part I*, L. D. Landau and E. M Lifshitz, Pergamon Press, Oxford (1980); gravitational collapse (leading to black holes) is discussed by the same authors in *Classical Theory of Fields*, Section 102, Pergamon (1985).

[36]At white dwarf mass densities, the Fermi energy calculated for the corresponding average electron density greatly exceeds the binding energies of these same electrons in the isolated atoms making up the star; i.e. the (repulsive) electron–electron and nuclear–nuclear and the (attractive) electron–nuclear Coulomb energies play a negligible role. Furthermore, one is also in a limit where the associated Fermi velocity is approaching that of light, so the energy of the gas must be calculated using the relativistic energy–momentum relation. This near total dominance of the electron kinetic energy makes BCS pairing at typical star temperatures unlikely.

The pressure balancing that arises from the quantum degeneracy of the electron gas is provided by the gravitational potential. But with increasing mass of the star a point is reached where quantum degeneracy can no longer resist the pull of gravity and the star becomes unstable; this happens at about 1.5 solar masses, the so-called Chandrasekhar limit.

2.7.1 Some systematics of neutron stars [120–124]

Leaving aside the black holes, neutron stars are objects having the highest densities in the universe, in the region of 10^{15} g/cc and corresponding to 5–10 times that of nuclear matter, $n_0 \approx 0.16\,\mathrm{fm}^{-3}$. Total masses are ≈ 1.5 solar masses while the radii are $\approx 12\,\mathrm{km}$; temperatures are in the range 10^6–10^9 K. In such stars local equilibrium requires that the global chemical potential be a constant, since this quantity measures the energy needed to transfer a particle from one region of the star to another. Restricting ourselves to the neutrons, their global chemical potential is given by the sum of the local chemical potential of the highly degenerate neutron liquid, which is positive and position dependent through the local density, and the local gravitational potential energy, which is negative and also position dependent; both calculated on a per neutron basis. As an approximation one might treat the neutrons as a non-relativistic free Fermi gas. In practice one needs the energy density of the interacting neutron liquid itself, which is not well known at densities beyond those encountered in nuclear physics. In addition to the neutron matter, which comprises most of the mass density (outside a possible inner core), one has an interpenetrating proton liquid and a charge compensating relativistic electron gas.

On approaching the star from afar one first encounters a stellar atmosphere which, although important for the emission properties (in the X-ray regime), we do not discuss. At the bottom of this atmosphere, the star has a solid crust composed of heavy nuclei and relativistic electrons (see Fig. 2.51.). Neglecting the (negligible) pressure from the atmosphere, one can take the pressure at the crust surface to be zero. As one proceeds inward, the rapidly increasing pressure from the crust matter lying above raises the local electron density and with it the chemical potential. When the sum of the electron rest mass and Fermi energy (chemical potential) match that of the neutron–proton mass difference in the nuclei, reactions occur in which those protons capture free electrons forming neutrons (simultaneously emitting a neutrino, an inverse beta decay); this leads to neutron-rich nuclei and as neutron numbers in the nuclei increase they become unstable and shed neutrons (called neutron drip), where the latter form a gas with a rising chemical potential; this in turn feeds back in a balanced way to allow neutron numbers in the nuclei to further increase and become quite large. All the while, as we proceed inward, the number density of the nuclei increases and at some point they contact and merge. Here a (likely) first-order transition occurs, beyond which we have a *quantum liquid solution* consisting of three interpenetrating materials: neutrons, electrons, and protons. Since these particles can change their identities through beta and inverse beta processes they can adjust their relative chemical potentials, and hence their concentrations, so as to create overall chemical equilibrium, both with themselves and with the other two species, via the reaction,[37,38]

$$\mu_n = \mu_e + \mu_p. \tag{2.170}$$

When this condition applies we can say that we are in the neutron star proper (called the outer core).

As we continue to approach the star center, and the density further increases, the chemical potentials of the three equilibrated ingredients (n, e, p) can rise to the point where other hyperons[39] may become stable, all but the Λ containing strange quarks; mesons, as we discuss below, may also be stable. Given more degrees of freedom the chemical potentials of existing species rise more slowly, and the equation of state becomes "softer" as additional populations enter.

It is widely believed that as we go closer to the center of the star an *inner core* is encountered involving different physics. One possibility is a deconfinement transition in which the neutrons "melt" into quark matter; in particular the u, d, and s quarks; the remaining quarks, c, b, and t, are not expected since they are too heavy to be stable relative to the other quarks. One can again think of this as a softening occurring in the equation of state, made possible by the availability of more particle species to slow the rise in chemical potentials with density.

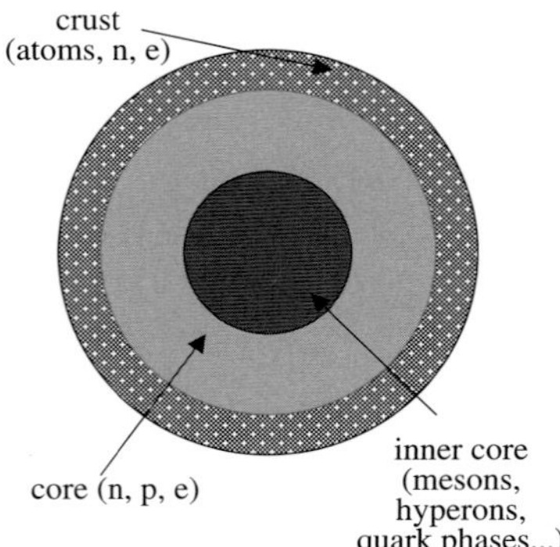

Fig. 2.51 Schematic representation of a neutron star with its crust (containing electrons atoms and progressively more neutrons), core (containing electrons, neutrons, and protons), and inner core (with possible meson and quark phases).

[37]More correctly the condition is $\mu_e + \mu_p = \mu_n + \mu_\nu$, and this is important in the earlier stages following collapse. However as the star matures the neutrinos leak out and since their number in then not constrained, $\mu_\nu \to 0$.

[38]When nuclear reactions can occur one must include the rest mass of the individual particles in evaluating the chemical potentials.

[39]The spin 1/2 and 3/2 hyperons, together with their charges, are denoted Λ, $\Sigma^{0,\pm1}$, $\Xi^{0,-1}$ and $\Delta^{2,1,0,-1}$, $\Sigma^{*+,0,-}$, $\Xi^{*0,-1}$, Ω^{-1}, respectively, in the order of increasing mass within each multiplet.

2.7.2 Quark nuggets and quark stars

Witten [125] has suggested that a (first-order) QCD phase transition may have occurred in the early universe that, among other things, might concentrate most of the quark matter into "nuggets" which, if they survived to the present epoch, might conceivably explain the dark matter problem without the need to invoke any new particles.

On a larger scale, *strange quark matter* (SQM) stars might exist where deconfined $u\ d\ s$ quark matter exists without requiring a neutron liquid intermediary (as in the neutron inner core scenario above) [126]. It is conjectured that such stars could have either a bare quark-matter surface, with zero pressure (like the neutron star crust) and a large super-nuclear density, or start with a thin layer of normal-matter crust (like the neutron star) but then transition to quark matter below this surface. Note that if this system had a greater binding energy per baryon at *zero pressure* than, say, an iron nucleus, it would be the *ultimate ground state of matter;* i.e. normal matter would be metastable, and if compressed to a high enough density would convert into de-confined quark matter. So unlike the neutron stars, SQM stars would not require gravity to bind them. Stars with a free quark surface might be identifiable by their emission characteristics which would involve electron–positron pairs and X-rays arising from their recombination [127].

2.7.3 BCS superfluidity in the neutron liquid

Based on strong theoretical arguments, neutron stars are expected to display superfluid phases, the nature of which varies in the different regions of the star (crust, core, and inner core);[40] it is the subject of Chapter 20 of Volume 2 of this book. Pairing, it is argued, is already present for both neutrons and protons in ordinary nuclei on the basis of the so-called odd–even contribution to the semi-empirical mass formula, as well as their excitation spectrum. On the basis of such data, A. Bohr et al. estimated that a BCS-like energy gap with $\Delta \approx 12A^{-1/2}$ Mev was present for both proton and neutron superfluids [129] where A is the nucleon number. Further evidence for superfluidity came from the rotational properties of nuclei (which is an analog of the Meissner effect) [130].

[40] For an early review see [128].

On the basis of neutron scattering experiments the energy dependence of the phase shifts can be deduced [131]. Since pairing involves particles at the Fermi surface, and assuming the Fermi energy scales with the mass density as $\rho^{2/3}$, one can use the energy-dependent phase shift data to deduce an effective pair potential and to predict the nature of the pairing in a neutron star. On the basis of such an analysis, Hoffberg et al. concluded [132] that the triplet $^3P_2$ state is dominant in the outer core.[41] Single $^1S_0$ pairing is expected for neutrons in the crust [133] while protons in the core are also thought to be paired in a $^1S_0$ state. The predicted form of the neutron $^3P_2$ superfluid state (together with normalization factors) is

[41] In this sense one can say that unconventional superfluidity was first discovered in nuclear matter, since the phase shifts were known empirically and those for ^{3}He were essentially unknown.

$$\underset{\sim}{\psi} = \begin{pmatrix} \sqrt{2}\Delta_0 Y_{1-1} + \sqrt{6}\Delta_1 Y_{10} + \sqrt{12}\Delta_2 Y_{11} & -\sqrt{3}\Delta_1^* Y_{1-1} + 2\Delta_0 Y_{10} + \sqrt{3}\Delta_1 Y_{11} \\ -\sqrt{3}\Delta_1^* Y_{1-1} + 2\Delta_0 Y_{10} + \sqrt{3} Y_{11} & \sqrt{2}\Delta_0 Y_{11} - \sqrt{6}\Delta_1^* Y_{10} + \sqrt{12}\Delta_2^* Y_{1-1} \end{pmatrix}, \tag{2.171}$$

where the ratios of the parameters Δ_1/Δ_0 and Δ_2/Δ_0 are not fixed at the transition temperature; some of this degeneracy is retained near the transition temperature when strong coupling corrections are added [134], and more is removed by a magnetic field [135]. The form (2.171) is to be compared with Eqs. (1.329) for the B–W and (1.335) for the ABM states of superfluid ^{3}He; i.e. it differs from both. Superfluid vortices in the neutron fluid and superconducting fluxons in the proton superfluid, which we do not discuss here, play major roles in the rotational dynamics and magnetic flux retention properties of neutron stars.

One can ask what experimental evidence exists for superfluidity in neutron stars. We start by noting that as these stars evolves they gradually slow down due to external electromagnetic "braking torques". But occasionally one observes "glitches", sudden increases, spin-ups, in their rotation rates (see Fig. 2.52) [136, 137]. These events are thought to result from angular momentum transfer between the (solid) crust, which is what is astronomically observed, and a co-rotating neutron superfluid component within the crust. One model is that patches of pinned vorticity in the neutron superfluid are suddenly released, similarly to flux depinning in a superconductor (which can cause quenching in a superconducting magnet) [138]. As a result the superfluid transfers angular momentum to the core causing it to spin up. Other models have been proposed including sudden structural changes in the crust, such as crack formation (star quakes), and their aftermath; in any case the connection between superfluidity and spin-ups remains unsettled.

An emerging method for infering the presence of superfluidity in a neutron star is from the observed cooling rate in an early stage of its evolution when it is transitioning to the superfluid state. Based on theoretical modeling of their atmospheres, the observed rate of the recently formed neutron star in the supernova[42] remnant in Cassiopeia A [139] is judged to be anomalously large. To explain this discrepancy it is proposed that enhanced neutrino

[42] Some uncertainty exists as to when this event actually occurred, but an extrapolation yields AD 1667.

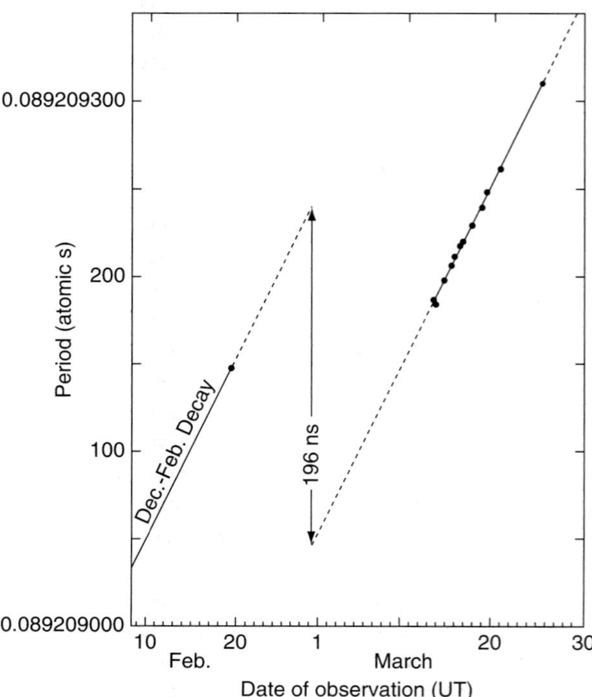

Fig. 2.52 A plot of the period versus time for pulsar PSA 0833-45, showing a glitch on 02/28/09.

emission occurs, associated with the rapid breaking and forming of neutron Cooper pairs (in the $^3P_2$ channel) in the vicinity of the transition [140, 141] via Urca processes[43] They further conclude that for this to happen the protons had to already be in a superconducting state with a larger critical temperature, and that all together this implies that both superfluidity *and* superconductivity are present within this star. They further predict that the cooling will continue for several decades at the present rate.

[43]The process was first discussed by G. Gamow and M. Schoenberg, see [142]. It refers to any astrophysical weak reaction by which a baryon changes its identity; e.g. $p + e \rightarrow n + v$.

With the planned upgrades to the two U.S. gravitational wave interferometers (LIGO), and various newly proposed instruments, signals resulting from the merger of two neutron stars, or a neutron star/black hole combination, may be observable at the level of many events per year. One then has the possibility that the gravitational wave signatures detected by such instruments will contain information that can be utilized to infer the make-up of the merging entities [143–146].

2.7.4 Kaon condensation

Neutron star interiors may also contain macroscopic condensates of bosons, with the K^- meson being the most likely [147, 148]. This particle, with a mass of 494 MeV and quark content ($\bar{u}s$), forms a bound state within nuclear matter, although it subsequently decays by reactions such as $K^- + N \rightarrow \Sigma^- + \pi$, where $N = (n, p)$. Friedman et al. [149] have made systematic measurements of the binding energies involving many atoms; for Ni, which has a density we can take as that of nuclear matter, denoted as ρ_0, the binding energy is 200 MeV. Now in a neutron star, where $\rho = (2\text{–}4)\rho_0$, we can argue that the binding is enhanced by a ratio ρ/ρ_0; i.e. we write the binding energy as $-200(\rho/\rho_0)$ MeV and, allowing for the K^- rest mass, the energy of a kaon in nuclear matter would be

$$\varepsilon_{K^-} = [494 - 200(\rho/\rho_0)]. \tag{2.172}$$

At high enough densities, particles in the core of the star can undergo reactions through the weak interaction provided that they conserve energy. We consider $e \rightarrow K^- + \upsilon$ and for densities between $(2\text{–}4)\rho_0$, Brown estimates the electron chemical potential μ_e lies in the range $200\,\text{MeV} < \mu_e < 300\,\text{MeV}$. Thus at some critical density, ρ_c, where

$$\varepsilon_{K^-}(\rho_c) = \mu_e, \tag{2.173}$$

kaons will spontaneously enter that region of the star. Being bosons, which can accumulate in a zero momentum state, their density for $\rho > \rho_c$ would involve interaction effects or concentration shifts of the fermion constituents (e, n, p).

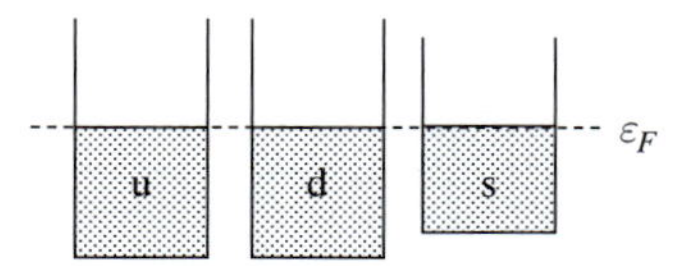

Fig. 2.53 Equilibrated Fermi levels of a deconfined quark liquid of u, d, and s quarks. The s quark well is slight shallower since its mass is larger.

2.7.5 Superfluidity in a quark liquid

The possibility that the inner core of a neutron star consists of a deconfined quark liquid (see Fig. 2.53) begs the question of whether it is a superfluid and, if so, what state (or states) we might encounter. This topic is discussed in Chapter 10 of the present volume and by Schwenk in the second volume, and here we restrict ourselves to a few qualitative remarks on some characteristics of proposed states. The topic is of fundamental interest and refines the above question to read: what is the ultimate ground state of matter (short of gravitational collapse) and is it a superfluid?

In the conventional BCS superfluids that we have discussed up to this point (those with Cooper pairs of electrons, ^{3}He atoms, neutrons, or protons) we had order parameters of the form $\Psi_{\alpha\beta}(k) = \langle \hat{c}_{\mathbf{k}\alpha}\hat{c}_{-\mathbf{k}\beta}\rangle$ (see Eq. 2.172). But quarks have flavor *and* color degrees of freedom. Assuming that quarks and gluons are the dominant degrees of freedom in the problem, pairing will involve gluon exchange; this allows quark pairs to *interchange color* by exchanging the appropriate gluon. Through weak interactions they can also interchange flavor. Thus we should anticipate pairings involving *different flavors*. The most general kind of pairing amplitudes then have the form $\left\langle \hat{q}_a^\alpha\, \hat{q}_b^\beta \right\rangle$, where $\hat{q}$ is a quark operator and the Roman and Greek indices refer to the three flavors and colors respectively; note, we have suppressed the spin and momentum quantum numbers for simplicity.

We can write the overall wave function in the form,

$$\Psi = \psi_{\text{orbital}}\psi_{\text{spin}}\psi_{\text{color}}\psi_{\text{flavor}}, \tag{2.174}$$

and as discussed in Section 2.7.2 it must be antisymmetric. The simplest form for the combined orbital and spin states is a symmetric $^1\text{S}_0$ state; if the color and flavor pair states are both antisymmetric overall, antisymmetry of the pair amplitude will be satisfied. A problem that immediately arises is that, unlike baryons $(n, p, \ldots)$ pairs of quarks cannot be color singlets. Therefore two-flavor Cooper pairs formed from massless u and d quarks must form in some other representation of the color group.

As the baryon density is increased, and for sufficiently large m_s, a first-order transition to a superfluid quark phase occurs, which is designated 2SC(u,d) [150] and consists of $\langle \hat{u}\hat{d}\rangle - \langle \hat{d}\hat{u}\rangle$ pairs; this particular combination can simultaneously be both a color and flavor singlet (with the charge imbalance taken up by electrons). In addition five of the gluons acquire mass; a mass for a gluon is the analog of the Meissner effect in that it sets a length scale for the penetration of the gluon field into the superfluid.

It turns out that Cooper pairs combining three flavors of massless quarks cannot be flavor singlets, and both color and flavor symmetries are necessarily broken. One of the proposed phases involves a superposition of three u, d, and s combinations with equal amplitudes,

$$\langle \hat{u}_R\hat{d}_G\rangle = \langle \hat{d}_G\hat{s}_B\rangle = \langle \hat{s}_B\hat{u}_R\rangle\,, \tag{2.175}$$

where R, G, and B denote the quark colors; note we have assigned a particular color to each flavor and the resulting superfluid state is therefore called the *color–flavor locked* phase. Since it involves equal amounts of the three flavors, the total charge sums to zero ($u = s = -1/3$; $d = +2/3$) the resulting state turns out to an insulating superfluid. Here all eight gluons acquire a mass. A $T_c \approx 10^{12}$ K has been estimated for such stars [151]. Figure 2.54 shows a proposed phase diagram for quark matter, as presented by Alford, Schmitt, Rajagopal, and Schafer [152], as discussed later in this book.

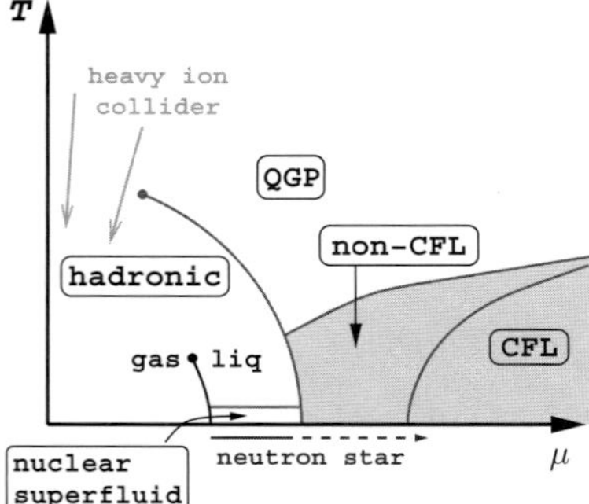

Fig. 2.54 A schematic phase diagram of quark matter as a function of the temperature (T) and chemical potential (μ). The highest temperatures correspond to the quark–gluon plasma (QGP). Lower T and μ yield hydronic matter. With increasing chemical potential non-color–flavor-locked (non-CFL) and color–flavor-locked (CFL) phases are proposed. (After Alford et al. (2008).)

References

[1] Klaers, J. Schmitt, J. Vewinger, F. and Weitz, M. Bose–Einstein condensation of photons in an optical microcavity, *Nature* **468**, 45 (2010).

[2] Klaers, J. Vewinger, F. and Weitz, M. Thermalization of a two-dimensional photonic gas in a 'white wall' photon box, *Nature Physics* **6**, 512 (2010).

[3] Saleh, B. E. A. and Teich, M. C. *Fundamentals of Photonics*, Wiley Interscience, New York, Section 3.3, Pgs. 100–109 (1991).

[4] Hohenberg, P. C. *Phys. Rev.* **158**, 383 (1967).

[5] Bagnato, V. and Kleppner, D. Bose–Einstein condensation in low-dimensional traps, *Phys. Rev.* A**44**, 7439 (1991).

[6] Mullin, W. J. Bose–Einstein condensation in a harmonic potential, *J. of Low Temp. Physics*, **106**, 615 (1997).

[7] Wannier, G. The structure of electronic excitation levels in insulating crystals, *Phys. Rev.* **52**, 191 (1937).

[8] Frenkel, I. J. On the Transformation of light into heat in solids. *Phys. Rev.* **37**, 17 (1931).

[9] Moskalenko, S. and Snoke, D. *Bose–Einstein Condensation of Excitons and Biexcitons*, Cambridge University Press (2000).

[10] Apfel, J. H. and Hadey, L. N. Exciton absorption in cuprous oxide, *Phys. Rev.* **100**, 1689 (1955).

[11] Baumeister, P. W. Optical absorption of cuprous oxide, *Phys. Rev.* **121**, 359 (1961).

[12] Ito, T. Yamaguchi, H. Okabes, K. and Masumi, T. Single-crystal growth and characterization of Cu_2O and CuO, *J. Mat. Sci.* **33**, 3555 (1998).

[13] Agranovich, V. M. and Ginsburg, V. *Crystal Optics with Spatial Dispersion and Excitons,* Springer-Verlag Series in Solid State Physics, Vol. 42, Heidelerg (1942).

[14] Frohlich, D. Kulik, A. Uebbing, B. Mysyrowicz, A. Langer, V. Stolz, H. and von der Osten, W. Coherent propagation of quantum beats of qudrupole polaritons in Cu_2O, *Phys. Rev. Lett.* **67**, 2341 (1991).

[15] Wolfe, J. P. Lin, J. L. and Snoke, D. W. in *Bose–Einstein Condensation*, edited by Griffin, A. Snoke, D. W. and Stringari, S. Cambridge University Press, Cambridge, England, 1995.

[16] Snoke, D. Wolfe, J. and Mysyrowicz, A. Evidence for Bose–Einstein condensation of a two-component exciton gas, *Phys. Rev. Lett.* **59**, 827 (1987); *Phys. Rev.* B **41**, 11171 (1990).

[17] O'Hara, K. E., Suilleabhain, L. O. and Wolfe, J. Strong nonradiative recombination of excitons in Cu_2O and its impact on Bose–Einstein statistics *Phys. Rev.* B **60**, 10565 (1999).

[18] Warren, J. T. O'Hara, K. and Wolfe, J. Two-body decay of thermalized excitons in Cu_2O, *Phys. Rev.* B **61**, 8215 (2000).

[19] Jang, J. I. and Wolfe, J. P. Biexcitons in the semiconductor Cu_2O: An explanation of the rapid decay of excitons, *Phys. Rev.* B **72**, 241201 (2005).

[20] Jang, J. I. and Wolfe, J. P. Auger recombination and biexcitons in Cu_2O: A case for dark excitonic matter, *Phys. Rev.* B **74**, 045211 (2006).

[21] Trauernicht, P. Mysyrowicz, A. and Wolfe, J. P. Thermodynamics of strain-confined paraexcitons in Cu_2O, *Phys. Rev.* B **34**, 2561 (1986).

[22] Hertz, H. Math (Crelle's) J. **92**, 156 (1881).

[23] Landau, L. D. and Lifshitz, E. M. *Theory of Elasticity*, 3rd Edition, Sec. 9, Pergamon Press, Oxford, 1986.

[24] Sun, Y. Wong, G. K. L. and Ketterson, J. B. Production of 1s quadrupole-orthoexciton polaritons in Cu_2O by two-photon pumping, *Phys. Rev.* B **63**, 5323 (2001).

[25] Elliot, R. J. Symmetry of Excitons in Cu_2O, *Phys. Rev.* **124**, 340 (1961).

[26] Inoue, M. and Toyozawa, Y. 2-Photon absorption and energy band structure, *J. Phys. Soc. Jpn.* **20**, 363 (1965).

[27] Naka, N. and Nagasawa, N. Bosonic stimulation of cold 1s excitons into a harmonic potential minimum in Cu_2O, Solid State Communications **126**, 523 (2003).

[28] Liu, Y. and Snoke, D. Resonant two-photon excitation of 1s paraexcitons in cuprous oxide, *Solid State Communications* **134**, 159 (2005).

[29] Beloussov, I. V. Ketterson, J. B. and Sun, Y. Four-wave-mixing theory for two-photon generation of excitons in Cu_2O, *Phys. Rev.* B **80**, 245213 (2009).

[30] Beloussov, I. V. Ketterson, J. B. and Sun, Y. Four-wave mixing theory for two-photon generation of excitons in thin films of Cu_2O, *Phys. Rev.* B **81**, 205208 (2010).

[31] Jang, J. I. and Ketterson, J. B. Suppression of molecule formation for orthoexciton–polaritons in Cu_2O, *Solid State Communications* **146**, 128 (2008).

[32] Yoshioka, K. Ideguchi, T. Mysyrowicz, A. and Kuwata-Gonokami, M. Quantum inelastic collisions between paraexcitons in Cu_2O, *Phys. Rev.* B **82**, 041201(R) (2010).

[33] Yoshioka, K. Chae, E. and Kuwata-Gonokami, M. Transition to a Bose–Einstein condensate and relaxation explosion of excitons at sub-Kelvin temperatures, *Nature Comm.*, **328**, 1 (2011).

[34] Weisbuch, C. Nishioka, M. Ishikawa, A. and Arakawa, Y. Observation of the coupled exciton-photon mode splitting in a semiconductor quantum microcavity, *Phys. Rev. Lett.* **69**, 3314 (1992).

[35] Kavokin, A. Exciton–polaritons in microcavities: present and future, *Appl. Phys.* A **89**, 241 (2007).

[36] Deng, H. Weihs, G. Santori, C. Bloch, J. and Yamamoto, Y. Condensation of semiconductor microcavity exciton polaritons, *Science* **298**, 199 (2002).

[37] Kasprzak, J. Richard, M. Kundermann, S. Baas, A. Jeambrun, P. Keeling, J. M. J. Marchetti, F. M. Szymanska, M. H. Andre, R. Staehli, J. L. Savona, V. Littlewood, P. B. Deveaud, B. and Dang, L. S. Bose–Einstein condensation of exciton polaritons, *Nature* **443**, 409 |(2006).

[38] Balili, R. Hartwell, V. Snoke, D. Pfeiffer, L. and West, K. Bose–Einstein condensation of microcavity polaritons in a trap, *Science* **316**, 1007 (2007).

[39] Deng, H. Haug, H. and Yamamoto, Y. Exciton–polariton Bose–Einstein condensation, *Rev. of Mod. Phys.*, **82**, (2010).

[40] Demokritov, S. O. Demidov, V. E. Dzyapko, O. Melkov, G. A. Serga, A. A. Hillebrands, B. and Slavin, A. N. Bose–Einstein condensation of quasi-equilibrium magnons at room temperature under pumping, *Nature Letters* **443**, 430 (2006).

[41] Demidov, V. E. Dzyapko, O. Demokritov, S. O. Melkov, G. A. and Slavin, A. N. Thermalization of a Parametrically Driven Magnon Gas Leading to Bose–Einstein Condensation, *Phys. Rev. Lett.* **99**, 037205 (2007).

[42] Dzyapko, O. Demidov, V. E. Melkov, G. A. and Demokritov, S. O. Bose–Einstein condensation of spin wave quanta at room temperature, Phil. Trans. Royal Society A: Mathematical Physical and Engineering Sciences **369**, 3575 (2011).

[43] Deman, R. W. and Eshbach, J. R. Magnetostatic modes in a ferromagnetic slab, *J. Chem. Phys. Sol.* **19**, 308 (1961).

[44] Fletcher, P. C. and Kittel, C. Considerations on the propagation of magnetostatic waves and spin waves, *Phys. Rev.* **120**, 2004 (1960).

[45] Cottam, M. G. and Lockwood, D. J. *Light Scattering in Magnetic Solids*, Ch. 8, Wiley (1986).

[46] Sparks, M. Ferromagnetic resonance in thin films I: Normal mode frequencies, *Phys. Rev.* B **1** 3831 (1970).

[47] Wolfram, T. and De Wames, R. E. Magnetoexchange branches and spin-wave resonance in conducting and insulting films: perpendicular resonance, *Phys. Rev.* B **4**, 3125 (1971).

[48] Akhiezer, A. I. Bar'Yakhtar, V. G. and Peletimnskii, S. *Spin Waves*, North-Holland, Amsterdam (1968).

[49] Kalinikos, B. A. and Slavin, A. N. Theory of dipole-exchange spin wave spectrum for ferromagnetic films with mixed exchange boundary conditions, *J. Phys. C: Solid State Phys.* **19** (1986).

[50] Kreisel, A. Sauli, F. Bartosch, L. and Kopietz, P. Microscopic spin-wave theory for yttrium-iron garnet films, *Eur. Phys. J.* **B71**, 59 (2009).

[51] Demidov, V. E. Dzyapko, O. Demokritov, S. O. Melkov, G. A. and Slavin, A. N. Observation of spontaneous coherence in Bose–Einstein condensate of magnons, *Phys. Rev. Lett.* **100**, 047205 (2008).

[52] Chumak, A. V. Melkov, G. A. Demidov, V. E. Dzyapko, O. Safonov, V. L. and Demokritov, S. O. Bose–Einstein Condensation of Magnons under Incoherent Pumping, *Phys. Rev. Lett.* **102**, 187205 (2009).

[53] Dzyapko, O. Demidov, V. E. Buchmeier, M. Stockhoff, T. Schmitz, G. Melkov, G. A. and Demokritov, S. O. Excitation of two spatially separated Bose–Einstein condensates of magnons, *Phys. Rev.* B **80**, 060401(R) (2009).

[54] Anderson, M. H. Ensher, J. R. Matthews, M. R. Wieman, C. E. and Cornell, E. A. Observation of Bose–Einstein condensation in a dilute atomic vapor, *Science* **269**, 198 (1995).

[55] Davis, K. B. Mewes, M. O. Andrews, M. R. van Druten, N. J. Durfee, D. S. Kurn, D. M. and Ketterle, W. Bose–Einstein condensation in a gas of sodium atoms, *Phys. Rev. Lett.* **75**, 3969 (1995).

[56] Metcaff, H. J. and van der Straten, P. *Laser Cooling and Trapping*, Springer-Verlag, New York (1999).

[57] Pethick, C. J. and Smith, H. *Bose–Einstein Condensation in Dilute Gases*, Cambridge University Press, Cambridge, 2001.

[58] Pitaevskii, L. P. and Stringari, S. *Bose–Einstein Condensation*, Clarendon Press, Oxford, 2003.

[59] Duine, R. A. and Stoof, H. T. C. *Atom–molecule coherence in Bose gases*, Physics Reports **396**, 115 (2004).

[60] Ashkin, A. Acceleration and trapping of particles by radiation pressure, *Phys. Rev. Lett.* **24**, 156 (1970).

[61] Ashkin, A. and Dziedzic, J. M. Optical trapping and manipulation of viruses and bacteria, *Science* **235**, 1517 (1987).

[62] Jaksch, D. Bruder, C. Cirac, J. I. Gardiner, C. W. and Zoller, P. Cold bosonic atoms in optical lattices, *Phys. Rev. Lett.* **81**, 3108 (1998).

[63] Fisher, M. P. A. Weichman, P. B. Grinstein, G. and Fisher, D. S. Boson localization and the superfluid – insulator transition, *Phys. Rev.* B **40**, 546 (1989).

[64] Sheshadri, K. Krishnamurthy, H. R. Pandit, R. and Ramakrishnan, T. V. Superfluid and insulating phases in an interacting-boson model: mean-field theory and the RPA, *Europhys. Lett.* **22**, 257 (1993).

[65] Freericks, J. K. and Monien, H. Phase diagram of the Bose Hubbard model, *Europhys. Lett.* **26**, 545 (1995).

[66] van Oosten, D. van der Straten, P. and Stoof, H. T. C. Quantum phases in an optical lattice, *Phys. Rev.* A**63**, 053601 (2001).

[67] Hänsch, T. and Schawlow, A. Cooling of gases by laser radiation, *Opt. Comm.* **13**, 68 (1975).

[68] Wineland, D. and Dehmelt, H. Proposed 1014 delta upsilon less than upsilon laser fluorescence spectroscopy on $T1^+$ mono-ion oscillator III, *Bull. Am. Phys. Soc.* **20**, 637, (1975).

[69] Ketterle, W. Davis, K. B. Joff'e, M. A. Martin, A. and Pritchard, D. E. High densities of cold atoms in a dark spontaneous-force optical trap, *Phys. Rev. Lett.* **70**, 2253 (1993)

[70] Dalibard, J. and Cohen-Tannoudji, C. Laser cooling below the Doppler limit by polarization gradients: simple theoretical models, *J. Optical Society of America*, B **6**, 2023 (1989).

[71] Solomon, C. Dalibar, J. Phillips, W. D. Clairon, A. and Guellanti, S. Laser cooling of cesium atoms below 3 microKelvin, *Europhys. Lett.* **12**, 683 (1990).

[72] Wing, W. H. On neutral particle trapping in quasistatic electromagnetic-fields, *Prog. in Quantum Electronics* **8**, 181 (1984).

[73] Gott, Y. V. Ioffe, M. S. and Tel'kovskii, V. G. *Nuc. Fusion*, 1962 Suppl. Part 3, Pg. 1045 (1962).

[74] Pritchard, D. E. Cooling neutral atoms in a magnetic trap for precision spectroscopy, *Phys. Rev. Lett.* **51**, 1336 (1983).

[75] Stamper-Kurn, D. M. Andrews, M. R. Chikkatur, A. P. Inouye, S. Miesner, H.-J. Stenger, J. and Ketterle, W. Optical confinement of a Bose–Einstein condensate, *Phys. Rev. Lett.* **80**, 2027 (1998).

[76] Petrich, W. Anderson, M. H. Ensher, J. R. and Cornell, E. A. Stable, tightly confining magnetic trap for evaporative cooling of neutral atoms, *Phys. Rev. Lett.* **74**, 3352 (1995).

[77] Barrett, M. D. Sauer, J. A. and Chapman, M. S. All-optical formation of an atomic Bose–Einstein condensate, *Phys. Rev. Lett.* **87**, 010404 (2001).

[78] Takasu, Y. Honda, K. Komori, K. Kuwamoto, T. Kumakura, M. Takahashi, Y. and Yabuzaki, T. High-density trapping of cold ytterbium atoms by an optical dipole force, *Phys. Rev. Lett.* **90** 023003 (2003).

[79] Rychtarik, D. Engeser, B. Nägerl, H.-C. and Grimm, R. Two-dimensional Bose–Einstein condensate in an optical surface trap, *Phys. Rev. Lett.* **92**, 173003 (2004).

[80] C. Chin, private communication.

[81] Inouye, S. Andrews, M. R. Stenger, J. Miesner, H.-J. Stamper-Kurn, D. M. and Ketterle, W. Observation of Feshbach resonances in a Bose–Einstein condensate, *Nature* **392**, 151 (1998).

[82] Roberts, J. L. Claussen, N. R. Burke, Jr., J. P. Greene, C. H. Cornell, E. A. and Wieman, C. E. Resonant magnetic field control of elastic scattering in cold ^{85}Rb, *Phys. Rev. Lett.* **81**, 5109 (1998).

[83] Roberts, J. L. Claussen, N. R. Cornish, S. L. Donley, E. A. Cornell, E. A. and Wieman, C. E. Controlled collapse of a Bose–Einstein condensate, *Phys. Rev. Lett.* **86**, 4211 (2001).

[84] Donley, E. A. Claussen, N. R. Cornish, S. L. Roberts, J. L. Cornell, E. A. and Wieman, C. E. Dynamics of collapsing and exploding Bose–Einstein condensates, *Nature* **412**, 295 (2001).

[85] Ruprecht, P. A. Holland, M. J. and Burnett, K. Time-dependent solution of the nonlinear Schrödinger equation for Bose-condensed trapped neutral atoms, *Phys. Rev.* A **51**, 4704 (1995).

[86] Anderson, P. W. Considerations on the flow of superfluid helium, *Rev. Mod. Phys.* **38**, 298 (1966).

[87] Andrews, M. R. Townsend, C. G. Miesner, H.-J. Durfee, D. S. Kurn, D. M. and Ketterle, W. Observation of interference between two bose condensates, *Science*, **275**, 637 (1997).

[88] Pethick, C. J. and Smith, H. *Bose–Einstein Condensation in Dilute Gases*, Cambridge Univ. Press, 2002, Chapter13.

[89] Raman, C. Köhl, M. Onofrio, R. Durfee, D. S. Kuklewicz, C. E. Hadzibabic, Z. and Ketterle, W. Evidence for a critical velocity in a Bose–Einstein condensed gas, *Phys. Rev. Lett.* **83**, 2502 (2002).

[90] Andrews, M. R. Kurn, D. M. Miesner, H.-J. Durfee, D. S. Townsend, C. G. Inouye, S. and Ketterle, W. Propagation of sound in a Bose–Einstein condensate, *Phys. Rev. Lett.* **79**, 553 (1997).

[91] DeMarco, B. and Jin, D. S. Onset of Fermi degeneracy in a trapped atomic gas, *Science* **285**, 1703 (1999).

[92] Quantum behavior of an atomic Fermi gas, Brian DeMarco, Thesis, Department of Physics, University of Colorado, 2001.

[93] Truscott, A. G. Strecker, K. E. McAlexander, W. I. Partridge, G. B. and Hulet, R. G. Observation of Fermi pressure in a gas of trapped atoms, *Science* **291**, 2570 (2001).

[94] Regal, C. A. Greiner, M. and Jin, D. S. Lifetime of molecule–atom mixtures near a Feshbach resonance in ^{40}K, *Phys. Rev. Lett.* **92**, 083201 (2004).

[95] Regal, C. A. Ticknor, C. Bohn, J. L. and Jin, D. S. Creation of ultracold molecules from a Fermi gas of atoms, *Nature* **427**, 47 (2003).

[96] Strecker, K. E. Partridge, G. B. and Hulet, R. G. Conversion of an atomic Fermi gas to a long-lived molecular Bose gas, *Phys. Rev. Lett.* **91**, 080406 (2003).

[97] Cubizolles, J. Bourdel, T. Kokkelmans, S. J. J. M. F. Shlyapnikov, G. V. and Salomon, C. Production of long-lived ultracold Li2 molecules from a Fermi gas, *Phys. Rev. Lett.*, **91**, 240401 (2003).

[98] Greiner, M. Regal, C. A. and Jin, D. Emergence of a molecular Bose–Einstein condensate from a Fermi gas, *Nature* **426**, 537 (2003).

[99] Jochim, S. Bartenstein, M. Altmeyer, A. Hendl, G. Riedl, S. Chin, C. Denschlag, J. H. and Grimm, R. Bose–Einstein condensation of molecules, *Science* **302**, 2101 (2003).

[100] Zwierlein, M. W. Stan, C. A. Schunck, C. H. Raupach, S. M. F. Gupta, S. Hadzibabic, Z. and Ketterle, W. Observation of Bose–Einstein condensation of molecules, *Phys. Rev. Lett.* **91**, 250401 (2003).

[101] Regal, C. A. Greiner, M. and Jin, D. S. Observation of resonance condensation of fermionic Atom Pairs, *Phys. Rev. Lett.* **92**, 040403 (2004).

[102] A. J. Leggett, in *Modern Trends in the Theory of Condensed Matter*, edited by Pekalski, A. and Przystawa, R. (Springer-Verlag, Berlin, 1980); J. Phys. (Paris), Colloq. **41**, C7-19 (1980).

[103] Heiselberg, H. Pethick, C. J. Smith, H. and Viverit, L. Influence of induced interactions on the superfluid transition in dilute Fermi gases, *Phys. Rev. Lett.* **85**, 2418 (2000).

[104] Bourdel, T. Khaykovich, L. Cubizolles, J. Zhang, J. Chevy, F. Teichmann, M. Tarruell, L. Kokkelmans, S. J. J. M. F. and Salomon, C. Experimental study of the BEC–BCS crossover region in lithium 6, *Phys. Rev. Lett.* **93**, 050401 (2004).

[105] Partridge, G. B. Strecker, K. E. Kamar, R. I. Jack, M. W. and Hulet, R. G. Molecular probe of pairing in the BEC–BCS crossover, *Phys. Rev. Lett.* **95**, 020404 (2005).

[106] Holland, M. Kokkelmans, S. J. J. M. F. Chiofalo, M. L. and Walser, R. Resonance superfluidity in a quantum degenerate Fermi gas, *Phys Rev. Lett.* **87**, 120406 (2001).

[107] Timmermans, E. Furuya, K. Milonni, P. W. and Kerman, A. K. Prospect of creating a composite Fermi–Bose superfluid, *Phys. Lett.* A **285**, 228 (2001).

[108] Zwierlein, M. W. Abo-Shaeer, J. R. Schirotzek, A. Schunck, C. H. and Ketterle, W. Vortices and superfluidity in a strongly interacting Fermi gas, *Nature* **435**, 1047 (2005).

[109] Clogston, A. M. Upper limit for the critical field in hard superconductors, *Phys. Rev. Lett.* **9**, 266 (1962).

[110] Fulde, P. and Ferrell, R. A. Superconductivity in a strong spin-exchange Field, *Phys. Rev.* **135**, A550 (1964).

[111] Larkin, A. I. and Ovchinnikov, Yu. N. Inhomogeneous state of superconductors, *Sov. Phys. JETP* **20**, 762, (1965).

[112] Bedaque, P. F. Caldas, H. and Rupak, G. Phase separation in asymmetrical fermion superfluids, *Phys. Rev. Lett.* **91** 247002 (2003).

[113] Partridge, G. B. Li, W. Kamar, R. I. Liao, Y. and Hulet, R. G. Pairing and phase separation in a polarized Fermi gas, *Science* **311**, 503 (2006).

[114] Shin, Y. Zwierlein, M. W. Schunck, C. H. Schirotzek, A. and Ketterle, W. Observation of phase separation in a strongly interacting imbalanced Fermi gas, *Phys. Rev. Lett.* **97**, 030401 (2006).

[115] Zwierlein, M. W. Schirotzek, A. Schunck, C. H. and Ketterle, W. Fermionic superfluidity with imbalanced spin populations, *Science* **311**, 492 (2006).

[116] Schunck, C. H. Shin, Y. Schirotzek, A. Zwierlein, M. W. and Ketterle, W. Pairing without superfluidity:the ground state of an imbalanced Fermi mixture, *Science* **316**, 867 (2007).

[117] Shin, Y. Schunck, C. H. Schirotzek, A. and Ketterle, W. Phase diagram of a two-component Fermi gas with resonant interactions, *Nature* **451**, 689 (2008).

[118] Joseph, J. Clancy, B. Luo, L. Kinast, J. Turlapov, A. and Thomas, J. E. Measurement of sound velocity in a Fermi gas near a Feshbach resonance, *Phys. Rev. Lett.* **98**, 170401 (2007).

[119] Greiner, M. Mandel, O. Esslinger, T. Hänsch, T. W. and Bloch, I. Quantum phase transition from a super fluid to a Mott insulator in a gas of ultracold atoms, *Nature* **415**, 39 (2002).

[120] Lattimer, J. M. and Prakash, M. The Physics of neutron stars, *Science* **D 304**, 536 (2004).

[121] Baym, G. and Pethick, C. The physics of neutron stars, *Ann. Rev. Astron. Astrophys.* **17**, 415 (1979).

[122] Heiselberg, H. and Pandharipande, V. Recent progress in neutron star theory, *Annu. Rev. Nucl. Part. Sci.* **50**, 481 (2000).

[123] Blaschke, D. Glendenning, N. K. and Sedrakian, A. *Physics of Neutron Star Interiors*, Edited by Springer Verlag, Berlin (2001).

[124] Haensel, P. Haensel, Pawe Potekhin, A. Y. *Neutron stars: Equation of State and Structure*, Springer-Verlag, Berlin (2007).

[125] Witten, E. Cosmic separation of phases, *Phys. Rev.* **30**, 272 (1984).

[126] Alcock, C. and Olinto, A. Exotic phases of hadronic matter and their astrophysical application, *Rev. Nucl. Part. Sci.* **38**, 161 (1988).

[127] Page, D. and Usov, V. V. Thermal Evolution and Light Curves of Young Bare Strange Stars, *Phys. Rev. Lett.* 131101, **89** (2002).

[128] Sauls, J. A. *Superfluidity in the Interiors of Neutron Stars*, NATO ASI Series C, **262**, 457, Kluwer Academic Press, 1989.

[129] Bohr, A. Mottelson, B. R. and Pines, D. Possible Analogy between the excitarion spectra of nuclei and those of the superconducting metallic state, *Phys. Rev.* **110**, 936 (1958).

[130] Migdal, A. B. Superfluidity and moments of inertia of nuclei, *Nuclear Physics* **13**, 655 (1959).

[131] Tabakin, F. An effective interaction for nuclear Hartree–Fock calculations, *Ann. Phys.* **30**, 51 (1964).

[132] Hoffberg, M. Glassgold, A. E. Richardson, R. W. and Ruderman, M. Anisotropic superfluidity in neutron star matter, *Phys. Rev. Lett.* **24**, 777 (1970).

[133] Fabrocini, A. Fantoni, S. Illarionov, A. Yu. and Schmidt, K. E. $^1S_0$ superfluid phase transition in neutron matter with realistic nuclear potentials and modern many-body theories, *Phys. Rev. Lett.* **95**, 192501 (2005).

[134] Sauls, J. A. and Serene, J. W. P_2 pairing near the transition temperature in neutron-star matter, *Phys. Rev.* **D 17**, 1524 (1978).

[135] Muzikar, P. Sauls, J. A. and Serene, J. W. $^3P_2$ pairing in neutron-star matter: magnetic field effects and vortices, *Phys. Rev.* **21**, 1494 (1980).

[136] Radhakrishnan, V. and Manchester, R. N. Detection of a change of state in the pulsar PSR 0833-45, *Nature* **222**, 228 (1969).

[137] Reichley, P. E. and Downs, G. S. Observed decrease in the periods of pulsar PSR 0833–45P, *Nature* **222**, 229 (1969).

[138] Anderson, P. W. and Ito, N. Pulsar glitches and restlessness as a hard superfluidity phenomenon, *Nature* **256**, 25 (1975).

[139] Heinke, C. O. and Ho, W. C. G. Direct observation of the cooling of the Cassiopeia A neutron star, *The Astrophysical Journal Letters*, **719** L167 (2010).

[140] Page, D. Prakash, M. Lattimer, J. M. and Steiner, A. W. Rapid cooling of the neutron star in Cassiopeia a triggered by neutron superfluidity in dense matter, *Phys. Rev. Lett* **106**, 081101 (2011).

[141] Yakovlev, D. G. Ho, W. C. G. Shternin, P. S. Heinke, C. O. and Potekhin, A. Y. Cooling rates of neutron stars and the young neutron star in the Cassiopeia A supernova remnant, *Monthly Notices of the Royal Astronomical Society* **411**, 1977 (2011).

[142] Nadyozhin, D. K. Gamow and the physics and evolution of stars, *Space Science Reviews* **74**, 455 (1995).

[143] Michele, V. Prospects for gravitational-wave observations of neutron-star tidal disruption in neutron-star-black-hole binaries, *Phys. Rev. Lett.* **84**, 3519 (2000).

[144] Faber, J. A. Grandclément, P. Rasio, F. A. and Taniguchi, K. Measuring neutron-star radii with gravitational-wave detectors, *Phys. Rev. Lett.* **8**, 231102 (2002).

[145] Hinderer, T. Lackey, B. D. Lang, R. N. and Read, S. J. Tidal deformability of neutron stars with realistic equations of state and their gravitational wave signatures in binary inspiral, *Phys. Rev.* D **8**, 123016 (2010).

[146] Bauswein, A. and Janka, H.-T. Measuring neutron-star properties via gravitational waves from neutron-star mergers, *Phys. Rev. Lett.* **108**, 011101 (2012).

[147] Brown, G. E. Kaon condensation in dense matter, in *Bose-Einstein Condensation*, Edited by Griffin, A. Snoke, D. W. and Stringari, S. Cambridge University Press. Cambridge (1995) Pg. 438.

[148] Thorsson, V. Prakash, M. and Lattimer, J. M. Composition, structure and evolution of neutron stars with kaon condensates, *Nucl. Phys.* A **572**, 693 (1994).

[149] Friedman, E. Gal, A. and Batty, C. J. Density dependence in kaonic atoms, *Phys. Lett.* B **308**, 6 (1993).

[150] Alford, M. Berges, J. and Rajagopal, K. Unlocking color and flavor in superconducting strange quark matter, *Nuclear Physics* B **558**, 219 (1999).

[151] Alford, M. Bowers, J. A. and Rajagopal, K. Color superconductivity in compact stars, *J. Phys.* G **27**, 541 (2001).

[152] Alford, M. G. Schmitt, A. Rajagopal, K. and Schäfer, T. Color superconductivity in dense quark matter, *Rev. Mod. Phys.* **80**, 1455 (2008).

3 Quantized vortices in superfluid helium and atomic Bose–Einstein condensates

Makoto Tsubota, Kenichi Kasamatsu, and Michikazu Kobayashi

This chapter reviews recent developments in the physics of quantized vortices in superfluid helium and atomic Bose–Einstein condensates. Quantized vortices appear in low-temperature quantum condensed systems as the direct product of Bose–Einstein condensation. Quantized vortices were first discovered in superfluid ^{4}He in the 1950s, and have since been studied with a primary focus on the quantum hydrodynamics of this system. Since the discovery of superfluid ^{3}He in 1972, quantized vortices characteristic of an anisotropic superfluid have been studied theoretically and observed experimentally using rotating cryostats. The realization of atomic Bose–Einstein condensation in 1995 has opened up new possibilities, because it then became possible to control and to directly visualize condensates and quantized vortices. Historically, many ideas developed in superfluid ^{4}He and ^{3}He have been imported to the field of cold atoms and utilized effectively. Here, we review and summarize our current understanding of quantized vortices, bridging superfluid helium and atomic Bose–Einstein condensates. This review begins with a basic introduction, which is followed by discussion of modern topics such as quantum turbulence and vortices in unusual cold atom condensates.

This field is currently very active. However, many unresolved problems remain. It is our hope that this chapter will attract the interest of scientists, not only those working in this field but also in other fields, especially young scientists, and that it will contribute to further development and progress in physics.

3.1 Introduction

Bose–Einstein condensation is often considered to be a macroscopic quantum phenomenon. This is because bosons occupy the same single-particle ground state below a critical temperature through Bose–Einstein condensation to form a macroscopic wave function (order parameter) extending over the entire system. As a direct result of the formation of a macroscopic wave function, quantized vortices appear in the Bose-condensed system. A quantized vortex is a vortex of inviscid superflow, and any rotational motion of a superfluid is sustained by quantized vortices. A quantized vortex is a stable and well-defined topological defect, very different from ordinary vortices in a conventional fluid. Hydrodynamics dominated by quantized vortices is called *quantum hydrodynamics*, and turbulence comprised of quantized vortices is known as *quantum turbulence* (QT). Studies of quantized vortices originally began in the 1950s using superfluid ^{4}He, and much theoretical, numerical, and experimental effort has been devoted to the field. Superfluid ^{3}He, discovered in 1972, presented a system with a variety of quantized vortices characteristic of p-wave superfluids. However, quantum hydrodynamics has become very important again in recent years, for two reasons. The first reason was the appearance of new research activity on QT [1, 2]. Before 1990, most studies of QT were limited to thermal counterflow in ^{4}He in which the normal fluid and the superfluid flow oppositely through the injection of heat current. Since counterflow turbulence has no classical analog, understanding the relationship between QT and traditional classical turbulence (CT) has been deserted. However, some experiments without counterflow appeared in the 1990s, demonstrating the Kolmogorov law of energy spectra, which is one of the most important statistical laws in turbulence, thus opening a new stage of QT research. The second reason is that Bose–Einstein condensation of cold atoms was realized in 1995. This system enables the control and direct visualization of condensates, which was impossible in other quantum condensed systems, such as superfluid helium and superconductors. The present article will review these recent studies on quantized vortices in both superfluid helium and cold atoms.

The contents of this article are as follows. Section 3.2 deals with quantized vortices in superfluid helium. A brief review of the research history and the dynamics of quantized vortices is followed by an important modern topic, QT, which includes the energy spectra issue, dissipation at very low temperatures, QT created by vibrating structures, and the visualization of quantized vortices and turbulence. In Section 3.3, which begins with the basics on atomic Bose–Einstein condensates (BECs), we review the topics of vortices in single-component, two-component, spinor, and dipolar BECs, and then discuss quantum turbulence. Section 3.4 is devoted to a summary and conclusions.

3.2 Quantized vortices in superfluid helium

Quantum hydrodynamics and quantized vortices have long been studied in superfluid ^{4}He. This section begins by describing the research history and some basic dynamics of quantized vortices, and then discusses the modern topic of QT.

3.2.1 Research history

Liquid ^{4}He enters a superfluid state below the λ point ($T_\lambda = 2.17\,\mathrm{K}$) with Bose–Einstein condensation of the ^{4}He atoms. The characteristic phenomena of superfluidity were discovered experimentally in the 1930s by Kapitza [3] and Allen et al. [4]. The hydrodynamics of superfluid helium are well described by the two-fluid model proposed by Landau [5] and Tisza [6]. According to the two-fluid model, the system consists of an inviscid superfluid (density ρ_s) and a viscous normal fluid (density ρ_n) with two independent velocity fields $\mathbf{v}_s$ and $\mathbf{v}_n$. The mixing ratio of the two fluids depends on the temperature. As the temperature is reduced below the λ point, the ratio of the superfluid component increases, and the fluid becomes entirely superfluid below about 1 K. The two-fluid model successfully explained the phenomenon of superfluidity, while it was known in 1940s that superfluidity breaks down when it flows fast [7] and this phenomenon was not explained through the two-fluid model. This was later found to be caused by turbulence of the superfluid component due to random motion of quantized vortices.

The λ transition is closely related to the Bose–Einstein condensation of ^{4}He atoms, as first proposed by London [8]. The Bose-condensed system exhibits the macroscopic wave function $\Psi(\mathbf{x},t) = |\Psi(\mathbf{x},t)|e^{i\theta(\mathbf{x},t)}$ as an order parameter. The superfluid velocity field is given by $\mathbf{v}_s = (\hbar/m)\nabla\theta$, with boson mass m, representing the potential flow. Since the macroscopic wave function should be single-valued for the space coordinate $\mathbf{x}$, the circulation $\Gamma = \oint \mathbf{v} \cdot d\boldsymbol{\ell}$ for an arbitrary closed loop in the fluid is quantized by the quantum $\kappa = h/m$. A vortex with quantized circulation is called a quantized vortex. Any rotational motion of a superfluid is sustained only by quantized vortices.

A quantized vortex is a topological defect characteristic of a Bose–Einstein condensate, and is different from a vortex in a classical viscous fluid. First, the circulation is quantized, which is contrary to a classical vortex which can have any circulation value. Second, a quantized vortex is a vortex of inviscid superflow. Thus, it cannot decay by the viscous diffusion of vorticity that occurs in a classical fluid. Third, the core of a quantized vortex is very thin, of the order of the coherence length, which is only a few angstroms in superfluid ^{4}He. Because the vortex core is very thin and does not decay by diffusion, it is always possible to identify the position of a quantized vortex in the fluid. These properties make a quantized vortex more stable and definite than a classical vortex.

The idea of quantized circulation was first proposed by Onsager, for a series of annular rings in a rotating superfluid [9]. Feynman considered that a vortex in a superfluid can take the form of a vortex filament, with quantized circulation κ and a core of atomic dimensions [10]. Early experimental studies on superfluid hydrodynamics focused primarily on thermal counterflow. The flow is driven by an injected heat current, and the normal fluid and superfluid flow in opposite directions. The superflow was found to become dissipative when the relative velocity between the two fluids exceeds a critical value [7]. Gorter and Mellink attributed the dissipation to mutual friction between the two fluids, and considered the possibility of superfluid turbulence. Feynman proposed a turbulent superfluid state consisting of a tangle of quantized vortices [10]. Hall and Vinen performed the experiments of second sound attenuation in rotating ^{4}He, and found that mutual friction arises from interaction between the normal fluid and quantized vortices [11, 12]; second sound refers to an entropy wave in which superfluid and normal fluid oscillate oppositely, and its propagation and attenuation give information on the vortex density in the fluid. Vinen confirmed

Feynman's findings experimentally, by showing that the dissipation arises from mutual friction between vortices and the normal flow [13–16]. Vinen also succeeded in observing quantized circulation using vibrating wires in rotating superfluid ^{4}He [17]. Subsequently, many experimental studies have examined superfluid turbulence (ST) in thermal counterflow systems, and have revealed a variety of physical phenomena [18]. Since the dynamics of quantized vortices are nonlinear and non-local, it has not been easy to quantitatively understand these observations on the basis of vortex dynamics. Schwarz clarified the picture of ST based on tangled vortices by numerical simulation of the quantized vortex filament model in the thermal counterflow [19, 20]. However, since the thermal counterflow has no analogy in conventional fluid dynamics, this study was not helpful in clarifying the relationship between ST and classical turbulence (CT). Superfluid turbulence is often called quantum turbulence (QT), which emphasizes the belief that it is comprised of quantized vortices.

Turbulence has long been one of the great mysteries in nature, with discussion dating back to the era of Leonardo da Vinci. Turbulence has been intensely studied in a number of fields, but it is still far from being completely understood. This is primarily because turbulence is a complicated dynamical phenomenon with strong nonlinearity. Comparing QT and CT demonstrates the importance of studying QT. Turbulence in a classical viscous fluid appears to be comprised of vortices, as pointed out by Da Vinci. However, these vortices are unstable, and appear and disappear repeatedly. Moreover, circulation is not conserved, nor is it identical for each vortex. QT consists of a tangle of quantized vortices that have the same conserved circulation. Thus, QT may present an easier system for study than CT because the elements are more definite and clear, which is confirmed at relatively large scales by many works described later. However, the situation is not so simple. The severe constraint of quantum mechanics will be effective at small scales and leads to some definite difference from CT, and what happens there is not so trivial.

Based on these considerations, QT research has headed in a new direction since the mid 1990s. One primary interest has been to understand the relationship between QT and CT [1]. The energy spectrum of fully developed CT is known to obey the Kolmogorov law in an inertial range. The energy transfer in an inertial range is believed to be sustained by the Richardson cascade process, in which large eddies are broken up self-similarly into smaller eddies. Recent experimental and numerical studies have supported the Kolmogorov spectrum, even in QT. Another important problem is the dissipative mechanism in QT. At a finite temperature, mutual friction works as a dissipative mechanism. However, it is not so easy to understand how the energy cascade is for the full range of scales and what mechanism causes dissipation at very low temperatures, at which the normal fluid component is negligible.

3.2.2 Dynamics of quantized vortices

Quantum hydrodynamics, including QT, is reduced to the motion of quantized vortices. Hence, understanding the dynamics of quantized vortices is a key issue in quantum hydrodynamics. Two formulations are generally available for studying the dynamics of quantized vortices. One is the vortex filament model, and the other is the Gross–Pitaevskii (GP) model. We will briefly describe these two formulations.

3.2.2.1 *Vortex filament model*

As described in Section 3.2.1, a quantized vortex has quantized circulation. The vortex core is extremely thin, usually much smaller than other characteristic scales of vortex motion. These properties allow a quantized vortex to be represented as a vortex filament. In classical fluid dynamics [21], the vortex filament model is just a convenient idealization; the vorticity in a realistic classical fluid flow rarely takes the form of clearly discrete vorticity filaments. However, the vortex filament model is accurate and realistic for a quantized vortex in superfluid helium.

The vortex filament formulation represents a quantized vortex as a filament passing through the fluid, having a definite direction corresponding to its vorticity. Except for the thin core region, the superflow velocity field has a classically well-defined meaning, and can be described by ideal fluid dynamics. The velocity at a point $\mathbf{r}$ due to a filament is given by the Biot–Savart expression

$$\mathbf{v}_s(\mathbf{r}) = \frac{\kappa}{4\pi} \int_{\mathcal{L}} \frac{(\mathbf{s}_1 - \mathbf{r}) \times d\mathbf{s}_1}{|\mathbf{s}_1 - \mathbf{r}|^3}, \tag{3.1}$$

where κ is the quantum of circulation. The filament is represented by the parametric form $\mathbf{s} = \mathbf{s}(\varsigma, t)$ with the one-dimensional coordinate ς along the filament. The vector $\mathbf{s}_1$ refers to a point on the filament, and the integration is taken along the filament. Helmholtz's theorem for a perfect fluid states that the vortex moves at the superfluid velocity. Calculating the velocity $\mathbf{v}_s$ at a point $\mathbf{r} = \mathbf{s}$ on the filament causes the integral to diverge as $\mathbf{s}_1 \to \mathbf{s}$. To avoid this divergence, we separate the velocity $\dot{\mathbf{s}}$ of the filament at the point $\mathbf{s}$ into two components [19],

$$\dot{\mathbf{s}} = \frac{\kappa}{4\pi} \mathbf{s}' \times \mathbf{s}'' \ln \left(\frac{2(\ell_+ \ell_-)^{1/2}}{e^{1/4} a_0} \right) + \frac{\kappa}{4\pi} \int_{\mathcal{L}}' \frac{(\mathbf{s}_1 - \mathbf{r}) \times d\mathbf{s}_1}{|\mathbf{s}_1 - \mathbf{r}|^3}. \tag{3.2}$$

The first term is the localized induction field arising from a curved line element acting on itself, and ℓ_+ and ℓ_- are the lengths of the two adjacent line elements after discretization, separated by the point $\mathbf{s}$. The prime denotes differentiation with respect to the arc length ς. The mutually perpendicular vectors $\mathbf{s}'$, $\mathbf{s}''$, and $\mathbf{s}' \times \mathbf{s}''$ are directed along the tangent, the principal normal, and the binormal, respectively, at the point $\mathbf{s}$, and their respective magnitudes are 1, R^{-1}, and R^{-1}, where R is the local radius of curvature. The parameter a_0 is the cutoff, corresponding to the core radius. Thus, the first term represents the tendency to move the local point $\mathbf{s}$ in the binormal direction with a velocity inversely proportional to R. The second term represents the non-local field obtained by integrating the integral of Eq. (3.1) along the rest of the filament, except in the neighborhood of $\mathbf{s}$.

Neglecting the non-local terms and replacing Eq. (3.2) by $\dot{\mathbf{s}} = \beta \mathbf{s}' \times \mathbf{s}''$ is referred to as the localized induction approximation (LIA). Here, the coefficient β is defined by $\beta = (\kappa/4\pi) \ln (c\langle R \rangle / a_0)$, where c is a constant of order 1 and $(\ell_+ \ell_-)^{1/2}$ is replaced by the mean radius of curvature $\langle R \rangle$ along the length of the filament. This approximation is believed to be effective for analyzing isotropic dense tangles due to cancellations between non-local contributions. However, the ILA lacks the interaction between vortices, not suitable for the description of a realistic vortex tangle.

A better understanding of vortices in a real system is obtained when boundaries are included in the analysis. For this purpose, a boundary-induced velocity field $\mathbf{v}_{s,b}$ is added to $\mathbf{v}_s$, so that the superflow can satisfy the boundary condition of an inviscid flow, that is, the normal component of the velocity should disappear at the boundaries. To allow for another, presently unspecified, applied field, we include $\mathbf{v}_{s,a}$. Hence, the total velocity $\dot{\mathbf{s}}_0$ of the vortex filament without dissipation is

$$\dot{\mathbf{s}}_0 = \frac{\kappa}{4\pi}\mathbf{s}' \times \mathbf{s}'' \ln\left(\frac{2(\ell_+\ell_-)^{1/2}}{e^{1/4}a_0}\right) + \frac{\kappa}{4\pi}\int_{\mathcal{L}}' \frac{(\mathbf{s}_1 - \mathbf{r}) \times d\mathbf{s}_1}{|\mathbf{s}_1 - \mathbf{r}|^3} + \mathbf{v}_{s,b}(\mathbf{s}) + \mathbf{v}_{s,a}(\mathbf{s}). \tag{3.3}$$

At finite temperatures, it is necessary to take into account the mutual friction between the vortex core and the normal flow $\mathbf{v}_n$. Including this term, the velocity of $\mathbf{s}$ is given by

$$\dot{\mathbf{s}} = \dot{\mathbf{s}}_0 + \alpha\mathbf{s}' \times (\mathbf{v}_n - \dot{\mathbf{s}}_0) - \alpha'\mathbf{s}' \times [\mathbf{s}' \times (\mathbf{v}_n - \dot{\mathbf{s}}_0)], \tag{3.4}$$

where α and α' are temperature-dependent friction coefficients [19], and $\dot{\mathbf{s}}_0$ is calculated from Eq. (3.3).

The numerical simulation method based on this model has been described in detail elsewhere [19, 20, 22, 23]. A vortex filament is represented by a single string of points separated by a distance $\Delta\varsigma$. The vortex configuration at a given time determines the velocity field in the fluid, thus moving the vortex filaments according to Eqs. (3.3) and (3.4). Vortex reconnection should be properly included when simulating vortex dynamics. A numerical study of a classical fluid shows that the close interaction of two vortices leads to their reconnection, primarily because of viscous diffusion of the vorticity [24]. Schwarz assumed that two vortex filaments reconnect when they come within a critical distance of one another, and showed that statistical quantities such as the vortex line density were not sensitive to how these reconnections occur [19, 20]. Even after Schwarz's study, it remained unclear as to whether quantized vortices can actually reconnect. However, Koplik and Levine directly solved the GP equation to show that two closely quantized vortices reconnect, even in an inviscid superfluid [25]. More recent simulations have shown that reconnections are accompanied by emissions of sound waves having wavelengths on the order of the healing length [26, 27].

Starting with several remnant vortices under thermal counterflow, Schwarz studied numerically how these vortices developed into a vortex tangle [20]. The tangle was self-sustained by the competition between excitation due to the applied flow and dissipation through mutual friction. The numerical results were quantitatively consistent with typical experimental results. Since Schwarz used the LIA, he could not obtain the statistical steady state without an artificial procedure. Performing the full non-local calculation, Adachi et al. succeeded in obtaining a steady state in counterflow turbulence that was consistent with typical observations. This was a significant accomplishment in numerical research [23].

Thus, the vortex filament model is very useful for QT, although it cannot describe phenomena directly related to vortex cores, such as reconnection, nucleation, and annihilation. These phenomena can be analyzed only by the GP model.

3.2.2.2 *The Gross–Pitaevskii model*

In a weakly interacting Bose system, the macroscopic wave function $\Psi(\mathbf{r}, t)$ appears as the order parameter of Bose–Einstein condensation, obeying the Gross–Pitaevskii (GP) equation [28, 29],

$$i\hbar\frac{\partial\Psi(\mathbf{r},t)}{\partial t} = \left(-\frac{\hbar^2}{2m}\nabla^2 + g|\Psi(\mathbf{r},t)|^2 - \mu\right)\Psi(\mathbf{r},t). \tag{3.5}$$

Here $g = 4\pi\hbar^2 m/a$ represents the strength of the interaction characterized by the s-wave scattering length a, m is the mass of each particle, and μ is the chemical potential. In $\Psi = |\Psi|\exp(i\theta)$, the squared amplitude $|\Psi|^2$ is the condensate, density, and the gradient of the phase θ gives the superfluid velocity $\mathbf{v}_s = (\hbar/m)\nabla\theta$, a frictionless flow of the condensate. This relation causes quantized vortices to appear with quantized circulation. The only characteristic scale of the GP model is the coherence length, defined by $\xi = \hbar/(\sqrt{2mg}|\Psi|)$, which gives the vortex core size.

The GP model can explain not only the vortex dynamics but also vortex core phenomena such as reconnection and nucleation. However, strictly speaking, the GP equation is not applicable to superfluid ^{4}He, which is not a weakly interacting Bose system. The GP model does not feature, for example, short-wavelength excitations such as rotons, which are present in superfluid ^{4}He. The GP equation is applicable to Bose–Einstein condensation of a dilute atomic Bose gas [30].

3.2.3 Quantum turbulence

In this section, we review recent developments in the study of QT. "Quantum turbulence" is defined as the turbulent state of a quantum fluid, and this term is often used to emphasize that the state is dominated by the behavior of quantized vortices at low temperatures, with little thermal effect [1]. From a theoretical point of view, the properties of QT are often discussed in reference to statistical quantities such as the energy spectrum, which is important in turbulence. Since quantized vortices in QT are definite and stable topological defects, and can be positively identified, the energy spectrum of QT is usually related to the dynamics of vortices, such as vortex cascades [31].

The energy spectrum and the vortex cascade process are usually divided into two regions in wavenumber space. The first region is called the classical region, and exists at wavenumbers below the inverse of the mean intervortex spacing. The dynamics of vortices in the classical region are dominated by the Richardson cascade, in which large vortices are broken up self-similarly into smaller ones, or collective dynamics of aggregated quantized vortices at scales larger than the intervortex spacing. Such behavior of vortices supports the analogy of QT to CT, namely, the Kolmogorov energy spectrum, which has been confirmed by several theoretical and experimental efforts [32]. The second region is called the quantum region, in which vortex dynamics are dominated by the effects of the quantized circulation, specifically the Kelvin wave cascade of vortices [33, 34], which does not appear in CT. The Kelvin wave cascade is also a very important concept in understanding the dissipation mechanism of QT at very low temperatures.

At finite temperatures, turbulent flow is carried by not only quantized vortices, but also by normal fluid. On larger scales, superfluids and normal fluids are coupled together by the mutual friction between them, and behave as a classical fluid to show the Kolmogorov energy spectrum. On small scales, turbulent flow is dissipated by the viscosity of the normal fluid, and Kelvin waves do not exist.

Another topic in QT is the decay mechanism of quantized vortices [1]. The total vortex line length, or the vortex line density, can be observed experimentally, and the decay of vortices provides important information regarding QT. Since the decay of vortices depends strongly on the initial QT state, we can indirectly determine the statistical properties, such as the energy spectrum, of vortices in the initial state.

This section is organized as follows. First, we review the energy spectrum of QT, from the classical region to the quantum region, and the connection between these two regions, which remains an open question. Next, we discuss the decay of quantized vortices in QT, recent experimental developments, QT created by vibrating structures, and visualization of quantized vortices.

3.2.3.1 *Energy spectra in classical turbulence*

To discuss energy spectra of QT, we begin with a brief review of the energy spectrum in CT. The dynamics of classical fluids are usually described by the Navier–Stokes equation [35],

$$\frac{\partial \mathbf{v}}{\partial t} + \mathbf{v} \cdot \nabla \mathbf{v} = -\frac{1}{\rho} \nabla P + \nu \nabla^2 \mathbf{v}. \tag{3.6}$$

Here, $\mathbf{v} = \mathbf{v}(\mathbf{x}, t)$ is the fluid velocity, $P = P(\mathbf{x}, t)$ is the pressure, ρ is the fluid density, and ν is the kinematic viscosity. The flow of this fluid can be characterized by the ratio of the second term of the left-hand side of Eq. (3.6), hereafter the inertial term, to the second term of the right-hand side, hereafter the viscous term. This ratio is the Reynolds number: $R = \bar{v}D/\nu$. Here $\bar{v}$ and D are the characteristic velocity of flow and the characteristic scale, respectively. When $\bar{v}$ increases to the point that the Reynolds number exceeds a critical value, the system enters a turbulent state in which the flow is highly complicated, with many eddies.

The energy spectrum can be obtained from the spatial Fourier transformation of the equal-time two-point velocity correlation,

$$F(\mathbf{k}, t) = \frac{1}{2} \int d\mathbf{x}\, e^{-i\mathbf{k}\cdot\mathbf{x}} \int d\mathbf{y}\, \mathbf{v}(\mathbf{y}, t) \cdot \mathbf{v}(\mathbf{y} + \mathbf{x}, t). \tag{3.7}$$

Here k is the wavenumber from the Fourier transformation. Kolmogorov proposed the concept of a globally steady state of fully developed turbulence [36, 37] in which energy is injected into the fluid at scales comparable to the system size in the energy-containing range. In the inertial range, this energy is transferred to smaller scales without being dissipated. In the inertial range, the system is assumed to be locally homogeneous and isotropic, and the angular dependence of $E(\mathbf{k}, t)$ in wavenumber space becomes less important. The integration over the angle is defined as the energy spectrum,

$$E(k,t) = \frac{1}{(2\pi)^3} \int d\phi_{\mathbf{k}} d\theta_{\mathbf{k}}\, k^2 F(\mathbf{k}, t). \tag{3.8}$$

The energy spectrum holds the following relation with the spatial integration of the kinetic energy:

$$\int dk\, E(k,t) = \frac{1}{2(2\pi)^3} \int d\mathbf{k}\, |\mathbf{v}(\mathbf{k}, t)|^2 = \frac{1}{2} \int d\mathbf{x}\, \mathbf{v}(\mathbf{x}, t)^2 = E(t). \tag{3.9}$$

Here, $E(t)$ is the total kinetic energy per unit mass, and $\mathbf{v}(\mathbf{k}, t)$ is the Fourier transformation of the fluid velocity. In the steady state, the averaged value of $E(k, t)$ over t is regarded as the statistical value, together with a statistical law known as the Kolmogorov law,

$$E(k) = C\varepsilon^{2/3} k^{-5/3}. \tag{3.10}$$

The energy transferred to smaller scales in the energy-dissipative range is dissipated at the Kolmogorov wavenumber $k_K = (\varepsilon/\nu)^{1/4}$ through the viscosity of the fluid, at the dissipation rate ε in Eq. (3.10). This spectrum of Eq. (3.10) is easily derived by assuming that $E(k)$ is locally determined by only ε and k. The Kolmogorov constant C is a dimensionless parameter of order unity.

The inertial range is thought to be sustained by the self-similar Richardson cascade, in which large eddies are broken up into smaller ones with many reconnections [38]. In CT, however, the Richardson cascade is not completely understood because it is impossible to definitively identify each individual eddy.

The Kolmogorov law is based on the assumption that the turbulence is locally homogeneous and isotropic. However, actual turbulence is neither homogeneous nor isotropic and hence the energy spectrum deviates from Eq. (3.10). This phenomenon is called "intermittency", which is closely related to the coherent dynamics of eddies [35].

3.2.3.2 *Energy spectra in quantum turbulence: overall picture*

In QT, it is important to consider the energy spectrum for the following two reasons. First, like CT, QT is a complicated dynamical phenomenon with many degrees of freedom, and it is very useful to consider the energy spectrum as the statistical law which extracts the important imformation for the dynamical properties of the turbulence. Secondly, in QT, unlike in CT, vortices are definite and stable topological defects, which enables the study of actual vortex dynamics. In particular, when there is no normal fluid component at near-zero temperatures, any rotational motion is carried by quantized vortices and their dynamics are not affected by thermal components, so the inherent dynamics of vortices in turbulence are revealed. If QT has an analogy to CT, vortex dynamics in QT should show the real vortex dynamics in the Richardson cascade. Therefore, QT is an ideal prototype for the study of the statistics of turbulence, such as the relationship between the Kolmogorov law in wavenumber space and the Richardson cascade in real space [2]. Whether this scenario is true can be confirmed by investigating the energy spectrum and energy transfer in QT.

As in CT, the energy spectra in QT can also be obtained from the spatial Fourier transformation of the equal-time two-point superfluid velocity correlation: Eq. (3.8), where $\mathbf{v}(\mathbf{x}, t)$ is replaced with the superfluid velocity $\mathbf{v}_s(\mathbf{x}, t)$.

Here, we summarize the overall picture of the energy spectrum of QT at zero temperature on the basis of theoretical and numerical studies (Fig. 3.1) [1]. In QT, vortices form a highly complicated tangled structure. If the tangle is assumed to be homogeneous and isotropic, there are two characteristic length scales: one is the mean intervortex spacing $l = L^{-1/2}$ with a vortex line length density L (the vortex line length per unit volume), and the other is the coherence length ξ corresponding to the size of the vortex core. Generally, l is much larger than ξ; $l \gg \xi$. Using the length scales l and ξ, we can define the characteristic wavenumbers $k_l = 2\pi/l$ and $k_\xi = 2\pi/\xi$.

At length scales larger than l, the dynamics of QT are dominated by a tangled structure of many vortices. Because vortex dynamics becomes collective at large scales, quantization of the circulation is not relevant, and the dynamics are similar to those of eddies in CT. Therefore, this region can be referred to as the classical region. As a result, the energy spectrum $E(k)$ in the range of $k < k_l$ obeys the Kolmogorov law (3.10). Vortices in QT sustain a Richardson cascade that transfers energy from smaller wavenumbers to larger ones without dissipation. The Richardson cascade can be understood as large vortices breaking up into smaller ones in real space. In the Richardson cascade process, the key vortex dynamic is reconnection; when two vortex lines approach each other, they become locally antiparallel and reconnect [25–27]. Reconnection causes topological changes of the vortex lines in the tangle, formation of distortion waves on the vortex lines (Kelvin waves), or fission of vortex loops through self-reconnections, all of which are responsible for the cascade of energy towards smaller scales [22, 33, 39]. The Richardson cascade and the Kolmogorov energy spectrum have been confirmed by several numerical studies [40–45].

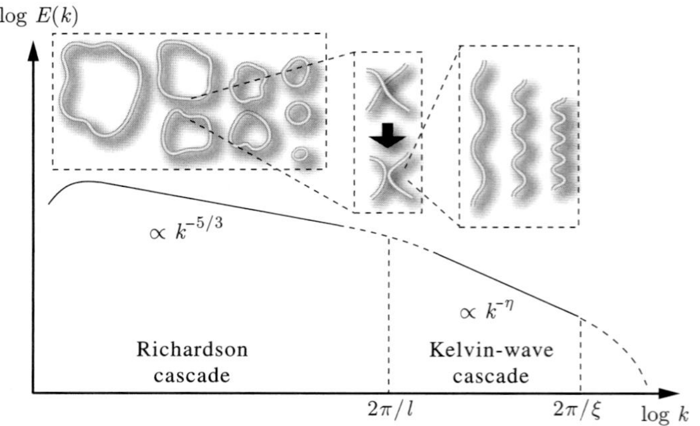

Fig. 3.1 Overall picture of the energy spectrum of QT at zero temperature. The energy spectrum depends on the scale, and its properties change at about the scale of the mean intervortex spacing l. When $k < k_l = 2\pi/l$, a Richardson cascade of quantized vortices transfers energy from large to small scales, maintaining the Kolmogorov spectrum $E(k) = C\varepsilon^{2/3}k^{-5/3}$. When $k > k_l$, energy is transferred by the Kelvin-wave cascade, which is a nonlinear interaction between Kelvin waves of different wavenumbers. Eventually, energy is dissipated at scales of ξ by the radiation of phonons.

At length scales comparable to l, reconnection is the dominant dynamic. At these scales, two vortex lines separated by a distance $\sim l$ reconnect, forming small cusps on the vortex lines, which are regarded as the primary source of Kelvin waves in QT. The wavelength of the created Kelvin waves is in the order of l.

At length scales smaller than l, which is referred to as the quantum region, the Richardson cascade is no longer dominant, and the quantized circulation and the motion of each vortex line become significant [33, 46–51, 66]. In this range, vortex dynamics are characterized by the cascade process of the Kelvin waves which are formed by reconnection. The nonlinear interaction of the Kelvin waves is the origin of the cascade from smaller to larger wavenumbers. The cascade dynamics of the Kelvin waves along a single vortex line has been confirmed, both numerically and theoretically. The Kelvin-wave cascade in QT has also been studied. The energy spectrum in the quantum region $k_l < k < k_\xi$ is theoretically predicted to obey a Kolmogorov-like power law: $E(k) \propto k^\eta$. Several values have been proposed for the exponent η, which originates from both the configuration of the Kelvin waves and their cascade process. The power-law behavior of the energy spectrum in this region has been confirmed numerically [45, 47–51].

In the region $k \sim k_\xi$, the Kelvin waves with wavelength ξ change to elementary excitations, such as phonons and rotons, in the primary decay process of QT near zero temperature [34].

There is one open question regarding the energy spectrum in the region $k \sim k_l$, which is the transitional region between the Richardson cascade and the Kelvin-wave cascade. A theoretical study proposed a bottleneck effect connecting the spectrum (3.10) of the classical region to that of the quantum region [52]. Another theoretical prediction was based on vortex reconnection dynamics on the scale $\sim l$, in which the transitional cascade process was predicted to occur by reconnection of the vortex bundle [53].

3.2.3.3 *Energy spectra in quantum turbulence: classical region*

Energy spectra in the classical region $k < k_l$ have been a major problem in terms of the analogy of QT to CT. In this region, the energy spectrum is determined by the collective behavior of many vortices, such as the vortex tangle and the aggregated bundle structure at scales larger than l. Several numerical studies have calculated the energy spectrum in this region by simulating QT at zero temperature.

There are two formulations for studying the vortex dynamics in QT, as described in Sec. 3.2.2. The first is the vortex filament model, in which the time evolution of each element $\mathbf{s}$ of a vortex filament follows Eq. (3.4). At zero temperature there is no mutual friction, and we obtain $\alpha = \alpha' = 0$ in (2.4) and $\dot{\mathbf{s}} = \dot{\mathbf{s}}_0$. In this model, the energy spectrum can be calculated directly from the configuration of the vortex filaments, but not through the superfluid velocity $\mathbf{v}_s(\mathbf{f})$ of Eq. (3.1). The vorticity $\boldsymbol{\omega}(\mathbf{r}) = \nabla \times \mathbf{v}_s(\mathbf{r})$ is concentrated on the vortex filament,

$$\boldsymbol{\omega}(\mathbf{r}) = \kappa \int d\varsigma \; \mathbf{s}'(\varsigma)\delta(\mathbf{s}(\varsigma) - \mathbf{r}). \tag{3.11}$$

Here, ς is the arc length along the vortex filament. The Fourier transformation of the vorticity $\boldsymbol{\omega}(\mathbf{k}) = i\mathbf{k} \times \mathbf{v}_s(\mathbf{k})$ becomes

$$\boldsymbol{\omega}(\mathbf{k}) = \kappa \int d\varsigma \ \exp[-i\mathbf{k} \cdot \mathbf{s}(\varsigma)]\mathbf{s}'(\varsigma). \tag{3.12}$$

Assuming fluid incompressibility $\nabla \cdot \mathbf{v}_s(\mathbf{r}) = 0$ and $\mathbf{v}_s(\mathbf{k}) = i\mathbf{k} \times \boldsymbol{\omega}(\mathbf{k})/|\mathbf{k}|^2$, we obtain the energy spectrum as the integrated form over ς:

$$E(k) = \frac{\kappa^2}{2(2\pi)^3} \int d\phi_{\mathbf{k}} d\theta_{\mathbf{k}} \int d\varsigma_1 d\varsigma_2 \ \exp[-i\mathbf{k} \cdot (\mathbf{s}(\varsigma_1) - \mathbf{s}(\varsigma_2))]\mathbf{s}'(\varsigma_1)\mathbf{s}'(\varsigma_2). \tag{3.13}$$

This formula enables us to calculate the energy spectrum directly from the configuration of the vortex filament [42].

Using the vortex filament model, Araki et al. simulated QT with no mutual friction, starting from a Taylor Green vortex in a cube of size $D = 1.0\,\text{cm}$ [42]. A cutoff was introduced for the smallest vortices, the size of which was comparable to the space resolution $\Delta\varsigma = 1.83 \times 10^{-2}$, which was the only dissipation mechanism in the simulation. The energy spectrum $E(k)$ below k_l had a peak at the smallest k, and depended strongly on the initial configuration. After about 70 sec, the system exhibited a nearly homogeneous and isotropic turbulent state, and lost its dependence on the initial configuration. The energy dissipation rate $\varepsilon = -dE/dt$ became less dependent on time, and the system behaved as a quasi-steady turbulent state. The energy spectrum obtained from Eq. (3.13) agreed with the Kolmogorov law (3.10) at $k < k_l$ (Fig. 3.2). They also calculated the vortex length distribution $n(x)$, where $n(x)\Delta x$ is the number of vortex loops, the length of which ranged from x to $x + \Delta x$. In the range from l to D, $n(x)$ obeyed the scaling property $n(x) \propto x^{-1.34 \pm 0.18}$, which supports the self-similar Richardson cascade from larger to smaller scales. The energy transfer rate through the Richardson cascade is equal to the energy dissipation rate $\varepsilon = -dE/dt$ in Eq. (3.10). Furthermore, the energy spectrum $E(k)$ in the region $k < k_l$ is determined by the energy dissipation rate, although the dissipation mechanism is effective only at wavenumbers corresponding to the cutoff $2\pi/\Delta\xi$. In the region of $k > k_l$, the energy spectrum deviates from the Kolmogorov law, showing a different power-law behavior. Although this behavior may come from the velocity field near each vortex line, the simulation was not appropriate to examine the spectrum in this region because it did not have sufficient spatial resolution.

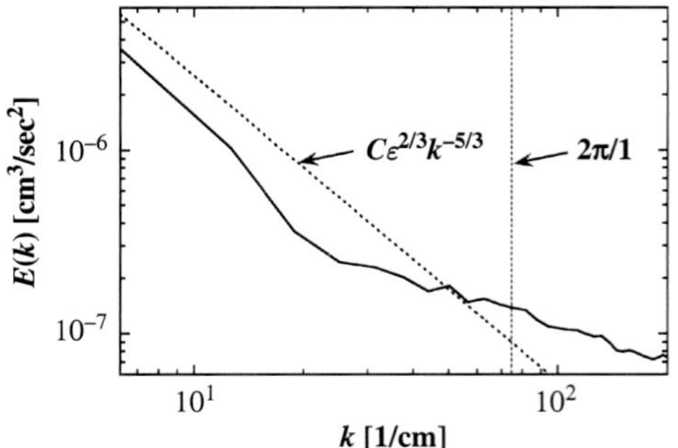

Fig. 3.2 Comparison of the energy spectrum (solid line) at $t = 70$ sec after the Taylor Green vortex with the Kolmogorov law (3.10) (dotted line) with $C = 1$ and $\varepsilon = 1.287 \times 10^{-6}\,\text{cm}^2/\text{sec}^3$ [Araki, Tsubota, and Nemirovskii: *Phys. Rev. Lett.* **89** (2002) 145301, reproduced with permission. Copyright 2002 by the American Physical Society].

The other formulation is the GP model. The GP equation (3.5) can be obtained from the energy functional,

$$H = \int d\mathbf{x} \left[\frac{\hbar^2}{2m} |\nabla \Psi(\mathbf{x}, t)|^2 + \frac{g}{2} |\Psi(\mathbf{x}, t)|^4 - \mu |\Psi(\mathbf{x}, t)|^2 \right]. \tag{3.14}$$

Because the vorticity $\boldsymbol{\omega}(\mathbf{x}, t) = \nabla \times \mathbf{v}_s(\mathbf{x}, t)$ vanishes everywhere in a singly connected region of the fluid, all rotational flow is carried by quantized vortices. In the core of each vortex, $\Psi(\mathbf{x}, t)$ vanishes so that the circulation $\oint \mathbf{v}_s(\mathbf{x}, t) \cdot d\mathbf{s}$ around the core is quantized by h/m. The vortex core size is given by the healing length $\xi = \hbar/\sqrt{2mg\rho_s}$, where the superfluid density ρ_s is defined as the spatially averaged condensate density $|\Psi(\mathbf{x}, t)|^2$. When considering the energy spectrum, it should be noted that the hydrodynamics are compressible in the GP model. The total energy of the GP equation per unit mass,

$$E(t) = \frac{1}{mN} \int d\mathbf{x} \left[\frac{\hbar^2}{2m} |\nabla \Psi(\mathbf{x}, t)|^2 + \frac{g}{2} |\Psi(\mathbf{x}, t)|^4 \right] \tag{3.15}$$

can be separated into the interaction energy $E_{\mathrm{int}}(t)$, the quantum energy $E_{\mathrm{q}}(t)$, and the kinetic energy $E_{\mathrm{kin}}(t)$:

$$E_{\mathrm{int}}(t) = \frac{g}{2mN} \int d\mathbf{x}\, |\Psi(\mathbf{x}, t)|^4 \tag{3.16}$$

$$E_{\mathrm{q}}(t) = \frac{(2\pi)^2 \kappa^2}{2N} \int d\mathbf{x}\, (\nabla |\Psi(\mathbf{x}, t)|)^2 \tag{3.17}$$

$$E_{\mathrm{kin}}(t) = \frac{(2\pi)^2 \kappa^2}{2N} \int d\mathbf{x}\, (|\Psi(\mathbf{x}, t)| \nabla \theta(\mathbf{x}, t))^2. \tag{3.18}$$

Here, $N \equiv \int d\mathbf{x}\, |\Psi(\mathbf{x}, t)|^2$ is the total number of particles. The kinetic energy can be further divided into a compressible part $E^{\mathrm{c}}_{\mathrm{kin}}(t)$ due to compressible excitations, and an incompressible part $E^{\mathrm{i}}_{\mathrm{kin}}(t)$ due to vortices [40, 41],

$$E^{\mathrm{c,i}}_{\mathrm{kin}}(t) = \frac{(2\pi)^2 \kappa^2}{2N} \int d\mathbf{x}\, [(|\Psi(\mathbf{x}, t)| \nabla \theta(\mathbf{x}, t))^{\mathrm{c,i}}]^2. \tag{3.19}$$

Here, $[\cdots]^{\mathrm{c}}$ denotes the compressible part and $\nabla \times [\cdots]^{\mathrm{c}} = 0$, and $[\cdots]^{\mathrm{i}}$ denotes the incompressible part $\nabla \cdot [\cdots]^{\mathrm{i}} = 0$. Corresponding to each energy, there are several kinds of energy spectra. The most important of these is the energy spectrum of the incompressible kinetic energy,

$$E^{\mathrm{i}}_{\mathrm{kin}}(k, t) = \frac{\kappa^2}{2(2\pi)N} \int d\phi_{\mathbf{k}} d\theta_{\mathbf{k}}\, [\mathbf{p}^{\mathrm{i}}(\mathbf{k}, t)]^2, \tag{3.20}$$

because it should obey the Kolmogorov law with the Richardson cascade of quantized vortices. Here $\mathbf{p}^{\mathrm{i}}(\mathbf{k}, t)$ is the Fourier transformation of the momentum density $\{|\Psi(\mathbf{x}, t)| \nabla \theta(\mathbf{x}, t)\}^{\mathrm{i}}$.

Starting from a Taylor Green vortex, Nore et al. simulated QT by numerically solving the GP model [40, 41]. After some time, the initial vortices became tangled, and the calculated spectrum obeyed the power-law behavior $E^{\mathrm{i}}_{\mathrm{kin}}(k, t) \propto k^{-\eta(t)}$. When vortices formed a tangle, the exponent $\eta(t)$ was about $5/3$, but this value did not hold for long because the turbulence was decaying with the conservation of total energy. Because of energy conservation, the energy of the vortices $E^{\mathrm{i}}_{\mathrm{kin}}(t)$ was transferred to compressible excitations $E^{\mathrm{i}}_{\mathrm{kin}}(t)$ through repeated reconnections and disappearances of small vortex loops with sizes in the order of ξ. This thermalization process hindered the vortex cascade process through interactions between vortices and compressible excitations, which caused $E^{\mathrm{i}}_{\mathrm{kin}}(k,t)$ to deviate from the Kolmogorov law.

To avoid this thermalization, Kobayashi and Tsubota proposed a modified GP model, in which a dissipation term was introduced to remove the compressible excitations [43, 44]. The characteristic wavelength of the compressible excitations is in the order of ξ, so the introduced dissipation was set to act only at scales smaller than ξ and to not affect vortices. The resulting GP equation in wavenumber space becomes

$$\hbar[i-\gamma(k)]\frac{\partial\Psi(\mathbf{k},t)}{\partial t} = \left(\frac{\hbar^2k^2}{2m} - \mu(t)\right)\Psi(\mathbf{k},t) + \frac{g}{(2\pi)^6}\int d\mathbf{k}_1 d\mathbf{k}_2\, \Psi(\mathbf{k}_1,t)\Psi^*(\mathbf{k}_2,t)\Psi(\mathbf{k}-\mathbf{k}_1+\mathbf{k}_2,t). \tag{3.21}$$

Here, $\Psi(\mathbf{k},t)$ is the Fourier transformation of $\Psi(\mathbf{x},t)$. The dissipation term has the form $\gamma(k) = \gamma_0\theta(k-2\pi/\xi)$ with the step function θ, being effective only at $k > k_\xi$. Introduction of $\gamma(k)$ conserves neither the energy E nor the number of particles N. However, when studying the hydrodynamics of turbulence, it is realistic to assume that the number of particles is conserved. Hence, the time dependence of the chemical potential $\mu(t)$ was introduced to conserve the total number of particles N. By numerically analyzing Eq. (3.21), they confirmed the Kolmogorov spectrum in two kinds of QT.

The first analysis was of decaying turbulence [43]. To obtain a turbulent state, an initial configuration was set to have a uniform density $|\Psi(\mathbf{x},t=0)|^2 = 1$, and the phase had a random spatial distribution. Because the initial superfluid velocity $\mathbf{v}_s(\mathbf{x},t=0) = (\hbar/m)\nabla\theta(\mathbf{x},t=0)$ was also random, the initial wave function was dynamically unstable and soon produced fully developed turbulence with many quantized vortex loops. They confirmed that only the compressible kinetic energy was decreased by the dissipation term, and that the incompressible kinetic energy $E^{\mathrm{i}}_{\mathrm{kin}}(t)$ dominated the total kinetic energy $E_{\mathrm{kin}}(t)$, demonstrating that the thermalization was effectively suppressed. The obtained energy spectrum of the incompressible kinetic energy was consistent with the Kolmogorov law in the period during which the vortex tangle was formed, and this consistency lasted longer than it did without the dissipation term. In this period, the energy dissipation rate $\varepsilon = -dE^{\mathrm{i}}_{\mathrm{kin}}(t)/dt$ became nearly constant, and the system can be considered to occupy a quasi-steady turbulent state with a stationary Kolmogorov spectrum of $\varepsilon^{2/3}k^{-5/3}$.

The second analysis was of steady turbulence, with not only energy dissipation but also large-scale energy injection [44]. The energy injection was accomplished by introducing the stochastic external potential $V(\mathbf{x},t)$ into the GP equation (3.21). $V(\mathbf{x},t)$ has a Gaussian two-point correlation,

$$\langle V(\mathbf{x},t)V(\mathbf{x}',t')\rangle = V_0^2\exp\left[-\frac{|\mathbf{x}-\mathbf{x}'|^2}{2X_0^2} - \frac{(t-t')^2}{2T_0^2}\right], \tag{3.22}$$

where X_0 and T_0 are the characteristic spatial and time scales of the potential. By introducing $V(\mathbf{x}, t)$, energy was injected at the spatial scale of X_0, and quantized vortices of radius X_0 were effectively nucleated. Starting from the uniform state $\Psi(\mathbf{x}, t = 0) = 1$, they developed the GP equation (3.21) with the potential (3.22), and obtained a steady QT in which $E(t)$, $E_{\text{kin}}(t)$, $E^{\text{c}}_{\text{kin}}(t)$, and $E^{\text{i}}_{\text{kin}}(t)$ were nearly constant. In the simulation, the incompressible kinetic energy $E^{\text{i}}_{\text{kin}}(t)$ was always dominant in the total kinetic energy $E_{\text{kin}}(t)$; the introduced potential contributed to the nucleation of vortices rather than that of compressible waves. The obtained energy spectrum was consistent with the Kolmogorov law in the range $2\pi/X_0 < k_\xi$, which is regarded as the inertial range of Richardson cascade of vortices. (Fig. 3.3). They also calculated the flux $\Pi(k, t)$ of the incompressible kinetic energy from smaller to larger wavenumbers,

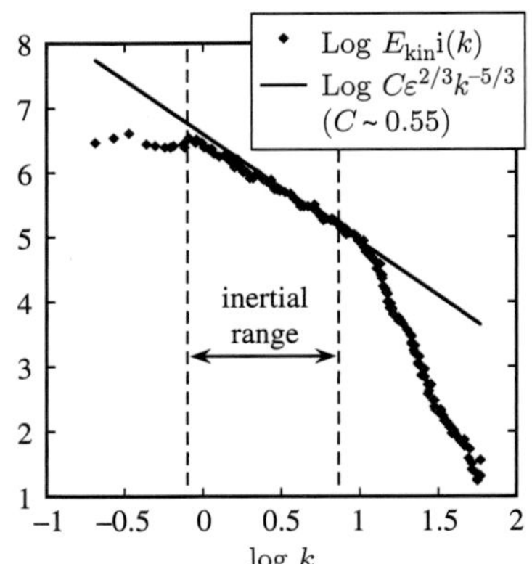

Fig. 3.3 Incompressible energy spectrum $E^{\text{i}}_{\text{kin}}(k)$ obtained by Kobayashi and Tsubota [44]. The plotted points are from an ensemble average of 50 randomly selected states. The solid line is the Kolmogorov law [Kobayashi and Tsubota: *J. Phys. Soc. Jpn.* **74** (2005) 3248, reproduced with permission].

$$\Pi(k, t) = \frac{(2\pi)^2\kappa^2}{N} \int d\mathbf{x}\, L_k[\{\mathbf{p}(\mathbf{x}, t) \cdot \nabla \mathbf{v}_s(\mathbf{x}, t)\}^{\text{i}}] L_k[\{\mathbf{p}(\mathbf{x}, t)\}^{\text{i}}], \tag{3.23}$$

where L_k is the operator for the low-pass filter,

$$L_k[s(\mathbf{x})] = \frac{1}{(2\pi)^3} \int_{|\mathbf{k}|<k} d\mathbf{k} \int d\mathbf{x}'\, e^{i\mathbf{k}\cdot(\mathbf{x}-\mathbf{x}')} s(\mathbf{x}), \tag{3.24}$$

for an arbitrary function $s(\mathbf{x})$. $\Pi(k, t)$ can be obtained from the scale-by-scale energy budget equation derived from the GP equation (3.21). $\Pi(k, t)$ was nearly constant in the inertial range, and almost the same as the energy dissipation rate $\varepsilon = -dE^{\text{i}}_{\text{kin}}(t)/dt$ after switching off the moving random potential. This confirmed the picture of the inertial range and the Richardson cascade of quantized vortices in QT; the incompressible kinetic energy steadily flows in wavenumber space through the Richardson cascade at the constant rate Π, and finally dissipates at the rate $\varepsilon \simeq \Pi$.

In the above three numerical studies, the system size was not so large that the inertial range was less than one order in wavenumber space. Furthermore, the mean intervortex spacing l was close to the healing length ξ, and was too short to study the Kelvin-wave cascade. To obtain the energy spectrum of a wider range of wavenumber space, Yepez et al. performed a large-scale simulation of the GP model, the size of which was $(11.25)^3$ times larger than the above three simulations [45]. They also used a highly accurate numerical code for the time development; the unitary quantum lattice gas algorithm, in which the time development of the order parameter $\Psi(\mathbf{x}, t)$ in Eq. (3.5) could be performed as the unitary time evolution of qubits on a cubic lattice. Starting from the initial state with 12 straight vortices, which consisted of three groups of four vortices aligned along x, y, and z axes, they obtained QT with highly tangled vortices. In this simulation, two length scales were defined; the inner radius of a vortex core ξ, and its outer radius $\pi\xi$. On an L^3 grid, the wavenumber corresponding to the core's inner radius was $k_{\text{inner}} \simeq (\sqrt{3}/2)L/\xi$, while the outer radius wavenumber was $k_{\text{outer}} \simeq k_{\text{inner}}/\pi$. They found that the incompressible kinetic energy spectrum (3.20) had three distinct power-law $k^{-\alpha}$ regions that ranged from the classical turbulent regime of Kolmogorov $\alpha = -5/3$ at large scales $k < k_{\text{outer}}$ to the quantum Kelvin-wave cascades $\alpha = 3$ at small scales $k > k_{\text{inner}}$. There was a semiclassical region $6.34 \lesssim \alpha \lesssim 7.11$ connecting the

Kolmogorov and Kelvin-wave spectra $k_{\mathrm{outer}} < k < k_{\mathrm{inner}}$. Compared to previous simulations, this simulation supplied the Kolmogorov spectrum over a much wider inertial range, with about two orders in wavenumber space, and the Kelvin-wave spectrum in the quantum region. Although they related the k^{-3} spectrum at small scales $k > k_{\mathrm{inner}}$ to the Kelvin-wave cascade, the length scale in this region is smaller than the vortex core size. The one-dimensional picture of a Kelvin wave is relevant in scales much larger than the vortex core, and a nature of short waves propagating in three-dimensional space is considered to be dominant in the length scale smaller than the vortex core size. They also did not discuss the mean intervortex spacing l, which is the key length scale to distinguish the classical and quantum regions. Therefore, it is not so clear that the k^{-3} spectrum reflects the Kelvin-wave cascade.

It should be noted that the Richardson cascade of quantized vortices is genuine in QT, and does not completely imitate that of CT. The Kolmogorov constant C in Eq. (3.10) in QT is not necessarily equal to that of CT. The obtained Kolmogorov constant was $C \simeq 0.7$ for the vortex filament model [42], and $C \simeq 0.32$ [43] and 0.55 [44] for the decaying and steady QT in the GP model, which was smaller than that of CT, in which $C \simeq 1.5$. These smaller Kolmogorov constants may be characteristic of QT.

3.2.3.4 *Energy spectra in quantum turbulence: quantum region*

In the region of $k > k_l$, the picture of aggregated vortices is no longer effective, and the motion of each vortex line becomes essential. The most probable dominant dynamic of vortices is Kelvin waves, which originate from distortion waves on the vortex lines after their reconnection. A Kelvin wave is a transverse, circularly polarized wave motion, with the approximate dispersion relation for a rectilinear vortex:

$$\omega_k = \frac{\kappa k^2}{4\pi}\left[\ln\left(\frac{1}{k\xi}\right) + c\right], \tag{3.25}$$

with a dimensionless constant $c \sim 1$. k is the wavenumber of the Kelvin wave, and was different from that used for the energy spectrum. Kelvin waves were theoretically proposed to exist in inviscid fluids [54], and were first observed by inducing torsional oscillations in a rotating superfluid ^{4}He [55, 56].

At finite temperatures where there is a significant fraction of normal fluid, Kelvin waves are damped by mutual friction. On the other hand, at very low temperatures they can be damped only by the radiation of phonons. Vinen estimated the rate of radiation, and found that it is extremely low unless the frequency is very high, typically in the order of 4 GHz for superfluid ^{4}He [34]. In these circumstances, low-frequency Kelvin waves can lose energy only by nonlinear coupling to waves of a different frequency, which is the Kelvin wave cascade from small to large wavenumbers.

An earlier numerical work showed the nonlinear interaction of Kelvin waves in the framework of the vortex filament model. Samuels and Donnelly performed a simulation of a full Biot–Savart law (3.2) and the local induction approximation (LIA), in which the second Biot–Savart term in (3.2) was neglected, and they found sideband instability in both cases above a critical amplitude of the Kelvin wave [46]: $A_0 > \lambda/(2\pi n)$, where A_0 is the amplitude of the main helical wave, λ is the wavelength, and n is the number of half-waves on the vortex. The occurrence of sideband instability indicates the mixing of Kelvin waves of the main wavelength with those of smaller wavelengths via nonlinear interactions, and the transfer of energy from small to large wavenumbers.

Although the Kelvin-wave cascade seems to be a very important mechanism in QT at scales smaller than l, it is a non-trivial problem about the actual cascade process. Vinen et al. performed a numerical simulation of a Kelvin wave excited along a single vortex line using the vortex filament model [47]. They considered a model system in which the helium was contained in a space between two parallel sheets separated by a distance $\ell_B = 1$ cm. A single, initially rectilinear (along the z axis), vortex was stretched between the two sheets. The allowed wavenumbers of the Kelvin waves were $k = 2\pi n/\ell_B$, where n is a positive integer. A Kelvin wave with a small wavenumber of n_0 was continuously driven, and all modes with n exceeding a large critical value n_c were strongly damped. The simulation was based on the full Biot–Savart law, and the force driving one mode was of the form $V\rho\kappa \sin(k_0 z - \omega_0)$, where $k_0 = 2\pi n_0/\ell_B$, ρ is the density of helium, and ω_0 is related to the frequency k_0 by the dispersion relation (3.25). Damping was applied by a periodic smoothing process. Starting from a straight vortex line, the total length of the vortex line evolved to reach a steady average value after application of the driving force, which suggests the existence of a steady state. In the typical simulation, with $V = 2.5 \times 10^{-5}$ cm s^{-1}, $k_0 = 10\pi$ cm^{-1}, and $k_c = 2\pi n_c/\ell_B = 120\pi$ cm^{-1}, they calculated the root mean square amplitudes $n_k(t) = \langle \zeta_k^* \zeta_k \rangle^{1/2}$ of the Fourier components of the displacement of the vortex. Initially, only the mode that resonated with the drive was excited. However, as time passed, nonlinear interactions led to the excitation of all modes. The spectrum $n_k(t)$ also reached a steady state, and in this stage energy was injected at a certain rate at wavenumber k_0 and dissipated at the same rate at wavenumber k_c. For large k, where the modes nearly formed a continuum, the steady state was observed to have, to a good approximation, a spectrum of the form

$$n_k = A\ell_B^{-1}k^{-3}, \tag{3.26}$$

with the dimensionless parameter A of order unity. They continued their simulations, changing the driving amplitude V and the drive wavenumber k_0 to show that there was no effect on the steady state, which suggests that the scaling property of the spectrum (3.26) is universal. The mean energy per unit length of a vortex in mode k is related to n_k by the equation [34]

$$E_K(k) = \epsilon_K k^2 n_k, \tag{3.27}$$

where ϵ_K is an effective energy per unit length of vortex, given by

$$\epsilon_K = \frac{\rho\kappa^2}{4\pi}\left[\ln\left(\frac{1}{ka}\right) + c_1\right]. \tag{3.28}$$

From Eqs. (3.26) and (3.27), the energy spectrum of the single Kelvin wave becomes $E_K(k) = A\epsilon_K(k\ell_B)^{-1}$.

When two vortices reconnect, they twist to become locally antiparallel at the reconnection point and create small cusps or kinks after the reconnection, initially confirmed numerically by Schwarz [19, 20], and later by Tsubota et al. in detail [22] using the vortex filament model. Svistunov suggested that the relaxation process of these cusps or kinks causes the emission of Kelvin waves, and plays an important role in the decay of QT at low temperatures [33]. Following Svistunov, Vinen analyzed the energy spectrum of the Kelvin-wave cascade by introducing a "smoothed" length of vortex line per unit volume after all the Kelvin waves were removed, and considered $E_{\rm K}(k)dk$, the energy per unit length of the smoothed vortex lines associated with Kelvin waves in the range k to $k + dk$ [34]. By dimensional analysis, $E_{\rm K}(k)$ was estimated as

$$E_{\rm K}(k) = A\rho\kappa^2 k^{-1} \tag{3.29}$$

with a constant A.

Kivotides et al. confirmed numerically the generation of Kelvin waves through reconnections using the vortex filament model [39]. In their simulation, four vortex rings were placed symmetrically on opposite sides and oriented so that they all moved toward the center. The four rings approached each other and underwent reconnections, which induced cusp relaxation and the generation of large-amplitude Kelvin waves. They calculated the energy spectrum $E(k)$ defined by Eq. (3.8) for the superfluid velocity, and found that $E(k)$ developed approximately a k^{-1} form after the reconnections occurred and that the reconnected vortex lines began to move farther apart. This supports the nonlinear transport of energy between different wavenumbers. Considering that the fluctuations of the velocity field were induced by the Kelvin waves on the filament, this result is consistent with Vinen's analysis of Eq. (3.29) with $E_{\rm K}(k) \sim E(k)$.

Kozik and Svistunov analyzed the Kelvin-wave cascade using weak-turbulence theory [48]. They employed the Hamiltonian representation of the vortex line motion; a Kelvin wave was excited along the z direction and the position of the vortex line was specified in the parametric form $x = x(z)$, $y = y(z)$. In terms of the complex canonical variable $w(z,t) = x(z,t) + iy(z,t)$, the Biot–Savart dynamics equation (3.2), acquires the Hamiltonian form $i\dot{w} = \delta H[w, w^*]/\delta w^*$ with

$$H[w, w^*] = \frac{\kappa}{4\pi} \int dz_1 dz_2 \, \frac{1 + \mathrm{Re}[w'^*(z_1)w'(z_2)]}{\sqrt{(z_1 - z_2)^2 + |w(z_1) - w(z_2)|^2}}, \tag{3.30}$$

where $w'(z) = dw(z)/dz$. Under the assumption that the amplitude of the Kelvin wave is small compared to its wavelength, the Hamiltonian (3.30) can be expanded in powers of $|w(z_1) - w(z_2)|/|z_1 - z_2|$: $H = E_0 + H_0 + H_1 + \cdots$ (E_0 is just a number). H_0 can be diagonalized by the Fourier transformation $w_k = \int dz \, w(z)e^{-ikz}$ to

$$H_0 = \frac{\kappa}{4\pi} \sum_k \omega_k w_k^* w_k, \tag{3.31}$$

where ω_k is the dispersion given in Eq. (3.25). By introducing kelvons, which are the quanta $\hat{w}_k$ for the amplitude of Kelvin waves w_k, the Hamiltonian can be canonically quantized with the kelvon annihilation operator $\hat{a}_k = \sqrt{\kappa\rho/2\hbar}\hat{w}_k$.

Because of the conservation of momentum and energy, the two-kelvon scattering channel is suppressed and three-kelvon scattering becomes the dominant process. The effective vertex $V_{1,2,3}^{4,5,6}$ for the three-kelvon scattering process consists of two parts; a two-kelvon vertex in the second order of perturbation associated with H_1, and the bare three-kelvon vertex associated with H_2. In the classical-field limit, the kinetic equation of the averaged kelvon occupation number $n_k = \langle a_k^\dagger a_k \rangle$ over the statistical ensemble is

$$\dot{n}_1 = 216\pi \sum_{2,\cdots,6} \left| V_{1,2,3}^{4,5,6} \right|^2 \delta(\Delta\omega)\delta(\Delta k) \left(f_{4,5,6}^{1,2,3} - f_{1,2,3}^{4,5,6} \right), \tag{3.32}$$

with $\Delta k = k_1 + k_2 + k_3 - k_4 - k_5 - k_6$, $\Delta\omega = \omega_1 + \omega_2 + \omega_3 - \omega_4 - \omega_5 - \omega_6$, and $f_{1,2,3}^{4,5,6} = n_1 n_2 n_3 (n_4 n_5 + n_5 n_6 + n_6 n_4)$. Here, $(1, \cdots, 6)$ denotes the set of the wavenumber $(k_1, \cdots, k_6)$. The kinetic equation (3.32) supports the energy cascade, if two conditions are satisfied. The first is that kinetic time grows progressively smaller in the limit of large wavenumbers, which can be checked by a dimensional estimate. The second is that the collision term is local in wavenumber space, which has been verified numerically. By the dimensional analysis, Eq. (3.32) yields

$$\dot{n}_k \propto \omega_k^{-1} n_k^5 k^{16}, \tag{3.33}$$

with $k_1 \sim \cdots \sim k_6 \sim k$ and $|V| \sim k^6$. The energy flux θ_k per unit vortex line length at momentum k is defined as $\theta_k = L^{-1} \sum_{k'<k} \omega_{k'} \dot{n}_{k'}$, where L is the system size. This implies $\theta_k \sim k \dot{n}_k \omega_k \sim n_k^5 k^{17}$ with Eq. (3.33). The cascade requirement that θ_k is k independent, without energy dissipation in the inertial range, leads to the spectrum: $n_k \propto k^{-17/5}$ and $H_0 \propto \sum_k \omega_k n_k \sim \sum_k k^{-7/5}$. To investigate the above spectrum in more detail, Kozik and Svistunov performed a numerical simulation of the full Biot–Savart law of Eq. (3.2) over a wide range of scale [49]. Starting from a single vortex line with the initial distribution $n_k \propto k^{-3}$, they found that the waves converted into a new power-law distribution $n_k \propto k^{-17/5}$.

For the spectrum of Kelvin waves, Nazarenko presented a nonlinear differential equation model, pointing out that turbulence displays a dual cascade behavior of both the direct energy cascade and an inverse cascade of wave action, which correspond to two constants of motion:

$$E = \iint \frac{d\mathbf{s} d\mathbf{s}_0}{|\mathbf{s} - \mathbf{s}_0|} \tag{3.34}$$

and

$$\mathbf{P} = \int \mathbf{s} \times d\mathbf{s} \tag{3.35}$$

for the full Biot–Savart equation (3.2) [50]. The nonlinear differential equation was constructed to preserve the main scaling of the original closure $n_k \propto k^{-17/5}$ as

$$\dot{n} = \frac{C}{\kappa^{10}} \omega^{1/2} \frac{\partial^2}{\partial\omega^2} \left(n^6 \omega^{21/2} \frac{\partial^2}{\partial\omega^2} \frac{1}{n} \right), \tag{3.36}$$

where C is a dimensionless constant, $\omega = \omega(k) = \kappa k^2/4\pi$ is the approximate dispersion of the Kelvin wave. This equation preserves the energy $E = \int d\omega\, n\omega^{1/2}$, and the wave action $N = \int d\omega\, n\omega^{-1/2}$. Equation (3.36) has both the direct cascade solution $n_k \propto k^{-17/5}$ and the inverse cascade solution $n_k \propto k^{-3}$. It also has the same thermodynamic Rayleigh–Jeans solutions $n = T/(\omega + \mu)$ with constants representing temperature T and chemical potential μ. This model was developed to include the effect of generation of Kelvin waves by reconnection, dissipation caused by mutual friction, and phonon radiation.

Boffetta et al. suggested a simpler model for the Kelvin-wave cascade than that of Svistunov et al. [48], using a truncated expansion of the LIA [51]. As is the case in Eq. (3.30), a Hamiltonian form for the LIA can be obtained as

$$H[w] = 2\frac{2\kappa}{4\pi} \ln\left(\frac{\ell}{\xi}\right) \int dz \sqrt{1 + |w'(z)|^2} = 2\beta L[w], \tag{3.37}$$

where ℓ is a length in the order of the curvature radius, and $\beta = (\kappa/4\pi)\ln(\ell/\xi)$. The Hamiltonian is proportional to the vortex length $L[w] = \int dz \sqrt{1 + |w'(z)|^2}$. The equation of motion is

$$\dot{w} = \frac{i}{2}\left(\frac{w'}{\sqrt{1 + |w'|^2}}\right)' \tag{3.38}$$

with $\beta = 1/2$. As a consequence of the invariance under phase transformation, the total wave action $N[w] = \int dz\, |w|^2$ is also conserved. In the framework of the LIA, the system becomes integrable with infinite numbers of conserved quantities, so that the energy and the wave action cannot cascade. The integrability is broken by considering a truncated expansion of the Hamiltonian in the power of the wave amplitude $w'(z)$ such as

$$H_{\mathrm{exp}}[w] = H_0 + H_1 + H_2 = \int dz \left(1 + \frac{|w'|^2}{2} - \frac{|w'|^4}{8}\right). \tag{3.39}$$

Taking into account only the six-wave resonant condition, and taking the Taylor expansion of $w(k, t)$ around $w(k, t) = c(k, 0)$ ($w(k, t) = \int dz e^{-ikz} w(z, t)$ as the Fourier transformation of $w(z, t)$), one can obtain the resulting Hamiltonian expressed in c_k and the equation of motion for $n_k = \langle |c_k|^2 \rangle$,

$$H_c = \int dk\, \omega_k |c_k|^2 + \int dk_{123456}\, C_{123456} \delta^{123}_{456} c_1^* c_2^* c_3^* c_4 c_5 c_6, \tag{3.40}$$

$$\dot{n}_k = 18\pi \int dk_{23456}\, |C_{k23456}|^2 \delta^{k23}_{456} \delta(\omega^{k234}_{456}) f_{k23456}, \tag{3.41}$$

where C_{123456} is the interaction coefficient, and

$$f_{k23456} = n_k n_2 n_3 n_4 n_5 n_6 \left[\frac{1}{n_k} + \frac{1}{n_2} + \frac{1}{n_3} - \frac{1}{n_4} - \frac{1}{n_5} - \frac{1}{n_6}\right], \tag{3.42}$$

$$\delta(\omega^{k23}_{456}) = \delta(\omega_k + \omega_2 + \omega_3 - \omega_4 - \omega_5 - \omega_6). \tag{3.43}$$

The kinetic equation (3.41) is almost the same as that of the Biot–Savart formulation (3.32) by Svistunov, and a simple dimensional analysis of Eq. (3.41) gives $\dot{n}_k \propto k^{14} n_k^5$. Besides the equilibrium Rayleigh–Jeans solution $n_k = T/(\omega_k + \mu)$, there are non-equilibrium steady state solutions of the kinetic equation (3.41) in wave turbulence theory, which rely on a constant flux in some inertial range and are known as Kolmogorov–Zakharov solutions [57]. Using a dimensional analysis, one can obtain the energy flux $\Pi_k^{(\mathrm{H})} = \int dk' \dot{n}_{k'} \omega_{k'} \sim k^{17} n_k^5$ and the wave action flux $\Pi_k^{(\mathrm{N})} = \int dk' \dot{n}_{k'} \sim k^{15} n_k^5$. Requiring the existence of two ranges of scales in which $\Pi_k^{(\mathrm{H})}$ and $\Pi_k^{(\mathrm{N})}$ are k-independent leads to the spectra,

$$n_k \sim k^{-17/5}, \tag{3.44a}$$

$$n_k \sim k^{-3}, \tag{3.44b}$$

respectively. As in the analysis by Nazarenko [50], the first spectrum is expected to be the direct cascade of energy flowing to large k, and the second is expected to be the inverse cascade of wave action flowing to small k. Finally, we mention the energy spectra estimated from the spectra in (3.44). By using Eq. (3.27) and the relation $E_{\mathrm{K}}(k) \sim E(k)$ proposed by Kivotides et al., we can obtain $E(k) \propto k^{-7/5}$ for the direct cascade and $E(k) \propto k^{-1}$ for the inverse cascade.

Yepez et al. investigated numerically the energy spectrum in the quantum region [45] using the Gross–Pitaevskii model. The obtained spectrum $E(k) \propto k^{-3}$ is inconsistent with those of the above analytical works, and the energy spectra in the quantum region remain as an open question. As described in the previous subsection, however, the relationship between the Kelvin-wave spectrum and the mean intervortex spacing l was not mentioned, and it is not clear whether the obtained spectrum can appropriately be regarded as the spectrum of the Kelvin-wave cascade.

3.2.3.5 *Energy spectra in classical–quantum crossover*

As discussed in the previous sections, there are two different types of energy spectra in the classical ($k < k_l$) and quantum ($k_l < k < k_\xi$) regions. An important question arises: how do these two energy spectra connect to each other at the length scale l? Although there are several theoretical and numerical reports on this region, consistency amongst this work has not yet been obtained, and the problem remains controversial. In analysis of classical–quantum crossover, $\Lambda = \ln(l/\xi)$ appears to be an important parameter. In typical ^{4}He experiments, Λ is about 15.

L'vov et al. suggested a bottleneck crossover between the two regions [52]. For a given mean intervortex spacing l, the mean vorticity in the system $\langle|\omega|\rangle$ is given by $\langle|\omega|\rangle \simeq \kappa l^{-2}$. The energy spectrum in the classical region is of the Kolmogorov form (Eq. (3.10)): $E_{\mathrm{cl}}(k) \simeq \varepsilon^{2/3} k^{-5/3}$, where the Kolmogorov constant C is approximately unity. Assuming that $\langle|\omega|\rangle$ is dominated by the classical–quantum crossover scale, we obtain

$$\langle|\omega|\rangle^2 \simeq \int^{1/l} k^2 E_{\mathrm{cl}}(k) \simeq \varepsilon^{2/3} l^{-4/3}. \tag{3.45}$$

In the quantum region, the reformulated kelvon occupation number n_k from the analysis by Kozik and Svistunov is $n_k \simeq (l\varepsilon)^{1/5}\kappa^{2/5}k^{-17/5}$ from dimensional analysis, and the corresponding energy spectrum becomes

$$E_{\rm qu}(k) \simeq \Lambda \left(\frac{\kappa^7 \varepsilon}{l^8}\right)^{1/5} k^{-7/5}, \tag{3.46}$$

which is consistent with Eq. (3.44a). This equation also gives the mean vorticity as

$$\langle|\omega|\rangle^2 \simeq \frac{E_{\rm qu}(k=1/l)}{l^3} \simeq \Lambda \left(\frac{\kappa^7 \varepsilon}{l^{16}}\right)^{1/5}. \tag{3.47}$$

If the energy flux ε is the same in both the classical and the quantum regions, the ratio between $E_{\rm cl}(1/l)$ and $E_{\rm qu}(1/l)$ at the crossover of $k \simeq 1/l$ is $E_{\rm qu}(1/l)/E_{\rm cl}(1/l) \simeq \Lambda^{10/3} \gg 1$. This large mismatch indicates that the energy flux carried by classical hydrodynamic turbulence cannot fully propagate through the crossover region, and that larger scale hydrodynamic motion will increase in energy up to the level $E_{\rm qu}(1/l)$; this is called the bottleneck effect (Fig. 3.4). To get a qualitative picture of the bottleneck, L'vov et al. used the warm cascade solutions following from the model

$$\varepsilon = -\frac{1}{8}\sqrt{k^{13}F_k}\frac{dF_k}{dk}, \quad F_k = \frac{E(k)}{k^2}, \tag{3.48}$$

for the energy flux of hydrodynamic turbulence [58]. Here F_k is the three-dimensional spectrum of turbulence. For a constant energy flux, the solution of Eq. (3.48) is

$$F_k = \left[\frac{24\varepsilon}{11k^{11/2}} + \left(\frac{T}{\pi\rho}\right)^{3/2}\right]^{2/3}, \tag{3.49}$$

where ρ is the fluid density. While (3.49) coincides with the Kolmogorov energy spectrum at low k, the large k range comprises a thermalized portion of the spectrum with equipartition of energy, characterized by an effective temperature T.

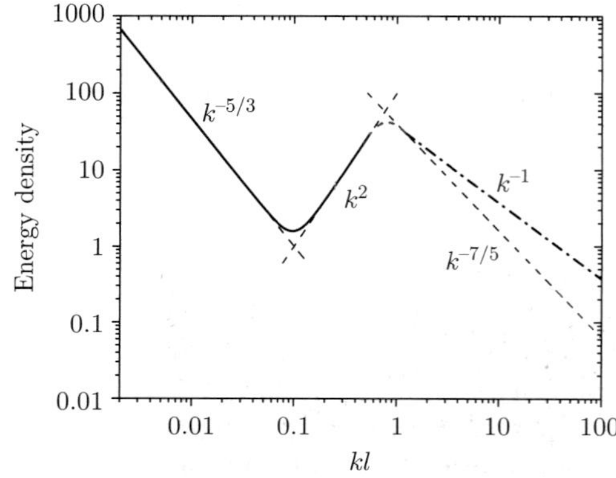

Fig. 3.4 The energy spectra $E(k)$ in the classical, $k < 1/l$, and quantum $k > 1/l$ ranges, proposed by L'vov et al. [52]. The two straight solid lines in the classical range indicate pure Kolmogorov scaling $E(k) \propto k^{-5/3}$ and pure thermodynamic scaling $E(k) \propto k^2$. In the quantum range, the dashed line indicates the Kelvin wave cascade spectrum (slope $-7/5$, whereas the dash-dotted line marks the spectrum corresponding to the noncascading part of the vortex tangle energy (slope -1) [L'vov, Nazarenko, and Rudenko: *Phys. Rev.* B **76** (2007) 024520, reproduced with permission. Copyright 2007 by the American Physical Society].

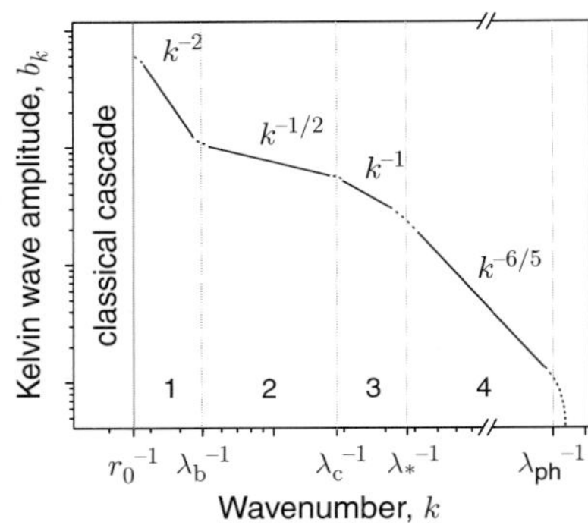

Fig. 3.5 Spectrum of Kelvin waves in the quantized regime proposed by Kozik et al. [53]. The inertial range consists of a chain of cascades driven by different mechanisms: (1) reconnections of vortex-line bundles, (2) reconnections between nearest-neighbor vortex lines in a bundle, (3) self-reconnections on single vortex lines, and (4) nonlinear dynamics of single vortex lines without reconnections [Kozik and Svistunov: *Phys. Rev.* B **77** (2008) 060502(R), reproduced with permission. Copyright 2008 by the American Physical Society].

The analysis by L'vov et al. was based on the assumption that the coarse-grained macroscopic description of quantized vorticity is effective down to the scale of l. Based on the LIA, Kozik and Svistunov suggested a different picture for the crossover region, in which the locally induced motion of the vortex lines emerges at the scale of $r_0 \sim \Lambda^{1/2} l$, and the crossover range is divided into three subranges, $r_0^{-1} < k < \lambda_b^{-1}$, $\lambda_b^{-1} < k < \lambda_c^{-1}$, and $\lambda_c^{-1} < k < \lambda_*^{-1}$, where $\lambda_b \sim \Lambda^{1/4} l$, $\lambda_c \sim l/\Lambda^{1/4}$, and $\lambda_* \sim l/\Lambda^{1/2}$ (Fig. 3.5) [53]. In the first region, $r_0^{-1} < k < \lambda_b$, polarized vortex lines are organized in bundles and reconnect with other bundles to form Kelvin waves with amplitude $b_k \sim r_0^{-1} k^{-2}$, where b_k is defined as $b_k^2 \sim L^{-1} \sum_{q \sim k} \langle \hat{a}_q^\dagger \hat{a}_q \rangle = L^{-1} \sum_{q \sim k} n_q \sim k n_k$. Here, $\hat{a}_q$ is the kelvon annihilation operator and n_q is the kelvon occupation number. In the second region, $\lambda_b^{-1} < k < \lambda_c^{-1}$, the cascade is supported by nearest-neighbor reconnections in a bundle, and $b_k \sim l(\lambda_b k)^{-1/2}$. In the third range, $\lambda_c^{-1} < k < \lambda_*^{-1}$, the cascade is driven by self-reconnection of vortex lines, giving $b_k \sim k^{-1}$. The quantum region for the Kelvin-wave cascade of a single vortex line begins from λ_*^{-1}, giving $b_k \propto k^{-6/5}$. The Kelvin-wave spectrum b_k smoothly connects these ranges, and there is no bottleneck effect in the model. Although they emphasized that the energy spectrum $E(k)$ is practically meaningful only in the classical region, we can estimate $E(k)$ from their model: $E(k) \sim k^{-3}$ in $r_0^{-1} < k < \lambda_b^{-1}$, $E(k) \sim k^0$ in $\lambda_b^{-1} < k < \lambda_c^{-1}$, and $E(k) \sim k^{-1}$ in $\lambda_c^{-1} < k < \lambda_*^{-1}$.

The theoretical works of both L'vov et al. and Kozik et al. were based on the idea that quantized vortices are locally polarized in a tangle, forming bundles. However, the nature of their polarization is not clear. Furthermore, the energy spectra estimated from their model were inconsistent with those given from the numerical work of the GP model by Yepez et al. [45] which also remained a problem in this wavenumber region, as we commented in subsection 3.2.3.3. More quantitative analysis and detailed numerical work is needed to clarify the nature of the crossover region.

3.2.3.6 *Energy spectra in quantum turbulence at finite temperatures*

Experimental studies have measured the energy spectrum of QT at finite temperatures, and supported the Kolmogorov spectrum directly or indirectly [59–65]. These experiments were also consistent, in the sense that they showed similarities between QT and CT. Vinen considered this similarity theoretically, and proposed that the superfluid and normal fluid are likely to be coupled together by mutual friction at scales larger than the intervortex spacing l, and would thus behave like a classical fluid and show a Kolmogorov energy spectrum where the mutual friction does not cause dissipation [66]. This idea was confirmed by Kivotides et al. through numerical simulation of coupled dynamics of a vortex filament and a normal fluid [67], and by L'vov et al. through theoretical analysis of the two fluid equations [68]. Although they also showed a decoupling of the two fluids at small scales, this remains a controversial topic.

3.2.3.7 *Experimental study of quantum turbulence*

Early experimental studies of QT focused on thermal counterflow, in which the normal fluid and superfluid flow in opposite directions [69]. However, as thermal counterflow has no analogy with conventional fluid dynamics, it has not enabled an understanding of the relationship between QT and CT. In the

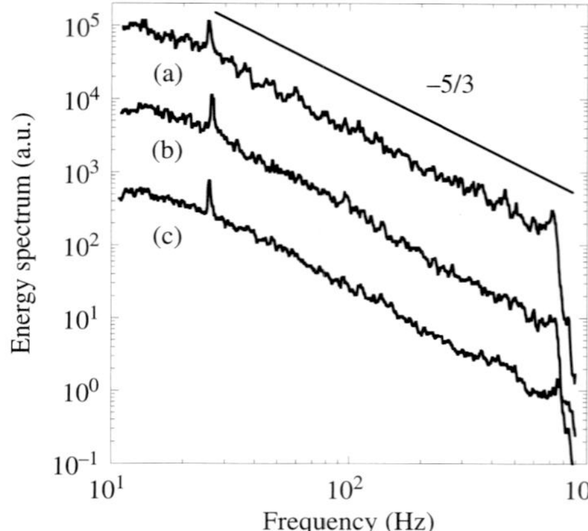

Fig. 3.6 Energy spectra obtained using counter-rotating disks at (a) 2.3 K, (b) 2.08 K, and (c) 1.4 K [60].

mid 1990s, QT experiments were performed that did not involve thermal counterflow. Maurer and Tabeling studied QT of superfluid ^{4}He that was produced in a cylinder 8 cm in diameter and 20 cm high and driven by two counter-rotating disks [60]. They observed local pressure fluctuations, and converted these into an energy spectrum. The experiments were done at three temperatures: 2.3 K ($\rho_s/\rho = 0$), 2.08 K ($\rho_s/\rho = 0.05$), and 1.4 K ($\rho_s/\rho = 0.9$), which are both above and below the λ point. In all cases, a Kolmogorov energy spectrum was observed (Fig. 3.6). Above the λ point, the obtained spectrum is reasonable because the system is a classical viscous fluid. Below the λ point, however, it is debatable whether the obtained spectra are consistent with the Kolmogorov law and independent of the ratio between ρ_s and ρ_n, namely, the dissipative mechanism. Their observations were understood on the basis of the idea that the two fluids were probably coupled together by mutual friction, and behaved as a one-component fluid [66].

Afterwards, a series of experiments with superfluid ^{4}He above 1 K were performed by the Oregon group [59, 61–65]. The helium was contained in a channel with a square (1 cm $\times$ 1 cm) cross-section, along which a grid was pulled at a constant velocity. This method of creating QT is similar to a method of generating homogeneous, isotropic turbulence in a classical fluid [35]. A pair of second-sound transducers placed in the walls of the channel detected the appearance of a vortex tangle by second-sound attenuation [32]. The decay of the vorticity of the tangle behind the grid was observed. To obtain the energy spectrum from this result, the authors made some assumptions. In a fully developed CT, the energy dissipation rate becomes $\varepsilon = \nu\langle\omega^2\rangle$, where $\langle\omega^2\rangle$ is the mean square vorticity [70]. A similar formulation could be satisfied in fully developed QT above 1 K, in which $\kappa^2 L^2$ is a measure of $\langle\omega^2\rangle$. They assumed that the dissipation rate in the experiment became

$$\varepsilon = \nu'\kappa^2 L^2, \tag{3.50}$$

with an effective kinematic viscosity ν'. They also assumed that the flows of the superfluid and normal fluid were coupled together to behave as a one-component fluid by mutual friction at scales larger than l, and Eq. (3.50) could be directly connected with the observation of second-sound attenuation by choosing a suitable value of ν' as a function of temperature [64]. The result suggests that κL decays as $\kappa L \propto t^{-3/2}$, which is indirectly consistent with the Kolmogorov energy spectrum in QT.

The relationship between $\kappa L \propto t^{-3/2}$ and the Kolmogorov energy spectrum was reviewed by Vinen and Niemela [32]. When we assume that the flow motion in the inertial range $D^{-1} < k < k_{\mathrm{d}}$ with $D^{-1} \ll k_{\mathrm{d}}$ dominates the overall fluid dynamics and gives a Kolmogorov energy spectrum (3.10), the total energy can be approximated by

$$E = \int_{D^{-1}}^{k_{\mathrm{d}}} C\varepsilon^{2/3} k^{-5/3} \sim \frac{3}{2} C\varepsilon^{2/3} D^{2/3}. \tag{3.51}$$

If the turbulence decays slowly, the dissipation rate ε can be written as

$$\varepsilon = -\frac{dE}{dt} = -C\varepsilon^{-1/3} D^{2/3} \frac{d\varepsilon}{dt}. \tag{3.52}$$

The solution of this differential equation is $\varepsilon = 27C^3 D^2/(t+t_0)^3$, with a constant t_0. This result and Eq. (3.50) give

$$L = \frac{(3C)^{3/2} D}{\kappa \nu'^{1/2}} (t+t_0)^{-3/2}. \tag{3.53}$$

The behavior $L \sim t^{-3/2}$ for large t suggests that the Kolmogorov energy spectrum in QT is independent of the value of ν', namely the temperature.

This type of decay $L \sim t^{-3/2}$ was also observed in turbulence by an impulsive spin down for superfluid ^{4}He [71], in turbulence by injecting negative ions into superfluid ^{4}He [72], and for grid turbulence in superfluid ^{3}He-B [73], with little normal fluid at low temperatures, which supports the classical analog of QT with no normal fluid component.

The Helsinki group studied experimentally vortex dynamics of propagation into a region of vortex-free flow in a rotating superfluid ^{3}He-B [74]. They measured the velocity of the vortex front toward the metastable region, and determined the rate of dissipation as a function of temperature. The results showed a transition from laminar through quasiclassical turbulent to quantum turbulent flow with decreasing temperature. Below $0.25\,T_{\mathrm{c}}$ with a superfluid critical temperature T_{c}, there was a peculiar decrease of dissipation. As one possibility, the authors suggested that the energy flux toward small scales propagates to the scale l, and vortex discreteness and quantization effects become important. The dominant part of the energy loss was the Kelvin-wave cascade below the scale l. As proposed by L'vov et al., Kelvin waves are much less efficient in downscale energy transfer than classical turbulence, which leads to a bottleneck accumulation of kinetic energy [52].

3.2.3.8 *Decay of vortices in quantum turbulence at low temperatures*

In this section, we consider the decay process of QT at low temperatures, at which the normal fluid component is negligible and mutual friction does not occur. In this case, there is no dissipation in vortex lines at large scales and dissipation occurs only at scales comparable to the core radius ξ. Energy at large scales cannot dissipate, but flows to smaller scales via the Richardson cascade and the Kelvin-wave cascade [32].

Feynman first proposed a dissipation mechanism of QT at zero temperature [10]. In this mechanism, large vortex loops are broken up into smaller loops via the Richardson cascade, and the smallest vortex rings, with radii comparable to the atomic scale, decay into excitations, such as rotons. This mechanism is not currently accepted. Later, Vinen considered the decay of superfluid turbulence at finite temperatures in an examination of his experimental results from thermal counterflow, and proposed Vinen's equation [15]. At finite temperatures, vortices in turbulence are randomly spaced with no polarization, and there is only one length scale l. The energy and the decay of vortices spreads over a wide range of scales. The decay of the total energy is expressed in terms of a characteristic velocity $v_{\mathrm{s}} = \kappa/2\pi l$ and a characteristic time constant $\tau = l/v_{\mathrm{s}}$ as

$$\frac{dv_{\mathrm{s}}^2}{dt} = -\chi\frac{v_{\mathrm{s}}^2}{\tau} = -\chi\frac{v_{\mathrm{s}}^3}{l}, \tag{3.54}$$

where χ is a temperature-dependent dimensionless parameter [15]. Incorporating the vortex line density $L = l^{-2}$, this equation becomes

$$\frac{dL}{dt} = -\chi\frac{\kappa}{2\pi}L^2. \tag{3.55}$$

This is a type of Vinen's equation without a generation term proportional to $(v_{\mathrm{s}} - v_{\mathrm{n}})L^{3/2}$ due to thermal counterflow. The solution of this equation is

$$\frac{1}{L} = \frac{1}{L_0} + \chi\frac{\kappa}{2\pi}t, \tag{3.56}$$

where $L_0 = L(t = 0)$. At large t, L behaves as $L \propto t^{-1}$. The experimental observations of thermal counterflow can be explained by this solution, and χ can be obtained as a function of temperature. Although the decay of QT at finite temperatures is due to mutual friction, this solution can describe QT at zero temperature without mutual friction. Using a vortex filament model, Tsubota et al. performed numerical simulations of QT without mutual friction, and estimated the value of χ [22]. Equation (3.56) can also be formulated by assuming that the motion of the quantized vortices dominates the fluid dynamics, and their energy $\Lambda/(4\pi)\kappa^2 L$ is their primary contribution to the total energy [72]. By using Eq. (3.50), we obtain

$$\frac{dE}{dt} = \frac{\Lambda}{4\pi}\kappa^2\frac{dL}{dt} = -\nu'\kappa^2L^2. \tag{3.57}$$

The solution of this differential equation is

$$\frac{1}{L} = \frac{1}{L_0} + \frac{\Lambda}{4\pi\nu'}t^{-1}, \tag{3.58}$$

which is the same result as Eq. (3.56). The effective kinematic viscosity ν' in Eq. (3.53) and that in Eq. (3.58) are naturally different. We denote them as ν'_{K} for Eq. (3.53) and ν'_{V} for Eq. (3.58) [72].

Two important questions arise. What causes the difference in decay between Eqs. (3.58) ($L \propto t^{-1}$) and (3.53) ($L \propto t^{-3/2}$), and what is the origin of the decay mechanism at zero temperature? The answer to the first question comes from the structural differences of vortex tangles in QT. When a vortex tangle supports the Kolmogorov energy spectrum, vortices form an inertial range in which the tangle is self-similar, creating polarized vortex bundle structures that are different from a completely random distribution of vortex lines in QT at finite temperatures. In this case, QT decays as $L \propto t^{-3/2}$. However, when a vortex tangle is dilute and random, with no correlation, there is only one length scale l in the tangle and QT decays as $L \propto t^{-1}$, even at zero temperature. This different behavior of L enables us to understand the structure of a vortex tangle in QT. The behavior of $L \propto t^{-1}$ has been observed in turbulence by injecting negative ions into ^{4}He [72] and by a vibrating grid in ^{3}He-B [73]. As discussed in the previous subsection, the behavior of $L \propto t^{-3/2}$ has been observed in turbulence by an impulsive spin down of superfluid ^{4}He, by injecting negative ions into superfluid ^{4}He [71], and by a vibrating grid in superfluid ^{3}He-B [73]. Bradley et al. also observed the crossover from $L \propto t^{-3/2}$ to $L \propto t^{-1}$ behavior of grid turbulence in ^{3}He [73]. The structure of a generated vortex tangle depends on the velocity of the grid, and the decay of the turbulence becomes $L \propto t^{-3/2}$ for a dense vortex tangle with a rapidly oscillating grid, and $L \propto t^{-1}$ for a dilute vortex tangle with a slowly oscillating grid. By using Eqs. (3.53) and (3.58), Walmsley and Golov estimated the effective kinematic viscosity $\nu'_{\rm K}$ and $\nu'_{\rm V}$ for turbulence in ^{4}He and showed that they are almost the same at high temperatures $T > 1.0\,\mathrm{K}$ for a normal fluid, and that $\nu'_{\rm K}$ becomes much smaller than $\nu'_{\rm V}$ at low temperatures with no normal fluid [72].

For the second question, there are several possible answers. The first is acoustic emissions at vortex reconnections, similar to that of eddies in a classical fluid. Numerical simulations of the GP model support acoustic emissions at every reconnection [26, 27]. The energy dissipation for each reconnection is about 3ξ times the vortex line energy per unit length: $\sim \Lambda\kappa^2/4\pi$. In the case of superfluid ^{4}He, however, ξ is quite small and the dissipation due to reconnections can be negligible. This dissipation mechanism may emerge for the turbulent state of atomic Bose–Einstein condensates [30]. By using the GP model, Kobayashi and Tsubota investigated the decay from fully developed turbulence and reported both $L \propto t^{-3/2}$ to $L \propto t^{-1}$ behaviors [75]. In their simulation, ξ is close to the system size, and acoustic emission upon reconnection is the main dissipation mechanism. Another possible answer is the radiation of phonons from high-frequency Kelvin waves [34]. Vortex reconnections excite Kelvin waves whose wavelength is of the order of l [19, 33, 22, 39]. Although Kelvin waves with wavelength $\sim l$ cannot cause effective radiation, the Kelvin-wave cascade creates waves with much shorter wavelengths [34, 47–51].

In actual experiments, one must also consider vortex diffusion as an origin of the decrease of L. When a vortex tangle is inhomogeneous or local and the probe observing the turbulence is also local, a vortex tangle may escape from the observable region. Using the vortex filament model, Tsubota et al. studied numerically the diffusion of an inhomogeneous vortex tangle [76]. The effects of diffusion can be quantitatively evaluated by the equation

$$\frac{\partial L(\mathbf{x},t)}{\partial t} = -\chi\frac{\kappa}{2\pi}L(\mathbf{x},t)^2 + D\nabla^2 L(\mathbf{x},t). \tag{3.59}$$

Here, $L(\mathbf{x},t)$ is the space-dependent line length density, and D is the diffusion constant. The numerical simulation indicated that $D \sim 0.1\kappa$

3.2.3.9 *Quantum turbulence created by vibrating structures*

Recently, vibrating structures, such as discs, spheres, grids, wires, and tuning forks, have been widely used for research into QT [77]. In spite of detailed differences between the structures, the experiments have shown some surprisingly common phenomena. This trend started with the pioneering observation of QT on an oscillating microsphere by Jäger et al. [78]. Subsequently, many groups have investigated experimentally the transition to turbulence in superfluid ^{4}He and ^{3}He-B by using grids [79–83, 73], wires [84–88], and tuning forks [89, 90]. The details of these observations were described in a review article [77].

Here, we will describe briefly the essence of the observations by referring to a typical result from Yano et al. [91]. A thin superconducting (typically NbTi) wire with a micron-size radius is formed into a semicircle, and its two edges are attached to the wall of a vessel. The wire vibrates in resonance with a Lorentz force due to an alternating electric current under a static magnetic field. Figure 3.7 shows the response velocity of the wire as a function of the driving force. When the driving force is relatively low, the wire moves smoothly. However, if the driving force exceeds some value, the velocity suddenly drops, indicating that the wire does not vibrate so much despite an increase of the driving force. The energy injected from the drive must go somewhere, and the only possible escape is the excitation of quantum turbulence. The response of the wire shows clearly a hysteresis between the upward and downward sweeps of the driving force.

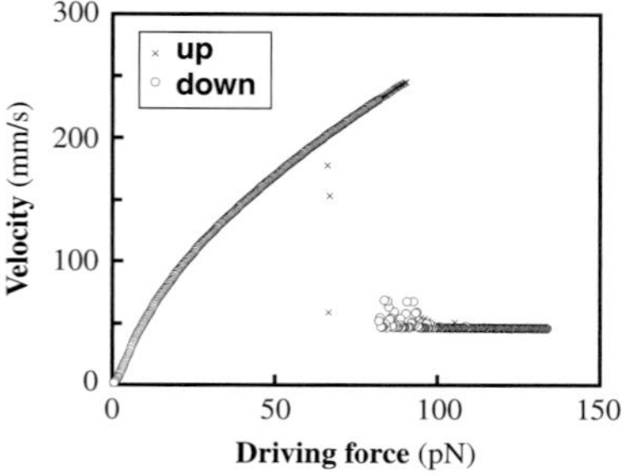

Fig. 3.7 Typical response of a vibrating wire in superfluid ^{4}He at 30 mK. Courtesy of H. Yano.

Experimental studies conducted by many groups have reported common behaviors, independent of the structural details, such as type, shape, and surface roughness. However, there were also some differences. The key points are the critical velocity for the onset of turbulence, the shift of the resonance frequency, and the hysteresis and the drag coefficient.

The observed critical velocities were in the range 1 mm/s to approximately 200 mm/s, much lower than the velocity necessary for intrinsic nucleation of quantized vortices, which is in the order of 10 m/s. It is known that there are initially some remnant vortices in superfluid ^{4}He [92], and the transition to turbulence should come from their extension or amplification. This scenario was confirmed by numerical simulations using the vortex filament model [94] and an experiment using a wire free from remnant vortices [88], which will be described later. However, it should be noted that this scenario is applicable only to ^{4}He. The transition to turbulence observed in the B phase of superfluid ^{3}He [83, 73] can be related to the intrinsic nucleation of vortices, which is known to occur near a solid surface at very small velocities, of order 4 mm/s, depending on the surface roughness [93]. This issue was discussed in detail in a review article [77]. The dependence of the critical velocity v_c on the oscillation ω is very important. Most experiments have provided evidence for a universal scaling property $v_c \sim \sqrt{\kappa\omega}$, independent of the geometry of the vibrating structure. This relation is trivial from dimensional analysis, but it is necessary to investigate its physical origin. Hänninen and Schoepe discussed the scaling property [95]. This relation was obtained from the "superfluid Reynolds number" $Re = v\ell/\kappa$ with some characteristic length scale ℓ [96]. It is also possible to understand the scaling by extending the scenario of the dynamical behavior of a vortex tangle in counterflow turbulence at constant velocity [97] to the case of oscillating flows. However, more detailed investigation of the issue is required.

The transition to turbulence is accompanied by a characteristic shift of the resonance frequency. In the wire experiments Yano et al. observed that the resonance frequency decreased with increasing drive force in the laminar regime, but increased in the turbulent regime [86]. Bradley et al. also observed an upward shift of the resonance frequency at the onset of turbulence [82]. Such an increase in the resonance frequency is surprising, because it indicates either an increased stiffness or a decreased effective mass of the wire. If vortices were amplified and a tangle surrounds the wire, the effective mass should increase. The reasons for the frequency shift are not yet resolved, and more study is required.

Many groups have reported some hysteresis in the transition to turbulence, which arises from the stability of the laminar and turbulent states. By using an oscillating microsphere, Jäger et al. observed that the transition from laminar to turbulent response was accompanied by significant hysteresis at low temperatures [78]. Below 0.5 K, turbulence was observed to be unstable and the flow switched intermittently between turbulent and laminar flow. Schoepe analyzed the lifetime of turbulence in connection with the statistical fluctuations of the vortex line length density [98]. Yano et al. [91] and Bradley et al. [82] also observed hysteresis when using vibrating wires.

The net drag force is another important quantity when we consider the transition to QT. The drag force F is often expressed in terms of a drag coefficient C_{D}, defined by the equation [99]

$$F = \frac{1}{2} C_{\mathrm{D}} \rho A U^2, \tag{3.60}$$

where ρ is the density of the fluid, A is the projected area of the object normal to the flow, and U is the flow velocity. The behavior of C_{D} is well known in a steady classical flow past an obstacle. In the case of laminar potential flow of an ideal fluid past a cylinder, the flow is symmetric about the plane through the center of the cylinder and normal to the flow. As a result, the net force vanishes with $C_{\mathrm{D}} = 0$ (the d'Alembert paradox). For laminar viscous flow, the drag force is approximately proportional to U so that $C_{\mathrm{D}} \sim U^{-1}$ (the Stokes formula). In the case of flow at a high Reynolds number, the flow behind the obstacle accompanies a wake, which destroys the symmetry and leads to the d'Alembert paradox, and C_{D} becomes of order unity. The value of C_{D} depends on the geometry of the obstacle [99]; C_{D} is approximately unity for a disc or cylinder, and about 0.3 for a sphere. Also, in quantum turbulence created by vibrating structures, it is possible to obtain the drag coefficient from the dependence of velocity on the driving force, as in Fig. 3.7. Much of the data on microspheres [78], mm-scale spheres [100], grids [79, 80], and quartz forks [90] show the classical analog behavior in which the drag coefficient is about U^{-1} in the laminar regime and of order unity in the turbulent regime. This is another classical analog of QT. Although the experimental results are trivial, we need a clearer understanding of the phenomena. The drag coefficient characteristic of the turbulent regime was numerically confirmed by simulation of the vortex filament model [101].

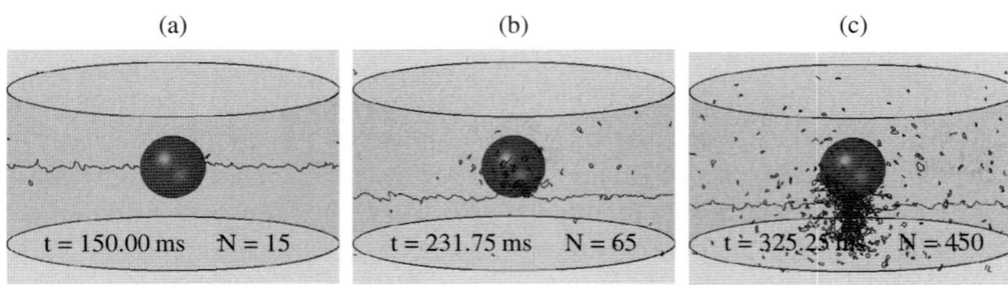

Fig. 3.8 Evolution of the vortex line near a sphere of radius 100 μm in an oscillating superflow of 150 mm^{-1} at 200 Hz [Hänninen, Tsubota and Vinen: *Phys. Rev.* B **75** (2007) 064502, reproduced with permission. Copyright 2007 by the American Physical Society].

The remnant vortices play several important roles in these phenomena, which are discussed in the remainder of this subsection.

Hänninen et al. performed a numerical simulation using the vortex filament model, and described how remnant vortices develop into a tangle under an AC superflow, as shown in Fig. 3.8 [94]. A smooth solid sphere of radius 100 μm was placed in a cylindrical vessel filled with superfluid ^{4}He. Generally, we do not know how remnant vortices remain in a given geometry, but here they were assumed initially to extend between the sphere and the vessel wall. When an oscillating superflow of 150 mm^{-1} at 200 Hz was applied, Kelvin waves resonant with the flow were gradually excited along the remnant vortices. The amplitude of the Kelvin waves grew large enough to lead to self-reconnection and the emission of small vortex rings. These vortices gathered around the stagnation points, repeatedly reconnected, and were amplified by the flow, eventually developing into a vortex tangle. This simulation likely captured the essence of what occurred in the experiments. However, the simulation was not necessarily satisfactory, and some problems remain. The first is the critical velocity. The situation of Fig. 3.8 is quite similar to the experiment using a microsphere [78] and gives a critical velocity of about 120 mm^{-1}, which is much larger than the observed value of about 40 mm^{-1}. The second problem is hysteresis. The simulation does not produce any hysteresis between upward and downward sweeps. If the velocity is taken above the critical value for the formation of a tangle, and then reduced below this value, the vortices immediately decay, eliminating the turbulent state. These two difficulties may arise from the artificially smooth surface of the sphere. A solid surface is generally rough, and even tiny bumps can act as pinning sites for quantized vortices with such a thin core [19, 102]. If such pinning effects on the surface are taken into account, they should induce additional disturbance of vortices and help the transition to turbulence, thereby reducing the critical velocity. Regarding the hysteresis, some vortices may be trapped by pinning sites through the downward sweep, which can make the physics different from the upward sweep. However, we do not know how to consider the effect of surface roughness, namely multi-pinning sites for numerical vortex dynamics, and development of applicable methods is urgently needed.

The roles of remnant vortices were revealed by several interesting experiments by the Osaka City University group. First, Hashimoto et al. succeeded in setting a vibrating wire free from remnant vortices using a unique idea [87]. They added a small chamber with a pinhole in a sample cell, and filled the chamber extremely slowly with superfluid ^{4}He. The vortices were likely filtered out by the pinhole, providing a vibrating wire free from remnant vortices. The wire never caused a transition to turbulence, even when the velocity exceeded 1 m/s. Next, Goto et al. set two wires in the small chamber. By using the same procedure, they obtained one wire (labeled A) free from remnant vortices and another (labeled B) with remnant vortices [88]. Wire B can create turbulence by itself from remnant vortices and the resulting tangle emits vortex rings. Although wire A alone never caused turbulence, it could create turbulence if it received seed vortices from wire B. This observation clearly demonstrated that the turbulence arises from remnant vortices. This behavior was confirmed by numerical simulation of the vortex filament model [88, 101].

3.2.3.10 *Visualization of quantized vortices and turbulence*

There has been little direct experimental information regarding the flow in superfluid ^{4}He. This is mainly because usual flow visualization techniques are not applicable to cryogenic superfluids. However, this situation is changing rapidly [103]. For QT, one can seed the fluid with tracer particles in order to visualize the flow field and possibly quantized vortices, which are observable by appropriate optical techniques.

A significant contribution was made by Zhang and Van Sciver [104]. Using a particle image velocimetry (PIV) technique with 1.7 μm-diameter polymer particles, they visualized a large-scale turbulent flow both in front of and behind a cylinder in a counterflow in superfluid ^{4}He at finite temperatures. In classical fluids, such large-scale turbulent structures are seen downstream of objects such as cylinders, with the structures periodically detaching to form a vortex street. In the present case of ^{4}He counterflow, the locations of the large-scale turbulent structures were relatively stable, and they did not detach and move downstream, although local fluctuations in the turbulence were evident.

Another significant contribution was the visualization of quantized vortices by Bewley et al. [105]. In their experiments, the liquid helium was seeded with solid hydrogen particles smaller than 2.7 μm at a temperature slightly above T_λ, after which the fluid was cooled to below T_λ. When the temperature was above T_λ, the particles were seen to form a homogeneous cloud that dispersed throughout the fluid. However, on passing through T_λ, the particles coalesced into web-like structures, as shown in Fig. 3.9. Bewley et al. suggested that these structures represent decorated quantized vortex lines. They reported that the vortex lines appear to form connections rather than remaining separated, and were homogeneously distributed throughout the fluid. The fork-like structures may indicate that several vortices were attached to the same particle, as indicated by numerical simulations of vortex pinning [106]. Using the same technique, Paoletti et al. obtained the trajectories of tracer particles and visualized thermal counterflow [107]. The observed trajectories showed two distinct types of behavior. One group consisted of trajectories that moved in the direction of the heat flux, while the other consisted of those that opposed this motion. The former trajectories were smooth and uniform, but the latter could be quite erratic. Particles of the former trajectories were probably dragged by the normal fluid, while those of the latter were trapped in vortex tangles.

Here, it is necessary to understand whether such tracer particles follow the normal flow or the superflow, or an even more complex flow. Poole et al. studied this problem theoretically and numerically, and showed that the situation

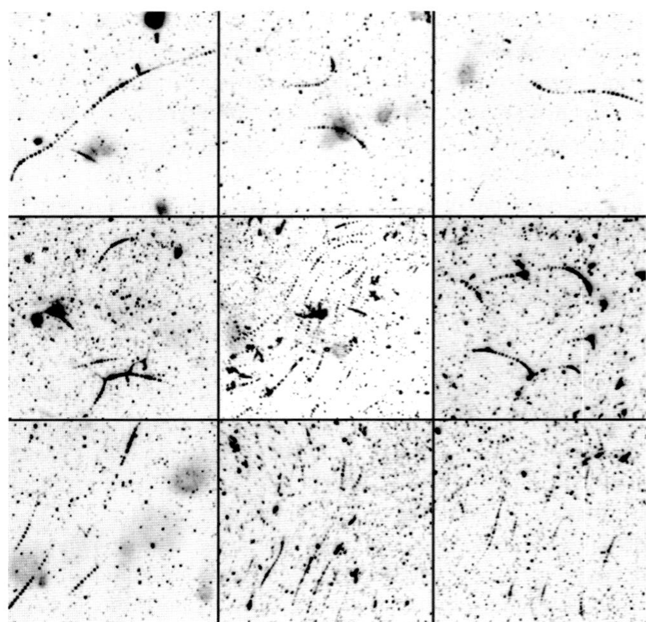

Fig. 3.9 Visualization of quantized vortices by intensity-inverted images. Each image shows the full field of view of 6.78 × 6.78 mm^2. Multiple particles are often trapped in each visible vortex, and trapped particles tend to be uniformly spaced along the vortex core [107]. Courtesy of E. Fonda, K. Gaff, M. S. Paoletti, and D. P. Lathrop.

changes depending on the size and mass of the tracer particles [108]. However, the situation is so complicated that many issues remain unanswered [109–112].

3.3 Quantized vortices in atomic Bose–Einstein condensates

3.3.1 Introduction

The achievement of Bose–Einstein condensation in trapped atomic gases at ultra-low temperatures [113–115] has stimulated intense experimental and theoretical activity in modern physics. In this section, we review the physics of quantized vortices in atomic-gas BECs. While quantized vortices have been thoroughly studied in superfluid helium [69], there has been a resurgence of interest in vortices in atomic BECs because of the peculiar features of this system:

(1) *Small gas parameter*
Because the gas is dilute, the GP equation (3.5) gives a quantitatively accurate description of the static and dynamic properties of the atomic condensate [30, 116]. Therefore, the vortex structure and dynamics can be discussed by a more fundamental approach than with superfluid helium, in which the interaction cannot be described in such a simple local form. Furthermore, the diluteness of a gas leads to a relatively large healing length that characterizes the vortex core size, thus enabling direct experimental visualization of vortex cores [117–121].

(2) *Trapping potential*
The finite size effect of the trapping potential and the associated density inhomogeneity yield new characteristics of vortex physics. This also provides a new spectroscopy to study characteristic vortex dynamics through collective motion of the inhomogeneous condensate [122–125].

(3) *Laser manipulation and rotation*
The manipulation of a condensate wave function via external (magnetic or optical) fields provides a versatile scheme to control the vortex states. Laser beams can yield effective repulsive or attractive forces on atoms, creating confining potentials, periodic lattice potentials, and impurity/obstacle potentials for atomic condensates. Vortex formation can be achieved by the mechanical rotation of a laser-created optical spoon [118–120] or the transfer of orbital angular momentum from a Laguerre–Gaussian laser beam [126]. The tunability of the rotation over a wide range is useful for studying various vortex phases and their dynamics in trapped BECs. This provides an opportunity to study a rapidly rotating condensate with a dense vortex lattice [119, 127, 128], which is not easy to accomplish in superfluid helium systems. Such rapidly rotating BECs present a so-called mean-field quantum-Hall regime described by the orbitals of the lowest Landau levels (LLL), and also a regime dominated by a strong many-body correlation [129–131]. Moreover, laser-induced coherent transition of atoms between different hyperfine levels can control the phase profile in such a way that the condensate has a circulation [117]. Recently, the laser field has played the role of an artificial vector potential [132, 133], which could be an alternative means of manipulating the rotational properties of condensates.

(4) *Detection*
The vortex cores can be visualized directly through the observed density profile by a time-of-flight (TOF) technique [117–120]. The TOF technique involves switching off the trapping field (magnetic or optical) at time $t = 0$ and taking an image of the BEC several (typically 5 to 25) milliseconds later. Switching off the trap allows the atomic gas to expand until a laser beam probe becomes available to observe the density profile. Images are taken by shining a resonant laser beam into the atomic gas and using a CCD camera to observe the shadow cast by the absorption of photons, from which the integrated atomic density can be determined.

The matter–wave interference technique is also applicable to vortex detection [134, 135]. The presence of vortices can be revealed by finding the dislocations in the interference fringes between expanding condensates [136]. In addition, the Bragg scattering of matter waves from laser beams can be used to detect the vortex state, because the behavior of the scattered condensate is sensitive to the spatial phase distribution of the initial state [137, 138].

(5) *Multicomponent condensates*
Since it is possible to load and cool atoms in more than one hyperfine spin state or more than one atomic element in the same trap,

multicomponent BECs with internal degrees of freedom can be created experimentally [139–148]. Multicomponent condensates allow the formation of various unconventional topological defects with complex properties that arise from interactions between different components, providing a new platform for the study of unconventional vortices. This content relates closely to other fields of physics, such as superfluids ^{3}He [149], unconventional superconductors [150], quantum Hall systems [151], as well as high-energy physics and cosmology [152]. Atomic BEC is an extremely flexible system for the study of such topological defects, since optical techniques allow the control of the condensate wave functions. An external field can couple the internal sublevels of the condensate and cause a coherent transition of the population. This coherent transition can be used to control the spatial variation of the condensate wave functions, resulting in an "imprinting" of a phase pattern onto the condensate. In most schemes, the spatial configuration of the field, the intensity and detuning of the laser fields, and the phase relationship between the different fields must be carefully controlled to create the correct phase pattern of topological excitations through the complex internal dynamics.

(6) *Feshbach resonance*
The most salient feature of the cold atom system is that a field-induced Feshbach resonance can tune the s-wave scattering length between atoms [153], which determines the strength of atom–atom interaction. A Feshbach resonance occurs when a quasi-bound molecular state in a closed channel has an energy equal to that of two colliding atoms in an open channel. This technique can change the scattering length over a wide range, from negative to positive values, thus creating condensates with strongly attractive interactions [154], strongly repulsive interactions [155], or dominant long-range dipole–dipole interactions [156]. It also provides new physics associated with the formation of quasi-bound molecules, provided by the coherent transition between an atomic condensate and a molecular one [157]. In the case of fermions, this has been used to induce pairing of two fermions, which results in the formation of fermion condensates by controlling the character of the two-body interactions from strongly bound bosonic molecules to weakly interacting fermion (Cooper) pairs [158]. Vortices in these exotic condensates are expected to show quite intriguing properties. A review of Feshbach resonance was recently given by Chin et al., [159].

There are several excellent reviews of quantized vortices in ultracold atomic BECs. The basics of the theoretical formulation and early research on the physics of quantized vortices were reviewed by Fetter and Svidzinsky [160]. Recent progress of the extensive study of quantized vortices is described in Refs. [161, 162]. These reviews were mainly concerned with the properties of a conventional single-component BEC, whose order parameter is scalar and interaction is characterized only by an s-wave scattering length. In this section, therefore, we will also discuss vortex physics in unconventional BECs, which was not covered by these reviews. We begin with the basic theory and experiments on quantized vortices in conventional BECs. Then, we describe vortices in some unconventional condensates realized in ultra-cold atomic systems, such as spinor condensates and dipolar condensates.

3.3.2 Vortices in single-component Bose–Einstein condensates

First, we review vortices in a single-component BEC in a trapping potential from both theoretical and experimental points of view. A more detailed review and references can be found in Refs. [161, 162]. The contents of this section will give the basic information necessary for subsequent sections. After reviewing the theoretical formulation for trapped BECs, we describe the properties of a single vortex and a vortex lattice in trapped BECs, together with various interesting current topics.

3.3.2.1 *Theoretical description of ultra-cold atomic-gas BECs*

GROSS–PITAEVSKII EQUATION

The many-body Hamiltonian for bosons in a trapping potential $V_{\rm ex}(\mathbf{r})$ is given by

$$\hat{H} = \int d\mathbf{r}\hat{\psi}^{\dagger}(\mathbf{r})\left[-\frac{\hbar^2\nabla^2}{2m} + V_{\rm ex}\right]\hat{\psi}(\mathbf{r}) + \frac{1}{2}\int d\mathbf{r}\int d\mathbf{r}'\hat{\psi}^{\dagger}(\mathbf{r})\hat{\psi}^{\dagger}(\mathbf{r}')V(\mathbf{r}-\mathbf{r}')\hat{\psi}(\mathbf{r}')\hat{\psi}(\mathbf{r}) \tag{3.61}$$

with the bosonic field operator $\hat{\psi}(\mathbf{r})$ and the two-body interaction $V(\mathbf{r}-\mathbf{r}')$. The dynamics of $\hat{\psi}$ are governed by the Heisenberg equation $i\hbar\partial\hat{\psi}/\partial t = [\hat{\psi}, \hat{H}]$. In the low temperature limit, the field operator can be approximated by a classical field known as the "condensate wave function" $\Psi = \langle\hat{\psi}\rangle$, and the trapped BECs can thus be described by the GP equations:

$$i\hbar\frac{\partial\Psi(\mathbf{r},t)}{\partial t} = \left[-\frac{\hbar^2\nabla^2}{2m} + V_{\rm ex} + g|\Psi(\mathbf{r},t)|^2\right]\Psi(\mathbf{r},t). \tag{3.62}$$

The trapping potential typically has the form of a harmonic oscillator $V_{\rm ex} = (m/2)(\omega_x^2x^2 + \omega_y^2y^2 + \omega_z^2z^2)$. Comparison of the kinetic and trap energies introduces a characteristic length scale $a_{\rm ho} = \sqrt{\hbar/m\omega}$ with $\omega = (\omega_x\omega_y\omega_z)^{1/3}$. The time-dependent GP equation (3.62) can be used to explore the dynamic behavior of the condensate, which is characterized by variations of the order parameter over distances larger than the mean distance between atoms. This equation is valid when the s-wave scattering length is much smaller than the average distance between atoms, and the number of atoms in the condensate is much larger than unity. The success of the quantitative description of the trapped BECs is described in books by Pethick and Smith [30], and Pitaevskii and Stringari [116].

To determine the ground state of a trapped BEC, we can write the condensate wave function as $\Psi(\mathbf{r},t) = \Phi(\mathbf{r})e^{-i\mu t/\hbar}$, where $\Phi(\mathbf{r})$ obeys the time-independent GP equation

$$\left[-\frac{\hbar^2\nabla^2}{2m} + V_{\rm ex} + g|\Phi(\mathbf{r})|^2\right]\Phi(\mathbf{r}) = \mu\Phi(\mathbf{r}). \tag{3.63}$$

Here, Φ is normalized to the number of condensed particles $\int d\mathbf{r}|\Phi(\mathbf{r})|^2 = N_0$, which determines the chemical potential μ. In the dilute limit (the gas parameter $\bar{n}|a|^3$ is typically less than 10^{-3}, where $\bar{n}$ is the average density of the gas) the condensed particles N_0 can be approximated as the total number N, because depletion of the condensate is small as $N' = N - N_0 \propto \sqrt{\bar{n}|a|^3}N \ll N$. The time-independent GP equation (3.63) is also derived by minimizing the GP energy functional

$$E[\Phi, \Phi^*] = \int d\mathbf{r}\Phi^*\left(-\frac{\hbar^2\nabla^2}{2m} + V_{\rm ex} + \frac{g}{2}|\Phi|^2\right)\Phi \equiv E_{\rm kin} + E_{\rm tr} + E_{\rm int}, \tag{3.64}$$

subject to the constraint of a fixed particle number N. This constraint is taken into account by the Lagrange multiplier μ as $\delta(E - \mu N)/\delta\Phi^* = 0$, where μ is the chemical potential that ensures a fixed N.

Although the exact solutions of the ground state can generally be obtained only by solving Eq. (3.63) numerically, an approximate analytic solution can be found when the interaction energy $E_{\rm int}$ is much larger than $E_{\rm kin}$ [163]. To see this, let us neglect the anisotropy of the harmonic potential and assume that the cloud occupies a region of radius $\sim R$, so that $n \sim N/R^3$. Thus, the scale of the harmonic oscillator energy per particle is $\sim m\omega^2R^2/2$, while each particle experiences an interaction with the other particles of energy $\sim gN/R^3$. By equating these energies, the radius is found to be $R \sim a_{\rm ho}(8\pi Na/a_{\rm ho})^{1/5}$. Since the kinetic energy is of order $\hbar^2/2mR^2$, the ratio of the kinetic to interaction (or trap) energies is $\sim (Na/a_{\rm ho})^{-4/5}$. In the limit $Na/a_{\rm ho} \gg 1$, which is relevant to most of the experiments, the repulsive interactions significantly expand the condensate, so that the kinetic energy associated with the density variation becomes negligible compared to the trap and interaction energies. As a result, the kinetic-energy operator can be omitted in Eq. (3.63), which gives the Thomas–Fermi (TF) parabolic profile for the ground-state density

$$n(\mathbf{r}) \simeq |\Phi_{\rm TF}(\mathbf{r})|^2 = \frac{\mu - V_{\rm ex}(\mathbf{r})}{g}\Theta\left[\mu - V_{\rm ex}(\mathbf{r})\right], \tag{3.65}$$

where $\Theta(x)$ denotes the step function. The resulting ellipsoidal density in three-dimensional (3D) space is characterized by two types of parameters: the central density $n_0 = \mu/g$ and the three condensate radii $R_j^2 = 2\mu/m\omega_j^2$ $(j = x, y, z)$. The chemical potential μ is determined by the normalization $\int d\mathbf{r}n(\mathbf{r}) = N$ as $\mu = (\hbar\omega/2)(15Na/a_{\rm ho})^{2/5}$. In the TF regime, the time-dependent GP equation (3.62) reduces to the incompressible hydrodynamic equation for superfluids [116].

THE BOGOLIUBOV–DE GENNE EQUATION

The spectrum of elementary excitations of the condensate is an essential ingredient in calculations of thermodynamic properties. To study the low-lying collective-excitation spectrum of trapped BECs, the Bogoliubov–de Gennes (BdG) equation coupled with the GP equation is a general formalism. Let us consider the equation of motion for a small perturbation around the stationary state Φ, which is a solution of Eq. (3.63). The wave function may take the form $\Psi(\mathbf{r},t) = [\Phi(\mathbf{r}) + u_j(\mathbf{r})e^{-i\omega_j t} - v_j^*(\mathbf{r})e^{i\omega_j t}]e^{-i\mu t}$. Inserting this ansatz into Eq. (3.62), and retaining terms up to the first order in u and v, we obtain the BdG equation:

$$\begin{pmatrix} L(\mathbf{r}) & -g\Phi(\mathbf{r})^2 \\ g\Phi^*(\mathbf{r})^2 & -L(\mathbf{r}) \end{pmatrix} \begin{pmatrix} u_j(\mathbf{r}) \\ v_j(\mathbf{r}) \end{pmatrix} = \hbar\omega_j \begin{pmatrix} u_j(\mathbf{r}) \\ v_j(\mathbf{r}) \end{pmatrix}, \tag{3.66}$$

where $L(\mathbf{r}) = -\hbar^2\nabla^2/2m + V_{\text{ex}} - \mu + 2g|\Phi(\mathbf{r})|^2$, and ω_j are the eigenfrequencies related to the quasiparticle normal-mode functions $u_j(\mathbf{r})$ and $v_j(\mathbf{r})$. The mode functions are subject to the orthogonality and symmetry relations $\int d\mathbf{r}[u_i u_j^* - v_i v_j^*] = \delta_{ij}$ and $\int d\mathbf{r}[u_i v_j^* - v_i u_j^*] = 0$.

Since the energy $\hbar\omega_j$ of this quasiparticle is defined with respect to the condensate energy (Eq. (3.64) with the stationary solution Φ), the presence of quasiparticles with negative frequencies implies a "thermodynamic" instability for the solution Φ; if there is energy dissipation in the system, the excitation of negative energy modes lowers the total energy, and the modes grow spontaneously to relax Φ into a more stable state. In other words, there is a path in configuration space along which the energy decreases. This argument is closely related to the Landau criterion for superfluidity (Landau instability). On the other hand, one can note that, since the matrix element of Eq. (3.66) is non-Hermitian, the eigenfrequencies can assume complex values. Therefore, the small-amplitude fluctuations of the corresponding eigenmodes grow exponentially, even for the energy-conserving time development. This is known as "dynamical instability". These instabilities play a key role in vortex dynamics.

3.3.2.2 *Vortex experiments in rotating BECs*

The first experimental detection of a vortex in an atomic BEC was made by Matthews et al. in 1999 [117]. Their study involved condensates of ^{87}Rb atoms residing in two hyperfine states. We describe this experiment in the next section, and summarize here the experiments on vortices in single-component BECs under rotation.

An intuitive method of creating vortices is to stir a condensate mechanically with a rotating "bucket". Using this scheme, Madison et al. observed the formation of vortices in a single-component elongated BEC [118] (see Fig. 3.10). The condensate was trapped in a static axisymmetric magnetic trap which was deformed by a nonaxisymmetric attractive dipole potential created with stirring laser beams. This combined potential produced a cigar-shaped harmonic trap with a slightly anisotropic transverse profile. By rotating the orientation of the transverse anisotropy at a frequency Ω, they observed the dynamic formation

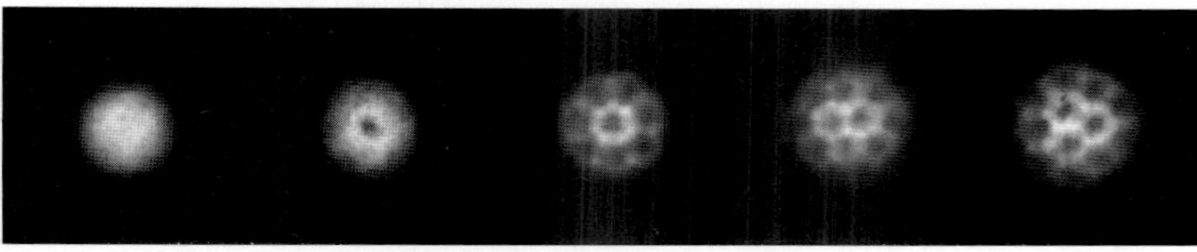

Fig. 3.10 Typical density profiles of a rotating condensate in the plane perpendicular to the rotation axis, taken by TOF measurement. From left to right, the rotation frequency increases. Courtesy of J. Dalibard.

of a vortex above a certain critical value of Ω. When Ω was increased further, multiple vortices appeared, forming a triangular lattice. The quantized vortices could be directly visualized as "dips" in the transverse density profile of the TOF image. Following the experiments of the ENS group, other groups have also observed quantized vortices using slightly different methods based on the concept of the rotating bucket. Abo-Shaeer et al. observed a vortex lattice consisting of up to 100 vortices in a ^{23}Na BEC [119]. Hodby et al. created a vortex lattice by rotating the anisotropic magnetic trap directly without using a laser beam, which is similar to the rotating bucket experiment [121]. Rotating an optical spoon made by multiple-spot laser beams or narrow focusing beams has also been used to nucleate vortices [164].

Haljan et al. created a vortex state by cooling an initially rotating thermal gas in a static confining potential [127]. Thermal gas above the transition temperature was rotated by a slightly anisotropic trap. After recovering the anisotropy of the potential, the rotating thermal gas was evaporatively cooled until most of the atoms were condensed. Although the number of atoms decreased through the evaporative cooling, the condensate continued to rotate because the angular momentum per atom did not change, and thus the vortex lattice was created. Since atoms can be selectively removed during evaporation, spinning up of the condensate can be efficiently achieved by removing atoms extending in the axial direction, as opposed to the radial direction, allowing a BEC with a high rotation rate to be obtained. Various properties of rapidly rotating BECs have been studied, e.g. equilibrium properties [165], collective dynamics of a vortex lattice [166, 167], vortex aggregation [168], and the lowest Landau level regime [169].

3.3.2.3 *Structure of a single vortex*

As a simple example, let us consider the structure of a single vortex in a condensate trapped by an axisymmetric harmonic potential $V_{\text{ex}}(r,z) = m\omega_\perp^2(r^2 + \lambda^2 z^2)/2$ with transverse trapping frequency $\omega_\perp$ and aspect ratio $\lambda = \omega_z/\omega_\perp$. The condensate wave function with a straight vortex line located along the z-axis takes the form $\Phi(\mathbf{r}) = \phi(r,z)e^{iq\theta}$ with a winding number q and cylindrical coordinate (r,θ,z). The velocity field around the vortex line is $\mathbf{v}_s = (q\hbar/mr)\hat{\theta}$. Equation (33.3) for a real function ϕ becomes

$$\left[-\frac{\hbar^2}{2m}\left(\frac{\partial^2}{\partial r^2} + \frac{1}{r}\frac{\partial}{\partial r} + \frac{\partial^2}{\partial z^2}\right) + \frac{q^2\hbar^2}{2mr^2} + V_{\text{ex}}(r,z) + gn\right]\phi = \mu\phi. \quad (3.67)$$

Here, the centrifugal term $q^2\hbar^2/2mr^2$ arises from the azimuthal motion of the condensate. Equation (3.67), which must be solved numerically, gives the structure of the vortex state. Asymptotically, the solution takes $n(r,z) \simeq n_0(r/\xi)^{2q^2}$ for $r \ll \xi$, and $n(r,z) = n_0(1 - r^2/R_\perp^2 - z^2/R_z^2 - q^2\xi^2/r^2)$ for $\xi \ll r < R_\perp$. The latter can be obtained in the TF limit $Na/a_{\text{ho}} \gg 1$ for Eq. (3.67), where $n_0 = \mu/g$ is the density at the center of the vortex-free TF profile. We have defined the TF radii $R_\perp^2 = 2\mu/m\omega_\perp^2$, $R_z^2 = 2\mu/m\omega_z^2$ and the healing length $\xi = (\hbar^2/2mgn_0)^{1/2}$. The condensate density vanishes at the center, out to a distance of order ξ, whereas the density in the outer region has the form of an upward-oriented parabola. Hence, the healing length ξ characterizes the vortex core size; for typical BEC parameters, $\xi \sim 0.2$ μm. In the TF limit, the core size is very small because $\xi/R_\perp = \hbar\omega_\perp/2\mu = (15Na/a_{\mathrm{ho}})^{-2/5} \ll 1$. Increasing the winding number q widens the core radius due to centrifugal effects.

This axisymmetric solution holds for a BEC in a pancake trap with $\lambda \gg 1$, but for a cigar-shaped trap with $\lambda \ll 1$ the axisymmetry is spontaneously broken as a result of *vortex bending*. Numerical simulation of the 3D GP equation revealed that the vortex bending is held stationary for the ground state of a rotating cigar-shaped condensate [170–172]. This bending is a symmetry breaking effect, which happens even in a completely axisymmetric setup. Evidence of vortex bending in the ground state was observed by the ENS group [173].

3.3.2.4 *Vortex stability*

Stability of the vortex state should be ensured to demonstrate that a BEC exhibits the superfluidity characterized by a "persistent current". At zero temperature, the BdG analysis shows that the excitation spectrum of a condensate with a centered vortex in a cylindrical system has at least one negative energy mode with a positive norm $\int d\mathbf{r}(|u|^2 - |v|^2) > 0$, localized at the core (so-called "anomalous modes") [174]. This implies that the single-vortex state is thermodynamically unstable [175]. In fact, in the absence of rotation, the centered vortex would decay by spiraling outward due to dissipation by the thermal atoms.

As demonstrated in the experiments, imposing rotation on the system is a direct way to achieve vortex stabilization. If the system is under rotation, it is convenient to consider the corresponding rotating frame; for a rotation frequency $\mathbf{\Omega} = \Omega\hat{\mathbf{z}}$, the integrand of the GP energy functional (3.64) acquires an additional term,

$$E' = \int d\mathbf{r}\Psi^* \left(-\frac{\hbar^2\nabla^2}{2m} + V_{\mathrm{ex}} + \frac{g}{2}|\Psi|^2 - \Omega L_z \right) \Psi, \tag{3.68}$$

where $L_z = -i\hbar(x\partial_y - y\partial_x)$ is the angular momentum operator along the z-axis. The corresponding GP equation becomes

$$i\hbar\frac{\partial\Psi(\mathbf{r},t)}{\partial t} = \left[-\frac{\hbar^2\nabla^2}{2m} + V_{\mathrm{ex}} + g|\Psi(\mathbf{r},t)|^2 - \Omega L_z \right] \Psi(\mathbf{r},t). \tag{3.69}$$

The energy associated with a single vortex line is predominantly contributed by the kinetic energy of the superfluid flow in the vortex, which can be estimated as

$$E_1 = \int \frac{1}{2}mnv_s^2 d\mathbf{r} \simeq \frac{m\bar{n}}{2}2R_z \int_\xi^{R_\perp} v_s^2 2\pi r dr = q^2 R_z \frac{2\pi\hbar^2\bar{n}}{m} \ln\left(\frac{R_\perp}{\xi}\right), \tag{3.70}$$

where $\bar{n}$ is the mean uniform density. Since $E_1 \propto q^2$, the energy cost to create one $q = 2$ vortex is higher than that to create two $q = 1$ vortices, thus vortices with $q > 1$ are energetically unfavorable. Therefore, a stable quantized vortex usually has $q = 1$, and we will mainly concentrate on the $q = 1$ vortex in the following discussion.

If there is a quantized vortex along the trap axis, $\langle L_z \rangle = N\hbar$, the corresponding energy of the system in the rotating frame is $E_1' = E_1 - N\hbar\Omega$. The difference between E'_1 and the vortex-free energy E'_0 determines the energetic favorability of a vortex entering the condensate. Since E'_0 is equal to the energy E_0 in the laboratory frame, the difference is given by $\Delta E' = E'_1 - E'_0 = E_1 - E_0 - N\hbar\Omega$. Thus, the critical rotation frequency Ω_c for the energetic stability of a vortex line is given by $\Omega_c = (E_1 - E_0)/N\hbar$. When Ω exceeds Ω_c, the single vortex state is *thermodynamically* stable.

To calculate E_1 more precisely, it is necessary to take into account the inhomogeneous effects of condensate density [176]. In the TF limit, for a condensate in a cylindrical trap $\omega_z = 0$ (an effective 2D condensate), the critical frequency is given by $\Omega_c = (2\hbar/mR_\perp^2)\ln(0.888R_\perp/\xi)$. For an axisymmetric trap $V_{\rm ex}(r,z)$, the critical frequency is $\Omega_c = (5\hbar/2mR_\perp^2)\ln(0.671R_\perp/\xi)$. For a nonaxisymmetric trap, Ω_c is slightly modified by a small numerical factor [177, 178].

3.3.2.5 *Vortex nucleation*

The critical rotation frequency Ω_c only indicates the energetic stability of the central vortex state. Vortex nucleation in a non-rotating condensate occurs when the trap is rotated at a higher frequency than Ω_c to overcome the energy barrier that stops the transition from the nonvortex state to the vortex state [179]. The threshold of the rotation frequency for instability, leading to vortex nucleation, is related to the excitation of surface modes of the condensate [180–184]. According to the Landau criterion for rotationally invariant systems, the critical rotation frequency is given by $\Omega_v = \min(\omega_l/l)$, where ω_l is the frequency of a surface mode with the profile $\sim e^{il\theta}$ and the quantum number l of the azimuthal (θ-) direction. Above Ω_v, some surface modes with negative energy appear in the spectrum of a non-rotating condensate [179, 181], and their growth may trigger vortex nucleation. The negative-energy modes can grow exponentially in the presence of dissipation, caused, for example, by interactions with thermal atoms [183].

The frequency Ω_v can explain vortex nucleation by a rotating thermal gas [127, 182], but cannot yet explain the results of external stirring potentials [118–121]. For example, in the case of the ENS group [122], the number of nucleated vortices had a peak near $\Omega = 0.7\omega_\perp$. Their experiments demonstrated that instability occurs when a particular surface mode is resonantly excited by a deformed rotating potential. The rotating potential of the ENS group mainly excited the surface mode with $l = 2$ (quadrupole mode). In a rotating frame with frequency Ω, the frequency of the surface mode is increased by $-l\Omega$. The resonance thereby occurs close to the rotation frequency $\Omega = \omega_l/l$. In the TF limit, the dispersion relation for the surface mode is given by $\omega_l = \sqrt{l}\omega_\perp$ with the trapping frequency $\omega_\perp$ [185]. Hence, it is expected that the quadrupole mode with $l = 2$ is resonantly excited at $\Omega = \omega_\perp/\sqrt{2} \simeq 0.707\omega_\perp$. A theoretical study revealed that when the quadrupole mode is resonantly excited, an imaginary component in the excitation frequency appears in surface modes with high multipolarity [186]. This indicates that dynamic instability can trigger vortex nucleation even at zero temperature. This picture was further supported by an experiment of the MIT group [164], in which surface modes with higher multipolarity ($l = 3, 4$) were resonantly excited using multiple laser-beam spots. The largest number of vortices were generated at frequencies close to the expected values $\Omega = \omega_\perp/\sqrt{l}$. However, a recent theory proposed an alternative mechanism, noting that the single vortex state can be regarded as the first Zeeman-like excited state, and its resonance frequency is quantitatively consistent with all of the above experiments [187].

The observation of vortex nucleation and lattice formation [120] has been well reproduced by numerical simulations of the time-dependent GP equation (3.69) [188–194]. The simulation results clarified the following dynamics. When a sufficiently high rotation is applied to an initially non-rotating condensate, it becomes elliptical and undergoes a quadrupole oscillation (Fig. 3.11(a)). Then, the boundary surface of the condensate becomes unstable, and generates ripples that propagate along the surface (Fig. 3.11(b)). It is also possible to identify quantized vortices in the phase profile. As soon as the rotation starts, many vortices appear in the low-density region outside of the condensate (Fig. 3.11(a)). Since quantized vortices are excitations, their nucleation increases the energy of the system. Because of the low density in the outskirts of the condensate, however, their nucleation contributes little to the energy or angular momentum. Since these vortices outside of the condensate are not observed in the density profile, they are called "ghost vortices". Their movement toward the Thomas–Fermi surface excites ripples (Fig. 3.11(c)). It is not easy for these ghost vortices to enter the condensate, because that would increase both the energy and angular momentum. Only some vortices enter the condensate cloud to become "real vortices" with the usual density profile of quantized vortices (Fig. 3.11(d)), eventually forming a vortex lattice (Fig. 3.11(e) and (f)).

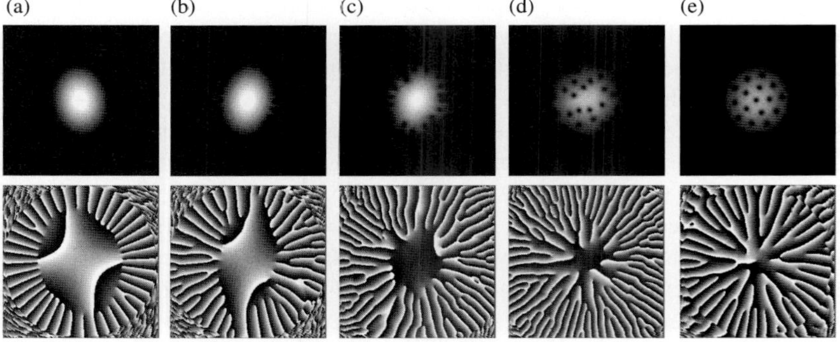

Fig. 3.11 Dynamics of vortex lattice formation in a rotating BEC. The graphs show both the condensate density (upper panels) and phase (lower panels), being at (a) t = 300 ms, (b) 370 ms, (c) 385 ms, (d) 410 ms, and (e) 550 ms after the start of rotation. The phase varies continuously from 0 (white) to 2π (black). The simulations were performed using the 2D GP equation with experimentally accessible parameters; see Ref. [188] for details.

3.3.2.6 *Dynamics of a single vortex*

PRECESSION OF AN OFF-CENTERED VORTEX

Precession of a vortex core upon the off-center of the condensate is a simple example of vortex motion. Core precession can be described in terms of a Magnus force effect. A net force on a quantized vortex core creates a pressure imbalance, resulting in core motion perpendicular to both the force and the vortex quantization axis. In the case of trapped BECs, these net forces can be caused by either condensate density gradients [177, 195, 196] or drag due to thermal atoms [197]. The former may effectively act as a buoyancy of the vortex. Typically, the total buoyant force is towards the condensate surface, and the net effect is a precession of the core around the condensate center via the Magnus effect. The latter causes radial drag and spiraling of the core towards the condensate surface due to energy dissipation.

Core precession was investigated experimentally by Anderson et al. [198]. Initially, a vortex was prepared to be off-centered, and its precession frequency was determined from the vortex position subtracted directly from the snapshots of the density profile. The vortex core precessed in the same direction as the vortex fluid flow around the core. These features can be understood as a result of the anomalous mode (with negative energy and positive norm); when the energy is negative, the precession has the same direction as the vortex flow. The observed excitation frequency also agrees with theory [177] and numerical simulations [195, 196, 199].

VORTEX WAVE

As in the studies of superfluid helium, vortex waves are also an interesting subject in trapped BECs. The dispersion relation of the vortex wave can be obtained by BdG analysis of a single vortex state, as shown in Fig. 3.12(a). In this calculation, the perturbation can be taken as $\delta\Psi = e^{i\theta}[u_{k_z,l}e^{i(k_z z+l\theta-\omega t)} - v^*_{k_z,l}e^{-i(k_z z+l\theta-\omega t)}]$ in the cylindrical system, where k_z and l refer to the wave number and the angular quantum number along the z-axis. The lowest modes along the radial direction can be classified into three groups: the Kelvin wave ($l = -1$) (Fig. 3.12(b)), the varicose wave ($l = 0$) (Fig. 3.12(c)), and the surface waves ($l \neq 0, -1$). Because of the finite size of the system, the dispersion of the Kelvin wave is well described by [200]:

$$\omega(k_0 + k_z) = \omega_0 + \frac{\hbar k_z^2}{2m}\ln\left(\frac{1}{r_c k_z}\right) \quad (kr_c \ll 1), \tag{3.71}$$

where r_c is of the order of the core size, $k_0 \sim 2\pi/R_z$ the smallest wave number, and $\omega_0 < 0$ the frequency of the anomalous mode described above. The varicose wave with $l = 0$ is known in classical fluids as the axisymmetrically propagating mode, in which the core diameter of a vortex oscillates along the vortex line.

The collective modes of the condensate have a strong influence on the excitation of vortex waves due to the finite size effect. Bretin et al. observed that

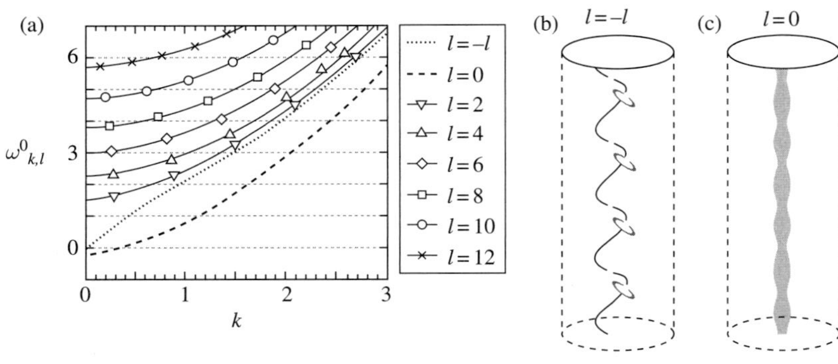

Fig. 3.12 (a) Dispersion relation of the axisymmetric single-vortex state of Kelvin waves with $l = -1$ (dashed line), varicose waves with $l = 0$ (dotted line), and several surface waves with $l > 0$ (solid lines), calculated for a cylindrically symmetric condensate with a vortex (see Ref. [203]). (b) and (c) show schematic illustrations of the Kelvin and varicose waves of the vortex line, respectively.

when superimposed $l = \pm 2$ quadrupole modes were excited with equal amplitudes in the single vortex state, the oscillation of the $l = -2$ mode decayed faster than that of the $l = +2$ mode [124]. This indicates that the $l = -2$ mode decays to Kelvin modes through a nonlinear Beliaev process, which is supported by theoretical analysis based on the BdG equation [201] and by numerical simulations [200].

When an additional velocity V is applied along the z-axis, the dispersion of the vortex waves behaves as $\omega - Vk_z$, and the frequency can become negative above the critical velocity. This is known as the Donnely–Glaberson instability in superfluid helium [202], and it results in the amplification of Kelvin waves [203]. This could induce reconnections of adjoining vortex lines, and eventually a turbulent state.

SPLITTING OF A MULTIPLY CHARGED VORTEX

As seen in Eq. (3.70), the energy cost to create a $q > 1$ multiply charged vortex is unfavorable compared to q singly charged vortices. This raises an interesting question: what happens when such an unstable vortex is created? A multiply charged vortex can be created experimentally in atomic BECs by using topological phase imprinting [204], and causes interesting disintegration dynamics [205] (the physics of topological phase imprinting will be described in Section 3.3.4).

The stability of a multiply charged vortex in a trapped BEC shows an interesting interaction dependence, which originates from the finite size effect [206]. BdG analysis for the cylindrical system revealed the appearance of complex eigenvalue modes, which implies that the multiply charged vortex is dynamically unstable. For a vortex with a winding number q, the conservation of angular momentum leads to constraint of the normal mode functions $u_j(\mathbf{r}) = u_j(r)e^{i(q+2)\theta}$ and $v_j(\mathbf{r}) = v_j(r)e^{i(q-2)\theta}$. For $q = 2$ there are alternating stable and unstable regions with respect to the interaction parameter $an_z = a\int |\Psi|^2 dxdy$; the first (second) region appears for $0 < an_z < 3$ $(11.4 < an_z < 16)$. Numerical simulations demonstrated that, when the system is in the unstable region, a doubly charged vortex decays into two singly charged vortices [207].

The experiment by Shin et al. investigated the splitting process of a doubly charged vortex and its characteristic time scale as a function of $an_{z=0}$ [205]. The results showed that the doubly charged vortex decayed, but that its lifetime increased monotonically with $an_{z=0}$, showing no periodic behavior. Numerical simulation of the 3D GP equations also displayed this mysterious phenomenon, demonstrating that the detailed dynamical behavior of a vortex along the entire z-axis is relevant in characterization of the splitting process in an axially elongated condensate [208, 209]. The vortex begins to split near both condensate edges in the z-direction, and then the splitting propagates to the center.

Isoshima et al. studied the splitting dynamics of quadruply charged ($q = 4$) vortices, both theoretically and experimentally [210]. In this case, a rich variety of splitting patterns with l-fold symmetry was expected [211]. Initially, they prepared an off-centered $q = 4$ vortex, and observed that the vortex split into an array of four linearly aligned $q = 1$ vortices. This is because the perturbation with two-fold rotational symmetry ($l = 2$) is the most unstable mode under the experimentally relevant parameters. These observations were well reproduced by numerical simulation of the GP equation, where the initial displacement of the vortex position was crucial to causing splitting instability, due to the conservation of angular momentum.

3.3.2.7 *Vortex lattices*

At very high frequencies Ω, a rotating superfluid mimics rigid body rotation with $\nabla \times \mathbf{v}_s = 2\boldsymbol{\Omega}$, by forming a triangular vortex lattice which arranges itself parallel to the rotation axis [212]. These properties closely resemble the magnetic flux-line lattice in type-II superconductors predicted by Abrikosov [213]. Here, we summarize the physics of vortex lattices in trapped BECs. Since various features of the rapidly rotating BECs have been reviewed thoroughly by Cooper [129] and Fetter [162], in this section we describe only the theoretical basics that will be useful in the following sections.

BASICS

For vortex lines parallel to the z-axis and located at $\mathbf{r}_i = (x_i, y_i)$, the vorticity is given by the form $\nabla \times \mathbf{v}_s = \sum \kappa \delta^{(2)}(\mathbf{r}_i)\hat{\mathbf{z}}$. Therefore, the average vorticity per unit area is given by $\nabla \times \mathbf{v}_s = \kappa n_v \hat{\mathbf{z}}$, where n_v is the number of vortices per unit area. Hence, the density of the vortices is related to the rotational frequency Ω as $n_v = 2\Omega/\kappa$ ("Feynman's rule") [10]. This relation can be used to estimate the maximum possible number of vortices in a given area as a function of Ω. The vortex lattice can be characterized by the nearest-neighbor lattice spacing $\sim b = (\hbar/m\Omega)^{1/2}$ defined by the area per vortex $n_v^{-1} = \pi b^2$, and by the radius of each vortex core $r_c \sim \xi$.

THOMAS–FERMI REGIME

The GP energy functional of Eq. (3.68) in a rotating frame can be rewritten as

$$E' = \int d\mathbf{r} \left(\frac{\hbar^2}{2m} \left| \left(-i\nabla - \frac{m}{\hbar} \boldsymbol{\Omega} \times \mathbf{r} \right) \Psi \right|^2 + V_{\text{eff}} |\Psi|^2 + \frac{g}{2} |\Psi|^4 \right), \tag{3.72}$$

where $V_{\text{eff}} = m(\omega_\perp^2 - \Omega^2) r^2/2 + m\omega_z^2 z^2/2$ is the effective trapping potential combined with the centrifugal potential; the rotation effectively releases the radial potential, and causes it to vanish at $\Omega = \omega_\perp$. Because the first term in Eq. (3.72) reads $\hbar^2 (\nabla |\Psi|)^2/2m + m(\mathbf{v}_s - \boldsymbol{\Omega} \times \mathbf{r})^2 |\Psi|^2/2$, it can be neglected when both the TF limit and the rigid-body rotation limit $\mathbf{v}_s = \boldsymbol{\Omega} \times \mathbf{r}$ are satisfied. This TF approximation is valid when the lattice spacing b is much larger than the vortex core size r_c, giving an effective coarse-grained description of the rotating BECs. The TF radius is given by $R_\perp(\Omega) = R_\perp/[1 - (\Omega/\omega_\perp)^2]^{3/10}$ with $R_\perp$ for a non-rotating condensate, providing an aspect ratio $\lambda_{\text{rb}} = R_\perp(\Omega)/R_z = \lambda/[1 - (\Omega/\omega_\perp)^2]^{1/2}$. Thus, measurement of λ_{rb} can give the rotation rate of the condensate [127, 164]. Also, in the high rotation limit $\Omega \to \omega_\perp$, the condensate flattens out and reaches an interesting quasi-2D regime; current experiments have reached $\Omega/\omega_\perp \approx 0.995$ [165].

MEAN-FIELD QUANTUM HALL REGIME

Note that the first term in Eq. (3.72) can be identified as the Hamiltonian $H_L = (-i\hbar\nabla - e\mathbf{A}/c)^2/2m$ of a charge $-e$ particle moving in the xy plane under a magnetic field $B\hat{\mathbf{z}}$ with a vector potential $\mathbf{A} = (mc/e)\boldsymbol{\Omega} \times \mathbf{r}$. If interaction is neglected ($g = 0$), the eigenvalues of the Hamiltonian of Eq. (3.72) forms Landau levels as $\epsilon_{n,m,n_z}/\hbar = \omega_\perp + n(\omega_\perp + \Omega) + m(\omega_\perp - \Omega) + (n_z + 1/2)\omega_z$, where n is the Landau level index, m indexes the substates within the n-th Landau level, and n_z is the index of the states along the z-axis. The lowest energy states of two adjacent Landau levels are separated by $\hbar(\omega_\perp + \Omega)$, whereas the distance between two adjacent substates in a given Landau level is $\hbar(\omega_\perp - \Omega)$; when $\Omega = \omega_\perp$, all states in a given Landau level are degenerate. This corresponds to a situation where the centrifugal force exactly balances the trapping force in the x-y plane, and only the Coriolis force remains. The system is then invariant under translation, and therefore exhibits macroscopic degeneracy. This formal analogy has led to the prediction that quantum Hall-like properties would emerge in rapidly rotating BECs; see [129, 130] and references therein for the details of this strongly correlated phase in rapidly rotating bosons.

The quantum Hall formalism provides a useful mean-field description of rotating BECs with vortex lattices [214]. Interaction will lead to the mixing of different (n, m, n_z) states. However, because the averaged density $\bar{n}$ of the system drops as $\Omega \to \omega_\perp$, the interaction energy $\sim g\bar{n}$ can become low compared to $2\hbar\omega_\perp$ and $\hbar\omega_z$. In this limit, particles should condense into the lowest Landau levels (LLL) with $n = 0$. The system then enters the "mean field" quantum Hall regime, where the wave function can be described by only the LLL orbitals with the form

$$\Psi_{\mathrm{LLL}} = \sum_{m \geq 0} a_m \psi_m(\mathbf{r}) = A \prod_j (z - z_j) e^{-r^2/2a_{\mathrm{ho}}^2}, \tag{3.73}$$

where $z = x + iy$, z_j are the positions of vortices (zeros), and A is a normalization constant. By minimizing Eq. (3.72) using the ansatz Ψ_{LLL} with respect to z_j, one can analytically describe the vortex structure [215–220], although this method is effective only near the limit $\Omega \simeq \omega_\perp$. However, this would be impractical for numerical calculations because of the time and accuracy required. A similar approach was applied to investigate the ground state of (not rapidly) rotating BECs with very weak interaction [221–223], where the coefficients a_m of the harmonic oscillator basis ψ_m were minimized.

Schwaikhard et al. created rapidly rotating BECs by spinning up the condensates to $\Omega/\omega_\perp > 0.99$ [169], and found some evidence that the condensate entered the LLL regime: (1) The 2D signature of a rapidly rotating BEC was confirmed by the excitation of an axial breathing mode ($m_z = 0$). For a BEC in the axial TF regime, an axial breathing frequency $\omega_{\mathrm{B}} = \sqrt{3}\omega_z$ has been predicted in the limit $\Omega/\omega_\perp \to 1$ [224], whereas $\omega_{\mathrm{B}} = 2\omega_z$ is expected for a noninteracting gas with $\mu < \hbar\omega_z$. Schweikhard et al. observed a crossover of ω_{B} from $\sqrt{3}\omega_z$ to $2\omega_z$ with increasing Ω ($\mu \sim 3\hbar\omega_z$). (2) A signature of the LLL regime is that the vortex core size r_c is similar to the vortex separation distance $b = (\hbar/m\Omega)^{1/2}$. The theory [220, 225–227] predicted that the vortex cores begin to shrink as the intervortex spacing becomes comparable to the healing length ξ, and eventually the core radius becomes proportional to the intervortex spacing. The saturation of the fractional area $A = r_c^2/b^2$

with increasing Ω was actually observed, indicating that the system was in the LLL regime. (3) The observed global density profile remained a parabolic TF profile even for $\Omega \to \omega_\perp$, although it was expected to be a Gaussian [214] (see Eq. (3.73)) in the LLL regime. However, very small distortions of the vortex lattice from a perfect triangle could result in large changes in the global density distribution, such as from a Gaussian to a TF form [215–217].

TKACHENKO OSCILLATION OF A VORTEX LATTICE

It should be possible to propagate collective waves transverse to the vortex lattice in the superfluid, in so-called Tkachenko (TK) modes [228]. For incompressible superfluids, the dispersion law is given by $\omega_{\rm TK}(k) = \sqrt{\hbar\Omega/4m}\, k$. The TK mode of a vortex lattice in a trapped BEC was observed experimentally [167], and was identified by the sinusoidal displacement of vortex cores with an origin at the center of the condensate.

To explain the observed frequency of the TK mode $\omega_{(n,m)}$, where the quantum number (n, m) refers to the radial and angular nodes, the effects of compressibility should be taken into account. According to the elastohydrodynamic approach developed by Baym [229], the TK frequency is described by the compressional modulus C_1 and the shear modulus C_2 of the vortex lattice included in the elastic energy:

$$E_{\rm el} = \int d\mathbf{r} \left\{ 2C_1(\nabla\cdot\boldsymbol{\epsilon})^2 + C_2\left[\left(\frac{\partial\epsilon_x}{\partial x} - \frac{\partial\epsilon_y}{\partial y}\right)^2 + \left(\frac{\partial\epsilon_x}{\partial y} + \frac{\partial\epsilon_y}{\partial x}\right)^2\right]\right\}, \tag{3.74}$$

where $\boldsymbol{\epsilon}(\mathbf{r}, t)$ is the continuum displacement field of the vortices from their home positions. In the incompressible TF regime, $C_2 = -C_1 = n\hbar\Omega/8$. Compressibility yields two branches in the energy spectrum: The upper branch follows the dispersion law $\omega_+^2 = 4\Omega^2 + c^2k^2$, where $c = \sqrt{gn/m}$ is the velocity of sound, which is the standard inertial mode of a rotating fluid with a gap at $k = 0$. The lower frequency branch corresponds to the TK mode with dispersion $\omega_-^2 = (\hbar\Omega/4m)[c^2k^4/(4\Omega^2 + c^2k^2)]$. For large k, this reproduces the original TK frequency $\omega_{\rm TK}$, but for small k it exhibits quadratic behavior $\omega_- \simeq \sqrt{\hbar/16m\Omega}ck^2$. The transition between k^2 and k dependence takes place at $k \sim \Omega/c > R_\perp^{-1}$. This suggests that the effects of compressibility, which characterize the k^2 dependence, play a crucial role in the TK mode. Thus, this regime is distinguished from the usual incompressible TF regime, and is referred to as the "soft" TF regime. Including finite compressibility, this can explain the observed values of $\omega_{(1,0)}$ [229–231]. First-principles simulations based on the GP and BdG formalism also agreed excellently with the experimental data [232, 233].

3.3.2.8 *Various topics on vortices in single-component BECs*

VORTICES IN AN ANHARMONIC POTENTIAL

For a rotating condensate with frequency Ω in a harmonic potential $(1/2)m\omega_\perp^2 r^2$, the centrifugal potential effectively reduces the confinement, preventing a BEC from rotating at Ω beyond $\omega_\perp$. This restriction can be avoided

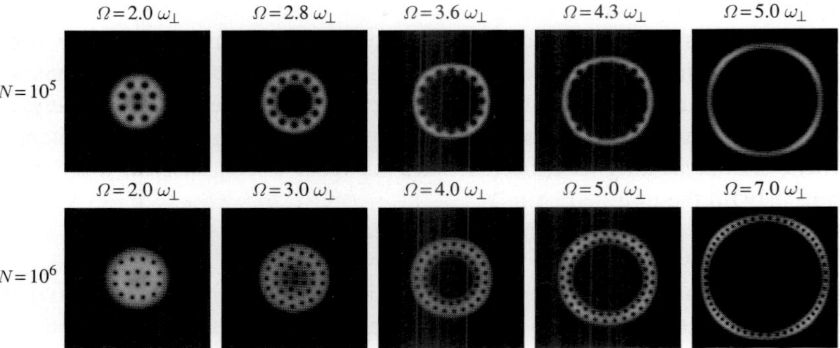

Fig. 3.13 Vortex structure of a rapidly rotating BEC in a combined harmonic–quadratic potential. This calculation was done by 2D simulations of the GP equation with the parameter values $m = m_{^{87}\mathrm{Rb}}$, $\omega_\perp = 10 \times 2\pi$ Hz, $a = 5.61$ nm and $\lambda = 0.5$. The upper and lower panels correspond to $N = 10^5$ and $N = 10^6$, respectively.

by introducing an additional quartic potential [234–236], so that the combined trapping potential in the xy plane becomes $V_{\mathrm{ex}}(r) = (1/2)m\omega_\perp^2(r^2 + \lambda r^4/a_{\mathrm{ho}}^2)$, where the dimensionless parameter λ represents the relative strength of the quartic potential. The properties of a rotating condensate in an anharmonic potential have attracted considerable theoretical attention [238–241]. The vortex phases in an anharmonic trap are quite different from those in a harmonic trap, as shown in Fig. 3.13, because it is possible to rotate the system arbitrarily fast. For small Ω, the equilibrium state is the usual vortex lattice state. As Ω increases, the vortices begin to merge in the central region, and the centrifugal force pushes the particles towards the edge of the trap. This results in a new vortex state consisting of a uniform lattice (multiple circular arrays of vortices) with a central density hole. The central hole becomes larger with increasing Ω, and the condensate forms an annular structure with a single circular array of vortices. A further increase of Ω stabilizes a *giant vortex*, where all vortices are concentrated in the single hole.

A combined harmonic–quartic potential was constructed by Bretin et al., who superimposed a blue detuned laser with the Gaussian profile [128]. Since the waist w of the beam propagating along the z-axis was larger than the condensate radius, the potential created by the laser $U_0 \exp(-2r^2/w^2)$ can be written as $U(r) \simeq U_0(1 - 2r^2/w^2 + 2r^4/w^4)$. The second term leads to a reduction of the transverse trapping frequency $\omega_\perp$, and the third term provides the desired quartic confinement, giving $\omega_\perp/2\pi = 65$ Hz and $\lambda \simeq 10^{-3}$ for $V_{\mathrm{ex}}(r)$. For $\Omega < \omega_\perp$, a vortex lattice was clearly observed. When $\Omega > \omega_\perp$, however, the vortices became gradually more difficult to observe and the images became less clear for $\Omega = 1.05\omega_\perp$. The most plausible explanation of this observation is that the vortices were not in equilibrium, and were strongly bent [239].

VORTEX PINNING IN AN OPTICAL LATTICE

Rotating BECs combined with a co-rotating optical lattice are an interesting system, which has two competing length scales: vortex separation and the periodicity of the optical lattice. The structure of the vortex lattice is strongly dependent on the externally applied optical lattice. Various vortex phases appear, depending on the number of vortices per pinning center, i.e. the filling factor. Tung et al. created a rotating optical lattice using a rotating mask [242], which provided a periodic pinning potential that was stationary in the corresponding rotating frame of vortices. The authors observed a structural crossover from a triangular to a square lattice with increasing potential amplitude of the optical lattice, which was consistent with theoretical predictions [243, 244]. In a deep optical lattice, the condensates are well localized at each potential minimum, so that the system can be regarded as a Josephson-junction array (JJA). A rotating Josephson-junction array can have characteristic vortex patterns with the unit-cell structure [245], and was realized experimentally by Williams et al. [246].

THERMALLY-ACTIVATED VORTEX GENERATION

In 2D systems with continuous symmetry, true long-range order is destroyed by thermal fluctuations at any finite temperature. For 2D Bose systems, a quasi-condensate can be formed with a correlation decaying algebraically in space, where superfluidity is still expected below a certain critical temperature. This 2D phase transition is closely connected with the emergence of thermally activated vortex–antivortex pairs, known as the Berezinskii–Kosterlitz–Thouless (BKT) phase transition occurring at $T = T_{\rm BKT}$ [247, 248]. For $T < T_{\rm BKT}$, there are no isolated free vortices; vortices only exist as bound pairs, consisting of two vortices with opposite circulations. The contribution of these vortex pairs to the decay of the correlation is negligible, and the algebraic decay is dominated by phonons. For $T > T_{\rm BKT}$, the vortex pairs are separated, and the free vortices form a disordered gas of phase defects and give rise to an exponential decay of the correlation.

Recently, a BKT transition was observed experimentally in ultracold atomic gases [249, 250]. In the experiment, a 1D optical lattice was applied to an elongated condensate, splitting the 3D condensate into an array of independent quasi-2D BECs. The interference technique revealed the temperature dependence of an exponent of the first-order correlation function of the fluctuating 2D bosonic field [251]. A universal jump in the superfluid density characteristic of the BKT transition was confirmed by observing a sudden change in the exponent, in which a finite size effect caused a finite-width crossover, rather than a sharp transition. The microscopic origin of this transition was clarified from the image of the interference of the two 2D condensates. If isolated free vortices were present in either of the two condensates, the interference fringes would exhibit dislocations. Such a dislocation was observed in the high-T region of the crossover, as predicted theoretically by classical field simulations [252].

VORTEX FORMATION DURING A RAPID QUENCH

One universal element of many continuous phase transitions is the spontaneous formation of topological defects during a quench through the critical point, known as the Kibble–Zurek mechanism [253, 254]. The rapid symmetry-breaking phase transition in the early universe is expected to have produced causally disconnected domains of a true vacuum state, and their phase mismatch would leave behind topological defects like cosmic strings or monopoles. Since the cosmological phase transition may be second order, it is especially significant to perform experiments in a superfluid system whose transition is also second order [255, 256].

Ultracold atomic BECs offer an ideal opportunity for studying the Kibble–Zurek mechanism because the temperature, strength of interaction, and external parameters such as the magnetic field and trapping potentials, can be changed in a time shorter than the characteristic time scale of topological defect formation. A quench through the BEC transition temperature was realized by using a rapid ramping of the rf-field for evaporative cooling. Weiler et al. studied the spontaneous formation of vortices in BEC, both experimentally and theoretically [257]. Vortices appeared spontaneously during condensate formation, without the application of any external rotation. This experiment was compared to a simulation of BEC growth dynamics, calculated using the stochastic projected GP equation [258],

$$d\Psi(\mathbf{r},t) = \mathcal{P}\left\{-\frac{i}{\hbar}L_{\mathrm{GP}}\Psi(\mathbf{r},t)dt + \frac{\gamma(\mathbf{r})}{k_{\mathrm{B}}T}(\mu - L_{\mathrm{GP}})\Psi(\mathbf{r},t)dt + dW_{\gamma}(\mathbf{r},t)\right\}, \tag{3.75}$$

with $L_{\mathrm{GP}} = -\hbar^2\nabla^2/2m + V_{\mathrm{ex}} + g|\Psi|^2$. This equation was derived from first principles using the Wigner phase-space representation [259]. Here, the states of the trapped system are divided into the "coherent" region and the "incoherent" region by the high-energy cut-off of the harmonic oscillator mode. The first term on the right describes unitary evolution of the classical field in the coherent region according to the GP equation. The second term represents growth processes, that is, collisions that transfer atoms from the thermal bath to the classical field and vice versa, and the form of $\gamma(\mathbf{r})$ may be determined from kinetic theory [260]. The third term dW_{γ} is the complex-valued noise associated with condensate growth, which is consistent with the fluctuation–dissipation theorem. The projection operator $\mathcal{P}$ restricts the dynamics to the coherent region. The simulation results show that, after the temperature quench, the condensate domains begin to grow in various regions of the fluctuating classical field, and several vortices and antivortices (like a quantum turbulence) are formed. Eventually, a single vortex remains in the well-developed condensate.

VORTEX FORMATION DURING THE INTERFERENCE OF BECS

In a related work, vortices were formed by merging three uncorrelated BECs that were initially separated by a triple-well potential [261]. This may verify one side of the KZ mechanism: can the causally disconnected domains generate a net circular flow, i.e. vortices, through a merging process? When the potential barrier created by a repulsive laser beam was removed, the three BECs merged. Depending on the relative phases between the condensates and the rate of barrier removal, vortices formed stochastically even without an applied rotation. If the condensates have random initial phases, a vortex should appear with a probability of 0.25 [261, 262], which was observed with a slow removal rate of the barrier. With quick barrier removal, the resulting interference fringes could generate pairs of vortices and antivortices via "snake instability" [263, 264], but these soon disappear due to pair annihilation. These properties were studied by extensive numerical simulations of the GP equation [265, 266].

3.3.3 Vortices in two-component condensates

We now address the properties of vortices in multicomponent BECs. Although there have been many theoretical predictions of exotic vortices in multicomponent BECs, much experimental work remains to be performed. Here, we primarily describe theoretical studies of the topic, along with a few related experimental results [117, 267–270]. In this section, we focus on the properties of vortices in two-component BECs.

3.3.3.1 *Multicomponent Bose–Einstein condensates*

Before discussing vortices, we summarize the types of multicomponent BECs and describe how they are obtained. In this system, one can load and cool atoms in more than one hyperfine spin state or more than one atomic element in the same trap, thereby realizing experimentally multicomponent condensates. The simplest and most intuitive example is a mixture of different kinds of atoms. However, because of the effect of strong inelastic collisions between different kinds of atoms, this example is generally difficult to realize. One successful example is the ^{87}Rb–^{41}K mixture [143]. Recently, a BEC mixture of isotopes of Rb atoms (^{87}Rb–^{85}Rb) was realized by two groups [147, 148]. These experiments demonstrated that the interacting properties of two atomic species were tuned via a Feshbach resonance, opening a new avenue to study superfluid mixtures in a well-controlled manner.

A common method used to realize a condensate mixture of ultra-cold atoms is to utilize the hyperfine spin states of atoms. The resulting BECs can have internal degrees of freedom, due to these hyperfine states. A hyperfine-Zeeman sublevel of an atom with total electronic angular momentum $\mathbf{J}$ and nuclear spin $\mathbf{I}$ may be labeled by the projection m_F of total atomic spin $\mathbf{F} = \mathbf{I} + \mathbf{J}$ on the axis of the field $\mathbf{B}$ and by the total F, which can take a value from $|I - J|$ to $|I + J|$. This is because the hyperfine coupling, which is proportional to $\mathbf{I} \cdot \mathbf{J}$, is much larger than the typical temperature of an ultra-cold atomic system. The hyperfine state is denoted by $|F, m_F\rangle$ with $m_F = -F, -F+1, \cdots, F-1, F$. The simultaneous trapping of atoms with different hyperfine sublevels makes it possible to create multicomponent (often called "spinor") BECs with internal degrees of freedom, characterized by multiple order parameters [140–142, 144–146]. The order parameter of a BEC with hyperfine spin F has $2F+1$ component Ψ_{m_F} with respect to the basis vectors $|m_F\rangle$ defined by $\hat{F}_z|m_F\rangle = m_F|m_F\rangle$, which can be expanded as $|\Psi\rangle = \sum_{m_F=0,\pm1,\cdots\pm F} \Psi_{m_F}|m_F\rangle$.

Two-component mixtures of BECs with different hyperfine spin states have been created in the laboratory for the systems of ^{87}Rb [140, 271, 272] and ^{23}Na [273]. In Ref. [140], the two-component BECs of ^{87}Rb atoms consisted of the hyperfine spin states $|F = 1, m_F = -1\rangle \equiv |1\rangle$ and $|F = 2, m_F = 1\rangle \equiv |2\rangle$. The MIT group used an optical trap, all spin manifold $F = 1$ spin states of ^{23}Na ($m_F = 0, \pm1$) were simultaneously trapped in a single trap. In this case, the spin exchange interactions (see Sec. 3.3.4) gave rise to spin mixing dynamics among the different components. They made use of a quadratic Zeeman effect for atoms by applying a magnetic field to prevent the $|m_F = -1\rangle$ component from appearing. As a result, the system can be regarded as a two-component BEC.

When the trapping potential $V^i_{\rm ex}(\mathbf{r})$ $(i = 1, 2)$ is produced by a magnetic field, the atoms of each hyperfine state have their own potential energy because the potential energy depends on the magnetic moment of the atom. For example, the trapping potential for atoms in the $|F_i, m_{F_i}\rangle$ state is expressed as a function of the magnetic field $|B(\mathbf{r})|$ as [274]

$$V^i_{\rm ex}(\mathbf{r}) = g_{F_i}\mu_B m_{F_i}|B(\mathbf{r})| \simeq g_{F_i}\mu_B m_{F_i}\left[B_0 + \frac{1}{2}(K_x x^2 + K_y y^2 + K_z z^2)\right]$$
$$\equiv V^i_0 + \frac{1}{2}m_i(\omega^2_{ix}x^2 + \omega^2_{iy}y^2 + \omega^2_{iz}z^2), \tag{3.76}$$

where g_{F_i} is the g-factor of the i-th atom, μ_B is the Bohr magneton, $V^i_0 = g_{F_i}\mu_B m_{F_i}B_0$, and ω_{ik} $(k = x, y, z)$ is the trapping frequency satisfying the relation $m_i\omega^2_{ik} = g_{F_i}\mu_B m_{F_i}K_k$. The potential minima V^i_0 for BECs with different values of $g_{F_i}m_{F_i}$ do not coincide. This positional shift is also influenced by a gravitational effect, the difference in nuclear magnetic moments, and nonlinearity of the Zeeman shift [139, 140]. If the potential is created by a polarized optical laser [141, 142, 144, 145], atoms in all hyperfine states share the same confining potential.

3.3.3.2 *Coupled Gross–Pitaevskii equations*

First, we describe the mean-field theory for two-component BECs, which is a natural extension from that of single-component BECs. The condensates are confined by trapping potentials $V^i_{\rm ex}(\mathbf{r})$ $(i = 1, 2)$, which are assumed to rotate at a rotation frequency Ω about the z axis as $\mathbf{\Omega} = \Omega\hat{\mathbf{z}}$. The GP energy functional of the two-component BEC in the rotating frame is

$$E = \int d\mathbf{r}\left[\sum_{i=1,2}\Psi^*_i\left(-\frac{\hbar^2\nabla^2}{2m_i} + V^i_{\rm ex}(\mathbf{r}) - \Omega L_z + \frac{g_i}{2}|\Psi_i|^2\right)\Psi_i + g_{12}|\Psi_1|^2|\Psi_2|^2\right]. \tag{3.77}$$

Here, m_i is the mass of the i-th atom. The coefficients g_i and g_{12} represent the atom–atom interactions between atoms in the same component, and atoms in different components, respectively. These are expressed in terms of the corresponding s-wave scattering lengths a_1, a_2, and a_{12} as $g_i = 4\pi\hbar^2 a_i/m_i$, and $g_{12} = 2\pi\hbar^2 a_{12}/m_{12}$, where $m^{-1}_{12} = m^{-1}_1 + m^{-1}_2$ is the reduced mass. The time-dependent coupled GP equations for two-component BECs can be obtained using a variational procedure $i\hbar\partial\Psi_i = \delta E/\delta\Psi^*_i$ as

$$i\hbar\frac{\partial\Psi_i}{\partial t} = \left(-\frac{\hbar^2\nabla^2}{2m_i} + V^i_{\rm ex}(\mathbf{r}) + g_i|\Psi_i|^2 + g_{12}|\Psi_{3-i}|^2 - \Omega L_z\right)\Psi_i. \tag{3.78}$$

Substituting $\Psi_i(\mathbf{r}, t) = \Phi_i(\mathbf{r})e^{-i\mu_i t/\hbar}$ yields the time-independent coupled GP equations

$$\left(-\frac{\hbar^2\nabla^2}{2m_i} + V^i_{\rm ex}(\mathbf{r}) + g_i|\Phi_i|^2 + g_{12}|\Phi_{3-i}|^2 - \Omega L_z\right)\Phi_i = \mu_i\Phi_i, \tag{3.79}$$

where we have introduced the Lagrange multiplier μ_i, which represents the chemical potential and was determined so as to satisfy the conservation of particle number $N_i = \int d\mathbf{r}|\Phi_i|^2$ for each component.

It is instructive to examine the ground state without rotation ($\Omega = 0$). Compared to single-component BECs, two-component BECs exhibit a rich variety of ground state structures, depending on the various parameters of the system [274]. The intercomponent interaction g_{12} plays an especially important role; phase separation of homogeneous two-component BECs occurs when $g_{12}^2 > \sqrt{g_1 g_2}$ [275, 276], where strong intercomponent repulsion overcomes the intracomponent repulsion. In ^{87}Rb, the scattering lengths are nearly equal ($a_1 = 5.67$ nm, $a_2 = 5.34$ nm, and $a_{12} = 5.50$ nm), thus $g_{12}^2 \simeq \sqrt{g_1 g_2}$, but the small relative displacement of the trapping minima leads to a spatial separation between the two components [140]. Even without the trap displacement, the two components phase separate into oscillating ring structures [272]. In contrast, for the ^{23}Na-mixture, $a_1 = 2.75$ nm, $a_2 = 2.65$ nm, and $a_{12} = 2.75$ nm, thus $g_{12}^2 > \sqrt{g_1 g_2}$, causing a well-separated spin domain to form [141, 273]. A recent experiment demonstrated the control of the miscibility of a ^{85}Rb–^{87}Rb mixture via Feshbach resonance [148].

3.3.3.3 *A single vortex in a two-component BEC*

EXPERIMENTAL CREATION

The first observation of a quantized vortex in an atomic BEC was achieved in a two-component BEC consisting of ^{87}Rb atoms with hyperfine spin states $|F = 1, m_F = -1\rangle \equiv |1\rangle$ and $|F = 2, m_F = 1\rangle \equiv |2\rangle$ [117], which were confined simultaneously in almost identical magnetic potentials. Initially, condensed atoms were trapped in one state, such as the $|1\rangle$ state. Then, a two-photon microwave field was applied, inducing coherent Rabi transitions of atomic populations between the $|1\rangle$ state and the $|2\rangle$ state. For a homogenous system in which both components have uniform phases, the interconversion takes place at the same rate everywhere. However, time variation of the spatially inhomogeneous potential changes the nature of the interconversion. Let us consider a co-rotating frame of an off-centered perturbation potential at the rotation frequency Ω'. In this frame, the energy of a vortex with one unit of angular momentum is shifted by $\hbar\Omega'$ relative to its value in the laboratory frame. When this energy shift is compensated for by the sum of the detuning energy of an applied microwave field and the small chemical potential difference between the vortex and non-vortex states, a resonant transfer of population can occur between the non-vortex $|1\rangle$ state and the vortex $|2\rangle$ state. The rotating perturbation created by a spatially inhomogeneous laser beam was rotated around the initial non-rotating component $|1\rangle$. By adjusting Ω' and the detuning, the $|2\rangle$ component was resonantly transferred to a state with unit angular momentum by controlling precisely the time at which the coupling drive was turned off [277]. This procedure resulted in a "composite" vortex, where the $|2\rangle$ component had a vortex at the center, while the non-rotating $|1\rangle$ component occupied the center and acted as a pinning potential to stabilize the vortex core.

It is possible to prepare the initial condensate into either the $|1\rangle$ or $|2\rangle$ state, and to then form a vortex in the $|2\rangle$ or $|1\rangle$ state. It was observed that the stability of the vortex states was strongly dependent on the atom–atom interactions [117]. The vortex was stable only when the vortex was in the $|1\rangle$ state and the core was in the $|2\rangle$ state. The other state, with the vortex in the $|2\rangle$ state, exhibited an instability, with the $|2\rangle$ vortex sinking inward toward the trap center and breaking up. The details are discussed below.

EQUILIBRIUM STRUCTURE

Here, we present theoretical studies on the structures of a single vortex state in two-component BECs. The analysis is based on the coupled GP equations (3.79). As observed in Ref. [117], the vortex structure consists of one circulating component that surrounds the other non-rotating component, where both wave functions have axisymmetric profiles. We assume that the condensate wave functions have one quantized vortex at the center as $\Phi_i(\mathbf{r}) = \phi_i(r,z)e^{iq_i\theta}$, where $n_i = \phi_i^2$ is the condensate density and the trapping potential is given by $V_{\text{ex}}^i(\mathbf{r}) = m_i(\omega_{i\perp}^2 r^2 + \omega_{iz}^2 z^2)/2$. Thus, the axisymmetric (singly charged) vortex states are characterized by $(q_1,q_2) = (1,0)$, $(0,1)$, $(1,1)$.

For ^{87}Rb atoms in non-rotating traps, in order to decrease the intracomponent mean field energy, the Ψ_1 component with the larger intracomponent scattering length a_1 forms a shell outside the Ψ_2 component occupying the central part, as shown in Fig. 3.14(a). For $(q_1,q_2) = (1,0)$ [Fig. 3.14(b)], a density depletion associated with the vortex core appears in the density of Ψ_1 along the z-axis. The Ψ_2 component then fills the vortex core, because of intercomponent repulsion. As a result, the characteristic size of the core becomes larger than the healing length $\xi = \hbar/\sqrt{2m_1g_1n_1}$ of the single component. In this case, the centrifugal force associated with the vortex causes the Ψ_1 component to expand radially, which allows a decrease in the intracomponent mean-field energy of the Ψ_1 component, rather than that of the non-vortex state.

On the other hand, the structure of $(q_1,q_2) = (0,1)$ is different from the $(q_1,q_2) = (1,0)$ solution [see Fig. 3.14(c)]. The core size becomes smaller than that of the $(1,0)$ solution. While a fraction of the Ψ_1 component fills the vortex core, the excessive Ψ_1 component extends outward, resulting in an increase in size of the region of coexistence with the rotating Ψ_2 component. As discussed below, this configuration is *dynamically* unstable [278, 279]. Also, the $(q_1,q_2) = (1,1)$ state is unstable because the two components fully overlap,

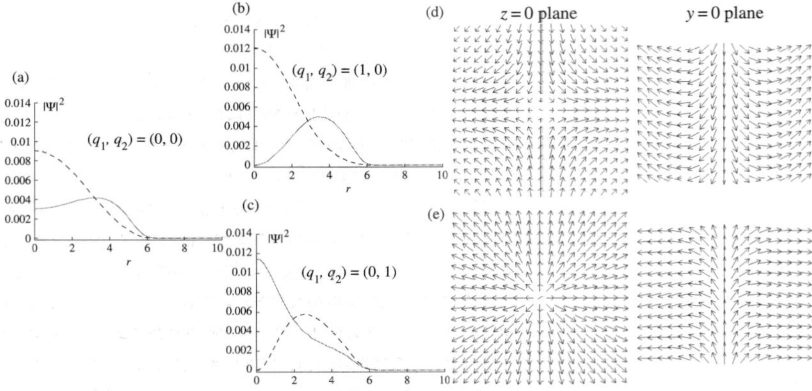

Fig. 3.14 Stationary solutions of two-component BECs with winding number $(q_1,q_2) = (0,0)$, $(1,0)$, and $(0,1)$. The parameters of the ^{87}Rb atoms in $|1,-1\rangle$ and $|2,1\rangle$ states were used. The trapping potential was oblate as $\omega_\perp = 5 \times 2\pi$ Hz and $\omega_z = 10 \times 2\pi$ and the particle number $N_1 = N_2 = 10^5$. In (a)–(c), the radial density profile is shown. The corresponding (pseudo)spin textures in the $z = 0$ and $y = 0$ planes are shown for (d) $(1,0)$ and (e) $(0,1)$ states.

which is energetically unfavorable. In the presence of a slow rotation, the thermodynamically stable configuration of $(1, 1)$ is that in which the two vortices in each component are displaced from the center, which decreases the size of the overlapping region.

It is instructive to represent the spinor order parameter of two-component BECs using the "pseudospin" [280–282], which allows us to analyze this system as a spin-1/2 BEC. We introduce a normalized spinor $\boldsymbol{\zeta}$ to represent the two-component wave functions as $[\Psi_1, \Psi_2]^T = \sqrt{n_{\rm T}} e^{i\Theta/2} \boldsymbol{\zeta}$, where

$$\boldsymbol{\zeta} = \begin{pmatrix} \zeta_1 \\ \zeta_2 \end{pmatrix} = \begin{bmatrix} \cos(\theta/2) e^{-i\varphi/2} \\ \sin(\theta/2) e^{i\varphi/2} \end{bmatrix} \tag{3.80}$$

and $|\zeta_1|^2 + |\zeta_2|^2 = 1$. Here, we assume that Ψ_1 and Ψ_2 represent the up and down components of the spin-1/2 spinor, respectively. Hence, the four degrees of freedom of the original wave functions $\Psi_j = \sqrt{n_j} e^{i\theta_j}$ (their amplitudes n_j and phases θ_j) are expressed in terms of the total density $n_{\rm T} = n_1 + n_2$, the total phase $\Theta = \theta_1 + \theta_2$, and the polar and azimuthal angles (θ, φ) of the local pseudospin $\mathbf{S} = (S_x, S_y, S_z)$, defined as

$$\mathbf{S} = \boldsymbol{\zeta}^\dagger \boldsymbol{\sigma} \boldsymbol{\zeta} = \begin{bmatrix} \zeta_1^* \zeta_2 + \zeta_2^* \zeta_1 \\ -i(\zeta_1^* \zeta_2 - \zeta_2^* \zeta_1) \\ |\zeta_1|^2 - |\zeta_2|^2 \end{bmatrix} = \begin{bmatrix} \sin\theta \cos\varphi \\ \sin\theta \sin\varphi \\ \cos\theta \end{bmatrix}, \tag{3.81}$$

where $\boldsymbol{\sigma}$ is the Pauli matrix, $\cos\theta = (n_1 - n_2)/n_{\rm T}$, $\varphi = \theta_2 - \theta_1$, and $|\mathbf{S}|^2 = 1$.

In the pseudospin picture, the axisymmetric vortex is analogous to a spin texture, extensively studied in superfluid ^{3}He. When the wave function for the $(1, 0)$ case is parameterized as $(\Psi_1, \Psi_2) = \sqrt{n_{\rm T}}(e^{i\theta} \cos(\beta(r)/2), \sin(\beta(r)/2))$, the configuration satisfying the boundary condition $\beta(0) = \pi$ and $\beta(\infty) = 0$ is referred to as an Anderson–Toulouse vortex [283], and $\beta(\infty) = \pi/2$ is a Mermin–Ho vortex [284]. The spin texture for the $(1, 0)$ solution is shown in Fig. 3.14(d). Here, the Ψ_1 component vanishes at the center, so that the pseudospin points down at $r = 0$ $(\beta(0) = \pi)$ according to the definition of Eq. (3.81). The spin aligns with a hyperbolic distribution as $(S_x, S_y) \propto (-x, y)$ around the center in the x-y planes. At the edge of the atomic cloud, the Ψ_2 component vanishes, and the pseudospin points up. In between, the bending angle $\beta(r) = \cos^{-1} S_z$ decreases smoothly from π at $r = 0$ as r increases, as seen in the profile in the $y = 0$ plane. For the $(0, 1)$ state, the texture exhibits a radial-disgyration as shown in Fig. 3.14(e), where the spin at $x = y = 0$ points up and aligns as $(S_x, S_y) \propto (x, y)$ in the x–y plane. Note that the spin rotates $\pi/2$ as going radially outward from the center. This configuration corresponds to a Mermin–Ho vortex. In the case of superfluid ^{3}He, an MH vortex is stabilized by the boundary condition of the $\hat{\mathbf{l}}$-vector imposed by a cylindrical vessel. However, in an atomic-BEC system there is no constraint at the boundary; the value $\beta(r)$ at the boundary $r = R$ should be determined self-consistently. The value $\beta(R)$ can change arbitrarily with Ω by varying the ratio N_1/N_2 or g_1/g_2 [282], which implies that an intermediate configuration between a Mermin–Ho vortex $(\beta(R) = \pi/2)$ and an Anderson–Toulouse vortex $(\beta(R) = 0)$ can be thermodynamically stable.

In terms of the spin-1/2 BEC, since the density $n_{\rm T}$ does not vanish at the vortex core, these vortices can be called *coreless* vortices. We can define an effective velocity field for the spinor BEC as

$$\mathbf{v}_{\rm eff} = \frac{\hbar}{2im}\sum_{j=1,2}\left(\zeta_j^*\nabla\zeta_j - \zeta_j\nabla\zeta_j^*\right) = \frac{\hbar}{2m}\left(\nabla\Theta - \cos\theta\nabla\varphi\right), \qquad (3.82)$$

which depends on the gradient of the total phase $\Theta = \theta_1 + \theta_2$ and that of the angle of the pseudospin. By neglecting the singular contribution of the velocity due to the total phase Θ, we can obtain the vorticity,

$$\boldsymbol{\omega}_{\rm eff} = \nabla\times\mathbf{v}_{\rm eff} = \frac{\hbar}{2m}(\nabla\theta)\times(\sin\theta\nabla\varphi) = \frac{\hbar}{4m}\epsilon_{\alpha\beta\gamma}S_\alpha\nabla S_\beta\times\nabla S_\gamma \qquad (3.83)$$

with the Levi-Civita symbol $\epsilon_{\alpha\beta\gamma}$. The last equality is known as the Mermin–Ho relation [284]. The vortex was characterized by the topological charge (or the Pontryagian index) [285],

$$Q = \int d^2r q(\mathbf{r}) = \frac{m}{h}\int d^2r\omega_{{\rm eff}\,z} = \frac{1}{8\pi}\int d^2r\left(\epsilon_{\alpha\beta\gamma}S_\alpha\nabla S_\beta\times\nabla S_\gamma\right)_z, \quad (3.84)$$

where $q(\mathbf{r})$ is the topological charge density characterizing the spatial distribution of the vorticity of the coreless vortex. In the coreless vortex, the topological charge density $q(\mathbf{r})$ is continuously distributed around the vortex core at which $|\mathbf{v}_{\rm eff}|$ vanishes, contrary to the case of a conventional vortex in a single-component condensate.

It is interesting to consider the nonaxisymmetric (1, 1) state in terms of the pseudospin. Then, the pair of coreless vortices can be identified as a pair of Mermin–Ho vortices (meron-pair) [284], where the spins are oriented along the x–y plane far from the two vortices. An example is shown in Fig. 3.15. Note that the boundary condition for the pseudospin is not well defined in our system. If we introduce the *internal Josephson coupling* between the two components, the spin should be oriented along the x-axis, far from the defect. This can be achieved by applying an external driving field that couples the two hyperfine levels of the atoms. When the atoms are condensed, coherent Rabi-oscillation between the two components occurs [286]. If the field strength is increased gradually from zero and its frequency is gradually ramped to resonance, a stationary nearly-weight superposition of the two components can be obtained. This Josephson coupling energy, given by $E_{\rm J} = -\hbar\omega_{\rm R}(\Psi_1^*\Psi_2 + {\rm c.c}) = -2\hbar\omega_{\rm R}\sqrt{n_1 n_2}\cos\varphi = -2\hbar\omega_{\rm R}n_{\rm T}S_x$ with the Rabi frequency $\omega_{\rm R}$, acts as an external field that aligns the spin along the x-axis. Therefore, the meron-pair is stabilized [287]. In the presence of the Josephson coupling, the two

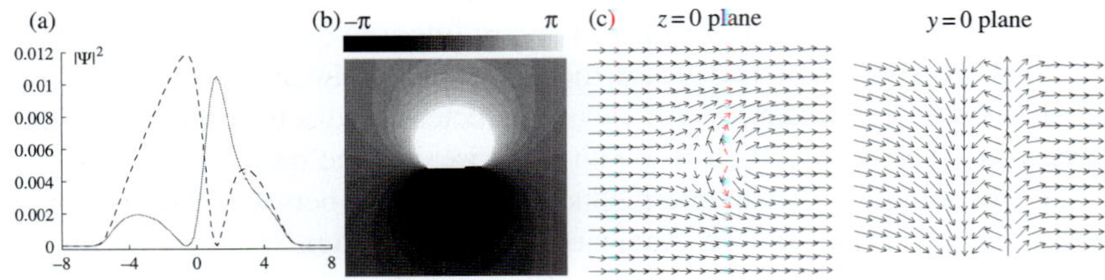

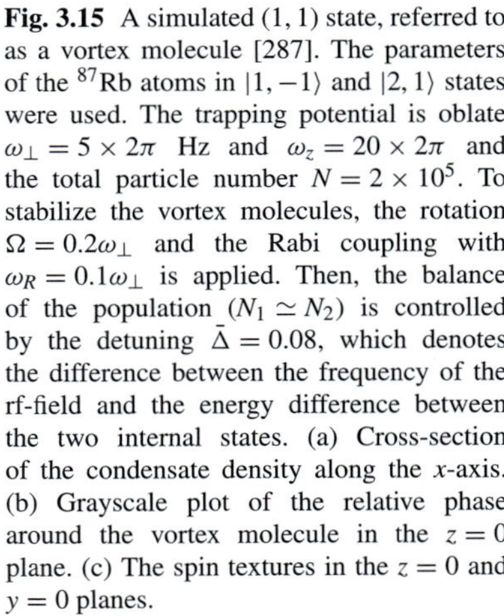
Fig. 3.15 A simulated (1, 1) state, referred to as a vortex molecule [287]. The parameters of the ^{87}Rb atoms in $|1,-1\rangle$ and $|2,1\rangle$ states were used. The trapping potential is oblate $\omega_\perp = 5\times 2\pi$ Hz and $\omega_z = 20\times 2\pi$ and the total particle number $N = 2\times 10^5$. To stabilize the vortex molecules, the rotation $\Omega = 0.2\omega_\perp$ and the Rabi coupling with $\omega_R = 0.1\omega_\perp$ is applied. Then, the balance of the population $(N_1 \simeq N_2)$ is controlled by the detuning $\bar{\Delta} = 0.08$, which denotes the difference between the frequency of the rf-field and the energy difference between the two internal states. (a) Cross-section of the condensate density along the x-axis. (b) Grayscale plot of the relative phase around the vortex molecule in the $z = 0$ plane. (c) The spin textures in the $z = 0$ and $y = 0$ planes.

vortices are bounded by the domain wall of the *relative phase* [288], shown in Fig. 3.15(b), whose binding energy can be controlled by the Josephson coupling strength.

VORTEX STABILITY AND DYNAMICS

Unstable vortex dynamics were experimentally observed to depend on which component had a vortex [117]. As seen in the ground state configuration of Fig. 3.14(a), if a component with smaller g_i is displaced from the trap center, it will tend to return to the trap center. This effect is sometimes called the "buoyancy" effect in trapped two-component BECs [196], and plays an important role in vortex dynamics.

Quantitatively, the instability can be understood from the BdG analysis of the vortex state [278, 279]. Numerical simulations showed that for equal populations $N_1 = N_2 = N$, and with the coupling constants $g_1 > g_{12} > g_2$, the axisymmetric vortex $(1, 0)$ state is stable, while the $(0, 1)$ state is *dynamically* unstable. Among the normal modes of the $(1, 0)$ case, there is a negative eigenvalue. This mode, which is a dipole-type mode, belongs to a perturbation that causes an initial displacement of the vortex from the trap center. However, the drift instability is affected by dissipation; the configuration is stable dynamically without dissipation, and furthermore, it can be stabilized energetically by rotation [282]. In the $(0, 1)$ case, on the other hand, there are normal modes with complex frequencies. The unstable modes have a similar shape to the negative-eigenvalue modes of the $(1, 0)$ case, which means that the perturbations also displace the vortex from the center. The imaginary part of the eigenvalues implies that vortices with unit charge in $|2\rangle$ are dynamically unstable under a generic perturbation of the initial configuration, and this instability grows without dissipation. This result is consistent with the JILA experiments [117].

To study the dynamics beyond linear stability, one begins with the time-dependent coupled GP equations (3.78). Numerical simulations of nonlinear vortex dynamics caused by large perturbations were performed by García-Ripoll and Pérez-García [278]. The linearly stable state $(1, 0)$ is robust and survives even under a wide range of perturbations; the vortex only shows a precession around the center. In contrast, the unstable configuration $(0, 1)$ gives rise to recurrent dynamics: the density of the $|1\rangle$ component and the vortex in $|2\rangle$ oscillate synchronously. These oscillations grow in amplitude until the vortex spirals outward. Since Eq. (3.18) conserves the total angular momentum $\langle L_z \rangle = \int d\mathbf{r}(\Psi_1^* L_z \Psi_1 + \Psi_2^* L_z \Psi_2)$, the vortex is transferred from $|2\rangle$ to $|1\rangle$. Although the dynamics are not completely periodic, this mechanism exhibits some recurrence, and the vortex eventually returns to $|2\rangle$. The unstable dynamics of $|2\rangle$ with a multiply quantized vortex such as $(q_1, q_2) = (0, 2), (0, 3)$ were investigated by Skryabin [279].

The effect of intercomponent interaction was also demonstrated by the precession of an off-centered vortex [198]. The experiment observed that the precession frequency of the *filled*-core (coreless) vortex was slower than that of the empty core. The slower precession of filled cores can be understood in terms of the buoyancy effect. Because of its slightly smaller scattering length, the $|2\rangle$ component has a negative buoyancy with respect to the $|1\rangle$ component, and consequently tends to sink inward toward the center of the condensate. With increasing amounts of the $|2\rangle$ component in the core, the inward force on the core begins to counteract the effective outward buoyancy of the vortex (described in Section 3.3.2.6), resulting in a reduced precession velocity. It was predicted that with a filling component of sufficiently negative buoyancy in the core, the core precession may stop, or even precess in a direction opposite to that of the fluid flow [196].

3.3.3.4 *Vortex lattices in two-component BECs*

Vortex lattices in rotating two-component BECS are more complicated than those in single-component condensates, because the systems have two-component order parameters and are coupled by intercomponent interactions. To simplify the problem, each component has an equal number of bosons, and the trapping potentials of the two components are identical. Then, the two components will have the same size and the same density of vortices. In this case, one would expect that each component will contain identical but *staggered* vortex lattices, with one lattice displaced relative to the other because of the intercomponent repulsive interaction. We will focus on the case of $g_1 = g_2 \neq g_{12}$, and vary the parameter $\delta = g_{12}/\sqrt{g_1 g_2}$. As described above, the condition of phase separation for non-rotating condensates is given as $\delta > 1$. The presence of a vortex lattice naturally modulates the density of each component, with the high density regions of one component overlapping with the low density regions of the other (the vortex core is filled by the other component). Thus, the system is effectively phase-separated whenever staggered vortex lattices are present, even for $\delta < 1$.

The first theoretical study of the lattice structure was done by Mueller and Ho within the mean-field quantum Hall regime [289], extending the analysis in Section 3.3.2.7 to a two-component system. In this analysis, (i) the rotation frequency was assumed to be very close to the radial trapping frequency ($\Omega \simeq \omega_\perp$), (ii) the vortex lattice was assumed to be perfectly regular, (iii) the range $\delta > 1$ could not be described, because a non-periodic structure appears as shown below. For $\delta \leq 0$, the two components prefer to overlap and form a single triangular lattice. As δ increases, the lattices undergo several structural transitions. There are five distinct phases of the lattice: triangular, displaced triangular, rhombic, square, and rectangular with increasing δ. Keçeli and Oktal calculated the dispersion relation of the TK mode for each lattice type, and discussed its relation to the structural phase transition [290].

Beyond the above restrictions (i)–(iii), one needs full numerical calculations of the coupled GP equations (3.79) [291]. Figure 3.16 shows the numerically obtained phase diagram of vortex states in the (δ, Ω) plane. In the overlapping region $\delta < 1$, two types of regular vortex lattices were obtained. For $\delta = 0$, the two components did not interact. Therefore, triangular vortex lattices formed, as in a single component BEC. As δ increased, the positions of the vortex cores in one component gradually shifted away from those of the other component and the triangular lattices became distorted. Eventually, the vortices in each component formed a square lattice rather than a triangular one. The two vortex lattices were interlaced in such a manner that a peak in the density of one component was located at the density minimum of the other.

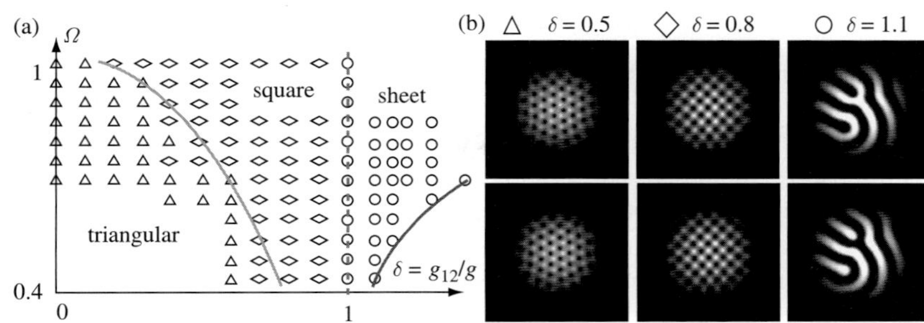

Fig. 3.16 (a) Ω-δ phase diagram for vortex states in rotating two-component BECs, obtained by 2D numerical simulations of Eq. (3.79): $\triangle$: triangular lattice, $\diamond$: square lattice, $\circ$: vortex sheet; see Ref. [291] for details. Because of the continuous change between triangular and square lattices, their boundary is shown by a combination of $\triangle$ and $\diamond$. The plots at $\Omega = 1$ show the results obtained by Ref. [289] based on the LLL approximation. (b) Typical vortex states in each phase for $\Omega = 0.8\omega_\perp$ and $\delta = 0.5, 0.8, 1.1$.

The stable region of the square lattice depended not only on δ, but also on Ω. While an increase in δ did cause deformation of the lattices from triangular to square, the transition occurred at a significantly higher value of δ than that of the LLL result ($\delta = 0.373$ [289]); for example, the transition occurred at $\delta \simeq 0.65$ for $\Omega = 0.7$. This implies that an increase in rotation frequency causes a transition from triangular to square lattices. The excitation spectra of these equilibrium states were calculated numerically by Woo et al. [292], showing complicated behavior consisting of TK modes, hydrodynamic modes, and surface modes.

When δ exceeds unity, the condensates undergo phase separation to spontaneously form domains having the same spin component. Concurrently, vortices begin to overlap. In the strongly phase-separated region $\delta > 1$ ($c_2 < 0$), the domains of the same spin component, at which the other-component vortices are located, merge further, resulting in the formation of "serpentine" vortex sheets [291]. A typical example is shown in Fig. 3.16. Singly quantized vortices line up in sheets, and the sheets of components 1 and 2 are interwoven. The sheet distance can be estimated from the competition of the kinetic energy of the superflow and the surface tension of the domain wall. A detailed discussion of the vortex sheet was reported in Ref. [293].

Schweikhard et al. observed an interlaced vortex lattice in a two-component BEC consisting of two hyperfine levels of the ^{87}Rb atom: $|1\rangle$ and $|2\rangle$ [268]. They initially created a regular triangular vortex lattice in a BEC in the $|1\rangle$ state. Then, a short pulse of the coupling drive transferred a 80–85% fraction of the population into the $|2\rangle$ state. After a variable wait time, they took images of the condensates, revealing interesting dynamics of the structural transition of the lattices. For the first period 0.1–0.25 s, there was little dynamical behavior in either component, and certainly no structural transition in the vortex lattice. From 0.25–2 s, turbulent behavior appeared in both components in which vortex visibility degraded significantly, exhibiting a transition from overlapping hexagonal vortex lattices to interlaced square lattices. From 2–3 s, square lattices emerged from the turbulent state. From 3–5.5 s stable square lattices were observed in both components. At this stage, in spite of the large (80–85 %) initial population transfer to state $|2\rangle$, the number of $|2\rangle$ atoms decreased rapidly because of a larger inelastic scattering loss of the $|2\rangle$ state than of the $|1\rangle$ state. As the $|2\rangle$ state population continued to decay, the vortex lattice reverted to a hexagonal lattice in the $|1\rangle$ state.

3.3.4 Vortices in spinor condensates

An optical trap removes the restriction of the confinable hyperfine spin states of atoms, thus realizing new multicomponent BECs with internal degrees of freedom, referred to as "spinor BECs" [141, 142, 144–146]. In contrast with the two-component BECs, interatomic interactions allow for a coherent transfer of population between different hyperfine spin states (spin-exchange collisions), which yields a fascinating physics of both ground-state properties and spin dynamics; a review addressing various problems on spinor BECs is described in Ref. [294]. In this section, we focus on the studies of quantized vortices in such spinor BECs with $F = 1$ and $F = 2$ hyperfine spin states.

3.3.4.1 *General formulation of spin-F Bose systems*

To describe the zero-temperature properties of spinor BECs, a generalized GP mean-field model was introduced by Ohmi and Machida [295] and Ho [296] for $F = 1$ spinor BECs. Here, we consider the most general case: spin-F bosons under a magnetic field $\mathbf{B}$. The second quantized Hamiltonian of the system is $\hat{H} = \hat{H}_0 + \hat{H}_{\text{int}}$. The single-particle part $\hat{H}_0$ can be written as

$$\hat{H}_0 = \int d\mathbf{r} \sum_{\alpha,\beta=-F}^{F} \hat{\psi}_\alpha^\dagger(\mathbf{r})\Big(\hat{h} + g_F \mu_B \mathbf{B} \cdot \mathbf{F}_{\alpha\beta}\Big)\hat{\psi}_\beta(\mathbf{r}), \tag{3.85}$$

where $\hat{\psi}_\alpha(\mathbf{r})$ is the field annihilation operator for an atom in the hyperfine state $|F,\alpha\rangle$ $(\alpha = -F, -F+1, \cdots, F-1, F)$, satisfying the commutation relation $[\hat{\psi}_\alpha(\mathbf{r}), \hat{\psi}_\beta^\dagger(\mathbf{r}')] = \delta_{\alpha\beta}\delta(\mathbf{r}-\mathbf{r}')$ and $[\hat{\psi}_\alpha(\mathbf{r}), \hat{\psi}_\beta(\mathbf{r}')] = 0$, and $\hat{h} = -\hbar^2\nabla^2/2m + V_{\text{ex}}(\mathbf{r})$ with an optical potential $V_{\text{ex}}(\mathbf{r})$ that can confine the atoms independently of their hyperfine spin states. The spin matrix $\mathbf{F}_{\alpha\beta} = [(F_x)_{\alpha\beta}, (F_y)_{\alpha\beta}, (F_z)_{\alpha\beta}]$ is the spin-F representation of the $SU(2)$ Lie algebra.

The interaction Hamiltonian $\hat{H}_{\text{int}} = \int d\mathbf{r}_1 d\mathbf{r}_2 V_F(\mathbf{r}_1 - \mathbf{r}_2)$ is governed by a two-body interaction $V_F(\mathbf{r}_1 - \mathbf{r}_2)$ that is invariant under spin rotation and preserves the hyperfine spin of the individual atoms. In a low-energy limit, this is of the form

$$V_F(\mathbf{r}_1 - \mathbf{r}_2) = \delta(\mathbf{r}_1 - \mathbf{r}_2)\sum_{f=0}^{2F} g_f \mathcal{P}_f, \tag{3.86}$$

where $g_f = 4\pi\hbar^2 a_f/m$ for atoms with an s-wave scattering length a_f corresponding to a channel with a total hyperfine spin f of two colliding atoms. The operator $\mathcal{P}_f$ projects the pair of atoms to a state with total hyperfine spin f, written in the second quantized form as $\mathcal{P}_f = \sum_{\alpha=-f}^{f} \hat{O}_{f,\alpha}^\dagger \hat{O}_{f,\alpha}$, where $\hat{O}_{f,\alpha} = \sum_{\alpha_1,\alpha_2} \langle f\alpha|F,\alpha_1;F,\alpha_2\rangle \hat{\psi}_{\alpha_1}^\dagger \hat{\psi}_{\alpha_2}$ with the Clebsch–Gordon coefficient $\langle f\alpha|F,\alpha_1;F,\alpha_2\rangle$, forming a total spin f state from two spin-F particles. The sum is taken for even numbers of f, because only symmetric spin channels are allowed for bosons.

3.3.4.2 *Mean-field theory for spin-1 BECs*

GENERALIZED GP EQUATION

For the spin-1 case, the field operator has three components: $(\hat{\psi}_1, \hat{\psi}_0, \hat{\psi}_{-1})$. The interaction can be expressed as

$$V_{F=1} = \frac{g_n}{2} \sum_{\alpha,\beta=-1}^{1} \hat{\psi}_\alpha^\dagger \hat{\psi}_\beta^\dagger \hat{\psi}_\beta \hat{\psi}_\alpha + \frac{g_s}{2} \sum_{\alpha,\alpha',\beta,\beta'=-1}^{1} \hat{\psi}_\alpha^\dagger \hat{\psi}_\beta^\dagger \mathbf{F}_{\alpha\alpha'} \cdot \mathbf{F}_{\beta\beta'} \hat{\psi}_{\beta'} \hat{\psi}_{\alpha'}, \tag{3.87}$$

where we have omitted $\delta(\mathbf{r}_1 - \mathbf{r}_2)$. The collision coefficients are written as $g_n = (g_0 + 2g_2)/3$ and $g_s = (g_2 - g_0)/3$. The term proportional to g_n is symmetric in the spin indices and represents the spin-independent contact interaction. The term proportional to g_s, on the other hand, is spin dependent and represents the short-range spin-exchange interaction. The sign of g_s determines the nature of the spin-exchange coupling: a negative (positive) g_s represents ferromagnetic (antiferromagnetic) coupling. The spinor BEC of $F = 1$ ^{23}Na atoms was demonstrated to have an antiferromagnetic interaction [141], while the $F = 1$ ^{87}Rb condensate has a ferromagnetic interaction [142]. Although $g_n \gg g_s$ in most systems, the g_s-term plays a crucial role in determining the properties of spinor BECs.

In the zero-temperature limit, the total Hamiltonian reduces to an energy functional by taking the expectation value $E = \langle \hat{H} \rangle$ and introducing the condensate wave function $\Psi_\alpha = \langle \hat{\psi}_\alpha \rangle$, which amounts to neglecting the effects of the non-condensed gas. For the condensates trapped by an optical potential and in a uniform magnetic field $\mathbf{B} = B\hat{\mathbf{z}}$, we can obtain

$$E[\boldsymbol{\Psi}] = \int d\mathbf{r} \left[\sum_{\alpha=-1}^{1} \Psi_\alpha^* \hat{h} \Psi_\alpha - p\langle F_z \rangle + q\langle F_z^2 \rangle + \frac{g_n}{2} n^2 + \frac{g_s}{2} |\mathbf{S}|^2 \right], \tag{3.88}$$

where $n_{\mathrm{T}}(\mathbf{r}) = \sum_{\alpha=-1}^{1} |\Psi_\alpha(\mathbf{r})|^2$ is the total particle density, and $\mathbf{S} = (\langle F_x \rangle, \langle F_y \rangle, \langle F_z \rangle)$ is the spin density vector defined by $\langle F_i(\mathbf{r}) \rangle = \sum_{\alpha,\beta=-1}^{1} \Psi_\alpha^*(\mathbf{r}) (F_i)_{\alpha\beta} \Psi_\beta(\mathbf{r})$ $(i = x, y, z)$. In addition to the linear Zeeman term with $p = -g_F \mu_B B$, we have introduced an additional quadratic Zeeman term characterized by $q = (g_F \mu_B B)^2 / E_{\mathrm{hf}}$ with the hyperfine splitting E_{hf}; these coefficients can be varied arbitrarily in experiments [141], and play an important role in determining the properties of spinor BECs.

The time evolution of the wave function is governed by $i\hbar \partial \Psi_\alpha(\mathbf{r})/\partial t = \delta E / \delta \Psi_\alpha^*$; substituting Eq. (3.88) into the right-hand side gives

$$i\hbar \frac{\partial \Psi_\alpha}{\partial t} = \hat{h} \Psi_\alpha + g_n n_{\mathrm{T}} \Psi_\alpha + \sum_{\beta=-1}^{1} \left[-p(F_z)_{\alpha\beta} + q(F_z^2)_{\alpha\beta} + g_s \mathbf{S} \cdot \mathbf{F}_{\alpha\beta} \right] \Psi_\beta, \tag{3.89}$$

which are the multicomponent GP equations that describe the mean-field properties of spin-1 BECs [295, 296]. In a stationary state, we substitute $\Psi_\alpha(\mathbf{r}, t) = \Phi_\alpha(\mathbf{r}) e^{-i\mu t/\hbar}$ into Eq. (3.89) to obtain

$$\hat{h}\Phi_\alpha + g_n n_{\rm T}\Phi_\alpha + \sum_{\beta=-1}^{1}\left[-p(F_z)_{\alpha\beta} + q(F_z^2)_{\alpha\beta} + g_s\mathbf{S}\cdot\mathbf{F}_{\alpha\beta}\right]\Phi_\beta = \mu\Phi_\alpha. \tag{3.90}$$

The chemical potential μ is determined so as to conserve the total particle number $N = \int d\mathbf{r} n_{\rm T}(\mathbf{r})$. In addition, the spin dynamics governed by Eq. (3.89) conserves the total magnetization $M = \int d\mathbf{r}\langle F_z(\mathbf{r})\rangle$, because the interactions conserve the total spin of two colliding atoms. Therefore, when we search for the ground state at a given magnetization M, we must replace E with $E - \lambda M$, where λ is a Lagrange multiplier. p is then replaced with $\tilde{p} = p + \lambda$, which is determined as a function of M.

GROUND STATE PROPERTIES

It is easy to identify the ground state of a uniform system when the Zeeman terms are negligible ($p = q = 0$), because we can set the kinetic energy and $V_{\rm ex}(\mathbf{r})$ to zero. From the energy functional (3.88), the ground state is ferromagnetic ($|\mathbf{S}| = n_{\rm T}$) for $g_s < 0$, and polar or antiferromagnetic ($|\mathbf{S}| = 0$) for $g_s > 0$. Introducing the normalized spinor $\boldsymbol{\zeta}$ as $\boldsymbol{\Psi}(\mathbf{r}) = \sqrt{n_{\rm T}(\mathbf{r})}\boldsymbol{\zeta}(\mathbf{r}) = \sqrt{n_{\rm T}(\mathbf{r})}(\zeta_+(\mathbf{r}), \zeta_0(\mathbf{r}), \zeta_-(\mathbf{r}))^T$, where $|\boldsymbol{\zeta}| = 1$, one can write the order parameter of the ferromagnetic phase as $\sqrt{n_{\rm T}}(1,0,0)^T$, while that of the polar (antiferromagnetic) phase is given by $\sqrt{n}(0,1,0)^T$ ($\sqrt{n_{\rm T}/2}(1,0,1)^T$) [295, 296]. Here, the polar and antiferromagnetic states are related to the $\pi/2$-rotation of the spin vector about the y-axis, which is degenerate in the absence of a magnetic field. Because of the symmetry of the Hamiltonian, all states obtained by the gauge transformation $e^{i\chi}$ and the spin rotation $U(\alpha,\beta,\gamma) = e^{-iF_z\alpha}e^{-iF_y\beta}e^{-iF_z\gamma}$ of these ground states are degenerate, where (α,β,γ) are the Euler angles. Thus, the general forms of the order parameter can be written as

$$\boldsymbol{\Psi}^{\rm ferro} = \sqrt{n_{\rm T}}e^{i(\chi-\gamma)}\begin{pmatrix} e^{-i\alpha}\cos^2\frac{\beta}{2} \\ \sqrt{2}\cos\frac{\beta}{2}\sin\frac{\beta}{2} \\ e^{i\alpha}\sin^2\frac{\beta}{2}\end{pmatrix}, \quad \boldsymbol{\Psi}^{\rm polar} = \sqrt{n_{\rm T}}e^{i\chi}\begin{pmatrix} -\frac{e^{-i\alpha}\sin\beta}{\sqrt{2}} \\ \cos\beta \\ \frac{e^{i\alpha}\sin\beta}{\sqrt{2}}\end{pmatrix} \tag{3.91}$$

for the ferromagnetic and polar states, respectively. The symmetry group of the ferromagnetic state is $SO(3)$, since the distinct configurations of $\boldsymbol{\zeta}$ are given by the full range of Euler angles. Note that the phase change by χ and the spin rotation by $-\gamma$ play an equivalent role, which represents a "spin-gauge" symmetry [297]. That of the polar state is $U(1) \times S^2/Z_2$, where $U(1)$ denotes the phase angle χ and S^2 is a surface of a unit sphere denoting all orientations (α,β) of the spin quantization axis. Moreover, a discrete symmetry Z_2 indicates that $\boldsymbol{\Psi}^{\rm polar}$ is invariant under π gauge transformation combined with π spin rotation around an axis perpendicular to the quantization axis $\hat{\mathbf{n}} = (\cos\alpha\sin\beta, \sin\alpha\sin\beta, \cos\beta)$, i.e., $\chi \to \chi + \pi$ and $\hat{\mathbf{n}} \to -\hat{\mathbf{n}}$ [298].

In the presence of an external magnetic field, the ground state becomes more complicated due to linear and quadratic Zeeman effects. The ground-state phase diagram in a parameter space of (p, q) was studied in Ref. [141]. Since $g_n n_{\rm T}$ can be absorbed into μ by defining $\tilde{\mu} \equiv \mu - g_n n_{\rm T}$, the ground state corresponds to the spinor, which minimizes the spin-dependent portion of

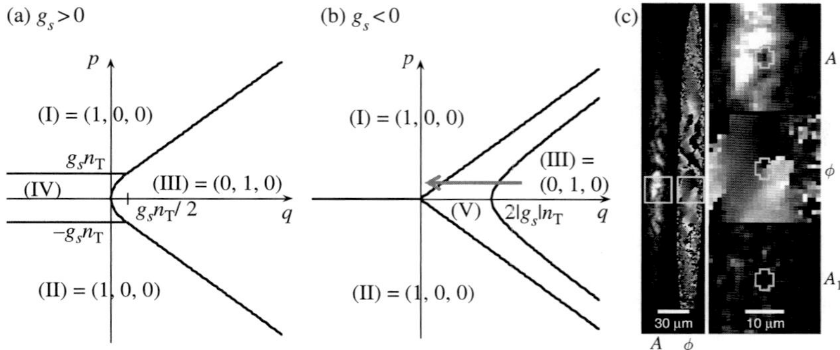

Fig. 3.17 p–q phase diagram of a uniform spin-1 BEC with (a) $g_s > 0$ and (b) $g_s < 0$. For (a) $g_s > 0$, the ferromagnetic phase and the polar phase are separated by a clear boundary $|p| = q + g_s$, which indicates that the Ψ_0 and $\Psi_{\pm 1}$ components are immiscible. For small bias fields, with $q < g_s n_T/2$ and $|p| < g_s n_T$, there is a region (IV), where the $\Psi_{\pm 1}$ components are mixed and the ratio of the populations is given by $|\Psi_+|^2/|\Psi_-^2| = (2g_s + p)/(2g_s - p)$. The boundary between (III) and (IV) lies at $|p| = 2\sqrt{cq}$. For (b) $g_s < 0$, The polar state occurs for $|p| \leq \sqrt{q(q - 4|g_s|)}$, and the ferromagnetic state occurs for $|p| > q$. Here, all three components are generally miscible, because the polar state is skirted by a region of a broken-axisymmetry phase (V) where all components have non-zero population. In (c), the in-situ observation of spontaneously generated spin vortices in a ferromagnetic spinor BEC [269] is shown, where a magnetic field was quenched along the arrow in (b). The picture represents the profile of the magnitude of the transverse magnetization A, its orientation ϕ, and the magnitude of the longitudinal magnetization A_l [Sadler et al.: *Nature* (London) **443** (2006) 312, reproduced with permission. Copyright 2006 by the Nature Publishing Group].

the energy density $E_s/V = g_s|\mathbf{S}|^2/2 - p\langle F_z\rangle + q\langle F_z^2\rangle$. A phase diagram of the ground state is shown in Fig. 3.17(a) and (b); the details of which are described in Refs. [294, 299]. Briefly, there are five distinct states (I)–(V) depending on the values of (g_s, p, q): (I) and (II) are ferromagnetic phases, whereas (III) is polar. In (IV), the longitudinal magnetization depends on p as $S_z = p/g_s$. In (V), the magnetization is tilted from the z-axis and its azimuthal angle is chosen spontaneously. This is referred to as the broken-axisymmetry phase [299]. The rotational symmetry of the order parameter about the magnetic-field axis is broken in states (III) and (V).

3.3.4.3 *Topological vortex formation in spinor BECs*

The internal degrees of freedom in BECs provide a peculiar way to generate vortices, which is called "topological phase imprinting" [300–302]. The MIT group [204, 205, 267] and the Kyoto university group [210, 304] used this method to create a vortex in a trapped BEC. In a magnetic field, atoms can be trapped by the Zeeman interaction of the electron spin with an inhomogeneous magnetic field. Atoms with electron spins parallel to the magnetic field are attracted to the minimum of the magnetic field (weak-field seeking state), while those with an electron spin antiparallel are repelled (strong-field seeking state). In the experiment at MIT, ^{23}Na BECs were prepared in weak-field seeking states, a $|F, m_F\rangle = |1, -1\rangle$ or $|2, +2\rangle$ hyperfine state, and confined in a Ioffe–Pritchard magnetic trap, described by $\mathbf{B} = B'_\perp(x\hat{\mathbf{x}} - y\hat{\mathbf{y}}) + B_z(t)\hat{\mathbf{z}}$. An essential idea in topological phase imprinting is azimuthally dependent adiabatic inversion of the bias field $B_z(t)$.

To interpret the mechanism, let us consider a BEC with $F = 1$. The basis vectors in the representation $(\Psi_1, \Psi_0, \Psi_{-1})$ are $\{|\pm\rangle, |0\rangle\}$. We introduce another set of basis vectors $|x\rangle, |y\rangle$ and $|z\rangle$, which are defined by $F_x|x\rangle = F_y|y\rangle = F_z|z\rangle = 0$; these vectors are related to the previous vectors as $|\pm 1\rangle = \mp(1/\sqrt{2})\,(|x\rangle \pm i|y\rangle)$ and $|0\rangle = |z\rangle$. When the z-axis is taken to be parallel to the uniform magnetic field, the order parameter of the weak-field seeking state takes the form $\Psi_{-1} = \psi$ and $\Psi_0 = \Psi_1 = 0$, or in vectorial form as $\boldsymbol{\Psi} = (\psi/\sqrt{2})\,(\hat{\mathbf{x}} - i\hat{\mathbf{y}})$. When the magnetic field points in the direction $\mathbf{B} = B(\cos\alpha\sin\beta, \sin\alpha\sin\beta, \cos\beta)$, a rotational transformation with respect to the Euler angle (α, β, γ) gives $\boldsymbol{\Psi} = (\psi/\sqrt{2})e^{-i\gamma}(\hat{\mathbf{m}} + i\hat{\mathbf{n}})$, where $\hat{\mathbf{m}} = (\cos\alpha\cos\beta, \sin\alpha\cos\beta, -\sin\beta)$ and $\hat{\mathbf{n}} = (\sin\alpha, -\cos\alpha, 0)$. The unit vector $\hat{\mathbf{l}} = \hat{\mathbf{m}} \times \hat{\mathbf{n}} = (-\cos\alpha \sin\beta, -\sin\alpha \sin\beta, -\cos\beta)$ corresponds to the direction of the spin $\mathbf{S}$, which is antiparallel to $\mathbf{B}$. The three real vectors $\{\hat{\mathbf{l}}, \hat{\mathbf{m}}, \hat{\mathbf{n}}\}$ form a triad, analogous to the order parameter of the orbital part of superfluid ^{3}He. The same amplitudes in the basis $\{|0\rangle, |\pm\rangle\}$ are $\Psi_1 = (\psi/2)(1 - \cos\beta)e^{-i\alpha - i\gamma}$, $\Psi_0 = -(\psi/\sqrt{2})\sin\beta e^{-i\gamma}$, and $\Psi_{-1} = (\psi/2)(1 + \cos\beta)e^{i\alpha - i\gamma}$.

When the field B_z is strong compared to the transverse quadrupole field $B'_\perp r$, the trapped condensate initially has an order parameter $\boldsymbol{\Psi} = (\psi/\sqrt{2})(\hat{\mathbf{x}} - i\hat{\mathbf{y}})$ without vorticity, i.e. $\Psi_{-1} = \psi$ and $\Psi_1 = \Psi_0 = 0$. This configuration corresponds to $\beta = 0$ and $\gamma = \alpha = \theta$, where θ is the azimuthal angle. The adiabatic condition with respect to the change of B_z is required for atoms to remain in the weak-field seeking state, so that $\hat{\mathbf{l}}$ is always antiparallel to $\mathbf{B}$. Thus, when B_z is gradually changed in the opposite $-\hat{\mathbf{z}}$ direction, the $\hat{\mathbf{l}}$-vector points $+\hat{\mathbf{z}}$ so that $\beta = \pi$, but $\gamma = \alpha = \theta$ is maintained by the effect of $B'_\perp$. Substituting these angles into $\Psi_{\pm 1}$ and Ψ_0, we obtain $\Psi_{-1} = \Psi_0 = 0$ and $\Psi_1 = \psi e^{-2i\theta}$, which corresponds to a vortex with a winding number $q = 2$. This result can be reinterpreted in terms of Berry's phase [204, 303]. In general, when the hyperfine spin is F, we obtain a vortex with a winding number $2F$ since Ψ_{-F} and Ψ_F acquire phases $F(\alpha - \gamma)$ and $F(-\alpha - \gamma)$, respectively [205, 304].

When the bias field is turned off at $B_z = 0$, a spin texture known as cross disgyration appears in the Ioffe–Pritchard trap. Here, the angle β increases from 0 to $\pi/2$, whereas $\gamma = \alpha = \theta$. The $\hat{\mathbf{l}}$-vector aligns with a hyperbolic distribution around the singularity at the center by following the direction of a local magnetic field of the Ioffe–Pritchard trap. This spin texture has been observed as a coreless vortex composed of three-component order parameters $\Psi_{\pm 1,0}$ [267], and its theoretical account will be given below.

If one inverts the bias field back to the initial state after vortex creation, the induced vortex unwinds itself, and the resulting state is non-rotating. On the other hand, if higher-order magnetic fields are employed, it is possible to gain more than $2F\hbar$ of angular momentum per particle through a single inversion of the bias field. Hence, by using, e.g. a hexapole field in the radial direction and inverting the bias field B_z, a $4F$-quantum vortex is produced. Switching to a quadrupole field and inverting the bias field back to the original value, one loses only $2F\hbar$ of angular momentum and the final state has a total angular momentum of $2F\hbar$. This process can be repeated, and each cycle increases the angular momentum of the system by $2F\hbar$ per particle. The idea of cyclically pumping vortices into a condensate was proposed theoretically in Refs. [305, 306].

3.3.4.4 *A single vortex (a few vortices) in spin-1 BECs*

Next, we consider the vortex states in slowly rotating spin-1 BECs. The characteristics of the vortex states are strongly dependent on the sign of g_s, i.e. ferromagnetic ($g_s > 0$) or antiferromagnetic ($g_s < 0$). This can be seen by their superfluid velocity $\mathbf{v} = (\hbar/m)\zeta^\dagger\nabla\zeta$; from Eq. (3.91), they are

$$\mathbf{v}^{\text{ferro}} = \frac{\hbar}{m}\left[\nabla(\chi - \gamma) - \cos\beta\nabla\alpha\right], \qquad \mathbf{v}^{\text{polar}} = \frac{\hbar}{m}\nabla\chi, \tag{3.92}$$

where the angles are assumed to be spatially varying functions. For the ferromagnetic case, the supercurrent can be induced by not only a spatial variation of the phase χ, but also by that of the spin, similarly to the pseudospin interpretation of two-component BECs described in Sec. 3.3.3.3. On the other hand, the polar case resembles that of scalar BECs. However, it allows half of the unit of quantization due to order parameter symmetry, as shown below.

The structure and stability of the axisymmetric vortex states for various Ω and total magnetizations M were thoroughly investigated in Refs. [307, 308]. The equilibrium state can be obtained by minimizing the energy in the rotating frame: $E - \lambda M - \Omega\langle L_z\rangle$. The axisymmetric vortex configuration is classified by a combination of the winding number q_j of the condensate wave function $\Psi_j = \sqrt{n_j(r)}e^{i(\chi_j + q_j\theta)}$ with the cylindrical coordinate (r, θ, z), where the homogeneity of the wave functions along the z-axis is assumed, and χ_j is the overall phase of the j-th component. The phases are constrained to minimize the spin-exchange interaction energy in Eq. (3.88):

$$E_s = \frac{g_s}{2}|\mathbf{S}|^2 = \frac{g_s}{2}\left\{(n_1 - n_{-1})^2 + 2n_0\left[n_1 + n_{-1} + 2\sqrt{n_1 n_{-1}}\cos(\bar{\chi} + \bar{q}\theta)\right]\right\}, \tag{3.93}$$

where $\bar{\chi} = \chi_1 + \chi_{-1} - 2\chi_0$ and $\bar{q} = q_1 + q_{-1} - 2q_0$. To minimize E_s, χ_j and q_j should satisfy $2\chi_0 = \chi_1 + \chi_{-1} + n'\pi$ and $2q_0 = q_1 + q_{-1}$, where n' is an odd (even) integer for antiferromagnetic (ferromagnetic) interaction. The global phase $\bar{\chi}$ has no effect on the vortex structure, and was therefore set to zero. The possible combination of q_j satisfying the above relation gives this system a characteristic vortex structure. If the values of q_j are restricted to $|q_j| \leq 1$, we have $(q_1, q_0, q_{-1}) = \pm(1, 1, 1)$, $\pm(1, 0, -1)$ and $\pm(1, 1/2, 0)$, where cases with negative sign are omitted in the following. For $(1, 1/2, 0)$, the value $1/2$ is not allowed in an axisymmetric system, so the Ψ_0-component vanishes; we denote this state as $(1, \times, 0)$.

Unfortunately, the assumption of axisymmetry restricts the possible ground states, limiting stable vortex configurations. The vortex states without assuming axisymmetry have been considered by some authors [309–314], who constructed phase diagrams of vortex states for various parameter spaces in various situations. The details for each vortex state are summarized below.

$(1, 1, 1)$ VORTEX

In this case, all components have singly quantized vortices at the center, and the densities of the three components are fully overlapped. As a result, the total density is equivalent to that of a conventional BEC with an empty vortex core.

With a fixed M, this axisymmetric vortex state cannot be stabilized in any parameter region for both $g_s < 0$ and $g_s > 0$ [308]. However, as in the (1,1) vortex state in two-component BECs shown in Fig. 3.14, by displacing the vortex cores of each component from the center of the trap, the (1,1,1) vortex can be stabilized in a nonaxisymmetric coreless configuration [311]. For $g_s > 0$, the vortex cores are displaced such that the condensates have two singularities through an overlap of the Ψ_1 and Ψ_{-1} components, or they have three singularities that form a triangular configuration. For $g_s < 0$, two singularities of Ψ_1 and Ψ_{-1} are displaced from the center, while Ψ_0 with a singularity at the center prevents phase separation. For all of these cases, the breaking of axisymmetry leads to a smooth variation of the total density by decreasing the overlap area of $\Psi_{\pm 1}$ and Ψ_0 for $g_s > 0$, and Ψ_1 and Ψ_{-1} for $g_s < 0$.

$(1, 0, -1)$ VORTEX

In this configuration, the non-rotating Ψ_0 component occupies the central region of the vortex core, which is made up of the Ψ_1 and Ψ_{-1} components with opposite circulations. This vortex state is referred to as a "polar-core" vortex. Since the condensate at the center of the trap consists only of the polar state, the spin texture has a singularity, although it can vary continuously around the vortex core. This vortex state is thermodynamically stable in the high magnetization region, which is independent of the strength and sign of the spin exchange interaction [311]. Also, it becomes stable in a non-rotating harmonic trap with an applied Ioffe–Pritchard magnetic field [313].

Interestingly, this vortex state can be generated by the intrinsic dynamical instability of a ferromagnetic spinor BEC. Saito et al. demonstrated that an initially prepared Ψ_0 without vorticity can generate transverse magnetization due to ferromagnetic interaction [315]. Since spin conservation prohibits the appearance of longitudinal magnetization, the dynamical instability results in a polar–core vortex by spontaneously breaking the chiral symmetry of the initial state, where the $\Psi_{\pm 1}$ component can take either $q = \pm 1$ circulation. Since oppositely rotating $\Psi_{\pm 1}$ components have the same amplitude due to $M = 0$, the vortex carries no net mass current but a spin current with one quantum of circulation. This spontaneous formation of such polar–core "spin" vortices was demonstrated experimentally by the Berkeley group, using an instantaneous quench from the polar phase to the ferromagnetic phase [269] [see Fig. 3.17].

$(1, \times, 0)$ VORTEX

This vortex state consists of a non-rotating Ψ_{-1} component at the center filling the vortex core of the Ψ_1 component. This is similar to the (1,0) vortex of two-component BECs, but its interpretation is quite different. In the $(1,\times,0)$ vortex, the spin component is suddenly reversed near the vortex core of the Ψ_1 component because the absence of the Ψ_0 component means that $S_x = S_y = 0$, which cannot yield a continuous variation of the spin texture. This vortex state is thermodynamically stable only for the polar case ($g_s > 0$) [308, 311].

This vortex state is called the half-quantum vortex state ("Alice state") [280, 308]. To see this, let us consider $\boldsymbol{\Psi}^{\rm polar}$ and a loop that encircles a vortex with a fixed radius. Each point on the loop can be specified by its azimuthal angle θ. The single-valuedness of the order parameter $\boldsymbol{\Psi}^{\rm polar}$ in Eq. (3.91) is met if we take, for example, $-\alpha = \chi = q\theta/2$ and $\beta = \pi/2$. The order parameter at $r \to \infty$ is given by $\boldsymbol{\Psi}^{\rm polar} = \sqrt{n_{\rm T}/2}(-e^{iq\theta}, 0, 1)$. Thus, from Eq. (3.92), the circulation becomes $\oint \mathbf{v}^{\rm polar} \cdot d\boldsymbol{\ell} = (h/2m)q$, quantized in units of $h/(2m)$ rather than the usual h/m. The underlying physics for this half-quantum number are based on the discrete Z_2 symmetry of $\boldsymbol{\Psi}^{\rm polar}$, which is invariant under π gauge transformation ($\chi \to \chi + \pi$) combined with a π spin rotation around an axis perpendicular to $\hat{\mathbf{n}} = (\cos\alpha \sin\beta, \sin\alpha \sin\beta, \cos\beta)$: $\hat{\mathbf{n}} \to -\hat{\mathbf{n}}$. Thus, the polar phase of a spin-1 BEC can host a half-quantum vortex [298, 280]. The dynamical creation of half-quantum vortices under an external rotating potential was demonstrated in numerical simulations [316, 317].

The half-quantum vortex state is unstable for ferromagnetic spinor BECs. The axisymmetry of the transverse magnetization is broken spontaneously to form three-fold domains [318]. This originates from the topological structure of the half-quantum vortex and spin conservation.

(0, 1, 2) VORTEX

In ferromagnetic interactions, a more stable vortex configuration occurs when the winding number q_j exceeds unity. Mizushima et al. found that ferromagnetic interaction supports the thermodynamic stability of the $(0, 1, 2)$ axisymmetric vortex state over wide parameter regions [310, 311]. The central region of the harmonic trap is occupied by a non-rotating Ψ_1 component. The Ψ_0 component with $q_0 = 1$ is pushed outward, while the Ψ_{-1} component with $q_{-1} = 2$ occupies the outermost region. This configuration is favorable for the ferromagnetic case, because it is more favorable than spatial phase separation of Ψ_1 and Ψ_{-1}; the presence of vortices with different q_j effectively causes phase separation in the radial direction. Only a (0,1,2) vortex can have a nonsingular continuous spin texture under slow rotation. Mizushima et al., performed extensive numerical studies to determine the Ω-M phase diagram of the vortex states, including the $(2, 1, 0)$ state referred to as the mixed-twist texture [314].

When the wave function $\boldsymbol{\Psi}^{\text{ferro}}$ is parameterized by the spatially dependent bending angle $\beta(r)$ and $\chi = 0$ and $\alpha = -\gamma = \theta$ in Eq. (3.91), we have $\boldsymbol{\Psi}^{\text{ferro}} = \sqrt{n_{\text{T}}(r)}\ (\cos^2 \frac{\beta(r)}{2},\ e^{i\theta}\sqrt{2}\sin\frac{\beta(r)}{2}\cos\frac{\beta(r)}{2},\ e^{2i\theta}\sin^2\frac{\beta(r)}{2})$, where $\beta(r)$ is an increasing function of r from $\beta(0) = 0$ and $0 < \beta(r) < \pi$. The spin direction is related to the $\hat{\mathbf{l}}$-vector, and is given as $\hat{\mathbf{l}}(r) = \hat{\mathbf{z}}\cos\beta(r) + \sin\beta(r)(\cos\theta\hat{\mathbf{x}} + \sin\theta\hat{\mathbf{y}})$, where β varies from $\beta(0) = 0$ to $\beta(R) = \pi/2\ (= \pi)$ for a Mermin–Ho (Anderson–Toulouse) vortex. As in two-component BECs [see Sec. 3.3.3.3], the $(0, 1, 2)$ vortex can be interpreted as a vortex with an intermediate boundary condition ($\pi/2 < \beta(R) < \pi$); we can control the value of $\beta(R)$ from the Mermin–Ho condition to the Anderson–Toulouse condition by merely changing the total magnetization M [310]. In another numerical study, Martikainen et al. analyzed the coreless vortex state as a function of rotational frequency, without fixing the total magnetization [312], finding that $\beta(R)$ increases with increasing Ω and that the upper value of $\beta(R)$ is $3\pi/4$, above which additional vortices nucleate. This implies that an Anderson–Toulouse vortex can never be the ground state of the system.

Because this coreless vortex contains a doubly charged vortex in one of the components, it is interesting to study whether the coreless vortex state inherits the dynamical instability of doubly charged vortices in scalar condensates, as shown in Section 3.3.2.6. Pietilä et al. studied the low-lying excitations of the $(0, 1, 2)$ state for $g_s < 0$ in the presence of an Ioffe–Pritchard trap based on the BdG approach. The phase diagram of the dynamical stability was explored for the bias field B_z and the transverse quadrupole field $B'_\perp$ [319]. Takahashi et al. considered the same problem for both $g_s < 0$ and $g_s > 0$ without the Ioffe–Pritchard trap, finding that the dynamical instability is suppressed (enhanced) for $g_s < 0$ ($g_s > 0$) [320]. The unstable dynamics are associated with vortex splitting of the $q = 2$ vortex, and phase separation in the $g_s > 0$ case.

3.3.4.5 *Vortex lattices in $F = 1$ spinor BECs*

Vortex lattices in rapidly rotating spin-1 BECs can possess richer vortex phases than those in two-component BECs, but their classification is very complicated. Kita et al. studied vortex-lattice structures of antiferromagnetic spin-1 BECs using the phenomenological Ginzburg–Landau equation [321], which is formally similar to the mean-field quantum Hall approach. This study revealed that the conventional Abrikosov lattice with hard vortex cores is unstable, and the vortex cores shift their locations. The system has many metastable configurations of vortices, depending sensitively on the ratio g_s/g_n. The vortices are characterized by the distribution of magnetization and the difference in the number of circulations of quanta per unit cell, all of which makes the characterization of the vortex lattice structure complicated.

Further discussion on the phase diagram of the ground state in rotating spin-1 bosons was reported in Reijnders et al. [322]. This paper also describes a study of vortex-lattice structures with the mean-field quantum Hall approach over wide parameter regions and the exact diagonalization study of the quantum liquid phase. A similar analysis was done by Mueller for slowly rotating spin-1 BECs [281]. In addition, Mizushima et al. performed numerical simulations of the GP equations (3.90) for a ferromagnetic spinor BEC in a fast-rotating regime and also in an external magnetic field [314]. For $M = 0$, the equilibrium state is a square lattice constructed from two sublattices of the Mermin–Ho vortex and the mixed-twisted vortex, and the local spin on the two sublattice sites are locked in alternate directions $\hat{\mathbf{z}}$ and $-\hat{\mathbf{z}}$. For $M \neq 0$, the composite lattice of a coreless vortex and a polar–core vortex was found.

3.3.4.6 *Quenched spinor BEC: Kibble–Zurek mechanism*

Topological defect formation via the Kibble–Zurek mechanism can occur in a ferromagnetic phase of a spinor BEC. In an experiment by the Berkeley group [269], ^{87}Rb atoms (with $g_s < 0$) in the $m = 0$ state were prepared by applying a magnetic field along the z-axis. This initial state corresponds to the point in region (III) of the phase diagram in Fig 3.17(b). The strength of the magnetic field was then suddenly decreased toward region (I) along a constant-p line of the phase diagram. The $m = 0$ state was no longer the ground state, and a spontaneous transverse magnetization emerged due to the conservation of longitudinal magnetization. Since this phase transition was triggered by a sudden change in a magnetic field, this can be regarded as a *quantum* quench at $T = 0$. The magnetization dynamics were observed by a spin-sensitive *in situ* measurement to form complicated ferromagnetic domains. Remarkably, polar–core spin vortices were identified in some snapshots of the spin distribution, as shown in Fig. 3.17(c), consistent with the prediction [315].

Theoretical analysis of this experiment was done by several authors [323–327], in relation to the Kibble–Zurek mechanism. The numerical simulations revealed that the creation of the spin vortices was caused by instability of the initially generated solitons, whose pattern and characteristic length scales were strongly dependent, not only on the quench time, but also the properties of the initial noise.

3.3.4.7 *Vortices in $F = 2$ spinor condensates*

For the $F = 2$ case, the field operator has five components ($\hat{\psi}_2, \hat{\psi}_1, \hat{\psi}_0, \hat{\psi}_{-1}, \hat{\psi}_{-2}$). The interaction becomes

$$
\begin{aligned}
V_{F=2} = \frac{c_0}{2} \sum_{\alpha,\beta=-2}^{2} \hat{\psi}_\alpha^\dagger \hat{\psi}_\beta^\dagger \hat{\psi}_\beta \hat{\psi}_\alpha \\
+ \frac{1}{2} \sum_{\alpha,\alpha',\beta,\beta'=-2}^{2} \Bigg(c_1 \hat{\psi}_\alpha^\dagger \hat{\psi}_\beta^\dagger \mathbf{F}_{\alpha\alpha'} \cdot \mathbf{F}_{\beta\beta'} \hat{\psi}_{\beta'} \hat{\psi}_{\alpha'} \\
+ 5c_2 \hat{\psi}_\alpha^\dagger \hat{\psi}_\beta^\dagger \langle 2,\alpha;2,\alpha'|0,0\rangle \langle 0,0|2,\beta;2,\beta'\rangle \hat{\psi}_{\beta'} \hat{\psi}_{\alpha'} \Bigg).
\end{aligned}
\tag{3.94}
$$

The collision coefficients are given by $c_0 = (4g_2 + 3g_4)/7$, $c_1 = -(g_2 - g_4)/7$, and $c_2 = (g_0 - g_4)/5 - 2(g_2 - g_4)/7$. The energy functional $E = \langle \hat{H} \rangle$ for spin-2 BECs can be written as

$$
E[\mathbf{\Psi}] = \int d\mathbf{r} \left(\sum_{\alpha=-2}^{2} \Psi_\alpha^* \hat{h} \Psi_\alpha + \frac{c_0}{2} n_{\mathrm{T}}^2 + \frac{c_1}{2} |\mathbf{S}|^2 + \frac{c_2}{2} |A_{20}|^2 \right), \tag{3.95}
$$

where $n_{\mathrm{T}}(\mathbf{r}) = \sum_{\alpha=-2}^{2} |\Psi_\alpha(\mathbf{r})|^2$ is the total particle density and $\mathbf{S} = (\langle F_x \rangle, \langle F_y \rangle, \langle F_z \rangle)$ is the spin density vector defined by $\langle F_i(\mathbf{r}) \rangle = \sum_{\alpha,\beta=-2}^{2} \Psi_\alpha^*(\mathbf{r}) (F_i)_{\alpha\beta} \Psi_\beta(\mathbf{r})$ $(i = x, y, z)$. The new feature of spin-2 BECs, not found in spin-1 BECs, is the presence of the spin-singlet pair amplitude $A_{20} = (2\Psi_2\Psi_{-2} - 2\Psi_1\Psi_{-1} + \Psi_0^2)/\sqrt{5}$. Figure 3.18(a) shows a mean-field phase diagram in the absence of a magnetic field. There are four distinct phases: ferromagnetic, uniaxial nematic, biaxial nematic, and cyclic, where the order parameters of the last three phases are characterized by a discrete symmetry. According to mean-field theory, the uniaxial nematic phase and biaxial phase are degenerate in the absence of a magnetic field [328], however, quantum fluctuations induce a quantum phase transition between the two phases, lifting the degeneracy [329, 330]. For $c_1 > 0$ and $c_2 > 0$, both $\mathbf{S}$ and A_{20} vanish, which yields the cyclic phase concentration characterized by the spin-singlet "trios" peculiar to spin-2 BECs [331]; one representation for the cyclic phase is given by $\mathbf{\Psi}^{\mathrm{cyc}} = \sqrt{n_{\mathrm{T}}}(i/2, 0, 1/\sqrt{2}, 0, i/2)^T$.

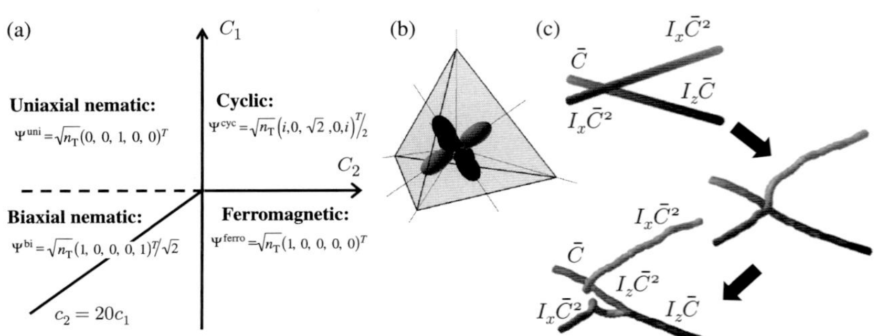

Fig. 3.18 (a) Mean-field phase diagram of spin-2 BECs. (b) Schematic illustration showing that the symmetry axes of the cyclic order parameter constitute a tetrahedron. (c) Numerical simulation of the collision dynamics of non-Abelian vortices. After the collision, two vortices get connected and a rung appears between them. The topological charges for each vortex line are denoted [Kobayashi et al.: *Phys. Rev. Lett.* **103** (2009) 115301, reproduced with permission. Copyright 2009 by the American Physical Society].

Although several experiments have demonstrated that spinor BECs with an $F = 2$ manifold of ^{87}Rb atoms were characterized by biaxial-nematic spinor dynamics, [144, 146], the possibility of the cyclic phase has not yet been excluded due to complications arising from quadratic Zeeman effects and hyperfine-spin-exchange relaxations [332].

The vortex states in spin-2 BECs are affected strongly by the symmetry of the order parameters in each phase. For ferromagnetic phases, the properties of the vortex states are not so different from those of spin-1 BECs. However, the symmetry of the biaxial nematic phase and the cyclic phase can yield exotic vortices governed by the non-Abelian group [333–337]. Here, we focus on the vortex states in the cyclic phase, and briefly summarize the salient results.

VORTICES IN CYCLIC CONDENSATES

As in the spin-1 BEC, through the $U(1)$ gauge transformation and the $SO(3)$ spin rotation, it is possible to transform from one to another cyclic state, $\Psi^{\mathrm{cyc}\prime} = e^{i\chi} e^{-i\mathbf{F}\cdot\hat{\mathbf{n}}\theta} \Psi^{\mathrm{cyc}}$, where $\hat{\mathbf{n}}$ and θ are the unit vector of the rotational axis and angle of the spin rotation, respectively. The order parameter Ψ^{cyc} is invariant under the following 12 transformations: $\mathbf{1}, I_x = e^{iF_x\pi}, I_y = e^{iF_y\pi}, I_z = e^{iF_z\pi}$, $\bar{C} = e^{2\pi i/3} e^{-2\pi i(F_x+F_y+F_z)/3\sqrt{3}}$, $\bar{C}^2, I_x\bar{C}, I_y\bar{C}, I_z\bar{C}, I_x\bar{C}^2, I_y\bar{C}^2$, and $I_z\bar{C}^2$ [333]. The overbar was added to emphasize that the operation includes not only a rotation in spin space but also a gauge transformation. These 12 transformations form the non-Abelian tetrahedral group T, shown in Fig. 3.18(b).

Topological charges of vortices can be classified by 12 generators $\mathbf{1}$, I_x, I_y, $\cdots$ of the non-Abelian group T. Vortices are also classified into four conjugacy classes: (I): integer vortex; $\mathbf{1}$, (II): 1/2 - spin vortex; I_x, I_y, and I_z, (III): 1/3 vortex; $\bar{C}$, $I_x\bar{C}$, $I_y\bar{C}$, and $I_z\bar{C}$, and (IV): 2/3 vortex; $\bar{C}^2$, $I_x\bar{C}^2$, $I_y\bar{C}^2$, and $I_z\bar{C}^2$. Topological charges in the same conjugacy class transform into one another under the arbitrary global gauge transformation and the spin rotation, denoted by $\hat{S}$. The order parameters for straight vortices along the z-axis in each conjugacy class can be written in cylindrical coordinates (r, θ, z) as

$$\Psi = \frac{1}{2}\sqrt{n_{\mathrm{T}}(r)} e^{in_1\theta} \hat{S} \times \begin{cases} \left(if(r)e^{2i(n_2+1)\theta}, 0, \sqrt{2}h(r), 0, if(r)e^{-2i(n_2+1)\theta}\right)^T \\ \left(if(r)e^{i(2n_2+1)\theta}, 0, \sqrt{2}h(r), 0, if(r)e^{-i(2n_2+1)\theta}\right)^T \\ \frac{2}{\sqrt{3}}\left(f(r)e^{i(2n_2+1)\theta}, 0, 0, \sqrt{2}g(r)e^{-in_2\theta}, 0\right)^T \\ \frac{2}{\sqrt{3}}\left(f(r)e^{i(2n_2-1)\theta}, 0, 0, \sqrt{2}g(r)e^{-in_2\theta}, 0\right)^T \end{cases} \tag{3.96}$$

for (I), (II), (III), and (IV), respectively, where the vortex is located at $r = 0$. Here, n_1 and n_2 are the integer winding numbers, and f, g, and h are real functions that satisfy $[f^2 + h^2]/2 = [f^2 + 2g^2]/3 = 1$ and $f(r \to \infty) = g(r \to \infty) = h(r \to \infty) = 1$. At the vortex core, the cyclic order parameter changes to that of a different phase. With the minimum windings ($n_1 = n_2 = 0$), we obtain the core structure of each conjugacy class by taking $f(r = 0) = 0$ as: (I)(II) $\Psi \propto \hat{S}(0, 0, 1, 0, 0)^T$ and (III)(IV) $\Psi \propto \hat{S}(0, 0, 0, 1, 0)^T$, i.e. the core of (I) and (II) vortices have a finite spin-singlet pair amplitude ($A_{20} \neq 0$), and that of (III) and (IV) vortices have a finite magnetization ($S_z \neq 0$). The 1/3 and 2/3 vortices are a hallmark of the cyclic state with tetrahedral symmetry. The energetic stability of these fractional vortices in rotating potentials was examined in Ref. [337]

A salient feature of non-Abelian vortices emerges in their collision dynamics [336]. When two Abelian vortices collide, there are three possibilities: (i) they reconnect themselves, (ii) they pass through each other, or (iii) they form a rung that bridges the two vortices, depending on the kinematic parameters and initial conditions. Usually, reconnection occurs due to the energetic constraint. However, when two non-Abelian vortices collide, only a rung can be formed; reconnection and passing through are topologically forbidden because the corresponding generators do not commute with each other. In fact, the nonzero commutator of the two generators gives the generator of the rung vortex. Figure 3.18(c) illustrates a typical rung formation.

In other studies, vortex molecules consisting of a bound pair of fractional vortices were discussed in Refs. [335, 337]. The structure of vortex lattices in cyclic condensates was discussed in Ref. [334], where the transitions of the various lattice geometries were demonstrated as a function of applied magnetic field and temperature.

3.3.5 Vortices in dipolar condensates

A dipolar condensate refers to a condensate in which atoms interact via dipole–dipole interactions (DDIs), in addition to the usual s-wave contact interactions. For two particles 1 and 2 with dipole moments along the unit vector $\mathbf{e}_1$ and $\mathbf{e}_2$, and relative position $\mathbf{r}$, the DDI has the form

$$V_{\rm dd}(\mathbf{r}) = \frac{c_{\rm dd}}{4\pi}\frac{\mathbf{e}_1\cdot\mathbf{e}_2 - 3(\mathbf{e}_1\cdot\hat{\mathbf{r}})(\mathbf{e}_2\cdot\hat{\mathbf{r}})}{r^3}. \tag{3.97}$$

The coupling constant c_{dd} is given as $\mu_0\mu_{\rm d}^2$ for particles with a permanent magnetic dipole moment $\mu_{\rm d}$ (μ_0 is the magnetic permeability of vacuum) and d^2/ϵ_0 for a permanent electric dipole moment d (ϵ_0 is the permittivity of vacuum). Recently, BECs of chromium atoms have been created [338], in which the atoms exhibited a larger magnetic-dipole moment ($\mu_{\rm d} = 6\mu_{\rm B}$) than typical alkali atoms ($\mu_{\rm d} \simeq \mu_{\rm B}$). This enables the study of the effects of *anisotropic long-range* interactions in BECs. The rapid progress in studies of dipolar BECs can be seen in the review article of Ref. [342].

In particular, when the dipole moments are polarized under an external field along the z-axis, the interaction potential between two magnetic dipoles with $\mathbf{e}_z$ separated by $\mathbf{r}$ is given by $V_{\rm dd}(\mathbf{r}) = (\mu_0\mu_d^2/4\pi)(1-3\cos^2\theta)/r^3$, where $\hat{\mathbf{e}}_z\cdot\hat{\mathbf{r}} = \cos\theta$. Thus, the system is described by the scalar condensate wave function, and the DDIs contribute to the GP equation as a non-local mean-field potential

$$i\hbar\frac{\partial\Psi}{\partial t} = \left[-\frac{\hbar^2\nabla^2}{2m} + V_{\rm ex} + g|\Psi|^2 + \int d\mathbf{r}' V_{\rm dd}(\mathbf{r}'-\mathbf{r})|\Psi(\mathbf{r}')|^2\right]\Psi. \tag{3.98}$$

Since the contact interaction $g = 4\pi\hbar^2 a/m$ can be tuned to zero by the Feshbach resonance technique, we can obtain novel quantum *ferrofluids* dominated by the dipole–dipole interaction [156].

Before studying the vortex states, it is useful to examine the properties of non-rotating dipolar BECs. The principal effect of $V_{\rm dd}$ on the equilibrium shape of a trapped condensate is to cause distortion of its aspect ratio [339, 340]. Since the dipoles are aligned along the z-axis, the attractive interaction of dipoles becomes significant in a cigar-shaped trap, and the atomic cloud is elongated along the z-axis. Although this may appear counterintuitive, it can be understood by considering the energy density of the DDI. It has the anisotropic form of a saddle, with negative curvature along the direction of the dipoles [341]. Because of attractive interactions, the condensate will eventually collapse, once the particle number exceeds some critical value N_c, similar to condensates with attractive contact interactions [154]. For a pancake trap, the condensate is elongated similarly along the z-axis, but stable against collapse because the DDI is mostly repulsive. The instability condition also depends on the strength of the contact interaction. These collapse dynamics were demonstrated by tuning the s-wave scattering length into the unstable regime, showing an anisotropic explosion characterized by d-wave symmetry [343].

Another interesting feature of a dipolar BEC is that the Bogoliubov excitation spectrum exhibits a "roton–maxon" behavior, caused by the attractive nature of the DDI. Santos et al., found that, for a quasi-2D system, harmonically confined in the z direction and free in the x and y directions, the excitation energy has a local maximum and minimum dependent upon the in-plane momentum q [344]. Although this behavior is reminiscent of that found in the excitation spectrum of liquid helium II, its physical origin is somewhat different. This can intuitively be understood as follows. When the in-plane momenta q are much smaller than the inverse size L of the condensate in the z direction, excitations have a 2D character. Because the dipoles are oriented along the z-axis, the interaction is effectively repulsive and the in-plane excitations become phonons. For $q \ll 1/L$, excitations begin to have a 3D character and the interparticle repulsion is reduced due to the attractive force in the z direction. This decreases the excitation energy with an increase in q. The excitation energy reaches a minimum (roton) and then begins to increase with any further increase in q, eventually leading to a quadratic behavior $\propto q^2$.

As the dipole interaction becomes stronger, the energy at the roton minimum decreases, and then reaches zero, representing an onset of instability. This "roton instability" is probably towards the formation of a density wave. Ronen et al. studied numerically the equilibrium solutions for a pancake geometry, finding that the condensate always has an unstable criterion of the dipolar interaction strength, even for the pancake-trap limit [345]. They observed that, near the instability threshold in a certain range of the trap aspect ratio, the equilibrium density possesses "biconcave" shape, where the density modulates radially and its local minimum appears at the center. This density modulation is associated with roton instability. Note that this physical origin of the instability is different from that discussed in the dipolar collapse experiment; the roton instability exists even for a 2D pancake geometry. A more detailed discussion and references can be found in the review article [342].

3.3.5.1 *Vortices in spin-polarized dipolar BECs*

Here, we summarize some of the theoretical results on vortices in a spin-polarized dipolar BEC, in which the above features of the excitation spectrum have a strong influence on vortex states.

STRUCTURE OF A SINGLE VORTEX

The effect of dipolar interactions on the single vortex state was first studied by Yi and Pu by numerically solving Eq. (3.98) [346]. While there was no significant difference of the vortex structure for the repulsive contact interaction, the density profile exhibited a "biconcave" structure around the vortex core when the condensate had attractive contact interactions ($a < 0$) and was trapped in a highly prolate trap. More detailed numerical study was performed by Wilson et al. [347], who found that these biconcave density ripples around the vortex core emerge at milder trap aspect ratios and with purely dipolar interactions ($a = 0$). By using perturbation theory, they related these density oscillations to the roton mode near the instability.

VORTEX STABILITY AND DYNAMICS

The thermodynamic critical rotation frequency Ω_c for dipolar BECs in the TF limit was investigated by O'Dell and Eberlein [348]. The value of Ω_c decreased for a condensate in a pancake-shaped trap, while it increased in a cigar-shaped trap, compared to that of a conventional BEC. This is because, for the pancake trap, the DDI is repulsive on average, and results in a slight increase of both the axial size R_z and the radial size $R_\perp$. Thus, Ω_c decreases since $\Omega_c \propto R_\perp^{-2}$. For the cigar-shaped trap, the DDI caused a slight increase in the axial size R_z but decreased the radial size $R_\perp$, which increased Ω_c. The TF approximation is invalid for pure dipolar BECs. The full numerical calculations for the single-vortex state were done by Abad et al. [349], and their results were in good agreement with Ref. [348].

Another important issue is the critical rotation frequency for vortex nucleation in rotating dipolar BECs. Following the analysis for a conventional BEC [186], Bijnen et al. studied the dynamical instability of a rotating dipolar BEC in the TF limit [350, 351]. The dynamically unstable region of the rotation frequency Ω versus trap aspect ratio $\omega_z/\omega_\perp$ non-trivially depends on the dipole interaction strength. Interestingly, the critical frequency Ω_c for a cigar-shaped trap can become larger than the onset of the dynamical instability of a rotating condensate. This is an intriguing regime, in which a rotating dipolar BEC is dynamically unstable, but vortices will not enter.

The BdG analysis for a single vortex state in dipolar BECs was done by Wilson et al. [352]. As discussed in Section 3.3.2.4, the single vortex state in a conventional BEC (without rotation) is thermodynamically unstable, but dynamically stable. However, the vortex states in dipolar BECs can be dynamically unstable, depending on the dipolar strength and the aspect ratio of the trap potential. For a pancake trap, the instability is associated with the radial and angular roton modes, which induce a density wave and local collapse. For a cigar-shaped trap, the 3D character of the DDI becomes important. Thus, the vortex wave (Kelvin mode) is strongly affected by the DDI. Klawunn et al. showed that, when applying a periodic potential along the vortex line, which changes the effective mass along this direction, dispersion of the Kelvin mode has a roton-like minimum [353]. As the dipolar interaction increases, this minimum can reach zero, forming an instability similar to the Donnelly–Glaberson instability (see Section 3.3.2.6 for discussion of vortex waves).

VORTEX LATTICES

Rapidly rotating dipolar BECs exhibit a rich variety of vortex phases characterized by different symmetries of the lattice structure [354–356]. Cooper et al. treated this problem within the LLL approximation [354]. In particular, with increasing dipole interaction strength or with a high filling factor, the vortex lattice may undergo transitions between different symmetries: triangular, square, striped vortex crystal, and bubble state. In addition, for vortex lattices in double-well potentials, the competition between tunneling and interlayer DDI should lead to a quantum phase transition from a coincident phase to a staggered phase [355].

3.3.5.2 *Vortices in spinor dipolar condensates*

When the spin degree of freedom is taken into account for dipolar BECs, extremely rich physics appear. Let us consider the $F = 1$ case. Typically, the energy associated with spin-exchange interactions is much smaller than that of contact (spin-preserving) interactions, i.e. $g_n \gg g_s$. Hence, the spin-exchange interaction may become comparable to the DDI. As a consequence, even for alkali spinor BECs (in particular ^{87}Rb) the DDI plays a significant role [357].

Under a uniform magnetic field $\mathbf{B}$, the second quantized Hamiltonian for spin-1 dipolar BECs is $\hat{H} = \hat{H}_0 + \hat{H}_{\rm int} + \hat{H}_{\rm dd}$, where $\hat{H}_0$ and $\hat{H}_{\rm int} = \int d\mathbf{r}_1 d\mathbf{r}_2 V_{F=1}(\mathbf{r}_1 - \mathbf{r}_2)$ is given by Eqs. (3.85) and (3.87), respectively. The dipolar contribution becomes

$$\begin{aligned}\hat{H}_{\rm dd} = \frac{c_{\rm dd}}{8\pi}\int\int \frac{d\mathbf{r}d\mathbf{r}'}{|\mathbf{r}-\mathbf{r}'|^3} \\ \times \sum_{\alpha,\alpha',\beta,\beta'=-1}^{1} \Big[\hat{\psi}^\dagger_\alpha(\mathbf{r})\hat{\psi}^\dagger_\beta(\mathbf{r}')\mathbf{F}_{\alpha\alpha'}\cdot\mathbf{F}_{\beta\beta'}\hat{\psi}_{\beta'}(\mathbf{r}')\hat{\psi}_{\alpha'}(\mathbf{r}) \\ - 3\hat{\psi}^\dagger_\alpha(\mathbf{r})\hat{\psi}^\dagger_\beta(\mathbf{r}')(\mathbf{F}_{\alpha\alpha'}\cdot\hat{\mathbf{r}}_{12})(\mathbf{F}_{\beta\beta'}\cdot\hat{\mathbf{r}}_{12})\hat{\psi}_{\beta'}(\mathbf{r}')\hat{\psi}_{\alpha'}(\mathbf{r})\Big],\end{aligned} \tag{3.99}$$

where $\hat{\mathbf{r}}_{12} = (\mathbf{r}-\mathbf{r}')/|\mathbf{r}-\mathbf{r}'|$ is a unit vector. The spin-exchange term and the dipolar term describe two types of spin-dependent interactions. Contrary to the contact interaction, the DDI does not necessarily conserve the spin projection along the quantization axis due to the anisotropic character of the interaction. Note that the direction of spin of the order parameter is not polarized by an external field. In such a situation, the system can develop a non-trivial spin texture due to the interplay and competition between these two terms.

EINSTEIN–DE HAAS EFFECT

If atoms are initially prepared into a maximally polarized state, e.g. $m_F = -F$, contact interactions cannot induce any spinor dynamics due to the conservation of total magnetization M. DDIs, on the contrary, may induce a transfer into an $m_F + 1$ state. Here, the DDI conserves the total (spin + orbital) angular momentum of the system. If the system preserves cylindrical symmetry around the quantization axis, this violation of the spin conservation is accompanied by a transfer of angular momentum from the spin to the orbital part. Because of this transfer, an initially spin-polarized dipolar condensate can dynamically generate vortices in a process resembling the Einstein–de Haas effect [358, 359].

Unfortunately, the Einstein–de Haas effect is prevented in the presence of even a weak magnetic field, which favors spin polarization. However, for a ferromagnetic spinor BEC such as $F = 1$ ^{87}Rb BECs, a significant Einstein–de Haas effect may be expected under specific resonant conditions [360]. Population transfer away from the initially prepared $m_F = 1$ state is typically very small, but can be significantly enhanced by applying a resonant magnetic field, such that the linear Zeeman energy of an atom in the $m_F = 1$ state is totally transferred into the kinetic energy of the rotating atom in $m_F = 0$.

SPONTANEOUS CIRCULATION IN THE GROUND STATE

The ground state structure of the dipolar spinor condensate was first studied by Yi et al. [361] under single mode approximation (SMA), which assumes that all components share the same spatial wave function. For spinor BECs, SMA is valid when the spin-dependent interaction is sufficiently weaker than the spin-independent interaction ($g_n \gg g_s$ for spin-1 case). However, in the presence of even weak DDIs, dipole moments are spatially modulated due to the long-range anisotropic nature of the interactions, thus the application of SMA is questionable. It is interesting that, in this system, the spatial spin variation, caused by the DDI, induces the supercurrent due to spin-gauge symmetry, as seen in Eq. (3.92).

Systematic numerical studies of the ground state are described in Refs. [362, 363], revealing the spontaneous emergence of spin textures in the ground state. Yi and Pu constructed a phase diagram of the ground state of a spin-1 dipolar BEC with respect to the trap aspect ratio λ and the dipolar interaction strength $c_{\rm dd}$, for both the ferromagnetic case and the antiferromagnetic case [362]. For $c_{\rm dd}$ smaller than g_n, the SMA is valid. Otherwise, three distinct spin textures appear, which can be classified as: (a) for a pancake geometry ($\lambda > 1$), the polar–core spin vortex with $(-1, 0, 1)$. (b) for a cigar-shaped geometry ($\lambda < 1$), the coreless vortex with $(0, 1, 2)$, where the spins twist around the z-axis. (c) for a spherical geometry ($\lambda \sim 1$), the non-axisymmetric state. Kawaguchi et al. studied ferromagnetic spin-1 BECs by varying g_s and $c_{\rm dd}$, finding three distinct phases: a polar–core vortex, a flower phase, and a chiral spin-vortex phase [363]. Hence, dipolar spinor BECs can have spontaneous circulation in the ground state, even without rotation.

3.3.6 Quantum turbulence

As discussed previously, the study of the turbulent state in quantum fluids and its relation to classical turbulence (CT) is an intriguing physical problem. Although the study of quantum turbulence (QT) has a long history, only superfluid ^{4}He and ^{3}He systems have been used to realize QT until recently.

Recently, atomic Bose–Einstein condensates have become another candidate for QT research, since a turbulent state was realized in this system.

In this section, we briefly review some of the theoretical studies and important experimental work for QT in atomic BECs.

3.3.6.1 *Theoretical work on quantum turbulence in atomic BECs*

A major problem in the study of QT in atomic BEC is the difficulty of applying a velocity field to generate the turbulent state. Thus, alternative methods are required to realize QT.

Berloff and Svistunov suggested the realization of QT in the dynamics of the formation of a BEC from a strongly degenerate non-equilibrium gas [364, 365]. Such dynamics can also be described by the GP equation in the framework of a classical field description.

Parker and Adams suggest the emergence and decay of turbulence in an atomic BEC under a simple rotation, starting from a vortex-free equilibrium BEC [193]. Starting from the vortex-free steady state with a constant potential, they performed a numerical simulation of the GP model with realistic experimental parameters for ^{87}Rb BECs, and showed that the total dynamics can be divided into four distinct stages as follows: (i) Fragmentation—quadrupolar oscillation of the condensate breaks down, ejecting energetic atoms to form an outer cloud, (ii) symmetry breaking—the two-fold rotational symmetry of the system is broken, allowing the rotation to couple to additional modes, thereby rapidly injecting energy into the system, (iii) turbulence—a turbulent cloud containing vortices and high energy density fluctuations (sound field) is formed, and (iv) crystallization—the loss of energy at short length scales coupled with vortex–sound interactions allows the system to relax into an ordered lattice. In the third region of turbulence, the system produces a Kolmogorov energy spectrum (3.10) demonstrating a classical analog of QT.

To realize steady QT, Kobayashi and Tsubota suggested combining rotations around two axes [366]. Starting from a vortex-free initial state, they performed numerical simulations of the GP model with realistic parameters for ^{87}Rb BEC, and obtained a steady turbulent state with no crystallization, but with highly tangled quantized vortices. The incompressible kinetic energy spectrum satisfies the Kolmogorov law in the inertial range.

White et al. investigated QT numerically in atomic BECs, and found non-classical velocity statistics [367]. The probability density function (PDF) of velocity is another important statistical measure to study turbulence, and for the case of CT, the PDF obeys a Gaussian distribution [368–371]. They performed numerical simulations of the GP model (3.5) using realistic experimental parameters for a ^{23}Na condensate. The method to create turbulence was "phase imprinting" in the condensate. They calculated the PDF of the superfluid velocity, and found that the velocity statistics were non-Gaussian and had a power-law dependence. Their result is consistent with the high-velocity tails found experimentally by Paoletti et al. in turbulent superfluid ^{4}He [107].

3.3.6.2 *Experimental work on quantum turbulence in atomic BECs*

Recently, a turbulent state was realized in atomic BECs by two methods. Weiler et al. performed a rapid quench of an ^{87}Rb gas through the BEC transition temperature [257], which is a similar situation to that examined by Berloff and Svistunov [365]. Through the high density fluctuation regime (weak turbulence) in a short period, several vortices and anti-vortices were formed to make the system turbulent. This experiment was interpreted in the previous section (subsection 3.3.2.8).

The turbulent state created in the above method depended strongly on the initial uncontrollable thermal state. Furthermore, the turbulent state is merely intermediate, and the final state that contains only a few vortices is not QT. For a method with better control of the turbulence, Henn et al. introduced an external oscillatory perturbation to an ^{87}Rb BEC [372, 373]. This oscillating magnetic field was produced by a pair of anti-Helmholtz coils which were not perfectly aligned to the vertical axis of the cigar-shaped condensate. Additionally, the components along the two equal directions that result in the radial symmetry of the trap were slightly different. This oscillatory field induced a coherent mode excitation in a BEC. For small amplitudes of the oscillating field and short excitation periods, dipolar modes, quadrupolar modes, and scissors modes of the BEC were observed, but no vortices appeared. Increasing both parameters, vortices grew in number, eventually leading to the turbulent state. In the turbulent regime, they observed a rapid increase in the number of vortices followed by proliferation of vortex lines in all directions, where many vortices with no preferred orientation formed a vortex tangle (Fig. 3.19). Another remarkable feature is that a completely different hydrodynamic regime was followed: the suppression of aspect ratio inversion during free expansion, despite the asymmetric expansion (from a cigar-shaped to pancake-shaped) of the usual quantum gas of bosons, or the isotropic expansion of a thermal cloud. Although theoretical understanding of this effect remains incomplete, it represents a remarkable new effect in atomic superfluids. The emergence of QT with vortex reconnections, excitation of Kelvin waves, and the existence of cascade phenomena are expected to be observed directly in this well-controllable QT system in the near future.

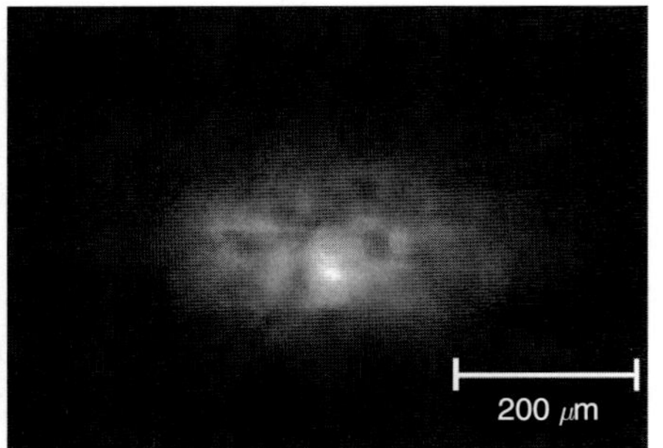

Fig. 3.19 Atomic optical density after 15 ms of free expansion (left) and a corresponding schematic diagram showing the inferred distribution of vortices obtained from the image (right) [373] [Henn, Seman, Roati, Magalhães, and Bagnato: *Phys. Rev. Lett.* **103** (2009) 045301, reproduced with permission. Copyright 2009 by the American Physical Society].

3.4 Summary and conclusions

In this chapter, we have reviewed recent topics on quantized vortices in superfluid helium and atomic Bose–Einstein condensates. Quantized vortices were discovered in superfluid ^{4}He in the 1990s. However, they have recently grown in importance, for two reasons. The first reason is that research on QT has entered a new era since the mid 1950s. A primary interest is to understand the relationship between CT and QT. The second reason is the realization of atomic BECs. Modern optical techniques have enabled direct visualization of quantized vortices, and multi-component BECs have further enriched the world of quantized vortices.

Research on vortices in classical fluids has a very long history. When Leonardo da Vinci drew his famous sketches of water turbulence in the 16th century, he certainly recognized that turbulence consists of many eddies of different sizes. It was in 1858 that Helmholtz proposed the idea of a vortex filament. However, it is not straightforward to understand the role of vortices in the classical hydrodynamics of turbulence. One reason is that vortices in classical fluids are unstable and not very well defined. Quantum condensed systems create more stable and well-defined vortices, namely quantized vortices.

This field is currently very active, attracting the interest of scientists from many different fields. We hope that this chapter will contribute to further development and progress in this fascinating field of physics.

Acknowledgments

The authors are grateful to the following colleagues and friends for their collaboration and encouragement: H. Adachi, T. Araki, V. S. Bagnato, C. F. Barenghi, R. J. Donnelly, V. B. Eltsov, S. N. Fisher, S. Fujiyama, A. I. Golov, R. Hänninen, T. Hata, M. Krusius, D. P. Lathrop, V. S. L'vov, P. V. E. McClintock, A. Mitani, S. K. Nemirovskii, S. Ogawa, G. R. Pickett, S. W. Van Sciver, L. Skrbek, H. Takeuchi, M. Ueda, W. F. Vinen, G. E. Volovik, and H. Yano.

References

[1] Halperin, W. P. and Tsubota, M. eds. *Progress in Low Temperature Physics* Vol. 16. Elsevier, Amsterdam (2009).

[2] Tsubota, M. Quantum turbulence. *J. Phys. Soc. Jpn.*, **77**, 111006(1–12) (2008).

[3] Kapitza, P. Viscosity of liquid helium below the λ point. *Nature*, **141**, 74 (1938).

[4] Allen, J. F. and Misener, A. D. Flow of liquid helium II. *Nature*, **141**, 75 (1938).

[5] Landau, L. The theory of superfluidity of helium II. *J. Phys. USSR*, **5**, 71–90 (1941).

[6] Tisza, L. Transport phenomena in helium II. *Nature*, **141**, 913 (1938).

[7] Gorter, C. J. and Mellink, J. H. On the irreversible processes in liquid helium II. *Physica*, **15**, 285–304 (1949).

[8] London, F. On the Bose–Einstein condensation. *Phys. Rev.*, **54**, 947–954 (1938).

[9] Onsager, L. *Nuovo Cimento Suppl.*, **6**, 249–250 (1949).

[10] Feynman, R. P. Application of quantum mechanics to liquid helium. *Progress in Low Temperature Physics* Vol. 1 (Gorter, C. J. ed.). Amsterdam. North-Holland, 17–53 (1955).

[11] Hall, H. E. and Vinen, W. F. The rotation of liquid helium II I. Experiments on the propagation of second sound in uniformly rotating helium II. *Proc. Roy. Soc. London*, **A238**, 204–214 (1956).

[12] Hall, H. E. and Vinen, W. F. The rotation of liquid helium II II. The theory of mutual friction in uniformly rotating helium II. *Proc. Roy. Soc. London*, **A238**, 215–234 (1956).

[13] Vinen, W.F. Mutual friction in a heat current in liquid helium II I. Experiments on steady heat currents, *Proc. Roy. Soc. London*, **A240**, 114–127 (1957).

[14] Vinen, W. F. Mutual friction in a heat current in liquid helium II. II. Experiments on transient effects. *Proc. Roy. Soc. London*, **A240**, 128–143 (1957).

[15] Vinen, W. F. Mutual friction in a heat current in liquid helium II III. Theory of mutual friction. *Proc. Roy. Soc. London*, **A242**, 493–515 (1957).

[16] Vinen, W. F. Mutual friction in a heat current in liquid helium II IV. Critical heat currents in wide channels. *Proc. Roy. Soc. London*, **A243**, 400–413 (1957).

[17] Vinen, W. F. The detection of single quanta circulation in liquid helium II. *Proc. Roy. Soc. London*, **A260**, 218–236 (1961).

[18] Tough, J. T. Superfluid turbulence. *Progress in Low Temperature Physics* Vol. 8 (Gorter, C. J. ed.). Amsterdam. North-Holland, 133–220 (1982).

[19] Schwarz, K. W. Three-dimensional vortex dynamics in superfluid ^{4}He: Line–line and line–boundary interactions. *Phys. Rev. B*, **31**, 5782–5803 (1985).

[20] Schwarz, K. W. Three-dimensional vortex dynamics in superfluid ^{4}He: Homogeneous superfluid turbulence. *Phys. Rev. B*, **38**, 2398–2417 (1988).

[21] Saffman, P. G. *Vortex Dynamics*. Cambridge University Press, Cambridge (1992).

[22] Tsubota, M., Araki T. and Nemirovskii, S. K. Dynamics of vortex tangle without mutual friction in superfluid ^{4}He. *Phys. Rev. B*, **62**, 11751–11762 (2000).

[23] Adachi, H., Fujiyama, S. and Tsubota, M. Steady-state counterflow quantum turbulence: Simulation of vortex filaments using the full Biot–Savart law. *Phys. Rev. B*, **81**, 104511(1–7) (2010).

[24] Boratav, O. N., Pelz, R. B. and Zabusky, N. J. Reconnection in orthogonally interacting vortex tubes: direct numerical simulations and quantifications. *Phys. Fluids A*, **4**, 581–605 (1992).

[25] Koplik, J. and Levine, H. Vortex reconnection in superfluid helium. *Phys. Rev. Lett.*, **71**, 1375–1378 (1993).

[26] Leadbeater, M., Winiecki, T., Samuels, D. C., Barenghi, C. F. and Adams, C. S. Sound emission due to superfluid vortex reconnection. *Phys. Rev. Lett.*, **86**, 1410–1413 (2001).

[27] Ogawa, S., Tsubota, M. and Hattori, Y. Study of reconnection and acoustic emission of quantized vortices in superfluid by the numerical analysis of the Gross-Pitaevskii equation. *J. Phys. Soc. Jpn.*, **71**, 813–821 (2002).

[28] Gross, E. P. Structure of a quantized vortex in boson systems. *Nuovo Cimento*, **20**, 454–457 (1961).

[29] Pitaevskii, L. P. Vortex lines in an imperfect Bose gas. *Zh. Eksp. Teor. Fiz.*, **40**, 646–651 [*Sov. Phys. JETP*, **13**, 451–454] (1961).

[30] Pethick, C. J. and Smith, H. *Bose–Einstein Condensation in Dilute Gases* (2nd edn.). Cambridge University Press, Cambridge (2008).

[31] Vinen, W. F. and Donnelly, R. J. Quantum turbulence. *Phys. Today*, **60**, 43–48 (2007).

[32] Vinen, W. F. and Niemela, J. J. Quantum turbulence. *J. Low Temp. Phys.*, **128**, 167–231 (2002).

[33] Svistunov, B. V. Superfluid turbulence in the low-temperature limit. *Phys. Rev. B*, **52**, 3647–3653 (1995).

[34] Vinen, V. F. Decay of superfluid turbulence at a very low temperature: The radiation of sound from a Kelvin wave on a quantized vortex. *Phys. Rev. B*, **64**, 134520(1–4) (2001).

[35] Frisch, U. *Turbulence*. Cambridge University Press, Cambridge (1995).

[36] Kolmogorov, A. N. The local structure of turbulence in incompressible viscous fluid for very large Reynolds numbers. *Dokl. Akad. Nauk SSSR*, **30**, 299–303 (reprinted in 1991, *Proc. Roy. Soc. A*, **434**, 9–13) (1941).

[37] Kolmogorov, A. N. Dissipation of energy in the locally isotropic turbulence. *Dokl. Akad. Nauk SSSR*, **32**, 16–18 (reprinted in 1991, *Proc. Roy. Soc. A*, **434**, 15–17) (1941).

[38] Richardson, L. F. *Weather Prediction by Numerical Process* (2nd edn.). Cambridge University Press, Cambridge (2007).

[39] Kivotides, D., Vassilicos, J. C., Samuels, D. C., and Barenghi, C. F. Kelvin waves cascade in superfluid turbulence. *Phys. Rev. Lett.*, **86**, 3080–3083 (2001).

[40] Nore, C., Abid, M., and Brachet, M. E. Kolmogorov turbulence in low-temperature superflows. *Phys. Rev. Lett.*, **78**, 3896–3899 (1997).

[41] Nore, C., Abid, M., and Brachet, M. E. Decaying Kolmogorov turbulence in a model of superflow. *Phys. Fluids*, **9**, 2644–2669 (1997).

[42] Araki, T., Tsubota, M., and Nemirovskii, S. K. Energy spectrum of superfluid turbulence with no normal-fluid component. *Phys. Rev. Lett.*, **89**, 145301(1–4) (2002).

[43] Kobayashi, M. and Tsubota, M. Kolmogorov spectrum of superfluid turbulence: numerical analysis of the Gross–Pitaevskii equation with a small-scale dissipation. *Phys. Rev. Lett.*, **94**, 065302(1–4) (2005).

[44] Kobayashi, M. and Tsubota, M. Kolmogorov spectrum of quantum turbulence. *J. Phys. Soc. Jpn.*, **74**, 3248–3258 (2005).

[45] Yepez, J., Vahala, G., Vahala, L., and Soe, M. Superfluid turbulence from quantum Kelvin wave to classical Kolmogorov cascade. *Phys. Rev. Lett.*, **103**, 084501(1–4) (2009).

[46] Samuels, D. C. and Donnelly, R. J. Sideband instability and recurrence of Kelvin waves on vortex cores. *Phys. Rev. Lett.*, **64**, 1385–1388 (1990).

[47] Vinen, W. F., Tsubota, M., and Mitani, A. Kelvin-wave cascade on a vortex in superfluid ^{4}He at a very low temperature. *Phys. Rev. Lett.*, **91**, 135301(1–4) (2003).

[48] Kozik, E. and Svistunov, B. Kelvin-wave cascade and decay of superfluid turbulence. *Phys. Rev. Lett.*, **92**, 035301(1–4) (2004).

[49] Kozik, E. and Svistunov, B. Scale-separation scheme for simulating superfluid turbulence: Kelvin-wave cascade. *Phys. Rev. Lett.*, **94**, 025301(1–4) (2005).

[50] Nazarenko, S. Differential approximation for Kelvin wave turbulence. *JETP Lett.*, **83**, 198–200 (2006).

[51] Boffetta, G., Celani, A., Dezzani, D., Laurie, J., and Nazarenko, S. Modeling Kelvin wave cascades in superfluid helium. *J. Low Temp. Phys.*, **156**, 193–214 (2009).

[52] L'vov, V. S., Nazarenko, S. V., and Rudenko, O. Bottleneck crossover between classical and quantum superfluid turbulence. *Phys. Rev. B*, **76**, 024520(1–9) (2007).

[53] Kozik, E. and Svistunov, B. Kolmogorov and Kelvin-wave cascades of superfluid turbulence at $T = 0$: what lies between. *Phys. Rev. B*, **77**, 060502(R)(1–4) (2008).

[54] Thomson, W. Vibrations of a columnar vortex. *Phil. Mag.*, **10**, 155–168 (1880).

[55] Hall, H. E. An experimental and theoretical study of torsional oscillations in uniformly rotating liquid helium II. *Proc. Roy. Soc. London*, **A245**, 546–561 (1958).

[56] Hall, H. E. The rotation of liquid helium II. *Phil. Mag. Suppl.*, **9**, 89–146 (1960).

[57] Zakharov, V. E., L'vov, V. S., and Fal'kovich, G. *Kolmogorov Spectra of Turbulence*, Springer, Berlin (1992).

[58] Connaughton, C. and Nazarenko, S. Warm cascades and anomalous scaling in a diffusion model of turbulence. *Phys. Rev. Lett.*, **92**, 044501(1–4) (2004).

[59] Smith, M. R., Donnelly, R. J., Goldenfeld, N., and Vinen, W. F. Decay of vorticity in homogeneous turbulence. *Phys. Rev. Lett.*, **71**, 2583–2586 (1993).

[60] Maurer, J. and Tabeling, P. Local investigation of superfluid turbulence. *Europhys. Lett.*, **43**, 29–34 (1998).

[61] Stalp, S. R., Skrbek, L., and Donnelly, R. J. Decay of grid turbulence in a finite channel. *Phys. Rev. Lett.*, **82**, 4831–4834 (1999).

[62] Skrbek, L., Niemela, J. J., and Donnelly, R. Four regimes of decaying grid turbulence in a finite channel. *Phys. Rev. Lett.*, **85**, 2973–2976 (2000).

[63] Skrbek, L. and Stalp, S. R. On the decay of homogeneous isotropic turbulence. *Phys. Fluids*, **12**, 1997–2019 (2000).

[64] Stalp, S. R., Niemela, J. J., Vinen, W. F., and Donnelly, R. J. Dissipation of grid turbulence in helium II. *Phys. Fluids*, **14**, 1377–1379 (2002).

[65] Skrbek, L., Gordeev, A. V., and Soukup, F. Decay of counterflow He II turbulence in a finite channel: possibility of missing links between classical and quantum turbulence. *Phys. Rev. E*, **67**, 047302(1–4) (2003).

[66] Vinen, W. F. Classical character of turbulence in a quantum liquid. *Phys. Rev. B*, **61**, 1410–1420 (2000).

[67] Kivotides, D. Relaxation of superfluid vortex bundles via energy transfer to the normal fluid. *Phys. Rev. B*, **76**, 054503(1–12) (2007).

[68] L'vov, V. S., Nazarenko, S. V., and Skrbek, L. Energy spectra of developed turbulence in helium superfluids. *J. Low Temp. Phys.*, **145**, 125–142 (2006).

[69] Donnelly, R. J. *Quantized Vortices in Helium II*, Cambridge University Press, Cambridge (1991).

[70] Hinze, J. O. *Turbulence* (2nd edn.) McGraw Hill, New York (1975).

[71] Walmsley, P. M., Golov, A. I., Hall, H. E., Levchenko, A. A., and Vinen, W. F. Dissipation of quantum turbulence in the zero temperature limit. *Phys. Rev. Lett.*, **99**, 265302(1–4) (2007).

[72] Walmsley, P. M. and Golov, A. I. Quantum and quasiclassical types of superfluid turbulence. *Phys. Rev. Lett.*, **100**, 245301(1–4) (2008).

[73] Bradley, D. I., Clubb, D. O., Fisher, S. N., Guenault, A. M., Halay, R. P., Matthews, C. J., Pickett, G. R., Tsepelin, V. and Zaki, K. Decay of pure quantum turbulence in superfluid ^{3}He-B. *Phys. Rev. Lett.*, **96**, 035301(1–4) (2006).

[74] Eltsov, V. B., Golov, A. I., de Graaf, R., Hänninen, R., Krusius, M., L'vov, V. S., and Solntsev, R. E. Quantum turbulence in a propagating superfluid vortex front. *Phys. Rev. Lett.*, **99**, 265301(1–4) (2007).

[75] Kobayashi, M. and Tsubota, M. Decay of quantized vortices in quantum turbulence. *J. Low Temp. Phys.*, **145**, 209–218 (2006).

[76] Tsubota, M., Araki, T., and Vinen, W. F. Diffusion of an inhomogeneous vortex tangle. *Physica B*, **329–333**, 224–225 (2003).

[77] Skrbek, L. and Vinen, W. F. The use of vibrating structures in the study of quantum turbulence. *Progress in Low Temperature Physics* Vol. 16 (Halperin, W.P. and Tsubota, M. edn.). Amsterdam. Elsevier, 195–246 (2009).

[78] Jäger, J., Schuderer, B. and Schoepe, W. Turbulent and laminar drag of superfluid helium on an oscillating microsphere. *Phys. Rev. Lett.*, **74**, 566–569 (1995).

[79] Nichol, H. A., Skrbek, L., Hendry, P. C. and McClintock, P. V. E. Flow of He II due to an oscillating grid in the low-temperature limit. *Phys. Rev. Lett.*, **92**, 244501(1–4) (2004).

[80] Nichol, H. A., Skrbek, L., Hendry, P. C. and McClintock, P. V. E. Experimental investigation of the macroscopic flow of He II due to an oscillating grid in the zero temperature limit. *Phys. Rev. E*, **70**, 056307(1–14) (2004).

[81] Charalambous, D., Hendry, P. C., McClintock, P. V. E., Skrbek, L. and Vinen, W. F. Experimental investigation of the dynamics of a vibrating grid in superfluid ^{4}He over a range of temperatures and pressures, *Phys. Rev. E*, **74**, 036307(1–10) (2006).

[82] Bradley, D. I., Clubb, D. O., Fisher, S. N., Guénault, A. M., Haley, R. P., Matthews, C. J., Pickett, G. R. and Zaki, K. L. Turbulence generated by vibrating wire resonators in superfluid ^{4}He at low temperatures. *J. Low Temp. Phys.*, **138**, 493–498 (2005).

[83] Bradley, D. I., Fisher, S. N., Guénault, A. M., Lowe, M. R., Pickett, G. R., Rahm, A. and Whitehead, R. C. V. Emission of discrete vortex rings by a vibrating grid In superfluid 3He-B: A precursor to quantum turbulence. *Phys. Rev. Lett.*, **95**, 035302(1–4) (2004).

[84] Fisher, S. N., Hale, A.J., Guénault, A. M. and Pickett, G. R. Generation and detection of quantum turbulence in superfluid ^{3}He-B. *Phys. Rev. Lett.*, **86**, 244–247 (2001).

[85] Bradley, D. I., Fisher, S. N., Guénault, A. M., Lowe, M. R., Pickett, G. R., Rahm, A. and Whitehead, R. C. V. Quantum turbulence in superfluid 3He illuminated by a beam of quasiparticle excitations. *Phys. Rev. Lett.*, **93**, 235302(1–4) (2004).

[86] Yano, H., Hashimoto, N., Handa, A., Nakagawa, M., Obara, K., Ishikawa, O. and Hata, T. Motion of quantized vortices attached to a boundary in alternating currents of superfluid ^{4}He. *Phys. Rev. B*, **75**, 012502(1–4) (2007).

[87] Hashimoto, N., Goto, R., Yano, H., Obara, K., Ishikawa, O. and Hata, T. Control of turbulence in boundary layers of superfluid ^{4}He by filtering out remnant vortices. *Phys. Rev. B*, **76**, 020504(R)(1–4) (2007).

[88] Goto, R., Fujiyama, S., Yano, H., Nago, Y., Hashimoto, N., Obara, K., Ishikawa, O., Tsubota, M. and Hata, T. Turbulence in boundary flow of superfluid ^{4}He triggered by free vortex rings. *Phys. Rev. Lett.*, **100**, 045301(1–4) (2008).

[89] M. Blažková, M. Človečko, V. B. Eltsov, E. Gažo, R. de Graaf, J. J. Hosio, M. Krusius, D. Schmoranzer, W. Schoepe, L. Skrbek, P. Skyba, R. E. Solntsev and W. F. Vinen: *J. Low Temp. Phys.* **150** 525–535 (2008).

[90] Blažková, M., Schmoranzer, D., Skrbek, L. and Vinen, W. F. Generation of turbulence by vibrating forks and other structures in superfluid ^{4}He. *Phys. Rev. B*, **79**, 054522(1–11) (2009).

[91] Yano. H., Ogawa, T., Mori, A., Miura, Y., Nago, Y., Obara, K., Ishikawa, O. and Hata, T. Transition to quantum turbulence generated by thin vibrating wires in superfluid ^{4}He. *J. Low Temp. Phys.*, **156**, 132–144 (2009).

[92] Awshalom, D. D. and Schwarz, K. W. Observation of a remnant vortex-line density in superfluid helium. *Phys. Rev. Lett.*, **52**, 49–52 (1984).

[93] Ruutu, V. M. H., Parts, Ü, Koivuniemi, J. H., Kopnin, N. B. and Krusius, M. Intrinsic and extrinsic mechanisms of vortex formation in superfluid ^{3}He-B. *J. Low Temp. Phys.*, **106**, 93–164 (1997).

[94] Hänninen, R., Tsubota, M. and Vinen, W. F. Generation of turbulence by oscillating structures in superfluid helium at very low temperatures. *Phys. Rev. B*, **75**, 064502(1–12) (2007).

[95] Hänninen, R. and Schoepe, W. Universal onset of quantum turbulence in oscillating flows and crossover to steady flows. *J. Low Temp. Phys.*, **158**, 410–414 (2010).

[96] Volovik, G. E. Classical and quantum regimes of the superfluid turbulence. *JETP Lett.*, **78**, 533–537 (2003).

[97] Kopnin, N. B. Vortex instability and the onset of superfluid turbulence. *Phys. Rev. Lett.*, **92**, 135301(1–4) (2004).

[98] Schoepe, W. Fluctuations and stability of superfluid turbulence at mK temperatures. *Phys. Rev. Lett.*, **92**, 095301(1–3) (2004).

[99] Batchelor, G. K. *An Introduction to Fluid Mechanics*. Cambridge University Press, Cambridge (1967).

[100] Hemmati, A. M., Fuzier, S., Bosque, E. and Van Sciver, S. W. Drag measurement on an oscillating sphere in Helium II. *J. Low Temp. Phys.*, **156**, 71–83 (2009).

[101] Fujiyama, S. and Tsubota, M. Drag force on an oscillating object in quantum turbulence. *Phys. Rev. B*, **79**, 094513(1–7) (2009).

[102] Tsubota, M. and Maekawa, S. Pinning and depinning of two quantized vortices in superfluid ^{4}He. *Phys. Rev. B*, **47**, 12040–12050 (1993).

[103] Van Sciver, S. W. and Barenghi, C. F. Visualization of quantum turbulence. *Progress in Low Temperature Physics* Vol. 16 (Halperin, W.P. and Tsubota, M.eds.). Amsterdam. Elsevier, 247–303 (2009).

[104] Zhang, T. and Van Sciver, S. W. Large-scale turbulent flow around a cylinder in counterflow superfluid ^{4}He (He(II)), *Nature Phys.*, **1**, 36–38 (2005).

[105] Bewley, G. P., Lathrop, D. P. and Sreenivasan, K. R. Visualization of quantized vortices. *Nature*. **441**, 588 (2006).

[106] Tsubota, M. Capacity of a pinning site for trapping quantized vortices in superfluid ^{4}He. *Phys. Rev.B*, **50**, 579–581 (1994).

[107] Paoletti, M. S., Fioroto, R. B., Sreenivasan, K. R. and Lathrop, D. P. Visualization of superfluid helium flow. *J. Phys. Soc. Jpn.*, **77**, 111007(1–7) (2008).

[108] Poole, D. R., Barenghi, C. F., Sergeev, Y. A. and Vinen, W. F. Motion of tracer particles in He II. *Phys. Rev. B*, **71**, 064514(1–16) (2005).

[109] Sergeev, Y. A., Barenghi, C. F. and Kivotides, D. Motion of micron-size particles in turbulent He II. *Phys. Rev. B*, **74**, 184506(1–5) (2006).

[110] Kivotides, D., Barenghi, C. F. and Sergeev, Y. A. Collision of a tracer particle and a quantized vortex in superfluid helium: self-consistent calculation. *Phys. Rev. B*, **75**, 212502(1–4) (2007).

[111] Kivotides, D., Barenghi, C. F. and Sergeev, Y. A. Interaction between particles and quantized vortices in superfluid helium. *Phys. Rev. B*, **77**, 014527(1–13) (2008).

[112] Barenghi, C. F. and Sergeev, Y. A. Motion of vortex ring with tracer particles in superfluid helium. *Phys. Rev. B*, **80**, 024514(1–5) (2009).

[113] Anderson, M.H., Ensher, J.R., Matthews, M.R., Wieman, C.E., and Cornell, E.A. Observation of Bose–Einstein condensation in a dilute atomic vapor. *Science*, **269**, 198–201 (1995).

[114] Bradley, C.C., Sackett, C.A., Tollett, J.J., and Hulet, R.G. Evidence of Bose–Einstein condensation in an atomic gas with attractive interactions. *Phys. Rev. Lett.*, **75**, 1687–1690 (1995).

[115] Davis, K.B., Mewes, M.-O., Andrews, M.R., van Druten, N.J., Durfee, D.S., Kurn, D.M., and Ketterle, W. Bose–Einstein condensation in a gas of sodium atoms. *Phys. Rev. Lett.*, **75**, 3969–3973 (1995).

[116] Pitaevskii, L. and Stringari, S. *Bose–Einstein Condensation*. Oxford University Press, Oxford (2003).

[117] Matthews, M.R., Anderson, B.P., Haljan, P.C., Hall, D.S., Wieman, C.E., and Cornell, E.A. Vortices in a Bose–Einstein condensate. *Phys. Rev. Lett.*, **83**, 2498–2501 (1999).

[118] Madison, K.W., Chevy, F., Wohlleben, W., and Dalibard, J. Vortex formation in a stirred Bose–Einstein condensate. *Phys. Rev. Lett.*, **84**, 806–809 (2000).

[119] Abo-Shaeer, J.R., Raman, C., Vogels, J. M., and Ketterle, W. Observation of vortex lattices in Bose–Einstein condensates. *Science*, **292**, 476–479 (2001).

[120] Madison, K. W., Chevy, F., Bretin, V., Dalibard, J. Stationary states of a rotating Bose–Einstein condensate: Routes to vortex nucleation. *Phys. Rev. Lett.*, **86**, 4443–4446 (2001).

[121] Hodby, E., Hechenblaikner, G., Hopkins, S. A., Maragó, O. M., and Foot, C. J. Vortex nucleation in Bose–Einstein condensates in an oblate, purely magnetic potential. *Phys. Rev. Lett.*, **88**, 010405(1–4) (2002).

[122] Chevy, F., Madison, K.W., and Dalibard, J. Measurement of the angular momentum of a rotating Bose–Einstein condensate. *Phys. Rev. Lett.*, **85**, 2223–2227 (2000).

[123] Haljan, P. C., Anderson, B. P., Coddington, I., and Cornell, E. A. Use of surface-wave spectroscopy to characterize tilt modes of a vortex in a Bose–Einstein condensate. *Phys. Rev. Lett.*, **86**, 2922–2925 (2001).

[124] Bretin, V., Rosenbusch, P., Chevy, F., Shlyapnikov, G.V., and Dalibard, J. Quadrupole oscillation of a single-vortex Bose–Einstein condensate: Evidence for Kelvin modes. *Phys. Rev. Lett.*, **90**, 100403(1–4) (2003).

[125] Hodby, E., Hopkins, S.A., Hechenblaikner, G., Smith, N.L., and Foot, C.J. Experimental observation of a superfluid gyroscope in a dilute Bose–Einstein condensate. *Phys. Rev. Lett.*, **91**, 090403(1–4) (2003).

[126] Anderson, M.F., Ryu, C., Cladé, P., Natarajan, V., Vaziri, A., Helmerson, K., and Phillips, W.D. Quantized rotation of atoms from photons with orbital angular momentum. *Phys. Rev. Lett.*, **97**, 170406(1–4) (2006).

[127] Haljan, P.C., Coddington, I., Engels, P., and Cornell, E.A. Driving Bose–Einstein-condensate vorticity with a rotating normal cloud. *Phys. Rev. Lett.*, **87**, 210403(1–4) (2001).

[128] Bretin, V., Stock, S., Seurin, Y., and Dalibard, J. Fast rotation of a Bose–Einstein condensate. *Phys. Rev. Lett.*, **92**, 050403(1–4) (2004).

[129] Cooper, N. R. Rapidly rotating atomic gases. *Adv. Phys.*, **57**, 539–616 (2008).

[130] Viefers, S. Quantum Hall physics in rotating Bose–Einstein condensates. *J. Phys.: Condens. Matter*, **20**, 123202(1–14) (2008).

[131] Bloch, I., Dalibard, J. and Zwerger W. Many-body physics with ultracold gases. *Rev. Mod. Phys.*, **80**, 885–964 (2008).

[132] Lin, Y.-J., Compton, R. L., Perry, A. R., Phillips, W. D., Porto, J. V., and Spielman I. B. Bose–Einstein condensate in a uniform light-induced vector potential. *Phys. Rev. Lett.*, **102**, 130401(1–4) (2009).

[133] Lin, Y.-J., Compton, R. L., Jiménez-García, K., Porto, J. V., and Spielman, I. B. Synthetic magnetic fields for ultracold neutral atoms. *Nature (London)*, **462**, 628–632 (2009).

[134] Inouye, S., Gupta, S., Rosenband, T., Chikkatur, A. P., Görlitz, A., Gustavson, T. L., Leanhardt, A. E., Pritchard, D. E., and Ketterle W. Observation of vortex phase singularities in Bose–Einstein condensates. *Phys. Rev. Lett.* 87, 080402 (1–4) (2001).

[135] Chevy, F., Madison, K. W., Bretin, V., and Dalibard, J. Interferometric detection of a single vortex in a dilute Bose–Einstein condensate. *Phys. Rev. A* 64, 031601(R)(1–4) (2001).

[136] Bolda, E. L., and Walls, D. F. Detection of vorticity in Bose–Einstein condensed gases by matter-wave interference. *Phys. Rev. Lett.* 81, 5477–5480 (1998).

[137] Blakie, P. B. and Ballagh, R. J. Spatially selective Bragg scattering: A signature for vortices in Bose–Einstein condensates. *Phys. Rev. Lett.* 86, 3930–3933 (2001).

[138] Muniz, S. R., Naik, D. S., and Raman, C. Bragg spectroscopy of vortex lattices in Bose–Einstein condensates. *Phys. Rev. A* 73, 041605(R)(1–4) (2006).

[139] Myatt, C. J., Burt, E. A., Ghrist, R. W., Cornell, E. A., and Wieman, C. E. Production of two overlapping Bose–Einstein condensates by sympathetic cooling. *Phys. Rev. Lett.*, **78**, 586–589 (1997).

[140] Hall, D. S., Matthews, M. R., Ensher, J. R., Wieman, C. E., and Cornell, E. A. Dynamics of component separation in a binary mixture of Bose–Einstein condensates. *Phys. Rev. Lett.*, **81**, 1539–1542 (1998).

[141] Stenger, J., Inouye, S., Stamper-Kurn, D. M., Miesner, H. -J., Chikkatur, A. P., and Ketterle, W. Spin domains in ground-state Bose–Einstein condensates. *Nature (London)*, **396**, 345–348 (1998).

[142] Barrett, M. D., Sauer, J. A., and Chapman, M. S. All-optical formation of an atomic Bose–Einstein condensate. *Phys. Rev. Lett.*, **87**, 010404(1–4) (2001).

[143] Modugno, G., Modugno, M., Riboli, F., Roati, G., and Inguscio, M. Two atomic species superfluid. *Phys. Rev. Lett.*, **89**, 190404(1–4) (2002).

[144] Schmaljohann, H., Erhard, M., Kronjäger, J., Kottke, M., Staa, S. van, Cacciapuoti, L., Arlt, J. J., Bongs, K., and Sengstock, K. Dynamics of F = 2 spinor Bose–Einstein condensates. *Phys. Rev. Lett.*, **92**, 040402(1–4) (2004).

[145] Chang, M.-S., Hamley, C. D., Barrett, M.D., Sauer, J. A., Fortier, K. M., Zhang, W., You, L., and Chapman, M. S. Observation of spinor dynamics in optically trapped Rb-87 Bose–Einstein condensates. *Phys. Rev. Lett.*, **92**, 140403(1–4) (2004).

[146] Kuwamoto, T., Araki, K., Eno, T., and Hirano, T. Magnetic field dependence of the dynamics of Rb-87 spin-2 Bose–Einstein condensates. *Phys. Rev. A*, **69**, 063604(1–6) (2004).

[147] Thalhammer, G., Barontini, G., De Sarlo, L., Catani, J., Minardi, F., and Inguscio, M. Double species Bose–Einstein condensate with tunable interspecies interactions. *Phys. Rev. Lett*, **100**, 210402(1–4) (2008).

[148] Papp, S. B., Pino, J.M., and Wieman, C.E. Tunable miscibility in a dual-species Bose–Einstein condensate. *Phys. Rev. Lett*, **101**, 040402(1–4) (2008).

[149] Vollhardt D., and Woelfle, P. *The Superfluid Phases of Helium 3*. Taylor and Francis, London (1990).

[150] Bennemann, K. H. and Ketterson, J. B. eds. *Superconductivity* Volume 1: Conventional and High Temperature Superconductors. Springer-Verlag (2008).

[151] Girvin, S.M. *The Quantum Hall effect: Novel Excitations and Broken Symmetries, Proceedings of the Les Houches Summer School of Theoretical Physics*, Springer Verlag, Berlin, and Les Editions de Physique, Paris (1998).

[152] Volovik, G.E. *The Universe in a Helium Droplet*. Oxford University Press, Oxford (2003).

[153] Inouye, S., Andrews, M. R., Stenger, J., Miesner, H.-J., Stamper-Kurn, D. M., and Ketterle, W. Observation of Feshbach resonances in a Bose–Einstein condensate. *Nature (London)*, **392**, 151–154 (1998).

[154] Donley, E. A., Claussen, R. R., Cornish, S. L., Roberts, J. L., Cornell, E. A., and Wieman, C. E. Dynamics of collapsing and exploding Bose–Einstein condensates. *Nature (London)*, **412**, 295–299 (2001).

[155] Papp, S. B., Pino, J. M., Wild, R. J., Ronen, S., Wieman, C. E., Jin, D. E., and Cornell, E. A. Bragg spectroscopy of a strongly interacting Rb-85 Bose–Einstein condensate. *Phys. Rev. Lett.*, **101**, 135301(1–4) (2008).

[156] Lahaye, T., Koch, T., Frohlich, B., Fattori, M., Metz, J., Griesmaier, A., Giovanazzi, S. and Pfau, T. Strong dipolar effects in a quantum ferrofluid. *Nature (London)*, **448**, 672–675 (2007).

[157] Donley, E.A., Claussen, N.R., Thompson, S.T. and Wieman, C.E. Atom-molecule coherence in a Bose–Einstein condensate. *Nature (London)*, **417**, 529–533 (2002).

[158] Inguscio, M., Ketterle, W., and Salomon, C. Eds, *Ultra-cold Fermi Gases*, Proc. Int. School "Enrico Fermi," Course CLXIV (IOS Press, Amsterdam) (2007).

[159] Chin, C., Grimm, R., Julienne, P. and Tiesinga, E. Feshbach resonances in ultracold gases. *Rev. Mod. Phys.*, to be published (2010).

[160] Fetter, A. L. and Svidzinsky, A. A. Vortices in a trapped dilute Bose–Einstein condensate. *J. Phys. Condens. Matter*, **13**, R135–R194 (2001).

[161] Kasamatsu, K. and Tsubota, M. Quantized vortices in atomic Bose–Einstein condensates. *Prog. Low Temp. Phys.*, **16**, (Halperin, W. P. and Tsubota, M. eds.) Elsevier, Amsterdam, 351–403 (2009).

[162] Fetter, A. L. Rotating trapped Bose–Einstein condensates. *Rev. Mod. Phys.*, **81**, 647–691 (2009).

[163] Baym, G., Pethick, C. J. Ground-state properties of magnetically trapped Bose-condensed rubidium gas. *Phys. Rev. Lett.*, **76**, 6–9 (1995).

[164] Raman, C., Abo-Shaeer, J. R., Vogels, J. M., Xu, K., and Ketterle, W. Vortex nucleation in a stirred Bose–Einstein condensate. *Phys. Rev. Lett.*, **87**, 210402(1–4) (2001).

[165] Coddington, I., Haljan, P. C., Engels, P., Schweikhard, V., Tung, S., and Cornell, E. A. Experimental studies of equilibrium vortex properties in a Bose-condensed gas. *Phys. Rev. A*, **70**, 063607(1–11) (2004).

[166] Engels, P., Coddington, I., Haljan, P. C., and Cornell, E. A. Nonequilibrium effects of anisotropic compression applied to vortex lattices in Bose–Einstein condensates. *Phys. Rev. Lett.*, **89**, 100403(1–4) (2002).

[167] Coddington, I., Engels, P., Schweikhard, V., and Cornell, E. A. Observation of Tkachenko oscillations in rapidly rotating Bose–Einstein condensates. *Phys. Rev. Lett.*, **91**, 100402(1–4) (2003).

[168] Engels, P., Coddington, I., Haljan, P. C., Schweikhard, V., and Cornell, E. A. Observation of long-lived vortex aggregates in rapidly rotating Bose–Einstein condensates. *Phys. Rev. Lett.*, **90**, 170405(1–4) (2003).

[169] Schweikhard, V., Coddington, I., Engels, P., Mogendorff, V. P., and Cornell, E. A. Rapidly rotating Bose–Einstein condensates in and near the lowest Landau level. *Phys. Rev. Lett.*, **90**, 170405(1–4) (2004).

[170] García-Ripoll, J. J., and Pérez-García, V. M. Vortex bending and tightly packed vortex lattices in Bose–Einstein condensates. *Phys. Rev. A*, **64**, 053611(1–7) (2001).

[171] Modugno, M., Pricoupenko, L., and Castin, Y. Bose–Einstein condensates with a bent vortex in rotating traps. *Euro. Phys. J. D*, **22**, 235–257 (2003).

[172] Aftalion, A., and Danaila, I. Three-dimensional vortex configurations in a rotating Bose–Einstein condensate. *Phys. Rev. A*, **68**, 023603(1–6) (2003).

[173] Rosenbusch, P., Bretin, V., and Dalibard, J. Dynamics of a single vortex line in a Bose–Einstein condensate. *Phys. Rev. Lett.*, **89**, 200403(1–4) (2002).

[174] Dodd, R. J., Burnett, K., Edwards, M., and Clark, C. W. Excitation spectroscopy of vortex states in dilute Bose–Einstein condensed gases. *Phys. Rev. A*, **56**, 587–590 (1997).

[175] Rokhsar, D. S. Vortex stability and persistent currents in trapped Bose gases. *Phys. Rev. Lett.*, **79**, 2164–2167 (1997).

[176] Lundh, E., Pethick, C. J., and Smith, H. Zero-temperature properties of a trapped Bose-condensed gas: beyond the Thomas–Fermi approximation. *Phys. Rev. A*, **55**, 2126–2131 (1997).

[177] Svidzinsky, A. A., and Fetter, A. L. Stability of a vortex in a trapped Bose–Einstein condensate. *Phys. Rev. Lett.*, **84**, 5919–5923 (2000).

[178] Feder, D. L., Clark, C. W., and Schneider, B. I. Vortex stability of interacting Bose–Einstein condensates confined in anisotropic harmonic traps. *Phys. Rev. Lett.*, **82**, 4956–4959 (1999).

[179] Isoshima, T., and Machida, K. Instability of the nonvortex state toward a quantized vortex in a Bose–Einstein condensate under external rotation. *Phys. Rev. A*, **60**, 3313–3316 (1999).

[180] Feder, D. L., Clark, C. W., and Schneider, B. I. Nucleation of vortex arrays in rotating anisotropic Bose–Einstein condensates. *Phys. Rev. A*, **61**, 011601(R) (1–4) (1999).

[181] Dalfovo, F., and Stringari, S. Shape deformations and angular-momentum transfer in trapped Bose–Einstein condensates. *Phys. Rev. A*, **63**, 011601(R)(1–4) (2000).

[182] Anglin, J. R. Local vortex generation and the surface mode spectrum of large Bose–Einstein condensates. *Phys. Rev. Lett.*, **87**, 240401(1–4) (2001).

[183] Williams, J.E., Zaremba, E., Jackson, B., Nikuni, T., and Griffin, A. Dynamical instability of a condensate induced by a rotating thermal gas. *Phys. Rev. Lett.*, **88**, 070401(1–4) (2002).

[184] Simula, T. P., Virtanen, S. M. M., and Salomaa, M. M. Surface modes and vortex formation in dilute Bose–Einstein condensates at finite temperatures. *Phys. Rev. A*, **66**, 035601(1–4) (2002).

[185] Stringari, S. Collective excitations of a trapped Bose-condensed gas. *Phys. Rev. Lett.*, **77**, 2360–2363 (1996).

[186] Sinha, S., and Castin, Y. Dynamic instability of a rotating Bose–Einstein condensate. *Phys. Rev. Lett.*, **87**, 190402(1–4) (2001).

[187] Reinisch, G. Vortex-nucleating Zeeman resonance in axisymmetric rotating Bose–Einstein condensates. *Phys. Rev. Lett*, **99**, 120402(1–4) (2007).

[188] Tsubota, M., Kasamatsu, K., and Ueda, M. Vortex lattice formation in a rotating Bose–Einstein condensate. *Phys. Rev. A*, **65**, 023603(1–4) (2002).

[189] Kasamatsu, K., Tsubota, M., and Ueda, M. Nonlinear dynamics of vortex lattice formation in a rotating Bose–Einstein condensate. *Phys. Rev. A*, **67**, 033610 (1–14) (2003).

[190] Penckwitt, A. A., Ballagh, R. J., and Gardiner, C. W. Nucleation, growth, and stabilization of Bose–Einstein condensate vortex lattices. *Phys. Rev. Lett.*, **89**, 260402(1–4) (2002).

[191] Lobo, C., Sinatra, A., and Castin, Y. Vortex lattice formation in Bose–Einstein condensates. *Phys. Rev. Lett.*, **92**, 020403(1–4) (2004).

[192] Kasamatsu, K., Machida, M., Sasa, N., and Tsubota, M. Three-dimensional dynamics of vortex-lattice formation in Bose–Einstein condensates. *Phys. Rev. A*, **71**, 063616(1–5) (2005).

[193] Parker, N. G., and Adams, C. S. Emergence and decay of turbulence in stirred atomic Bose–Einstein condensates. *Phys. Rev. Lett.*, **95**, 145301(1–4) (2005).

[194] Wright, T. M., Ballagh, R. J., Bradley, A. S., Blakie, P. B., and Gardiner, C. W. Dynamical thermalization and vortex formation in stirred two-dimensional Bose–Einstein condensates. *Phys. Rev. A*, **78**, 063601(1–22) (2008).

[195] Jackson, B., McCann, J. F., and Adams, C. S. Vortex line and ring dynamics in trapped Bose–Einstein condensates. *Phys. Rev. A*, **61**, 013604(1–7) (1999).

[196] McGee, S. A., and Holland, M. J. Rotational dynamics of vortices in confined Bose–Einstein condensates. *Phys. Rev. A*, **63**, 043608(1–6) (2001).

[197] Fedichev, P. O., and Shlyapnikov, G. V. Dissipative dynamics of a vortex state in a trapped Bose-condensed gas. *Phys. Rev. A*, **60**, R1779–R1782 (1999).

[198] Anderson, B. P., Haljan, P. C., Wieman, C. E., and Cornell, E. A. Vortex precession in Bose–Einstein condensates: Observations with filled and empty cores. *Phys. Rev. Lett.*, **85**, 2857(1–4) (2000).

[199] Feder, D. L., Svidzinsky, A. A. Fetter, A. L., and Clark, C. W. Anomalous modes drive vortex dynamics in confined Bose–Einstein condensates. *Phys. Rev. Lett.*, **86**, 564(1–4) (2001).

[200] Simula, T. P., Mizushima, T., and Machida, K. Kelvin waves of quantized vortex lines in trapped Bose–Einstein condensates. *Phys. Rev. Lett.*, **101**, 020402(1–4) (2008).

[201] Mizushima, T., Ichioka, M., and Machida, K. Beliaev damping and Kelvin mode spectroscopy of a Bose–Einstein condensate in the presence of a vortex line. *Phys. Rev. Lett.*, **90**, 180401(1–4) (2003).

[202] Tsubota, M., Araki, T., and Barenghi, C. F. Rotating superfluid turbulence. *Phys. Rev. Lett.*, **90**, 205301(1–4) (2003).

[203] Takeuchi, H., Kasamatsu, K., and Tsubota, M. Spontaneous radiation and amplification of Kelvin waves on quantized vortices in Bose–Einstein condensates. *Phys. Rev. A*, **79**, 033619(1–5) (2009).

[204] Leanhardt, A. E., Görlitz, A., Chikkatur, A. P., Kielpinski, D., Shin, Y., Pritchard, D. E., and Ketterle, W. Imprinting vortices in a Bose–Einstein condensate using topological phases. *Phys. Rev. Lett.*, **89**, 190403(1–4) (2002).

[205] Shin, Y., Saba, M., Vengalattore, M., Pasquini, T. A., Sanner, C., Leanhardt, A. E., Prentiss, M., Pritchard, D. E., and Ketterle, W. Dynamical instability of a doubly quantized vortex in a Bose–Einstein condensate. *Phys. Rev. Lett.*, **93**, 160406(1–4) (2004).

[206] Pu, H., Law, C. K., Eberly, J. H., and Bigelow, N. P. Coherent disintegration and stability of vortices in trapped Bose condensates. *Phys. Rev. A*, **59**, 1533–1537 (1999).

[207] Möttönen, M., Mizushima, T., Isoshima, T., Salomaa, M. M., and Machida, K. Splitting of a doubly quantized vortex through intertwining in Bose–Einstein condensates. *Phys. Rev. A*, **68**, 023611(1–4) (2003).

[208] Huhtamäki, J. A. M., Möttönen, M., Isoshima, T., Pietilä, V., and Virtanen, S. M. M. Splitting times of doubly quantized vortices in dilute Bose–Einstein condensates. *Phys. Rev. Lett.*, **97**, 110406(1–4) (2006).

[209] Muñoz Mateo, A., and Delgado, V. Dynamical evolution of a doubly quantized vortex imprinted in a Bose–Einstein condensate. *Phys. Rev. Lett.*, **97**, 180409 (1–4) (2006).

[210] Isoshima, T., Okano, M., Yasuda, H., Kasa, K., Huhtamäki, J. A. M., Kumakura, M., Takahashi, Y. Spontaneous splitting of a quadruply charged vortex. *Phys. Rev. Lett.*, **99**, 200403(1–4) (2007).

[211] Kawaguchi, Y., and Ohmi, T. Splitting instability of a multiply charged vortex in a Bose–Einstein condensate. *Phys. Rev. A*, **70**, 043610(1–7) (2004).

[212] Yarmchuk, E. J., Gordon, M. J. V., and Packard, R. E. Observation of stationary vortex arrays in rotating superfluid helium. *Phys. Rev. Lett.*, **43**, 214–217 (1979).

[213] Abrikosov, A. A. On the magnetic properties of superconductors of the second group. *Sov. Phys. JETP*, **5**, 1174–1182 (1957).

[214] Ho, T. L. Bose–Einstein condensates with large number of vortices. *Phys. Rev. Lett.*, **87**, 060403(1–4) (2001).

[215] Watanabe, G., Baym, G., and Pethick, C. J. Landau levels and the Thomas-Fermi structure of rapidly rotating Bose–Einstein condensates. *Phys. Rev. Lett.*, **93**, 190401(1–4) (2004).

[216] Cooper, N. R., Komineas, S., and Read, N. Vortex lattices in the lowest Landau level for confined Bose–Einstein condensates. *Phys. Rev. A*, **70**, 033604(1–5) (2004).

[217] Aftalion, A., Blanc, X., and Dalibard, J. Vortex patterns in a fast rotating Bose–Einstein condensate. *Phys. Rev. A*, **71**, 023611(1–11) (2005).

[218] Sonin, E. B. Ground state and Tkachenko modes of a rapidly rotating Bose–Einstein condensate in the lowest-Landau-level state. *Phys. Rev. A*, **72**, 021606(R)(1–4) (2005).

[219] Aftalion, A., Blanc, X., and Nier, F. Vortex distribution in the lowest Landau level. *Phys. Rev. A*, **73**, 011601(R)(1–4) (2006).

[220] Cozzini, M., Stringari, S., and Tozzo, C. Vortex lattices in Bose–Einstein condensates: From the Thomas–Fermi regime to the lowest-Landau-level regime. *Phys. Rev. A*, **73**, 023615(1–4) (2006).

[221] Butts, D. A. and Rokhsar, D. S. Predicted signatures of rotating Bose–Einstein condensates. *Nature (London)*, **397**, 327–329 (1999).

[222] Kavoulakis, G. M., Mottelson, B., and Pethick, C. J. Weakly interacting Bose–Einstein condensates under rotation. *Phys. Rev. A*, **62**, 063605(1–10) (2000).

[223] Vorov, O. K., Isacker, P. V., Hussein, M. S., and Bartschat, K. Nucleation and growth of vortices in a rotating Bose–Einstein condensate. *Phys. Rev. Lett.*, **95**, 023406(1–4) (2005).

[224] 2003 Cozzini, M., and Stringari, S. Macroscopic dynamics of a Bose–Einstein condensate containing a vortex lattice. *Phys. Rev. A*, **67**, 041602(R)(1–4) (2003).

[225] Fischer, U. R., and Baym, G. Vortex states of rapidly rotating dilute Bose–Einstein condensates. *Phys. Rev. Lett.*, **90**, 140402(1–4) (2003).

[226] Baym, G., and Pethick, C. J. Vortex core structure and global properties of rapidly rotating Bose–Einstein condensates. *Phys. Rev. A*, **69**, 043619(1–9) (2004).

[227] Watanabe, G., Gifford, S. A., Baym, G., and Pethick, C. J. Structure of vortices in rotating Bose–Einstein condensates. *Phys. Rev. A*, **74**, 063621(1–9) (2006).

[228] Tkachenko, V. K. Stability of vortex lattices. *Sov. Phys. JETP*, **23**, 1049–1056 (1966).

[229] Baym, G. Tkachenko modes of vortex lattices in rapidly rotating Bose–Einstein condensates. *Phys. Rev. Lett.*, **91**, 110402(1–4) (2003).

[230] Cozzini, M., Pitaevskii, L. P., and Stringari, S. Tkachenko oscillations and the compressibility of a rotating Bose–Einstein condensate. *Phys. Rev. Lett.*, **92**, 220401(1–4) (2004).

[231] Sonin, E. B. Continuum theory of Tkachenko modes in rotating Bose–Einstein condensate. *Phys. Rev. A*, **71**, 011603(R)(1–4) (2005).

[232] Mizushima, T., Kawaguchi, Y., Machida, K., Ohmi, T., Isoshima, T., and Salomaa, M. M. Collective oscillations of vortex lattices in rotating Bose–Einstein condensates. *Phys. Rev. Lett.*, **92**, 060407(1–4) (2004).

[233] Baksmaty, L. O., Woo, S. J., Choi, S., and Bigelow, N. P. Tkachenko waves in rapidly rotating Bose–Einstein condensates. *Phys. Rev. Lett.*, **92**, 160405(1–4) (2004).

[234] Fetter, A. L. Rotating vortex lattice in a Bose–Einstein condensate trapped in combined quadratic and quartic radial potentials. *Phys. Rev. A*, **64**, 063608(1–6) (2001).

[235] Lundh, E. Multiply quantized vortices in trapped Bose–Einstein condensates. *Phys. Rev. A*, **65**, 043604(1–6) (2002).

[236] Kasamatsu, K., Tsubota, M., and Ueda, M. Giant hole and circular superflow in a fast rotating Bose–Einstein condensate. *Phys. Rev. A*, **66**, 053606(1–4) (2002).

[237] Jackson, A. D., Kavoulakis, G. M., and Lundh, E. Phase diagram of a rotating Bose–Einstein condensate with anharmonic confinement. *Phys. Rev. A*, **69**, 053619(1–7) (2004).

[238] Fetter, A. L., Jackson, B., and Stringari, S. Rapid rotation of a Bose–Einstein condensate in a harmonic plus quartic trap. *Phys. Rev. A*, **71**, 013605(1–9) (2005).

[239] Danaila, I. Three-dimensional vortex structure of a fast rotating Bose–Einstein condensate with harmonic-plus-quartic confinement. *Phys. Rev. A*, **72**, 013605(1–6) (2005).

[240] Fu, H., and Zaremba, E. Transition to the giant vortex state in a harmonic-plus-quartic trap. *Phys. Rev. A*, **73**, 013614(1–14) (2006).

[241] Blanc, X., and Rougerie, N. Lowest-Landau-level vortex structure of a Bose–Einstein condensate rotating in a harmonic plus quartic trap. *Phys. Rev. A*, **77**, 053615(1–8) (2008).

[242] Tung, S., Schweikhard, V., and Cornell, E. A. Observation of vortex pinning in Bose–Einstein condensates. *Phys. Rev. Lett.*, **97**, 240402(1–4) (2006).

[243] Reijnders J. W., and Duine, R. A. Pinning of vortices in a Bose–Einstein condensate by an optical lattice. *Phys. Rev. Lett.*, **93**, 060401(1–4) (2004).

[244] Pu, H., Baksmaty, L. O., Yi, S., and Bigelow, N. P. Structural phase transitions of vortex matter in an optical lattice. *Phys. Rev. Lett.*, **94**, 190401(1–4) (2005).

[245] Kasamatsu, K. Uniformly frustrated bosonic Josephson-junction arrays. *Phys. Rev. A*, **79**, 021604(R)(1–4) (2009).

[246] Williams, R. A., Al-Assam, S., and Foot, C. J. Observation of vortex nucleation in a rotating two-dimensional lattice of Bose–Einstein condensates. *Phys. Rev. Lett.*, **104**, 050404(1–4) (2010).

[247] Berezinskii, V. L. Destruction of long-range order in one-dimensional and two-dimensional systems possessing a continuous symmetry group. II. quantum systems *Sov. Phys. JETP*, **34**, 610–616 (1972).

[248] Kosterlitz, J. M., and Thouless, D. J. Ordering, metastability and phase transitions in two-dimensional systems. *J. Phys. C*, **6**, 1181–1203 (1973).

[249] Stock, S., Hadzibabic, Z., Battelier, B., Cheneau, M., and Dalibard, J. Observation of phase defects in quasi-two-dimensional Bose–Einstein condensates. *Phys. Rev. Lett.*, **95**, 190403(1–4) (2005).

[250] Hadzibabic, Z., Krüger, P., Cheneau, M., Battelier, B., and Dalibard, J. Berezinskii–Kosterlitz–Thouless crossover in a trapped atomic gas. *Nature (London)*, **441**, 1118–1121 (2006).

[251] Polkovnikov, A., Altman, E., and Demler, E. Interference between independent fluctuating condensates. *Proc. Natl. Acad. Sci. USA*, **103**, 6125–6129 (2006).

[252] Simula, T. P., and Blakie, P. B. Thermal activation of vortex–antivortex pairs in quasi-two-dimensional Bose–Einstein condensates. *Phys. Rev. Lett.*, **96**, 020404(1–4) (2006).

[253] Kibble, T. W. B. Topology of cosmic domains and strings. *J. Phys. A*, **9**, 1387–1398 (1976).

[254] Zurek, W. H. Cosmological experiments in superfluid helium? *Nature (London)*, **317**, 505–508 (1985).

[255] Bäuerle, C., Bunkov, Y. M. Fisher, S. N., Godfrin, H., and Pickett, G. R. Laboratory simulation of cosmic string formation in the early universe using superfluid ^{3}He. *Nature (London)*, **382**, 332–334 (1996).

[256] Ruutu, V. M. H., Eltsov, V. B., Gill, A. J., Kibble, T. W. B., Krusius, M., Makhlin, Y. G., Plaçais, B., Volovik, G. E., Xu, W. Vortex formation in neutron-irradiated superfluid ^{3}He as an analogue of cosmological defect formation. *Nature (London)*, **382**, 334–336 (1996).

[257] Weiler, C. N., Neely, T. W., Scherer, D. R., Bradley, A. S., Davis, M. J., and Anderson, B. P. Spontaneous vortices in the formation of Bose–Einstein condensates. *Nature (London)*, **455**, 948–951 (2008).

[258] Gardiner, C. W., Anglin, J. R., and Fudge, T. I. A. The stochastic Gross–Pitaevskii equation. *J. Phys. B*, **35**, 1555–1582 (2002).

[259] Gardiner, C. W. and Davis, M. J. The stochastic Gross–Pitaevskii equation: II. *J. Phys. B*, **36**, 4731–4753 (2003).

[260] Bradley, A. S., Gardiner, C. W. Davis, M. J. Bose–Einstein condensation from a rotating thermal cloud: Vortex nucleation and lattice formation. *Phys. Rev. A*, **77**, 033616(1–14) (2008).

[261] Scherer, D. R., Weiler, C. N., Neely, T. W., and Anderson, B. P. Vortex formation by merging of multiple trapped Bose–Einstein condensates. *Phys. Rev. Lett.*, **98**, 110402(1–4) (2007).

[262] Kasamatsu, K., and Tsubota, M. Vortex generation in cyclically coupled superfluids and the Kibble–Zurek mechanism. *J. Low Temp. Phys.*, **126**, 315–320 (2002).

[263] Feder, D. L., Pindzola, M. S., Collins, L. A., Schneider, B. I., and Clark, C. W. Dark-soliton states of Bose–Einstein condensates in anisotropic traps. *Phys. Rev. A*, **62**, 053606(1–11) (2000).

[264] Carretero-González, R., Whitaker, N., Kevrekidis, P. G., and Frantzeskakis, D. J. Vortex structures formed by the interference of sliced condensates. *Phys. Rev. A*, **77**, 023605(1–8) (2008).

[265] Carretero-González, R., Anderson, B. P., Kevrekidis, P. G., Frantzeskakis, D. J., and Weiler, C.N. Dynamics of vortex formation in merging Bose–Einstein condensate fragments. *Phys. Rev. A*, **77**, 033625(1–11) (2008).

[266] Ruben, G., Paganin, D. M., and Morgan, M. J. Vortex-lattice formation and melting in a nonrotating Bose–Einstein condensate. *Phys. Rev. A*, **78**, 013631(1–9) (2008).

[267] Leanhardt, A.E., Shin, Y., Kielpinski, D., Pritchard, D. E., and Ketterle, W. Coreless vortex formation in a spinor Bose–Einstein condensate. *Phys. Rev. Lett.*, **90**, 140403(1–4) (2003).

[268] Schweikhard, V., Coddington, I., Engels, P., Tung, S., and Cornell, E. A. Vortex-lattice dynamics in rotating spinor Bose–Einstein condensates. *Phys. Rev. Lett.*, **93**, 210403(1–4) (2004).

[269] Sadler, L. E., Higbie, J. M., Leslie, S. R., Vengalattore, M., and Stamper-Kurn, D. M. Spontaneous symmetry breaking in a quenched ferromagnetic spinor Bose–Einstein condensate. *Nature (London)*, **443**, 312–315 (2006).

[270] Leslie, L. S., Hansen, A., Wright, K. C., Deutsch, B. M., and Bigelow, N. P. Creation and detection of skyrmions in a Bose–Einstein condensate. *Phys. Rev. Lett.*, **103**, 250401(1–4) (2009).

[271] Maddaloni, P., Modugno, M., Fort, C., Minardi, F., and Inguscio, M. Collective oscillations of two colliding Bose–Einstein condensates. *Phys. Rev. Lett.*, **85**, 2413–2417 (2000).

[272] Mertes, K. M., Merrill, J. W., Carretero-Gonzalez, R., Frantzeskakis, D. J., Kevrekidis, P. G., and Hall, D. S. Nonequilibrium dynamics and superfluid ring excitations in binary Bose–Einstein condensates. *Phys. Rev. Lett.*, **99**, 190402(1–4) (2007).

[273] Miesner, H.-J., Stamper-Kurn, D. M., Stenger, J., Inouye, S., Chikkatur, A. P., and Ketterle W. Observation of metastable states in spinor Bose–Einstein condensates. *Phys. Rev. Lett.*, **82**, 2228–2231 (1999).

[274] Ho, T.-L., and Shenoy, V. B. Binary mixtures of bose condensates of alkali atoms. *Phys. Rev. Lett.*, **77**, 3276–3279 (1996).

[275] Timmermans, E. Phase separation of Bose–Einstein condensates. *Phys. Rev. Lett.*, **81**, 5718–5721 (1998).

[276] Ao, P., and Chui, S. T. Binary Bose–Einstein condensate mixtures in weakly and strongly segregated phases. *Phys. Rev. A*, **58**, 4836–4840 (1998).

[277] Williams, J. E. and Holland, M. J. Preparing topological states of a Bose–Einstein condensate. *Nature (London)*, **401**, 568–572 (1999).

[278] García-Ripoll J. J., and Pérez-García, V. M. Stable and unstable vortices in multicomponent Bose–Einstein condensates. *Phys. Rev. Lett.*, **84**, 4264–4267 (2000).

[279] Skryabin, D.V. Instabilities of vortices in a binary mixture of trapped Bose–Einstein condensates: Role of collective excitations with positive and negative energies. *Phys. Rev. A*, **63**, 013602(1–10) (2000).

[280] Leonhardt, U. and Volovik, G. E. How to create an Alice string (half-quantum vortex) in a vector Bose–Einstein condensate. *JETP Letters*, **72**, 46–48 (2000).

[281] Mueller, E. J. Spin textures in slowly rotating Bose–Einstein condensates. *Phys. Rev. A*, **69**, 033606(1–14) (2004).

[282] Kasamatsu, K., Tsubota, M., and Ueda, M., Spin textures in rotating two-component Bose–Einstein condensates *Phys. Rev. A*, **71**, 043611(1–14) (2005).

[283] Anderson, P. W., and Toulouse, G. Phase slippage without vortex cores: vortex textures in superfluid ^{3}He. *Phys. Rev. Lett.*, **38**, 508–511 (1977).

[284] Mermin, N. D., and Ho, T. L. Circulation and angular momentum in the a phase of superfluid Helium-3. *Phys. Rev. Lett.*, **36**, 594–597 (1976).

[285] Rajaraman, R. *Soliton and Instantons*, North-Holland, Amsterdam (1989).

[286] Matthews, M. R., Anderson, B. P., Haljan, P. C., Hall, D. S., Holland, M. J., Williams, J. E., Wieman, C. E., and Cornell, E. A. Watching a superfluid untwist itself: recurrence of Rabi oscillations in a Bose–Einstein condensate. *Phys. Rev. Lett.*, **83**, 3358–3361 (1999).

[287] Kasamatsu, K., Tsubota, M., and Ueda, M. Vortex molecules in coherently coupled two-component Bose–Einstein condensates. *Phys. Rev. Lett.*, **93**, 250406(1–4) (2004).

[288] Son, D. T. and Stephanov, M. A. Domain walls of relative phase in two-component Bose–Einstein condensates. *Phys. Rev. A*, **65**, 063621(1–10) (2002).

[289] Mueller, E. J., and Ho, T.-L. Two-component Bose–Einstein condensates with a large number of vortices. *Phys. Rev. Lett.*, **88**, 180403(1–4) (2002).

[290] Keçeli, M., and Oktel, M. Ö. Tkachenko modes and structural phase transitions of the vortex lattice of a two-component Bose–Einstein condensate. *Phys. Rev. A*, **73**, 023611(1–17) (2006).

[291] Kasamatsu, K., Tsubota, M., and Ueda, M. Vortex phase diagram in rotating two-component Bose–Einstein condensates. *Phys. Rev. Lett.*, **91**, 150406(1–4) (2003).

[292] Woo, S. J., Choi, S., Baksmaty, L. O., and Bigelow, N. P. Dynamics of vortex matter in rotating two-species Bose–Einstein condensates. *Phys. Rev. A*, **75**, 031604(R)(1–4) (2007).

[293] Kasamatsu, K., and Tsubota, M. Vortex sheet in rotating two-component Bose–Einstein condensates. *Phys. Rev. A*, **79**, 023606(1–7) (2009).

[294] Ueda, M., and Kawaguchi, Y. Spinor Bose–Einstein condensates. *arXiv: 1001.2072* (2010).

[295] Ohmi, T. and Machida, K. Bose–Einstein condensation with internal degrees of freedom in alkali atom gases. *J. Phys. Soc. Jpn.*, **67**, 1822–1825 (1998).

[296] Ho, T. L. Spinor Bose condensates in optical traps. *Phys. Rev. Lett.*, **81**, 742–745 (1998).

[297] Ho, T. L., and Shenoy, V. B. Local spin-gauge symmetry of the Bose–Einstein condensates in atomic gases. *Phys. Rev. Lett.*, **77**, 2595–2599 (1996).

[298] Zhou, F. Spin correlation and discrete symmetry in spinor Bose–Einstein condensates. *Phys. Rev. Lett.*, **87**, 080401(1–4) (2001).

[299] Murata, K., Saito, H., and Ueda, M. Broken-axisymmetry phase of a spin-1 ferromagnetic Bose–Einstein condensate. *Phys. Rev. A*, **75**, 013607(1–10) (2007).

[300] Nakahara, M., Isoshima, T., Machida, K., Ogawa, S., and Ohmi, T. A simple method to create a vortex in Bose–Einstein condensate of alkali atoms. *Phys. B*, **284**, 17–18 (2000).

[301] Isoshima, T., Nakahara, M., Ohmi, T., and Machida, K. Creation of a persistent current and vortex in a Bose–Einstein condensate of alkali-metal atoms. *Phys. Rev. A*, **61**, 063610(1–10) (2000).

[302] Pietilä, V., Möttönen, M., and Nakahara, M. Topological vortex creation in spinor Bose–Einstein condensates. *Electromagnetic, Magnetostatic, and Exchange-Interaction Vortices in Confined Magnetic Structures* (Kamenetskii, E. O. eds.). Kerala. Transworld Research Network, 297–329 (2008).

[303] Ogawa, S. -I., Möttönen, M., Nakahara, M., Ohmi, T., and Shimada, H. Method to create a vortex in a Bose–Einstein condensate. *Phys. Rev. A*, **66**, 013617(1–7) (2002).

[304] Kumakura, M., Hirotani, T., Okano, M., Takahashi, Y., and Yabuzaki, T. Topological formation of a multiply charged vortex in the Rb Bose–Einstein condensate: Effectiveness of the gravity compensation. *Phys. Rev. A*, **73**, 063605(1–7) (2006).

[305] Möttönen, M., Pietilä, V., Virtanen, S. M. M. Vortex pump for dilute Bose–Einstein condensates. *Phys. Rev. Lett.*, **99**, 250406(1–4) (2007).

[306] Xu, Z. F., Zhang, P., Raman, C., and You, L. Continuous vortex pumping into a spinor condensate with magnetic fields. *Phys. Rev. A*, **78**, 043606(1–8) (2008).

[307] Isoshima, T., Machida, K., and Ohmi, T. Quantum vortex in a spinor Bose–Einstein condensate. *J. Phys. Soc. Jpn.*, **70**, 1604–1610 (2001).

[308] Isoshima, T. and Machida, K. Axisymmetric vortices in spinor Bose–Einstein condensates under rotation. *Phys. Rev. A*, **66**, 023602(1–8) (2002).

[309] Yip, S. K. Internal vortex structure of a trapped spinor Bose–Einstein condensate. *Phys. Rev. Lett.*, **83**, 4677–4681 (1999).

[310] Mizushima, T., Machida, K., and Kita, T. Mermin-Ho vortex in ferromagnetic spinor Bose–Einstein condensates. *Phys. Rev. Lett.*, **89**, 030401(1–4) (2002).

[311] Mizushima, T., Machida, K., and Kita, T. Axisymmetric versus nonaxisymmetric vortices in spinor Bose–Einstein condensates. *Phys. Rev. A*, **66**, 053610(1–8) (2002).

[312] Martikainen, J. P., Collin, A., and Suominen, K. A. Coreless vortex ground state of the rotating spinor condensate. *Phys. Rev. A*, **66**, 053604(1–6) (2002).

[313] Bulgakov, E. N. and Sadreev, A. F. Vortex phase diagram of F = 1 spinor Bose–Einstein condensates. *Phys. Rev. Lett.*, **90**, 200401(1–4) (2003).

[314] Mizushima, T., Kobayashi, N., and Machida, K. Coreless and singular vortex lattices in rotating spinor Bose–Einstein condensates. *Phys. Rev. A*, **70**, 043613(1–10) (2004).

[315] Saito, H., Kawaguchi, Y., and Ueda, M. Breaking of chiral symmetry and spontaneous rotation in a spinor Bose–Einstein condensate. *Phys. Rev. Lett.*, **96**, 065302(1–4) (2006).

[316] Ji, A.-C., Liu, W. M., Song, J. L., and Zhou, F. Dynamical creation of fractionalized vortices and vortex lattices. *Phys. Rev. Lett.*, **101**, 010402(1–4) (2008).

[317] Chiba, H. and Saito, H. Spin-vortex nucleation in a Bose–Einstein condensate by a spin-dependent rotating trap. *Phys. Rev. A*, **78**, 043602(1–5) (2008).

[318] Hoshi, S. and Saito, H. Magnetization of a half-quantum vortex in a spinor Bose–Einstein condensate. *Phys. Rev. A*, **78**, 053618(1–6) (2008).

[319] Pietilä, V., Möttönen, M., and Virtanen, S. M. M. Stability of coreless vortices in ferromagnetic spinor Bose–Einstein condensates. *Phys. Rev. A*, **76**, 023610(1–7) (2007).

[320] Takahashi, M., Pietilä, V., Möttönen, M., Mizushima, T., and Machida, K. Vortex-splitting and phase-separating instabilities of coreless vortices in F = 1 spinor Bose–Einstein condensates. *Phys. Rev. A*, **79**, 023618(1–10) (2009).

[321] Kita, T., Mizushima, T., and Machida, K. Spinor Bose–Einstein condensates with many vortices. *Phys. Rev. A*, **66**, 061601(R)(1–4) (2002).

[322] Reijnders, J. W., van Lankvelt, F. J. M., Schoutens, K., and Read, N. Rotating spin-1 bosons in the lowest Landau level. *Phys. Rev. A*, **69**, 023612(1–23) (2002).

[323] Saito, H., Kawaguchi, Y., Ueda, M. Topological defect formation in a quenched ferromagnetic Bose–Einstein condensates. *Phys. Rev. A*, **75**, 013621(1–10) (2007).

[324] Lamacraft, A. Quantum quenches in a spinor condensate. *Phys. Rev. Lett.*, **98**, 160404(1–4) (2007).

[325] Uhlmann, M., Schutzhold, R., Fischer, U. R. Vortex quantum creation and winding number scaling in a quenched spinor Bose gas. *Phys. Rev. Lett.*, **99**, 120407(1–4) (2007).

[326] Damski, B. and Zurek, W. H. Dynamics of a quantum phase transition in a ferromagnetic Bose–Einstein condensate. *Phys. Rev. Lett.*, **99**, 130402(1–4) (2007).

[327] Saito, H., Kawaguchi, Y., Ueda, M. Kibble–Zurek mechanism in a quenched ferromagnetic Bose–Einstein condensate. *Phys. Rev. A*, **76**, 043613(1–10) (2007).

[328] Ciobanu, C. V., Yip, S.-K., and Ho, T. L. Phase diagram of F = 2 spinor Bose–Einstein condensates. *Phys. Rev. A*, **61**, 033607(1–5) (2000).

[329] Turner, A. M., Barnett, R., Demler, E., and Vishwanath, A. Nematic order by disorder in spin-2 Bose–Einstein condensates. *Phys. Rev. Lett.*, **98**, 190404(1–4) (2007).

[330] Song, J. L., Semenoff, G. W., and Zhou, F. Uniaxial and biaxial spin nematic phases induced by quantum fluctuations. *Phys. Rev. Lett.*, **98**, 160408(1–4) (2007).

[331] Ueda, M. and Koashi, M. Theory of spin-2 Bose–Einstein condensates: Spin correlations, magnetic response, and excitation spectra. *Phys. Rev. A*, **65**, 063602(1–22) (2002).

[332] Tojo, S., Hayashi, T., Tanabe, T., Hirano, T., Kawaguchi, Y., Saito, H., and Ueda, M. Spin-dependent inelastic collisions in spin-2 Bose–Einstein condensates. *Phys. Rev. A*, **80**, 042704(1–7) (2009).

[333] Semenoff, G. W., and Zhou, F. Discrete symmetries and 1/3-quantum vortices in condensates of F = 2 cold atoms. *Phys. Rev. Lett.*, **98**, 100401(1–4) (2007).

[334] Barnett, R., Mukerjee, S., and Moore, J. E. Vortex lattice transition in cyclic spinor condensates. *Phys. Rev. Lett.*, **100**, 240405(1–4) (2008).

[335] Turner, A. M. and Demler, E. Vortex molecules in spinor condensates. *Phys. Rev. B*, **79**, 214522(1–24) (2009).

[336] Kobayashi, M., Kawaguchi, Y., Nitta, M., and Ueda, M. Collision dynamics and rung formation of non-Abelian vortices. *Phys. Rev. Lett.*, **103**, 115301(1–4) (2009).

[337] Huhtamäki, J. A. M., Simula, T. P., Kobayashi, M., and Machida, K. Stable fractional vortices in the cyclic states of Bose–Einstein condensates. *Phys. Rev. A*, **80**, 051601(R)(1–4) (2009).

[338] Griesmaier, A., Werner, J., Hensler, S., Stuhler, J., and Pfau, T. Bose–Einstein condensation of chromium. *Phys. Rev. Lett.*, **61**, 160401(1–4) (2005).

[339] Santos, L., Shlyapnikov, G. V., Zoller, P., and Lewenstein M. Bose–Einstein condensation in trapped dipolar gases. *Phys. Rev. Lett.*, **85**, 1791–1794 (2000).

[340] Yi, S. and You. L. Trapped condensates of atoms with dipole interactions. *Phys. Rev. A*, **63**, 053607(1–14) (2001).

[341] Stuhler, J., Griesmaier, A., Koch, T., Fattori, M., Pfau, T., Giovanazzi, S., Pedri, P., and Santos, L. Observation of dipole–dipole interaction in a degenerate quantum gas. *Phys. Rev. Lett.*, **95**, 150406(1–4) (2005).

[342] Lahaye, T., Menotti, C., Santos, L., Lewenstein, M., and Pfau, T. The physics of dipolar bosonic quantum gases. *Rep. Prog. Phys.*, **72**, 126401(1–41) (2009).

[343] Lahaye, T., Metz, J., Fröhlich, B., Koch, T., Meister, M., Griesmaier, A., Pfau, T., Saito, H., Kawaguchi, Y., and Ueda M. d-wave collapse and explosion of a dipolar Bose–Einstein condensate. *Phys. Rev. Lett.*, **101**, 080401(1–4) (2008).

[344] Santos, L., Shlyapnikov, G. V., and Lewenstein M. Roton-maxon spectrum and stability of trapped dipolar Bose–Einstein condensates. *Phys. Rev. Lett.*, **90**, 250403(1–4) (2003).

[345] Ronen, S., Bortolotti, D. C. E., and Bohn, J. L. Radial and angular rotons in trapped dipolar gases. *Phys. Rev. Lett.*, **98**, 030406(1–4) (2007).

[346] Yi, S., and Pu, H. Vortex structures in dipolar condensates. *Phys. Rev. A*, **73**, 061602(R)(1–4) (2006).

[347] Wilson, R. M., Ronen, S., Bohn, J. L., and Pu, H. Manifestations of the roton mode in dipolar Bose–Einstein condensates. *Phys. Rev. Lett.*, **100**, 245302(1–4) (2008).

[348] O'Dell, D. H. J., and Eberlein, C. Vortex in a trapped Bose–Einstein condensate with dipole–dipole interactions. *Phys. Rev. A*, **75**, 013604(1–11) (2007).

[349] Abad, M., Guilleumas, M., Mayol, R., Pi, M., and Jezek, D. M. Vortices in Bose–Einstein condensates with dominant dipolar interactions. *Phys. Rev. A*, **79**, 063622(1–9) (2009).

[350] van Bijnen, R. M. W., O'Dell, D. H. J., Parker, N. G., Martin, A. M. Dynamical instability of a rotating dipolar Bose–Einstein condensate. *Phys. Rev. Lett.*, **98**, 150401(1–4) (2007).

[351] van Bijnen, R. M. W., Dow, A. J., O'Dell, D. H. J., Parker, N. G., Martin, A. M. Exact solutions and stability of rotating dipolar Bose–Einstein condensates in the Thomas–Fermi limit. *Phys. Rev. A*, **80**, 033617(1–11) (2009).

[352] Wilson, R. M., Ronen, S., and Bohn, J. L. Stability and excitations of a dipolar Bose–Einstein condensate with a vortex. *Phys. Rev. A*, **79**, 013621(1–7) (2009).

[353] Klawunn, M., Nath, R., Pedri, P., and Santos, L. Transverse instability of straight vortex lines in dipolar Bose–Einstein condensates. *Phys. Rev. Lett.*, **100**, 240403(1–4) (2008).

[354] Cooper, N. R., Rezayi, E. H., and Simon, S. H. Vortex lattices in rotating atomic Bose gases with dipolar interactions. *Phys. Rev. Lett.*, **95**, 200402(1–4) (2005).

[355] Zhang, J., and Zhai, H. Vortex lattices in planar Bose–Einstein condensates with dipolar interactions. *Phys. Rev. Lett.*, **95**, 200403(1–4) (2005).

[356] Komineas, S., and Cooper, N. R. Vortex lattices in Bose–Einstein condensates with dipolar interactions beyond the weak-interaction limit. *Phys. Rev. A*, **75**, 023623(1–5) (2007).

[357] Vengalattore, M., Leslie, S. R., Guzman, J., Stamper-Kurn, D. M. Spontaneously modulated spin textures in a dipolar spinor Bose–Einstein condensate. *Phys. Rev. Lett.*, **100**, 170403(1–4) (2008).

[358] Kawaguchi, Y., Saito, H., Ueda, M. Einstein–de Haas effect in dipolar Bose–Einstein condensates. *Phys. Rev. Lett.*, **96**, 080405(1–4) (2006).

[359] Santos, L. and Pfau, T. Spin-3 chromium Bose–Einstein condensates. *Phys. Rev. Lett.*, **96**, 190404(1–4) (2006).

[360] Gawryluk, K., Brewczyk, M., Bongs, K., Gajda, M. Resonant Einstein–de Haas effect in a rubidium condensate. *Phys. Rev. Lett.*, **99**, 130401(1–4) (2007).

[361] Yi, S., You, L., and Pu, H. Quantum phases of dipolar spinor condensates. *Phys. Rev. Lett.*, **93**, 040403(1–4) (2004).

[362] Yi, S. and Pu, H. Spontaneous spin textures in dipolar spinor condensates. *Phys. Rev. Lett.*, **97**, 130404(1–4) (2006).

[363] Kawaguchi, Y., Saito, H., Ueda, M. Spontaneous circulation in ground-state spinor dipolar Bose–Einstein condensates. *Phys. Rev. Lett.*, **97**, 130404(1–4) (2006).

[364] Svistunov, B. in *Quantized Vortex Dynamics and Superfluid Turbulence* Vol. 571 (Barenghi, C. F., Donnelly, R. J., and Vinen, W. F. ed.). Springer-Verlag, Berlin, 327–333 (2001).

[365] Berloff, N. G. and Svistunov, B. Scenario of strongly nonequilibrated Bose–Einstein condensation. *Phys. Rev. A*, **66**, 013603(1–7) (2002).

[366] Kobayashi, M. and Tsubota, M. Quantum turbulence in a trapped Bose–Einstein condensate. *Phys. Rev. A*, **76**, 045603(1–4) (2007).

[367] White, A. C., Barenghi, C. F., Proukakis, N. P., Youd, A. J., and Wacks, D. H. Nonclassical velocity statistics in a turbulent atomic Bose–Einstein condensate. *Phys. Rev. Lett.*, **104**, 075301(1–4) (2010).

[368] Vincent, A. and Meneguzzi, M. The spatial structure and statistical properties of homogeneous turbulence. *J. Fluid Mech.*, **225**, 1–20 (1991).

[369] Noullez, A., Wallace, G., Lempert, W., Miles, R. B., and Frisch, U. Transverse velocity increments in turbulent flow using the RELIEF technique. *J. Fluid Mec.*, **339**, 287–307 (1997).

[370] Gotoh, T., Fukayama, D., and Nakano, T. Velocity field statistics in homogeneous steady turbulence obtained using a high-resolution direct numerical simulation. *Phys. Fluids*, **14**, 1065–1081 (2002).

[371] Min, I. A., Mezić, I., and Leonard, A. Lévy stable distributions for velocity and velocity difference in systems of vortex elements. *Phys. Fluids*, **8**, 1169–1180 (1996).

[372] Henn, E. A. L., Seman, J. A., Ramos, E. R. F., Caracanhas, M., Castilho, P., Olímpio, E. P., Roati, G., Magalhães, D. V., Magalhães, K. M. F., and Bagnato, V. S. Observation of vortex formation in an oscillating trapped Bose–Einstein condensate. *Phys. Rev. A*, **79**, 043618(1–5) (2009).

[373] Henn, E. A. L., Seman, J. A., Roati, G., Magalhães, K. M. F., and Bagnato, V. S. Emergence of turbulence in an oscillating Bose–Einstein condensate. *Phys. Rev. Lett.*, **103**, 045301(1–4) (2009).

Spin superfluidity and magnon Bose–Einstein condensation

4

Yu. M. Bunkov and G. E. Volovik

The superfluid current of spins—spin supercurrent—is one more representative of superfluid currents, such as the superfluid current of mass and atoms in superfluid ^{4}He; superfluid current of electric charge in superconductors; superfluid current of hypercharge in the Standard Model of particle physics; superfluid baryonic current and current of chiral charge in quark matter; etc. Spin superfluidity manifests as the spontaneous phase-coherent precession of spins first discovered in 1984 in ^{3}He-B, and can be described in terms of the Bose condensation of spin waves—magnons. We discuss different phases of magnon superfluidity, including those in a magnetic trap; and signatures of magnon superfluidity: (i) spin supercurrent, which transports the magnetization for a macroscopic distance (up to 1 cm); (ii) spin current Josephson effect, which shows the interference between two condensates; (iii) spin current vortex—a topological defect which is an analog of a quantized vortex in superfluids, of an Abrikosov vortex in superconductors, and cosmic strings in relativistic theories; (iv) Goldstone modes related to broken $U(1)$ symmetry—phonons in the spin-superfluid magnon gas; etc. We also touch the topic of spin supercurrent in general, including spin-Hall and intrinsic quantum spin-Hall effects.

4.1 Introduction

Nature knows different types of ordered states. One major class is represented by equilibrium macroscopic ordered states exhibiting spontaneous breaking of symmetry. This class contains crystals; nematic, cholesteric, and other liquid crystals; different types of ordered magnets (antiferromagnets, ferromagnets, etc.); superfluids, superconductors, and Bose condensates; all types of Higgs fields in high energy physics; etc. The important subclasses of this class contain systems with macroscopic quantum coherence exhibiting off-diagonal long-range order (ODLRO), and/or non-dissipative superfluid currents (mass current, spin current, electric current, hypercharge current, etc.). The class of ordered systems is characterized by rigidity, stable gradients of order parameter (non-dissipative currents in quantum coherent systems), and topologically stable defects (vortices, solitons, cosmic strings, monopoles, etc.).

A second large class is presented by dynamical systems out of equilibrium. Ordered states may emerge under external flux of energy. Examples are the coherent emission from lasers; water flow in a draining bathtub; pattern formation in dissipative systems; etc.

Some of the latter dynamic systems can be close to stationary equilibrium systems of the first class. For example, ultra-cold gases in optical traps are not fully equilibrium states since the number of atoms in the trap is not conserved, and thus the steady state requires pumping. However, if the decay is small then the system is close to an equilibrium Bose condensate, and experiences all of the corresponding superfluid properties.

4.1.1 BEC of quasiparticles

Bose–Einstein condensations (BEC) of quasiparticles whose numbers are not conserved is presently one of the debated phenomena of condensed matter physics. In thermal equilibrium the chemical potential of excitations vanishes and, as a result, their condensate does not form. The only way to overcome this situation is to create a non-equilibrium but dynamically steady state, in which the number of excitations is conserved, since the loss of quasiparticles owing to their decay is compensated by pumping of energy. Thus the Bose condensation of quasiparticles belongs to the phenomenon of second class, when the emerging steady state of the system is not in a full thermodynamic equilibrium.

Formally BEC requires conservation of charge or particle number. However, condensation can still be extended to systems with weakly violated conservation. For sufficiently long-lived quasiparticles their distribution may be close to the thermodynamic equilibrium with a well-defined finite chemical potential, which follows from the quasi-conservation of numbers of quasiparticles, and the Bose condensation becomes possible. Several examples of Bose condensation of quasiparticles have been observed or suggested, including phonons [1], excitons [2], exciton–polaritons [3], photons [4], and rotons [5]. The BEC of quasi-equilibrium magnons—spin waves—in ferromagnets has been discussed in Ref. [6] and investigated in [7–9].

In this review we consider the BEC of magnons and their spin superfluidity. Magnons are magnetic excitations in magnetic materials, such as magnetically ordered systems, like ferromagnets, antiferromagnets, etc., and paramagnetic systems with external magnetic ordering such as Fermi liquids. The most suitable systems for studying the phenomenon of magnon BEC are superfluid phases of helium-3. The absolute purity, long lifetime of magnons, different types of magnon–magnon interactions and well controlled magnetic anisotropy make antiferromagnetic superfluid phases of ^{3}He a basic laboratory of magnon BEC. The first BEC state of magnons was discovered in 1984 in ^{3}He-B as a coherent spin precession, and it was baptized as Homogeneously Precessing Domain (HPD) [10, 11]. This is the spontaneously emerging steady state of precession, which preserves the phase coherence across the whole sample even in an inhomogeneous external magnetic field and even in the absence of energy pumping. This is equivalent to the appearance of a coherent superfluid Bose–Einstein condensate.

In the absence of energy pumping this HPD state slowly decays, but during the decay the system remains in the coherent state of BEC: the volume of the Bose condensate (the volume of HPD) gradually decreases with time without violation of the observed properties of the spin-superfluid phase-coherent state. A steady state of phase-coherent precession can be supported by pumping. In particular, the coherence of electron spins induced by periodic pumping has been observed in ensemble of (In,Ga)As/GaAs quantum dots [12]. But in the case of magnon BEC, the pumping need not be coherent—it can be chaotic: the system chooses its own (eigen) frequency of coherent precession, which emphasizes the spontaneous emergence of coherence from chaos.

HPD is very close to the thermodynamic equilibrium of the magnon Bose condensate and exhibits all the superfluid properties which follow from the off-diagonal long-range order (ODLRO) of the coherent precession. Following the discovery of HPD, several other states of magnon BEC have been observed in superfluid phases of ^{3}He, which we discuss in this review, including finite magnon BEC states in magnetic traps. Very similar BEC states were recently observed in antiferromagnets with the so-called Suhl–Nacamura interaction, the long-range nuclear–nuclear interaction via a magnetically ordered electronic subsystem [13, 14].

4.1.2 Spin superfluidity versus superfluidity of mass and charge

The past decade was marked by fundamental studies of mesoscopic quantum states of dilute ultra-cold atomic gases in the regime where the de Broglie wavelength of the atoms is comparable with their spacing, giving rise to the phenomenon of Bose–Einstein condensation (see reviews [15, 16]). Formation of the Bose–Einstein condensate (BEC)—accumulation of the macroscopic number of particles in the lowest energy state—was predicted by Einstein in 1925 [17]. In an ideal gas, all atoms are in the lowest energy state in the zero temperature limit. In dilute atomic gases, weak interactions between atoms produce a small fraction of non-condensed atoms.

In the only known bosonic liquid ^{4}He, which remains liquid at zero temperature, BEC is strongly modified by interactions. Depletion of the condensate due to interactions is very strong: in the limit of zero temperature only about 10% of particles occupy the state with zero momentum. Nevertheless, BEC still remains the key mechanism for the phenomenon of superfluidity in liquid ^{4}He: because of BEC the whole liquid (100% of ^{4}He atoms) forms a coherent quantum state at $T = 0$ and participates in non-dissipative superfluid flow.

Superfluidity is a very general quantum property of matter at low temperatures, with a variety of mechanisms and possible non-dissipative superfluid currents. These include the supercurrent of electric charge in superconductors and mass supercurrent in superfluid ^{3}He, where the mechanism of superfluidity is the Cooper pairing; hypercharge supercurrent in the vacuum of the Standard Model of elementary particle physics, which comes from the Higgs mechanism; supercurrent of color charge in a dense quark matter in quantum chromo-dynamics; etc. All these supercurrents have the same origin: spontaneous breaking of the $U(1)$ or higher symmetry related to conservation of the corresponding charge or particle number, which leads to so-called off-diagonal long-range order.

This spin supercurrent—the superfluid current of spins—is one more representative of superfluid currents. Here the $U(1)$ symmetry is the approximate symmetry of spin rotation, which is related to the quasi-conservation of spin. It appears that the finite lifetime of magnons, and non-conservation of spin due to spin–orbital coupling do not prevent coherence and superfluidity of magnon BEC in ^{3}He-B. The non-conservation leads to a decrease in the number of magnons until the HPD disappears completely, but during this relaxation, the coherence of magnon BEC is preserved with all the signatures of spin superfluidity: (i) the spin supercurrent, which transports the magnetization over a macroscopic distance of more than 1 cm; (ii) the spin current Josephson effect, which shows interference between two condensates; (iii) phase-slip processes at the critical current; (iv) the spin current vortex—a topological defect which is an analog of a quantized vortex in superfluids, of an Abrikosov vortex in superconductors, and cosmic strings in relativistic theories; (v) Goldstone modes related to the broken $U(1)$ symmetry—phonons in the spin-superfluid magnon gas; etc.

4.1.3 Magnon BEC versus equilibrium magnets

The magnetic $U(1)$ symmetry is also broken spontaneously in some static magnetic systems. Sometimes this symmetry breaking is described in terms of BEC of magnons [18–22]. Let us stress from the beginning that there is a principal difference between the magnetic ordering in equilibrium and the BEC of quasiparticles which we are discussing in this review.

In the equilibrium static magnetic systems, symmetry breaking phase transition may start when the system becomes softly unstable towards the growth of one of the magnon modes. Condensation of this mode can be used for a description of the soft mechanism of formation of ferromagnetic and antiferromagnetic states (see e.g. [23]). However, the final outcome of the condensation is the true equilibrium ordered state. In the same manner, the Bose condensation of phonon modes may serve as a soft mechanism of formation for the equilibrium solid crystals [24]. But this does not mean that the final crystal state is the Bose condensate of phonons. See also the paper [25] and the Comment on this paper [26].

On the contrary, BEC of quasiparticles is in principle a non-equilibrium phenomenon, since quasiparticles (magnons) have a finite lifetime. In our case, magnons live long enough to form a state very close to thermodynamic equilibrium BEC, but still it is not an equilibrium. In the final equilibrium state at $T = 0$ all of the magnons will die out. In this respect, the growth of a single mode in the nonlinear process after a hydrodynamic instability [27], which has been discussed in terms of the Bose condensation of classical sound or surface waves [28], is more close to magnon BEC than equilibrium magnets with spontaneously broken $U(1)$ symmetry.

The other difference is that the ordered magnetic states are states with diagonal long-range order. The magnon BEC is a dynamic state characterized by off-diagonal long-range order (see Section 4.8.3 below), which is the main signature of spin superfluidity in this nonequilibrium dynamical state.

4.2 Coherent precession as magnon superfluid

4.2.1 Spin precession

The magnetic subsystem that we discuss is precessing magnetization. In full correspondence with atomic systems, the precessing spins can either be in the normal state or in the ordered spin-superfluid state. In the normal state, spins of atoms are precessing, with the local frequency being determined by the local magnetic field and interactions. In the ordered state the precession of all spins is coherent: they spontaneously develop a common global frequency and global phase of precession.

In pulsed NMR experiments in ^{3}He-B, magnetization is created by an applied static magnetic field: $\mathbf{M} = \chi \mathbf{H}$, where χ is the magnetic susceptibility. Then a pulse of radio-frequency (RF) field $\mathbf{H}_{\rm RF} \perp \mathbf{H}$ deflects the magnetization by an angle β, and after that the induction signal from the free precession is measured. In the state of disordered precession, spins almost immediately lose information on the original common phase and frequency induced by the RF field, and due to this decoherence the measured induction signal is very short, of order $1/\Delta\nu$ where $\Delta\nu$ is the line-width coming from the inhomogeneity of the magnetic field in the sample. In the BEC state, all spins precess coherently, which means that the whole macroscopic magnetization of the sample of volume V is precessing,

$$\mathcal{M}_x + i\mathcal{M}_y = \mathcal{M}_\perp e^{i\omega t + i\alpha}, \ \mathcal{M}_\perp = \chi H V \sin\beta. \tag{4.1}$$

This coherent precession is manifested as a huge and long-lived induction signal (Fig. 4.1). It is important that the coherence of precession is spontaneous: the global frequency and the global phase of precession are formed by the system itself and do not depend on the frequency and phase of the initial RF pulse.

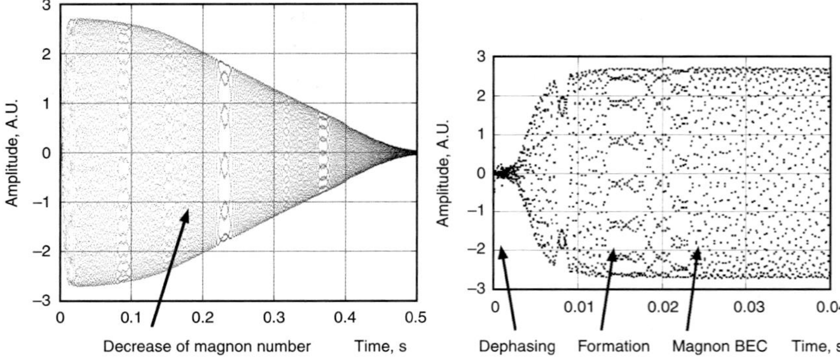

Fig. 4.1 The stroboscopic record of an induction decay signal at a frequency about 1 MHz. *Left*: During the first stage of about 0.002 s the induction signal completely disappears due to dephasing. Then, over about 0.02 s, the spin supercurrent redistributes the magnetization and creates a phase coherent precession, which is equivalent to the magnon BEC state. As a result of a small magnetic relaxation, the number of magnons slowly decreases but the precession remains coherent. *Right*: The initial part of the magnon BEC signal.

4.2.2 Off-diagonal long-range order

Superfluid atomic systems are characterized by off-diagonal long-range order (ODLRO) [29]. In superfluid ^{4}He and in coherent atomic systems the operator of annihilation of atoms with momentum $\mathbf{p} = 0$ has a non-zero vacuum expectation value:

$$\langle \hat{a}_0 \rangle = \mathcal{N}_0^{1/2} e^{i\mu t + i\alpha}, \tag{4.2}$$

where $\mathcal{N}_0$ is the number of particles in the Bose condensate, which in the limit of weak interactions between the atoms coincides, at $T = 0$, with the total number of atoms $\mathcal{N}$.

Equation (4.1) demonstrates that in coherent precession the ODLRO is manifested by a non-zero vacuum expectation value of the operator of creation of spin:

$$\langle \hat{S}_+ \rangle = \mathcal{S}_x + i\mathcal{S}_y = \frac{\mathcal{M}_\perp}{\gamma} e^{i\omega t + i\alpha}, \tag{4.3}$$

where γ is the gyromagnetic ratio, which relates magnetic moment and spin. This analogy suggests that in coherent spin precession the role of particle number $\mathcal{N}$ is played by the projection of total spin on the direction of the magnetic field $\mathcal{S}_z$. The corresponding symmetry group $U(1)$ in magnetic systems is the group of $O(2)$ rotations about the direction of the magnetic field. This quantity $\mathcal{S}_z$ is conserved in the absence of spin–orbit interactions.

Spin–orbit interactions transform the spin angular momentum of a magnetic subsystem to orbital angular momentum, which causes the loss of spin $\mathcal{S}_z$ during precession. In superfluid ^{3}He, spin–orbit coupling is relatively rather small, and thus $\mathcal{S}_z$ is quasi-conserved. Because of the loss of spin, precession will finally decay, but, during its long lifetime the precession remains coherent (Fig. 4.1). This is similar to non-conservation of the number of atoms in laser traps: although the number of atoms decreases with time due to evaporation, this does not destroy the coherence of the remaining atomic BEC.

4.2.3 ODLRO and magnon BEC

The ODLRO in (4.3) can be represented in terms of magnon condensation. To view that, let us use the Holstein–Primakoff transformation, which relates the spin operators with the operators of creation and annihilation of magnons,

$$\hat{a}_0 \sqrt{1 - \frac{\hbar a_0^\dagger a_0}{2\mathcal{S}}} = \frac{\hat{\mathcal{S}}_+}{\sqrt{2\mathcal{S}\hbar}}, \tag{4.4}$$

$$\sqrt{1 - \frac{\hbar a_0^\dagger a_0}{2\mathcal{S}}}\, \hat{a}_0^\dagger = \frac{\hat{\mathcal{S}}_-}{\sqrt{2\mathcal{S}\hbar}}, \tag{4.5}$$

$$\hat{\mathcal{N}} = \hat{a}_0^\dagger \hat{a}_0 = \frac{\mathcal{S} - \hat{\mathcal{S}}_z}{\hbar}. \tag{4.6}$$

Equation (4.6) relates the number of magnons $\mathcal{N}$ to the deviation of spin $\mathcal{S}_z$ from its equilibrium value $\mathcal{S}_z^{(\text{equilibrium})} = \mathcal{S} = \chi H V/\gamma$. In the full thermodynamic equilibrium, magnons are absent (in ^{3}He-B thermal magnons can be ignored, see Section 4.3.1). Each magnon has spin $-\hbar$, and thus the total spin projection after pumping of $\mathcal{N}$ magnons into the system by the RF pulse is reduced by the number of magnons, $\mathcal{S}_z = \mathcal{S} - \hbar\mathcal{N}$. The ODLRO in magnon BEC is given by Eq. (4.2), where $\mathcal{N}_0 = \mathcal{N}$ is the total number of magnons (4.6) in the BEC:

$$\langle \hat{a}_0 \rangle = \mathcal{N}^{1/2} e^{i\omega t + i\alpha} = \sqrt{\frac{2\mathcal{S}}{\hbar}} \sin\frac{\beta}{2}\, e^{i\omega t + i\alpha}. \tag{4.7}$$

Comparing (4.7) and (4.2), one can see that the role of the global chemical potential in atomic systems μ is played by the global frequency of the coherent precession ω, i.e. $\mu \equiv \omega$. This demonstrates that this analogy with the phenomenon of BEC in atomic gases takes place only for the dynamic states of a magnetic subsystem—the states of precession. The ordered magnetic systems discussed in Refs. [18–21] are static, and for them the chemical potential of magnons is always zero.

There are two approaches to studying the thermodynamics of atomic systems: at fixed particle number $\mathcal{N}$ or at fixed chemical potential μ. For the magnon BEC, these two approaches correspond to two different experimental arrangements: pulsed NMR and continuous wave (CW) NMR, respectively. In the case of free precession after the pulse, the number of magnons pumped into the system is conserved (if one neglects the loss of spin). This corresponds to the situation with fixed $\mathcal{N}$, in which the system itself will choose the global frequency of the coherent precession (the magnon chemical potential). The opposite case is CW NMR, when a small RF field is applied continuously to compensate the losses. In this case the frequency of precession is fixed by the frequency of the RF field, $\mu \equiv \omega = \omega_{\mathrm{RF}}$, and now the number of magnons will be adjusted to this frequency to match the resonance condition.

Finally, let us mention that in the approach in which $\mathcal{N}$ is strictly conserved and has quantized integer values, the quantity $< \hat{a}_0 >= 0$ in Eq. (4.2). In the same way, the quantity $\langle \hat{S}_+ \rangle = 0$ in Eq. (4.3), if spin $\mathcal{S}_z$ is strictly conserved and takes quantized values. This means that formally there is no precession if the system is in the quantum state with fixed spin quantum number $\mathcal{S}_z$. However, this does not lead to any paradox in the thermodynamic limit: in the limit of infinite $\mathcal{N}$ and $\mathcal{S}_z$, the description in terms of fixed $\mathcal{N}$ (or $\mathcal{S}_z$) is equivalent to the description in terms of fixed chemical potential μ (or frequency ω).

4.3 Phenomenology of magnon superfluidity

We consider this phenomenology using ^{3}He-B as an example.

4.3.1 Magnon spectrum and magnon mass

Let us neglect for a moment the anisotropy of spin wave velocity c and spin–orbit interaction. Then the magnon spectrum in ^{3}He-B has the following form:

$$\omega(k) = \frac{\omega_L}{2} + \sqrt{\frac{\omega_L^2}{4} + k^2c^2}\,, \tag{4.8}$$

where $\omega_L = \gamma H$. At large momentum, $ck \gg \omega$, this spectrum transforms to the linear spectrum $\omega = ck$ of spin waves propagating with velocity c, which is in the order of the Fermi velocity v_F. At small k, $ck \ll \omega$, this is the spectrum of a massive particle

$$E_k = \hbar\omega(k),\ \omega(k) = \omega_L + \frac{\hbar k^2}{2m_M}, \tag{4.9}$$

where the magnon mass is

$$m_M = \frac{\hbar\omega_L}{2c^2}. \tag{4.10}$$

Since in ^{3}He-B one has $c \sim v_F$, the relative magnitude of the magnon mass compared with the bare mass m_3 of the ^{3}He atom is

$$\frac{m_M}{m_3} \sim \frac{\hbar\omega_L}{E_F}, \tag{4.11}$$

where the Fermi energy $E_F \sim m_3 v_F^2 \sim p_F^2/m_3$. With the magnon gap $\hbar\omega_L \sim 50\,\mu K$ at $\omega_L \sim 1$ MHz, and $E_F \sim 1\,K$, one has $m_M \sim 10^{-4} m_3$. A small mass of these bosons favors Bose condensation. The opposite factor is the small density n of the bosons. From (4.6) it follows that the magnon density is

$$n = \frac{S - S_z}{\hbar} = \frac{\chi H}{\hbar\gamma}(1 - \cos\beta). \tag{4.12}$$

In the typical precessing state of ^{3}He-B the magnon density n is by the same factor $\hbar\omega_L/E_F$ smaller than the density of ^{3}He atoms $n_3 = p_F^3/3\pi^2\hbar^3$ in the liquid

$$\frac{n}{n_3} \sim \frac{\hbar\omega_L}{E_F}. \tag{4.13}$$

In ^{3}He-B, the typical temperature $T \sim 10^{-3}E_F$. As we shall see below it is small compared with the temperature of Bose condensation, $T < T_{\mathrm{BEC}}$. However, T is big compared to the magnon gap, $T \gg \hbar\omega_L$. But this does not cause a problem, it simply means that according to (4.8), thermal magnons are mostly the spin waves with linear spectrum $\omega(k) = ck$ and with characteristic momenta $k_T \sim T/\hbar c$. The density of thermal magnons is $n_T \sim k_T^3$. At ^{3}He-B temperatures, this density is much smaller then the density of condensed magnons and can be neglected, $n_T/n \sim T^3/E_F^2\omega_L \ll 1$.

The smallness of $\omega_L \ll T$ modifies the estimate of the temperature of the Bose condensation, compared with the atomic gases. Before we start pumping magnons, we have an equilibrium system of thermal magnons with $\mu = 0$. After pumping of extra magnons with density n we obtain the quasi-equilibrium state in which the number of magnons is temporarily conserved and thus the magnon system acquires a non-zero chemical potential, $\mu \neq 0$. The number of extra magnons that can be absorbed by thermal distribution without formation of BEC is thus the difference between the distribution functions at $\mu = 0$ and $\mu \neq 0$ at the same temperature:

$$n = \sum_{\mathbf{k}} (f(E_k) - f(E_k - \mu)). \tag{4.14}$$

This quantity reaches its maximum value when $\mu = \omega_L$. Since $\omega_L \ll T$ one has

$$n_{\max} = \omega_L \sum_{\mathbf{k}} \frac{df}{dE} \sim \frac{T^2\omega_L}{c^3}. \tag{4.15}$$

This gives the dependence of BEC transition temperature on the number of pumped magnons,

$$T_{\rm BEC} \sim \left(\frac{nc^3}{\omega_L}\right)^{1/2}. \tag{4.16}$$

At $T < T_{\rm BEC}$ the magnon BEC with $k = 0$ must be formed. In ^{3}He-B, $T_{\rm BEC} \sim E_F$ which is higher than the temperature at which superfluid ^{3}He exists by three to four orders of magnitude. As a result, in the coherently precessing state of ^{3}He-B practically all the magnons are condensed in the ground state with $k = 0$, with a negligible amount of thermal magnons, i.e. the Bose condensation of magnons in ^{3}He-B is almost perfect.

The above estimate also demonstrates that in solid state magnetic systems the BEC may occur even at room temperature, see Refs. [7, 8].

4.3.2 Order parameter and the Gross–Pitaevskii equation

As in the case of atomic Bose condensates, the main physics of the magnon BEC can be found from consideration of the Gross–Pitaevskii equation for the complex order parameter. The local order parameter is obtained by extension of Eq. (4.7) to the inhomogeneous case and is determined as the vacuum expectation value of the magnon field operator:

$$\Psi(\mathbf{r},t) = \left\langle \hat{\Psi}(\mathbf{r},t)\right\rangle, \quad n = |\Psi|^2, \quad \mathcal{N} = \int d^3r\, |\Psi|^2, \tag{4.17}$$

where n is magnon density. To avoid confusion, let us mention that this order parameter (4.17) describes coherent precession in any system, superfluid or non-superfluid. It has nothing to do with the multi-component order parameter which describes the underlying systems—superfluid phases of ^{3}He [30]. In other words, the mass superfluidity of ^{3}He is accompanied by antiferromagnetic ordering in the subsystem of nuclear spins. All of the magnetic properties that we are discussing here are the properties of this magnetically ordered subsystem, and are not connected directly to the mass superfluidity of ^{3}He.

If dissipation and pumping of magnons are ignored (for relaxation and pumping terms in magnon BEC see Ref. [31]), the corresponding Gross–Pitaevskii equation has the conventional form ($\hbar = 1$),

$$-i\frac{\partial \Psi}{\partial t} = \frac{\delta \mathcal{F}}{\delta \Psi^*}, \tag{4.18}$$

where $\mathcal{F}\{\Psi\}$ is the free energy functional, which plays the role of the effective Hamiltonian of the spin subsystem. In coherent precession, the global frequency is constant in space and time (if dissipation is neglected),

$$\Psi(\mathbf{r},t) = \Psi(\mathbf{r})e^{i\omega t}, \tag{4.19}$$

and the Gross–Pitaevskii equation transforms into the Ginzburg–Landau equation with $\omega = \mu$:

$$\frac{\delta \mathcal{F}}{\delta \Psi^*} - \mu\Psi = 0. \tag{4.20}$$

The important feature of the magnon systems is that their number density is limited:

$$n < n_{\max} = \frac{2S}{\hbar}. \tag{4.21}$$

For small $n \ll n_{\max}$ the Ginzburg–Landau free energy functional has the conventional form,

$$\mathcal{F} - \mu\mathcal{N} = \int d^3r \left(\frac{|\nabla\Psi|^2}{2m} + (\omega_L(\mathbf{r}) - \omega)|\Psi|^2 + F_{\rm so}(|\Psi|^2) \right). \tag{4.22}$$

Here, $\omega_L(\mathbf{r}) = \gamma H(\mathbf{r})$ is the local Larmor frequency, which plays the role of external potential $U(\mathbf{r})$ in atomic condensates. The last term $F_{\rm so}(|\Psi|^2)$ contains non-linearity which comes from spin–orbit interaction. It is analogous to the fourth-order term in the atomic BEC, which describes the interaction between the atoms.

4.3.3 Spin–orbit interaction as interaction between magnons

In the magnetic subsystem of superfluid ^{3}He, the interaction term in the Ginzburg–Landau free energy is provided by spin–orbit interaction—interaction between the spin and orbital degrees of freedom. Although the structure of superfluid phases of ^{3}He is rather complicated and is described by the multi-component superfluid order parameter [30], the only output needed for investigation of coherent precession is the structure of the spin–orbit interaction term $F_{\rm so}(|\Psi|^2)$, which appears to be rather simple. The spin–orbit interaction provides the effective interaction between magnons, which can be attractive or repulsive, depending on the orientation of spin and orbital degrees with respect to each other and with respect to the magnetic field. The orbital degrees of freedom in the superfluid phases of ^{3}He are characterized by the direction of the orbital momentum of the Cooper pair $\hat{\mathbf{l}}$, which also marks the axis of the spatial anisotropy of these superfluid liquids. By changing the orientation of $\hat{\mathbf{l}}$ with respect to the magnetic field one is able to regulate the interaction term in experiments.

In superfluid ^{3}He-B, the spin–orbit interaction has very peculiar properties. The microscopic derivation (see (4.98)) leads to the following form [32]:

$$F_{\rm so}(s,l,\gamma) = \frac{2}{15}\frac{\chi}{\gamma^2}\Omega_L^2 \left[\left(sl - \frac{1}{2} + \frac{1}{2}\cos\gamma(1+s)(1+l) \right)^2 + \frac{1}{8}(1-s)^2(1-l)^2 + (1-s^2)(1-l^2)(1+\cos\gamma) \right]. \tag{4.23}$$

It is obtained by averaging the spin–orbit energy over the fast precession of spins. Here $s = \cos\beta$, while $l = \hat{\mathbf{l}}\cdot\hat{\mathbf{H}}$ describes the orientation of the unit vector $\hat{\mathbf{l}}$ with respect to the direction $\hat{\mathbf{H}}$ of the magnetic field. The parameter Ω_L is the so-called Leggett frequency, which characterizes the magnitude of the spin–orbit interaction, and thus the shift of the resonance frequency from the Larmor value caused by spin–orbit interaction. In typical experimental situations, $\Omega_L^2 \ll \omega^2$, which means that the frequency shift is relatively small. Finally γ is another angle, which characterizes the mutual orientation of spin and orbital degrees of freedom. At not extremely low temperatures ($T \geq 0.2\,T_c$), it is a passive quantity: it takes a value corresponding to the minimum of $F_{\rm so}$ for given s and l, i.e. $\gamma = \gamma(s, l)$.

To obtain $F_{\rm so}(|\Psi|^2)$ in (4.22) at fixed $\hat{\mathbf{l}}$, one must express s via $|\Psi|^2$:

$$1 - s = 1 - \cos\beta = \frac{\hbar|\Psi|^2}{S}, \tag{4.24}$$

where $S = \chi H/\gamma$ is the spin density. Since Eq. (4.23) is quadratic in s, the spin–orbit interaction contains quadratic and quartic terms in $|\Psi|$. While the quadratic term modifies the potential U in the Ginzburg–Landau free energy, the quartic term simulates the interaction between magnons.

The profile of the spin–orbit interaction $F_{\rm so}(s, l, \gamma(s, l))$ shown in Fig. 4.2, determines different states of coherent precession and thus different types of magnon BEC in ^{3}He-B, which depend on the orientation of the orbital vector $\hat{\mathbf{l}}$. The most important of these, which has the name HPD, was discovered about 30 years ago [10, 11].

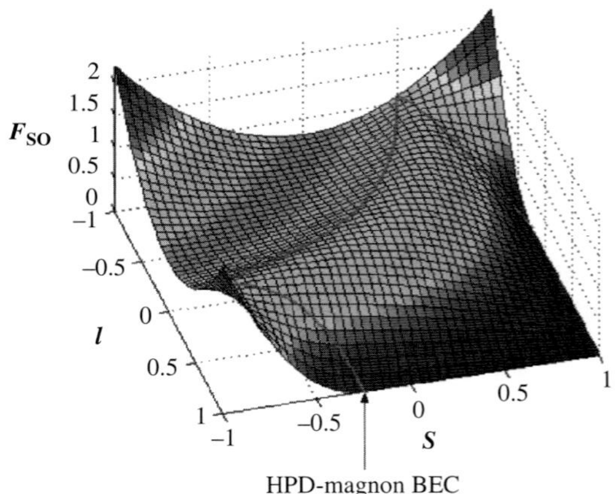

Fig. 4.2 The profile of the spin–orbit energy as a function of $s = \cos\beta$ and orbital variable l, where β is the tipping angle of precession and l is the projection of the orbital angular momentum of a Cooper pair in the direction of the magnetic field. Spontaneous phase-coherent precession emerging at $l = 1$ and $s \approx -1/4$ is called Homogeneously Precessing Domain (HPD). HPD is one of the representatives of magnon BEC in ^{3}He-B.

4.4 Magnon superfluids in ^{3}He-B

In this section we shall describe the main properties of magnon superfluidity in ^{3}He-B. More detailed descriptions can be found in original papers as well as in some review articles [33–38].

4.4.1 HPD as unconventional magnon superfluid

In the bottom-right corner of Fig. 4.2, the minimum of the free energy occurs for $l = 1$, i.e. for the orbital vector $\hat{\mathbf{l}}$ oriented along the magnetic field. This means that if there is no other orientational effect on the orbital vector $\hat{\mathbf{l}}$, the spin–orbit interaction orients it along the magnetic field, and one automatically obtains $l = \cos\beta_L = 1$. The most surprising property emerging at such an orientation is the existence of the completely flat region in Fig. 4.2. The spin–orbit interaction is identically zero in the large range of the tipping angle β of precession, for $1 > s = \cos\beta > -\frac{1}{4}$ [30]:

$$F_{\rm so}(\beta)_{l=1} = 0\,,\ \cos\beta > -\frac{1}{4}, \tag{4.25}$$

$$F_{\rm so}(\beta)_{l=1} = \frac{8}{15}\frac{\chi}{\gamma^2}\Omega_L^2\left(\cos\beta + \frac{1}{4}\right)^2,\ \cos\beta < -\frac{1}{4}. \tag{4.26}$$

Using Eq. (4.24) one obtains the Ginzburg–Landau potential in (4.22) with,

$$F_{\rm so}\left(|\Psi|^2\right) = 0\,,\ |\Psi|^2 < n_c = \frac{5}{4}S, \tag{4.27}$$

$$F_{\rm so}\left(|\Psi|^2\right) = \frac{8}{15}\frac{\chi}{\gamma^2}\Omega_L^2\left(\frac{|\Psi|^2}{S} - \frac{5}{4}\right)^2,\ |\Psi|^2 > n_c = \frac{5}{4}S. \tag{4.28}$$

Eqs. (4.27) and (4.28) demonstrate that when the orbital momentum is oriented along the magnetic field, magnons are non-interacting for all densities n below the threshold value $n_c = (5/4)S/\hbar$. This is a very unconventional gas.

The energy profile of $F - \mu n$ is shown in Fig. 4.3 for different values of the chemical potential $\mu \equiv \omega$. For μ below the external potential U, i.e. for $\omega < \omega_L$, the minimum of $F - \mu n$ corresponds to zero magnons, $n = 0$. It is the static state of ^{3}He-B without precession. For $\mu > U$, i.e. for $\omega > \omega_L$, the minimum of $F - \mu n$ corresponds to the finite value of the magnon density:

$$n = n_c\left(1 + \frac{3}{4}\frac{(\omega - \omega_L)\omega_L}{\Omega_L^2}\right). \tag{4.29}$$

This shows that the formation of HPD starts with a discontinuous jump from zero density of magnons to finite density $n_c = 5S/4\hbar$, which corresponds to coherent precession with a large tipping angle—the so-called magic Leggett angle, $\beta_c \approx 104°$ ($\cos\beta_c = -1/4$). This is distinct from the standard Ginzburg–Landau energy functional (see Eq. (4.60) for magnon BEC in ^{3}He-A-phase below), where the Bose condensate density starts growing smoothly from zero and is proportional to $\mu - U$ for $\mu > U$.

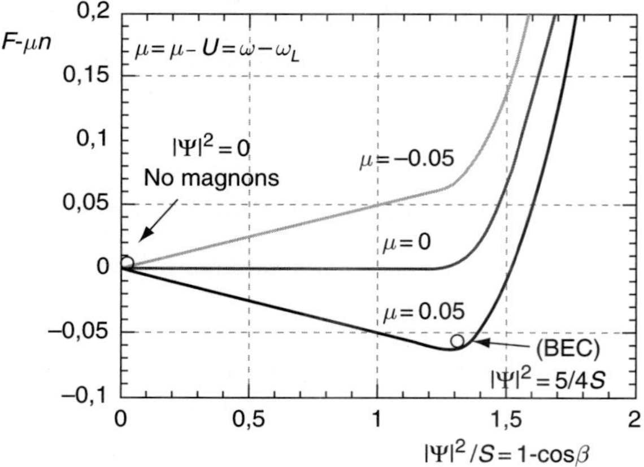

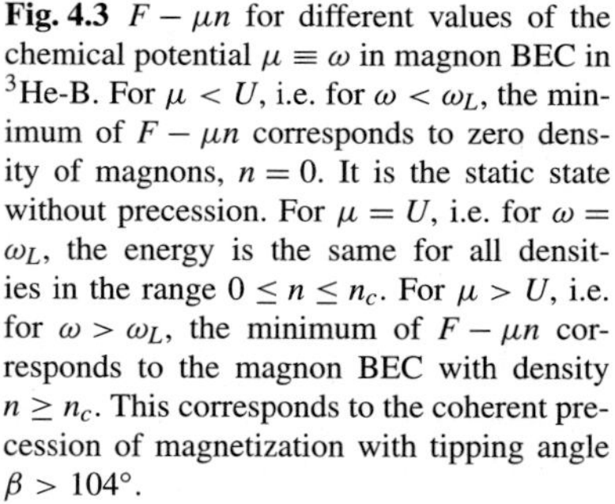
Fig. 4.3 $F - \mu n$ for different values of the chemical potential $\mu \equiv \omega$ in magnon BEC in ^{3}He-B. For $\mu < U$, i.e. for $\omega < \omega_L$, the minimum of $F - \mu n$ corresponds to zero density of magnons, $n = 0$. It is the static state without precession. For $\mu = U$, i.e. for $\omega = \omega_L$, the energy is the same for all densities in the range $0 \le n \le n_c$. For $\mu > U$, i.e. for $\omega > \omega_L$, the minimum of $F - \mu n$ corresponds to the magnon BEC with density $n \ge n_c$. This corresponds to the coherent precession of magnetization with tipping angle $\beta > 104°$.

4.4.2 Two-domain precession

The two states with zero and finite density of magnons, resemble the low-density gas state and the high-density liquid state, respectively. Gas and liquid can be separated in the gravitational field: the heavier liquid state will be concentrated in the lower part of the vessel. For the magnon BEC, the role of the gravitational field is played by the gradient of the magnetic field:

$$\nabla U \equiv \nabla \omega_L = \gamma \nabla H. \tag{4.30}$$

Thus applying the gradient of the magnetic field along axis z, one enforces phase separation (Fig. 4.4). The static thermodynamic equilibrium state with no magnons is concentrated in the region of higher field, where $\omega_L(z) > \omega$, i.e. $U(z) > \mu$. The magnon superfluid—the coherently precessing state—occupies the low-field region, where $\omega_L(z) < \omega$, i.e. $U(z) < \mu$. This is the HPD state, in which all spins precess with the same frequency ω and the same phase α. In typical experiments the gradient is small, and magnon density is close to the threshold value n_c.

The interface between the two domains is situated at the position z_0 where $\omega_L(z_0) = \omega$, i.e. $U(z_0) = \mu$. In continuous NMR, the chemical potential is fixed by the frequency of the RF field: $\mu = \omega_{\text{RF}}$, this determines the position of the interface in the experimental cell.

In pulsed NMR, the two-domain structure emerges spontaneously after the magnetization is deflected by the RF pulse (Fig. 4.4, *left* and *middle*). The position of the interface between the domains is determined by the number of magnons pumped into the system: $\mathcal{N} = (\mathcal{S} - \mathcal{S}_z)/\hbar$. The number of magnons is quasi-conserved, i.e. it is well conserved during the time of the formation of the two-domain state of precession. That is why the volume of the domain occupied by the magnon BEC after its formation is $V = \mathcal{N}/n_c$. This determines

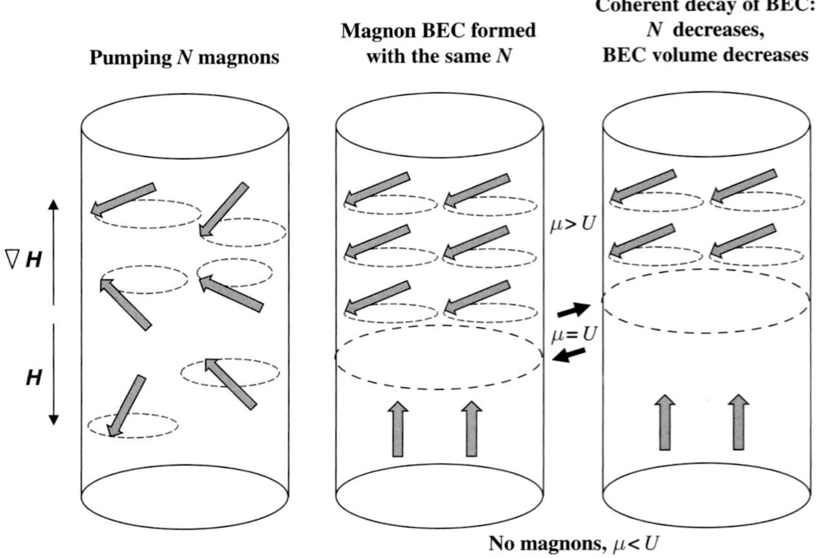

Fig. 4.4 Two domains emerging in ^{3}He-B in pulsed NMR experiments. *Left*: Incoherent spin precession after the pulse of the RF field deflects magnetization from its equilibrium value. The total number of magnons pumped into the system is $\mathcal{N} = (\mathcal{S} - \mathcal{S}_z)/\hbar$. *Middle*: Formation of two domains. All $\mathcal{N}$ magnons are concentrated in the part of the cell with lower magnetic field, forming the BEC state there. The volume of this state is determined by the magnon density in BEC, $V = \mathcal{N}/n$, where $n \approx n_c$. This volume determines the position z_0 of the domain boundary $z_0 = V/A$, where A is the area of the cross-section of the cylindrical cell. The position of the interface in turn determines the global frequency of precession, which is equal to the local Larmor frequency at the phase boundary, $\mu \equiv \omega = \omega_L(z_0)$. *Right*: Decay of magnon BEC. The number of magnons decreases due to spin and energy losses. Since the magnon density in BEC is fixed (it is always close to n_c), the relaxation leads to a decrease in the volume of the BEC domain. However, within this domain the precession remains fully coherent. While the phase boundary moves slowly down, the frequency of the global precession gradually decreases (Fig. 4.5 *left*).

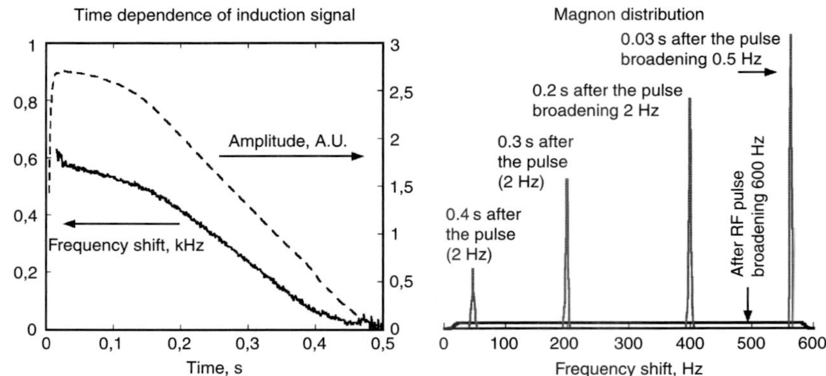

Fig. 4.5 The amplitude and frequency of the induction decay signal from magnon BEC. *Left*: The condensate occupies the domain where the chemical potential $\mu > U$ and radiates the signal corresponding to the Larmor frequency at the domain boundary of the condensate. With relaxation the number of magnons decreases, and the chemical potential moves to the region with lower Larmor frequency. *Right*: The spectroscopic distribution of magnons. Immediately after the RF pulse each spin precesses with the local Larmor frequency. After BEC formation, all of the spins precess with a common frequency ω and spontaneously emergent common phase α. Because of relaxation the number of magnons decreases, leading to a continuously decreasing frequency. The small broadening of the BEC state is due to relaxation. By comparing the initial broadening of the NMR line of about 600 Hz and final broadening of about 0.5 Hz we can estimate that about 99.9% of the pumped magnons are in the condensate.

the position z_0 of the interface, and the chemical potential μ will be adjusted to this position: $\mu = \omega_L(z_0)$.

In the absence of the RF field, i.e. without continuous pumping of magnons, the magnon BEC decays due to loss of magnons. But the precessing domain remains in the fully coherent Bose condensate state, while the volume of the magnon superfluid gradually decreases due to losses and the domain boundary slowly moves down (Fig. 4.4 *right*). The frequency ω of spontaneous coherence, as well as the phase of precession, remain homogeneous across the whole domain of magnon BEC, but the frequency changes with time, since it is determined by the Larmor frequency at the position of the interface, $\mu(t) \equiv \omega_L(z_0(t))$. The change in frequency during the decay is shown in Fig. 4.5 *left*. This frequency change during relaxation was the main observational fact that led Fomin to construct the theory of two-domain precession [11].

Details of the formation of the magnon BEC are shown in Fig. 4.1, where the stroboscopic record of the induction decay signal is shown. During the first stage of about 0.002 s, the induction signal completely disappears due to dephasing. Then, over about 0.02 s, the phase coherent precession emerges spontaneously, which is equivalent to the magnon BEC state. Because of a weak magnetic relaxation, the number of magnons slowly decreases but the precession remains coherent during the whole process of relaxation. The time of formation of magnon BEC is essentially shorter than the relaxation time, as is clearly shown in Fig. 4.1 *right*.

4.4.3 Mass and spin supercurrents in magnon BEC

As we have already discussed, superfluidity is a phenomenon arising due to spontaneous breaking of $U(1)$ symmetry, which in our case is represented by the symmetry group $SO(2)$ of spin rotations about the direction of the magnetic field. In atomic BEC and in helium superfluids such symmetry breaking leads to a non-zero value of superfluid rigidity—the superfluid density ρ_s which enters the non-dissipative supercurrent of particles and thus to mass supercurrent. The same takes place for magnon BEC. But since magnons have both mass m_M and spin $-\hbar$, they carry both the mass and spin supercurrents. The corresponding Goldstone phonon mode of the magnon BEC has been observed experimentally: it manifests as twist oscillations of the precessing domain in ^{3}He-B [39] (see Section 4.4.5 below).

For a small density of magnon condensate $n \ll n_c$ the mass current of magnons is given by the traditional equation,

$$\mathbf{J} = \rho_s \mathbf{v}_s \ , \ \mathbf{v}_s = \frac{\hbar}{m_M} \nabla \alpha \ , \ \rho_s(T=0) = n m_M \ . \tag{4.31}$$

In translationally invariant systems, where the mass current coincides with the density of linear momentum, Eq. (4.31) can be obtained directly from the definition of linear momentum density in spin systems:

$$\mathbf{P} = (S - S_z)\nabla\alpha = n\hbar\nabla\alpha \ , \tag{4.32}$$

where we have used the fact that $S - S_z$ and α are canonically conjugated variables. As in conventional superfluids, the superfluid density of the magnon liquid is determined by the magnon density n and magnon mass m_M.

To avoid confusion, let us mention that in magnon BEC the superfluid density describes the coherent precession in a magnetic subsystem and has nothing to do with the superfluid density of the underlying system—the superfluid ^{3}He. In magnon BEC the superfluid mass current (4.31) carries magnons with mass m_M, while in atomic superfluids the superfluid mass current carries atoms. The mass current generated by precessing magnetization in magnon BEC is similar to the electric current generated by precessing magnetization in ferromagnets [40], which is now used in spintronics (see e.g. the recent review [41]).

It is important that the proper atomic analog of magnon BEC is actually the A_1 phase of ^{3}He. Both systems are spin polarized: magnons have spin $-\hbar$ while in ^{3}He-A_1 only the atoms with one spin polarization experience superfluidity [30]. As a result, both the superfluid current of magnons in magnon BEC and the superfluid current of atoms in ^{3}He-A_1 are necessarily accompanied by a superfluid spin current. Since each magnon carries spin $-\hbar$, the magnon mass supercurrent is accompanied by the magnetization supercurrent—the supercurrent of the z-component of spin:

$$J_i^z = -\frac{\hbar}{m_M} J_i = -n\frac{\hbar^2}{m_M}\nabla_i\alpha \ . \tag{4.33}$$

The same is valid for ^{3}He-A_1 where the spin current is $J_i^z = -\frac{\hbar}{2m_3} J_i$.

Let us note here that the density of linear momentum of the spin subsystem is not well defined globally. While the total momentum of the system is conserved, the canonical momenta of the spin and orbital subsystems are not conserved separately [40, 42, 43]. For a particular choice of linear momentum density in (4.32), $\mathbf{P}$ is not defined at the points where the tilting angle of magnetization $\beta = \pi$, because at $\beta = \pi$ spins are stationary and thus the spin precession angle α is ill defined. This is another interesting feature of magnon superfluids, which becomes important for the magnon BEC emerging in normal (non-superfluid) ^{3}He. The latter represents a coherently precessing structure at the interface between the equilibrium domain with $\beta = 0$ and the domain with reversed magnetization, i.e. with $\beta = \pi$ [44].

4.4.4 London limit: hydrodynamics of magnon BEC

In the HPD state of magnon BEC, the magnon density is comparable with the limiting value, $n \approx (5/8)n_{\rm max}$, as a result the kinetic term in the Ginzburg–Landau energy becomes complicated. However, the theory of HPD becomes simple in the London limit, where the magnon superfluidity is described by the hydrodynamic energy functional written in terms of density n and superfluid velocity $\mathbf{v}_s$. This hydrodynamic energy functional is similar to that in superfluid liquids and atomic BEC, but with some important differences. One of these is the anisotropy of magnon mass, which leads to anisotropy of superfluid density, even at $T = 0$. Another difference is the presence of the symmetry breaking term which depends explicitly on α, and which gives rise to the mass of the Goldstone boson.

The hydrodynamic energy functional is expressed in terms of the canonically conjugated variables—number density n (magnon density) and superfluid velocity $\mathbf{v}_{\rm s}$, which is expressed via the gradient of the phase of the Bose condensate α (the phase of the coherent precession of magnetization). It has the following general form:

$$F = \frac{1}{2}\rho_{{\rm s}ij}(n) v_{{\rm s}i} v_{{\rm s}j} + \varepsilon(n) - \mu n + F_{\rm sb}(\alpha, n). \tag{4.34}$$

Here μ, as before, is the chemical potential; $\rho_{{\rm s}ij}$ is the tensor of anisotropic superfluid density, and $v_{{\rm s}i}$ is the superfluid velocity of magnon superfluid:

$$\rho_{{\rm s}ij}(n) = n m_{ij}\,, \quad v_{{\rm s}i} = \hbar \left(m^{-1}\right)_{ij} \nabla_j \alpha\,, \tag{4.35}$$

where the matrix of magnon masses $m_{ij}(n)$ depends on magnon density n and tilting angle of precession. For magnons propagating along the field and in transverse directions, their mass depends on the tilting angle in the following way:

$$\frac{1}{m^{\parallel}(n)} = 2\frac{c_{\parallel}^2 \cos\beta + c_{\perp}^2(1 - \cos\beta)}{\hbar\omega_L}\,, \tag{4.36}$$

$$\frac{1}{m^{\perp}(n)} = \frac{c_{\parallel}^2(1 + \cos\beta) + c_{\perp}^2(1 - \cos\beta)}{\hbar\omega_L}\,, \tag{4.37}$$

where the parameters $c_{\parallel}$ and $c_{\perp}$ are in the order of the Fermi velocity v_F. The mass supercurrent is however isotropic, when it is expressed via α:

$$J_i = \frac{dF}{dv_{{\rm s}i}} = \hbar n \nabla_i \alpha\,, \tag{4.38}$$

which is in agreement with Eq. (4.32) for linear momentum.

Finally $F_{\rm sb}$ in Eq. (4.34) is the symmetry breaking term which depends explicitly on α. It arises only in the case of continuous wave NMR, where it comes from interaction of the condensate with an applied RF field, which is needed to compensate for the relaxation of magnons, see Eq. (4.48). This term explicitly violates the $U(1)$ symmetry which is why it gives rise to the mass of the Goldstone boson, which we discuss later.

The hydrodynamic equations for a magnon superfluid are the Hamilton equations for the canonically conjugated variables n and α:

$$\dot{\alpha} = \frac{\delta F}{\delta n}, \quad \dot{n} = -\frac{\delta F}{\delta \alpha}. \tag{4.39}$$

The superfluid spin current is, as before, determined by the spin to mass ratio for the magnon. But because the magnon mass is anisotropic, the spin current transferred by the coherent spin precession is anisotropic too:

$$J_z^z = -\frac{\hbar^2}{m^{\parallel}(n)} n \nabla_z \alpha, \tag{4.40}$$

$$\mathbf{J}_\perp^z = -\frac{\hbar^2}{m^{\perp}(n)} n \boldsymbol{\nabla}_\perp \alpha. \tag{4.41}$$

The anisotropy of the current in Eqs. (4.40–4.41) is an important modification of conventional Bose condensation, since it is absent in the atomic Bose condensates. The spin supercurrent becomes isotropic, when it is expressed in terms of superfluid velocity:

$$J_i^z = \hbar \left(m^{-1}\right)_{ij} J_j = \hbar n v_{\mathrm{s}i}. \tag{4.42}$$

4.4.5 Goldstone mode of coherent precession—sound in magnon BEC

In atomic superfluids, sound is the Goldstone mode of the spontaneously broken $U(1)$ symmetry. The same sound mode exists in magnon BEC.

The HPD formation corresponds to an energy minimum under the condition of conservation of total longitudinal magnetization of the sample. There is a powerful feedback mechanism that returns the system to the homogeneous precession state after any disturbance. This is the excitation of spin supercurrent transport between non-homogeneous states of the precessing magnetization. Therefore oscillations of the magnetization distribution near the equilibrium HPD state can take place. The frequencies of two modes of such oscillation were calculated first by Fomin [45]. There are torsional oscillations in bulk and surface oscillations (Fig. 4.6). Under experimental conditions both bulk and surface oscillation modes have been observed [39, 46, 47]. A complete review of these experiments can be found in Ref. [33].

The surface oscillations at the domain boundary are analogous to gravity waves on the surface of liquids, whose spectrum is $\omega^2 = gk$, where g is the gravitational field. The role of the gravitational field is played by the gradient of the Zeeman energy, while the kinetic energy of the spin supercurrent plays the role of the kinetic energy of the flow of the liquid. For a

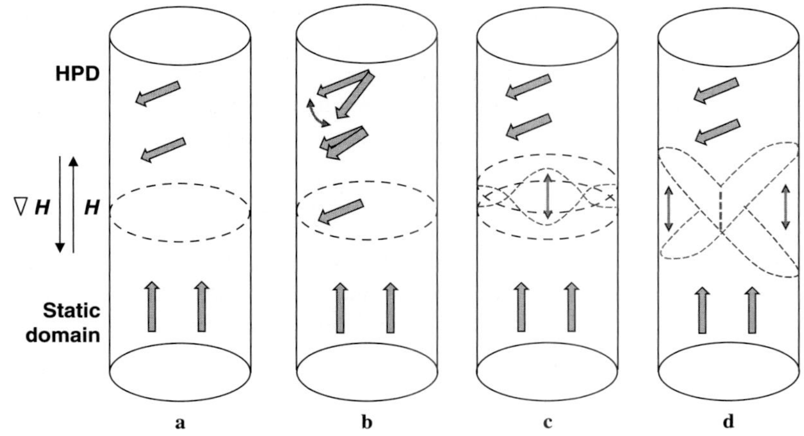

Fig. 4.6 Schematic representation of precessing magnetization in the rotating frame for an equilibrium HPD (a); Goldstone mode of coherent precession—analog of the sound wave in atomic BEC—is the mode of twist oscillations (b); surface oscillations—analogs of gravity waves on the surface of a liquid (c, d).

cylindrical cell, we can visualize these oscillations as the surface waves of water in a glass.

Torsional oscillations originate from the degeneracy of the precessing states with respect to the phase of the precession α. This mode, called the twisting mode, corresponds to spatial oscillations of the phase of the magnetization precession inside the HPD with spin supercurrent feedback response, and thus represents the Goldstone mode of spontaneously broken $U(1)$ symmetry. It is analogous to a sound wave in atomic superfluids. The sound mode in a magnon subsystem is obtained from linearization of the hydrodynamic equations (4.39). It has been calculated in Ref. [45] and identified experimentally in Ref. [39].

There are several conditions required for the existence and stability of magnon BEC. Some of them are the same as for conventional atomic BEC, but there are also important differences, which we discuss later. One of the conditions is that the compressibility β_M of the magnon gas must be positive:

$$\beta_M^{-1} = n\frac{dP}{dn} = n^2\frac{d^2\varepsilon}{dn^2} > 0. \tag{4.43}$$

This condition means that the fourth-order term in the Ginzburg–Landau free energy should be positive, i.e. the interaction between magnons should be repulsive. The magnon interaction energy $\varepsilon(n)$ is provided by spin–orbit (dipole–dipole) interaction. It has a very peculiar form for HPD in ^{3}He-B:

$$\epsilon(n) \equiv E_{\mathrm{so}}(n) = \frac{8\chi\Omega_L^2}{15\gamma^2}\left(\frac{\hbar n}{S} - \frac{5}{4}\right)^2 \Theta\left(\frac{\hbar n}{S} - \frac{5}{4}\right), \tag{4.44}$$

where $\Theta(x)$ is the Heaviside step function; Ω_L is the Leggett frequency (we assume that $\Omega_L \ll \omega_L$). This means that in this state of magnon BEC, a stable coherent precession occurs only at large enough magnon density $n > 5S/4\hbar$, where $d^2E_{\mathrm{so}}/dn^2 > 0$. This corresponds to $\cos\beta < -1/4$. The magnon BEC state in (4.56) also satisfies the condition (4.43), while the magnon condensates in (4.55) and in bulk ^{3}He-A are unstable.

The compressibility of the magnon gas determines the speed of sound propagating in the magnon gas. Since the magnon's mass is anisotropic the phonon spectrum is also anisotropic:

$$\left(c_s^2\right)^{ij} = \left(m^{-1}\right)^{ij} \frac{dP}{dn} = n\frac{d^2 E_{\rm so}}{dn^2} \left(m^{-1}\right)^{ij} . \tag{4.45}$$

In typical experiments with HPD, $\cos\beta$ is close to $-1/4$, i.e. $\cos\beta = -1/4 - 0$. For such a β one has,

$$c_{s\parallel}^2 = \frac{n}{m_\parallel}\frac{d^2 E_{\rm so}}{dn^2} = \frac{2}{3}\frac{\Omega_L^2}{\omega_L^2}\left(5c_\perp^2 - c_\parallel^2\right) , \tag{4.46}$$

$$c_{s\perp}^2 = \frac{n}{m_\perp}\frac{d^2 E_{\rm so}}{dn^2} = \frac{1}{3}\frac{\Omega_L^2}{\omega_L^2}\left(5c_\perp^2 + 3c_\parallel^2\right) . \tag{4.47}$$

Owing to the anisotropy of phonon spectra, the spin wave velocities appear differently in the modes of oscillation in Fig. 4.6. While the frequency of twist oscillations is proportional to $c_{s\parallel}$, the frequency of surface waves is proportional to $\sqrt{c_{s\parallel}c_{s\perp}}$. By experimental investigation of these two modes of oscillation, the experimental group in Kapitza Institute was able to measure the spin wave velocities $c_\parallel$ and $c_\perp$ [33, 47].

4.4.6 Mass of phonons in magnon superfluid

As distinct from conventional superfluids, in magnon BEC one may experimentally introduce the symmetry breaking field which smoothly violates $U(1)$ symmetry and induces a small gap (mass) in the phonon spectrum. This mass has been measured.

The symmetry breaking term appears in continuous wave NMR, when the relaxation of magnon BEC is compensated by the RF field. It describes the interaction $F_{\rm sb}(\alpha, n) = -\gamma \mathbf{H}_{\rm RF} \cdot \mathbf{S}$ of the precessing magnetization with the RF field $\mathbf{H}_{\rm RF}$, which is transverse to the applied constant field $\mathbf{H}$. In continuous wave NMR experiments the RF field prescribes the frequency of precession, $\omega = \omega_{\rm RF}$, and thus fixes the chemical potential μ; while in the state of free precession, the chemical potential μ is determined by the number of pumped magnons. The symmetry breaking term depends explicitly on the phase of precession α with respect to the direction of the RF field in the precessing frame:

$$F_{\rm sb} = -\gamma H_{\rm RF} S_\perp \cos\alpha = -\gamma H_{\rm RF} S \sin\beta\left(1 - \frac{\alpha^2}{2}\right) . \tag{4.48}$$

Due to explicit dependence on α, this term generates the mass of the Goldstone boson (phonon) [48]. For $\cos\beta = -1/4$ the phonon spectrum becomes:

$$\omega_s^2(\mathbf{k}) = \left(c_s^2\right)^{ij} k_i k_j + m_s^2 \ , \ m_s^2 = \frac{4}{\sqrt{15}}\gamma H_{\rm RF}\frac{\Omega_L^2}{\omega_L} . \tag{4.49}$$

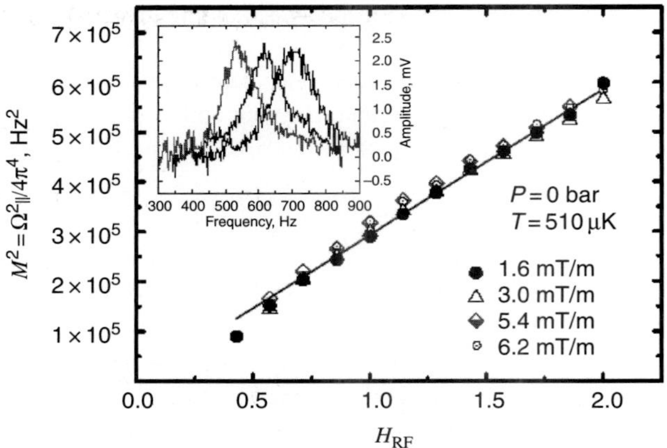

Fig. 4.7 Phonon mass in magnon BEC as a function of the symmetry breaking field. From [50]. (See Plate 11 for color image).

Two experiments with HPD [49, 50] reported the gap in the spectrum of the collective mode of the coherent precession. The measured gap is proportional to $H_{\rm RF}^{1/2}$, in agreement with (4.49) (see Fig. 4.7).

4.4.7 Spin vortex—topological defect of magnon BEC

The phase coherent precession of magnetization in superfluid ^{3}He has all the properties of the coherent Bose condensate of magnons. The main spin-superfluid properties of HPD had already been verified in earlier experiments around 30 years ago, including spin supercurrent, which transports the magnetization (analog of the mass current in conventional superfluids); the spin current Josephson effect, and phase-slip processes at the critical current [51, 55], which we shall discuss later.

The spin current vortex was then observed [57]—a topological defect which is an analog of a quantized vortex in superfluids and of an Abrikosov vortex in superconductors. The precession angle α has 2π winding around the vortex core. In the magnon BEC description, where α is the phase of the magnon condensate, this is the mass current vortex, and since magnons are spin polarized this gives rise to spin current in Eq. (4.41) circulating around the vortex core (see Fig. 4.8).

Since in the central part of a cylindrical cell the phase α changes by 2π around the center, the transverse magnetization $\mathbf{M}_\perp$ is opposite on opposite sides of the cell. In the central part of the cell, i.e. in the vortex core, the magnetization remains vertical and does not precess. The magnon BEC with a spin vortex is created by applying a quadrupole RF field. For this purpose two parts of the saddle NMR coil are connected in opposite directions, so that the phase of the RF field (and consequently the phase α) is opposite at the opposite sides of the cell. Using these NMR coils, practically the same HPD signal was observed as in the conventional arrangement with parallel connection of the coils, although with a slightly reduced amplitude. This shows that HPD is created with opposite α on opposite sides of the cell. To verify this a pair of small pick-up coils were installed at the top of the cell connected in the usual

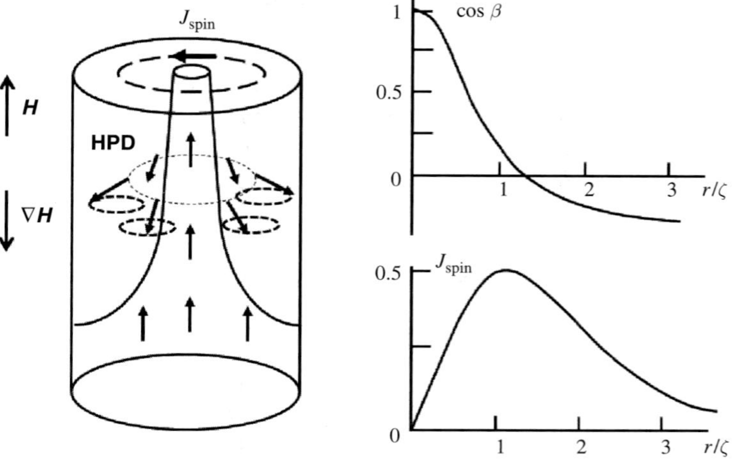

Fig. 4.8 Spin supercurrent vortex in magnon BEC in ^{3}He-B. As in the case of the mass supercurrent vortex in ^{3}He-A, the core of the spin vortex does not have a singularity. The density of spin supercurrent in Eq. (4.41), which is $J_{\text{spin}} \propto (1 - \cos\beta)\nabla\alpha$, goes virtually to zero near the core, as was calculated by Fomin. According to Eq. (4.52), the magnetic coherence length ξ and the size of the vortex core diverge when the HPD domain boundary is approached where the local Larmor frequency $\omega_L = \omega$.

way. When the RF field was switched off, the pick-up coils received a very small RF signal from HPD, while the frequency of this signal corresponded to the full HPD signal. This means that HPD generated the signal with the opposite phase at the two sides of the pick-up coils, which nearly compensated each other (see Fig. 4.9). This corresponds to HPD with a circular gradient of α, as shown in Fig. 4.8. The magnetization is oriented vertically in the vortex core. On the periphery of the cell it precesses with tipping angle 104°, and with 2π phase winding around the center. This type of HPD should radiate at a frequency, which corresponds to the Larmor field on the boundary of HPD, but should not produce any signal in the pick-up coil. A small signal appears due to asymmetry of the pick-up coil; oscillations of this signal may correspond to nutations of the vortex core.

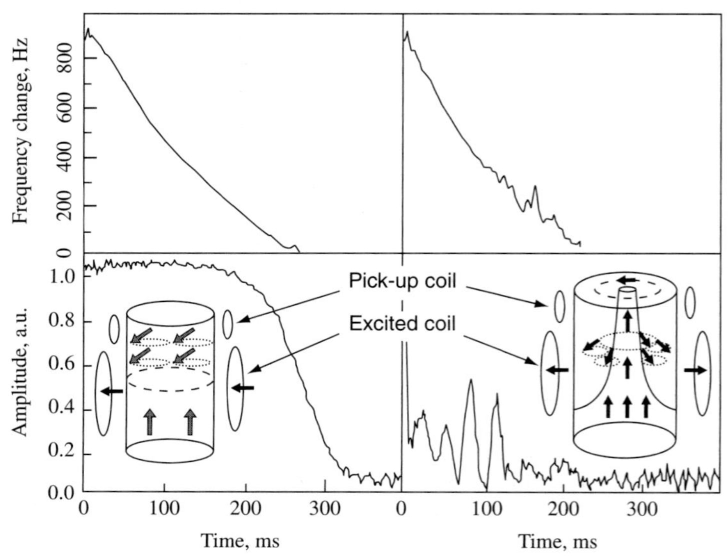

Fig. 4.9 NMR signature of spin vortex in magnon BEC in ^{3}He-B. Frequency (top) and amplitude (bottom) of HPD induction decay measured by a pick-up coil. The HPD was maintained with an RF field from parallel connected coils (left) and oppositely (quadrupole) connected coils (right) at $P = 29.3$ bar, $T = 0.5\,T_c$. After the quadrupole excitation the frequency decays with time in the same way as in conventional vortex-free precession, while the amplitude of the HPD signal is nearly zero because of compensation of signals from opposite sides of the cells, where the phase α of precession differs by π.

4.4.8 Critical velocities, coherence length, and the vortex core radius

In atomic BEC the speed of sound determines the coherence length, the size of the vortex core, and the Landau critical velocity of flow at which phonons are created:

$$v_{\mathrm{L}} = c_s \ , \ \ r_{\mathrm{core}} \sim \xi \sim \frac{\hbar}{mc_s} . \tag{4.50}$$

The extension to the magnon BEC would suggest that the coherence length and the size of the vortex core in the HPD state should be in the order of,

$$r_{\mathrm{core}} \sim \frac{\hbar}{m_M c_s} \sim \frac{c}{\Omega_L} . \tag{4.51}$$

However, this naive extension does not work, and Eq. (4.27) gives only the lower bound on the core size. The core is larger due to the specific profile of the Ginzburg–Landau (dipole) energy in Eq. (4.44) which is strictly zero for $\cos\beta > -1/4$. This leads to the special topological properties of coherent precession (see Ref. [58]). As a result the spin vortex created and observed in Ref. [57] has a continuous core with broken symmetry, similar to vortices in superfluid ^{3}He-A [59]. The size of the continuous core is determined by the proper coherence length [60], which can be found from the competition between the first two terms in the Ginzburg–Landau free energy in Eq. (4.22):

$$r_{\mathrm{core}} \sim \xi \sim \frac{\hbar}{\sqrt{m_M(\mu - \omega_L)}} \sim \frac{c}{\sqrt{\omega_L(\omega - \omega_L)}} . \tag{4.52}$$

This coherence length also determines the critical velocity for creation of vortices:

$$v_c \sim \frac{\hbar}{m_M r_{\mathrm{core}}} . \tag{4.53}$$

It is smaller than the Landau critical velocity for creation of phonons, which in the case of isotropic sound is

$$v_{\mathrm{L}} = c_s \, . \tag{4.54}$$

The Landau criterion for the onset of phonon radiation in the case of anisotropic speed of sound has been derived in Ref. [61].

For large tipping angles of precession the symmetry of the vortex core is restored: the vortex becomes singular with the core radius $r_{\mathrm{core}} \sim c/\Omega_L$ in (4.51) [62].

Other topological defects possible in the coherent precession beyond the Ginzburg–Landau model of magnon BEC are discussed in [58]. In BEC of excitations, the topological defects have been detected in exciton–polariton condensates [63]. Among these are the half-quantum vortices—vortices with half of the circulation quantum. Half-quantum vortices are topologically stable in the superfluid ^{3}He-A [64], but they still remain elusive there.

4.4.9 Spin supercurrent transport

The next step in investigations of the magnon BEC was experimental studies of spin supercurrent between two independent HPD states, connected by a channel which was either perpendicular [33, 51, 52] or parallel to the magnetic field [53].

In the first case the steady state spin supercurrent was created between two magnon condensates formed in two different cells. The two cells were connected by a channel of diameter 1.4 mm. The HPD states were formed in both cells by a CW NMR. The frequency of the RF field ω was chosen to be slightly above the local Larmor frequency ω_L in the channel; this determines the coherence length ξ of magnon BEC in (4.52). The HPD penetrates in a channel as shown in Fig. (4.10). Then, one slightly changes the frequency in one of the cells. The gradient of phase of precession appears which leads to a growing spin supercurrent in the channel, which is proportional to the difference between the phases of the two HPDs. This is the current of magnons, which transports the magnetization and consequently the Zeeman energy from one cell to another. As a result, in one of the cells the density of magnons decreases and the HPD starts to absorb more RF energy to compensate the extra losses. Contrarily, in the other cell more magnons appear and thus the absorbed energy decreases. At some conditions the transport of magnons becomes so large, that absorption transforms to radiation from the HPD. This means that the apparatus starts to operate as a spin supercurrent transformer, which transports the RF signal from one coil to another. By changing the amplitude of the RF field one can measure, directly, the value of the spin supercurrent in the channel. If one removes the difference in frequencies of the two HPD states (the difference in chemical potentials of the two magnon condensates), the spin supercurrent

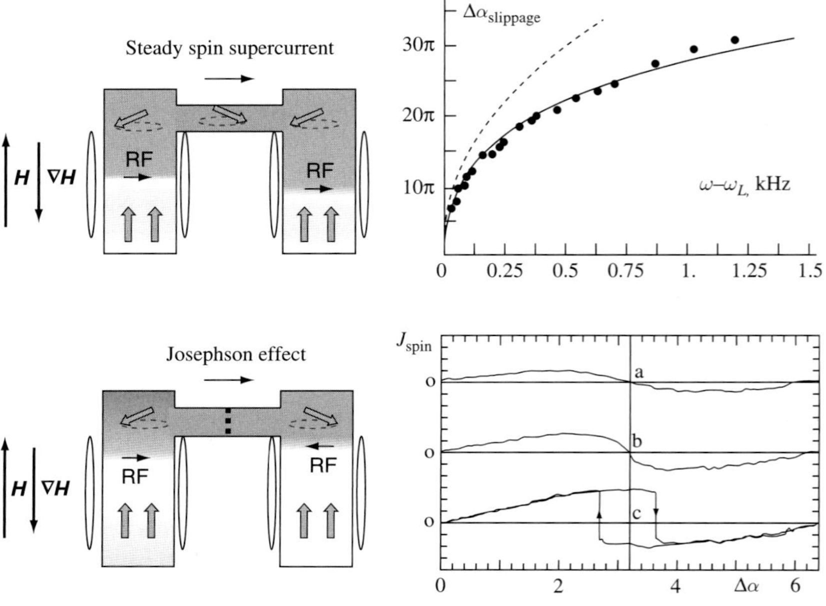

Fig. 4.10 Illustration of the experimental observation of spin supercurrent between two HPD states (*top left*) and the Josephson effect (*bottom left*). The measured critical spin current in the channel as a function of $\omega - \omega_L$ and thus as a function of the coherence length ξ in (4.52) (*top right*). For observation of the dc and ac Josephson effects an orifice of diameter about 0.4 mm was installed (*bottom left*). The Josephson effect for magnon BEC demonstrates the interference between two magnon condensates. The spin current as a function of the phase difference across the junction, $\alpha_2 - \alpha_1$, where α_1 and α_2 are phases of precession in two coherently precessing domains (*bottom right*). Different experimental records correspond to a different ratio between the diameter of the orifice and the coherence length ξ of magnon BEC. The pure dc Josephson phenomenon was observed for magnetic coherent length $\xi = 1.3$ mm (a), and the distorted one for $\xi = 0.8$ mm (b). Phase slippage processes were observed for $\xi = 0.7$ mm (c).

remains stable and is proportional to the difference of phase between the condensates. By increasing the phase difference one is able to reach a critical current, at which the spin supercurrent starts losing the 2π windings. These phase slippages were measured as a function of $\omega - \omega_L$ in the channel and thus as a function of the coherence length ξ, in good agreement with theory, see Fig. 4.10 *top right*. Both 2π slippage cases as well as multi 2π slippage were observed. In some cases the 2π slippage was separated by two independent slippages, which can be explained by the formation of a spin-current vortex [33] or π-soliton inside the channel as the intermediate state.

4.4.10 Spin-current Josephson effect

The Josephson effect is the response of the current to the phase between two weakly connected regions of coherent quantum states. It was described by Josephson [54] for the case of two quantum states, separated by a potential barrier. This phenomenon is usually studied for the case of quantum states connected by a conducting bridge with dimensions smaller than the coherence length. In this case the coherent state in the bridge cannot be established so there is no phase memory, which determines the direction of the phase gradient. As a result the supercurrent is determined only by the phase difference between the two states. As the dimensions of the conducting bridge increase, a more complex current–phase relation is observed. For bridge dimensions of the order of the coherence length, a transition to a hysteretic scenario with phase slippage appears.

In the case of mass and electronic supercurrents the coherence length is a function of the temperature. In the case of spin supercurrents, however, the Ginzburg–Landau coherence length ξ is not only a function of temperature, but also a function of the difference between the HPD precession frequency and the local Larmor frequency, according to Eq. (4.52). This quantity can be varied experimentally with a magnetic field gradient or with the position of the domain boundary. As a result one is able to change the coherence length in the region of the orifice in the channel and observe the change from canonical current–phase relation to phase slip behavior. This experiment performed in the Kapitza Institute [33, 55, 56] is presented schematically in Fig. 4.10. The orifice, of diameter 0.48 mm, was placed in the central part of the channel. The current–phase characteristics, observed in this experiment are represented for different positions of the domain boundary related to the orifice. One can easily see that the current in Fig. 4.10 (*bottom right*) (a) corresponds to the canonical current–phase relation, which transforms to the non-linear relation in (b) and then to a phase slip phenomenon in (c).

4.4.11 Other states of magnon BEC in ^{3}He-B

Recent experiments in ^{3}He-B allowed us to probe the BEC states that emerge in the valley on the other side of the energy barrier in Fig. 4.2. This became possible by immersing the superfluid ^{3}He in a very porous material called aerogel. By squeezing or stretching the aerogel sample, one creates a global anisotropy which captures the orbital vector $\hat{\mathbf{l}}$. This allows orientation of the orbital vector $\hat{\mathbf{l}}$ in the desirable direction with respect to the magnetic field [65, 66].

For the transverse orientation of $\hat{\mathbf{l}}$, i.e. for $l = 0$, two new BEC states have been identified. One of them exists at $|\Psi|^2 < S/\hbar$ and has the following form of spin–orbit interaction obtained from Eq. (4.23) (we omit the constant term for simplicity):

$$F_{\mathrm{so}}(\Psi)_{l=0} = -\frac{\chi}{4\gamma^2}\Omega_L^2\left(\frac{|\Psi|^2}{S} - \frac{4}{5}\right)^2, \quad |\Psi|^2 < \frac{S}{\hbar}. \tag{4.55}$$

This state has an attractive interaction between magnons, and is unstable, since the compressibility β_M of the magnon gas in (4.43) is negative: $d^2\varepsilon/dn^2 < 0$.

The other state exists at $|\Psi|^2 > S/\hbar$ and has the following form of spin–orbit interaction:

$$F_{\mathrm{so}}(\Psi)_{l=0} = \frac{\chi}{20\gamma^2}\Omega_L^2\left(\frac{|\Psi|^2}{S} - 2\right)^2, \quad |\Psi|^2 > \frac{S}{\hbar}. \tag{4.56}$$

This state has a repulsive interaction between magnons and is stable. The magnon BEC formation under these conditions has been observed [67].

4.5 Magnon BEC in ^{3}He-A

As in the case of ^{3}He-B, all the information on the ^{3}He-A order parameter that is needed to study the coherent precession is encoded in the spin–orbit interaction.

4.5.1 Instability of magnon BEC in bulk ^{3}He-A

For ^{3}He-A, the spin–orbit interaction averaged over the fast precession has the following form [70]:

$$F_{\mathrm{so}}(|\Psi|) = \frac{\chi\Omega_L^2}{4\gamma^2}\left[-2\frac{|\Psi|^2}{S} + \frac{|\Psi|^4}{S^2} + \left(-2 + 4\frac{|\Psi|^2}{S} - \frac{7}{4}\frac{|\Psi|^4}{S^2}\right)(1 - l^2)\right] \tag{4.57}$$

In a static bulk ^{3}He-A, when $\Psi = 0$, the spin–orbit energy F_{so} in Eq. (4.57) is minimized when the orbital vector $\hat{\mathbf{l}}$ is perpendicular to the magnetic field, i.e. for $l = 0$. Then one has

$$F_{\mathrm{so}}(|\Psi|, l = 0) = \frac{\chi\Omega_L^2}{4\gamma^2}\left[-2 + 2\frac{|\Psi|^2}{S} - \frac{3}{4}\frac{|\Psi|^4}{S^2}\right], \tag{4.58}$$

with a negative quartic term. The attractive interaction between magnons destabilizes the BEC, which means that homogeneous precession of magnetization in ^{3}He-A becomes unstable. This instability, predicted by Fomin [68], was confirmed experimentally [69].

However, as follows from (4.57), at sufficiently large magnon density $n = |\Psi|^2$,

$$\frac{8+\sqrt{8}}{7}S > n > \frac{8-\sqrt{8}}{7}S, \tag{4.59}$$

the factor in front of $1 - l^2$ becomes negative. Therefore it becomes energetically favorable to orient the orbital momentum $\hat{\mathbf{l}}$ along the magnetic field, $l = 1$. For this orientation one obtains the Ginzburg–Landau free energy with

$$F_{\rm so}\left(|\Psi|, l = 1\right) = \frac{\chi\Omega_L^2}{4\gamma^2}\left[-2\frac{|\Psi|^2}{S} + \frac{|\Psi|^4}{S^2}\right]. \tag{4.60}$$

This corresponds to the conventional Ginzburg–Landau free energy in atomic BEC. The quadratic term modifies the potential U; the quartic term is now positive.

In the language of BEC, this means that, with increasing density of Bose condensate, the originally attractive interaction between magnons should spontaneously become repulsive when the critical magnon density $n_c = S(8 - \sqrt{8})/7$ is reached. If this happens, the magnon BEC becomes stable, and in this way the state with spontaneous coherent precession could be formed [70]. This self-stabilization effect is similar to the effect of Q-ball, where bosons create the potential well in which they condense (we shall discuss the Q-ball phenomenon in magnon BEC later in Section 4.6.1). However, such a self-sustaining BEC with originally attractive boson interaction has not been achieved experimentally in bulk ^{3}He-A, most probably because of the large dissipation, as a result of which the threshold value of the condensate density has not been reached.

4.5.2 Magnon BEC of ^{3}He-A in deformed aerogel

Finally, the fixed orientation of the orbital vector $\hat{\mathbf{l}}$ has been achieved in ^{3}He-A confined in aerogel—a material with high porosity (about 98% of volume). Silicon strands of aerogel play the role of impurities with local anisotropy along the strands. According to the Larkin–Imry–Ma effect, the random anisotropy suppresses the orientational long-range order of the orbital vector $\hat{\mathbf{l}}$; however, when the aerogel sample is deformed the long-range order of $\hat{\mathbf{l}}$ is restored [71]. Experiments with globally squeezed aerogel [65] demonstrated that a uni-axial deformation by about 1% is sufficient for global orientation of the vector $\hat{\mathbf{l}}$ along the anisotropy axis. When a magnetic field is also oriented along the anisotropy axis one obtains the required geometry with $l = 1$, at which the magnon BEC in ^{3}He-A becomes stable. The first indication of coherent precession in ^{3}He-A was reported in [72, 73] and confirmed in [74]. Contrary to the unconventional magnon BEC in the form of HPD in ^{3}He-B, the magnon BEC emerging in superfluid ^{3}He-A is in one-to-one correspondence with atomic BEC, see Fig. 4.11. For $\mu > U$, the condensate density determined from equation $dF/dn = \mu$ grows continuously from zero as $n \propto \mu - U$.

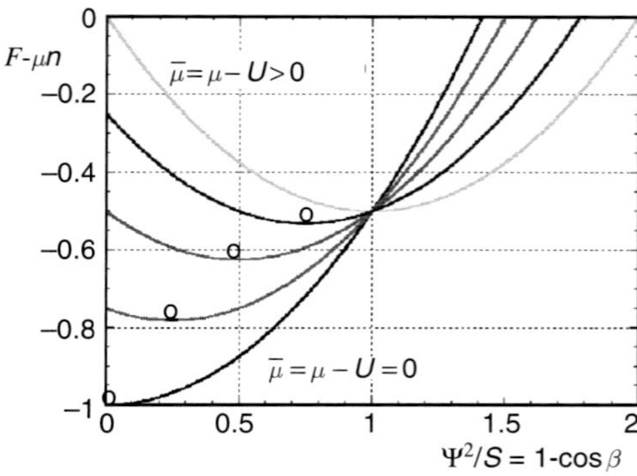

Fig. 4.11 $F - \mu n$ for different values of the chemical potential $\mu \geq U$ in magnon BEC in ^{3}He-A. Magnon BEC in ^{3}He-A is similar to BEC in atomic gases.

For $l = 1$ the Ginzburg–Landau free energy acquires the standard form,

$$F = \int d^3r \left(\frac{|\nabla\Psi|^2}{2m} + (\omega_L(\mathbf{r}) - \mu)|\Psi|^2 + \frac{1}{2}b|\Psi|^4 \right), \tag{4.61}$$

where we have modified the chemical potential by the constant frequency shift

$$\mu = \omega + \frac{\Omega_L^2}{2\omega}, \tag{4.62}$$

and the parameter b of repulsive magnon interaction is

$$b = \frac{\Omega_L^2}{2\omega S}. \tag{4.63}$$

At $\mu > \omega_L$, magnon BEC must be formed with density

$$|\Psi|^2 = \frac{\mu - \omega_L}{b}. \tag{4.64}$$

This is distinct from ^{3}He-B, where condensation starts with finite condensate density. Equation (4.64) corresponds to the following dependence of the frequency shift on tipping angle β of coherence precession:

$$\omega - \omega_L = -\frac{\Omega_L^2}{2\omega} \cos\beta. \tag{4.65}$$

The final proof of the coherence of precession in ^{3}He-A in aerogel was the observation of free precession after a pulsed NMR and also after a switch-off of the CW NMR [74]. In conclusion, in contrast to the homogeneously precessing domain (HPD) in ^{3}He-B, the magnon Bose condensation in ^{3}He-A obeys the standard Gross–Pitaevskii equation. In bulk ^{3}He-A, the Bose condensate of magnons is unstable because of the attractive interaction between magnons. In ^{3}He-A confined in aerogel, a repulsive interaction is achieved by proper deformation of the aerogel sample, and the Bose condensate becomes stable.

4.6 Magnon BEC in magnetic trap and MIT bag

4.6.1 Magnon BEC in the form of Q-ball

There are many new physical phenomena related to the Bose condensation of magnons, which have been observed following the discovery of HPD. These include in particular compact objects—coherently precessing states trapped by orbital texture [75]. At a small number N of pumped magnons, the system is similar to the Bose condensate of the ultracold atoms in harmonic traps, while at larger N the analog of the Q-ball in particle physics develops [76].

A Q-ball is a non-topological soliton solution in field theories containing a complex scalar field Ψ. Q-balls are stabilized through conservation of the global $U(1)$ charge Q [77]. They are formed as a result of suitable attractive interaction that binds the quanta of the Ψ-field into a large compact object. In some modern SUSY scenarios Q-balls are considered as heavy particle-like objects, with Q being the baryon and/or lepton number. For many conceivable alternatives, Q-balls may contribute significantly to the dark matter and baryon contents of the universe, as described in review [78]. Stable cosmological Q-balls can be searched for in existing and planned experiments [79].

The Q-ball is a rather general physical object, which in principle can be formed in condensed matter systems. In particular, Q-balls were suggested in the atomic Bose–Einstein condensates [80]. In ^{3}He-B, Q-balls are formed as special states of phase coherent precession of magnetization. The role of the Q-charge is played by the projection $\mathcal{S}_z$ of the total spin of the system on the axis of the magnetic field, which is a rather well conserved quantity at low temperature, or which is the same the magnon number $\mathcal{N}$. At the quantum level, this Q-ball is a compact object formed by magnons—quanta of the corresponding Ψ-field.

In ^{3}He-B the Q-balls are formed at low temperatures, when homogeneous magnon BEC in the form of a Homogeneously Precessing Domain (HPD) becomes unstable due to parametric Suhl instability [81–83], which we shall discuss later in Section 4.9.4. At low temperatures the condensate can only be formed in a trap, similar to that in atomic gases [16], and the Q-balls are either formed in these traps or dig their own trap.

Experimentally, the Q-ball in ^{3}He-B [76] manifests as a long-lived ringing (of up to an hour!) of the free induction decay after a NMR tipping pulse [84, 85]. In a steady state it can be maintained by CW RF pumping [87, 88] (Fig. 4.14), and even by off-resonance excitation [75, 86]. Detailed experimental investigations of Q-balls formed in specially prepared and well-controlled traps were made in [89].

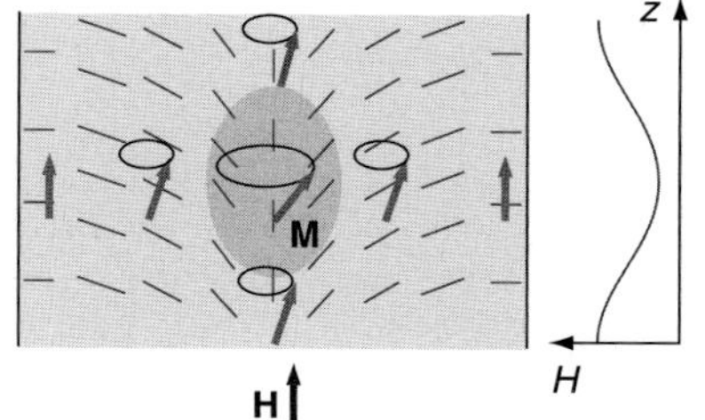

Fig. 4.12 The trapping potential Eq. (4.67) used in [89] is formed in the cylindrically symmetric "flare-out" texture of the orbital angular momentum **l** (dashed lines) in a shallow minimum of the vertical magnetic field (*right*). The arrows represent the precessing magnetization $\mathcal{M}$, which precesses coherently within the condensate droplet (darker gray). The radial texture is manipulated by rotation.

4.6.2 Magneto-textural trap for magnon BEC in ^{3}He-B

A cylindrically symmetric magnetic trap for magnon BEC in ^{3}He-B is shown schematically in Fig. 4.12 [89]. The confinement potential $U_{\parallel}(z) = \gamma H(z)$ in the axial direction is produced by local perturbation of the magnetic field with a small pinch coil. In the radial direction, the well $U_{\perp}(r)$ is formed by the cylindrically symmetric flare-out texture of the orbital vector **l**. This comes from the spin–orbit interaction energy (4.23), which enters the Ginzburg–Landau functional. The relevant term in Eq. (4.23), which is responsible for the radial potential, is

$$U_{\perp}(r)|\Psi|^2 = \frac{2\Omega_{\mathrm{L}}^2}{5\omega_{\mathrm{L}}}\left(1 - l(r)\right)|\Psi|^2 . \tag{4.66}$$

Here, as before, $l = \hat{\mathbf{l}} \cdot \hat{\mathbf{H}} \equiv \cos\beta_L$ describes the orientation of the unit vector $\hat{\mathbf{l}}$ with respect to the direction $\hat{\mathbf{H}}$ of the magnetic field. On the side wall of the cylindrical container the orbital momentum $\hat{\mathbf{l}}$ is normal to the wall, while in the center it is parallel to the axially oriented applied magnetic field. This produces a minimum of the potential $U_{\perp}(r)$ on the cylinder axis. So the total confinement potential is

$$U(\mathbf{r}) = U_{\parallel}(z) + U_{\perp}(r) = \omega_{\mathrm{L}}(z) + \frac{2\Omega_{\mathrm{L}}^2}{5\omega_{\mathrm{L}}}\left(1 - l(r)\right) . \tag{4.67}$$

Close to the axis the polar angle β_L of the $\hat{\mathbf{l}}$-vector varies linearly with distance r from the axis [59]. As a result, the potential $U(\mathbf{r})$ reduces to that of a normal harmonic trap used for the confinement of dilute Bose gases [16]:

$$U(\mathbf{r}) = U(0) + \frac{m_{\mathrm{M}}}{2}\left(\omega_z^2 z^2 + \omega_r^2 r^2\right), \tag{4.68}$$

where we also take into account that the axial trap is close to harmonic. This is checked by measurement of the spectrum of magnons—standing spin waves in the trap, which has equidistant levels:

$$\omega_{\mathrm{nm}} = \omega_{\mathrm{L}}(0) + \omega_r(\mathrm{n} + 1) + \omega_z(\mathrm{m} + 1/2) . \tag{4.69}$$

The oscillator frequency ω_z of the well in the axial direction can be regulated by the field in the pinch coil (the equidistant levels of spin precession localized around the minimum of the external magnetic field have been derived in Ref. [90] using the full set of equations for spin dynamics in ^{3}He-B). The frequency ω_r in the radial direction can be adjusted by applying rotation, since the vortex-free superfluid flow or the array of rectilinear vortex lines created by rotation modifies the flare-out texture $\hat{\mathbf{l}}(r)$.

4.6.3 Ground-state condensate and self-localization

In atomic BEC the condensate is formed in the ground state (0,0) of the trap. When the number of atoms $\mathcal{N}$ in the ground state increases, the interactions between atoms become important and the condensate wave function starts deviating from the Gaussian form of an ideal gas. However, as distinct from a system of cold atoms, the peculiarity of the magnon Ginzburg–Landau functional in Eq. (4.23) is that the prefactor of the quartic term, which describes the interactions between the magnons, is not a constant, but $\propto (1 - l(r))^2|\Psi|^4 \propto r^4|\Psi|^4$. It is small in the region of the trap and can be neglected.

Under conditions of experiment, the main effect is caused not by the magnon-magnon interactions, but by interaction of magnons with $\hat{\mathbf{l}}$-field, which leads to the self-localization discussed in [76]. At a high density of magnons they start to influence the radial $\hat{\mathbf{l}}$-texture. According to (4.66) the condensation of Ψ in the trap leads to the preferable orientation of $\hat{\mathbf{l}}$ parallel to the magnetic field, $l = 1$, in the region of the trap. As a result the potential well becomes wider and the energy of the level in the trap decreases, and at large $\mathcal{N}$ the harmonic trap gradually transforms to a box with $\beta_{\rm L} \approx 0$ within which magnons are localized. This allows incorporation of more magnons at this same level by sweeping up the frequency. This is equivalent to effective attractive interaction between magnons induced by exchange of the quanta of the $\hat{\mathbf{l}}$-field.

In the language of relativistic quantum fields, this is a particular representation of the Q-ball [91], in which self-localization is caused by interaction between the charged field (magnon field Ψ) and the neutral field ($\hat{\mathbf{l}}$-field), where the neutral field provides the potential for the charged one. In the process of self-localization the charged field locally modifies the neutral field so that a potential well is formed in which the charge is condensed. We remember that the charge Q corresponds to the spin $S - S_z$ or, equivalently, to the magnon number $\mathcal{N}$.

4.6.4 Localization with formation of a box: analog of electron bubble and MIT bag

The phenomenon of self-localization with formation of a box in [89] is not unique in nature. Other examples of self-formation of a box-like trapping potential are the electron bubble in liquid helium and the MIT bag model of a hadron [92], where the asymptotically free quarks are confined within a cavity surrounded by the QCD vacuum.

The MIT bag model has been used for construction of different hadrons, including mesons, baryons, and even multiquark hadrons, such as tetraquarks [93] and pentaquarks [94]. In the MIT bag model, free quarks are forced to move only inside a given spatial region, within which they occupy single-particle orbitals. The MIT bag is described by the following energy whose minimization determines the equilibrium radius R of a given hadron:

$$E(R) = \sum_a N_a \sqrt{m_a^2 c^4 + \frac{\hbar^2 c^2 x_a^2}{R^2}} + F(R), \quad F(R) = B\frac{4\pi R^3}{3}. \tag{4.70}$$

Here the first term is the kinetic energy of quarks with mass m_a in the cavity of radius R, where the dimensionless parameters x_a are determined by the boundary conditions for fermions on the boundary of the bag and radial quantum numbers. For the fermion in the ground state in the box, and in the ultra-relativistic limit of vanishing fermionic masses, $m_a \to 0$, the parameter $x = 2.04$. The second term is the potential energy, B is the so-called bag constant which reflects the bag pressure. At zero temperature, the bag constant B is the difference in energy density between the false vacuum inside the bag (the deconfinement phase) and the true QCD vacuum outside (the confinement phase). In the non-relativistic limit, ignoring the term which does not depend on R, one gets

$$E(R) = \sum_a N_a \frac{\hbar^2 x_a^2}{2m_a R^2} + F(R). \tag{4.71}$$

The same equation describes the electron bubble in superfluid ^{4}He, where m is the electron mass; $x = \pi^2$ for the ground state level; the potential energy $F(R) = (4\pi/3)R^3P + 4\pi\sigma R^2$, with P being the external pressure and σ the surface tension. For extension to the multi-electron bubbles in superfluid ^{4}He, see Ref. [95].

This consideration is applicable for magnon BEC. In the harmonic trap for magnons presented in Fig. 4.12 the flexible texture of the orbital momentum $\hat{\mathbf{l}}$ of Cooper pairs in our analogy either plays the role of the pion field or that of the non-perturbative gluonic field, depending on the microscopic structure of the confinement phase. The trap is modified by pumped magnons due to spin–orbit interaction in (4.66), which repels the $\hat{\mathbf{l}}$-field from the region, where magnons are localized. At a large number $\mathcal{N}$ of magnons in the trap the system becomes similar to a MIT bag with cavity, free from the orbital field, which is occupied by magnons. So magnons, like quarks, dig a hole, pushing the orbital field away due to repulsive interaction, see Fig. 4.13. The main difference from the MIT bag model is that magnons are bosons and may macroscopically occupy the same energy state in the trap, forming a Bose-condensate, while in the MIT bag, the number of fermions on the same energy level is limited by the Pauli principle. The bosonic bag becomes equivalent to the fermionic bag in the limit of a large number of quark flavors, when $\mathcal{N} \gg 1$ quarks may occupy the same level.

In experiments, the trap is elongated, so, without losing generality, we may consider the 2D approximation, i.e. the 2D cylindrical trap. In the limit of large $\mathcal{N}$, the radius R of the cavity filled with $\mathcal{N}$ magnons occupying the quantum state with radial number n_r is determined by a balance of two terms in the total energy of the bag [89]:

$$E(R, n_r) = \mathcal{N}\varepsilon_{n_r}(R) + F(R), \quad \varepsilon_{n_r}(R) = \frac{\hbar^2\lambda^2_{n_r+1}}{2m_{\rm M}R^2} . \tag{4.72}$$

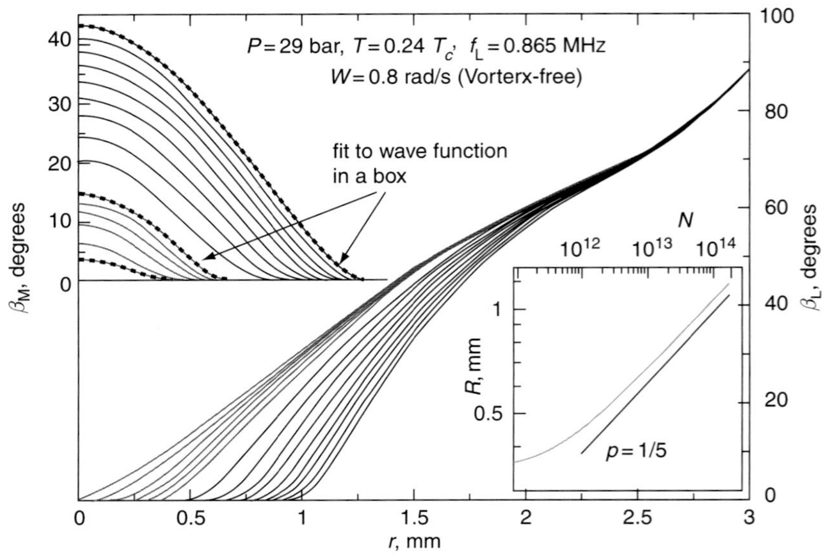

Fig. 4.13 Calculations of the multi-magnon bubble in a 2D textural trap using the magnon BEC approximation (see Ref. [89]). The condensate is formed in the ground state, $n_r = 0$. The deflection angles of the magnetization, $\beta_{\rm M}$ and of the textural anisotropy axis, $\beta_{\rm L}$ are plotted as a function of radius for different condensate populations. The population increases from bottom to top in the upper plot and from left to right in the lower plot. When magnon occupation increases, the magnon wave function suppresses the orbital texture $\beta_{\rm L}$ and the potential well transforms towards a box with impenetrable walls. Fit to the wave function of the condensate in a box, which is nullified at the box boundary, is shown in the upper plot. The effective radius of the box obtained from such fits is shown in the *insert* as a function of the magnon occupation number $\mathcal{N}$. Slope $p = 1/(k+2)$ from Eq. (4.74) with $k = 3$ is shown for comparison.

The first term on the right-hand side is the magnon zero-point energy in the radial cavity. This is the magnon number $\mathcal{N}$ in the Bose condensate times the energy ε_{n_r} of a single magnon on the radial level n_r in the cylindrical box with impenetrable walls. In NMR experiments, where a homogeneous RF field is used, only the energy levels with zero azimuthal quantum number are excited, which corresponds to $n = 2n_r$ in Eq. (4.69). These are measured as the shift of the frequency of the NMR peak, corresponding to excitation of a magnon, with respect to the Larmor frequency ω_L:

$$\Delta\omega = \omega - \omega_L = \frac{\varepsilon_{n_r}}{\hbar} = \frac{\hbar\lambda^2_{\mathrm{n_r}+1}}{2m_M R^2} \,. \tag{4.73}$$

Here, m_M is the magnon mass in Eq. (4.10), and the parameter x in (4.76) equals the $n_r + 1$-th root of the Bessel function, $x = \lambda_{\mathrm{n_r}+1}$, which corresponds to the proper boundary condition for magnons populating the radial level n_r in the impenetrable box. The potential energy $F(R)$ in (4.72) corresponds to the pressure exerted to the bag by the field of the orbital texture, which is expelled from the bag. It is the difference in energy of the orbital field texture with and without the cavity.

Experimental results for the ground state magnon condensate in [89] demonstrated that they can be reproduced by the phenomenological equation (4.72) if one assumes that there is a scaling law $F(R) \propto R^k$. Minimization of the phenomenological equation Eq. (4.72) with respect to R suggests that at large $\mathcal{N}$ the radius of localization approaches the asymptote:

$$R(\mathcal{N}) \sim a_r \,(\mathcal{N}/\mathcal{N}_{\mathrm{c}})^{1/(k+2)}, \;\; \mathcal{N} \gg \mathcal{N}_{\mathrm{c}} \,, \tag{4.74}$$

where a_r is the harmonic oscillator length in the original radial trap (at $\mathcal{N} \ll \mathcal{N}_{\mathrm{c}}$), $\mathcal{N}_{\mathrm{c}}$ is the characteristic number at which the scaling starts. In experiments, the dependence of the transverse magnetization $\mathcal{M}_\perp$ on the frequency shift $\Delta\omega$ is measured. As distinct from the magnon number $\mathcal{N} = \int d^2r|\Psi|^2$, where Ψ is the wave function of the magnon condensate, the transverse magnetization density represents the order parameter and is proportional to Ψ, see (4.3). The total magnetization is thus $\mathcal{M}_\perp \propto \int d^2r|\Psi| \propto \mathcal{N}^{1/2}R$. Since $\Delta\omega \propto 1/R^2$ according to (4.73), one obtains $\mathcal{M}_\perp \propto (\Delta\omega)^{-1-k/4}$. The measured transverse magnetization suggests that for large $\mathcal{N}$ the scaling law is approached with $k \approx 3$, see insert in Fig. 4.14. The magnetization estimated in numerical simulations also suggests that, under experimental conditions, the $k = 3$ scaling is the reasonable fit [89], see insert in Fig. 4.13, which is just the scaling corresponding to the MIT bag model.

In the above approach the energy consideration has been used, where the energy potential is obtained by averaging over fast precession. Numerical simulations of the Q-ball, using the full dynamical equations, can be found in [75].

4.6.5 Comparison with atomic BEC in a trap

Incidentally, for an atomic condensate in a harmonic trap the radius R as a function of number of atoms at large N also approaches the scaling in Eq. (4.74)

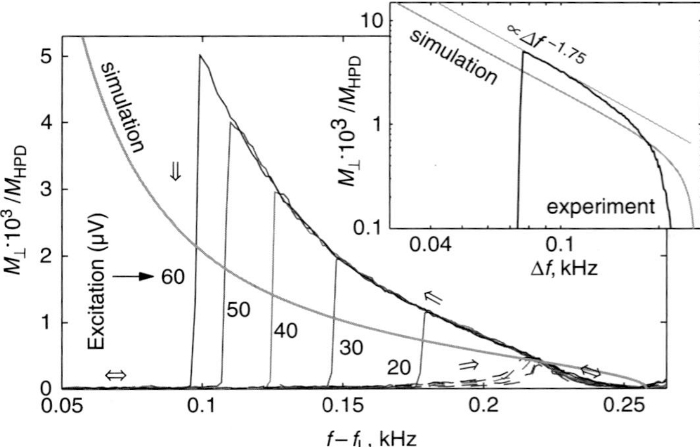

Fig. 4.14 Formation of a magnon condensate droplet in the ground state $n_r = 0$, which corresponds to $n = 0$ in the original trap in Fig. 4.12, in cw NMR measurement. The condensate magnetization $M_\perp$ precessing in the transverse plane is plotted on the vertical axis, normalized to that when the *homogeneously precessing domain* (HPD) fills the volume within the detector coil. The arrows indicate the sweep direction of the applied rf frequency $f = \omega/(2\pi)$. $M_\perp$ grows when the frequency is swept down. Only a tiny response is obtained on sweeping in the opposite direction. During the downward frequency sweep the condensate is destroyed (vertical lines), when energy dissipation exceeds rf pumping. This point depends on the applied rf excitation amplitude, marked on each line. The lower line denoted as simulation represents the result of calculations from Fig. 4.13. The calculations have no fitting parameters and the difference in the measurements can be attributed to the experimental uncertainty in determining $M_\perp$ for the vertical scaling. *(Insert)* The experimental curve for the largest excitation and the numerical curve from the main panel are replotted in double-logarithmic coordinates to demonstrate the asymptotic limit for large magnon number: $M_\perp \propto (f - f_L(0))^{-1.75}$, which corresponds to the condensate in a box in Eq. (4.74) with $k = 3$.

with $k = 3$ [16]. This behavior results from repulsive inter-particle interactions in the Thomas–Fermi limit. However, this similarity in the scaling law, $k = 3$, both with atomic condensate and with the MIT bag model, is accidental. Moreover, the $k = 3$ scaling is actually in disagreement with typical bag models. If the main contribution to the pressure comes from the bulk, as happens for the hadron model, then the energy $F(R)$ should be proportional to the volume of the bag; and then for the two-dimensional radial trap one would expect $F(R) = \pi R^2 P$, i.e. $k = 2$. On the other hand, if the main contribution to the pressure comes from the surface tension, as happens for electron bubble in liquid helium at $P = 0$, then the energy $F(R)$ should be proportional to the surface area; and for our 2D case this would give $F(R) = 2\pi\sigma R$, i.e. $k = 1$. The observed softer behavior with approximate scaling law $k \approx 3$ in the 2D case reflects the flexibility of the orbital field, which is inhomogeneous outside the cavity. On general grounds, $F(R)$ depends on several length scales: radius R of the bubble; radius R_c of the cylindrical container, where the boundary conditions on the $\hat{\mathbf{l}}$-texture are imposed; and the textural healing lengths: magnetic length ξ_H (the thickness of the layer near the wall of the container in which the orientation of $\hat{\mathbf{l}}$ by the magnetic field is restored), and the lengths related to the orientational effects of rotation and vortices on $\hat{\mathbf{l}}$.

In both experiments and in numerical simulations in [89], all of the length scales were of the same order, and thus no really small parameter was available, which could justify the scaling law. The true scaling behavior may only appear in some limit cases. For example, in a vessel rotating with angular velocity Ω in a vortex-free state in the regime $R_0 \gg R \gg \xi_v \gg \xi_H$ (where ξ_v is the healing length related to counterflow $|\mathbf{v}_s - \mathbf{v}_n| = \Omega R$) one may expect that the main contribution comes from the orientational effect of the counterflow, which is removed from the cavity $F(R) \propto R^2(\Omega R)^2$. The obtained scaling law with $k = 4$ gives $R \sim \mathcal{N}^{1/6}$ and $\mathcal{M}_\perp \propto (\omega - \omega_L)^{-2}$. Such an asymptotic regime, which can be approached in a large vessel, was probed in numerical simulations and is in reasonable agreement [97].

Under the conditions of the experiment the exponent k in Eq. (4.74) is close to $k = 3$ for atomic condensates in a harmonic trap. However, the physics of the formation of this exponent is different (formation of a box inside a flexible texture versus atom–atom interaction). As a result one has completely opposite behavior for analogous quantities—the frequency shift $\Delta\omega(\mathcal{N})$ in magnon BEC and the chemical potential $\mu(\mathcal{N})$ in an atomic condensate:

$$\omega - \omega_{\rm L}(0) \equiv \mu - U(0) \sim \omega_r\,(\mathcal{N}/\mathcal{N}_{\rm c})^{-2/5}, \quad \text{magnon BEC,} \qquad (4.75)$$

$$\mu - U(0) \sim \omega_r\,(\mathcal{N}/\mathcal{N}_{\rm c})^{2/5}, \quad \text{atomic BEC.} \qquad (4.76)$$

As a result, in contrast to atomic condensates, the magnon condensate droplet has negative derivative $d\mu/d\mathcal{N} < 0$. This means that with a growing Q-ball, its frequency ω decreases approaching the Larmor frequency asymptotically, and this behavior determines the way in which the magnon condensate is grown in a CW NMR measurement, as seen in Fig. 4.14 for the formation of ground-state BEC in the trap. The magnons are created when the frequency ω of the applied RF field is swept down and crosses the ground state level ω_{00}. When ω is reduced further, the number of magnons follows asymptotically Eq. (4.75), i.e. $\mathcal{N} \sim (\omega - \omega_{\rm L})^{-5/2}$.

The negative value of $d\mu/d\mathcal{N} < 0$ also allows formation of the condensates on excited levels under conditions where the ground-state condensate still does not exist—a situation which is impossible for atomic condensates with repulsive interaction, where $d\mu/d\mathcal{N} > 0$.

4.6.6 Formation of magnon BEC on an excited state in the trap

A condensate can be formed when one starts filling magnons to one of the levels (n,m) in Eq. (4.69). Then one obtains a non-ground-state condensate [89, 97]. The formation of non-ground-state condensates has been proposed for cold atoms [98], but as a dynamic mixture of the ground state and an excited level. It was suggested to use resonant modulation of either the trap potential or the atomic scattering length, e.g. by applying a temporal modulation of the atomic interactions via the Feshbach resonance technique. In contrast with such schemes, the excited states (n, m) of a magnon condensate can be populated directly without the original ground-state condensate. This is because the frequency $\omega_{\rm nm}(\mathcal{N})$ of excited state condensate also decreases with increasing magnon number $\mathcal{N}$. The condensate in the state (n, m) grows when the frequency of the RF field is swept down and crosses the level $\omega_{\rm nm}(0)$ from above, while at such a frequency the ground-state condensate does not exist. The numerical simulation of formation of the multi-magnon bubble with magnon condensate on the first excited level in a trap [97] is shown in Fig. 4.15.

The condensate in the excited state is metastable: it is supported by continuous pumping at $\omega_{\rm mn}(\mathcal{N})$, i.e. it exists in the regime of the controlled chemical potential μ. After switching off the pumping, i.e. in the regime of the controlled magnon number $\mathcal{N}$, the excited state condensate decays to the magnon BEC in the ground state. This quantum process of formation of the ground-state BEC from the pumping of magnons to an excited level is similar to the

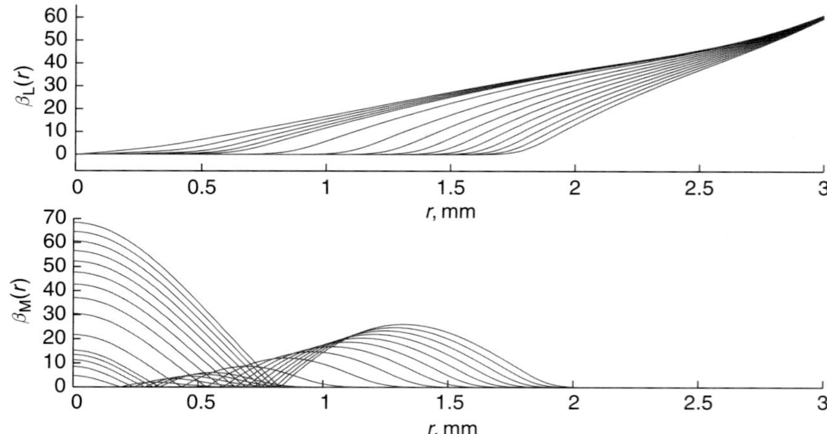

Fig. 4.15 Simulation of formation of a multi-magnon bubble with magnon condensate on the first excited level $n_r = 1$ (i.e. $n = 2$ in the initially harmonic 2D trap) [97]. At a large number of magnons, the harmonic trap gradually transforms to the box (*top*), while the wave function of the condensate gradually transforms to the Bessel function (*bottom*).

formation of magnon BEC by incoherent pumping in yttrium-iron garnet [9] (see Section 4.8.2) and also to the off-resonance excitation of the ground-state magnon condensate, observed in ^{3}He-B in Ref. [86]. This demonstrates that formation of magnon BEC is a spontaneous process, in which the precession frequency emerges spontaneously, and is not produced by an external rf field, i.e. the magnon condensate is not the "driven condensate" [99]. An external r.f. field if applied is needed for compensation of losses.

As distinct from the excited-state magnon BEC, the condensate in the ground state can be formed in pulsed NMR measurements after a large number of magnons, $\mathcal{N}$, are pumped to the cell. This clearly demonstrates the effect of self-localization: the main part of the pumped "charge" relaxes but the rest of $\mathcal{N}$ starts to concentrate at some place on the axis, digging a potential well there and attracting the "charge" from the other parts of the container. In earlier experiments [96] without the confinement potential $U_{\parallel}(z)$ in the axial direction, Q-balls were typically formed at the bottom of the cell. However, Q-balls were often formed on the axis of the flared-out texture, away from the horizontal walls. This shows that a Q-ball may dig the potential well in different places in the cell. In the formation of a Q-ball with off-resonance excitation [75, 86], the effect of self-localization also plays a crucial role.

In conclusion, Q-balls represent a new phase-coherent state of Larmor precession. They emerge at low T, when the homogeneous bulk BEC of magnons (HPD) becomes unstable. These Q-balls are compact objects which exist due to the conservation of the global $U(1)$ charge $Q = \mathcal{N}$. At small Q they are stabilized in the potential well, while at large Q the effect of self-localization is observed. In terms of relativistic quantum fields, the localization is caused by the peculiar interaction between the charged and neutral fields [91]. The neutral field $\hat{\mathbf{l}}$ provides the potential for the charged field Ψ; the charged field modifies locally the neutral field so that the potential well is deformed, and forms a box in which the charge Q is further condensed. In this limit, the magnon BEC becomes the bosonic analog of the MIT bag with trapped quarks in QCD or of an electron trapped in the cavity formed in liquid ^{4}He.

4.7 Exploiting Bose condensation of magnons

Bose condensation of magnons in superfluid ^{3}He-B has many practical applications. In Helsinki, owing to the extreme sensitivity of the Bose condensate to textural inhomogeneity, the phenomenon of Bose condensation has been applied to studies of supercurrents and topological defects in ^{3}He-B in a rotating cryostat. The measurement technique was HPD spectroscopy [100, 101].

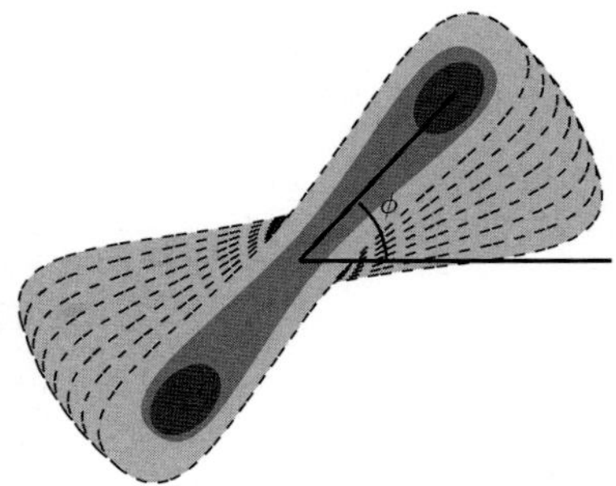

Fig. 4.16 Twisted core of non-axisymmetric vortex in ^{3}He-B. The gradient of the Goldstone field $\nabla\phi$ along the string corresponds to the superconducting current along the superconducting cosmic string. Such a twisted core has been obtained and detected using coherent precession of magnetization [102].

4.7.1 Observation of Witten string in ^{3}He-B

In particular, HPD spectroscopy has provided direct experimental evidence for broken axial symmetry in the core of quantized vortices in ^{3}He-B [102].

The dominating area of the phase diagram of the vortex states in ^{3}He-B is occupied by vortices with non-axisymmetric cores, i.e. vortices with the spontaneously broken rotational $SO(2)$ of the core, calculated in Refs. [103, 104]. The core with broken rotational symmetry can be considered as a pair of half-quantum vortices, connected by a non-topological soliton wall (see Fig. 4.16). The separation of the half-quantum vortices increases with decreasing pressure and thus the double-core structure is most pronounced at zero pressure.

In the physics of cosmic strings, an analogous breaking of continuous symmetry in the core was first discussed by Witten [105], who considered the spontaneous breaking of electromagnetic gauge symmetry $U(1)$. Since the same symmetry group is broken in condensed matter superconductors, one can say that in the core of the cosmic string there appears the superconductivity of the electric charges, hence the name "superconducting cosmic strings."

For the ^{3}He-B vortices, the spontaneous breaking of the $SO(2)$ symmetry in the core leads to Goldstone bosons—the mode in which the degeneracy parameter, the axis of anisotropy of the vortex core, is oscillating. The homogeneous magnon condensate, the HPD state, has been used to study the structure and twisting dynamics of the non-axisymmetric core of the low-temperature vortex in ^{3}He-B [102]. This is because the coherent precession of magnetization excites the vibrational Goldstone mode via spin–orbit interaction. Moreover, because of spin–orbit interaction the precessing magnetization rotates the core around its axis with constant angular velocity. In addition, since the core was pinned to the top and the bottom of the container, it was even possible to screw the core (see Fig. 4.16). Such a twisted core corresponds to a Witten superconducting string with electric supercurrent along the core. The rigidity of a twisted core differs from that of a straight core. This has allowed a detailed experimental study of the Goldstone mode of the vortex core resulting from the spontaneous violation of rotational $U(1)$ symmetry in the core [102].

4.7.2 Observation of the spin–mass vortex in ^{3}He-B

There are different types of topological defects in (non-precessing) ^{3}He-B. Among these is a Z_2 spin vortex—a topological defect of the order parameter matrix $R_{\alpha i}$ (see Ref. [30]). Because of spin–orbit coupling this defect serves as the termination line of the topological soliton wall, and because of soliton

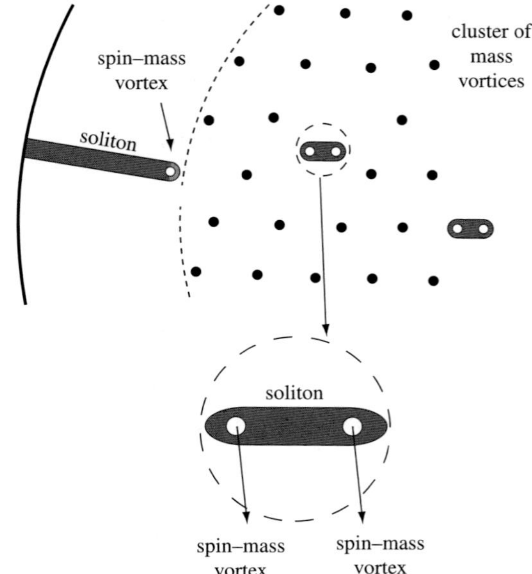

Fig. 4.17 Vortices in rotating ^{3}He-B. Mass vortices form a regular structure like Abrikosov vortices in an applied magnetic field. If the number of vortices is less than the equilibrium number for a given rotation velocity, vortices are collected in the vortex cluster. Within the cluster the average superfluid velocity $\langle \mathbf{v}_s \rangle = \mathbf{v}_n$. On the periphery there is a region devoid of vortices—the counterflow region, where $\langle \mathbf{v}_s \rangle \neq \mathbf{v}_n$. Spin–mass vortices can be created and stabilized in the rotating vessel. The confining potential produced by the soliton wall is compensated by the logarithmic repulsion of vortices forming the vortex pair—the doubly quantized vortex inside the cluster. A single spin–mass vortex is stabilized at the periphery of the cluster by the combined effect of soliton tension and Magnus force acting on the mass part of the vortex.

tension it cannot be stabilized in the rotating vessel. However, spin and mass vortices attract each other and form a combined defect with common core, the so-called spin–mass vortex, which can be stabilized under rotation (see Fig. 4.17). Spin–mass vortices also form molecules where the soliton serves as the chemical bond. These defects—a spin–mass vortex connected to the wall by a soliton and bound pairs of spin–mass vortices—have been observed and studied using HPD spectroscopy [106].

4.7.3 Magnon condensates in aerogel

HPD spectroscopy has proved to be extremely useful for the investigation of the superfluid order parameter in a novel system—superfluid ^{3}He confined in aerogel [65, 66, 72, 107–109].

4.7.4 Towards observation of Majorana fermions

The condensate in a magnetic trap can be used to continue NMR measurements in ^{3}He-B down to 0.1–0.2 T_c, where HPD does not exist. Of great current interest are fermionic states bound to the core of mass vortices and to the surface of ^{3}He-B, especially the still elusive Majorana fermions with zero energy (see the review on Majorana fermions [110]). Recently, the topologically non-trivial gapless and gapped phases of matter—topological insulators, semimetals, superconductors, and superfluids—have attracted attention [111, 112]. ^{3}He-B is the best representative of a three-dimensional coherent quantum system with time reversal symmetry. Its non-trivial topology gives rise to gapless Andreev–Majorana fermions as surface states [113–115]. There is now experimental evidence for Andreev surface states in ^{3}He-B at a solid wall [116, 117], but the Majorana signature of these fermions—the linear "relativistic" energy spectrum at low energy—can only be observed at extremely low temperatures, when the thermal quasiparticles in bulk ^{3}He-B are exponentially depleted. Majorana fermions, both in the vortex core and at surfaces, are then expected to give the main contribution to thermodynamics and dissipation, with a power-law dependence of the physical quantities on T. In rotation, or by moving the magnon condensate droplet next to the wall, one will be able to probe Majorana fermion states in ^{3}He-B.

4.8 Magnon BEC in other systems

4.8.1 Magnon BEC in normal ^{3}He

A very long-lived induction signal has been observed in normal Fermi liquids: in spin-polarized ^{3}He-^{4}He solutions [118] and in normal liquid ^{3}He [119]. This was explained as a coherently precessing structure at the interface between the equilibrium domain and the domain with reversed magnetization [44]. It would be interesting to treat this type of dynamic magnetic ordering as a new mode of magnon BEC.

4.8.2 Magnon condensation in solid materials

Recently, indications of magnon BEC in terms of coherent spin precession have been reported in a solid state material $CsMnF_3$ [13]. This condensate is similar to the magnon BEC in ^{3}He-A in aerogel. The magnon BEC obtained by parametric pumping of magnons has been investigated in yttrium–iron garnet (YIG) films [7–9]. Let us discuss the latter.

Magnons in yttrium–iron garnet have a quasi 2D spectrum:

$$\omega_n(k_x, k_y) = \Delta_n + \frac{k_y^2}{2m_y} + \frac{(k_x \pm k_0)^2}{2m_x}, \tag{4.77}$$

where the magnetic field is along x; the gap in the lowest branch $\Delta_0 = 2.1\,\text{GHz} \equiv 101\,\text{mK}$ at $H = 700\,\text{Oe}$ [7] and $\Delta_0 = 2.9\,\text{GHz}$ at $H = 1000\,\text{Oe}$ [8]; there are two minima with $k_0 = 5 \times 10^4$ 1/cm [7]; the anisotropic magnon mass can probably be estimated as $m_x \sim k_0^2/\Delta$, with m_y being somewhat bigger, both are of the order of electron mass.

In two-dimensional systems the number of extra magnons—the difference in the distribution function at $\mu = 0$ and $\mu \neq 0$—is determined by the low energy Rayleigh–Jeans part of the spectrum:

$$n = \sum_{\mathbf{k}} \left(\frac{T}{E_{\mathbf{k}} - \mu} - \frac{T}{E_{\mathbf{k}}} \right). \tag{4.78}$$

If one neglects the contribution of the higher levels and considers the 2D gas, then Eq. (4.78) becomes

$$n_{\text{extra}} = \frac{T}{2\pi\hbar}\sqrt{m_x m_y}\, \ln \frac{\Delta_0}{\Delta_0 - \mu}. \tag{4.79}$$

In two dimensions, all extra magnons can be absorbed by thermal distribution at any temperature without formation of a Bose condensate. The larger the number n of pumped magnons the closer is μ to Δ_0, but μ never crosses Δ_0. At large n the chemical potential exponentially approaches Δ_0 from below and the width of the distribution becomes exponentially narrow:

$$\frac{(\delta k_y)^2}{m_y \Delta_0} \sim \frac{(\delta k_x)^2}{m_x \Delta_0} \sim \frac{\Delta_0 - \mu}{\Delta_0} \sim \exp\left(-\frac{2\pi \hbar n_M}{T\sqrt{m_x m_y}}\right). \tag{4.80}$$

If one uses the 2D number density $n = \delta N\, d$ with film thickness $d = 5\ \mu\text{m}$ and 3D number density $\delta N \sim 5 \times 10^{18}\ \text{cm}^{-3}$, one obtains that at room temperature the exponent is

$$\frac{2\pi \hbar n_M}{T\sqrt{m_x m_y}} \sim \frac{2\pi\, \Delta_0}{T} \frac{\delta N\, d}{10 k_0^2} \sim 10^2\,. \tag{4.81}$$

If this estimation is correct, the peak should be extremely narrow, so that all extra magnons are concentrated at the lowest level of the discrete spectrum. However, there are other contributions to the width of the peak due to: finite resolution of the spectrometer, magnon interaction, finite lifetime of magnons, and the influence of the higher discrete levels $n \neq 0$.

In any case, the process of the concentration of extra magnons in the states very close to the lowest energy is the signature of the BEC of magnons. The main property of room temperature BEC in YIG is that the transition temperature T_c is only slightly higher than temperature, $T_c - T \ll T$; as a result, the number of condensed magnons is small compared to the number of thermal magnons: $n \ll n_T$. The situation with magnon BEC in ^{3}He is the opposite, one has $T \ll T_c$ and thus $n \gg n_T$. In ^{3}He-B, the typical temperature is high compared to the magnon gap, $T \gg \hbar\omega_L$, and thus according to (4.3) the thermal magnons are spin waves with linear spectrum $\omega(k) = ck$, with characteristic momenta $k_T \sim T/\hbar c$. The density of such thermal magnons $n_T \sim k_T^3$ is much smaller than the density of the Bose-condensed magnons and in ^{3}He-B they can be neglected.

4.8.3 Magnon BEC versus planar ferromagnet

Let us compare the HPD state (4.82) in ^{3}He-B, the coherent precession in a solid antiferromangnet, and BEC in YIG on one side, with the equilibrium magnetic states discussed in [18–21, 23] on the other side. In both groups the $U(1)$ symmetry is spontaneously broken, and thus they both belong to the same symmetry class as atomic BEC. At first glance, both groups can be described in terms of the ODLRO. The spin density in the coherently precessing HPD state of ^{3}He-B and in YIG are correspondingly

$$\langle S_+ \rangle = S_x + iS_y = S_\perp e^{i\omega t + i\alpha}, \tag{4.82}$$

and

$$\langle S_+ \rangle = S_x + iS_y = S_\perp \cos(k_0 x) e^{i\omega t + i\alpha}. \tag{4.83}$$

For the equilibrium planar ferromagnets one can also express the broken symmetry state in terms of a vacuum expectation value of the spin creation operator [23, 120],

$$\langle S_+ \rangle = S_x + iS_y = S_\perp e^{i\alpha}. \tag{4.84}$$

However, as distinct from Eqs. (4.82) and (4.83), Eq. (4.84) is time independent, and as a result it can be described by ordinary diagonal long-range order $\langle \mathbf{S} \rangle$ instead of the off-diagonal vacuum expectation value $\langle S_+ \rangle$. The phenomenon of ODLRO, which results in time dependence of the magnetic state manifested by a coherent precession, is the major feature that distinguishes between these two phenomena.

For the precessing states the $U(1)$ symmetry is related to the quasi-conservation of charge Q, which is the analog of the number of atoms in atomic BEC, and of the number of electrons in superconductors. In magnetic materials, charge Q is played by the spin projection S_z, or by the related number $\mathcal{N}$ of magnons. This approximate conservation law gives rise to a non-equilibrium chemical potential $\mu = dE/d\mathcal{N}$, which is non-zero only in dynamic states where it coincides with the precession frequency ω. On the contrary, in a static state one has $\omega = 0$, and thus Eq. (4.84) does not contain the analog of chemical potential. This means that the conservation law is not in the origin of the formation of the static equilibrium state. While the magnetic field may play the role of external potential, it cannot play the role of magnon chemical potential, since in a fully equilibrium state the chemical potential of magnons is always strictly zero, $\mu = 0$, which results in $\omega = 0$.

For both groups, the underlying $U(1)$ symmetry is approximate due to spin–orbit interaction, which violates the conservation of S_z and makes the lifetime for magnons finite. For the precessing states (4.82) and (4.83) this leads to a finite lifetime for the coherent precession. To support the steady state of precession, pumping of spin and energy is required. Contrary to this, spin–orbit interaction does not destroy the diagonal long-range magnetic order of the static states: these are the fully equilibrium states which do not decay and thus do not require pumping. That is why the approximate $U(1)$ symmetry and its spontaneous violation are not necessary conditions for the existence of equilibrium magnetic systems. A planar ferromagnet (4.84) is just one more equilibrium state of matter with broken time reversal symmetry, in addition to the easy axis ferromagnetic or antiferromagnetic state, rather than the magnon condensate. Formally, in the limit of small spin–orbit interaction the symmetry breaking scheme in these materials belongs to the same $U(1)$ class as conventional superfluids and superconductors, and thus they share many properties of this class, except for the ODLRO and related phenomena.

The property of (quasi) conservation of the $U(1)$ charge Q distinguishes the coherent precession from other coherent phenomena, such as optical lasers and standing waves. For the real BEC one needs conservation of particle number or charge Q during the time of equilibration. BEC occurs due to thermodynamics, when the number of particles (or charge Q) cannot be accommodated by thermal distribution, and as a result, the extra part must be accumulated in the lowest energy state. This is the essence of BEC.

Photons and phonons can also form BEC under pumping, if the lifetime of these excitations is larger than the thermalization time. For photons this condition has been realized, and photon BEC has been observed [4]. These thermodynamic BEC states are certainly different from other coherent states such as optical lasers and also from the equilibrium deformations of solids.

4.9 Beyond magnon BEC: Suhl instability of coherent precession

4.9.1 Catastrophic relaxation of magnon BEC

The instability of BEC which we discuss here is applicable only to the BEC of magnon quasiparticles, and is irrelevant for atomic condensates. For quasiparticles the $U(1)$ symmetry is not strictly conserved. For a magnetic subsystem, this is the $SO(2)$ symmetry with respect to spin rotations in the plane perpendicular to the magnetic field, and it is violated by spin–orbit interactions. The magnon BEC is a time-dependent process, and it may experience instabilities which do not occur in equilibrium condensates of stable particles. In 1989, it was found that the original magnon condensate—the HPD state—abruptly loses its stability below about 0.4 T_c [81]. This was called catastrophic relaxation. This phenomenon was left unexplained for a long time.

One of the earliest suggestions was very interesting. It was suggested that the instability appears due to crossover between NMR in the external field and NMR in the Landau field [121]. The molecular Landau field was introduced in phenomenological Landau theory of a Fermi liquid. According to this concept, the magnetization in superfluid ^{3}He can be presented as a sum of two components, the superfluid one which comes from the superfluid order parameter, and the normal one which comes from thermal quasiparticles. This is similar to the two-fluid model of superfluidity. In analogy with the second sound, in NMR both components precess around the common direction—the Landau field direction, which in turn is directed along the total magnetization and thus does not contribute to the frequency of NMR (see Fig. 4.18). This two-component precession is described by Leggett–Takagi equations (see the book Ref. [30]), with each component having its own spin–orbit interaction, phase of precession, and relaxation. The value of the Landau field is proportional to the magnetization and changes in superfluid ^{3}He with temperature, so that in the range of pressures between 0 and 10 bar there is a crossover temperature when the value of the Landau field is equal to the value of the external magnetic field. At this crossover temperature the fast relaxation of the conventional NMR signal was found [122]. This results from the two-mode precession around the Landau field, which is highly dissipative and pumps out the energy from the usual mode of NMR [123]. This was the first evidence that the Landau field is indeed a real molecular field and not just a mathematical construction. However, the Landau field does not explain the catastrophic relaxation phenomenon, since the crossover temperature has an opposite dependence on pressure [124].

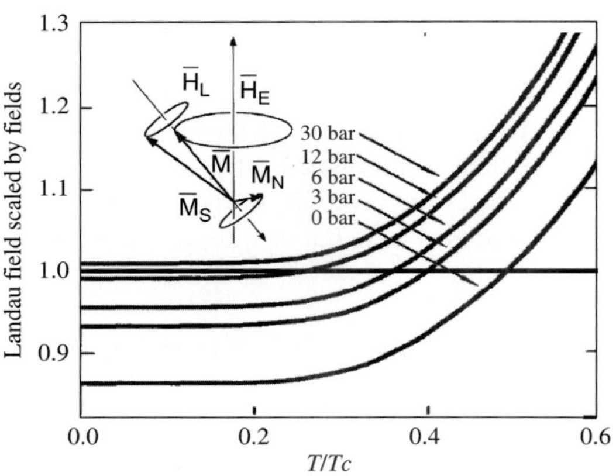

Fig. 4.18 Illustration of the concept of the "molecular" Landau field: NMR in terms of the two-component precession. The normal $\mathbf{M}_{\mathrm{N}}$ and superfluid $\mathbf{M}_{\mathrm{S}}$ components of magnetization process around the direction of the Landau field, which in turn is directed along the full magnetization $\mathbf{M} = \mathbf{M}_{\mathrm{N}} + \mathbf{M}_{\mathrm{S}}$, which precesses around the external magnetic field. In some range of pressures, there is the transition temperature, at which the magnitude of the Landau field crosses the magnitude of the external field. A rapid decay of the NMR signal is observed at this temperature [122], which justifies the phenomenology of the Landau field. However, the concept of the Landau field does not explain catastrophic relaxation. The origin of the latter is Suhl instability.

Finally, the reason for catastrophic relaxation was established: in the low-temperature regime, where dissipation becomes sufficiently small, the Suhl instability destroys homogeneous precession [82, 83, 125]. This is a parametric instability, which leads to decay of HPD due to parametric amplification of an intrinsic mode, not previously discussed. The latter modes are different from the magnons which we discuss here: they represent another branch of the collective modes of superfluid ^{3}He-B which appears in the regions where the orbital momentum is not parallel to the magnetic field [126, 127]. In particular they are excited near the walls of the cell which are oriented along the magnetic field [128]. Instability occurs because the spin–orbit interactions violate $U(1)$ symmetry.

For magnetically ordered systems the instability of homogeneous precession is a well-known phenomenon. Suhl [129] explained it in terms of parametric instability of the mode of precession with respect to excitations of pairs of spin waves satisfying the condition of resonance:

$$n\omega_L = \omega_s(\mathbf{k}) + \omega_s(-\mathbf{k}), \tag{4.85}$$

where ω_L is the precession frequency and n is an integer (see also review [130]). All the magnetic systems, where the Suhl instability has been observed, are anisotropic. In particular, as quantum solids and liquids are concerned, Suhl instability has been observed in anti-ferromagnetic solid ^{3}He [131], and was predicted by Fomin for anisotropic superfluid liquid ^{3}He-A [132], where it was later observed [69]. Because of the extreme isotropy of ^{3}He-B and the unique symmetry of the spin–orbit interaction, the Suhl instability was not expected there.

However, under the conditions of the experiment, the boundary conditions on the wall of the container induce the texture of the order parameter in which the orbital vector $\mathbf{l}$ deviates from its symmetric orientation along the magnetic field $\mathbf{H}$. The symmetry of the spin–orbit interaction is violated, providing the additional term in the interaction between the modes, which is dominating in typical experiments with catastrophic relaxation [81, 128, 133–135].

4.9.2 Precessing states and their symmetry

To describe the interaction of magnon condensate with other modes of superfluid ^{3}He-B, we must go beyond the magnon BEC description and consider all of the degrees of freedom of homogeneous free precession in external magnetic field $\mathbf{H}$. In liquid ^{3}He the spin–orbit (dipole–dipole) interaction is weak. If it is neglected, we can apply the powerful Larmor theorem, according to which, in the spin-space coordinate frame rotating with the Larmor frequency the effect of the magnetic field on the spins of the ^{3}He atoms is completely compensated. This follows from the observation that, in dynamics the time derivative of a spin vector enters equations together with the Larmor frequency vector $\boldsymbol{\omega}_L = \gamma \mathbf{H}$:

$$D_t \mathbf{f} = \partial_t \mathbf{f} - \boldsymbol{\omega}_L \times \mathbf{f}. \tag{4.86}$$

In other words, the Pauli magnetic field acts on spin vectors as a time component of the effective $SO(3)$ gauge field,

$$\mathbf{A}_0 = \gamma \mathbf{H}. \tag{4.87}$$

We shall use this equation later in Section 4.10 for discussion of spin currents and the spin-quantum Hall effect.

Since the magnetic field becomes irrelevant, the symmetry group of the physical laws in the precessing frame becomes the same as in the absence of the field. If the spin–orbit interaction is neglected, it is the product of the SO_3^L group of orbital rotations and the SO_3^S group of spin rotations:

$$G = SO_3^L \times SO_3^S. \tag{4.88}$$

The difference from the symmetry group $G = SO_3^L \times SO_3^S$ in the static case, is that the SO_3^S rotations are now considered in the precessing frame rather than in the laboratory frame. The elements of the latter group $\tilde{\mathbf{g}}(\mathbf{t})$ are constructed from the elements $\mathbf{g}$ of conventional spin rotations in the laboratory frame:

$$\tilde{\mathbf{g}}(\mathbf{t}) = \mathbf{O}^{-1}(\hat{\mathbf{z}}, \omega_L t)\, \mathbf{g}\, \mathbf{O}(\hat{\mathbf{z}}, \omega_L t). \tag{4.89}$$

Here, matrix $O_{\alpha\beta}(\hat{\mathbf{z}}, \omega t)$ describes the transformation from the laboratory frame into the rotating frame—this is the rotation about the magnetic field axis $\hat{\mathbf{z}}$ by angle $\omega_L t$. Now we can find all of the degenerate coherent states of the Larmor precession, applying the symmetry group G to the order parameter in the simplest equilibrium state of the given superfluid phase. The order parameter in superfluid ^{3}He is 3×3 matrix, which corresponds to the Cooper pairing with orbital and spin momenta $L = S = 1$ [30]. Thus, choosing the simplest 3×3 static matrix $\mathbf{A}^{(0)}$ in a given superfluid phase, one obtains all the precessing states in this phase if spin–orbit coupling is neglected:

$$\mathbf{A}(t) = \mathbf{O}^{-1}(t)\mathbf{R}^{(1)}\mathbf{O}(t)\mathbf{A}^{(0)}(\mathbf{R}^{(2)})^{-1}. \tag{4.90}$$

Here $\mathbf{R}^{(1)}$ is the arbitrary matrix describing spin rotations in the precessing frame and $\mathbf{R}^{(2)}$ is another arbitrary matrix which describes the orbital rotations in the laboratory frame. In the case of ^{3}He-B, the simplest state corresponds to the total angular momentum of the Cooper pair $J = 0$, which is described by the isotropic matrix [30]:

$$A^{(0)}_{\alpha i} = \Delta_B\, \delta_{\alpha i}, \tag{4.91}$$

where Δ_B is the gap in the fermionic spectrum. The action of elements of the group G on this stationary state leads to the following general precession of ^{3}He-B with the Larmor frequency (if spin–orbit interaction is neglected):

$$A_{\alpha i}(t) = \Delta_B R_{\alpha i}(t), \tag{4.92}$$

$$R_{\alpha i}(t) = O_{\alpha\beta}(\hat{\mathbf{z}}, -\omega t) R^{(1)}_{\beta\gamma} O_{\gamma\mu}(\hat{\mathbf{z}}, \omega t)(R^{(2)})^{-1}_{\mu i}. \tag{4.93}$$

Matrix $\mathbf{R}^{(1)}$ determines the direction of spin density in the precessing frame:

$$S_\alpha = \chi R^{(1)}_{\alpha\beta} H_\beta, \tag{4.94}$$

where χ is the spin susceptibility of ^{3}He-B. This corresponds to the precession of spin magnetization with the tipping angle $\cos\beta_M = R^{(1)}_{zz}$. Matrix $\mathbf{R}^{(2)}$ determines the direction of orbital momentum density in the laboratory frame:

$$L_i = -R_{\alpha i}(t) S_\alpha(t) = -\chi R^{(2)}_{i\alpha} H_\alpha, \tag{4.95}$$

with the tipping angle $\cos\beta_L = R^{(2)}_{zz}$.

4.9.3 Spin–orbit interaction as perturbation

Spin–orbit interaction couples the spin and orbital components of matrix $A_{\alpha i}$. For ^{3}He-B in (4.92) one obtains [30],

$$F_D = \frac{2}{15}\chi\Omega_L^2\left(R_{ii}(t) - \frac{1}{2}\right)^2 = \frac{8}{15}\chi\Omega_L^2\left(\cos\theta(t) + \frac{1}{4}\right)^2, \tag{4.96}$$

where Ω_L is the so called Leggett frequency—the frequency of the longitudinal NMR; θ is the angle of rotation in the parameterization of the matrix $R_{\alpha i}$ in terms of the angle and axis of rotation [30]; here, we shall use the system of units in which the gyromagnetic ratio γ for the ^{3}He atom is 1, hence the magnetic field and the frequency will have the same physical dimension.

In the general state of the Larmor precession (4.92), the spin–orbit interaction contains the time-independent part and rapidly oscillating terms with frequencies ω_L, $2\omega_L$, $3\omega_L$, and $4\omega_L$:

$$F_D(\gamma) = F_0 + \sum_{n=1}^{4} F_n \cos(n\omega_L t). \tag{4.97}$$

The time-independent part—the average over fast oscillations—gives the spin–orbit potential in (4.23), which determines the phases of magnon BEC in ^{3}He-B:

$$F_0 = F_{\rm so}(s,l,\gamma) = \frac{2}{15}\chi\Omega_L^2\left[\left(sl - \frac{1}{2} + \frac{1}{2}(1+s)(1+l)\cos\gamma\right)^2 + \frac{1}{8}(1-s)^2(1-l)^2 + (1-s^2)(1-l^2)(1+\cos\gamma)\right]. \tag{4.98}$$

Here $s = \cos\beta_M$ (or simply $\cos\beta$) and $l = \cos\beta_L$ are z projections of unit vectors $\hat{\mathbf{s}} = \mathbf{S}/S$ and $\hat{\mathbf{l}} = -\mathbf{L}/L$; and γ is another free parameter of the general precession. Altogether the free precession is characterized by five independent parameters coming from two matrices $\mathbf{R}^{(1)}$ and $\mathbf{R}^{(2)}$ [58]: two angles of spin $\mathbf{S}$, two angles of the orbital momentum $\hat{\mathbf{l}}$, and the relative rotation of matrices by angle γ. In the case of the stationary (non-precessing) magnetization, the γ-mode corresponds to the longitudinal NMR mode.

4.9.4 Parametric instability of HPD

In the simplest description, the dynamics of the γ-mode is determined by the following Lagrangian:

$$\mathcal{L} = -\frac{1}{2}\chi\left(\dot{\gamma}^2 - c^2(\nabla\gamma)^2\right) + F_D(\gamma). \tag{4.99}$$

Here we have used the approximation of an isotropic speed of spin waves c. In the time-dependent part of F_D in (4.97) we only consider the first harmonic, i.e. according to Eq. (4.85) we discuss the parametric excitation of two γ-modes with $ck \approx \omega_L/2$. The amplitude of the first harmonic is:

$$\begin{aligned} F_1 = {}& \frac{4}{15}\chi\Omega_L^2 \sin\beta \sin\beta_L \cos(\gamma/2) \\ & \times \left(2sl - 1 + \frac{(1-s)(1-l)}{2} + (1+s)(1+l)\cos\gamma\right). \end{aligned} \tag{4.100}$$

Further we assume that the system is in the minimum of the dipole energy F_0 as a function of γ. The equilibrium value $\gamma = \gamma_0$ is

$$\cos\gamma_0 = -\frac{(2sl-1) + 2(1-s)(1-l)}{(1+s)(1+l)}, \tag{4.101}$$

which is valid if the right-hand side of Eq. (4.101) does not exceed unity, i.e. when $s + l - 5sl < 2$.

For discussion of Suhl instability we need the time-dependent term which is quadratic in $\gamma - \gamma_0$. Then the Lagrangian (4.99) which describes the parametric instability towards decay of the Larmor precession to two γ-modes with $kc \approx \omega_L/2$ is (after the shift $\gamma - \gamma_0 \to \gamma$; neglecting Ω_L compared to ω_L; and neglecting the anisotropy of the spin-wave velocity)

$$\mathcal{L} = \frac{1}{2}\chi\left(-\dot{\gamma}^2 + c^2(\nabla\gamma)^2 + a\Omega_L^2\gamma^2\cos\omega_L t\right), \tag{4.102}$$

where, if $s + l - 5sl < 2$, the parameter a is

$$\begin{aligned} a = {}& \frac{4}{15}\sin\beta\sin\beta_L\left[\frac{3(s+l-sl)}{2(1+s)(1+l)}\right]^{1/2} \\ & \times\left[(1+s)(1+l) + 2(2sl-1) + \frac{35}{8}(1-s)(1-l)\right]. \end{aligned} \tag{4.103}$$

Let us rewrite the Lagrangian (4.102) in terms of a Hamiltonian as a function of creation and annihilation operators $b_{\mathbf{k}}$ and $b^*_{\mathbf{k}}$:

$$\gamma_{\mathbf{k}} = \frac{i}{\sqrt{2\chi\omega_s(k)}}(b_{\mathbf{k}} - b^*_{\mathbf{k}}) \ , \ \omega_s^2(k) = c^2k^2 \quad (4.104)$$

$$p_{\mathbf{k}} = \chi\dot{\gamma}_{\mathbf{k}} = \sqrt{\chi\omega_s(k)/2}(b_{\mathbf{k}} + b^*_{\mathbf{k}}), \quad (4.105)$$

$$\mathcal{H} = \sum_{\mathbf{k}} \omega_s(k) b^*_{\mathbf{k}} b_{\mathbf{k}} + \sum_{\mathbf{k}} \frac{a\Omega_L^2}{2\omega_s(k)} \left(e^{-i\omega_L t} b_{\mathbf{k}} b_{-\mathbf{k}} + e^{i\omega_L t} b^*_{\mathbf{k}} b^*_{-\mathbf{k}}\right), \quad (4.106)$$

where we have neglected Ω_L compared to ω_L. The spectrum of the excited mode is

$$b_{\mathbf{k}}(t) = \tilde{b}_{\mathbf{k}} e^{-i\omega_L/2t + i\nu_{\mathbf{k}} t} \ .$$

$$\nu_{\mathbf{k}} = \sqrt{(\omega_s(k) - \omega_L/2)^2 - a^2\Omega_L^4/\omega_s^2(k)} \quad (4.107)$$

At the resonance, i.e. when $\omega_s(k) = \omega_L/2$, the mode grows exponentially:

$$b_{\mathbf{k}}(t) \propto e^{\lambda t}, \ \ \lambda = 2a\Omega_L^2/\omega_L. \quad (4.108)$$

At finite temperatures this growth is damped by dissipation, but at low temperature the dissipation becomes small and catastrophic relaxation occurs. Following Ref. [125] one may assume the spin diffusion mechanism of dissipation. In this case the equation for the temperature $T_{\rm cat}$ below which the instability of the homogeneous precession towards radiation of spin waves with $\omega_s(k) = ck = \omega_L/2$ starts to develop, is [82, 83]:

$$D(T_{\rm cat}) = 2\lambda c^2/\omega_L^2. \quad (4.109)$$

Here $D(T)$ is the spin diffusion coefficient, which depends on temperature and which decreases with decreasing T. This agrees with observations.

4.10 Beyond magnon BEC: spin supercurrents and spin-Hall effects

Spin current has many faces. Spin can be transferred by convective, diffusive, or ballistic motion of particles; it can also be transferred from particle to particle without any particle motion. In real systems these mechanisms compete with each other, so that in the phenomenological description only in very particular cases is it possible to resolve between them. In a ferromagnetically ordered A_1-phase of superfluid ^{3}He, the nuclear spin is transferred by superfluid mass current. In magnon BEC in ^{3}He-B, the nuclear spin is transferred by the superfluid current of magnons (Section 4.4.4); in particular, this spin current transfers spin from one experimental cell to another in the spin-current Josephson effect (Section 4.4.10). In magnetically ordered systems the spin current can also be represented in terms of the rigidity of the order parameter to inhomogeneous spin rotations. An example of the non-dissipative spin current arising in inhomogeneous states of these materials is the spin current circulating around the core of the spin-mass vortex in ^{3}He-B and inside the soliton which terminates on a spin-mass vortex (Section 4.7.2). As its electric counterpart—the charge current, the spin current can be dissipative and non-dissipative, with or without spin accumulation. It may give rise to a spin-Josephson effect and also to a spin-Hall effect with and without external magnetic field, which may be ordinary (Section 4.10.2) and quantized (Section 4.10.4).

4.10.1 Microscopic theory of spin supercurrent in ^{3}He-B

Here we give the "microscopic" derivation for the spin supercurrent, which has been discussed on the phenomenological level of magnon BEC. The underlying microscopic physics is the BCS theory of p-wave spin-triplet superfluid ^{3}He. Superfluid spin currents in ^{3}He-B exist even in the absence of magnon BEC. They come from the spontaneous breaking of spin-rotation symmetry. The spin supercurrent in ^{3}He-B is expressed in terms of the spin superfluid velocities:

$$\omega_{\alpha i} = \nabla_i \theta_\alpha = \frac{1}{2} e_{\alpha\beta\gamma} R_{\beta j} \nabla_i R_{\gamma j}. \tag{4.110}$$

The corresponding gradient energy is

$$F_{\text{grad}} = \frac{1}{2} \rho_{\alpha i,\beta j} \omega_{\alpha i} \omega_{\beta j}, \tag{4.111}$$

where $\rho_{\alpha i,\beta j}$ is the spin rigidity tensor with spin rigidity parameters

$$\rho_{\alpha i,\beta j} = \frac{\chi_B}{\gamma^2} \left[\tilde{c}_{\parallel}^2 \delta_{\alpha\beta} \delta_{ij} - (\tilde{c}_{\parallel}^2 - \tilde{c}_{\perp}^2)(R_{\alpha i} R_{\beta j} + R_{\alpha j} R_{\beta i}) \right]. \tag{4.112}$$

From these equations one obtains the spin supercurrent in ^{3}He-B,

$$J_{\alpha i} = -\frac{\partial F_{\text{grad}}}{\partial \omega_{\alpha i}} = -\rho_{\alpha i,\beta j} \omega_{\beta j}. \tag{4.113}$$

This spin current averaged over the fast precession determines the parameters of the phenomenological equations (4.40) and (4.41) for the spin supercurrent emerging in magnon BEC.

4.10.2 Spin-Hall effect in ^{3}He-B

The symmetry properties of the spin superfluid velocity, and thus of the spin supercurrent, allows linear coupling of the spin current with the electric field, even in the absence of the spin–orbital interaction. The following term in this action is possible [136]:

$$F = -\beta e_{ijk} \omega_{\alpha i} R_{\alpha j} E_k. \tag{4.114}$$

The parameter β is not well defined from the microscopic theory due to the unknown Fermi-liquid corrections involved. The estimate for β reported in Ref. [137] is $\beta \sim 10^{-4}$ e/cm. Variation with respect to $\omega_{\alpha i}$ demonstrates that there exists a linear response of the spin supercurrent on the electric field:

$$J_{\alpha i} = -\frac{\partial F}{\partial \omega_{\alpha i}} = \beta e_{ijk} R_{\alpha j} E_k. \tag{4.115}$$

This spin current is transverse to the electric field and thus represents the spin current-Hall effect. As distinct from the spin-Hall effect predicted by Dyakonov and Perel [138], this spin-Hall effect occurs in the absence of spin–orbit interaction. The bridge, which connects spin and orbital motion, is provided by the order parameter matrix $R_{\alpha j}$.

4.10.3 Electric and magnetic fields as $SU(2)$ gauge fields

The interaction of the electric and magnetic fields with the order parameter in superfluid phases of ^{3}He, can also be found using the observation that **H** and **E** may be considered as temporal and spatial components of the $SU(2)$ gauge field, where $SU(2)$ is the group of spin rotations. The auxiliary $SU(2)$ gauge field A^α_μ is convenient for description of the effects related to the spin current. In the spinor representation one has the following covariant derivatives coming from the auxiliary $SU(2)$ gauge field [136, 139, 140]

$$\mathbf{D} = \nabla - i\mathbf{A}^i \frac{\sigma^i}{2}, \quad D_0 = \partial_t + iA^i_0 \frac{\sigma^i}{2}. \tag{4.116}$$

Some components of the field A^α_μ are physical, being represented by the real physical quantities which couple to the fermionic charges. An example is provided by the Pauli magnetic field H^i, which plays the role of component A^i_0 of the $SU(2)$ gauge field, see (4.87) [58, 141], while the spatial components are played by the electric field, which enters the gradient energy via the covariant spatial derivative [136, 139, 140, 142]:

$$A^i_0 = B^i, \quad A^i_j = e_{jik} E_k\,. \tag{4.117}$$

The electric field enters due to the relativistic spin–orbit interaction of the spin of the ^{3}He atom with the electric field **E**.

The spin current is obtained as a variation of the action with respect to the fictitious $SU(2)$ gauge field:

$$\mathbf{J}^i = \frac{\delta S}{\delta \mathbf{A}^i}. \tag{4.118}$$

After the spin current is calculated the values of the auxiliary fields are made equal to zero or to the values of the corresponding physical fields which simulate the gauge fields.

For example, in the presence of an electric field, equation (4.111) becomes

$$F_{\rm grad} = \frac{1}{2}\rho_{\alpha i,\beta j}\left(\omega_{\alpha i} - \frac{\gamma}{c} e_{\alpha ik} E_k\right)\left(\omega_{\beta j} - \frac{\gamma}{c} e_{\beta jl} E_l\right), \tag{4.119}$$

which demonstrates that the electric field enters as the $SU(2)$ gauge field forming the covariant derivative. This gives another response of the spin supercurrent on the external electric field,

$$J_{\alpha i} = -\frac{\partial F_{\rm grad}}{\partial \omega_{\alpha i}} = -\rho_{\alpha i,\beta j}\left(\omega_{\beta j} - \frac{\gamma}{c} e_{\beta jk} E_k\right), \tag{4.120}$$

As distinct from (4.11.6) this spin-Hall effect is governed by the spin–orbit interaction. According to Ref. [56] both spin-Hall effects in ^{3}He-B should modify the spin current-Josephson effect in magnon BEC. The supercurrent, induced by the electric field, leads to an additional phase shift proportional to the electric field, which is to be measured.

4.10.4 Quantum spin-Hall effect

There are several types of responses of spin and electric currents to transverse forces which are quantized in 2+1 systems under appropriate conditions. The most familiar is the conventional quantum Hall effect (QHE). This is a quantized response of the particle current to a transverse force, say to a transverse gradient of chemical potential, $\mathbf{J} = \sigma_{xy}\hat{\mathbf{z}} \times \nabla\mu$. In electrically charged systems it is the quantized response of the electric current $\mathbf{J}^e$ to transverse electric field $\mathbf{J}^e = e^2\sigma_{xy}\hat{\mathbf{z}} \times \mathbf{E}$.

The other effects involve the spin degrees of freedom. An example is the mixed spin quantum Hall effect: a quantized response of the particle current $\mathbf{J}$ (or electric current $\mathbf{J}^e$) to a transverse gradient of a magnetic field interacting with Pauli spins (Pauli field in short) [143, 144]:

$$\mathbf{J} = \sigma_{xy}^{\text{mixed}}\hat{\mathbf{z}} \times \nabla(\gamma H^z), \quad \mathbf{J}^e = e\mathbf{J}. \tag{4.121}$$

The related effect, which is determined by the same quantized parameter $\sigma_{xy}^{\text{mixed}}$, is the quantized response of the spin current, say current $\mathbf{J}^z$ of the z component of spin, to the gradient of a chemical potential [145]. In electrically charged systems this corresponds to the quantized response of the spin current to a transverse electric field:

$$\mathbf{J}^z = \sigma_{xy}^{\text{mixed}}\hat{\mathbf{z}} \times \nabla\mu = e\sigma_{xy}^{\text{mixed}}\hat{\mathbf{z}} \times \mathbf{E}. \tag{4.122}$$

This kind of mixed Hall effect is now used in spintronics [146], which exploits the coupling between spin and charge transport in condensed matter.

Finally, there is a pure spin-Hall effect—the quantized response of the spin current to a transverse gradient of a magnetic field [143, 144, 147, 148]:

$$\mathbf{J}^z = \sigma_{xy}^{\text{spin/spin}}\hat{\mathbf{z}} \times \nabla(\gamma H^z). \tag{4.123}$$

Let us consider the mixed spin-Hall effects in (4.122) and (4.123). These two effects are related, since they are described by the same topological Chern–Simons action [143] and thus by the same parameter $\sigma_{xy}^{\text{mixed}}$. To see this, let us remember that the spin current is obtained as a variation of the action over the fictitious $SU(2)$ or $SO(3)$ gauge field, see (4.118). For example, the current of the z-projection of spin is

$$\mathbf{J}^z = \frac{\delta S}{\delta \mathbf{A}^z}. \tag{4.124}$$

The corresponding Chern–Simons term in the action is given by [143]

$$F_{\text{CS}} = e\sigma_{xy}^{\text{mixed}}e^{\nu\alpha\beta}\int d^2x dt A_\nu^z \nabla_\alpha A_\beta\,, \tag{4.125}$$

where A_β is the vector potential of the conventional electromagnetic field, and $A^z_\nu = (A^z_i, A^z_0)$ represent components of an auxiliary (fictitious) $SU(2)$ gauge field. Variation of the action with respect to the field A^z_i gives the spin current in (4.122). On the other hand, the variation of the action with respect to field A_i gives electric current in terms of the gradients of an auxiliary gauge field. However, from equation (4.87) or (4.117) it follows that the role of the auxiliary gauge field A^z_0 is played by magnetic field H^z. As a result one obtains equation (4.121) for an electric current.

Equation (4.125) was originally introduced for a thin film of the so-called planar phase of superfluid ^{3}He [143]. However, it is better suited for two-dimensional topological insulators with time reversal invariance (for topological insulators, see reviews [111, 112]). These materials have the same topological structure as the planar phase, which is also time reversal invariant, but the advantage of these materials is that they are insulating and thus the superconductivity does not mask the spin-Hall quantization.

Discussion of the mixed Chern–Simons term can be found in Ref. [149]. For the related phenomenon of axial anomaly in particle physics, the mixed action in terms of different (real and fictitious) gauge fields has been introduced in Ref. [150].

4.11 Conclusion

The superfluid phases of liquid ^{3}He at extreme low temperatures are unique states of condensed matter with physical properties which can be compared to the vacuum of relativistic quantum field theories (see Chapter 11 "The Superfluid Universe" in this book). The A phase (^{3}He-A) belongs to the same symmetry and topology class as the vacuum of the Standard Model of particle physics in its massless (i.e. gapless) phase and can also be described as a semi-metal-like system with non-trivial topology. The B phase (^{3}He-B), in contrast, is similar to the vacuum of the Standard Model in its massive (or gapped) phase and to three-dimensional topological insulators with time reversal symmetry. In addition to fermionic quasiparticle excitations, superfluid ^{3}He also has bosonic quasiparticles, such as magnons—quanta of excitations of the magnetic subsystem. These magnon excitations can form long-lived Bose–Einstein condensates both in ^{3}He-A and ^{3}He-B, and these condensates experience their own superfluidity, which is not related to the superfluidity of the underlying system.

Formally, the phenomenon of superfluidity requires the conservation of charge or particle number. However, consideration can be extended to systems with a weakly violated conservation law, including a system of sufficiently long-lived quasiparticles—discrete quanta of energy that can be treated as real particles in condensed matter. The spin superfluidity—superfluidity in the magnetic subsystem of condensed matter—is manifested in magnon BEC as the spontaneous phase-coherent precession of spins first discovered in 1984 [10, 11]. This superfluid current of spins is one more representative of superfluid currents known or discussed in other systems, such as the superfluid current of mass and atoms in superfluid ^{4}He; superfluid current of electric charge in superconductors; superfluid current of hypercharge in the Standard Model; superfluid baryonic current and current of chiral charge in quark matter; etc. The analogy of the dynamical superfluid state of coherent precession with the non-perturbative dynamics of the physical vacuum has been discussed in [151].

Different condensates and thus different states of magnon superfluidity have been created by choosing different experimental arrangements. At low temperatures the condensate is confined in a magnetic trap which is formed by the order parameter texture of the superfluid state. This produces the analog of atomic BEC in laser traps, but adds some new features, such as formation of the non-ground-state condensate; self-localization, and formation of the multi-boson bubble which is the analog of the MIT bag model of hadrons; non-zero mass of the Goldstone bosons, etc. The magnon condensates can be used to probe the quantum vacuum of ^{3}He in the limit $T \to 0$, where conventional measuring signals become insensitive.

Acknowledgments

This work is supported in part by the Academy of Finland, Centers of excellence program 2006-2011, by the EU's 7th Framework Programme (FP7/2007-2013: grant agreement # 228464 MICROKELVIN), and by collaboration between CNRS and the Russian Academy of Sciences (project # N16569).

References

[1] Kagan, Yu., et al. Condensation of phonons in an ultracold Bose gas, *Phys. Lett.* A **361**, 401 (2007).

[2] Butov, L.V., Ivanov, A. L., Imamoglu, A., Littlewood, P. B., Shashkin, A. A., Dolgopolov, V. T., Campman, K. L., and Gossard, A. C. Stimulated scattering of indirect excitons in coupled quantum wells: Signature of a degenerate Bose-gas of excitons, *Phys. Rev. Lett.* **86**, 5608 (2001).

[3] Kasprzak, J., Richard, M., Kundermann, S., Baas, A., Jeambrun, P., Keeling, J.M.J., Marchetti, F.M., Szymaska, M.H., Andre, R., Staehli, J.L., Savona, V., Littlewood, P.B., Deveaud, B., and Dang, Le Si. Bose–Einstein condensation of exciton polaritons, *Nature* **443**, 409–414 (2006); B. Deveaud-Pledran, On the condensation of polaritons *J. Opt. Soc. Am.* B **29**, A138–A145 (2012).

[4] Klaers, J., Schmitt, J., Vewinger, F., and Weitz, M. Bose–Einstein condensation of photons in an optical microcavity *Nature* **468**, 545 (2010); Schmitt, J., Damm, T., Vewinger, F., and Weitz, M. Bose–Einstein condensation of paraxial light, *Appl. Phys.* B Laser Opt. **105**, 17–33 (2011); arXiv:1109.4023.

[5] Melnikovsky, L.A. Bose–Einstein condensation of rotons, *Phys. Rev.* B **84**, 024525 (2011).

[6] Kalafati, Yu. D., and Safonov, V.L. Possibility of Bose condensation of magnons excited by incoherent pump, *JETP Lett.* **50**, 149–151 (1989).

[7] Demokritov, S.O., Demidov, V.E., Dzyapko, O., Melkov, G.A., Serga, A.A., Hillebrands, B., and Slavin, A.N. Bose–Einstein condensation of quasi-equilibrium magnons at room temperature under pumping, *Nature* **443**, 430–433 (2006).

[8] Demidov, V.E., et al. Observation of spontaneous coherence in Bose–Einstein condensate of magnons, *Phys. Rev. Lett.* **100**, 047205 (2008).

[9] Chumak, A.V., Melkov, G.A., Demidov, V.E., Dzyapko, O., Safonov, V.L., and Demokritov, S.O. Bose–Einstein condensation of magnons under incoherent pumping, *Phys. Rev. Lett.* **102**, 187205 (2009).

[10] Borovik-Romanov, A.S., Bunkov, Yu.M., Dmitriev, V.V., and Mukharskiy, Yu.M. Long-lived induction decay signal investigations in ^{3}He, *JETP Lett.* **40**, 1033, (1984); Stratification of ^{3}He spin precession in two magnetic domains, *Sov. Phys. JETP* **61**, 1199, (1985).

[11] Fomin, I.A. Long-lived induction signal and spatially nonuniform spin precession in ^{3}He-B, *JETP Lett.* **40**, 1037 (1984).

[12] Greilich, A., Shabaev, A., Yakovlev, D.R., Efros, Al.L., Yugova, I.A., Reuter, D., Wieck, A.D., and Bayer, M. Nuclei-induced frequency focusing of electron spin coherence, *Science* **317**, 1896–1899 (2007); Glazov, M.M. A. Yugova and Al.L. Efros, Electron spin synchronization induced by optical nuclear magnetic resonance feedback, arXiv:1103.3249.

[13] Bunkov, Yu.M., Alakshin, E.M., Gazizulin, R.R., Klochkov, A.V., Kuzmin, V.V., Safin, T.R., and Tagirov, M.S. Discovery of the classical Bose–Einstein condensation of magnons in solid antiferromagnets, *JETP Lett.*, **94**, 68–72 (2011).

[14] Bunkov, Yu. M., Alakshin, E. M., Gazizulin, R. R., Klochkov, A.V., Kuzmin, V.V., L'vov, V.S., and Tagirov, M.S. High T_c spin superfluidity in antiferromagnets, *Phys. Rev. Lett.* **108**, 177002 (2012).

[15] Pethick, C.J., and Smith, H. *Bose–Einstein Condensation in Dilute Gases*, Ed. by Cambridge University Press (2002); Leggett, A.J. *Rev. Mod. Phys.* **73**, 307 (2001).

[16] Pitaevskii, L., and Stringari, S. *Bose–Einstein Condensation* (Clarendon Press, Oxford, 2003).

[17] Einstein, A. Quantentheorie des einatomigen idealen Gases, Sitzungsberichte der Preussischen Akademie der Wissenschaften **1**, 3 (1925).

[18] Ruegg, C. *Nature* **423**, 63 (2003).

[19] Della Torre, E., Bennett, L.H., and Watson, R.E. Extension of the Bloch $T^{3/2}$ law to magnetic nanostructures: Bose–Einstein condensation, *Phys. Rev. Lett.*, **94**, 147210 (2005).

[20] Radu, T., Wilhelm, H., Yushankhai, V., Kovrizhin, D., Coldea, R., Tylczynski, Z., Lhmann, T., and Steglich, F. Bose–Einstein condensation of magnons in Cs_2CuCl_4, *Phys. Rev. Lett.*, **95**, 127202 (2005).

[21] Giamarchi, T., Rüegg, Ch., and Tchernyshyov, O. Bose–Einstein condensation in magnetic insulators, *Nature Phys.* **4**, 198–204 (2008).

[22] Kaul, S.N., and Mathew, S.P. Magnons as a Bose–Einstein condensate in nanocrystalline gadolinium, *Phys. Rev. Lett.* 106, 247204 (2011).

[23] Nikuni, T., Oshikawa, M., Oosawa, A., and Tanaka, H. Bose–Einstein condensation of dilute magnons in $TlCuCl_3$, *Phys. Rev. Lett.* **84**, 5868–5871 (2000).

[24] Kohn, W., and Sherrington, D. Two kinds of bosons and Bose condensates, *Rev. Mod. Phys.* **42**, 1–11 (1970).

[25] Radu, T., Wilhelm, H., Yushankhai, V., Kovrizhin, D., Coldea, R., Tylczynski, Z., Lühmann, T., and Steglich, F., Radu et al. reply, *Phys. Rev. Lett.* **98**, 039702 (2007).

[26] Mills, D.L. A comment on the letter by Radu T. et al., *Phys. Rev. Lett.* **98**, 039701 (2007).

[27] Landau, L.D., and Lifshitz, E.M. *Fluid Mechanics*, Pergamon, Oxford (1959).

[28] Zakharov, V.E., and Nazarenko, S.V. Dynamics of the Bose–Einstein condensation, *Physica* **D 201**, 203–211 (2005)

[29] Yang, C.N. Concept of off-diagonal long-range order and the quantum phases of liquid He and of superconductors, *Rev. Mod. Phys.* **34**, 694–704 (1962).

[30] Vollhardt, D., and Wölfle, P. *The Superfluid Phases of Helium 3*, Taylor and Francis, London (1990).

[31] Malomed, B.A., Dzyapko, O., Demidov, V.E., and Demokritov, S.O. The Ginzburg–Landau model of Bose–Einstein condensation of magnons, *Phys. Rev.* B **81**, 024418 (2010).

[32] Bunkov, Yu. M., and Volovik, G. E. Homogeneously precessing domains in ^{3}He-B, JETP, **76**, 794–801 (1993).

[33] Bunkov, Yu. M. Spin supercurrent and novel properties of NMR in superfluid ^{3}He, in: *Prog. Low Temp. Phys.* Vol XIV, p. 69, ed. W. Halperin, Elsevier, Amsterdam (1995).

[34] Fomin, I.A. Pulsed NMR in superfluid phases of ^{3}He, in: *Helium Three* ed. by Halperin, W.P. and Pitaevsky, L.P. Elsevier (1995).

[35] Bunkov, Yu.M., and Volovik, G.E. Bose–Einstein condensation of magnons in superfluid ^{3}He, *J. Low Temp. Phys.* **150**, 135–144 (2008).

[36] Volovik, G.E. Twenty years of magnon Bose condensation and spin current superfluidity in ^{3}He-B, *J. Low Temp. Phys.* **153**, 266–284 (2008).

[37] Bunkov, Yu.M. Spin supercurrent and coherent spin precession, London prize lecture, *J. Phys.: Condens. Matter* **21**, 164201 (2009).

[38] Bunkov, Yu. M., and Volovik, G. E. Magnon Bose–Einstein condensation and spin superfluidity, *J. Phys.: Condens. Matter* **22**, 164210 (2010).

[39] Bunkov, Yu.M., Dmitriev, V.V., and Mukharskiy, Yu.M. Twist oscillations of homogeneous precession domain in ^{3}He-B, *JETP Lett.* **43**, 168–171 (1986).

[40] Volovik, G.E. Linear momentum in ferromagnets, *J. Phys.* C **20**, L83–L87 (1987).

[41] Brataas, A., Kent, A.D., and Ohno, H. Current-induced torques in magnetic materials, *Nature Materials* **10**, 372–381 (2012).

[42] Stone, M. Magnus force on skyrmions in ferromagnets and quantum Hall systems, *Phys. Rev.* B **53**, 16573–16578 (1996).

[43] Wong, C.H., and Tserkovnyak, Y. Hydrodynamic theory of coupled current and magnetization dynamics in spin-textured ferromagnets *Phys. Rev.* B **80**, 184411 (2009).

[44] Dmitriev, V.V., and Fomin, I.A. Coherently precessing spin structure in a normal Fermi liquid, *JETP Lett.* **59**, 378–384 (1994).

[45] Fomin, I. A. Low-frequency oscillations of a precessing magnetic domain in ^{3}He-B, *JETP Lett.* **43**, 171–174 (1986); Spin waves in pulsed NMR experiments in the B phase of ^{3}He, JETP Lett. **28**, 334–336 (1978).

[46] Bunkov, Yu.M., Dmitriev, V.V., and Mukharskiy, Yu.M. Low frequency oscillations of the homogeneously precessing domain in ^{3}He-B, *Physica* B **178**, 196–201 (1992).

[47] Lokner, L., Feher, A., Kupka, M., Harakly, R., Scheibel, R., Bunkov, Yu.M., and Skyba, P. Surface oscillations of homogeneously precessing domain with axial symmetry, *Europhys. Lett.* **40**, 539, (1997).

[48] Volovik, G.E. Phonons in magnon superfluid and symmetry breaking field, *JETP Lett.* **87**, 639–640 (2008).

[49] Dmitriev, V.V., Zavjalov, V.V., and Zmeev, D.Ye. Spatially homogeneous oscillations of homogeneously precessing domain in ^{3}He-B, *J. Low Temp. Phys.* **138**, 765–770 (2005).

[50] Človečko, M., Gažo, E., Kupka, M., and Skyba, P. New non-Goldstone collective mode of BEC of magnons in superfluid ^{3}He-B, *Phys. Rev. Lett.* **100**, 155301 (2008).

[51] Borovik-Romanov, A.S., Bunkov, Yu.M., Dmitriev, V.V., and Mukharskiy, Yu.M. Phase slippage observations of spin supercurrent in ^{3}He-B, *JETP Lett.* **45**, 124 (1987).

[52] Borovik-Romanov, A.S., Bunkov, Yu.M., Dmitriev, V.V., Mukharskiy, Yu.M., and Sergatskov, D.A. Investigation of spin supercurrent in ^{3}He-B, *Phys. Rev. Lett.* **62**, 1631–1634 (1989).

[53] Bunkov, Yu.M., Dmitriev, V.V., Mukharskiy, Yu.M., and Tvalashvily, G.K. Superfluid spin current in a channel parallel to the magnetic field, *Sov. Phys. JETP* **67**, 300 (1988).

[54] Josephson, B.D. Possible new effects in superconducting tunnelling, *Phys. Lett.* **1**, 251–253 (1962).

[55] Borovik-Romanov, A.S., Bunkov, Yu.M., de Waard, A., Dmitriev, V.V. Makrotsieva, V., Mukharskiy, Yu.M., and Sergatskov, D.A. Observation of a spin supercurrent analog of the Josephson effect, *JETP Lett.* **47**, 478–482 (1988).

[56] Borovik-Romanov, A.S., Bunkov, Yu.M., Dmitriev, V.V. Mukharskiy, Yu.M., and Sergatskov, D.A. Josephson effect in spin supercurrent in ^{3}He-B, *AIP Conf. Proc.* **194**, 27 (1989).

[57] Borovik-Romanov, A.S., Bunkov, Yu.M., Dmitriev, V.V., Mukharskiy Yu.M., and Sergatskov, D.A. Observation of vortex-like spin supercurrent in ^{3}He-B, *Physica* B **165**, 649–650 (1990).

[58] Misirpashaev, T.Sh., and Volovik, G.E. Topology of coherent precession in superfluid ^{3}He-B, *JETP* **75**, 650–665 (1992).

[59] Salomaa, M.M., and Volovik, G.E. Quantized vortices in superfluid ^{3}He, *Rev. Mod. Phys.* **59**, 533–613 (1987).

[60] Fomin, I.A. Critical superfluid spin current in ^{3}He-B, *JETP Lett.* **45**, 106–108 (1987).

[61] Sonin, E.B. Superfluid transport of precession in ^{3}He-B, *JETP Lett.* **45**, 747–751 (1987).

[62] Sonin, E.B. Spin-precession vortex and spin-precession supercurrent stability in ^{3}He-B, *JETP Lett.* **88**, 238–242 (2008).

[63] Lagoudakis, K. G. et al. Quantized vortices in an exciton–polariton condensate, *Nature Phys.* **4**, 706–710 (2008).

[64] Volovik, G.E. and Mineev, V.P. Line and point singularities in superfluid ^{3}He, *JETP Lett.* **24**, 561–563 (1976).

[65] Kunimatsu, T., Sato, T., Izumina, K., Matsubara, A., Sasaki, Y., Kubota, M., Ishikawa, O., Mizusaki, T., and Bunkov, Yu.M. The orientation effect on superfluid ^{3}3He in anisotropic aerogel, *JETP Lett.* **86**, 216–220 (2007).

[66] Elbs, J., Bunkov, Yu. M., Collin, E., Godfrin, H., and Volovik, G. E. Strong orientational effect of stretched aerogel on the ^{3}He order parameter, *Phys. Rev. Lett.* **100**, 215304 (2008).

[67] Hunger, P., Bunkov, Y. M., Collin, E., and Godfrin, H. Evidence for magnon BEC in superfluid ^{3}He-B in squeezed aerogel, to be published.

[68] Fomin, I.A. Instability of homogeneous precession of magnetization in the superfluid A phase of He^3, *JETP Lett.* **30**, 164–166 (1979).

[69] Borovik-Romanov, A.S., Bunkov, Yu.M., Dmitriev, V.V., and Mukharskiy, Yu.M. Instability of homogeneous spin precession in superfluid ^{3}He-A, *JETP Lett.* **39**, 469–473 (1984).

[70] Bunkov, Yu.M., and Volovik, G.E. On the possibility of the homogeneously precessing domain in bulk ^{3}He-A, *Europhys. Lett.* **21**, 837–843 (1993).

[71] Volovik, G.E. Random anisotropy disorder in superfluid ^{3}He-A in aerogel, *Pis'ma ZhETF* **84**, 533–538 (2006), *JETP Lett.* **84**, 455 (2006).

[72] Sato, T., Kunimatsu, T., Izumina, K., Matsubara, A., Kubota, M., Mizusaki, T., and Bunkov, Yu.M. Coherent precession of magnetization in the superfluid ^{3}He A-phase, *Phys. Rev. Lett.* **101**, 055301 (2008).

[73] Bunkov, Yu.M., and Volovik, G.E. Magnon BEC in superfluid ^{3}He-A, *JETP Lett.* **89**, 306–310 (2009).

[74] Hunger, P., Bunkov, Y. M., Collin, E., and Godfrin, H. Evidence for magnon BEC in superfluid ^{3}He-A, *J. Low Temp. Phys.* **158**, 129–134 (2010).

[75] Bunkov, Yu.M. Persistent signal; coherent NMR state trapped by orbital texture, *J. Low Temp. Phys.* **138**, 753–758 (2005).

[76] Bunkov, Yu.M., and Volovik, G.E. Magnon condensation into a Q-ball in ^{3}He-B, *Phys. Rev. Lett.* **98**, 265302 (2007).

[77] Coleman, S.R. Q-balls, *Nucl. Phys.* B **262**, 263–283 (1985).

[78] Enqvist, K., and Mazumdar, A. Cosmological consequences of MSSM flat directions, *Phys. Rept.* **380**, 99–234 (2003).

[79] Kusenko, A., Kuzmin, V., Shaposhnikov, M., and Tinyakov, P.G. Experimental signatures of supersymmetric dark-matter Q-balls, *Phys. Rev. Lett.* **80**, 3185–3188 (1998).

[80] Enqvist, K., and Laine, M. Q-ball dynamics from atomic Bose–Einstein condensates, *JCAP* 0308 (2003) 003.

[81] Bunkov, Yu.M., Dmitriev, V.V., Mukharskiy, Yu.M., Nyeki, J., and Sergatskov, D.A. Catastrophic relaxation in 3He-B at $0.4\,T_c$, *Europhys. Lett.* **8**, 645–649 (1989).

[82] Bunkov, Yu.M., Lvov, V.S., and Volovik, G.E. Solution of the problem of catastrophic relaxation of homogeneous spin precession in superfluid ^{3}He-B, *JETP Lett.* **83**, 530–535 (2006).

[83] Bunkov, Yu.M., Lvov, V.S., and Volovik, G.E. On the problem of catastrophic relaxation in superfluid ^{3}He-B, *JETP Lett.* **84**, 289–293 (2006).

[84] Bunkov, Yu.M., Fisher, S.N., Guenault, A.M., and Pickett, G.R. Persistent spin precession in ^{3}He-B in the regime of vanishing quasiparticle density, *Phys. Rev. Lett.* **69**, 3092–3095 (1992).

[85] Bunkov, Yu.M., Fisher, S.N., Guenault, A.M., Pickett, G.R., and Zakazov, S.R. Persistent spin precession in superfluid ^{3}He-B, *Physica* B **194**, 827–828 (1994).

[86] Cousins, D. J., Fisher, S. N., Gregory, A. I., Pickett, G. R., and Shaw, N. S. Persistent coherent spin precession in superfluid ^{3}He-B driven by off-resonant excitation, *Phys. Rev. Lett.* **82**, 4484–4487 (1999).

[87] Chen, A.S., Bunkov, Yu.M., Godfrin, H., Schanen, R., and Scheffer, F. Nonlinear spin waves on topological defects of texture in superfluid 3He, *J. Low Temp. Phys,* **110**, 51–56 (1998).

[88] Chen, A.S., Bunkov, Yu. M., Godfrin, H., Schanen, R., and Scheffler, F. Nonlinear stationary spin-waves in superfluid 3He-B at ultra-low temperatures, *J. Low Temp. Phys.*, **113**, 693–698 (1998).

[89] Autti, S., Bunkov, Yu.M., Eltsov, V.B., Heikkinen, P.J., Hosio, J.J., Hunger, P., Krusius, M., and Volovik, G.E. Self-trapping of magnon Bose–Einstein condensates in the ground state and on excited levels: From harmonic to box confinement, *Phys. Rev. Lett.* **108**, 145303 (2012).

[90] Kupka, M., and Skyba, P. On the spin dynamics of ^{3}HeB at the minimum of spatially nonuniform external magnetic field, *Phys. Lett.* A **317**, 324–328 (2003).

[91] Friedberg, R., Lee, T. D., and Sirlin, A. Class of scalar-field soliton solutions in three space dimensions, *Phys. Rev.* D **13**, 2739–2761 (1976).

[92] Chodos, A., Jaffe, R.L., Johnson, K., Thorn, C.B., and Weisskopf, V. F. New extended model of hadrons, *Phys. Rev.* D **9**, 3471–3495 (1974); Chodos, A. Jaffe, R.L. Johnson, K. and Thorn, C.B. Baryon structure in the bag theory, *Phys. Rev.* D **10**, 2599–2604 (1974).

[93] Jaffe, R. L. Multiquark hadrons. I. Phenomenology of $Q^2\bar{Q}^2$ mesons *Phys. Rev.* D **15**, 267–280 (1977); Multiquark hadrons. II. Methods, *Phys. Rev.* D **15**, 281–289 (1977).

[94] Strottman, D. Multiquark baryons and the MIT bag model, *Phys. Rev.* D **20**, 748–767 (1979).

[95] Salomaa, M.M., and Williams, G.A. Structure and stability of multielectron bubbles in liquid helium, *Phys. Rev. Lett.* **47**, 1730–1733 (1981); Silvera, I.F. Blanchfield, J. and Tempere, J. Stability of multielectron bubbles against single-electron split-off, *Phys. Stat. Sol.* B **237**, 274–279 (2003).

[96] Bradley, D.I., Clubb, D.O., Fisher, S.N., Guenault, A.M., Matthews, C.J., Pickett, G.R., and Skyba, P. Spatial manipulation of the persistent precessing spin domain in superfluid ^{3}He-B, *J. Low Temp. Phys.* **134**, 351–356 (1998).

[97] Autti, S., Eltsov, V.B., and Volovik, G.E. Bose analogs of MIT bag model of hadrons in coherent precession, *Pis'ma ZhETF* **95**, 610–614 (2012); *JETP Lett.* **95**, (2012); arXiv:1204.3423.

[98] Ramos, E.R.F., Henn, E.A.L., Seman, J.A., Caracanhas, M.A., Magalhaes, K.M.F., Helmerson, K., Yukalov, V.I., and Bagnato, V.S. Generation of nonground-state Bose–Einstein condensates by modulating atomic interactions, *Phys. Rev.* A **78**, 063412 (2008).

[99] Snoke, D. Coherent questions, *Nature* **443**, 403–404 (2006).

[100] Bunkov, Yu.M. Principles of HPD NMR spectroscopy of ^{3}He-B, *Physica* B**178**, 187–195 (1992).

[101] Korhonen, J.S., Bunkov, Yu.M., Dmitriev, V.V., Kondo, Y., Krusius, M., Mukharskiy, Yu.M., Parts, U., and Thuneberg, E.V. Homogeneous spin precession in rotating vortex-free ^{3}He-B; measurements of the superfluid density anisotropy, *Phys. Rev.* B **46**, 13983–13990 (1992).

[102] Kondo, Y., Korhonen, J.S., Krusius, M., Dmitriev, V.V., Mukharsky, Yu.M., Sonin, E.B., and Volovik, G.E. Direct observation of the nonaxisymmetric vortex in superfluid ^{3}He-B, *Phys. Rev. Lett.* **67**, 81–84 (1991).

[103] Thuneberg, E.V. Identification of vortices in superfluid ^{3}He-B, *Phys. Rev. Lett.* **56**, 359–362 (1986).

[104] Salomaa, M.M., and Volovik, G.E. Vortices with spontaneously broken axisymmetry in ^{3}He-B, *Phys. Rev. Lett.* **56**, 363–366 (1986).

[105] Witten, E. Superconducting strings, *Nucl. Phys.* **B249**, 557–592 (1985).

[106] Kondo, Y., Korhonen, J.S., Krusius, M., Dmitriev, V.V., Thuneberg, E.V., and Volovik, G.E. Combined spin-mass vortex with soliton tail in superfluid ^{3}He-B, *Phys. Rev. Lett.* **68**, 3331–3334 (1992).

[107] Dmitriev, V.V., Zavjalov, V.V., Zmeev, D.E., Kosarev, I.V., and Mulders, N. Nonlinear NMR in a superfluid B phase of ^{3}He in aerogel, *JETP Lett.* **76**, 321 (2002); Dmitriev, V.V., Zavjalov, V.V., and Zmeev, D.Ye. Measurements of the Leggett frequency in ^{3}He-B in aerogel, *JETP Lett.* **76**, 499 (2004).

[108] Bunkov, Yu.M., Collin, E., Godfrin, H., and Harakaly, R. Topological defects and coherent magnetization precession of ^{3}He in aerogel, *Physica* B **329-333**, 305–306 (2003).

[109] Kunimatsu, T., Matsubara, A., Izumina, K., Sato, T., Kubota, M., Takagi, T., Bunkov, Yu.M., and Mizusaki, T. Quantum fluid dynamics of rotating superfluid ^{3}He in aerogel, *J. Low Temp. Phys.* **150**, 435–444 (2008).

[110] Beenakker, C. W. J. Search for Majorana fermions in superconductors, arXiv:1112.1950.

[111] Hasan, M.Z., and Kane, C.L. Topological Insulators, *Rev. Mod. Phys.* **82**, 3045–3067 (2010).

[112] Qi, Xiao-Liang, and Zhang, Shou-Cheng. Topological insulators and superconductors, *Rev. Mod. Phys.* **83**, 1057–1110 (2011).

[113] Schnyder, A.P., Ryu, S., Furusaki, A., and Ludwig, A.W.W. Classification of topological insulators and superconductors, *AIP Conf. Proc.* **1134**, 10–21 (2009); arXiv:0905.2029.

[114] Chung, Suk Bum, and Zhang, Shou-Cheng. Detecting the Majorana fermion surface state of ^{3}He-B through spin relaxation, *Phys. Rev. Lett.* **103**, 235301 (2009).

[115] Volovik, G.E. Topological invariant for superfluid ^{3}He-B and quantum phase transitions, *JETP Lett.* **90**, 587 (2009).

[116] Davis, J.P., Pollanen, J., Choi, H., Sauls, J.A., Halperin, W.P., and Vorontsov, A.B. Anomalous attenuation of transverse sound in ^{3}He, *Phys. Rev. Lett.* **101**, 085301 (2008).

[117] Murakawa, S., Tamura, Y., Wada, Y., Wasai, M., Saitoh, M., Aoki, Y., Nomura, R., Okuda, Y., Nagato, Y., Yamamoto, M., Higashitani, S., and Nagai, K. New

anomaly in transverse acoustic impedance of superfluid ^{3}He-B with a wall coated by several layers of ^{4}He, *Phys. Rev. Lett.* **103**, 155301 (2009).

[118] Nunes, Jr. G., Jin, C., Hawthorne, D.L., Putnam, A.M., and Lee, D.M. Spin-polarized ^{3}He-^{4}He solutions: Longitudinal spin diffusion and nonlinear spin dynamics, *Phys. Rev.* B **46**, 9082–9103 (1992).

[119] Dmitriev, V.V., Zakazov, S.R., and Moroz, V.V. Coherently precessing magnetization structure in normal ^{3}He in pulsed NMR, *JETP Lett.* **61**, 309–315 (1995).

[120] Hohenberg, P.C. in: Critical Phenomena. *Proc. Int. School Phys.* "Enrico Fermi". Course LI., Acad. Press, NY (1971).

[121] Markelov, A.V. Josephson effect on a spin current, *JETP* **67**, 520–523 (1988).

[122] Bunkov, Yu.M., Fisher, S.N., Guenault, A.M., and Pickett, G.R. Resonant observation of the Landau field in superfluid ^{3}He by NMR, *Phys. Rev. Lett.* **68**, 600–603 (1992).

[123] Bunkov, Yu.M., and Golo, V.L. A chaotic regime of internal precession in ^{3}He, *J. Low Temp. Phys.* **90**, 167–179 (1993).

[124] Bunkov, Yu.M., Fisher, S.N., Guenault, A.M., Kennedy, C.J., and Pickett, G.R. A new NMR mode in the Landau field in superfluid ^{3}He-B, *J. Low Temp. Phys.* **89**, 27–36 (1992).

[125] Sourovtsev, E.V., and Fomin, I.A. Parametric instability of uniform spin precession in superfluid ^{3}He-B, *JETP Lett.* **83**, 410 (2006).

[126] Bunkov, Yu. M. NMR in superfluid helium-3 in the non-hydrodynamic regime, *J. of Low Temp. Phys.* **135**, 337–359 (2004).

[127] Bunkov, Yu. M., and Golo, V.L. Spin-orbit dynamics in the B-phase of superfluid helium-3, *J. Low Temp. Phys.* **137**, 625–654 (2004).

[128] Bunkov, Yu.M., Timofeevskaya, O.D., Volovik, G.E. Nonwetting conditions for coherent quantum precession in superfluid ^{3}He-B, *Phys. Rev. Lett.* **73**, 1817–1820 (1994).

[129] Suhl, H. The theory of ferromagnetic resonance at high signal powers, *J. Phys. Chem. Solids,* **1**, 209–227 (1959).

[130] L'vov, V.S. Solitons and nonlinear phenomena in parametrically excited spin waves, in Chap. 5, *Solitons* eds. Trullinger, S.E. Zakharov, V.E. and Pokrovsky, V.L. pp. 243–298, North-Holland Physics Publishing, Amsterdam 1986.

[131] Matsushita, T., Nomura, R., Hensley, H. H., Shiga, H., and Mizusaki, T. Spin dynamics and onset of Suhl instability in bcc solid ^{3}He in the nuclear-ordered U2D2 phase, *J. Low Temp. Phys.* **105**, 67–92 (1996).

[132] Fomin, I.A. Instability of homogeneous precession of magnetization in the superfluid A phase of He^3, *JETP Lett.* **30**, 164–166 (1979).

[133] Bunkov, Yu. M., and Golo, V.L. Spin-orbit dynamics in the B-phase of superfluid helium-3, *J. Low Temp. Phys.* **137**, 625–654 (2004).

[134] Borovik-Romanov, A.S., Bunkov, Yu.M., Dmitriev, V.V., and Mukharsky, Yu.M. AIP conference proceedings **194**, 15 (1989).

[135] Bunkov, Yu.M., Dmitriev, V.V., Mukharsky, Yu.M., Nyeki, J., Sergatskov, D.A., and Fomin, I.A. Instability of the homogeneous precession in ^{3}He- B (catastrophic relaxation), *Physica* B, **165**, 675–676 (1990).

[136] Mineev, V.P., and Volovik, G.E. Electric dipole moment and spin supercurrent in superfluid ^{3}He-B, *J. Low Temp. Phys.* **89**, 823–830 (1992).

[137] Volovik, G.E. Spontaneous electric polarization of vortices in superfluid ^{3}He, *JETP Lett.* **39**, 200–203 (1984).

[138] Dyakonov, M.I., and Perel, V.I. Possibility of orienting electron spins with current, *JETP Lett.* **13**, 467–469 (1971).

[139] Leurs, B.W.A., Nazario, Z., Santiago, D.I., and Zaanen, J. Non-Abelian hydrodynamics and the flow of spin in spin orbit coupled substances, *Annals of Physics,* **323**, 907–945 (2008).

[140] Berche, B., Medina, E., and Lopez, A. Spin superfluidity and spin–orbit gauge symmetry fixing, *EPL* **97**, 67007 (2012).

[141] Misirpashaev, T.Sh., and Volovik, G.E. Collective modes of Larmor precession: transverse NMR on HPD in superfluid ^{3}He-B, *J. Low Temp. Phys.*, 885–895 (1992).

[142] Goldhaber, A. S. Comment on "Topological quantum effects for neutral particles", *Phys. Rev. Lett.* **62**, 482 (1989); J. Fröhlich and Studer, U. M. $U(1) \times SU(2)$-gauge invariance of non-relativistic quantum mechanics, and generalized Hall effects, *Comm. Math. Phys.* **148**, 553–600 (1992).

[143] Volovik, G.E., and Yakovenko, V.M. Fractional charge, spin and statistics of solitons in superfluid ^{3}He film, *J. Phys.: Cond. Matter* **1**, 5263–5974 (1989).

[144] Volovik, G.E. *Exotic Properties of Superfluid ^{3}He*, World Scientific, Singapore, 1992.

[145] Volovik, G.E. Fractional statistics and analogs of quantum Hall effect in superfluid ^{3}He films, AIP Conference Proceedings **194**, 136–146 (1989).

[146] Awschalom, D., and Samarth, N. Spintronics without magnetism, *Physics* **2**, 50 (2009).

[147] Haldane, F.D.M., and Arovas, D.P. Quantized spin currents in 2-dimensional chiral magnets, *Phys. Rev.* B **52**, 4223–4225 (1995).

[148] Senthil, T., Marston, J.B., and Fisher, M.P.A. The spin quantum Hall effect in unconventional superconductors, *Phys. Rev.* B **60**, 4245–4254 (1999).

[149] Kou, Su-Peng, Qi, Xiao-Liang, and Weng, Zheng-Yu. Spin Hall effect in a doped Mott insulator, *Phys. Rev.* B **72**, 165114 (2005).

[150] Son, D.T., and Zhitnitsky, A.R.: Quantum anomalies in dense matter, *Phys. Rev.* D **70**, 074018 (2004).

[151] Klinkhamer, F.R. Dynamically broken Lorentz invariance from the Higgs sector? arXiv:1202.0531.

5 Superfluid helium three in aerogel: experiment

Yoonseok Lee and Richard P. Haley

5.1 Introduction

Impurities, and in a broader sense disorder, might be considered a nuisance for understanding the intrinsic properties of a system. Indeed, a tremendous amount of effort has been expended on eliminating unwanted disorder. However, as our appreciation of the role of disorder grew, scientists started to embrace it from a different perspective, and even began to introduce impurities as a strategy to investigate specific processes. Nowadays the physics of disorder forms a common theme that underlies large areas of research, and its impact ranges from the highly technical to the purely academic. The presence of disorder is crucial in the manufacture of practical superconducting wires, and doping controlled species and amounts of impurities into semiconducting materials makes modern electronic magic a reality. Furthermore, the influence of disorder on ordered states, principally magnetic or superconducting phases, has been the subject of intense research, displaying diverse and fascinating phenomena such as quantum phase transitions, universal behavior in the critical region, and the emergence of novel phases.

The early theoretical works of Anderson [6], Abrikosov and Gor'kov [2], and Larkin [108] have revealed fundamental insights into the role played by an impurity as a pair-breaking agent in conventional as well as unconventional superconductors. The response of Cooper pairs to various types of impurity depends on the symmetry of the order parameter. The strong influence of a small concentration of paramagnetic impurities on a low temperature superconductor is in contrast to its insensitivity to non-magnetic impurities. Since the early 1980s many different classes of novel superconductors have been discovered and a majority of them are known to have unconventional pairing, whose extreme sensitivity to any impurities or disorder has been well documented and even utilized as the first indicator of their unconventional nature [11]. In these systems it is imperative to understand the role of disorder, not only to characterize the superconducting states, but also to identify the nature of the pairing mechanism.

At first sight liquid ^{3}He might seem an unusual choice of material for a systematic investigation of this topic. The extreme low temperature environment that is necessary to study the liquid phase excludes any possibility of introducing static impurities or quenched disorder in a conventional way. Furthermore, their distinct quantum nature makes ^{3}He and ^{4}He the only isotopic system that spontaneously phase separates at low temperatures. The extremely low solubility of ^{4}He in liquid ^{3}He near the superfluid transition is estimated to be on the order of one mole of ^{4}He in a volume of ^{3}He equal to that of the universe. Therefore liquid ^{3}He presents us with a practically disorder free, homogeneous, and isotropic neutral fermionic system, which is arguably one of the best understood physical systems on a quantitative level. The superfluid phases of ^{3}He emerge below ≈ 2 mK as a consequence of unconventional BCS pairing with *p-wave* symmetry. Unlike other unconventional superconductors, the structure of the order parameters of the superfluid phases of ^{3}He have been clearly identified and confirmed through numerous theoretical and experimental studies [117, 169, 54]. In spite of our clear understanding of the symmetry and the intrinsic properties of the phases, efforts to probe disordered superfluid ^{3}He had been hampered until the advent of high porosity aerogel.

In 1995, groups at Cornell University [140] and Northwestern University [160] conducted experiments on liquid ^{3}He filling 98.2% porosity silica aerogel samples, provided by Moses Chan's group at Pennsylvania State University. Their experiments were originally motivated by the discovery by Chan's group [101] of the striking influence of the same porosity aerogel on the phase diagram of ^{3}He–^{4}He mixtures: the λ-transition line is completely detached from the boundary of the coexistence region, giving rise to the intriguing possibility of a miscible superfluid mixture at high ^{3}He concentration and low temperatures. Both groups initiated experiments to investigate the effects of aerogel on pure liquid ^{3}He, and their first steps would evolve into a lasting endeavor. In their measurements, using torsional oscillators at Cornell and nuclear magnetic resonance (NMR) at Northwestern, surprisingly sharp superfluid transition features were observed at substantially lower temperatures than expected in pure bulk liquid. The results of these experiments triggered immediate interest, because the observed behavior was markedly different from what had been seen previously in confined geometries using conventional porous materials [92, 155] or parallel plates [66]. In spite of the rather wide range of pore size—to be more precise, length scales—in aerogel, the superfluid transition could be determined within a percent and the onset of the suppressed transition showed a strong pressure dependence. In contrast, the superfluid transition in packed powders exhibits a generic behavior with a severely rounded tail reaching very close to the bulk transition. For example, the heat capacity measurement by Schrenk and König performed in a silver sinter of ≈ 100 nm pores shows a rather broad peak about 0.5 mK below the onset temperature that is near the bulk transition [157], as shown in Fig. 5.1. This type of behavior has been understood in terms of a healing length arising from diffusive surface scattering [92]. The onset of superfluidity in a pore of radius R occurs at the temperature where $R \approx \xi(T)$, the temperature-dependent coherence length. Therefore a broad distribution of pore sizes results in a large variance in T_c since $\xi(T) \propto (1 - T/T_c)^{-1/2}$. However, the superfluid transition

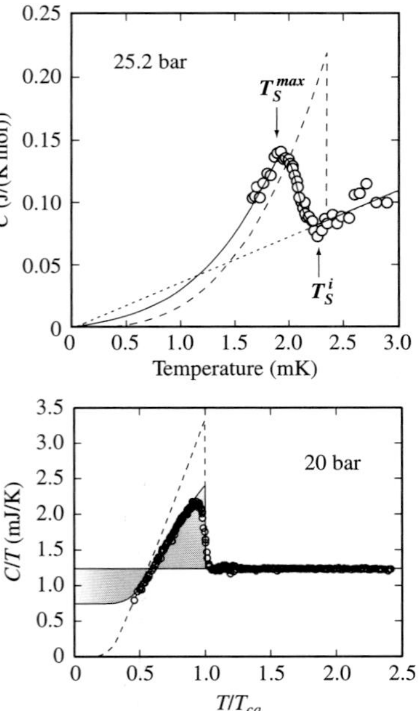

Fig. 5.1 Top: Temperature dependence of the specific heat of liquid ^{3}He in a silver sinter at 25.2 bar, from [157]. Bottom: Temperature dependence of the specific heat of liquid ^{3}He in 98% aerogel at 20 bar, from [37]. T_S^i indicates the onset of the superfluid and T_S^{max} marks the temperature at which the specific heat has its maximum value. The specific heat of bulk ^{3}He is shown by the dashed lines for comparison. The bulk superfluid transition at 20 bar is at a temperature 30% higher than the superfluid transition in aerogel.

in aerogel exhibited distinct features that did not fit in with the healing length scenario [140, 160]. The unique nature of aerogel, differing from conventional porous materials, had also been recognized in studies on liquid ^{4}He and ^{3}He-^{4}He mixtures [34]. Its tenuous fibrillar structure and large porosity facilitate an open geometry. Thus the observation of superfluid transitions in aerogel has furnished us with a new technique for introducing static disorder/impurities rather than confinement. With a great deal of quantitative understanding of the intrinsic properties of superfluid ^{3}He, the ^{3}He-aerogel system has become an exemplar for a systematic investigation of the effects of static disorder in unconventional superfluids.

There are several fascinating aspects to ^{3}He-aerogel that cannot easily be realized in other condensed matter systems. First, a wide range of impurity pair-breaking behavior can be explored by continuously varying the sample pressure without altering the degree of disorder, i.e. the porosity of aerogel. The bulk coherence length, $\xi_o = \hbar v_F / \pi \Delta_o$, varies from 17 to 85 nm in the pressure range 34–0 bar, which allows for the investigation of a significant range of the pair-breaking parameter, $x \equiv \xi_o/\ell$, for a given aerogel porosity. Here v_F, Δ_o, and ℓ are the Fermi velocity, the zero temperature superfluid gap, and the mean free path, respectively. Second, the nature of impurity scattering can be adjusted by modifying the composition of the surface layers. The ^{3}He atoms adsorbed on the surface of aerogel form localized moments and can behave as magnetic scattering centers, in addition to the potential scattering from the aerogel strand itself. The scattering through the spin channel can be turned off by systematically replacing the ^{3}He solid layers with non-magnetic ^{4}He atoms. Finally, the structure of aerogel, being formed from an entangled network of nanometer scale SiO_2 strands, presents more than just conventional randomly distributed isotropic scattering centers, and the correlated structure inevitably introduces anisotropic disorder. Thus the effect of disorder is not simply limited to the suppression of superfluidity by pair-breaking, and provides a rare situation in which one can study the role of anisotropic disorder on an anisotropic superfluid with continuous symmetry.

Numerous experimental and theoretical efforts have been dedicated to understanding this exciting system. To date, a few hundred papers have been published on this subject. Although there are issues that remain to be resolved or verified experimentally, the subject is mature enough to warrant an extensive review. There is already a small section in a monograph by Dobbs [54] and a brief summary by Halperin and Sauls [83] in which earlier results are summarized. A recent concise but informative review by Halperin et al. [82] highlights a selected number of experimental results with an extensive discussion of Ginzburg–Landau (G–L) theory. In this chapter we have attempted to provide a comprehensive survey of the experiments conducted in the low temperature phases of liquid ^{3}He in aerogel. The companion chapter by Sauls in Volume 2 provides a full review of theoretical results and ideas including the G–L theory of disordered superfluid ^{3}He and the various impurity scattering models. We refer readers to many wonderful monographs and reviews [117, 169, 54] for general background material on normal and superfluid ^{3}He.

In the following section the structure and the relevant physical properties of aerogel are discussed, with a brief description of the aerogel growth process.

An overview of the current status is outlined in Section 5.3, which is followed by the main sections of experimental results. There are about a dozen experimental groups that are actively involved in this research. Each of them employs one or two specific measurement techniques and we chose to review the experimental results categorized by measurement technique. An investigation combining more than one technique on the same aerogel sample is a rare event. Therefore, we focus on the evolution and development of experimental studies using common techniques. The first two experiments seem to have set the preferred sample porosity used by most groups, and so a majority of the research has been performed using nominally 98% porosity aerogels. Throughout this chapter, we will use 98% as the default porosity, and we will make note of any different porosities where necessary for clarity.

5.2 Physical properties of silica aerogel

5.2.1 Preparation of aerogel

Aerogel is a highly porous material composed of a cross-linked network of chains that are a few nanometers in diameter. This unusual structure is established through a sol–gel process in which nanometer-sized silica clusters are polymerized into the final formation in a solvent. Aerogel is then produced by replacing the solvent with air. However, if the liquid component is extracted from the gel simply by drying out the solvent, this inevitably causes a significant amount of shrinkage and/or structural damage owing to capillary action. In 1931, Kistler [102], who coined the term aerogel, found a clever way of circumventing this problem by employing the so-called supercritical extraction (SCE) process. Using this method, he first produced high porosity silica aerogel with little or no shrinkage. He was certainly intrigued by this new material and described [102] that "The silica aerogels are highly opalescent, although quite transparent; they display a glassy fracture and small pieces emit a metallic ring when dropped." His visual observation is wonderfully reflected in Fig. 5.2.

The standard method of making silica aerogel [67] starts with the hydrolysis of tetramethoxysilane (TMOS), $Si(OCH_3)_4$. With the addition of either basic or acidic catalyst to the mixture of water and TMOS, orthosilicic acid $Si(OH)_4$ and methanol are produced. In this solution $Si(OH)_4$ molecules start to break down, aggregate into silica clusters 3–5 nm in diameter, and finally coalesce into a gel structure:

$$Si(OCH_3)_4 + 4H_2O \longrightarrow Si(OH)_4 + 4CH_3OH \tag{5.1}$$

$$n Si(OH)_4 \longrightarrow (SiO_2)_n + 2nH_2O. \tag{5.2}$$

The supercritical condition is achieved in an autoclave by raising the temperature and pressure above the critical point of methanol at 512 K and 81 bar. Then the autoclave is depressurized slowly, extracting the hypercritical gas without destroying the structure. There are several variations adopted by various groups in the gelation and SCE processes. All of the aerogel samples used in the very early experiments were provided by Chan's group and a majority of the later samples were grown in Mulders' group at the University of

Fig. 5.2 A picture of a 98% porosity silica aerogel sample produced by Mulders. As described by Kistler, it is opalescent but transparent. The effect of Rayleigh scattering from the nanometer-sized aerogel strands is displayed by the hues of scattered light. (See Plate 12 for color image).

Delaware. Both of them used a two-step method in gelation [167] over the one-step method described in Eqs. (5.1) and (5.2), followed by the standard supercritical drying process. The aerogel samples used in the experiments in Osheroff's group at Stanford were manufactured by Poco and Hrubesh at Lawrence Livermore National Laboratory. They used both one-step and two-step methods and rapid supercritical extraction (RSCE) to shorten the length of drying time and to minimize possible shrinkage during the drying process. The Osaka City University group used exclusively aerogel samples produced by Yokoyama and Yokogawa at Matsushita Electric Works Ltd. using a method similar to that of Mulders [97]. Recently, Halperin's group at Northwestern started to produce their own aerogels using both one and two-step processes and a slightly modified RSCE [138].

It is known that aerogels of the same porosity but with different growth environments possess slightly different microscopic structures. This was linked to the variances in superfluid density and superfluid transition temperatures measured in the very early years in samples of the same porosity [142], notwithstanding that most of the aerogels were provided by a single source or grown under similar conditions. The most widely used 98% porosity aerogels produced by Mulders followed the two-step method using acetonitrile as the basic solvent to set the final aerogel density ($\approx 0.04\,\mathrm{g\,cm^{-3}}$). The gels are typically aged for two weeks and then supercritically dried at 290 °C. The resulting aerogels are hydrophobic and stable in air for several years. The specific surface area, as determined from helium adsorption isotherms, is about $1000\,\mathrm{m^2 g^{-1}}$.

5.2.2 Structure of aerogel: important length scales

Most of the physical properties of pure liquid ^{3}He are related to the Fermi wavelength λ_F, the coherence length ξ_o, and inelastic scattering mean free path ℓ_i. It is the interplay between these length scales and those introduced by the aerogel that gives rise to the observed disordered liquid ^{3}He phenomena. Therefore it is essential to have a clear understanding of the structure of aerogel. The percentage porosity P is defined by $P = 100(1 - \rho_a/\rho_{SiO_2})$, where $\rho_{a(SiO_2)}$ is the density of aerogel (solid amorphous SiO_2). With $\rho_{SiO_2} = 2.20\,\mathrm{g\,cm^{-3}}$, 98% aerogel has $\rho_a = 0.044\,\mathrm{g\,cm^{-3}}$, and for comparison the density of ^{3}He is $\approx 0.1\,\mathrm{g\,cm^{-3}}$. Aerogel is, in fact, the lightest solid material. An evacuated 99.95% aerogel piece actually has lower density than air. The fact that such a small solid fraction can maintain a three-dimensional structure indicates the existence of strong structural correlations. The first experimental confirmation of the aerogel structure was performed on a 96% porosity aerogel by Schaefer and Keefer [154] using small angle X-ray scattering (SAXS). Their primary goal was to establish the existence of a fractal porous structure, which had been postulated to exist in some porous media. They observed three distinct regimes in SAXS intensity $I(q)$, and identified two fractal regimes as a function of the momentum transfer q. Since then, SAXS has been the primary tool for the quantitative analysis of aerogel structure.

Detailed SAXS measurements have been conducted by the Cornell group on various samples of 98% and 99.5% porosity in which actual torsional oscillator measurements have been performed in liquid ^{3}He [142, 111]. Their measurements provided important insight into relating the structure to the physical properties in the superfluid phase. The aerogel structure can be described by four parameters determined from SAXS. Two length scales, the correlation length ζ_a and the diameter of silica cluster δ_a, separate the three distinct regimes in $I(q)$, and the two fractal dimensions, D_v and D_s, characterize the fractal regions defined by $I(q) \sim q^{-D_v, D_s}$. For $q^{-1} \gg \zeta_a$, $I(q)$ is q-independent, indicating that above the length scale ζ_a, the system appears homogeneous. On the other hand, for $q^{-1} \ll \delta_a$, aerogel is in the surface fractal regime where the scattered intensity arises predominantly from surface scattering. If one measures the surface area S within a sphere of radius R, $S \sim R^{6-D_s}$. In the

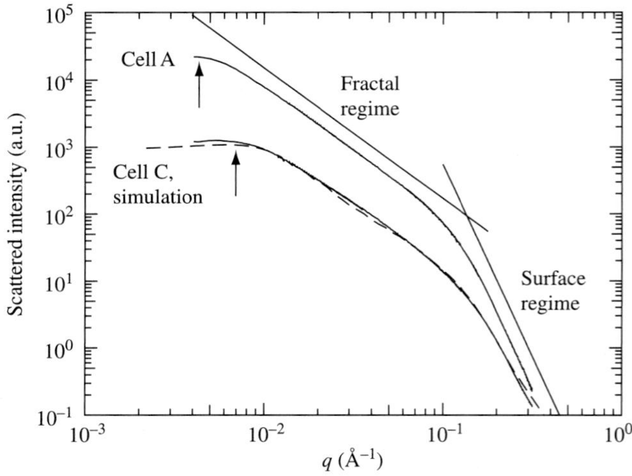

Fig. 5.3 Small angle X-ray scattering from two different aerogel samples of 98% porosity grown by Chan's group, from [142]. Although they have the same porosity, the scattered intensity shows a quantitative difference between the two samples. The dashed line is calculated based on the DLCA simulation described in the text. The arrows indicate the position of $q \simeq 2\pi/\zeta_a$.

intermediate length scale, $\delta_a \ll q^{-1} \ll \zeta_a$, $I(q) \sim q^{-D_v}$ in the volume fractal regime in which the volume or mass within a sphere of radius R follows $V \sim R^{D_v}$. They found $D_v \approx 1.7$–1.9 and $D_s \approx 4.5$–6 for all samples studied.

It is reasonable to expect ζ_a to be dependent on porosity. However, an interesting finding from the Cornell measurements is that ζ_a differs substantially even in the same porosity samples if they were prepared under slightly different conditions. Figure 5.3 shows SAXS results for two such 98% samples. Sample A is the first sample studied [141] and has $\zeta_a \approx 130$ nm. Unfortunately, the growth parameters for this sample were not well documented. Sample C, prepared with a protocol that most of the ensuing productions have adopted, has much shorter $\zeta_a \approx 84$ nm, although both samples exhibit qualitatively consistent behavior in $I(q)$. The length scale ζ_a is comparable to ξ_o, and significant differences in ζ_a are likely to strongly affect the superfluid properties. Indeed, in these two samples, substantial differences in superfluid transition temperature T_{ca} and superfluid density ρ_s^a have been observed [141, 142, 119] (and as shown later in Fig. 5.30). It is, however, worth noting that they found evidence for a scaling of both T_{ca} and ρ_s^a to ζ_a [111]. A SAXS study on aerogel samples with 97–98% porosity grown at Northwestern found a consistent $D_v \approx 1.55 - 1.80$ but significantly smaller $\zeta_a \approx 20$ nm [138]. Therefore, it seems clear that at least in base-catalyzed aerogels the correlation length depends strongly on the porosity as well as the growth conditions.

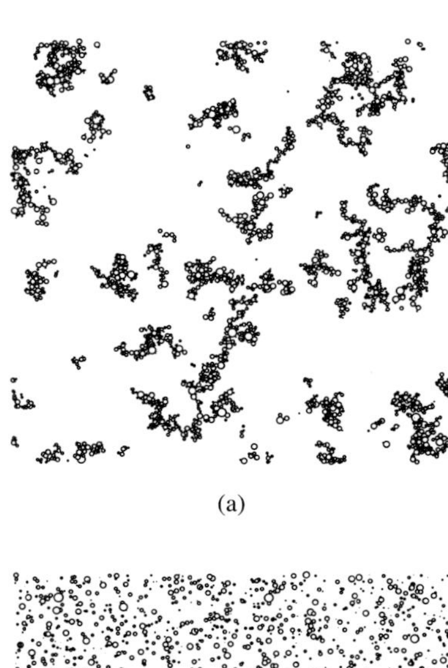

Fig. 5.4 (a) Projection of a 30 nm-thick slice of 98% aerogel, 350 nm on a side, simulated with the DLCA model. (b) Projection of a similar slice with a random arrangement of the same density. The particles represented by circles have an average diameter of 3 nm [142]. The juxtaposition of the two images clearly illustrates the open structure in aerogel and the presence of strong correlation.

To our knowledge, all aerogel samples used in ^{3}He have been base-catalyzed, and here the sol–gel process is known to be diffusion-limited and tends to produce smaller clusters [67]. For such aerogels, numerical simulations based on a diffusion-limited cluster aggregation (DLCA) model produce realistic structure representations that are consistent with SAXS [142, 79], as shown in Fig. 5.3. Unlike SAXS, the DLCA simulation provides real space information on the structure from which additional length scales can be extracted, shown in Fig. 5.4. The simulation of 98% aerogel [142] shows that half of the open volume is more than 10 nm away from the silica, but none of it is further away than 35 nm. This implies that on average a Cooper pair will encounter at least one silica strand [142]. One remaining important length scale is the geometrical mean free path of aerogel, ℓ_a, which sets the scattering length in the system at low temperatures. The observed value of D_v in aerogel indicates that a longer ℓ_a should be observed compared to a randomly distributed structure of the same porosity. This aspect was corroborated by the DLCA simulation performed by Porto and Parpia [142], from which a simple relation for the porosity dependent geometrical mean free path was found, given by

$$\ell_a(P) = 1.76\left(1 - \frac{P}{100}\right)^{-1.1} \text{ (nm)}. \tag{5.3}$$

For 98% aerogel this model estimates $\ell_a = 130$ nm, which is in excellent agreement with the values determined by spin diffusion (130 nm) [151] and sound attenuation (120 nm) [42], but somewhat different from the values obtained from thermal conductivity (92 nm) [146, 58] and heat capacity measurements (150–180 nm) [37].

In sum, the structure of aerogel, resulting from correlated aggregates of silica clusters, presents multiple length scales, namely δ_a, ζ_a, and ℓ_a. In the limit of $\delta_a, \zeta_a \to 0$, aerogel can in effect be viewed as a collection of randomly distributed point-like scattering centers. However, in liquid ^{3}He, where $\delta_a \approx 10\,\lambda_F$ and $\zeta_a \sim \xi_o < \ell_a$, this hierarchy in length scales leads directly to the observation of fascinating phenomena. For example, the pair-breaking parameter appearing in theoretical models [165, 166, 88, 84] should incorporate the correlation effect, in order to fully understand the suppression of the superfluid transition in the full pressure range [152]. Various experimental and theoretical analyses have narrowed down the values of the parameters: $\delta_a \approx$ 3–5 nm, $\zeta_a \approx$ 40–80 nm, and $\ell_a \approx$ 120–170 nm for 98% aerogel.

5.2.3 Physical properties of high porosity aerogel

The singular structure and extreme high porosity of these aerogels have led to a number of exceptional physical properties, which have prompted a wide variety of applications in science and technology [3]. Silica aerogels have been used extensively in Cerenkov counters in particle physics since the 1970s [33] and have recently been utilized to capture cosmic dust in outer space [31]. Their low thermal conductivity and high solar transmittance offer possibilities for use as superinsulating fillers for window systems and as a solar energy collector [147]. Here, we survey some of the properties pertinent to the ^{3}He work under discussion.

Aerogel is a compliant material with very low elastic constants, since elastic strain comes mainly from bending and wedging of silica strands rather than from the deformation of bulk silica. It has been reported that growth conditions and heat treatment alter the elastic properties of aerogel [78, 77, 156]. In the linear regime, both the Young's and bulk moduli follow a power-law dependence on density with an exponent ≈ 2.6 for higher porosity $P > 95\%$ [77]. The Young's modulus of 98% aerogel, ≈ 0.1 MPa, is many orders of magnitude smaller than ≈ 70 GPa of fused silica at room temperature. However, the Poisson's ratio of high porosity aerogel, $\approx$ 0.15–0.22, does not depend on porosity and is close to the value for fused silica of ≈ 0.23 [78, 138, 23]. Small elastic moduli result in low sound speeds, reported to be as low as $20\,\mathrm{m\,s^{-1}}$ [77]. The speed of transverse sound for 98% aerogel has been measured down to 400 mK, with $c_t \approx 90\,\mathrm{m\,s^{-1}}$ at the lowest temperature [44]. The speed of longitudinal sound for the same porosity is $c_a = c_\ell \approx 150\,\mathrm{m\,s^{-1}}$.

On average, aerogel looks homogeneous and isotropic on length scales larger than ζ_a. However, local fluctuations of the structure could create an intriguing situation where local anisotropy should come into play. There has been continued interest in the consequences of the interaction between anisotropic disorder presented by aerogel and the anisotropic order parameter of superfluid ^{3}He [170, 166, 64, 7, 168]. It has been proposed that this effect could be studied systematically by introducing controlled global anisotropy through uniaxial deformation of aerogel [168]. This scheme is particularly attractive considering its unusual mechanical properties. Two groups, at Northwestern and the University of Florida, conducted systematic studies to confirm and quantify

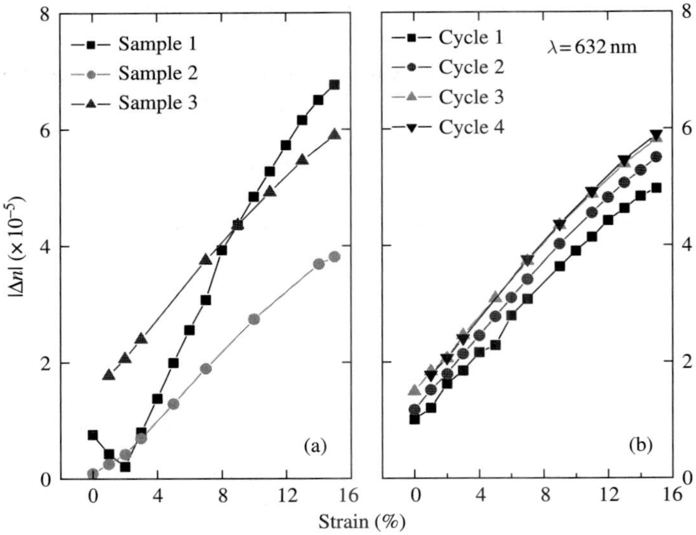

Fig. 5.5 (a) Optical birefringence Δn versus strain at 632 nm for three different samples of 98% porosity, from [23], showing the presence of anisotropy in uniaxially compressed aerogels. Samples 1 and 3 exhibit weak built-in anisotropy caused by multiple compressions or during the drying process. (b) Δn versus strain at 632 nm for Sample 3 at different cycles of compression from its virgin condition.

the global anisotropy induced in uniaxially compressed aerogels [138, 22, 23]. The SAXS measurements on compressed aerogels by the Northwestern group clearly showed anisotropy in ζ_a, shorter in the axial direction by up to 7 nm at 30% strain [138]. They also observed optical birefringence in the compressed samples, a hallmark of anisotropic dielectric materials. Optical birefringence, defined by $\Delta n = n_e - n_o$, where $n_{o(e)}$ is the index of refraction of the (extra)ordinary ray, showed a linear dependence on strain in the low compression limit [22, 23]. However, a significant level of hysteresis was observed when compressed beyond 5% strain, suggesting possible structural damage at higher strain levels [23], as shown in Fig. 5.5. An interesting aspect of these experiments is that they observed clear evidence of weak anisotropy present in some of the uncompressed samples [138, 22, 23]. This weak built-in anisotropy was prominent in a series of samples which exhibited preferential radial shrinkage that was probably caused during the supercritical extraction process [138].

It is evident that aerogel is susceptible to even small stresses. Significant deformation was observed during capillary condensation of liquid ^{4}He in high porosity aerogels [87, 100], despite its small surface tension, $\approx 90\,\mu\text{Nm}^{-1}$ at 4.2 K [87], being three orders of magnitude smaller than that of water at room temperature. These observations could provide a clue in understanding early experiments that showed different behavior even in same porosity samples of the same origin [142] and which surprisingly reported large bulk contributions in their measurements [160, 161]. Stress from differential thermal contraction between aerogel and the surrounding structure could also cause unwanted strain in aerogel. Although complications arising from capillary condensation can be minimized by introducing ^{3}He hypercritically, it is imperative to conduct structure characterization using SAXS and/or optical birefringence.

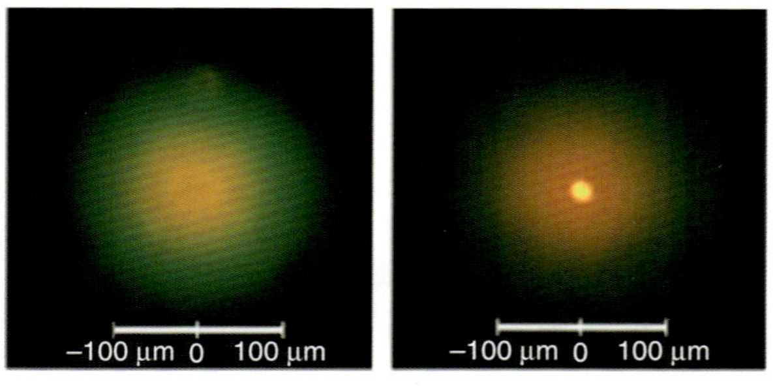

Plate 1 Images of the lateral intensity distribution in the cavity. The image to the left is recorded at a temperature $T > T_c$, while that to the right is for $T < T_c$. Note the distinct narrowing and that excitations progressively fall into the ground state harmonic oscillator level as the temperature falls. (After Klaers et al. (2010).)

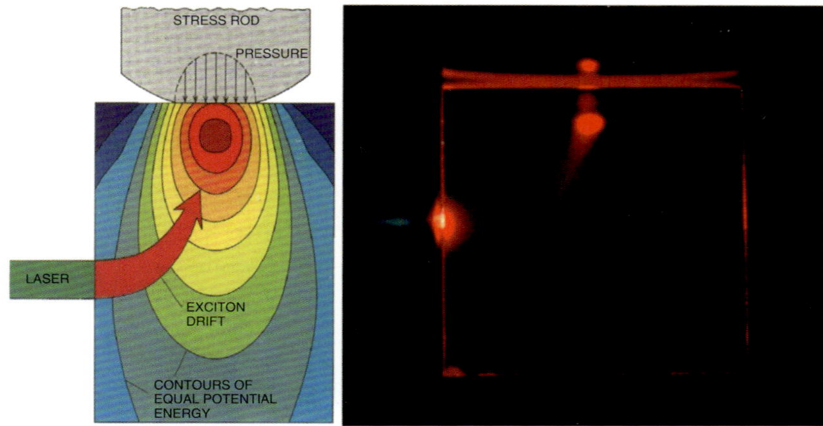

Plate 2 In the schematic drawing to the left we see a depiction of the strain contours arising from a curved piston which is pressed against the upper surface of the crystal; the maximum lies slightly below this surface. Also shown is an incoming laser beam (green) which produces excitons (red) which drift normal to the strain contours and are ultimately trapped in the vicinity of the maximum strain point. To the right we see the actual luminescence profile; the two very bright regions are where the laser beam enters the crystal and where the excitons accumulate respectively. (Courtesy of J. Wolfe.)

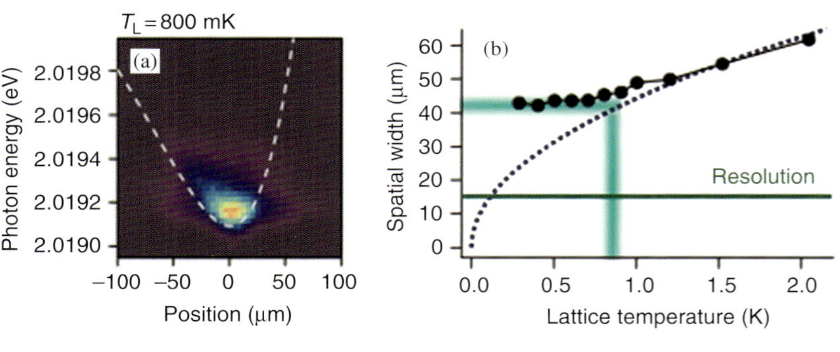

Plate 3 (a) A typical cloud luminescence profile. The dashed line shows the position dependence of the potential energy of the strain trap. (b) The temperature dependence of the spatial width of the cloud luminescence profile. Note the observed width tracks the lattice temperature down to about 800 mK, below which it is larger, implying an elevated temperature. The dashed line shows the trajectory for thermal equilibrium. (After Yoshioka et al. (2011).)

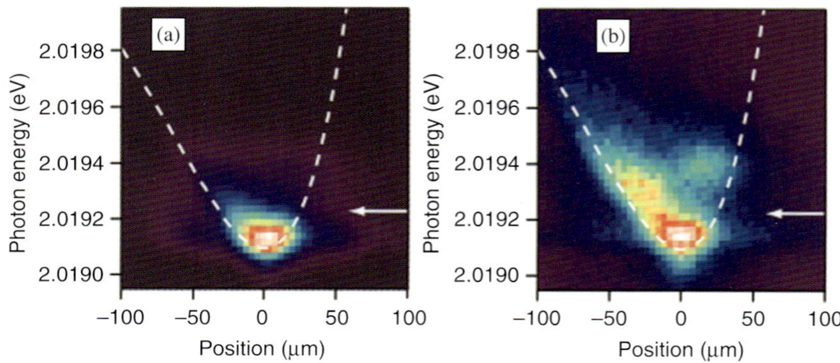

Plate 4 Luminescence signals taken at a lattice temperature of 354 mK, where the estimated number N of paraexcitons in the trap is (a) 2×10^7 and (b) 2×10^9. The dashed line shows the potential profile of the strain well, while the arrow shows the energy chosen to partition high- and low-energy excitations. (After Yoshioka et al. (2011).)

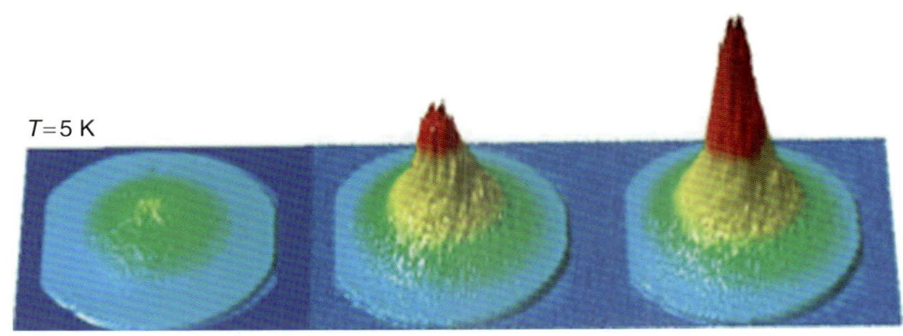

Plate 5 Far-field emission measured at 5 K for three excitaton intensities which are increasing to the right. Shown as pseudo-3D images of the farfield emission within an angular cone of $\pm 23^\circ$ with the emission intensity displayed on the vertical axis (in arbitrary units). With increasing excitation power, a sharp and intense peak is formed in the center of the emission distribution corresponding to emission angles $\theta_x = \theta_y = 0$, implying that they are associated with $k_\perp = 0$. (After Kasprzak et al. (2006).)

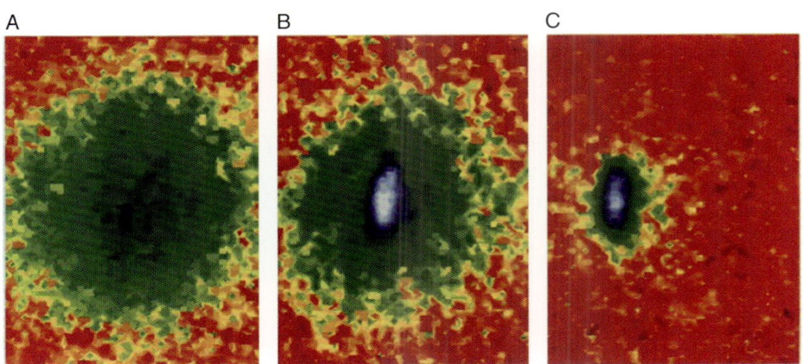

Plate 6 False color images of an expanded atom cloud. (A) Just before the appearance of a condensate; note the essentially isotropic (green & yellow) profile expected for a Maxwell distribution. (B) Slightly after the formation of a condensate; the elliptical blue-white region arises from the probability density of the condensate wave function emerging from the green & yellow background associated with the excited atoms. (C) A case where the condensate occupies a much larger fraction of the cloud (note the magnification is lower here). (After Anderson et al. (1995).)

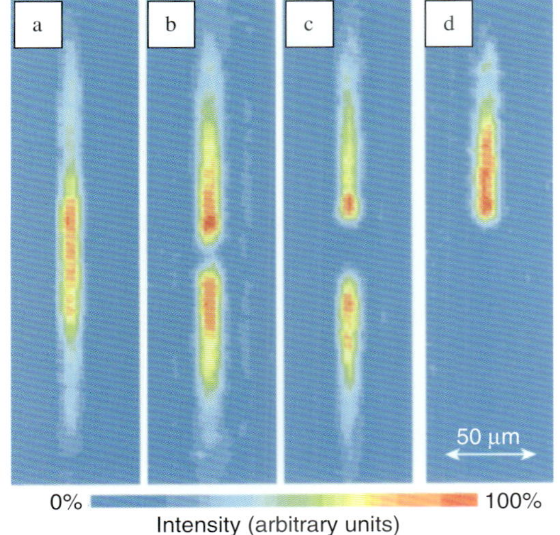

Plate 7 Phase-contrast images of: (a) a single Bose condensate; (b) and (c) double Bose condensates with different separations (achieved by altering the argon laser power); and (d) an initial double condensate after sweeping one half away. (Anderson et al. (1997).)

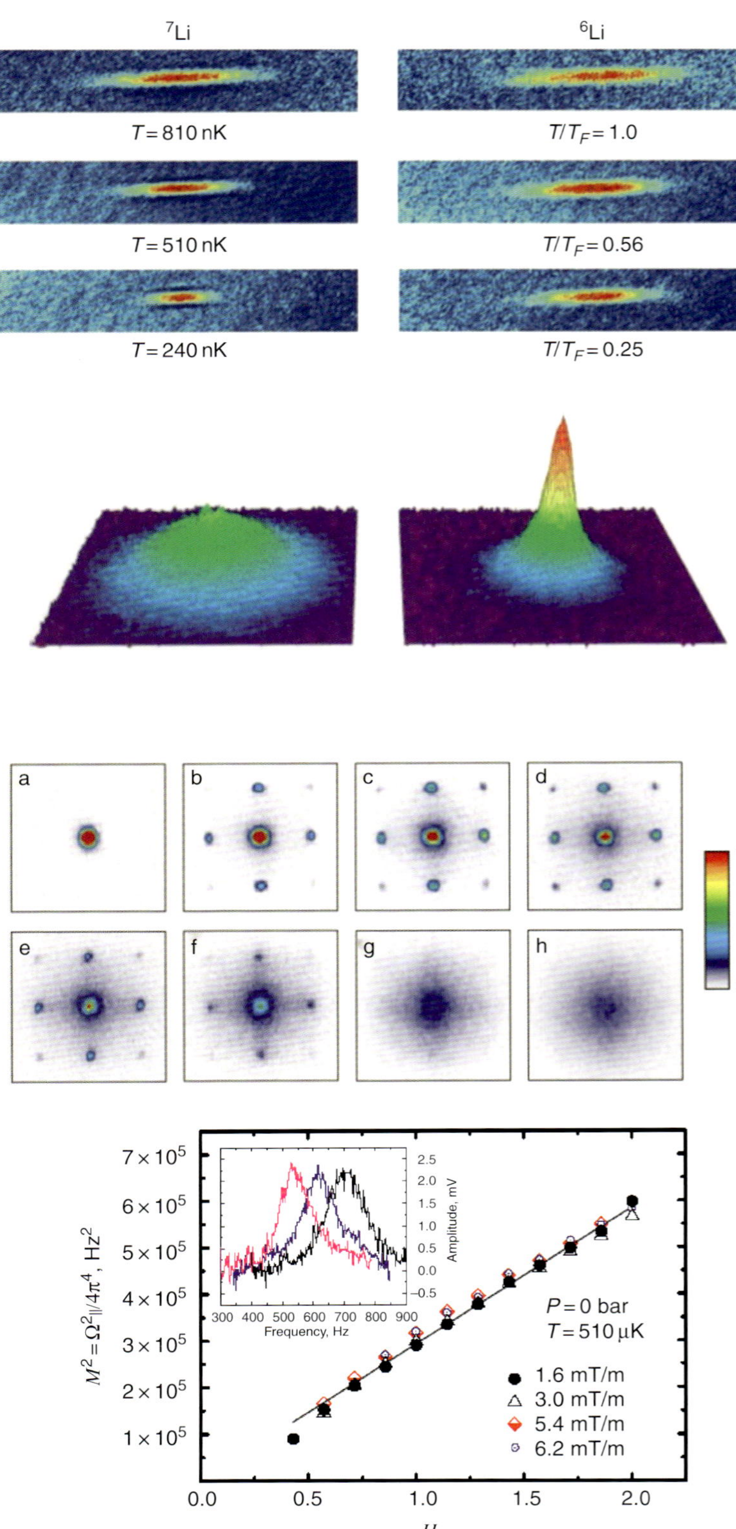

Plate 8 Absorption profiles of expanded clouds on ^{7}Li and ^{6}Li for various starting temperatures. Points to note are: (i) because of the Fermi pressure, the Fermi cloud (right) undergoes much less of a contraction relative to the Bose cloud (left) as the temperature is reduced, and (ii) the boson cloud undergoes a greater change in shape with falling temperatures. (After Truscott et al. (2001).)

Plate 9 Surface plot of the optical density for a molecular cloud created by applying a magnetic-field sweep to a Fermi gas of ^{40}K atoms. The two experiments start with fermionic atoms at different temperatures: $T = 0.19T_F$ (left) and $T = 0.06T_F$ (right). A molecular cloud is then created by passing through the Feshbach resonance. For the higher starting temperature, shown on the left, a condensate does not form, while for the lower starting temperature, shown to the right, a condensate clearly forms. (After Greiner et al. (2003).)

Plate 10 Absorption images of expanded cloud profiles for various values of the trapping potential V_0 after a time of 15 ms. In units of V_0/E_R the various frames correspond to: (a) 0, (b) 3, (c) 7, (d) 10, (e) 13, (f) 14, (g) 16, (h) 20. (After Greiner et al. (2002).)

Plate 11 Phonon mass in magnon BEC as a function of the symmetry breaking field. From [50].

Plate 12 A picture of a 98% porosity silica aerogel sample produced by Mulders. As described by Kistler, it is opalescent but transparent. The effect of Rayleigh scattering from the nanometer-sized aerogel strands is displayed by the hues of scattered light.

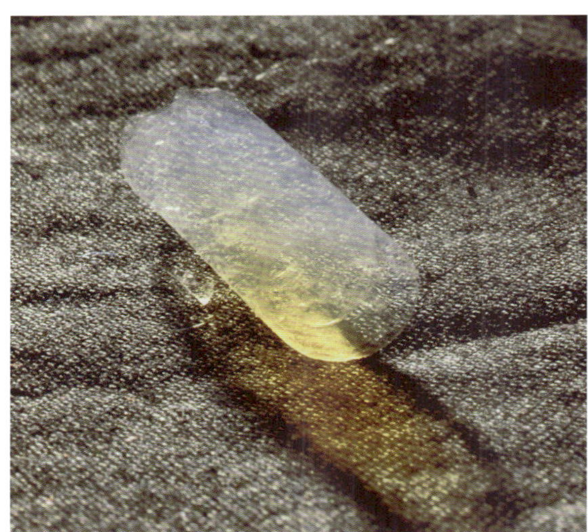

Plate 13 (a) The connected circles give measured spectral intensity distributions for different optical pump powers. The legend gives the intracavity optical power, determining the photon number, in units of $P_{\mathrm{c,exp}} = (1.55 \pm 0.6)$ W. A spectrally sharp BEC peak at the position of the cavity cutoff is visible above the critical power. The observed peak width is limited by spectrometer resolution. The inset gives theoretical spectra. (b) Images of the radiation emitted along the cavity axis, below (top) and above (bottom) the critical power. In the latter case, a condensate peak is visible in the center.

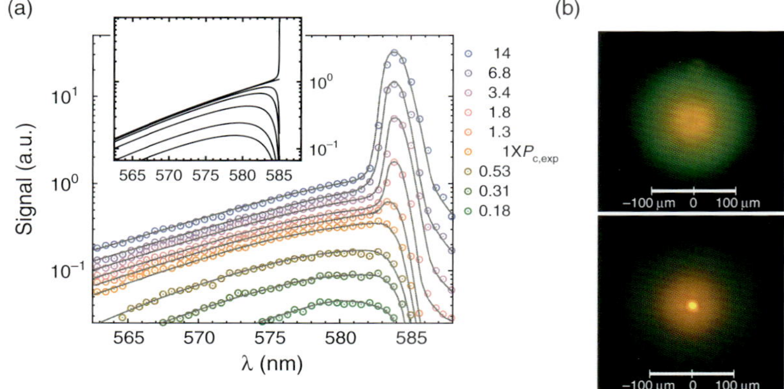

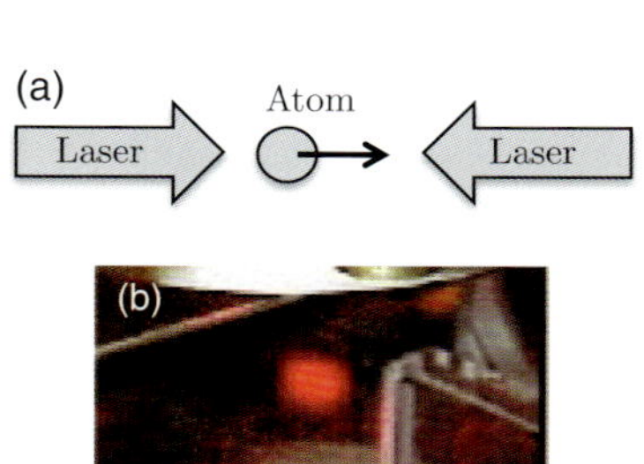

Plate 14 (a) Principle of Doppler cooling in a one-dimensional configuration. The sum of the radiation pressures of two counter-propagating laser beams creates a damping force on a moving atom. (b) Photograph of a lithium magneto-optical trap (photo from Laboratoire Kastler Brossel, ENS Paris). The glowing ball corresponds to $\sim 10^{10}$ atoms trapped at 1 mK.

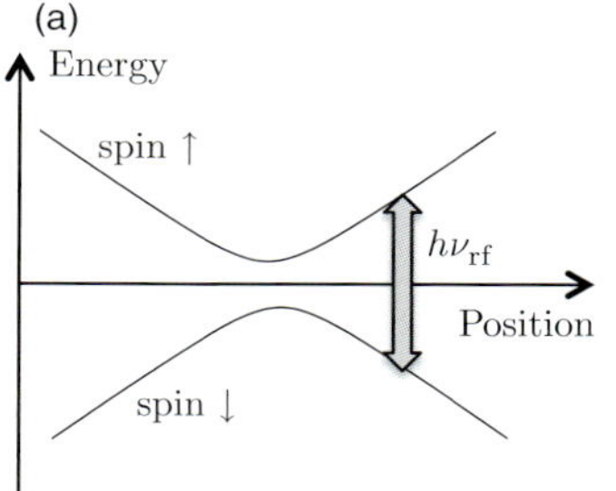

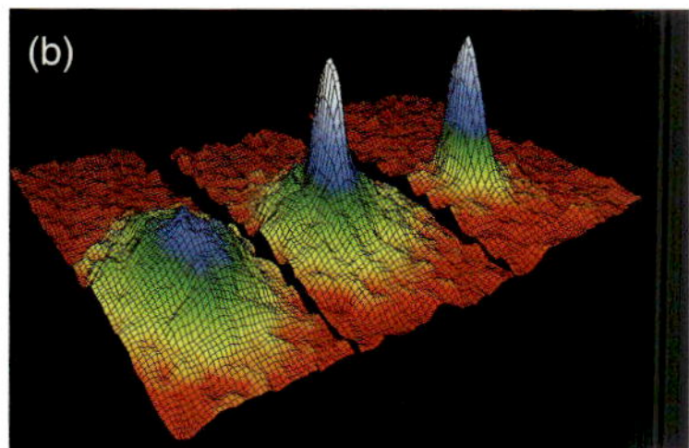

Plate 15 (a) Evaporative cooling in a magnetic trap, using a radio-frequency that flips the magnetic moment of atoms at a given position in the trap. (b) First observation of a gaseous Bose–Einstein condensate (photos: courtesy of Eric Cornell, NIST Boulder). From the left to the right, the three density profiles correspond to decreasing temperatures. The first one is still in the classical regime, where the density distribution is given by the classical Boltzmann law. The last one corresponds to a quasi-pure condensate.

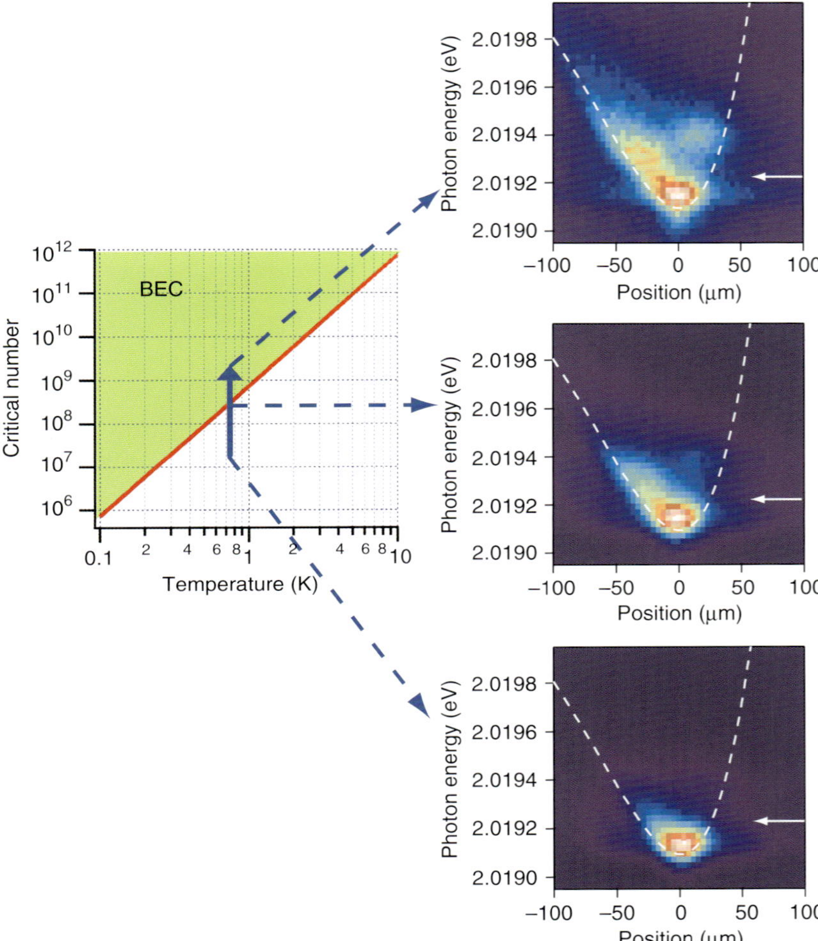

Plate 16 Direct luminescence of the trapped paraexcitons across the BEC phase boundary.

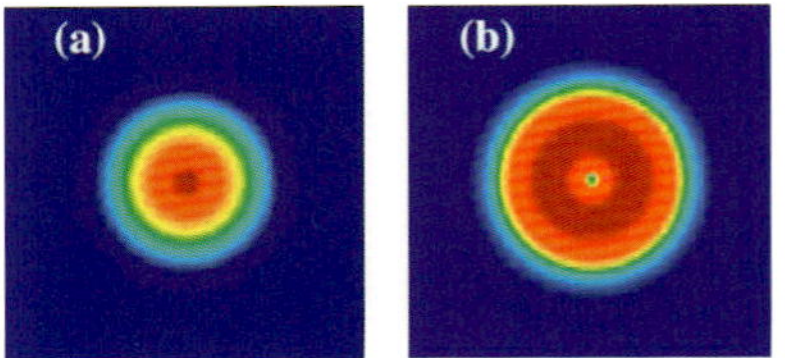

Plate 17 Condensate density for the exciton–polaritons. (a) Condensate density for non-interacting polaritons. (b) Condensate density for interacting polaritons with $n = 3 \times 10^5$ and $P/P_{th} = 5$.

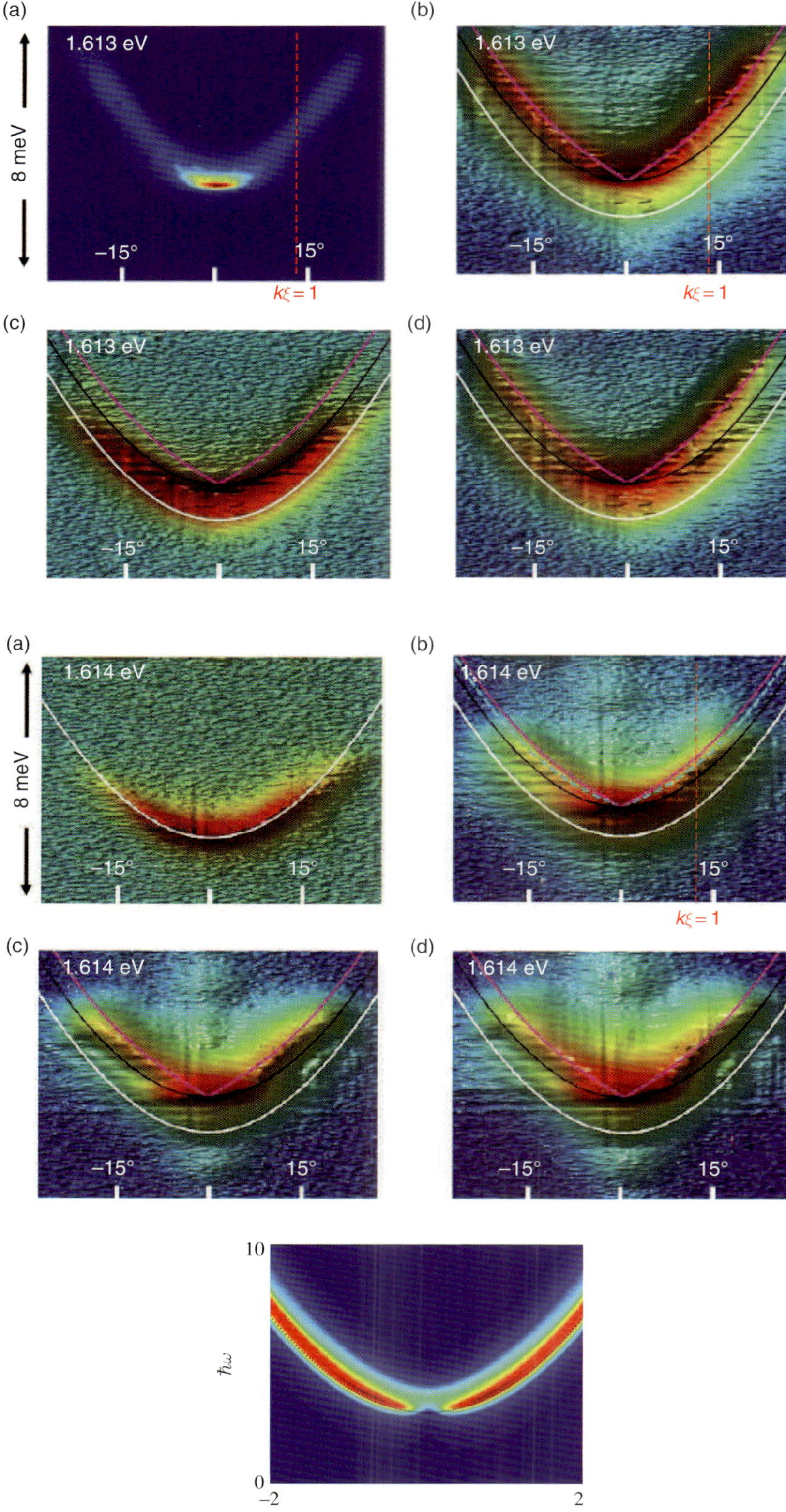

Plate 18 Polarization dependence of the excitation spectrum for an untrapped condensate system. Here (a) represents a linear plot of the intensity, while (b)–(d) employ three-dimensional logarithmic plots of the intensity to magnify the excitation spectra. Time integrated dispersion relation between the LP energy (in the range of 8 meV centered at 1.609 eV) versus in-plane wavenumber for the untrapped condensate system, where the detuning parameter is $\Delta = 1.4\,\text{meV}$ and the pump rate is $P = 3P_{th}$ ($P_{th} = 17\,\text{mW}$). A circularly polarized pump beam was incident with an angle of 60°. Three detection schemes: (a) and (b) detection of the leakage photons with co-circular polarization as the pump beam, (c) detection of the cross–circular polarization (d) detection of a small amount of a mixture of the co-circular polarization with the cross-circular polarization. The theoretical curves represent the Bogoliubov excitation energy E_B (pink line), the quadratic dispersion relations E'_{LP} (black line) which starts from the condensate energy and the non-interacting free polariton dispersion relation E_{LP} (white line), which is determined experimentally by the data taken far below the threshold $P = 0.001P_{th}$.

Plate 19 Pump rate dependence of the excitation spectrum for a trapped condensate system. Time integrated dispersion relations between the LP energy (in the range of 8 meV centered at 1.61 eV) versus in-plane wavenumber. The circularly polarized pump beam is injected into a trap with 8 μm diameter, where the detuning parameter is $\Delta =$ 1.6 meV. Pump rates are (a) $P = 0.05P_{th}$, (b) $P = 1.2P_{th}$, (c) $P = 4P_{th}$, and (d) $P = 6P_{th}$, where $Pth = 4$ mW. Three theoretical curves represent the Bogoliubov excitation energy E_B based on the homogeneous model (pink line), the quadratic dispersion curve E'_{LP} starting from the condensate energy (black line), and the non-interacting free polariton quadratic dispersion curve E_{LP} (white line) that is determined by the experimental data shown in Fig. 9.5(a). In Fig. 9.5(b), the light blue dotted line shows the Bogoliubov excitation curve based on the local density approximation.

Plate 20 $S(k, \omega)$ for $P/P_{th} = 5$ with $\xi = 0.6\,\mu\text{m}$ and the unit for $\hbar\omega$ is meV. The dotted line is the free particle dispersion starting from the condensate energy.

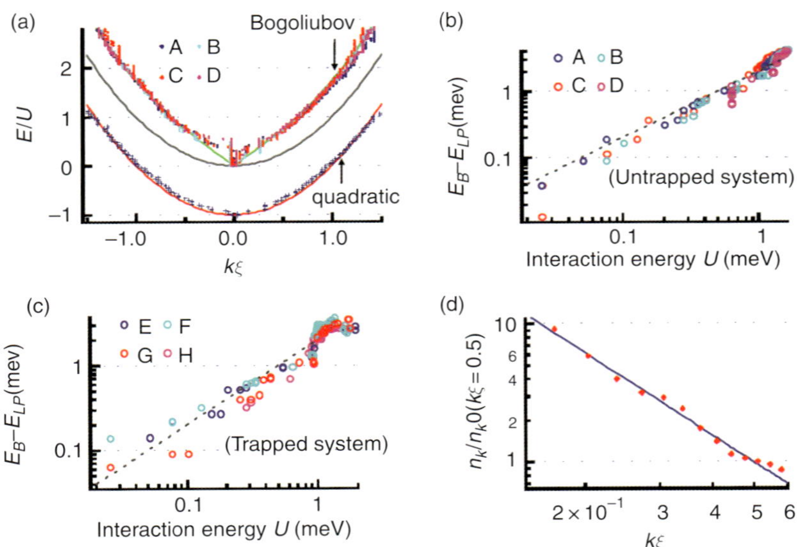

Plate 21 Dispersion relation, energy shift in free particle regime, and population distribution in excitation spectrum. (a) Numerically searched excitation energy normalized by the interaction energy $E/U(n)$ as a function of normalized wavenumber $k\xi$ for four different untrapped condensate systems. A(blue dot): $\Delta = 1.41$ meV, $P = 4P_{th}$ ($P_{th} = 6.3$ mW), B(light blue dot): $\Delta = 0.82$ meV, $P = 8P_{th}$ ($P_{th} = 8.2$ mW), C(red dot): $\Delta = 4.2$ meV, $P = 4P_{th}$ ($P_{th} = 6.4$ mW), D(pink dot): $\Delta = -0.23$ meV, $P = 24P_{th}$ ($P_{th} = 8.2$ mW). The experimental data far below threshold is also plotted by blue crosses for the system A. Three theoretical dispersion curves normalized by the interaction energy are plotted; the Bogoliubov excitation energy $E_B/U(n)$ starting from the condensate energy (green solid line), the quadratic dispersion curves $E'_{LP}/U(n)$ (gray solid line) and free polariton dispersion $E_{LP}/U(n)$ (red solid line). (b) and (c) The energy shift $E_B - E_{LP}$ in the free particle regime ($|k\xi = 1$) is plotted as a function of the interaction energy $U(n)$ for the same four different untrapped systems as in Fig. 9.9(a)–(b) and four different trapped systems (c), where trapped condensate systems are labeled as below; E(blue circle): d(diameter)$= 7\,\mu$m, $\Delta = 3.3$ meV, F(light blue circle): $d = 7\,\mu$m, $\Delta = 2.9$ meV, G(red circle): $d = 8\,\mu$m, $\Delta = 1.6$ meV, H(pink circle): $d = 8\,\mu$m, $\Delta = 2.5$ meV. The dashed line represents the theoretical prediction based on the homogeneous model $E_B - E_{LP} = 2U(n)$. (d) LP population distribution normalized by the value at $k\xi = 0.3$ for the trapped system G (in Fig. 9.9(c)) at the pump rate; $P = 2P_{th}(P_{th} = 4$ mW). Theoretical $1/k^2$ dependence for the thermal depletion is shown by the blue line and $1/k$ dependence for the quantum depletion is shown by the blue dotted line. The experimental data are plotted at the range, $k\xi = 0.2 \sim 0.6$ because the experimental data at $|k\xi < 0.2$ is dominated by the condensate with a finite Δk and the Bogoliubov excitation is suppressed.

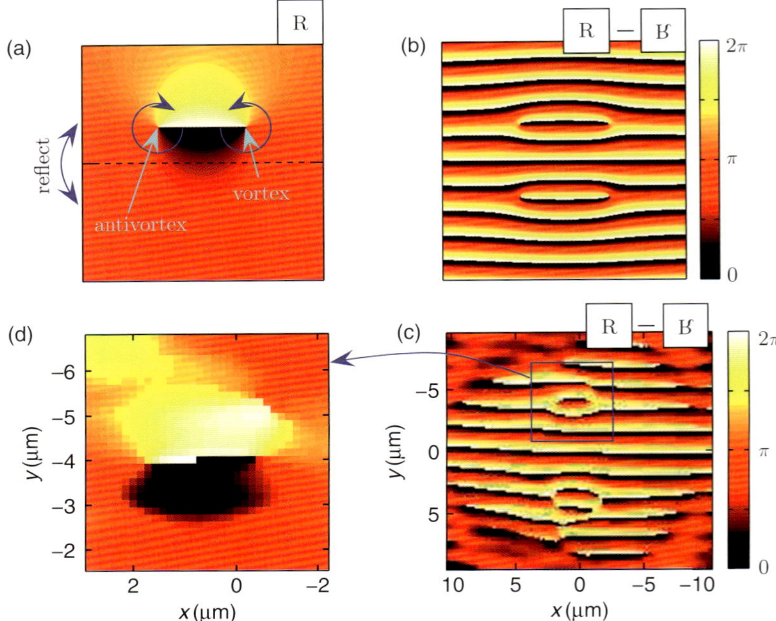

Plate 22 Phase map of a pinned pair. (a) Expected phase map of a condensate including a single vortex–antivortex pair. Arrows show the direction of the phase increase around the vortex and antivortex. (b) Simulation of the experimentally measured phase map when a interferes with its reflection along the horizontal (dashed) line. A global phase slope along the vertical direction is added. (c) Experimentally measured phase map at 55 mW above the condensation threshold of 20 mW. The blue square marks the position of a double dislocation pattern. (d) Expanded view of the blue square in c, where the global slope along the horizontal direction is subtracted. (c) and (d) are rotated by 90° with respect to all other experimental data.

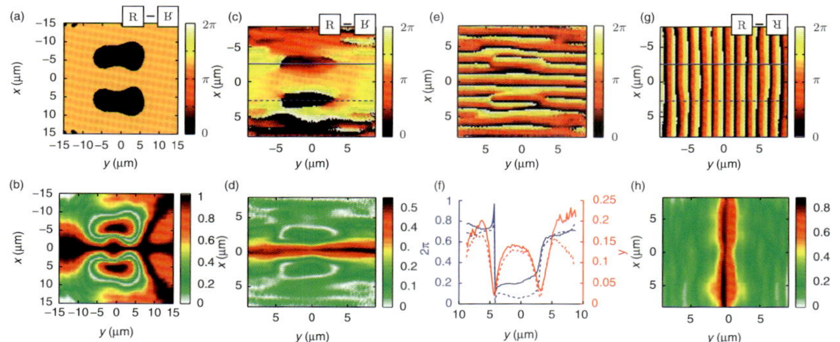

Plate 23 Phase and fringe visibility map of a free pair. (a) Theoretical time-integrated phase map for a vortex–antivortex pair imprinted into a condensate and evolving according to the dissipative GP equation. (b) Corresponding fringe visibility map. (c) and (d) Measured phase and fringe visibility maps. (e) Same as in (c) but now the global phase slope is not subtracted. (f) Blue: phase cross-section along the continuous and dashed lines in (c). Red: fringe visibility cross-section along the same lines. (g) Measured phase map when the prism is rotated by 90°, along with a schematic showing the orientation of the interfering images. (h) Corresponding fringe visibility map. Experimental data are taken at 55 mW, above the condensation threshold of 20 mW.

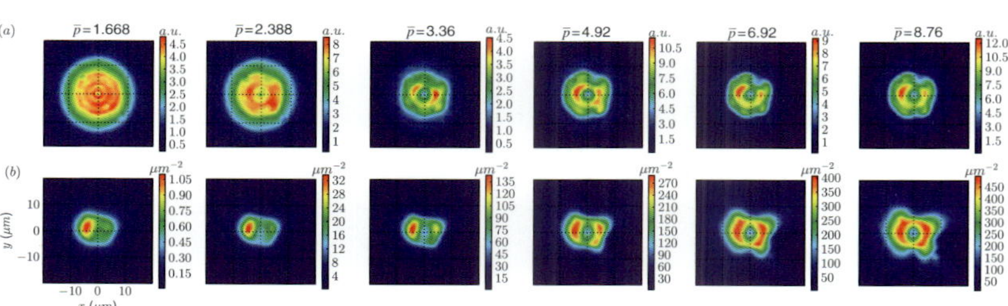

Plate 24 Steady-state condensate density profiles as a function of the normalized pumping power for (a) experimental measurements and (b) numerical simulations, both using the same measured experimental pump profile.

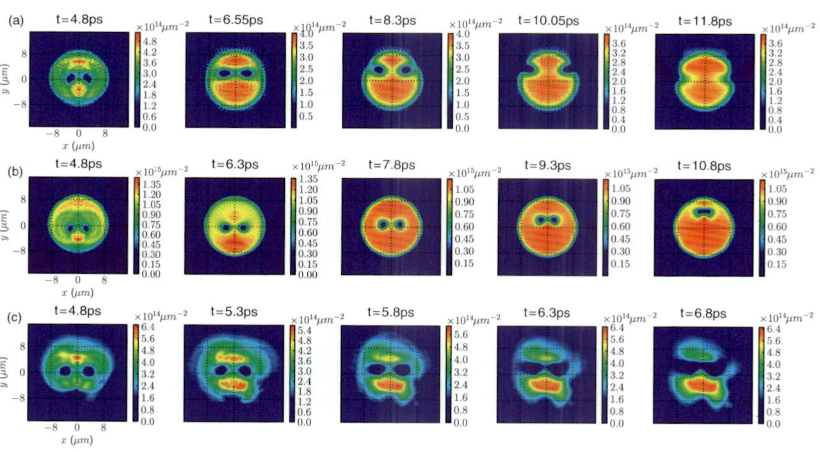

Plate 25 Numerical simulations giving the time-dependent dynamics of (a) a radially splitting vortex pair and (b) a slowly recombining vortex pair and (c) a rapidly recombining vortex pair. The first two scenarios use a perfect top-hat pumping profile, while the third uses the measurement of the experimental profile. The parameter space of (c) is also chosen to closely match that of the experiment, notably that this parameter space dictates both the presence of a central density dip in the steady-state profile as well as a recombining vortex pair trajectory.

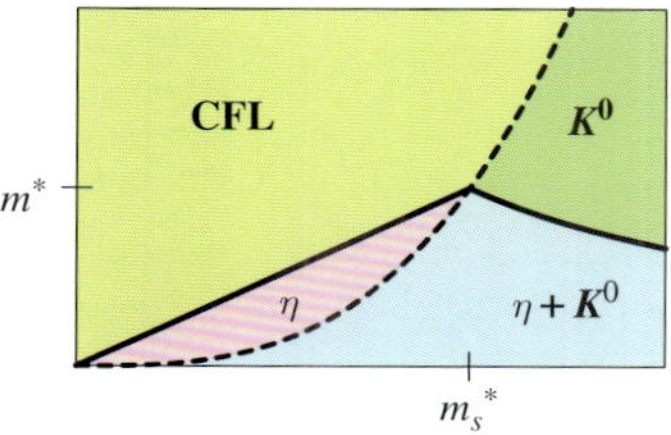

Plate 26 Phase structure of CFL matter as a function of the light quark mass m and the strange quark mass m_s, from (Kryjevski et al., 2005). CFL denotes pure CFL matter, while K^0 and η denote CFL phases with K^0 and/or η condensation. Solid lines are first-order transitions, dashed lines are second order.

5.3 Overview of the current status

5.3.1 Dirty Fermi liquid

In this section we outline the current status of the field, centered around the low temperature phase diagram that has emerged from experimental efforts that began in 1995. This section provides an overview and should also serve as a guide leading to the subsequent sections. Although this field was initiated and has been driven by keen interest in the experimental realization of a *dirty p-wave superfluid*, there are important and interesting phenomena in the normal fluid, which we will first discuss.

Normal ^{3}He liquid below about 100 mK is an ideal example of a Fermi liquid [117], and its physical properties can be described by a set of Fermi liquid parameters $\{F_l^s, F_l^a\}$ and the inelastic scattering time, τ_i, arising from binary collisions between quasiparticles [137]. Thermodynamic properties in a Fermi liquid are essentially identical to those of a Fermi gas with proper renormalization through the Fermi liquid parameters as

$$C_N = \frac{m^*}{m} C_g = \frac{m^* k_F k_B^2}{3\hbar^2} T, \tag{5.4}$$

$$\frac{m^*}{m} = 1 + \frac{1}{3} F_1^s, \tag{5.5}$$

$$\chi_N = \chi_g \frac{m^*}{m} (1 + F_0^a)^{-1} = \frac{\hbar^2 \gamma^2}{4} N_o (1 + F_0^a)^{-1}, \tag{5.6}$$

where γ is the gyromagnetic ratio of ^{3}He nuclei and $N_o = m^* k_F/(\pi^2\hbar^2)$ is the density of states per volume for both spins with effective mass m^*. $C_{N(g)}$ and $\chi_{N(g)}$ represent the heat capacity and magnetic susceptibility of the Fermi liquid (Fermi gas), respectively. The transport properties such as thermal conductivity κ, viscosity η, and spin diffusion coefficient D_M are directly related to the relevant transport scattering time $\tau_t \sim \tau_i$ through

$$\kappa = \frac{1}{3} C_N v_f^2 \tau_\kappa \tag{5.7}$$

$$\eta = \frac{1}{5} \left(1 + \frac{F_1^s}{3}\right) \left(1 + \frac{F_2^s}{5}\right) \rho v_F^2 \tau_\eta, \tag{5.8}$$

$$D_M = \frac{1}{3} \left(1 + \frac{F_1^s}{3}\right) v_f^2 \tau_D. \tag{5.9}$$

In a Fermi liquid, the number of thermally excited quasiparticles decreases with temperature. As a result, $\tau_i \sim \tau_t \propto 1/T^2$ and inelastic scattering becomes rare at low temperatures. However, the presence of aerogel introduces an additional scattering mechanism, elastic scattering of quasiparticles with silica strands. This is characterized by a temperature-independent scattering time, $\tau_e = \ell_a/v_F$ in association with the length presented by the aerogel structure. Therefore, the total scattering rate, $1/\tau$, should incorporate both the elastic and the inelastic scattering, following Matthiessen's rule, $1/\tau = 1/\tau_e + 1/\tau_i$. At low temperatures the temperature-independent elastic scattering becomes dominant, preventing the usual divergent growth of the mean free path that is expected for a Fermi liquid. The scattering mechanism goes through a crossover from inelastic collision at high temperatures to the elastic collision regime at low temperatures. Thus the temperature T^*, defined by $\tau_i(T^*) = \tau_e$, separates two regimes of distinct scattering processes [153]. For $\ell_a = 130\,\text{nm}$, $T^* \approx 17\,\text{mK}$ at 29 bar. This property is of particular importance in understanding transport properties in aerogel.

For high porosities, where $\ell_a \gg \lambda_F$, it is expected that aerogel does not significantly modify the bulk properties of normal liquid [144]. Specifically, the Fermi liquid parameters and consequently the thermodynamic properties should not be affected. However, experimental confirmation of this is not trivial. For example, in heat capacity measurements, the primary means of ascertaining the most fundamental Fermi liquid parameter F_1^s, it is essential to make sophisticated background corrections for paramagnetic solid ^{3}He adsorbed on the aerogel surface and any bulk liquid residing in the calorimeter. The early heat capacity measurement by He et al. [85, 86] confirmed the linear temperature dependence expected for the normal fluid in aerogel at 22.5 bar, after properly subtracting the bulk contribution, but their value of m^* in aerogel was 30% larger than the bulk. Alternatively Choi et al. [37] analyzed their heat capacity measurement by assuming that the liquid in aerogel had the same heat capacity as the bulk. The notion that Fermi liquid properties would not be affected by the presence of aerogel has been taken for granted without direct experimental validation. A very recent NMR experiment finally provided the most convincing confirmation [43].

The first NMR experiment on the ^{3}He-aerogel system observed a Curie-type contribution to magnetization likely attributable to the solid layers adsorbed on aerogel strands [160, 161]. The extremely large surface area in aerogel makes this contribution a dominant component at low temperatures and also facilitates fast exchange between the nuclear spins in solid and liquid, thus producing a narrow single NMR line. Therefore the total magnetization M in aerogel should have two components from liquid M_l and solid M_s given by

$$M = M_l + M_s, \tag{5.10}$$

$$M_l = C_o \frac{n_l(p)}{T_F^{**}}, \tag{5.11}$$

$$M_s = C_o \frac{n_s(p)}{T - \theta_{cw}(p)}, \tag{5.12}$$

where C_o is the Curie constant per spin, $n_{l(s)}$ is the spin density in liquid (solid), and finally $\theta_{cw}(p)$ is the pressure-dependent Curie–Weiss temperature [43]. This effect caused complications in interpreting NMR results in superfluid phases, and most of the NMR studies avoided this nuisance by preplating the aerogel surface with non-magnetic ^{4}He. However, a detailed study of the nature of the solid layers is relevant to the investigation of a dirty superfluid and to a better understanding of the phenomena occurring in a normal fluid. The spin exchange interaction between localized paramagnetic spins in the solid and itinerant spins in the liquid adds an additional scattering channel and raises interesting questions. For example, what is the effect of spin-flip scattering on the transport properties? Will it influence the onset of superfluidity or the superfluid phases in aerogel? There have been several theoretical studies that have investigated the implications of spin exchange interactions, mainly on the superfluid phases [13, 15, 122, 152], but only a few experiments have addressed this issue.

Recently, an extensive NMR study [43] was conducted in a normal fluid for the full pressure range to achieve a better grasp on the nature of the solid layers and the interactions of these with liquid. They found that the magnetic degeneracy temperature T_F^{**} in Eq. (5.11), a measure of Fermi liquid interaction, was neither sensitive to the presence of aerogel nor to the solid layers, that is, the Fermi liquid properties remained unchanged. Their improved experimental sensitivity allowed for quantitative analysis, and they were able to show that the number of localized spins in solid layers grew linearly with pressure while θ_{cw} decreased, which is characteristic of a disordered solid, as shown in Fig. 5.6. The Lancaster group has used NMR and heat capacity measurements to study the properties of solid layers in contact with normal fluid at extremely low temperatures, well below 1 mK [27]. They reported the unexpected discovery of an apparent magnetic phase transition, and suggest that the solid layers may undergo antiferromagnetic ordering.

Unlike thermodynamic properties, the influence of aerogel on transport properties is significant and gives rise to intriguing phenomena in the normal fluid. The temperature dependencies of thermal conductivity [57, 146], the spin diffusion coefficient [151], and ultrasound attenuation [134, 42] reflect the crossover behavior from inelastic quasiparticle scattering to the regime of elastic scattering from aerogel strands at lower temperatures. In thermal conductivity, Eq. 5.7 predicts $\kappa \propto 1/T$ for $T >> T^*$ but $\kappa \propto T$ for $T << T^*$ in aerogel, while one expects a $1/T$-dependence in bulk at all temperatures

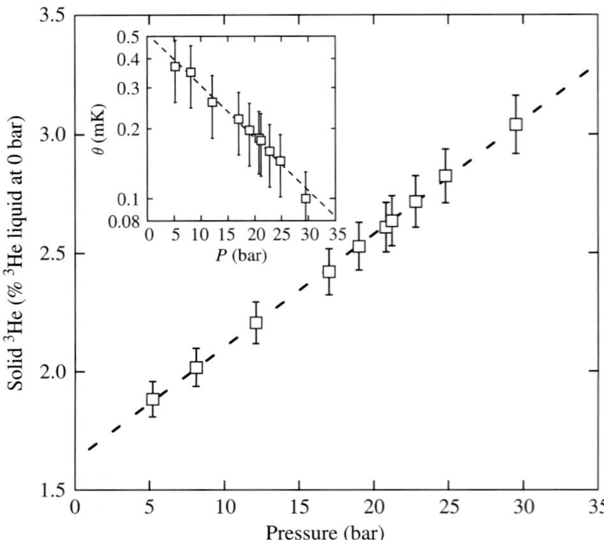

Fig. 5.6 The number of localized spins adsorbed on the 98% aerogel surface as a function of pressure, from [43]. The inset shows the pressure-dependent Curie–Weiss temperature.

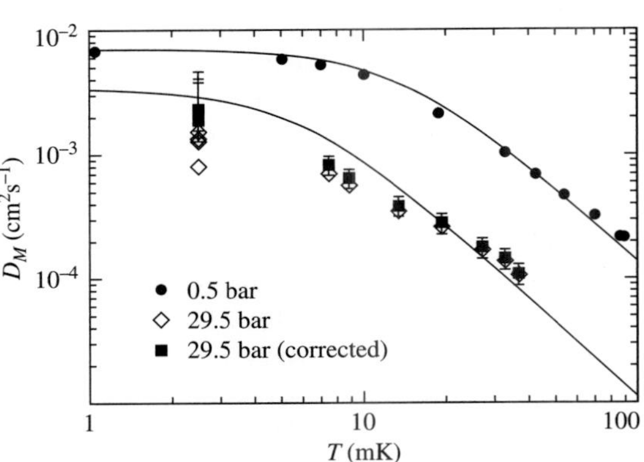

Fig. 5.7 Spin diffusion coefficient as a function of temperature at 0.5 and 29.5 bar, from [151]. The experimental data are consistent with the theoretical results shown as solid lines with $\ell_a = 130$ nm. From Eq. (5.9), one expects $D_M \propto T^{-2}$ for $T \gg T^*$, and $D_M = \text{const.}$ for $T \ll T^*$. The squares represent the data corrected for the component of solid magnetization.

above T_c. Reeves et al. [146] showed that the thermal conductivity in 95% and 98% aerogel below 10 mK was reduced by a factor of ten below that of bulk liquid, and observed the expected temperature dependence ($\sim T$) for an additional temperature-independent scattering mechanism. Recent theoretical analysis based on two-channel scattering predicts scaling behavior for the normalized thermal conductivity in terms of a reduced temperature, T/T^* [153]. Measurements of spin diffusion [30, 151] have provided the most convincing demonstration of crossover behavior, which was consistent with theoretical predictions based on scattering from both the aerogel and binary quasiparticle collisions, shown in Fig. 5.7.

One of the most fascinating properties of a Fermi liquid is the existence of zero sound [107]. This emerges as a well-defined mode at low temperatures where the conventional hydrodynamic condition breaks down. As temperature decreases, the crossover from the hydrodynamic region ($\omega\tau \ll 1$) to the collisionless region ($\omega\tau \gg 1$) occurs at a fixed sound frequency ω. In 98% aerogel, $1/\tau_e \sim 100$ MHz, and consequently, the first to zero sound crossover in normal ^{3}He should be effectively inhibited in aerogel for $\omega < 100$ MHz [144]. Therefore first sound attenuation $\alpha_1 = \frac{2\omega^2}{3\rho c_\ell^3}\eta$ becomes temperature-independent for $T \ll T^*$, as verified in the experiments by Nomura et al. [134] and Choi et al. [42]. A more sophisticated theoretical model [93, 89] was proposed to explain the observed behavior near T^*. This model, based on visco-elastic theory, considers the motion of both liquid and aerogel. A quasiparticle impinging on an aerogel strand transfers momentum and thus causes a collisional drag effect. When this process generates relative motion between the liquid and aerogel, it gives rise to an additional damping mechanism. Although this model is in qualitative agreement with the experimental observation [134], measured attenuation for $T > T^*$ shows rather unusual behavior, following a $T^{0.7}$-dependence at high temperatures after going through a broad minimum at around 40 mK [35], shown in Fig. 5.8. Although the rise in attenuation at high temperatures is probably related to decoupling of liquid from the aerogel, this phenomenon, especially the power-law T-dependence, is not fully understood. It is interesting to note that the minimum in α_1 occurs where $\zeta_a \approx \ell$.

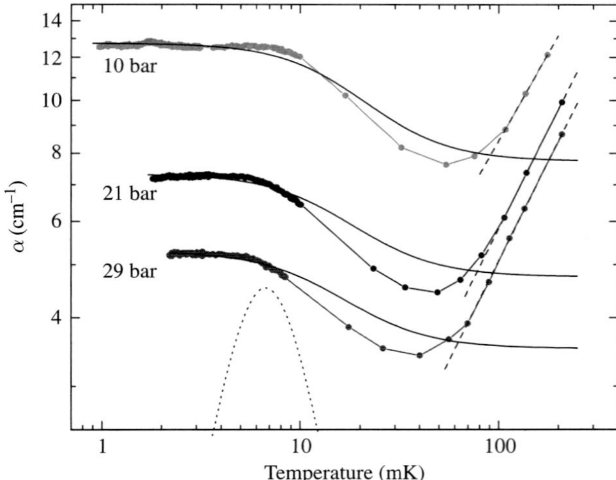

Fig. 5.8 Absolute attenuation in a normal liquid at 10, 21, and 29 bar in 98% aerogel, determined using 9.5 MHz longitudinal sound, from [42, 38]. The solid lines are the result of a fit based on the theory of Higashitani et al. [89] (see text). The dotted line indicates attenuation in bulk at 29 bar. The dashed lines represent a $T^{0.7}$-dependence.

5.3.2 Superfluid phases in aerogel

At this point we would like to introduce the order parameters of the superfluid phases in pure ^{3}He and to lay out a brief discussion of their pertinent properties. In bulk, the A and B phases appear in zero magnetic field and the A_1 phase appears only in the presence of magnetic fields by splitting the second-order superfluid transition into two second-order transitions. Having *p-wave spin-triplet* pairing, the order parameter possesses internal structure in spin and orbital spaces. Because the spin–orbit coupling from the nuclear dipole–dipole interaction is weak compared with other characteristic energy scales in this system, the order parameter, $A_{\mu j}$, is represented by two independent components for the spin μ and the orbital j spaces. The order parameters of these phases [169] are given by

$$A_{\mu j} = \frac{1}{\sqrt{2}}\hat{d}_\mu(\hat{m}_j + i\hat{n}_j) \quad \text{(A phase)}, \tag{5.13}$$

$$A_{\mu j} = \frac{1}{\sqrt{3}}R_{\mu j}(\hat{n}, \theta) \quad \text{(B phase)}, \tag{5.14}$$

$$A_{\mu j} = \frac{1}{\sqrt{2}}(\hat{d}_\mu + i\hat{e}_\mu)(\hat{m}_j + i\hat{n}_j) \quad (\text{A}_1 \text{ phase}), \tag{5.15}$$

where the overall phase factor is omitted for clarity. The Balian–Werthamer (BW) state, identified with the B phase, is unique among the possible states with p-wave spin-triplet pairing in the sense that it has an isotropic gap and is composed of an equal admixture of the three possible spin states. However, the Anderson–Brinkman–Morel (ABM) state for the A and A_1 phases carries two nodes in the superfluid gap along the common angular momentum axis shared by all pairs, and is an equal spin pairing (ESP) state, being composed of $(\uparrow\uparrow)$ and/or $(\downarrow\downarrow)$ spins only.

The order parameters of these phases reveal their anisotropic nature in the form of $\hat{n}$, $\hat{d}$, $\hat{l} = \hat{m} \times \hat{n}$, and $\hat{f} = \hat{d} \times \hat{e}$. Vector $\hat{d}$ represents the preferred direction in spin space, while $\hat{l}$ plays the corresponding role in orbital space in the direction of the angular momentum of a Cooper pair. $R_{\mu j}(\hat{n}, \theta)$ describes a relative rotation of spin and orbital spaces around $\hat{n}$ by an angle θ. The appearance of these vectors indicates that the system chooses a preferred orientation. Degeneracy in the actual choice of direction is lifted by both internal perturbation from the dipole–dipole interaction, and external perturbations that couple to the order parameter, such as magnetic and electric fields, container walls and boundaries, and superflow. When only one perturbation is considered, the order parameter aligns uniformly throughout the system giving rise to a uniform texture of these representative vectors. In reality, multiple sources of perturbation compete with each other, and the system will find the lowest energy configuration incorporating spatial variations of the preferred directions, thus producing a non-uniform texture.

The dipole energy scale is many orders of magnitude smaller ($\approx 10^{-7}$ K) than thermal energy, but its effect is amplified due to the coherent nature of the condensate [117]. The dipole energy densities of the A and B phases are given by [169]

$$\Delta f_D^A = -\frac{3}{5} g_D(T)(\hat{d} \cdot \hat{l})^2 \quad \text{(A phase)}, \tag{5.16}$$

$$\Delta f_D^B = \frac{8}{5} g_D(T) \left(\cos\theta + \frac{1}{4} \right)^2 \quad \text{(B phase)}, \tag{5.17}$$

where $g_D(T) \equiv \lambda_D N_o \Delta^2(T)$ is the temperature-dependent dipole coupling strength, and λ_D is the dimensionless dipole coupling strength [169]. Therefore the minimum energy configuration is achieved by $\hat{d} \parallel \hat{l}$ for the A phase and the relative spin-orbit rotation angle $\theta_L = \cos^{-1}(-\frac{1}{4}) \approx 104\,°$, known as the "Leggett angle" for the B phase. However, the dipole energy alone cannot lift the degeneracy completely. In the A phase, degeneracy remains in the plane perpendicular to either $\hat{l}$ or $\hat{d}$. In the B phase, all possible orientations of $\hat{n}$ share the same energy for a fixed value of θ. The effect of a magnetic field ($\vec{H}$) is represented through the magnetic energy density given by [169]

$$\Delta f_H^A \propto (\hat{d} \cdot \vec{H})^2, \tag{5.18}$$

$$\Delta f_H^B \propto -(\hat{n} \cdot \vec{H})^2, \tag{5.19}$$

so that the preferred orientations are $\hat{d} \perp \vec{H}$ and $\hat{n} \parallel \vec{H}$. Walls or surface boundaries also have a significant effect on the texture and favor both vectors, $\hat{l}$ and $\hat{n}$, aligned to the surface normal $\hat{s}$. In this case, although the magnitude of the order parameter recovers its bulk value within a few coherence lengths, the direction of the order parameter conforms to the bulk configuration over a longer length scale, called the healing length [117]. Healing lengths are determined for the various orienting forces by equating the energy gains to the bending energy cost. The healing lengths for three of the orienting forces are summarized in Table 5.1. Understanding the texture in a sample cell with a specific

Table 5.1 Healing lengths for various orienting forces [169].

Orienting force	Healing length	Notes
Dipole	$\xi_D^A \approx 8\,\mu\mathrm{m}$ $\xi_D^B \approx 7\,\mu\mathrm{m}$	In the G–L regime at melting pressure
Magnetic field	$\xi_H^A \propto \frac{1}{H}\xi_0$ $\xi_H^B \propto \frac{1}{H}\xi_0$	$\xi_H^A \approx \xi_D$ at ≈ 28 G $\xi_H^B \gg \xi_H^A$ and ξ_D
Surface	$\xi_S^A \sim 10\,\mu\mathrm{m}$ $\xi_S^B \approx 2(1-T/T_c)^{1/2}$ mm	Weak T-dependence $\xi_S^B \approx 1$ mm at $T = 0.7\,T_c$

geometry is crucial in interpreting results, in particular from NMR measurements. The surface healing length in the B phase is quite long and comparable to a typical sample cell size, and therefore these B-phase textures should be evident in NMR spectra.

Although the first two measurements on ^{3}He in aerogel clearly demonstrated the fragile disposition of the p-wave superfluid toward impurities, early experiments exhibited varying degrees of suppression, even in samples of the same porosity [140, 160, 119, 16]. The obvious task was to construct a dependable phase diagram for a given porosity. Within a few years a consistent pressure-dependent superfluid transition line started to emerge from a number of experiments using diverse experimental techniques [119, 5, 13, 70, 152]. Figure 5.9 shows the most widely accepted zero field phase diagram in 98% aerogel. One striking feature is the appearance of the quantum phase transition at fluid density $\rho_c \approx 94\,\mathrm{mg\,cm^{-3}}$ ($p_c \approx 6.5$ bar) [119], below which no evidence of a superfluid transition was observed. It is stimulating to contemplate the effect of quantum fluctuations on the normal as well as the superfluid state [121, 12], but this remains unexplored territory.

The phase diagram exhibits a remarkable resemblance to the theoretical predictions of Abrikosov and Gor'kov [2]: an initial linear suppression of T_c with

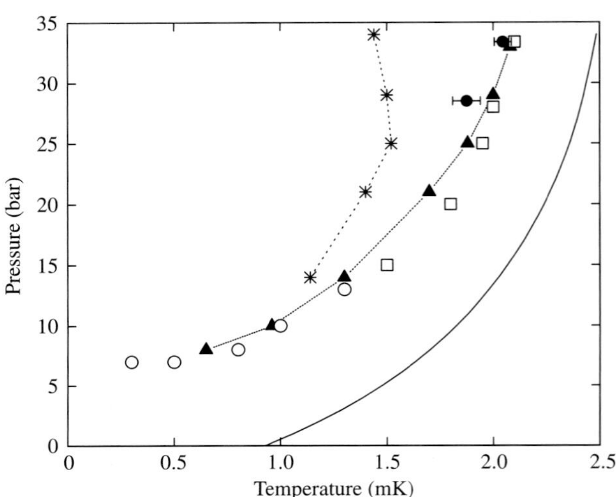

Fig. 5.9 Low temperature phase diagram of ^{3}He in 98% aerogel in zero magnetic field from various measurements: transverse acoustic impedance by Gervais et al. [71] (open squares); torsional oscillator by Matsumoto et al. [119] (open circles); longitudinal sound by Choi et al. [41, 38] (solid triangles). The solid line represents the superfluid transition in bulk. The two solid circles are the A–B-like phase transition observed on warming from Vicente et al. [168], and the stars are the A–B-like transition observed on cooling, from [38], which demonstrates the existence of the metastable A-like phase over a wide range of pressures.

increasing pair-breaking parameter x and the extinction of superconductivity at x_c. Thuneberg et al. [166] first proposed a theoretical model in which aerogel was considered as a collection of isotropic and uniformly but randomly distributed pair-breaking scattering centers. This so-called homogeneous isotropic scattering model (HISM) predicts a suppression of T_{ca} in the direction compatible with experimental observations, given by

$$\ln\left(\frac{T_c}{T_{ca}}\right) = 2\sum_{n=1}^{\infty}\left(\frac{1}{2n-1} - \frac{1}{2n-1+x}\right), \tag{5.20}$$

but it fell short in accounting for experimental results over the entire pressure range, with a single pair-breaking parameter x. Furthermore, numerous experiments [140, 160, 16, 37] consistently estimated much more severe suppression of the superfluid gap and superfluid density than predicted by HISM, as summarized by Halperin et al. [82]. Several variations of HISM models have been developed to resolve the discrepancies by considering the detailed structure of aerogel [166, 84, 152]. As already mentioned, aerogel appears homogeneous and uniform only on length scales larger than ζ_a. Therefore, when $\xi_o < \zeta_a$, as is the case at high pressures, it is reasonable to consider the inhomogeneity and anisotropy of aerogel. Hänninen and Thuneberg [84] studied inhomogeneous but isotropic scattering models (IISM), which give better agreement with some experimental aspects, but predict a significant temperature dependence of the suppression factor for the superfluid gap, which is not consistent with observations. Sauls and Sharma [152] proposed a phenomenological IISM by redefining the pair-breaking parameter $\tilde{x} = x/(1 + \tilde{\zeta}_a^2/x)$, where $\tilde{\zeta}_a = \zeta_a/\ell$. This two-parameter model (ℓ_a and ζ_a) produced an excellent fit of the superfluid transition in 98% aerogel for the whole pressure range. Recently, the effects of anisotropic scattering have attracted attention. Thuneberg et al. [166] and Thuneberg [164] have incorporated the effect of anisotropy into the Ginzburg–Landau free energy. A series of calculations by Aoyama and Ikeda [7, 8, 10, 94] revealed intriguing effects of anisotropic scattering on the phase diagram, although not all of them have been verified experimentally.

To date, three distinct superfluid phases have been observed in aerogel, and it is believed that these correspond to their counterparts in bulk: the A phase, the B phase, and the A_1 phase. The phases in aerogel are conventionally called the A-like [160], the B-like [5], and the A_1-like phase [39] in the literature. Identification of the symmetry of these phases is not as complete or unequivocal as in bulk. However, persistent experimental efforts have been building confidence in the current identification of the observed phases. In addition to the work of Alles et al. [5], Dmitriev et al. unambiguously showed that the B-like phase had the same spin structure as the bulk B phase through the dependence of frequency shift on tipping angle [53] and the existence of the homogeneous spin precession domain (HPD) [46]. The A_1-like phase, identified only in high magnetic fields, behaves as the bulk A_1 phase in terms of its response to a magnetic field [39].

The A-like phase has puzzled researchers since its discovery. Early NMR work [160] identified its spin structure as an equal spin pairing (ESP) state.

However, the orbital component of the order parameter has been elusive, mainly because there are many possible orbital structures associated with ESP and no reliable experimental tool is available to probe the orbital structure directly. In bulk superfluid, ultrasound spectroscopy of the order parameter collective modes has been the best tool for this task, providing unambiguous verification of the symmetry of the order parameter [169, 54]. In aerogel these modes are buried in the background of the impurity states induced by pair-breaking scattering and their detection is extremely difficult, if not impossible [144, 134, 41]. A few exotic states besides the ABM state of the bulk A phase have been put forward as candidates for the A-like phase. Volovik [170–172] proposed a globally isotropic state without long-range orientational order, called the glass or Larkin–Imry–Ma (LIM) state. Fomin [62–64] suggested a completely different class of states, the so-called robust phase. The possible existence of the planar state has also been mooted [18]. Although the final verdict has yet to be pronounced, an accumulating amount of experimental [52, 49, 106, 150, 56, 91] and theoretical [171, 172, 49] analyses suggest that the A-like phase has axial symmetry, albeit with unusual textures associated with the anisotropic disorder presented by aerogel, and the proposed LIM state seems to be realized in uncompressed aerogel samples with no built-in anisotropy.

The first experimental confirmation of the first order A-like to B-like transition was made by Barker et al. [16] in a low magnetic field of 28.4 mT using NMR, and later by Vicente et al. [168] in zero field using a transverse acoustic impedance technique. Aerogel has a profound effect on this transition. As shown in Fig. 5.9, the width of the A-like phase is drastically diminished from that in bulk. However, the A-like phase should exist even below the polycritical pressure of bulk, $p = 21$ bar, in the absence of a magnetic field, judging from the supercooled metastable A-like phase [69, 133, 132, 126]. Furthermore, a recent acoustic study revealed a drastically different pressure dependence of the field-suppressed transition line [126] as reproduced in Fig. 5.10. Unlike the

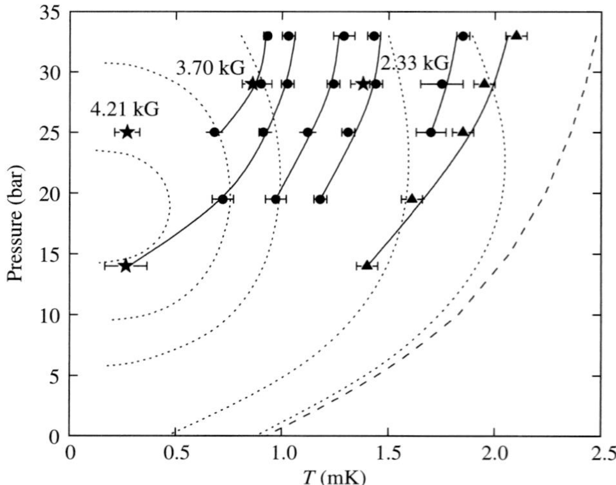

Fig. 5.10 The A-like to B-like transition of superfluid ^{3}He in 98% aerogel in various magnetic fields, from [126]. The solid triangles represent the aerogel superfluid transition determined by ultrasound transmission. The solid circles are the A-like to B-like transition identified on warming at a constant field. The solid lines going through the data points are guides for the eye but conform to the constant field phase boundaries for 1.11, 2.22, 2.75, 3.33, and 3.85 kG, respectively from right to left. For comparison, the constant field A-B phase boundaries for the bulk liquid are shown by dotted lines for 1, 3, 5, 5.5, and 5.8 kG, respectively. The numbers right next to the star symbols indicate the mid-field strength of the transition determined by a field sweep at constant temperature.

bulk liquid where the field-driven extinction of the B phase occurs near the polycritical pressure, as represented by the dotted line in the figure, the B-like phase appears to retract its territory toward the melting pressure. The key features of the phase diagram can be understood on the basis of two fundamental points: first, strong-coupling effects are significantly reduced in this system by impurity scattering; and second, the anisotropic disorder introduced by the aerogel strands emulates the effect of a magnetic field [168, 126, 127].

One consequence of impurity scattering in superconductors is the accumulation of impurity bound states inside the gap. Abrikosov and Gor'kov [2], Buchholtz and Zwicknagl [29], and Maki and Puchkaryov [118] demonstrated this effect theoretically in various types of superconducting states in the Born (phase shift $\delta_o \ll 1$) and the unitary ($\delta_o \rightarrow \pi/2$) scattering limits. Although the detailed spectrum of the bound states depends on the symmetry and type of scattering, a general feature of this phenomenon is the appearance of gapless superconductivity beyond a certain level of disorder or impurity concentration. These theoretical results are applicable to superfluid ^{3}He in aerogel, and qualitatively similar results were obtained by Sharma and Sauls [158] and Higashitani et al. [90]. The results of low temperature thermal conductivity [58], heat capacity [39], and ultrasound attenuation [36] measurements were consistent with the presence of impurity states. Here the density of states at the Fermi energy, $N(0)$, has been extracted from these measurements, shown in Fig. 5.11. There is reasonable agreement, at least providing indirect evidence of gapless superfluidity at lower pressures.

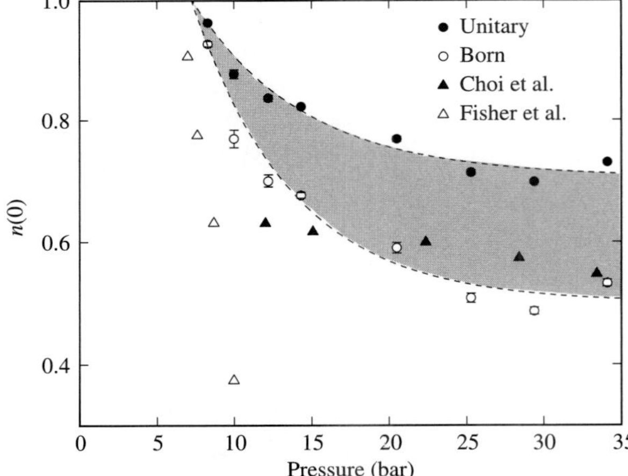

Fig. 5.11 The density of states normalized by the value at T_{ca} as a function of pressure. The circles are extracted from [36] in the Born (open circles) and unitary (solid circles) limits based on the theory of Higashitani et al. [90]. The dashed lines are guides for the eye, and the shaded region represents the range of $n(0)$ bound by the unitary and Born limits. Results from thermal conductivity (open triangles) [58] and heat capacity measurements (solid triangles) [39] are also presented.

5.4 Experimental properties

5.4.1 NMR experiments

Without a doubt, nuclear magnetic resonance [1] and ultrasound spectroscopy [81] have been proven to be the most valuable spectroscopic tools in the study of superfluid ^{3}He. Both experimental techniques played determining roles in uncovering the symmetry of the superfluid phases [54]. Although each technique primarily probes a specific component of the order parameter, they are far from mutually exclusive because of the important spin-orbit coupling present in the superfluid phases of ^{3}He. A wealth of information on the nature of a superfluid phase is contained in magnetization, frequency shifts of NMR resonance, and various relaxation times. In liquid ^{3}He, both continuous wave (cw) and pulsed NMR have been used. Unlike solid state NMR, cw NMR has been extremely valuable in the study of superfluid ^{3}He because of its ability to produce extensive NMR line behavior that can be used to elucidate the textures. Pulsed NMR, a time-domain probe, has produced unusual results in the superfluid phases originating from nonlinear effects, specifically in tipping angle dependencies in frequency shifts. The theory of Leggett [116] provided the fundamental framework for understanding NMR experiments in p-wave superfluids. We will briefly recollect some of the important results of the theory and the experiments in bulk liquid [1, 116].

In his comprehensive study, Leggett [116] demonstrated that the dipole interaction should be considered and that it produces an additional torque in the Bloch equation governing spin dynamics. Consequently, the resonance of transverse NMR is shifted from its Larmor frequency, $\omega_L \equiv \gamma H_o$, where H_o is the strength of the applied magnetic field. Furthermore, oscillation of the longitudinal magnetization, longitudinal resonance, was also predicted to exist at a frequency named the Leggett frequency, $\Omega_{A,B}(T)$, different for each phase. In the G–L limit, the ratio of these quantities has a simple but useful ratio,

$$\left(\frac{\Omega_B}{\Omega_A}\right)^2 \approx \frac{5\chi_A}{2\chi_B}, \tag{5.21}$$

where $\chi_{A(B)}$ is the susceptibility of the A(B) phase. The amount of transverse shift depends on various factors: the symmetry of the superfluid phase; the configuration of relevant order parameters; and in pulsed NMR, the tipping angle β. Typically, $\Omega_{A,B}/\gamma \sim 1$ mT, and for applied fields larger than this the frequency shifts from ω_L in the A and the B phases can be represented by

$$\Delta\omega = \begin{cases} \dfrac{\Omega_A^2}{2\omega_L}\cos 2\phi & \text{for A phase (cw NMR)}, \\[2ex] \dfrac{\Omega_B^2}{2\omega_L}\sin^2\phi & \text{for B phase (cw NMR)}, \end{cases} \tag{5.22}$$

$$\Delta\omega = \begin{cases} \dfrac{\Omega_A^2}{8\omega_L}(1 + 3\cos\beta) & \text{for A phase}, \\[2ex] -\dfrac{4\Omega_B^2}{15\omega_L}(1 + 4\cos\beta) & \text{for B phase and } \beta > \theta_L, \end{cases} \tag{5.23}$$

where ϕ is the angle between $\hat{l}$ and $\hat{d}$ for the A phase or $\hat{n}$ and $\vec{H}$ for the B phase. It should be noted that in the bulk equilibrium B phase there is no dependence of frequency shift on tipping angle until the Leggett angle

θ_L is exceeded (Eq. 5.23). Continuous wave NMR is performed practically at $\beta \approx 0$. Therefore in B-phase cw NMR it is impossible to have a negative frequency shift, regardless of the configuration of the order parameter. However, the A phase could have a negative frequency shift for $\pi/4 < \phi < 3\pi/4$. This property is of special importance in interpreting NMR results in aerogel.

In actual experiments, the situation is more complicated due to inhomogeneous textural broadening of NMR lines. For example, in a cylindrical NMR cell of a few millimeters in diameter, oriented perpendicular to the field, a typical NMR line in the B phase has a large weight at zero frequency shift, with quite a broad tail only in the positive direction to the maximum shift bounded by $\Omega_B^2(T)/2\omega_L$. On the other hand, the A phase usually exhibits a rather narrow line shifted by $\Omega_A^2(T)/2\omega_L$, because the short dipole healing length means that most of the texture is unaffected by the cell surface. All of these features have been verified experimentally in bulk ^{3}He [54, 1] and provide the background for understanding NMR results in aerogel.

The first round of NMR experiments [160, 159, 162] by the Northwestern group revealed an impressive amount of intriguing features in ^{3}He-aerogel. Their experiments and the early torsional oscillator work [140, 119] by the Cornell group laid the groundwork by providing a cogent physical picture as well as experimental protocols that have been followed by other groups. The pulsed NMR measurements at 112 mT ($\beta \approx 10°$) observed a Curie–Weiss contribution from the solid layers in fast exchange with the spins in liquid as described by Eq. (5.10). They observed a rather sharp onset of a large frequency shift at T_{ca}, signaling the superfluid transition, but no appreciable deviation in magnetization from Curie–Weiss behavior (Fig. 5.12). Their pressure dependence of the transition temperature—significantly depressed from the bulk transition temperature—was consistent with the torsional oscillator measurements by Porto and Parpia [140]. By taking into account the fast

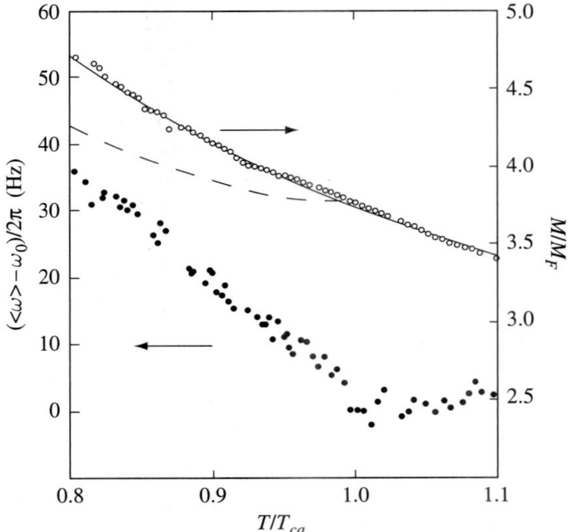

Fig. 5.12 Magnetization (open circles) and frequency shift as a function of reduced temperature, from [160]. The transition is well defined by the sharp increase in the shift. The magnetization is normalized by its Fermi liquid value M_F and the Curie–Weiss behavior is apparent. The solid (dashed) line represents the expected total magnetization assuming an ESP pairing state (the B phase) below the superfluid transition.

exchange between spins in liquid and solid, they were able to extract the frequency shift associated with the superfluid component using the relation

$$\langle\omega\rangle = \frac{\langle\omega_l\rangle M_l + \langle\omega_s\rangle M_s}{M_l + M_s}, \tag{5.24}$$

where $\langle\omega\rangle$ represents the average precession frequency of the observed NMR line and $\langle\omega_{l(s)}\rangle$ is the corresponding quantity for the liquid (solid) component. The effects of internuclear dipolar fields that produce frequency shifts in the solid layers are averaged to zero, assuming random orientation of the aerogel strands, $\langle\omega_s\rangle = 0$. The magnetization, the large frequency shift, and the symmetric lineshape of the superfluid indicated that the superfluid phase observed was an ESP state similar to the A phase in bulk. This observation is in contradiction to what is currently accepted, but it is now understood that their experiment was probably done in the metastable A-like phase. On the other hand, the unusual tipping angle dependence in $\Delta\omega$, showing abrupt disappearance for $\beta > 40°$, was completely new and inconsistent with either the A or B phases in bulk. The sharp superfluid transition and strongly pressure-dependent suppression prompted an interpretation of aerogel as a collection of pair-breaking impurities. However, they found inconsistency with the Abrikosov–Gor'kov model in that the suppression of the gap estimated from the measured longitudinal frequency was consistently larger than that estimated from the transition temperature. This has been verified repeatedly by other NMR [159, 16, 46, 20, 49] and superfluid density measurements [140, 119, 142, 73], calling for a theoretical description beyond the isotropic scattering model.

Sprague et al. [159, 162] studied the effect of a magnetic field on the superfluid transition temperature up to 200 mT, which was a logical step knowing the presence of localized moments in solid layers. They found that T_{ca} decreased quadratically with field and made an attempt to explain this behavior in terms of spin exchange scattering. However, later experiments from the same group [71] and others [143, 130, 39, 40] could not confirm their results up to fairly high fields (≈ 0.8 T). They also showed that preplating with non-magnetic ^{4}He layers effectively eliminated the Curie–Weiss contribution and shut off the spin exchange scattering. One surprising observation of their experiment was that a new phase, a non-ESP state, emerged after the addition of 3.4 layers of ^{4}He. Again, the superfluid gap of this phase was suppressed more than the transition temperature, under the assumption that the non-EPS state was described by the same order parameter as the bulk B phase.

For various technical reasons, data acquisition in an experiment studying superfluid ^{3}He is usually done during the warm-up sequence after cooling by nuclear demagnetization. This conventional protocol has delayed observation of the A-like to B-like transition in aerogel, because the transition on warming appears subtly, and very close to the superfluid transition. Barker et al. [16] managed to take data on cooling and were able to identify clear transition features in both magnetization and frequency shift. They performed cw NMR at 28.4 mT with and without ^{3}He solid layers. Unlike the Northwestern experiment, they did not find any changes in the nature of the superfluid phase between the two configurations [17], and the main part of their experiment

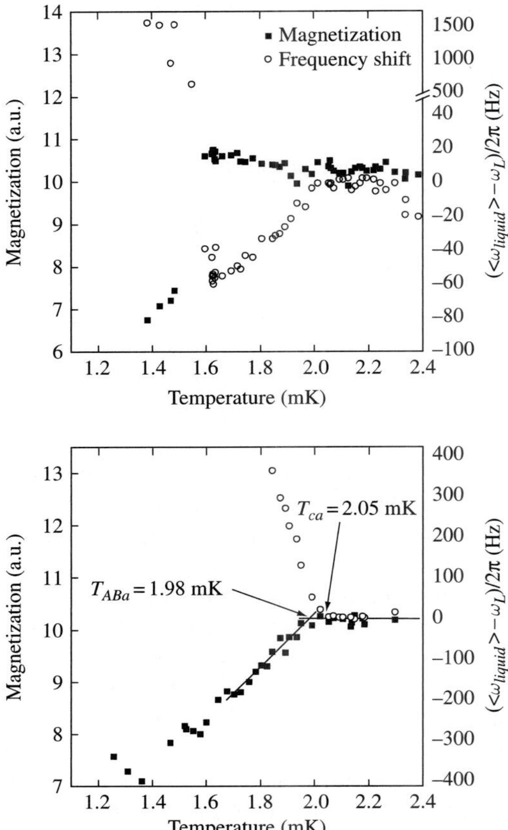

Fig. 5.13 Magnetization and frequency shift in 98% aerogel as a function of temperature on cooling (top) and warming (bottom), from [16]. Two layers of solid ^{4}He were preplated on the aerogel surface. The effect of supercooling is evident. The narrow window in temperature between 1.89 mK and 2.05 mK could be the A-like phase on warming.

was conducted without the magnetic solid layers. On cooling, discontinuities in the magnetization and the frequency shift were observed, suggesting the transition was of first order. The magnetization exhibited a sudden drop at the transition. The NMR line in the higher temperature phase was relatively narrow and *negatively* shifted, but went through a sudden broadening to the high frequency side at the transition into the low temperature phase. However, on warming, the magnetization continuously reached the full normal fluid value at a slightly lower temperature than the superfluid transition identified by the disappearance of the frequency shift, as shown in Fig. 5.13. Therefore they unambiguously demonstrated the existence of two distinct superfluid phases separated by a first-order transition: the A-like and the B-like phases. Alles et al. [5] confirmed that the NMR line of the B-like phase was indeed what was expected in the texture broadened bulk B phase. A few years later, a series of pulsed and cw NMR measurements provided unequivocal evidence for this identification [53, 48, 47, 46]. A signature of the BW state, the Leggett angle θ_L was found to be identical to the value of the bulk B phase from the tipping angle dependence measurement, shown in Fig. 5.14, and a homogeneous precession domain (HPD), proprietary to the bulk B phase [25, 61], was observed through nonlinear cw NMR in a field gradient [46].

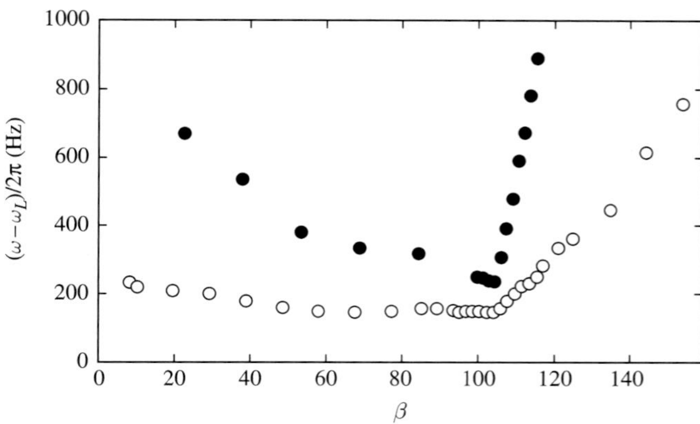

Fig. 5.14 Tipping angle dependence of transverse NMR frequency shift, from [53]. Data were taken at $0.83\,T_{ca}$ and 101 mT (open circles) and at $0.78\,T_{ca}$ at 28.5 mT (solid circles). The non-zero shift and the weak tipping angle dependence for $\beta < \theta_L$ indicate that the texture is determined by the inhomogeneous aerogel density, rather than by the geometry of the sample cell.

In a continuation of Barker's experiment, Baumgardner et al. [19] extended their measurements to higher porosity aerogels and observed similar behavior to the 98% sample. In 99.3% aerogel, they were able to clearly identify a wide region of supercooled A-like phase on cooling, but a very narrow A-like phase ($\approx 150\,\mu$K) on warming. However, unlike in Barker's experiment, the average frequency shift was positive on both cooling and warming, despite a small portion of the spectrum appearing in the negative side. Their detailed analysis of the spectrum acquired through a so-called *tracking experiment*, a method employed earlier by Gervais et al. [70], revealed coexistence of both the A-like and B-like phases in the warming A-like phase region, with the fraction of the A-like component gradually increasing toward T_{ca}. A typical tracking experiment is done in the following manner. The liquid is initially cooled from the normal phase to below the supercooled A-like to B-like transition. This establishes a wide range of A-like phase behavior in a specific measurement. Then the liquid is warmed slowly to the desired turnaround or trajectory temperature, T_t. After a few hours of waiting for equilibrium, the temperature is lowered again. This procedure is repeated for strategically selected values of T_t. One expects to see different cooling traces and the reappearance of the supercooled transition only when $T_t > T_{ABa}$, the warming B-like to A-like transition temperature, because the conversion between the two phases is supposed to occur only in that temperature range. In Baumgardner's experiment, the spectra were taken at 1.82 mK on cooling after going through tracking. Each spectrum was then fitted with the known spectra of the A-like and B-like phases with their relative composition as a fitting parameter. This routine produced remarkably good fits to the actual spectra and, convincingly, determined the fraction of the A-like phase. Figure 5.15 shows the results of this experiment. One can clearly see the onset of a non-zero A-like fraction, concurrent with the kink in the magnetization, and the gradual growth of the A-like fraction toward T_{ca}. This behavior was interpreted as a consequence of the pinning of the A-B interface by aerogel. A similar interpretation was made by Brussard et al. [28] in high magnetic fields and significantly lower temperatures. However, a different scenario [168, 126] was proposed based on acoustic experiments performed in zero magnetic field, which will be discussed in the following section.

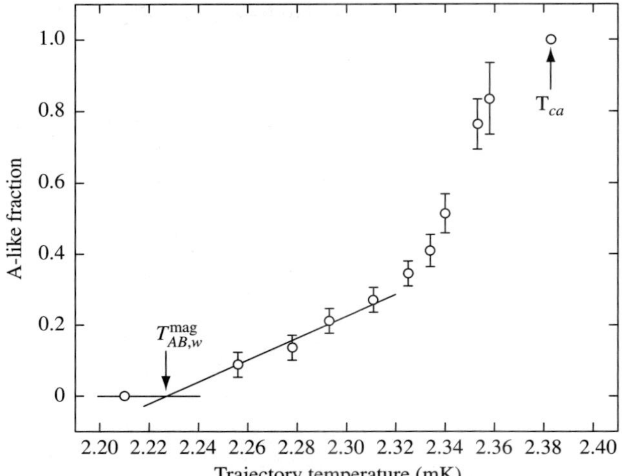

Fig. 5.15 The fraction of A-like phase versus the trajectory (or turnaround) temperature, see text, from [19]. The arrow indicates the temperature where a kink in magnetization appeared in Fig. 2 in [19].

These observations and interpretations raised an interesting question on the spatial distribution of these phases in the coexistence region. Is there a single boundary between the phases? Or do multiple domains exist? A group at Osaka City University conducted cw NMR with a radial gradient of an axial field (28 mT) on a cylindrical aerogel sample of 97.5% porosity, aligned in the direction of the static field [98]. Their experiment indicated that the A to B conversion on cooling started from the central region of the cylindrical sample in a columnar shape. However, this could be driven by the field gradient and still remains an open question.

The hysteretic behavior of the A-like to B-like transition and the possible pinning of the interface could make determination of the thermodynamic first-order transition rather dicey. Fisher et al. [60] argued that the pinning of the interface could cause both supercooling and superheating of the transition, and hence the true thermodynamic transition could not be determined by a temperature sweep. Baumgardner et al. [20] developed a method for determining the equilibrium A-like to B-like transition temperature through a thermodynamic analysis of precise measurements of longitudinal frequencies and magnetization. The thermodynamic transition temperature obtained was within 30 μK below the observed transition on warming, indicating negligible superheating in aerogel, as is also the case in bulk.

Other than the fact that the A-like phase is an ESP state, the experimental results from this phase were ostensibly inconsistent and puzzling. The frequency shifts obtained from cw NMR and small tipping angle pulsed NMR exhibited wildly different behaviors: large positive shifts [160, 21, 52, 129]; large negative shifts [16, 106]; and also small negative shifts [48]. Furthermore, various types of tipping angle dependencies have been found in the A-like phase. In addition to the abrupt disappearance of $\Delta\omega$ for $\beta > 40°$ observed by the Northwestern group, at least three other tipping angle dependencies have been identified: $\Delta\omega \propto (1 + \cos\beta)$ [96]; $\Delta\omega \propto (\beta - \pi/2)$ [48]; and $\Delta\omega \propto \cos\beta$ [52]. In an experiment by Dmitriev et al. [52], two simultaneous absorption lines, both positively shifted but with drastically different properties, were observed in the metastable A-like phase of a single aerogel sample. One peak called the "c-state" appeared close to ω_L, while the "f-state" peak appeared far away from ω_L. They found that the c- and f-states came from different regions of the ^{3}He in the aerogel. The ratio of the shifts was about 4, independent of temperature. They were able to stabilize the c-state only when a large RF excitation was applied during cooling through T_{ca}. In the c-state it was found that $\Delta\omega \propto \cos\beta$, but in the $(f + c)$-state, $\Delta\omega \propto (1 + \cos\beta)$.

How do we understand all of these behaviors occurring in the same superfluid phase with ESP? As briefly mentioned earlier, *anisotropy* is the key to understanding the nature of the A-like phase and the first-order transition between the two phases observed in aerogel. Soon after the discovery of superfluid ^{3}He in aerogel, Volovik pointed out the significance of the quenched random anisotropic disorder presented by the strand-like aerogel structure and its interaction with the anisotropic order parameter [170]. This coupling was thought to be particularly important in the A phase whose order parameter is doubly anisotropic, in the sense that the rotational symmetries in the spin and the orbital space are broken separately. In consideration of the interaction between the anisotropic disorder and the orbital vector $\hat{l}$, Volovik argued that the long-range order established in pure superfluid ^{3}He was unstable against the presence of any degree of disorder, and therefore a strong candidate for the A-like phase in aerogel should be the A phase with short-range orientational order of $\hat{l}$, effectively a random texture [171, 172]. This glass-like state is called the LIM state, acknowledging the earlier work of Larkin [109] and Imry and Ma [95]. Detailed free energy considerations [145] indicate that a cylindrical object with axis $\hat{a}$ and a radius smaller than ξ_o would favor the $\hat{a} \perp \hat{l}$ configuration in the A phase within a length $\sim \xi_o$ around the object. An aerogel strand should have a similar effect on the order parameter of the A-like phase if it has the same order parameter structure as the bulk A phase. Then, the LIM state in the A-like phase is characterized by the characteristic length scale of the short-range orientational order of $\hat{l}$, L_{LIM}, determined by balancing the energy cost of the random anisotropy against the gain from gradient energy. Volovik [172] provided an order of magnitude estimation using reasonable length scales associated with 98% aerogel and superfluid ^{3}He to find $L_{LIM} \sim 1\,\mu$m. Therefore, if the LIM state were realized in the A-like phase, it should lead to the inhomogeneous broadening of NMR lines with a substantial spectral weight below ω_L, since the orientation of $\hat{d}$ is confined in the plane perpendicular to $\vec{H}$ for $\omega_L \gg \Omega_A$. However, actual experimental results showed much more diverse behavior, as previously discussed.

Other possibilities such as the robust phase and the planar state have been proposed as candidates for the A-like phase in aerogel. Baumgardner [18] argued that the slope of the pressure-dependent thermodynamic first-order transition line extracted from his experiment was much more consistent with the value calculated under an assumption of the A-like phase being the two-dimensional planar state. On the other hand, a brand new class of equal spin pairing state, the robust phase, which is inert to arbitrary spatial rotation in the presence of random anisotropic disorder, has been proposed by Fomin [63], and some of the NMR results were interpreted to support this idea [96, 124, 65]. However, a growing number of theoretical and experimental results have weakened these arguments, and the LIM state has emerged as the strongest candidate for the A-like phase in isotropic aerogels.

It is the NMR experiments in the past 5–6 years that have significantly advanced our understanding of the role of anisotropic disorder in this system. The suggestion by Vicente et al. for the use of uniaxially deformed aerogels to introduce global anisotropy, and confirmation of built-in anisotropy in supposedly isotropic undeformed aerogels have added an additional dimension to the design of new experiments, as well as the interpretation of results. In the rest of this section, we will review the results of NMR experiments conducted using uniaxially deformed aerogels. In particular, we allocate space for the recent experiments by Dmitriev et al. [50, 49] which, in our opinion, provide the most comprehensive and coherent picture on this subject.

When a piece of isotropic aerogel is compressed uniaxially, say in the z-direction, then the randomly oriented aerogel strands tend to rotate into the x–y plane. This mechanical deformation introduces a small degree of global anisotropy in the sample, as verified by optical birefringence measurements [138, 22, 23]. This promotes $\hat{l}$ orienting towards the z-direction from presumably the random configuration of the LIM state. One finds that an ordered state with $\hat{l} \parallel \hat{z}$ becomes energetically favorable when the strain exceeds a critical value, $\sigma_c \approx (\zeta_a/L_{LIM})^{3/2}$ [171, 172]. Volovik estimated $\sigma_c \approx 10^{-3}$–10^{-2} in the weak pinning limit. On the other hand, in a stretched aerogel, $\hat{l}$ is expected to lie preferentially in the x–y plane. The direction of a magnetic field, which fixes the orientation of $\hat{d} \perp \vec{H}$, will eventually dictate the dipole configuration between $\hat{l}$ and $\hat{d}$. This suggests that even tiny deformations of aerogel with various origins can give rise to large effects.

The first experimental confirmation of this scenario was made in the experiment of Kunimatsu et al. [106]. They conducted cw NMR on aerogel samples under 2–3% compressive strain at 29.3 bar of sample pressure in a field of 29 mT. The magnetic field, the axis of compression, and the cylindrical axis of the sample were all in the z-direction. They found *negatively* shifted, but relatively *narrow*, NMR lines in the A-like phase. The B-like phase observed on cooling also had a narrow, positively shifted line. These results are shown in Fig. 5.16. This behavior is consistent with having most of $\hat{l}$ aligned along the z-direction, but $\hat{d}$ in the x–y plane, rendering $\phi \approx 90°$ in Eq. (5.16). Their quantitative analysis showed that the maximally shifted value in the A-like phase indeed corresponded to $\hat{l} \parallel \hat{z} \perp \hat{d}$. Dmitriev et al. [50] observed similar behavior in a rather different setup. In their experiment, three aerogel samples of 98% porosity were prepared: two supposedly isotropic and one under $\approx 1\%$ radial compression through differential thermal contraction during cooling. The geometry of the experimental setup was identical to Kunimatsu's, except that they were able to rotate the static field and implemented an additional coil for direct longitudinal resonance measurement. They made a remarkable observation from one of the uncompressed—supposedly isotropic—aerogel samples. With $\vec{H} \parallel \hat{z}$, relatively *narrow, negatively* shifted NMR lines were observed, but when the field was rotated to the y-direction, the lines shifted symmetrically to the positive side. This behavior was exactly what was expected for the A phase in bulk with a homogeneous texture, as described by

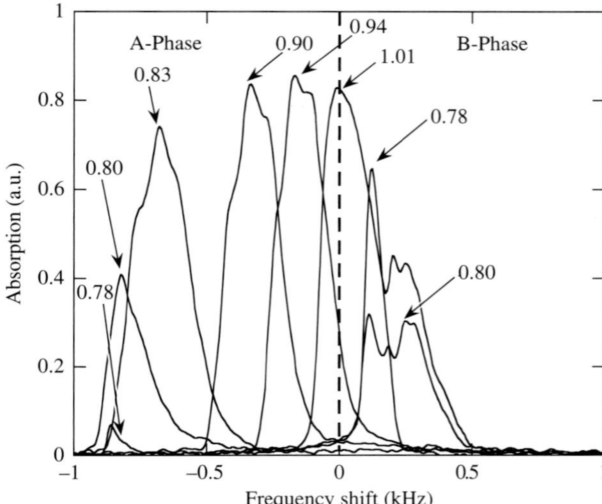

Fig. 5.16 NMR spectra in uniaxially compressed aerogel for various reduced temperatures T/T_{ca}, from [106]. The lines appearing in the negative (positive) side are from the A(B)-like phase. The A-like to B-like conversion starts below $0.82\,T_{ca}$.

Eq. (5.22). For $\vec{H} \parallel \hat{y}$, a reasonably homogeneous $\hat{l} \parallel \hat{d} \parallel \hat{z}$ configuration was realized—dipole-locked bulk A phase—which was corroborated by their measured tipping angle dependence being consistent with Eq. (5.23). This meant that the uncompressed aerogel actually possessed built-in global anisotropy equivalent to the compressed case of Kunimatsu et al. [106]. Their B-like phase results were also consistent with this interpretation. The ratio of the Leggett frequencies in the A-like and the B-like phase aerogel were close to the bulk values. However, they found that the behavior from the third sample under radial compression was drastically different, which led them to speculate that the built-in anisotropy might be compensated by external deformation. The experiments by Sato et al. [150], using uniaxially compressed aerogel, and Elbs et al. [56], with a radially compressed (effectively stretched) aerogel, also produced results in line with this picture. Sato's experiment provided strong confirmation of the $\hat{l} \parallel \hat{z} \perp \hat{d}$ structure in a 2% compressed aerogel, by demonstrating coherent precession in the A-like phase, which does not exist for $\hat{l} \parallel \hat{d}$.

A coherent picture encompassing a wide range of phenomena in the A-like phase has emerged from these experiments: (i) the A-like phase has the same symmetry as the ABM state; (ii) in isotropic aerogels, the A-like phase might be in a state in which the long-range orientational order in $\hat{l}$ is destroyed by the quenched random anisotropy provided by aerogel strands; (iii) a small degree of global anisotropy can restore the long-range order. Depending on the type of anisotropy induced or built-in, a variety of possible order parameter structures can be realized and give rise to fascinating NMR behavior.

Here we would like to summarize this picture by borrowing the theoretical argument developed by Volovik [171] and expanded by Dmitriev et al. [49] with a few additional experimental results. In an isotropic aerogel with strong disorder, the orbital angular momentum $\hat{l}$ is randomized and

$$\langle \hat{l} \rangle = 0, \quad \langle \hat{l}_i^2 \rangle = 1/3 \quad (i = x, y, z), \tag{5.25}$$

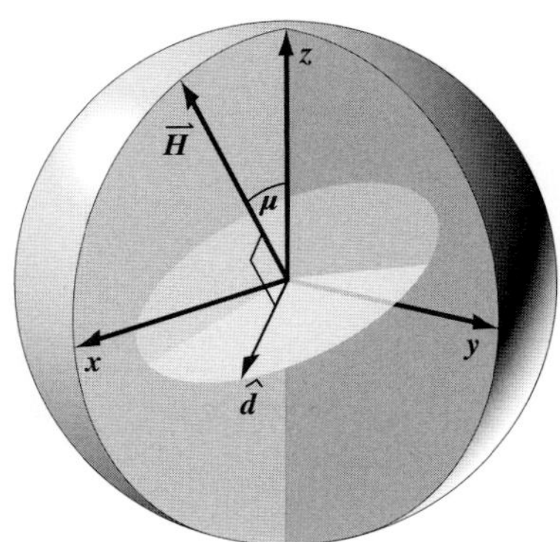

Fig. 5.17 Vector diagram of $\vec{H}$ and $\hat{d}$. By convention the direction of uniaxial deformation is along z. The angle μ determines the relative orientation of a magnetic field from $\hat{z}$. $\hat{d}$ is confined in the plane normal to the magnetic field.

where $\langle .. \rangle$ indicates the spatial average. This configuration represents the orbital-glass and spin-glass (OG-SG) state corresponding to the LIM state. In the presence of a large enough magnetic field, $\hat{d}$ is randomly oriented in the plane normal to $\vec{H}$ due to the spin-orbit interaction from random $\hat{l}$. Among these energy scales the dipole energy, responsible for fixing the relative orientation between $\hat{d}$ and $\hat{l}$, is the weakest. Figure 5.17 illustrates the configuration of $\hat{d}$ and $\vec{H}$. Uniaxially deformed aerogels can be characterized by introducing a parameter, $\tilde{a}$, representing the degree of global anisotropy as

$$\langle \hat{l} \rangle = 0, \quad \langle \hat{l}_z^{\,2} \rangle = \frac{1+2\tilde{a}}{3}, \quad \langle \hat{l}_x^{\,2} \rangle = \langle \hat{l}_y^{\,2} \rangle = \frac{1-\tilde{a}}{3}. \tag{5.26}$$

Here, the axis of deformation is assumed to be in the z-direction. For $0 < \tilde{a} < 1$, aerogel is compressed and $\langle \hat{l}_z^{\,2} \rangle$ will gradually increase with the compressive strain. However, it is expected that $\langle \hat{l} \rangle \neq 0$ for compression beyond the critical value, σ_c. This state is equivalent to an orbital ferromagnet (OF). Stretching gives the easy x–y plane anisotropy with $-1/2 < \tilde{a} < 0$. Figure 5.18 shows cartoons of the various configurations in aerogel under uniaxial deformation.

Volovik [171] and Dmitriev et al. [49] derived transverse NMR frequency shifts for various cases in the presence of a magnetic field oriented in an arbitrary direction. For the orbital-glass and spin-nematic (OG-SN) state in a compressed aerogel, the frequency shift is given by

$$\Delta\omega = \tilde{a}\Delta\omega_A \left\{ -\cos\beta + \sin^2\mu \left(\frac{7\cos\beta + 1}{4} \right) \right\}, \tag{5.27}$$

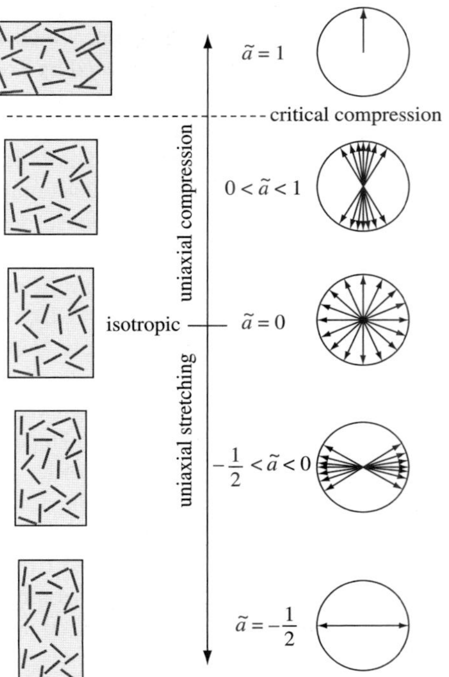

Fig. 5.18 Cartoon of uniaxially deformed aerogel (left) and schematic diagram (right) of the angular momentum configuration. The lines in the cartoons depict aerogel strands. Deformation is along the vertical direction. The arrows in the schematic diagrams represent $\hat{l}$ in aerogel. See text for details.

where μ is the angle between $\vec{H}$ and $\hat{z}$, and $\Delta\omega_A = \Omega_A^2/2\omega_L$, the maximum transverse shift in the bulk A phase. This equation is also applicable to the OF state with $\tilde{a} \approx 1$. For the OG-SN state in a stretched aerogel,

$$\Delta\omega = \tilde{a}\Delta\omega_A \left\{ -\cos\beta + \sin^2\mu \left(\frac{5\cos\beta - 1}{4} \right) \right\}, \tag{5.28}$$

and for the OG-SG state in both compressed and stretched aerogel,

$$\Delta\omega = \tilde{a}\Delta\omega_A \cos\beta \left(\frac{3}{2}\sin^2\mu - 1 \right). \tag{5.29}$$

For example, in aerogel that is uniaxially compressed beyond the critical strain, and with $\vec{H} \perp \hat{z}$, the frequency shift given by Eq. (5.27) recovers the expression for the A phase in the dipole minimum configuration (Eq. 5.22). Interestingly, this also predicts zero frequency shift in the LIM state, for which $\tilde{a} = 0$.

Recent experiments by Dmitriev et al. [49] have verified many features predicted by this theory. Their experimental setup was practically identical to the previous one [50]. Continuous wave NMR on a 9%-compressed aerogel (sample 1) showed that the frequency shifts in the A-like phase followed the behavior expected from Eq. (5.22). Therefore, the Leggett frequency could be extracted from the shift. They compared their data directly with the results from previous experiments [106, 50, 51, 56] by properly adjusting the sample pressures to 26 bar, using the known pressure dependence of Ω_A in the bulk A phase. They found that all the data fell on a common temperature dependence, as shown in Fig. 5.19. This suggests that all the samples employed in these experiments were in the OF state with $\tilde{a} \approx 1$. However, the frequency shifts observed in the 5%-compressed (sample 2) and the uncompressed (sample 3) aerogels showed smaller but positive transverse shifts and their Leggett frequencies were distinct from the other samples, also

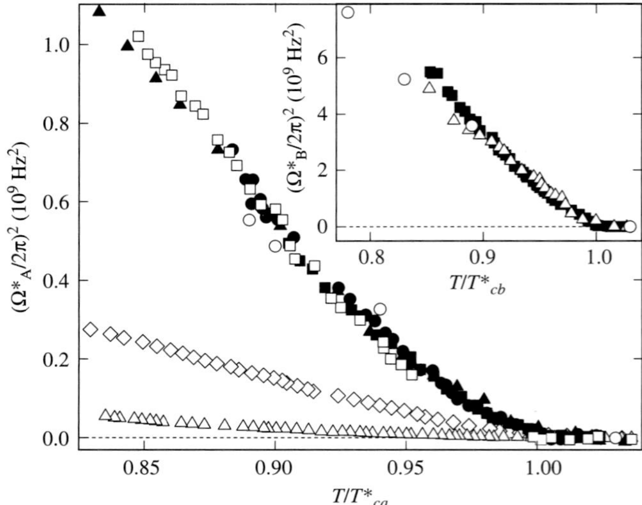

Fig. 5.19 Temperature dependence of the Leggett frequency in the A-like phase and in the B-like phase (inset), from [49]. $(\Omega^*_{A,B})^2 = 2\omega_L|\Delta\omega|$. The data from various experiments are rescaled to 26 bar, as described in the text. Open squares - sample 1 9% compressed at 26 bar, solid triangles - 4% compressed at 29.3 bar [106], solid circles - intrinsically anisotropic at 28.6 bar [51], solid squares - intrinsically anisotropic at 26 bar, open circles - 20% radially compressed at 25 bar, open diamonds - sample 2 at 27.2 bar, open triangles - sample 3 at 27.2 bar.

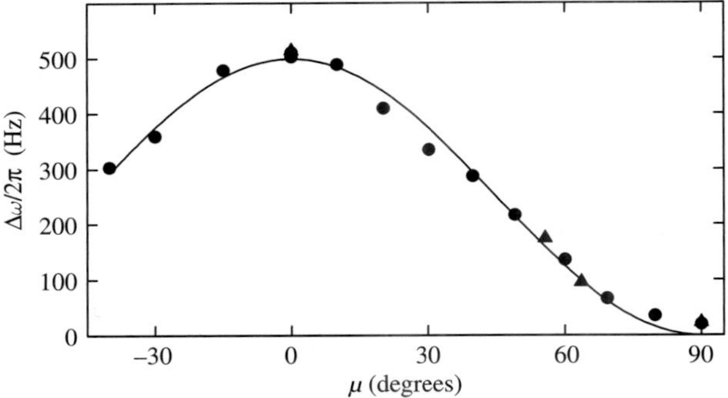

Fig. 5.20 Field orientation dependence of cw NMR frequency shift in the A-like phase from sample 2, from [49]. Data were taken at $0.81T_{ca}$ in 100 mT (circles) and 104 mT (triangles), but are rescaled at 100 mT in the plot. The solid line represents the best fit to the data using $\Delta\omega = -\tilde{a}\Delta\omega_A \cos^2\mu$ and $\tilde{a} = -0.25$.

shown in Fig. 5.19. It was more consistent with what was expected in the stretched OG-SN state. If this is the case, the cw NMR shift should also follow the distinct field orientation dependence, $\Delta\omega = -\tilde{a}\Delta\omega_A \cos^2\mu$. As shown in Fig. 5.20, their field orientation dependent cw NMR shift in the 5%-compressed aerogel showed excellent agreement with the prediction for $\tilde{a} = -0.25$ (stretched). Their tipping angle dependencies observed in the A-like phase contained in these samples under various experimental conditions were internally consistent, and were in quantitative agreement with the theoretical predictions.

The results from NMR experiments have been exceptionally valuable in understanding the nature of superfluid phases in aerogel. In particular, the symmetry of the A-like phase, which has been elusive for an extended period, is now thought to be the same as the bulk A phase. Although not completely consistent with all measurements, a large proportion of the results can be understood in terms of the LIM state realized in isotropic aerogel, and orientational effects arising from global anisotropy induced in deformed aerogels. There are a few other interesting NMR results that await further investigation. Experiments in a rotating cryostat [173, 174, 106, 104, 105] have revealed intriguing phenomena such as pinning of vortices by aerogel [174]. The controllable presence of localized spins makes relaxation mechanisms in the normal and superfluid phases interesting. Only a few experiments have been dedicated to probing relaxation in the normal fluid [32, 76] and in the superfluid phases [53, 48, 150]. This remains one of the lesser investigated topics.

5.4.2 Sound experiments

The acoustic properties of a superfluid are unique because novel sound modes arise from the motions of the quantum condensate and the normal fluid [54, 169]. In addition to ordinary longitudinal sound (first sound), there are three additional (second, third, and fourth) sound modes present in superfluids whose properties are quite well understood in terms of the two-fluid model. The in-phase motion of the two fluid components generates first sound, and its speed is directly related to the thermodynamic quantities of the system, $c_1^2 = (\partial p/\partial\rho)_T$. For second sound, the normal and superfluid components oscillate in an out-of-phase fashion. Therefore, this mode propagates at a low speed, expressed by $c_2 = S(T\rho_s/C_p\rho_n)^{1/2}$, where S and C_p represent entropy and isobaric heat capacity, respectively. It is of note that there are no elastic quantities appearing in the expression of the speed of second sound. Fourth sound only exists in a confined geometry such as porous media or narrow channels, where the normal fluid component is viscously clamped by the structure. If the characteristic size of the pore or channel is much smaller than the viscous penetration depth, $\delta_v = (2\eta/\rho_n\omega)^{1/2}$, the normal component is completely locked to the immobile matrix, and fourth sound propagates at speed $c_4 = c_1(\rho_s/\rho_n)^{1/2}$ with low damping. Therefore, the superfluid density can be independently determined from c_4. Third sound is a property of thin superfluid films and is not relevant to the ^{3}He-aerogel system.

In liquid ^{3}He, high frequency zero sound is expected to disappear in the superfluid state as the superfluid gap develops at the Fermi level. However, this mode is revived as a collective mode of coherent phase oscillation resulting from broken gauge symmetry. Below T_c, zero sound exhibits broad absorption due to the pair-breaking process for $\omega > 2\Delta(T)$. Coupling to some of the order parameter collective modes results in strong damping of the mode at resonance, accompanied by significant modification of its dispersion due to repulsive mode coupling. All of these features have been studied extensively in bulk superfluid [81], and gap spectroscopy using high frequency zero sound has been instrumental in identifying the microscopic structure of the order parameter [81]. Unfortunately, in aerogel, none of these features have been observed experimentally [134, 41, 126].

As a result of its singular structure and mechanical properties, acoustic properties in aerogel are expected to show significant changes and quite unusual behavior. Aerogel is compliant and by no means immobile. Therefore, a proper description of this system should incorporate the motion of aerogel strands clamped to the normal component of the fluid. This stipulates renormalization of the inertia as well as the elastic moduli, and the sound modes of liquid ^{3}He are effectively coupled to the mode of the aerogel matrix, which has a relatively low sound speed. As a result, the first sound mode is expected to split into two modes: fast and slow sound. In the fast sound mode, the fluid moves together in phase with aerogel, close to conventional longitudinal sound. It is also possible to have a mode where the superfluid component oscillates out of phase with the aerogel matrix and the clamped normal component. This mode, reminiscent of second sound in bulk superfluid, has a much lower speed and is thus termed slow sound. McKenna et al. [120] formulated a phenomenological theory to describe the acoustic properties of superfluid ^{4}He in aerogel, and this theory is equally applicable to the ^{3}He-aerogel system. Two-fluid hydrodynamic equations were modified to incorporate the motion of the aerogel matrix, and the secular equation for the sound velocity was obtained as

$$\left(c_x^2 - c_1^2\right)\left(c_x^2 - c_2^2\right) + \frac{\rho_a}{\rho_n}\left(c_x^2 - c_a^2\right)\left(c_x^2 - c_4^2\right) = 0. \qquad (5.30)$$

Here $\rho_{n(s)}$ is the normal fluid (superfluid) density in aerogel and $\rho = \rho_n + \rho_s$. The solutions of this equation, the speeds of slow and fast sound modes, $c_{s,f}$, can be simplified to give [73, 74]

$$c_f^2 = c_1^2 \left(1 + \frac{\rho_a}{\rho} \frac{\rho_s(T)}{\rho_n(T)}\right) / \left(1 + \frac{\rho_a}{\rho_n(T)}\right) \quad (c_2^2 \ll c_f^2 \text{ and } c_a^2 \ll c_f^2), \quad (5.31)$$

$$c_s^2 = c_a^2 \frac{\rho_a \rho_s(T)}{\rho^2} \quad (c_2^2 \ll c_s^2 \ll c_a^2 \ll c_1^2). \quad (5.32)$$

The slow sound mode is of particular importance since it provides an additional experimental tool to determine the superfluid density. Although one can relate slow sound in aerogel to second sound in bulk superfluid, the source of the restoring force on the normal component in slow sound is mainly in the elasticity of aerogel, rather than in thermal energy as in second sound [73]. Therefore, the scale of c_s is set by c_a and is much higher than c_2. On the other hand, slow sound also shares some characteristics of fourth sound in a porous medium, although there the speed is scaled by c_1 rather than c_a. The slow sound mode has been studied extensively by the Cornell group [68, 73, 74, 130, 110, 133, 132, 131].

The Osaka City group successfully observed fourth sound in aerogel grown in a sintered silver sponge [103, 128, 135]. This technique allowed their experiment to be performed in a more ideal situation by limiting the motion of aerogel strands and the normal component, and also eliminated potential complications in detection by boosting the intrinsic resonance frequency of the aerogel matrix well above the fourth sound resonances.

Unlike sound speed, the damping of the various sound modes in aerogel is much more complicated. In general, the relative motion of the aerogel strand to normal fluid introduces an additional damping mechanism. Ichikawa et al. [93] considered this effect in order to understand the damping mechanism in high frequency hydrodynamic sound. Low mass density and the compliant nature of aerogel necessitate consideration of the effective momentum transfer upon quasiparticle scattering off strands, which generates dragged motion of the aerogel. This so-called collision drag effect was included in calculating the dispersion relation in the normal fluid. Higashitani et al. [90] and Miura et al. [123] extended this model to the longitudinal fast sound mode in the superfluid within the framework of the two-fluid description. The drag effect was introduced phenomenologically by a frictional drag force density, $\vec{F}_d$, originating from the relative motion of the normal fluid and aerogel,

$$\vec{F}_d = \frac{1}{\tau_f} \rho_n (\vec{v}_n - \vec{v}_a), \quad (5.33)$$

where τ_f is the frictional relaxation time, determined from a microscopic calculation, and $\vec{v}_{n(a)}$ is the velocity of the normal fluid (aerogel strand). By incorporating the drag effect into the equation of McKenna et al. [120], Miura et al. [123] obtained a generalized dispersion relation,

$$\left(z_x^2 - c_1^2\right)\left(z_x^2 - c_2^2\right) + \left(i\frac{4\eta\omega}{3\rho_n} + \frac{i}{\omega\tau_f} \frac{z^2}{1 + \dfrac{i}{\omega\tau_f} \dfrac{\rho_n}{\rho_a} \tilde{\omega}_a} \right) \left(z_x^2 - c_4^2\right) = 0, \quad (5.34)$$

where $z = \omega/k$, $\tilde{\omega}_a = \omega^2/(\omega^2 - \omega_a^2)$, and $\omega_a = c_a k$. From this dispersion relation, the attenuation for fast sound was derived [90] to be

$$\alpha_f = \frac{\omega^2/2c_f}{1+\rho_a\rho_s/\rho_n\rho}\left(\frac{\rho_a^2\tau_f/\rho\rho_n}{1+\rho_a/\rho_n} + \frac{4\eta/3\rho c_1^2}{1+\rho_a\rho_s/\rho_n\rho}\right). \tag{5.35}$$

The same mechanism could also be a source of damping for fourth sound in aerogel.

High frequency transverse sound takes an interesting position in the study of superfluid ^{3}He. In a conventional liquid, transverse sound is an over-damped mode and does not propagate. However, in a Fermi liquid its presence as a propagating mode has been predicted in the collisionless limit [107]. This prediction was finally verified experimentally many decades later in the normal state of ^{3}He [149] and in the B phase of the superfluid [99, 115]. In the B phase, this mode can only be stabilized as a propagating mode through off-resonant coupling to an order parameter collective mode (OPCM), the imaginary squashing mode. With a lack of OPCMs and the difficulty of reaching the collisionless limit in aerogel, the utility of this mode does not seem to be very promising. However, high frequency transverse sound has been employed in ^{3}He-aerogel, though its efficacy has been limited to detecting various phase transitions.

Experiments involving propagating modes use either a resonant cavity method with cw excitation or a pulse propagation technique. In the resonant cavity method, which has been employed in slow sound, fourth sound, and high frequency longitudinal sound experiments, a ^{3}He-aerogel sample is enclosed in a cylindrical cavity of specific dimensions that are related to the wavelength of the sound mode. The two ends of the cavity are composed of sound transducers used as a speaker (generator) and a microphone (detector). Since this method involves sweeping the frequency, broadband piezoelectric ceramic transducers have been the popular choice. As the speaker sweeps the frequency of cw excitation, the microphone detects its cavity resonance features. Then, the speed of the mode is determined from the *m-th* harmonic resonance, f_m, by $c_x = 2A_x L f_m/m$, where L is the size of the cavity, and A_x is a numerical factor which depends on the mode, boundary conditions, and the type of medium. For a compressional wave in a pure liquid, A_x is unity. The inverse quality factor, determined from the frequency width of the cavity resonance, is related to the attenuation of the sound mode as $Q_m^{-1} \propto \alpha/\omega$.

In a pulse propagation experiment, the speed of sound is determined from the time of flight or through phase analysis, and absolute attenuation can be extracted in principle from the amplitudes of the pulse-echoes if the path length is known. High-Q piezoelectric crystals such as quartz or $LiNbO_3$ are commonly used for this scheme. Because of the drastic acoustic mismatch[1] at the boundary between the transducer and ^{3}He-aerogel, less than $\approx 1\%$ of loss is estimated at the reflection [114], but this weak coupling is also a cause of long ringing in the transducer, the main drawback of this technique.

Transverse sound experiments rely on acoustic impedance. Excitation and detection of transverse sound is done by a single shear mode high-Q quartz

[1] The acoustic impedance of liquid ^{3}He, $Z_a \approx 3 \times 10^3$ g s^{-1} cm^{-2}, while Z_a for $LiNbO_3$ is 3.4×10^6 g s^{-1} cm^{-2}.

transducer. A sensitive cw impedance bridge can detect a minute change in electrical impedance of a transducer in direct contact with the sample [38]. When there is no reflected wave, the acoustic response is determined by the bulk acoustic impedance, $Z_a = \rho\omega/\tilde{q}$, where $\tilde{q} = k + i\alpha$ is the complex wavenumber. A change in either the attenuation or the phase velocity produces a change in the impedance, Z_a, and as a consequence changes the electrical impedance of the transducer. Although it is not trivial to extract physical quantities from such measurements, this method has been used successfully owing to its high resolution and its capability for collecting data continuously.

Acoustic measurements can be conducted with or without a magnetic field, and do not require a laborious tuning process at a given field. Therefore, they have proven to be useful in mapping the phase diagram, in particular as a function of field. However, unlike NMR, the transducers in acoustic measurements have to be in direct contact with the sample. At least for high frequency experiments, the quality of the contact is of utmost importance to the outcome of the experiment. Best results have been produced when aerogel samples were grown directly onto the transducer or inside a cavity [70, 39, 41, 126]. Most of the acoustic measurements have used pure ^{3}He. Therefore, unless mentioned otherwise, experiments described below were performed with pure ^{3}He.

Geller et al. [68] reported the first observation of slow and fast sound in aerogel, which was followed by a comprehensive study by Golov et al. [73]. Employing an acoustic cavity technique, they detected multiple resonance features over a wide range of frequencies in zero magnetic field for several pressures. The superfluid transitions in aerogel were clearly identified by the sharp onset of positive shifts in both fast and slow sound resonances. The speed of the fast sound mode in the normal fluid was about 80% of the bulk ^{3}He first sound speed. It increased in the superfluid phase by only $\sim$1%. This observation is consistent with Eq. (5.31). Figure 5.21 shows the spectra near the slow sound mode and the temperature dependencies of the speed of the slow sound mode at various pressures. The sharp positive peak starts to grow and move to a higher frequency below the aerogel superfluid transition (top panel). Note that $c_s \sim 10\,\mathrm{m\ s^{-1}}$ at low temperatures. The superfluid densities extracted using Eq. (5.32) are comparable to those obtained in the torsional oscillator experiments [140, 119]. The data acquisition in this experiment was done on warming in the B-like phase and no signature of the first-order transition to the A-like phase was seen. In their follow-up experiment in small magnetic fields [131], the superfluid density in the metastable A-like phase was measured for the first time at 27.9 bar. While no significant field dependence was observed in the superfluid density in either phase, they found a significantly lower superfluid density in the A-like phase, $\rho_s^A/\rho \approx 0.5\rho_s^B/\rho$ compared with that in bulk, which can hardly be accounted for by any reasonable textural configurations possible in the ABM state.

Observation of the first-order transition between the A-like and B-like phases in the NMR experiments of Barker et al. [16] called for independent and thorough investigation on this transition with and without a magnetic field. The Northwestern group conducted a detailed field dependence study of the

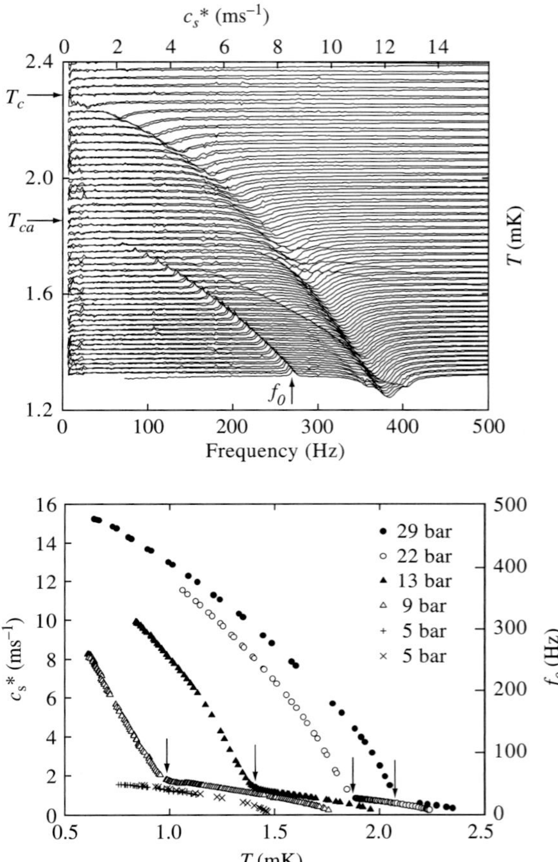

Fig. 5.21 Top: Spectra of the slow sound mode at 21.7 bar, from [73]. The sharp positive peak at f_0 is the fundamental slow sound resonance. The broader negative peak is the Helmholtz resonance from the acoustic cavity and the narrow filling tube. The slow sound resonance appears below the aerogel superfluid transition but the Helmholtz mode is extended to the bulk superfluid transition. A few other higher harmonics of the slow mode are observable in this plot. Bottom: Temperature dependence of the slow sound speed, c_s^* determined assuming $A_s = 1$, from [73]. The onset of superfluidity in aerogel is marked by arrows. The superfluid density obtained using Eq. (5.32) is comparable to the results obtained from torsional oscillator experiments.

first-order transition on both cooling and warming using transverse acoustic impedance [69, 71]. In zero magnetic field they confirmed the transition from the metastable A-like phase to the B-like phase on cooling, but no evidence of a transition on warming was found. Their tracking experiments (explained in the NMR section) could not show any evidence of B to A conversion up to $\sim 20\,\mu$K below T_{ca}. However, the metastable A-like phase appeared at pressures even below the bulk polycritical point, $p_{pc} = 21.2$ bar, where the bulk A phase does not exist. The transition on warming could only be identified in magnetic fields above ~ 0.14 T.

Almost identical behavior was observed in the slow sound experiments by Nazaretski et al. [132, 131]. In zero field they identified the stable coexistence of the metastable A-like and B-like phases akin to the Stanford NMR experiment in a low magnetic field. They also found bands in temperature approximately $20\,\mu$K in width where partial conversions between the phases occurred well below T_{ca} on cooling, but at around T_{ca} on warming, shown in the bottom panel of Fig. 5.22. The top panel of Fig. 5.22 shows the metastable phase region, denoted A*. The supercooled first-order transition was found to

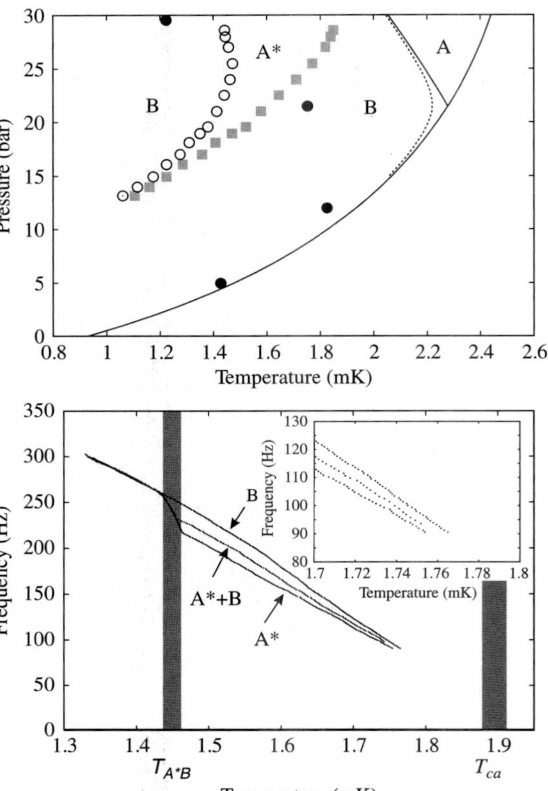

Fig. 5.22 Top: The region of the metastable A-like phase in 98% aerogel, from [132]. A* represents the metastable A-like phase. The solid lines show the bulk phase diagram in zero field. The dotted line is the thermodynamic A–B transition line in bulk at 28.4 mT. Bottom: Traces of the resonance peak of the slow sound mode from [132]. The lower trace taken on cooling shows the extended region of the metastable A-like phase. The intermediate curve is a warming trace taken on warming after going through a partial conversion in the shaded band. This trace suggests that the system maintains a stable mixture of the two phases. The two shaded bands are where A* to B (lower temperature) or B to A* conversion occurs.

be unaffected by the cooling rate and history. These observations by the Cornell and Northwestern groups show that the A-like phase should be thermodynamically stable at least in close proximity to T_{ca}, even below the polycritical point in zero field. Finally, B-like to A-like transitions were observed on warming at pressures of 28.4 and 33.5 bar by Vicente et al. [168], as shown in Fig. 5.9.

The field dependence study of the Northwestern group unveiled a couple of intriguing features [69, 71]. First of all, the superfluid transition did not show any shift or hint of splitting in response to increasing magnetic field up to ~ 0.8 T. In pure superfluid ^{3}He, a magnetic field induces a narrow sliver of A_1 phase between the normal and the A phase by splitting the superfluid transition. This phenomenon is caused by Zeeman coupling in the presence of minute particle–hole asymmetry, and the width of the A_1 phase increases almost linearly with field at the rate of $60\,\mu\text{K}\text{T}^{-1}$ at melting pressure. If the A-like phase in aerogel behaves like the bulk A phase, a similar field-dependent splitting must exist. The splitting in temperature was estimated to be large enough to be detected in their experiment, if the same degree of splitting in bulk was assumed in aerogel, and therefore their observation warranted further investigation.

Choi et al. [39] at the University of Florida performed the same measurements on the same sample used in the study of Gervais et al. in strong magnetic fields up to 15 T. They indeed observed splitting of the second-order transition above ~ 3 T. The rate of the observed splitting and its field dependence are consistent with the A_1 phase observed in pure liquid, as demonstrated in Fig. 5.23. However, the width of the A_1-like phase tends to fall below the linear splitting behavior for magnetic fields below ~ 4 T, which seems to be in line with the observation by the Northwestern group. A theoretical idea first put forward by Baramidze and Kharadze [13] and strengthened by detailed calculations [152, 15] predicts that the spin-exchange scattering between the ^{3}He spins in liquid and solid layers on the aerogel surface could give rise to an independent mechanism for the splitting of the transition. Antiferromagnetic (ferromagnetic) exchange reduces (enhances) the total splitting in low fields, but the rate of the particle–hole asymmetry contribution is recovered in high fields as the polarization of the localized spins saturates. An obvious next step would be to repeat the experiment with the magnetic layers deactivated by preplating with ^{4}He. However, as yet no further experiments have been performed, the main obstacle being thermometry in the relevant field and temperature range. On the other hand, the asymmetry ratio of the splitting, defined by $r = (T_{A_1} - T_c)/(T_c - T_{A_2})$, is a direct measure of strong coupling effects, as $r \to 1$ in the weak coupling limit. Choi et al. found that the asymmetry ratio in aerogel was consistently lower by about 20% than that of bulk at the same pressure, suggesting weakening of the effects. Furthermore, it is also of note that the observed ratio is an order of magnitude larger than expected for the robust phase [39, 15].

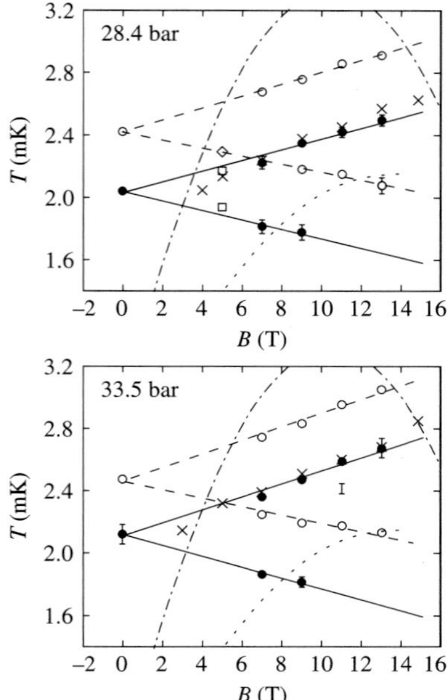

Fig. 5.23 Magnetic field dependence of A_1–A_2 splitting at 28.4 and 33.5 bar, from [39]. Solid circles and crosses are for the transitions in aerogel and open circles represent transitions in bulk. Other features in the plot are described in [39].

The bulk A–B transition has a solid theoretical description in the G–L limit for $p < p_{pc}$ [163, 169]. Relative to the normal state, the G–L free energy of the A(B) phase is $f_{A(B)} = -\tilde{\alpha}^2/2\beta_{A(B)}$ where $\tilde{\alpha} = N_o(T/T_c - 1)$, and $\beta_{A(B)}$ is the appropriate combination of the β-parameters, the coefficients of the fourth-order terms in the G–L expansion [169]; $\beta_A = \beta_{245}, \beta_B = \beta_{12} + \beta_{345}/3$.[2] For $p < p_{pc}, f_A > f_B$, forbidding the A phase below T_c, but at $p = p_{pc}, f_A = f_B$, and consequently $\beta_A = \beta_B$. The presence of a magnetic field requires an additional term in the G–L expansion given by

$$f_z = g_z B_\mu A_{\mu i} A^*_{\nu i} B_\nu. \tag{5.36}$$

With two distinct symmetries in the A and B phase order parameters, this quadratic contribution lifts the degeneracy in the superfluid transition temperature, thereby pushing the A phase T_c slightly above that of the B phase. As a result, a narrow region of the A phase must be wedged between the normal and the B phase, even for an infinitesimally small magnetic field. The suppression of the B-like phase in finite magnetic fields can be expressed as

$$1 - T_{AB}(T)/T_c = g(\beta)(B/B_c)^2 + \mathcal{O}(B/B_c)^4, \tag{5.37}$$

where B_c represents a characteristic field scale directly related to the transition temperature (see [163] for the full expression). The coefficient of the quadratic field dependence, $g(\beta)$, is another measure of strong coupling effects,

[2] The conventional notation, $\beta_{ijk} = \beta_i + \beta_j + \beta_k$, is used.

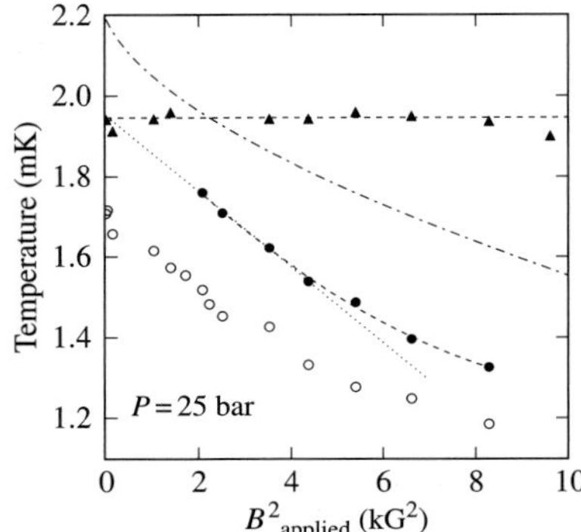

Fig. 5.24 Field dependencies of various transition features at 25 bar, from [69]. The triangles represent the superfluid transition. The circles are for the first-order transition observed on cooling (open) and warming (solid). The dotted line represents the result of a low field fit to a quadratic field dependence. The dot-dashed line is the field dependence of the bulk A–B transition at the same pressure.

increasing from 1 in the weak coupling limit, and is a function of the pressure-dependent β-parameters, $g(\beta) \propto (\beta_A - \beta_B)^{-1}$. Therefore, in bulk superfluid, $g(\beta)$ displays divergent behavior as $p \rightarrow p_{pc}$.

The systematic field and pressure dependence study of Gervais et al. [71] uncovered puzzling behavior in the first-order A–B transition. They observed that the warming B-like to A-like transition was suppressed quadratically in the field at all pressures from 15 to 33 bar. One of their measurements at 25 bar is shown in Fig. 5.24. This behavior allowed them to apply Eq. 5.37 to extract values of $g(\beta)$ by properly rescaling the superfluid transitions in aerogel, under the assumption that the warming transition was the thermodynamic transition. They found that in contrast to the bulk, $g(\beta)$ in aerogel increased monotonically with pressure without showing any anomalies (Fig. 5.25), which was later confirmed by Moon et al. [126]. This observation naturally raised questions about the position or the existence of the polycritical point in aerogel, bearing in mind the observations of metastable A-like phases at all pressures in zero field. Vicente et al. [168] offered an interesting scenario that could encompass the seemingly contradicting experimental facts. Their argument stands on the basis of two fundamental points: (i) the strong coupling effect is significantly reduced in aerogel, and as a result the polycritical point is moved up beyond the melting pressure and; (ii) the anisotropic disorder presented by aerogel has an effect analogous to that of an applied magnetic field. The coupling of anisotropic disorder to the order parameter introduces an additional term in the G–L free energy [166, 164, 63],

$$f_a = g_a \hat{a}_i A_{\mu i} A^*_{\mu j} \hat{a}_j, \tag{5.38}$$

where $\hat{a}$ represents the direction of local anisotropic disorder. This contribution, isomorphic to f_z (Eq. 5.36), emulates the role, via the orbital channel, played by a magnetic field through the spin channel. The inhomogeneous distribution of local anisotropy on length scales larger than ξ_o could be responsible for the region of mixed state. Based on this argument, they proposed an

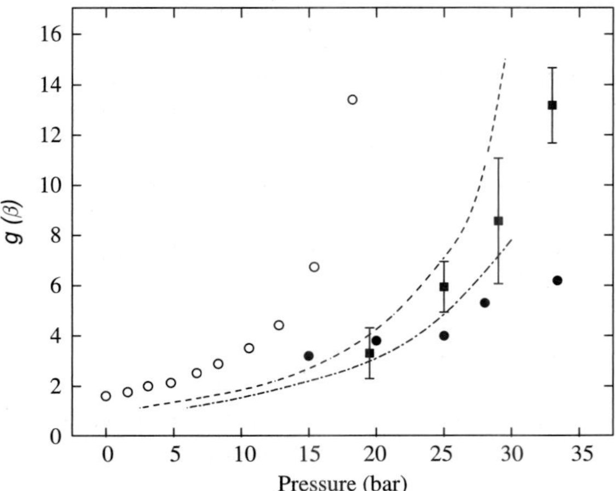

Fig. 5.25 Pressure dependence of $g(\beta)$ from [126]. The solid circles (squares) are the results in aerogel by Gervais et al. [71] (Moon et al. [126]). The open circles are for the bulk liquid [163]. The dashed and dot-dashed lines are theoretical results based on HISM with $\ell = 200$ nm and 150 nm, respectively.

experiment using a uniaxially compressed aerogel in order to study the role of anisotropic disorder in a more systematic way.

The importance of anisotropic disorder and the dramatic effects of global anisotropy have been demonstrated by a number of NMR experiments reviewed in the previous section. A series of calculations supports these ideas and predicts some intriguing phenomena. The impurity scattering necessitates various corrections to T_c and the β-parameters [166, 164, 14, 9, 10]. The most extensive calculation including vertex corrections to the β-parameters was done by Aoyama and Ikeda [10]. Their theoretical phase diagram based on those corrections indeed resembles the bulk phase diagram that has, in effect, been shifted to lower temperature and, simultaneously, to higher pressure, resulting in relocation of the polycritical point to an increased pressure. Furthermore, in a uniaxially deformed aerogel, the calculation of Aoyama and Ikeda [7] shows unambiguously that global anisotropy has an effect similar to that of a magnetic field. The effect of the orbital field was estimated to be comparable to ~ 0.1 T in the case of complete alignment [168]. One fascinating aspect of their calculation is the appearance of a new phase, the polar phase, in a uniaxially stretched aerogel, see Fig. 5.26.

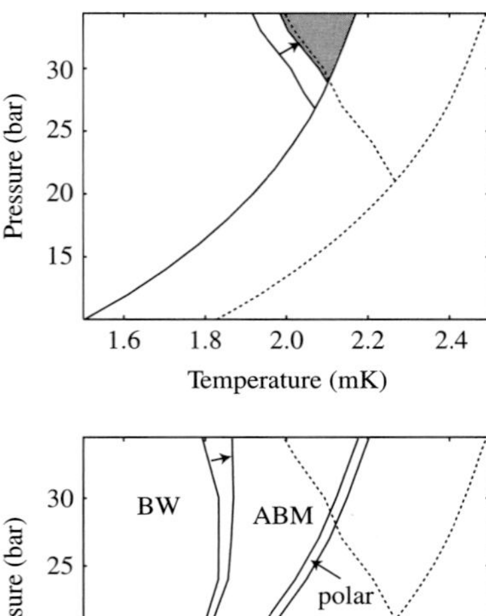

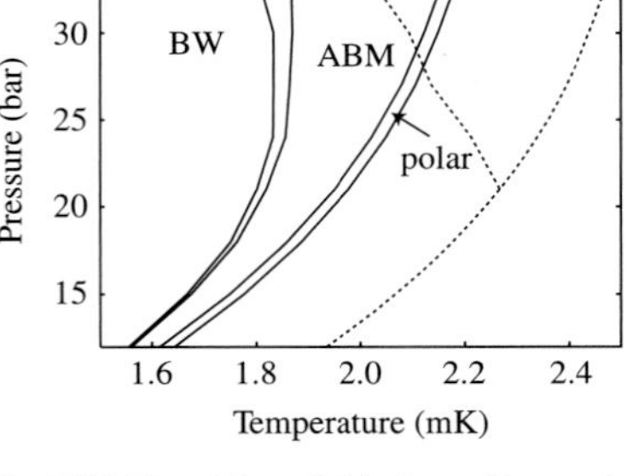

Fig. 5.26 Top: Mean field phase diagram in an isotropic aerogel with $1/2\pi\tau = 0.13$ mK and $\lambda_c = 0.821$, the dimensionless coupling constant for two-body interaction in the theory, see [10] for details. Bottom: Phase diagram in a uniaxially stretched aerogel. The degree of deformation is entered as the difference in the mean free path parallel (perpendicular) to the axis of deformation, $\ell_{\parallel(\perp)}$. Here, $(\ell_\parallel - \ell_\perp)/\ell_a = -0.06$ and the other conditions are identical to the left panel. The arrows represent the shift due to vertex corrections to the β-parameters. Note the appearance of A-like phase at all pressures and the narrow band of the polar state in the stretched case.

Davis et al. [45] and Bhupathi et al. [24] performed transverse acoustic impedance measurements to investigate the effect of global anisotropy on the phase diagram. While the Davis et al. experiment did not show any difference in the first-order transition between the uncompressed and 17% axially compressed aerogels, Bhupathi et al. did observe a clear difference in the field-dependent suppression of the first-order transition, which might suggest a widened A-like region in their 5%-compressed aerogel. However, no evidence of a new phase has been seen in an NMR experiment on a radially squeezed aerogel, even though it is thought to be equivalent to a uniaxially stretched aerogel [56].

The University of Florida group embarked on completing the *p-B-T* phase diagram with the intention of verifying the scenario proposed by Vicente et al. [126]. They conducted high frequency longitudinal sound attenuation by directly propagating sound pulses in magnetic fields up to 0.44 T. At the highest field, the metastable A-like phase was extended to the lowest temperature reached ($\approx 200\,\mu$K), which allowed them to establish the full temperature dependence of attenuation in this phase. With the B-like phase measurements obtained in zero field they were able to identify warming transitions between the phases at pressures from 14 to 33 bar. Most of the transitions were identified by temperature sweeps on warming at a constant field. However, their secondary method, an isothermal field sweep, produced consistent results within their experimental uncertainty with no hysteresis between both directions of the field sweep, suggesting that the superheating effect was negligible. The phase diagram that emerged from this experiment, Fig. 5.10, is in evident contrast to that of the bulk. First, the slope of the constant-field phase boundary is positive in aerogel but negative in bulk for most of the corresponding pressure range. Second, the phase boundary in aerogel recedes toward the corner of the melting pressure and zero temperature, rather than $p \sim p_{pc}$ as in bulk. This phase diagram is fully compatible with the behavior of $g(\beta)$ and the interpretation put forward based on various sound experiments.

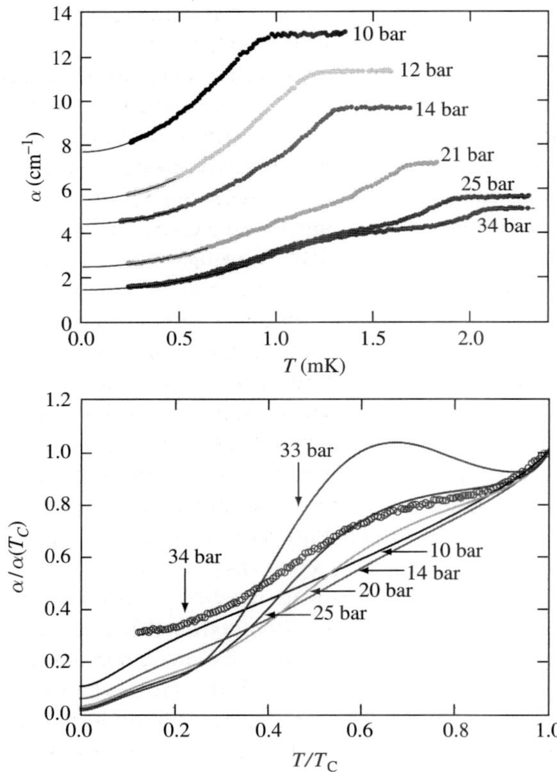

Fig. 5.27 Top: Absolute attenuation for various pressures versus temperature at various pressures. Bottom: Normalized sound attenuation versus normalized temperature. The results of theoretical calculation (solid lines) are plotted along with the experimental results at 34 bar for comparison from [41]. As pressure decreases $\alpha(0)$ increases.

A comprehensive study of the attenuation of fast sound was conducted in the B-like phase using 9.5 MHz sound pulses by Choi et al. [41, 42]. They were able to detect multiple echoes at low temperatures and determined the absolute attenuation over a wide range of pressure. As shown in Fig. 5.27, the overall temperature dependence is drastically different from what is expected in the bulk. In the bulk B phase, the attenuation follows $\alpha \propto e^{-\Delta(T)/k_BT}$ below the attenuation peak associated with the order parameter collective modes and the pair-breaking process, and reaches practically zero attenuation below $0.6\,T_c$. In contrast, the attenuation in aerogel exhibits a weaker temperature dependence and remains high even at $0.2\,T_{ca}$. Furthermore, a peculiar shoulder feature appears around $0.6\,T_{ca}$ for higher pressures. This feature weakens gradually and eventually disappears at lower pressures. This behavior is in qualitative agreement with the theory of Higashitani [90] introduced earlier. The conspicuous hump structure is intimately related to the temperature dependence of τ_f in Eq. 5.35 and the weakening of this structure at lower pressures is reproduced by their theory, also shown in Fig. 5.27. As $T \to 0$, $\alpha \propto \omega^2\eta(0)$ and therefore the density of states at zero energy (E_f) could be extracted using $\alpha(0)/\alpha(T_{ca}) \propto n^z(0)$, where $n(0)$ is the normalized density of states and $z = 2, 4$ in the Born and Unitary limits, respectively [90]. The pressure dependence of $n(0)$ is in qualitative agreement with the combined results of Fisher et al. [58] and Choi et al. [37], shown in Fig. 5.11.

Recently Moon et al. [125] extended the experiment of Choi et al. [42] to investigate the frequency dependence of attenuation. They performed sound attenuation measurements in the B-like phase at four frequencies ranging from 3.6 to 11.3 MHz. At all of the pressures studied (14, 25, and 33 bar), the attenuation exhibits non-trivial frequency dependencies that progressively deviate from the expected hydrodynamic ω^2 behavior as the temperature decreases. It is interesting to note that in the presence of aerogel, it is possible to have an unusual circumstance, $\omega \sim \Delta/\hbar$ and $\omega\tau < 1$, in a wide range of temperature, a regime which is more difficult to establish and explore in the bulk. In this regime, various mechanisms are responsible for sound absorption such as resonant absorption between various combinations of impurity states and gap edge states. They suggested that the non-trivial frequency dependence might be related to the structure of the impurity states inside the gap; these would have a non-trivial spectrum originating from the structure of aerogel, for instance due to finite size effects since $k_F \delta_a \approx 10$–100. It should be noted that the density of impurity states extracted from sound attenuation (Fig. 5.11) might provide an upper bound, considering this unusual frequency dependence.

The Osaka City University group successfully observed fourth sound in 99% aerogel directly grown in an acoustic cavity filled with a silver sinter of 70 μm particles. The effective pore size, in the order of 10 μm, is large enough that no superfluid suppression was observed in pure ^{3}He. Therefore any suppression in the presence of aerogel could be attributed to pair-breaking from the strands. In this geometry, the eigenmode of the aerogel strands held by the porous structure is at a much higher frequency than the fourth sound resonances, and consequently the slow sound mode is effectively suppressed. The superfluid density determined from the resonance frequency is in good agreement with the theoretical calculation based on the HISM. The energy loss obtained from multiple harmonics displays typical hydrodynamic behavior, no frequency dependence in Q^{-1}/ω, as shown in Fig. 5.28 (recall $Q_m^{-1} \propto \alpha/\omega$ and $\alpha \propto \omega^2$). But the loss in aerogel almost vanishes around $0.3\,T_{ca}$ and is substantially smaller than in bulk. This is rather surprising since ρ_s is significantly

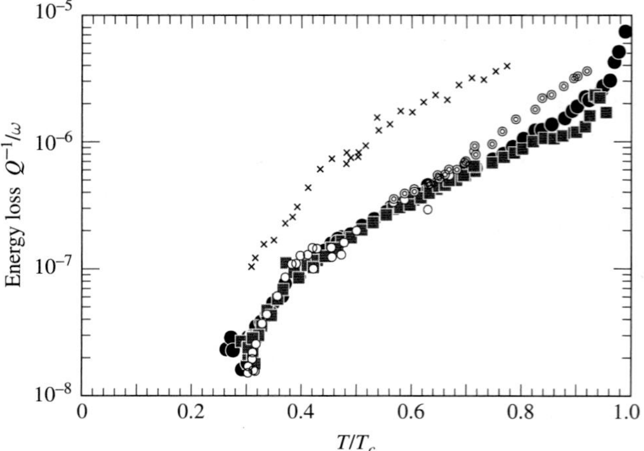

Fig. 5.28 Temperature dependencies of the energy loss of fourth sound scaled by the resonance frequency, from [135]. Solid circles, solid squares, double circles, and empty circles represent the 2nd, 3rd, 4th, and 5th harmonics in aerogel, respectively. The crosses are from the 2nd harmonic resonance in pure ^{3}He.

suppressed in aerogel and a substantial amount of ρ_n is present below 0.3 T_{ca}. They were able to explain this behavior by introducing collision drag effects. From the generalized dispersion relation, Eq. (5.34), the speed of sound and the energy loss were calculated recognizing $v_a \to 0$ in their experimental setup, giving $Q^{-1} \approx \frac{\rho_n}{\rho_s}\omega\tau_f$. This expression corresponds to Eq. (5.35) for the damping of fast sound. As discussed earlier, the broad hump structure in fast sound attenuation was understood to be a direct consequence of the frictional relaxation time τ_f. It is important to notice the appearance of a similar structure around 0.5 T_{ca} in Q^{-1}/ω in Fig. 5.28. Based on this analysis they concluded that the main dissipation of fourth sound in aerogel is frictional rather than viscous in origin. This phenomenon can also be understood as a crossover in the flow properties of the normal component in aerogel from the viscous Hagen–Poiseuille to the frictional Drude regime [55].

5.4.3 Mechanical/dynamical measurements

TORSIONAL OSCILLATORS

In 1995 Porto and Parpia [140] at Cornell reported the first observation of a superfluid transition in ^{3}He confined within aerogel using a torsion pendulum. This powerful and sensitive technique has a long history in superfluid helium experiments [148], essentially measuring the reduction in rotational inertia as helium becomes superfluid, loses its viscosity, and thus decouples from the torsion bob. This leads to a measurable increase in the resonant frequency (or decrease in period Γ) of the pendulum that is temperature dependent and related to the relative superfluid density as

$$\frac{\rho_s}{\rho} = \frac{\Gamma(T_{ca}) - \Gamma(T)}{\Gamma_0 - \Gamma_{\text{empty}}}, \tag{5.39}$$

where $\Gamma_0 - \Gamma_{\text{empty}}$ is the period shift from filling the empty cell with ^{3}He. The strand spacing in aerogel is so small compared with the viscous penetration depth of ^{3}He that it is reasonable to assume that all normal fluid is coupled to the oscillator at all temperatures.

The first measurements by the Cornell group were made with a 943 Hz oscillator containing 0.29 cm^3 of 98.2% open aerogel that was grown directly into the torsion bob. They used purified ^{3}He to reduce the ^{4}He content below 10 ppm, so that it would not preferentially coat the aerogel strands and potentially alter the surface interactions. From their measurements of ρ_s^a/ρ, shown in Fig. 5.29, they were able to infer the pressure dependence of T_{ca}, the onset of superfluidity in the aerogel, from 2.7 to 29 bar, shown as triangles in Fig. 5.30. There is strong suppression of the superfluid transition temperature and, as one can see from Fig. 5.29, the transitions are well defined. These findings were in surprising contrast to what one might expect based on previous measurements of ^{3}He in confined geometries, which had been generally understood in terms of a healing length, as discussed in Section 5.1. Now, sharp transition features are observed in well-defined regular geometries such as parallel plates where the onset of superfluidity occurs at a fixed value of bulk coherence length $\xi(T)$, which can be tuned by changing the pressure. Nevertheless, the Cornell group

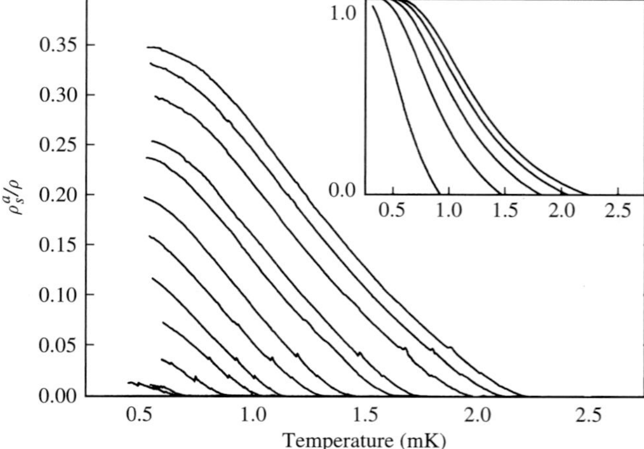

Fig. 5.29 The superfluid fraction in a 98.2% aerogel at various pressures as a function of pressure, from [140]. From left to right the curves correspond to pressures of 3.4, 4.0, 5.0, 6.1, 7.0, 8.5, 10, 13, 15, 20, 25, and 29 bar. The small spikes are due to nuisance acoustic resonances in the cell. The inset shows the bulk superfluid fraction for pressures 0, 5, 10, 15, and 20 bar for comparison.

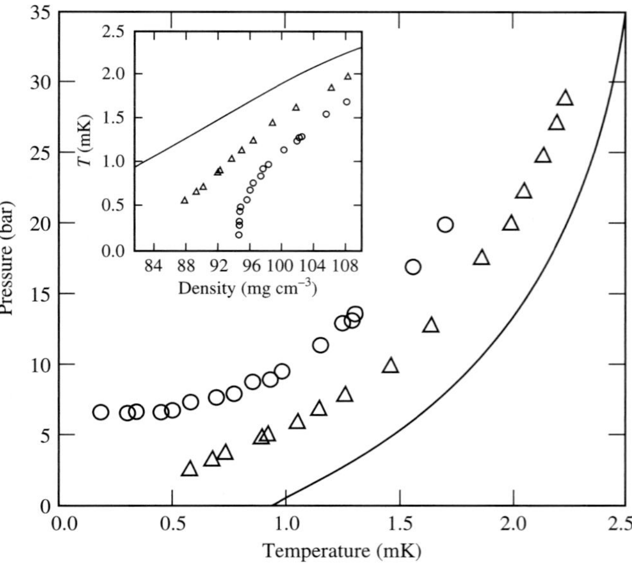

Fig. 5.30 Pressure–temperature phase diagram measurements for ^{3}He in 98.2% porosity aerogel, from [119]. Circles show the data from this later work, and triangles from the earlier work of [140]. The solid line is the bulk transition for comparison. The inset shows T_{ca} as a function of density.

showed that this could not be used here as an argument for a similar single well-defined length scale in aerogel; as they increased the pressure, their transitions did not occur at the same value of bulk $\xi(T)$. These observations led Porto and Parpia to suggest that aerogel in ^{3}He may be best described as an impurity that suppresses pairing of the p-wave BCS state.

Intriguingly, in Porto and Parpia's initial measurements, they did not detect any superfluid at pressures below 2.7 bar, even though it appeared that T_{ca} had not gone to zero. The Cornell group returned to this corner of the phase diagram with a different 98.2% aerogel sample [119] and made the remarkable discovery of a quantum phase transition (QPT), a continuous transition between two zero entropy states (normal and superfluid ^{3}He at $T = 0$). Their measurements are summarized in Fig. 5.30. In order to track the transitions below 0.93 mK, where the pressure dependence starts to flatten, they varied pressure rather than temperature as they monitored the period shift of the oscillator. Thus, changes in the density of the system, rather than temperature, were used to control the continuous transition from normal to superfluid. The values of critical density that indicated the start of the transition are shown in the inset of Fig. 5.30 and demonstrate a strikingly steep drop as $T = 0$ is approached. This figure also includes the transition data from Porto and Parpia's first measurements and points out one of the problems with the early aerogel work; although both samples are nominally the same 98.2% porosity, they show very different behavior. The Cornell group made a third torsional oscillator cell, again with 98.2% aerogel, to study the influence of adding ^{4}He to the ^{3}He [75], discussed below. Care was taken to use the same gelation process as the QPT cell, with the result that the transition temperatures for pure ^{3}He were now in approximate agreement [139], at least for these two cells.

At around the same time Alles et al. in Manchester constructed a combined torsional oscillator and NMR cell [4, 5]. They succeeded in their aim of demonstrating that the superfluid transition is seen simultaneously by both techniques in a ~98% aerogel. This could be taken as verifying that the torsional oscillator is measuring superfluidity that extends throughout the whole sample, rather than a series of more local transitions, and it thus lent weight to the view that the ^{3}He-aerogel system is a "dirty" superfluid whose properties are governed by impurity scattering. The Manchester group was careful to have their aerogel grown in situ, and to use the same source as the Northwestern group, but the values of T_{ca} were not identical. Although the Northwestern values agreed at least approximately with the later Cornell measurements, those of the Manchester group fell between these and the first Cornell measurements. Variability in samples was still an unresolved issue.

Returning to the early work of the Cornell group [140], in addition to the unexpected behavior of T_{ca}, the temperature dependence of ρ_s^a/ρ in aerogel was distinct from that of the bulk, shown in Fig. 5.29. In bulk, at all pressures, ρ_s/ρ approaches 1 at $T = 0$, whereas in aerogel the superfluid fraction increases with pressure but never approaches unity. Furthermore, close to T_c the bulk superfluid fraction varies linearly as a function of reduced temperature $(1 - T/T_c)$, while ^{3}He in aerogel follows a power law $(1 - T/T_{ca})^n$ where $n \sim 1.5$ over a large temperature range. (In later work n is given as 1.45 above 15 bar and 1.33 for lower pressures [136, 142].) The Manchester group measured slightly lower values of ρ_s^a/ρ in their 98% aerogel, though their data still fits a similar power-law dependence well, with $n \sim 1.5$ at higher pressures. In bulk ^{3}He the temperature dependence of ρ_s/ρ is principally governed by the energy gap $\Delta(T)$ whose scale is set by the pressure-dependent T_c. The Cornell group demonstrated that the important energy scale for ^{3}He in aerogel was one that appeared to depend on how close T was to T_{ca}, so that one could collapse curves of ρ_s^a/ρ at different pressures onto universal power-law curves as a function of $(T_{ca} - T)$ [142]. Explanation of this behavior remains an open question [136].

The Cornell group went on to make measurements on a fourth aerogel sample, this time a very open 99.5% aerogel that was grown inside the pores of a coarse silver sinter, to improve thermal contact and suppress parasitic

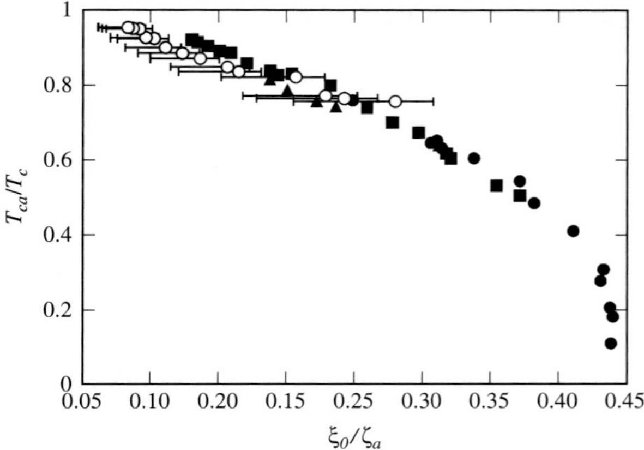

Fig. 5.31 Relative suppression of the aerogel transition temperature T_{ca} scaled by the bulk T_c, against the bulk coherence length ξ_o scaled by the correlation length ζ_a, from [112]. The correlation length of the 99.5% sample (open circles) was taken to be 220 nm with the error bars corresponding to the spread between low and high values of 200–300 nm. Filled symbols are for 98% aerogels with correlation lengths of 130 nm (squares), 90 nm (triangles), and 84 nm (circles).

resonances [112]. This aerogel was approximately four times less dense than their previous samples but, more importantly, it had a very different correlation length ζ_a, the length scale measured independently by SAXS [142]. The earlier 98.2% aerogels each had slightly different values of ζ_a, attributed to differences in their detailed growth history, but they were all roughly in the order of 100 nm. The 99.5% aerogel had a correlation length in excess of 200 nm. As one might expect, the 99.5% samples, being more dilute, showed transition temperature and ρ_s^a behavior that approached that of the bulk. However, what was remarkable was that the measurements could be used to show that both T_{ca} and ρ_s^a could be scaled so that they only depended on the bulk T_c and the ratio ξ/ζ_a. Their T_{ca} data are reproduced in Fig. 5.31. This demonstrated that the behavior of ^{3}He is independent of the aerogel density, but heavily dependent on its microstructure, and goes some way towards explaining the problems with variability between samples that plagued some of the early work. Their measurements of ρ_s^a do not fit the homogeneous scattering model well, but the alternative isotropic inhomogeneous scattering model does not fare much better on a quantitative level.

The Cornell group went further and used the oscillator measurements of ρ_s^a to derive values of the *bare* superfluid density,

$$\frac{\rho_s^b}{\rho} = \frac{\left(1 + \dfrac{1}{3}F_1\right)\left(\rho_s^a/\rho\right)}{1 + \dfrac{1}{3}F_1\left(\rho_s^a/\rho\right)}, \tag{5.40}$$

thereby removing pressure-dependent Fermi liquid corrections to the effective mass. This bare density is related to the temperature-dependent Yosida function $Y(T)$, and so to the gap parameter $\Delta(T)$. Thus, they were able to extract estimates of the gap parameter for ^{3}He in aerogel as a function of T [113]. They found that impurity scattering by the aerogel reduced the gap by a factor ~ 0.4 to 0.76, depending on pressure.

The Cornell group investigated the effects of ^{4}He added to the ^{3}He using the third 98.2% aerogel cell [75]. Measurements were made with ^{4}He concentrations of 2%, 13%, 21%, and 34%. For all these samples, T_{ca} was slightly enhanced but the superfluid density was suppressed for the samples with higher ^{4}He contents. The authors suggest that as the ^{4}He content increases, coating and enveloping the aerogel strands, the structure experienced by the confined ^{3}He evolves towards one of interconnected voids, and is thus better described as being analogous to a granular superconductor.

It should also be mentioned that in all torsional oscillator measurements, dissipation was measured as the inverse quality factor of the resonator, Q^{-1}. In general these measurements were difficult to interpret, owing to parasitic resonances and level crossing from sound modes inside the oscillators. The effect of these modes can be seen in the small spikes in Fig. 5.29. The Cornell group found that these were reduced significantly when the aerogel was grown in the pores of a silver sinter [112]. The Manchester group indicated that adding a small amount of ^{4}He to their ^{3}He in 99% aerogel led to a massive reduction in the influence of parasitic resonances. The Cornell group attempted to model the dissipation behavior of their oscillator for various ^{4}He concentrations [72].

VIBRATING WIRES

The Lancaster group has made a series of investigations by attaching aerogel to vibrating wire resonators and oscillating the samples within a volume of ^{3}He. Although conceptually similar to torsional oscillator measurements, there is an added complication in that one must account for the fluid outside the aerogel that is displaced as it moves. There are thus three contributions to the effective mass: the vacuum mass of the resonator with empty aerogel; the mass of the normal fluid entrained within the aerogel, and assumed to move with it; and the fluid backflow around the aerogel, which must also be related to how transparent the sample is to superflow through it. Assuming that the spring constant of the wire is unchanged, these three cases can be combined to find the aerogel superfluid fraction through measurement of the resonant frequency f using

$$\frac{\rho_s^a}{\rho} = 1 - \frac{(f_0/f)^2 - 1}{(f_0/f_n)^2 - 1}, \tag{5.41}$$

where f_0 and f_n are the measured resonant frequencies in vacuum and when the ^{3}He in the aerogel is completely normal, respectively.

Typical data from their early work [28], and a sketch of the resonator with attached aerogel disk are shown in Fig. 5.32. Note that at all temperatures and pressures measured here, the bulk ^{3}He outside the aerogel is in the B phase, and the labels on the figure refer to the phase contained inside the aerogel. The ρ_s^a/ρ behavior in low field B phase is similar to that measured in torsional oscillators [140], albeit quantitatively larger, and the measured transition temperatures are consistent. However, what is really striking here is that the demagnetization technique employed by the Lancaster group enabled them to clearly observe the transition from the A-like phase to the B-like phase and map out a magnetic phase diagram. This transition has not been observed in any of the torsional oscillator measurements, even though the A-like phase is thought to be metastable on cooling, even in zero magnetic field. Their aerogel A–B phase boundary lines, measured at 4.8 and 7.4 bar, appear to be simply consistent with

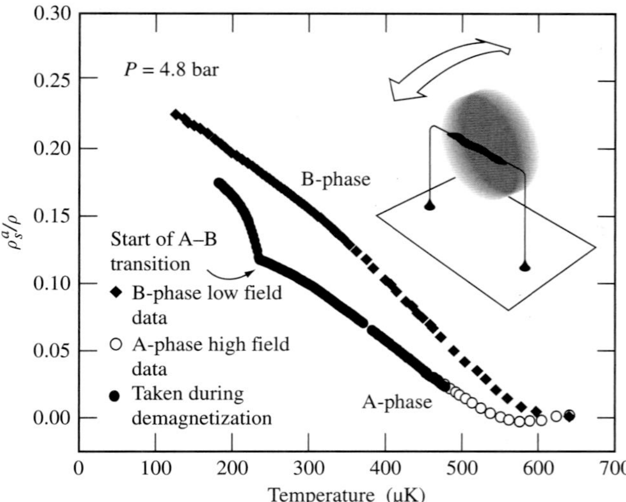

Fig. 5.32 Inferred values of ρ_s^a/ρ as a function of temperature in 98% aerogel, from [28]. The upper B-phase curve was obtained during a slow warm-up in a low magnetic field. The open circles were measured in a high field, with the A phase inside the aerogel. The filled circles were taken while cooling under demagnetization, and there is a clear signature of the onset of the A–B transition in the aerogel at ~230 μK. Inset shows the aerogel resonator, a disk 5 mm diameter, 1.1 mm thick, glued with epoxy resin to a superconducting tantalum wire.

an equal scaling of the bulk phase line down both the magnetic field and the temperature axes [80]. The authors point out that this line should be sensitive to any difference in the influence of the aerogel on the stability of the A or B phase, and this result implies that any scattering model should affect the free energy of the two phases in the same way. One can also see from Fig. 5.32 that the transition from A to B is not a discontinuous jump, as one would expect for a first-order phase transition. The authors explain that the magnetic field driving the transition is not actually constant over the cell, and there is a gradient in field across the aerogel disk. This means that the equilibrium position of the A–B interface moves through the aerogel as the field is ramped, and so the transition is forced to be gradual. In fact, the authors were able to measure that the transition is hysteretic, and argued that the AB interface is actually pinned by the aerogel, lagging behind its equilibrium position. This implies that the interface has a surface tension, a characteristic of a first-order transition.

Later, the group made measurements on a second aerogel resonator, again 98% porosity, but this time in the shape of a cylinder 4 mm long and 2 mm in diameter, rather than a disk [26]. The cylinder was made to oscillate through the bulk ^{3}He along a direction perpendicular to its symmetry axis. They made measurements of the velocity dependence of the superfluid fraction and also of the damping of the resonator. When the aerogel was full of B-like phase, surrounded by bulk B phase, the measured ρ_s^a/ρ was virtually independent of velocity, and reproducible. However, the behavior when aerogel contained A-like phase (again surrounded by bulk B phase) was seen to be markedly different, and history dependent. When driven to higher velocities, the apparent values of ρ_s^a/ρ decreased, but on returning to low velocities ρ_s^a/ρ was observed to have increased. This increase in ρ_s^a/ρ at low velocity became larger when the resonator was pushed to higher and higher velocities. (The damping for the A-like phase was similarly hysteretic, also in contrast to that of the B-like phase.) The authors argued that this is strong evidence that the A-like phase order parameter in aerogel is highly anisotropic, like the bulk ABM state, and the behavior is due to a smoothing of its texture and removal of defects. If, like the bulk, the A phase in aerogel has an $\hat{l}$-direction where the energy gap goes to zero, then this will form a texture. Owing to the convoluted geometry and high magnetic fields, when this texture first forms, it is likely to be highly irregular and full of defects. This is reflected in the low initial values of ρ_s^a/ρ. When the aerogel is moved at higher velocities, superflow tries to align the $\hat{l}$-texture along the flow direction. This would reduce the apparent ρ_s^a at high velocities, but would also remove and re-orient defects. This disturbance effectively anneals the texture, creating a cleaner, more uniform A phase which thus has a higher effective superfluid fraction when the velocity is reduced.

5.4.4 Thermodynamic measurements

HEAT CAPACITY

In order to prove that the onset of superfluidity in aerogel is a true thermodynamic phase transition, it was necessary to measure the heat capacity of the "dirty" ^{3}He and demonstrate that a change in heat capacity accompanies the onset behavior measured by other techniques. The heat capacity contribution from the ^{3}He in aerogel alone can only be measured correctly after a careful correction for the background addendum heat capacity from solid ^{3}He on the aerogel surface, any bulk liquid in the cell, and the cell itself. The first effort was made by the Cornell group who reported measurements on a 97.6% aerogel at a single pressure of 22.5 bar. They employed a heat pulse technique in a cell that was designed to function as a torsional oscillator as well, allowing for simultaneous measurement of the superfluid transition [85, 86]. They confirmed the concurrent onset of superfluidity in both measurements. After accounting for an unexpectedly large bulk contribution, as well as a small component presumably from the ^{3}He solid layers, they confirmed that the heat capacity above T_{ca} is linear, as expected for a Fermi liquid. This allowed them to extract a value for the effective mass m^* in aerogel that is 30% larger than the bulk value (see Section 5.3.1, Eqs. 5.4 and 5.5). This is unexpected, since one would not expect aerogel to make a difference to Fermi liquid parameters, where the relevant length scale is atomic, and in fact other workers take this to be a valid assumption.

It was also anticipated that the heat capacity change at the transition should be sharp, whereas the data here implies a distribution of critical temperatures in the order of 100 μK centered around T_{ca}. In their analysis this rounding is accounted for by assuming that a Gaussian spread is convoluted with the expected BCS form. The data can then be fitted in two ways: either a superfluid with a full BCS gap where ~70% of the liquid in the aerogel remains in the normal state as T→0; or 100% superfluid as T→0, albeit with a suppressed gap, consistent with their torsional oscillator measurements [113]. However, the contribution from impurity states, whose presence is now well established, was not included in their analysis.

Later, the Northwestern group reported heat capacity measurements on ^{3}He contained in a 98.2% aerogel for pressures ranging from 11 to 29 bar, shown

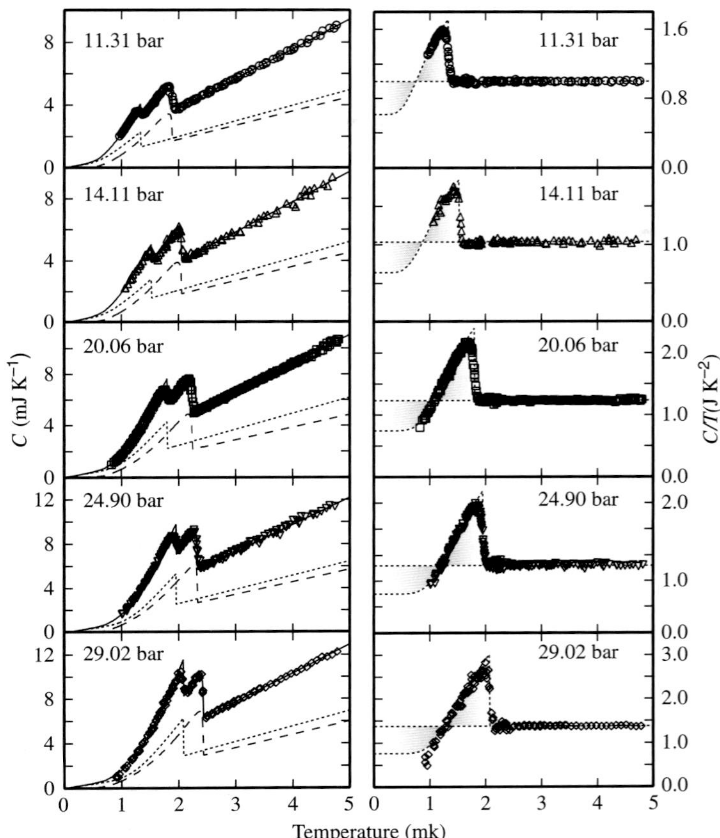

Fig. 5.33 Measurements of heat capacity for ^{3}He in a 98.2% aerogel over a range of pressures, from [37]. The left-hand side shows the contributions from the bulk ^{3}He in the cell, as well as that contained in the aerogel, after the heat capacity from solid ^{3}He adsorbed on the aerogel strands has been subtracted. The dashed lines represent the calculated temperature dependence of the bulk, which can then be subtracted to leave only the heat capacity from the ^{3}He in aerogel. This is plotted on the right-hand side as C/T versus T. The dotted lines represent a fit to the data that invokes entropy conservation to determine the $T \to 0$ limiting behavior.

in Fig. 5.33 [37]. For this figure, the extra heat capacity owing to the paramagnetic surface solid layers has already been removed. The group confirmed this addendum by replacing the ^{3}He on the surface with non-magnetic ^{4}He. They also accounted for an addendum heat capacity arising from a region of normal state fluid at the surface of their heat exchanger. They were then left with contributions from the ^{3}He inside the aerogel, and bulk outside. No evidence was found for any large voids inside the aerogel, so they were able confidently to ascertain the volume of bulk ^{3}He and subtract this from their measurements. Finally they were left with the contribution to the heat capacity from the ^{3}He in the aerogel, shown on the right-hand side of Fig. 5.33, as C/T against T.

These data show the expected sharp increase at the transition, though the values of $\Delta C/C$ are approximately half those of bulk ^{3}He at all the pressures that they measured. Since the superfluid energy gap is proportional to $(\Delta C/C)^{\frac{1}{2}}$ in the Ginzburg–Landau limit, this measurement also provides evidence that the amplitude of the ^{3}He order parameter must be reduced in aerogel. They were also able to show that the heat capacity jump fitted the homogeneous isotropic scattering model well, quantitatively, for a mean free path of 180 nm, which is consistent with the value of 150 nm estimated from transport measurements and the phase diagram. The Northwestern group also extended analysis of their heat capacity measurements by invoking the third law of thermodynamics to deduce the limiting behavior of the heat capacity as T approached zero. They showed that the $T = 0$ values of C/T at all pressures must be non-zero, and argued that this limiting linear dependence of C on T was evidence for the existence of quasiparticle bound states, owing to order parameter suppression close to the aerogel strands. This implies a finite density of states centered around the Fermi energy, within the energy gap, and thus the ^{3}He in the aerogel is a gapless superfluid. This confirmed the interpretation of thermal conductivity measurements by the Lancaster group [58]. The derived values for the density of states agree only reasonably well with the predictions of the homogeneous isotropic scattering model, and are displayed along with the results from other measurements in Fig. 5.11.

The early Cornell measurements hinted at the existence of a magnetic ordering in the layers of solid ^{3}He adsorbed on the aerogel strands [86]. Recent work by the Lancaster group at temperatures below 150 μK appears to show that there is indeed a magnetic phase transition in the solid that manifests itself in a heat capacity peak [27]. Their measurements were made on a 98% aerogel sample placed inside a larger volume of bulk ^{3}He. At zero pressure the ^{3}He inside the aerogel remains in the normal state, while that outside is in the pure condensate limit. They were not able to measure the temperature of the liquid inside the aerogel directly, but could estimate it from the temperature of the connected superfluid. At such low temperatures the liquid heat capacity should be negligible, and they assumed this in order to infer an effective heat capacity of the solid from warm-up curves. Measurements were made at different magnetic fields, and their data show a peak in heat capacity that is pushed to lower temperatures as the field increases, disappearing above ~5 mT. Since their measurements of magnetization above 1 mK indicated a Curie–Weiss temperature of ~0.5 mK, in agreement with other groups, they suggest that this low temperature heat capacity peak may signal a transition from a ferromagnetic-like high-field phase to a low-field antiferromagnetic phase.

THERMAL CONDUCTIVITY

At Lancaster, the thermal conductivity κ of ^{3}He in aerogel was measured in two distinct regimes: in the normal fluid at temperatures above 5 mK [146]; and in the superfluid state at temperatures below 1 mK [57, 58]. Both sets of experiments were performed using a bolometric technique, where a small enclosed volume of bulk ^{3}He was connected to the larger open bulk volume through a tube (or disk) of aerogel. A known amount of heat power $\dot{Q}$ was applied to the enclosed volume, and the resulting temperature difference ΔT between the enclosure and the outside bulk was used to measure the thermal conductivity of the ^{3}He inside the piece of aerogel. In the normal fluid, as described in Section 5.3.1., their analysis showed a tendency towards $\kappa \propto T$ at lower temperatures, rather than the T^{-1} dependence that one would expect for a pure Fermi liquid. From these measurements they thus derived mean free path values of 56 nm and 92 nm for 95% and 98% porosity aerogels, respectively.

The first experiments in the superfluid were carried out using a 98% aerogel [57] at pressures of 4.8 and 7.4 bar. Fits to the inferred normal state conductivity limit returned values for the mean free path of $\sim 200\,\text{nm}$ at both pressures. This is twice as large as the 92 nm measured above for nominally the same aerogel. A second 98% aerogel sample used in a later superfluid experiment, described below, gave a value of 90 nm for ℓ_a, and also had a higher critical pressure. These differences must be ascribed to variability in the aerogel samples. Below T_{ca} the conductivity dropped below that of the normal state, as one would expect when an energy gap opens up and thus reduces the number density of quasiparticle carriers. There appeared to be no measurable difference in conductivity whether the aerogel contained A-like phase or B-like phase. The data were fitted to a simple impurity scattering model where the only free parameters were ℓ_a (already set by the limiting normal state conductivity) and the aerogel energy gap. The data were fitted well an energy gap that was close to the BCS value given by T_{ca} and thus appeared to show no further suppression in aerogel.

However, in the analysis of the second round of superfluid experiments it was realized that a more sophisticated model was necessary, since one must take into account the role of the thermal boundary resistance between the ^{3}He inside and outside the aerogel [58, 59]. In effect, because the energy gap in bulk B phase is always larger than that inside the aerogel, whether the ^{3}He is normal or superfluid, quasiparticles cannot cross the boundary between the two unless they have energy larger than the B phase gap. Thus, in analyzing their measurements, the Lancaster group had to treat the aerogel thermal link as a series of resistances: the boundary resistance between ^{3}He in the small enclosed volume and in the aerogel sample; the aerogel ^{3}He itself; and a further boundary resistance between the aerogel ^{3}He and that in the outside open volume. Their measurements of total thermal conductance K at various pressures are shown in Fig. 5.34 as plots K/T versus T. The dotted line represents the expected conductance if all the liquid inside the aerogel remains normal,

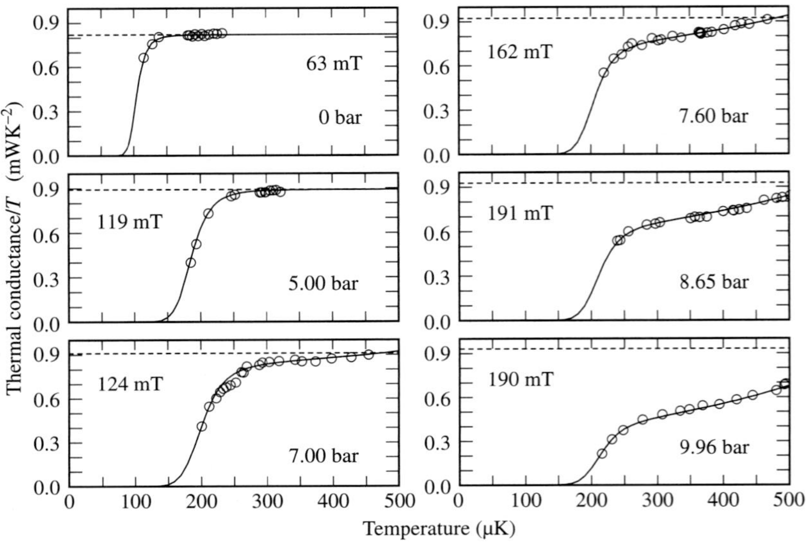

Fig. 5.34 Measured thermal conductance K for ^{3}He contained in a 98% aerogel, from [58], plotted as K/T versus T to remove the linear temperature dependence. The dotted line represents the expected conductance if all the liquid in the aerogel were normal. The solid line is a fit to the data that accounts for the onset of superfluidity in the aerogel, and also includes the boundary resistance between the bulk ^{3}He and that inside the aerogel which dominates at lower temperatures and is responsible for the fall in conductance. Thus no superfluidity is seen inside the aerogel below 7 bar. Above this pressure, the ^{3}He in the aerogel becomes superfluid, its thermal conductivity reduces, and its contribution to the total conductance drops.

which it does at pressures of 0 and 5 bar. The drop in K at the low T end is a consequence of the boundary resistance as the number of quasiparticles in the bulk falls exponentially. The interesting behavior occurs as the pressure is increased, the ^{3}He inside the aerogel goes superfluid, and the conductance falls below the normal state value before the boundary resistance dominates. It was possible to calculate the boundary resistance exactly and then subtract it from the measured data, leaving only the conductance of the ^{3}He in aerogel, from which the conductivity was derived. It was found that, rather than extrapolating to zero as $T \to 0$, as one would expect for a superfluid with a well-defined gap, there was a pressure range of 6.5 to 11 bar where the conductivity had a finite limiting value. The group argued that this was compelling evidence for a *gapless* superfluid inside the aerogel, one in which there is a finite density of states at the Fermi energy that arises because of impurity scattering. Here these data have been used to infer the normalized density of states and appear in Fig. 5.11.

5.5 Summary

Since the realization that aerogel acts as a static impurity which introduces controlled disorder into superfluid ^{3}He, many experimental investigations have been made to elucidate the properties of this dirty superfluid. It is now clear that aerogel cannot simply be considered as a collection of randomly distributed scattering centers. Its unique structure proffers quenched anisotropic disorder with various length scales which interact with the anisotropic superfluid order parameters. Most experiments have used a nominal 98% aerogel, and after the early days of non-reproducibility between different samples, there is now an accepted phase diagram for this porosity. The strong pair-breaking effect of the aerogel manifests itself with the appearance of a disorder-driven zero temperature transition and the existence of gapless superfluidity. The profound influence of aerogel on the AB transition shows striking similarities with the effect of a magnetic field on the pure phases; orbital fields associated with aerogel's anisotropic structure mimic the effect of an applied magnetic field on the order parameter. This is of particular importance in understanding the nature of the A-like phase which has puzzled researchers over an extended period. It now appears that the A-like phase in isotropic aerogel is best described by the glassy Larkin–Imry–Ma state. The community has just started effectively to take advantage of global anisotropy induced by deformation and is now making steady progress towards a better understanding of the consequences of anisotropic disorder. For example, textural transitions observed in globally anisotropic aerogels are extremely fascinating, and the theoretical prediction of a novel superfluid phase in a uniaxially stretched aerogel provides impetus for future experiments.

While experimentalists have answered many questions on the nature of ^{3}He in aerogel, others are yet to be addressed, and research has inevitably led to the posing of more. Here we propose several possible avenues for further work that range from tying up important loose ends to exploiting the unique properties of the ^{3}He-aerogel system.

The possibility of a quantum phase transition at low pressures and very low temperatures was first observed in torsional oscillator studies in the 1990s. This could be an exemplar QPT between normal fluid and either the B-like or A-like phases. Furthermore, it will be extremely interesting to investigate theoretically, as well as experimentally, the effects of quantum fluctuations at a finite temperature.

The aerogel strands provide scattering centers for quasiparticles that are in an interesting mesoscopic regime between point-like impurities and surface scattering. The effects of finite-size correlated scattering are not well understood. The ^{3}He-aerogel system also offers an opportunity to investigate various phenomena caused by spin-exchange interactions between localized spins in the solid layers and the itinerant spins in the surrounding liquid. We do not have clear experimental evidence as to whether the interaction is antiferromagnetic or ferromagnetic. Detailed investigation of an A_1-like phase with and without the magnetic solid layers could be an important step here. One might also wonder what the effect of magnetic scattering has on transport properties.

The solid ^{3}He itself provides a unique environment for investigating magnetic behavior, and might be a good test-bed for nano-scale spin ordering. Do the solid layers order? If they do, do they order ferromagnetically, antiferromagnetically, or are both transitions present, as in bulk solid? Furthermore, at temperatures above any ordering, it has been possible to demagnetize the paramagnetic solid and thus cool the surrounding ^{3}He liquid [27]. Exploiting this effect as a new cooling technique could open up hitherto unexplored temperature regimes.

An important aspect of bulk superfluid behavior has been its response to rotation, and the properties of the vortex defects that this can create. While some work has been done on ^{3}He-aerogel under rotation, this could provide a rich area of prospective new research. For example, there is an extensive phase diagram of different types of vorticity in the bulk superfluids, and one may question whether these vortices can also exist in ^{3}He-aerogel and/or whether there are new types favored by order parameter suppression and pinning on strands.

The interface between bulk ^{3}He and that contained inside the aerogel has barely been considered at an experimental level. This could be an extremely fruitful area of future research as the interplay between bulk and ^{3}He aerogel phase diagrams can be used to investigate the interfaces between the bulk superfluids and normal fluid inside the aerogel. One could also probe the interface between bulk B phase and the A-like phase. This affords many possibilities, from nucleation behavior to potential proximity effects. Could the aerogel even be exploited as a weak link between two volumes of bulk fluid and thus lead to Josephson effects?

Finally, the existence of anisotropy has been shown to be very important in controlling the properties of the A-like phase and its stability relative to other phases. Much work has been done, but little to date on isotropic samples that are then deformed in situ and in a controlled manner. It is especially important to investigate the behavior of strain-induced textural transitions, starting from the isotropic state, and to search for novel phases induced by anisotropy. Such techniques could also be used to investigate the analogy between strain and magnetic field and to further manipulate the phase diagram.

Acknowledgments

We would like to express our gratitude to the many authors who have allowed us to reproduce their data and figures, and who have discussed their work with us. We would like to thank Byunghee Moon and Hyunchang Choi for allowing us to make use of their unpublished data and for their assistance in producing the related figures used in this chapter. George Pickett's assistance in the preparation of figures was invaluable. We also acknowledge the financial support of the Royal Society and the UK EPSRC (RPH) and the National Science Foundation (YL). Finally, we thank Mark Meisel and Bill Halperin for their helpful suggestions and comments on our manuscript.

References

[1] Abragam, A. and Goldman, M. *Nuclear Magnetism: Order and Disorder*. Clarendon Press, Oxford (1982).

[2] Abrikosov, A. A. and Gor'kov, L. P. Contribution to the theory with superconducting alloys with paramagnetic impurities. *Sov. Phys. JETP*, **12**, 1243–1253 (1961).

[3] Akimov, Y. K. Fields of application of aerogels (review). *Instrum. Exp. Tech.*, **46**(3), 287–299 (2003).

[4] Alles, H., Kaplinsky, J. J., Wootton, P. S., Reppy, J. D., and Hook, J. R. Torsional oscillator studies of the superfluidity of ^{3}He in aerogel. *Physica* B, **255**, 1–10 (1998).

[5] Alles, H., Kaplinsky, J. J., Wootton, P. S., Reppy, J. D., Naish, J. H., and Hook, J. R. Evidence for superfluid B phase of ^{3}He in aerogel. *Phys. Rev. Lett.*, **83**(7), 1367–1370 (1999).

[6] Anderson, P. W. Theory of dirty superconductors. *J. Phys. Chem. Solids*, **11**(1–2), 26–30 (1959).

[7] Aoyama, K. and Ikeda, R. Equal-spin pairing state of superfluid ^{3}He in aerogel. *Phys. Rev.* B, **72**(1), 012515 (2005).

[8] Aoyama, K. and Ikeda, R. Pairing states of superfluid ^{3}He in uniaxially anisotropic aerogel. *Phys. Rev.* B, **73**(6), 06054 (2006).

[9] Aoyama, K. and Ikeda, R. Perturbative study of strong coupling correction in superfluid ^{3}He in aerogel. *J. Low Temp. Phys.*, **148**(5–6), 573–577 (2007).

[10] Aoyama, K. and Ikeda, R. Strong-coupling effects in superfluid ^{3}He in aerogel. *Phys. Rev.* B, **76**(10), 104512 (2007).

[11] Balatsky, A. V., Vekhter, I., and Zhu, Jian-Xin. Impurity-induced states in conventional and unconventional superconductors. *Rev. Mod. Phys.*, **78**(2), 373–433 (2006).

[12] Baramidze, G. A. and Kharadze, G. A. Superfluidity of ^{3}He in aerogel at T = 0 in a magnetic field. *JETP*, **88**(2), 415–419 (1999).

[13] Baramidze, G. A. and Kharadze, G. A. Cooper pairing in ^{3}He in the presence of spin-polarized scattering centers. *Physica* B, **284**(Part 1), 305–306 (2000).

[14] Baramidze, G. A. and Kharadze, G. A. Strong-coupling effects in 'dirty' superfluid ^{3}He. *J. Phys.: Condens. Matter*, **14**(32), 7471–7477 (2002).

[15] Baramidze, G. A. and Kharadze, G. A. Superfluid ^{3}He in presence of spin-polarized scattering centers. *J. Low Temp. Phys.*, **135**(5–6), 399–409 (2004).

[16] Barker, B. I., Lee, Y., Polukhina, L., Osheroff, D. D., Hrubesh, L. W., and Poco, J. F. Observation of a superfluid ^{3}He A-B phase transition in silica aerogel. *Phys. Rev. Lett.*, **85**(10), 2148–2151 (2000).

[17] Barker, B. I., Polukhina, L., Poco, J. F., Hrubesh, L. W., and Osheroff, D. D. Spin dynamics of ^{3}He in aerogel. *J. Low Temp. Phys.*, **113**(5-6), 635–644 (1998).

[18] Baumgardner, J. E. *Impurity Scattering in Superfluid ^{3}He: A New Phase*. Ph.D. thesis, Stanford University (2004).

[19] Baumgardner, J. E., Lee, Y., Osheroff, D. D., Hrubesh, L. W., and Poco, J. F. Interfacial pinning in the superfluid ^{3}He A–B transition in aerogel. *Phys. Rev. Lett.*, **93**(5), 055301 (2004).

[20] Baumgardner, J. E. and Osheroff, D. D. Phase diagram of superfluid ^{3}He in 99.3% porosity aerogel. *Phys. Rev. Lett.*, **93**(15), 155301 (2004).

[21] Baumgardner, J. E., Polukhina, L. V., Lee, Y., Poco, J. F., Hrubesh, L. W., and Osheroff, D. D. NMR studies of superfluid ^{3}He in low density silica aerogels. *Physica* B, **329**, 292–295 (2003).

[22] Bhupathi, P., Hwang, J., Martin, R. M., Blankstein, J., Jaworski, L., Mulders, N., Tanner, D. B., and Lee, Y. Aerogel waveplates. *Opt. Express*, **17**(13), 10599–10605 (2009).

[23] Bhupathi, P., Jaworski, L., Hwang, J., Tanner, D. B., Obukov, S., Lee, Y., and Mulders, N. Optical birefringence in uniaxially compressed aerogels. *New J. Phys.*, **12**, 103016–1–20 (2010).

[24] Bhupathi, P., Moon, B. H., Gonzalez, M., and Lee, Y. Transverse acoustic spectroscopy of superfluid ^{3}He in compressed aerogel. *J. Low Temp. Phys.*, **158**(1–2, Sp. Iss. SI), 176–181 (2010).

[25] Borovik-Romanov, A. S., Bunkov, Y. M., Dmitriev, V. V., and Mukharskiy, M. (1984). Instability of homogeneous spin precession in superfluid ^{3}He-A. *JETP Lett.*, **40**, 469–473 (1984).

[26] Bradley, D. I., Fisher, S. N., Guénault, A. M., Haley, R. P., Mulders, N., O'Sullivan, S., Pickett, G. R., Roberts, J. E., and Tsepelin, V. Contrasting mechanical anisotropies of the superfluid ^{3}He phases in aerogel. *Phys. Rev. Lett.*, **98**(7), 075302 (2007).

[27] Bradley, D. I., Fisher, S. N., Guénault, A. M., Haley, R. P., Mulders, N., Pickett, G. R., Potts, D., Skyba, P., Smith, J., Tsepelin, V., and Whitehead, R. C. V. Magnetic phase transition in a nanonetwork of solid ^{3}He in aerogel. *Phys. Rev. Lett.*, **105**(12), 125303 (2010).

[28] Brussard, P., Fisher, S. N., Guénault, A. M., Hale, A. J., Mulders, N., and Pickett, G. R. Phase diagram of the A and B phases of superfluid ^{3}He in aerogel. *Phys. Rev. Lett.*, **86**(20), 4580–4583 (2001).

[29] Buchholtz, L. J. and Zwicknagl, G. Identification of p-wave superconductors. *Phys. Rev.* B, **23**(11), 5788–5796 (1981).

[30] Bunkov, Y. M., Collin, E., and Godfrin, H. ^{3}He NMR in aerogel. *J. Phys. Chem. Solids*, **66**(8–9), 1325–1329 (2005).

[31] Burchell, M. J., Graham, G., and Kearsley, A. Cosmic dust collection in aerogel. *Annu. Rev. Earth Planet. Sci.*, **34**, 385–418 (2006).

[32] Candela, D. and Kalechofsky, N. Nuclear magnetism of normal ^{3}He and ^{3}He-^{4}He mixtures in aerogel. *J. Low Temp. Phys.*, **113**(3–4), 351–356 (1998).

[33] Cantin, M., Casse, M., Koch, L., Jouan, R., Mestreau, P., Roussel, D., Bonnin, F., Moutel, J., and Teichner, S. J. Silica aerogels used as Cherenkov radiators. *Nucl. Instrum. Methods*, **118**(1), 177–182 (1974).

[34] Chan, M. H. W., Mulders, N., and Reppy, J. D. Helium in aerogel. *Physics Today*, **49**(8, Part 1), 30–37 (1996).

[35] Choi, H., Davis, J. P., Pollanen, J., Haard, T. M., and Halperin, W. P. Analysis of strong-coupling parameters for superfluid ^{3}He. *J. Low Temp. Phys.*, **148**(5–6), 507–511 (2007).

[36] Choi, H., Davis, J. P., Pollanen, J., Haard, T. M., and Halperin, W. P. Strong coupling corrections to the Ginzburg–Landau theory of superfluid ^{3}He. *Phys. Rev.* B, **75**(17), 174503 (2007).

[37] Choi, H., Yawata, K., Haard, T. M., Davis, J. P., Gervais, G., Mulders, N., Sharma, P., Sauls, J. A., and Halperin, W. P. Specific heat of disordered superfluid ^{3}He. *Phys. Rev. Lett.*, **93**(14), 145301 (2004).

[38] Choi, H. C. *Study of disordered liquid ^{3}He in high porosity silica aerogel*. Ph.D. thesis, University of Florida (2007).

[39] Choi, H. C., Gray, A. J., Vicente, C. L., Xia, J. S., Gervais, G., Halperin, W. P., Mulders, N., and Lee, Y. A_1 and A_2 transitions in superfluid ^{3}He in 98% porosity aerogel. *Phys. Rev. Lett.*, **93**(14), 145302 (2004).

[40] Choi, H. C., Gray, A. J., Vicente, C. L., Xia, J. S., Gervais, G., Halperin, W. P., Mulders, N., and Lee, Y. Effect of strong magnetic fields on superfluid ^{3}He in 98% porosity aerogel. *J. Low Temp. Phys.*, **138**(1–2, Sp. Iss. SI), 107–115 (2005).

[41] Choi, H. C., Masuhara, N., Moon, B. H., Bhupathi, P., Meisel, M. W., Lee, Y., Mulders, N., Higashitani, S., Miura, M., and Nagai, K. Ultrasound attenuation of superfluid ^{3}He in aerogel. *Phys. Rev. Lett.*, **98**(22), 225301 (2007).

[42] Choi, H. C., Masuhara, N., Moon, B. H., Bhupathi, P., Mulders, N., Meisel, M. W., and Lee, Y. Acoustic properties of normal liquid ^{3}He in 98% aerogel. *J. Low Temp. Phys.*, **148**(5–6), 609–613 (2007).

[43] Colin, E., Triqueneaux, S., Bunkov, Yu. M., and Godfrin, H. Fast-exchange model visualized with ^{3}He confined in aerogel: A Fermi liquid in contact with a ferromagnetic solid. *Phys. Rev.* B, **80**, 094422 (2009).

[44] Daughton, D. R., MacDonald, J., and Mulders, N. Acoustic properties of silica aerogels from 400 mK to 400 K. *Physica* B, **329**(Part 2), 1233–1234 (2003).

[45] Davis, J. P., Pollanen, J., Reddy, B., Shirer, K. R., Choi, H., and Halperin, W. P. Stability of the axial phase of superfluid ^{3}He in aerogel with globally anisotropic scattering. *Phys. Rev.* B, **77**(14), 140502 (2008).

[46] Dmitriev, V. V., Kosarev, I. V., Mulders, N., Zavjalov, V. V., and Zmeev, D. E. Homogeneous spin precession in superfluid ^{3}He confined to aerogel. *Physica* B, **329**, 324–326 (2003).

[47] Dmitriev, V. V., Kosarev, I. V., Mulders, N., Zavjalov, V. V., and Zmeev, D. E. Measurements of longitudinal and transverse magnetic relaxation in superfluid ^{3}He confined to aerogel. *Physica* B, **329**, 322–323 (2003).

[48] Dmitriev, V. V., Kosarev, I. V., Mulders, N., Zavjalov, V. V., and Zmeev, D. E. Pulsed NMR experiments in superfluid ^{3}He confined in aerogel. *Physica* B, **329**, 296–298 (2003).

[49] Dmitriev, V. V., Krasnikhin, D. A., Mulders, N., Senin, A. A., Volovik, G. E., and Yudin, A. N. Orbital glass and spin glass states of ^{3}He-A in aerogel. *JETP Lett.*, **91**(11), 599–606 (2010).

[50] Dmitriev, V. V., Krasnikhin, D. A., Mulders, N., Zavjalov, V. V., and Zmeev, D. E. Transverse and longitudinal nuclear magnetic resonance in superfluid ^{3}He in anisotropic aerogel. *JETP Lett.*, **86**(9), 594–599 (2007).

[51] Dmitriev, V. V., Krasnikhin, D. A., Mulders, N., and Zmeev, D. E. Soliton-like spin state in the A-like phase of ^{3}He in anisotropic aerogel. *J. Low Temp. Phys.*, **150**(3–4), 493–498 (2008).

[52] Dmitriev, V. V., Levitin, L. V., Mulders, N., and Zmeev, D. E. Longitudinal NMR and spin states in the A-like phase of ^{3}He in aerogel. *JETP Lett.*, **84**(8), 461–465 (2006).

[53] Dmitriev, V. V., Zavjalov, V. V., Zmeev, D. E., Kosarev, I. V., and Mulders, N. Nonlinear NMR in a superfluid B phase of ^{3}He in aerogel. *JETP Lett.*, **76**(5), 312–317 (2002).

[54] Dobbs, E. R. *Helium Three*. Oxford University Press, New York (2000).

[55] Einzel, D. and Parpia, J. M. Liquid ^{3}He in aerogel: Crossover from Drude's to Hagen–Poiseuille's law. *Phys. Rev. Lett.*, **81**(18), 3896–3899 (1998).

[56] Elbs, J., Bunkov, Y. M., Collin, E., Godfrin, H., and Volovik, G. E. Strong orientational effect of stretched aerogel on the ^{3}He order parameter. *Phys. Rev. Lett.*, **100**(21), 215304 (2008).

[57] Fisher, S. N., Guénault, A. M., Hale, A. J., and Pickett, G. R. The thermal conductivity of superfluid ^{3}He in aerogel: a measurement of the energy gap. *J. Low Temp. Phys.*, **126**(1–2), 673–678 (2002).

[58] Fisher, S. N., Guénault, A. M., Mulders, N., and Pickett, G. R. Thermal conductivity of liquid ^{3}He in aerogel: a gapless superfluid. *Phys. Rev. Lett.*, **91**(10), 105303 (2003).

[59] Fisher, S. N., Guénault, A. M., Pickett, G. R., and Sauer, F. The interfacial thermal resistance between bulk superfluid ^{3}He and liquid ^{3}He in aerogel at ultralow temperatures. *Physica* B, **329**, 311–312 (2003).

[60] Fisher, S. N., Haley, R. P., and Pickett, G. R. Comment on "Nucleation and interfacial coupling between pure and dirty superfluid phases of ^{3}He". *Phys. Rev. Lett.*, **88**, 209601 (2002).

[61] Fomin, I. A. Long-lived induction signal and spatially nonuniform spin precession in ^{3}He-B. *JETP Lett.*, **40**, 1037–1040 (1984).

[62] Fomin, I. A. Random textures of the order parameter of superfluid ^{3}He-B in aerogel. *JETP Lett.*, **75**(4), 187–189 (2002).

[63] Fomin, I. A. Order parameter of A-like ^{3}He phase in aerogel. *JETP Lett.*, **77**(5), 240–242 (2003).

[64] Fomin, I. A. Superfluid ^{3}He phases in aerogel. *JETP*, **98**(5), 974–980 (2004).

[65] Fomin, I. A. Spin nutation in the quasi-isotropic A-like superfluid phase of ^{3}He. *J. Exp. Theor. Phys.*, **102**(6), 966–971 (2006).

[66] Freeman, M. R., Germain, R. S., Thuneberg, E. V., and Richardson, R. C. Size effects in thin-films of superfluid ^{3}He. *Phys. Rev. Lett.*, **60**(7), 596–599 (1988).

[67] Fricke, J. and Emerling, A. Aerogels – preparation, properties, applications. *Structure and Bonding*, **77**, 37–87 (1992).

[68] Geller, D. A., Golov, A. I., Mulders, N., Chan, M. H. W., and Parpia, J. M. Sound modes of superfluid ^{3}He in aerogel. *J. Low Temp. Phys.*, **113**(3-4), 339–344 (1998).

[69] Gervais, G., Haard, T. M., Nomura, R., Mulders, N., and Halperin, W. P. Modification of the superfluid ^{3}He phase diagram by impurity scattering. *Phys. Rev. Lett.*, **87**(3), 035701 (2001).

[70] Gervais, G., Yawata, K., Mulders, N., and Halperin, W. P. Nucleation and interfacial coupling between pure and dirty superfluid phases of ^{3}He. *Phys. Rev. Lett.*, **88**(4), 045505 (2002).

[71] Gervais, G., Yawata, K., Mulders, N., and Halperin, W. P. Phase diagram of the superfluid phases of ^{3}He in 98% aerogel. *Phys. Rev.* B, **66**(5), 054528 (2002).

[72] Golov, A. I., Einzel, D., Lawes, G. J., Matsumoto, K., and Parpia, J. M. Dissipation mechanisms near the superfluid ^{3}He transition in aerogel. *Phys. Rev. Lett.*, **92**(19), 195301 (2004).

[73] Golov, A. I., Geller, D. A., Parpia, J. M., and Mulders, N. Acoustic spectroscopy of superfluid ^{3}He in aerogel. *Phys. Rev. Lett.*, **82**(17), 3492–3495 (1999).

[74] Golov, A. I., Porto, J. V., Geller, D. A., Mulders, N., Lawes, G. J., and Parpia, J. M. ^{3}He superfluidity in the presence of aerogel. *Physica* B, **280**(1–4), 134–139 (2000).

[75] Golov, A. I., Porto, J. V., and Parpia, J. M. Superfluidity of ^{3}He in aerogel covered with a thick ^{4}He film. *Phys. Rev. Lett.*, **80**(20), 4486–4489 (1998).

[76] Gotz, H. and Eska, G. Spin diffusion in normal fluid ^{3}He in 97% porous silica-aerogel. *Physica* B, **329**, 307–308 (2003).

[77] Gross, J., Fricke, J., and Hrubesh, L. W. Sound-propagation in SiO_2 aerogels. *J. Acoust. Soc. Am.*, **91**(4, Part 1), 2004–2006 (1992).

[78] Gross, J., Reichnauer, G., and Fricke, J. Mechanical properties of SiO_2 aerogels. *J. Phys.* D, **21**(9), 1447–1451 (1988).

[79] Haard, T. M., Gervais, G., Nomura, R., and Halperin, W. P. The pathlength distribution of simulated aerogels. *Physica* B, **284**(Part 1), 289–290 (2000).

[80] Hahn, I., Boyd, S. T. P., Bozler, H. M., and Gould, C. M. Thermodynamic magnetization discontinuity between the A-phase and B-phase of superfluid ^{3}He. *J. Low Temp. Phys.*, **101**(3–4), 781–786 (1995).

[81] Halperin, W. P. and Varoquauax, E. *Helium Three*. North-Holland, Amsterdam (1990).

[82] Halperin, W. P., Choi, H., Davis, J. P., and Pollanen, J. Impurity effects of aerogel in superfluid ^{3}He. *J. Phys. Soc. Jpn.*, **77**(11), 111002 (2008).

[83] Halperin, W. P. and Sauls, J. A. Helium-three in aerogel. *arXiv:cond-mat/0408593v1* (2004).

[84] Hänninen, R. and Thuneberg, E. V. Model of inhomogeneous impurity distribution in Fermi superfluids. *Phys. Rev.* B, **67**(21), 214507 (2003).

[85] He, J., Corwin, A. D., Mulders, N., Parpia, J. M., Reppy, J. D., and Chan, M. H. W. An experiment to measure heat capacity of ^{3}He in aerogel. *J. Low Temp. Phys.*, **126**(1–2), 679–684 (2002).

[86] He, J., Corwin, A. D., Parpia, J. M., and Reppy, J. D. Heat capacity of ^{3}He in aerogel. *Phys. Rev. Lett.*, **89**(11), 115301 (2002).

[87] Herman, T., Day, J., and Beamish, J. R. Deformation of silica aerogel during fluid adsorption. *Phys. Rev.* B, **73**(9), 094127 (2006).

[88] Higashitani, S. Impurity scattering effect on superfluidity of ^{3}He in aerogel. *J. Low Temp. Phys.*, **114**(1–2), 161–172 (1999).

[89] Higashitani, S., Ichikawa, T., Yamamoto, M., Miura, M., and Nagai, K. Viscoelastic theory of liquid ^{3}He in aerogel. *Physica* B, **329**, 299–300 (2003).

[90] Higashitani, S., Miura, M., Yamamoto, M., and Nagai, K. Microscopic theory of sound propagation in the superfluid ^{3}He-aerogel system. *Phys. Rev.* B, **71**(13), 134508 (2005).

[91] Hunger, P., Bunkov, Y. M., Collin, E., and Godfrin, H. Evidence for magnon BEC in superfluid ^{3}He-A. *J. Low Temp. Phys.*, **158**(1–2, Sp. Iss. SI), 129–134 (2010).

[92] Ichikawa, K., Yamasaki, S., Akimoto, H., Kodama, T., Shigi, T., and Kojima, H. Healing length of superfluid ^{3}He. *Phys. Rev. Lett.*, **58**(19), 1949–1952 (1987).

[93] Ichikawa, T., Yamamoto, M., Higashitani, S., and Nagai, K. Collision drag effect on propagation of sound in liquid ^{3}He in aerogel. *J. Phys. Soc. Jpn.*, **70**(12), 3483–3486 (2001).

[94] Ikeda, R. and Aoyama, K. Superfluid ^{3}He in globally isotropic random media. *Phys. Rev.* B, **79**(6), 064527 (2009).

[95] Imry, Y. and Ma, S. K. Random-field stability of the ordered state of continuous symmetry. *Phys. Rev. Lett.*, **35**, 1399–1401 (1975).

[96] Ishikawa, O., Kado, R., Nakagawa, H., Obara, K., Yano, H., Hata, T., Yokogawa, H., and Yokoyama, M. Pulsed NMR measurements in superfluid ^{3}He in aerogel of 97.5%. *AIP Conference Proceedings*, **850**, 235–236 (2006).

[97] Ishino, M., Chiba, J., En'yo, H., Funahashi, H., Ichikawa, A., Ieiri, M., Kanda, H., Masaike, A., Mihara, S., Miyashita, T., Murakami, T., Nakamura, A., .Naruki, M., Muto, R., Ozawa, K., Sato, H. D., Sekimoto, M., Tabaru, T., Tanaka, K. H., Yoshimura, Y., Yokkaichi, S., Yokoyama, M., and Yokogawa, H. Mass production of hydrophobic silica aerogel and readout optics of Cherenkov light. *Nucl. Instrum. Methods*, **A457**, 581–587 (2001).

[98] Kado, R., Nakagawa, H., Obara, K., Yano, H., Ishikawa, O., and Hata, T. A-B phase conversion and coexistence of superfluid ^{3}He in aerogel. *J. Low Temp. Phys.*, **150**(3–4), 472–475 (2008).

[99] Kalbfeld, S., Kucera, D. M., and Ketterson, J. B. Observation of an evolving standing-wave pattern involving a transverse disturbance in superfluid ^{3}He-B. *Phys. Rev. Lett.*, **71**(14), 2264–2267 (1993).

[100] Kato, H., Miyashita, W., Nomura, R., and Okuda, Y. Observation of shrinkage of silica aerogel during capillary condensation of ^{4}He. *J. Low Temp. Phys.*, **148** (5–6), 621–625 (2007).

[101] Kim, S. B., Ma, J., and Chan, M. H. W. Phase-diagram of ^{3}He-^{4}He mixture in aerogel. *Phys. Rev. Lett.*, **71**(14), 2268–2271 (1993).

[102] Kistler, S. S. Coherent expanded aerogels and jellies. *Nature*, **127**, 741 (1931).

[103] Kotera, K., Hatate, T., Nakagawa, H., Yano, H., Ishikawa, O., Hata, T., Yokogawa, H., and Yokoyama, M. Observation of superfluidity of ^{3}He in aerogel by fourth sound technique. *Physica* B, **329**, 316–317 (2003).

[104] Kunimatsu, T., Matsubara, A., Izumina, K., Sato, T., Kubota, M., Bunkov, Y. M., and Mizusaki, T. Observation of vortex-creep in superfluid ^{3}He B-like phase in aerogel by the hpd. *Physica* C, **468**(7–10), 605–608 (2008).

[105] Kunimatsu, T., Matsubara, A., Izumina, K., Sato, T., Kubota, M., Takagi, T., Bunkov, Y. M., and Mizusaki, T. Quantum fluid dynamics of rotating superfluid ^{3}He in aerogel. *J. Low Temp. Phys.*, **150**(3–4), 435–444 (2008).

[106] Kunimatsu, T., Sato, T., Izumina, K., Matsubara, A., Sasaki, Y., Kubota, M., Ishikawa, O., Mizusaki, T., and Bunkov, Y. M. Orientation effect on superfluid ^{3}He in anisotropic aerogel. *JETP Lett.*, **86**(3), 216–220 (2007).

[107] Landau, L. D. Oscillations in a Fermi liquid. *JETP*, **5**, 101–108 (1957).

[108] Larkin, A. I. Vector pairing of superconductors of small dimension. *JETP Lett.*, **2**, 130 (1965).

[109] Larkin, A. I. Effect of inhomogeneity on structure of mixed state of superconductors. *JETP*, **31**, 784 (1970).

[110] Lawes, G. J., Golov, A. I., Nazaretski, E., Mulders, N., and Parpia, J. M. Sound propagation in coexistent Bose and Fermi superfluids in aerogel. *Phys. Rev. Lett.*, **90**(19), 195301 (2003).

[111] Lawes, G. J., Kingsley, S. C. J., Mulders, N., Beamish, J. R., and Parpia, J. M. Torsional oscillator measurements on superfluid ^{3}He in 99.5% porous aerogel. *Physica* B, **284**(Part 1), 299–300 (2000).

[112] Lawes, G. J., Kingsley, S. C. J., Mulders, N., and Parpia, J. M. Scaling of the superfluid fraction and T_c of ^{3}He in aerogel. *Phys. Rev. Lett.*, **84**(18), 4148–4151 (2000).

[113] Lawes, G. J. and Parpia, J. M. Estimate of the gap parameter for superfluid ^{3}He in aerogel. *Phys. Rev.* B, **65**(9), 092511 (2002).

[114] Lee, Y., Choi, H. C., Masuhara, N., Moon, B. H., Bhupathi, P., Meisel, M. W., and Mulders, N. Absolute ultrasound attenuation measurements in superfluid ^{3}He in 98% aerogel by direct transmission. *J. Low Temp. Phys.*, **148**(5–6), 565–572 (2007).

[115] Lee, Y., Haard, T. M., Halperin, W. P., and Sauls, J. A. Discovery of acoustic Faraday effect in superfluid ^{3}He-B. *Nature*, **400**, 431–433 (1999).

[116] Leggett, A. J. Spin dynamics of anisotropic Fermi superfluid ^{3}He. *Ann. Phys.*, **85**(1), 11–55 (1974).

[117] Leggett, A. J. Theoretical description of new phases of liquid ^{3}He. *Rev. Mod. Phys.*, **47**(2), 331–414 (1975).

[118] Maki, K. and Puchkaryov, E. Impurity scattering in isotropic p-wave superconductors. *Europhys. Lett.*, **45**(2), 263–268 (1999).

[119] Matsumoto, K., Porto, J. V., Pollack, L., Smith, E. N., Ho, T. L., and Parpia, J. M. Quantum phase transition of ^{3}He in aerogel at a nonzero pressure. *Phys. Rev. Lett.*, **79**(2), 253–256 (1997).

[120] McKenna, M. J., Slawecki, T., and Maynard, J. D. Observation of a second-sound-like mode in superfluid-filled aerogel. *Phys. Rev. Lett.*, **66**(14), 1878–1881 (1991).

[121] Mineev, V. P. Phase transition in liquid ^{3}He in an aerogel at zero temperature. *JETP Lett.*, **66**(10), 693–698 (1997).

[122] Mineev, V. P. and Krotkov, P. L. Spin susceptibility of the superfluid ^{3}He-B in aerogel. *Phys. Rev.* B, **65**(2), 024501 (2002).

[123] Miura, M., Higashitani, S., Yamamoto, M., and Nagai, K. Collision drag effect on two-fluid hydrodynamics of superfluid ^{3}He in aerogel. *J. Low Temp. Phys.*, **134**(3–4), 843–850 (2004).

[124] Miura, M., Higashitani, S., Yamamoto, M., and Nagai, K. NMR properties of a possible non-unitary state of superfluid ^{3}He in aerogel. *J. Low Temp. Phys.*, **138**(1–2, Sp. Iss. SI), 153–157 (2005).

[125] Moon, B. H., Masuhara, N., Bhupathi, P., Gonzalez, M., Meisel, M. W., Lee, Y., and Mulders, N. Frequency-dependent ultrasound attenuation in superfluid ^{3}He in aerogel. *Phys. Rev.* B, **82**(6), 060501(R) (2010).

[126] Moon, B. H., Masuhara, N., Bhupathi, P., Gonzalez, M., Meisel, M. W., Lee, Y., and Mulders, N. Ultrasound attenuation and a P-B-T phase diagram of superfluid ^{3}He in 98% aerogel. *Phys. Rev.* B, **81**(13), 134526 (2010).

[127] Moon, B. H., Masuhara, N., Bhupathi, P., Gonzalez, M., Meisel, M. W., and Mulders, N. and Lee, Y. Magnetic field dependence of the A-like to B-like transition of superfluid ^{3}He in aerogel. *J. Low Temp. Phys.*, **158**(1–2, Sp. Iss. SI), 170–175 (2010).

[128] Nago, Y., Obara, K., Kado, R., Yano, H., Ishikawa, O., and Hata, T. Fourth sound measurement of superfluid ^{3}He in aerogel. *J. Low Temp. Phys.*, **148**(5–6), 597–601 (2007).

[129] Nakagawa, H., Kado, R., Obara, K., Yano, H., Ishikawa, O., Hata, T., Yokogawa, H., and Yokoyama, M. Equal-spin-pairing superfluid phase of ^{3}He in an aerogel acting as an impurity. *Phys. Rev.* B, **76**(17), 172504 (2007).

[130] Nazaretski, E., Lawes, G. J., Lee, D. M., Mulders, N., Ponarin, D., and Parpia, J. M. Acoustic spectroscopy of superfluid ^{3}He in aerogel in the presence of a magnetic field. *J. Low Temp. Phys.*, **126**(1–2), 685–690 (2002).

[131] Nazaretski, E., Lee, D. M., and Parpia, J. M. Superfluid density of ^{3}He in 98% aerogel in small magnetic fields. *Phys. Rev.* B, **71**(14), 144506 (2005).

[132] Nazaretski, E., Mulders, N., and Parpia, J. M. Metastability and superfluid fraction of the A-like and B phases of ^{3}He in aerogel in zero magnetic field. *JETP Lett.*, **79**(8), 383–387 (2004).

[133] Nazaretski, E., Mulders, N., and Parpia, J. M. Sound spectroscopy of the superfluid phases of ^{3}He in aerogel in zero magnetic field. *J. Low Temp. Phys.*, **134**(1–2), 763–768 (2004).

[134] Nomura, R., Gervais, G., Haard, T. M., Lee, Y., Mulders, N., and Halperin, W. P. High-frequency acoustics of ^{3}He in aerogel. *Phys. Rev. Lett.*, **85**(20), 4325–4328 (2000).

[135] Obara, K., Kato, C., Matsukura, T., Nago, Y., Kado, R., Yano, H., Ishikawa, O., Hata, T., Higashitani, S., and Nagai, K. Frictional motion of normal-fluid component of superfluid ^{3}He in aerogel. *Phys. Rev.* B, **82**(5), 054521 (2010).

[136] Parpia, J. M., Fefferman, A. D., Porto, J. V., Dmitriev, V. V., Levitin, L. V., and Zmeev, D. E. Scaling results for superfluid ^{3}He in 98% open aerogel. *J. Low Temp. Phys.*, **150**(3–4), 482–486 (2008).

[137] Pines, D. and Nosieres, P. *The Theory of Quantum Liquids*. W.A. Benjamin, New York (1966).

[138] Pollanen, J., Shirer, K. R., Blinstein, S., Davis, J. P., Choi, H., Lippman, T. M., Halperin, W. P., and Lurio, L. B. Globally anisotropic high porosity silica aerogels. *J. Non-Cryst. Solids*, **354**(40–41), 4668–4674 (2008).

[139] Porto, J. V., Golov, A. I., Matsumoto, K., Biggar, R. D., and Parpia, J. M. Suppression of ^{3}He superfluidity by aerogel. *J. Non-Cryst. Solids*, **225**(1–3), 205–209 (1998).

[140] Porto, J. V. and Parpia, J. M. Superfluid ^{3}He in aerogel. *Phys. Rev. Lett.*, **74**(23), 4667–4670 (1995).

[141] Porto, J. V. and Parpia, J. M. The effect of surface ^{4}He on superfluid ^{3}He in aerogel. *Czech. J. Phys.*, **46**(Suppl. 1), 123–124 (1996).

[142] Porto, J. V. and Parpia, J. M. Correlated disorder in a p-wave superfluid. *Phys. Rev.* B, **59**(22), 14583–14592 (1999).

[143] Porto, J. V., Pollack, L., Matsumoto, K., Smith, E. N., and Parpia, J. M. An experiment to measure the effect of magnetic fields on the superfluid fraction and transition temperature of ^{3}He in aerogel. *Czech. J. Phys.*, **46**(Suppl. 1), 125–126 (1996).

[144] Rainer, D. and Sauls, J. A. Sound propagation and transport properties of liquid ^{3}He in aerogel. *J. Low Temp. Phys.*, **110**(1–2), 525–531 (1998).

[145] Rainer, D. and Vuorio, M. Small objects in superfluid ^{3}He. *J. Phys. C*, **10**, 3093–3106 (1977).

[146] Reeves, P. A., Tvalashvili, G., Fisher, S. N., Guénault, A. M., and Pickett, G. R. Thermal conductivity of normal liquid ^{3}He in aerogel. *J. Low Temp. Phys.*, **129**(3–4), 185–193 (2002).

[147] Reim, M., Korner, W., Manara, J., Korder, S., Arduini-Schuster, M., Ebert, H. P., and Fricke, J. Silica aerogel granulate material for thermal insulation and daylighting. *Solar Energy*, **79**(2), 131–139 (2005).

[148] Richardson, R. C. and Smith, E. N. *Experimental Techniques in Condensed Matter Physics at Low Temperatures*. Addison-Wesley, Redwood City (1988).

[149] Roach, Pat R. and Ketterson, J. B. Observation of transverse zero sound in normal ^{3}He. *Phys. Rev. Lett.*, **36**(13), 736–740 (1976).

[150] Sato, T., Kunimatsu, T., Izumina, K., Matsubara, A., Kubota, M., Mizusaki, T., and Bunkov, Y. M. Coherent precession of magnetization in the superfluid ^{3}He-A-phase. *Phys. Rev. Lett.*, **101**(5), 055301 (2008).

[151] Sauls, J. A., Bunkov, Y. M., Collin, E., Godfrin, H., and Sharma, P. Magnetization and spin diffusion of liquid ^{3}He in aerogel. *Phys. Rev.* B, **72**(2), 024507 (2005).

[152] Sauls, J. A. and Sharma, P. Impurity effects on the A_1-A_2 splitting of superfluid ^{3}He in aerogel. *Phys. Rev.* B, **68**(22), 224502 (2003).

[153] Sauls, J. A. and Sharma, P. Theory of heat transport of normal liquid ^{3}He in aerogel. *New J. Phys.*, **12**, 983056–1–35 (2010).

[154] Schaefer, D. W. and Keefer, K. D. Structure of random porous materials–silica aerogel. *Phys. Rev. Lett.*, **56**(20), 2199–2202 (1986).

[155] Schechter, A. M. R., Simmonds, R. W., Packard, R. E., and Davis, J. C. Observation of 'third sound' in superfluid ^{3}He. *Nature*, **396**(6711), 554–557 (1998).

[156] Scherer, G. W., Smith, D. M., Qiu, X. M., and Anderson, J. M. Compression of aerogel. *J. Non-Cryst. Solids*, **186**, 316–320 (1995).

[157] Schrenk, R. and König, R. Specific heat of liquid ^{3}He under pressure in a restricted geometry. *Phys. Rev.* B, **57**(14), 8518–8525 (1998).

[158] Sharma, P. and Sauls, J. A. Thermal conductivity of superfluid ^{3}He in aerogel. *Physica* B, **329**, 313–315 (2003).

[159] Sprague, D. T., Haard, T. M., Kycia, J. B., Rand, M. R., Lee, Y., and Halperin, W. P. Magnetic field suppression of ^{3}He superfluidity in aerogel. *Czech. J. Phys.*, **46**(Suppl. 1), 119–120 (1996).

[160] Sprague, D. T., Haard, T. M., Kycia, J. B., Rand, M. R., Lee, Y., Hamot, P. J., and Halperin, W. P. Homogeneous equal-spin pairing superfluid state of ^{3}He in aerogel. *Phys. Rev. Lett.*, **75**(4), 661–664 (1995).

[161] Sprague, D. T., Haard, T. M., Kycia, J. B., Rand, M. R., Lee, Y., Hamot, P. J., and Halperin, W. P. Nuclear magnetic resonance of ^{3}He superfluid confined in high-porosity aerogel. *J. Low Temp. Phys.*, **101**(1–2), 185–194 (1995).

[162] Sprague, D. T., Haard, T. M., Kycia, J. B., Rand, M. R., Lee, Y., Hamot, P. J., and Halperin, W. P. Effect of magnetic scattering on the ^{3}He superfluid state in aerogel. *Phys. Rev. Lett.*, **77**(22), 4568–4571 (1996).

[163] Tang, Y. H., Hahn, I., Bozler, H. M., and Gould, C. M. Magnetic suppression of the B phase of superfluid ^{3}He. *Phys. Rev. Lett.*, **67**(13), 1775–1778 (1991).

[164] Thuneberg, E. V. Ginzburg–Landau theory for impure superfluid ^{3}He. *arXiv:cond-mat/9802044* (1998).

[165] Thuneberg, E. V., Fogelstrom, M., Yip, S. K., and Sauls, J. A. Localized vs. delocalized scattering in superfluid ^{3}He-aerogel. *Czech. J. Phys.*, **46**(Suppl. 1), 113–114 (1996).

[166] Thuneberg, E. V., Yip, S. K., Fogelstrom, M., and Sauls, J. A. Models for superfluid ^{3}He in aerogel. *Phys. Rev. Lett.*, **80**(13), 2861–2864 (1998).

[167] Tillotson, T. M. and Hrubesh, L. W. Transparent ultralow-density silica aerogels prepared by a 2-step sol–gel process. *J. Non-Cryst. Solids*, **145**, 44 (1992).

[168] Vicente, C. L., Choi, H. C., Xia, J. S., Halperin, W. P., Mulders, N., and Lee, Y. A-B transition of superfluid ^{3}He in aerogel and the effect of anisotropic scattering. *Phys. Rev.* B, **72**(9), 094519 (2005).

[169] Vollardt, D. and Wölfle, P. *The Superfluid Phases of Helium Three*. Taylor and Francis, London (1990).

[170] Volovik, G. E. Glass state of superfluid ^{3}He-A in an aerogel. *JETP Lett.*, **63**(4), 301–304 (1996).

[171] Volovik, G. E. Random anisotropy disorder in superfluid ^{3}He-A in aerogel. *JETP Lett.*, **84**(8), 455–460 (2006).

[172] Volovik, G. E. On Larkin-Imry-Ma state of ^{3}He-A in aerogel. *J. Low Temp. Phys.*, **150**(3–4), 453–463 (2008).

[173] Yamashita, M., Matsubara, A., Ishiguro, R., Sasaki, Y., Ishikawa, O., Kubota, M., Bunkov, Y. M., and Mizusaki, T. Rotating superfluid ^{3}He in aerogel. *J. Low Temp. Phys.*, **134**(1–2), 749–755 (2004).

[174] Yamashita, M., Matsubara, A., Ishiguro, R., Sasaki, Y., Kataoka, Y., Kubota, M., Ishikawa, O., Bunkov, Y. M., Ohmi, T., Takagi, T., and Mizusaki, T. Pinning of texture and vortices of the rotating B-like phase of superfluid ^{3}He confined in a 98% aerogel. *Phys. Rev. Lett.*, **94**(7), 075301 (2005).

6 Bose–Einstein condensation of photons

Jan Klaers and Martin Weitz

We review recent work on the Bose–Einstein condensation of photons in a dye microcavity environment. Other than for material particles, such as cold atomic Bose gases, photons usually do not condense at low temperatures. For Planck's blackbody radiation, the most ubiquitous Bose gas, photon number, and temperature are not independently tunable and at low temperatures the photons simply disappear in the system's walls, instead of massively occupying the cavity ground mode. In the approach described here, this obstacle is overcome by a fluorescence-induced thermalization mechanism in a dye-filled microcavity. Experimentally, both the thermalization of the photon gas and, at high photon densities, Bose–Einstein condensation have been observed. This article describes the thermalization mechanism of the photon gas in detail and summarizes experimental work performed so far.

6.1 Introduction

When a gas of particles with given density is cooled to such low temperatures that the associated de Broglie wavepackets spatially overlap, quantum statistical effects come into play. Specifically, for a gas of particles with integer spin (bosons), Bose–Einstein condensation into the ground state sets in above a critical phase space density. For dilute atomic gases, this effect was first observed in 1995 by means of laser and subsequent evaporative cooling of alkali atoms [2, 7, 11], see also Chapter 7 in this volume. The signature of Bose–Einstein condensation has also been observed for several solid-state quasiparticles, such as exciton–polaritons and magnons, see e.g. the contributions by Y. Yamamoto in this volume, and by D. Demokritov and A. Slavin in the second volume.

Photons, quantized particles of light, are also bosons, but usually show no Bose–Einstein condensation. In a blackbody radiator, the chemical potential of photons vanishes, i.e. the average particle number does not follow a given conservation law, but adjusts itself to the available thermal energy [19]. This is the essence of the Stefan–Boltzmann law, linking total radiation energy U to the fourth power of the temperature, $U \propto T^4$. At low temperatures, the photon number simply decreases and no macroscopic occupation of the cavity ground state occurs. Thus, a necessary precondition for a Bose–Einstein condensation of photons is to find a thermalization process that allows for independent adjustment of both photon number and temperature. The lack of such a mechanism has long prevented the realization of light sources that are capable of generating single mode light, without the necessity to be driven out of thermal equilibrium, as e.g. with a laser. Note that in a laser both the state of the light field and that of the active medium are far removed from thermal equilibrium [49]. To some extent, lasing is even a prime example of a non-equilibrium process, as only the absence of thermal equilibrium allows for inversion and optical gain. Early theoretical work proposed achievement of Bose–Einstein condensation of photons by Compton scattering off a thermal electron gas [56]. Later, Chiao et al. proposed a two-dimensional photon quantum fluid in a nonlinear Fabry–Perot resonator, where thermalization was sought from photon–photon scattering [6, 8, 9]. This concept is similar to atom–atom scattering processes in atomic physics BEC experiments, although the limited nonlinearity has so far prevented a thermalization of the photon gas [42]. In other work, demonstration of (quasi-)equilibrium Bose–Einstein condensation of exciton–polaritons, mixed states of matter and light, has been reported [5, 12, 21]. Here interparticle collisions of the excitons, i.e. the material parts of the polaritons, act as a thermalization mechanism. In other experiments, superfluidity of polaritons has been observed [1, 32].

In recent experiments by our group, photon Bose–Einstein condensation is achieved in a dye-solution-filled optical microresonator [26, 28, 29]. Thermalization of the photon gas with the dye is achieved by repeated absorption/emission cycles. For such systems it is known that frequent collisions ($\sim$ 10 fs time scale) between solvent and dye molecules causes rapid transverse decoherence at room temperature, so that the condition of strong light–matter coupling is not met [3, 54]. The distance between the two spherically curved resonator mirrors is in the micrometer region, which causes a large frequency spacing between the longitudinal resonator modes. The latter is of the order of the emission width of the dye molecules. In combination with an intracavity modification of the spontaneous emission, limiting the emission to small volume modes (low transversal excitation), a regime is reached where, to good approximation, the resonator is populated only by photons of a single longitudinal mode number, see Fig. 6.1a. The longitudinal modal quantum number is frozen out and the photon gas effectively becomes two-dimensional. As is indicated in Fig. 6.1b, the photon dispersion relation acquires quadratic, i.e. particle-like, character with the frequency of the transverse TEM_{00} mode acting as a low frequency cutoff frequency. Furthermore, a harmonic trapping potential for the photon gas is induced by the mirrors' curvature. Thermal equilibrium of the photon gas with its environment (at room temperature) is achieved as the photons are repeatedly absorbed and emitted by the dye molecules. The photon frequencies will accumulate within a spectral range of order $\sim k_B T/\hbar$ above the low frequency cutoff. Other than in a blackbody radiator, the thermalization process allows for independent adjustment of temperature and photon number. This becomes clear by noting that

(a)

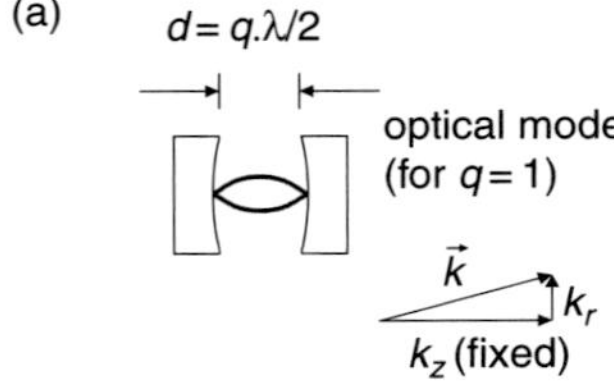

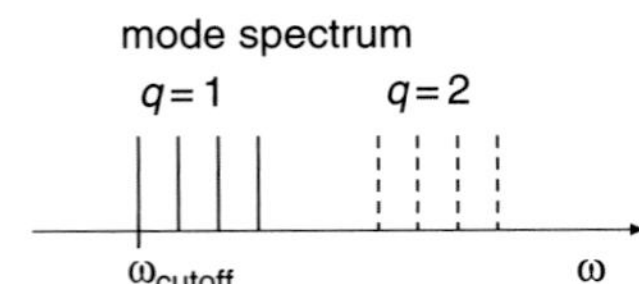

(b)

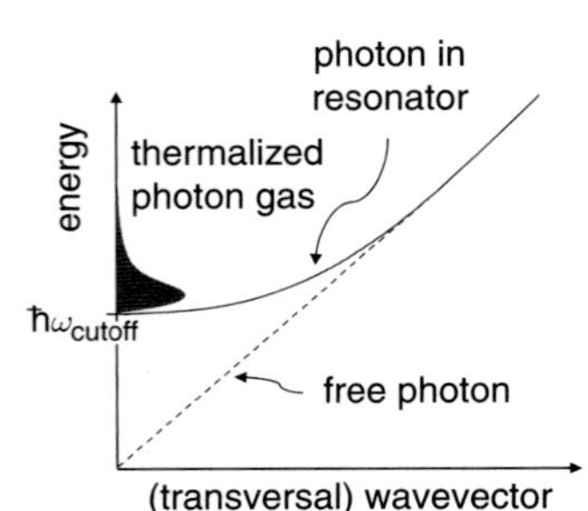

Fig. 6.1 (a) Scheme of optical resonator (top) and the cavity modes (bottom) for the case of a spacing between resonator mirrors of half an optical wavelength of the lowest optical mode. The resonator is filled with dye solution. The photon gas thermalizes to the temperature of the dye solution. (b) Photon dispersion in resonator (solid line) and the dispersion of a free photon (dashed line).

the energy of fluorescence photons, which takes values above the cutoff frequency of the resonator in energy units ($\hbar\omega_{\text{cutoff}} \simeq 2.1\,\text{eV}$), is far above the thermal energy, $k_{\text{B}}T \simeq 1/40\,\text{eV}$. Thus, purely thermal excitation of (optical) photons is negligible. Instead, the number of photons will be determined by the strength of the optical pumping. One can show that the photon gas confined in the resonator is formally equivalent to a harmonically trapped two-dimensional gas of massive bosons with effective mass $m_{\text{eff}} = \hbar\omega_{\text{cutoff}}n_0^2/c^2$, where n_0 denotes the refractive index of the medium and c the vacuum speed of light. For such a system it is well known that a Bose–Einstein condensate exists at a finite temperature [4, 43]. In recent experiments, we have observed both thermalization of the photon gas in the dye-filled microcavity system [29] as well as Bose–Einstein condensation [28]. So far, the observed properties of photon Bose–Einstein condensates in many respects resemble those of atomic gases, while the approximately 10 orders of magnitude smaller effective photon mass allows for transition temperatures in the room temperature regime.

Bose–Einstein condensation and superfluidity are two closely related phenomena [16]. While the former is connected to equilibrium properties, the latter deals with transport properties. Bose–Einstein condensation is, in principle, possible with an ideal gas, while the presence of superfluidity requires interparticle interactions. In future, it remains to be verified experimentally that the photon Bose–Einstein condensate also exhibits superfluidity.

In the following, Section 6.2 describes the fluorescence induced thermalization mechanism of the photon gas and Section 6.3 the statistical theory of the trapped photon gas. Further, Section 6.4 reviews experiments on the thermalization process and Section 6.5, corresponding results on Bose–Einstein condensation. Finally, Section 6.6 gives conclusions.

6.2 Fluorescence induced thermalization

6.2.1 Kennard–Stepanov theory of dye spectra

Early experimental work has shown that spectra of dye molecules in liquid solution show several universal properties [13, 33, 47, 48]. These are, for example, the mirror rule which states that the spectral absorption profile is a mirror image of the fluorescence profile; the Stokes rule, stating that the spectral centroid of fluorescence occurs at a higher wavelength than that of absorption; and Kasha's rule [20], which expresses that the fluorescence does not depend on the wavelength of the exciting light. These common properties, which to good accuracy are fulfilled in many dye species, suggest that absorption and fluorescence in such systems follow a general mechanism. Most of these properties can be understood as being the consequence of a collisionally induced thermalization mechanism [33, 48]. For a corresponding model, consider an idealized dye molecule with an electronic ground state S_0 and an electronically excited state S_1, each of which are subject to additional rovibronic level splitting, as shown in Fig. 6.2a. The vibrational and rotational state of a dye molecule in liquid solution is permanently altered by collisions

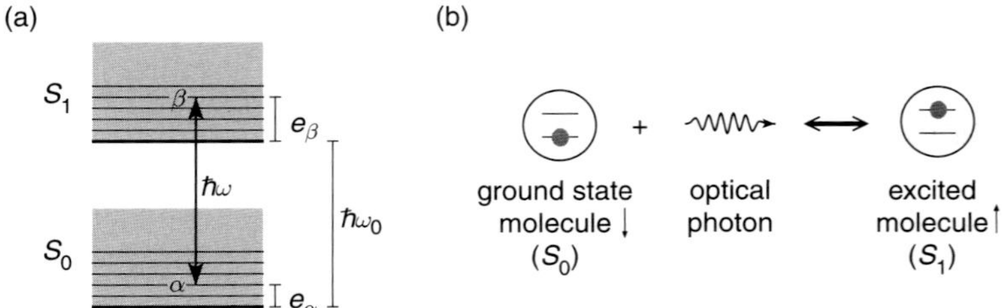

Fig. 6.2 (a) Jablonski diagram of a dye molecule with two electronic levels. Both lower and upper electronic states, S_0 and S_1, are split into a rovibronic substructure. (b) The process of absorption and fluorescence of a photon can be seen as a photochemical reaction.

with solvent molecules, e.g. on a femtosecond time scale at room temperature. The frequent collisions lead to a thermalization of the rovibrational state within a sub-picosecond time scale, which is much faster than the electronic lifetime of the excited state, being typically of the order of nanoseconds. When a photon is absorbed, the dye molecule is likely to be transferred to a highly rovibronically excited substate of the S_1 manifold, but the excessive rovibronic energy will quickly be dissipated into the solvent bath. The fluorescence photon will be emitted from a dye state that is in thermal equilibrium with the solvent bath, with typically lower rovibrational quantum number. This thermalization process, which occurs both in the electronically excited and in the ground state level, explains why fluorescence can be dissipative (Stokes shift) and why there is typically no correlation between the wavelength of the absorbed and the emitted photon (Kasha's rule).

Thermalization of the rovibronic degrees of freedom has another, though closely related, consequence: the Einstein coefficients of absorption and emission at a certain photon energy $\hbar\omega$ are connected by the Boltzmann factor of that energy. This relation is known as the Kennard–Stepanov law, and can be written in the form

$$\frac{B_{21}(\omega)}{B_{12}(\omega)} = \frac{w_\downarrow}{w_\uparrow} e^{-\frac{\hbar(\omega-\omega_0)}{k_B T}}, \tag{6.1}$$

where $B_{12}(\omega)$ and $B_{21}(\omega)$ are the Einstein coefficients of absorption and stimulated emission respectively, ω_0 is the frequency of the zero-phonon line of the dye, and $w_{\downarrow,\uparrow}$ are statistical weights related to the rovibronic density of states, which will be defined subsequently in this text. We note that the Kennard–Stepanov relation can also be stated in terms of $A(\omega)/B_{12}(\omega)$, where $A(\omega)$ is the Einstein coefficient for spontaneous emission—with the difference to the above definition being essentially the density of states. This relation was discovered at the beginning of the last century [23, 24], and has been "rediscovered" several times. A short historical outline can be found in [45]. Both theoretical and experimental investigations can be found in the literature [22, 39, 45, 50, 52].

In the following, we give a short derivation of the Kennard–Stepanov relation. As described above, we model the dye molecule by an electronic two-level system with levels S_0 and S_1, also denoted by ↓ and ↑, each of which is subject to additional rovibronic level splitting. It is important to note that the Einstein coefficients of such a medium, at a given frequency ω, are an average over all pairs of individual rovibronic substates (α, β), with $\alpha \in S_0$, $\beta \in S_1$ that match the transition frequency:

$$e_\alpha + \hbar\omega = \hbar\omega_0 + e_\beta. \tag{6.2}$$

The latter equation expresses energy conservation, see also the Jablonski diagram of Fig. 6.2a. We assume that the population of rovibronic states in both lower and upper electronic states is fully thermalized from frequent collisions with solvent molecules. This assumption will be valid as long as the radiative lifetime of the electronically excited state S_1 remains clearly longer than the thermalization time. We thus expect that the corresponding substates within the lower and upper electronic manifolds will be occupied with a probability given by the Boltzmann factors

$$p_\alpha = e^{-\frac{e_\alpha}{k_B T}}/w_\downarrow \qquad \text{and} \qquad p_\beta = e^{-\frac{e_\beta}{k_B T}}/w_\uparrow, \tag{6.3}$$

with the normalization factors

$$w_\downarrow = \sum_{\alpha \in S_0} e^{-\frac{e_\alpha}{k_B T}} \qquad \text{and} \qquad w_\uparrow = \sum_{\beta \in S_1} e^{-\frac{e_\beta}{k_B T}}. \tag{6.4}$$

From Eqs. (6.2) and (6.3) one immediately obtains

$$p_\beta = \frac{w_\downarrow}{w_\uparrow} e^{-\frac{\hbar(\omega-\omega_0)}{k_B T}} p_\alpha. \tag{6.5}$$

We can now write

$$\frac{B_{21}(\omega)}{B_{12}(\omega)} = \frac{\sum_{(\alpha,\beta)} p_\beta \, B(\beta \to \alpha)}{\sum_{(\alpha,\beta)} p_\alpha \, B(\alpha \to \beta)}, \tag{6.6}$$

where we have introduced the Einstein coefficients $B(\alpha \to \beta)$ for transitions between the corresponding individual rovibronic states. Applying $B(\alpha \to \beta) = B(\beta \to \alpha)$ and using in addition (6.5), we finally obtain

$$\begin{aligned} \frac{B_{21}(\omega)}{B_{12}(\omega)} &= \frac{w_\downarrow}{w_\uparrow} e^{-\frac{\hbar(\omega-\omega_0)}{k_B T}} \frac{\sum_{(\alpha,\beta)} p_\alpha \, B(\alpha \to \beta)}{\sum_{(\alpha,\beta)} p_\alpha \, B(\alpha \to \beta)} \\ &= \frac{w_\downarrow}{w_\uparrow} e^{-\frac{\hbar(\omega-\omega_0)}{k_B T}}, \end{aligned} \tag{6.7}$$

which is the Kennard–Stepanov law of Eq. (6.1). As described above, the Kennard–Stepanov relation is experimentally well established. However, many dye species show smaller or even larger deviations. For a discussion of these issues, see Section 6.2.5.

6.2.2 Chemical equilibrium between photons and molecules

Molecules in the electronic ground state can be transferred to the upper electronic level by the absorption of a photon, while photon emission leads to de-excitation. These processes can be seen as a photochemical reaction of the type

$$\gamma + \downarrow \leftrightharpoons \uparrow, \tag{6.8}$$

see also the illustration in Fig. 6.2b. Here γ stands for a photon, $\downarrow$ for a molecule in the electronic ground state, and $\uparrow$ for a molecule in the electronically excited state. In chemical equilibrium the corresponding chemical potentials, with $\mu_\downarrow$ ($\mu_\uparrow$) denoting the chemical potential of a ground (excited) state molecule and μ_γ indicating the photon chemical potential, satisfy the relation

$$\mu_\gamma + \mu_\downarrow = \mu_\uparrow. \tag{6.9}$$

If we express this relation in terms of the photon fugacity z, the equation reads

$$z = e^{\frac{\mu_\gamma}{k_B T}} = e^{\frac{\mu_\uparrow}{k_B T}} / e^{\frac{\mu_\downarrow}{k_B T}}. \tag{6.10}$$

Next, let the partition function of a dye molecule be

$$\mathcal{F} = w_\downarrow \, e^{\frac{\mu_\downarrow}{k_B T}} + w_\uparrow \, e^{-\frac{\hbar\omega_0 - \mu_\uparrow}{k_B T}}, \tag{6.11}$$

where $w_{\downarrow,\uparrow}$ are the statistical weights of Eq. (6.4). Further, ω_0 is the frequency of the zero-phonon line of the dye. We can then identify

$$w_\uparrow \, e^{-\frac{\hbar\omega_0 - \mu_\uparrow}{k_B T}} / \mathcal{F} = \frac{\rho_\uparrow}{\rho}, \tag{6.12}$$

$$w_\downarrow \, e^{\frac{\mu_\downarrow}{k_B T}} / \mathcal{F} = \frac{\rho_\downarrow}{\rho}, \tag{6.13}$$

as the probability of finding a molecule in the excited (ground) state, where $\rho_\uparrow$ ($\rho_\downarrow$) is the spatial density of the excited (ground) state dye molecules respectively, and ρ is the total dye molecular density. Using Eq. (6.10), we find that the photon chemical potential is determined by the relation

$$z = e^{\frac{\mu_\gamma}{k_B T}} = \frac{w_\downarrow}{w_\uparrow} \frac{\rho_\uparrow}{\rho_\downarrow} e^{\frac{\hbar\omega_0}{k_B T}}. \tag{6.14}$$

In chemical equilibrium, the photon chemical potential is thus determined by the excitation ratio $\rho_\uparrow/\rho_\downarrow$, which corresponds to the relative number of dye molecules in the electronically excited state [26, 27].

6.2.3 Thermal equilibrium and Markov processes

To begin discussion of the thermalization process, let us characterize thermal equilibrium in terms of a Markov process. In general, the time evolution of a physical system coupled to a heat bath can be regarded as a random walk in configuration space [34, 40]. Such a random walk is fully characterized by the present state of the system and fixed transition rates between configurations (process "without memory"). Not all sets of transition rates are meaningful physically. A necessary requirement is that for a sufficiently long time evolution of the system, each configuration has to occur with the statistical weight of its Boltzmann factor. In the following, we derive a condition from which one can easily decide whether a given set of transition rates leads to thermal equilibrium. In the next section this will then be applied to the fluorescence induced thermalization process of a photon gas.

Let K denote a certain state of a physical system (for a photon gas this will be given by a set of mode occupation numbers). Furthermore, $p_K(t)$ denotes the probability of finding state K at time t. The temporal evolution of $p_K(t)$ is given by the master equation

$$p_K(t+1) - p_K(t) = \sum_{K'} p_{K'}(t) R(K' \to K) - \sum_{K'} p_K(t) R(K \to K'), \quad (6.15)$$

where the coefficients $R(K \to K')$ denote the transition rates between configurations. We are interested in transition rates that asymptotically bring the system into thermal equilibrium, i.e. which yield $p_K(t \to \infty) = \exp(-E_K/k_B T)/Z$, with Z as the partition function. The asymptotic master equation for this case is given by

$$0 = \sum_{K'} \exp(-E_{K'}/k_B T) R(K' \to K) - \sum_{K'} \exp(-E_K/k_B T) R(K \to K'). \quad (6.16)$$

Equation (6.16) has many solutions. One solution stands out by showing no net probability flow between two given states (detailed balance):

$$0 = \exp(-E_{K'}/k_B T) R(K' \to K) - \exp(-E_K/k_B T) R(K \to K') \;\forall K, K', \quad (6.17)$$

or

$$\frac{R(K \to K')}{R(K' \to K)} = \exp(-\Delta E/k_B T) \quad \forall K, K', \quad (6.18)$$

with an energy difference $\Delta E = E_{K'} - E_K$. If the transition rates $R(K \to K')$ fulfill Eq. (6.18), this gives a sufficient condition for the Markov process to drive the system into thermal equilibrium.

6.2.4 Light–matter thermalization process

To begin with, consider an experimental scenario, as depicted in Fig. 6.3, consisting of a macroscopic box with reflecting walls, that is filled with a dye solution. Light, emitted e.g. by a spectrally narrow optical source, is irradiated at point A. What is the spectrum of the radiation leaving the dye container at point B? If the optical density of the dye is sufficiently large, multiple absorption–emission fluorescence cycles will occur. One then finds experimentally that the spectrum at point B is red shifted with respect to that obtained for a single fluorescence event. In the context of fluorescence spectroscopy, this effect is known as the inner filter effect [33, 48]. Interestingly, this red shift of the fluorescence can be understood in terms of a partial thermalization process. By fluorescence and reabsorption, the photons exchange energy with the dye solution. This energy exchange establishes a thermal contact between the light and the dye medium. In principle, one would expect that the state of the light field relaxes asymptotically towards a thermal state at the temperature of the dye solution, i.e. after many complete cycles the spectrum of blackbody radiation at room temperature would be obtained. However, owing to the succesive red shift, along with the limited spectral bandwith of the dye,

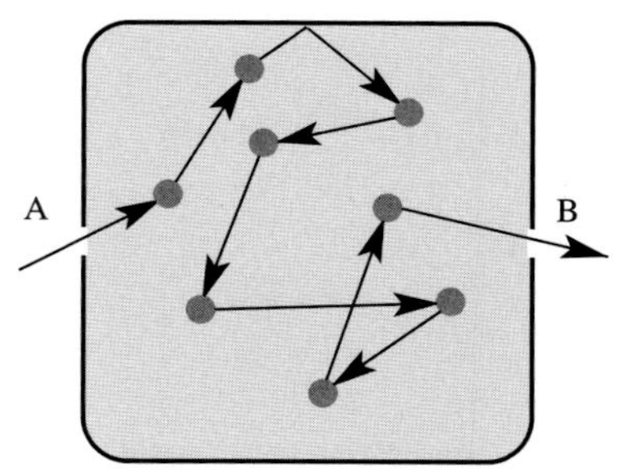

Fig. 6.3 Multiple absorption–fluorescence cycles in a macroscopic dye-filled photon box. Asymptotically, the photons entering at point A would be transferred to blackbody radiation at the temperature of the environment. However, without a low-frequency cutoff the thermalization process remains incomplete, as photons leave the absorption bandwidth of the dye.

the reabsorption probability rapidly decreases, and the dye solution gradually becomes transparent. The thermal contact soon breaks down and the process remains incomplete. Such an incomplete thermalization occurs e.g. in some solar light concentrators [41].

Regarding photon number conservation, this, on the other hand, is not desirable for transferring irradiated light into blackbody radiation. As soon as the photon energy became comparable to thermal energy, one would lose the ability to experimentally tune the photon number. In equilibrium, the photon number would then readjust following the available thermal energy, independently of the optical pumping power irradiated at point A (provided that the pumping does not significantly heat up the dye solution). However, in a microcavity environment, as used in our experiment, both a full thermalization and an experimentally tunable photon number can be achieved. This becomes clear by noting that the low-frequency cutoff, imposed by the resonator, prevents a succesive red shift of the photon gas. Thermalization by repeated absorption–emission cycles can now proceed until the photon gas is fully thermalized, i.e. thermal contact to the dye solution is not lost. Owing to the low-frequency cutoff, the photon energy can remain far above the thermal energy. With $\hbar\omega_{\text{cutoff}} \simeq 2.1\,\text{eV} \gg 1/40\,\text{eV} \simeq k_{\text{B}}T$ this limit is fulfilled in our experiment. In this situation, no photons are created or destroyed on average. Here, the thermalization process conserves the photon number, with the latter being determined by the strength of the optical pumping.

In the remainder of this section, we present a more rigorous treatment of light–matter thermalization [27, 29]. The thermalization process is considered as a random walk in the configuration space of all allowed light field states. Here a state K is given by the cavity mode occupation numbers $K = (n_0^K, n_1^K, n_2^K, \ldots)$. The mode occupation numbers are permanently altered by photon absorption and emission processes. In first-order perturbation theory, the rates (per volume) for absorption and emission of one photon in mode i at cavity position $\mathbf{r}$, denoted by $R_{12}^{K,i}(\mathbf{r})$ and $R_{21}^{K,i}(\mathbf{r})$, have the form

$$R_{12}^{K,i}(\mathbf{r}) = B_{12}(\omega_i)\, u^i(\mathbf{r})\, \rho_\downarrow\, n_i^K \tag{6.19}$$

$$R_{21}^{K,i}(\mathbf{r}) = B_{21}(\omega_i)\, u^i(\mathbf{r})\, \rho_\uparrow\, (n_i^K + 1), \tag{6.20}$$

where $u^i(\mathbf{r})$ is the spectral energy density of one photon in mode i. We assume that the number of dye molecules is sufficiently large, that photon absorption and emission leave the densities of ground state ($\rho_\downarrow$) and excited state ($\rho_\uparrow$) molecules unchanged, i.e. $\rho_{\uparrow,\downarrow}$ can be treated as fixed parameters. This assumption corresponds to a grandcanonical limit, see also Section 6.3.3.

Suppose that a state K' emerges from state K by the absorption of a photon in mode i, with $n_i^{K'} = n_i^K - 1$, and accordingly, that K emerges from K' by an emission process into this mode. The corresponding rates are $R_{12}^{K,i}(\mathbf{r})$ and $R_{21}^{K',i}(\mathbf{r})$, and with Eqs. (6.19) and (6.20), their ratio is given by $R_{12}^{K,i}(\mathbf{r})/R_{21}^{K',i}(\mathbf{r}) = B_{12}(\omega_i)\,\rho_\downarrow/B_{21}(\omega_i)\,\rho_\uparrow$. In Section 6.2.3 it was shown that the random walk given by the transition rates of Eq. (6.19) and (6.20) will lead to thermal equilibrium, if the ratio $R_{12}^{K,i}(\mathbf{r})/R_{21}^{K',i}(\mathbf{r})$ is given by the Boltzmann factor of the energy difference between K and K',

$$\frac{B_{12}(\omega_i)}{B_{21}(\omega_i)} \frac{\rho_\downarrow}{\rho_\uparrow} \stackrel{!}{=} e^{\frac{\hbar\omega_i - \mu_\gamma}{k_B T}}, \tag{6.21}$$

see also [25, 27, 44]. If we now apply the Kennard–Stepanov relation, Eq. (6.1), and assume chemical equilibrium, Eq. (6.14), the detailed balance condition of Eq. (6.21) is indeed verified. This means that multiple absorption–emission cycles drive the photon gas into thermal equilibrium with the dye solution at temperature T, and with a photon chemical potential μ_γ determined by the molecular excitation ratio. The average occupation number $\bar{n}_i$ of mode i can be determined by balancing the average absorption and emission rates at a given cavity position. As expected, this gives a Bose–Einstein distribution for the average occupation number, with $\bar{n}_i = \left(\exp\left[(\hbar\omega_i - \mu_\gamma)/k_B T\right] - 1\right)^{-1}$.

6.2.5 Spectral temperature

From the Kennard–Stepanov theory of dye spectra, as presented in Section 6.2.1, one expects that the spectra of absorption and emission are interlinked by a Boltzmann factor. In general, this law is well confirmed empirically. However, it is known that many real dyes show more or less pronounced deviations. Before choosing a particular dye medium, it is thus necessary to check whether it fulfills the Kennard–Stepanov relation. For this it proves helpful to introduce a spectral temperature, which is derived using a formal solution of Eq. (6.1). Consider the Kennard–Stepanov relation for two different frequencies ω and $\omega + \delta\omega$. From Eq. (6.1), we obtain

$$\frac{B_{21}(\omega)\, B_{12}(\omega + \delta\omega)}{B_{12}(\omega)\, B_{21}(\omega + \delta\omega)} = e^{\frac{\hbar\delta\omega}{k_B T}}. \tag{6.22}$$

We can now solve this equation for the frequency-dependent spectral temperature, $T_{\text{spec}}(\omega)$,

$$T_{\text{spec}}(\omega) = \frac{\hbar\delta\omega}{k_B \ln\left(\frac{B_{21}(\omega)\, B_{12}(\omega+\delta\omega)}{B_{12}(\omega)\, B_{21}(\omega+\delta\omega)}\right)} \tag{6.23}$$

$$\stackrel{\delta\omega\to 0}{=} \frac{\hbar}{k_B \frac{\partial}{\partial\omega} \ln \frac{B_{12}(\omega)}{B_{21}(\omega)}}, \tag{6.24}$$

and eventually perform the limit $\delta\omega \to 0$. Thus, if the spectral profiles of absorption and emission are known, one can easily calculate the corresponding spectral temperature. We note that the particular form of T_{spec} given by Eqs. (6.23) and (6.24) has the advantage of being independent of the statistical weights $w_{\downarrow,\uparrow}$ and the frequency of the zero-phonon line ω_0, which might not be known in all cases. Moreover, in this form it is sufficient to know the coefficients $B_{12,21}(\omega)$ only up to a proportionality factor.

The Kennard–Stepanov relation holds for a given dye, if its spectral temperature coincides with the thermodynamic temperature of the solution, $T_{\text{spec}}(\omega) = T$, independent of the frequency ω. Figure 6.4 shows the spectral temperatures, here as a function of wavelength, for eight different dye

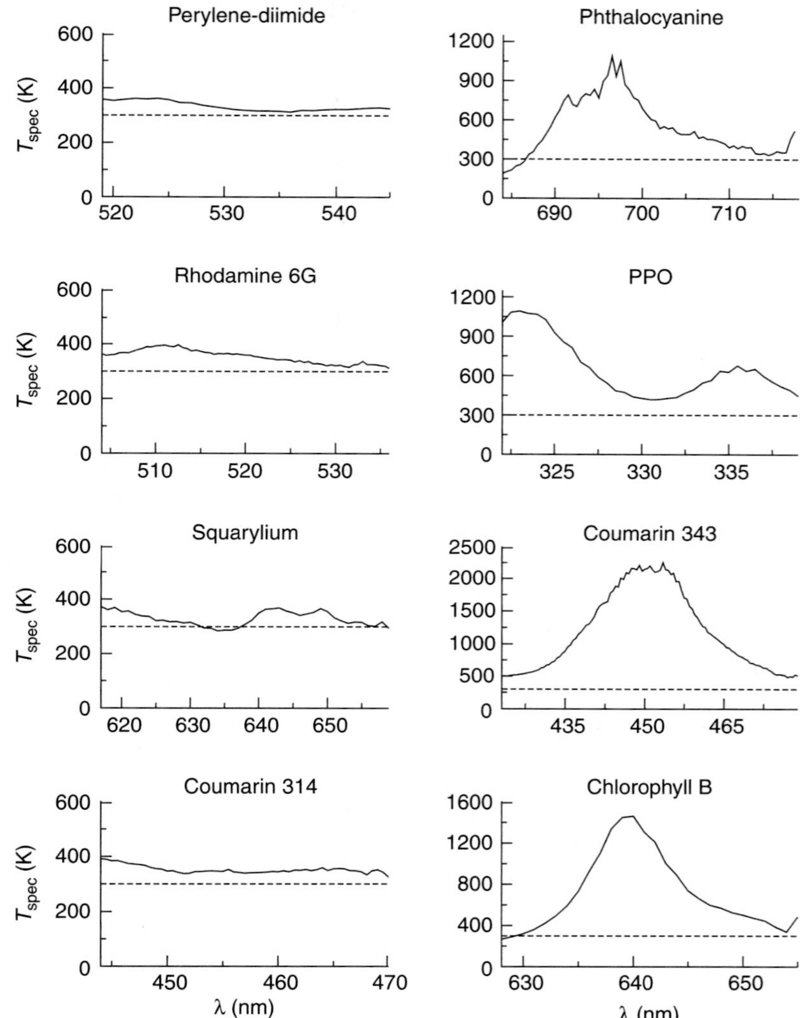

Fig. 6.4 Spectral temperatures $T_{\text{spec}}(\lambda)$ for different dyes at room temperature, derived from the spectra of the database of Ref. [14]. This numerical determination is most precise in the Stokes region, where both absorption and fluorescence are sufficiently strong and thus most precisely known. We nevertheless estimate an error of the given spectral temperatures of at least 15 %. The corresponding quantum efficiencies are: perylene-diimide: $\Phi = 0.97$, rhodamine 6G: $\Phi = 0.95$, squarylium: $\Phi = 0.65$, coumarin 314: $\Phi = 0.68$, phthalocyanine: $\Phi = 0.60$, PPO: $\Phi \approx 1$, coumarin 343: $\Phi = 0.63$, chlorophyll B: $\Phi = 0.117$.

species. They were calculated with Eq. (6.23) using the corresponding spectra of the database of Ref. [14]. The dyes on the left-hand side of the figure fulfill the condition $T_{\text{spec}}(\omega) \simeq T$ to good approximation, while the dyes on the right-hand side show significant deviations. It is noteworthy that the spectral temperature here has a tendency to be higher than the ambient temperature, see also [31, 52]. In explanation for the observed deviations, one mainly finds two lines of reasoning in the literature, either based on incomplete rovibronic thermalization or on inhomogeneous broadening [37, 38, 46, 52]. Our experiment uses either perylene-dimide or rhodamine 6G dye, for both of which the quantum efficiency is 95% or above, and the Kennard–Stepanov relation is well fulfilled.

6.3 Thermodynamics of the two-dimensional photon gas

Formally, the photon gas confined in the resonator is equivalent to a two-dimensional gas of massive bosonic particles. This can be seen by investigating the energy–momentum relation. We will derive the equilibrium properties of such a gas, and discuss the expected condensate fluctuations.

6.3.1 Cavity photon dispersion

We begin by expressing the energy of a photon in the microcavity as a function of longitudinal (i.e. along the optical axis) wavenumber k_z and transverse wavenumber k_r,

$$E = \frac{\hbar c}{n_0}|k| = \frac{\hbar c}{n_0}\sqrt{k_z^2 + k_r^2}\,, \tag{6.25}$$

where n_0 is the refractive index of the medium. Let us denote the spacing between the two cavity mirrors on the optical axis as D_0 ($\simeq 1.46\,\mu$m in our experiment), and the radii of the two curved mirrors, which we assume to be identical, as R ($\simeq 1$ m typically). The boundary conditions generated by the mirrors imply a resonance condition,

$$k_z(r) = q\pi/D(r)\,, \tag{6.26}$$

with q as the longitudinal mode number and $D(r) = D_0 - 2(R - \sqrt{R^2 - r^2})$ as the mirror spacing at a distance r from the optical axis. In a paraxial limit, with $k_r \ll k_z$ and $r \ll R$, we can approximate the photon energy by the expansion

$$E \simeq \frac{m_{\text{eff}}c^2}{n_0^2} + \frac{(\hbar k_r)^2}{2m_{\text{eff}}} + \frac{1}{2}m_{\text{eff}}\Omega^2 r^2, \tag{6.27}$$

with $m_{\text{eff}} = \hbar k_z(0)n_0/c = \hbar\omega_{\text{cutoff}}n_0^2/c^2$ and $\Omega = (c/n_0)/\sqrt{D_0 R/2}$. This yields the dispersion relation of a particle with mass m_{eff} moving in the transverse resonator that is subject to additional harmonic confinement with trapping frequency Ω. If we in addition account for a nonlinear self-interaction of photons that modifies the refractive index following $n_0 \to n_0 + n_2 I(r)$, where n_2 is a nonlinear index of refraction, and $I(r)$ is the optical intensity, we find that the photon energy is shifted by an amount

$$E_{\text{int}} \simeq -m_{\text{eff}}c^2\frac{n_2}{n_0^3}I(r)\,. \tag{6.28}$$

A non-zero value of n_2 can arise e.g. from a Kerr nonlinearity or thermal lensing. Since the confined photons have the quadratic dispersion relation of massive particles, we can define a thermal de Broglie wavelength by $\lambda_{\text{th}} = h/\sqrt{2\pi m_{\text{eff}}k_B T}$, in direct analogy with e.g. a gas of atoms. Essentially, λ_{th} is inversely proportional to the average transversal wavenumber $\sqrt{\langle k_r^2\rangle}$ of the photons, and its value at room temperature is $\simeq 1.58\,\mu$m. Bose–Einstein condensation is expected when the phase space density $n\lambda_{\text{th}}^2$ exceeds a value near unity, where n denotes the two-dimensional number density.

6.3.2 Statistical theory of two-dimensional Bose gas in trap

For an exact determination of the phase transition, a statistical multimode treatment is necessary. The average photon number at a transversal energy $u = \hbar(\omega - \omega_{\text{cutoff}})$ is given by a Bose–Einstein distribution

$$n_{T,\mu}(u) = \frac{g(u)}{e^{\frac{u-\mu}{k_B T}} - 1}, \tag{6.29}$$

where $g(u) = 2(u/\hbar\Omega + 1)$ is the degeneracy factor (which here increases linearly with energy), and the factor 2 accounts for the two possible polarizations. A macroscopic occupation of the ground state mode at $u = 0$ sets in when the particle number N reaches a critical value of $N_c = \sum_{u>0} n_{T,\mu=0}(u)$. We find,

$$N_c = \frac{\pi^2}{3}\left(\frac{k_B T}{\hbar\Omega}\right)^2. \tag{6.30}$$

The typical photon trapping frequency in our setup is $\Omega \simeq 2\pi \cdot 4.1 \times 10^{10}$ Hz. For room temperature, $T = 300$ K, we arrive at $N_c \simeq 77\,000$, which is experimentally feasible. If this critical value is reached, the occupation of transversally excited modes is expected to saturate and the ground mode starts to become macroscopically populated. The possibility of observing a Bose–Einstein condensation at room temperature can be understood from the extremely small effective photon mass $m_{\text{eff}} = \hbar\omega_{\text{cutoff}} n_0^2/c^2$, which in our case corresponds to $2.1\,\text{eV}\, n_0^2/c^2 \simeq 7 \cdot 10^{-36}$ kg. This is 10 orders of magnitude below the mass of alkali atoms, which enhances the phase transition temperature extremely with respect to atomic systems.

6.3.3 Condensate fluctuations

In our experiment, photons are initially brought into the system by pumping the dye medium with a laser. Pumping is maintained throughout the measurement to compensate for photon losses due to unconfined optical modes, finite dye quantum efficiency, and mirror losses. In spite of pumping, the photon gas is expected to both spectrally and spatially relax to thermal equilibrium, provided that a photon scatters several times off a dye molecule before being lost [29]. By optical, pumping, we generate and uphold a reservoir of electronic excitations in the dye medium that can exchange particles with the photon gas. This experimental situation is well described by a grandcanonical statistical ensemble, see Fig. 6.5, in which the photon gas acquires the temperature of the dye solution and the photon chemical potential is directly related to the excitation level in the medium, as given by Eq. (6.14).

The particle exchange with the reservoir of excited state dye molecules is of special relevance for the expected photon statistics and second-order coherence of the condensate. It is well known that there is no ensemble equivalence for Bose–Einstein condensation [15, 30, 57]. In particular, the particle number fluctuations of the condensate depend strongly on the underlying statistical ensemble. For a (micro-)canonical ensemble, which is typically realized in atomic BEC experiments, the particle number distribution of the ground state changes from Bose–Einstein to Poisson-like when the condensation sets in. This is accompanied by a damping of the number fluctuations. On the other

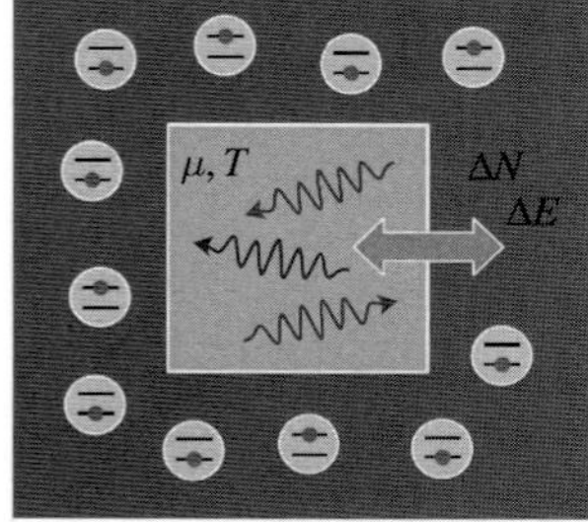

Fig. 6.5 Schematic illustration of the excitation exchange between the photon gas and a reservoir of electronically excited dye molecules. The photon gas can be seen as an open system, in the sense of grandcanonical experimental conditions.

hand, the particle number distribution for a grandcanonical ensemble remains Bose–Einstein-like and the number fluctuations remain of the order of the average occupation number, $\Delta n_0 = \sqrt{\langle n_0^2 \rangle - \bar{n}_0^2} \simeq \bar{n}_0$. This can also be seen in the second-order coherence of the condensate. The zero-delay autocorrelation function $g^{(2)}(0) = \langle n_0(n_0 - 1) \rangle \,/\, \langle n_0 \rangle^2$ is not expected to drop off to $g^{(2)}(0) = 1$, as it does for (micro-)canonical BECs. Instead, a bunching behavior with $g^{(2)}(0) = 2$ is expected, even below the critical temperature [27]. In the future, it will be very interesting to test for such unusually large condensate fluctuations. This regime is not observed in Bose–Einstein condensates of ultra-cold atomic gases, but is likely to occur for photonic BECs.

6.4 Experiments on photon gas thermalization

6.4.1 Experimental setup

We begin by describing measurements on the thermalization of the photon gas in a dye-filled microresonator, which have been carried out at comparatively small photon numbers, i.e. far below the critical photon number. A scheme of the experimental setup used in the experiments of Refs. [28, 29] by our group

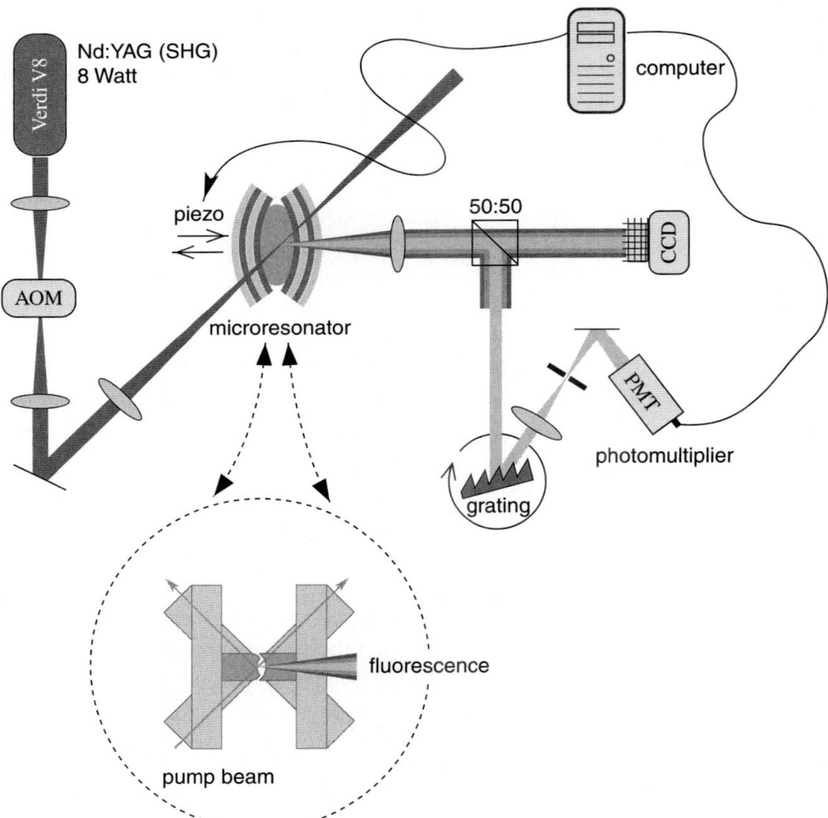

Fig. 6.6 Schematic setup of the microcavity experiment. The dye-filled microresonator is pumped at an angle of 45° to the optical axis. The radiation emerging from the microresonator is examined both spatially and spectrally.

is shown in Fig. 6.6. The optical resonator consists of two highly reflecting mirrors with reflectivity above 99.997% in the wavelength region 500–590 nm. This reflectivity translates to a finesse of >100 000 for the empty cavity. The mirrors are spherically curved with a typical radius of curvature of $R = 1$ m. To allow for a cavity length in the micrometer regime despite the curvature, one of the mirrors is cut to 1×1 mm surface size. Both resonator mirrors are glued onto glass substrates. Additional glass prisms are attached to the side to allow for pumping under an angle of 45°, see Fig. 6.6. The typical (effective) cavity length of $D_0 \simeq 1.45\,\mu$m is determined from the resonator free spectral range, and corresponds to the $q = 7$ longitudinal mode. The used dyes are either rhodamine 6G or perylenedimide (PDI), whose absorption and fluorescence spectra are shown in Fig. 6.7. The dyes have a high quantum efficiency between 95% and 97% [35], and spectral temperatures close to the thermodynamic temperature of the dye solution. We use organic solvent, e.g. methanol and ethylene glycol, and typical dye concentrations of 1.5×10^{-3} mol/l for rhodamine 6G. A careful filtering of the dye solution is necessary before placing it between the mirrors. The microcavity is pumped with a laser beam at a wavelength of 532 nm, derived from a frequency doubled Nd:YAG laser. The light transmitted through one of the cavity mirrors is split into two partial beams by a non-polarizing beamsplitter. One beam is imaged onto a CCD-camera for spatial analysis, while the other is sent into a spectrometer for monitoring of the emitted spectrum. We use different types of spectrometers, both commercially available and self-built devices. Care has to be taken when the spectrometer makes use of an entrance slit. The coupling efficiency to the spectrometer will be different for the various transversally excited cavity modes. This problem can be overcome by placing a diffusing plate in front of the entrance slit. However, a more advantageous method is to remove the entrance slit altogether. This is possible because the microresonator fluorescence can be sufficiently well collimated even without an additional spatial filtering. In Fig. 6.6, a self-built version of such a device is shown. This spectrometer uses a rotating grating and a photomultiplier tube.

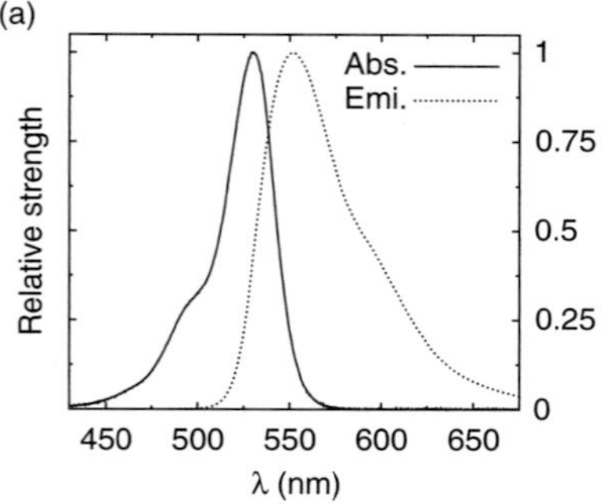

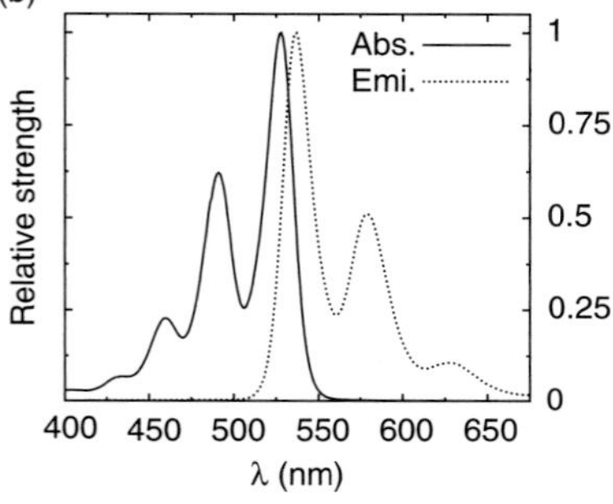

Fig. 6.7 Relative strengths of absorption, $B_{12}(\lambda)/B_{12}^{\max}$, and emission, $B_{21}(\lambda)/B_{21}^{\max}$, for the dyes (a) rhodamine 6G and (b) perylene diimide.

6.4.2 Spectral and spatial intensity distribution

In initial experiments, we tested for the thermalization of the two-dimensional photon gas in the dye-filled resonator [29]. Figure 6.8a shows spectra of light transmitted through one of the cavity mirrors for two different temperatures of the resonator setup (top: room temperature, $T = 300$ K, bottom: $T = 365$ K). The output power for these measurements was (50 ± 5 nW), which corresponds to an average photon number in the cavity of $N_{\mathrm{ph}} = 60 \pm 10$. From this, the chemical potential can be determined by numerical solving $\sum_{u\geq 0} n_{T,\mu}(u) = N_{\mathrm{ph}}$, with $n_{T,\mu}(u)$ given by Eq. (6.29). For the two datasets in Fig. 6.8a, we obtain $\mu/k_{\mathrm{B}}T = -6.76 \pm 0.17$ ($T = 300$ K) and $\mu/k_{\mathrm{B}}T = -7.16 \pm 0.17$ ($T = 365$ K), respectively. Clearly, both measurements are performed far below the phase transition, which for this finite size system is expected to occur once the chemical potential becomes comparable to the trap level spacing (i.e. $\mu \simeq -\hbar\Omega$). Because of $-\mu/k_{\mathrm{B}}T \gg 1$, the term -1 in the denominator of Eq. (6.29) can be neglected and the distribution becomes Boltzmann-like. As can be

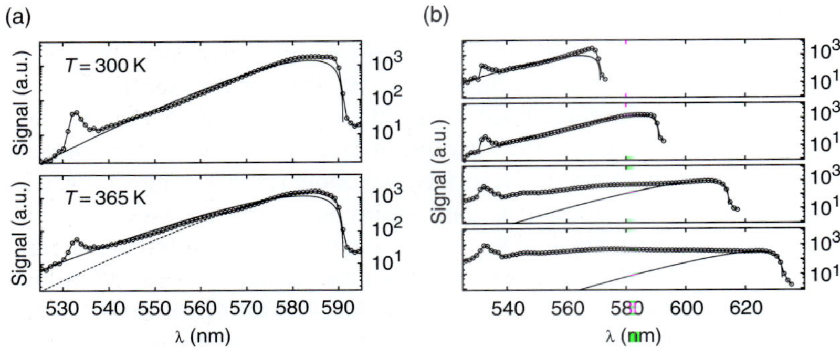

Fig. 6.8 (a) Spectral distribution of the radiation emitted from the microcavity far below the critical photon number at $T = 300\,\text{K}$ and $T = 365\,\text{K}$ (circles). The spectra are in good agreement with the Boltzmann-distributed photon energies (lines). For comparison, a $T = 300\,\text{K}$ Boltzmann distribution is additionally plotted in the bottom graph (dashed line). (Rhodamine 6G in ethylene glycol, concentration $5 \times 10^{-4}\,\text{mol/l}$, mirror curvatures $R_1 = R_2 = 1\,\text{m}$, cavity order $q = 7$) (b) Normalized spectral distribution of the radiation emitted from the microcavity (connected circles) for four different cutoff wavelengths $\lambda_c \simeq \{570, 590, 615, 630\}\,\text{nm}$, from top to bottom. In addition, Boltzmann-distributed photon energies for $T = 300\,\text{K}$ are plotted (lines).

seen from the figure, besides a derivation near 532 nm from residual pump light, the measured spectra are in good agreement with theoretical expectations over a wide spectral range for both temperatures. We interpret this as evidence for the photon gas to be in thermal equilibrium with the dye solution. In another set of measurements, the cutoff wavelength $\lambda_{\text{cutoff}} = 2\pi c/\omega_{\text{cutoff}}$ was varied, which can easily be achieved in our setup by piezo tuning of the cavity length. Corresponding spectra are shown in Fig. 6.8b, along with theory spectra obtained by assuming a Boltzmann distribution of the transversally excited modes. Satisfactory agreement between theory and experiment is only obtained for the two upper spectra with cutoff wavelengths near 570 nm and 590 nm. For the two lower spectra with longer cutoff wavelength, no satisfactory thermalization is present, which can be attributed to the weak dye reabsorption in this spectral regime. These observations illustrate the importance of both emission and reabsorption for the thermalization process.

We have also monitored the spatial distribution of the photon gas at similarly low pumping power. As before, the experimental data can be well explained by assuming a Boltzmann-like population of the transversally excited cavity modes [29]. Furthermore, a spatial concentration of light into the center was observed, which can be seen as a consequence of minimizing the photon energy in the effective trapping potential induced by the mirror curvature. Technical applications of this light concentration effect could include the collection of diffuse solar light to a central spot [41, 53].

6.5 Experiments on photon Bose–Einstein condensation

In subsequent experiments, the dye microresonator was operated at higher pump powers, allowing for photon numbers near and above the critical regime. To avoid population of triplet states and excessive heat deposition, the pump beam in these measurements was acousto-optically chopped to 0.5 μs-long pulses with 8 ms repetition time. Since the pulses are two orders of magnitude longer than the lifetime of the excited dye molecules and four orders of magnitude longer than the lifetime of the photons (average time between emission and reabsorption), the experimental conditions can be considered as quasistatic.

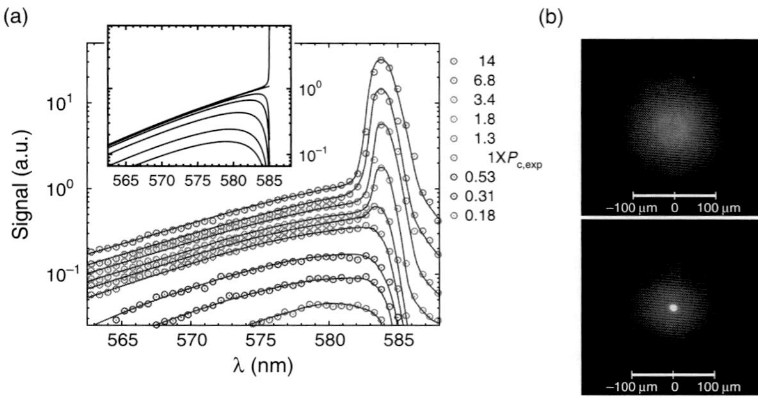

Fig. 6.9 (a) The connected circles give measured spectral intensity distributions for different optical pump powers. The legend gives the intracavity optical power, determining the photon number, in units of $P_{\mathrm{c,exp}} = (1.55 \pm 0.6)\,\mathrm{W}$. A spectrally sharp BEC peak at the position of the cavity cutoff is visible above the critical power. The observed peak width is limited by spectrometer resolution. The inset gives theoretical spectra. (b) Images of the radiation emitted along the cavity axis, below (top) and above (bottom) the critical power. In the latter case, a condensate peak is visible in the center. (See Plate 13 for color image).

Typical experimental spectra of the photon gas are shown in Fig. 6.9a for different optical pumping powers [28]. At low pumping and correspondingly low intracavity optical power, a broad spectral distribution resembling a Boltzmann-like occupation of modes above the cavity cutoff is seen. Near the phase transition, the spectral maximum shifts towards the cutoff. When exceeding the critical photon number, a spectrally sharp peak at the position of the cavity cutoff is visible. The observed spectral width is limited by the resolution of the spectrometer used. The experimental results are in good agreement with theoretical spectra based on a Bose–Einstein-like population of cavity modes, see the inset of the figure. Observed imperfect saturation of the transversally excited modes for even higher photon numbers is attributed to an interaction-induced deformation of the effective trapping potential [51]. At the phase transition, the optical intracavity power is $P_{\mathrm{c,exp}} = (1.55 \pm 0.6)\,\mathrm{W}$, which corresponds to a photon number of $(6.3 \pm 2.4) \times 10^4$. This value is in good agreement with the theoretical prediction of Eq. (6.30). Notably, a similar measurement with PDI gives the same result within experimental uncertainties. Thus, the obtained results seem to be largely independent of which dye is used.

Figure 6.9b shows spatial images of the radiation emitted by the dye microcavity (real image onto a color CCD camera) both below (top) and above (bottom) the critical power. Both pictures show a shift from the yellow spectral regime for the transversally low excited cavity modes located near the trap center to the green for transversally higher excited modes appearing at the outer trap regions. In the lower image, a bright spot is visible in the center with a measured FWHM diameter of $(14 \pm 2)\,\mu\mathrm{m}$. Within the quoted uncertainties, this agrees with the expected diameter of the TEM_{00} mode, $d = 2\sqrt{\hbar \ln 2 / m_{\mathrm{eff}}\Omega} \simeq 12.2\,\mu\mathrm{m}$, yielding clear evidence for a macroscopic population of the transversal ground mode. Further, we observe that not only does the height of the condensate peak increase for higher photon numbers, but also its width [28]. This mode diameter increase suggests a weak repulsive self-interaction mediated by the dye solution. The origin of this is most likely thermal lensing, but in principle it could also be due to a microscopic Kerr nonlinearity in the dye medium. Both effects contribute to the nonlinear index

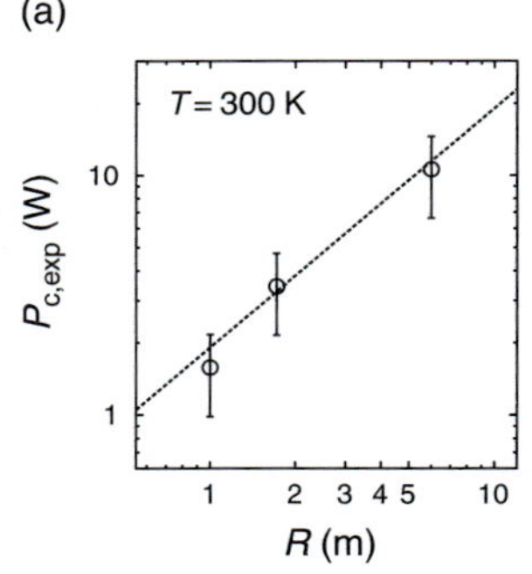

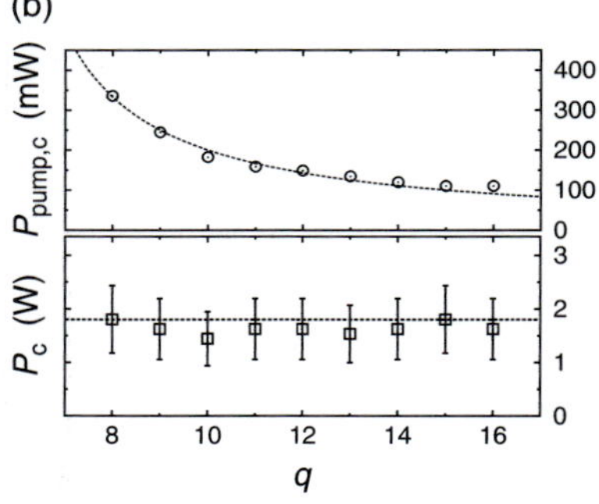

Fig. 6.10 (a) Critical circulating intracavity power as a function of the effective mirror curvature R. (Rhodamine 6G in methanol, concentration 1.5×10^{-3} mol/l, cavity order $q = 7$, pulse duration $0.5\,\mu$s) (b) Pump power at the phase transition (top) and critical intracavity power (bottom), respectively, as a function of the cavity order q. (Rhodamine 6G in methanol, concentration 1.5×10^{-3} mol/l, mirror curvatures $R_1 = R_2 = 1$ m, pulse duration $0.5\,\mu$s).

of refraction of Eq. (6.28), if they are treated on a mean-field level. By comparing the observed increase of mode diameter with numerical solutions of the two-dimensional Gross–Pitaevski equation, we estimate a dimensionless interaction parameter [17] of $\tilde{g} = (7 \pm 3) \times 10^{-4}$. This is significantly smaller than the values in the range $10^{-2} \ldots 10^{-1}$ reported for two-dimensional atomic gas experiments [10, 18], and also below the regime in which Kosterlitz–Thouless physics can be expected to be relevant [17]. An indication of the latter would be the loss of long-range order. Experimentally, when directing the condensate peak through a shearing interferometer, we have not seen any signatures for such a spatially varying phase. Thus, the observed condensation does not seem to deviate from the BEC scenario of an ideal or weakly interacting Bose gas.

In further experiments, we have investigated the influence of resonator geometry on the observed critical particle number. For that we measured the critical circulating power for various mirror radii and different mirror spacings. From Eq. (6.30) we expect that the critical power $P_c = N_c\hbar\omega^2/2\pi q = (\pi^2/12)(k_B T)^2(\omega/\hbar c)R$ should increase linearly with the mirror radius R with no dependence on the cavity order q (at least as long as the two-dimensionality of photon gas holds). Figure 6.10a and the lower graph of Fig. 6.10b give corresponding experimental results, which confirm the theoretical predictions, for both scaling and absolute values. The upper graph of Fig. 6.10b shows the pump power required to achieve criticality as a function of the longitudinal mode number q. Here, a decrease is observed for larger mirror spacings. This becomes clear by noting that the pump power absorption, which is of order 1% of the incident light, increases with the thickness of the dye film. Therefore, less pump power is necessary to reach a given circulating power. The experimental data can be well modeled by assuming that the absorbed pump power at criticality remains constant when increasing q (solid line). Note that a quite contrary scaling was observed in early work on "thresholdless" microlasers, for which an increase of the pump threshold with mirror spacing has been reported [36, 55].

As already discussed, for lower intracavity powers the thermalization process is accompanied by a spatial redistribution of photons towards the trap center. This effect is also observed for higher pump powers, and we expect it to provide a critical photon density in the center of the cavity, even for a displaced pumping spot. We note that this effect has been observed in polariton BEC experiments [5], but is unobserved in lasers. For a corresponding measurement, the pump beam was displaced $50\,\mu$m from the trap center. A set of spatial intensity profiles for different values of cavity cutoff wavelength is shown in Fig. 6.11. By tuning the cavity cutoff one can vary the reabsorption probability, which alters the degree of thermalization. The lowest profile displays results recorded with a cutoff wavelength near $610\,\mu$m. The low dye reabsorption in this wavelength range inhibits efficient thermalization, and spatial accumulation at the trap center. Thus, the maximum of the fluorescence remains at the location of the pump spot (shown by a dashed line). If the cutoff wavelength is decreased, the reabsorption probability increases and the intensity profiles get more symmetrically distributed around the trap center. The top profile shows data recorded for a cutoff near 570 nm. Here, the photon gas seems fully thermalized. The spatial intensity profile shows a cusp at the trap

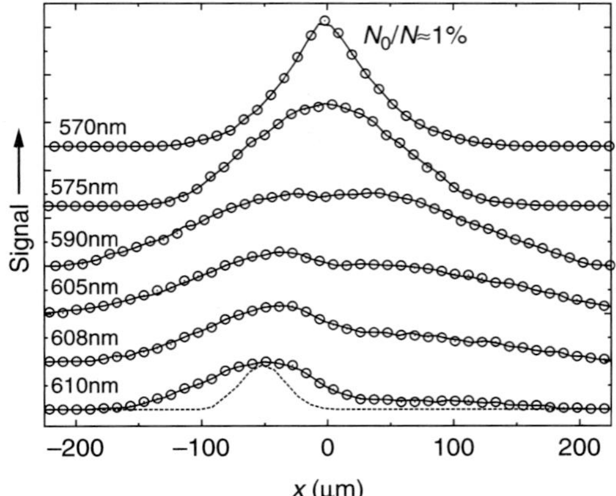

Fig. 6.11 Intensity distribution of the photon gas along an axis intersecting the trap center for different cutoff wavelengths λ_c (indicated on the left-hand side). The pump beam (dashed line) is located outside the trap center and its position as well as its power are kept fixed during the course of the measurement. The top curve shows a photon gas at criticality with a ground state population of $N_0/N \lesssim 1\%$ (Rhodamine 6G in methanol, concentration 1.5×10^{-3} mol/l, cavity order $q = 7$, mirror curvatures $R_1 = R_2 = 1$ m, pulse duration 0.5 μs).

center, which indicates criticality (condensate fraction $\lesssim 1\%$). As the pump beam intensity at the trap center is almost negligible, we conclude that the critical photon density is not directly generated by pumping, but is a consequence of spatial relaxation towards the cavity ground state connected to the thermalization.

6.6 Conclusions

We have investigated thermodynamic properties of a two-dimensional photon gas in a dye-filled optical microcavity. A fluorescence induced thermalization process leads to a thermal photon gas at room temperature with freely adjustable chemical potential. Evidence for a Bose–Einstein condensation of photons was obtained from Bose–Einstein distributed photon energies, including a massively populated ground state mode, the phase transition occurring at the expected critical photon number and exhibiting the predicted dependence on cavity geometry, and a spatial relaxation process leading to a condensation even for a displaced pump spot.

An interesting question is, in what respect does photon Bose–Einstein condensation relate to lasing. In general, the main borderline between a laser and a photonic Bose–Einstein condensate is that lasing is a non-equilibrium phenomenon, while Bose–Einstein condensation occurs in thermal equilibrium. Thermalization of the photon gas above a cavity low-frequency cutoff can be achieved in media, fulfilling the Kennard–Stepanov relation, when recapturing (spontaneous) emission in one of the many transversal cavity modes, and tuning to a regime with strong reabsorption. In such a system, the condition for the emergence of a macroscopically occupied photon mode, which for a laser is usually given by a (small-signal) gain condition in an inverted medium, is replaced by a Bose–Einstein criticality condition, as given in Eq. (6.30) for the case of a harmonically trapped two-dimensional photon gas. In essence, this is equivalent to the condition of the phase space density exceeding a value near unity. The thermalization process will select the cavity ground state to have the biggest statistical weight, which even becomes macroscopic when condensation sets in.

For the future, we plan to investigate condensate properties in more detail. For example, an intriguing question is whether the observed self-interaction is sufficient to cause superfluidity. Also, the coherence properties of the photon condensate should be further explored. Initial interferometric measurements have already delivered a lower bound for the first-order coherence length in the centimeter regime [25, 26]. It will also be important to investigate the second-order coherence of the photon Bose–Einstein condensate. This is of particular interest because of the unusually large intensity fluctuations that are expected for grandcanonical Bose–Einstein condensates.

References

[1] Amo, A., Lefrère, J., Pigeon, S., Adrados, C., Ciuti, C., Carusotto, I., Houdré, R., Giacobino, E., and Bramati, A. Superfluidity of polaritons in semiconductor microcavities. *Nat. Phys.* **5**, 805 (2009).

[2] Anderson, M. H., Ensher, J. R., Matthews, M. R., Wieman, C. E., and Cornell, E. A. Observation of Bose–Einstein condensation in a dilute atomic vapor. *Science*, **269**, 198 (1995).

[3] Angelis, E. De, Martini, F. De, and Mataloni, P. Microcavity quantum superradiance. *J. Optics* B: *Quantum Semiclassical Opt.*, **2**, 149 (2000).

[4] Bagnato, V. and Kleppner, D. Bose–Einstein condensation in low-dimensional traps. *Phys. Rev.* A, **44**, 7439 (1991).

[5] Balili, R., Hartwell, V., Snoke, D., and Pfeiffer, L. Bose–Einstein condensation of microcavity polaritons in a trap. *Science*, **316**, 1007 (2007).

[6] Bolda, E. L., Chiao, R. Y., and Zurek, W. H. Dissipative optical flow in a nonlinear Fabry-Pérot cavity. *Phys. Rev. Lett.* **86**, 416 (2001).

[7] Bradley, C. C., Sackett, C. A., and Hulet, R. G. Bose–Einstein condensation of lithium: Observation of limited condensate number. *Phys. Rev. Lett.* **78**, 985 (1997).

[8] Chiao, R. Y. Bogoliubov dispersion relation and the possibility of superfluidity for weakly interacting photons in a two-dimensional photon fluid. *Phys. Rev.* A, **60**, 4114 (1999).

[9] Chiao, R. Y. Bogoliubov dispersion relation for a 'photon fluid': Is this a superfluid? *Opt. Comm.* **179**, 157 (2000).

[10] Clade, P., Ryu, C., Ramanathan, A., Helmerson, K., and Phillips, W. D. Observation of a 2d Bose gas: From thermal to quasicondensate to superfluid. *Phys. Rev. Lett.* **102**, 170401 (2009).

[11] Davis, K. B., Mewes, M.-O., Andrews, M. R., van Druten, N. J., Durfee, D. M., Kurn, D. M., and Ketterle, W. Bose–Einstein condensation in a gas of sodium atoms. *Phys. Rev. Lett.* **75**, 3969 (1995).

[12] Deng, H., Haug, H., and Yamamoto, Y. Exciton–polariton Bose–Einstein condensation. *Rev. Mod. Phys.* **82**, 1489 (2010).

[13] Drexhage, K. H. Structure and properties of laser dyes. In *Dye Lasers*. Springer-Verlag, Berlin (1990).

[14] Du, H., Fuh, R.-C. A., Li, J., Corkan, L. A., and Lindsey, J. S. PhotochemCAD. A computer-aided design and research tool in photochemistry and photobiology. *Photochem. Photobiol.* **68**, 141 (1998).

[15] Fujiwara, I., Haar, D. Ter, and Wergeland, H. Fluctuations in the population of the ground state of Bose systems. *J. Stat. Phys.* **2**, 329 (1970).

[16] Griffin, A., Snoke, D. W., and Stringari, S. *Bose–Einstein Condensation*. Cambridge University Press (1995).

[17] Hadzibabic, Z. and Dalibard, J. Two-dimensional Bose fluids: An atomic physics perspective. *arXiv:0912.1490* (2009).

[18] Hadzibabic, Z., Krüger, P., Cheneau, M., Battelier, B., and Dalibard, J. Berezinskii–Kosterlitz–Thouless crossover in a trapped atomic gas. *Nature*, **441**, 1111 (2006).

[19] Huang, K. *Statistical Mechanics*. Wiley, New York (1987).

[20] Kasha, M. Characterization of electronic transitions in complex molecules. *Discussions of the Faraday Society*, **9**, 14 (1950).

[21] Kasprzak, J., Richard, M., Kundermann, S., Baas, A., Jeambrun, P., Keeling, J. M. J., Marchetti, F. M., Szymanska, M. H., Andre, R., Staehli, J. L., Savona, V., Littlewood, P. B., Deveaud, B., and Dang, Le Si. Bose–Einstein condensation of exciton polaritons. *Nature*, **443**, 409 (2006).

[22] Kazachenko, L. P. and Stepanov, B. I. Mirror symmetry and the shape of absorption and luminescence bands of complex molecules. *Optika i Spektroskopiya*, **2**, 339 (1957).

[23] Kennard, E. H. On the thermodynamics of fluorescence. *Phys. Rev.* **11**, 29 (1918).

[24] Kennard, E. H. The excitation of fluorescence in fluorescein. *Phys. Rev.* **29**, 466 (1927).

[25] Klaers, J. *Bose-Einstein-Kondensation von paraxialem Licht*. PhD thesis, University of Bonn (2011).

[26] Klaers, J., Schmitt, J., Damm, T., Vewinger, F., and Weitz, M. Bose-einstein condensation of paraxial light. *App. Phys.* B, **105**, 17 (2011).

[27] Klaers, J., Schmitt, J., Damm, T., Vewinger, F., and Weitz, M. Statistical physics of Bose-Einstein-condensed light in a dye microcavity. *Phys. Rev. Lett.* **108**, 160403 (2012).

[28] Klaers, J., Schmitt, J., Vewinger, F., and Weitz, M. Bose–Einstein condensation of photons in an optical microcavity. *Nature*, **468**, 545 (2010).

[29] Klaers, J., Vewinger, F., and Weitz, M. Thermalization of a two-dimensional photonic gas in a 'white-wall' photon box. *Nature Phys.* **6**, 512 (2010).

[30] Kocharovsky, V. V., Kocharovsky, V. V., Holthaus, M., Ooi, C. H. Raymond, Svidzinsky, A. A., Ketterle, W., and Scully, M. O. Fluctuations in ideal and interacting Bose–Einstein condensates: From the laser phase transition analogy to squeezed states and Bogoliubov quasiparticles. *Adv. At. Mol. Opt. Phys.* **53**, 291 (2006).

[31] Kozma, L., Szalay, L., and Hevesi, J. Influence of environment on luminescence of dissolved dye molecules. *Acta Phys. Chem.* **10**, 67 (1964).

[32] Lagoudakis, K. G., Wouters, M., Richard, M., Baas, A., Carusotto, I., André, R., Dang, L. S., and Deveaud-Plédran, B. Quantized vortices in an exciton–polariton condensate. *Nature Phys.* **4**, 706 (2008).

[33] Lakowicz, J. R. *Principles of Fluorescence Spectroscopy*. Kluwer Academic, New York (1999).

[34] Landau, D. P. and Binder, K. *A Guide to Monte-Carlo Simulations in Statistical Physics*. Cambridge Univerity Press, Cambridge (2000).

[35] Magde, D., Wong, R., and Seybold, P. G. Fluorescence quantum yields and their relation to lifetimes of rhodamine 6G and fluorescein in nine solvents: Improved absolute standards for quantum yields. *Photochem. Photobiol.* **75**, 327 (2002).

[36] Martini, F. De and Jacobovitz, G. Anomalous spontaneous-stimulated-decay phase transition and zero-threshold laser action in a microscopic cavity. *Phys. Rev. Lett.* **60**, 1711 (1988).

[37] Mazurenko, Y. T. Broadening of electronic-spectra of complex molecules in a polar medium. *Optika i Spektroskopiya*, **33**, 22 (1972).

[38] Mazurenko, Y. T. Kinetics of luminescence of polar solutions. *Optika i Spektroskopiya*, **36**, 491 (1974).

[39] McCumber, D. E. Einstein relations connecting broadband emission and absorption spectra. *Phys. Rev.* **136**, A954 (1964).

[40] Metropolis, N., Rosenbluth, A., Rosenbluth, M., Teller, A. H., and Teller, E. Equation of state calculations by fast computing machines. *J. Chem. Phys.* **21**, 1087 (1953).

[41] Meyer, T. J. J. and Markvart, T. The chemical potential of light in fluorescent solar collectors. *J. App. Phys.* **105**, 063110 (2009).

[42] Mitchell, M. W., Hancox, C. I., and Chiao, R. Y. Dynamics of atom-mediated photon–photon scattering. *Phys. Rev.* A, **62**, 043819 (2000).

[43] Mullin, W. J. Bose–Einstein condensation in a harmonic potential. *J. Low Temp. Phys.* **106**, 615 (1997).

[44] Norman, G.E., Filinov, V.S., and Mataloni, P. *High Temp.*, **7**, 216 (1969).

[45] Ross, R. Some thermodynamics of photochemical systems. *J. Chem. Phys.* **46**, 4590 (1967).

[46] Sawicki, D. A. and Knox, R. S. Universal relationship between optical emission and absorption of complex systems: An alternative approach. *Phys. Rev.* A, **54**, 4837 (1996).

[47] Schäfer, F. P. Principles of dye laser operation. In *Dye Lasers*. Springer-Verlag, Berlin (1990).

[48] Sharma, A. and Schulman, S. G. *Introduction to Fluorescence Spectroscopy*. Wiley Interscience (1999).

[49] Siegman, A. E. *Lasers*. University Science Books (1986).

[50] Stepanov, B. I. Universal relation between the absorption spectra and luminescence spectra of complex molecules. *Doklady Akademii Nauk SSSR*, **112**, 839 (1957).

[51] Tammuz, N., Smith, R. P., Campbell, R. L. D., Beattie, S., Moulder, S., Dalibard, J., and Hadzibabic, Z. Can a Bose gas be saturated? *Phys. Rev. Lett.* **106**, 230401 (2011).

[52] van Metter, R. L. and Knox, R. S. On the relation between absorption and emission spectra of molecules in solution. *Chem. Phys.* **12** (1976).

[53] van Sark W. et al. Luminescent solar concentrators – A review of recent results. *Optics Express*, **16**, 21773 (2008).

[54] Yokoyama, H. and Brorson, S. D. Rate equation analysis of microcavity lasers. *J. App. Phys.* **66**, 4801 (1989).

[55] Yokoyama, H., Nishi, K., Anan, T., Nambu, Y., Brorson, S. D., Ippen, E. P., and Suzuki, M. Controlling spontaneous emission and threshold-less laser oscillation with optical microcavities. *Opt. Quant. Electr.* **24**, 245 (1992).

[56] Zel'dovich, Y. B. and Levich, E. V. Bose condensation and shock waves in photon spectra. *Sov. Phys. JETP*, **28**, 1287 (1969).

[57] Ziff, R. M., Uhlenbeck, G. E., and Kac, M. Ideal Bose–Einstein gas, revisited. *Phys. Rep.* **32**, 169 (1977).

7 Bose–Einstein condensation of atomic gases

Frédéric Chevy and Jean Dalibard

Discovery of the superfluid transition of liquid helium [1, 2] marked the first achievement of Bose–Einstein condensation in the laboratory, more than a decade after Einstein's prediction for an ideal gas [3, 4]. Together with superconductivity, they offered the first examples of macroscopic quantum phenomena and as such constituted a milestone in the history of Physics. The quest for an understanding of liquid helium superfluidity was the source of major advances in quantum many-body physics, such as the development of techniques inspired from quantum field theory and Landau's phenomenological two-fluid model. The latter was in particular very successful for describing the hydrodynamics of this quantum liquid.

However, interactions between atoms in liquid helium are strong and make the comparison between experiments and *ab initio* theories a difficult task. A striking example is calculation of the critical temperature for superfluid transition, which was determined only recently using Quantum Monte Carlo simulations [5]. By contrast, gaseous Bose–Einstein condensates (BECs), discovered in 1995 after the development of laser cooling and trapping techniques, constitute weakly interacting systems much closer to Einstein's original idea. The condensation temperature is usually close to the prediction for an ideal gas and more generally, gaseous BECs offer the opportunity to test quantitatively the theoretical ideas elaborated over the past fifty years.

Bose–Einstein condensation has been achieved in dilute gaseous systems either with bosonic atoms [6–8], or with molecules made with pairs of fermionic atoms [9–11]. In these dilute systems, the weakness of interactions allows one to adopt a mean-field description in which the many-body wave function $\Psi(\boldsymbol{r}_1, \ldots, \boldsymbol{r}_N)$ is approximated by a factorized state $\varphi(\boldsymbol{r}_1)\ldots\varphi(\boldsymbol{r}_N)$. The macroscopic matter wave $\varphi(\boldsymbol{r})$, which was introduced phenomenologically for liquid helium, provides for atomic gases an accurate description of the microscopic degrees of freedom [12–14]. As a consequence, ultra-cold atoms allow for a large variety of spectacular phenomena, such as interference between independent condensates [15] and long-range phase coherence in an *atom laser* [16].

In this chapter, we present an overview of the specific experimental tools developed in the field of ultra-cold gases to achieve and probe superfluidity in vapors of bosonic atoms. First, we describe the main cooling techniques: the magneto-optical trap, which brings an atomic vapor from room temperature down to the sub-mK range, and the evaporative cooling scheme, which bridges the gap to the superfluid regime. We then proceed to the study of interaction effects. We show that these play a central role in the understanding of both static and dynamic properties of trapped BECs, despite the dilute character of these gases. As a consequence, gaseous BECs and superfluid helium obey the same laws of hydrodynamics and exhibit similar dynamical properties, although their densities differ by several orders of magnitude. The third section is devoted to the coherence properties of gaseous BECs. We show that, by contrast with liquid helium, specific features of ultra-cold gases make them suitable for a direct probing of the quantum coherence associated with Bose–Einstein condensation. Finally, we discuss the possibility of tailoring trapping potentials and realizing low-dimensional systems where one or two directions of motion are frozen.

7.1 Production of a gaseous atomic condensate

Every quantum gas experiment starts with the cooling of a vapor of atoms from room temperature (or even higher) down to the milliKelvin range. The standard tool for this spectacular freezing of atomic motion is laser cooling, which was developed during the 1980s [17–19]. In this section, we present the basic principle of the method and we show how the very same spontaneous emission processes responsible for cooling also set intrinsic limits preventing one from reaching quantum degeneracy. We then discuss how evaporative cooling strategies, based on selective elimination of the most energetic atoms of the gas, overcame this fundamental barrier and led to observation of the first Bose–Einstein condensates.

7.1.1 Laser cooling of atomic vapors

Most laser cooling schemes use the radiative forces exerted on atoms by continuous laser beams, with a frequency that is quasi-resonant with an electronic transition of the species of interest. The simplest scheme, conceptually is *Doppler cooling*, whose basic principle is recalled in Fig. 7.1a, in the case of a one-dimensional system. Two counterpropagating light beams of frequency ω_L are shone onto atoms with a resonance frequency $\omega_A > \omega_L$. For an atom at rest, the radiation pressure forces exerted by the two beams are balanced and the overall force is zero. For a moving atom this balance is broken. Consider for example an atom moving to the right as in Fig. 7.1a; the Doppler effect shifts the apparent frequency of the right beam upwards and that of the left beam downwards. Being closer to resonance, the radiation pressure force from the right beam is stronger than that from the left. The atom thus feels a force opposite to its velocity that damps its motion and which provides the cooling of

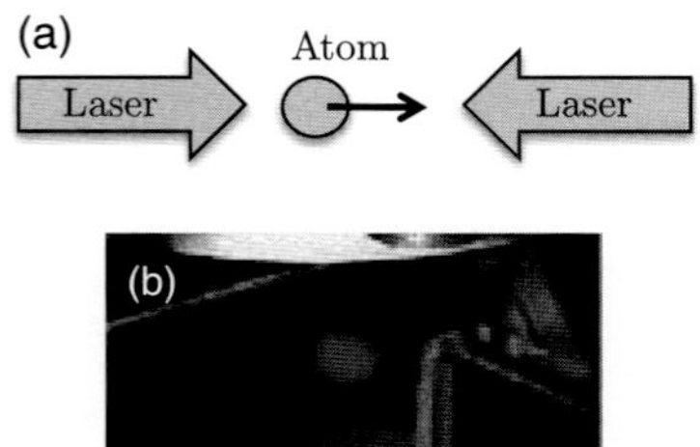

Fig. 7.1 (a) Principle of Doppler cooling in a one-dimensional configuration. The sum of the radiation pressures of two counter-propagating laser beams creates a damping force on a moving atom. (b) Photograph of a lithium magneto-optical trap (photo from Laboratoire Kastler Brossel, ENS Paris). The glowing ball corresponds to $\sim 10^{10}$ atoms trapped at 1 mK. (See Plate 14 for color image).

its translational degree of freedom. Using three pairs of laser beams propagating in independent directions, one can extend this scheme to all three spatial directions, thus creating an *optical molasses* for the atoms. The volume of the optical molasses delimited by intersection of the laser beams is typically a few centimeters cubed.

The equilibrium energy of atoms in an optical molasses results from the balance between the cooling effect that we have just described and the heating associated with the random character of spontaneous emission processes. The Brownian motion of the atoms in the molasses can be characterized by a temperature T whose minimal value is $\hbar\Gamma/2k_B$ for Doppler cooling; here Γ is the natural width of the electronic transition excited by the cooling lasers. This temperature is in the range of several hundred microkelvins for alkali atoms, which are the most frequently used species in these experiments. More subtle cooling processes, like *Sisyphus cooling*, also take place in optical molasses and they can lower the temperature to a few E_R/k_B. Here E_R is the *recoil energy* of an atom when it emits or absorbs a single photon: $E_R = \hbar^2 k_L^2/2m$, where k_L is the wavevector of the laser beams and m the atomic mass. The temperature obtained with Sisyphus cooling is usually in the range 10–100 microkelvins. Note that there also exist subrecoil cooling mechanisms [18], but these are not generally used in quantum gases experiments because of their relatively complex implementation.

7.1.2 The magneto-optical trap

An optical molasses only provides a damping force on the atoms, but it does not trap them. Because of their Brownian motion in the light beams, the atoms eventually leave the region of the optical molasses, in a fraction of a second. This severely limits the number of atoms in the cold gas and its spatial density. To solve this issue, one superimposes onto the laser beams a static quadrupolar magnetic field, with the zero of the field in the center of the molasses, thus creating a *magneto-optical trap* (MOT). Because of the spatially varying Zeeman shift of the atomic levels, the radiation pressure force now depends on position. For a proper choice of polarization of the molasses beams with respect to the local direction of the magnetic field, the radiative pressure not only damps the atomic motion, but also creates a restoring force towards the zero of the magnetic field [20]. The number of atoms at equilibrium in a MOT is in the range 10^8–10^{10}; this depends essentially on the available laser power at the desired wavelength.

For an ideal gas of density n and temperature T, Einstein's criterion for condensation is $n\lambda^3 = \zeta(3/2) \approx 2.6$, where $\lambda = (2\pi\hbar^2/mk_BT)^{1/2}$ is the thermal wavelength and ζ is the Riemann function [21]. This requires "large" densities and/or low temperatures. A typical target for the density of an ultra-cold atomic vapour is 10^{14} cm^{-3}. Above this value, the rate of three-body recombination processes leading to the formation of molecules exceeds 1 s^{-1} and the sample decays before having reached thermal equilibrium. This density is lower than that of liquid helium by 8 orders of magnitude, which brings the degeneracy temperature from the kelvin region for liquid helium down to

0.1–1 microkelvin for atomic gases. Although the magneto-optical trap is a very powerful tool to bring an atomic vapor into the sub-milliKelvin range, it is not suited to directly reach quantum degeneracy. The spatial density in a MOT is indeed limited to values much smaller than the target density of 10^{14} cm^{-3} mentioned above, because of the permanent scattering of photons by the trapped atoms. More precisely a fluorescence photon emitted by a given atom can be reabsorbed by another nearby atom, leading to an effective repulsive force between these two atoms. This laser-induced repulsion limits the spatial density in a MOT to values in the range 10^{10}–10^{12} cm^{-3}, corresponding to a phase space density 10^{-4}–10^{-6} for alkali vapors [24].[1]

[1] Larger phase space densities—though not yet at quantum degeneracy—have been achieved in ytterbium and strontium MOTs, thanks to the much narrower linewidth of the cooling transition [22, 23].

7.1.3 Pure magnetic and pure optical confinements

To circumvent the limit on spatial density imposed in a MOT by light-induced repulsion, and also to further lower the temperature of the gas, it is necessary to use a non-dissipative confinement. Up to now two kinds of traps, magnetic and optical, have been used successfully in the quest for Bose–Einstein condensation. These two families of traps correspond to relatively shallow potential depths, which are insufficient to trap atoms at room temperature. Therefore a first stage of laser cooling is necessary in all quantum gas experiments, in order to capture room temperature atoms and to reduce their energy down to a level where they can be transferred efficiently to a non-dissipative potential.

MAGNETIC TRAPS

The first class of non-dissipative trap is based on a spatially varying magnetic field $\boldsymbol{B}(\boldsymbol{r})$, and can be used for any atom possessing a non-zero magnetic moment $\boldsymbol{\mu}$. In the presence of a magnetic field, the atomic potential energy is $V(\boldsymbol{r}) = -\boldsymbol{\mu} \cdot \boldsymbol{B}(\boldsymbol{r})$. If the motion of the atom is slow enough, the projection $\mu_{\parallel}$ of the magnetic moment along the direction of the magnetic field remains constant. The trapping potential is thus simply $V(\boldsymbol{r}) = -\mu_{\parallel} B(\boldsymbol{r})$. Depending on the sign of $\mu_{\parallel}$, the atom is attracted towards the minima of B ($\mu_{\parallel} < 0$) or towards its maxima ($\mu_{\parallel} > 0$). Because of the structure of Maxwell's equations, only local minima of the modulus of a static magnetic field can exist in a region with no current, which means that only low-field seeking spin states corresponding to $\mu_{\parallel} < 0$ can be trapped by a static magnetic field.

LASER DIPOLE TRAPS

The second class of non-dissipative trap uses a laser beam whose frequency ω_{L} is chosen far from the atomic resonance line [25]. Dissipation processes associated with spontaneous emission are then negligible and only virtual scattering of photons is permitted. The electric–dipole interaction of the atom with the laser electric field $E(\boldsymbol{r})$ gives rise to potential energy $V(\boldsymbol{r}) = -\alpha(\omega_{\mathrm{L}})E^2(\boldsymbol{r})/2$, where α is the dynamical polarizability of the atom. Just like a classical driven oscillator, the sign of α depends on detuning of the laser with respect to the atomic resonance. For red-detuned light ($\omega_{\mathrm{L}} < \omega_{\mathrm{A}}$), α is positive, while it is negative for blue-detuned light ($\omega_{\mathrm{L}} > \omega_{\mathrm{A}}$), a result that can be recovered easily in Thomson's classical model of the atom. For red-detuned light, the potential energy is a minimum at the point(s) where the laser intensity is the highest: a focused laser beam acts as a potential well that keeps the atoms trapped around the focal point. Dipole traps have a large variety of applications, since one can produce laser beams with intensity profiles, hence potential landscapes, of nearly arbitrary shapes. They can for instance be used to engineer periodic potentials, the so-called *optical lattices* (see chapter 17 volume 2 by Chin and Gemelke), or to strongly confine atoms in one or two dimensions to a point where the atomic motion is frozen in these directions, thus realizing a quasi two-dimensional or one-dimensional system (see Section 7.4).

7.1.4 Evaporative cooling to the degenerate regime

Up to now, evaporative cooling is the only way to cool an atomic vapor down to the quantum degenerate regime. Although several implementations exist, the principle of evaporative cooling is always the same: one removes atoms carrying large energy from the trap, so as to decrease the mean energy—hence the temperature—of the remaining particles. This is achieved by truncating the trapping potential at some energy U_0. A key ingredient in the process is the elastic collision rate that fixes the speed at which atoms are removed from the trap, hence the cooling rate dT/dt [26, 27].

Basic evaporative cooling suffers from the fact that as the temperature decreases, fewer particles reach an energy larger than U_0 and the evaporation process slows down. To circumvent this problem, one turns to *forced evaporative cooling*, obtained by continuously decreasing U_0 in order to keep the ratio $\eta = U_0/k_BT$ constant. When the initial collision rate is large enough, one reaches a runaway regime in which the collision rate increases as the gas gets colder and denser. The quantum degenerate regime is reached after an evaporation time corresponding to a few hundred elastic collisions per atom. During the evaporation process the number of atoms is divided by a factor of 100 to 1000. The number of atoms at the condensation point ranges between 10^3 and 10^8, depending on the initial loading of the magnetic trap, on the optimization of the evaporation process, and on the subsequent goal of the experiment.

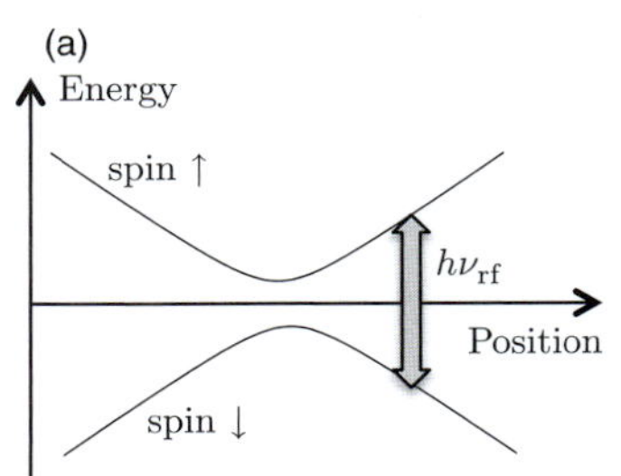

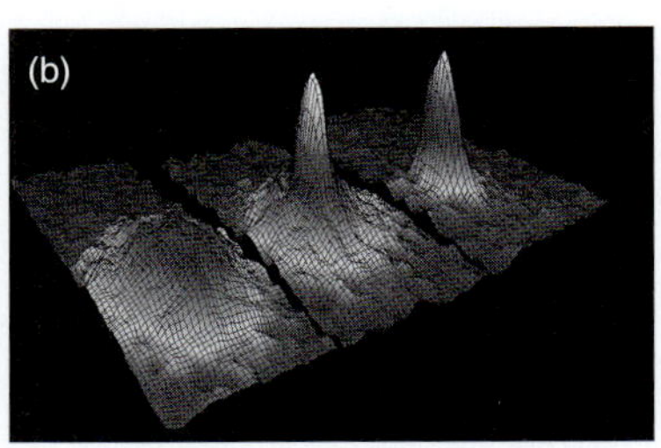

Fig. 7.2 (a) Evaporative cooling in a magnetic trap, using a radio-frequency that flips the magnetic moment of atoms at a given position in the trap. (b) First observation of a gaseous Bose–Einstein condensate (photos: courtesy of Eric Cornell, NIST Boulder). From the left to the right, the three density profiles correspond to decreasing temperatures. The first one is still in the classical regime, where the density distribution is given by the classical Boltzmann law. The last one corresponds to a quasi-pure condensate. (See Plate 15 for color image).

In a magnetic trap, evaporation is performed using a radio-frequency (rf) electromagnetic field of frequency ν_{rf} that flips the atom spin to a non-trapped state. Since the trapping magnetic field $\boldsymbol{B}$ depends on position, the expulsion only takes place at positions $\boldsymbol{r}$ where the resonance condition $h\nu_{rf} = g_L\mu_\parallel B(\boldsymbol{r})$ is satisfied, where g_L is the Landé factor of the relevant internal atomic state (Figure 7.2a). By sweeping ν_{rf} one can thus produce the desired change of U_0 without changing the characteristics of the trap itself.

In a laser dipole trap, all spin states feel the same potential and rf evaporation cannot be used. In this case, one takes advantage of the relation between the depth of the trapping potential and the intensity of the laser light shining on the atoms: forced evaporation is obtained simply by decreasing the laser power. One drawback of this scheme is that the trapping strength also decreases during the evaporation, which diminishes the collision rate that is so crucial for the success of evaporation. The criterion on the initial collision rate for a successful evaporation is therefore more stringent in a dipole trap than in a magnetic trap.[2]

[2]Note that it is also possible to recover a fully efficient evaporation in a dipole trap by refocusing the laser beam on the atoms as the evaporation proceeds, so as to maintain a constant trapping strength as in a magnetic trap.

Evaporative cooling of atoms in a magnetic trap led, in 1995, to the first observation of a Bose–Einstein condensate of rubidium atoms by the group of E. A. Cornell and C. Wieman, in Boulder [6, 28], soon followed by the groups of W. Ketterle, at MIT, with sodium atoms [7, 29], and of R. Hulet, in Houston, with lithium atoms [8]. Evaporation down to the degenerate regime in a pure optical trap was achieved in 2001 by the group of M. S. Chapman [30]. In Fig. 7.2b we show images of the momentum distribution of the first Bose–Einstein condensates obtained at JILA in 1995. The momentum distribution was obtained by taking a picture after releasing the atoms from their trap. For low atom numbers, interactions are negligible and the gas expands freely.[3] In this case, the density profile after time of flight is proportional to the initial momentum distribution. The onset of Bose–Einstein condensation is clearly observed by the appearance of a narrow peak in the momentum distribution, corresponding to the macroscopic accumulation of atoms in the single-particle ground state of the trap.

[3]Although this situation was indeed achieved for the first gaseous BECs, we will see below that in most cases, interactions actually play a crucial role in the equilibrium shape and in the dynamics of the condensate.

In practice, the atoms are observed by absorption imaging, a process in which one shines a resonant beam on the cloud and images the shadow cast on a CCD camera. The drawback of this method is that it destroys the cloud and forbids real time imaging of its dynamics. An alternative method uses a non-resonant probe beam, which practically is not absorbed by the cloud of atoms, but simply dephased. By using a phase contrast technique, also used in standard microscopy to image transparent objects, it is possible to reconstruct the refractive index profile of the cloud, hence its density profile.

7.1.5 Bose–Einstein condensation in a trap

In order to calculate the critical temperature for a Bose gas confined by a potential $V(\boldsymbol{r})$, one often starts with the semi-classical approximation, which states that the phase space density of an ideal gas of chemical potential μ and temperature T is,

$$f(\boldsymbol{p},\boldsymbol{r}) \approx \frac{1}{(2\pi\hbar)^3} \frac{1}{e^{\beta(h(\boldsymbol{p},\boldsymbol{r})-\mu)}-1}. \tag{7.1}$$

Here, $\beta = 1/k_B T$ and $h(\boldsymbol{p},\boldsymbol{r}) = p^2/2m + V(\boldsymbol{r})$ is the classical Hamiltonian for a particle of mass m trapped in the potential V. The semi-classical approximation is valid as soon as the size of the cloud is larger than other characteristic length scales, such as the spatial extent of the ground state wave function in the trap or the thermal wavelength λ.

For simplicity we restrict ourselves from now on to the case of a harmonic potential with frequencies ω_i, $i = x, y, z$ and $V(0) = 0$. As in free space, the chemical potential can take only non-positive values, and the phase space density is at any place smaller than its value for $\mu = 0$. For a given temperature, the number of atoms that can be accounted for with the semi-classical result (7.1) is bounded from above by

$$N_{\max} = \int \frac{1}{e^{\beta h(\boldsymbol{p},\boldsymbol{r})}-1} \frac{d^3r\, d^3p}{(2\pi\hbar)^3} = \zeta(3)\left(\frac{k_B T}{\hbar\bar{\omega}}\right)^3, \tag{7.2}$$

where $\zeta(3) \approx 1.2$ and $\bar{\omega}^3 = \omega_x\omega_y\omega_z$. When the atom number is larger than N_{max}, atoms in excess accumulate in the ground state of the trap, following the general Bose–Einstein condensation scenario. Conversely, for a given atom number N, condensation occurs when the temperature passes below the critical value T_{c} such that $k_{\text{B}}T_{\text{c}} = \hbar\bar{\omega}(N/\zeta(3))^{1/3}$.

Within the semi-classical approximation, the threshold for condensation in a trap is directly related to the critical point of a homogeneous system. To prove this point we first note that the density at the center of the trap $n(0)$ can be calculated from expression (7.1) for the phase space density. Suppose now that the number of atoms in the trap is equal to the maximal value given in (7.2). One readily finds that $n(0) = n_{\text{c}}$ where $n_{\text{c}} = \zeta(3/2)\lambda^{-3}$ is the critical density for Bose–Einstein condensation in a homogeneous system.

So far we have only considered the case of an ideal gas. To go further, one needs a proper modeling of the interaction potential $U(\boldsymbol{r})$ between a pair of atoms separated by distance $\boldsymbol{r}$. This modeling can be written in a simple form, thanks to the fact that at low temperature, the thermal wavelength is larger than the range of the interatomic potential. One can therefore replace the (complicated) true potential by a contact interaction, with a strength g, proportional to the two-body scattering length a_{s}:

$$U(\boldsymbol{r}) = g\,\delta(\boldsymbol{r}) \qquad \text{with} \qquad g = \frac{4\pi\hbar^2 a_{\text{s}}}{m}. \tag{7.3}$$

In the rest of this chapter we will use this simple modeling of atomic interactions, except in Section 7.2.5, where we will address the case of long-range (dipolar) forces.

Trapped atomic gases being very dilute, at least for $T > T_{\text{c}}$, means that the influence of interactions on the critical point is relatively weak. The main effect is that because of repulsive (respectively attractive) interactions, the density at the center of the trap for a given number of atoms is lower (resp. higher) than its value in the absence of interactions. Therefore one needs to place more (resp. less) atoms in the trap in order to reach the threshold $n(0)\,\lambda^3 = \zeta(3/2)$. This can be expressed as a shift of the critical temperature [31],

$$\frac{\delta T_{\text{c}}}{T_{\text{c}}} \approx -1.33\,N^{1/6}\,\frac{a_{\text{s}}}{a_{\text{ho}}}, \tag{7.4}$$

where $a_{\text{ho}} = \sqrt{\hbar/m\bar{\omega}}$ is the extension of the ground state wave function in a harmonic potential of frequency $\bar{\omega}$. For typical situations, the scattering length a_{s} is at the nanometer scale, whereas a_{ho} is at the micrometer scale. Therefore, the relative shift of T_{c} is usually a few percent. Its measurement is described in [32], where corrections to T_{c} due to atom correlations are also reviewed and discussed.

7.2 Probing a condensate: the hydrodynamic approach

Typical densities in quantum degenerate vapors are in the range 10^{13}–10^{14} atoms/cm^3, five to six orders of magnitude more dilute than air. However, in spite of this extreme dilution, the dynamics of Bose–Einstein condensates follow the classical laws of hydrodynamics for inviscid fluids. We investigate the low energy modes arising from this hydrodynamic behavior and show how they were confirmed with remarkable accuracy by experiment.

7.2.1 Hartree approximation and Gross–Pitaevskii equation

Compared with liquid helium, ultra-cold gases have a very small density and they can often be considered as systems with independent particles, with quasi-negligible correlations between the atoms. In this respect, these gases are close to the situation discussed by Einstein in his seminal work [3, 4]. The quasi-independence between the atoms in a cold gas is illustrated by the fact that the condensed fraction—defined as the largest eigenvalue of the one-body density matrix—can be close to 100% at very low temperature, whereas it never exceeds 10% in superfluid liquid helium.

To exploit this absence of correlation between the atoms of the gas, a simple and useful approach is the *Hartree approximation*, which consists in describing the many-body wave function of a condensate containing N particles by the product

$$\Psi(\boldsymbol{r}_1,\boldsymbol{r}_2,\ldots,\boldsymbol{r}_N,t)=\prod_{i=1}^{N}\varphi(\boldsymbol{r}_i,t), \tag{7.5}$$

where φ is the macroscopic wave function describing the behavior of the system. For particles of mass m interacting through a two-body potential $U(\boldsymbol{r}_1,\boldsymbol{r}_2)$ and trapped in an external potential $V(\boldsymbol{r}_1)$, the evolution of φ can be obtained by minimization of the action associated with the Lagrangian density,

$$\mathcal{L}=-i\hbar\Psi^*\partial_t\Psi+\sum_{i=1}^{N}\left(\frac{\hbar^2}{2m}|\nabla_i\Psi|^2+V(\boldsymbol{r}_i)|\Psi|^2\right)+\frac{1}{2}\sum_{i,j=1}^{N}U(\boldsymbol{r}_i,\boldsymbol{r}_j)|\Psi|^2. \tag{7.6}$$

Writing the Euler–Lagrange equations with the Hartree anzatz (7.5) finally yields the Gross–Pitaevskii equation [33, 34],

$$i\hbar\partial_t\varphi=-\frac{\hbar^2}{2m}\nabla^2\varphi+V(\boldsymbol{r})\varphi(\boldsymbol{r},t)+(N-1)\int U(\boldsymbol{r},\boldsymbol{r}')\left|\varphi(\boldsymbol{r}',t)\right|^2\varphi(\boldsymbol{r},t)\,d^3r'. \tag{7.7}$$

This equation has a clear physical interpretation: each particle in the state $\varphi(\boldsymbol{r},t)$ evolves in a potential that is the sum of the external trapping potential $V(\boldsymbol{r})$ and the mean-field interaction energy due to the $N-1$ remaining particles. In the following, we will assume $N\gg 1$, hence we will replace $N-1$ by N.

In most experimental situations,[4] the range of the interatomic potential U is much shorter that other relevant length scales, like the interatomic distance or the thermal wavelength. Then, as explained in Section 7.1.5, we can replace U by a contact potential $g\,\delta(\boldsymbol{r}-\boldsymbol{r}')$, and the Gross–Pitaevskii equation turns into the nonlinear Schrödinger equation,

$$i\hbar\partial_t\varphi=-\frac{\hbar^2}{2m}\nabla^2\varphi+V(\boldsymbol{r})\varphi(\boldsymbol{r},t)+Ng\left|\varphi(\boldsymbol{r},t)\right|^2\varphi(\boldsymbol{r},t). \tag{7.8}$$

[4]The most notable exception being dipolar gases discussed in Section 7.2.5.

In the case where the gas is kept in a flat ($V = 0$) cubic box of size L, one can look for stationary solutions of the Gross–Pitaevskii equation with the form $\varphi_0(t) = e^{-i\mu t/\hbar}/L^{3/2}$, where μ is the chemical potential of the system.[5] The resolution is straightforward and yields

[5]One can check that the value of μ obtained with this definition coincides with the energy required to add an Nth particle to the gas already containing $N - 1$ atoms.

$$\mu = gn_0, \tag{7.9}$$

where $n_0 = N/L^3$ is the particle density in the condensate.

7.2.2 The Bogoliubov spectrum of collective excitations

Because of the nonlinear nature of the Gross–Pitaevskii equation, its resolution for arbitrary initial conditions can only be obtained numerically. However, when the system is weakly perturbed, a first-order expansion can be performed. In this section we restrict ourselves to the case of a gas confined in a cubic box of size L, for which the linearization of the Gross–Pitaevskii and the research of its eigenmodes is relatively simple.

We start by writing $\varphi(\boldsymbol{r}, t) = e^{-i\mu t/\hbar}[1 + \delta\varphi(\boldsymbol{r}, t)]/L^{3/2}$, so that we obtain at first order in $\delta\varphi$ the linear system

$$i\hbar\partial_t \begin{pmatrix} \delta\varphi \\ \delta\varphi^* \end{pmatrix} = \mathcal{L}_{GP} \begin{pmatrix} \delta\varphi \\ \delta\varphi^* \end{pmatrix}, \tag{7.10}$$

with the linear operator $\mathcal{L}_{GP}$ given by

$$\mathcal{L}_{GP} = \begin{pmatrix} -\hbar^2\nabla^2/2m + gn_0 & gn_0 \\ -gn_0 & \hbar^2\nabla^2/2m - gn_0 \end{pmatrix}. \tag{7.11}$$

Using translational invariance, solutions of Eq. (7.10) can be expanded on a set of plane waves $(u_k, v_k)e^{i(\boldsymbol{k}\cdot\boldsymbol{r}-\omega t)}$ diagonalizing the operator $\mathcal{L}_{GP}$ (the so-called Bogoliubov modes). A simple algebra then yields

$$\begin{pmatrix} u_k \\ v_k \end{pmatrix} \propto \begin{pmatrix} \cosh\theta_k \\ \sinh\theta_k \end{pmatrix}, \qquad \tanh 2\theta_k = \frac{1}{k^2\xi^2 + 1}, \tag{7.12}$$

with the Bogoliubov dispersion relation

$$\omega_k = \sqrt{gn_0k^2/m + (\hbar k^2/2m)^2}. \tag{7.13}$$

Here we have introduced the *healing length* $\xi = (8\pi n_0 a_s)^{-1/2}$, which characterizes the distance over which the condensate recovers its homogeneous density when one applies a local perturbation. Note that we have implicitly assumed that the scattering length a_s is positive, corresponding to an effective repulsive interaction between atoms. In this case, ω is real for any value of k.

The Bogoliubov approximation is valid in the dilute limit $n_0 a_s^3 \ll 1$, or equivalently $n_0\xi^3 \gg 1$. The Bogoliubov modes describe the low-energy excitations of a Bose-condensed gas, and they can be used to study its low-temperature thermodynamic properties. For instance, the quantum fluctuations of the Bogoliubov modes give access to quantum depletion of the condensate, i.e. the difference at zero temperature between the total density n and the condensed density n_0. They also provide the first beyond-mean-field corrections of the zero temperature equation of state (7.9) [35].

The dispersion relation (7.13) displays two different asymptotic regimes. For $k\xi \gg 1$, one recovers the single-particle dispersion $\omega \sim \hbar k^2/2m$, as for a non-interacting gas. For $k\xi \ll 1$, the dispersion relation is linear, $\omega \sim k\sqrt{gn_0/m}$, and in this regime we can identify the eigenmodes as acoustic waves with sound velocity $c_s = \sqrt{gn_0/m}$.

In gaseous BECs, the Bogoliubov spectrum can be studied by Raman scattering experiments in which two laser beams (labeled 1 and 2) with different frequencies and wavevectors are shone onto the atoms. Let us consider a process where one photon of beam 1 is transferred to beam 2 in an "absorption-stimulated emission" cycle (Fig. 7.3a). From energy–momentum conservation, this process is associated with the creation of a Bogoliubov excitation of momentum $\boldsymbol{k} = \boldsymbol{q}_1 - \boldsymbol{q}_2$ and frequency $\omega_k = \omega_1 - \omega_2$. This can only happen when the condition $\omega_{\boldsymbol{q}_1-\boldsymbol{q}_2} = \omega_1 - \omega_2$ is satisfied. Therefore, for a given set of directions $(\boldsymbol{q}_1, \boldsymbol{q}_2)$, the rate of Raman scattering varies resonantly with $\omega_1 - \omega_2$, providing one point on the dispersion curve ω_k. The experiment is then repeated for other directions $(\boldsymbol{q}_1, \boldsymbol{q}_2)$, to map the complete dispersion curve. Strictly speaking, the formalism developed above does not apply as such to the case of trapped gases, because the presence of the confining potential breaks the translational symmetry that led to the modes given in (7.12). However, the gas can be considered as quasi-homogeneous on length scales much smaller than the cloud size R, and the Bogoliubov spectrum (7.13) is therefore relevant if one uses only short wavelength excitations satisfying $kR \gg 1$.

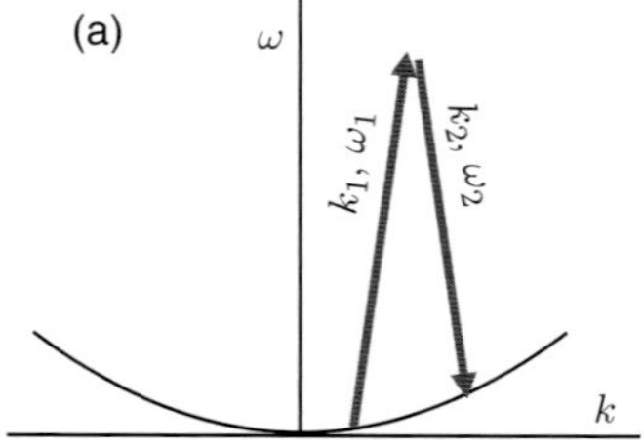

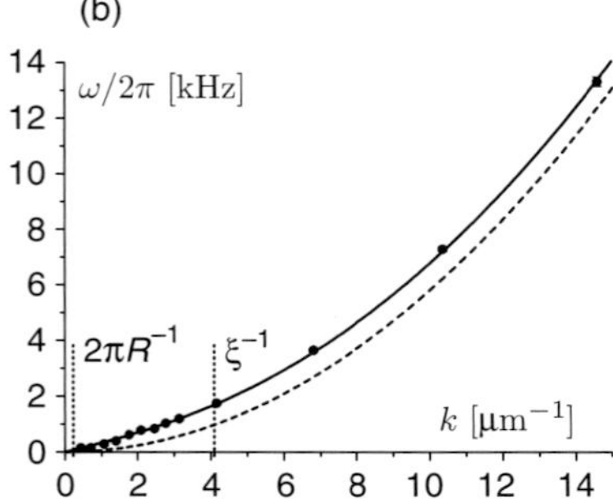

Fig. 7.3 (a) Example of a Raman scattering process, in which a photon of wavevector $\boldsymbol{k}_1$ and frequency ω_1 disappears, and a photon of wavevector $\boldsymbol{k}_2$ and frequency ω_2 is created. (b) Dots: Experimental measurement of the Bogoliubov spectrum in a rubidium condensate using Raman spectroscopy. The solid and dashed lines correspond to the Bogoliubov and free-particle spectra, respectively (data from [36]).

We show in Fig. 7.3b results obtained with this technique by the group of N. Davidson, at the Weizmann Institute [36]. The experiment was performed with rubidium atoms and led to results in excellent agreement with the Bogoliubov dispersion relation (7.13). Interestingly, the excitation spectrum of a gaseous Bose–Einstein is simpler than that of liquid helium, which exhibits a *roton branch* associated with a local minimum of ω_k. This is a result of the fact that a cloud of ultra-cold atoms is a weakly interacting system, for which the mean-field approximation adopted here is quite accurate. In recent experiments, the group of E. Cornell at JILA have started to explore the strongly interacting regime $na_s^3 \sim 1$ [37] and could observe deviations from the Bogoliubov dispersion relation (7.13), indicating a breakdown of the mean-field approximation.

THE CASE OF ATTRACTIVE INTERACTIONS

The scattering length a_s describing atomic interaction at low energy is negative for some atomic species, like ^{7}Li atoms in their lowest energy state. When this occurs, the Bogoliubov dispersion relation (7.13) leads to imaginary values for ω_k. This feature is the signature of an instability of the gas which collapses under the effect of attractive interactions. For a gas confined in a harmonic potential this collapse was indeed observed, when the number of atoms exceeded a threshold value [38]. In a one-dimensional geometry, this instability is connected to the existence of solitonic solutions of the stationary Gross–Pitaevskii equation,[6] which were also observed experimentally with Bose–Einstein condensates of ^{7}Li [40, 41]. These coherent atomic "packets" propagate without deformation, due to a balance between the interaction-induced collapse, and the broadening of their wave function due

[6]Note that solitonic solutions exist also for $g > 0$. In this case, they correspond to a dip in the density profile, and are therefore called dark (or gray when the contrast is not 100%) solitons [39].

to the dispersive nature of the single-particle dispersion relation $\omega = \hbar k^2/2m$. This phenomenon is also well known in other domains of physics (in particular hydrodynamics and nonlinear optics [42]).

7.2.3 Equilibrium shape and eigenmodes of a trapped condensate

In the presence of a harmonic trapping potential $V(\boldsymbol{r})$, the resolution of the Gross–Pitaevskii equation is more involved than for a homogeneous system. First, we consider the equilibrium state of the condensate, setting $\varphi(r,t) = \varphi_0(r)\, e^{-i\mu t/\hbar}$ in (7.2.4). In the absence of interactions ($g = 0$), the lowest energy solution $\varphi_0(r)$ is the single-particle ground state in the harmonic trap, i.e. a Gaussian function with an extension $a_{\text{ho}} = (\hbar/m\omega)^{1/2}$ and a chemical potential $\mu = 3\hbar\omega/2$ (for simplicity, here we assume an isotropic confinement of frequency ω). In the case where the scattering length is positive, repulsive interactions increase the size R of the cloud. In the limit of large atom numbers, the kinetic energy $\sim \hbar^2/mR^2$ can be neglected with respect to the trapping energy $\sim m\omega^2R^2$. In this so-called *Thomas–Fermi* regime, the stationary Gross–Pitaevskii equation leads to [43]

$$\mu = gn_0(\boldsymbol{r}) + V(\boldsymbol{r}), \tag{7.14}$$

where $n_0(\boldsymbol{r}) = N|\varphi_0(\boldsymbol{r})|^2$ is the atom density. This relation readily yields the density profile of the cloud in the presence of the external potential. It can be recovered from (7.9) using the *local density approximation*, where one considers that the system is locally homogeneous, with a space-dependent chemical potential $\mu_{\text{loc}}(\boldsymbol{r}) = \mu - V(\boldsymbol{r})$. For a condensate with N atoms in a harmonic potential, (7.14) entails that the density distribution be an inverted parabola. The radius of distribution is given by $R = a_{\text{ho}}\,\eta^{1/5}$ and the chemical potential is $\mu = \hbar\bar{\omega}\,\eta^{2/5}/2$, where we set $\eta = 15Na_{\text{s}}/a_{\text{ho}}$; the Thomas–Fermi approximation is valid if $R \gg a_{\text{ho}}$, i.e. $\eta \gg 1$.

A similar approach can be followed in the dynamical regime, in which one recovers equations analogous to Euler's equation in classical hydrodynamics. We start by writing the condensate wave function as $\varphi(\boldsymbol{r},t) = \sqrt{n(\boldsymbol{r},t)/N}\, e^{i\theta(\boldsymbol{r},t)}$ (Madelung transform). Expressing the Gross–Pitaevskii equation in terms of the real variables θ and n, we obtain the set of equations:

$$\partial_t n = -\nabla\cdot(n\boldsymbol{v}) \tag{7.15}$$

$$m\,\partial_t\boldsymbol{v} = -\nabla\left(gn + V + \frac{mv^2}{2} - \frac{\hbar^2}{2m\sqrt{n}}\nabla^2\sqrt{n}\right), \tag{7.16}$$

where $\boldsymbol{v}(\boldsymbol{r},t) = \hbar\nabla\theta/m$ is the velocity field of the condensate. Physical interpretation of these two equations is straightforward. The first one expresses the mass conservation, the second is the Euler equation for an inviscid fluid with an irrotational flow, with θ playing the role of the velocity potential. The term proportional to $\hbar^2$ arises from quantum fluctuations and is called *quantum pressure*. In the semi-classical limit $\hbar \to 0$, this can be neglected, and the above set of equations becomes:

$$\partial_t n = -\nabla \cdot (n\mathbf{v}) \quad (7.17)$$

$$m\,\partial_t \mathbf{v} = -\nabla\left(gn + V + mv^2/2\right), \quad (7.18)$$

which can identified with the Euler equations for a gas of pressure P characterized by the simple equation of state $P = gn$.

A quick inspection shows that the hydrodynamic regime described by Eqs. (7.17–7.18) is valid in the Thomas–Fermi regime $g\bar{n}/\hbar\omega \gg 1$, where $\bar{n}$ is the typical atomic density in the trap. Interestingly, this criterion is much less stringent than the condition for reaching the hydrodynamic regime for a *classical* trapped gas: $\gamma = \bar{n}\sigma\bar{v}/\omega \gg 1$. The latter condition compares the trap oscillation frequency ω to the collision rate $\bar{n}\sigma\bar{v}$, where $\sigma = 8\pi a_s^2$ is the s-wave scattering cross-section and $\bar{v}$ is the characteristic atomic velocity. In a Bose–Einstein condensate, the velocity is small and is Fourier-limited with $\bar{v} \sim \hbar/mR$, where R is the cloud size. We then find that

$$\gamma \sim \frac{a_s}{R}\left(\frac{g\bar{n}}{\hbar\omega}\right). \quad (7.19)$$

In typical experimental conditions, $g\bar{n}/\hbar\omega \sim 10$ and the Thomas–Fermi condition is fulfilled. Contrary to this, although scattering lengths can be rather large compared with typical atomic lengths (for rubidium it is ~ 5 nm, i.e. 100 times larger than the Bohr radius), their values are still much lower than the cloud radius $R \sim 10\,\mu$m, and the validity condition $\gamma \gg 1$ for reaching the classical hydrodynamic regime is not fulfilled. We thus see that, contrary to classical fluids, hydrodynamicity in quantum gases is not driven by collisions, but by quantum coherence, entailing the existence of a one-body wave function that encapsulates the macroscopic properties of the system.

The resolution of this set of equations in the case of low-lying excitation modes in an harmonic trap has been the subject of a large amount of both theoretical and experimental work [13, 14]. For instance, the quadrupolar mode associated with the oscillation of the aspect ratio of the cloud could be related to the formation of vortices in a BEC stirred by the rotation of an anisotropic harmonic potential [44]. The same quadrupolar mode was used to probe the angular momentum of a rotating condensate [45].

One specific mode, called the *scissor mode*, is more specifically connected with the issue of superfluidity. It is related to a reduction of the moment of inertia that is itself characteristic of a non-classical fluid behavior. In cold gases, the scissor mode is excited by the sudden tilting of one of the axes of an anisotropic trap (Fig. 7.4a). One can show that the quenching of the moment of inertia of the superfluid is associated with the existence of a single high frequency mode in the quadrupolar response of the cloud [46, 47]. By contrast, the response of a non-condensed (i.e. non-superfluid) Bose gas is characterized by two frequencies, one of these being proportional to the anisotropy of the trap, and therefore vanishingly small for weakly anisotropic potentials. Experimental observation of a single, high frequency scissor mode, hence proving superfluidity, was performed in Oxford by the group of C. J. Foot (Fig. 7.4b) [48, 49].

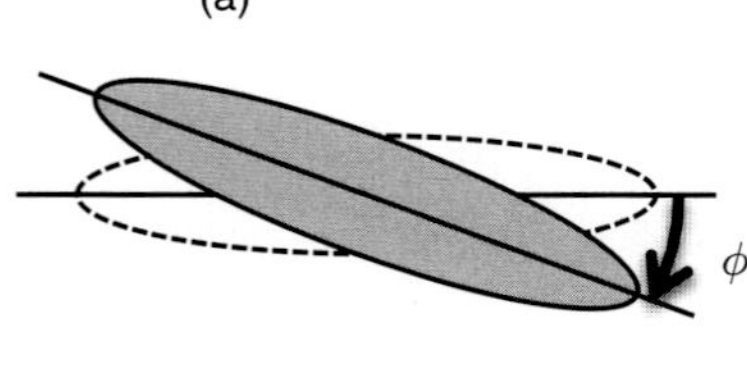

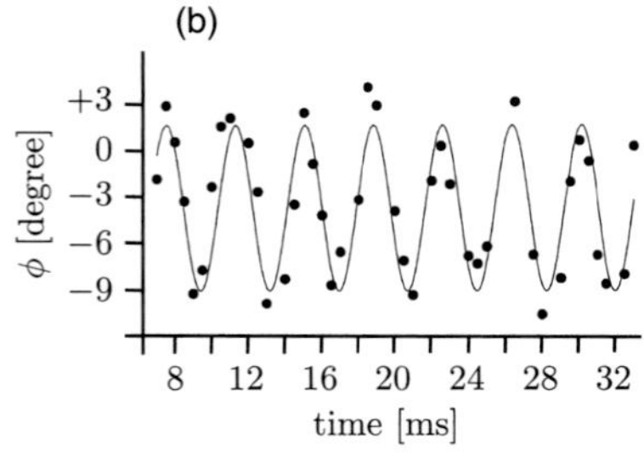

Fig. 7.4 Scissor mode and superfluidity. (a) Schematic representation of the scissor mode: the axes of an anistropic trap are suddenly rotated by a small amount and one observes the subsequent dynamics of the cloud. (b) Oscillation of the condensate axis (scissor mode). Only one frequency appears in the oscillatory motion, which is a consequence of superfluidity (figure extracted from [48]).

7.2.4 Probing superfluidity with a moving impurity

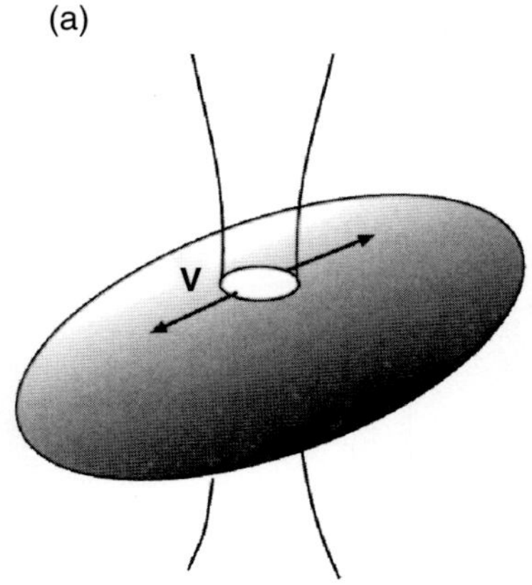

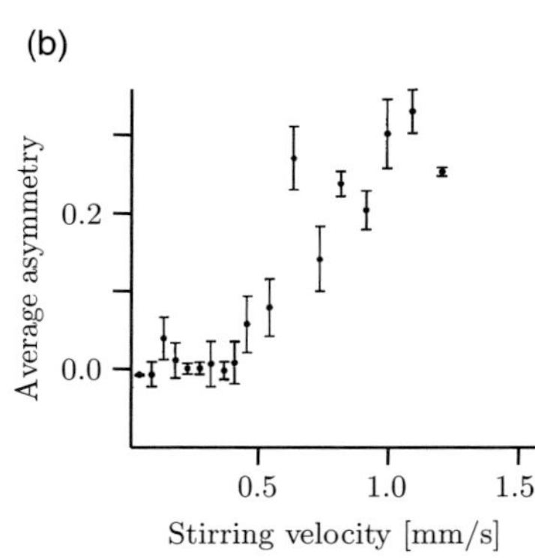

Fig. 7.5 (a) Probing superfluidity: a blue-detuned laser creating a repulsive potential is moved at constant velocity along a Bose–Einstein condensate (figure from [52]). (b) Drag force on the condensate, derived as asymmetry in the density profile. The central sound velocity is 4.8 mm/s (data from [53]).

Apart from measurement of the moment of inertia, another historical characterization of superfluidity in liquid helium is the absence of drag when an obstacle is moved along the superfluid, below a certain critical velocity. A simple explanation of this property, based on the structure of the excitation spectrum, was proposed by Landau. He noted that when the perturbation imparted by the obstacle is small, the energy transfer due to the drag can be described by the formation of elementary excitations in the fluid. Using a simple energy–momentum balance, one can show readily that a viscous drag only happens when the relative velocity of the obstacle with respect to the superfluid is larger than the critical velocity V_c given by the Landau criterion,

$$V_c = \min_k \left(\frac{\omega_k}{k}\right). \tag{7.20}$$

For the Bogoliubov spectrum (7.13), the critical velocity is simply the sound velocity. The drag can then be interpreted as an acoustic version of the Cerenkov radiation, associated in electromagnetism with the emission of an electromagnetic wake when a charged particle moves faster than the light velocity of the surrounding medium.

The absence of drag for a slowly moving microscopic impurity was tested quantitatively at MIT with a sodium Bose–Einstein condensate. The impurities were also sodium atoms that were transferred to an untrapped internal spin state using a stimulated Raman transition. In accordance with Landau's scenario, the scattering cross-section of the impurity with the condensed atoms dramatically decreased when the velocity of the impurity atom was smaller than the sound velocity in the BEC [50].

When the obstacle creating the perturbation has a larger size, the viscous drag is still negligible below a certain critical velocity, but the mechanism for the energy transfer is different. In this case, vortex shedding in the wake of the obstacle is the main source of dissipation in the system [51]. The onset of macroscopic dissipation was also studied by the group of W. Ketterle at MIT [52, 53]. They demonstrated that when a blue-detuned laser creating a hole inside the condensate was moved, heating was observed only above a certain critical velocity (Fig. 7.5). Using matter–wave interferometric techniques, the superfluidity breakdown mechanism was later on attributed to the nucleation of a vortex wake, in agreement with the large object scenario [54].

7.2.5 The case of long-range forces: dipolar condensates

In some sense, the physics of dilute Bose–Einstein condensates with short-range interactions constitutes an extension of the phenomena observed in liquid helium to the weak interaction regime $na_s^3 \ll 1$. By contrast, novel phenomena are expected when long-range forces, like dipole–dipole interactions, are dominant. This explains why in recent years, much attention has been devoted to the realization of polar Bose–Einstein condensates [55].

At the time of writing (June 2010), Bose–Einstein condensation of chromium is the only successful attempt in this direction [56, 57]. Chromium is an atom with a rather large magnetic moment (six times that of an alkali) for which the ratio between long-range and short-range interactions can be modified using a Feshbach resonance. It is thus possible to achieve a situation where interactions are dominated by dipolar effects, in which case dramatic phenomena can be observed [55]. In addition to their long-range character, dipolar forces are strongly anisotropic and their attractive or repulsive overall nature will depend on the geometry of the trapping potential. For cigar-shaped potentials and a dipole aligned with the trap axis, dipole forces are essentially attractive. In a pancake geometry and a dipole orientation perpendicular to the plane of the condensate, dipole forces are repulsive. In the first case, the cloud collapses due to an instability akin to the Rosensweig instability in classical ferrofluids [58]. By contrast, the repulsive interactions in a pancake geometry can overcome the instability of a Bose–Einstein with short-range attractive interactions, as observed in [59].

In parallel with attempts to manipulate atomic species with larger magnetic moments (erbium, dysprosium [60]), other lines of research are currently exploring the possibility of producing quantum gases with electric dipole interactions, which exceed magnetic interaction by several orders of magnitude. One possibility is to take advantage of the large electric dipole moment that can exist for Rydberg atoms, i.e. atoms where one electron is excited to a high energy level [61]. Another promising option aims at producing a cold gas of heteronuclear molecules, using for example the photo-association of a mixture of two atomic gases [62–64].

7.3 Probing a condensate: the quantum approach

With the possibility of manipulating the confining potential in space and time, original probing schemes have been developed for atomic gases. Using interference experiments one can access the phase distribution of the fluid and its one-body distribution function. One can also measure the spatial distribution of particles in the gas with single-atom resolution, and determine the density–density correlation function. In this section we present these probing schemes and illustrate them with some spectacular examples, such as the interference between independent condensates, the beat between two atom lasers, and the atomic Hanbury Brown and Twiss effect.

7.3.1 Interference of condensates

The prime feature of Bose–Einstein condensation is the accumulation of many particles in a single quantum state. This is reminiscent of the laser operation principle, where a macroscopic number of photons accumulate in the same mode of an electomagnetic cavity. The first probe that we describe here is direct proof of this macroscopic population of a single level.

We start with an experiment that W. Ketterle and his group performed in 1997 [15]. This constituted an experimental answer to the question raised by

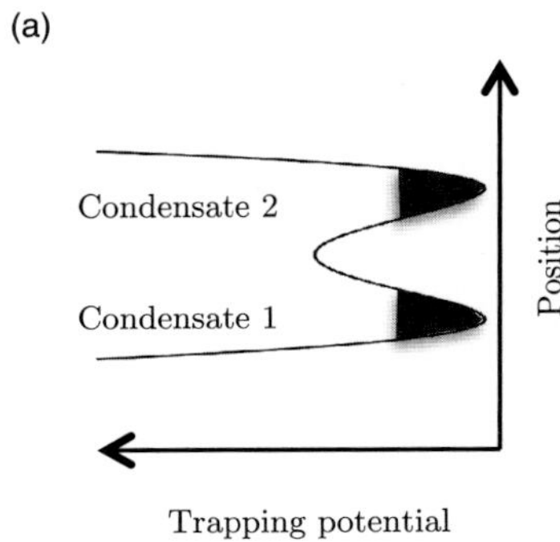

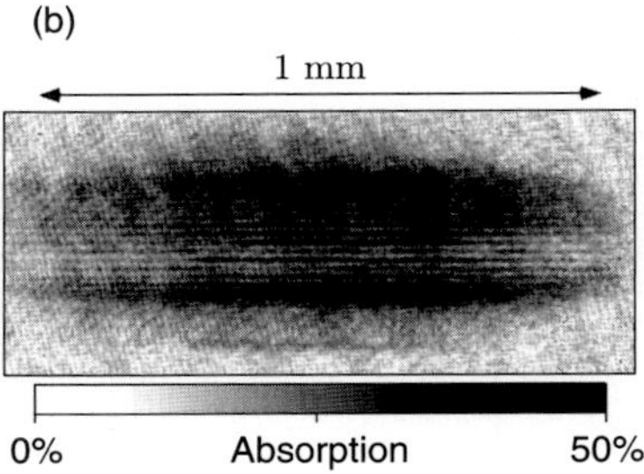

Fig. 7.6 (a) Double well potential obtained by shining a laser beam on the center of the magnetic trap. After evaporation one obtains two independent condensates. (b) After release from the magnetic + optical potential the two condensates expand and overlap. The spatial distribution in the overlap region shows interference fringes with a large contrast (photograph: courtesy of W. Ketterle, MIT).

P.W. Anderson [65]: "Do two superfluids that have never seen one another possess a definite relative phase?". The center of a magnetic trap was irradiated by a light sheet, creating a large repulsive barrier to produce a double well potential (Fig. 7.6a). Using evaporative cooling, a condensate was prepared around each potential minimum. The magnetic trap was then switched off, as well as the light sheet. The two atom clouds expanded and overlapped, and an image of the resulting spatial distribution was taken. The density profile exhibited interference fringes with a large contrast ($> 70\%$), which proved the coherence of each initial cloud (Fig. 7.6b).

To give a quantitative account of the interference pattern, the simplest approach consists of associating a classical field with a random phase φ_j ($j = 1, 2$) to each condensate [66]. Here we assume that the condensates are centered on the points $\pm\boldsymbol{a}/2$ and, for simplicity, we neglect the role of atomic interaction during the time-of-flight (TOF) expansion. This approximation may be questionable for short expansion times t, but eventually becomes correct, since the atomic density drops as t^{-3} when t increases. Expansion of each condensate wave function is thus obtained using the single-particle propagator associated with the Schrödinger equation. Assuming that the initial distance a between the condensates is much larger than their initial size, the density at a point $\boldsymbol{r}$ after a TOF duration t is approximately proportional to

$$\begin{aligned} \rho(\boldsymbol{r}, t) &\propto \left|\exp\left[\varphi_1 + im(\boldsymbol{r} + \boldsymbol{a}/2)^2/2\hbar t\right] + \exp\left[\varphi_2 + im(\boldsymbol{r} - \boldsymbol{a}/2)^2/2\hbar t\right]\right|^2 \\ &\propto \cos^2(\Delta\varphi + m\boldsymbol{r}\cdot\boldsymbol{a}/2\hbar t), \end{aligned} \tag{7.21}$$

where $\Delta\varphi = \varphi_1 - \varphi_2$. The interference pattern consists of straight fringes perpendicular to the line joining the condensate centers, with a fringe spacing equal to ht/ma. The contrast of the interference reaches 100% in this simple model, as a result of the macroscopic occupation of a single quantum state. For a given experimental shot, the positions of the bright fringes give access to the relative phase $\Delta\varphi$ between the two condensates. The relative phase fluctuates randomly from shot to shot, and the superposition of many interference patterns recorded in the same experimental condition leads to a uniformly gray image.

One can also explain the emergence of the interference pattern by assuming that the initial state of the two condensates is of the form $|N_1, N_2\rangle$, with a well-defined number of particles N_1 and N_2 in each subsystem. In this case phases φ_1 and φ_2 are initially not defined and the probability distribution for the relative phase $\varphi_1 - \varphi_2$ is a uniform function between 0 and 2π. The phase distribution evolves towards a narrower distribution as the number of detected atoms increases. From this point of view, the emergence of a relative phase is a consequence of the information acquired on the system via atomic position measurements [67–69]. Such interfering independent condensates can be used to investigate possible violations of local realism, using generalized Bell-type inequalities [70].

So far we have restricted our discussion to the case of independent condensates, assuming that the tunneling between the central barrier was negligible on the time scale of the experiment. When coupling between the two sides of the barrier is significant, the situation is reminiscent of a Josephson junction and it can give rise to a wealth of quantum phenomena, such as quantum self-trapping [71] and a.c./d.c. Josephson effects [72]. Repulsive interactions between particles in this double well geometry can also lead to a reduction in fluctuations of $N_1 - N_2$. This so-called *number squeezing* was demonstrated in [73] and can lead to significant improvement in atom interferometry methods [74, 75].

7.3.2 One-body correlation function

The most direct tool for investigating formation of a Bose–Einstein condensate is the one-body correlation function,

$$G_1(\boldsymbol{r}, \boldsymbol{r}') = \langle \hat{\psi}^\dagger(\boldsymbol{r}) \hat{\psi}(\boldsymbol{r}') \rangle, \qquad (7.22)$$

where the operator $\hat{\psi}^\dagger(\boldsymbol{r})$ creates a particle in $\boldsymbol{r}$. For a uniform fluid, G_1 depends only on the distance $|\boldsymbol{r} - \boldsymbol{r}'|$ and the Penrose–Onsager criterion relates Bose–Einstein condensation with a non-zero limit of G_1 when $|\boldsymbol{r} - \boldsymbol{r}'|$ tends to infinity [76]. For a non-degenerate ideal atomic gas, G_1 is a Gaussian function that decays to zero over a distance of the order of the thermal wavelength λ.

Several strategies have been developed to access G_1. One can take advantage of the fact that the momentum distribution $\mathcal{P}(\boldsymbol{p})$, which can be measured using the Bragg spectroscopy method presented in the previous section, is the Fourier transform with respect to variable $\boldsymbol{u}$ of $\int G_1(\boldsymbol{R} + \boldsymbol{u}/2, \boldsymbol{R} - \boldsymbol{u}/2)\, d\boldsymbol{R}$. Here we concentrate on a direct measurement of G_1, which can be obtained by looking at the interference of one part of the gas located around $\boldsymbol{r}$ with another part located around $\boldsymbol{r}'$.

The NIST group have developed a procedure that consists of measuring the interference between two spatially displaced copies of an original condensate [77]. In the NIST experiment, each copy was produced using a light pulse with a laser standing wave along a given direction z, which transferred a small fraction of the atoms of the BEC to a state with momentum $\boldsymbol{p}_0 = 2\hbar k\hat{z}$, where k is the wavevector of the photons and where $\hat{z}$ is a unit vector along the z-axis. This momentum kick, $\boldsymbol{p}_0$, resulted from the absorption of a photon in one of the beams creating the standing wave, and from the stimulated emission of a photon in the other beam. The kick, p_0, was much larger than the typical momentum of an atom in a trapped BEC. The total number of atoms with momentum p_0 was measured as a function of the time t between the two light pulses. For a pure condensate, one can show that this number is related to the overlap between the initial condensate wave function and the same wave function displaced by a distance $\boldsymbol{\rho} = \boldsymbol{p}_0 t/m$. More generally this method gives access to the integral over $\boldsymbol{r}$ of $G_1(\boldsymbol{r}, \boldsymbol{r} + \boldsymbol{\rho})$. It has been used by several groups to study the emergence of coherence in atomic gases, in particular in low-dimensional systems.

A second procedure consists in generating two continuous atomic beams out of an atom cloud, and in looking at the spatial interference between these beams. We show in Fig. 7.7 a result obtained by the Munich group. The atoms were confined in a magnetic trap and each beam was extracted using

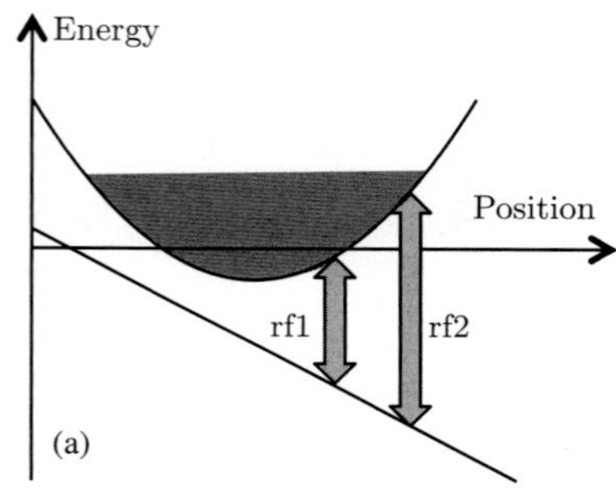

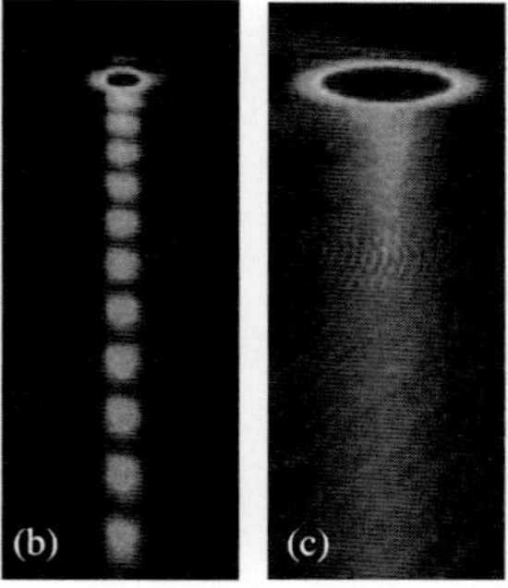

Fig. 7.7 (a) Extraction of two atomic beams ("atom lasers") from a cloud of rubidium atoms. Two radio-frequency waves, rf1 and rf2, flip the magnetic moment of the atoms at well-defined positions in the trap. After the spin flip the atoms are in an internal state that is not confined in the magnetic trap, and they fall under gravity. (b) The atom cloud is a quasi-pure condensate and a strong interference contrast is observed between the two beams, which reveals the phase coherence of the sample. (c) For a cloud above the critical temperature, no phase coherence is measured if the distance between the extraction points exceeds 200 nm (photographs: courtesy of Immanuel Bloch, Munich).

a radio-frequency (rf) electromagnetic field. As for evaporative cooling (see Section 7.1.4), the rf flipped the magnetic moments of the atoms at some definite locations. After the flip these atoms were no longer trapped and fell under the influence of gravity. The choice of the rf value determined the precise location in the trap from which the atoms were extracted [78]. When a single rf wave is applied, the atomic beam produced in this way is often referred to as an *atom laser* [79]. By simultaneously applying two different rf values, the Munich group obtained two atomic beams emerging from two different points of the atom cloud (see Fig. 7.7a) [16]. When temperature T was chosen well below the critical temperature T_c for BEC, they observed an interference with a large contrast in the region where the two atomic beams overlapped (Fig. 7.7b). The interference pattern remained visible, even for a large difference between the two radio-frequencies, corresponding to a distance between the two point sources of the order of the size of the cloud. Contrary to this, when $T > T_c$, no detectable interference was visible in the zone where the two beams overlap (Fig. 7.7c), unless the distance between the two sources was below 200 nm, i.e. the coherence length of the gas $\ell \approx \lambda$ in these experimental conditions. Measurement of the visibility of the interference pattern between two matter-waves also provides a means to investigate the critical behavior of the gas at the Bose–Einstein condensation point [80].

7.3.3 Two-body correlation function and Hanbury Brown and Twiss effect

In general the one-body correlation function does not capture all the physics of a many-body system and one needs correlation functions involving an arbitrary number of particles to characterize fully the state of a quantum fluid. Measurement of the two-body correlation function,

$$G_2(\boldsymbol{r},\boldsymbol{r}') = \langle \hat{\psi}^\dagger(\boldsymbol{r})\, \hat{\psi}^\dagger(\boldsymbol{r}')\, \hat{\psi}(\boldsymbol{r}')\hat{\psi}(\boldsymbol{r})\rangle \tag{7.23}$$

is a particularly important step. Indeed G_2 corresponds to the probability of detecting one atom in $\boldsymbol{r}$ and another atom in $\boldsymbol{r}'$, and it gives access to the density fluctuations in the superfluid, by contrast to G_1 which characterizes its phase fluctuations.

An efficient tool for measuring G_2 is a position-resolved single atom counter. Here we briefly describe results obtained in a collaboration between the Amsterdam and Orsay groups, using a microchannel plate to detect metastable helium atoms [81, 82]. The plate was placed below the atom cloud; when released from the trap, the atoms fell onto the plate and the position of each detected atom in the horizontal plane was recorded, as well as its arrival time. Assuming ballistic expansion one could then reconstruct the in situ G_2 function. This experiment provided a nice illustration of the Hanbury Brown and Twiss (HBT) effect, i.e. the bunching of bosonic particles in a thermal source, corresponding to a maximum of G_2 for $\boldsymbol{r} = \boldsymbol{r}'$. The HBT effect can be interpreted in simple terms, by considering the detection in $\boldsymbol{r}$ and $\boldsymbol{r}'$ of two atoms that were in different initial states a and b. There are two quantum paths $\{a \to \boldsymbol{r}, b \to \boldsymbol{r}'\}$ and $\{a \to \boldsymbol{r}', b \to \boldsymbol{r}\}$ that correspond to this process and that can interfere. For bosonic particles, thanks to the symmetry of the global wave function by exchange of the two particles, the interference is constructive in $\boldsymbol{r} = \boldsymbol{r}'$ and provides the HBT bunching. For thermal Bose gases observation of this bunching was reported in [83, 81]. The bunching is not present for a pure condensate, because all particles then occupy the same initial state [81]. If the experiment is performed with a Fermi gas instead of a Bose gas (^{3}He instead of ^{4}He for example), one expects from the Pauli principle an antibunching of particles at $\boldsymbol{r} = \boldsymbol{r}'$. This was indeed observed in [82].

A technique that is also directly inspired by the HBT effect is quantum noise interferometry [84]. This method, which does not require detection at the single atom level, is based on the autocorrelation function of individual images of a quantum gas. Many images are taken under the same experimental conditions (temperature, chemical potential), and each image differs from the others only in its atomic shot noise. The average of the autocorrelation function over these many images provides the desired density–density correlation function. Spectacular illustrations of this technique are evidence of the spatial order of the Mott-insulator state of a gas in an optical lattice [85], and the detection of correlations between pairs of atoms produced in the dissociation of a weakly bound molecule [86].

Finally, let us mention that one can access higher-order correlation functions (at least their values at short distances) by looking at the loss rate from the gas. As mentioned above, the main loss process in cold atomic gases is usually three-body recombination. This process occurs when three atoms are close to each other; two of them can form a dimer bound state, and the third atom carries away the released energy. The corresponding rate is approximately proportional to $\langle n^3(\boldsymbol{r})\rangle$, and its measurement as a function of temperature and density allows one to follow the entrance of the gas into the quantum degenerate regime [87]. In particular for a constant density $\langle n\rangle$, one can observe the reduction of $\langle n^3\rangle$ by a factor $3! = 6$ between a thermal state and a pure condensate [88]. The strong decrease in the three-body recombination rate in a quasi-one-dimensional gas was also used as a signature of the entrance in the strongly correlated Tonks–Girardeau regime [89].

7.4 Low-dimensional aspects: BEC versus superfluidity

Dimensionality has a strong influence on the type of phase transitions that can take place in a physical system [90]. Indeed, phase transitions result from competition between cooperativity effects and quantum or thermal fluctuations. Because a particle in a 1D or 2D geometry has fewer neighbors than in 3D, the role of interactions is weakened, and disordered states are favored. More precisely, the Mermin–Wagner theorem states that long-range order cannot occur at non-zero temperature in a 1D or 2D system with short-ranged interactions and a continuous symmetry [91]. An illustration of this result is the absence of Bose–Einstein condensation in an infinite, homogeneous Bose gas in one or two dimensions [92], which holds for both the ideal and interacting cases.

The absence of true Bose–Einstein condensation in a low-dimensional gas is still compatible with the presence of a superfluid component. The proper definition of superfluidity is based on the sensitivity of the N-body wave function $\Psi(\boldsymbol{r}_1, \ldots, \boldsymbol{r}_N)$ with respect to a boost modeled by a change in the boundary conditions. Instead of choosing the usual periodic boundary conditions in a box of size L, one can consider twisted boundary conditions such that Ψ is multiplied by $e^{i\theta}$ when the coordinates $\boldsymbol{r}_i$ are increased by $L\boldsymbol{e}$, where $\boldsymbol{e}$ is a unit vector along one of the directions of space. If the system is normal (non-superfluid) its free energy is unaffected by the phase twist. On the contrary, a superfluid system possesses some phase rigidity, and its free energy increases by an amount ΔF proportional to θ^2 for small θ. The rigorous definition of superfluid density is then based on the non-zero value of the ratio $\Delta F/\theta^2$. From a practical point of view the twist in the boundary conditions is provided by a slow rotation of the system, and the angle θ is the Sagnac phase appearing in the frame rotating with the gas. The non-zero value of ΔF corresponds to a reduction of the moment of inertia of the gas, with respect to the value expected for a classical fluid. Twisted boundary conditions can also be imposed by taking advantage of the geometric Berry's phase [93].

In the following, we discuss the case of the superfluid transition in a two-dimensional Bose gas. We first recall why no Bose–Einstein condensation occurs in an infinite, ideal gas, and we briefly describe the Berezinski–Kosterlitz–Thouless mechanism which is at the origin of the superfluid transition in this system. We then turn to trapped two-dimensional atomic gases, for which the finite size of the system makes possible the emergence of a significant condensed fraction at a non-zero temperature. To keep this section to a reasonable length, we do not address here the case of the superfluidity of one-dimensional systems and we refer the reader to [94] and references therein for a discussion of this problem.

7.4.1 The superfluid transition in a uniform 2D gas

For an ideal 2D gas, the absence of Bose–Einstein condensation at any non-zero temperature is a direct consequence of the equation of state,

$$n_2\lambda^2 = -\ln(1 - Z), \tag{7.24}$$

where n_2 is the surface density of the gas and Z the fugacity, defined as $Z = \exp(\mu/k_\mathrm{B}T)$. From this result it is clear that one can associate a value of the chemical potential to an arbitrary large phase space density $n_2\lambda^2$. This is very different from the 3D situation where, as mentioned in the first section, the equation of state for the ideal gas no longer possesses any solution for $n_3\lambda^3 > \zeta(3/2)$.

The two-dimensional situation is marginal in the sense that although thermal fluctuations prevent the appearance of a true Bose–Einstein condensate, a superfluid transition at a non-zero temperature is still possible. This transition has been investigated with great precision in helium films [95]. Its key feature, first described by Berezinskii [96] and by Kosterlitz and Thouless [97] (BKT), is well captured by the decay at large distances $|\boldsymbol{r} - \boldsymbol{r}'|$ of the one-body correlation function $G_1(\boldsymbol{r}-\boldsymbol{r}')$, defined in (22). For a Bose gas with repulsive interactions, three regimes can be identified when the temperature is decreased while maintaining a fixed spatial density n_2. At high temperature, the interaction energy is negligible compared with $k_B T$; in this case G_1 is a Gaussian function that tends to zero over a distance given by the thermal wavelength λ. When T is lowered, the interaction energy becomes significant and density fluctuations are gradually suppressed. The decay of G_1 then becomes exponential with a characteristic length ℓ which increases when the temperature decreases. At a critical temperature T_c, the length ℓ diverges and a fraction of the gas becomes superfluid. For $T < T_c$, the one-body correlation function still decays at infinity (otherwise a true BEC would be present) but the decay is only algebraic: $G_1 \propto |\boldsymbol{r}-\boldsymbol{r}'|^{-\alpha}$. Remarkably the 2D superfluid density $n_{2,s}$ is related to exponent α by the simple law $\alpha = 1/n_{2,s}\lambda^2$. Just below T_c, exponent α takes the universal value 1/4, irrespective of the strength of the interactions, so that $n_{2,s}^{(c)} = 4/\lambda^2$ [98]. Note that the universal relation $n_{2,s}^{(c)}\lambda^2 = 4$ giving the critical point is implicit, since $n_{2,s}$ is itself a function of temperature. The relation between the total density n_2 and the temperature at the critical point has been determined numerically in [99] in the regime of weak interactions. It can be written as

$$n_2^{(c)}\lambda^2 = \ln(C/g), \tag{7.25}$$

where $g \ll 1$ is the dimensionless parameter characterizing the interactions in the 2D fluid and $C \approx 380$ is a constant.

The microscopic mechanism at the origin of the 2D superfluid transition is the breaking of vortex pairs. In this context a vortex is a point in space where the superfluid density vanishes, and around which the phase rotates by $\pm 2\pi$ (vortices corresponding to multiples of $\pm 2\pi$ play a negligible role in practice). In the domain of the temperature of interest, the relevant excitations of the fluid are either vortices or phonons, both corresponding to phase fluctuations. Density fluctuations play a minor role, at least at the qualitative level. In the low temperature superfluid phase, vortices can only exist in the form of bound pairs, formed by the association of two vortices with opposite circulation. Indeed, the free-energy cost of a single isolated vortex is large and even diverges in the thermodynamic limit. When the temperature increases and $n_{2,s}\lambda^2$ reaches the value 4, the free energy for a single vortex decreases and finally vanishes. Isolated vortices can then appear, which further reduces the value of $n_{2,s}$ and makes the emergence of other free vortices even more likely. This avalanche effect means that the properly renormalized superfluid density vanishes [97, 98].

7.4.2 The 2D trapped Bose gas

Cold atom experiments are usually performed in a harmonic potential and the confinement significantly modifies the results obtained for the infinite homogeneous case. We consider first an ideal gas in an isotropic potential of frequency ω. The single-particle density of states is proportional to the energy, as for a particle moving freely in a 3D space. One recovers therefore a genuine

Bose–Einstein condensation when the number of particles exceeds the critical number [100],

$$N_c = \frac{\pi^2}{6}\left(\frac{k_B T}{\hbar\omega}\right)^2. \tag{7.26}$$

This transition survives in the thermodynamic limit which corresponds, for a two-dimensional harmonic trap, to letting the number of particles N go to infinity, the trap frequency ω to 0, while keeping the product $N\omega^2$ constant.

The next question to ask is whether this result still holds in the presence of interactions. The answer is quite subtle (see [101] for a review) and here we outline only the main results. The simplest approach is based on the mean-field Hartree–Fock approximation, in which repulsive interactions are accounted for by adding a term proportional to the density $n_2(\boldsymbol{r})$ to the trapping potential. The mean-field energy reduces the strength of the confinement at the bottom of the trap, where the density is the largest, and makes the situation very similar to the case of a 2D gas in a flat potential. A quantitative analysis shows that this flattening of the confinement has a dramatic effect: within the mean-field description, the singularity for $N > N_c$ that was signaling the condensation for an ideal gas disappears [102].

The Hartree–Fock approach is however not the end of the story. Indeed, it does not predict the superfluid BKT transition when it is applied to the uniform infinite system. To go further it is convenient to turn to the local density approximation, and make use of the known results for the infinite uniform 2D gas [103]. Within this approximation, one predicts that the superfluid transition takes place in the trapped system when the phase space density at the center of the trap reaches a critical value (7.25). This approximation has been accurately checked using a quantum Monte Carlo calculation [104] as well as using semi-classical field simulations [105].

Let us now focus on the case where a superfluid with a size L has formed at the bottom of the trap. Is there also a significant condensed fraction in this case? We recall that the answer would be negative in an infinite uniform system, because the one-body correlation function, G_1 decays algebraically with distance in a low temperature two-dimensional gas. However, for a system of finite size L this argument does not apply. The condensed fraction f, which is defined as the largest eigenvalue of the one-body density operator, is approximately given in this case by $G_1(L) \sim (\xi/L)^\alpha$, where ξ is the healing length. We have mentioned previously that the exponent α is smaller than 1/4; taking as a typical value $L \sim 100\,\xi$ we find $f \geq 0.3$, which is confirmed by Monte Carlo calculations [104, 106]. Here, we reach a conclusion that is well known in the domain of two-dimensional magnetism [107]: Because of finite size effects, the two phenomena of superfluidity and condensation cannot be dissociated for any practical case.

7.4.3 Making a 2D atomic fluid in practice

In order to realize experimentally a 2D atomic gas in the xy plane, one usually freezes the third degree of freedom with a laser beam which provides a strong confinement $m\omega_z^2 z^2/2$ along the z-direction. If the temperature and the chemical potential are both smaller than $\hbar\omega_z$, the gas can be considered as a 2D system from the thermodynamic point of view, with a thickness given by the size of the ground state of the harmonic oscillator along z, $a_z = \sqrt{\hbar/m\omega_z}$. The 3D scattering length a_s is usually much smaller than a_z, so that atom interactions can still be described by the usual 3D scattering formalism [108]. The interaction energy of the gas E_{int} can then be written in good approximation as, $E_{\text{int}} = (\hbar^2 g/2m) \int n_2^2(\boldsymbol{r})\, d^2r$, with $g = \sqrt{8\pi}\, a_s/a_z$.

The simplest laser scheme that produces the required confinement along z is a single Gaussian beam. It must be red-detuned with respect to the atomic resonance, so that the atoms are attracted by the dipole force towards the high intensity region [109]. The beam is focused with a cylindrical lens in order to produce a horizontal light sheet with a thickness of a few microns. More elaborate schemes involve evanescent waves at the surface of a dielectric medium, optical lattices, and holographic wave plates (see [94] for a review). The first diagnosis of a 2D gas consists in measuring its density profile in the xy plane, either in situ or after a time of flight. Experiments with rubidium atoms, corresponding to $g = 0.15$, have shown the existence of a sharp transition [110, 111]. For a small phase space density at the center, the density profile is smooth and relatively well described by a single Gaussian function. Above a critical value (which is in good agreement with the BKT prediction (25)), a narrow feature appears at the center of the trap, on top of a broader distribution, in good agreement with a two-fluid model. For experiments performed with sodium, a good description of the data around the critical point is obtained only if one introduces a third intermediate component, which is interpreted as a "non-superfluid" quasi-condensate [112]. A possible hint towards explaining the differences between the sodium and rubidium cases is the notably smaller interaction parameter for the Na experiment ($g = 0.02$), making it closer to the ideal gas case.

Further insight on 2D gases is provided by measurement of the function G_1, which characterizes the phase distribution in the fluid. In order to access this, one can interfere two independent planes of atoms prepared under the same conditions (Fig. 7.8a) and study the distribution of contrasts of the interference patterns [113, 114]. Such a measurement has been performed with rubidium atoms and gave evidence for the rapid increase of coherence of the gas below the critical point (Figs. 7.8b–d) [115]. In addition, some interference patterns revealed the presence of isolated vortices, which appear as dislocations of the fringe system (Fig. 7.8d). The number of dislocations increased with temperature, until the critical point was reached and the interference disappeared. The link between fringe dislocations and vortices was confirmed by numerical simulations based on a classical field stochastic evolution [116]. Another way for measuring G_1 is a homodyning method, in which one interferes the 2D gas with its own copy, after it has been displaced by an adjustable distance [112]. This method is reminiscent of that presented in Section 7.3.2 for 3D samples, and it gives access to the coherence length ℓ of the gas as a function of temperature above and below the critical point. The authors of [112] could in this way observe the transition between the "hot" regime where $\ell \sim \lambda$ and the colder one where $\ell \gg \lambda$.

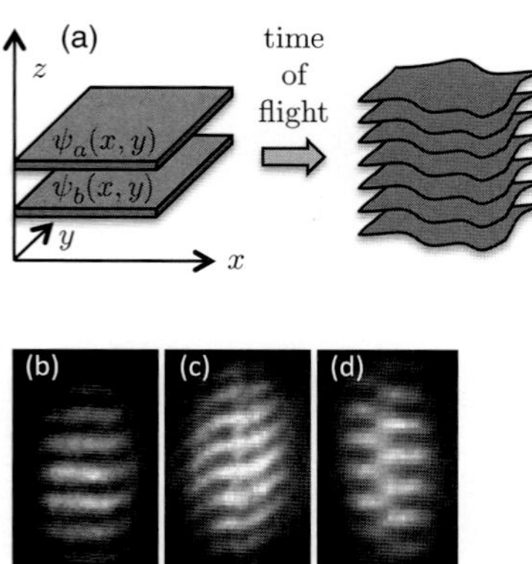

Fig. 7.8 (a) Interference between two independent planar gases, observed after time of flight. (b–d): Examples of interference patterns measured with the experimental setup described in [115]. The imaging beam is propagating along the y-axis. Pattern (b) is obtained with very cold gases, whereas (c) corresponds to a higher temperature. The dislocation in (d) is a signature for the presence of a vortex in one of the two gases.

Research on 2D atomic superfluids is still a very open field and, at the time of writing, many issues remain to be investigated. Here, we will mention only two of them. The first series of problems deals with "out of equilibrium" questions: how do two independent planes dephase with respect to each other, if their phases were locked at initial time? This question has already been answered experimentally for 1D gases [117] and it would be interesting to revisit this for 2d fluids, where the loss of coherence is predicted to behave as a power-law function of the evolution time [118]. A second crucial aspect is the direct determination of the superfluid component, by measuring, for example, the moment of inertia of the gas. Up to now, investigations on 2D fluids have mostly focused on phase coherence. Looking directly at superfluid properties would allow one to construct a bridge to the physics of helium films, where transport measurement is a natural diagnostic [95].

7.5 Summary and outlook

We have presented in this chapter the methods that allow one to prepare and probe a quantum atomic gas. These gases provide a practical realization of a paradigm of many-body theory: the weakly interacting Bose gas. This model was developed over the period 1950–1970 but could not be applied in a quantitative manner to the only Bose superfluid (liquid ^{4}He) that existed at that time. Indeed in superfluid ^{4}He the interactions are strong and lead to a severe depletion of the condensate, even at zero temperature. Contrary to this, the interparticle distance in atomic gases is usually much larger than the scattering length characterizing the interactions, and the microscopic description of the fluid in terms of the Gross–Pitaevskii equation (GPE) is an excellent approximation. The Bogoliubov linearization of the GPE provides an excitation spectrum that is in very good agreement with experimental observations, and the GPE can even be used to study the turbulent dynamics of these gases ([119] and Chapter 3 in the present book).

By contrast with "conventional" superfluids like liquid ^{4}He or superconductors, transport measurements are not well suited for atomic vapors. The required "circuitry" would in most cases be difficult to implement with the standard tools of atomic physics: magnetic fields and laser beams. Therefore investigation of atomic gases has required the development of novel probing techniques: time-of-flight expansion, Bragg spectroscopy, interference between independent samples, etc. These tools give access in particular to the momentum distribution of the particles or to its Fourier transform, the one-body correlation function G_1. Recently, the achievement of detection methods at the single atom level (see e.g. [120]) has provided the opportunity to take a direct "snapshot" of a many-body wave function.

The possibilities presented by the achievement of degenerate atomic gases go well beyond the mean-field physics described by the GPE. A celebrated example is transition from superfluid to Mott-insulator states, which can be observed with a gas confined in an optical lattice (see [121] and see chapter 17 volume 2 by Chin and Gemelke). Another class of strongly correlated states that are actively being looked for, are the analogs of those appearing in the fractional quantum-Hall effect. Such states are expected to emerge in an atomic gas submitted to a strong gauge field [122]. This gauge field can originate from a fast rotation of the system, or from the geometrical phase accumulated by an atom when it follows, adiabatically, one of its internal states [94]. The study of the BEC–BCS crossover also belongs to "beyond mean-field" physics, using fermionic species instead of bosonic ones (see the chapter by M. Zwierlein in Volume Z).

To keep this chapter to a reasonable length we have had to omit several important developments of many-body physics with cold atomic gases. We have restricted ourselves in particular to single component superfluids, but we should mention that the research on multi-component gases is also very active. One can take advantage of the spin degeneracy and prepare a spinor gas with 3 (5) components for a $F = 1$ (2) atomic state, which leads for example to a spectacular texture dynamics [123]. One can also mix two different atomic species (or even three species as in [124]) with different masses and possibly different statistical natures (Bose–Bose, Bose–Fermi and Fermi–Fermi). Finally, we mention a promising emerging subject, devoted to the effect of disorder on superfluidity. Recent experiments have studied the localization of atoms in a disordered potential created by laser light in a one-dimensional geometry [125, 126]. The transition to two and three dimensions is now underway [127], and these studies could help to clarify the subtle interplay between the single particle Anderson localization phenomenon and superfluidity of an interacting gas.

References

[1] Kapitza, P. Viscosity of liquid helium below the λ-point. *Nature*, **141**:74 (1938).

[2] Allen, J. F., and Misener, A. D. Flow of liquid helium II. *Nature*, **141**:75 (1938).

[3] Einstein, A. Quantentheorie des einatomigen idealen gases. *Sitzungsberichte/ Physikalische Klasse, Preussische Akademie der Wissenschaften*, **22**:261 (1924).

[4] Einstein, A. Quantentheorie des einatomigen idealen gases. ii. *Sitzungsberichte/ Physikalische Klasse, Preussische Akademie der Wissenschaften*, **1**:3 1925.

[5] Grüter, P., Ceperley, D., and Laloë, F. Critical temperature of Bose–Einstein condensation of hard-sphere gases. *Phys. Rev. Lett.*, **79**(19):3549–3552 (Nov 1997).

[6] Anderson, M. H., Ensher, J. R., Matthews, M. R., Wieman, C. E., and Cornell, E. A. Observation of Bose–Einstein condensation in a dilute atomic vapor. *Science*, **269**:198 (1995).

[7] Davis, K. B., Mewes, M.-O., Andrews, M. R., van Druten, N. J., Durfee, D. S., Kurn, D. M., and Ketterle, W. Bose–Einstein condensation in a gas of sodium atoms. *Phys. Rev. Lett.*, **75**:3969 (1995).

[8] Bradley, C. C., Sackett, C. A., and Hulet, R. G. Bose–Einstein condensation of lithium: Observation of limited condensate number. *Phys. Rev. Lett.*, **78**:985 (1997).

[9] Greiner, M., Regal, C. A., and Jin, D. S. Emergence of a molecular Bose–Einstein condensate from a Fermi gas. *Nature*, **426**:537 (2003).

[10] Jochim, S., Bartenstein, M., Altmeyer, A., Hendl, G., Riedl, S., Chin, C., Hecker-Denschlag, J., and Grimm, R. Bose–Einstein condensation of molecules. *Science*, **302**:2101 (2003).

[11] Zwierlein, M. W., Stan, C. A., Schunk, C. H., Raupach, S. M. F., Gupta, S., Hadzibabic, Z., and Ketterle, W. Observation of Bose–Einstein condensation of molecules. *Phys. Rev. Lett.*, **91**:250401 (2003).

[12] Leggett, A. J. Bose–Einstein condensation in the alkali gases. *Rev. Mod. Phys.*, **73**:333 (2001).

[13] Pitaevskii, L., and Stringari, S. *Bose–Einstein Condensation*. Oxford University Press, Oxford (2003).

[14] Pethick, C. J., and Smith, H. *Bose–Einstein Condensation in Dilute Gases*. Cambridge University Press (2002).

[15] Andrews, M. R., Townsend, C. G., Miesner, H. J., Durfee, D. S., Kurn, D. M., and Ketterle W. Observation of interference between two Bose condensates. *Science*, **275**:637 (1997).

[16] Bloch, I., Hänsch, T. W., and Esslinger T. Measurement of the spatial coherence of a trapped Bose gas at the phase transition. *Nature*, **403**:166 (2000).

[17] Chu, S. Nobel Lecture: The manipulation of neutral particles. *Rev. Mod. Phys.*, **70**:685 (1998).

[18] Cohen-Tannoudji, C. Nobel Lecture: Manipulating atoms with photons. *Rev. Mod. Phys.*, **70**:707 (1998).

[19] Phillips, W. D. Nobel Lecture: Laser cooling and trapping of neutral atoms. *Rev. Mod. Phys.*, **70**:721 (1998).

[20] Raab, E. L., Prentiss, M., Cable, A., Chu, S., and Pritchard, D. E. Trapping of neutral sodium with radiation pressure. *Phys. Rev. Lett.*, **59**:2631 (1987).

[21] Huang, K. *Statistical Mechanics*. Wiley, New York (1987).

[22] Katori, H., Ido, T., Isoya, Y., and Kuwata-Gonokami, M. Magneto-optical trapping and cooling of strontium atoms down to the photon recoil temperature. *Phys. Rev. Lett.*, **82**:116 (1999).

[23] Kuwamoto, T., Honda, K., Takahashi, Y., and Yabuzaki, T. Magneto-optical trapping of Yb atoms using an intercombination transition. *Phys. Rev.* A, **60**:745–748 (1999).

[24] Townsend, C. G., Edwards, N. H., Cooper, C. J., Zetie, K. P., Foot, C. J., Steane, A. M., Szriftgiser, P., Perrin, H., and Dalibard, J. Phase-space density in the magneto-optical trap. *Phys. Rev.* A, **52**:1423 (1995).

[25] Grimm, R., Weidemüller, M., and Ovchinnikov, Y. B. Optical dipole traps for neutral atoms. *Adv. At. Mol. Opt. Phys.*, **42**:95 (2000).

[26] Luiten, O. J., Reynolds, M. W., and Walraven, J. T. M. Kinetic theory of evaporative cooling of a trapped gas. *Phys. Rev.* A, **53**:381 (1996).

[27] Ketterle, W., and van Druten, N. J. Evaporative cooling of trapped atoms. In Bederson, B. and Walther, H. editors, *Advances in Atomic, Molecular, and Optical Physics*, volume 37, pages 181–236. Academic Press, San Diego (1996).

[28] Cornell, E. A., and Wieman, C. E. Nobel Lecture: Bose–Einstein condensation in a dilute gas, the first 70 years and some recent experiments. *Rev. Mod. Phys.*, **74**(3):875–893 (2002).

[29] Ketterle, W. Nobel Lecture: When atoms behave as waves: Bose–Einstein condensation and the atom laser. *Rev. Mod. Phys.*, **74**(4):1131–1151 (2002).

[30] Barrett, M. D., Sauer, J. A., and Chapman, M. S. All-optical formation of an atomic Bose–Einstein condensate. *Phys. Rev. Lett.*, **87**(1):10404 (2001).

[31] Dalfovo, F. S., Pitaevkii, L. P., Stringari, S., and Giorgini, S. Theory of Bose–Einstein condensation in trapped gases. *Rev. Mod. Phys.*, **71**:463 (1999).

[32] Gerbier, F., Thywissen, J. H., Richard, S., Hugbart, M., Bouyer, P., and Aspect, A. Critical temperature of a trapped, weakly interacting Bose gas. *Phys. Rev. Lett.*, **92**(3):030405 (2004).

[33] Gross, E. P. Structure of a quantized vortex in Boson systems. *Il Nuovo Cimento*, **20**(3):454–477 (1961).

[34] Pitaevskii, L. P. Vortex lines in an imperfect Bose gas. *Sov. Phys. JETP*, **13**(2):451–454 (1961).

[35] Lee, T. D., Huang, K., and Yang, C. N. Eigenvalues and eigenfunctions of a Bose system of hard spheres and its low-temperature properties. *Phys. Rev. Lett.*, **106**:1135 (1957).

[36] Steinhauer, J., Ozeri, R., Katz, N., and Davidson, N. Excitation spectrum of a Bose–Einstein condensate. *Phys. Phys. Lett.*, **88**(12):120407 (2002).

[37] Papp, S. B., Pino, J. M., Wild, R. J., Ronen, S., Wieman, C. E., Jin, D. S., and Cornell, E. A. Bragg spectroscopy of a strongly interacting Rb85 Bose–Einstein condensate. *Phys. Rev. Lett.*, **101**:135301 (2008).

[38] Sackett, C. A., Gerton, J. M., Welling, M., and Hulet, R. G. Measurement of collective collapse in a Bose–Einstein condensate with attractive interactions. *Phys. Rev. Lett.*, **82**:876 (1999).

[39] Burger, S., Bongs, K., Dettmer, S., Ertmer, W., Sengstock, K., Sanpera, A., Shlyapnikov, G. V., and Lewenstein, M. Dark solitons in Bose–Einstein condensates. *Phys. Rev. Lett.*, **83**(25):5198–5201 (1999).

[40] Strecker, K. E., Partridge, G. B., Truscott, A. G., and Hulet, R. G. Formation and propagation of matter–wave soliton trains. *Nature*, **417**(6885):150–153 (2002).

[41] Khaykovich, L., Schreck, F., Ferrari, G., Bourdel, T., Cubizolles, J., Carr, L. D., Castin, Y., and Salomon, C. Formation of a matter–wave bright soliton. *Science*, **296**(5571):1290 (2002).

[42] Drazin, P. G., and Johnson, R. S. *Solitons: An Introduction*. Cambridge University Press (1989).

[43] Baym, G., and Pethick, C. J. Ground-state properties of magnetically trapped Bose-condensed rubidium gas. *Phys. Rev. Lett.*, **76**:6 (1996).

[44] Madison, K. W., Chevy, F., Bretin, V., and Dalibard, J. Stationary states of a rotating Bose–Einstein condensate: Routes to vortex nucleation. *Phys. Rev. Lett.*, **86**:4443 (2001).

[45] Chevy, F., Madison, K. W., and Dalibard, J. Measurement of the angular momentum of a rotating Bose–Einstein condensate. *Phys. Rev. Lett.*, **85**:2223 (2000).

[46] Guéry-Odelin, D., and Stringari, S. Scissor mode and superfluidity of a trapped Bose–Einstein condensed gas. *Phys. Rev. Lett.*, **83**(22):4452–4455 (1999).

[47] Zambelli, F., and Stringari, S. Moment of inertia and quadrupole response function of a trapped superfluid. *Phys. Rev.* A, **63**(3):33602 (2001).

[48] Maragó, O. M., Hopkins, S. A., Arlt, J., Hodby, E., Hechenblaikner, G., and Foot, C. J. Observation of the scissor mode and evidence for superfluidity of a trapped Bose–Einstein condensed gas. *Phys. Rev. Lett.*, **84**:2056 (2000).

[49] Maragó, O. M., Hechenblaikner, G., Hodby, E., and Foot, C. J. Temperature dependence and frequency shifts of the scissor mode of a trapped Bose–Einstein condensate. *Phys. Rev. Lett.*, **86**:3938 (2001).

[50] Chikkatur, A. P., Görlitz, A., Stamper-Kurn, D. M., Inouye, S., Gupta, S., and Ketterle, W. Suppression and enhancement of impurity scattering in a Bose–Einstein condensate. *Phys. Rev. Lett.*, **85**:483 (2000).

[51] Frisch, T., Pomeau, Y., and Rica, S. Transition to dissipation in a model of superflow. *Phys. Rev. Lett.*, **69**:1644 (1992).

[52] Raman, C., Köhl, M., Onofrio, R., Durfee, D. S., Kuklewicz, C. E., Hadzibabic, Z., and Ketterle, W. Evidence for a critical velocity in a Bose–Einstein condensed gas. *Phys. Rev. Lett.*, **83**:2502 (1999).

[53] Onofrio, R., Raman, C., Vogels, J. M., Abo-Shaeer, J. R., Chikkatur, A. P., and Ketterle, W. Observation of superfluid flow in a Bose–Einstein condensed gas. *Phys. Rev. Lett.*, **85**:2228–2231 (2000).

[54] Inouye, S., Gupta, S., Rosenband, T., Chikkatur, A. P., Görlitz, A., Gustavson, T. L., Leanhardt, A. E., Pritchard, D. E., and Ketter, W. Observation of vortex phase singularities in Bose–Einstein condensates. *Phys. Rev. Lett.*, **87**:080402 (2001).

[55] Lahaye, T., Menotti, C., Santos, L., Lewenstein M., and Pfau, T. The physics of dipolar bosonic quantum gases. *Rep. Prog. Phys.*, **72**(12):126401–126442 (2009).

[56] Griesmaier, A., Werner, J., Hensler, S., Stuhler, J., and Pfau, T. Bose–Einstein condensation of chromium. *Phys. Rev. Lett.*, **94**:160401 (2005).

[57] Beaufils, Q., Chicireanu, R., Zanon, T., Laburthe-Tolra, B., Maréchal, E., Vernac, L., Keller, J. C., and Gorceix, O. All-optical production of chromium Bose–Einstein condensates. *Phys. Rev.* A, **77**(6):061601, Jun (2008).

[58] Rosensweig, R. E. *Ferrohydrodynamics*. Cambridge University Press (1985).

[59] Koch, T., Lahaye, T., Metz, J., Fröhlich, B., Griesmaier, A., and Pfau, T. Stabilizing a purely dipolar quantum gas against collapse. *Nat. Phys.*, **4**:218–222 (2008).

[60] Lu, M., Youn, S. H., and Lev, B. L. Trapping ultracold dysprosium: A highly magnetic gas for dipolar physics. *Phys. Rev. Lett.*, **104**(6):63001 (2010).

[61] Heidemann, R., Raitzsch, U., Bendkowsky, V., Butscher, B., Löw, R., and Pfau, T. Rydberg excitation of Bose–Einstein condensates. *Phys. Rev. Lett.*, **100**:33601 (2008).

[62] Deiglmayr, J., Grochola, A., Repp, M., Mörtlbauer, K., Glück, C., Lange, J., Dulieu, O., Wester, R., and Weidemüller, M. Formation of ultracold polar molecules in the rovibrational ground state. *Phys. Rev. Lett.*, **101**(13):133004 (2008).

[63] Ni, K. K., Ospelkaus, S., de Miranda, M. H. G., Pe'er, A., Neyenhuis, B., Zirbel, J. J., Kotochigova, S., Julienne, P. S., Jin, D. S., and Ye, J. A high phase-space-density gas of polar molecules. *Science*, **322**(5899):231 (2008).

[64] Danzl, J. G., Haller, E., Gustavsson, M., Mark, M. J., Hart, R., Bouloufa, N., Dulieu, O., Ritsch, H., and Nagerl, H. C. Quantum gas of deeply bound ground state molecules. *Science*, **321**(5892):1062 (2008).

[65] Anderson, P. W. Measurement in quantum theory and the problem of complex systems. In Boer, J. d., Dal, E., and Ulfbeck, O., editors, *The Lesson of Quantum Theory*, page 23, Elsevier, Amsterdam (1986).

[66] Leggett, A. J., and Sols, F. On the concept of spontaneously broken gauge symmetry in condensed matter physics. *Found. Phys.*, **21**:353 (1991).

[67] Javanainen, J., and Yoo, S. M. Quantum phase of a Bose–Einstein condensate with an arbitrary number of atoms. *Phys. Rev. Lett.*, **76**:161 (1996).

[68] Castin, Y., and Dalibard, J. Relative phase of two Bose–Einstein condensates. *Phys. Rev.* A, **55**:4330 (1997).

[69] Cirac, J. I., Gardiner, C. W., Naraschewski, M., and Zoller, P. Continuous observation of interference fringes from Bose condensates. *Phys. Rev.* A, **54**:R3714 (1996).

[70] Mullin W. J., and Laloë, F. Interference of Bose–Einstein condensates: Quantum nonlocal effects. *Phys. Rev. A*, **78**(6):061605 (2008).

[71] Albiez, M., Gati, R., Fölling, J., Hunsmann, S., Cristiani, M., and Oberthaler, M. K. Direct observation of tunneling and nonlinear self-trapping in a single bosonic Josephson junction. *Phys. Rev. Lett.*, **95**(1):10402 (2005).

[72] Levy, S., Lahoud, E., Shomroni, I., and Steinhauer, J. The ac and dc Josephson effects in a Bose–Einstein condensate. *Nature*, **449**(7162):579 (2007).

[73] Esteve, J., Gross, C., Weller, A., Giovanazzi, S., and Oberthaler, M. K. Squeezing and entanglement in a Bose–Einstein condensate. *Nature*, **455**(7217):1216 (2008).

[74] Riedel, M. F., Böhi, Pascal, Li, Y., Hänsch, T. W., Sinatra, A., and Treutlein, P. Atom-chip-based generation of entanglement for quantum metrology. *Nature*, **464**:1170 (2010).

[75] Gross, C., Zibold, T., Nicklas, E., Estève, J., and Oberthaler, M. K. Nonlinear atom interferometer surpasses classical precision limit. *Nature*, **464**:1165 (2010).

[76] Penrose, O., and Onsager, L. Bose–Einstein condensation and liquid helium. *Phys. Rev.*, **104**:576 (1956).

[77] Hagley, E. W., Deng, L., Kozuma, M., Trippenbach, M., Band, Y. B., Edwards, M., Doery, M., Julienne, P. S., Helmerson, K., Rolston, S. L., and Phillips, W. D. Measurement of the coherence of a Bose–Einstein condensate. *Phys. Rev. Lett.*, **83**:3112 (1999).

[78] Bloch, I., Hänsch, T. W., and Esslinger, T. An atom laser with a cw output coupler. *Phys. Rev. Lett.*, **82**:3008 (1999).

[79] Helmerson, K., Hutchinson, D., Burnett, K., and Phillips, W. D. Atom lasers. *Physics World*, page 31, August (1999).

[80] Donner, T., Ritter, S., Bourdel, T., Ottl, A., Köhl, M., and Esslinger, T. Critical behavior of a trapped interacting Bose gas. *Science*, **315**(5818):1556–1558 (2007).

[81] Schellekens, M., Hoppeler, R., Perrin, A., Viana Gomes, J., Boiron, D., Aspect, A., and Westbrook, C. I. Hanbury Brown Twiss effect for ultracold quantum gases. *Science*, **310**:648 (2005).

[82] Jeltes, T., McNamara, J. M., Hogervorst, W., Vassen, W., Krachmalnicoff, V., Schellekens, M., Perrin, A., Chang, H., Boiron, D., Aspect, A., and Westbrook, C. Hanbury Brown Twiss effect for bosons versus fermions. *Nature*, **445**:402–405 (2007).

[83] Yasuda, M., and Shimizu, F. Observation of two-atom correlation of an ultracold neon atomic beam. *Phys. Rev. Lett.*, **77**:3090 (1996).

[84] Altman, E., Demler, E., and Lukin, M. D. Probing many-body states of ultracold atoms via noise correlations. *Phys. Rev.* A, **70**:013603 (2004).

[85] Fölling, S., Gerbier, F., Widera, A., Mandel, O., Gericke, T., and Bloch, I. Spatial quantum noise interferometry in expanding ultracold atom clouds. *Nature*, **434**:481–484 (2005).

[86] Greiner, M., Regal, C. A., Stewart, J. T., and Jin, D. S. Probing pair-correlated Fermionic atoms through correlations in atom shot noise. *Phys. Rev. Lett.*, **94**:110401 (2005).

[87] Kagan, Y., Svistunov, B. V., and Shlyapnikov, G. V. Effect of Bose condensation on inelastic processes in gases. *JETP Lett.*, **642**:209 (1985).

[88] Burt, E. A., Ghrist, R. W., Myatt, C. J., Holland, M. J., Cornell, E. A., and Wieman, C. E. Coherence, correlations, and collisions: What one learns about Bose–Einstein condensates from their decay. *Phys. Rev. Lett.*, **79**:337 (1997).

[89] Laburthe Tolra, B., O'Hara, K. M., Huckans, J. H., Phillips, W. D., Rolston, S. L., and Porto, J. V. Observation of reduced three-body recombination in a correlated 1d degenerate Bose gas. *Phys. Rev. Lett.*, **92**(19):190401, May (2004).

[90] Peierls, R. E. Quelques propriétés typiques des corps solides. *Ann. Inst. Henri Poincaré*, **5**:177 (1935).

[91] Mermin, N. D., and Wagner, H. Absence of ferromagnetism or antiferromagnetism in one- or two-dimensional isotropic Heisenberg models. *Phys. Rev. Lett.*, **17**:1133 (1966).

[92] Hohenberg, P. C. Existence of long-range order in one and two dimensions. *Phys. Rev.*, **158**:383 (1967).

[93] Cooper, N. R., and Hadzibabic, Z. Measuring the superfluid fraction of an ultracold atomic gas. *Phys. Rev. Lett.*, **104**(3):030401 (2010).

[94] Bloch, I., Dalibard, J., and Zwerger, W. Many-body physics with ultracold gases. *Rev. Mod. Phys*, **80**(3):885 (2008).

[95] Bishop D. J., and Reppy, J. D. Study of the superfluid transition in two-dimensional ^{4}He films. *Phys. Rev. Lett.*, **40**(26):1727–1730 (1978).

[96] Berezinskii, V. L. Destruction of long-range order in one-dimensional and two-dimensional system possessing a continuous symmetry group - ii. Quantum systems. *Sov. Phys. JETP*, **34**:610 (1971).

[97] Kosterlitz, J. M., and Thouless, D. J. Ordering, metastability and phase transitions in two dimensional systems. *J. Phys.* C *Solid State Phys.*, **6**:1181 (1973).

[98] Nelson, D. R., and Kosterlitz, J. M. Universal jump in the superfluid density of two-dimensional superfluids. *Phys. Rev. Lett.*, **39**:1201 (1977).

[99] Prokof'ev, N. V., Ruebenacker, O., and Svistunov, B. V. Critical point of a weakly interacting two-dimensional Bose gas. *Phys. Rev. Lett.*, **87**:270402 (2001).

[100] Bagnato V. S., and Kleppner, D. Bose–Einstein condensation in low-dimensional traps. *Phys. Rev.* A, **44**(11):7439–7441 (1991).

[101] Hadzibabic Z., and Dalibard, J. Two-dimensional Bose fluids: An atomic physics perspective. In Kaiser, R., and Wiersma, D., editors, *Nano Optics and Atomics: Transport of Light and Matter Waves*, volume CLXXIII of *Proceedings of the International School of Physics Enrico Fermi, 2009*. IOS Press (2010). arXiv:0912.1490.

[102] Bhaduri, R. K., Reimann, S. M., Viefers, S., Ghose Choudhury, A., and Srivastava, M. K. The effect of interactions on Bose–Einstein condensation in

a quasi two-dimensional harmonic trap. *J. Phys.* B. *At. Mol. Opt. Phys.*, **33**:3895 (2000).

[103] Holzmann, M., Baym, G., Blaizot, J. P., and Laloë, F. Superfluid transition of homogeneous and trapped two-dimensional Bose gases. *Proc. Natl. Acad. Sci. USA*, **104**:1476 (2007).

[104] Holzmann, M., and Krauth, W. Kosterlitz–Thouless transition of the quasi two-dimensional trapped Bose gas. *Phys. Rev. Lett.*, **100**(19):190402 (2008).

[105] Bisset, R. N., Davis, M. J., Simula, T. P., and Blakie, P. B. Quasicondensation and coherence in the quasi-two-dimensional trapped bose gas. *Phys. Rev.* A, **79**(3):033626 (2009).

[106] Giorgetti, L., Carusotto, I., and Castin, Y. Semiclassical field method for the equilibrium Bose gas and application to thermal vortices in two dimensions. *Phys. Rev.* A, **76**(1):013613 (2007).

[107] Bramwell S. T., and Holdsworth, P. C. W. Magnetization: A characteristic of the Kosterlit–Thouless–Berezinskii transition. *Phys. Rev.* B, **49**(13):8811–8814 (1994).

[108] Petrov D. S., and Shlyapnikov, G. V. Interatomic collisions in a tightly confined Bose gas. *Phys. Rev.* A, **64**:012706 (2001).

[109] Görlitz, A., Vogels, J. M., Leanhardt, A. E., Raman, C., Gustavson, T. L., Abo-Shaeer, J. R., Chikkatur, A. P., Gupta, S., Inouye, S., Rosenband, T., and Ketterle, W. Realization of Bose–Einstein condensates in lower dimensions. *Phys. Rev. Lett.*, **87**:130402 (2001).

[110] Krüger, P., Hadzibabic, Z., and Dalibard, J. Critical point of an interacting two-dimensional atomic Bose gas. *Phys. Rev. Lett.*, **99**(4):040402 (2007).

[111] Hadzibabic, Z., Krüger, P., Cheneau, M., Rath, S. P., and Dalibard, J. The trapped two-dimensional Bose gas: from Bose–Einstein condensation to Berezinskii–Kosterlitz–Thouless physics. *New J. Phys.*, **10**(4):045006 (2008).

[112] Cladé, P., Ryu, C., Ramanathan, A., Helmerson, K., and Phillips, W. D. Observation of a 2d Bose gas: From thermal to quasicondensate to superfluid. *Phys. Rev. Lett.*, **102**(17):170401 (2009).

[113] Polkovnikov, A., Altman, E., and Demler, E. Interference between independent fluctuating condensates. *Proc. Natl. Acad. Sci. USA*, **103**:6125 (2006).

[114] Gritsev, V., Altman, E., Demler, E., and Polkovnikov, A. Full quantum distribution of contrast in interference experiments between interacting one-dimensional Bose liquids. *Nat. Phys.*, **2**:705–709 (2006).

[115] Hadzibabic, Z., Krüger, P., Cheneau, M., Battelier, B., and Dalibard, J. Berezinskii–Kosterlitz–Thouless crossover in a trapped atomic gas. *Nature*, **441**:1118–1121 (2006).

[116] Simula, T. P., and Blakie, P. B. Thermal activation of vortex-antivortex pairs in quasi-two-dimensional Bose–Einstein condensates. *Phys. Rev. Lett.*, **96**:020404 (2006).

[117] Hofferberth, S., Lesanovsky, I., Fischer, B., Schumm, T., and Schmiedmayer, J. Non-equilibrium coherence dynamics in one-dimensional Bose gases. *Nature*, **449**(7160):324–327 (2007).

[118] Burkov, A. A., Lukin, M. D., and Demler Eugene. Decoherence dynamics in low-dimensional cold atom interferometers. *Phys. Rev. Lett.*, **98**(20):200404 (2007).

[119] Henn, E. A. L., Seman, J. A., Roati, G., Magalhães, K. M. F., and Bagnato, V. S. Emergence of turbulence in an oscillating Bose–Einstein condensate. *Phys. Rev. Lett.*, **103**(4):045301, Jul (2009).

[120] Bakr, W. S., Peng, A., Folling, S., and Greiner, M. A quantum gas microscope for detecting single atoms in a Hubbard-regime optical lattice. *Nature*, **462**:74–77 (2009).

[121] Greiner, M., Mandel, M. O., Esslinger, T., Hänsch, T., and Bloch, I. Quantum phase transition from a superfluid to a mott insulator in a gas of ultracold atoms. *Nature*, **415**:39 (2002).

[122] Cooper, N. R. Rapidly rotating atomic gases. *Adv. Phys.*, **57**(6):539–616 (2008).

[123] Sadler, L. E., Higbie, J. M., Leslie, S. R., Vengalattore, M., and Stamper-Kurn, D. M. Spontaneous symmetry breaking in a quenched ferromagnetic spinor Bose–Einstein condensate. *Nature*, **443**:312–315 (2006).

[124] Taglieber, M., Voigt, A.-C., Aoki, T., Hänsch, T. W., and Dieckmann, K. Quantum degenerate two-species Fermi–Fermi mixture coexisting with a Bose–Einstein condensate. *Phys. Rev. Lett.*, **100**(1):010401, Jan (2008).

[125] Billy, J., Josse, V., Zuo, Z., Bernard, A., Hambrecht, B., Lugan, P., Clément, D., Sanchez-Palencia, L., Bouyer, P., and Aspect, A. Direct observation of Anderson localization of matter waves in a controlled disorder. *Nature*, **453**:891 (2008).

[126] Roati, G., D'Errico, C., Fallani, L., Fattori, M., Fort, C., Zaccanti, M., Modugno, G., Modugno, M., and Inguscio, M. Anderson localization of a non-interacting Bose–Einstein condensate. *Nature*, **453**:895 (2008).

[127] Sanchez-Palencia, L. and Lewenstein, M. Disordered quantum gases under control. *Nat. Phys.*, **6**(87) (2010).

Experiments on excitons in bulk semiconductors

M. Kuwata-Gonokami

8.1 Introduction: excitonic matter in semiconductors

Collective quantum mechanical phenomena such as superfluidity and superconductivity appear in ensembles of bosonic particles at high densities and low temperatures. An ensemble of electrons and holes in a semiconductor provides an opportunity to study the collective quantum mechanical phenomena involving carriers with strong Coulomb correlation. A small effective mass of bipolar carriers causes pronounced quantum effects [1, 2].

Figure 8.1 is a schematic view of the forms that electron and hole ensembles take at various densities and temperatures. In the dilute region, electrons and holes combine to form a bound state known as an exciton. An exciton is a quantum consisting of an electronic excitation wave that does not carry electronic current, but can carry energy via the correlated motions of electrons and holes. This concept of excitons as an elementary excitation of dielectric materials was introduced by Frenkel (1931a) [3] and Peierls et al. (1932) [4]. According to the band theory, the behavior of electrons and holes is fully controlled by the band structure. The electrons and holes created by an absorbed photon are close to one another, so Coulomb attraction results in the formation of bound states or excitons.

The electronic structure of excitons involves many-body interactions of electrons in the ground and excited states, and calculations of their electronic structure are rather complicated and non-trivial. However, in the limit where the distance between an electron and a hole is much larger than the lattice spacing, we can apply the effective mass approximation (EMA), as first discussed by Wannier et al. (1937) [5]. In conventional semiconductor systems, the EMA appropriately describes the motions of electrons and holes. Within this approximation, an electron and a hole form a hydrogen-like bound state having a Bohr radius much larger than the lattice spacing. This type of exciton is called a Wannier–Mott exciton. An exciton is the excited state of the many-body

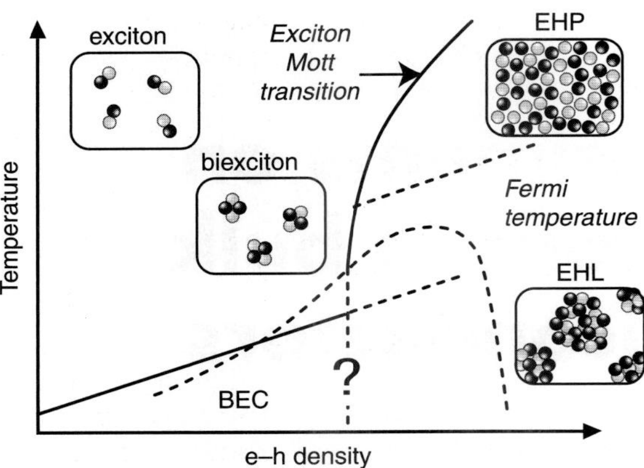

Fig. 8.1 Schematic phase diagram of electron–hole systems in semiconductors.

system that constitutes a semiconductor crystal; it can be seen as a quantum of electronic excitation that can carry energy in a crystal lattice with a finite momentum.

Although this description of Wannier–Mott excitons successfully explains the properties of excitons in various semiconductors, an exciton is a much more complex and rich system than simply a pair of electrostatically bound particles (positive holes and negative electrons) immersed in a dielectric matrix. Its consistent description should be based on the many-body theory, which enables the physics of relevant processes to be visualized. Pioneering work by Sham and Rice in terms of the many-body theory showed that an exciton can still be regarded as a bound pair of quasiparticles with positive and negative charges [6]. The simple picture of an exciton composed of coupled electrons and holes may fail in the limit of a weakly bound or large exciton Bohr radius. The coupling between the quasi-electron and the quasi-hole that form the exciton may involve both static Coulomb and retarded interactions, which should be taken into account for strong electron–hole coupling. Nevertheless, in the limit of weak binding and large spatial extent, the Coulomb interaction screened by the macroscopic dielectric constant dominates.

As the exciton density increases, a bound state of two excitons, an excitonic molecule referred to as a biexciton, appears. Since both excitons and biexcitons contain an even number of fermions, these quasiparticles obey Bose statistics. The possibility of Bose–Einstein condensation (BEC) of excitons or biexcitons has been extensively studied. As noted above, ensembles of electrons and holes are complex quantum systems; thus, it is a non-trivial question as to whether nature chooses a form of BEC of excitons or biexcitons.

At higher densities, an exciton ensemble produces screening of Coulomb attraction and finally the ionization of excitons. The ionized electron–hole ensemble is called an electron–hole plasma (EHP). This transformation from an insulator to a metallic phase is referred to as the exciton Mott transition. At a sufficiently low temperature, the EHP condenses inhomogeneously and forms electron–hole droplets (EHDs), which resemble spatial phase separation at the first-order phase transition. In this phase, which is called the electron–hole liquid phase, the mean interparticle distance is less than the Bohr radius of an exciton; therefore, the electron–hole ensemble shows high Fermi degeneracy. EHDs have been investigated extensively in indirect-gap semiconductors with large conduction band degeneracy, and EHD formation has been well established both experimentally and theoretically [7]. EHDs in diamond crystals show a very high critical temperature of 165 K [8, 9].

On the other hand, in direct-gap semiconductors, the lifetimes of photogenerated carriers and excitons are short, typically less than one nanosecond. Under intense optical excitation, the initial temperature of an electron–hole ensemble is usually very high. Thus, it is difficult for an electron–hole ensemble to reach a low-temperature state within its short lifetime. For example, in the direct-gap semiconductor CuCl, high-density electrons and holes have been shown to be generated at a low temperature under the resonant excitation of excitons with short pulses. Above the Mott transition density, cold high-density electrons and holes form a droplet-like phase [10].

The transformation from an excitonic Bose–Einstein condensate (BEC) to a fermionic liquid in the quantum degenerate regime at low temperatures is of particular interest. An exciton ensemble might evolve continuously from a dilute BEC into a dielectric superfluid consisting of a Bardeen–Cooper–Schrieffer (BCS)-like degenerate two-component Fermi liquid with Coulomb attraction [11–14].

It is important to develop experimental methods for probing the behavior of high-density carriers. Conventionally, luminescence spectroscopy has been widely used to examine the behavior of photogenerated carriers. The recent development in ultrashort-pulse laser technology has expanded the optical probing range in both frequency and time domains. In particular, developments in far-infrared (THz) spectroscopy provide a new opportunity to probe the electronic properties of photogenerated carriers. The theoretical bases of such methods have also been established [10, 15, 16]. In the insulating phase, electrons and holes form ensembles of excitons or biexcitons with optical responses characteristic of the internal motion of bound electrons and holes. In the ionic phase of electrons and holes, metal-like optical responses appear (Fig. 8.2). Time- and space-resolved low-frequency electromagnetic response measurements reveal the signature of an insulating-to-metallic-phase transformation.

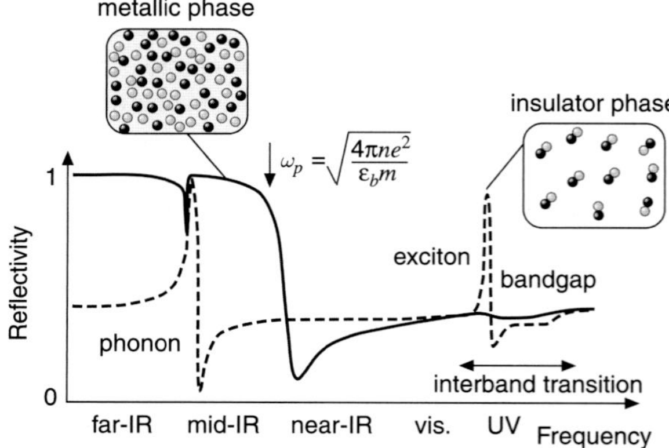

Fig. 8.2 Schematic reflection spectra of a photoexcited semiconductor (solid curve) and a non-excited, undoped semiconductor (dashed curve).

8.2 Direct optical manipulation of supercooled excitonic particles

This section describes experimental methods for generating highly degenerate excitonic particles by using energy- and momentum-selective optical excitation processes for biexcitons in CuCl and orthoexcitons in Cu_2O. Phase-sensitive interference of particles demonstrates the macroscopic coherence of such ensembles of cold particles prepared in the quantum degenerate regime.

8.2.1 Generation of highly degenerate biexcitons

Particle–wave duality is well documented for photons as well as massive particles in the dilute limit, where a particle interferes with itself. Interference of BECs of atoms demonstrates the wave nature of a large number of massive particles in a direct and spectacular way. Excitonic particles in semiconductors offer an alternative system for probing and exploiting the dual aspect of a wave and a massive particle at high densities and low temperatures. In laser cooling of atoms, highly collimated incident laser light with almost zero entropy interacts with thermal atoms and extracts their entropy [17, 18]. On the other hand, we can dramatically reduce the entropy of excitonic particles by directly transferring the small entropy of laser light to matter through state-selective generation of excitonic particles, via the energy and momentum selection rules of optical transition.

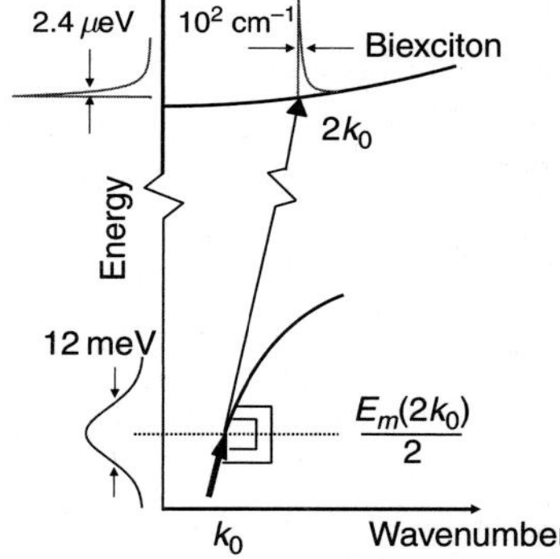

Fig. 8.3 Phase space compression scheme for the creation of supercooled biexcitons (a coherent biexcitonic matter wave) by two-photon excitation [23]. A photon with excess energy ΔE can find a partner photon having an equivalent energy deficit $-\Delta E$ to complete a two-photon transition and thus convert into a biexciton while still satisfying energy and momentum conservation. The calculated distribution in momentum space of biexcitons created by a femtosecond laser pulse is also shown. The initial momentum spread of the optical pulses $\Delta k = 1.0 \times 10^5\,\text{cm}^{-1}$ is compressed by three orders of magnitude. The corresponding initial biexcitonic effective temperature is estimated to be $T \sim 3 \times 10^{-2}$ K.

First, we focus on biexcitons in a CuCl crystal. The biexciton state is a bound state of two excitons, an exciton analog of a hydrogen molecule. In CuCl, a biexciton state with a total energy of 6.372 eV and binding energy of 30 meV has been observed in various experiments and is known to be very stable. Its wave function exhibits total s-like symmetry (Γ_1). The translational mass of biexcitons has been determined to be $m = 5.29\,m_e$ using momentum-resolved resonant two-photon Raman spectroscopy [19]. Ensembles of biexcitons show a characteristic luminescence band, i.e. the M band, and their spectral profiles reflect the thermal distribution function of biexcitons. The observed spectra indicate that a collection of biexcitons behaves like a gas of independent Bose particles that are spread in momentum space according to the Maxwell distribution at low densities and evolve toward a Bose distribution at higher densities and lower temperatures [20].

Γ_1 biexcitons can be created directly by light in a two-photon process with an extremely large cross-section [21, 22]. A pulsed coherent light source tuned to half of the biexciton energy, 3.186 eV, permits the creation of a large number of biexcitons in a very small number of momentum states, thereby effectively producing a supercooled gas of biexcitons [23]. Figure 8.3 illustrates the strong phase space compression that occurs even when we use a spectrally broad femtosecond laser pulse of large bandwidth $\Delta k = 1 \times 10^5\,\text{cm}^{-1}$ around $k_0 = 4.45 \times 10^5\,\text{cm}^{-1}$ for excitation. The high-frequency component of the incident pulse couples with the low-frequency component to obey the two-photon transition selection rule, while satisfying energy and momentum conservation. As a result, most of the created particles occupy a few biexciton modes around $2k_0$, with an initial occupation number per mode of $N_k > 10^4$. The rest is spread in a tail with an estimated effective temperature of $T \sim 3 \times 10^{-2}$ K.

In terms of the wave picture, this high occupancy in a mode implies that we can prepare a large-amplitude phase-coherent biexciton wave. Two such waves exhibit interference; therefore, the phase and amplitude of the biexciton ensemble can be manipulated by adjusting the relative phase of the two biexciton waves. In contrast with the conventional scheme of coherent control in semiconductors [24], the present method of biexciton ensemble manipulation relies on second-order two-photon coherence [25] rather than on interband polarization. In CuCl, biexciton–acoustic phonon coupling is weak at low temperatures [26]; thus, biexcitons can preserve their initial phase coherence for a long time. At low temperatures, the dephasing time of biexcitons is in the order of 100 ps, which is limited by spontaneous two-photon decay [26]. The long-lived coherence of biexcitons enables the macroscopic occupation of biexcitons in a small phase space region with high quantum degeneracy. This type of biexciton ensemble can be regarded as a "forced condensate." [27].

Figure 8.4 shows the schematic of an experiment demonstrating biexciton wave interference. The sample is exposed to linearly polarized femtosecond laser pulses at times $t_1 = 0$ and $t_2 = \tau$, creating biexcitonic waves $A = A_0(t)\exp\{-i\omega_b t + ik_b z\} + \text{c.c.}$ and $B = B_0(t)\exp\{-i\omega_b(t-\tau) + ik_b z\} + \text{c.c.}$ with $\omega_b = 9.69 \times 10^{15}\,\text{s}^{-1}$ and $k_b = 2k_0 = 8.9 \times 10^5\,\text{cm}^{-1}$. To eliminate direct optical interference, the pulses are set to be orthogonally polarized. An ultra-short femtosecond diffracted laser pulse is proportional to the intensity of the coherent biexciton wave $|A+B|^2$ at t_3. The variation in the diffracted probe pulse intensity as a function of τ corresponds to fringes in the interference pattern of biexciton waves A and B. A polarization beam splitter is placed after the sample to separate the first and second pulses so that the transmission measurement of the second pulse impinges on the sample at time t.

To optimize the fringe contrast, we adjusted the amplitudes of waves A and B so that they were equal and then introduced a delay τ of 2 ps. The interference pattern was detected by a weak probe pulse from the same laser impinging on

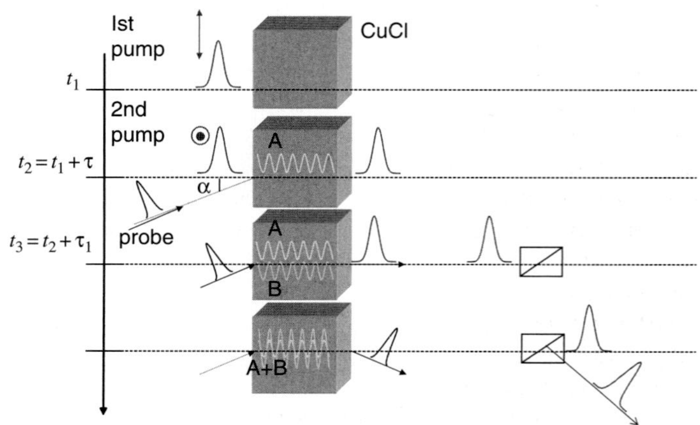

Fig. 8.4 Schematic description of the experiment [23]. At time t_1, the first linearly polarized laser pulse is partially converted into biexciton wave A through a two-photon absorption process. At time $t_2 = t_1 + \tau$, the second laser pulse, which is linearly polarized orthogonal to the first pulse, produces a second coherent biexciton wave B. The superposition wave is measured by a third probe pulse at time t_3. The intensity of the diffracted pulse is proportional to that of the coherent biexciton wave $|A+B|^2$ at t_3. The variation in the diffracted probe pulse intensity as a function of τ corresponds to fringes in the interference pattern of biexciton waves A and B. A polarization beam splitter is placed after the sample to separate the first and second pulses for the transmission measurement of the second pulse.

the sample under a small angle α at time t_3, with $t_3 - t_2 > 300\,\text{fs}$. The probe pulse produced a scattered signal proportional to the square amplitude of the superposed biexciton wave at time $t = t_3$.

We used a carefully selected high-quality 33-μm-thick single crystal platelet, mounted on a stress-free holder, and cooled to 10 K. Figure 8.5(a) shows the biexciton wave interference pattern and transmitted intensity of the second pulse recorded by finely adjusting the delay τ using a piezoelectric element. The fringe period $2\pi/\omega = 0.65\,\text{fs}$ corresponds exactly with the resonance frequency of the biexciton. The recorded waveform shown in Fig. 8.5(a) demonstrates an almost perfectly sine-wave-shaped interference pattern, indicating that biexcitons in CuCl behave as almost ideal Bose particles at a density of $10^{13}\,\text{cm}^{-3}$, with as many as 10^4 particles accommodated in a single mode.

Energy conservation in a system of biexcitons and photons requires that at the fringe minima, the energy of biexcitons provided by the first pulse is transferred to photons, producing two-photon amplification of the second pulse. This amplification is clearly observed as a phase-dependent modulation of the second pulse transmission with a phase shift of π relative to the biexciton interference pattern, as shown in Fig. 8.5(b).

A coherent biexciton wave prepared by the phase space compression of photons shown in Fig. 8.3 acts in reverse to generate a photon pair.

For a strong first pulse intensity of 0.7 MW/cm^2, 40% of the light energy is transferred to the crystal in the form of a coherent biexciton wave with an initial biexciton density of $10^{14}\,\text{cm}^{-3}$ [28]. We amplified the second optical

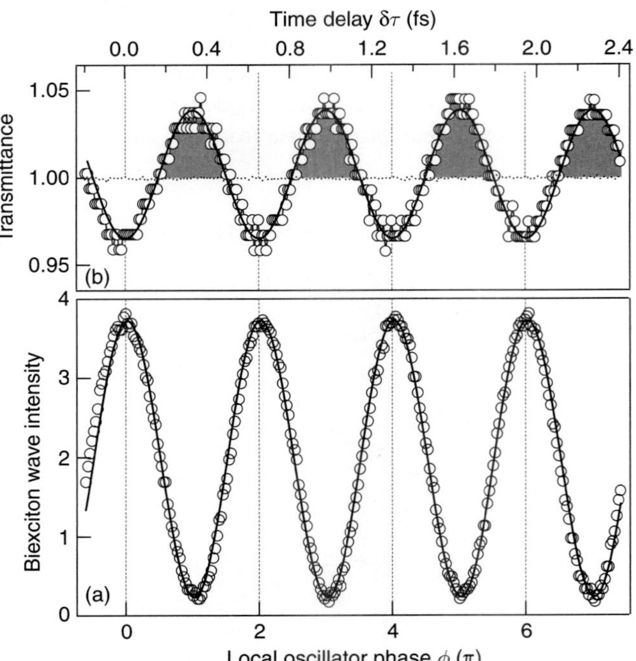

Fig. 8.5 (a) Interference pattern created by two biexciton waves having density of $4 \times 10^{11}\,\text{cm}^{-3}$. The intensity of the scattered probe pulse at $t_3 = t_2 + 0.3\,\text{ps}$ is plotted as a function of the fine tuning of τ around 2 ps. (b) Transmitted intensity of the second pulse at $t_2 = t_1 + \tau$ recorded under the same conditions as in (a) [23].

pulse of lower amplitude by up to three times, with a corresponding gain of $G = 350\,\text{cm}^{-1}$ (Figs. 8.6 and 8.7). The efficient interaction between light and matter waves implies that strong squeezing of the generated light can be achieved. The amplification coefficient is a function of the product of pulse duration and intensity when the biexciton dephasing time is longer than the pulse

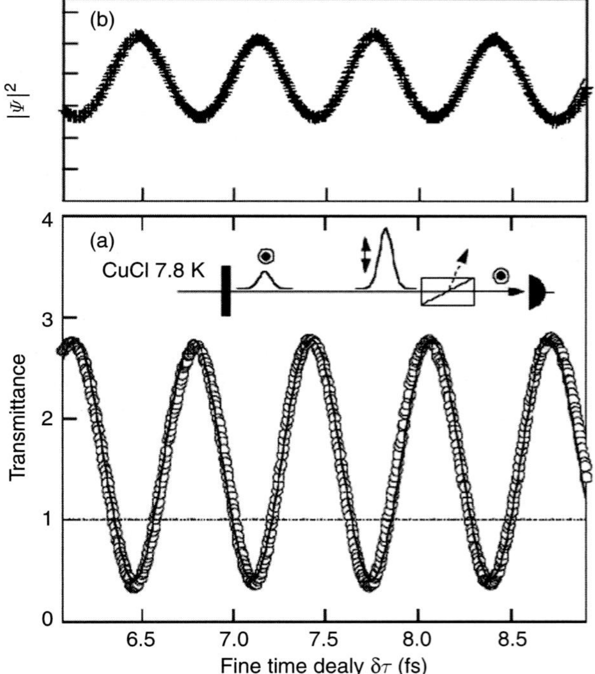

Fig. 8.6 (a) Transmittance of the second pulse as a function of fine time delay between the first (700 kW/cm^2) and second (46 kW/cm^2) pulses. The transmittance shows phase-sensitive amplification and de-amplification at an oscillation period corresponding to the biexciton frequency ω_b. (b) Simultaneously measured scattered probe pulse and biexciton-mediated four-wave mixing signal intensity, showing the squared biexciton wave amplitude $|\psi|^2$ [28].

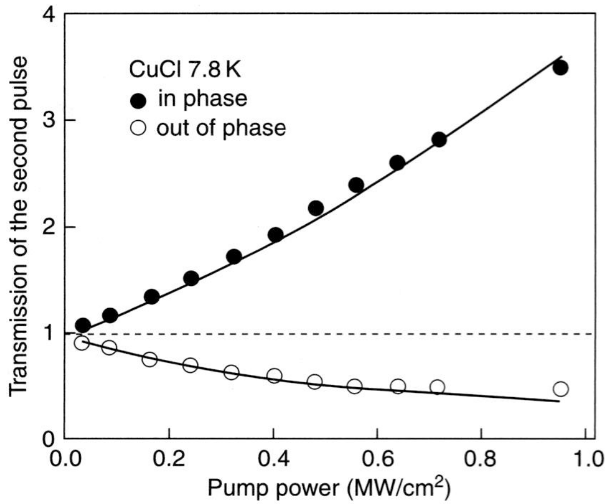

Fig. 8.7 Dependence of the maximum and minimum of second pulse transmittance on first pulse intensity. Second pulse intensity is kept constant at 46 kW/cm^2 [28].

duration. From the obtained value of the two-photon parametric amplification coefficient, we can estimate the required pump intensity for strong squeezing. For instance, 10 dB squeezing would require 11 MW/cm^2 for a 10 ps pump pulse.

Ultra-fast time-gated operation of squeezed light generation and entangled photon pair generation is another intriguing possibility [29]. A phase jump of $\pi/2$ is sufficient to switch the system from an amplifying to an absorbing state. The biexciton gain can be extinguished by using a phase-controlled succession of optical pulses and by inducing a $\pi/2$ shift.

8.2.2 Two-photon coherence storage in dark excitons in Cu_2O

Long-lived phase coherence of biexcitons provides a reservoir of photons exhibiting two-photon coherence. In this subsection, we present another example of two-photon coherence storage in excitonic particles. The yellow series of exciton states in Cu_2O has been discussed as a model system of Wannier–Mott excitons. Since both the valence and conduction bands have the same parity, the interband transition is forbidden at $k = 0$ in dipole approximation. A 4s-like conduction electron and a 3d-like valence hole combine to form hydrogen-like envelope functions. The yellow exciton series, which appears in the 2.1–2.2 eV spectral range, is rather weak, in the order of 10^3–10^4 cm^{-1}, which is one or two orders weaker than typical direct-gap semiconductors. An electron–hole exchange interaction splits the fourfold 1s exciton states into threefold degenerate orthoexciton states and a lower non-degenerate paraexciton state. The 1s orthoexcitons are electric dipole forbidden but active for the electric quadrupole one-photon transition and two-photon dipole transition. Owing to these selection rules, resonant second-harmonic coherent emission can occur. The electron–hole exchange interaction in crystalline solids has a K-dependent component arising from the non-local nature of the interaction. Hence, the threefold degeneracy of orthoexciton states $|O_{yz}\rangle$, $|O_{zx}\rangle$, and $|O_{xy}\rangle$ is lifted for non-zero K states. Usually such splitting is very small. Recently, high-resolution linear absorption spectroscopy has revealed a fine splitting in the order of a few microelectronvolts [30].

When the excitation beam is set parallel to the [110] crystal axis, two dark exciton states, $|O_1\rangle = \left(-|O_{yz}\rangle + |O_{zx}\rangle\right)/\sqrt{2}$ and $|O_3\rangle = |O_{xy}\rangle$, and one bright state, $|O_2\rangle = \left(|O_{yz}\rangle + |O_{zx}\rangle\right)/\sqrt{2}$, are obtained, as shown in Fig. 8.8. For the dark states, one-photon transitions are forbidden, but two-photon dipole transitions are allowed. For the bright state, only one-photon quadrupole transitions are allowed, and two-photon transitions are forbidden. Because of this complementarity in the selection rules, resonant coherent two-photon emission processes, namely resonant second-harmonic generation, are inhibited. Emission from the bright state $|O_2\rangle$ occurs via incoherent repopulation processes from the coherent dark exciton states $|O_1\rangle$ and $|O_3\rangle$. As a result, the quadrupole emission occurring in response to pulsed two-photon excitation of dark states has a finite rise time.

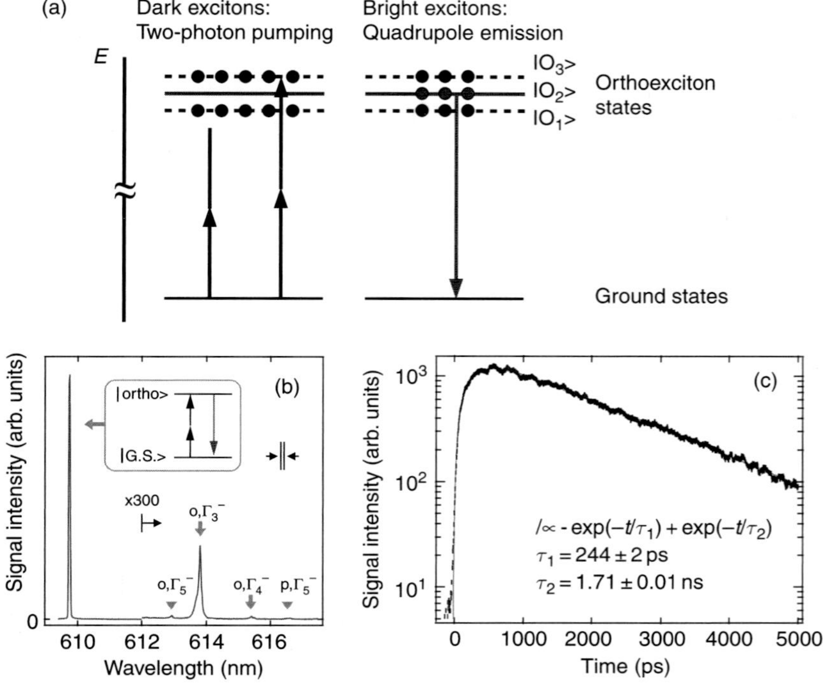

Fig. 8.8 (a) 1s orthoexciton energy levels in Cu_2O when a linearly polarized pump beam with a wavevector parallel to the [110] crystal axis is used. Solid line and dashed line indicate bright and dark orthoexciton states, respectively. The population ratio of the $|O_1\rangle$ and $|O_3\rangle$ dark exciton states can be tuned by rotating the pump polarization. Two-photon pumping is allowed only for the dark exciton states. The exciton population relaxes to a uniform distribution among the three states, which lose their initial phase coherence. Radiation occurs when excitons in the $|O_2\rangle$ bright state recombine. (b) Time-integrated photoluminescence spectrum under two-photon resonant excitation of 1s orthoexcitons in a 440-μm-thick Cu_2O crystal at 1.8 K. The duration of the excitation pulses is 2 ps, and they are nearly transform-limited, with a central wavelength of 1219.5 nm and a repetition rate of 76 MHz. A significant two-photon resonant emission signal appears at 609.74 nm. The four lines on the right are the phonon-assisted luminescence of ortho (o) and para (p) excitons, which are not discussed here. Each Γ_i^- represents the symmetry of the optical phonon mode involved in the transition. (c) Time-resolved two-photon resonant emission (609.74 nm) at 1.8 K detected by a synchroscan streak camera. A double-exponential fit convoluted with a 40-ps resolution was used to obtain a 244 ± 2 ps rise time and a 1.71 ± 0.01 ns fall time [31].

High-quality naturally grown single Cu_2O crystal samples with 440 and 800 μm thickness were set in a sample holder in a nearly strain-free state and maintained at 1.8 K in a liquid helium cryostat. Figure 8.8(b) shows the emission spectrum under two-photon excitation, with the wavevector along the [110] crystal axis. A very sharp emission line appears at the orthoexciton resonance ($\lambda = 609.74$ nm, Fig. 8.8(b)). The temporal evolution (Fig. 8.8(c)) shows a finite rise time of 244 ps, followed by a long decay time of 1.7 ns, in strong contrast to the case of coherent second-harmonic generation, in which the pulsed signal follows the excitation pulse closely. This implies that orthoexciton levels are split into bright and dark exciton states, as discussed above.

Since the dark excitons are decoupled from the radiation field, they can preserve phase information for a long time. To measure their coherence time, we performed a two-photon coherent control experiment, similar to that for biexcitons of CuCl in the previous subsection. Figure 8.9(a) shows the experimental setup. A sequence of phase-locked laser pulse pairs is obtained using a Michelson interferometer, and the two-photon resonant quadrupole emission intensity is measured as a function of the pulse separation time. The lower part of Fig. 8.9(b) shows the optical interference pattern of the two pump pulses, which have a central wavelength $\lambda_p = 1219.5$ nm, when they overlap each other nearly completely in time ($\tau_c = 0.0$ ps). We observed an oscillation period $\Delta\tau = (c/\lambda_p)^{-1} = 4.07$ fs, where c is the velocity of light.

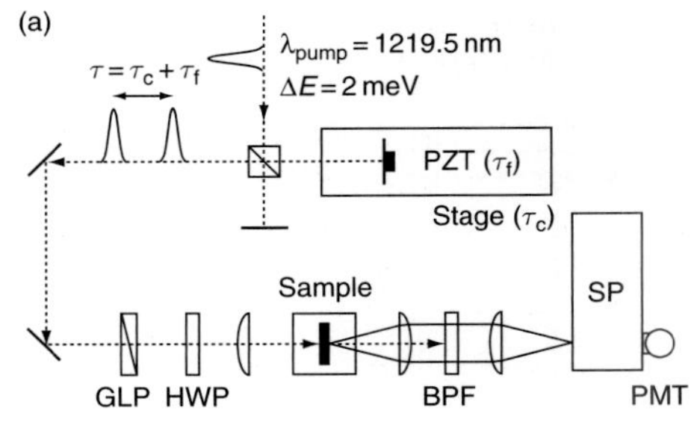

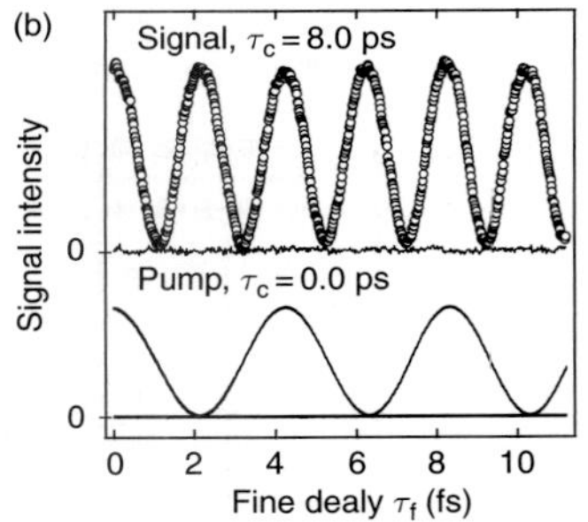

Fig. 8.9 (a) Experimental setup for two-photon coherent control of orthoexcitons. The time-integrated resonant emission signal is detected with a photomultiplier tube. The excitation power density is $6.1\,\mathrm{MW/cm^2}$, and the resulting estimated initial orthoexciton density per pulse is $2.5 \times 10^{11}\,\mathrm{cm^{-3}}$. (b) (Upper) Resonant emission intensity of orthoexcitons as a function of the fine delay τ_f (coarse delay was $\tau_c = 8.0\,\mathrm{ps}$) for $T = 1.7\,\mathrm{K}$. (Lower) Optical interference pattern of pump pulses observed for $\tau_c = 0.0\,\mathrm{ps}$. The linear polarization of the excitation light is 45° relative to the [001] crystal axis. Flat lines in the two graphs represent the background signal intensity obtained when the pump light beam is cut [31].

We set a coarse delay $\tau_c = 8.0\,\mathrm{ps}$, which is longer than the 2.0 ps pulse duration, for the pump–pulse pair, and varied the fine pulse interval τ_f. An interference pattern with almost perfect contrast was obtained. The oscillation frequency was twice that of the pump pulse frequency, as shown in the upper part of Fig. 8.9(b). This demonstrates that phase-coherent ensembles of dark excitons are created via two-photon absorption of pulses and behave as waves [23].

The dephasing time of the coherent dark exciton waves was measured directly by observing the maxima and minima of the interference patterns as a function of the coarse delay (τ_c); the first-order field autocorrelation functions of dark exciton waves were thus obtained. According to the two-photon polarization selection rules, the ratio of the generated dark exciton population and the square of the exciton wave amplitude is given by $N_1/N_3 = \sin^2 2\theta / \sin^4 \theta$, where N_1 and N_3 are the number of dark excitons created in states $|O_1\rangle$ and $|O_3\rangle$, respectively, and θ is the linear polarization angle of the pump pulse relative to the crystal [001] axis. When $\theta \approx 50°$, the total efficiency of two-photon excitation of both the $|O_1\rangle$ and $|O_3\rangle$ states is maximized. We observed a clear beat signal with a decay time of more than 200 ps (Fig. 8.10(a)), which originated from the quantum beat of excitons with fine structures. Further proof was obtained when θ was set to 90° and we confirmed that the $|O_3\rangle$ state was almost selectively excited so that the beats were suppressed, as shown in Fig. 8.10(b).

A dark exciton wave is a coherent superposition of the $|O_1\rangle$ and $|O_3\rangle$ states with eigenenergies E_1 and E_3, respectively. Their dephasing time T_{relax} is limited by the repopulation of the bright orthoexciton level. We can describe the wave function of dark excitons coherently driven by pulsed two-photon excitation with the i-th pulse at $t = 0$ as

$$\left|\phi^{(i)}(t)\right\rangle = c_1 e^{-i(E_1/\hbar)t - t/T_{relax}} \,|O_1\rangle + c_3 e^{-i(E_3/\hbar)t - t/T_{relax} + \delta} \,|O_3\rangle, \tag{8.1}$$

where δ is the relative phase between the two pure states, and both components are assumed to have the same lifetime T_{relax}. The total wave function generated by double pulses ($i = 1, 2$) separated from each other by a delay $\tau = \tau_c + \tau_f$, is the sum of the wave functions $|\Psi(t,\tau)\rangle = \left|\phi^{(1)}(t)\right\rangle + \left|\phi^{(2)}(t-\tau)\right\rangle$. The quadruple emission signal I_{sig} from the bright state that we detected is proportional to the square amplitude of the total exciton waves, which is,

$$I_{sig} \propto \int_0^\infty \langle \Psi(t,\tau)|\Psi(t,\tau)\rangle dt. \tag{8.2}$$

By fitting the data in Fig. 8.10 with Eq. (8.2), we obtained the dephasing time, energy splitting, and created the population ratio. The measured T_{relax} of I_{sig} yields an estimated lower limit of the dephasing time of dark excitons of 235 ± 10 ps, which corresponds to a dephasing rate of $\Gamma_{relax} = 2\hbar/T_{relax} = 5.60 \pm 0.24\,\mu$eV. From the beat period, we obtained an energy splitting between the dark exciton levels $\Delta_{31} = E_3 - E_1 = 17.8 \pm 0.2\ \mu$eV, which is about four times larger than the value obtained by off-angle-beam one-photon absorption measurements [30]. This discrepancy can be explained by the residual strain in our sample, especially along the [110] axis. The obtained population ratio is $N_3/N_1 = 0.349 \pm 0.015$, which agrees well with the ratio $\sin^4(50°)/\sin^2(2 \times 50°) = 0.355$, calculated using the polarization selection rule.

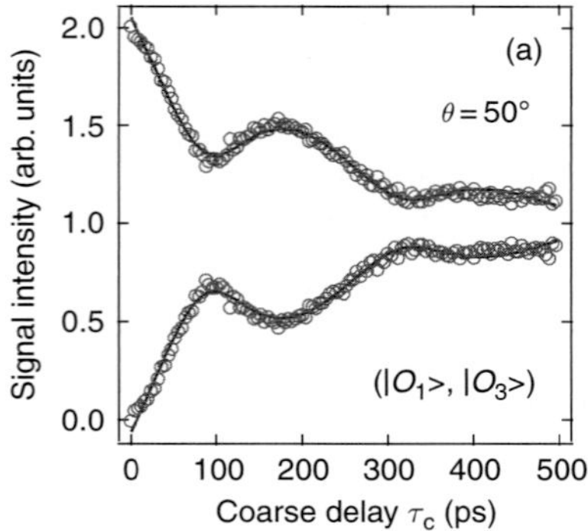

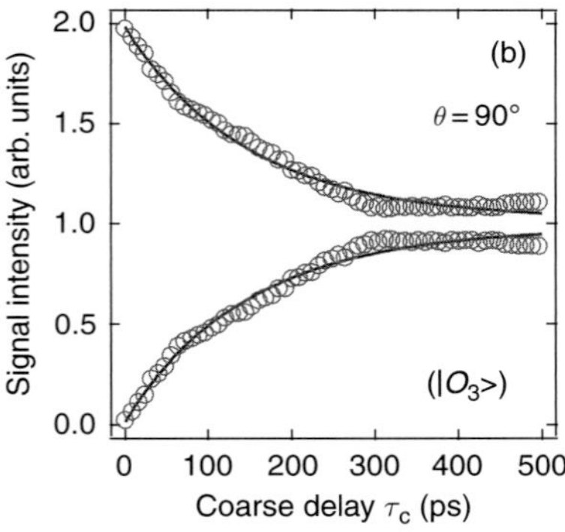

Fig. 8.10 First-order autocorrelation functions of dark exciton waves at $T = 1.7$ K. (a) Maxima and minima of the interference pattern as a function of the coarse delay τ_c (open circles). Each maximum and minimum is normalized with respect to the average of the two. The linear polarization angle of the excitation light relative to the [001] crystal axis is set to $\theta \approx 50°$ so that both states $|O_1\rangle$ and $|O_3\rangle$ are initially populated. The long-term oscillation corresponds to the energy difference between the two dark orthoexciton states. (b) Only the state $|O_3\rangle$ is populated ($\theta = 90°$). The signal intensity is about half of that in (a) [31].

8.3 Search for spontaneous Bose–Einstein condensation of excitons in a bulk crystal of Cu_2O

Creation of ultra-cold atomic ensembles with controllable density, temperature, and interaction strength has enabled the observation of new aspects of many-body quantum phenomena. These achievements have also renewed interest in BEC of excitons. A long-term quest to confirm this phenomenon has been underway since its theoretical prediction in the 1960s [11, 32]. However, electron–hole ensembles are complex quantum systems with strong Coulomb correlations, and thus, the question of whether nature chooses a form of exciton BEC is non-trivial. Various bulk [33–35] and two-dimensional semiconductor [36, 37] systems and exciton–photon hybrid systems [38–40] have been examined.

In Cu_2O, which is a cubic semiconductor crystal, the valence and conduction bands have positive parity and extrema at the center of the Brillouin zone. Because radiative recombination is forbidden, the $n = 1$ exciton state has a long lifetime; it is split by electron–hole exchange interaction into a triply degenerate orthoexciton and a lower-lying singly degenerate paraexciton state.

Quantum mechanical wave–particle duality is essential for BEC. Since the wave nature of an exciton ensemble as a macroscopic polarization wave is well established, it is important to observe an experimental signature of the particle nature of excitons. The corpuscular nature of excitons is well visualized in longitudinal optical (LO) phonon-assisted emission spectra, whose shapes reveal the thermal distribution functions of excitons. Photoluminescence of a naturally grown single crystal of very high structural quality was examined by Mysyrowicz and co-workers in 1979 [41]. Figure 8.11 shows photoluminescence spectra at various lattice temperatures. Optical phonon sidebands of 1s orthoexcitons indicate the quasi-thermal equilibrium distribution functions for the lattice. At low temperatures, narrowing of the distribution and enhanced

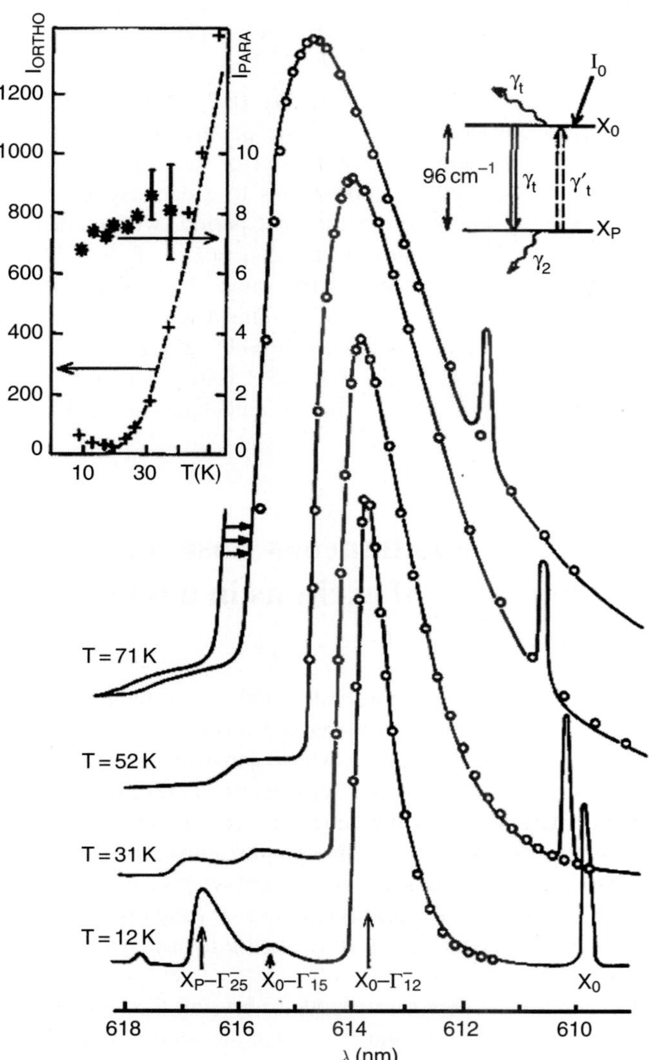

Fig. 8.11 Photoluminescence spectra of orthoexciton (X_O-Γ_{12}^-) and paraexciton (X_p-Γ_{25}^-) at various lattice temperatures. The inset shows the intensity of ortho- and para-luminescence versus temperatures. The exponential increase of the orthoexciton above $T = 20\,\mathrm{K}$ is well explained by the thermal activation from the lower lying paraexciton states. Circles show the fitted results using a Maxwell–Boltzman distribution function [42].

lower-lying 1s paraexciton emission were clearly observed. Luminescence decay measurements revealed a paraexciton emission decay time of 13 μs and an orthoexciton emission time of 3 ns.

The quality of the crystal strongly affects the relaxation kinetics and lifetime of excitons. Synthesized crystals have high-density structural defects and exhibit efficient non-radiative recombination processes [42]. Thus, the emission spectra showed non-equilibrium features.

For high-quality crystals, the signature of BE exciton statistics has been observed in the spectral shape of the luminescence at high-density excitation [43]. Figure 8.12 shows the LO phonon sideband emission spectra of 1s orthoexcitons at low-density excitation (a) and high-density excitation

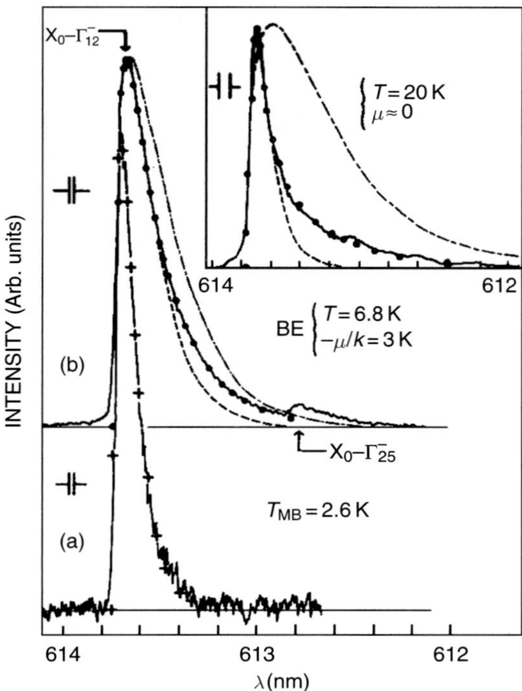

Fig. 8.12 The LO phonon sideband emission spectra of 1s orthoexcitons under cw excitation at low-density excitation (a) and high-density excitation (b) showing gradual evolution from classical to Bose distribution, measured at a bath temperature of 1.5 K. The inset shows a spectrum obtained with pulsed excitation (peak intensity of 10 MW/cm^2) [43].

(b) measured at a bath temperature of 1.5 K. At high-density excitation, deviations from the Maxwell–Boltzmann distribution functions appear. The density of orthoexcitons is almost linearly related to the excitation density, whereas the paraexciton emission saturates at high density. This saturation effect is the signature of collision-induced exciton decay, which is known as Auger decay.

The line shape of luminescence caused by high-density excitons has been measured and analyzed systematically [44]. Excitons were generated by excitation with a pulsed Ar-ion laser with a wavelength of 514.5 nm and a pulse duration of 10 ns, which was focused on a sample 30 μm in diameter. At high-density excitation, the time-resolved luminescence spectra of orthoexcitons were well fitted by BE distribution functions having a time-dependent temperature and chemical potential, as shown in Fig. 8.13 [44]. The chemical potential values reveal the exciton density. The temporal profiles of the incident pulses show that the exciton density increases, accompanied by heating. In a density versus temperature plot, the experimental points follow the line just below the phase transition boundary of BEC. This behavior is called quantum saturation of orthoexcitons. Spontaneous decay into paraexcitons prevents BEC of orthoexcitons.

The lowest energy 1s paraexciton state is the most important target for the realization of an exciton BEC. This is a pure spin-triplet state, so it is decoupled from the radiation field. In high-quality single crystals, 1s paraexcitons show a very long lifetime of several hundred nanoseconds [41, 45], which is enough to reach thermal equilibrium with the lattice. Taking into account their effective

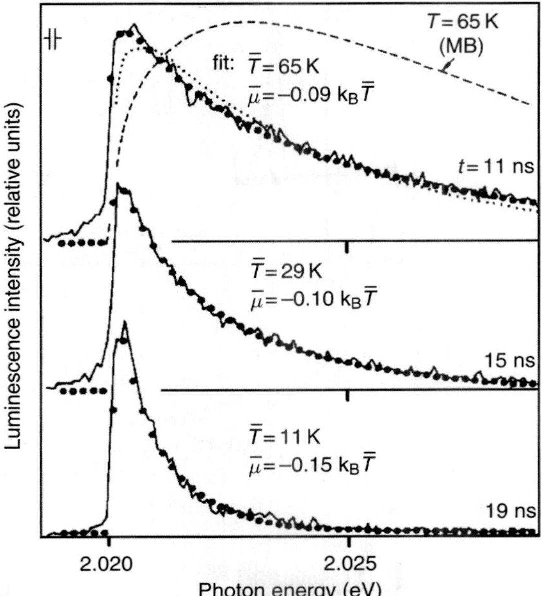

Fig. 8.13 Time-resolved spectra of the orthoexciton phonon-assisted luminescence in Cu_2O. Maximum absorbed power $\simeq 5 \times 10^6$ W/cm. Solid circles show best fits results with Bose–Einstein distribution function. The dotted line is a least squares fit to the first spectrum, imposing a lower degeneracy $\mu = 0.3\,k_B T$. (The corresponding T is 43 K.) The dashed curve is a classical ($\mu < -2\,k_B T$) distribution for the same T as the first spectrum [44].

mass of $m_{ex} = 2.6\,m_0$, where m_0 is the electron mass at rest [46], BEC can be realized at moderate density ($n \sim 10^{17}\,\text{cm}^{-3}$) at the superfluid helium temperature of 2 K. It is difficult to study paraexcitons because they are optically inactive. A very weak LO phonon sideband emission of 1s paraexcitons, which is typically 500 times less radiatively efficient than that of the orthoexciton band, has been examined [33, 47]. Figure 8.14 shows LO phonon sideband emission of 1s orthoexcitons (a) and 1s paraexcitons (b). At low densities, the emission lines of both ortho- and paraexcitons follow those of Maxwell–Boltzmann distribution functions. At moderate densities, the spectral profiles are fitted well by BE distribution functions. The obtained chemical potentials

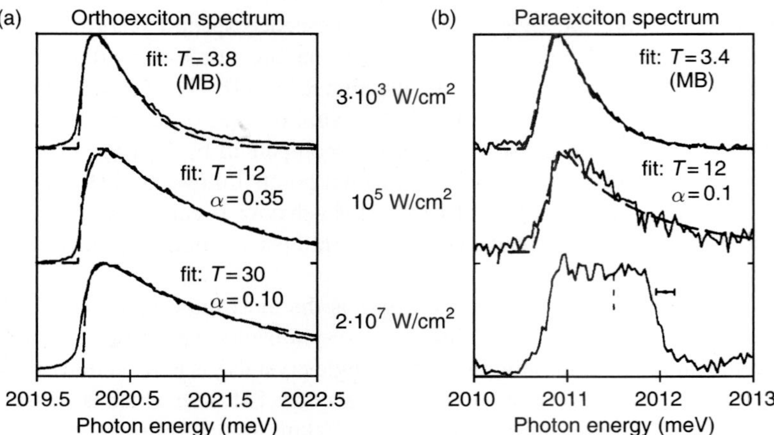

Fig. 8.14 LO phonon sideband emission of 1s orthoexcitons (a) and 1s paraexcitons (b) under the pulsed Ar-ion laser excitation for three incident powers.

of paraexcitons are smaller than those of orthoexcitons, indicating a higher occupation of paraexcitons. At high-density excitation, a bump appears at 1 meV above the ground state in paraexciton spectra. These peculiar features suggest Bogoliubov excitation of superfluid excitons.

The paraexciton spectrum at high density takes on an anomalous shape. The dashed line on this spectrum indicates the energy of an exciton which couples directly to the ground state via a longitudinal-acoustic phonon.

To visualize the optically inactive paraexcitons, Fortin and co-workers performed photovoltaic measurements [48]. Figure 8.15 shows their experimental configuration and results. A bulk Cu_2O crystal was excited with an intense nanosecond-pulsed laser (wavelength, 532 nm; pulse duration, 10 ns). Gold and copper electrodes were fabricated on the other side of the crystal. A strong electric field at the Schottky barrier of Cu and Cu_2O contact dissociated excitons and generated pulsed photovoltaic signals. At low densities or high temperatures, the signals showed broad profiles corresponding to diffusive motion of excitons. At high densities and low temperatures, the signal exhibited a pulse shape. This implies that the BEC transition changes exciton transport from diffusive to ballistic.

These intriguing results suggest the existence of a degenerate quantum fluid of paraexcitons. However, an analysis of the absolute quantum efficiency of luminescence indicates that collision-induced annihilation of excitons, which is known as the Auger effect, can have an extremely large cross-section [49]. In addition, the revised analyses raised rather serious objections to the earlier experimental claims of an exciton BEC in Cu_2O.

The lack of a quantitative method for monitoring optically inactive paraexcitons prevents clear evaluation of an exciton BEC. This difficulty can be

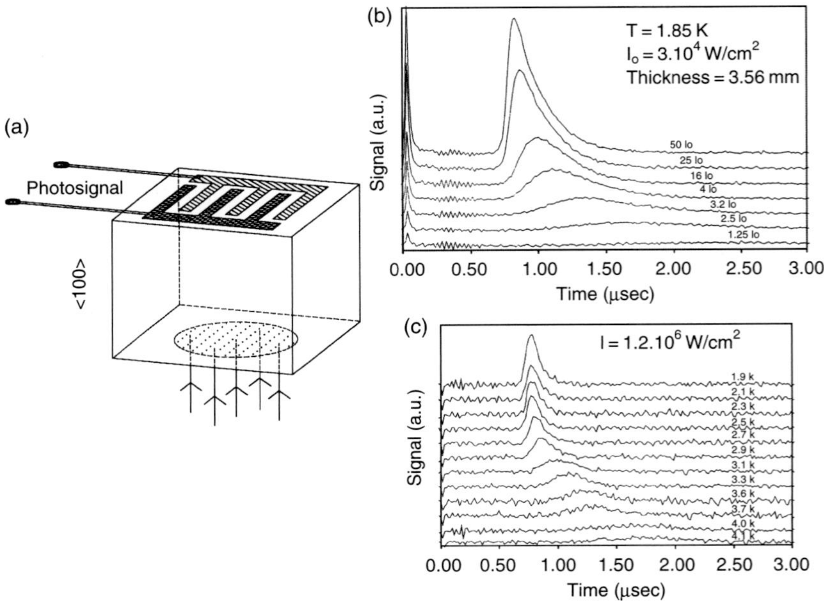

Fig. 8.15 Time-resolved photovoltaic voltage measurements. (a) Schematic representation of the sample with the electrodes to detect the photovoltaic signals. The detection electrodes consist of evaporated gold and copper with 200 μm width and a separation of 200 μm. Arrows represent the incoming laser beam. (b) Time-resolved photovoltaic voltage measured across an external 50 Ω resistance for different incident laser intensities. The fast signal at $t = 0$ results from direct illumination of the surface containing the electrodes and gives the detector response time. The sample, of thickness 3.56 mm, is immersed in liquid He at $T = 1.85$ K. (c) Time-resolved photovoltaic voltage in a 3.56-mm-thick sample as a function of sample temperature for a fixed incident laser intensity 1.2×10^6 W/cm^2.

overcome by using excitonic Lyman spectroscopy; this detection method probes the internal transitions of excitons, which occur in the mid-infrared spectral region [50–54]. In the next subsection, we present the results of our systematic measurements using excitonic Lyman spectroscopy.

8.3.1 Excitonic Lyman spectroscopy

The yellow series excitons in Cu_2O are a model system of Wannier excitons with hydrogen-like energy structures. Cu_2O has a particularly simple band structure, with non-degenerate parabolic valence (symmetry Γ_7^+) and conduction bands (symmetry Γ_6^+), and extrema located at the center of the Brillouin zone. The same parity implies that the yellow series starts with $n = 2$ and corresponds to the so-called second class (weakly allowed) dipole transition. The $n = 1$ term of the yellow series appears as a weak quadrupole absorption line. Electron–hole exchange interactions further split the $n = 1$ exciton into a highly forbidden singlet paraexciton of symmetry Γ_2^+ with an internal energy of 2.020 eV at 2 K and a triply degenerate Γ_5^+ orthoexciton (2.032 eV at 2 K). Both the $n = 1$ ortho- and paraexcitons have a long radiative lifetime.

Although photoluminescence measurements are the most sensitive method for obtaining information on excitons, the measured emission properties are very sensitive to the quality of the sample. An alternative method for exciton detection is to probe the transition between the $n = 1$ and higher principal quantum numbers [55, 56] by changing the angular momentum of the excitons from the $l = 0$ (s-like) state to the $l = 1$ (p-like) state, similar to the case of Lyman-series absorption lines in atomic hydrogen. In Cu_2O, these lines appear in the mid-infrared region. Because of the large difference in the electron–hole exchange energy for the 1s and 2p excitons (about 12 meV and 0 meV, respectively; see Fig. 8.16), the Lyman-α lines for orthoexcitons and paraexcitons can be resolved. Because 1s excitons have a small Bohr radius, their mass is greater; thus, the 1s–2p transition energy depends on the wavevector. Therefore, the line profile for the Lyman-α absorption band should undergo a characteristic narrowing when excitons become a BEC [57].

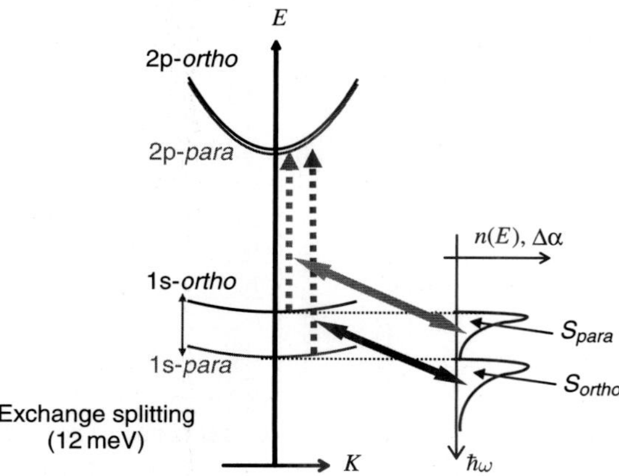

Fig. 8.16 Schematic energy diagram of excitonic Lyman spectroscopy.

Several groups have attempted to measure Lyman transitions in Cu_2O experimentally [50–54, 58–60].

First, we performed mid-infrared pump–probe spectroscopy with femtosecond pulses to confirm the Lyman transitions. With this technique, we can examine the temporal evolution of ortho- and paraexcitons under pulsed optical excitation.

Figure 8.17 shows the pump-induced change in absorption at the paraexciton 1s–2p resonance measured as a function of crystal temperature when paraexcitons are expected to be in equilibrium with respect to the lattice (at a probe–probe delay of 4 ns). The induced absorption band broadens at higher temperatures according to the thermal distribution of excitons. Solid lines show the results of fitting with the Maxwell–Boltzmann distribution of excitons using an orthoexciton mass of $m = 2.7\ \mathrm{m_e}$ [42]. Its value differs from the sum of the electron and hole masses (1.68 $\mathrm{m_e}$ [61]) because of the non-parabolic dispersion of 1s excitons with a Bohr radius comparable to the lattice constant of $a_l = 0.426$ nm. In contrast, the Bohr radius of the 2p state is as large as 4.4 nm; hence, the wave functions of electrons and holes are confined around the center of the Brillouin zone. The $n = 2$ exciton mass is thus close to the sum of the individual electron and hole masses. A precise line shape analysis of the paraexciton 1s–2p band at equilibrium yields an effective mass of the $n = 1$ exciton of $(2.40 \pm 0.34)\ m_e$ [54].

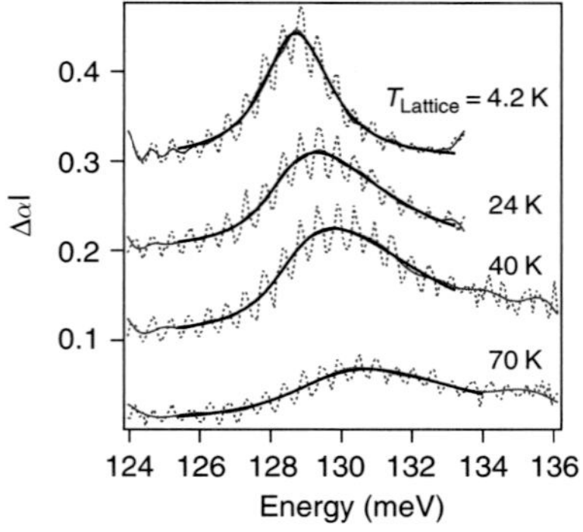

Fig. 8.17 Line shapes of the 1s–2p paraexciton-induced absorption band at lattice temperatures of 4.2, 24, 40, and 70 K [50].

As discussed in the previous section, we can generate orthoexcitons with a very narrow momentum distribution by using a pulsed resonant two-photon excitation scheme [52]. Figure 8.18 shows a schematic diagram of the Lyman transition from orthoexcitons to *n*p states under pulsed resonant two-photon excitation.

Figure 8.19(a) shows a linear absorption spectrum in the region of *n*p orthoexciton states. The induced absorption spectrum under resonant two-photon excitation of 1s orthoexcitons at a pump–probe delay of 5 ps is shown in Fig. 8.19(b); four lines appear at 116, 129, 133, and 136 meV, respectively. At a pump–probe delay of 4 ns (Fig. 8.19(c)), the orthoexcitons are converted into paraexcitons, and these resonances vanish, while a broader signal appears around 128 meV. This new line is assigned to the 1s–2p transition of paraexcitons [51].

The exciton Lyman series shows a systematic *n* dependence of the linewidths and absorption strengths. A similar behavior was observed in the *n*p absorption spectra (Fig. 8.19(a)) and has been discussed in terms of an *n*-dependent LO phonon scattering rate from the *n*p to the 1s state. The excitonic line widths in Fig. 8.19(a) are determined by the energy distribution of 1s excitons as well as the dephasing times of the 1s and *n*p states. This gives the following equation for the Lyman linewidths: $(\Delta E_{1s\text{-}np})^2 \sim (\Delta E_{np})^2 + [(m_{1s}/m_{2p} - 1)\Delta E_{1s}]^2$. The intrinsic linewidth of the 1s exciton state ($\gamma_{1s} < 1\ \mu$eV) [30] is negligible in our analysis, whereas the linewidths of the *n*p states are known to show an *n* dependence.

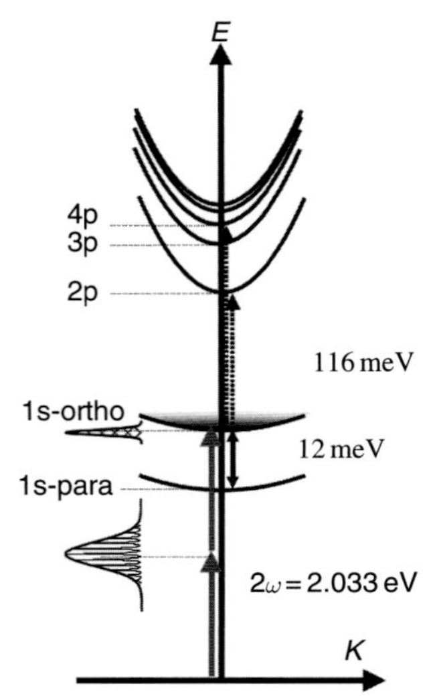

Fig. 8.18 Schematic diagram of Lyman series transition from supercooled 1s orthoexcitons to *n*p states under pulsed resonant two-photon excitation [52].

The upper limit of the energy spread of 1s orthoexcitons can be estimated from the width of the 1s–4p line, $\Delta E_{1s\text{-}4p} \approx 0.2$ meV, which gives $\Delta E_{1s} \approx 0.2\ \mathrm{meV}/(2.7\ \mathrm{m}_e/1.7\ \mathrm{m}_e - 1) \sim 0.34$ meV. Thus, the line is narrower than the lattice thermal energy at 4.2 K, $k_B T = 0.7$ meV. Figures 8.20(a) and (b) show

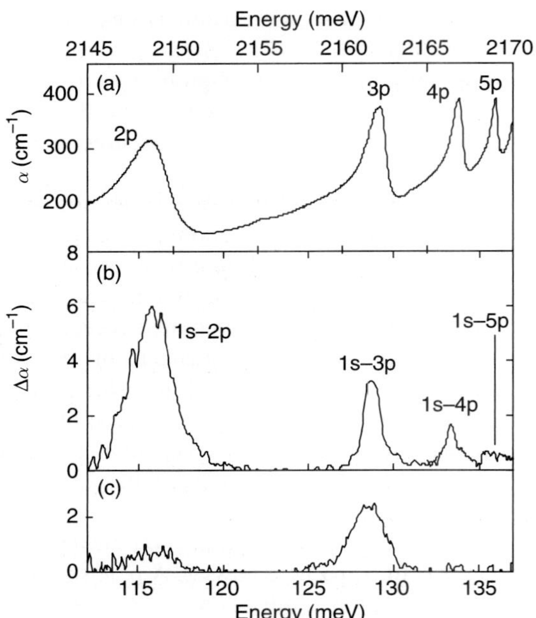

Fig. 8.19 (a) Linear absorption spectrum. (b) Induced absorption spectra of 1s orthoexcitons generated by a pulsed two-photon resonant excitation measured after a 5 ps pump–probe delay and (c) 1s–2p paraexciton signal measured 4 ns after excitation [52].

the temporal evolution of the 1s–4p line of orthoexcitons measured at 4.2 K and the extracted temporal evolution of the effective exciton temperature, respectively. The initial temperature of the created orthoexcitons is less than 1 K, and it increases gradually to the lattice temperature (4.2 K) in a 1 ns time interval.

Lyman spectra provide important information on the electronic structure of 1s excitons, the spectral position of which shows significant deviation from the Rydberg series [59, 62, 63]. Thus, systematic analysis of the exciton Lyman series enables us to determine the unknown 1s exciton wave function from the dipole matrix elements $\mu_{\text{1s-np}} = e < u_{1\text{s}}|z|u_{np} >$. This is because, in contrast with the 1s state, the spectral positions of np exciton states ($n \geq 2$) are well described by the Rydberg series, $E_{\text{n}} = 17525 - 786/n^2\ \text{cm}^{-1}$; that is, the wave functions of np states are hydrogen-like.

The area of the 1s–np exciton Lyman absorption line is given by

$$S_{1s-np} = \int_{1s-np} \Delta\alpha(E)dE = \frac{N_{1s}4\pi^2 E_{1s-np}|\mu_{1s-np}|^2}{\hbar c\sqrt{\varepsilon_b}}, \tag{8.3}$$

where $E_{\text{1s-np}}$, $\mu_{\text{1s-np}}$, and ε_{b} are the 1s–np transition energy, the dipole moment, and the background dielectric constant at the probe frequency, respectively. In the probe frequency region $\varepsilon_{\text{b}} \approx 6$, the dipole moment $\mu_{\text{1s-np}}$ can be obtained from Eq. (8.3) if the 1s exciton density is known. In practice, however, it is difficult to estimate the absolute value of $N_{1\text{s}}$ because in two-photon excitation, the number of absorbed photons depends strongly on the pumping conditions. Therefore, one may rely on relative rather than absolute values of the integrated absorption. This is because the radii $S_{\text{1s-np}}/S_{\text{1s-2p}}$ are density independent and insensitive to the pump conditions

$S_{1s\text{-}np}/S_{1s\text{-}2p} = |\mu_{1s\text{-}np}/\mu_{1s\text{-}2p}|^2 \times (E_{1s\text{-}np}/E_{1s\text{-}2p})$. Fitting the experimental data yields $S_{1s\text{-}3p}/S_{1s\text{-}2p} = 0.28 \pm 0.01$ and a 1s Bohr radius of $a_B = 7.3$ Å, which gives a dipole moment of $\mu_{1s\text{-}2p} = 3.5 \pm 0.3$ eÅ.

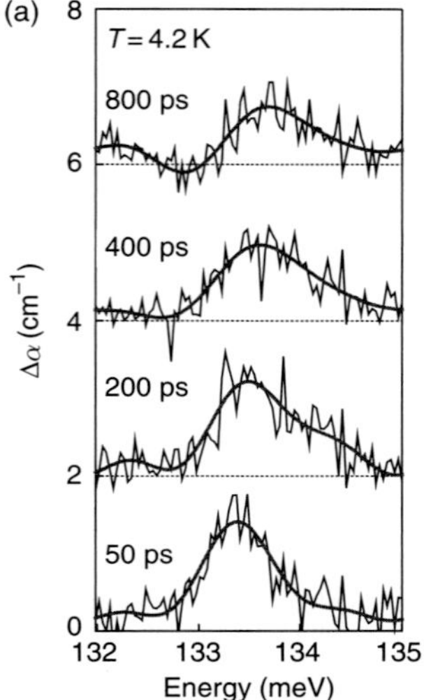

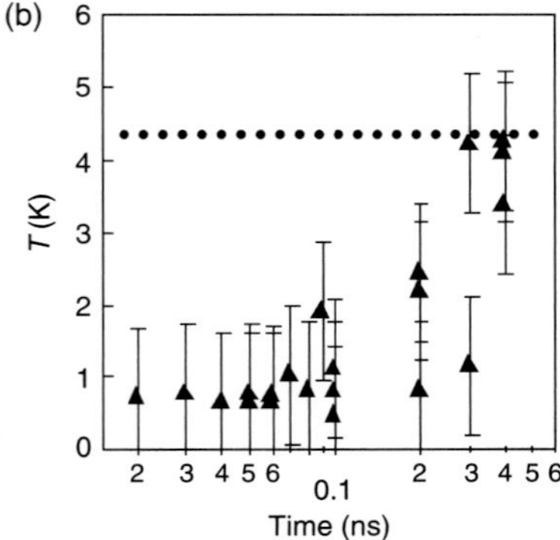

Fig. 8.20 (a) Temporal evolution of the 1s–4p lines from supercooled orthoexcitons generated by a pulsed two-photon resonant excitation measured at a lattice temperature of 4.2 K. Lines show fitting results. (b) Temporal evolution of effective temperature of orthoexcitons extracted from (a). Horizontal dashed line shows the lattice temperature ($T = 4.2$ K) [53].

8.3.2 Shaped pulse two-photon control of excitons

A pulsed two-photon excitation scheme has the strong potential to generate quantum-degenerate orthoexciton ensembles. At higher excitation levels, however, an excess heating mechanism exists, as depicted in Fig. 8.21 [64]. A three-photon band-to-band process can lead to heating via production of free electron–hole pairs with large excess energies. This heating mechanism becomes increasingly important at higher laser intensities I, since the rate of production of hot electron–hole pairs is proportional to I^3 [65].

By manipulating the phases of the ultra-short pulses employed for two-photon excitation using the experimental setup shown in Fig. 8.22, we can suppress this heating mechanism [64].

Frequency down-conversion with an optical parametric amplifier system produces tunable pulses of 100 fs duration in the required spectral region (~1.016 eV for the pump; ~116 meV for the probe). The pump pulses are introduced to a pulse shaper in a 4f configuration with paired gratings and a liquid crystal spatial light modulator [66].

To reduce undesirable three-photon heating without reducing the production rate of cold excitons by two-photon absorption, coherent control of the relevant multiphoton transitions can be employed by tailoring the phase and amplitude of the pulses. The two-photon transition probability in the absence of near-resonant intermediate states can be written as [67],

$$
\begin{aligned}
P^{2ph} &\propto \left| \int_{-\infty}^{\infty} E(\omega_0/2 + \Omega) E(\omega_0/2 - \Omega) d\Omega \right|^2 \\
&= \left| \int_{-\infty}^{\infty} A(\omega_0/2 + \Omega) A(\omega_0/2 - \Omega) \right. \\
&\qquad \left. \exp\left[i \{ \varphi(\omega_0/2 + \Omega) + \varphi(\omega_0/2 - \Omega) \} \right] d\Omega \right|^2 ,
\end{aligned}
\tag{8.4}
$$

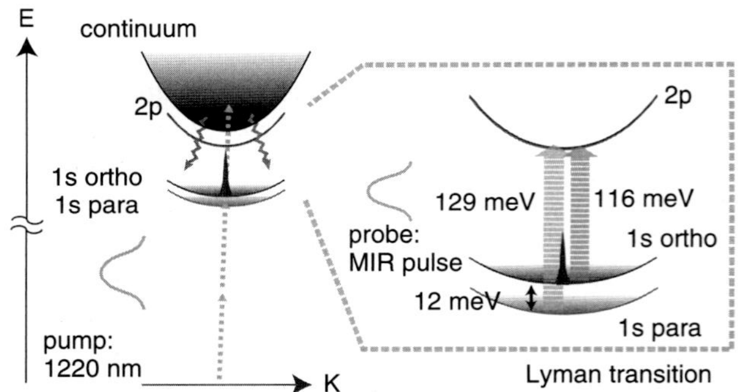

Fig. 8.21 Schematic description of optical processes involved in the experiments. Infrared pulses tuned to two-photon resonance of 1s orthoexcitons directly excite 1s excitons near zero kinetic energy and free electrons and holes via band-to-band continuum transitions. Inset shows the probing scheme of exciton Lyman transitions [64].

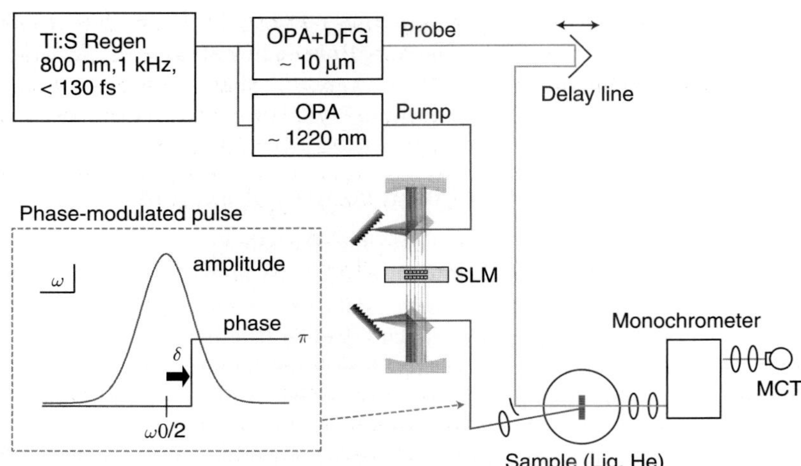

Fig. 8.22 Experimental setup of mid-infrared pump–probe measurement with phase-manipulated pulses [64].

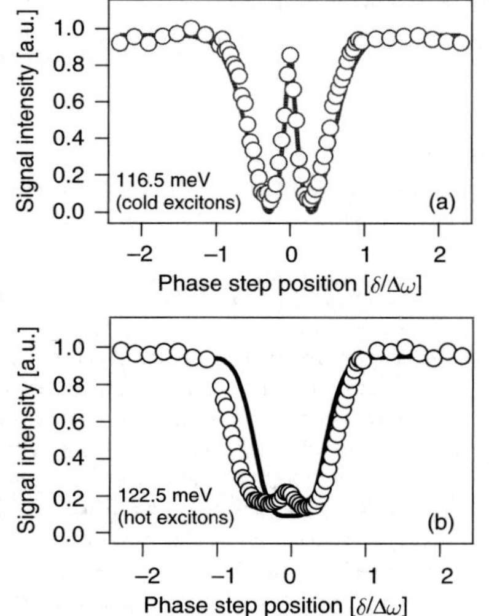

Fig. 8.23 Differential absorption signal as a function of the π-phase step frequency position for (a) cold excitons monitored at the peak of the 1s–2p orthoexciton transition energy (116.5 meV) and (b) hot excitons monitored on the high-energy side (122.5 meV) [64].

where $E(\omega) = A(\omega)\exp[i\varphi(\omega)]$ is the electric field of the excitation pulse in the frequency domain, and $A(\omega)$ and $\varphi(\omega)$ are the spectral amplitude and phase, respectively. According to Eq. (8.4), even if the bandwidth of the laser pulse having a central frequency of $\omega_0/2$ is much larger than the considered transition, all photons in the pulse can contribute to P^{2ph}. This is immediately apparent for a Fourier-transform-limited pulse (for which the spectral phase of the pulse is flat), because each photon with an energy deficit $\tilde{\omega}' - \omega_0/2 = \Delta$ can pair with a simultaneous photon of excess energy $-\Delta = \tilde{\omega}'' - \omega_0/2$ to satisfy the energy conservation law.

However, this is not usually true if we modify the pulse phase. For instance, in the presence of a linear chirp (linear phase shift), photons with energy $\omega_0/2 - \Delta$ have no simultaneous matching partner of frequency $\omega_0/2 + \Delta$. Nevertheless, there exist special phase-manipulated pulses, that have the same two-photon transition probability as that of the Fourier-transform-limited pulse, but a considerably reduced three-photon transition rate. One example is a pulse having a flat spectral phase with an abrupt phase shift of π at $\omega_0/\tilde{2}$.

To demonstrate the properties of the π-phase-shifted pulse in the two-photon process, we monitored the signal at the peak of the 1s–2p orthoexciton transition energy (116.5 meV), which is proportional to the cold 1s orthoexciton density generated by two-photon excitation as a function of the phase step frequency position (Fig. 8.23(a)). The two-photon transition rate is decreased by destructive interference between the different transition paths, except in a narrow frequency region around $\delta \sim 0$, where it reaches the value of a Fourier transform pulse with perfect constructive interference.

To examine the effect of the π-phase-shifted pulse on the three-photon process, we performed the same experiment to examine the generation of excitons via two-photon absorption by detecting a signal at the high-energy tail of the 1s–2p orthoexciton line at 122.5 meV (Fig. 8.23(b)). At this probe energy, only hot excitons generated by the three-photon process contribute to the signal. When the π-phase step shift position was tuned to the central frequency of the laser pulse ($\delta = 0$), the three-photon transition rate was sharply reduced.

Interference between transition paths is suppressed for a multiphoton transition with a broad final state distribution. In this case, the transition rate depends only on the temporal waveform of the excitation pulse. Because our tailored pulse had a double-humped temporal profile with a reduced peak intensity, it significantly reduced the three-photon transition rate.

As shown in Figs. 8.23(a) and (b), by changing the π-phase step position δ, we can easily control the relative contributions of the two types of transitions. At $\delta/\Delta\omega \sim 0$, (π-phase-shifted pulse) the three-photon transition is strongly suppressed, and the two-photon transition dominates. At $\delta/\Delta\omega \sim \pm 0.3$, the two-photon transition is quenched. (Here $\Delta\omega$ is the FWHM bandwidth of the pulse.)

One can compare differential absorption spectra at a probe delay of 5 ps, obtained under excitation with a π-phase-shifted pulse ($\delta/\Delta\omega \sim 0$) and with a transform-limited (TL) pulse (Fig. 8.24). At low-density excitation, both spectra show signals corresponding to cold 1s orthoexcitons directly generated via resonant two-photon absorption. At high-density excitation, the spectra of the TL pulses reveal hot excitons, whereas the spectra produced by the π-phase-shifted pulse reveal only cold excitons with a density of $10^{15}\,\mathrm{cm}^{-3}$. A higher density of cold excitons is expected to be obtained using more sophisticated phase modulation of the TL short pulses to eliminate the three-photon process.

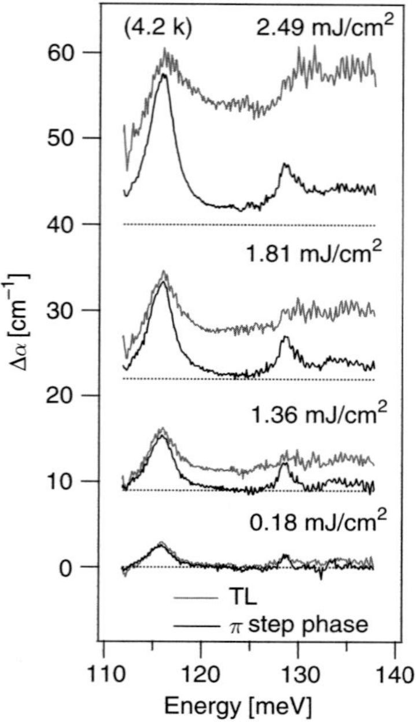

Fig. 8.24 Differential absorption spectra of the probe beam at a time delay of 5 ps with a π-phase step frequency of $\delta \sim 0$ (black curves) and with transform-limited pulses (gray curves) [64].

8.3.3 Collision-induced processes

In experiments on cold atoms, one carefully selects systems where interatomic interactions at sufficiently low temperatures are determined by a constant elastic-scattering length. On the other hand, interactions between excitons are only partially understood owing to many-body Coulomb correlations [68–71] and the added complication of strong coupling to the radiation field as well as the lattice environment. For excitons in Cu_2O, quantitative photon-counting emission measurements [72–76] revealed an efficient two-body collision-induced loss process. The density-dependent decay was first attributed to Auger recombination [77], a process in which a pair of excitons disappears in each inelastic collision, giving rise to a single pair of hot electron–hole free carriers, although a theoretical evaluation reports much smaller loss rates than those obtained experimentally [78]. A later theoretical estimate [79] showed that inelastic collisions between pairs of orthoexcitons, followed by conversion into paraexcitons, are more probable than the Auger process. These conversion processes were observed experimentally by time-resolved excitonic Lyman spectroscopy [51, 53].

In this section, we discuss measurements of the cross-section of collisions between 1s paraexcitons in Cu_2O at low temperatures. The density of optically inactive [80] 1s paraexcitons can be measured quantitatively by using excitonic Lyman spectroscopy [54]. In a steady state under continuous feeding with continuous wave (cw) laser excitation and cw probing, we can vary and probe their density over a wide range while maintaining quasi-thermal equilibrium with the lattice. Thus, we can extract the density-dependent loss rate on the basis of nonlinear rate equation analyses [76].

The sample was a high-quality, naturally single crystal 220 μm in thickness with the [100] crystal axis oriented perpendicular to the sample's main surface. The sample was irradiated with a cw single-mode dye laser set to 601 nm. Absorbed light generates 1s orthoexcitons homogeneously along the excitation direction via LO phonon-assisted absorption with a typical measured small absorption coefficient of $\alpha = 34\,\text{cm}^{-1}$ at 20 K. Orthoexcitons are photogenerated with a production efficiency of about 70% and are converted into paraexcitons within a few nanoseconds [81–83]. Under quasi-steady-state conditions, the ratio of paraexcitons to orthoexcitons is around 10^3. Thus, the population of excitons is dominated by that of paraexcitons, and the influence of orthoexcitons in exciton–exciton collisions can be safely neglected.

The mid-infrared probe beam for detecting the 1s to 2p internal transition of paraexcitons is obtained from a single-line cw CO_2 laser, which can be tuned in discrete steps in the region of interest, which covers most of the 1s–2p absorption band profile. Changes in the transmitted probe beam in the order of 10^{-6} could be detected. A detailed description can be found in Ref. [54]. Figure 8.25(a) shows typical induced absorption spectra taken at various lattice temperatures. The maximum generation rate of orthoexcitons was $1.2 \times 10^{12}\,\text{cm}^{-3}$/ns. At this excitation rate, no intensity-dependent change in the spectral profiles is observed, and we can exclude any excitation-induced heating effect. From the induced absorption spectra, we calculated the spatially averaged paraexciton densities. Figure 8.25(b) shows the density obtained as a function of the exciton generation rate at various lattice temperatures. Nonlinear saturation of paraexcitons is seen even at densities below $10^{14}\,\text{cm}^{-3}$, at which the corresponding interexcitonic distance is 290 nm (for comparison, the paraexciton Bohr radius is about 0.7 nm). This saturation in a dilute situation where 1s paraexcitons dominate the system clearly indicates the presence of a strong paraexciton–paraexciton inelastic collision process that is responsible for particle loss.

Because of the temperature dependence of parameters such as the exciton lifetime and diffusion, the curves in Fig. 8.25(b) show a peculiar temperature

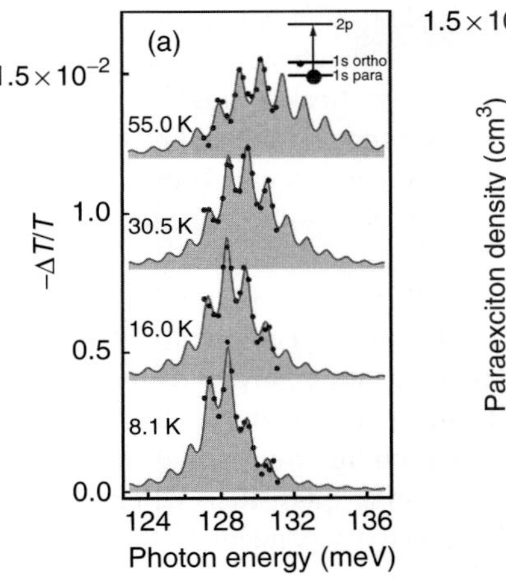

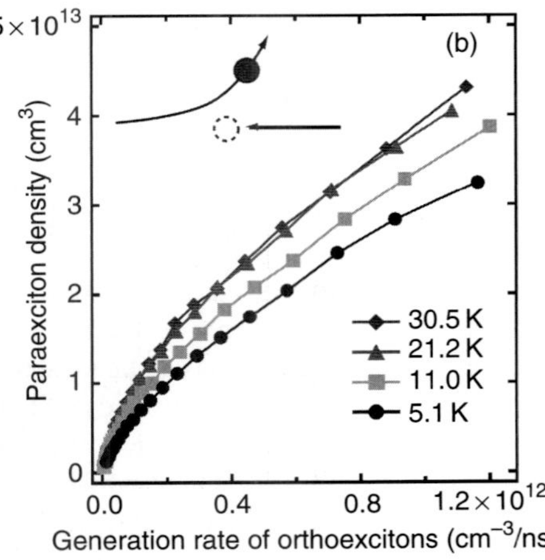

Fig. 8.25 (a) Typical 1s–2p induced absorption spectra of paraexcitons at various temperatures. Zero levels are offset for clarity. The absorbed excitation density is 7.4 W cm^2. Solid circles represent experimental data; discrete spectral values correspond to the 9P branch oscillation lines of the CO_2 laser. Filled area is the calculated spectrum taking into account the thermal distribution of the 1s state and the bandwidth of the 2p state. The oscillatory structure is due to the Fabry–Perot effect on the probe beam inside the sample. (Inset) Schematic of the corresponding transition. Size of the closed circles roughly indicates the population. (b) Paraexciton density at various lattice temperatures as a function of the generation rate of 1s orthoexcitons. The generation rate is known from the absorbed power density and the quantum efficiency for orthoexciton creation. (Inset) Schematic of the corresponding collision process.

dependence. To evaluate the two-body collision-induced loss rate from such experimental data taken at various temperatures, we performed three-dimensional explicit numerical modeling of the nonlinear dynamics of ortho- and paraexcitons using three-dimensional coupled rate equations. The effects of diffusion, temperature-dependent finite lifetime, collision-induced losses, and ortho–para transformations are taken into account. To find the value of the inelastic collision coefficient A, we parametrically changed the value from 1×10^{-19} to $1 \times 10^{-14}\,\mathrm{cm^3/ns}$ and compared the results with the data. When we set $A = 4 \times 10^{-16}\,\mathrm{cm^3/ns}$, we found good agreement between calculation and experiment. We also performed the calculation assuming a three-body collision process and confirmed that it produces much stronger saturation and does not fit the data. Figure 8.26 shows the temperature dependence of the inelastic collision coefficient A and the exciton–exciton inelastic collision cross-section. A is almost independent of temperature, with an average value $A = 3.7 \times 10^{-16}\,\mathrm{cm^3/ns}$.

With the obtained value of A, we can estimate the temperature dependence or momentum dependence of the inelastic collision cross-section σ_{inel}. The total collision rate is $\sigma_{inel} v_{th} n = An$, where $v_{th} = (3k_B T/m_{ex})^{1/2}$ is the mean thermal velocity of excitons and m_{ex} is the effective mass of a 1s paraexciton [46, 54, 84]. As shown in Fig. 8.26, a K_{th}^{-1} dependence (or $T^{-1/2}$ dependence) of the inelastic collision cross-section σ_{inel} appears as a function of the mean wavenumber $K_{th} = m_{ex} v_{ex}/\hbar$. This is a general feature of the quantum mechanical limit of inelastic cold collisions [85, 86]. This is in strong contrast with cold elastic collision processes, in which the cross-section becomes a constant, $\sigma_{el} = 4\pi a^2$, where a is the s-wave scattering length. Since an exciton is a very light particle with an effective mass comparable to the electron mass at rest, quantum effects are essential over a wide temperature range. The temperature at which the thermal de Broglie wavelength becomes comparable to the exciton Bohr radius a_B of the paraexciton is in the order of 10^3 K. We obtained a value for paraexcitons of $\sigma_{inel} = 85\,\mathrm{nm}^2$ at 5 K [87].

The magnitude of elastic scattering is the key parameter for BEC. Quantum Monte Carlo simulations yielded a value of $\sigma_{el} = 50\,\mathrm{nm}^2$ for 1s paraexcitons [68]. The measured value of σ_{inel} also provides the value of the elastic

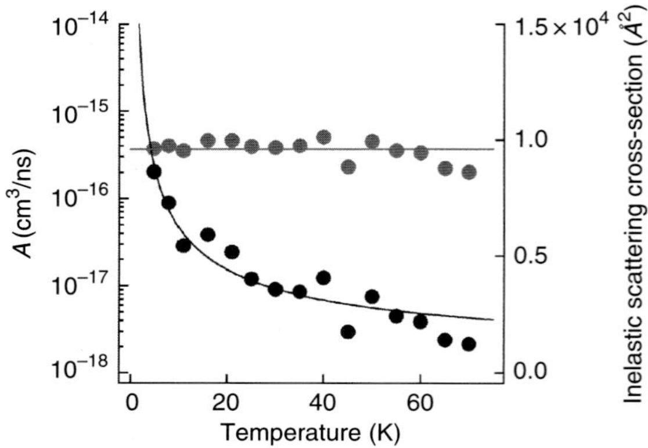

Fig. 8.26 Temperature dependence of the coefficient A (gray circles) and exciton–exciton inelastic collision cross-section (black circles). A is almost independent of temperature. The blue curve is a fitted curve assuming $\sigma_{inel} = aK_{th}^{-1}$, where a is a fitting parameter.

scattering collision cross-section, σ_{el}. The s-wave scattering matrix S_0 can be written as

$$S_0 \approx 1 - 2ik\alpha, \tag{8.5}$$

where $\alpha = \alpha' + i\alpha''$ is the complex scattering length and k is the wavenumber of the particle. The elastic (σ_{el}) and inelastic (σ_{inel}) cross-sections are expressed as

$$\sigma_{el} = 4\pi \left|\alpha\right|^2, \tag{8.6}$$

$$\sigma_{inel} = \frac{4\pi \left|\alpha''\right|}{k}. \tag{8.7}$$

At 5 K, the thermal velocity of excitons yields a thermal wavenumber of $K_{th} = 0.2\,\text{nm}^{-1} = k$. From Eq. (8.7), $\sigma_{inel} = 85\,\text{nm}^2$ yields $\alpha'' = -1.35\,\text{nm}$; therefore, this provides the lower boundary for the elastic cross-section: $\sigma_{el} \geq 4\pi \left|\alpha''\right|^2 = 23\,\text{nm}^2$. The upper boundary is given by the condition $|S_0| \leq 1$. As a result, the real part of the scattering length is subject to the following condition:

$$\alpha' \leq \frac{1}{2k}\left[1 - (1 + 2k\alpha'')^2\right].$$

Therefore, $\sigma_{el} = 4\pi\left(\left|\alpha'\right|^2 + \left|\alpha''\right|^2\right) \leq 30\,\text{nm}^2$.

Note that the condition $\sigma_{el} << \sigma_{inel}$ is satisfied over the temperature range considered in this experiment. Since the inelastic cross-section is inversely proportional to the square root of temperature, and the elastic cross-section is temperature independent, the inelastic collisional loss rate is always higher than the elastic scattering rate and therefore governs the properties of the system at low temperatures. To achieve BEC, we must set the critical density as low as possible by lowering the exciton temperature. Inefficient elastic collisions cannot redistribute the excess energy generated in inelastic collisions. A phonon bath ensures that the excitons are at thermal equilibrium. In principle, BEC is possible even when inelastic collisions dominate. Such a situation is inaccessible in atomic experiments; thus, the present system can offer a unique opportunity to study a new type of BEC.

8.3.4 Accumulation of paraexcitons in a three-dimensional trap at sub-Kelvin temperatures and observation of transition to a Bose–Einstein condensate

Excitonic Lyman spectroscopy is a powerful method for quantitative detection of optically inactive 1s paraexcitons. Using this method, we reconfirmed the basic parameters of 1s paraexcitons, which are crucial to finding the experimental conditions required to achieve BEC, including the Bohr radius, translational mass, and lifetime. The obtained small Bohr radius ensures the robustness of the excitons at high densities. The very long lifetime, in the order of one microsecond, enables us to prepare cold excitons in thermal equilibrium with the lattice. Importantly, we can extract the density and temperature of excitons independently, so we can determine the phase space density of paraexcitons directly. A newly obtained very large inelastic collisional rate [72, 87], however, severely limits the density that can be attained without heating the lattice via exciton recombination. Considering realistic experimental conditions, the attainable maximum density of paraexcitons is in the order of about $10^{16}\,\text{cm}^{-3}$. Thus, we must cool the excitons to less than 1 K. In this section, we describe our recent experimental study of the exciton–BEC transition in a three-dimensional potential trap in a lattice kept below 1 K.

We used a helium-3 refrigerator to cool the crystal to as low as 278 mK. Below 2 K, paraexcitons diffuse almost ballistically, so a trap potential is required to accumulate high-density excitons at low temperatures under continuous feeding. A three-dimensional parabolic potential well for excitons can be formed by introducing a non-uniform strain distribution with a Hertzian contact geometry [88, 89]. A Cu_2O single crystal was placed in the cold finger of the cryostat and pressed by a glass lens. The trap frequencies in the xy and z-directions (Fig. 8.27(a)) were $\omega_{x,y} = 2\pi \times 17\,\text{MHz}$ and $\omega_z = 2\pi \times 25\,\text{MHz}$,

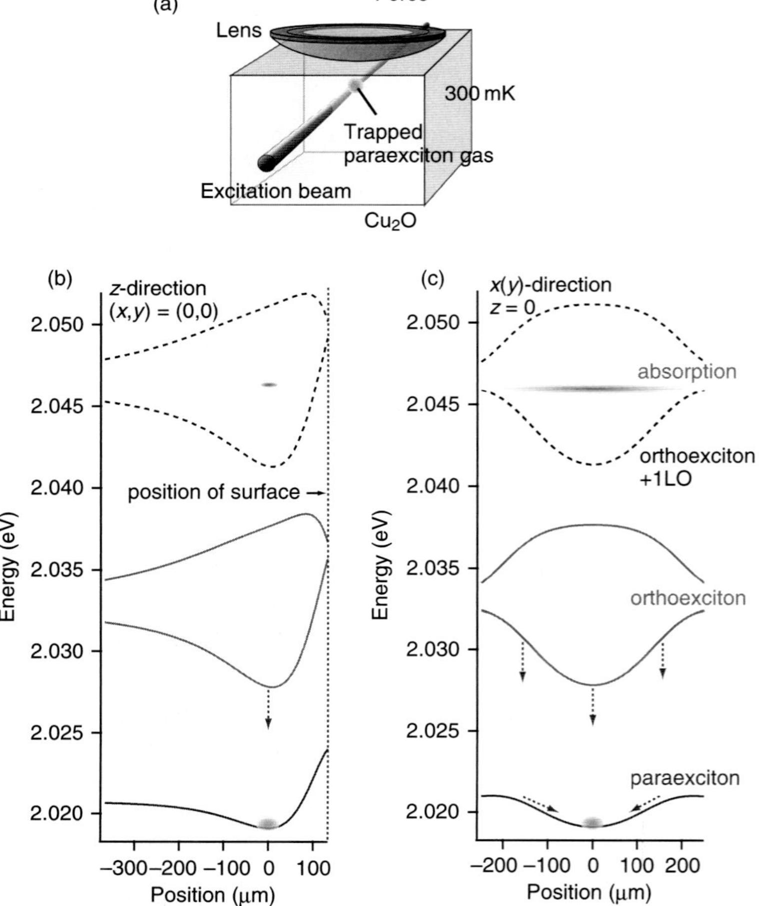

Fig. 8.27 Experimental configuration for 1s paraexciton trapping into a strain-induced three-dimensional trap potential in a single crystal of Cu_2O. (a) Schematic diagram of experimental configuration. (b, c) Position-dependent energy shift for ortho- and paraexcitons. Orthoexcitons are created via a phonon-assisted absorption process. Paraexcitons are generated via a rapid, typically within a few nanoseconds, ortho to para conversion process and accumulated around the bottom of the potential trap.

respectively. Note that the period of harmonic motion in the trap is shorter than the lifetime of 300 ns. This ensures a three-dimensional exciton trapping scheme. Paraexcitons are accumulated in the trap by continuously generating excitons near the trap by a cw ring dye laser tuned to the phonon-assisted absorption band of 1s orthoexcitons. The orthoexcitons are converted to 1s paraexcitons in nanoseconds, and they flow toward the bottom of the trap potential (Figs. 8.27(b), (c)). Spatially resolved spectra are then recorded by collecting the weak luminescence from the trapped paraexciton gas.

When the number of trapped paraexcitons is small, the lattice temperature dependence of the spatial and spectral width shows that the paraexciton gas is in thermal equilibrium and the gas temperature is about 800 mK (Fig. 8.28). The critical number for BEC of a non-interacting Bose gas at $T = 800\,\text{mK}$ is $N_c = 7.6 \times 10^8$ from the equation

$$N_c = \varsigma\,(3)\left(\frac{kT}{\hbar\overline{\omega}}\right)^3, \tag{8.8}$$

where $\varsigma(x)$ and $\overline{\omega}$ denote the Riemann zeta function and geometric angular frequency ($\overline{\omega} = \sqrt[3]{\omega_x\omega_y\omega_z} = 2\pi \times 19\,\text{MHz}$), respectively.

To increase the number of trapped paraexcitons while maintaining a constant lattice temperature ($T_L = 354\,\text{mK}$), we chopped the pump beam with appropriate duty cycles to achieve a constant average power. For ideal Bose particles well above the critical number, a large occupation number of the ground state is expected. Instead, as shown in Fig. 8.29, we observed anomalous spectra when the paraexciton number crossed the critical number. With increasing estimated paraexciton number, $N = 2 \times 10^7, 5 \times 10^8$, and 2×10^9, a high-energy signal suddenly appeared up to around 400 μeV above the potential minima. In Fig. 8.30(a), we plot the integrated emission intensity

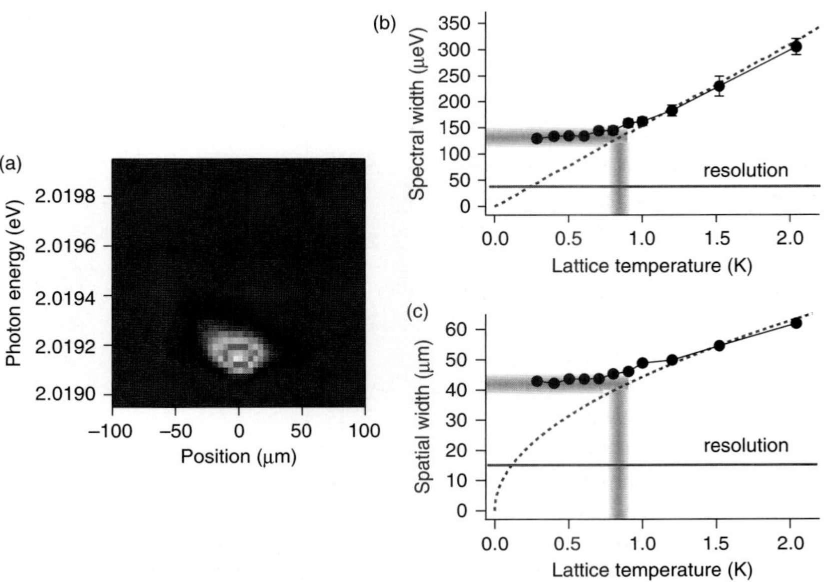

Fig. 8.28 Direct luminescence of the trapped paraexcitons at low density. (a) Spatially resolved luminescence of 1s paraexcitons at the lattice temperature of 287 mK. The dashed curve is the calculated potential profile of the trap. Spectral (b) and spatial (c) FWHM of the paraexciton gas.

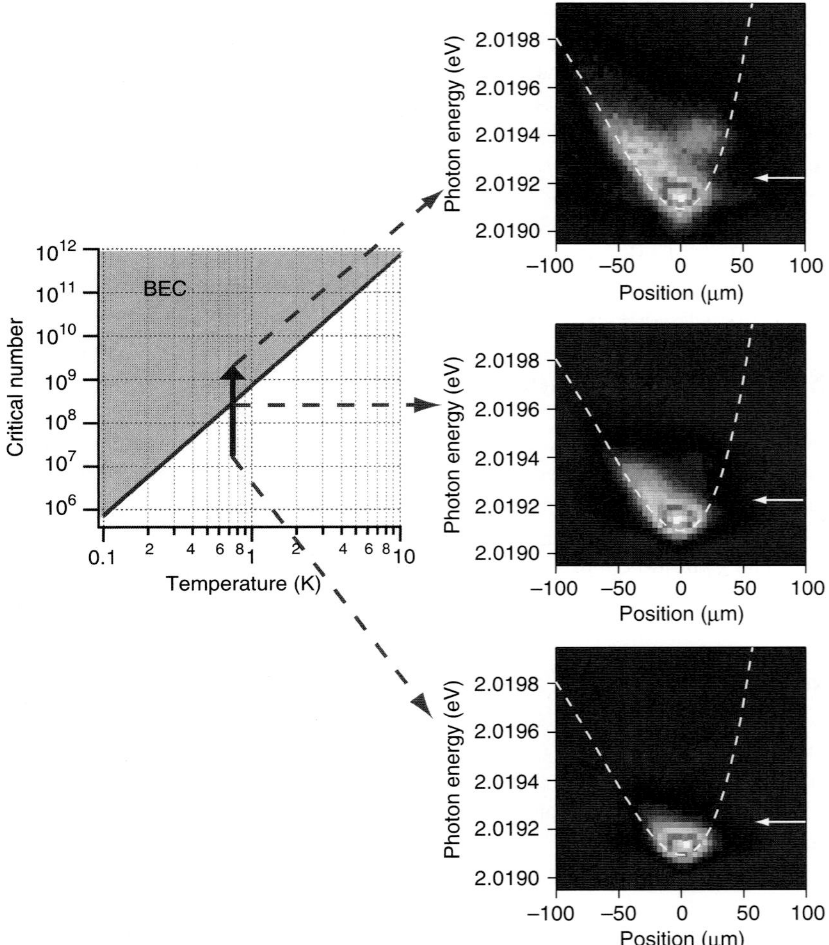

Fig. 8.29 Direct luminescence of the trapped paraexcitons across the BEC phase boundary. (See Plate 16 for color image).

ratio between the higher-energy part (2.01924–2.02004 eV) and the lower part (2.01902–2.01921 eV) versus the paraexciton number. A threshold-like rise in the ratio near N_c at an exciton temperature of 0.8 K is apparent, indicating that the BEC transition occurs under the conditions expected for ideal bosons. We also measured the temperature dependence under a constant excitation density. We set the excitation level to the maximum density shown in Fig. 8.30(b), $N = 2 \times 10^9$, at $T_L = 354$ mK. The lattice temperature dependence of the ratio also shows an abrupt increase at $T_L \sim 400$ mK, demonstrating the BEC transition of excitons (lowering T_L causes a slight exciton temperature drop and an increased exciton number).

Although the inelastic collision cross-section is very large, excitons are cooled by exciton–phonon interactions. A possible scenario for the peculiar spectra is as follows. Above the critical number, small but finite occupation of the ground state exists. Bosonic scattering into the ground state occurs, but the inelastic process is enhanced because of the small volume of the ground state

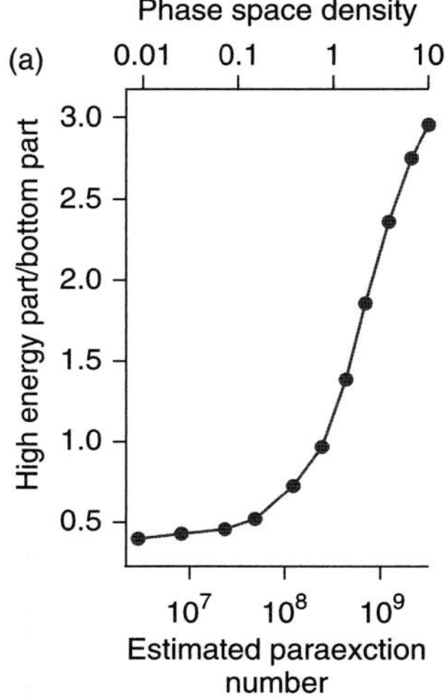

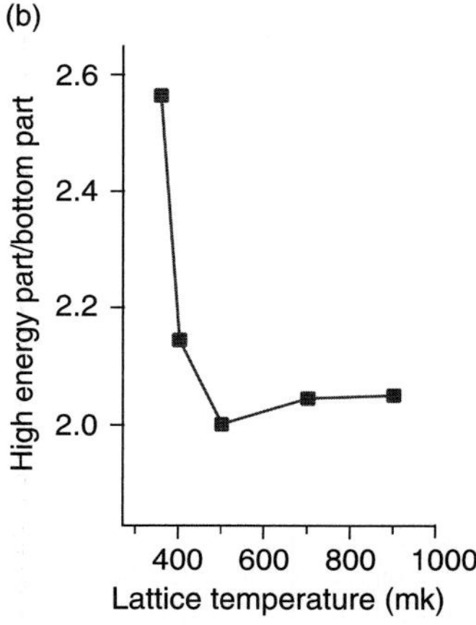

Fig. 8.30 Ratio between the higher energy part and bottom parts. (a) Hot/cold ratio versus the estimated paraexciton number at the effective exciton temperature of 0.8 K. Estimated phase space density is also shown. (b) Hot/cold ratio versus the lattice temperature under a constant excitation density ($N = 2 \times 10^9$ at 354 mK).

wave function. The collisions produce hot excitons, which eventually return to the bottom of the trap. Our analysis based on rate equations supported a condensate fraction of about 1%.

We conducted a simulation based on the direct simulation Monte Carlo method, considering all the processes involved (e.g. exciton–phonon interactions, lifetimes, collisions). The results confirmed the enhancement of the hot part in the presence of the condensate. This supports our conclusion that we observed a spontaneous BEC transition in a purely excitonic three-dimensional system. To make the condensate fraction much larger, we need to cool the gas further and achieve a dilute condensate to avoid frequent inelastic scattering.

8.4 Summary and perspective

In the past few decades, remarkable advances in laser technologies have produced various new spectroscopic techniques for probing matter from the far-infrared to the X-ray spectral region. In addition, precisely controlled light sources provided by advanced laser technologies enable us to manipulate matter under extreme conditions. A typical example is an ensemble of ultra-cold atoms prepared by laser cooling and trapping. Recent studies of ultra-cold fermionic atoms have opened a new area for exploring the nature of many-body quantum systems. For example, a continuous evolution from BEC to BCS-like limits could be observed experimentally [90, 91]. Such advanced laser technologies have also triggered renewed interest in the study of photoinduced phenomena in condensed matter systems. In particular, by using precisely controlled light, one can manipulate an electron–hole ensemble into a highly quantum degenerate state, where many-body quantum phenomena can be expected.

In this chapter, we have reviewed recent developments in laser-based spectroscopic studies of electron–hole systems in semiconductors. In particular, we focused on the search for BEC of excitons in bulk semiconductors. Excitonic Lyman spectroscopy is a powerful tool for investigating the space–time evolution and energy distribution of optically inactive excitons; it allows one to determine the basic parameters necessary to clarify the conditions for an exciton BEC. In particular, we developed cw-laser-based excitonic Lyman spectroscopy for sensitive, wide-dynamic-range detection of 1s paraexcitons in Cu_2O. With this method, we obtained the inelastic scattering cross-section of paraexcitons. The very large temperature-independent collision loss rate imposes a severe limit on the maximum reachable paraexciton density. At an ordinary cryogenic temperature, typically 2 K, nonlinear saturation makes it difficult to reach a phase transition density of 10^{17} cm^{-3}.

We have also presented our recent experiments on trapping of 1s paraexcitons at sub-Kelvin temperatures [92]. We can cool the paraexcitons in a cold phonon bath. Emission spectra from paraexcitons in a three-dimensional trap showed an anomalous threshold-like distribution at the critical number expected for BEC of ideal bosons. Bosonic stimulated scattering into the condensate and inelastic scattering processes in high-density condensates compete, limiting the condensate to a fraction of about 1%. This observation adds a new class of experimentally accessible BECs and provides opportunities for exploring a rich variety of matter phases of electron–hole ensembles.

To reach a large condensate fraction, we must reduce the local density. Therefore, it is essential to produce a colder exciton gas. Because inelastic scattering dominates elastic scattering, evaporative cooling does not work in this system. Thus, we need to develop new methods for efficient cooling of excitons. To understand the nature of an exciton condensate, it is evidently crucial to elucidate the mechanism for the large inelastic scattering cross-section. It is interesting to compare our results with those of recent experiments on cold molecules [93, 94]. To explore quantum many-body physics with excitonic matter, continuous efforts via systematic and quantitative experiments are necessary.

Acknowledgments

Experiments from Tokyo presented in this chapter were the results of long-term research at the Applied Physics department of the University of Tokyo.

The author thanks K. Yoshioka, R. Shimano, M. Kubouchi, T. Tayagaki, and A. Mysyrowicz for discussions and essential collaboration, and N. Naka for fruitful discussions.

This work was supported by Grant-in-Aid for Scientific Research on Innovative Area "Optical Science of dynamically correlated electrons (DYCE) 20104002", Photon Frontier Network Program of the Ministry of Education, Culture Sports, Science and Technology, Japan and Core Research for Evolutional Science and Technology (CREST) by Japan Science and Technology Agency.

References

[1] Klingshirn C.F., *Semiconductor Optics*, 2nd edition (Springer–Verlag, Berlin Heidelberg, 2005).

[2] Haug H. and Koch S.W., *Quantum Theory of the Optical and Electronic Properties of Semiconductors,* 3rd edition (World Scientific, Singapore, 1998).

[3] Frenkel J.I., *Phys. Rev.* **37**(17), 1276 (1931a).

[4] Peierls R.E., et al., *Ann. Phys. (Leipzig)* **13**, 905 (1932).

[5] Wannier G.H., et al., *Phys. Rev.* **52**, 191 (1937).

[6] Sham L.J., et al., *Phys. Rev*. **144**, 708 (1966).

[7] Keldysh L.V., *Electron-hole Droplets in Semiconductors,* edited by Jefferies C.D. and Keldysh L.V. (North Holland, Amsterdam, 1983).

[8] Shimano R., et al., *Phys. Rev. Lett.* **88**, 057404 (2002).

[9] Nagai M., et al., *Phys. Rev.* B **68**, 081202(R) (2003).

[10] Nagai M., et al., *Phys. Rev. Lett*. **86**, 5795 (2001).

[11] Keldysh V. and Koslov A.N., *Zh. Eksp. Teor. Fiz.* **54**, 978 [*Sov. Phys. JETP*. **27**, 521] (1968).

[12] Cosme C. and Nozieres P., *J. de Physique* **43,** 1069 (1982).

[13] Nozieres P. and Schmitt-Rink S., *J. Low Temp. Phys.* **59**, 195 (1984).

[14] Randeria M. in *Bose–Einstein Condensation* edited by Griffin A., Snoke D.W., and Stringari S. (Cambridge University Press, Cambridge, 1995) p. 355.

[15] Kiram., et al., *Phys. Rev. Lett.* **81**, 3263 (1998).

[16] Huber R., et al., *Nature* **414**, 286 (2001).

[17] Landau L., *J. Phys. U.S.S.R.* **10**, 503 (1946).

[18] Hänsh T.W. and Schawlow A., *Opt.Commun.* **13**, 68 (1975).

[19] Mita T., et al., *J. Phys. Soc. Jpn.* **48**, 496 (1980).

[20] Chase L.L., et al., *Phys. Rev. Lett.* **42**, 1231 (1979).

[21] Hanamura E., *Solid State Commun.* **12**, 951 (1973).

[22] Gale G.M. and Mysyrowicz A., *Phys. Rev. Lett.* **32**, 727 (1974).

[23] Kuwata-Gonokami M., et al., *J. Phys. Soc. Jpn.* **71**, 1257 (2002).

[24] Heberle A.P., et al., *Phys. Rev. Lett.* **75**, 2598 (1995).

[25] Blanchet V., et al., *Phys. Rev. Lett.* **78**, 2716 (1997).

[26] Akiyama H., et al., *Phys. Rev.* B **42**, 5621 (1990).

[27] Peyghanbarian N., et al., *Phys. Rev.* B **27**, 2325 (1983).

[28] Shimano R., et al., *Phys. Rev. Lett.* **23**, 233601 (2002).

[29] Edamatsu K., et al., *Nature* **431**, 167 (2004).

[30] Dasbach G., et al., *Phys. Rev. Lett.* **91**,107401 (2003).

[31] Yoshioka K. and Kuwata-Gonokami M., *Phys. Rev.* B **73**, 081202(R) (2006).

[32] Blatt J.M., et al., *Phys. Rev*. **126**, 1691 (1962).

[33] Snoke D.W., et al., *Phys. Rev.* B **41**, 11171 (1990).

[34] Lin J.L. and Wolfe J.P., *Phys. Rev. Lett.* **71**, 1222 (1993).

[35] Naka N. and Nagasawa N., *J. Lumin.* **112**, 11 (2005).

[36] Butov L.V., et al., *Nature* **418**, 751 (2002); **417**, 47 (2002).

[37] Snoke D.W., et al., *Nature* **418**, 754 (2002).

[38] Deng H., et al., *Science* **298**, 199 (2002).

[39] Kasprzak J., et al., *Nature* **443**, 409 (2006).

[40] Balili R., et al., *Science* **316**, 1007 (2007).

[41] Mysyrowicz A., *Phys. Rev. Lett.* **43**, 1123 (1979).

[42] Caswell N., et al., *Solid State Commun.* **40**, 843 (1981).

[43] Hulin D., et al., *Phys. Rev. Lett.* **45**, 1970 (1980).

[44] Snoke D.W., et al., *Phys. Rev. Lett.* **59**, 827 (1987).

[45] Yoshioka K., et al., *Phys. Rev. B* **76**, 033204 (2007).

[46] Brandt J., et al., *Phys. Rev. Lett.* **99**, 217403 (2007).

[47] Snoke D.W., et al., *Phys. Rev. Lett.* **64**, 2543 (1990).

[48] Fortin E., et al., *Phys. Rev. Lett.* **70**, 3951 (1993).

[49] O'Hara K.E., *Phys. Rev.* B **62**, 12909 (2000).

[50] Kuwata-Gonokami M., et al., *J. Phys. Soc. Jpn.* **73**, 1065 (2004).

[51] Kubouchi M., et al., *Phys. Rev. Lett.* **94**, 016403 (2005).

[52] Tayagaki T., et al., *J. Phys. Soc. Jpn.* **74**, 1423 (2005).

[53] Tayagaki T., et al., *Phys. Rev.* B **74**, 245127 (2006).

[54] Kuwata-Gonokami M., *Solid State Commun.* **134**, 127133 (2005).
[55] Haken H., Bericht in Fortschritte der Physik **6**, 271 (1958).
[56] Nikitine S., *J. Phys. Chem. Solids* **45**, 955 (1984).
[57] Johnsen K. and Kavoulakis G.M., *Phys. Rev. Lett.* **86**, 858 (2001).
[58] Goppert M., et al., *J. Modern Phys.* B **15**, 3615 (2001).
[59] Jörger M., et al., *Phys. Rev.* B **65**, 245209 (2005).
[60] Karpinska K., et al., *J. Lumin.* **112**, 17 (2005).
[61] Hodby J.W., et al., *J. Phys.* C **9**, 1429 (1976).
[62] Artoni M., et al., *Phys. Rev.* B **65**, 235422 (2002).
[63] Kavoulakis G.M., *Phys. Rev.* B **55**, 7593 (1997).
[64] Ideguchi T., et al., *Phys. Rev. Lett.* **100**, 233001 (2008).
[65] Mani S., et al., *Phys. Rev.* B **82**, 113203 (2010).
[66] Weiner A.M., *Rev. Sci. Instrum.* **71**, 1929 (2000).
[67] Meshulach D., et al., *Phys. Rev.* A **60**, 1287 (1999).
[68] Shumway J. and Ceperley D.M., *Phys. Rev.* B **63**, 165209 (2001).
[69] Okumura S. and Ogawa T., *Phys. Rev.* B **65**, 035105 (2001).
[70] Combescot M., et al., *Phys. Rev.* B **75**, 174305 (2007).
[71] Schindler C. and Zimmermann R., *Phys. Rev.* B **78**, 045313 (2008).
[72] O'Hara K.E., et al., *Phys. Rev.* B **60**, 10872 (1999).
[73] Snoke D.W. and Negoita V., *Phys. Rev.* B **61**, 2904 (2000).
[74] Warren J.T., et al., *Phys. Rev.* B **61**, 8215 (2000).
[75] Liu Y. and Snoke D.W., *Solid State Commun.* **140**, 208 (2006).
[76] Jang J.I. and Wolfe J.P., *Phys. Rev.* B **72**, 241201(R) (2005); **74**, 045211 (2006).
[77] Mysyrowicz A., et al., *J. Lumin*. **24**, 629 (1981).
[78] Kavoulakis G.M. and Baym G., *Phys. Rev.* B **54**, 16625 (1996).
[79] Kavoulakis G.M. and Mysyrowicz A., *Phys. Rev.* B **61**, 16619 (2000).
[80] Elliott R.J., *Phys. Rev.* **124**, 340 (1961).
[81] Jang J.I., et al., *Phys. Rev.* B **70**, 195205 (2004).
[82] Weiner J.S., et al., *Solid State Commun.* **46**, 105 (1983).
[83] Snoke D.W., et al., *Phys. Rev.* B **41**, 5266 (1990).
[84] Jörger M., et al., *Phys. Rev.* B **71**, 235210 (2005).
[85] Landau L.D. and Lifschitz E.M., *Quantum Mechanics*, 3rd ed. (Pergamon, New York, 1985).
[86] Wigner E.P., *Phys. Rev.* **73**, 1002 (1948).
[87] Yoshioka K., *Phys. Rev.* B **82**, 041201(R) (2010).
[88] Trauernicht D.P., et al., *Phys. Rev.* B **27**, 2562 (1983).
[89] Naka N. and Nagasawa N., *Phys. Rev.* B **70**, 155205 (2004).
[90] Regal C. A., et al., *Phys. Rev. Lett.* **92**, 040403 (2004).
[91] Zwierlein M.W., et al., *Phys. Rev. Lett.* **92**, 120403 (2004).
[92] Yoshsioka K., et al., *Nature Communications* **2**, 328(2011).
[93] Ospelkaus S., et al., *Science* **327**, 853 (2010).
[94] Ni K.K., et al., *Nature* **464**, 1324 (2010).

Superfluidity in exciton–polariton condensates

Y. Yamamoto

9.1 Introduction

Einstein's 1925 paper predicted the occurrence of Bose–Einstein condensation (BEC) in an ideal gas of non-interacting bosonic particles [1]. A quantum field-theoretical formulation for a weakly interacting Bose condensed system was developed by Bogoliubov in 1947, which predicted the phonon-like excitation spectrum [2] at low momentum regime. Experimental verification of the Bogoliubov theory on the quantitative level was performed for atomic BEC [3] using a two-photon Bragg scattering technique [4]. Exciton–polaritons in a semiconductor microcavity, which are elementary excitations created by strong coupling between quantum well excitons and microcavity photons, were proposed as a new BEC candidate in solid state systems [5]. Recent experiments with exciton–polaritons have demonstrated several interesting signatures from the viewpoint of polariton condensation, such as, quantum degeneracy at non-equilibrium condition [6–8], polariton bunching effect at condensation threshold [9], long spatial coherence [10–12], and quantum degeneracy at equilibrium condition [13], but the unique excitation spectra have only been studied theoretically for exciton–polaritons [14, 15]. In this chapter, we review the interaction effects on the exciton–polariton condensate and excitation spectra, which are in quantitative agreement with the Bogoliubov theory.

In a homogeneous two-dimensional system at non-zero temperature, although there can be no ordering of infinite range [16, 17], a superfluid phase is predicted [18–20]. Bound vortex–antivortex pairs dominate the thermodynamics and phase coherence properties in this superfluid regime. It is believed that several different systems share this common behavior, when the parameter describing their ordered state has two degrees of freedom [21], and the theory has been tested for some of these [22–27]. However, there has been no direct experimental observation of a quasi-condensate which includes a bound pair of vortex and antivortex. Here we present an experimental technique that can identify a single vortex–antivortex pair in a two-dimensional exciton–polariton condensate. The pair is generated by the inhomogeneous pumping spot profile, and is revealed in the time-integrated phase maps acquired using Michelson interferometry. Numerical modeling based on an open dissipative Gross–Pitaevskii equation suggests that pair evolution is quite different in this non-equilibrium system compared with atomic condensates [28].

9.2 Exciton–polariton condensation

9.2.1 Exciton–polariton trap

In a semiconductor microcavity with single or multiple quantum wells (QWs), eigenstates are altered to the new normal modes, called exciton–polaritons, when the cavity photon–QW exciton coupling rate exceeds the decay rates of the photon and exciton. The exciton–polariton is a promising solid state system for studying dynamical condensation phenomena in solids [29, 30]. Since its effective mass is eight orders of magnitude smaller than atomic mass and four orders of magnitude lighter than exciton mass, the critical temperature of polariton BEC transition is expected to be up to room temperature. The leakage photons carry the identical energy and in-plane momentum as the internal polaritons, thus it is possible to directly measure the energy–momentum dispersion relation and momentum distribution of the polaritons by an angle-resolved spectroscopy technique. This important information is available for liquid ^{4}He systems only through "quantum evaporation" [31] and for gaseous atoms only through "Feshbach resonance controlled free expansion".

The exciton–polariton trap used in our experiment is shown in Fig. 9.1a. Three stacks of 4 GaAs QWs are embedded at the central three antinode positions of an AlAs/AlGaAs distributed Bragg reflector (DBR) planar microcavity. The normal mode splitting is $2g_0 \sim 15$ meV and the cavity photon lifetime is 2 psec. This leads to a $k = 0$ lower polariton (LP) lifetime of ~ 4 psec at zero detuning, $\Delta \equiv E_C - E_X = 0$, where k is the in-plane wavenumber, E_C and E_X are the cavity photon and QW exciton energies at $k = 0$. The trap potential of $\sim 200\,\mu$eV is provided by a hole surrounded by a thin metal (Ti/Au) film [32]. The cavity resonant field normally has an antinode at the AlGaAs–air interface. However, the antinode position is shifted inside the AlGaAs layer with the metal film as shown in Fig. 9.1a, which results in the blue shift of the cavity resonance and the LP energy. In our experiment, the LPs were confined in circular holes of varying diameters from $5\,\mu$m to $100\,\mu$m. In a trap with $5 \sim 10\,\mu$m diameter, a single fundamental transverse mode dominated the condensation dynamics over other higher-order modes due to the relatively weak confining potential.

The total number n of injected LPs into a trap by the pump pulse and the number n_0 of the LP centered at $k = 0$ and within $|k| \leq 4 \times 10^3\,\text{cm}^{-1}$, approximately corresponding to the trapped ground state, are plotted in Fig. 9.1b as a function of normalized pump rate P/P_{th}. The fractional condensate n_0/n increases nonlinearity at condensation threshold and reaches a maximum of ~ 0.6 at $P/P_{th} \sim 6$.

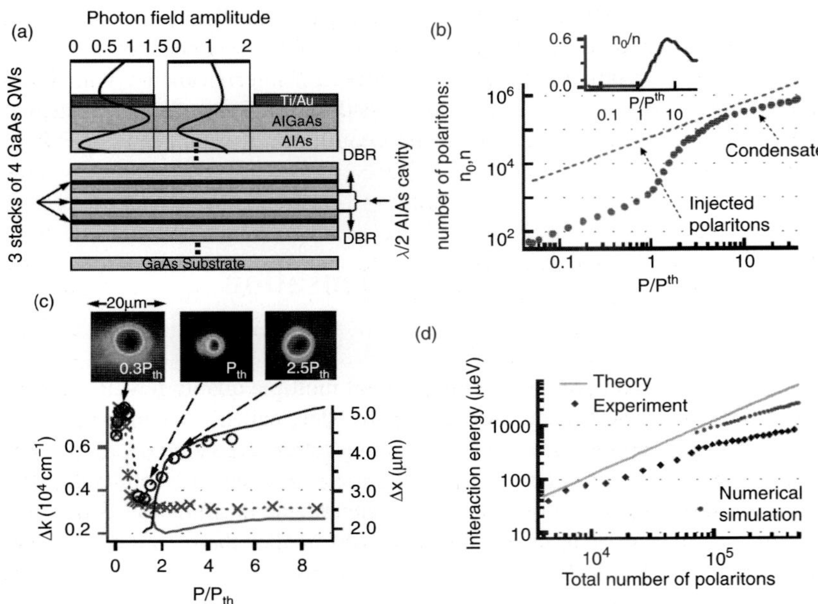

Fig. 9.1 An exciton–polariton condensate in a single mode trap. a, A schematic of a polariton trap formed by a thin metal film (Ti/Au: 4/20 nm) on top of a microcavity structure with circular holes (diameter: 4 ~ 100 μm). The microcavity consists of $\lambda/2$ AlAs optical cavity layer sandwiched by two distributed Bragg reflectors with alternating AlGaAs/AlAs $\lambda/4$ layers, where λ is the cavity resonance wavelength. Three stacks of 4 GaAs quantum wells (QWs) are placed at the central three antinodes of the microcavity photon field. A photon field amplitude has an antinode at an AlGaAs–air interface without a metal film, while a deposited thin metal film pushes an antinode position inside the AlGaAs layer, which results in a blue shift of the cavity resonance and the lower polariton (LP) energy. The blue shift of the LP energy under the metal film is typically $\sim 200\,\mu$eV. b, Total number of injected LPs n into a trap (pink dashed line) and number of the LP n_0 centered at $k = 0$ and within $|k| \leq 4 \times 10^3\,\text{cm}^{-1}$ (red dots), which approximately corresponds to the LP number of the ground state of the trap, are plotted as a function of P/P_{th}. [Inset] The fractional population n_0/n as a function of P/P_{th}. The linearly polarized laser beam is injected with an angle of $60°(k = 7 \times 10^4\,\text{cm}^{-1})$ into a trap with 8 m diameter, where the detuning parameter is $\Delta = 5.4$ meV. c, The measured standard deviations of LP distribution in coordinate Δx (blue circles) and in wavenumber Δk (red crosses) are plotted as a function of P/P_{th}. Theoretical values for Δx and Δk obtained by the GP equation are shown by blue and red solid lines. Top panels are the near-field images at three different pump levels; $0.3P_{th}$, P_{th}, and $2.5P_{th}$. d, The measured LP energy shift at $k = 0$ (blue diamonds) and calculated energy shift $U(n)$ by Eq. (9.5) (light blue solid line) are plotted as a function of total number of polaritons. The numerical results by the GP equation, including the effect of pump-dependent condensate size, are shown by red dots.

The top panels in Fig. 9.1c show the near-field emission patterns from a trap with 8 μm diameter at pump rates below, just above, and well above condensation threshold. The measured standard deviation for the polariton position Δx and the wavenumber Δk are plotted as a function of P/P_{th} in Fig. 9.1c. The sudden decrease in Δx and Δk were clearly observed at $P \gtrapprox P_{th}$. Just above threshold, the measured uncertainty product $\Delta x\ \Delta k$ is ~ 0.98, which is compared to the Heisenberg limit ($\Delta x \Delta k \sim 0.5$) for a minimum uncertainty wavepacket. The monotonic increase in Δx and x $\Delta x \Delta k$ at higher pump rates stems from the repulsive interaction among LPs in a condensate and is well reproduced by the theoretical analysis using the Gross–Pitaevskii (GP) equation as shown in Fig. 9.1c.

9.2.2 Gross–Pitaevskii equation

For weakly interacting bosons in a dilute limit [14], a single particle field $\Psi_0(\mathbf{r})e^{-i\frac{\varepsilon_0}{\hbar}t}$ can be introduced to describe the wave function of the condensate ground state,

$$\Psi_0(\mathbf{r}) = \prod_{I=1}^{N_0} \Psi_0(\mathbf{r}_I), \tag{9.1}$$

where $\mathbf{r}_I$ is the coordinate of the I-th particle in the two-dimensional system and N_0 is the number of particles in the ground state. At a temperature well below the critical temperature of condensation, $\Psi_0(\mathbf{r})$ obeys the following nonlinear Schrödinger equation (or Gross–Pitaevskii equation):

$$\left(-\frac{\hbar^2}{2m}\nabla^2 + V(\mathbf{r})\right)\Psi_0(\mathbf{r}) + g|\Psi_0(\mathbf{r})|^2\Psi_0(\mathbf{r}) = \varepsilon_0\Psi_0(\mathbf{r}), \tag{9.2}$$

where $V(\mathbf{r})$ is an external potential for the particles and ε_0 is the ground state energy per particle of the condensate. The nonlinear term in this equation results from the interaction between the particles with an interaction energy $g \equiv \frac{U(n')}{n'}$, where $U(n')$ is a density-dependent interaction energy introduced in the main text and a contact interaction with the form $\int d\mathbf{r}_1 d\mathbf{r}_2 g\delta(\mathbf{r}_1 - \mathbf{r}_2)\Psi^\dagger(\mathbf{r}_1)\Psi^\dagger(\mathbf{r}_2)\Psi(\mathbf{r}_2)\Psi(\mathbf{r}_1)$ has been assumed. (Please note that g is not an exciton–polariton splitting g_0.) Here n' is the local density of the condensate. The state $\Psi_0(\mathbf{r})$ describes the local condensate density as $n'(\mathbf{r}) = |\Psi_0(\mathbf{r})|^2$, and is renormalized to the total number of condensate particles as $\int d\mathbf{r}|\Psi_0(\mathbf{r})|^2 = n$. For a homogeneous system, the solution of Eq. (9.2) is approximated by $\Psi_0 = \sqrt{n'}$ with $n' = n/S$ and S is the size of the condensate, and $\varepsilon_0 = U(n') = gn'$.

For the exciton–polaritons in a microcavity, the system is not homogeneous either because of the Gaussian profile of the pumping laser beam or because of an external trapping potential. The effect of the Gaussian profile on the condensate has been thoroughly studied in our previous publication [4, 40]. First, we focus on the property of the condensate in an external trapping potential. In our system, a metallic layer provides a small cylindrical trapping potential of $V_0 \simeq 200\,\mu eV$ for the exciton–polaritons inside a trap of radii $R_0 = 3.5 - 4\,\mu$m. Meanwhile, it also reduces the absorbed pump power by $\sim 50\%$ for the area under the metallic layer.

We consider the solution of Eq. (9.2) in a two-dimensional system with the cylindrical coordinates (r, θ). The ground state wave function can be written as $\Psi_0(r) = \frac{\phi_0(r)}{\sqrt{r}}$ and $\phi_0(r)$ obeys the equation

$$\left\{-\frac{\hbar^2}{2m}\left[\frac{\partial^2}{\partial r^2} + \frac{1}{4r^2}\right] + V(r)\right\}\phi_0(r) + g\frac{\phi_0^3(r)}{r} = \varepsilon_0\phi_0(r), \qquad (9.3)$$

which can be solved numerically with the steepest descent method [14]. For non-interacting particles, the solution of Eq. (9.3) is given by the Bessel functions as

$$\phi_0(r) = \begin{cases} c_i J_0(k_0 r), & r \le R_0 \\ c_e K_0(k_0' r), & r > R_0 \end{cases}, \qquad (9.4)$$

with $k_0 = \sqrt{2m(V_0 + e_0)/\hbar^2}$, $k_0' = \sqrt{2m(-e_0)/\hbar^2}$ and $e_0\,(<0)$ being the ground state energy without interaction. Here, $c_{i,e}$ are the normalization coefficients determined by the boundary conditions. With $m = 10^{-4}m_e$, where m_e is the bare electron mass, we derive that $e_0 = -0.11$ meV for $R_0 = 3.5\,\mu$m and $V_0 = 200\,\mu eV$. The standard deviation of the ground state in coordinate space is $\Delta x = 1.86\,\mu$m and the deviation in momentum space is $\Delta k = 0.28\,(\mu\text{m})^{-1}$, with $\Delta x \Delta k \sim 0.52$ close to the limit imposed by the Heisenberg uncertainty principle. The density distribution is shown in Fig. 9.2 (a).

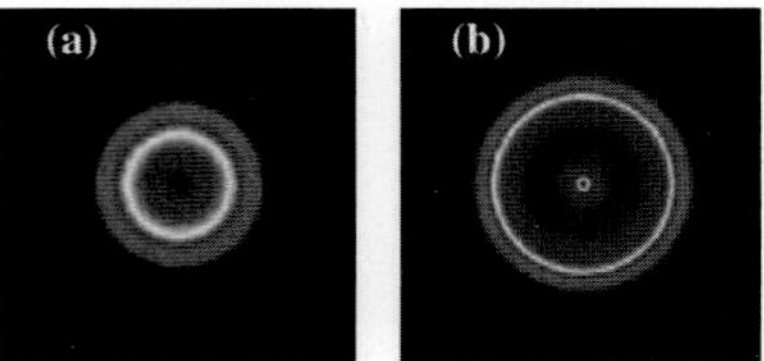

Fig. 9.2 Condensate density for the exciton–polaritons. (a) Condensate density for non-interacting polaritons. (b) Condensate density for interacting polaritons with $n = 3 \times 10^5$ and $P/P_{th} = 5$. (See Plate 17 for color image).

The interaction among the exciton–polaritons generates a nonlinear term in Eq. (9.3) and has a significant effect on the condensate properties. Originating from the interaction between the exciton components, the interaction energy in our model depends on the density of the exciton–polaritons as is described in the main text. When the density n increases, the interaction energy $U(n') = gn'$ increases with density, but saturates at large density. This means that the interaction constant g decreases with n'. In an inhomogeneous system, this causes a spatial dependence of the interaction energy, which introduces a second nonlinearity in the system besides the conventional nonlinearity in the GP equation.

The finite lifetime of exciton–polaritons affects measurement of the condensate and induces some complication for the calculation. First, only a finite time window is available for measurement of the leakage photons. During this time window, a substantial proportion of the non-condensed exciton–polaritons exist outside the trap. Second, the diffusion time of the injected exciton–polaritons is very long compared with the condensate lifetime. As a result, the exciton–polaritons outside the trap can be treated as a static background field that interacts with the condensate. We construct a simple model to describe these effects. The condensate is largely localized inside the trap. The exciton–polaritons outside the trap are treated as "idle" particles. The exciton–polaritons outside the trap, with a density of $n'_{ext} \simeq n'/2$, half of the density inside the trap due to absorption of pump photons in a metal film, are treated as an effective trapping potential.

We calculate the condensate properties for a wide range of pumping levels. In Fig. 9.2 (b), the spatial distribution of the condensate is shown for $P/P_{th} = 5$. Compared with the non-interacting case, the interaction flattens the density out into two sections. At $r \leq R_0$ (inside the trap), the condensate has a nearly uniform density; at $r > R_0$ (outside the trap), the density quickly decreases. The low density core at $r \sim 0$ is a numerical artifact due to the chosen boundary condition.

9.2.3 Interaction energy

The standard deviation of the condensate in coordinate space is calculated using a density-dependent interaction constant g. The calculated results agree well with the measurement results, except for the pump rate very close to the condensation threshold, as is shown in Fig. 9.1(c). As described in the next section, the condensation near threshold does not necessarily occur at the ground state so that the observed δx is larger than the theoretical value. At high pumping level, the size increase of the condensate saturates. Here, the increasing pump power increases the spatial density of the exciton–polaritons, and hence the interaction energy $gn'(r)$, which makes the condensate expand to a larger size. However, when the condensate expands too much, the interaction $gn'(r)$ decreases with the reduced density, which then acts as a back action to limit the growth of the condensate. Similarly, the standard deviation in momentum space can also be calculated. The condensate wave function in momentum space is $\psi_k \propto \int rdrJ_0(kr)\Psi_0(r)$, where $k = |\mathbf{k}|$ and $J_0(kr)$ is the zeroth order Bessel function. The results are presented in Fig. 9.1(c).

The condensate energy per particle ε_0 can also be calculated with the GP equation. For a homogeneous system, ε_0 is directly connected to the interaction energy $U(n') = g(n') \cdot n'$, and the measured ε_0 can be used to analyze the properties of the ground state as well as the excitations. We numerically calculate ε_0 for the trapped system. The results are shown in Fig. 9.3. In our system, ε_0 includes contributions from the interaction between the condensate particles $g|\Psi_0(r)|^2$, the effective potential, and the kinetic energy due to the change of

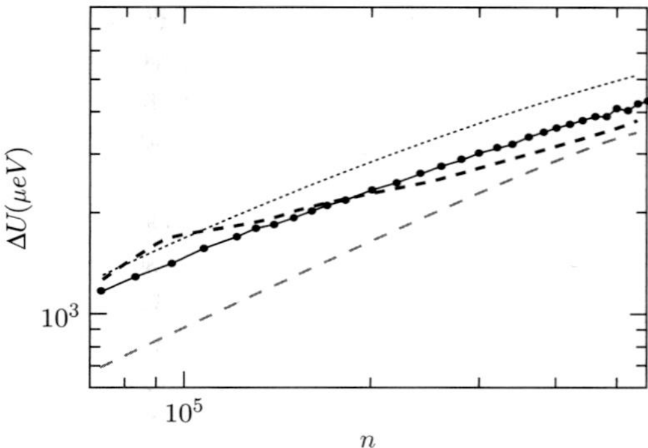

Fig. 9.3 Solid circle: the condensate energy per particle ε_0 versus total number of injected exciton–polaritons n inside a trap, calculated with the GP equation approach, which is shown in Fig. 9.1(d) by red dots. Blue dotted line: the interaction energy $U(n')$ calculated by Eq. (9.5). Black dashed line: the interaction energy $U_{in} = gn'(r < R_0)$ due to the polaritons inside the trap. Red dashed line: the interaction energy $U_{ext} = gn'_{ext}$ due to the polaritons outside the trap.

condensate geometry. Our results show that $\varepsilon_0 \sim U_{in}$ (interaction energy inside the trap) in a wide range of the pumping levels. This is because most of the condensate particles are in the central plateau, as shown in Fig. 9.2 (b), where we have a locally homogeneous system with the interaction U_{in}.

The $k = 0$ LP energy is blue shifted with the number of polaritons as shown in Fig. 9.1d. This is a direct manifestation of the aforementioned repulsive interaction among LPs in a condensate. The $k = 0$ LP energy shift $U(n)$ is calculated by the relation

$$U(n) = \frac{1}{2}\delta E_x + \sqrt{g_0^2 + \frac{\Delta^2}{4}} - \sqrt{g(n)^2 + \frac{(\Delta - \delta E_x)^2}{4}}, \tag{9.5}$$

where $\delta E_x = E_B \frac{n}{n_s}$ and $g(n) = g_0 \left(1 - \frac{n}{n_5}\right)$ represent the blue shift of the QW exciton energy due to fermionic exchange interaction [33, 34] and the reduced normal mode splitting due to phase space filling and fermionic exchange interaction [35], respectively. $n_s = \frac{N_{QW}S}{2.2\pi a_B^{*2}|X|^2}$ and $n'_s = \frac{N_{QW}S}{4\pi a_B^{*2}|X|^2}$ are the saturation polariton numbers for the above two nonlinear processes, respectively. a_B^* is the QW exciton Bohr radius, $N_{QW} = 12$ is the number of QWs, $g_0 \approx 7.5$ meV is the photon–exciton coupling strength, S is the cross-sectional area of the condensate, and $|X|^2 = \frac{1}{2}\left[1 + \Delta\Big/\sqrt{4g_0^2 + \Delta}\right]$ is the exciton fraction of the $k = 0$ LP.

If we neglect the pump rate dependence of the condensate cross-sectional area and assume a constant cross-sectional area determined by the trap area, $S = \pi\left(4 \times 10^{-4}\,\mathrm{cm}\right)^2$, the theoretical energy shift is shown by the light blue line in Fig. 9.1d. We can numerically solve the GP equation to incorporate the pump rate-dependent condensate size. The results are shown by red dots in Fig. 9.1d. These two theoretical predictions are compared to the experimental results (blue diamonds). We note that the above mentioned nonlinear model based on weakly interacting bosons [33–35] can reproduce the experimental data only in low polariton density regimes. Therefore, we use the experimental values (not theoretical values) for $U(n)$ as the interaction energy in subsequent discussions for the universal feature of the Bogoliubov excitations.

9.3 Bogoliubov excitation spectrum

9.3.1 Energy versus momentum dispersion relation

The dispersion relations between the LP energy E versus in-plane wavenumber k obtained by angle resolved spectroscopy are shown for the pump rate above threshold $P/P_{th} = 3$ in Fig. 9.4. Figure 9.4(a) represents a linear plot of the intensity, while Figs. 9.4(b)–(d) employ logarithmic plots of the intensity to magnify the excitation spectra. Here we show an "untrapped" case, where the pump spot size (diameter $\sim 30\,\mu$m) is considerably smaller than the trap size (diameter $\sim 90\,\mu$m). In such a case, the LP condensate is formed in an area determined by the pump spot size and the pump rate rather than the trap size, due to the limited lateral diffusion and varying spatial density of the LPs [12, 32]. Above threshold, two drastic changes are noticed compared with the standard quadratic dispersion observed far below threshold. One is the blue shift of the $k = 0$ LP energy and the other is the phonon-like linear dispersion relation at low momentum regime $|k\xi| < 1$, where $\xi = \hbar/\sqrt{2mU(n)}$ is the healing length. White and black lines in Fig. 9.4b represent the two quadratic dispersion relations, $E_{\rm LP} = -U(n) + \frac{(\hbar k)^2}{2m}$ and $E'_{LP} = \frac{(\hbar k)^2}{2m}$, where m is the effective mass of the $k = 0$ LP. Here we choose the zero energy as the condensate energy for convenience. Neither of the two theoretical curves can explain the measurement result. A solid pink line in Fig. 9.4b is the Bogoliubov excitation energy, given by [36, 37]

$$E_B = \sqrt{E'_{LP}\left(E'_{LP} + 2U(n)\right)} = U(n)\sqrt{(k\xi)^2\left[(k\xi)^2 + 2\right]}. \tag{9.6}$$

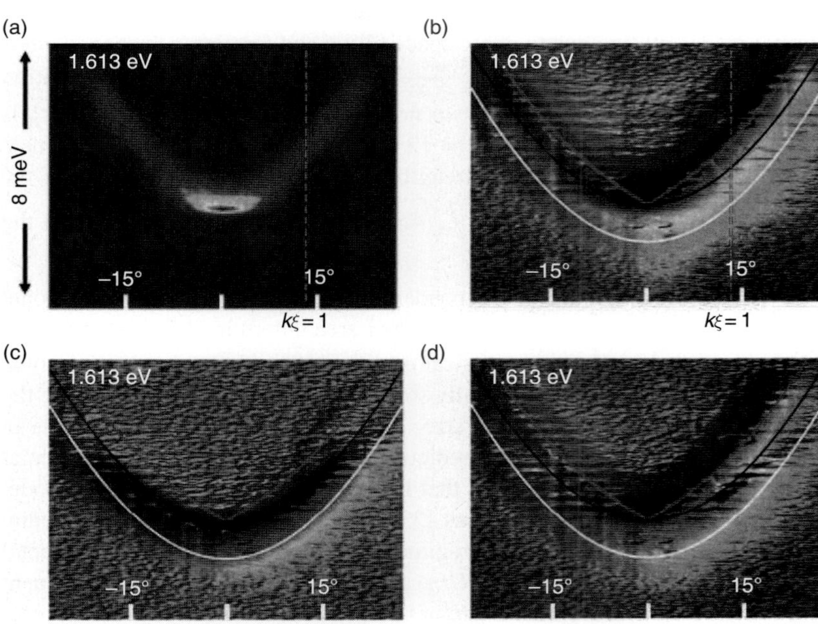

Fig. 9.4 Polarization dependence of the excitation spectrum for an untrapped condensate system. Here (a) represents a linear plot of the intensity, while (b)–(d) employ three-dimensional logarithmic plots of the intensity to magnify the excitation spectra. Time integrated dispersion relation between the LP energy (in the range of 8 meV centered at 1.609 eV) versus in-plane wavenumber for the untrapped condensate system, where the detuning parameter is $\Delta = 1.4$ meV and the pump rate is $P = 3P_{th}$ ($P_{th} = 17$ mW). A circularly polarized pump beam was incident with an angle of 60°. Three detection schemes: (a) and (b) detection of the leakage photons with co-circular polarization as the pump beam, (c) detection of the cross–circular polarization (d) detection of a small amount of a mixture of the co-circular polarization with the cross-circular polarization. The theoretical curves represent the Bogoliubov excitation energy E_B (pink line), the quadratic dispersion relations E'_{LP} (black line) which starts from the condensate energy and the non-interacting free polariton dispersion relation E_{LP} (white line), which is determined experimentally by the data taken far below the threshold $P = 0.001P_{th}$. (See Plate 18 for color image).

The measured dispersion relation for the excitation energy versus in-plane wavenumber is in good agreement with the Bogoliubov excitation spectrum without any fitting parameter.

A circularly polarized pump laser beam was used to inject spin polarized LPs in this experiment. In such a case, the LP condensate preserves the original spin polarization of optically injected polaritons [38]. The result shown in Fig. 9.4b was obtained for detection of the leakage photons with the same circular polarization as that of the pump (co-circular detection). If leakage photons with cross-circular polarization were detected, the standard quadratic dispersion was obtained but with slightly blue shifted energy, as shown in Fig. 9.4c. Even though the circularly polarized pump beam is injected, a small amount of cross-circularly polarized photons were detected because of spin-flip relaxation during the cooling process [38]. On increasing the pump rate, the number of LPs with opposite spin is also increased so that their energy is blue shifted compared to the single polariton energy E_{LP} (white line) in Fig. 9.4c. In Fig. 9.4d we intentionally mixed a small amount of the co-circular polarized photons to be detected with the cross-circular polarized photons. The difference between the Bogoliubov excitations with co-circular polarization and the standard quadratic excitation with cross-circular polarization is clearly seen.

The dispersion curves for the LP condensate in a trap with 8 μm diameter are shown in Fig. 9.5 for below and above threshold. The increasing blue shift of the condensate energy with the pump rate is seen from Fig. 9.5b to d. The Bogoliubov excitation energy is modified if the density of a condensate is not homogeneous in space due to trapping in a finite size [39]. Light blue dotted line in Fig. 9.5b is the modified Bogoliubov excitation energy based on the local density-dependent excitation energy $E_B(r) = \sqrt{E'_{LP}\left(E'_{LP} + 2U(n)\right)}$ and

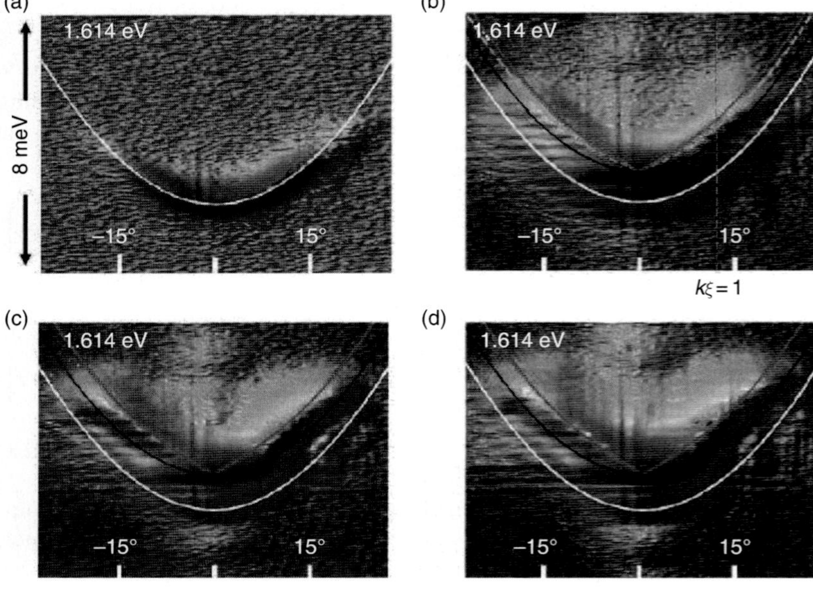

Fig. 9.5 Pump rate dependence of the excitation spectrum for a trapped condensate system. Time integrated dispersion relations between the LP energy (in the range of 8 meV centered at 1.61 eV) versus in-plane wavenumber. The circularly polarized pump beam is injected into a trap with 8 μm diameter, where the detuning parameter is $\Delta = 1.6$ meV. Pump rates are (a) $P = 0.05P_{th}$, (b) $P = 1.2P_{th}$, (c) $P = 4P_{th}$, and (d) $P = 6P_{th}$, where $Pth = 4$ mW. Three theoretical curves represent the Bogoliubov excitation energy E_B based on the homogeneous model (pink line), the quadratic dispersion curve E'_{LP} starting from the condensate energy (black line), and the non-interacting free polariton quadratic dispersion curve E_{LP} (white line) that is determined by the experimental data shown in Fig. 9.5(a). In Fig. 9.5(b), the light blue dotted line shows the Bogoliubov excitation curve based on the local density approximation. (See Plate 19 for color image).

the spatial average of $E_B(r)$, where $U(n)$ varies with position. In the phonon-like regime ($|k\xi| < 1$), the Bogoliubov dispersion is still linear $E \approx C'k$. However, the speed of sound is not equal to $C = \sqrt{\frac{U(n)}{m}}$ as in the homogeneous case, but is slightly reduced. In the free-particle regime ($|k\xi| > 1$), the dispersion is still quadratic, $E_B\,(|k\xi| > 1) \approx U'(n) + \frac{(\hbar k)^2}{2m}$. Here, the offset energy $U'(n)$ is also given by the spatial average of $U(n)$. Just above threshold, the experimental result agrees with such an inhomogeneous model (light blue dotted line) rather than a homogeneous model (pink solid line). However, as shown in Fig. 9.5c and d, the excitation spectra of the trapped condensates well above threshold are well reproduced by the homogeneous model. According to the local density approximation, at relatively high pump rates, the trapped condensate spreads over the entire trap due to the repulsive interaction among condensate polaritons and thus interacts with the excitations uniformly. Such a system can be well described by the homogeneous model.

9.3.2 Dynamic structure factor and excitations for a trapped polariton system

We assume the exciton–polaritons satisfiy the conditions of diluteness and small thermal depletion [41–47]. The diluteness condition and small thermal depletion condition are given by $U \ll k_BT_c$ and $T \ll T_c$ [14], where T_c is a BEC critical temperature and T is a polariton temperature. Both conditions are satisfied at pump rates well above threshold.

The GP equation can be used to study the properties of the excitations as well as the condensate if those conditions are satisfied. Below, we calculate the eigenfunctions and eigenenergies of the excitations. Assuming the excitation is in the form of a small oscillation with frequency ω, we have

$$\Psi(r,t) = \left(\Psi_0(r) + \frac{u(r)}{\sqrt{r}}e^{il\theta}e^{-i\omega t} + \frac{v^\star(r)}{\sqrt{r}}e^{-il\theta}e^{i\omega t}\right)e^{-i\frac{\varepsilon_0}{\hbar}t}, \tag{9.7}$$

where $e^{il\theta}$ is the angular part of the excitation with index l. By substituting Eq. (9.7) into Eq. (9.2) and collecting terms of $e^{\pm i\omega t}$ respectively, we obtain the following coupled equations

$$\begin{aligned}
\hbar\omega u(r) &= \left\{-\frac{\hbar^2}{2m}\left[\frac{\partial^2}{\partial^2 r} + \left(\frac{1}{4} - l^2\right)\frac{1}{r^2}\right] + 2g\frac{\phi_0^2(r)}{r} + V(r) - \varepsilon_0 + V_{eff}(r)\right\} \\
&\quad \times u(r) + g\frac{\phi_0^2(r)}{r}v(r) \\
-\hbar\omega v(r) &= \left\{-\frac{\hbar^2}{2m}\left[\frac{\partial^2}{\partial^2 r} + \left(\frac{1}{4} - l^2\right)\frac{1}{r^2}\right] + 2g\frac{\phi_0^2(r)}{r} + V(r) - \varepsilon_0 + V_{eff}(r)\right\} \\
&\quad \times v(r) + g\frac{\phi_0^2(r)}{r}u(r)
\end{aligned} \tag{9.8}$$

for the positive frequency component $u(r)$ and the negative frequency component $v(r)$, with the conditions $\hbar\omega > 0$ and $\int 2\pi dr(|u(r)|^2 - |v(r)|^2) = 1$. For a homogeneous system, the solution of Eq. (9.8) is a mixing between

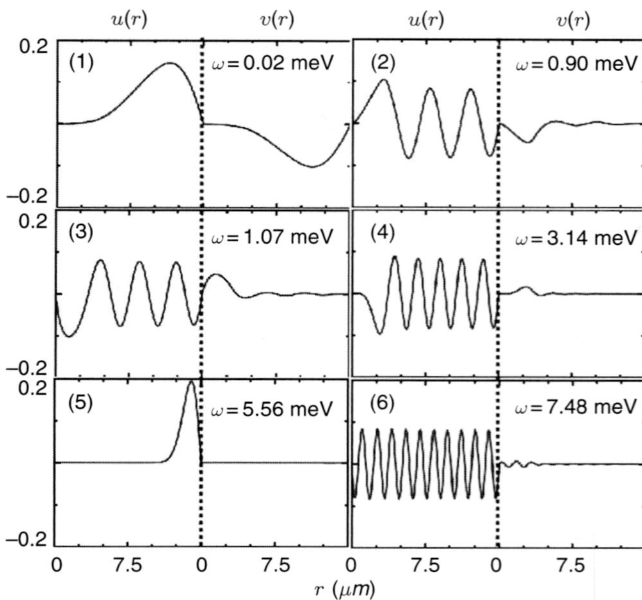

Fig. 9.6 Wave functions $(u(r), v(r))$ of the excitations at six different energies. The energy of each mode is labeled in the corresponding panel in the figure. In each panel, the positive frequency component $u(r)$ is plotted to the left of the dashed line and the negative frequency component $v(r)$ is plotted to the right of the dashed line. The index l in the angular component $e^{il\theta}$ of each mode is $l = 2, 1, 0, 5, 51, 0$ respectively for the six modes.

the plane wave modes of $e^{\pm i\mathbf{k}\cdot\mathbf{r}}$ with $u_k(\mathbf{r}) = e^{i\mathbf{k}\cdot\mathbf{r}}u_k$, $v_k(\mathbf{r}) = e^{-i\mathbf{k}\cdot\mathbf{r}}v_k$, and $\int 2\pi\, dr(u_k^2 - v_k^2) = 1$. The corresponding dispersion relation is the familiar form $E_k \equiv \hbar\omega_k = \sqrt{\epsilon_k^2 + 2\epsilon_k g n'}$ with $\varepsilon_k = \frac{(\hbar k)^2}{2m}$.

For the trapped system, the coupled equations in Eq. (9.8) need to be solved numerically. In Fig. 9.6, we plot the solutions of several excitation modes at various frequencies ω. For the low-frequency mode $\hbar\omega = 0.02\,\text{me}V$, the eigenfunction of the excitation is a nearly equal mixing between the positive frequency component and the negative frequency component, indicating that the interaction $g\phi_0^2/r$ plays a significant role. For the high-frequency mode $\hbar\omega = 7.48\,\text{me}V$, the kinetic energy ϵ_k dominates over the interaction energy, and the two components are almost decoupled with $v(r) \to 0$.

In an inhomogeneous condensate, modes with different momentum k become coupled and the plane wave solution is not the eigenmode any more. Only when $\epsilon_k \gg g\phi_0^2/r$, the k modes become eigenmodes again. To study the property of excitations, we calculate the dynamic structure factor defined as $S(k,\omega) = \sum_\alpha |\langle\Phi_\alpha|\hat{\rho}_{\mathbf{k}}|\Phi_g\rangle|^2$ at zero temperature [14], where the summation is over all the many body states Φ_α except for the ground state and $\hat{\rho}_k = \int dr e^{i\mathbf{k}\cdot\mathbf{r}}\hat{\rho}(\mathbf{r})$ with $\rho(\mathbf{r})$ being the density operator. The dynamic structure factor provides information on the excitations of the condensate. It can be shown that

$$S(k,\omega) = \sum_\alpha \left| \int r dr d\theta (u_\alpha^\star(r) + v_\alpha^\star(r)) e^{-il_\alpha\theta} e^{ikr\cos\theta} \frac{\phi_0(r)}{r} \right|^2 \delta\left(\omega - \frac{\epsilon_\alpha}{\hbar}\right), \tag{9.9}$$

where the condensate energy ε_0 is taken as the zero energy point, ϵ_α is the energy, and l_α is the angular index of the αth excitations calculated from Eq. (9.8). In the numerical calculation, the δ-function is treated as

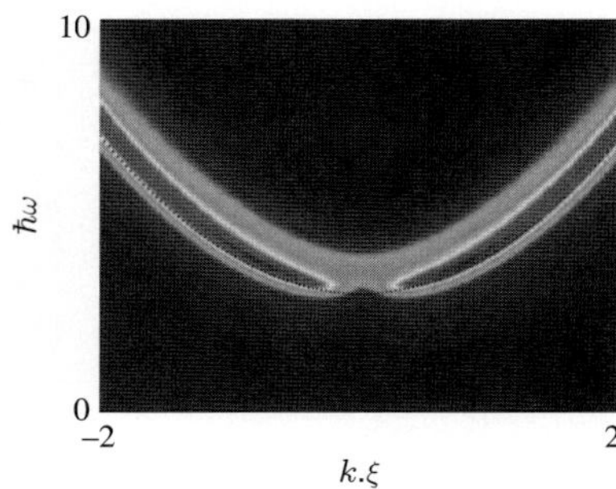

Fig. 9.7 $S(k,\omega)$ for $P/P_{th}=5$ with $\xi=0.6\,\mu\text{m}$ and the unit for $\hbar\omega$ is meV. The dotted line is the free particle dispersion starting from the condensate energy. (See Plate 20 for color image).

$$\delta(\omega-\omega_0)\to\frac{\delta_0}{\delta_0^2+(\omega-\omega_0)^2} \tag{9.10}$$

with a small energy, with δ_0 describing the broadening of the spectrum. In a homogeneous system, $S(k,\omega)=n\delta(\omega-\frac{\varepsilon_k}{\hbar})\frac{\epsilon_k}{\sqrt{\epsilon_k^2+2gn'\varepsilon_k}}$ has the form of a δ-function.

In Fig. 9.7, the calculated $S(k,\omega)$ for $P/P_{th}=5$ is shown, where we set $\delta_0=0.2\,\text{me}V$. At small momentum k with $|k\xi|\lesssim 0.2$, the intensity of $S(k,\omega)$ is very weak, which means that the quasiparticle population in this regime is small. In our model, the condensate ground state has a finite width of Δk in momentum space due to the finite spatial extent Δx. Hence, the excitations prefer to occupy phase space outside this regime, as can be seen in Fig. 9.6. Outside this regime, we can define an average dispersion relation ω_k by searching the momentum k of the maximum in $S(k,\omega)$ at fixed $\hbar\omega$. For $\epsilon_k<g\phi_0^2/r$, a linear behavior can be seen in $\hbar\omega$ versus $k\xi$, indicating the phonon-like behavior of the excitations. For $\epsilon_k>g\phi_0^2/r$ at large k, the average dispersion approaches $\epsilon_k+\Delta$, showing the behavior of a free particle, but with an energy shift of $\Delta=0.38\,\text{me}V$. This shift is much smaller than what is measured in the experiment. This difference stems from the fact that the theoretical result is the averaged interaction over the entire range of the system, while in the experiment the data were taken for the central plateau with radii of 4–5 μm. With a local density of $|\Psi_0(r)|^2$, the energy shift can be written as $\Delta=g|\Psi_0(r)|^2$, following the prediction of the Bogoliubov theory. In our trapped system, the density can be aproximated as $\sim|\Psi_0(r)(r=2.5\,\mu\text{m})|^2$ and the interaction constant is $g=8.9\times10^{-38}$ in units of $J\cdot m^2$, we have $\Delta=1.12\,\text{me}V$, which agrees with the experimental energy shift.

9.3.3 The solutions of the GP equation for an untrapped polariton system

The above estimation is based on the so-called local density approximation (LDA). In the following, we apply the LDA method to study the untrapped condensate. We assume that the density of the system varies smoothly over the space with density $n'(r)$. The dynamic structure factor under the LDA can be written as

$$S_{LDA}(k,\omega)=\int 2\pi r dr n'(r)\delta(\omega-\omega_k)\frac{\epsilon_k}{\hbar\omega_k}, \tag{9.11}$$

where $\hbar\omega_k=\sqrt{\epsilon_k^2+2\epsilon_k g n'(r)}$. The nth order moment of the structure factor is $M_n(k)=\int_{0^+}^{\infty}d\omega\omega^n S(k,\omega)$. The average excitation energy at momentum q can be derived as, $\bar{\omega}(k)=\hbar M_1(k)/M_0(k)$.

We consider a Gaussian profile for the pumping laser beam. The exciton–polariton density is $n'(r)=n_0'\exp(-r^2/\sigma)$, with n_0' proportional to the pumping power and $\sqrt{\frac{\sigma}{2}}$ being the variance of the Gaussian profile. For simplicity, we assume that when $n'(r)>n_{th}'$ (n_{th}' condensation threshold), the condensate appears within the area of $r\le R_0=\sqrt{\sigma\log(n_0'/n_{th}')}$. For $r>R_0$, there are only free particles with $\hbar\omega_k=\epsilon_k$.

It can be derived that the first moment of the dynamic structure factor is $\hbar M_1 = n_0' \sigma \pi \epsilon_k$ and the zeroth moment is

$$M_0 = n_0' \sigma \pi \left\{ e^{-\frac{R_0^2}{\sigma}} + \frac{\epsilon_k}{g n_0'} \left(\sqrt{1 + \frac{2 g n_0'}{\epsilon_k}} - \sqrt{1 + \frac{2 g n_0'}{\epsilon_k} e^{-\frac{R_0^2}{\sigma}}} \right) \right\}. \quad (9.12)$$

At small ϵ_k, we can expand the expression of M_0 and derive the average energy as

$$\hbar \bar{\omega}_k = \frac{\epsilon_k}{\sqrt{\frac{2\epsilon_k}{g n_0'}} \left(1 - e^{-\frac{R_0{}^2}{2\sigma}}\right) + e^{-\frac{R_0^2}{\sigma}}}, \quad (9.13)$$

which recovers the linear dispersion of the Bogoliubov excitations when $\exp(-R_0^2/\sigma) \to 0$. At large ϵ_k, we derive the average energy as

$$\hbar \bar{\omega}_k = \epsilon_k + \varepsilon_0 \left(1 - \left(\frac{P_{th}}{P}\right)^2\right). \quad (9.14)$$

The normalized excitation energy $\Delta E_{LDA} = (\hbar\omega(k) - \epsilon_k)/\varepsilon_0$ is plotted in Fig. 9.8. Note that $n_0'/n_{th}' = P/P_{th}$. It is interesting to note that $\hbar\bar{\omega}(k) - \varepsilon_k$ in Eq. (9.14) has the same expression as the averaged condensate energy in the LDA method. Here, the averaged condensate energy is $\langle \varepsilon_0 \rangle = \int 2\pi r dr n'(r) g |\Psi_0(r)|^2$. This result agrees with the experimental results shown in the next section.

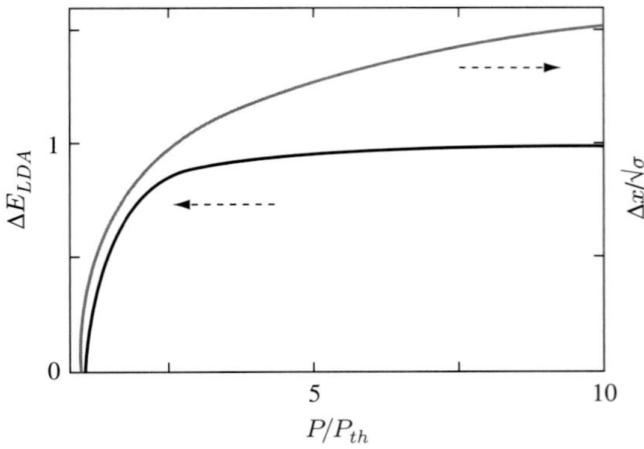

Fig. 9.8 Blue line: $\Delta E_{LDA} = (\hbar\bar{\omega}_k - \epsilon_k)/\varepsilon_0$ and red line: $\Delta x/\sqrt{\frac{\sigma}{2}}$ with LDA method versus the pumping level.

One interesting point to note is the phonon velocity at large pumping power. Here, we have $\langle \varepsilon_0 \rangle = \bar{U} = \frac{g n_0'}{2}$. From Eq. (9.13), the phonon velocity in the LDA method is $\bar{c} = \sqrt{\frac{\bar{U}}{2m}} \approx 0.7\sqrt{\frac{\bar{U}}{m}}$, which is in coincidence with the result for harmonically trapped condensates [14].

9.3.4 Universal scaling of the Bogoliubov spectrum

As indicated by Eq. (9.6), the Bogoliubov excitation energy normalized by the interaction energy $E_B/U(n)$ is a universal function of the wavenumber, normalized by the healing length $k\xi$. In Fig. 9.9a, this universal relation (green solid

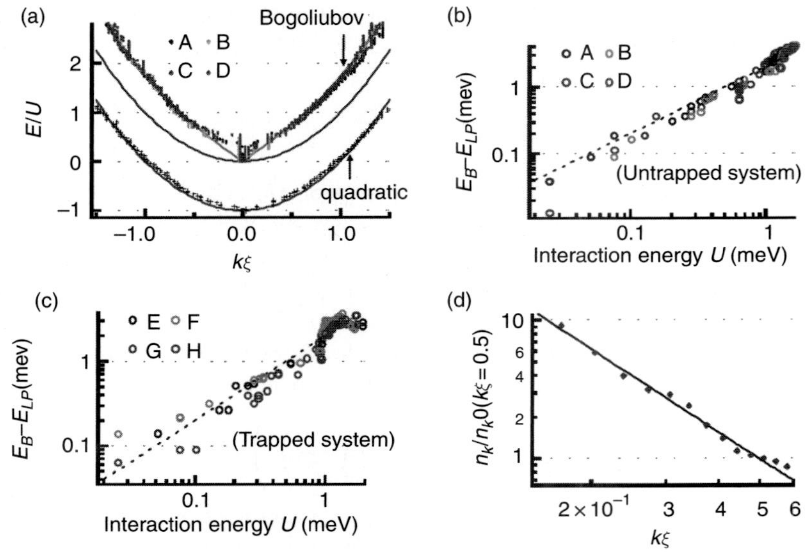

Fig. 9.9 Dispersion relation, energy shift in free particle regime, and population distribution in excitation spectrum. (a) Numerically searched excitation energy normalized by the interaction energy $E/U(n)$ as a function of normalized wavenumber $k\xi$ for four different untrapped condensate systems. A(blue dot): $\Delta = 1.41\,\text{meV}, P = 4P_{th}\,(P_{th} = 6.3\,\text{mW})$, B(light blue dot): $\Delta = 0.82\,\text{meV}$, $P = 8P_{th}$ $(P_{th} = 8.2\,\text{mW})$, C(red dot): $\Delta = 4.2\,\text{meV}$, $P = 4P_{th}\,(P_{th} = 6.4\,\text{mW})$, D(pink dot): $\Delta = -0.23\,\text{meV}$, $P = 24P_{th}\,(P_{th} = 8.2\,\text{mW})$. The experimental data far below threshold is also plotted by blue crosses for the system A. Three theoretical dispersion curves normalized by the interaction energy are plotted; the Bogoliubov excitation energy $E_B/U(n)$ starting from the condensate energy (green solid line), the quadratic dispersion curves $E'_{LP}/U(n)$ (gray solid line) and free polariton dispersion $E_{LP}/U(n)$ (red solid line). (b) and (c) The energy shift $E_B - E_{LP}$ in the free particle regime ($|k\xi = 1$) is plotted as a function of the interaction energy $U(n)$ for the same four different untrapped systems as in Fig. 9.9(a)–(b) and four different trapped systems (c), where trapped condensate systems are labeled as below; E(blue circle): d(diameter)$= 7\,\mu\text{m}$, $\Delta = 3.3\,\text{meV}$, F(light blue circle): $d = 7\,\mu\text{m}, \Delta = 2.9\,\text{meV}$, G(red circle): $d = 8\,\mu\text{m}, \Delta = 1.6\,\text{meV}$, H(pink circle): $d = 8\,\mu\text{m}, \Delta = 2.5\,\text{meV}$. The dashed line represents the theoretical prediction based on the homogeneous model $E_B - E_{LP} = 2U(n)$. (d) LP population distribution normalized by the value at $k\xi = 0.3$ for the trapped system G (in Fig. 9.9(c)) at the pump rate; $P = 2P_{th}(P_{th} = 4\,\text{mW})$. Theoretical $1/k^2$ dependence for the thermal depletion is shown by the blue line and $1/k$ dependence for the quantum depletion is shown by the blue dotted line. The experimental data are plotted at the range, $k\xi = 0.2 \sim 0.6$ because the experimental data at $|k\xi < 0.2$ is dominated by the condensate with a finite Δk and the Bogoliubov excitation is suppressed. (See Plate 21 for color image).

line) is compared with the experimental results for four different untrapped condensate systems. The experimental data shown in Fig. 9.9a are taken by numerical search for the intensity maximum wavenumber for varying E values. In both the phonon-like regime at $|k\xi| < 1$ and the free particle regime at $|k\xi| > 1$, the experimental results agree well with the universal curve. On the other hand, at a pump rate far below threshold, the measured dispersion relation is described completely by the single polariton energy E_{LP} (red solid line).

The sound velocity deduced from the phonon-like linear dispersion spectrum is of the order of $\sim 10^8$ cm/s. This value is eight orders of magnitude larger than that of atomic BEC. This enormous difference comes from the fact that the polariton mass is nine orders of magnitude smaller than the atomic mass and the polariton interaction energy is seven orders of magnitude larger than the atomic interaction energy. According to Landau's criterion [40], the observation of this linear dispersion at low momentum regime is an indication of superfluidity in the exciton–polariton system. However, we note that a polariton system is a dynamical system with a finite lifetime, so that Landau's criterion might be modified at a quantitative level.

The excitation spectrum is rounded near $k = 0$ and there is a gap between the condensate and the excitation spectrum in Fig. 9.9(a). This result is a consequence of the condensate–reservoir coupling (relaxation oscillation).

In the free particle regime ($|k\xi| > 1$), the excitation energy associated with the condensate is larger by $2U(n)$ than that of a single LP energy for the same wavenumber. In Figs. 9.9b and c, this important prediction of the Bogoliubov theory is compared with the experimental results for four different untrapped and trapped condensate systems, respectively. The experimental data were determined as the difference between the measured excitation energy with the presence of the condensate and the standard quadratic dispersion for a single LP state, which is determined by the experimental data obtained for a pump rate far below threshold ($P/P_{th} \ll 1$). The experimental data were taken for varying pump rates in the range of $P/P_{th} \gg 1$, so that the homogeneous model can be applied to both untrapped and trapped cases. The experimental data are in good agreement with the theoretical curve (gray dashed line) for both untrapped and trapped cases.

9.3.5 Thermal depletion versus quantum depletion

Figure 9.9d shows the normalized LP number n_k/n_k^0 versus the normalized wavenumber $|k\xi|$ for a trapped condensate, where n_k^0 is evaluated at $|k\xi| = 0.3$ for convenience. The LP occupation number n_k in the excitation spectrum can be calculated by applying the Bose–Einstein distribution for the Bogoliubov quasiparticles and subsequently taking the inverse Bogoliubov transformation [36],

$$n_k = |v_{-k}|^2 + \frac{|u_k|^2 + |v_{-k}|^2}{\exp(\beta E_B) - 1}, \tag{9.15}$$

where $u_k, v_{-k} = \pm\left[\frac{(\hbar k)^2/2m + U(n)}{2E_B} \pm \frac{1}{2}\right]^{1/2}$ and $\beta = \frac{1}{k_B T}$. The first and second terms of the right-hand side of Eq. (9.15) represent the real particles (exciton–polaritons) created by the quantum depletion and the thermal depletion, respectively. In the present polariton condensate system, the thermal depletion is much stronger than the quantum depletion, so that the second term of the right-hand side of Eq. (9.15) dominates over the first term. In such a case, the LP population is approximated by $n_k \approx \frac{mk_B T}{(\hbar k)^2}$ at a small $|k\xi|$ regime, while the LP population is given by $n_k \approx \frac{1}{2\sqrt{2}}\frac{1}{k\xi}$ if the quantum depletion is dominant [36]. This theoretical prediction of $1/k^2$ dependence of n_k for thermal depletion is compared with the experimental data in Fig. 9.9d and reasonable agreement was obtained.

The thermodynamic behavior of the excitations (Bogoliubov quasiparticles) can be obtained by setting the energy of the excitations, $E_k = [\frac{U}{m}(\hbar k)^2 + \frac{(\hbar k)^4}{4m^2}]^{\frac{1}{2}}$. The average occupation number of Bogoliubov quasiparticles with wavenumber k is given by

$$\langle \hat{b}_k^+ \hat{b}_k \rangle = \frac{1}{exp(\beta E_k) - 1}. \tag{9.16}$$

Even though Bogoliubov quasiparticles are the convenient theoretical concept, the experimentally accessible real particles in our experiment are the lower polaritons which have a one-to-one correspondence with the energy and the momentum of the photon leaking from the cavity. Using the Bogoliubov transformation,

$$\hat{a}_k = u_k \hat{b}_k + v_{-k}^* \hat{b}_{-k}^+, \quad \hat{a}_k^+ = u_k^* \hat{b}_k^+ + v_{-k} \hat{b}_{-k}, \tag{9.17}$$

$$u_k, v_{-k} = \pm\left[\frac{(\hbar k)^2/2m + U}{2E_k} \pm \frac{1}{2}\right]^{\frac{1}{2}}. \tag{9.18}$$

The real particle occupation number is given by

$$n_k \equiv \langle \hat{a}_k^+ \hat{a}_k \rangle = |u_k|^2 \langle \hat{b}_k^+ \hat{b}_k \rangle + |v_{-k}|^2 [1 + \langle \hat{b}_k^+ \hat{b}_k \rangle] \tag{9.19}$$

$$= |v_{-k}|^2 + \frac{|u_k|^2 + |v_{-k}|^2}{exp(\beta E_k) - 1}. \tag{9.20}$$

The interactions in the gas cause the presence of particles with finite k, even at absolute zero temperature where $\langle \hat{b}_k^+ \hat{b}_k \rangle = 0$. This is called quantum depletion and is described by the first term in the right-hand side of Eq. (9.20). The second term in the right-hand side of Eq. (9.20) represents the presence of particles at finite k due to thermal excitations at finite temperatures. We can distinguish the origin of the observed Bogoliubov excitations, whether quantum depletion or thermal depletion is responsible for the production of real particles, by examining the wavenumber dependence of the real particle population. Using Eq. (9.18) for v_{-k} we find the real particle occupation number at $T = 0$ and at small wavenumber $|k\xi| < 1$,

$$n_k \cong \frac{1}{2\sqrt{2}|k\xi|}, \tag{9.21}$$

where ξ is the healing length [14]. n_k decreases with $\frac{1}{k}$ dependence if the quantum depletion is dominant. As pointed out already in the previous section, Eq. (9.21) is valid in a k region larger than the momentum uncertainty Δk of the condensate.

Next, let us consider the thermal depletion. In the low energy regime where $E_k \ll k_B T$, Eq. (9.16) is reduced to, $\langle \hat{b}_k^+ \hat{b}_k \rangle \cong k_B T / E_k$. From Eq. (9.18) for u_k and v_{-k}, we find, at small wavenumber $|k\xi| < 1$, $|u_k|^2 + |v_{-k}|^2 = \frac{U}{E_k}$. Using the relations in the second term of the right-hand side of Eq. (9.20), the real particle occupation number at finite temperatures is given by

$$n_k \cong \frac{mk_B T}{(\hbar k)^2}, \tag{9.22}$$

n_k decreases with $\frac{1}{k^2}$ dependence if the thermal depletion is dominant.

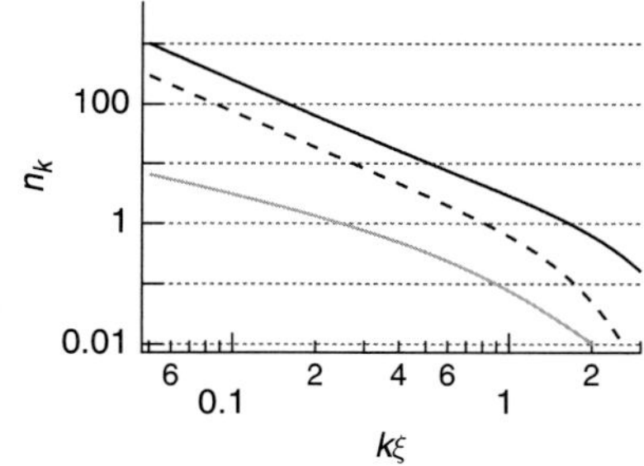

Fig. 9.10 Real particle (lower polariton) number n_k versus normalized wavenumber. Red solid line shows the real particle (lower polariton) number n_k using an effective temperature and polariton–polariton interaction energy of $k_B T = 10\,\text{meV}$ and $U = 2\,\text{meV}$, respectively. Red dotted line shows n_k for $k_B T = 3\,\text{meV}$ and $U = 2\,\text{meV}$. The component of quantum depletion $|v_{-k}|^2$ is plotted as the green solid line for $U = 2\,\text{meV}$, which are degenerate for the above two cases.

Figure 9.10 shows the real particle (lower polariton) number n_k versus the normalized wavenumber $k\xi$ using Eq. (9.20) for the typical experimental parameters in our system. In the exciton–polariton condensate, temperature T is a dynamical variable as a function of time after impulsive injection of high-energy exciton–polaritons [13]. The time resolved measurement of lower polariton population distribution $n_k(t)$ allows us to determine the time-dependent gas temperature $T(t)$, (Fig. 3 in [13]). By integrating those variables weighted by the emission intensity, we will obtain the time averaged temperature of the polariton gas. The averaged polariton temperature $k_B T$ ranges from 3–10 meV, depending on the detuning parameter and the pump rate. If we use those parameters with the observed interaction energy $U \simeq 2\,\text{meV}$, we confirm that the thermal depletion is indeed dominant over the quantum depletion in our experimental system, as shown in Fig. 9.10. The thermal depletion dominates over quantum depletion by one to two orders of magnitude in the experimental systems. The experimental results shown in Fig. 9.9(d) show the $\frac{1}{k^2}$ dependence, which suggests that Bogoliubov excitations are produced by the thermal depletion in our experiment.

9.4 Quantized vortex pair

Microcavity exciton polaritons [48] behave as a system of strictly two-dimensional bosons when their density is below the exciton saturation density. Because of their half-light/half-matter nature, their effective mass is extremely small, so that quantum many-body effects are important at relatively high temperatures, even up to room temperature [49]. However, the short lifetime only allows the formation of a quasi-equilibrium steady state, in which polaritons escaping from the cavity are continuously replenished by the external pump through an intermediate reservoir state.

Here, we show that single vortex–antivortex pairs can be observed in an exciton–polariton condensate. We have created a pumping spot which generates a minimum of the condensate density at the center. A zero in density can be thought of as a superposition of a vortex and antivortex that can be separated by an external perturbation [50]. Thus, the center of the condensate acts as a source of vortex–antivortex pairs. We have found that for a particular condensate size, there is on average one pair at any one time. A Michelson interferometer is used to reconstruct the time-integrated phase map of the system. When the sample disorder potential is stronger than the blue shift induced by polariton–polariton interactions, pinned pairs appear at certain locations. When the sample disorder potential is weak, on the other hand, they are mobile. They appear along a fixed axis, because of a small asymmetry in the pumping spot, and are created with a random polarization, namely, the vortex can appear on the right side of the spot and the antivortex on the left, or vice versa. In time-integrated measurement, two distinct characteristic phase defects appear in both cases.

We have applied an open-dissipative Gross–Pitaevskii (GP) equation model similar to that in Ref. [51], which reproduces the observed density minimum at the condensate center. We have also found that vortex–antivortex pair motion is significantly modified because of the dissipative nature of the polariton condensate and the repulsive interactions between condensate and reservoir particles. The vortex pair is found to migrate perpendicularly to its dipole moment and in our experimental parameter space, recombines before reaching the condensate edge. In spite of the short polariton lifetime, a vortex pair survives for long enough to be observed.

9.4.1 Experimental system

Our experiment is performed in a GaAs-based microcavity sample [52, 53]. The pump laser at above bandgap energy creates free electon–hole pairs, which form excitons and finally relax towards the lower polariton (LP) branch, as

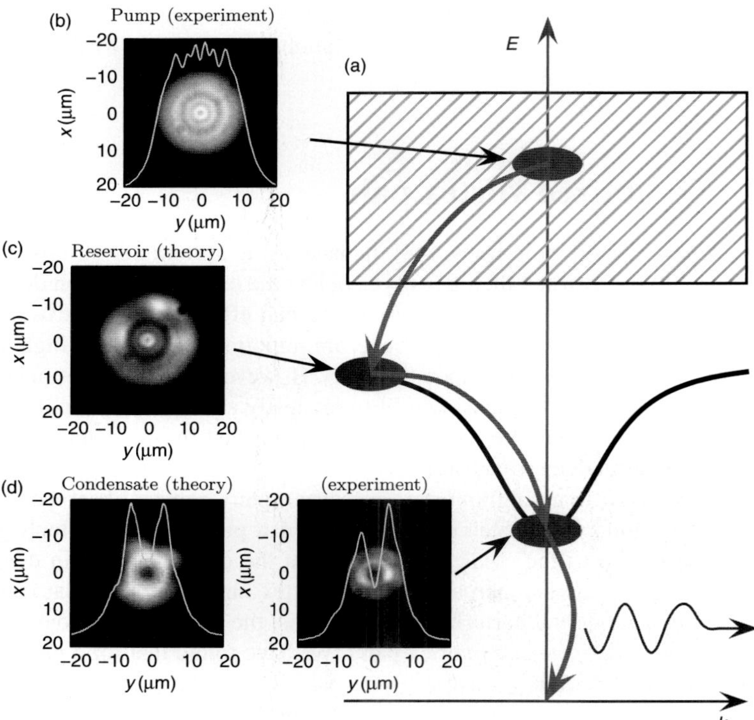

Fig. 9.11 Pump and condensate profile. a, Schematic of the pumping mechanism (see text). b, LP luminescence image at half the threshold power. The profile along $x = 0$ is shown in the yellow line. We assume that it has the shape of the laser pumping spot, and use it as an input to our numerical simulation. c, Numerically calculated reservoir density above threshold. d, Calculated and measured condensate density images for pumping power, three times above threshold.

shown in the schematic of Fig. 9.11a. When a LP decays, it emits one photon, which we observe. We employ a commercial beam shaper to create a flat-top pumping spot. Fig. 9.11b shows measured luminescence below the threshold, which features Airy-function-like patterns because of diffraction effects. We assume that the shape is the same as the laser pumping spot, and use it as an input to our numerical calculation. Above the condensation threshold, a population dip develops at the condensate centre, as shown in the measured luminescence image of Fig. 9.11d. Our numerical simulation reproduces the observed population dip (Fig. 9.11d), and suggests that the inhomogeneity of the pump spot profile, notably the ring near the center, generates this condensate shape. The condensate density minimum is stabilized by the repulsive reservoir–condensate interaction, since the reservoir shows a corresponding density maximum at the same point (Fig. 9.11c). This density minimum can be considered as an overlapping vortex and antivortex.

Phase fluctuations [54] of the condensate are responsible for the spontaneous vortex pair generation [50], and are locally maximized at the condensate center, where the condensate density is a minimum and the non-condensate density (reservoir polaritons) is a maximum. The mean-field interaction energy due to the reservoir is a maximum at the center, where the reservoir density is a maximum. Since power fluctuations of our multimode pump laser modulate the reservoir population strongly, this results in modulation of the local

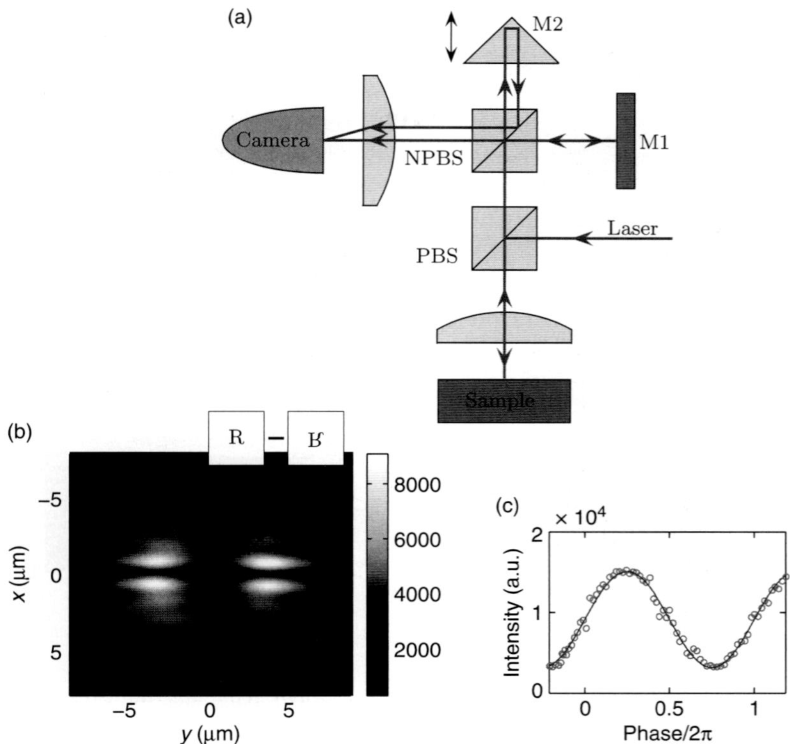

Fig. 9.12 Michelson interferometer. a, Schematic of the setup for measurement of phase and fringe visibility maps. It employs a mirror (M1) and a right-angle prism (M2), which creates the reflection of the original image along one axis, depending on prism orientation. A two-lens microscope setup overlaps the two real-space images of the polariton condensate on the camera. b, Typical interference pattern observed above the polariton condensation threshold (at 60 mW) along with a schematic showing the orientation of the two overlapping images. c, Blue circles: Intensity on one pixel of the camera as a function of the prism (M2) position in normalized units. Red line: Fitting to a sine function.

interaction energy and thus of the local phase of the condensate, which induces a separation of the vortex and antivortex. We have confirmed that a vortex–antivortex pair is not regularly generated when a quiet single-mode pump laser is used [53]. We also note that thermal effects are not strong enough to spontaneously create a vortex–antivortex pair in our system.

We use a Michelson interferometer to reconstruct the phase map of the condensate. A schematic of the setup is shown in Fig. 9.12a. We can overlap the real-space image with its reflected version on the camera, and because the two phase fronts are tilted with respect to each other, an interference pattern is observed (Fig. 9.12b). Changing the difference in the interferometer arm lengths shifts the fringes on the camera. Therefore, the intensity measured at one particular pixel shows a sinusoidal modulation (Fig. 9.12c). This reveals two pieces of information: the relative phase of the two overlapping images at this pixel point as well as the fringe visibility, which corresponds to the first-order spatial correlation function. Repeating this procedure for every pixel point, we can construct both phase and fringe visibility maps.

9.4.2 Observation of a pinned vortex–antivortex pair

The expected phase map for a pinned vortex–antivortex pair is shown in Fig. 9.13a. If we perform an interferometry measurement, in the way described above, with this phase map as an input, the expected result is Fig. 9.13b. The

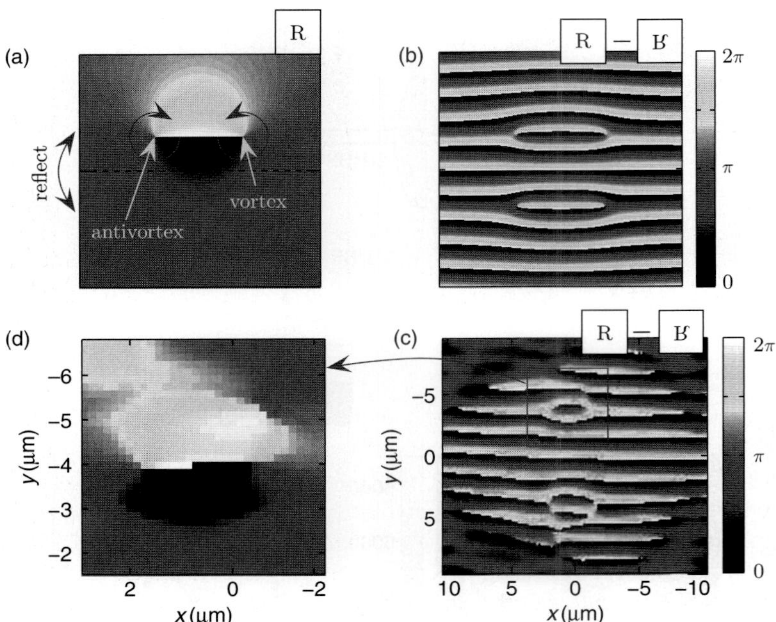

Fig. 9.13 Phase map of a pinned pair. (a) Expected phase map of a condensate including a single vortex–antivortex pair. Arrows show the direction of the phase increase around the vortex and antivortex. (b) Simulation of the experimentally measured phase map when a interferes with its reflection along the horizontal (dashed) line. A global phase slope along the vertical direction is added. (c) Experimentally measured phase map at 55 mW above the condensation threshold of 20 mW. The blue square marks the position of a double dislocation pattern. (d) Expanded view of the blue square in c, where the global slope along the horizontal direction is subtracted. (c) and (d) are rotated by 90° with respect to all other experimental data. (See Plate 22 for color image).

two double dislocation patterns, one at the upper part, and the other at the lower part of the figure, mark the position of the pair in the original and in the reflected images. Figure 9.13c shows a phase map measured experimentally. It features the double dislocation pattern characteristic of a pinned vortex–antivortex pair. In Fig. 9.13d, we have subtracted the global phase slope to reveal the actual phase map of a single vortex–antivortex pair, corresponding to that expected (Fig. 9.13a).

In this case (Fig. 9.13c–d), because of the sample disorder potential, the pair has a pinned position and polarization, defined as the vector pointing from the antivortex to the vortex. This is similar to the experiment in Ref. [51], where single vortices pinned by the disorder potential were observed. However, the disorder potential in our sample is very weak generally, much smaller than the blue shift due to polariton–polariton interactions, so the disorder is screened when the condensate is formed. Therefore, there are only a few locations where pinned vortex–antivortex pairs can be observed. If we mount the sample carefully, so as to minimize strain, we can no longer find any pinned vortex pair.

9.4.3 Observation of a mobile vortex–antivortex pair

We study numerically the evolution of a mobile vortex pair imprinted along the horizontal axis, using a time-dependent open-dissipative GP equation [55]. In a harmonically trapped conservative condensate, a single vortex pair will undergo linear motion with velocity $v_{v-av} = \hbar/md_v$, which is inversely proportional to the vortex–antivortex separation d_v, and upon interaction with the boundary, will wrap back upon itself with a cyclical motion. However,

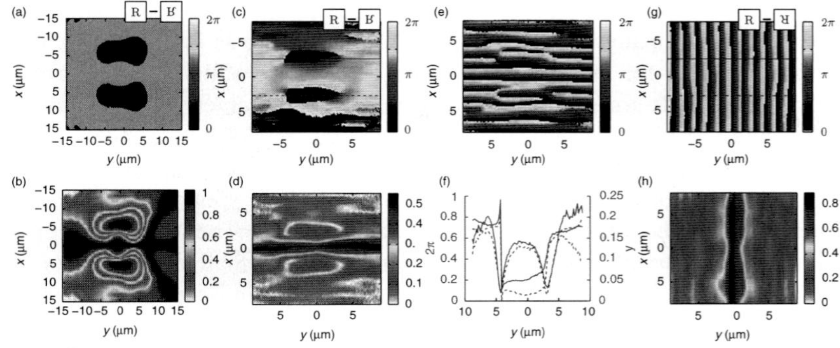

Fig. 9.14 Phase and fringe visibility map of a free pair. (a) Theoretical time-integrated phase map for a vortex–antivortex pair imprinted into a condensate and evolving according to the dissipative GP equation. (b) Corresponding fringe visibility map. (c) and (d) Measured phase and fringe visibility maps. (e) Same as in (c) but now the global phase slope is not subtracted. (f) Blue: phase cross-section along the continuous and dashed lines in (c). Red: fringe visibility cross-section along the same lines. (g) Measured phase map when the prism is rotated by 90°, along with a schematic showing the orientation of the interfering images. (h) Corresponding fringe visibility map. Experimental data are taken at 55 mW, above the condensation threshold of 20 mW. (See Plate 23 for color image).

a polariton condensate with considerable repulsive condensate–reservoir interaction (due to large reservoir population present in the experiment) experiences drag forces strongly perturbing the vortex pair motion [56], causing it to recombine after only a short travel distance. For these types of micro-motion, in spite of the vortex pair motion, a signature of the pair in a time-integrated interferogram is preserved. There are two areas where the phase is shifted by π, one at the top and the other at the bottom of the figure, and these are surrounded by minima in the fringe visibility. The measured positions of the phase defects depend on the time-averaged positions of the vortex and antivortex. The calculated time-integrated phase and visibility maps for such a moving pair are shown in Figs. 9.14a and 9.14b, respectively.

We indeed observed such unique patterns experimentally when the sample strain was removed, as shown in Fig. 9.14c,d, which illustrates typical experimental data. In Fig. 9.14e, we have not subtracted the global phase slope, so that the direct experimental phase map is given. In Fig. 9.14f, we plot the cross-section of the phase along two (continuous and dashed) lines passing through the two areas of π-phase shift, where this phase shift is shown quantitatively. We also plot the measured fringe visibility along the same lines, and find that the visibility minima coincide with the phase jumps. This result demonstrates that the vortex–antivortex pairs are created with a random polarization after removal of the strain. Moreover, when we rotate the prism by 90°, so that the reflected image is along the vertical axis, no phase defect is observed (Fig. 9.14g). The difference between Figs. 9.14e and 9.14g can be understood as follows. The vortex pair always remains on the horizontal axis and we fold the reflected image along the vertical axis. In this case, the phase rotation around the vortex by 2π and that around the antivortex by -2π cancel out, thus there is no phase defect in the interference pattern. The corresponding visibility map without any defect is shown in Fig. 9.14h.

The free pair of Fig. 9.14 shows a distinct signature compared with the pinned pair of Fig. 9.13. Indeed, in the interferogram of Fig. 9.13d there are two singularity points, one at the location of the vortex, and the other at the location of the antivortex, and the measured phase is continuous everywhere else. The experimental data in Fig. 9.14 suggests that such vortex–antivortex pairs are not pinned and travel a distance comparable to the spot size, in a process similar to the one described in Ref. [57]. When the pair can form with random polarization and also move, the measured time-integrated interferogram is a superposition of many interferograms of the same type as in Fig. 9.13d, with various positions for the vortex and antivortex. This results in a pattern with two π-phase shift areas (Fig. 9.14c), which is in agreement with our theory (Fig. 9.14a).

We consistently observe pairs along the same axis, even after we rotate the sample by 90°. However, when we rotate the pump laser spot by 90°, the polarization axis of the vortex pair is also rotated by 90°. These results suggest that the pair polarization direction is determined by a small asymmetry of the pump laser spot, rather than by the disorder potential landscape in the sample. Although the pair orientation follows the rotation of the laser spot, the random switching of pair polarization confirms that the pair generation mechanism is non-deterministic. We have found that there is no noticeable dependence of the measured phase maps on the laser pumping power, up to ~ 5 times the threshold power. Finally, we have created a Gaussian pumping spot and have not observed any phase defects. This observation supports our argument that the minimum of the condensate density at the center of the spot acts as a source of vortex–antivortex pairs. Given that the condensate size is not much larger than the size of the vortex pair it is not possible for more than one pair to enter into the system. Indeed, no phase defects are observed in a smaller condensate.

9.4.4 Open-dissipative Gross–Pitaevskii equation

The open-dissipative GP equation [51, 55] consists of two coupled equations describing the time evolution of the condensate order parameter $\psi(\mathbf{r},t)$ and reservoir polariton density $n_R(\mathbf{r},t)$,

$$\begin{aligned} i\hbar\frac{\partial\psi(\mathbf{r},t)}{\partial t} &= \left(-\frac{\hbar^2\nabla^2}{2m_{LP}} - \frac{i\hbar}{2}\left[\gamma_C - R(n_R(\mathbf{r},t))\right] + g_C|\psi(\mathbf{r},t)|^2 \right. \\ &\quad \left. + \; g_R n_R(\mathbf{r},t)\right)\psi(\mathbf{r},t) \qquad (9.23) \\ \frac{\partial_R(\mathbf{r},t)}{\partial t} &= P_1(\mathbf{r},t) - \gamma_R n_R(\mathbf{r},t) - R(r_R(\mathbf{r},t))|\psi(\mathbf{r},t)|^2; \end{aligned}$$

$n_R(\mathbf{r},t)$ is controlled by laser pumping gain $P_l(\mathbf{r})$ and reservoir loss γ_R. Interaction between condensate and reservoir is assigned a coupling constant $g_R = 2g_C$, where the condensate coupling constant is $g_C = 6\times 10^{-3}$ meV μm^2. Other parameters are the condensate polariton loss rate $\gamma_C = 0.33$ ps^{-1}, and stimulated scattering rate $R(n_R(\mathbf{r},t))$. The strong influence of the reservoir population on the condensate spatial profile allows us to probe indirectly the reservoir population density and uniquely determine the parameter space. In particular, we obtain best fits to experimental sequences of real-space data as a function of P_l for pumping threshold $P_{th} = 25$ mW with 24% pumping efficiency and $\gamma_R \approx 1.5\gamma_C$ and $R(n_R(\mathbf{r},t))/n_R(\mathbf{r},t) \approx \gamma_C\gamma_R/P_{th}$. To study the vortex pair dynamics, a vortex and antivortex are imprinted on a steady-state condensate directly via a phase factor $e^{i\theta}$ along the horizontal axis, and are subsequently left free to evolve in time. The temporal frames are stored and then, cumulatively, subjected to the same time-averaging mechanism as in the experiment.

9.4.5 Vortex nucleation and pair dynamics theory

We have described the observation of a single dynamic vortex–antivortex pair in a dissipative polariton condensate [58–61]. The time-integrated experiment results in clear interference fringes indicating the presence of a single vortex–antivortex pair. We have implemented a theoretical model and performed numerical simulations to understand the dynamics and nucleation mechanism. The model is found to reproduce the essential part of these observations.

9.4.5.1 *Vortex–antivortex pair dynamics*

We have analyzed the dynamics and stability of a vortex–antivortex pair in a polariton condensate in detail [56] and demonstrated significant deviations from the usual vortex pair motion [62] expected in a conservative atomic condensate. We found that the vortex pair will either recombine within the condensate or separate and dissipate from the condensate boundary. The choice between these two types of trajectory (in a largely homogeneous but confined condensate) is a competition between the force [63, 62] due to the radially directed superfluid flow of the unconfined repulsively interacting condensate, and the drag forces due to interaction with non-condensate (reservoir and thermal polaritons) [64]. The cross-over between these regimes essentially depends only on the magnitude and profile of $n_R(\mathbf{r})$.

The pumping profile used in this simulation is an experimentally measured top-hat profile (see Fig. 9.16b) and the condensate polariton lifetime is $\tau_r = 3\,\text{ps}$. The scattering rate $R(n_R) = R_{sc}n_R(\mathbf{r})$ is assigned a linear dependence on $n_R(\mathbf{r})$, and a measurement of the threshold pumping power P_{th} permits us to estimate the scattering rate via $R_{sc} = \gamma_C\gamma_R/P_{th}$, where we assume γ_R is comparable to γ_C, necessary to study this experimental parameter space. Thus, with these parameters fixed to correspond with experiments, a variation in the normalized pumping power $\bar{P} = P_l/P_{th}$ notably alters the relative fraction of reservoir n_R to condensate n_C particles according to $\dfrac{n_R}{n_C} = \dfrac{\gamma_C}{\gamma_R}\dfrac{1}{\bar{P}-1}$, and thus the pumping power should also control the choice between the two possible vortex pair trajectories outlined in reference [56].

Additionally, specific to these experiments, as the healing length of the condensate is not significantly smaller than the condensate size, the condensate boundaries are also expected to affect the vortex pair motion. Depending on the vortex pair initial energy and the local condensate environment, the process of the vortex pair either recombining within the condensate or splitting and leaving the condensate will happen on the order of the condensate polariton lifetime, due to the small condensate size. The next section aims to establish the position of these experiments in the possible parameter space and with this, determine vortex pair motions and attempts to reproduce time-integrated experimental results.

9.4.5.2 *Steady-state condensate profiles and time-integrated measurements*

The presence of a polariton reservoir modifies the polariton condensate spatial distribution and energy due to reservoir–condensate particle interactions. When a Gaussian pumping profile is used, this change is subtle given that there are no abrupt changes in density of the pump and therefore not in the condensate

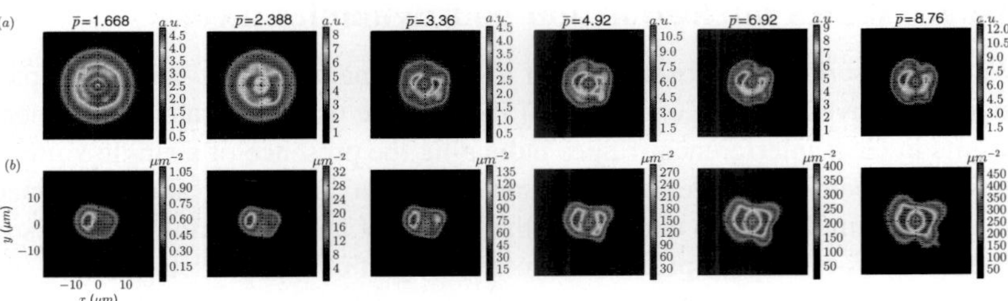

Fig. 9.15 Steady-state condensate density profiles as a function of the normalized pumping power for (a) experimental measurements and (b) numerical simulations, both using the same measured experimental pump profile. (See Plate 24 for color image).

either. However, under modified top-hat pumping potential in our experiments, the Airy-fringe-like diffraction patterns in the focused pump spot lead to spatial modulations in both reservoir and condensate populations. The amplitude of this spatial condensate modulation is determined by the system parameters γ_R/γ_C, P_{th} (which fixes R_{sc}), and to a lesser extent $\bar{P}$. With adjustment of these parameters, the resulting condensate distribution (under the experimental pumping profile in Fig. 9.1) ranges from a smooth Gaussian profile (when $n_C/n_R \gg 1$) to a strongly fragmented and unstable profile (when $n_C/n_R \sim 1$). The parameters can be chosen to carefully match experimental observations, such as the power-dependent steady-state condensate distribution. Figure 9.15 shows (a) the experimental condensate steady-state profile as a function of $\bar{P}$ (where the external threshold is $P_{th}^{ext} \approx 25\,\mathrm{mW}$ and (b) a numerical simulation of the open dissipative GPE with optimized parameters $\gamma_R \sim 1.5\gamma_C$ and $P_{th} = f_a \times P_{th}^{ext}$, where $f_a = 0.24$ is the fraction of light absorbed into useful polaritons.

The numerical simulation agrees well in predicting the appearance of the central condensate dip as a result of repulsive interactions between a large reservoir population at this parameter space location. The dip does not vanish for the values of $\bar{P}$ described, because the ratio n_C/n_R decreases only slowly with increasing $\bar{P}$, but if the pump power is increased sufficiently, this dip will eventually disappear. The discrepancy between experimental and numerical condensate profiles stems simply from the absence, in the theoretical simulation, of thermal polaritons. We expect a considerable thermal population to be present in the experiment for pumping powers close to threshold. This agreement improves as pumping power is increased, i.e. the condensate fraction approaches one.

9.4.6 Reconstructing time-dependent dynamics through interference patterns

Though the experimental pump profile (Fig. 9.11b) is roughly symmetric, existing asymmetries are enhanced through the presence of the large repulsive interaction between reservoir and condensate modes, resulting in an asymmetric condensate profile in the numerical simulation. This result biases the formation of vortex pairs preferentially dipole-aligned with one Cartesian axis, in this case the horizontal axis. Furthermore, the conclusion that an average of only one vortex pair is present is to be expected, based on our experimental results and from the observation that the minimum vortex pair size ($2d_v \sim 4\,\mu$m) is comparable to the size of the central dip in the condensate, which is believed to be responsible for vortex pair formation.

It is also desirable to know in more detail how the dynamics and life cycle of a single vortex–antivortex pair are reflected in the time-integrated measurements, and specifically why the defects in the fringes are not washed out by vortex motion. We assume that the vortex pair (imprinted in the calculation directly via a phase factor $e^{il\theta}$, where $\theta = \tan^{-1}\left(\dfrac{y}{x \mp d_v/2}\right)$ and $l = \pm 1$) is formed along the x-axis with core–core separation d_v, and is subsequently free to evolve in time. Essentially, given the mirror-symmetry of the problem across the y-axis, interference fringes will be observed due to the correlated motion of the vortex and antivortex. The system topology restricts pair motion to identical velocities in the y-direction and opposite velocities in the x-direction, so that the pair does not rotate.

In Fig. 9.16, we show time snapshots of the real time-evolution of an imprinted vortex pair, which subsequently moves along the $+y$-direction. The ratio of condensate density to reservoir density is approximated by

$$\frac{n_C}{n_R} \sim \frac{\gamma_R}{\gamma_C}(1 - \bar{P}(0)), \tag{9.24}$$

and the parameter space is adjusted such that the vortex pair trajectory (which is largely determined by this ratio) is modified to produce the scenarios of (a) a radially splitting pair (indicative of small reservoir population and lack of confining potential), and (b) a recombining vortex pair (indicative of a strong dissipation with large reservoir population). A perfect top-hat pumping profile is used for simplicity. In (c), we show the simulation result for the system parameters most relevant to our experiment. The pumping profile is the experimentally measured one (Fig. 9.11b), and the parameters are the same as for the phase and visibility maps shown in Figs. 9.14a–b, as well as the condensate shape images shown in Fig. 9.15b. The pair is found to recombine rapidly, as expected for a dissipative condensate with large reservoir population.

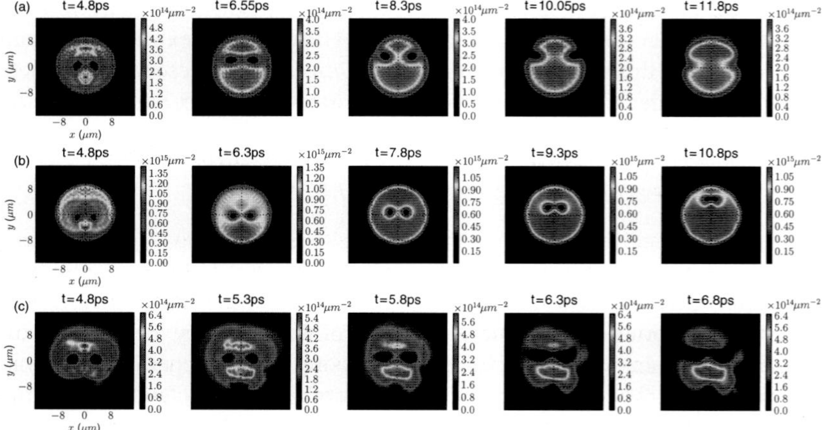

Fig. 9.16 Numerical simulations giving the time-dependent dynamics of (a) a radially splitting vortex pair and (b) a slowly recombining vortex pair and (c) a rapidly recombining vortex pair. The first two scenarios use a perfect top-hat pumping profile, while the third uses the measurement of the experimental profile. The parameter space of (c) is also chosen to closely match that of the experiment, notably that this parameter space dictates both the presence of a central density dip in the steady-state profile as well as a recombining vortex pair trajectory. (See Plate 25 for color image).

The dislocations in the interference measurement and visibility minima patterns are easily reproduced despite vortex pair motion given that (a) the mirror symmetry of the vortex pair about its midpoint and (b) that the condensate is occupied by a vortex pair most of the time. The shape of these π-phase shifted regions in the interference and fringe visibility experiments and simulations have a strong dependence on the vortex pair dynamics, specifically a correspondence between the area of the π-phase shifted region and the area mapped by the vortex pair trajectory. Small π-phase shifted areas (as in experiments) are only reproduced with recombining vortex pairs, while separating vortex pairs generally give distinct and significantly larger areas.

Thus, two requirements (a) strong modulation of condensate spatial profile by an inhomogeneous pump and (b) the vortex pair recombines rather than splits, both imply a similar parameter space position to successfully explain the experimental results. In this parameter region, the condensate and vortex dynamics are strongly influenced by interaction with reservoir populations.

9.4.7 Probability of spontaneous vortex formation

In two dimensions, the spontaneous formation of vortex pairs is usually the result of interaction of the condensate population with thermal phase fluctuations [36]. Knowing the energy E_v of a particular vortex pair configuration, the probability of thermal nucleation of this vortex pair configuration can be estimated. In three dimensions, the thermal excitation of a single vortex requires an energy proportional to the length of the vortex line, and is thus highly improbable in both atomic and polariton systems. In two dimensions (2D) though, the confinement along the axis of the vortex line results in a much smaller formation energy. In the present polariton condensate, the thermal energy is still insufficient to generate a single vortex which we can estimate from equation

$$E_v \sim \frac{m_{LP} n_C \kappa^2}{4\pi} \ln\left(\frac{L}{\xi}\right) \sim 10\,\text{eV} \gg k_B T \tag{9.25}$$

for the single on-axis vortex in an inhomogeneous condensate [66]. In this equation, L is the condensate radius and $\kappa = h/m_{LP}$. When dissipative effects are taken into account (through the dissipative GPE), the vortex energy is found to be lower than this estimate and nonlinear in n_C due to the reservoir presence, but is still of order eV. The energy of a 2D vortex–antivortex pair depends on its separation and thus can have a smaller energy for small separations.

$$E_p = \frac{m_{LP} n_C \kappa^2}{2\pi} \ln\left(\frac{d_v}{\xi}\right) \tag{9.26}$$

gives the energy for a uniform condensate, where d_v is the vortex–antivortex separation. Corrections to this equation for a finite size condensate are only important for vortex pairs within distances ξ of the boundary. Unless the vortex pair is given enough energy for the vortex cores to fully separate (core–core separation $d_v \gtrsim 2\xi$) it will immediately recombine rapidly and will not be observed in the current type of experiments.

The coefficient of the logarithm is still large, and thermal fluctuations ($k_B T \sim$ meV) of the phase cannot be solely responsible for vortex pair formation, for a reasonable large pair separation $d_v \gtrsim \xi$. Instead, density fluctuations in the reservoir, which has a population density comparable to or higher than the condensate, permit an alternative formation source. These fluctuations in the reservoir density result from the use of a noisy CW multi-mode laser with strong amplitude fluctuations expected at a time scale of the order of ps, which is similar to a vortex pair life time. In this case, the central dip in the condensate profile discussed previously presents ideal conditions for vortex pair formation.

Phase fluctuations [36, 65] are included in the operator of the condensate order parameter through $\hat{\psi} = \sqrt{n_C(\mathbf{r})}e^{i\hat{S}(\mathbf{r})}$, where $n_C(\mathbf{r})$ again represents the condensate density distribution at $T = 0$ and $\hat{S}(\mathbf{r})$ is the phase fluctuation operator. A phase correlation function can then be defined as $\chi(\mathbf{r} - \mathbf{r}') = \langle \hat{S}(\mathbf{r})\hat{S}(\mathbf{r}')\rangle$.

For a 2D finite area Bose gas, the phase fluctuation operator $\hat{S}(\mathbf{r})$ can be expressed in terms of the usual Bogoliubov coefficients $u_k(\mathbf{r})$ and $v_k(\mathbf{r})$, with

$$\hat{S}(\mathbf{r}) = \frac{1}{4\sqrt{n_C(\mathbf{r})}} \sum_k \left[f_k^+(\mathbf{r})\hat{a}_k + f_k^-(\mathbf{r})\hat{a}_k^\dagger \right], \tag{9.27}$$

where $\hat{a}_k$ is the annihilation operator of an excitation quantum from state k, and the amplitudes are $f_k(\mathbf{r})^\pm = u_k(\mathbf{r}) \pm v_k(\mathbf{r})$. A full treatment for the dissipative polariton condensate should also include density fluctuations and explicit contributions from interaction with reservoir populations, but the essential statement relevant here is that $\chi(\mathbf{r} - \mathbf{r}')$ is locally maximized when $n_C(\mathbf{r})$ is minimized and the reservoir population is maximized. Thus the local exchange of energy from condensate to reservoir is expected to be maximized at this position and the largest probability of phase fluctuations occurs, which are the specific conditions at the center of the excitation spot in present experiments.

9.5 Conclusions

We have presented the two experimental results related to superfluidity in exciton–polariton condensates: Bogoliubov excitation spectrum [52] and quantized vortex pair [67]. The BKT phase transition [18–21] in a uniform 2D Bose gas system is likely the underlying physics for the observed effects, but more direct experimental evidence such as power-law decay of a spatial correlation function has been obtained only recently [68] to identify the BKT physics as a real physical model for the observed effects.

The author wishes to thank S. Utsunomiya, G. Roumpos, L. Tian and M. Fraser for their experimental and theoretical contributions.

References

[1] Einstein, A. Quantentheorie des einatomigen idealen Gases: Zweite Abhandlung. *Sitzungber. Preuss. Akad. Wiss.*, **1**, 3–14 (1925).

[2] Bogoliubov, N. N. On the theory of superfluidity. *J. Phys. USSR*, **11**, 23–32 (1947).

[3] Anderson, M. H. et al. Observation of Bose–Einstein condensation in a dilute atomic vapor. *Science*, **269**, 198–201 (1995).

[4] Stamper-Kurn, D. M. et al. Excitation of phonons in a Bose–Einstein condensate by light scattering. *Phys. Rev. Lett.*, **83**, 2876–2879 (1999).

[5] Imamoglu, A. et al. Nonequilibrium condensates and lasers without inversion: Exciton-polariton lasers. *Phys. Rev.* A, **53**, 4250–4253 (1996).

[6] Dang, L. S. et al. Stimulation of polariton photoluminescence in semiconductor microcavity. *Phys. Rev. Lett.* **81**, 3920–3923 (1998).

[7] Senellart, P. and Bloch, J. Nonlinear emission of microcavity polaritons in the low density regime. *Phys. Rev. Lett.* **82**, 1233–1236 (1999).

[8] Savvidis, P. G. et al. Angle-resonant stimulated polariton amplifier. *Phys. Rev. Lett.* **84**, 1547–1550 (2000).

[9] Deng, H. et al. Condensation of semiconductor microcavity exciton polaritons. *Science*, **298**, 199–202 (2002).

[10] Kasprzak, J. et al. Bose–Einstein condensation of exciton polaritons. *Nature*, **443**, 409–414 (2006).

[11] Balili, R. et al. Bose–Einstein condensation of microcavity polaritons in a trap. *Science*, **316**, 1007–1010 (2007).

[12] Deng, H. et al. Spatial coherence of a polariton condensate. *Phys. Rev. Lett.* **99**, 126403 (2007).

[13] Deng, H. et al. Quantum degenerate exciton–polaritons in thermal equilibrium. *Phys. Rev. Lett.* **97**, 146402 (2006).

[14] Sarchi, D. and Savona, V. Spectrum and thermal fluctuations of a microcavity polariton Bose–Einstein condensate. *Phys. Rev.* B, **77**, 045304 (2008).

[15] Shelykh, I. A., Malpuech, G., and Kavokin, A. V. Bogoliubov theory of Bose-condensates of spinor exciton–polaritons. *Physica Status Solidi (a)*, **202**, 2614–2620 (2005).

[16] Hohenberg, P. C. Existence of long-range order in one and two dimensions. *Phys. Rev.*, **158**, 383–386 (1967).

[17] Mermin, N. D. and Wagner, H. Absence of ferromagnetism or antiferromagnetism in one- or two-dimensional isotropic Heisenberg models. *Phys. Rev. Lett.*, **17**, 1133–1136 (1966).

[18] Berezinskii, V. L. Destruction of long-range order in one-dimensional and two-dimensional systems possessing a continuous symmetry group. II. Quantum systems. *Sov. Phys. JETP*, **34**, 1144–1156 (1972).

[19] Kosterlitz, J. M. and Thouless, D. J. Ordering, metastability and phase transitions in two-dimensional systems. *J. Phys. C: Solid State Phys*, **6**, 1181–1203 (1973).

[20] Posazhennikova, A. Weakly interacting, dilute Bose gases in 2D. *Rev. Mod. Phys.*, **78**, 1111–1134 (2006).

[21] Kosterlitz, J. M. and Thouless, D. J. Two-dimensional physics. *Prog. Low Temp. Phys.*, **VII B**, 371–433 (1978).

[22] Bishop, D. J. and Reppy, J. D. Study of the superfluid transition in two-dimensional ^{4}He films. *Phys. Rev.* B, **22**, 5171–5185 (1980).

[23] Resnick, D. J. et al. Kosterlitz-Thouless transition in proximity-coupled superconducting arrays. *Phys. Rev. Lett.*, **47**, 1542–1545 (1981).

[24] van der Zant, H. S. J. et al. Phase transition of frustrated two-dimensional Josephson junction arrays. *J. Low Temp. Phys.*, **82**, 67–92 (1991).

[25] Safonov, A. I. et al. Observation of quasicondensate in two-dimensional atomic hydrogen. *Phys. Rev. Lett.*, **81**, 4545–4548 (1998).

[26] Hadzibabic, Z. et al. Berezinskii-Kosterlitz-Thouless crossover in a trapped atomic gas. *Nature*, **441**, 1118–1121 (2006).

[27] Cladé, P. et al. Observation of a 2D Bose gas: from thermal to quasicondensate to superfluid. *Phys. Rev. Lett.*, **102**, 170401 (2009).

[28] Neely, T. W. et al. Observation of vortex dipoles in an oblate Bose–Einstein condensate. *Phys. Rev. Lett.*, **104**, 160401 (2010).

[29] Keeling, J., Marchetti, F. M., Szymanska, M. H., and Littlewood, P. B. Collective coherence in planar semiconductor microcavities. *Semiconductor Sci. and Tech.*, **22**, R1–R26 (2007).

[30] Deng, H. et al. Exciton-polariton Bose–einstein condensation. *Rev. Mod. Phys.*, **82**, 1489–1537 (2010).

[31] Wyatt, Adrian F. G. Evidence for a Bose–Einstein condensate in liquid 4He from quantum evaporation. *Nature*, **391**, 56–59 (1997).

[32] Lai, C. W. et al. Spontaneous buildup of a coherent zero-state and π-state in an exciton-polariton condensate link. *Nature*, **450**, 529 (2007).

[33] Ciuti, C. et al. Role of the exchange of carriers in elastic exciton-exciton scattering in quantum wells. *Phys. Rev.* B, **58**, 7926–7933 (1998).

[34] Rochat, G. et al. Excitonic Bloch equations for a two-dimensional system of interacting excitons. *Phys. Rev.* B, **61**, 13856–13862 (2000).

[35] Schmitt-Rink, S. et al. Theory of transient excitonic optical nonlinearities in semiconductor quantum-well structures. *Phys. Rev.* B, **32**, 6601–6609 (1985).

[36] Pitaevskii, L. P. and Stringari, S. *Bose–Einstein condensation*. Clarendon Press, Oxford (2003).

[37] Ozeri, R. et al. Colloquium: Bulk Bogoliubov excitations in a Bose–Einstein condensate. *Rev. Mod. Phys.*, **77**, 187–205 (2005).

[38] Deng, H. et al. Polariton lasing vs. photon lasing in a semiconductor microcavity. *Proceedings of the National Academy of Sciences of the United States of America*, **100**, 15318–15323 (2003).

[39] Stenger, J. et al. Bragg spectroscopy of a Bose–Einstein condensate. *Phys. Rev. Lett.*, **82**, 4569–4573 (1999).

[40] Landau, L. D. and Lifshiëtis, E. M. *Fluid Mechanics*, 2nd edn. Pergamon Press, Oxford (1987).

[41] Lee, T. D., Huang, K. and Yang, C. N. *Phys. Rev.*, **106**, 1135–1145 (1957).

[42] Lee, T. D. and Yang, C. N. *Phys. Rev.*, **105**, 1119–1120 (1957).

[43] Nozieres, P. and Pines, D. *The Theory of Quantum Liquids*. Perseus Books, Cambridge, Mass (1999).

[44] Landau, L. D., Pitaevskii, L. P. and Lifshits, E. M. *Statistical Physics*, Part 2. Pergamon Press, Oxford, New York (1980).

[45] Leggett, A. J. *Quantum liquids: Bose Condensation and Cooper Pairing in Condensed Matter Systems*. Oxford University Press, Oxford, New York (2006).

[46] Leggett, A. J. *Rev. Mod. Phys.* **73**, 307–356 (2001).

[47] Hora, C. and Castin, Y. *Phys. Rev.* A **67**, 053615 (2003).

[48] Weisbuch, C. et al. Observation of the coupled exciton–photon mode splitting in a semiconductor quantum microcavity. *Phys. Rev. Lett.*, **69**, 3314–3317 (1992).

[49] Christopoulos, S. et al. Room-temperature polariton lasing in semiconductor microcavities. *Phys. Rev. Lett.*, **98**, 126405 (2007).

[50] Giorgetti, L. et al. Semiclassical field method for the equilibrium Bose gas and application to thermal vortices in two dimensions. *Phys. Rev.* A **76**, 013613 (2007).

[51] Lagoudakis, K. G. et al. Quantized vortices in an exciton–polariton condensate. *Nature Phys.*, **4**, 706–710 (2008).

[52] Utsunomiya, S. et al. Observation of Bogoliubov excitations in exciton–polariton condensates. *Nature Phys.*, **4**, 700–705 (2008).

[53] Roumpos, G. et al. Gain-induced trapping of microcavity exciton polariton condensates. *Phys. Rev. Lett.*, **104**, 126403 (2010).

[54] Amo, A. et al. Light engineering of the polariton landscape in semiconductor microcavities. *Phys. Rev.* B, **82**, 081301(R) (2010).

[55] Wouters, M. and Carusotto, I. Excitations in a nonequilibrium Bose–Einstein condensate of exciton polaritons. *Phys. Rev. Lett.*, **99**, 140402 (2007).

[56] Fraser, M. et al. Vortex–antivortex pair dynamics in an exciton polariton condensate. *New J. Phys.*, **11**, 113048 (2009).

[57] Pigeon, S. et al. Hydrodynamic nucleation of vortices and solitons in a resonantly excited polariton superfluid. arXiv:1006.4755v1 [cond-mat.other] (2010).

[58] Carusotto, I. and Ciuti, C. Probing microcavity polariton superfluidity through resonant Rayleigh scattering. *Phys. Rev. Lett.*, **93**, 166401 (2004).

[59] Amo, A. et al. Superfluidity of polaritons in semiconductor microcavities. *Nature Phys.*, **5**, 805–810 (2009).

[60] Wouters, M. and Savona, V. Superfluidity of a nonequilibrium Bose–Einstein condensate of polaritons. *Phys. Rev.* B, **81**, 054508 (2010).

[61] Sanvitto, D. et al. Persistent currents and quantized vortices in a polariton superfluid. *Nature Phys.*, **6**. 527–533 (2010).

[62] Donnelly, R. et al. *Quantized Vortices in Helium II*. Cambridge University Press (1991).

[63] Sonin, E. B. *Phys. Rev.* B. **55**. 485 (1997).

[64] Fedichev, P. O. and Shlyapnikov, G. V. *Phys. Rev.* A. **60**. R1779 (1999).

[65] Khawaja, U. A. et al. *Phys. Rev.* A. **66**. 013615 (2002).

[66] Pethick, C. J. and Smith, H. *Bose–Einstein Condensation in Dilute Gases*. Cambridge University Press (2001).

[67] Roumpos, G. et al. Observation of Bogoliubov excitation in exciton–polariton condensates. *Nature Phys.*, **7**, 126 (2011).

[68] Roumpos, G. Power-law decay of the spatial correlation function in exciton–polariton condensates. *Proceedings of the National Academy of Sciences of the United States of America*, **109**, 6467 (2012).

Color superconductivity in dense quark matter

10

Mark G. Alford, Krishna Rajagopal, Thomas Schäfer, and Andreas Schmitt

Matter at high density and low temperature is expected to be a color superconductor, which is a degenerate Fermi gas of quarks with a condensate of Cooper pairs near the Fermi surface that induces color Meissner effects. At the highest densities, where the QCD coupling is weak, rigorous calculations are possible, and the ground state is a particularly symmetric state, the color–flavor locked (CFL) phase. The CFL phase is a superfluid, an electromagnetic insulator, and breaks chiral symmetry. The effective theory of the low-energy excitations in the CFL phase is known and can be used, even at more moderate densities, to describe its physical properties. At lower densities the CFL phase may be disfavored by stresses that seek to separate the Fermi surfaces of the different flavors, and comparison with the competing alternative phases, which may break translation and/or rotation invariance, is done using phenomenological models. We review the calculations that underlie these results, and then discuss transport properties of several color-superconducting phases and their consequences for signatures of color superconductivity in neutron stars.

10.1 Introduction

10.1.1 General outline

In this chapter we review the current understanding of matter at ultra-high density and low temperature. One might call this field the "condensed matter physics of quantum chromodynamics". We describe how one can have build on our understanding of the strong interaction, derived from experimental observation of few-body processes, to predict the behavior of macroscopic quantities in many-body systems where the fundamental particles of the standard model—quarks and leptons—become the relevant degrees of freedom. As in conventional condensed-matter physics, we seek to map the phase diagram and calculate the properties of the phases. However, we are in the unusual position of having a sector of the phase diagram where we can calculate many properties of quark matter rigorously from first principles. This sector is the

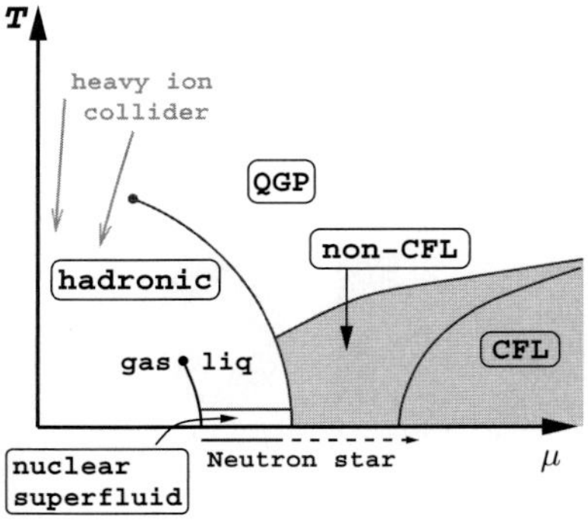

Fig. 10.1 A schematic outline for the phase diagram of matter at ultra-high density and temperature. The CFL phase is a superfluid (like cold nuclear matter) and has broken chiral symmetry (like the hadronic phase).

region of "asymptotically high" densities, where quantum chromodynamics is weakly coupled. We will review those rigorous results and describe the progress that has been made in building on this solid foundation to extend our understanding to lower and more phenomenologically relevant densities. Quark matter occurs in various forms, depending on the temperature T and quark chemical potential μ (see Fig. 10.1). At high temperatures ($T \gg \mu$) entropy precludes any pattern of order and there is only quark–gluon plasma (QGP), the phase of strongly interacting matter that has no spontaneous symmetry breaking, and which filled the universe for the first microseconds after the big bang. Quark–gluon plasma is also being created in small, very short-lived, droplets in ultrarelativistic heavy ion collisions at the Relativistic Heavy Ion Collider.

We will concentrate on the regime of relatively low temperatures, $T \ll \mu$, where we find a rich variety of spontaneous symmetry breaking phases. To create such material in nature requires a piston that can compress matter to super-nuclear densities and hold it while it cools. The only known context where this might happen is in the interior of neutron stars, where gravity squeezes the star to an ultra-high density state where it remains for millions of years. This gives time for weak interactions to equilibrate, and for the temperature of the star to drop far below the quark chemical potential. We do not currently know whether quark matter exists in the cores of neutron stars. One of the reasons for studying color superconductivity is to improve our understanding of how a quark matter core would affect the observable behavior of a neutron star, and thereby resolve this uncertainty.

In our discussion of matter at the highest densities, we shall always take the high density limit with up, down, and strange quarks only. We do so because neutron star cores are not dense enough (by more than an order of magnitude) to contain any charm or heavier quarks, and our ultimate goal is to gain insight into quark matter at densities that may be found in nature. For the same reason we focus on temperatures below about ten MeV, which are appropriate for neutron stars that are more than a few seconds old. At such high densities and low temepratures we expect to find a degenerate liquid of quarks, with Cooper pairing near the Fermi surface that spontaneously breaks the color gauge symmetry ("color superconductivity"). As recent reviews make clear, (Alford and Rajagopal, 2006; Alford et al., 2008b; Buballa, 2005a; Huang, 2005; Ren, 2004; Rischke, 2004; Schäfer, 2003b; Shovkovy, 2005) the last decade has seen dramatic progress in our understanding of dense matter. We can now obtain, directly from QCD, rigorous and quantitative answers to the basic question: "What happens to matter if you squeeze it to arbitrarily high density?". In Section 10.4 we will show how QCD then becomes analytically tractable: the coupling is weak and the physics of confinement never arises, since long-wavelength magnetic interactions are cut off, both by Landau damping and by the Meissner effect. As a result, matter at the highest densities is known to be in the CFL phase, whose properties (see Section 10.2) are understood rigorously from first principles. There is a well-developed effective field theory describing the low energy excitations of CFL matter (see Section 10.5), so that at any density at which the CFL phase occurs, even if this density is not high enough for a weak-coupling QCD calculation to be valid, many phenomena can nevertheless be described quantitatively in terms of a few parameters, via the effective field theory.

It should be emphasized that QCD at arbitrarily high density is more fully understood than in any other context. High energy scattering, for example, can be treated by perturbative QCD, but making contact with observables brings in poorly understood nonperturbative physics via structure functions and/or fragmentation functions. Or, in quark-gluon plasma in the high temperature limit much of the physics is weakly coupled but the lowest energy modes remain strongly coupled with nonperturbative physics arising in the nonabelian color-magnetic sector. We shall see that there are no analogous difficulties in the analysis of CFL matter at asymptotic densities.

If the CFL phase persists all the way down to the transition to nuclear matter then we have an exceptionally good theoretical understanding of the properties of quark matter in nature. However, less symmetrically paired phases of quark matter may well intervene in the intermediate density region between nuclear and CFL matter (Section 10.1.5). We enumerate some of the possibilities in Section 10.3. In principle this region could also be understood from first principles, using brute-force numerical methods (lattice QCD) to evaluate the QCD path integral, but unfortunately current lattice QCD algorithms are defeated by the fermion sign problem in the high-density low-temperature regime (Schmidt, 2006). This means we have to use models, or try to derive information from astrophysical observations. In Section 10.6 we sketch an example of a (Nambu–Jona-Lasinio) model analysis within which one can compare some of the possible intermediate-density phases suggested in Sec. 10.1.5. We finally discuss the observational approach, which involves elucidating the properties of the suggested phases of quark matter (Sections 10.6.3 and 10.7), and then finding astrophysical signatures by which their presence inside neutron stars might be established or ruled out using astronomical observations (Section 10.8).

10.1.2 Inevitability of color superconductivity

At sufficiently high density and low temperature it is a good starting point to imagine that quarks form a degenerate Fermi liquid. Because QCD is asymptotically free—the interaction becomes weaker as the momentum transferred grows—the quarks near the Fermi surface are almost free, with weak QCD interactions between them. (Small-angle quark-quark scattering via a low-momentum gluon is no problem because it is cut off by Landau damping, which, together with Debye screening, keeps perturbation theory at high density much better controlled than at high temperature (Pisarski and Rischke, 1999a; Son, 1999).) The quark–quark interaction is certainly attractive in some channels, since we know that quarks bind together to form baryons. As we will now argue, these conditions are sufficient to guarantee color superconductivity at sufficiently high density.

At zero temperature, the thermodynamic potential (which we will loosely refer to as the "free energy") is $\Omega = E - \mu N$, where E is the total energy of the system, μ is the chemical potential, and N is the number of fermions. If there were no interactions then the energy required to add a particle to the system would be the Fermi energy $E_F = \mu$, so adding or subtracting particles or holes near the Fermi surface would cost zero free energy. With a weak attractive interaction in any channel, if we add a pair of particles (or holes) with the quantum numbers of the attractive channel, the free energy is lowered by the potential energy of their attraction. Many such pairs will therefore be created in the modes near the Fermi surface, and these pairs, being bosonic, will form a condensate. The ground state will be a superposition of states with different numbers of pairs, breaking the fermion number symmetry. This argument, originally developed by Bardeen, Cooper, and Schrieffer (BCS) (Bardeen et al., 1957) is completely general, and can be applied to electrons in a metal, nucleons in nuclear matter, ^{3}He atoms, cold fermionic atoms in a trap, or quarks in quark matter.

The application of the BCS mechanism to pairing in dense quark matter is in a sense more direct than in its original setting. The dominant interaction between electrons in a metal is the repulsive Coulomb interaction, and it is only because this interaction is screened that the attraction mediated by phonons comes into play. This means that the effective interactions that govern superconductivity in a metal depend on band structure and other complications and are very difficult to determine accurately from first principles. In contrast, in QCD the "color Coulomb" interaction is attractive between quarks whose color wave function is antisymmetric, meaning that superconductivity arises as a direct consequence of the primary interaction in the theory. This has two important consequences. First, at asymptotic densities where the QCD interaction is weak we can derive the gap parameter and other properties of color superconducting quark matter rigorously from the underlying microscopic theory. Second, at accessible densities where the QCD interaction is stronger, the ratio of the gap parameter to the Fermi energy will be much larger than in conventional BCS superconducting metals. Thus, superconductivity in QCD is more robust, both in the theoretical sense and in the phenomenological sense, than superconductivity in metals.

It has long been known that, in the absence of pairing, an unscreened static magnetic interaction results in a "non-Fermi-liquid". However, in QCD the magnetic interaction is screened at non-zero frequency (Landau damping) and this produces a particularly mild form of non-Fermi-liquid behavior, as we describe in Section 10.5.1. In the absence of pairing but in the presence of interactions, there are still quark quasiparticles and there is still a "Fermi surface", and the BCS argument goes through. This argument is rigorous at high densities, where the QCD coupling g is small. The energy scale below which non-Fermi-liquid effects would become strong enough to modify the quasiparticle picture qualitatively is parametrically of order $\exp(-\text{const}/g^2)$ whereas the BCS gap that results from pairing is parametrically larger, of order $\exp(-\text{const}/g)$ as we shall see in Section 10.4. This means that pairing occurs in a regime where the basic logic of the BCS argument remains valid.

Since pairs of quarks cannot be color singlets, the Cooper pair condensate in quark matter will break the local color symmetry $SU(3)_c$, hence the term "color superconductivity". The quark pairs play the same role here as the Higgs particle does in the standard model: the color-superconducting phases can be thought of as Higgs phases of QCD. Here, the gauge bosons that acquire a mass through the process of spontaneous symmetry breaking are the gluons, giving rise to color Meissner effects. It is important to note that quarks, unlike electrons, have color and flavor as well as spin degrees of freedom, so many different patterns of pairing are possible. This leads us to expect a panoply of different possible color superconducting phases.

As we shall discuss in Section 10.2, at the highest densities we can achieve an *ab initio* understanding of the properties of dense matter, and we find that its preferred state is the CFL phase of three-flavor quark matter, which is unique in that *all* the quarks pair (all flavors, all colors, all spins, all momenta on the Fermi surfaces) and all the nonabelian gauge bosons are massive. The suppression of all of the infrared degrees of freedom of the types that typically indicate either instability toward further condensation or strongly coupled phenomena ensures that, at sufficiently high density, the CFL ground state, whose only infrared degrees of freedom are Goldstone bosons and an abelian photon, is stable. In this regime, quantitative calculations of observable properties of CFL matter can be done from first principles; there are no remaining non-perturbative gaps in our understanding.

As the density decreases, the effect of the strange quark mass becomes more noticeable, imposing stresses that may modify the Cooper pairing and the CFL phase may be replaced by other forms of color superconducting quark matter. Furthermore, as the attractive interaction between quarks becomes stronger at lower densities, correlations beyond the two-body correlation that yields Cooper pairing may become important, and at some point the ground state will no longer be a Cooper-paired state of quark matter, but something quite different. Indeed, by the time we decrease the density down to that of nuclear matter, the average separation between quarks has increased to the point that the interactions are strong enough to bind quarks into nucleons. It is worth noting that quark matter is in this regard different from Cooper-paired ultracold fermionic atoms (to be discussed in Section 10.3.9). For fermionic atoms, as the interaction strength increases there is a crossover from BCS-paired fermions to a Bose–Einstein condensate (BEC) of tightly bound, well-separated, weakly interacting di-atoms (molecules). In QCD, however, the color charge of a diquark is the same as that of an antiquark, so diquarks will interact with each other as strongly as quarks, and there will not be a literal analog of the BCS/BEC crossover seen in fermionic atoms. In QCD, the neutral bound states at low density that are (by QCD standards) weakly interacting are nucleons, containing three quarks not two.

We shall work with $N_c = 3$ colors throughout. In the limit $N_c \to \infty$ with fixed $\Lambda_{\rm QCD}$ (i.e. fixed $g^2 N_c$), Cooper pairing is not necessarily energetically preferred. A strong competitor for the large-N_c ground state is the chiral density wave (CDW), a condensate of quark–hole pairs, each with total momentum $2p_F$ (Deryagin et al., 1992). Quark–hole scattering is enhanced by a factor of N_c over quark–quark scattering, but, unlike Cooper pairing, it only uses a small fraction of the Fermi surface, and in the case of short-range forces the CDW phase is energetically favored in one-dimensional systems, but not in two or more spatial dimensions (Shankar, 1994). However, in QCD in the large N_c limit the equations governing the CDW state become effectively one-dimensional because the gluon propagator is not modified by the medium, so the quark–hole interaction is dominated by almost collinear scattering. Since pairing gaps are exponentially small in the coupling but medium effects only vanish as a power of N_c, the CDW state requires an exponentially large number of colors. It is estimated that for $\mu \sim 1$ GeV, quark–hole pairing becomes favored over Cooper pairing when $N_c \gtrsim 1000$ (Shuster and Son, 2000). Recent work (McLerran and Pisarski, 2007) discusses aspects of physics at large N_c at lower densities that may also be quite different from physics at $N_c = 3$.

Before turning to a description of CFL pairing in Section 10.2 and less-symmetrically paired forms of color superconducting quark matter in Section 10.3, we discuss some generic topics that arise in the analysis of color-superconducting phases: the gap equations, neutrality constraints, the resultant stresses on Cooper pairing, and the expected overall form of the phase diagram.

10.1.3 Quark Cooper pairing

The quark pair condensate can be characterized in a gauge-variant way by the expectation value of the one-particle-irreducible quark–quark two-point function, also known as the "anomalous self-energy",

$$\langle \psi^{\alpha}_{ia} \psi^{\beta}_{jb} \rangle = P^{\alpha\beta}_{ij\,ab} \Delta. \tag{10.1}$$

Here ψ is the quark field operator, color indices α, β range over red, green, and blue (r, g, b), flavor indices i, j range over up, down and strange (u, d, s), and a, b are the spinor Dirac indices. The angle brackets denote the one-particle-irreducible part of the quantum-mechanical ground-state expectation value. In general, both sides of this equation are functions of momentum. The color–flavor–spin matrix P characterizes a particular pairing channel, and Δ is the gap parameter which gives the strength of the pairing in this channel. A standard BCS condensate is position independent (so that in momentum space the pairing is between quarks with equal and opposite momentum) and a spin singlet (so that the gap is isotropic in momentum space). However, as we will see later, there is good reason to expect non-BCS condensates as well as BCS condensates in high-density quark matter.

Although (10.1) defines a gauge-variant quantity, it is still of physical relevance. Just as electroweak symmetry breaking is most straightforwardly understood in the unitary gauge where the Higgs vacuum expectation value is uniform in space, so color superconductivity is typically analyzed in the unitary gauge where the quark pair operator has a uniform color orientation in space. We then relate the gap parameter Δ to the spectrum of the quark-like excitations above the ground state ("quasiquarks"), which is gauge invariant.

In principle, a full analysis of the phase structure of quark matter in the μ-T plane would be performed by writing down the free energy Ω, which is a function of the temperature, the chemical potentials for all conserved quantities, and the gap parameters for all possible condensates, including the quark pair condensates, but also others such as chiral condensates of the form $\langle \bar{q}q \rangle$. We impose neutrality with respect to gauge charges (see Section 10.1.4 below) and then within the neutral subspace we minimize the free energy with respect to the strength of the condensate:

$$\frac{\partial\Omega}{\partial\Delta} = 0, \qquad \frac{\partial^2\Omega}{\partial\Delta^2} > 0. \tag{10.2}$$

We have written this gap equation and stability condition somewhat schematically since for many patterns of pairing there will be gap parameters with different magnitudes in different channels. The free energy must then be minimized with respect to each of the gap parameters, yielding a coupled set of gap equations. The solution to (10.2) with the lowest free energy that respects the neutrality constraints discussed below yields the favored phase.

10.1.4 Chemical potentials and neutrality constraints

Why do we describe "matter at high density" by introducing a large chemical potential μ for quark number but no chemical potentials for other quantities? The answer is that this reflects the physics of neutron stars, which are the main physical arena that we consider. Firstly, on the long time scales relevant to neutron stars, the only global charges that are conserved in the standard model are quark number and lepton number, so only these can be coupled to chemical potentials (we shall discuss gauged charges below). Secondly, a neutron star is permeable to lepton number because neutrinos are so light and weakly interacting that they can quickly escape from the star, so the chemical potential for lepton number is zero. Electrons are present because they carry electric charge, for which there is a non-zero potential. In the first few seconds of the life of a neutron star the neutrino mean free path may be short enough to sustain a non-zero lepton number chemical potential, but we will not discuss that scenario.

Stable bulk matter must be neutral under all gauged charges, whether they are spontaneously broken or not. Otherwise, the net charge density would create large electric fields, making the energy non-extensive. In the case of the electromagnetic gauge symmetry, this simply requires zero charge density, $Q = 0$. The correct formal requirement concerning the color charge of a large lump of matter is that it should be a color *singlet*, i.e. its wave function should be invariant under a general color gauge transformation. However, it is sufficient for us to impose color *neutrality*, meaning equality in the numbers of red, green, and blue quarks. This is a less stringent constraint (singlet $\Rightarrow$ neutral but neutral $\not\Rightarrow$ singlet) but the projection of a color neutral state onto a color singlet costs no extra free energy in the thermodynamic limit (Amore et al., 2002). (See also (Elze et al., 1983, 1984).) In general there are eight possible color charges, but because the Cartan subalgebra of $SU(3)_c$ is two-dimensional it is always possible to transform to a gauge where all are zero except Q_3 and Q_8, the charges associated with the diagonal generators $T_3 = \frac{1}{2}\,\mathrm{diag}(1, -1, 0)$ and $T_8 = \frac{1}{2\sqrt{3}}\,\mathrm{diag}(1, 1, -2)$ in (r, g, b) space (Buballa and Shovkovy, 2005; Rajagopal and Schmitt, 2006). In this review, we only discuss such gauges. So to impose color neutrality we just require $Q_3 = Q_8 = 0$.

In nature, electric and color neutrality are enforced by the dynamics of the electromagnetic and QCD gauge fields, whose zeroth components serve as chemical potentials coupled to the charges Q, Q_3, Q_8, and which are naturally driven to values that set these charges to zero (Alford and Rajagopal, 2002; Dietrich and Rischke, 2004; Gerhold and Rebhan, 2003; Iida and Baym, 2001; Kryjevski, 2003). In an NJL model with fermions but no gauge fields (see Section 10.6) one has to introduce the chemical potentials μ_e, μ_3, and μ_8 by hand in order to enforce color and electric neutrality. The neutrality conditions are then

$$\begin{aligned} Q &= \frac{\partial \Omega}{\partial \mu_e} = 0 \\ Q_3 &= -\frac{\partial \Omega}{\partial \mu_3} = 0 \\ Q_8 &= -\frac{\partial \Omega}{\partial \mu_8} = 0. \end{aligned} \tag{10.3}$$

(Note that we define an electrostatic potential μ_e that is coupled to the *negative* electric charge Q, so that in typical neutron star conditions, where there is a finite density of electrons rather than positrons, μ_e is positive.)

Finally we should note that enforcing local neutrality is appropriate for uniform phases, but there are also non-uniform charge-separated phases ("mixed phases"), consisting of positively and negatively charged domains which are neutral on average. These are discussed further in Section 10.3.8.

10.1.5 Stresses on BCS pairing

The free energy argument that we gave in Section 10.1.2 for the inevitability of BCS pairing in the presence of an attractive interaction relies on the assumption that the quarks that pair with equal and opposite momenta can each be arbitrarily close to their common Fermi surface. However, as we will see in Section 10.2, the neutrality constraint, combined with the mass of the strange quark and the requirement that matter be in beta equilibrium, tends to pull apart the Fermi momenta of the different flavors of quarks, imposing an extra energy cost ("stress") on the formation of Cooper pairs involving quarks of different flavors. This raises the possibility of non-BCS pairing in some regions of the phase diagram.

To set the stage here, let us discuss a simplified example: consider two massless species of fermions, labeled 1 and 2, with different chemical potentials μ_1 and μ_2, and an attractive interaction between them that favors cross-species BCS pairing with a gap parameter Δ. It will turn out that, to a good approximation, the color–flavor locked pairing pattern contains three such sectors, so this example captures the essential physics that we will encounter in later sections. We define the average chemical potential and the stress parameter

$$\begin{aligned} \bar{\mu} &= \tfrac{1}{2}(\mu_1 + \mu_2) \\ \delta\mu &= \tfrac{1}{2}(\mu_1 - \mu_2). \end{aligned} \tag{10.4}$$

As long as the stress $\delta\mu$ is small enough relative to Δ, BCS pairing between species 1 and 2 can occur, locking their Fermi surfaces together and ensuring that they occur in equal numbers. At the Chandrasekhar-Clogston point

(Chandrasekhar, 1962; Clogston, 1962), where $\delta\mu = \Delta/\sqrt{2}$, the two-species model undergoes a first-order transition to the unpaired phase. At this point BCS pairing still exists as a locally stable state, with a completely gapped spectrum of quasiparticles. When $\delta\mu$ reaches Δ the spectrum becomes gapless at momentum $p = \bar{\mu}$, indicating that cross-species BCS pairing is no longer favored at all momenta (Alford et al., 2004b). If the two species are part of a larger pairing pattern, the Chandrasekhar–Clogston transition can be shifted, and we shall see that in the two-species subsectors of the CFL pattern it is shifted to $\delta\mu > \Delta$. The onset of gaplessness is therefore the relevant threshold for our purposes, and it always occurs at $\delta\mu = \Delta$, independent of the larger context in which the two flavors pair. This follows from the fact that BCS pairing only occurs if the energy gained from turning a 1 quark into a 2 quark with the same momentum (namely $\mu_1 - \mu_2$) is less than the cost of breaking the Cooper pair formed by these quarks, which is 2Δ (Rajagopal and Wilczek, 2001). Thus the 1-2 Cooper pairs are energetically stable (or metastable) as long as $\delta\mu < \Delta$. A more detailed treatment of this illustrative example can be found in (Alford and Wang, 2005).

This example uses massless quarks, but it can easily be modified to include the leading effect of a quark mass M. A difference in the masses of the pairing quarks also stresses the pairing, because it gives them different Fermi momenta at the same chemical potential, so the quarks in a 1-2 Cooper pair, which have equal and opposite momenta, will not both be close to their Fermi energies. The leading-order effect is easily calculated, since for a quark near its Fermi surface it acts like a shift in the quark chemical potential by $-M^2/(2\bar{\mu})$ (given that Fermi momentum $p_F \approx \bar{\mu}$ to this order).

Returning from our toy model to realistic quark matter, the quark flavors that are potentially relevant at neutron-star densities are the light up (u) and down (d) quarks, with current masses m_u and m_d that are $\lesssim 5$ MeV, and a medium-weight flavor, the strange (s) quark, with current mass $m_s \sim 90$ MeV. Their effective "constituent" masses in the vacuum are hundreds of MeV larger, but are expected to decrease with increasing quark density. We shall refer to the density-dependent constituent masses as $M_{u,d,s}$ and shall typically neglect M_u and M_d. As our toy model has illustrated, however, the strange quark mass M_s will contribute to stresses on cross-flavor pairing, and those stresses will become more severe as the density (and hence $\bar{\mu}$) decreases. This will be a major theme of later sections.

10.1.6 Overview of the quark matter phase diagram

Figure 10.1 shows a schematic phase diagram for QCD that is consistent with what is currently known. Along the horizontal axis the temperature is zero, and the density is zero up to the onset transition where it jumps to nuclear density, and then rises with increasing μ. Neutron stars are in this region of the phase diagram, although it is not known whether their cores are dense enough to reach the quark matter phase. Along the vertical axis the temperature rises, taking us through the crossover from a hadronic gas to the quark–gluon plasma. This is the regime explored by high-energy heavy-ion colliders.

At the highest densities we find the color–flavor locked color-superconducting phase,[1] in which the strange quark participates symmetrically with the up and down quarks in Cooper pairing. This is described in more detail in Sections 10.2, 10.4, and 10.5. It is not yet clear what happens at intermediate density, and in Sections 10.3 and 10.6 we will discuss the factors that disfavor the CFL phase at intermediate densities, and survey the color superconducting phases that have been hypothesized to occur there.

[1]As explained in Section 10.1.1, we fix $N_f = 3$ at all densities, to maintain relevance to neutron star interiors. Pairing with arbitrary N_f has been studied (Schäfer, 2000a). For N_f a multiple of three, one finds multiple copies of the CFL pattern; for $N_f = 4, 5$ the pattern is more complicated.

Various aspects of color superconductivity at high temperatures have been studied, including the phase structure (see Section 10.6.1), spectral functions, pair-forming and -breaking fluctuations, possible precursors to condensation such as pseudogaps, and various collective phenomena (Fukushima and Iida, 2005; Hatsuda et al., 2006; Kitazawa et al., 2008). However, this chapter centers on quark matter at neutron star temperatures, and throughout Sections 10.2 and 10.3 we restrict ourselves to the phases of quark matter at zero temperature. This is because most of the phases that we discuss are expected to persist up to critical temperatures that are well above the core temperature of a typical neutron star, which drops below 1 MeV within seconds of its birth before cooling down through the keV range over millions of years.

10.2 Matter at the highest densities

10.2.1 Color-flavor locked (CFL) quark matter

Given that quarks form Cooper pairs, the next question is who pairs with whom? In quark matter at sufficiently high densities, where the up, down, and strange quarks can be treated on an equal footing and the disruptive effects of the strange quark mass can be neglected, the most symmetric and most attractive option is the color–flavor locked phase, where quarks of all three colors and all three flavors form conventional zero-momentum spinless Cooper pairs. This pattern is encoded in the quark–quark self-energy (Alford et al., 1999b)

$$\begin{aligned} \langle \psi_i^\alpha C\gamma_5 \psi_j^\beta \rangle &\propto \Delta_{\rm CFL}(\kappa+1)\delta_i^\alpha \delta_j^\beta + \Delta_{\rm CFL}(\kappa-1)\delta_j^\alpha \delta_i^\beta \\ &= \Delta_{\rm CFL}\epsilon^{\alpha\beta A}\epsilon_{ijA} + \Delta_{\rm CFL}\kappa(\delta_i^\alpha \delta_j^\beta + \delta_j^\alpha \delta_i^\beta). \end{aligned} \tag{10.5}$$

The symmetry breaking pattern is

$$[SU(3)_c] \times U(1)_B \times \underbrace{SU(3)_L \times SU(3)_R}_{\supset [U(1)_Q]} \rightarrow \underbrace{SU(3)_{c+L+R}}_{\supset [U(1)_{\tilde{Q}}]} \times \mathbb{Z}_2. \tag{10.6}$$

Color indices α, β and flavor indices i, j run from 1 to 3, Dirac indices are suppressed, and C is the Dirac charge-conjugation matrix. Gauge symmetries are in square brackets. $\Delta_{\rm CFL}$ is the CFL gap parameter. The Dirac structure $C\gamma_5$ is a Lorentz singlet, and corresponds to parity-even spin-singlet pairing, so it is antisymmetric in the Dirac indices. The two quarks in the Cooper pair are identical fermions, so the remaining color+flavor structure must be symmetric. The dominant color–flavor component in (10.5) transforms as $(\bar{\mathbf{3}}_A, \bar{\mathbf{3}}_A)$, antisymmetric in both. The subdominant term, multiplied by κ, transforms as

$(\mathbf{6}_S, \mathbf{6}_S)$. It is almost certainly not energetically favored on its own (all the arguments in Section 10.2.1.5 for the color triplet imply repulsion for the sextet), but in the presence of the dominant pairing it breaks no additional symmetries, so κ is in general small but not zero (Alford et al., 1999b; Pisarski and Rischke, 1999b; Schäfer, 2000a; Shovkovy and Wijewardhana, 1999).

10.2.1.1 *Color–flavor locking and chiral symmetry breaking*

A particularly striking feature of the CFL pairing pattern is that it breaks chiral symmetry. Because of color-flavor locking, chiral symmetry remains broken up to arbitrarily high densities in three-flavor quark matter. The mechanism is quite different from the formation of the $\langle \bar{\psi}\psi \rangle$ condensate which breaks chiral symmetry in the vacuum by pairing left-handed (L) quarks with right-handed (R) antiquarks. The CFL condensate pairs L quarks with each other and R quarks with each other (quarks in a Cooper pair have opposite momentum, and zero net spin, hence the same chirality) and so it might naively appear chirally symmetric. However, the Kronecker deltas in (10.5) connect color indices with flavor indices, so that the condensate is not invariant under color rotations, nor under flavor rotations, but only under simultaneous, equal and opposite, color and flavor rotations. Color is a vector symmetry, so the compensating flavor rotation must be the same for L and R quarks, so the axial part of the flavor group, which is the chiral symmetry, is broken by the locking of color and flavor rotations to each other (Alford et al., 1999b). Such locking is familiar from other contexts, including the QCD vacuum, where a condensate of quark–antiquark pairs locks $SU(3)_L$ to $SU(3)_R$ breaking chiral symmetry "directly", and the B phase of superfluid ^{3}He, where the condensate transforms nontrivially under rotations of spin and orbital angular momentum, but is invariant under simultaneous rotations of both.

The breaking of the chiral symmetry is associated with an expectation value for a gauge-invariant order parameter with the structure $\bar{\psi}\bar{\psi}\psi\psi$ (see Section 10.5). There is also a subdominant "conventional" chiral condensate $\langle \bar{\psi}\psi \rangle \ll \langle \psi C\gamma_5\psi \rangle$ (Schäfer, 2000a). These gauge-invariant observables distinguish the CFL phase from the QGP, and if a lattice QCD algorithm applicable at high density ever becomes available, they could be used to map the presence of color–flavor locking in the phase diagram.

We also expect massless Goldstone modes associated with chiral symmetry breaking (see Sections 10.2.1.4 and 10.5). In the real world there is small explicit breaking of chiral symmetry from the current quark masses, so the order parameters will not go to zero in the QGP, and the Goldstone bosons will be light but not massless.

10.2.1.2 *Superfluidity*

The CFL pairing pattern spontaneously breaks the exact global baryon number symmetry $U(1)_B$, leaving only a discrete $\mathbb{Z}_2$ symmetry under which all quark fields are multiplied by -1. There is an associated gauge-invariant 6-quark order parameter with the flavor and color structure of two Lambda baryons, $\langle \Lambda\Lambda \rangle$ where $\Lambda = \epsilon^{abc}\epsilon_{ijk}\psi_i^a\psi_j^b\psi_k^c$. This order parameter distinguishes the CFL phase from the QGP, and there is an associated massless Goldstone boson that makes the CFL phase a superfluid, see Section 10.5.3. The vortices that result when CFL quark matter is rotated have been studied in (Balachandran et al., 2006; Forbes and Zhitnitsky, 2002; Iida and Baym, 2002; Nakano et al., 2007).

10.2.1.3 *Gauge symmetry breaking and electromagnetism*

As explained above, the CFL condensate breaks the $SU(3)_c \times SU(3)_L \times SU(3)_R$ symmetry down to the diagonal group $SU(3)_{c+L+R}$ of simultaneous color and flavor rotations. Color is a gauge symmetry, and one of the generators of $SU(3)_{L+R}$ is the electric charge, which generates the $U(1)_Q$ gauge symmetry. This means that the unbroken $SU(3)_{c+L+R}$ contains one gauged generator, corresponding to an unbroken $U(1)_{\tilde{Q}}$ which consists of a simultaneous electromagnetic and color rotation. The rest of the color group is broken, so by the Higgs mechanism seven gluons and one gluon–photon linear combination become massive via the Meissner effect. The orthogonal gluon–photon generator $\tilde{Q}$ remains unbroken, because every diquark in the condensate has $\tilde{Q} = 0$. The mixing angle is $\cos\theta \equiv g/\sqrt{g^2 + 4e^2/3}$ where e and g are the QED and QCD couplings. Because $e \ll g$ the angle is close to zero, meaning that the $\tilde{Q}$ photon is mostly the original photon with a small admixture of gluon.

The $\tilde{Q}$ photon is massless. Given small but non-zero quark masses, there are no gapless $\tilde{Q}$-charged excitations; the lightest ones are the pseudoscalar pseudo-Goldstone bosons $\pi^\pm$ and $K^\pm$ (see Sections 10.2.1.4 and 10.5), so for temperatures well below their masses (and well below the electron mass (Shovkovy and Ellis, 2003)) the CFL phase is a transparent insulator, in which $\tilde{Q}$-electric and magnetic fields satisfy Maxwell's equations with a dielectric constant and index of refraction that can be calculated directly from QCD (Litim and Manuel, 2001),

$$n = 1 + \frac{e^2 \cos^2\theta}{9\pi^2}\,\frac{\mu^2}{\Delta_{\mathrm{CFL}}^2}. \tag{10.7}$$

(This result is valid as long as $n - 1 \ll 1$.) Apart from the fact that $n \neq 1$, the emergence of the $\tilde{Q}$ photon is an exact QCD-scale analog of the TeV-scale spontaneous symmetry breaking that gave rise to the photon as a linear combination of the W_3 and hypercharge gauge bosons, with the diquark condensate at the QCD scale playing the role of the Higgs condensate at the TeV scale.

If one could shine a beam of ordinary light on a lump of CFL matter in vacuum, some would be reflected and some would enter, refracted, as a beam of $\tilde{Q}$-light. The reflection and refraction coefficients are known (Manuel and Rajagopal, 2002) (see also (Alford and Good, 2004)). The static limit of this academic result is relevant: if a volume of CFL matter finds itself in a static magnetic field, as within a neutron star, surface currents are induced such that a fraction of this field is expelled via the Meissner effect for the non-$\tilde{Q}$ component of Q, while a fraction is admitted as $\tilde{Q}$-magnetic field (Alford et al., 2000). The magnetic field within the CFL volume is not confined to flux tubes, and is not frozen as in a conducting plasma: CFL quark matter is a color superconductor but it is an electromagnetic insulator.

All Cooper pairs have zero net $\tilde{Q}$-charge, but some have neutral constituents (both quarks $\tilde{Q}$-neutral) and some have charged constituents (the two quarks have opposite $\tilde{Q}$-charge). The $\tilde{Q}$-component of an external magnetic field will not affect the first type, but it will affect the pairing of the second type, so external magnetic fields can modify the CFL phase to the so-called magnetic CFL ("MCFL") phase. The MCFL phase has a different gap structure (Ferrer et al., 2005, 2006) and a different effective theory (Ferrer and de la Incera, 2007b). The original analyses of the MCFL phase were done for rotated magnetic fields $\tilde{B}$ large enough that all quarks are in the lowest Landau level; solving the gap equations at lower $\tilde{B}$ shows that the gap parameters in the MCFL phase exhibit de Haas-van Alphen oscillations, periodic in $1/\tilde{B}$ (Fukushima and Warringa, 2008; Noronha and Shovkovy, 2007).

10.2.1.4 *Low-energy excitations*

The low-energy excitations in the CFL phase are: the eight light pseudoscalars arising from broken chiral symmetry, the massless Goldstone boson associated with superfluidity, and the $\tilde{Q}$-photon. The pseudoscalars form an octet under the unbroken $SU(3)$ color+flavor symmetry, and can naturally be labeled according to their $\tilde{Q}$-charges as pions, kaons, and an η. The effective Lagrangian that describes their interactions, and the QCD calculation of their masses and decay constants will be discussed in Section 10.5. We shall find, in particular, that even though the quark–antiquark condensate is small, the pion decay constant is large, $f_\pi \sim \mu$.

The symmetry breaking pattern (10.6) does not include the spontaneous breaking of the $U(1)_A$ "symmetry" because it is explicitly broken by instanton effects. However, at large densities these effects become arbitrarily small, and the spontaneous breaking of $U(1)_A$ will have an associated order parameter and a ninth pseudo-Goldstone boson with the quantum numbers of the η'. This introduces the possibility of a second type of vortices (Forbes and Zhitnitsky, 2002; Son et al., 2001).

Among the gapped excitations, we find the quark-quasiparticles which fall into an $\mathbf{8} \oplus \mathbf{1}$ of the unbroken global $SU(3)_{c+L+R}$, so there are two gap parameters Δ_1 and Δ_8. The singlet has the larger gap $\Delta_1 = (2 + \mathcal{O}(\kappa))\Delta_8$. We also find an octet of massive vector mesons, which are the gluons that have acquired mass via the Higgs mechanism. The symmetries of the three-flavor CFL phase are the same as those one would expect for 3-flavor hypernuclear matter, and even the pattern of gapped excitations is remarkably similar, differing only in the absence of a ninth massive vector meson. It is therefore possible that there is no phase transition between hypernuclear matter and CFL quark matter (Schäfer and Wilczek, 1999b). This hadron–quark continuity can arise in nature only if the strange quark is so light that there is a hypernuclear phase, and this phase is characterized by proton-Ξ^-, neutron-Ξ^0, and Σ^+-Σ^- pairing, which can then continuously evolve into CFL quark matter upon further increase in density (Alford et al., 1999a).

10.2.1.5 *Why CFL is favored*

The dominant component of the CFL pairing pattern is the color $\bar{\mathbf{3}}_A$, flavor $\bar{\mathbf{3}}_A$, and Dirac $C\gamma_5$ (Lorentz scalar). There are many reasons to expect the color $\bar{\mathbf{3}}_A$ to be favored. First, this is the most attractive channel for quarks interacting via single-gluon exchange which is the dominant interaction at high densities where the QCD coupling is weak; second, it is also the most attractive channel for quarks interacting via the instanton-induced 't Hooft interaction, which is important at lower densities; third, qualitatively, combining two quarks that are each separately in the color-**3** representation to obtain a diquark that is a color-$\bar{\mathbf{3}}_A$ lowers the color-flux at large distances.

It is also easy to understand why pairing in the Lorentz-scalar channel is favorable: it leaves rotational invariance unbroken, allowing for quarks at all angles on the entire Fermi-sphere to participate coherently in the pairing. Many calculations have shown that pairing is weaker in channels that break rotational symmetry (Alford et al., 2003, 1998; Buballa et al., 2003; Iwasaki and Iwado, 1995; Schäfer, 2000b; Schmitt et al., 2002). There is also a rotationally invariant pairing channel with negative parity described by the order parameter $\langle \psi C \psi \rangle$. Perturbative gluon exchange interactions do not distinguish between positive and negative parity diquarks, but non-perturbative instanton induced interactions do, favoring the positive parity channel (Alford et al., 1998; Rapp et al., 1998, 2000).

Once we have antisymmetry in color and in Dirac indices, we are forced to antisymmetrize in flavor indices, and the most general color–flavor structure that the arguments above imply should be energetically favored is

$$\langle \psi_i^\alpha C\gamma_5 \psi_j^\beta \rangle \propto \epsilon^{\alpha\beta A} \epsilon_{ijB} \phi_B^A. \tag{10.8}$$

CFL pairing corresponds to $\phi_B^A = \delta_B^A$, and this is the only pattern that pairs all the quarks and leaves an entire $SU(3)$ global symmetry unbroken. The 2SC pattern is $\phi_B^A = \delta_3^A \delta_B^3$, in which only u and d quarks of two colors pair (Alford et al., 1998; Bailin and Love, 1984; Barrois, 1979; Rapp et al., 1998), see Section 10.3.1. As long as the strange quark mass can be neglected (the parametric criterion turns out to be $\Delta_{\mathrm{CFL}} \gg M_s^2/\mu$, see Section 10.3.2) calculations comparing patterns of the structure (10.8) always find the CFL phase to have the highest condensation energy, making it the favored pattern. This has been confirmed in weak-coupling QCD calculations valid at high density (Evans et al., 2000; Schäfer, 2000a; Shovkovy and Wijewardhana, 1999), in the Ginzburg–Landau approximation (Iida and Baym, 2001), and in many calculations using Nambu–Jona-Lasinio models (Alford et al., 1999a,b; Malekzadeh, 2006; Rapp et al., 2000; Schäfer and Wilczek, 1999b). In the high-density limit where $\Delta \gg M_s^2/\mu$ and $\Delta \ll \mu$ we can expand in powers of Δ/μ and explicitly compare CFL to 2SC pairing. The CFL condensation energy is $(8\Delta_8^2 + \Delta_1^2)\mu^2/(4\pi^2)$ which is $12\Delta_{\mathrm{CFL}}^2\mu^2/(4\pi^2)$ when $\kappa \ll 1$ (see Section 10.2.1.4) whereas the condensation energy in the 2SC phase is only $4\Delta_{\mathrm{2SC}}^2\mu^2/(4\pi^2)$. We shall see later that the 2SC gap parameter turns out to be larger than the CFL gap parameter by a factor of $2^{1/3}$, so up to corrections of order κ the CFL condensation energy is larger than that in the 2SC phase by a factor of $3 \times 2^{-2/3}$. At lower densities the condensation energies become smaller, and we cannot neglect negative M_s^4 terms which are energy penalties induced by the neutrality requirement. Their coefficient is larger for CFL than for 2SC, partly (but usually not completely) canceling the extra condensation energy—see Fig. 10.3 and Section 10.3.1.

10.2.2 Intermediate density: stresses on the CFL phase

As we noted in Section 10.1.5, BCS pairing between two species is suppressed if their chemical potentials are sufficiently different. In real-world quark matter such stresses arise from the strange quark mass, which gives the strange quark a lower Fermi momentum than the down quark at the same chemical potentials μ and μ_e, and from the neutrality requirement, which gives the up quark a different chemical potential from the down and strange quarks at the same μ and μ_e. Once flavor equilibrium under the weak interactions is reached, we find that all three flavors prefer to have different Fermi momenta at the same chemical potentials. This is illustrated in Fig. 10.2, which shows the Fermi momenta of the different species of quarks.

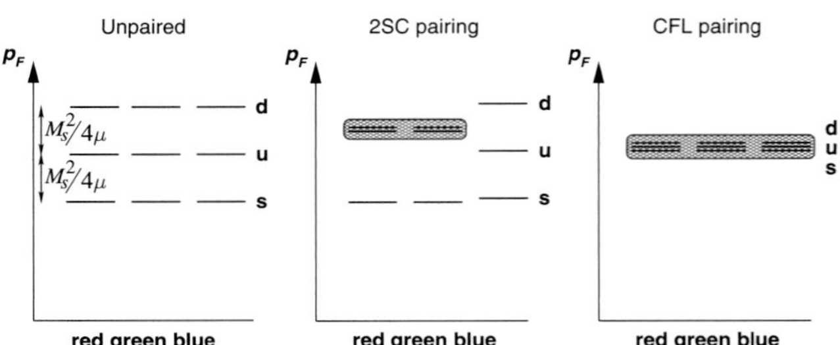

Fig. 10.2 Illustration of the splitting apart of the Fermi momenta of the various colors and flavors of quarks (exaggerated for easy visibility). In the unpaired phase, requirements of neutrality and weak interaction equilibration cause separation of the Fermi momenta of the various flavors. The splittings increase with decreasing density, as μ decreases and $M_s(\mu)$ increases. In the 2SC phase, up and down quarks of two colors pair, locking their Fermi momenta together. In the CFL phase, all colors and flavors pair and have a common Fermi momentum.

In the unpaired phase (Fig. 10.2, left panel), the strange quarks have a lower Fermi momentum because they are heavier, and to maintain electrical neutrality the number of down quarks is correspondingly increased. To lowest order in the strange quark mass, the separation between the Fermi momenta is $\delta p_F = M_s^2/(4\mu)$, so the splitting becomes larger as the density is reduced, and smaller as the density is increased. The phase space at the Fermi surface is proportional to μ^2, so the resultant difference in quark number densities is $n_d - n_u = n_u - n_s \propto \mu^2 \delta p_F \sim \mu M_s^2$. Electrons are also present in weak equilibrium, with $\mu_e = M_s^2/(4\mu)$, so their charge density is parametrically of order $\mu_e^3 \sim M_s^6/\mu^3 \ll \mu M_s^2$, meaning that they are unimportant in maintaining neutrality.

In the CFL phase all the colors and flavors pair with each other, locking all their Fermi momenta together at a common value (Fig. 10.2, right panel). This is possible as long as the energy cost of forcing all species to have the same Fermi momentum is compensated by the pairing energy that is released by the formation of the Cooper pairs. Still working to lowest order in M_s^2, we can say that parametrically the cost is $\mu^2 \delta p_F^2 \sim M_s^4$, and the pairing energy is $\mu^2 \Delta_{\mathrm{CFL}}^2$, so we expect CFL pairing to become disfavored when $\Delta_{\mathrm{CFL}} \lesssim M_s^2/\mu$. In fact, the CFL phase remains favored over the unpaired phase as long as $\Delta_{\mathrm{CFL}} > M_s^2/4\mu$ (Alford and Rajagopal, 2002), but already becomes unstable against unpairing when $\Delta_{\mathrm{CFL}} \gtrsim M_s^2/2\mu$ (see Sec. III.B). NJL model calculations (Abuki et al., 2005; Alford et al., 2005c; Alford and Rajagopal, 2002; Blaschke et al., 2005; Fukushima et al., 2005; Rüster et al., 2005) find that if the attractive interaction were strong enough to induce a 100 MeV CFL gap when $M_s = 0$, then the CFL phase would survive all the way down to the transition to nuclear matter. Otherwise, there must be a transition to some other quark matter phase: this is the "non-CFL" region shown schematically in Fig. 10.1. When the stress is small, the CFL pairing can bend rather than break, developing a condensate of K^0 mesons, described in Section 10.2.3 below. When the stress is larger, however, CFL pairing becomes disfavored. A comprehensive survey of possible BCS pairing patterns shows that all of them suffer from the stress of Fermi surface splitting (Rajagopal and Schmitt, 2006), so in the intermediate-density "non-CFL" region we expect more exotic non-BCS pairing patterns. In Section 10.3 we give a survey of possibilities that have been explored.

10.2.3 Kaon condensation: the CFL-K^0 phase

Bedaque and Schäfer (Bedaque and Schäfer, 2002) showed that when the stress is not too large (high density), it may simply modify the CFL pairing pattern by inducing a flavor rotation of the condensate. This modification can be interpreted as a condensate of "K^0" mesons. The K^0 meson carries negative strangeness (it has the same strangeness as a $\bar{s}$ quark), so forming a K^0 condensate relieves the stress on the CFL phase by reducing its strangeness content. At large density, kaon condensation occurs for $M_s \gtrsim m^{1/3}\Delta^{2/3}$, where m is the mass of the light (u and d) quarks. At moderate density, the critical strange quark mass is increased by instanton contribution to the kaon mass (Schäfer, 2002). Kaon condensation was initially demonstrated using an effective theory of the Goldstone bosons, but with some effort can also be seen in an NJL calculation (Buballa, 2005b; Forbes, 2005). The CFL-K^0 phase is a superfluid; it is a neutral insulator; all of its quark modes are gapped (as long as $M_s^2/(2\mu) < \Delta$); it breaks chiral symmetry. In all these respects it is similar to the CFL phase. Once we turn on small quark masses, different for all flavors, the $SU(3)_{c+L+R}$ symmetry of the CFL phase is reduced by explicit symmetry breaking to just $U(1)_{\tilde{Q}} \times U(1)_{\tilde{Y}}$, with $\tilde{Y}$ a linear combination of a diagonal color generator and hypercharge. In the CFL-K^0 phase, the kaon condensate breaks $U(1)_{\tilde{Y}}$ spontaneously. This modifies the spectrum of both quarks and Goldstone modes, and thus can affect transport properties.

10.3 Below CFL densities

As we discussed at the end of Section 10.1.1 and above (Section 10.2.2), at intermediate densities the CFL phase suffers from stresses induced by the strange quark mass, combined with beta-equilibration and neutrality requirements. It can only survive down to the transition to nuclear matter (occurring at quark chemical potential $\mu = \mu_{\text{nuc}}$) if the pairing is strong enough: roughly $\Delta_{\text{CFL}} > M_s(\mu_{\text{nuc}})^2/2\mu_{\text{nuc}}$, ignoring strong interaction corrections, which are presumably important in this regime. It is therefore quite possible that other pairing patterns occur at intermediate densities, and in this section we survey some of the possibilities that have been suggested.

Figure 10.3 shows a comparison of the free energies of some of these phases. We have chosen $\Delta_{\text{CFL}} = 25\,\text{MeV}$, so there is a window of non-CFL pairing

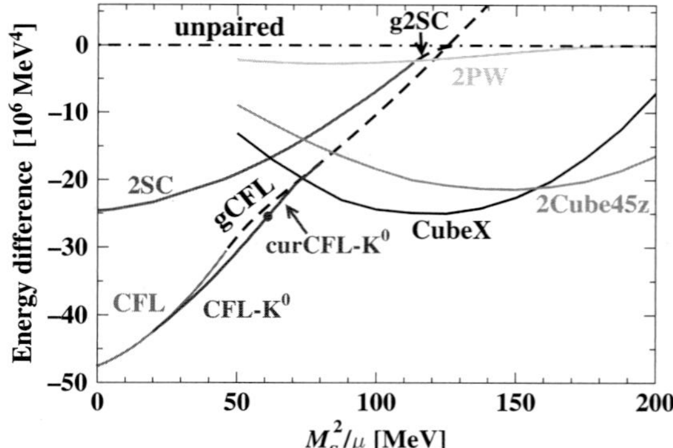

Fig. 10.3 Free energy of various phases of dense 3-flavor quark matter, assuming $\Delta_{\mathrm{CFL}} = 25$ MeV. The homogeneous phases are CFL and 2SC, their gapless analogs gCFL and g2SC, and the kaon-condensed phase CFL-K^0. The true ground state must have a free energy below that of the gCFL phase, which is known to be unstable. The inhomogeneous phases are curCFL-K^0, which is CFL-K^0 with meson supercurrents, and 2PW, CubeX, and 2Cube45z, which are crystalline color superconducting phases. The transition from CFL-K^0 to curCFL-K^0 is marked with a dot. In 2PW the condensate is a sum of only two plane waves. CubeX and 2Cube45z involve more plane waves, their condensation energies are larger but less reliably determined, so their curves should be assumed to have error bands comparable in size to the difference between them.

between nuclear density and the region where the CFL phase becomes stable. (For stronger pairing, $\Delta_{\mathrm{CFL}} \sim 100$ MeV, there would be no such window.) The curves for the CFL, 2SC, gCFL, g2SC, and crystalline phases (2PW, CubeX and 2Cube45z) are obtained from an NJL model as described in Section 10.6. The curves for the CFL-K^0 and meson supercurrent (curCFL-K^0) phases are calculated using the CFL effective theory with parameters chosen by matching to weak-coupling QCD, as described in Section 10.5, except that the gap was chosen to match $\Delta_{\mathrm{CFL}} = 25$ MeV. The phases displayed in Fig. 10.3 are discussed in the following sections.

10.3.1 Two-flavor pairing: the 2SC phase

After CFL, 2SC is the most straightforward less-symmetrically paired form of quark matter, and was one of the first patterns to be analyzed (Alford et al., 1998; Bailin and Love, 1979, 1984; Barrois, 1979; Rapp et al., 1998). In the 2SC phase, quarks with two out of three colors (red and green, say) and two out of three flavors, pair in the standard BCS fashion. The flavors with the most phase space near their Fermi surfaces, namely u and d, are the ones that pair, leaving the strange and blue quarks unpaired (middle panel of Fig. 10.2). According to NJL models, if the coupling is weak then there is no 2SC region in the phase diagram (Steiner et al., 2002). This can be understood by an expansion in powers of M_s, which finds that the CFL$\rightarrow$2SC transition occurs at the same point as the 2SC$\rightarrow$unpaired transition, leaving no 2SC window (Alford and Rajagopal, 2002) (this is the situation in Fig. 10.3). However, NJL models with stronger coupling leave open the possibility of a 2SC window in the "non-CFL" region of the phase diagram (Abuki and Kunihiro, 2006; Rüster et al., 2005). (These calculations have to date not included the possibility of meson current or crystalline color superconducting phases, discussed below, that may prove more favorable.)

The 2SC pairing pattern, corresponding to $\phi^A_B = \delta^A_3\delta^3_B$ in (10.8), is $\langle\psi^\alpha_i C\gamma_5\psi^\beta_j\rangle \propto \Delta_{2SC}\epsilon_{ij3}\epsilon^{\alpha\beta 3}$, where the symmetry breaking pattern, assuming massless up and down quarks, is

$$
\begin{aligned}
&[SU(3)_c] \times \underbrace{SU(2)_L \times SU(2)_R \times U(1)_B \times U(1)_S}_{\supset [U(1)_Q]} \\
&\to [SU(2)_{rg}] \times \underbrace{SU(2)_L \times SU(2)_R \times U(1)_{\tilde{B}} \times U(1)_S}_{\supset [U(1)_{\tilde{Q}}]}
\end{aligned}
\tag{10.9}
$$

using the same notation as in Eq. (10.6). The unpaired massive strange quarks introduce a $U(1)_S$ symmetry. The color $SU(3)_c$ gauge symmetry is broken down to an $SU(2)_{rg}$ red-green gauge symmetry, whose confinement distance rises exponentially with density, as $\Delta^{-1}\exp(\text{const}\,\mu/(g\Delta))$ (Rischke et al., 2001) (see also (Casalbuoni et al., 2002b; Ouyed and Sannino, 2001)). An interesting feature of 2SC pairing is that no global symmetries are broken. The condensate is a singlet of the $SU(2)_L \times SU(2)_R$ flavor symmetry, and baryon number survives as $\tilde{B}$, a linear combination of the original baryon number and the broken diagonal T_8 color generator. Electromagnetism, originally a linear combination of B, S, and I_3 (isospin), survives as an unbroken linear combination $\tilde{Q}$ of $\tilde{B}$, S, and I_3. 2SC quark matter is therefore a color superconductor but is neither a superfluid nor an electromagnetic superconductor, and there is no order parameter that distinguishes it from the unpaired phase or the QGP (Alford et al., 1998). With respect to the unbroken $U(1)_{\tilde{Q}}$ gauge symmetry, the 2SC phase is a conductor not an insulator, because some of the ungapped blue and strange quarks are $\tilde{Q}$-charged.

10.3.2 The unstable gapless phases

As was noted in Section 10.2.2, and can be seen in Fig. 10.3, the CFL phase becomes unstable when $\mu \approx \frac{1}{2}M_s^2/\Delta_{\text{CFL}}$. At this point the pairing in the *gs*-*bd* sector suffers the instability discussed in Sec. 10.1.5 and it becomes energetically favorable to convert *gs* quarks into *bd* quarks (both near their common Fermi momentum).[2] If we restrict ourselves to diquark condensates that are spatially homogeneous, the result is a modification of the pairing in which there is still pairing in all the color–flavor channels that characterize CFL, but *gs*-*bd* Cooper pairing ceases to occur in a range of momenta near the Fermi surface (Alford et al., 2004b, 2005c; Fukushima et al., 2005). In this range of momenta there are *bd* quarks but no *gs* quarks, and quark modes at the edges of this range are ungapped, hence this is called a gapless phase ("gCFL"). Such a phenomenon was first proposed for two flavor quark matter ("g2SC") (Shovkovy and Huang, 2003), see also (Gubankova et al., 2003). It has been confirmed in NJL analyses such as those in (Abuki et al., 2005; Abuki and Kunihiro, 2006; Alford et al., 2005b, 2004a,b, 2005c; Fukushima et al., 2005; Rüster et al., 2004, 2005), which predict that at densities too low for CFL pairing there will be gapless phases.

[2]The onset of gaplessness occurs at the μ at which $\frac{1}{2}(\mu_{bd}-\mu_{gs}) = \Delta_{\text{CFL}}$, as explained in Section 10.1.5. Note that in the CFL phase $(\mu_{bd}-\mu_{gs}) = M_s^2/\mu$, twice its value in unpaired quark matter because of the nonzero color chemical potential $\mu_8 \propto M_s^2/\mu$ required by color neutrality in the presence of CFL pairing (Alford and Rajagopal, 2002; Steiner et al., 2002).

In Fig. 10.3, where $\Delta_{\text{CFL}} = 25\,\text{MeV}$, we see the transition from CFL to gCFL at $M_s^2/\mu \approx 2\Delta_{\text{CFL}} = 50\,\text{MeV}$. (It is interesting to note that, whereas the CFL phase is a $\tilde{Q}$-insulator, the gCFL phase is a $\tilde{Q}$-conductor, because it has a small electron density, balanced by unpaired *bu* quarks from a very thin momentum shell of broken *bu-rs* pairing; the CFL→gCFL transition is the analogue of an insulator to metal transition at which a "band" that was unfilled in the insulating phase drops below the Fermi energy, making the material a metal.) The gCFL phase then remains favored beyond the value $M_s^2/\mu \approx 4\Delta_{\text{CFL}} = 100\,\text{MeV}$ at which the CFL phase would become unfavored relative to completely unpaired quark matter (Alford and Rajagopal, 2002).

However, it turns out that in QCD the gapless phases, both g2SC (Giannakis and Ren, 2005a; Huang and Shovkovy, 2004b) and gCFL (Casalbuoni et al., 2005b; Fukushima, 2005), are unstable at zero temperature. (Increasing the temperature above a critical value removes the instability; the critical value varies dramatically between phases, from a fraction of an MeV to of order 10 MeV (Fukushima, 2005).) The instability manifests itself in an imaginary Meissner mass m_M for some of the gluons. m_M^2 is the low-momentum current–current two-point function, and $m_M^2/(g^2\Delta^2)$ (where the strong interaction coupling is g) is the coefficient of the gradient term in the effective theory of small fluctuations around the ground-state condensate, so a negative value indicates an instability towards spontaneous breaking of translational invariance (Fukushima, 2006; Hashimoto, 2006; Huang, 2006; Iida and Fukushima, 2006; Reddy and Rupak, 2005). Calculations in a simple two-species model (Alford and Wang, 2005) show that gapless charged fermionic modes generically lead to imaginary m_M.

The instability of the gapless phases indicates that there must be other phases of even lower free energy, that occur in their place in the phase diagram. The nature of those phases is not reliably determined at present; likely candidates are discussed below.

10.3.3 Crystalline color superconductivity

The Meissner instability of the gCFL phase points to a breaking of translational invariance, and crystalline color superconductivity represents a possible resolution of that instability. The basic idea, first proposed in condensed matter physics (Fulde and Ferrell, 1964; Larkin and Ovchinnikov, 1965) and more recently analyzed in the context of color superconductivity (Alford et al., 2001a; Bowers and Rajagopal, 2002; Casalbuoni and Nardulli, 2004), is to allow the different quark flavors to have different Fermi momenta, thus accommodating the stress of the strange quark mass, and to form Cooper pairs with non-zero momentum, each quark lying close to its respective Fermi surface. The price one must pay for this arrangement is that only fermions in certain regions on the Fermi surface can pair. Pairs with non-zero momenta chosen from some set of wave vectors $\mathbf{q}_a$ yield condensates that vary in position space like $\sum_a \exp(i\mathbf{q}_a \cdot \mathbf{x})$, forming a crystalline pattern whose Bravais lattice is the set of $\mathbf{q}_a$.

Analyses to date have focused on *u-d* and *u-s* pairing, neglecting pairing of d and s because the separation of their Fermi momenta is twice as large (Fig. 10.2). If the $\langle ud\rangle$ condensate includes only pairs with a single non-zero momentum $\mathbf{q}$, this means that in position space the condensate is a single plane-wave and means that in momentum space pairing is allowed on a single ring on the u Fermi surface and a single ring on the opposite side of the d Fermi surface. The simplest "crystalline" phase of three-flavor quark matter that has been analyzed (Casalbuoni et al., 2005a; Mannarelli et al., 2006) includes two such single-plane wave condensates ("2PW"), one $\langle ud\rangle$ and one $\langle us\rangle$. The favored orientation of the two **q**s is parallel, keeping the two "pairing rings" on the u Fermi surface (from the $\langle us\rangle$ and $\langle ud\rangle$ condensates) as far apart as possible (Mannarelli et al., 2006). This simple pattern of pairing leaves much of the Fermi surfaces unpaired, and it is much more favorable to choose a pattern in which the $\langle us\rangle$ and $\langle ud\rangle$ condensates each include pairs with more than one **q**-vector, thus more than one ring and more than one plane wave. Among such more realistic pairing patterns, the two that appear most favorable have either four **q**s per condensate that together point at the eight corners of a cube in momentum space ("CubeX") or eight **q**s per condensate that each point at the corners of separate cubes, rotated relative to each other by 45 degrees ("2Cube45z") (Rajagopal and Sharma, 2006b). It has been shown that the chromomagnetic instability is no longer present in these phases (Ciminale et al., 2006). The free energies of the 2PW, CubeX, and 2Cube45z phases as calculated within an NJL model (see Section 10.6) are shown in Fig. 10.3. The calculation is an expansion in powers of $(\Delta/\delta p_F)^2$ which in the CubeX and 2Cube45z phases turns out to be of order a tenth to a quarter. According to results obtained in a calculation done to third order in this expansion parameter, the CubeX and 2Cube45z condensation energies are large enough that one or other of them is favored over a wide range of M_s^2/μ as illustrated in Fig. 10.3. The uncertainty in each is of the same order as the difference between them, so one cannot yet say which is favored, but the overall scale is plausible (one would expect condensation energies an order of magnitude bigger than that of the 2PW state). We discuss crystalline color superconductivity in greater detail in Section 10.6.

10.3.4 Meson supercurrent ("curCFL-K^0")

Kaon condensation alone does not remove the gapless modes that occur in the CFL phase when M_s becomes large enough, but it does affect the number of gapless modes and the onset value of M_s. In the CFL-K^0 phase, the electrically charged (bs) mode becomes gapless at $M_s^2/\mu \approx 8\Delta/3$ (compared with 2Δ in the CFL phase), and the electrically neutral (bd) mode becomes gapless for $M_s^2/\mu \approx 4\Delta$ (Kryjevski and Schäfer, 2005; Kryjevski and Yamada, 2005). (In an NJL model analysis (Forbes, 2005), the charged mode in the CFL-K^0 phase becomes gapless at $M_s^2/\mu \approx 2.44\Delta$ for $\Delta = 25$ MeV as in Fig. 10.3). The gapless CFL-K^0 phase has an instability which is similar to the instability of the gCFL phase. This instability can be viewed as a tendency towards spontaneous generation of Goldstone boson (kaon) currents (Kryjevski, 2008; Schäfer, 2006). The currents correspond to a spatial modulation of the kaon condensate. There is no net transfer of any charge because the Goldstone boson current is counterbalanced by a backflow of ungapped fermions. The meson supercurrent ground state is lower in energy than the CFL-K^0 state and the magnetic screening masses are real (Gerhold et al., 2007). Because the ungapped fermion mode is electrically charged, both the magnitude of the Goldstone boson current needed to stabilize the phase and the magnitude of the resulting energy gain relative to the phase without a current are very small. Goldstone boson currents can also be generated in the gCFL phase without K^0 condensation. In this case gauge invariance implies that the supercurrent state is equivalent to a single plane-wave LOFF state, but the analyses can be carried out in the limit that the gap is large compared to the magnitude of the current (Gerhold and Schäfer, 2006). This analysis is valid near the onset of the gCFL phase, but not for larger mismatches, where states with multiple currents are favored.

10.3.5 Single-flavor pairing

If the stress due to the strange quark mass is large enough then there may be a range of quark matter densities where no pairing between different flavors is possible, whether spatially uniform or inhomogeneous. From Fig. 10.3 we can estimate that this will occur when $M_s^2/(\mu\Delta_{\rm CFL}) \gtrsim 10$, so it requires a large effective strange quark mass and/or small CFL pairing gap. The best available option in this case is Cooper pairing of each flavor with itself. Single-flavor pairing may also arise among the strange quarks in a 2SC phase, since they are not involved in two-flavor pairing. We will discuss these cases separately below.

To maintain fermionic antisymmetry of the Cooper pair wave function, single-flavor pairing phases have to be either symmetric in color, which greatly weakens or eliminates the attractive interaction, or symmetric in Dirac indices, which compromises the uniform participation of the whole Fermi sphere. As a result, they have much lower critical temperatures than multi-flavor phases such as the CFL or 2SC phases, perhaps as large as a few MeV, more typically in the eV to many keV range (Alford et al., 2003, 1998; Buballa et al., 2003; Schäfer, 2000b; Schmitt, 2005; Schmitt et al., 2002).

Matter in which each flavor only pairs with itself has been studied using NJL models and weakly coupled QCD. These calculations agree that the energetically favored state is color-spin-locked (CSL) pairing for each flavor (Bailin and Love, 1979; Schäfer, 2000b; Schmitt, 2005). CSL pairing involves all three colors, with the color direction of each Cooper pair correlated with its spin direction, breaking $SU(3)_c \times SO(3)_{\rm rot} \to SO(3)_{c+{\rm rot}}$. The phase is isotropic, with rotational symmetry surviving as a group of simultaneous spatial and color rotations. Other possible phases exhibiting spin-one, single-flavor, pairing include the polar, planar, and A phases described in (Schäfer, 2000b; Schmitt, 2005) (for an NJL model treatment see (Alford et al., 2003)). Some of these phases exhibit point or line nodes in the energy gap at the Fermi surface, and hence do break rotational symmetry.

If 2SC pairing occurs with strange quarks present, one possible result is that the strange quarks of all three colors undergo CSL self-pairing, yielding an isotropic "2SC+CSL" pattern. However, NJL model calculations (Alford and Cowan, 2006) show that the color chemical potential induced by the breaking of color symmetry by the 2SC phases may well be large enough to disrupt the CSL pairing, leaving only spin-one pairing of the red and green strange quarks.

Because their gaps and critical temperatures can range as low as the eV scale, single-flavor pairing phases in compact stars would appear relatively late in the life of the star, and might cause dramatic changes in its behavior. For example, unlike the CFL and 2SC phases, many single-flavor-paired phases are electrical superconductors (Schmitt et al., 2003), so their appearance could significantly affect the magnetic field dynamics of the star.

10.3.6 Gluon condensation

In the 2SC phase (unlike in the CFL phase) the magnetic instability arises at a lower value of the stress on the BCS pairing than that at which the onset of gapless pairing occurs. In this 2SC regime, analyses done using a Ginzburg–Landau approach indicate that the instability can be cured by the appearance of a chromoelectric condensate (Gorbar et al., 2006a, 2007, 2006b; Hashimoto and Miransky, 2007). The 2SC condensate breaks the color group down to the $SU(2)_{rg}$ red-green subgroup, and five of the gluons become massive vector bosons via the Higgs mechanism. The new condensate involves some of these massive vector bosons, and because they transform non-trivially under $SU(2)_{rg}$ it now breaks that gauge symmetry. Because they are electrically charged vector particles, rotational symmetry is also broken, and the phase is an electrical superconductor. Alternatively, it has been suggested (Ferrer and de la Incera, 2007a) that the gluon condensate may be inhomogeneous with a large spontaneously induced $\tilde{Q}$ magnetic field.

10.3.7 Secondary pairing

Since the Meissner instability is generically associated with the presence of gapless fermionic modes, and the BCS mechanism implies that any gapless fermionic mode is unstable to Cooper pairing in the most attractive channel, one may ask whether the instability could be resolved without introducing spatial inhomogeneity simply by "secondary pairing" of the gapless quasi-particles, which would then acquire their own gap Δ_s (Hong, 2005; Huang and Shovkovy, 2003). Such pairing would be particularly favorable in the gCFL phase, where there is a mode in the gCFL phase whose dispersion relation is well approximated as quadratic, giving a greatly increased density of states at low energy (diverging as $(\text{energy})^{-1/2}$), and hence large pairing gaps ($\Delta_s \propto G^2$ for coupling strength G, as compared with the standard BCS result $\Delta \propto \exp(-\text{const}/G)$ (Hong, 2005). However, this does not generically resolve the magnetic instability because the Δ_s is still typically much smaller than the primary gap Δ_p, so secondary pairing is absent in the temperature range $\Delta_s \ll T \ll \Delta_p$ (Alford and Wang, 2006).

10.3.8 Mixed phases

Another way for a system to deal with a stress on its pairing pattern is to form a mixed phase, which is a charge-separated state consisting of positively and negatively charged domains which are neutral on average. The coexisting phases have a common pressure and a common value of the charge chemical potential which is not equal to the neutrality value for either phase (Glendenning, 1992; Ravenhall et al., 1983). The size of the domains is determined by a balance between surface tension (which favors large domains) and electric field energy (which favors small domains).

Separation of color charge (Neumann et al., 2003) is expected to be suppressed by the very high energy cost of color electric fields, and in quark matter it has been found that as long as we require local color neutrality such mixed phases are not the favored response to the stress imposed by the strange quark mass (Alford et al., 2004a,b).

However, electric charge separation is quite possible, and may occur at the interface between color-superconducting quark matter and nuclear matter (Alford et al., 2001b) and an interface between quark matter and the vacuum (Alford et al., 2006; Jaikumar et al., 2006), just as it occurs at interfaces between nuclear matter and a nucleon gas (Ravenhall et al., 1983). Mixed phases are a generic phenomenon, since, in the approximation where Coulomb energy costs are neglected, any phase can always lower its free energy density by becoming charged (this follows from the fact that free energies are concave functions of chemical potentials). In this approximation, if two phases A and B can coexist at the same pressure with opposite charge densities then such a mixture will always be favored over a uniform neutral phase of either A or B. For a pedagogical discussion, see (Alford et al., 2005c). Surface and Coulomb energy costs can cancel this energy advantage, however, and have to be calculated on a case-by-case basis.

10.3.9 Relation to cold atomic gases

Given the important role of stressed pairing in quark matter, it is extremely interesting that there is a class of condensed matter systems, namely trapped atomic gases, where stressed Cooper pairing can be studied experimentally. The experiments involve pairing between two different hyperfine states ("species") of the atom. The stress or "polarization" (relative number of the two species) and the interaction strength (scattering length) are both under experimental control, unlike quark matter where one physical variable (μ) controls both the coupling strength and the stress. It is therefore possible to vary the strength of the inter-atomic attraction from weak, where BCS pairing occurs, through the unitarity limit, where a bound state forms, to strong, where there is Bose–Einstein condensation of diatomic molecules.

The cold atom system is different from quark matter in some important ways, for example the relevant conserved charged are not gauged, so there are no neutrality constraints, which means that mixed phases, e.g. phase separation between an unpolarized superfluid and a polarized normal state, is expected to play a bigger role (Bedaque et al., 2003; Carlson and Reddy, 2005). There are already experimental indications of this behavior.

However, there are also intriguing similarities. One prospect is that crystalline superconducting (LOFF) phases, which appear to be a likely response of quark matter to the stress of the strange quark mass (Section 10.3.3), may be observed in cold atom systems. Another possibility is that magnetic instabilities, analogous to the chromomagnetic instability of the gapless color superconductors (Section 10.3.2) may be observed, in the form of a transition between a homogeneous gapless superfluid and a LOFF phase (Son and Stephanov, 2006). The LOFF phase is favored at weak coupling, and the homogeneous gapless superfluid (actually a uniform mixture of a an unpolarized superfluid and a fully polarized Fermi gas) is favored at strong coupling, raising the possibility of a phase transition between them at intermediate coupling.

10.4 Weak-coupling QCD calculations

We have asserted in Sections 10.1 and 10.2 that at sufficiently high densities it is possible to do controlled calculations of properties of CFL quark matter directly from the QCD Lagrangian. We describe how to do such calculations in this section. We shall focus on the calculation of the gap parameter, but also discuss the critical temperature T_c for the transition from the CFL phase to the quark–gluon plasma and the Meissner and Debye masses that control color-magnetic and color-electric effects in the CFL phase.

Weak-coupling calculations can be expected to be reliable only at densities for which $g(\mu) < 1$, which corresponds to densities many orders of magnitude greater than that at the centers of neutron stars. The reader may therefore be tempted to see this section as academic. From a theoretical point of view, it is exceptional to have an instance where the properties of a superconducting phase can be calculated rigorously from a fundamental short-distance theory, making this exploration a worthy pursuit even if academic. From a practical point of view, the quantitative understanding that we derive from calculations reviewed in this section provides a completely solid foundation from which we can extrapolate downwards in μ. The effective field theory described in Section 10.5 gives us a well-defined way of doing so as long as we stay within the CFL phase, meaning that we can come down from $\mu > 10^8$ MeV all the way down to $\mu \sim M_s^2/(2\Delta_{\text{CFL}})$. Finally, we shall gain qualitative insights into the CFL phase and other color superconducting phases, insights that guide our thinking at lower densities.

The QCD Lagrangian is given by

$$\mathcal{L} = \overline{\psi}(i\gamma^\mu D_\mu + \hat{\mu}\gamma_0 - \hat{m})\psi - \frac{1}{4}G_a^{\mu\nu}G_{\mu\nu}^a. \tag{10.10}$$

Here, ψ is the quark spinor in Dirac, color, and flavor space, i.e. a $4N_cN_f$-component spinor, and $\overline{\psi} \equiv \psi^\dagger\gamma_0$. The covariant derivative acting on the fermion field is $D_\mu = \partial_\mu + igT_aA_\mu^a$, where g is the strong coupling constant, A_μ^a are the gauge fields, $T^a = \lambda^a/2$ $(a = 1, \ldots, 8)$ are the generators of the gauge group $SU(3)_c$, and λ^a are the Gell–Mann matrices. The field strength tensor is $G_a^{\mu\nu} = \partial_\mu A_\nu^a - \partial_\nu A_\mu^a + gf^{abc}A_\mu^bA_\nu^c$ with the $SU(3)_c$ structure constants f^{abc}. The chemical potential $\hat{\mu}$ and the quark mass $\hat{m} = \text{diag}(m_u, m_d, m_s)$ are diagonal matrices in flavor space. If weak interactions are taken into account, flavor is no longer conserved and there are only two chemical potentials, one for quark (baryon) number, μ, and one for electric charge, μ_e. At the very high densities of interest in this section, the constituent quark masses are essentially the same as the current quark masses m_u, m_d, and m_s meaning that we need not distinguish between them. Furthermore, at asymptotic densities we can neglect even the strange quark mass, so throughout most of this section we shall set $m_u = m_d = m_s = 0$.

If the coupling is small then the natural starting point is a free Fermi gas of quarks. In a degenerate quark gas, all states with momenta $p < p_F = (\mu^2 - m_q^2)^{1/2}$ are occupied, and all states with $p > p_F$ are empty. Because of Pauli-blocking, interactions mainly modify states in the vicinity of the Fermi surface. Since the Fermi momentum is large, typical interactions between quarks near the Fermi surface involve large momentum transfer and are governed by the weak coupling $g(\mu)$. Interactions in which quarks scatter by only a small angle involve only a small momentum transfer and are therefore potentially dangerous. However, small momenta correspond to large distances, and medium modifications of the exchanged gluons are therefore important. In a dense medium, electric gluons are Debye screened at momenta $q \sim g\mu$. The dominant interaction for momenta below the screening scale is due to unscreened, almost static, magnetic gluons. In a hot quark–gluon gas, interactions between magnetic gluons become non-perturbative for momenta less than g^2T. This phenomenon does not take place in a very dense quark liquid, and gluon exchanges with arbitrarily small momenta remain perturbative. On a qualitative level this can be attributed to the absence of Bose enhancement factors in soft gluon propagators. A more detailed explanation will be given in Section 10.5.1. The unscreened magnetic interactions nevertheless make the fluid a "non-Fermi liquid" at temperatures above the critical temperature for color superconductivity.

10.4.1 Gap equation, quasiparticle excitations, and condensation energy

As discussed in Section 10.1.2, any attractive interaction in a many-fermion system leads to Cooper pairing. QCD at high density provides an attractive interaction via one-gluon exchange. In terms of quark representations of $SU(3)_c$, the attractive channel is the antisymmetric anti-triplet $\bar{\mathbf{3}}_A$, appearing by "pairing" two color triplets: $\mathbf{3} \otimes \mathbf{3} = \bar{\mathbf{3}}_A \oplus \mathbf{6}_S$. Consequently, only quarks of different colors form Cooper pairs. There is an induced pairing in the symmetric sextet channel $\mathbf{6}_S$. However, this pairing is much weaker (Alford et al., 1999b), and we shall largely neglect it in the following. As in an electronic superconductor, Cooper pairing results in an energy gap in the quasiparticle excitation spectrum. Its magnitude at zero temperature Δ is crucial for the phenomenology of a superconductor. In addition, it also sets the scale for the critical temperature T_c of the phase transition which can be expected to be of the same order as Δ (in BCS theory, $T_c = 0.57\Delta$).

Both Δ and T_c can be determined from the QCD gap equation

$$\Phi^+(K) = g^2 \int_Q \gamma^\mu \, T_a^T \, F^+(Q) \, \gamma^\nu \, T_b \, D_{\mu\nu}^{ab}(K - Q). \qquad (10.11)$$

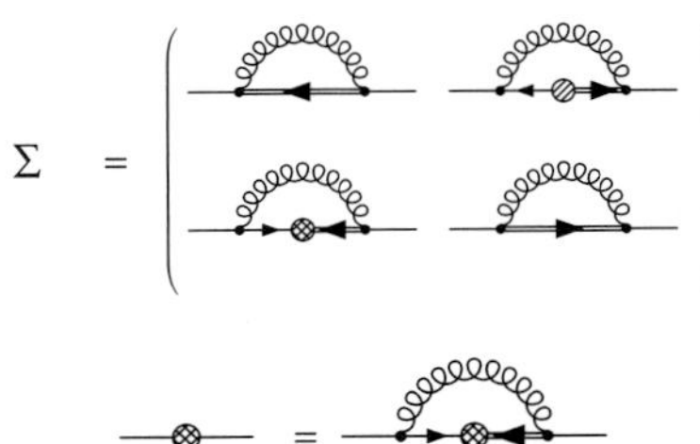

Fig. 10.4 Upper panel: Diagrammatic representation of the quark self-energy in Nambu–Gorkov space. Curly lines correspond to the gluon propagator D. The full quasiparticle and charge-conjugate quasiparticle propagators are denoted by double lines with an arrow pointing to the left and right, respectively. In the anomalous propagators (off-diagonal entries), thin lines correspond to the term $([G_0^{\mp}]^{-1} + \Sigma^{\mp})^{-1}$ (where $G_0^{\pm}$ are the tree-level propagators), while the cross-hatched and hatched circles denote the gap matrices Φ^+ and Φ^-, respectively. Lower panel: The QCD gap equation (10.11) is obtained by equating Φ^+ with the lower left entry of the self-energy depicted in the upper panel.

This gap equation can be derived from the QCD partition function which contains the QCD Lagrangian (10), for details and further references see Ref. (Alford et al., 2008b). In the lower panel of Fig. 10.4 we show the gap equation diagrammatically. In the upper panel, the quark self-energy in Nambu-Gorkov space is shown. The gap equation (and its charge-conjugate counterpart) are the two off-diagonal elements of the Dyson–Schwinger equation $S^{-1} = S_0^{-1} + \Sigma$ with the full quark propagator S and the tree-level propagator S_0, both 2×2 matrices in Nambu–Gorkov space. The off-diagonal elements F^+ and Φ^+ of S and Σ, respectively, appear in the gap equation (10.11). Both quantities are matrices in color, flavor, and Dirac space. The gluon propagator is denoted by $D^{ab}_{\mu\nu}$. In deriving the gap equation (10.11) we have assumed the system to be translationally invariant. This assumption fails for crystalline color superconductors, see Section 10.6.

Before we proceed with solving the gap equation, it is worthwhile discussing the dispersion relations for the fermionic quasiparticle excitations as well as the condensation energy in a color superconductor. That is, we suppose that the gap parameter(s) Δ have been obtained in the manner that we shall describe below, and ask what are the consequences for the quasiparticles and the resulting gain in free energy. The quasiparticle dispersion relations are encoded within the anomalous self-energy Φ^+ which we shall assume to have the form,

$$\Phi^+(K) = \sum_{e=\pm} \Delta^{(e)}(K)\, \mathcal{M}\, \Lambda_{\mathbf{k}}^{(e)}, \tag{10.12}$$

where $\mathcal{M}$ is a matrix in color, flavor, and Dirac space, and $\Lambda_{\mathbf{k}}^{(e)} \equiv (1 + e\gamma_0\boldsymbol{\gamma} \cdot \hat{\mathbf{k}})/2$ are projectors onto states of positive ($e = +$) or negative ($e = -$) energy. The corresponding gap functions are denoted as $\Delta^{(e)}(K)$ and will be determined by the gap equation. In our presentation we shall assume that $\mathcal{M}$ is momentum independent, corresponding to a condensate of Cooper pairs with angular momentum $J = 0$, but the formalism can easily be extended to allow a momentum-dependent $\mathcal{M}_{\mathbf{k}}$ as required for example in the analysis of the CSL phase and we shall quote results for this case also.

We shall analyze color superconducting phases whose color, flavor, and Dirac structure take the form

$$\mathcal{M}_{ij}^{\alpha\beta} = \phi_A^B \epsilon^{\alpha\beta A}\, \epsilon_{ijB}\, \gamma_5, \tag{10.13}$$

where the γ_5 Dirac structure selects a positive parity condensate, where, as described in Sections 10.1 and 10.2, the antisymmetric color matrix is favored since QCD is attractive in this channel and the antisymmetric flavor matrix is then required, and where ϕ is a 3×3 matrix.

The excitation spectrum is given by the poles of the full quark propagator S. It will turn out that the Hermitian matrix $\mathcal{M}\mathcal{M}^\dagger$ determines which quasiparticles are gapped and determines the ratios among the magnitudes of (possibly) different gaps. It is convenient to write this matrix via its spectral representation,

$$\mathcal{M}\mathcal{M}^\dagger = \sum_r \lambda_r \mathcal{P}_r, \tag{10.14}$$

where λ_r are the eigenvalues and $\mathcal{P}_r$ the projectors onto the corresponding eigenstates.

For the resulting explicit expressions for the propagators see Ref. (Alford et al., 2008b). Here we only quote the quasiparticle excitations,

$$\epsilon_{k,r}^{(e)} \equiv \sqrt{(ek-\mu)^2 + \lambda_r\,|\Delta^{(e)}|^2}, \tag{10.15}$$

which hold up to leading order in g; beyond this order, the wave function renormalization cannot be neglected and the quasiparticle and antiquasiparticle energies become more complicated. We see from (10.15) that the antiparticles have $\epsilon > \mu$ — in fact, for k near μ they have $\epsilon \sim 2\mu$. They therefore never play an important role at high density, and we may henceforth set the antiparticle gap to zero, $\Delta^{(-)} = 0$, and denote $\Delta \equiv \Delta^{(+)}$. We shall also use the notation $\epsilon_{k,r} \equiv \epsilon_{k,r}^{(+)}$. We then see that the minimum value of $\epsilon_{k,r}$ occurs at the Fermi surface, where $k = \mu$, and is given by $\sqrt{\lambda_r}\Delta$ which is conventionally referred to as the gap, again neglecting wave function renormalization. We see that although we must solve the gap equation in order to determine the magnitude of the gap parameter Δ, as we will do in Sections 10.4.2 and 10.4.3, the ratios among the actual gaps in the quasiparticle spectra that result are determined entirely by the λ_rs, namely the eigenvalues of $\mathcal{M}\mathcal{M}^\dagger$. The order parameters ϕ_A^B, eigenvalues λ_r, and corresponding projectors $\mathcal{P}_r$ for these CFL and 2SC phases are listed in Table 10.1.

The pressure P (equivalently, the thermodynamic potential since $\Omega = -P$) for a color superconductor can be obtained using the same formalism as for the derivation of the gap equation (10.11) (Alford et al., 2008b). At zero temperature and neglecting the effect of wave function renormalization, the pressure has the simple form,

$$P = N_c N_f \frac{\mu^4}{12\pi^2} + \delta P, \tag{10.16}$$

where we denote the pressure difference of the color-superconducting phase compared to the unpaired phase by δP. If we make the simplifying assumption (corrected in the next subsection) that the gap function is a constant in momentum space in the vicinity of the Fermi surface, we find the easily interpretable result

Table 10.1 Color–flavor structure of CFL and 2SC phases: Order parameters ϕ_A^B, eigenvalues λ_r of the matrix $\mathcal{M}\mathcal{M}^\dagger$, and corresponding projectors $\mathcal{P}_r$, derived from Eq. (10.13). Color (flavor) indices are denoted α, β, (i, j). In the CFL phase, there is an octet with gap Δ (and Cooper pairs *gu-rd*, *bd-gs*, *bu-rs* as well as two linear combinations of the three quarks *ru-gd-bs*) and a singlet with gap 2Δ (corresponding to the remaining orthogonal combination of *ru-gd-bs*). In the 2SC phase, there are four quasiparticles with gap Δ (and Cooper pairs *ru-gd*, *gu-rd*) and five unpaired quasiparticles.

Phase	ϕ_A^B	λ_1	λ_2	$(\mathcal{P}_1)_{\alpha\beta}^{ij}$	$(\mathcal{P}_2)_{\alpha\beta}^{ij}$
CFL	δ_A^B	4 (1–fold)	1 (8–fold)	$\delta_\alpha^i\delta_\beta^j/3$	$\delta_{\alpha\beta}\delta^{ij} - \delta_\alpha^i\delta_\beta^j/3$
2SC	$\delta_{A3}\delta^{B3}$	1 (4–fold)	0 (5–fold)	$(\delta_{\alpha\beta} - \delta_{\alpha3}\delta_{\beta3})(\delta^{ij} - \delta^{i3}\delta^{j3})$	$\delta_{\alpha3}\delta_{\beta3}\delta^{i3}\delta^{j3}$

$$\delta P = \frac{\mu^2}{4\pi^2} \sum_r \mathrm{Tr}[\mathcal{P}_r]\, \lambda_r \Delta^2. \tag{10.17}$$

At $T = 0$ this quantity is the condensation energy density of the color-superconducting state. The fact that $\delta P > 0$ implies that the superconducting state is favored relative to the normal phase. We observe that δP is proportional to the sum of the energy gap squared of the r-th branch, multiplied by the corresponding degeneracy $\mathrm{Tr}[\mathcal{P}_r]$.

We can use the result (10.17) to understand how to compare the favorability of different patterns of color superconducting pairing: the phase with lowest free energy (highest δP) is favored. As an example, in the CFL phase $\delta P = (\mu^2/(4\pi^2))(8 \cdot 1 + 1 \cdot 4)\Delta^2_{\mathrm{CFL}}$ while in the 2SC phase $\delta P = (\mu^2/(4\pi^2))$ $(4 \cdot 1 + 0 \cdot 5)\Delta^2_{\mathrm{2SC}}$ suggesting that the CFL phase is favored. (We shall make this conclusion firm in Section 10.4.3, where we shall find that Δ_{CFL} is smaller than Δ_{2SC} but only by a factor of $2^{1/3}$. This factor will also turn out to be determined entirely by the λ_rs and $\mathrm{Tr}[\mathcal{P}_r]$s.)

10.4.2 Weak-coupling solution of the gap equation

We can now solve the QCD gap equation (10.11) for an order parameter with a given matrix structure $\mathcal{M}$. The matrix structure of the equation is handled by multiplying both sides by $\mathcal{M}^\dagger \Lambda^{(+)}_{\mathbf{k}}$ and taking the trace over color, flavor, and Dirac indices.

The gap equation is sensitive to gluon modes with small momentum ($p \ll m_g$) and even smaller energy ($p_0 \sim p^3/m_g^2 \ll p$), where $m_g^2 = N_f g^2 \mu^2/(6\pi^2)$ is the square of the effective gluon mass at finite density. This means that medium effects in the gluon propagator have to be taken into account. In the low momentum limit, the gluon propagator takes on the standard hard-dense loop approximation form (Braaten and Pisarski, 1992), which we shall give below in Eqs. (10.20) and (10.21) upon simplifying it as appropriate for $p_0 \ll p$. In order to obtain $\log(\Delta/\mu)$ to order g^0, it suffices to keep only the leading terms in the propagator in the $p_0 \ll p$ limit. The gap equation reads

$$\begin{aligned} \Delta_{k,r} = \frac{g^2}{4} \int \frac{d^3q}{(2\pi)^3} \sum_s Z(\epsilon_{q,s}) \frac{\Delta_{q,s}}{\epsilon_{q,s}} \tanh\left(\frac{\epsilon_{q,s}}{2T}\right) \\ \times \left[D_\ell(p)\, \mathcal{T}^s_{00}(\mathbf{k},\mathbf{q}) + D_t(p, \epsilon_{q,s}, \epsilon_{k,r})\, \mathcal{T}^s_t(\mathbf{k},\mathbf{q}) \right], \end{aligned} \tag{10.18}$$

where we have abbreviated $P \equiv K - Q$ and have denoted the gap function on the quasiparticle mass shell by $\Delta_{k,r} \equiv \Delta(\epsilon_{k,r}, \mathbf{k})$. The wave function renormalization factor is denoted by $Z(\epsilon_{q,s})$, and we have denoted the traces over color, flavor, and Dirac space by

$$\mathcal{T}^s_{\mu\nu}(\mathbf{k},\mathbf{q}) \equiv -\frac{\mathrm{Tr}\left[\gamma_\mu T_a^T \gamma_0 \mathcal{M}_{\mathbf{q}} \gamma_0 \mathcal{P}_s \Lambda^{(-)}_{\mathbf{q}} \gamma_\nu T_a \mathcal{M}^\dagger_{\mathbf{k}} \Lambda^{(+)}_{\mathbf{k}}\right]}{\mathrm{Tr}\left[\mathcal{M}_{\mathbf{k}} \mathcal{M}^\dagger_{\mathbf{k}} \Lambda^{(+)}_{\mathbf{k}}\right]}, \tag{10.19}$$

and $\mathcal{T}^s_t(\mathbf{k},\mathbf{q}) \equiv -(\delta^{ij} - \hat{p}^i \hat{p}^j)\, \mathcal{T}^s_{ij}(\mathbf{k},\mathbf{q})$. The two terms inside the square bracket in Eq. (10.18) correspond to the contributions from electric and magnetic gluons. The dominant contribution comes from almost static gluons with $p_0 \ll p$. The static electric and almost static magnetic gluon propagators give,

$$D_\ell(p) \equiv \frac{2}{p^2 + 3m_g^2} \tag{10.20}$$

$$D_t(p, \epsilon, \epsilon') \equiv \frac{p^4}{p^6 + M_g^4(\epsilon + \epsilon')^2} + (\epsilon' \to -\epsilon'), \tag{10.21}$$

where $M_g^2 \equiv (3\pi/4)m_g^2$.

We can solve (10.18) for the zero temperature gap Δ on the Fermi surface. Or, we can solve for T in the $\Delta \to 0$ limit, thus obtaining the critical temperature T_c. Solving for Δ, we find that it has a weak coupling expansion of the form,

$$\log\left(\frac{\Delta}{\mu}\right) = -\frac{b_{-1}}{g} - \bar{b}_0 \log(g) - b_0 - \ldots \tag{10.22}$$

In our treatment of the fermion propagator, the gluon propagator, and in our truncation of the self-energy in Fig. 10.4 to one loop (for example neglecting vertex renormalization) we have been careful to keep all effects that contribute to b_0, but we have neglected many that contribute at order $g \log g$ and g. We shall describe the results for b_{-1}, $\bar{b}_0$, and b_0 in Section 10.4.3.

Before turning to quantitative results, it is worth highlighting the origin and the importance of the leading $-1/g$ behavior in (10.22), namely the fact that $(\Delta/\mu) \sim \exp(-\text{constant}/g)$. If in the gap equation of Fig. 10.4 we were to replace the exchanged gluon by a contact interaction, we would obtain a gap equation of the form,

$$\Delta \propto g^2 \int d\xi \frac{\Delta}{\sqrt{\xi^2 + \Delta^2}} \tag{10.23}$$

with $\xi \equiv k - \mu$. This always has the solution $\Delta = 0$; to seek non-zero solutions, we cancel Δ from both sides of the equation. Then, if Δ were 0, the remaining integral would diverge logarithmically at small ξ. Therefore, we find a non-zero Δ for any positive non-zero g no matter how small, with $\Delta \propto \exp(-\text{constant}/g^2)$. This is the original BCS argument for superconductivity as a consequence of an attractive interaction at a Fermi surface. However, once we restore the gluon propagator the argument is modified. The crucial point is that magnetic gluon exchange is an unscreened long-range interaction, meaning that the angular integral will diverge logarithmically at forward scattering in the absence of any mechanism that screens the magnetic interaction. The gap equation therefore takes the form

$$\Delta \propto g^2 \int d\xi \frac{\Delta}{\sqrt{\xi^2 + \Delta^2}} d\theta \frac{\mu^2}{\theta\mu^2 + \delta^2}, \tag{10.24}$$

where θ is the angle between the external momentum $\mathbf{k}$ and the loop momentum $\mathbf{q}$ and where δ is some quantity with dimensions of mass that cuts off the logarithmic collinear divergence of the angular integral. In the superconducting phase this divergence will at the least be cut off by the Meissner effect, which screens gluon modes with $p < \Delta$ (since the Cooper pairs have size

$1/\Delta$) giving $\delta \sim \Delta$. This yields $\Delta \sim g^2\Delta(\log\Delta)^2$ and hence a non-zero gap $\Delta \sim \exp(-\text{constant}/g)$. This consequence of the long-range nature of the magnetic gluon exchange was first discovered by Barrois (Barrois, 1979). However, pursuing the argument as just stated yields the wrong value of the constant b_{-1}; it was Son (Son, 1999) who realized that the collinear divergence is cut off by Landau damping at a larger value of angle θ than that at which the Meissner effect does. Loosely speaking, Landau damping leads to $\delta \sim (\Delta m_g^2)^{1/3} \gg \Delta$.

The $(\Delta/\mu) \propto \exp(-\text{constant}/g)$ behavior means that the color superconducting gap is parametrically larger at $\mu \to \infty$ than it would be for any four-fermion interaction. Furthermore, asymptotic freedom ensures that $1/g(\mu)^2$ increases logarithmically with μ, which means that $\exp[-\text{constant}/g(\mu)]$ decreases more slowly than $1/\mu$ at large μ. We can therefore conclude that Δ increases with increasing μ at asymptotically large μ, although of course Δ/μ decreases.

We conclude this subsection with a derivation of the correct value of the coefficient b_{-1}, namely the constant in $(\Delta/\mu) \propto \exp(-\text{constant}/g)$. This coefficient turns out to be independent of the spin–color–flavor structure $\mathcal{M}$, and it is therefore simplest to present its derivation in the 2SC phase, in which there is only one gap parameter $\Delta_k \equiv \Delta_{k,r=1}$, $\epsilon_k \equiv \epsilon_{k,r=1}$. The leading behavior of the gap is completely determined by magnetic gluon exchanges. We can also approximate the trace term by its value in the forward direction $\mathcal{T}_t(\mathbf{k},\mathbf{q}) \simeq \mathcal{T}_t(\mathbf{k},\mathbf{k}) = 2/3$ and set the wave function renormalization $Z(q_0) = 1$ (in the forward limit we also find $\mathcal{T}_{00}(\mathbf{k},\mathbf{q}) \simeq \mathcal{T}_t(\mathbf{k},\mathbf{q})$). Carrying out the angular integrals in the gap equation gives

$$\Delta_k = \frac{g^2}{18\pi^2}\int dq\, \frac{\Delta_q}{\epsilon_q}\,\frac{1}{2}\log\left(\frac{\mu^2}{|\epsilon_q^2 - \epsilon_k^2|}\right). \tag{10.25}$$

Son observed that at this order we can replace the logarithm by $\max\{\log(\mu/\epsilon_k), \log(\mu/\epsilon_q)\}$. Introducing logarithmic variables $x = \log[2\mu/(\xi_k + \epsilon_k)]$ with $\xi_k = |k - \mu|$, the integral equation (10.25) can be written as a differential equation,

$$\Delta''(x) = -\frac{g^2}{18\pi^2}\Delta(x), \tag{10.26}$$

with the boundary conditions $\Delta(0) = 0$ and $\Delta'(x_0) = 0$. Here, $x_0 = \log(2\mu/\Delta)$ determines the gap on the Fermi surface. The solution is

$$\Delta(x) = \Delta\sin\left(\frac{gx}{3\sqrt{2}\pi}\right), \quad \Delta = 2\mu\exp\left(-\frac{3\pi^2}{\sqrt{2}g}\right), \tag{10.27}$$

and thus $b_{-1} = 3\pi^2/\sqrt{2}$. We conclude that in the weak-coupling limit the gap function is peaked near the Fermi surface, with a width that is much smaller than μ but much larger than Δ. Had we not set $Z(q_0) = 1$, the x-dependence of $\Delta(x)$ would be more complicated than the simple sinusoid in (10.27), but the conclusion remains unchanged (Wang and Rischke, 2002).

10.4.3 Gap and critical temperature at weak coupling

The gap on the Fermi surface of a color superconductor at zero temperature can be written as

$$\Delta = \mu g^{-\bar{b}_0} e^{-b_0} \exp\left(-\frac{3\pi^2}{\sqrt{2}g}\right), \tag{10.28}$$

to order g^0 in the weak-coupling expansion of $\log(\Delta/\mu)$. We have derived the coefficient b_{-1} in the exponent above, starting from a simplified version of the gap equation (10.18), with no wave function renormalization and a simplified gluon propagator. Upon restoring these effects, analysis of the gap equation (10.18) yields

$$g^{-\bar{b}_0} e^{-b_0} = g^{-5} 512\pi^4 \left(\frac{2}{N_f}\right)^{5/2} e^{-b_0'} e^{-d} e^{-\zeta}. \tag{10.29}$$

In the following we shall define and explain the origin of each of the terms in this equation.

- The factor g^{-5} and the numerical factor in Eq. (10.29) are due to large angle magnetic as well as electric gluon exchanges and are independent of the pattern of pairing in the color superconducting phase, i.e. independent of $\mathcal{M}$.
- The factor

$$e^{-b_0'} = \exp\left(-\frac{\pi^2+4}{8}\right) \simeq 0.177 \tag{10.30}$$

  arises from the wave function renormalization factor $Z(q_0)$ (Brown et al., 2000b; Wang and Rischke, 2002) and is also independent of $\mathcal{M}$ and hence the same for all color superconducting phases.
- The factors that we have written as $e^{-d}e^{-\zeta}$ are different in different color superconducting phases. The factor e^{-d} is due to the angular structure of the gap. For the $J = 0$ condensates whose gap equation we have derived, $e^{-d} = 1$. Upon redoing the angular integrals for spin-1 condensates, we find that they are strongly suppressed (Schäfer, 2000b; Schmitt, 2005; Schmitt et al., 2002). For spin-1 pairing patterns in which quarks of the same chirality form Cooper pairs, $d = 6$. A smaller suppression occurs when quarks of opposite chirality pair, $d = 4.5$. Superpositions of these states yield values of d between these limits. Regardless, perturbative QCD predicts spin-1 gaps to be two to three orders of magnitude smaller than spin-0 gaps.
- The factor $e^{-\zeta}$ depends on $\mathcal{M}$, the color–flavor–spin matrix that describes the pattern of pairing in a particular color superconducting phase. In a phase in which $\mathcal{M}\mathcal{M}^\dagger$ has two different eigenvalues λ_1 and λ_2, describing $\mathrm{Tr}[\mathcal{P}_1]$ and $\mathrm{Tr}[\mathcal{P}_2]$ quasiparticles respectively, we find,

$$\zeta = \frac{1}{2} \frac{\langle \mathrm{Tr}[\mathcal{P}_1]\lambda_1 \log \lambda_1 + \mathrm{Tr}[\mathcal{P}_2]\lambda_2 \log \lambda_2 \rangle}{\langle \mathrm{Tr}[\mathcal{P}_1]\lambda_1 + \mathrm{Tr}[\mathcal{P}_2]\lambda_2 \rangle}, \tag{10.31}$$

where the angular brackets denote an angular average (trivial for $J = 0$ phases). From this result we obtain $\Delta_{\text{CFL}}/\Delta_{\text{2SC}} = 2^{-1/3}$, and we can conclude the discussion begun in Section 10.4.1, noting now that the condensation energy in the CFL phase is larger than that in the 2SC phase by a factor $3 \cdot 2^{-2/3}$.

It is instructive to extrapolate the perturbative results to lower baryon densities for which the running coupling constant is not small. Taking $\mu \simeq$ 400–500 MeV, and a strong coupling constant $g \simeq 3.5$ one obtains $\Delta \simeq 20$ MeV. This is comparable to (but on the small side of) the range of typical gaps $\Delta \sim$ (20–100) MeV (Rajagopal and Wilczek, 2000) obtained using an NJL model, to be discussed in Section 10.6, or numerical solutions of the Dyson-Schwinger equations (Marhauser et al., 2007; Nickel et al., 2006). This qualitative agreement between two completely different approaches gives us confidence that we understand the magnitude of Δ, the fundamental energy scale that characterizes color superconductivity.

Finally, we can use the gap equation (10.18) to extract the critical temperature T_c. The result is (Brown et al., 2000b,c; Pisarski and Rischke, 2000a,b; Schmitt et al., 2002)

$$\frac{T_c}{\Delta} = \frac{e^\gamma}{\pi} e^\zeta, \tag{10.32}$$

where $\gamma \simeq 0.577$ is the Euler–Mascheroni constant. This should be compared with the BCS result $T_c/\Delta = e^\gamma/\pi \simeq 0.57$. We observe that deviations from the BCS ratio occur in the case of two-gap structures and/or anisotropic gaps. Nevertheless, since e^ζ is of order one, the critical temperature is always of the same order of magnitude as the zero-temperature gap. We see that for the 2SC phase T_c/Δ is as in BCS theory, whereas in the CFL phase this ratio is larger by a factor of $2^{1/3}$. It therefore turns out that T_c is the same in the CFL and 2SC phases.

These estimates of T_c neglect gauge field fluctuations, making them valid only at asymptotic densities. We shall see in Section 10.5.2 that including the gauge field fluctuations turns the second order phase transition that we find by analyzing (10.18) into a first-order phase transition, and increases T_c by a factor $1 + \mathcal{O}(g)$, see Eq. (10.48).

10.4.4 Color and electromagnetic Meissner effect and chromomagnetic instability

One of the characteristic properties of a superconductor is the Meissner effect, the fact that an external magnetic field does not penetrate into the superconductor. The external field is shielded by supercurrents near the interface between the normal phase and the superconducting phase. The inverse penetration length defines a mass scale which can be viewed as an effective magnetic gauge boson mass.

This effect can also be described as the Anderson–Higgs phenomenon (Anderson, 1963; Higgs, 1964). The difermion condensate acts as a composite Higgs field which breaks all or part of the gauge symmetry of the theory. The gauge fields acquire a mass from the Higgs vacuum expectation value, and the would-be Goldstone bosons become the longitudinal components of the gauge fields. The gauge symmetry in QCD is $SU(3)_c \times U(1)_Q$. Different color superconducting order parameters realize different Higgs phases. The color gauge group may be partially or fully broken, and mixing between diagonal gluons and photons can occur.

The Meissner masses for the CFL phase (Rischke, 2000a; Son and Stephanov, 2000a; Zarembo, 2000), the 2SC phase (Rischke, 2000b) and the single-flavor CSL phase (Schmitt et al., 2004) are summarized in Table 10.2.

We observe that the chromomagnetic screening masses are of order $g\mu$. This means that the screening length is much shorter than the coherence length $\xi = 1/\Delta$, and color superconductivity is type I, see Section 10.5.2. The fact that the screening masses are independent of the gap does not contradict the fact that there is no magnetic screening in the normal phase. Magnetic screening disappears for energies and momenta larger than the gap. Therefore, if the $\Delta \to 0$ limit is taken before the limit $p \to 0$ then the magnetic screening vanishes, as expected. Of course, magnetic screening masses also vanish as the temperature approaches T_c.

In both 2SC and CFL phases, the off-diagonal masses vanish except for the eighth gluon and the photon, $m^2_{M,\gamma 8} = m^2_{M,8\gamma} \neq 0$. The two-by-two part of the gauge boson mass matrices that describe the eighth gluon and the photon has one vanishing eigenvalue and one non-zero eigenvalue. The eigenvectors are characterized by a mixing angle θ, given in the last column of Table 10.2. This angle defines the new gauge fields,

$$\tilde{A}^8_\mu = \cos\theta\, A^8_\mu + \sin\theta\, A_\mu, \tag{10.33a}$$

$$\tilde{A}_\mu = -\sin\theta\, A^8_\mu + \cos\theta\, A_\mu, \tag{10.33b}$$

where A^8_μ and A_μ denote the fields for the eighth gluon and the photon, respectively. The $\tilde{A}^8_\mu$ gauge boson feels a Meissner effect; it is the analog of the massive Z-boson in the electroweak standard model. The $\tilde{A}_\mu$ gauge boson, on the other hand, experiences no Meissner effect because the diquark condensate is $\tilde{Q}$-neutral. This is the photon of the unbroken Abelian $U(1)_{\tilde{Q}}$ gauge

Table 10.2 Zero-temperature Meissner masses m_M, rotated Meissner masses $\tilde{m}_M$, and gluon/photon mixing angle θ. The number a labels the gluons ($a = 1, \ldots, 8$) and the photon ($a = 9$). All masses are given in units of $N_f \mu^2/(6\pi^2)$, where $N_f = 3, 2, 1$ in the CFL, 2SC, CSL phases, respectively. We have abbreviated $\eta \equiv (21 - 8\log 2)/54$, $\alpha \equiv (3 + 4\log 2)/27$, $\beta \equiv (6 - 4\log 2)/9$. For the one-flavor CSL phase we have denoted the quark electric charge by q. While the rotated photon in the CFL and 2SC phases is massless, the photon acquires a Meissner mass in the CSL phase.

	$m^2_{M,aa}$								$m^2_{M,a\gamma} = m^2_{M,\gamma a}$		$m^2_{M,\gamma\gamma}$	$\tilde{m}^2_{M,88}$	$\tilde{m}^2_{M,\gamma\gamma}$	$\cos^2\theta$
a	1	2	3	4	5	6	7	8	1-7	8	9			
CFL	$\eta\, g^2$								0	$-\frac{2}{\sqrt{3}}\,\eta\, eg$	$\frac{4}{3}\,\eta\, e^2$	$\left(\frac{4}{3}e^2 + g^2\right)\eta$	0	$3g^2/(3g^2 + 4e^2)$
2SC	0			$\frac{1}{2}g^2$				$\frac{1}{3}g^2$	0	$\frac{1}{3\sqrt{3}}eg$	$\frac{1}{9}e^2$	$\frac{1}{3}g^2 + \frac{1}{9}e^2$	0	$3g^2/(3g^2 + e^2)$
CSL	βg^2	αg^2	βg^2	βg^2	αg^2	βg^2	αg^2	βg^2	0	0	$6q^2e^2$	βg^2	$6q^2e^2$	1

symmetry, consisting of simultaneous color and flavor (i.e. electromagnetic) rotations. The $\tilde{A}_\mu$ field satisfies Maxwell's equations. Because $g \gg e$, the mixing angle is very small and the $\tilde{A}_\mu$ photon contains only a small admixture of the original eighth gluon.

In contrast, $J = 1$ color superconductors show an electromagnetic Meissner effect (Schmitt et al., 2003, 2004). For example, in the CSL phase there is no mixing between the gluons and the photon, as can be seen in the last row of Table 10.2. The photon acquires a mass since the electromagnetic group is spontaneously broken. Other candidate spin-1 phases, such as the polar, planar, or *A* phase involve mixing but also (except for a one-flavor system) exhibit an electromagnetic Meissner effect. This difference in phenomenology of spin-0 versus spin-1 color superconductors may have consequences in compact stars (Aguilera, 2007).

As discussed in Section 10.3.2, if the CFL phase is stressed by a non-zero strange quark mass to the point that Cooper pairs break, the resulting gapless CFL (gCFL) phase found in analyses that presume a translationally invariant condensate exhibits *imaginary* Meissner masses (Casalbuoni et al., 2005b; Fukushima, 2005). This phenomenon was first discovered in the simpler gapless 2SC (g2SC) phase (Huang and Shovkovy, 2004a,b) and can be understood in either the gCFL or g2SC context via a simplified analysis involving two quark species only (Alford and Wang, 2005) that we introduced in Section 10.1.5. At zero temperature, in this simple context one finds the Meissner mass,

$$m_M^2 = m_0^2 \left[1 - \frac{\delta\mu\, \Theta(\delta\mu - \Delta)}{\sqrt{\delta\mu^2 - \Delta^2}} \right], \tag{10.34}$$

where m_0 is the Meissner mass obtained upon setting the mismatch in chemical potentials $\delta\mu = 0$. An instability occurs because a negative term $\propto \delta\mu/\sqrt{\delta\mu^2 - \Delta^2}$ appears for $\delta\mu > \Delta$. At $\delta\mu = \Delta$, which corresponds to the onset of gapless modes, this term diverges and thus it dominates the Meissner masses at least for $\delta\mu$ close to, but larger than, Δ. In essence, although more complicated, this is also what happens in the gCFL phase (Casalbuoni et al., 2005b; Fukushima, 2005).

The chromomagnetic instability of the gCFL phase only demonstrates that this phase is unstable; it does not determine the nature of the stable phase. However, the nature of the instability suggests that the stable phase should feature currents, which must be counterpropagating since in the ground state there can be no net current. Two of these phases have been enumerated in Section 10.3, the meson supercurrent phase and the crystalline color superconducting phases. In both phases the Meissner masses are real.

10.5 Effective theories of the CFL phase

At energies below the gap the response of superconducting quark matter is carried by collective excitations of the superfluid condensate. The lightest of these excitations are Goldstone bosons associated with broken global symmetries.

Effective theories for the Goldstone modes have a number of applications. They can be used to compute low temperature thermodynamic and transport properties, and to study the response to perturbations like non-zero quark masses and lepton chemical potentials. Other light degrees of freedom appear near special points in the phase diagram. Fermion modes become light near the CFL–gCFL transition, and fluctuations of the magnitude of the gap become light near T_c.

Effective field theories can be constructed "top down", by integrating out high energy degrees of freedom, or "bottom up", by writing down the most general effective Lagrangian consistent with the symmetries of a given phase. In QCD at moderate or low density the microscopic theory is non-perturbative, and the top down approach is not feasible. In this case the parameters of the effective Lagrangian can be estimated using dimensional analysis or models of QCD. If the density is very large then effective theories can be derived using the top down method. However, even in this case it is often easier to follow the bottom up approach, and determine the coefficients of the effective Lagrangian using matching arguments. Matching expresses the condition that low energy Green functions in the effective and fundamental theory have to agree.

Quark matter at very high density is characterized by several energy scales. In the limit of massless quarks the most important scales are the chemical potential μ, the screening scale m_g, and the pairing gap Δ. In the weak coupling limit we have $\mu \gg m_g \gg \Delta$. This hierarchy of scales can be exploited in order to simplify calculations of the properties of low energy degrees of freedom in the color superconducting phase. The corresponding effective theory is a non-Fermi liquid theory, which we shall describe in the following section. The Landau–Ginzburg theory is described in Section 10.5.2, and effective theories for Goldstone bosons are discussed in Sections 10.5.3 and 10.5.4.

10.5.1 Non-Fermi liquid theory

If the chemical potential is very large, $\mu \gg \Lambda_{QCD}$, then QCD reduces to an effective theory of particles and holes that has many similarities to Fermi liquid theory. The basic observation is that we can decompose the four-component quark field into low and high-energy components $\psi_\pm$ (Hong, 2000a,b),

$$\psi_\pm = e^{ip_F v_\mu x^\mu} \left(\frac{1 \pm \boldsymbol{\alpha} \cdot \hat{\mathbf{v}}_F}{2} \right) \psi, \tag{10.35}$$

where $\mathbf{v}_F$ is the Fermi velocity and $v_\mu = (1, \mathbf{v}_F)$. The prefactor removes the rapid phase variation common to all fermions in some patch on the Fermi surface specified by $\hat{\mathbf{v}}_F$. We can insert the decomposition Eq. (10.35) into the QCD Lagrangian and integrate out the ψ_- field as well as hard gluon exchanges. This generates an expansion of the QCD Lagrangian in powers of $1/p_F$. At leading order the high density effective Lagrangian for ψ_+ is

$$\mathcal{L}_{HDET} = \psi_+^\dagger \left(iv \cdot D - \frac{D_\perp^2}{2p_F} + \delta\mu \right) \psi_+ + \ldots, \tag{10.36}$$

where D_μ is the covariant derivative, $\delta\mu$ is the residual chemical potential, and the transverse component of D_μ is defined by $\mathbf{D}_\perp = \mathbf{D} - \mathbf{D}_{||}$ with $\mathbf{D}_{||} = \hat{\mathbf{v}}(\hat{\mathbf{v}} \cdot \mathbf{D})$. Higher order terms in the expansion contain four (six, eight, ...) fermion operators, and terms with more derivatives.

At energies below the screening scale particle–hole polarization diagrams have to be resummed. There is a simple generating functional for these effects, known as the hard dense loop (HDL) effective action (Braaten and Pisarski, 1992),

$$\mathcal{L}_{\rm HDL} = -\frac{m^2}{2} \sum_v G^a_{\mu\alpha} \frac{v^\alpha v^\beta}{(v^\lambda D_\lambda)^2} G^a_{\mu\beta}. \tag{10.37}$$

This is a gauge-invariant, but non-local, effective Lagrangian. Expanding $\mathcal{L}_{\rm HDL}$ in powers of the gauge field produces $2, 3, \ldots$ gluon vertices. The quadratic term describes dielectric screening of electric modes and Landau damping of magnetic modes. Higher-order terms contain corrections to the gluon self-interaction in a dense medium.

The low energy expansion defined by the Lagrangian $\mathcal{L}_{HDET} + \mathcal{L}_{HDL}$ was studied in (Schäfer and Schwenzer, 2006). Since electric fields are screened the interaction is dominated by the exchange of magnetic gluons. The transverse gauge boson propagator is

$$D_{ij}(K) = -\frac{i(\delta_{ij} - \hat{k}_i\hat{k}_j)}{k_0^2 - k^2 + iM_g^2 \dfrac{k_0}{k}}, \tag{10.38}$$

where $M_g^2 = (3\pi/4)m_g^2$ and we have assumed that $|k_0| < k$. We observe that the propagator becomes large in the regime $|k_0| \sim k^3/m_g^2$. If the energy is small, $|k_0| \ll m_g$, then the typical energy is much smaller than the typical momentum, $k \sim (m_g^2|k_0|)^{1/3} \gg |k_0|$. This implies that the gluon is very far off its energy shell and not a propagating state. In order for a quark to absorb the large momentum carried by a gluon and to stay close to the Fermi surface the gluon momentum has to be transverse to the momentum of the quark. This means that the term $k_\perp^2/(2\mu)$ in the quark propagator is relevant and has to be kept at leading order. Since $k_\perp \sim (k_0 m_g^2)^{1/3}$ we have $k_\perp^2/(2\mu) \gg k_0$ and the pole of the quark propagator is governed by the condition $k_{||} \sim k_\perp^2/(2\mu)$. This means that

$$k_\perp \sim m_g^{2/3} k_0^{1/3}, \qquad k_{||} \sim m_g^{4/3} \mu^{-1} k_0^{2/3}. \tag{10.39}$$

In this regime propagators and vertices can be simplified even further. The quark and gluon propagators are

$$S_{\alpha\beta}(K) = \frac{i\delta_{\alpha\beta}}{k_0 - k_{||} - \dfrac{k_\perp^2}{2\mu} + i\epsilon\,{\rm sgn}(k_0)}, \tag{10.40}$$

$$D_{ij}(K) = \frac{i\delta_{ij}}{k_\perp^2 - iM_g^2 \dfrac{k_0}{k_\perp}}, \tag{10.41}$$

and the quark gluon vertex is $gv_i(\lambda^a/2)$. Higher-order corrections can be found by expanding the quark and gluon propagators as well as the HDL vertices in powers of the small parameter $\epsilon \equiv (k_0/m)$.

The regime characterized by Eq. (10.39) is completely perturbative, i.e. graphs with extra loops are always suppressed by extra powers of $\epsilon^{1/3}$ (Schäfer and Schwenzer, 2006). The quark many-body system is a non-Fermi liquid. The excitations are quasiparticles with the quantum numbers of quarks, but Green functions scale with fractional powers and logarithms of the energy and the coupling constant (Gerhold and Rebhan, 2005; Ipp et al., 2004; Schäfer and Schwenzer, 2004b).

The corrections to Fermi liquid theory do not upset the logic that underlies the argument that leads to the BCS instability. For quark pairs with back-to-back momenta the basic one-gluon exchange interaction has to be summed to all orders, but all other interactions remain perturbative (Schäfer and Schwenzer, 2006). The gap equation that sums the leading order transverse gluon exchange in the color-antisymmetric channel is

$$\Delta(p_0) = -i\frac{2g^2}{3}\int\frac{dk_0}{2\pi}\int\frac{dk_\perp^2}{(2\pi)^2}\,\frac{k_\perp}{k_\perp^3 + iM_g^2(k_0 - p_0)} \times \int\frac{dk_{||}}{2\pi}\,\frac{\Delta(k_0)}{k_0^2 + k_{||}^2 + \Delta(k_0)^2}. \tag{10.42}$$

This equation is exactly equivalent to Eq. (10.25). In particular, all the kinematic approximations that were used to derive Eq. (10.25), like the low energy approximation to the HDL self-energies and the forward approximation to the Dirac traces, are built into the effective field theory vertices and propagators. The effective theory can now be used to study corrections to the leading order result. Higher-order corrections to the propagators and vertices of the effective theory modify the kernel of the integral equation Eq. (10.42), see (Brown et al., 2000a; Schäfer, 2003a).

The coefficients b_0 and $\bar{b}_0$ introduced in Section 10.4.2 can be determined by matching the four-fermion operators in the effective theory (Schäfer, 2003a). The b_0 term also receives contributions from the fermion wave function renormalization $Z \sim \log(k_0)$. All other terms give corrections beyond $\mathcal{O}(g^0)$ in $\log(\Delta_0/\mu)$. Vertex corrections scale as $\Gamma \sim p_0^{1/3}$ and are suppressed compared with the fermion wave function renormalization. The analogous statement in the case of phonon-induced electronic superconductors is known as Migdal's theorem. Gluon self-energy insertions beyond the $k_0/k_\perp$ term included in the leading order propagator are also suppressed by fractional powers of the coupling and the gluon energy.

10.5.2 Ginzburg–Landau theory

At zero temperature, fluctuations of the superconducting state are dominated by fluctuations of the phase of the order parameter. Near the critical temperature the gap becomes small and fluctuations of the magnitude of the gap are important, too. This regime can be described using the Ginzburg–Landau theory.

Ginzburg and Landau argued that in the vicinity of a second-order phase transition the thermodynamic potential of the system can be expanded in powers of the order parameter and its derivatives. This method was used very successfully in the study of superfluid phases of ^{3}He.

The Ginzburg–Landau approach was first applied to color superconductivity in (Bailin and Love, 1979). The problem was revisited by (Iida and Baym, 2001), who included the effects of unscreened gluon exchanges and charge neutrality. Consider the s-wave color anti-triplet condensate in QCD with three massless flavors. The order parameter can be written as

$$\langle \psi_i^\alpha C\gamma_5 \psi_j^\beta \rangle = \epsilon^{\alpha\beta A}\epsilon_{ijB}\phi_A^B, \tag{10.43}$$

where ϕ_A^B is a matrix in color–flavor space. Note that here we have included the energy gap into ϕ_A^B, in contrast with Eq. (10.13), where ϕ_A^B is dimensionless. We have fixed the orientations of left- and right-handed condensates. Fluctuations in the relative color–flavor orientation of the left- and right-handed fermions correspond to the Goldstone modes associated with chiral symmetry breaking, and will be considered in Section 10.5.4. Therefore, the ansatz (10.43) implies the assumption that chiral fluctuations near T_c are small compared with non-chiral gap fluctuations and fluctuations of the gauge field. The thermodynamic potential can be expanded as

$$\begin{aligned}\Omega = \Omega_0 + \alpha\,\mathrm{Tr}\left(\phi^\dagger\phi\right) + \beta_1\left[\mathrm{Tr}\left(\phi^\dagger\phi\right)\right]^2 \\ + \beta_2\,\mathrm{Tr}\left(\left[\phi^\dagger\phi\right]^2\right) + \kappa\,\mathrm{Tr}\left(\nabla\phi\nabla\phi^\dagger\right) + \ldots\end{aligned} \tag{10.44}$$

The coefficients α, β_i, κ can be treated as unknown parameters, or determined in QCD at weak coupling. The weak coupling QCD result is (Iida and Baym, 2001)

$$\alpha = 4N\frac{T - T_c}{T}, \quad \beta_1 = \beta_2 = \frac{7\zeta(3)}{8(\pi T_c)^2}N, \tag{10.45}$$

$$\kappa = \frac{7\zeta(3)}{8\pi^2 T_c^2}N \tag{10.46}$$

where $N = \mu^2/(2\pi^2)$ is the density of states on the Fermi surface. This result agrees with the BCS result. Using Eq. (10.45) we can verify that the ground state is in the CFL phase $\phi_A^B \sim \delta_A^B$.

From the study of electronic superconductors, it is known that the nature of the finite temperature phase transition depends on the ratio $\kappa = \lambda/\xi$ of the screening length λ and the correlation length ξ. If $\kappa > 1/\sqrt{2}$ the superconductor is type II, fluctuations of the order parameter are more important than fluctuations of the gauge field, and the transition is second order. In a type I superconductor the situation is reversed, and fluctuations of the gauge field drive the transition to first order (Halperin et al., 1974).

In the weak coupling limit, $\xi \sim 1/\Delta \gg \lambda \sim 1/(g\mu)$ and color superconductivity is strongly type I. The role of gauge field fluctuations was studied in (Bailin and Love, 1984; Giannakis et al., 2004; Giannakis and Ren,

2003; Matsuura et al., 2004; Noronha et al., 2006). The contribution to the thermodynamic potential is

$$\Omega_{fl} = 8T \int \frac{d^3k}{(2\pi)^3} \left\{ \log\left(1 + \frac{m_A^2(k)}{k^2}\right) - \frac{m_A^2(k)}{k^2} \right\}, \tag{10.47}$$

where $m_A(k)$ is the gauge field screening mass. In QCD the momentum dependence of m_A cannot be neglected. The contribution of the fluctuations Ω_{fl} induces a cubic term $\propto \phi^3$ in the thermodynamic potential which drives the transition first order. The first-order transition occurs at a critical temperature T_c^* (Giannakis et al., 2004),

$$\frac{T_c^* - T_c}{T_c} = \frac{\pi^2}{12\sqrt{2}} g, \tag{10.48}$$

where T_c is the critical temperature of the second-order transition obtained upon neglecting the cubic term. This result suggests that the phase transition between the color superconducting phase and the quark–gluon plasma will be strongly first order, in contrast to ordinary type I superconductors where this effect is very small.

10.5.3 $U(1)_B$ modes and superfluid hydrodynamics

Color flavor locked quark matter is a superfluid, and the low energy theory of the CFL phase must contain a Goldstone mode associated with the spontaneous breakdown of the $U(1)_B$ symmetry. We can identify this mode as the phase φ of the order parameter

$$\langle \psi_i^\alpha C\gamma_5 \psi_j^\beta \rangle = \epsilon^{\alpha\beta A} \epsilon_{ijB} e^{2i\varphi} \phi_A^B. \tag{10.49}$$

The field φ transforms as $\varphi \to \varphi + \alpha$ under $U(1)_B$ transformation of the quark fields $\psi \to \exp(i\alpha)\psi$. At order $\mathcal{O}((\partial\varphi)^2)$ the effective lagrangian for the $U(1)_B$ Goldstone mode is (Son and Stephanov, 2000a,b)

$$\mathcal{L} = \frac{f^2}{2} \left[(\partial_0\varphi)^2 - v^2(\nabla\varphi)^2 \right] + \ldots \tag{10.50}$$

The low-energy constants f and v can be determined in perturbative QCD. At leading order in the QCD coupling constant,

$$f^2 = \frac{6\mu^2}{\pi^2}, \qquad v^2 = \frac{1}{3}. \tag{10.51}$$

There is a simple symmetry argument that can be used to derive this result. This argument also fixes the form of the leading order Goldstone boson interaction (Greiter et al., 1989; Son, 2002). The main observations is that the baryon chemical potential appears in the QCD Lagrangian like the time component of a fictitious $U(1)_B$ gauge potential $B_\nu = (\mu, \mathbf{0})$. The $U(1)_B$ symmetry can be promoted to a local gauge symmetry is B_ν transforms as $B_\nu \to B_\nu + \partial_\nu\alpha$ under local gauge transformations $\psi \to \exp(i\alpha(x))\psi$. The local gauge symmetry can be implemented in the effective theory by introducing the covariant derivative $D_\nu\varphi = \partial_\nu\varphi + B_\nu$.

There are two constraints that fix the form of $\mathcal{L}(D_\nu\varphi)$: (1) Breaking of Lorentz invariance is only due to the expectation value of B_ν. The Lagrangian itself is Lorentz covariant. (2) For $\varphi = const$ and $B_\nu = (\mu, \mathbf{0})$ the dependence of $\mathcal{L}(B_\nu)$ on μ must match the dependence of the thermodynamic potential on μ. These two conditions imply that to leading (zeroth) order in the QCD coupling constant,

$$\mathcal{L} = \frac{3}{4\pi^2}\left[(\partial_0\varphi - \mu)^2 - (\nabla\varphi)^2\right]^2 + \ldots \tag{10.52}$$

Expanding Eq. (10.52) to second order in derivatives reproduces Eq. (10.50). In addition to that, Eq. (10.52) contains the leading three- and four-boson interactions.

The spontaneous breaking of $U(1)_B$ is related to superfluidity, and the $U(1)_B$ effective theory can be interpreted as superfluid hydrodynamics (Son, 2002). We can define the fluid velocity as

$$u_\alpha = -\frac{1}{\mu_0} D_\alpha\varphi, \tag{10.53}$$

where $D_\alpha\varphi \equiv \partial_\alpha\varphi + (\mu, \mathbf{0})$ and $\mu_0 \equiv (D_\alpha\varphi D^\alpha\varphi)^{1/2}$. Note that this definition ensures that the flow is irrotational, $\nabla \times \mathbf{u} = 0$. The identification (10.53) is motivated by the fact that the equation of motion for the $U(1)$ field φ can be written as a continuity equation,

$$\partial^\alpha(n_0 u_\alpha) = 0, \tag{10.54}$$

where $n_0 = 3\mu_0^3/\pi^2$ is the superfluid number density. At $T = 0$ the superfluid density is equal to the total density of the system, $n = dP/d\mu|_{\mu=\mu_0}$. The energy–momentum tensor has the ideal fluid form,

$$T_{\alpha\beta} = (\epsilon + P)u_\alpha u_\beta - P g_{\alpha\beta}, \tag{10.55}$$

and the conservation law $\partial^\alpha T_{\alpha\beta} = 0$ corresponds to the relativistic Euler equation of ideal fluid dynamics. We conclude that the effective theory for the $U(1)_B$ Goldstone mode accounts for the defining characteristics of a superfluid: irrotational, non-dissipative hydrodynamic flow.

10.5.4 Chiral Goldstone modes in the CFL phase

In the CFL phase the pattern of chiral symmetry breaking is identical to the one at $T = \mu = 0$. This implies that the effective Lagrangian has the same structure as chiral perturbation theory. The main difference is that Lorentz invariance is broken and only rotational invariance is a good symmetry. The effective Lagrangian for the Goldstone modes is given by (Casalbuoni and Gatto, 1999),

$$\mathcal{L}_{\text{eff}} = \frac{f_\pi^2}{4}\text{Tr}\left[\partial_0\Sigma\partial_0\Sigma^\dagger - v_\pi^2\partial_i\Sigma\partial_i\Sigma^\dagger\right] + \left[B\text{Tr}(M\Sigma^\dagger) + h.c.\right]$$
$$+ \left[A_1\text{Tr}(M\Sigma^\dagger)\text{Tr}(M\Sigma^\dagger) + A_2\text{Tr}(M\Sigma^\dagger M\Sigma^\dagger)\right.$$
$$\left. + A_3\text{Tr}(M\Sigma^\dagger)\text{Tr}(M^\dagger\Sigma) + h.c.\right] + \ldots \tag{10.56}$$

Here $\Sigma = \exp(i\phi^a\lambda^a/f_\pi)$ is the chiral field, f_π is the pion decay constant and M is a complex mass matrix. The fields ϕ^a describe the octet of Goldstone bosons $(\pi^\pm, \pi^0, K^\pm, K^0, \bar{K}^0, \eta)$. These Goldstone bosons are an octet under the unbroken $SU(3)_{c+L+R}$ symmetry of the CFL phase and their $\tilde{Q}$-charges under the unbroken gauge symmetry of the CFL phase are ± 1 and 0 as indicated by the superscripts, meaning that they have the same $\tilde{Q}$-charges as the Q-charges of the vacuum pseudoscalar mesons. The chiral field and the mass matrix transform as $\Sigma \to L\Sigma R^\dagger$ and $M \to LMR^\dagger$ under chiral transformations $(L, R) \in SU(3)_L \times SU(3)_R$.

The low energy constants f_π and v_π can be calculated in weak coupling QCD (Son and Stephanov, 2000a,b). We find,

$$f_\pi^2 = \frac{21 - 8\log 2}{18}\frac{\mu^2}{2\pi^2}, \qquad v_\pi^2 = \frac{1}{3}. \tag{10.57}$$

These results can be derived using matching arguments. The idea is that we can promote the $SU(3)_L \times SU(3)_R$ chiral symmetry of the effective Lagrangian to a local gauge symmetry. This gauge symmetry is broken by the vacuum expectation value $\langle\Sigma\rangle = 1$ of the chiral field in the CFL phase. As a consequence the chiral gauge bosons acquire electric and magnetic screening masses proportional to f_π^2 and $f_\pi^2 v_\pi^2$, respectively. The microscopic calculation of these screening masses is identical, up to factors of 2, to the calculation of the gluon screening masses reviewed in Section 10.4.4. Using the one-loop results given in Table 10.2 we can derive Eq. (10.57).

The structure of the mass terms in Eq. (10.56) is completely fixed by chiral symmetry. The low-energy constants B, A_i can be computed in the weak coupling limit. The basic observation is that B and A_i govern the dependence of the vacuum energy on the quark masses. In perturbative QCD $B = 0$ and

$$A_1 = -A_2 = \frac{3\Delta^2}{4\pi^2} \equiv A, \qquad A_3 = 0. \tag{10.58}$$

The coefficient B receives non-perturbative contributions from instantons. Instantons are topological gauge configurations that are related to the axial anomaly in QCD. Instanton effects can be computed reliably if the density is very large. The instanton contribution to B is (Schäfer, 2002)

$$B = c\left[\frac{3\sqrt{2}\pi}{g}\Delta\left(\frac{\mu^2}{2\pi^2}\right)\right]^2\left(\frac{8\pi^2}{g^2}\right)^6\frac{\Lambda_{QCD}^9}{\mu^{12}}, \tag{10.59}$$

where $c = 0.155$ and Λ_{QCD} is the QCD scale factor. We note that B is very small if $\mu \gg \Lambda_{QCD}$.

There is one additional effect related to the quark masses that has to be included in the effective Lagrangian. We explained in Section 10.2.2 that the quark mass leads to a shift $\delta p_F \simeq m_q^2/(2\mu)$ in the location of the Fermi surface. This is a an effect of order m_q^4 in the vacuum energy, formally of higher order than the B, A_i terms in the effective Lagrangian. The effect is nevertheless important, because B is highly suppressed and the A_i terms are proportional to products of the light and heavy quark masses.

Bedaque and Schäfer showed that $X_L \equiv MM^\dagger/(2p_F)$ and $X_R \equiv M^\dagger M/(2p_F)$ act as effective chemical potentials for left- and right-handed fermions, respectively. These effective chemical potentials can be included in the CFL chiral theory by promoting time derivatives to covariant derivatives (Bedaque and Schäfer, 2002),

$$\partial_0 \Sigma \to \nabla_0 \Sigma \equiv \partial_0 \Sigma + i\left(\frac{MM^\dagger}{2\mu}\right)\Sigma - i\Sigma\left(\frac{M^\dagger M}{2\mu}\right). \tag{10.60}$$

In the CFL vacuum $\Sigma = 1$ and the shift in the vacuum energy is indeed of order M^4.

Using Eqs. (10.60) and (10.58) we can compute the Goldstone boson dispersion relations. Neglecting terms proportional to B the energy of the flavored Goldstone modes is

$$\begin{aligned} E_{\pi^\pm} &= \mu_{\pi^\pm} + \left[v_\pi^2 p^2 + \frac{4A}{f_\pi^2}(m_u + m_d)m_s\right]^{1/2}, \\ E_{K^\pm} &= \mu_{K^\pm} + \left[v_\pi^2 p^2 + \frac{4A}{f_\pi^2}m_d(m_u + m_s)\right]^{1/2}, \\ E_{K^0,\bar{K}^0} &= \mu_{K^0,\bar{K}^0} + \left[v_\pi^2 p^2 + \frac{4A}{f_\pi^2}m_u(m_d + m_s)\right]^{1/2}, \end{aligned} \tag{10.61}$$

where

$$\begin{aligned} \mu_{\pi^\pm} &= \mp\frac{m_d^2 - m_u^2}{2\mu}, \qquad \mu_{K^\pm} = \mp\frac{m_s^2 - m_u^2}{2\mu}, \\ \mu_{K^0,\bar{K}^0} &= \mp\frac{m_s^2 - m_d^2}{2\mu}. \end{aligned} \tag{10.62}$$

The mass matrix for the remaining neutral Goldstone bosons, which mix, can be found in (Beane et al., 2000; Son and Stephanov, 2000a,b). We observe that the $\mathcal{O}(m)$ terms lead to an inverted mass spectrum with the kaons being lighter than the pions. This can be understood from a microscopic picture where we view the Goldstone modes as two-particle two-hole states. A mode with the quantum number of the pion is given by $\pi^+ \sim \epsilon^{abc}\epsilon^{ade}(\bar{d}_R^b \bar{s}_R^c)(u_L^d s_L^e)$. The structure of the field operators suggests that the mass of the π^+ is proportional to $(m_u + m_d)m_s$. By the same argument the mass of the K^+ is governed by $(m_u + m_s)m_d$, and $m_K < m_\pi$.

10.5.5 Kaon condensation

If the effective chemical potential in Eq. (10.62) becomes larger than the corresponding mass term in Eq. (10.61), then the energy of a Goldstone boson can become negative. In the physically relevant case $m_s \gg m_u \sim m_d$ this applies in particular to the K^0 and the K^+. When the Goldstone boson energy becomes negative the CFL ground state is reorganized and a Goldstone boson condensate is formed. The physical reason is that a non-zero m_s disfavors strange quarks relative to non-strange quarks. In normal quark matter the system responds to this stress by turning s quarks into (mostly) d quarks. In CFL matter this is difficult, since all quarks are gapped. Instead, the system can respond by populating mesons that contain d quarks and s holes.

The ground state can be determined from the effective potential,

$$V_{\text{eff}} = \frac{f_\pi^2}{4}\text{Tr}\left[2X_L\Sigma X_R\Sigma^\dagger - X_L^2 - X_R^2\right] - A_1\left\{\left[\text{Tr}(M\Sigma^\dagger)\right]^2 - \text{Tr}\left[(M\Sigma^\dagger)^2\right]\right\}, \tag{10.63}$$

where $X_L = MM^\dagger/(2p_F)$, $X_R = M^\dagger M/(2p_F)$ and $M = \text{diag}(m_u, m_d, m_s) = \text{diag}(m, m, m_s)$. Here we only discuss the $T = 0$ case and set $B = 0$. The first term on the right-hand side of Eq. (10.63) contains the effective chemical potential,

$$\mu_s \equiv -\mu_{K^0} \simeq \frac{m_s^2}{2p_F} \tag{10.64}$$

and favors states with a deficit of strange quarks. The second term favors the neutral ground state $\Sigma = 1$. The lightest excitation with positive strangeness is the K^0 meson. We consider the ansatz $\Sigma = \exp(i\alpha\lambda_4)$ which allows the order parameter to rotate in the K^0 direction. The vacuum energy is

$$V(\alpha) = f_\pi^2\left[-\frac{1}{2}\left(\frac{m_s^2 - m^2}{2p_F}\right)^2\sin^2\alpha + m_{K^0}^2(1-\cos\alpha)\right], \tag{10.65}$$

where $m_{K^0}^2 = (4A_1/f_\pi^2)m_u(m_d + m_s)$. Minimizing the vacuum energy we obtain

$$\cos(\alpha) = \begin{cases} 1 & \mu_s < m_{K^0} \\ \frac{m_{K^0}^2}{\mu_s^2} & \mu_s > m_{K^0} \end{cases} \tag{10.66}$$

We conclude that there is a second-order phase transition to a kaon condensed state at $\mu_s = m_{K^0}$. The strange quark mass breaks the $SU(3)$ flavor symmetry to $SU(2)_I \times U(1)_Y$. In the kaon condensed phase this symmetry is spontaneously broken to $U(1)_{\tilde{Q}}$. If $m_u = m_d$, isospin is an exact symmetry and there are two exact Goldstone modes (Miransky and Shovkovy, 2002; Schäfer et al., 2001) with zero energy gap, the K^0 and the K^+. Isospin breaking leads to a small energy gap for the K^+.

Using the perturbative result for A_1, and neglecting instanton effects by setting $B = 0$, we can get an estimate of the critical strange quark mass. The critical strange quark mass scales as $m_u^{1/3}\Delta^{2/3}$. Taking $\mu = 500\,\text{MeV}$, $\Delta = 50\,\text{MeV}$, $m_u = 4\,\text{MeV}$ and $m_d = 7\,\text{MeV}$, we find $m_s^{\text{crit}} \simeq 68\,\text{MeV}$. This is smaller than the physical mass of the strange quark mass, suggesting that the CFL phase at densities likely to occur in compact stars supports a condensate of neutral kaons.

The CFL phase also contains a very light flavor neutral mode which can potentially become unstable. This mode is a linear combination of the η and

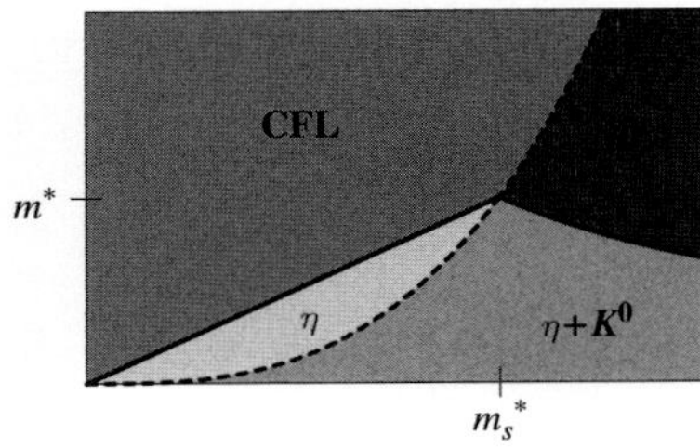

Fig. 10.5 Phase structure of CFL matter as a function of the light quark mass m and the strange quark mass m_s, from (Kryjevski et al., 2005). CFL denotes pure CFL matter, while K^0 and η denote CFL phases with K^0 and/or η condensation. Solid lines are first-order transitions, dashed lines are second order. (See Plate 26 for color image).

η' and its mass is proportional to $m_u m_d$. Because this mode has zero strangeness it is not affected by the μ_s term in the effective potential. However, since $m_u, m_d \ll m_s$ this state is sensitive to perturbative $\alpha_s m_s^2$ corrections (Kryjevski et al., 2005). The resulting phase diagram is shown in Fig. 10.5. The precise value of the tetra-critical point (m^*, m_s^*) depends sensitively on the value of the coupling constant. At very high density m^* is extremely small, but at moderate density m^* can become as large as 5 MeV, comparable to the physical values of the up and down quark mass.

10.5.6 Fermions in the CFL phase

A single quark excitation with energy close to Δ is long-lived and interacts only weakly with the Goldstone modes in the CFL phase. This means that it is possible to include quark fields in the chiral Lagrangian. This Lagrangian not only controls the interaction of quarks with pions and kaons, but it also constrains the dependence of the gap in the fermionic quasiparticle spectrum on the quark masses. This is of interest in connection with the existence and stability of the gapless CFL phase, see the discussion in Sections 10.1.5, 10.2.2, and 10.3.2.

The effective Lagrangian for fermions in the CFL phase is (Kryjevski and Schäfer, 2005)

$$\mathcal{L} = \mathrm{Tr}\left(N^\dagger i v^\mu D_\mu N\right) - D\mathrm{Tr}\left(N^\dagger v^\mu \gamma_5 \left\{\mathcal{A}_\mu, N\right\}\right) - F\mathrm{Tr}\left(N^\dagger v^\mu \gamma_5 \left[\mathcal{A}_\mu, N\right]\right)$$
$$+ \frac{\Delta}{2}\Big[\Big(\mathrm{Tr}\left(N_L N_L\right) - \left[\mathrm{Tr}\left(N_L\right)\right]^2\Big) - (L \leftrightarrow R) + h.c.\Big]. \quad (10.67)$$

$N_{L,R}$ are left- and right-handed baryon fields in the adjoint representation of flavor $SU(3)$. The baryon fields originate from quark–hadron complementarity (Alford et al., 1999a; Schäfer and Wilczek, 1999a). We can think of N as describing a quark which is surrounded by a diquark cloud, $N_L \sim q_L\langle q_L q_L\rangle$. The covariant derivative of the nucleon field is given by $D_\mu N = \partial_\mu N + i[\mathcal{V}_\mu, N]$. The vector and axial-vector currents

$$\mathcal{V}_\mu = -\frac{i}{2}\left(\xi\partial_\mu\xi^\dagger + \xi^\dagger\partial_\mu\xi\right), \quad \mathcal{A}_\mu = -\frac{i}{2}\xi\left(\nabla_\mu\Sigma^\dagger\right)\xi, \quad (10.68)$$

where ξ is defined by $\xi^2 = \Sigma$. It follows that ξ transforms as $\xi \to L\xi U^\dagger = U\xi R^\dagger$ with $U \in SU(3)_V$. The fermion field transforms as $N \to UNU^\dagger$. For pure $SU(3)$ flavor transformations $L = R = V$ we have $U = V$. F and D are low-energy constants that determine the baryon axial coupling. In QCD at weak coupling, we find $D = F = 1/2$ (Kryjevski and Schäfer, 2005).

Mass terms are strongly constrained by chiral symmetry. The effective chemical potentials (X_L, X_R) appear as left- and right-handed gauge potentials in the covariant derivative of the nucleon field. We have,

$$D_0 N = \partial_0 N + i[\Gamma_0, N], \quad (10.69)$$
$$\Gamma_0 = -\frac{i}{2}\left[\xi\left(\partial_0 + iX_R\right)\xi^\dagger + \xi^\dagger\left(\partial_0 + iX_L\right)\xi\right],$$

where $X_L = MM^\dagger/(2p_F)$ and $X_R = M^\dagger M/(2p_F)$ as before. (X_L, X_R) covariant derivatives also appears in the axial vector current given in Eq. (10.68).

We can now study how the fermion spectrum depends on the quark mass. In the CFL state we have $\xi = 1$. For $\mu_s = 0$ the baryon octet has an energy gap Δ and the singlet has gap 2Δ. The leading correction to this result comes from the commutator term in Eq. (10.69). We find that the gap of the proton and neutron is lowered, $\Delta_{p,n} = \Delta - \mu_s$, while the gap of the cascade particles Ξ^-, Ξ^0 is increased, $\Delta_\Xi = \Delta + \mu_s$. As a consequence we find gapless (p, n) excitations at $\mu_s = \Delta$. This result agrees with the spectrum discussed in Section 10.3.2 if the identification $p \equiv (bu)$ and $n \equiv (bd)$ is made.

The situation is more complicated when kaon condensation is taken into account. Numerical results for the eigenvalues are shown in Fig. 10.6. We observe that mixing within the charged and neutral baryon sectors leads to level repulsion. There are two modes that become light in the CFL window $\mu_s \leq 2\Delta$. One mode is a charged mode which is a linear combination of the proton and the Σ^+, while the other mode is a linear combination of the neutral baryons $(n, \Sigma^0, \Xi^0, \Lambda^8, \Lambda^0)$. The charged mode becomes gapless first, at $\mu_s = 4\Delta/3$, and the neutral modes becomes gapless at $\mu_s = 2\Delta$.

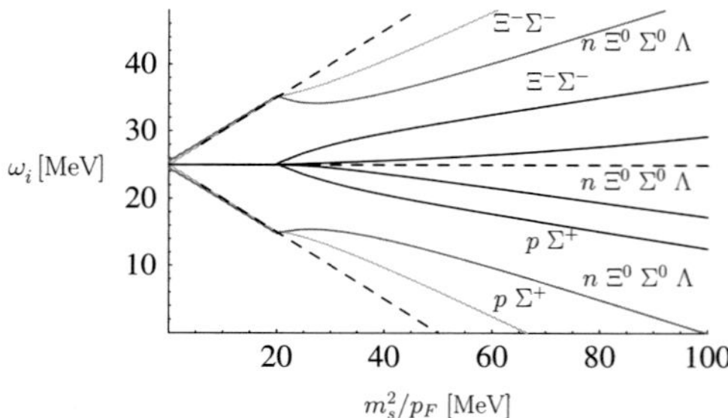

Fig. 10.6 This figure shows the fermion spectrum in the CFL phase. For $m_s = 0$ there are eight fermions with gap Δ (set to 25 MeV) and one fermion with gap 2Δ (not shown). As discussed in Section 10.3, the octet quasiparticles have the $SU(3)$ and $U(1)_{\tilde{Q}}$ quantum numbers of the octet baryons. Without kaon condensation, gapless fermion modes appear at $\mu_s = \Delta$ (dashed lines). With kaon condensation, gapless modes appear at $\mu_s = 4\Delta/3$. (Note that the scale on the horizontal axis is $2\mu_s$.)

10.5.7 Goldstone boson currents

In Section 10.4.4 we showed that gapless fermion modes lead to instabilities of the superfluid phase. Here we will discuss how these instabilities arise, and how they can be resolved, in the context of low-energy theories of the CFL state, by formation of the meson supercurrent state introduced in Section 10.3.4. The chromomagnetic instability is an instability towards the spontaneous generation of currents, that is to say the spontaneous generation of spatial variation in the phase of the diquark condensate. Consider a spatially varying $U(1)_Y$ rotation of the neutral kaon condensate,

$$\xi(\mathbf{x}) = U(\mathbf{x})\xi_K U^\dagger(\mathbf{x}), \tag{10.70}$$

where $\xi_K = \exp(i\pi\lambda_4)$ and $U(\mathbf{x}) = \exp(i\phi_K(\mathbf{x})\lambda_8)$. This state is characterized by nonzero vector and axial-vector currents, see Eq. (10.68). The dependence of the ground state energy on the kaon current $\jmath_K = \nabla\phi_K$ is determined by the effective Lagrangian. The gradient term in the meson sector gives a positive contribution $\mathcal{E} \sim v_\pi^2 f_\pi^2 \jmath_K^2$. A negative contribution can arise from gapless fermions. The kaon current shifts the energy of a fermion by $\Delta\omega \sim c_K \mathbf{v} \cdot \boldsymbol{j}_K$ where $\mathbf{v}$ is the Fermi velocity and the coefficient c_K depends on the quantum numbers of the fermion. If gapless states are present then this term can shift the energy of unoccupied states on one side of the Fermi sphere up, and lower the energy of occupied states on the opposite side. The net effect is that the groundstate energy is lowered.

The energy functional was analyzed in more detail in (Kryjevski, 2008; Schäfer, 2006). There is an instability near the point $\mu_s = 4\Delta/3$. The instability is resolved by the formation of a Goldstone boson current. If electric charge neutrality is enforced the magnitude of the current is very small, and there is no tendency towards the generation of multiple currents. It was also shown that all gluonic screening masses are real (Gerhold et al., 2007).

10.6 NJL model comparisons among candidate phases below CFL densities

As we have explained in Section 10.2, at sufficiently high densities, where the up, down, and strange quarks can be treated on an equal footing and the disruptive effects of the strange quark mass can be neglected, quark matter is in the CFL phase. At asymptotic densities, the CFL gap parameter Δ_{CFL} and indeed any property of CFL quark matter can be calculated in full QCD, as described in Section 10.4. At any density at which the CFL phase arises, its low energy excitations, and hence its properties and phenomenology, can be described by the effective field theory of Section 10.5, whose form is known and whose parameters can be systematically related to the CFL gap Δ_{CFL}. If we knew that the only form of color superconducting quark matter that arises in the QCD phase diagram were CFL, there would therefore be no need to resort to model analyses. However, as we have discussed in Section 10.3, $M_s^2/(\mu\Delta_{\text{CFL}})$ may not be small enough (at $\mu = \mu_{\text{nuc}}$ where the nuclear $\rightarrow$ quark matter transition occurs) for the QCD phase diagram to be this simple.

Even at the very center of a neutron star, μ cannot be larger than about 500 MeV, meaning that the (density-dependent) strange quark mass M_s cannot be neglected. In concert with the requirement that bulk matter must be neutral and must be in weak equilibrium, a non-zero M_s favors separation of the Fermi momenta of the three different flavors of quarks, and thus disfavors the cross-species BCS pairing that characterizes the CFL phase. If CFL pairing is disrupted by the heaviness of the strange quark at a higher μ than that at which color superconducting quark matter is superseded by baryonic matter, the CFL phase must be replaced by some phase of quark matter in which there is less, and less symmetric, pairing.

Within a spatially homogeneous ansatz, the next phase down in density is the gapless CFL (gCFL) phase described in Section 10.3.2. However, as we have described in Section 10.4.4, such gapless paired states suffer from a chromomagnetic instability: they can lower their energy by the formation of counter-propagating currents. It seems likely, therefore, that a ground state with counter-propagating currents is required. This could take the form of a crystalline color superconductor, that we have introduced in Section 10.3.3. Or, given that the CFL phase itself is likely augmented by kaon condensation as described in Sections 10.2.3 and 10.5.5, it could take the form of the phase we have described in Section 10.5.7 in which a CFL kaon condensate carries a current in one direction balanced by a counter-propagating current in the opposite direction carried by gapless quark quasiparticles.

Determining which phase or phases of quark matter occupy the regime of density between hadronic matter and CFL quark matter in the QCD phase diagram, if there is such a regime, remains an outstanding challenge. Barring a major breakthrough that would allow lattice QCD calculations to be brought to bear despite the fermion sign problem, a from-first-principles determination seems out of reach. This leaves two possible paths forward. First, as we describe in this section, we can analyze and compare many of the possible phases within a simplified few-parameter model, in so doing seeking qualitative insight into what phase(s) are favorable. Second, as we shall describe in Section 10.8, we can determine the observable consequences of the presence of various possible color superconducting phases in neutron stars, and then seek to use observational data to rule possibilities out or in.

10.6.1 Model, pairing ansatz, and homogeneous phases

We shall employ a Nambu–Jona-Lasinio (NJL) model in which the QCD interaction between quarks is replaced by a point-like four-quark interaction, with the quantum numbers of single-gluon exchange, analyzed in mean field theory. This is not a controlled approximation. However, it suffices for our purposes: because this model has attraction in the same channels as in QCD, its high-density phase is the CFL phase; and, the Fermi surface splitting effects whose qualitative consequences we wish to study can be built into the model. Note that we shall assume throughout that $\Delta_{\rm CFL} \ll \mu$. This weak coupling assumption means that the pairing is dominated by modes near the Fermi surfaces. Quantitatively, this means that results for the gaps and condensation energies of candidate phases are independent of the cutoff in the NJL model, when expressed in terms of the CFL gap $\Delta_{\rm CFL}$: if the cutoff is changed with the NJL coupling constant adjusted so that $\Delta_{\rm CFL}$ stays fixed, the gaps and condensation energies for the candidate crystalline phases also stay fixed. This makes the NJL model valuable for making the comparisons that are our goal. The NJL model has two parameters: the CFL gap $\Delta_{\rm CFL}$ which parameterizes the strength of the interaction and $M_s^2/(4\mu)$, the splitting between Fermi surfaces in neutral quark matter in the absence of pairing. The free energy of candidate patterns of pairing can be evaluated and compared as a function of these two parameters.

As a rather general pairing ansatz, we shall consider

$$\begin{aligned}
\langle ud\rangle &\sim \Delta_3 \sum_a \exp\left(2i\mathbf{q}_3^a\cdot\mathbf{r}\right)\\
\langle us\rangle &\sim \Delta_2 \sum_a \exp\left(2i\mathbf{q}_2^a\cdot\mathbf{r}\right)\\
\langle ds\rangle &\sim \Delta_1 \sum_a \exp\left(2i\mathbf{q}_1^a\cdot\mathbf{r}\right). \qquad (10.71)
\end{aligned}$$

If we set all the wavevectors $\mathbf{q}_I^a$ to zero, we can use this ansatz to compare spatially homogeneous phases including the CFL phase ($\Delta_1 = \Delta_2 = \Delta_3 \equiv \Delta_{\rm CFL}$), the gCFL phase ($\Delta_3 > \Delta_2 > \Delta_1 > 0$), and the 2SC phase ($\Delta_3 \equiv \Delta_{\rm 2SC}$; $\Delta_1 = \Delta_2 = 0$). Choosing different sets of wave vectors will allow us to analyze and compare different crystalline color superconducting phases of quark matter.

We shall analyze quark matter containing massless u and d quarks and s quarks with an effective mass M_s. The Lagrangian density describing this system in the absence of interactions is given by

$$\mathcal{L}_0 = \bar{\psi}_{i\alpha} \left(i \not{\partial} \delta^{\alpha\beta} \delta_{ij} - M_{ij}^{\alpha\beta} + \mu_{ij}^{\alpha\beta} \gamma_0 \right) \psi_{\beta j}, \tag{10.72}$$

where $i, j = 1, 2, 3$ are flavor indices and $\alpha, \beta = 1, 2, 3$ are color indices and we have suppressed the Dirac indices, where $M_{ij}^{\alpha\beta} = \delta^{\alpha\beta}\,\mathrm{diag}(0, 0, M_s)_{ij}$ is the mass matrix, and where the quark chemical potential matrix is given by

$$\mu_{ij}^{\alpha\beta} = (\mu\delta_{ij} - \mu_e Q_{ij})\delta^{\alpha\beta} + \delta_{ij}\left(\mu_3 T_3^{\alpha\beta} + \frac{2}{\sqrt{3}}\mu_8 T_8^{\alpha\beta}\right), \tag{10.73}$$

with $Q_{ij} = \mathrm{diag}(2/3, -1/3, -1/3)_{ij}$ the quark electric-charge matrix and T_3 and T_8 the diagonal color generators. In QCD, μ_e, μ_3 and μ_8 are the zeroth components of electromagnetic and color gauge fields, and the gauge field dynamics ensure that they take on values such that the matter is neutral (Alford and Rajagopal, 2002; Dietrich and Rischke, 2004; Gerhold and Rebhan, 2003; Kryjevski, 2003), satisfying the neutrality conditions (10.3). In the NJL model, quarks interact via four-fermion interactions and there are no gauge fields, so we introduce μ_e, μ_3, and μ_8 by hand, and choose them to satisfy the neutrality constraints (10.3). The assumption of weak equilibrium is built into the calculation via the fact that the only flavor-dependent chemical potential is μ_e, ensuring for example that the chemical potentials of d and s quarks with the same color must be equal. Because the strange quarks have greater mass, the equality of their chemical potentials implies that the s quarks have smaller Fermi momenta than the d quarks, in the absence of BCS pairing. In the absence of pairing, then, because weak equilibrium drives the massive strange quarks to be less numerous than the down quarks, electrical neutrality requires a $\mu_e > 0$, which makes the up quarks less numerous than the down quarks and introduces some electrons into the system. In the absence of pairing, color neutrality is obtained with $\mu_3 = \mu_8 = 0$.

As illustrated in Fig. 10.2, the Fermi momenta of the quarks and electrons in quark matter that is electrically and color neutral and in weak equilibrium are given in the absence of pairing by

$$\begin{aligned} p_F^u &= \mu - \frac{M_s^2}{6\mu} = p_F^d - \frac{M_s^2}{4\mu} = p_F^s + \frac{M_s^2}{4\mu} \\ p_F^e &= \mu_e = \frac{M_s^2}{4\mu}, \end{aligned} \tag{10.74}$$

where we have worked to linear order in μ_e and M_s^2. We see that the effect of the strange quark mass on unpaired quark matter is as though instead, one reduced the strange quark chemical potential by $M_s^2/(2\mu)$. We shall make this approximation throughout. Upon making this assumption, we need no longer be careful about the distinction between p_Fs and μs, as we can simply think of the three flavors of quarks as though they had chemical potentials,

$$\begin{aligned}\mu_d &= \mu_u + 2\delta\mu_3 \\ \mu_u &= p_F^u \\ \mu_s &= \mu_u - 2\delta\mu_2\end{aligned} \tag{10.75}$$

with

$$\delta\mu_3 = \delta\mu_2 = \frac{M_s^2}{8\mu} \equiv \delta\mu, \tag{10.76}$$

where the choice of subscripts indicates that $2\delta\mu_2$ is the splitting between the Fermi surfaces for quarks 1 and 3 and $2\delta\mu_3$ is that between the Fermi surfaces for quarks 1 and 2, identifying u, d, s with $1, 2, 3$.

As described in (Alford et al., 2004b; Alford and Rajagopal, 2002; Rajagopal and Wilczek, 2001; Steiner et al., 2002), BCS pairing introduces qualitative changes into the analysis of neutrality because wherever BCS pairing occurs between fermions whose Fermi surface would be split in the absence of pairing, the Fermi momenta of these fermions are locked together. This means that the μs required for neutrality can change qualitatively. For example, in the CFL phase $\mu_e = 0$ and μ_8 is non-zero and of order M_s^2/μ.

The NJL interaction term with the quantum numbers of single-gluon exchange that we add to the Lagrangian (10.72) is

$$\mathcal{L}_{\text{interaction}} = -\frac{3}{8}\lambda(\bar{\psi}\Gamma^{A\nu}\psi)(\bar{\psi}\Gamma_{A\nu}\psi), \tag{10.77}$$

where we have suppressed the color and flavor indices that we showed explicitly in (10.72), and have continued to suppress the Dirac indices. The full expression for $\Gamma^{A\nu}$ is $(\Gamma^{A\nu})_{\alpha i,\beta j} = \gamma^\nu (T^A)_{\alpha\beta}\delta_{ij}$. The NJL coupling constant λ has dimension -2, meaning that an ultraviolet cutoff Λ must be introduced as a second parameter in order to fully specify the interaction. We shall define Λ as restricting the momentum integrals to a shell around the Fermi surface, $\mu - \Lambda < |\mathbf{p}| < \mu + \Lambda$.

In the mean-field approximation, the interaction Lagrangian (10.77) takes on the form

$$\mathcal{L}_{\text{interaction}} = \frac{1}{2}\bar{\psi}\Delta(x)\bar{\psi}^T + \frac{1}{2}\psi^T\bar{\Delta}(x)\psi, \tag{10.78}$$

where $\Delta(x)$ is related to the diquark condensate by the relations

$$\begin{aligned}\Delta(x) &= \frac{3}{4}\lambda\Gamma^{A\nu}\langle\psi\psi^T\rangle(\Gamma_{A\nu})^T \\ \bar{\Delta}(x) &= \frac{3}{4}\lambda(\Gamma^{A\nu})^T\langle\bar{\psi}^T\bar{\psi}\rangle\Gamma_{A\nu} \\ &= \gamma^0\Delta^\dagger(x)\gamma^0 .\end{aligned} \tag{10.79}$$

The ansatz (10.71) can now be made precise: we take

$$\Delta(x) = \Delta_{CF}(x) \otimes C\gamma^5, \tag{10.80}$$

with the color–flavor part,

$$\Delta_{CF}(x)_{\alpha i,\beta j} = \sum_{I=1}^{3} \sum_{\mathbf{q}_I^a} \Delta(\mathbf{q}_I^a) e^{2i\mathbf{q}_I^a \cdot \mathbf{r}} \epsilon_{I\alpha\beta} \epsilon_{Iij}. \tag{10.81}$$

We have introduced notation that allows for the possibility of gap parameters $\Delta(\mathbf{q}_I^a)$ with different magnitudes for different I *and for different a*. In fact, we shall only consider circumstances in which $\Delta(\mathbf{q}_I^a) = \Delta_I$, as in (10.71).

The full Lagrangian, given by the sum of (10.72) and (10.78), is then quadratic and can be written very simply upon introducing two-component Nambu–Gorkov spinors, in terms of which

$$\mathcal{L} = \frac{1}{2} \bar{\Psi} \begin{pmatrix} i\partial\!\!\!/ + \mu\!\!\!/ & \Delta(x) \\ \bar{\Delta}(x) & (i\partial\!\!\!/ - \mu\!\!\!/)^T \end{pmatrix} \Psi. \tag{10.82}$$

Here, $\mu\!\!\!/ \equiv \mu\gamma_0$ and μ is the matrix (10.73).

The propagator corresponding to the Lagrangian (10.82) is given by,

$$\begin{aligned} \langle \Psi(x) \bar{\Psi}(x') \rangle &= \begin{pmatrix} \langle \psi(x) \bar{\psi}(x') \rangle & \langle \psi(x) \psi^T(x') \rangle \\ \langle \bar{\psi}^T(x) \bar{\psi}(x') \rangle & \langle \bar{\psi}^T(x) \psi^T(x') \rangle \end{pmatrix} \\ &= \begin{pmatrix} iG(x,x') & iF(x,x') \\ i\bar{F}(x,x') & i\bar{G}(x,x') \end{pmatrix}, \end{aligned} \tag{10.83}$$

where G and $\bar{G}$ are the "normal" components of the propagator and F and $\bar{F}$ are the "anomalous" components. They satisfy the coupled differential equations,

$$\begin{aligned} &\begin{pmatrix} i\partial\!\!\!/ + \mu\!\!\!/ & \Delta(x) \\ \bar{\Delta}(x) & (i\partial\!\!\!/ - \mu\!\!\!/)^T \end{pmatrix} \begin{pmatrix} G(x,x') & F(x,x') \\ \bar{F}(x,x') & \bar{G}(x,x') \end{pmatrix} \\ &\qquad = \begin{pmatrix} 1 & 0 \\ 0 & 1 \end{pmatrix} \delta^{(4)}(x - x'). \end{aligned} \tag{10.84}$$

We can now rewrite (10.79) as

$$\begin{aligned} \Delta(x) &= \frac{3i}{4} \lambda \Gamma^{A\nu} F(x,x) (\Gamma_{A\nu})^T \\ \bar{\Delta}(x) &= \frac{3i}{4} \lambda (\Gamma^{A\nu})^T \bar{F}(x,x) \Gamma_{A\nu}, \end{aligned} \tag{10.85}$$

either one of which is the self-consistency equation, or gap equation, that we must solve.

Without further approximation, (10.85) is not tractable. It yields an infinite set of coupled gap equations, one for each $\Delta(\mathbf{q}_I^a)$, because without further approximation it is not consistent to choose finite sets $\{\mathbf{q}_I\}$. When several plane waves are present in the condensate, they induce an infinite tower of higher momentum condensates (Bowers and Rajagopal, 2002). In the next subsection, we shall make a Ginzburg–Landau (i.e. small-Δ) approximation which eliminates these higher harmonics.

Of course, an even more dramatic simplification is obtained if we set all the wavevectors $\mathbf{q}_I^a$ to zero. Still, even in this case obtaining the general solution with $M_s \neq 0$ and $\Delta_1 \neq \Delta_2 \neq \Delta_3$ is somewhat involved (Alford et al., 2004b, 2005c; Fukushima et al., 2005). We shall not present the resulting analysis of the CFL→gCFL transition and the gCFL phase here. The free energies of these phases are depicted in Fig. 10.3, and their gap parameters are depicted below in Fig. 10.8.

If we simplify even further, by setting $M_s = 0$ and $\Delta_1 = \Delta_2 = \Delta_3 \equiv \Delta_{\rm CFL}$, the gap equation determining the CFL gap parameter $\Delta_{\rm CFL}$ can then be evaluated analytically, yielding (Bowers and Rajagopal, 2002)

$$\Delta_{\rm CFL} = 2^{\frac{2}{3}}\,\Lambda\,\exp\left[-\frac{\pi^2}{2\mu^2\lambda}\right]. \tag{10.86}$$

We shall see below that in the limit in which $\Delta \ll \Delta_{\rm CFL}, \delta\mu \ll \mu$, all results for the myriad possible crystalline phases can be expressed in terms of $\Delta_{\rm CFL}$; neither λ nor Λ shall appear. This reflects the fact that in this limit the physics of interest is dominated by quarks near the Fermi surfaces, not near Λ, and so once $\Delta_{\rm CFL}$ is used as the parameter describing the strength of the attraction between quarks, Λ is no longer visible; the cutoff Λ only appears in the relation between $\Delta_{\rm CFL}$ and λ, not in any comparison among different possible paired phases. We are using the NJL model in a specific, limited, fashion in which it serves as a two-parameter model allowing the comparison among different possible paired phases at a given $\Delta_{\rm CFL}$ and M_s. NJL models have also been employed to estimate the value of $\Delta_{\rm CFL}$ at a given μ (Alford et al., 1998, 1999b; Berges and Rajagopal, 1999; Carter and Diakonov, 1999; Rajagopal and Wilczek, 2000; Rapp et al., 1998); doing so requires normalization of the four-fermion interaction by calculating some zero density quantity like the vacuum chiral condensate, and in so doing introduces a dependence on the cutoff Λ. Such mean-field NJL analyses are important complements to extrapolation down from an analysis that is rigorous at high density and hence weak coupling, described in Section 10.4, and give us confidence that we understand the magnitude of $\Delta_{\rm CFL} \sim 10 - 100\,\text{MeV}$. This estimate receives further support from the lattice-NJL calculation of (Hands and Walters, 2004) which finds diquark condensation and a $\sim 60\,\text{MeV}$ gap in an NJL model whose parameters are normalized via calculation of f_π, m_π and a constituent quark mass in vacuum. With these as inputs, Δ is then calculated on the lattice, i.e. without making a mean-field approximation. With an understanding of its magnitude in hand, we shall treat $\Delta_{\rm CFL}$ as a parameter, thus making our results insensitive to Λ.

We shall focus below on the use of the NJL model that we have introduced to analyze and compare different possible crystalline phases, comparing their free energies to that of the CFL phase as a benchmark. The free energy of the 2SC phase is easily calculable in the same model, and the free energies of the unstable gapless CFL and gapless 2SC phases can also be obtained (Alford et al., 2005c). These free energies are all shown in Fig. 10.3. The free energies of phases with various patterns of single-flavor pairing have also been calculated in the same model (Alford et al., 2003). The NJL model is not a natural starting point for an analysis of the kaon condensate in the CFL-K^0 phase, but with considerable effort this has been accomplished in (Buballa, 2005b; Forbes, 2005; Kleinhaus et al., 2007; Warringa, 2006). The curCFL-K^0 phase of Sections 10.3.4 and 10.5.7, in which the K^0-condensate carries a current, has not been analyzed in an NJL model. But, because both the CFL-K^0 and curCFL-K^0 phases are continuously connected to the CFL phase, they can both be analyzed in a model-independent fashion using the effective field theory described in Section 10.5. The CFL-K^0 and curCFL-K^0 curves in Fig. 10.3 were obtained as described in Section 10.5. It remains a challenge for future work to do a calculation in which both curCFL-K^0 and crystalline phases are possible, allowing a direct comparison of their free energies within a single calculation and a study of whether they are distinct as current results seem to suggest, or are instead different limits of some more general inhomogeneous color superconducting phase.

10.6.2 Crystalline phases

Crystalline color superconductivity (Alford et al., 2001a; Bowers et al., 2001; Bowers and Rajagopal, 2002; Casalbuoni et al., 2006, 2002a, 2005a, 2001, 2002c; Casalbuoni and Nardulli, 2004; Casalbuoni et al., 2003, 2004; Ciminale et al., 2006; Giannakis et al., 2002; Kundu and Rajagopal, 2002; Leibovich et al., 2001; Mannarelli et al., 2006) naturally permits pairing between quarks living at split Fermi surfaces by allowing Cooper pairs with non-zero net momentum. In three-flavor quark matter, this allows pairing to occur even with the Fermi surfaces split in the free-energetically optimal way as in the absence of pairing, meaning that neutral crystalline phases are obtained in three-flavor quark matter with the chemical potential matrix (10.73) simplified to $\mu = \delta^{\alpha\beta} \otimes \text{diag}\,(\mu_u, \mu_d, \mu_s)$ with the flavor chemical potentials given simply by (10.75) (Casalbuoni et al., 2005a; Mannarelli et al., 2006; Rajagopal and Sharma, 2006b), up to higher-order corrections that have been investigated in (Casalbuoni et al., 2006). This is the origin of the advantage that crystalline color superconducting phases have over the CFL and gCFL phases at large values of the splitting $\delta\mu$. For example, by allowing u quarks with momentum $\mathbf{p} + \mathbf{q}_3$ to pair with d quarks with momentum $-\mathbf{p} + \mathbf{q}_3$, for any $\mathbf{p}$, we can pair u and d quarks along rings on their respective Fermi surfaces. In coordinate space, this corresponds to a condensate of the form $\langle ud \rangle \sim \Delta_3 \exp\big(2i\mathbf{q}_3 \cdot \mathbf{r}\big)$. The net free energy gained due to pairing is then a balance between increasing $|\mathbf{q}_3|$ yielding pairing on larger rings while exacting a greater kinetic energy cost. The optimum choice turns out to be $|\mathbf{q}_3| = \eta\delta\mu_3$ with $\eta = 1.1997$, corresponding to pairing rings on the Fermi surfaces with opening angle $67.1°$ (Alford et al., 2001a). Pairing with only a single $\mathbf{q}_3$ is disadvantaged because the only quarks on each Fermi surface that can then pair are those lying on a single ring. This disadvantage can be overcome in two ways. First, increasing Δ widens the pairing rings on the Fermi surfaces into pairing bands which fill in, forming pairing caps, at large enough Δ (Mannarelli et al., 2006). Second, it is possible to cover larger areas of the Fermi surfaces by allowing Cooper pairs with the same $|\mathbf{q}_3|$ but various $\hat{\mathbf{q}}_3$, yielding $\langle ud \rangle \sim \Delta_3 \sum_{\mathbf{q}_3^a} \exp\big(2\,i\,\mathbf{q}_3^a \cdot \mathbf{r}\big)$ with the $\mathbf{q}_3^a$ chosen from some specified set $\{\mathbf{q}_3^1, \mathbf{q}_3^2, \mathbf{q}_3^3, \ldots\} \equiv \{\mathbf{q}_3\}$. This is a condensate modulated in position space in some crystalline pattern, with the crystal structure defined by $\{\mathbf{q}_3\}$. In this two-flavor context, a Ginzburg–Landau analysis reveals that the best $\{\mathbf{q}_3\}$ contains eight vectors pointing at the corners of a cube, say in the $(\pm 1, \pm 1, \pm 1)$ directions in momentum space, yielding a face-centered cubic structure in position space (Bowers and Rajagopal, 2002).

This subsection describes the analysis of three-flavor crystalline phases in (Rajagopal and Sharma, 2006a). We use the ansatz given by (10.80) and (10.81) for the three-flavor crystalline color superconducting condensate. This is antisymmetric in color (α, β), spin, and flavor (i, j) indices and is a generalization of the CFL condensate to crystalline color superconductivity. We set $\Delta_1 = 0$, neglecting $\langle ds \rangle$ pairing because the d and s Fermi surfaces are twice as far apart from each other as each is from the intervening u Fermi surface. Hence, I can be taken to run over 2 and 3 only. $\{\mathbf{q}_2\}$ and $\{\mathbf{q}_3\}$ define the crystal structures of the $\langle us \rangle$ and $\langle ud \rangle$ condensates respectively. We only consider crystal structures in which all the vectors in $\{\mathbf{q}_2\}$ are equivalent to each other in the sense that any one can be transformed into any other by a symmetry operation of $\{\mathbf{q}_2\}$ and the same for $\{\mathbf{q}_3\}$. This justifies our simplifying assumption that the $\langle us \rangle$ and $\langle ud \rangle$ condensates are each specified by a single gap parameter (Δ_2 and Δ_3 respectively), avoiding having to introduce one gap parameter per $\mathbf{q}$. Furthermore, we only consider crystal structures which are exchange symmetric, meaning that $\{\mathbf{q}_2\}$ and $\{\mathbf{q}_3\}$ can be exchanged by some combination of rigid rotations and reflections applied simultaneously to all of the vectors in both sets. This simplification, together with $\delta\mu_2 = \delta\mu_3$ (an approximation corrected only at order M_s^4/μ^3), guarantees that we find solutions with $\Delta_2 = \Delta_3$.

We analyze and compare candidate crystal structures by evaluating the free energy $\Omega(\Delta_2, \Delta_3)$ for each crystal structure in a Ginzburg–Landau expansion in powers of the Δs. This approximation is controlled if $\Delta_2, \Delta_3 \ll \Delta_{\mathrm{CFL}}, \delta\mu$, with Δ_{CFL} the gap parameter in the CFL phase at $M_s^2/\mu = 0$. The terms in the Ginzburg–Landau expansion must respect the global $U(1)$ symmetry for each flavor, meaning that each Δ_I can only appear in the combination $|\Delta_I|^2$. (The $U(1)$ symmetries are spontaneously broken by the condensate, but not explicitly broken.) Therefore, $\Omega(\Delta_2, \Delta_3)$ is given to sextic order by,

$$
\begin{aligned}
\Omega(\Delta_2, \Delta_3) = \frac{2\mu^2}{\pi^2}\bigg[& P_2\alpha_2|\Delta_2|^2 + P_3\alpha_3|\Delta_3|^2 \\
& + \frac{1}{2}\Big(\beta_2|\Delta_2|^4 + \beta_3|\Delta_3|^4 + \beta_{32}|\Delta_2|^2|\Delta_3|^2\Big) \\
& + \frac{1}{3}\Big(\gamma_2|\Delta_2|^6 + \gamma_3|\Delta_3|^6 \\
& + \gamma_{322}|\Delta_3|^2|\Delta_2|^4 + \gamma_{233}|\Delta_3|^4|\Delta_2|^2\Big)\bigg],
\end{aligned}
\tag{10.87}
$$

where we have chosen notation consistent with that used in the two-flavor study of (Bowers and Rajagopal, 2002), which arises as a special case of (10.87) if we take Δ_2 or Δ_3 to be zero. P_I is the number of vectors in the set $\{\mathbf{q}_I\}$. The form of the Ginzburg–Landau expansion (10.87) is model independent, whereas the expressions for the coefficients α_I, β_I, β_{IJ}, γ_I, and γ_{IJJ} for a specific crystal structure are model dependent. We calculate these in the NJL model described in Section 10.6.1. For exchange symmetric crystal structures, $\alpha_2 = \alpha_3 \equiv \alpha$, $\beta_2 = \beta_3 \equiv \beta$, $\gamma_2 = \gamma_3 \equiv \gamma$ and $\gamma_{233} = \gamma_{322}$.

Because setting one of the Δ_I to zero reduces the problem to one with two-flavor pairing only, we can obtain α, β, and γ via applying the two-flavor analysis described in (Bowers and Rajagopal, 2002) to either $\{\mathbf{q}_2\}$ or $\{\mathbf{q}_3\}$ separately. Using α as an example, we learn that

$$\alpha_I = \alpha(q_I, \delta\mu_I) = -1 + \frac{\delta\mu_I}{2q_I}\log\left(\frac{q_I + \delta\mu_I}{q_I - \delta\mu_I}\right) - \frac{1}{2}\log\left(\frac{\Delta_{2SC}^2}{4(q_I^2 - \delta\mu_I^2)}\right). \tag{10.88}$$

Here, $q_I \equiv |\mathbf{q}_I|$ and Δ_{2SC} is the gap parameter for the 2SC (2-flavor, 2-color) BCS pairing obtained with $\delta\mu_I = 0$ and Δ_I non-zero with the other two gap parameters set to zero. Assuming that $\Delta_{CFL} \ll \mu$, the 2SC gap parameter is given by $\Delta_{2SC} = 2^{\frac{1}{3}}\Delta_{CFL}$ (Schäfer, 2000a), see Section 10.4. In the Ginzburg–Landau approximation, in which the Δ_I are assumed small, we must first minimize the quadratic contribution to the free energy, and only then investigate the quartic and sextic contributions. Minimizing α_I fixes the length of all the vectors in the set $\{\mathbf{q}_I\}$, and eliminates the possibility of waves at higher harmonics, yielding $q_I = \eta\,\delta\mu_I$ with $\eta = 1.1997$ the solution to $\frac{1}{2\eta}\log\left[(\eta+1)/(\eta-1)\right] = 1$ (Alford et al., 2001a). Upon setting $q_I = \eta\,\delta\mu_I$, (10.88) becomes

$$\alpha_I(\delta\mu_I) = -\frac{1}{2}\log\left(\frac{\Delta_{2SC}^2}{4\delta\mu_I^2(\eta^2 - 1)}\right). \tag{10.89}$$

Once the q_I have been fixed, the only dimensionful quantities on which the quartic and sextic coefficients can depend are the $\delta\mu_I$ (Bowers and Rajagopal, 2002; Rajagopal and Sharma, 2006b), meaning that for exchange symmetric crystal structures and with $\delta\mu_2 = \delta\mu_3 = \delta\mu$ we have $\beta = \bar{\beta}/\delta\mu^2$, $\beta_{32} = \bar{\beta}_{32}/\delta\mu^2$, $\gamma = \bar{\gamma}/\delta\mu^4$ and $\gamma_{322} = \bar{\gamma}_{322}/\delta\mu^4$ where the barred quantities are dimensionless numbers which depend only on $\{\hat{\mathbf{q}}_2\}$ and $\{\hat{\mathbf{q}}_3\}$ that must be evaluated for each crystal structure. Doing so requires evaluating one-loop Feynman diagrams with 4 or 6 insertions of Δ_Is. Each insertion of Δ_I (Δ_I^*) adds (subtracts) momentum $2\mathbf{q}_I^a$ for some a. The vector sum of all these external momenta inserted into a given one-loop diagram must vanish, meaning that the calculation consists of a bookkeeping task (determining which combinations of 4 or 6 $\mathbf{q}_I^a$s selected from the sets $\{\mathbf{q}_I\}$ satisfy this momentum-conservation constraint) that grows rapidly in complexity with the complexity of the crystal structure, and a loop integration that is non-trivial because the momentum in the propagator changes after each insertion. In (Rajagopal and Sharma, 2006b), this calculation is carried out explicitly for 11 crystal structures in the mean-field NJL model of Section 10.6.1 upon making the weak coupling ($\Delta_{\rm CFL}$ and $\delta\mu$ both much less than μ) approximation. Note that in this approximation neither the NJL cutoff nor the NJL coupling constant appear in any quartic or higher Ginzburg–Landau coefficient, and as we have seen above they appear in α only within $\Delta_{\rm CFL}$. Hence, the details of the model do not matter as long as one thinks of $\Delta_{\rm CFL}$ as a parameter, kept $\ll \mu$.

It is easy to show that for exchange symmetric crystal structures any extrema of $\Omega(\Delta_2, \Delta_3)$ in (Δ_2, Δ_3)-space must either have $\Delta_2 = \Delta_3 = \Delta$, or have one of Δ_2 and Δ_3 vanishing (Rajagopal and Sharma, 2006b). It is also possible to show that the three-flavor crystalline phases with $\Delta_2 = \Delta_3 = \Delta$ are electrically neutral, whereas two-flavor solutions in which only one of the Δs is non-zero are not (Rajagopal and Sharma, 2006b). We therefore analyze only solutions with $\Delta_2 = \Delta_3 = \Delta$. We find that $\Omega(\Delta, \Delta)$ is positive for large Δ for all the crystal structures that have been investigated to date (Rajagopal and Sharma, 2006b).[3] This allows us to minimize $\Omega(\Delta, \Delta)$ with respect to Δ, thus evaluating Δ and Ω.

[3]This is in marked contrast with what happens with only two flavors (and upon ignoring the requirement of neutrality.) in that context, many crystal structures have negative γ and hence sextic-order free energies that are unbounded from below (Bowers and Rajagopal, 2002).

We begin with the simplest three-flavor "crystal" structure in which $\{\mathbf{q}_2\}$ and $\{\mathbf{q}_3\}$ each contain only a single vector, making the $\langle us\rangle$ and $\langle ud\rangle$ condensates each a single plane wave (Casalbuoni et al., 2005a). We call this the 2PW phase. Unlike in the more realistic crystalline phases we describe below, in this "crystal" the magnitudes of the $\langle ud\rangle$ and $\langle us\rangle$ condensates are unmodulated. This simple condensate nevertheless yields a qualitative lesson which proves helpful in winnowing the space of multiple plane wave crystal structures (Rajagopal and Sharma, 2006b). For this simple "crystal" structure, all the coefficients in the Ginzburg–Landau free energy can be evaluated analytically (Casalbuoni et al., 2005a; Mannarelli et al., 2006; Rajagopal and Sharma, 2006b). The terms that occur in the three-flavor case but not in the two-flavor case, namely $\bar{\beta}_{32}$ and $\bar{\gamma}_{322}$, describe the interaction between the two condensates, and depend on the angle ϕ between $\mathbf{q}_2$ and $\mathbf{q}_3$. For any angle ϕ, both $\bar{\beta}_{32}$ and $\bar{\gamma}_{322}$ are positive. And, both increase monotonically with ϕ and diverge as $\phi \to \pi$. This divergence tells us that choosing $\mathbf{q}_2$ and $\mathbf{q}_3$ precisely antiparallel exacts an infinite free energy price in the combined Ginzburg–Landau and weak-coupling limit in which $\Delta \ll \delta\mu, \Delta_{\rm CFL} \ll \mu$, meaning that in this limit if we choose $\phi = \pi$ we find $\Delta = 0$. Away from the Ginzburg–Landau limit, when the pairing rings on the Fermi surfaces widen into bands, choosing $\phi = \pi$ exacts a finite price, meaning that Δ is non-zero but smaller than that for any other choice of ϕ. The high cost of choosing $\mathbf{q}_2$ and $\mathbf{q}_3$ precisely antiparallel can be understood qualitatively as arising from the fact that in this case the ring of states on the u-quark Fermi surface that "want to" pair with d-quarks coincides precisely with the ring that "wants to" pair with s-quarks (Mannarelli et al., 2006). This simple two plane wave ansatz has been analyzed upon making the weak-coupling approximation but without making the Ginzburg–Landau approximation (Mannarelli et al., 2006). All the qualitative lessons learned from the Ginzburg–Landau approximation remain valid and we learn further that the Ginzburg–Landau approximation always underestimates Δ (Mannarelli et al., 2006).

The analysis of the simple two plane wave "crystal" structure, together with the observation that in more complicated crystal structures with more than one vector in $\{\mathbf{q}_2\}$ and $\{\mathbf{q}_3\}$ the Ginzburg–Landau coefficient β_{32} (γ_{322}) is given in whole (in part) by a sum of many two plane wave contributions, yields one of two rules for constructing favorable crystal structures for three-flavor crystalline color superconductivity (Rajagopal and Sharma, 2006b): $\{\mathbf{q}_2\}$ and $\{\mathbf{q}_3\}$ should be rotated with respect to each other in a way that best keeps vectors in one set away from the antipodes of vectors in the other set. The second rule is that the sets $\{\mathbf{q}_2\}$ and $\{\mathbf{q}_3\}$ should each be chosen to yield crystal structures which, seen as separate two-flavor crystalline phases, are as favorable as possible. The 11 crystal structures analyzed in (Rajagopal and Sharma, 2006b) allow one to make several pairwise comparisons that test these two rules. These considerations, together with explicit calculations, indicate that two structures, which we denote "2Cube45z" and "CubeX", are particularly favorable.

In the 2Cube45z crystal, $\{\mathbf{q}_2\}$ and $\{\mathbf{q}_3\}$ each contain eight vectors pointing at the corners of a cube. If we orient $\{\mathbf{q}_2\}$ so that its vectors point in the $(\pm 1, \pm 1, \pm 1)$ directions in momentum space, then $\{\mathbf{q}_3\}$ is rotated relative to $\{\mathbf{q}_2\}$ by 45° about the z-axis. In this crystal structure, the $\langle ud \rangle$ and $\langle us \rangle$ condensates are each given by the most favored two-flavor crystal structure (Bowers and Rajagopal, 2002). The relative rotation maximizes the separation between any vector in $\{\mathbf{q}_2\}$ and the nearest antipodes of a vector in $\{\mathbf{q}_3\}$.

We arrive at the CubeX structure by reducing the number of vectors in $\{\mathbf{q}_2\}$ and $\{\mathbf{q}_3\}$. This worsens the two-flavor free energy of each condensate separately, but allows vectors in $\{\mathbf{q}_2\}$ to be kept farther away from the antipodes of vectors in $\{\mathbf{q}_3\}$. We have not analyzed all structures obtainable in this way, but we have found one and only one which has a condensation energy comparable to that of the 2Cube45z structure. In the CubeX structure, $\{\mathbf{q}_2\}$ and $\{\mathbf{q}_3\}$ each contain four vectors forming a rectangle. The eight vectors together point toward the corners of a cube. The two rectangles intersect to look like an "X" if viewed end-on. The color, flavor, and position space dependence of the CubeX condensate is given by,

$$\begin{aligned} &\epsilon_{2\alpha\beta}\epsilon_{2ij}\left[\cos\frac{2\pi}{a}(x+y+z)+\cos\frac{2\pi}{a}(-x-y+z)\right] \\ &+\epsilon_{3\alpha\beta}\epsilon_{3ij}\left[\cos\frac{2\pi}{a}(-x+y+z)+\cos\frac{2\pi}{a}(x-y+z)\right], \end{aligned} \tag{10.90}$$

where $a=\sqrt{3}\pi/q=4.536/\delta\mu=36.29\mu/M_s^2$ is the lattice spacing. For example, with $M_s^2/\mu=100, 150, 200\,\text{MeV}$ the lattice spacing is $a=72, 48, 36$ fm. We depict this condensate in Fig. 10.7.

In Figs. 10.8 and 10.3, we plot Δ and Ω versus M_s^2/μ for the most favorable crystal structures that we have found, namely the CubeX and 2Cube45z structures described above. We have taken the CFL gap parameter $\Delta_{\text{CFL}}=25$ MeV in these figures, but they can easily be rescaled to any value of $\Delta_{\text{CFL}} \ll \mu$ (Rajagopal and Sharma, 2006b): if the Δ and M_s^2/μ axes are rescaled by Δ_{CFL} and the energy axis is rescaled by Δ_{CFL}^2. Fig. 10.8 shows that the gap parameters are large enough that the Ginzburg–Landau approximation

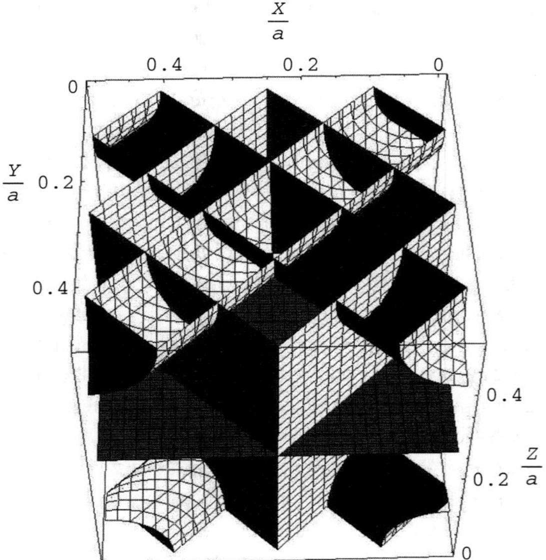

Fig. 10.7 The CubeX crystal structure of Eq. (10.90). The figure extends from 0 to $a/2$ in the x, y, and z directions. Both $\Delta_2(\mathbf{r})$ and $\Delta_3(\mathbf{r})$ vanish at the horizontal plane. $\Delta_2(\mathbf{r})$ vanishes on the darker vertical planes, and $\Delta_3(\mathbf{r})$ vanishes on the lighter vertical planes. On the upper (lower) dark cylinders and the lower (upper) two small corners of dark cylinders, $\Delta_2(\mathbf{r}) = +3.3\Delta$ $(\Delta_2(\mathbf{r}) = -3.3\Delta)$. On the upper (lower) lighter cylinders and the lower (upper) two small corners of lighter cylinders, $\Delta_3(\mathbf{r}) = -3.3\Delta$ $(\Delta_3(\mathbf{r}) = +3.3\Delta)$. The largest value of $|\Delta_I(\mathbf{r})|$ is 4Δ, occurring along lines at the centers of the cylinders. The lattice spacing is a when one takes into account the signs of the condensates; if one looks only at $|\Delta_I(\mathbf{r})|$, the lattice spacing is $a/2$.

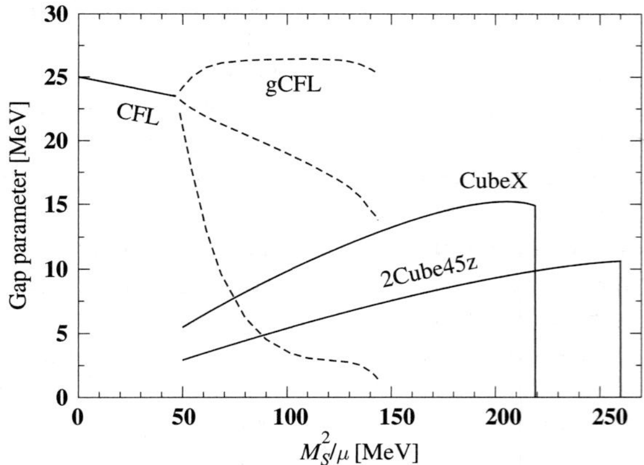

Fig. 10.8 Gap parameter Δ versus M_s^2/μ for: the CFL gap parameter (set to 25 MeV at $M_s^2/\mu = 0$), the three gap parameters $\Delta_1 < \Delta_2 < \Delta_3$ describing $\langle ds \rangle$, $\langle us \rangle$, and $\langle ud \rangle$ pairing in the gCFL phase, and the gap parameters in the crystalline color superconducting phases with CubeX and 2Cube45z crystal structures. Increasing M_s^2/μ corresponds to decreasing density.

is at the edge of its domain of reliability. However, results obtained for the simpler 2PW crystal structures suggest that the Ginzburg–Landau calculation underestimates Δ and the condensation energy and that, even when it breaks down, it is a good qualitative guide to the favorable structure (Mannarelli et al., 2006). We therefore trust the result, evident in Fig. 10.3, that these crystalline phases are both impressively robust, with one or other of them favored over a wide swath of M_s^2/μ and hence density. We do not trust the Ginzburg–Landau calculation to discriminate between these two structures, particularly given that although we have a qualitative understanding of why these two are favorable, we have no qualitative argument for why one should be favored over the other. Regardless of this, the 2Cube45z and CubeX crystalline phases together make the case that three-flavor crystalline color superconducting phases are the ground state of cold quark matter over a wide range of densities. If even better crystal structures can be found, this will only further strengthen this case.

Fig. 10.3 shows that over most of the range of M_s^2/μ where it was once considered a possibility, the gCFL phase can be replaced by a *much* more favorable three-flavor crystalline color superconducting phase. We find that the two most favorable crystal structures have large condensation energies, easily 1/3 to 1/2 of that in the CFL phase with $M_s = 0$, which is $3\Delta_{\text{CFL}}^2\mu^2/\pi^2$. This is at first surprising, given that the only quarks that pair are those lying on rings on the Fermi surfaces, whereas in the CFL phase with $M_s = 0$ pairing occurs over the entire u, d, and s Fermi surfaces. It can be understood qualitatively to a degree once we recall that there are in fact many rings, and note that as Δ increases, the pairing rings spread into bands on the Fermi surfaces, and for Δ as large as what we find to be favored these bands have expanded and filled in, becoming many "polar caps" on the Fermi surfaces (Mannarelli et al., 2006). In addition to being free-energetically favorable, these crystalline phases are, as far as we know, stable: they do not suffer from chromomagnetic instability (Ciminale et al., 2006; Gatto and Ruggieri, 2007; Giannakis et al., 2005; Giannakis and Ren, 2005b) and they are also stable with respect to kaon condensation (Anglani et al., 2007). In simplified analog contexts, it has even been possible to trace the path in configuration space from the unstable gapless phase (analog of gCFL) downward in free energy to the stable crystalline phase (Fukushima, 2006; Fukushima and Iida, 2007).

Figure 10.3 also shows that it is hard to find a crystalline phase with lower free energy than the gCFL phase at the lower values of M_s^2/μ (highest densities) within the "gCFL window". At these densities, however, the calculations described in Section 10.5 demonstrate that the gCFL phase is superseded by the stable CFL-K^0 and curCFL-K^0 phases, as shown in Fig. 10.3.

The three-flavor crystalline color superconducting phases with CubeX and 2Cube45z crystal structures are the lowest free-energy phases that we know of, and hence candidates for the ground state of QCD, over a wide range of densities. Within the Ginzburg–Landau approximation to the NJL model that we have employed, one or other is favored over the CFL, gCFL, and unpaired phases for $2.9\Delta_{\text{CFL}} < M_s^2/\mu < 10.4\Delta_{\text{CFL}}$, as shown in Fig. 10.3. For $\Delta_{\text{CFL}} = 25\,\text{MeV}$ and $M_s = 250\,\text{MeV}$, this translates to $240\,\text{MeV} < \mu < 847\,\text{MeV}$. With these choices of parameters, the lower part of this range of μ (higher part of the range of M_s^2/μ) is certainly superseded by nuclear matter. And, the high end of this range extends beyond the $\mu \sim 500\,\text{MeV}$ characteristic of the quark matter at the densities expected at the center of neutron stars. This qualitative feature persists in the analysis of (Ippolito et al., 2007) in which M_s is solved for rather than taken as a parameter. If neutron stars do have quark matter cores, then, it is possible that the entire quark matter core could be in a crystalline color superconducting phase.

10.6.3 Rigidity of crystalline color superconducting quark matter

The crystalline phases of color superconducting quark matter that we have described in the previous subsection are unique among all forms of dense matter that may arise within neutron star cores in one respect: they are rigid (Mannarelli et al., 2007). They are not solids in the usual sense: the quarks are not fixed in place at the vertices of some crystal structure. Instead, in fact, these phases are superfluid since the condensates all spontaneously break the $U(1)_B$ symmetry corresponding to quark number. We shall always write the condensates as real. This choice of overall phase breaks $U(1)_B$, and spatial gradients in this phase correspond to supercurrents. And yet, we shall see that crystalline color superconductors are rigid solids with large shear moduli. The diquark condensate, although spatially inhomogeneous, can carry supercurrents (Alford et al., 2001a; Mannarelli et al., 2007). It is the spatial modulation of the gap parameter that breaks translation invariance, as depicted for the CubeX phase in Fig. 10.7, and it is this pattern of modulation that is rigid.[4] This novel form of rigidity may sound tenuous upon first hearing, but we shall present the effective Lagrangian that describes the phonons in the CubeX and 2Cube45z crystalline phases, whose lowest-order coefficients have been calculated in the NJL model that we are employing (Mannarelli et al., 2007). We shall then extract the shear moduli from the phonon effective action, quantifying the rigidity and indicating the presence of transverse phonons. The fact that the crystalline phases are simultaneously rigid and superfluid means that their presence within neutron star cores has potentially observable consequences, as we shall describe in Section 10.8.5.

[4]Supersolids (Andreev and Lifshitz, 1969; Chester, 1970; Kim and Chan, 2004a,b; Leggett, 1970; Son, 2005) are another example of rigid superfluids, but they differ from crystalline color superconductors in that they are rigid by virtue of the presence of an underlying lattice of atoms.

The shear moduli of a crystal may be extracted from the effective Lagrangian that describes phonons in the crystal, namely space- and time-varying displacements of the crystalline pattern. Phonons in two-flavor crystalline phases were first investigated in (Casalbuoni et al., 2002a,c). In the present context, we introduce displacement fields for the $\langle ud\rangle$, $\langle us\rangle$, and $\langle ds\rangle$ condensates by making the replacement,

$$\Delta_I \sum_{\mathbf{q}_I^a} e^{2i\mathbf{q}_I^a\cdot\mathbf{r}} \rightarrow \Delta_I \sum_{\mathbf{q}_I^a} e^{2i\mathbf{q}_I^a\cdot(\mathbf{r}-\mathbf{u}_I(\mathbf{r}))} \tag{10.91}$$

in (10.81). One way to obtain an effective action describing the dynamics of the displacement fields $\mathbf{u}_I(\mathbf{r})$, including both its form and the values of its coefficients within the NJL model that we are employing, is to begin with the NJL model of Section 10.6.1 but with (10.91) and integrate out the fermion fields. After a lengthy calculation (Mannarelli et al., 2007), this yields

$$\begin{aligned} S[\mathbf{u}] = \frac{1}{2}\int d^4x \sum_I \kappa_I \\ \times \Bigg[\left(\sum_{\mathbf{q}_I^a}(\hat{q}_I^a)^m(\hat{q}_I^a)^n\right)(\partial_0 u_I^m)(\partial_0 u_I^n) \\ - \left(\sum_{\mathbf{q}_I^a}(\hat{q}_I^a)^m(\hat{q}_I^a)^v(\hat{q}_I^a)^n(\hat{q}_I^a)^w\right)(\partial_v u_I^m)(\partial_w u_I^n)\Bigg], \end{aligned} \tag{10.92}$$

where m, n, v, and w are spatial indices running over x, y, and z and where we have defined

$$\kappa_I \equiv \frac{2\mu^2|\Delta_I|^2\eta^2}{\pi^2(\eta^2-1)}. \tag{10.93}$$

Upon setting $\Delta_1 = 0$ and $\Delta_2 = \Delta_3 = \Delta$,

$$\kappa_2 = \kappa_3 \equiv \kappa = \frac{2\mu^2|\Delta|^2\eta^2}{\pi^2(\eta^2-1)} \simeq 0.664\,\mu^2|\Delta^2|. \tag{10.94}$$

$S[\mathbf{u}]$ is the low energy effective action for phonons in any crystalline color superconducting phase, valid to second order in derivatives, to second order in the gap parameters Δ_I, and to second order in the phonon fields $\mathbf{u}_I$. Because we are interested in long wavelength, small amplitude, phonon excitations, expanding to second order in derivatives and in the phonon fields is satisfactory. More complicated terms will arise at higher order, for example terms that couple the different $\mathbf{u}_I$s, but it is legitimate to neglect these complications (Mannarelli et al., 2007). Extending this calculation to higher order in the Ginzburg–Landau approximation would be worthwhile, however, since as we saw in Section 10.6.2 this approximation is at the edge of its domain of reliability.

In order to extract the shear moduli, we need to compare the phonon effective action to the general theory of elastic media (Landau and Lifshitz, 1981), which requires rewriting (10.92) in terms of the strain tensor,

$$s_I^{mv} \equiv \frac{1}{2}\left(\frac{\partial u_I^m}{\partial x^v} + \frac{\partial u_I^v}{\partial x^m}\right). \tag{10.95}$$

The elastic modulus tensor λ_I^{mvnw} is the coefficient multiplying the $s_I^{mv} s_I^{nw}$ term in the effective action. We shall not extract the bulk modulus because the contribution from the crystalline condensate is completely overwhelmed by that proportional to μ^4 that arises in the absence of any pairing. On the other hand, the response to shear stress arises only because of the presence of the crystalline condensate. The shear modulus in the mv plane (i.e. the strain in that plane induced in response to stress in that plane) is given by

$$\nu_I^{mv} = \frac{1}{2}\lambda_I^{mvmv}, \tag{10.96}$$

where the indices m and v are not summed. (The expression for ν is this simple only when the only non-zero entries in λ^{mvnw} with $m \neq v$ are the λ^{mvmv} entries, as is the case for all of the crystals we consider.)

For a given crystal structure, upon evaluating the sums in (10.92) and then using the definition (10.95) we can extract expressions for the λ tensor and thence for the shear moduli. This analysis, described in detail in (Mannarelli et al., 2007), shows that in the CubeX phase,

$$\nu_2 = \frac{16}{9}\kappa \begin{pmatrix} 0\,0\,1 \\ 0\,0\,0 \\ 1\,0\,0 \end{pmatrix}, \quad \nu_3 = \frac{16}{9}\kappa \begin{pmatrix} 0\,0\,0 \\ 0\,0\,1 \\ 0\,1\,0 \end{pmatrix}, \tag{10.97}$$

while in the 2Cube45z phase,

$$\nu_2 = \frac{16}{9}\kappa \begin{pmatrix} 0 & 1 & 1 \\ 1 & 0 & 1 \\ 1 & 1 & 0 \end{pmatrix}, \quad \nu_3 = \frac{16}{9}\kappa \begin{pmatrix} 0 & 0 & 1 \\ 0 & 0 & 1 \\ 1 & 1 & 0 \end{pmatrix}. \tag{10.98}$$

We shall see in Section 10.8.5 that it is relevant to check that both these crystals have enough non-zero entries in their shear moduli ν_I that if there are rotational vortices pinned within them, a force seeking to move such a vortex is opposed by the rigidity of the crystal structure described by one or more of the non-zero entries in the ν_I. This is demonstrated in (Mannarelli et al., 2007).

We see that all the non-zero shear moduli of both the CubeX and 2Cube45z crystalline color superconducting phases turn out to take on the same value,

$$\nu_{\rm CQM} = \frac{16}{9}\kappa, \tag{10.99}$$

with κ defined by (10.94). Evaluating κ yields

$$\begin{aligned} \nu_{\rm CQM} &= 1.18\,\mu^2\Delta^2 \\ &= 2.47\,\frac{\rm MeV}{\rm fm^3}\left(\frac{\Delta}{10\,\rm MeV}\right)^2\left(\frac{\mu}{400\,\rm MeV}\right)^2 \end{aligned} \tag{10.100}$$

From (10.100) we first of all see that the shear modulus is in no way suppressed relative to the scale $\mu^2\Delta^2$ that could have been guessed on dimensional grounds. And, second, we discover that a quark matter core in a crystalline color superconducting phase is 20 to 1000 times more rigid than the crust of a conventional neutron star (Mannarelli et al., 2007; Strohmayer et al., 1991). Finally, see (Mannarelli et al., 2007) for the extraction of the phonon dispersion relations from the effective action (10.92). The transverse phonons, whose restoring force is provided by the shear modulus and which correspond to propagating ripples in a condensation pattern like that in Fig. 10.7, turn out to have direction-dependent velocities that are typically a substantial fraction of the speed of light, in the specific instances evaluated in (Mannarelli et al., 2007), given by $\sqrt{1/3}$ and $\sqrt{2/3}$. This is yet a third way of seeing that this superfluid phase of matter is indeed rigid.

10.7 Transport properties and neutrino processes

In Section 10.8 we shall discuss how the observation of neutron star properties constrains the phase structure of dense quark matter. A crucial ingredient in these analyses are the transport properties, as well as neutrino emissivities and opacities of different phases of quark matter.

Using the methods introduced in Section 10.5 it is possible to perform rigorous calculations of transport properties of the CFL phase. The results are parameter-free predictions of QCD at asymptotically large density, and rigorous consequences of QCD expressed in terms of a few phenomenological parameters ($f_\pi, m_\pi, \ldots$) at lower density.

In the case of other color superconducting phases we perform calculations using perturbative QCD or models of QCD. For many quantities the results depend mainly on the spectrum of quark modes, and not on details of the quark–quark interaction.

10.7.1 Viscosity and thermal conductivity

Viscosity and thermal conductivity determine the dissipated energy $\dot{E}$ in a fluid with non-zero gradients of the velocity $\mathbf{v}$ and the temperature T,

$$\dot{E} = -\frac{\eta}{2} \int d^3x \left(\partial_i v_j + \partial_j v_i - \frac{2}{3} \delta_{ij} \partial_k v_k \right)^2 \\ -\zeta \int d^3x \left(\partial_i v_i \right)^2 - \frac{\kappa}{T} \int d^3x \left(\partial_i T \right)^2. \tag{10.101}$$

The transport coefficients η, ζ, and κ are the shear and bulk viscosity and the thermal conductivity, respectively. Eq. (10.101) is strictly valid only for non-relativistic fluids. In the case of relativistic fluids there is an extra contribution to the dissipated energy which is proportional to κ and the gradient of μ (Landau and Lifshitz, 1987). In the following we shall also neglect effects from stresses in the superfluid flow relative to the normal flow and interpret v_i in Eq. (10.101) as the normal fluid velocity.

In neutron stars an important contribution to the bulk viscosity arises from electroweak effects. In a bulk compression mode the density changes periodically and electroweak interactions may not be sufficiently fast to re-establish weak equilibrium. Weak effects occur on the same time scale as the oscillation period of the neutron star and the frequency dependence of the bulk viscosity is important. We define

$$\zeta(\omega) = 2 \langle \dot{E} \rangle \left(\frac{V_0}{\delta V_0} \right)^2 \frac{1}{\omega^2}, \tag{10.102}$$

where ω is the oscillation frequency, $\langle \ldots \rangle$ is a time average, and $\delta V_0 / V_0$ is the fractional change in the volume. The coefficient ζ in Eq. (10.101) is the $\omega \to 0$ limit of $\zeta(\omega)$.

Let us first discuss the CFL and CFL-K^0 phases. In these cases, all quark modes are gapped and the relevant excitations are Goldstone bosons. At very low temperature, transport properties are dominated by the massless Goldstone boson φ associated with the breaking of the $U(1)_B$ symmetry. Using the results in Section 10.5.3, we can compute the mean free path l_φ of the φ due to $\varphi \leftrightarrow \varphi + \varphi$ and $\varphi + \varphi \leftrightarrow \varphi + \varphi$ scattering. Small-angle scattering contributions give rise to $l_\varphi \propto \mu^4 / T^5$ (Manuel et al., 2005) and $l_\varphi \simeq 1$ km at $T = 0.1$ MeV, while large-angle scattering contributions yield an even longer $l_\varphi \propto \mu^8 / T^9$ (Shovkovy and Ellis, 2002). The thermal conductivity κ due to the φ is given by (Shovkovy and Ellis, 2002),

$$\kappa = \frac{2\pi^2 T^3}{45 v^2} l_\varphi, \tag{10.103}$$

where ℓ_φ is the φ mean-free path between large-angle scatterings and v is the φ velocity from Eqs. (10.50) and (10.51). For temperatures below ~ 1 MeV the thermal conductivity is very large and macroscopic amounts of CFL matter are expected to be isothermal. The electric conductivity in CFL matter is dominated by thermal electrons and positrons and was estimated in (Shovkovy and Ellis, 2003).

At low temperatures, the shear viscosity of the CFL phase is dominated by the φ contribution, which was computed in (Manuel et al., 2005) and is given by,

$$\eta = 1.3 \times 10^{-4} \frac{\mu^8}{T^5}. \tag{10.104}$$

The bulk viscosity ζ vanishes in an exactly scale invariant system. For realistic quark masses the dominant source of scale breaking is the strange quark mass. The contribution from the process $\varphi \leftrightarrow \varphi + \varphi$ is (Manuel and Llanes-Estrada, 2007)

$$\zeta = 0.011 \frac{M_s^4}{T}. \tag{10.105}$$

We show this contribution in Fig. 10.9. The other contribution to the CFL and CFL-K^0 bulk viscosities presented in the figure comes from the process $K^0 \leftrightarrow \varphi + \varphi$ (Alford et al., 2007, 2008a). We observe that at $T \simeq 10$ MeV the bulk viscosity of CFL matter is comparable to that of unpaired quark matter. For $T < 1$ MeV, ζ is strongly suppressed; kaon condensation leads, despite the additional light Goldstone mode, to an even stronger suppression. The reason is a smaller phase space for the relevant process at small temperatures. Depending on the poorly known value for $\delta m \equiv m_{K^0} - \mu_{K^0}$, the pure φ contribution given in Eq. (10.105) may dominate over the $K^0 \leftrightarrow \varphi + \varphi$ reaction at low enough temperatures. However, for $T < 0.1$ MeV the φ mean free path is of the order of the size of the star, i.e. the system is in the collisionless rather than in the hydrodynamic regime, and the result ceases to be meaningful.

For unpaired, ultrarelativistic three-flavor quark matter, thermal and electric conductivity as well as shear viscosity have been computed in (Heiselberg and

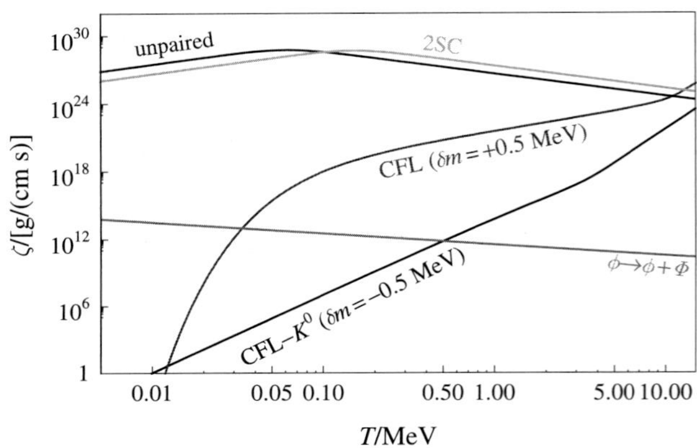

Fig. 10.9 Bulk viscosities as functions of temperature for an oscillation period $\tau = 2\pi/\omega = 1$ ms (and $\mu = 400$ MeV, $M_s = 120$ MeV). CFL/CFL-K^0 phases: contribution from the process $K^0 \leftrightarrow \varphi + \varphi$ for negative (CFL-K^0) and positive (CFL) values of $\delta m \equiv m_{K^0} - \mu_{K^0}$ and contribution from $\varphi \leftrightarrow \varphi + \varphi$, see Eq. (10.105). 2SC phase and unpaired quark matter: contribution from the process $u + d \leftrightarrow u + s$.

Pethick, 1993). In the low-temperature limit (in particular $T \ll m_D$ with the electric screening mass $m_D^2 = N_f g^2 \mu^2/(2\pi^2)$) they are

$$\kappa \simeq 0.5 \frac{m_D^2}{\alpha_s^2}, \qquad \sigma_{\rm el} \simeq 0.01 \frac{e^2 \mu^2 m_D^{2/3}}{\alpha_s^2 T^{5/3}}, \tag{10.106}$$

and

$$\eta \simeq 4.4 \times 10^{-3} \frac{\mu^4 m_D^{2/3}}{\alpha_s^2 T^{5/3}}. \tag{10.107}$$

These quantities have not yet been computed for partially gapped color superconductors such as the 2SC phase. The presence of ungapped modes, however, suggests that the results only differ by a numerical factor from the unpaired phase results.

The dominant flavor changing process that contributes to the bulk viscosity in unpaired quark matter is the reaction (Anand et al., 2000; Madsen, 1992)

$$u + d \leftrightarrow u + s. \tag{10.108}$$

Other relevant processes are the semi-leptonic processes $u + e \leftrightarrow d + \nu_e$ and $u + e \leftrightarrow s + \nu_e$ (Dong et al., 2007; Sa'd et al., 2007). In a partially gapped phase, such as the 2SC phase the bulk viscosity is also dominated by the process (10.108). We show the results for the 2SC phase (Alford and Schmitt, 2007) and unpaired quark matter in Fig. 10.9.

10.7.2 Neutrino emissivity and specific heat

Neutrino emissivity determines the rate at which quark matter can lose heat via neutrino emission. For the purpose of studying how neutron stars with ages ranging from tens of seconds to millions of years cool, as we shall discuss in Section 10.8.3, it is appropriate to treat the matter as completely transparent to the neutrinos that it emits.

In CFL quark matter, all quasifermion modes are gapped and neutrino emissivity is dominated by reactions involving (pseudo)-Goldstone modes such as,

$$\pi^{\pm}, K^{\pm} \to e^{\pm} + \bar{\nu}_e, \tag{10.109a}$$

$$\pi^0 \to \nu_e + \bar{\nu}_e, \tag{10.109b}$$

$$\varphi + \varphi \to \varphi + \nu_e + \bar{\nu}_e. \tag{10.109c}$$

These processes were studied in (Jaikumar et al., 2002; Reddy et al., 2003b). The decay rates of the massive mesons $\pi^{\pm}$, $K^{\pm}$, and π^0 are proportional to their number densities and are suppressed by Boltzmann factors $\exp(-E/T)$, where E is the energy gap of the meson. Since the pseudo-Goldstone boson energy gaps are in the order of a few MeV, the emissivities are strongly suppressed as compared with unpaired quark matter for temperatures below this scale. Neutrino emission from processes involving φ is not exponentially suppressed, but it involves a very large power of T,

$$\epsilon_\nu \sim \frac{G_F^2 T^{15}}{f^2 \mu^4}, \tag{10.110}$$

and is numerically very small. Reddy et al. also studied the neutrino mean free path l_ν. For $T \sim 30\,\text{MeV}$ the mean free path is in the order of 1 m, but for $T < 1\,\text{MeV}$, $l_\nu > 10$ km (Reddy et al., 2003b). In the CFL-K^0 phase, l_ν is almost the same as in the CFL phase, while the neutrino emissivity is larger (Reddy et al., 2003a).

The specific heat of CFL matter is also dominated by φ, yielding,

$$c_V = \frac{2\pi^2}{15 v^3} T^3. \tag{10.111}$$

In the CFL-K^0 phase, the Goldstone mode due to kaon condensation gives an additional contribution with the same T^3 dependence. This specific heat from Goldstone bosons is much smaller than the specific heat of any phase containing unpaired quarks, because their density is proportional to $\mu^2 T$, while that of ungapped bosons is T^3. This means that in any degenerate system ($T \ll \mu$) ungapped fermion modes, if they exist, will dominate the specific heat and the neutrino rates. In unpaired quark matter, neutrino emissivity is dominated by the direct Urca processes,

$$u + e \to d + \nu_e \qquad \text{(electron capture)}, \tag{10.112a}$$

$$d \to u + e + \bar{\nu}_e \qquad (\beta\text{-decay}). \tag{10.112b}$$

The radiated energy per unit of time and volume is (Iwamoto, 1980)

$$\epsilon_\nu \simeq \frac{457}{630} \alpha_s G_F^2 T^6 \mu_e \mu_u \mu_d. \tag{10.113}$$

Note that this result is proportional to the strong coupling constant α_s. The tree-level processes for massless quarks are approximately collinear and the weak matrix element vanishes in the forward direction. A non-zero emissivity arises from strong interaction corrections, which depress quark Fermi momenta relative to their chemical potentials. Because they do not at the same time depress the electron Fermi momentum, this opens up phase space for the reactions (10.112). A non-zero emissivity can also arise from quark mass effects, or higher-order corrections in T/μ. Since strange quark decays are Cabbibo suppressed and T/μ is small, the dominant contribution is likely to be that proportional to α_s, namely (10.113). Note that we have not included non-Fermi liquid corrections of $O(\alpha_s \log(T))$ (Schäfer and Schwenzer, 2004a).

The specific heat of unpaired quark matter is

$$c_V = \frac{N_c N_f}{3} \mu^2 T, \tag{10.114}$$

where we have again neglected terms of $O(\alpha_s \log(T))$ (Ipp et al., 2004) and assumed the flavor chemical potentials to be equal.

The presence of ungapped modes in most non-CFL color superconductors renders their specific heat and neutrino emissivity due to direct Urca processes virtually indistinguishable from the unpaired phase. For specific results and further references see Ref. (Alford et al., 2008b).

10.8 Color superconductivity in neutron stars

Neutron stars are the densest material objects in the universe, with masses of a similar to order that of the sun ($M_\odot$) and radii of order ten km. Depending on their mass and on the stiffness of the equation of state of the material of which they are composed, their central density lies between ~ 3 and ~ 12 times nuclear saturation density ($n_0 = 0.16$ nucleons/fm^3) (Lattimer and Prakash, 2001, 2007). Neutron stars consist of an outer crust made of a rigid lattice of positive ions embedded within a fluid of electrons and (in the inner layer of the crust) superfluid neutrons (Negele and Vautherin, 1973). Inside this crust, one finds a fluid "mantle" consisting of neutrons and protons, both likely superfluid, and electrons. Determining the composition of neutron star cores, namely of the densest matter in the universe, remains an outstanding challenge.[5] If the nuclear equation of state is stiff enough, neutron stars are made of neutrons, protons, and electrons all the way down to their centers. If higher densities are reached, other phases of baryonic matter (including either a pion condensate (Bahcall and Wolf, 1965; Baym and Campbell, 1978; Migdal, 1971; Sawyer, 1972; Scalapino, 1972), a kaon condensate (Brown, 1995; Kaplan and Nelson, 1986), or a non-zero density of one or several hyperons (Glendenning, 1985)) may result. Or, neutron star cores may be made of color superconducting quark matter.

[5]For review articles on neutron stars as laboratories for understanding dense matter, see for instance (Lattimer and Prakash, 2004, 2007; Page and Reddy, 2006; Prakash et al., 2001; Schmitt, 2010; Weber, 1999; Weber et al., 2006; Yakovlev and Pethick, 2004).

The density at which the transition from baryonic matter to quark matter occurs is not known; this depends on a comparison between the equations of state for both, which is not well determined for either. Very roughly, we expect this transition to occur when the density exceeds one nucleon per nucleon volume, a criterion which suggests a transition to quark matter at densities $\gtrsim 3n_0$. The question we shall pose in this section is how astrophysical observation of neutron stars could determine whether they do or do not contain quark matter within their cores. We have seen throughout the earlier sections of this review that quark matter at potentially accessible densities may be in the CFL phase, with all quarks paired, or may be in one of a number of possible phases in which there are some unpaired quarks, some of which are spatially inhomogeneous. If quark matter does exist within neutron stars, with their temperatures far below the critical temperatures for these paired phases, it will be in some color superconducting phase. We shall see in this section that these different phases have different observational consequences, making it possible for a combination of different types of observational data to cast light upon the question of which phase of color superconducting quark matter is favored in the QCD phase diagram, if neutron stars do in fact feature quark matter cores.

We shall discuss signatures of quark matter in neutron star cores, not the more radical possibility that nuclear matter in bulk is metastable at zero pressure, with the true ground state of strongly interacting matter in the infinite volume limit being color superconducting three-flavor quark matter.

10.8.1 Mass–radius relation

It has long been a central goal of neutron star astrophysics to measure the masses M and radii R of many neutron stars to a reasonable accuracy. Mapping out the curve in the mass–radius plane along which neutron stars are found would yield a strong constraint on the equation of state of dense matter. As this program represents such a large fraction of the effort to use observations of neutron stars to constrain dense matter physics, we begin by considering its implications for the presence of quark matter within neutron star cores.

The larger the maximum mass that can be attained by a neutron star, the stiffer the equation of state of dense matter. The recent observation of a neutron star with mass $1.97 \pm 0.04\,\mathrm{M}_\odot$ (Demorest et al., 2010) casts doubt on the existence of phases with a soft equation of state, such as baryonic matter with kaon or pion condensation. However, although the quark matter equation of state is not known from first principles, it may easily be as stiff as the stiffer equations of state posited for ordinary nuclear matter, and neutron stars with quark matter cores can in fact reach masses of order $2\,M_\odot$ (Alford et al., 2005a; Baldo et al., 2003; Blaschke et al., 2007; Fraga et al., 2001; Kurkela et al., 2010; Ozel et al., 2010; Rüster and Rischke, 2004).

The equation of state for CFL quark matter can be parameterized to a good approximation as (Alford et al., 2005a)

$$\Omega = -P = -\frac{3}{4\pi^2}(1-c)\mu^4 + \frac{3}{4\pi^2}(M_s^2 - 4\Delta^2)\mu^2 + B_{\text{eff}}. \tag{10.115}$$

If c were zero, the μ^4 term would be that for non-interacting quarks; c parameterizes the leading effect of interactions, modifying the relation between p_F and μ. At high densities, $c = 2\alpha_s/\pi$ to leading order in the strong coupling constant (Baym and Chin, 1976; Freedman and McLerran, 1977). Analysis of higher-order corrections suggests that $c \gtrsim 0.3$ at accessible densities (Fraga et al., 2001; Kurkela et al., 2009). B_{eff} can be thought of as parameterizing our ignorance of the μ at which the nuclear matter to quark matter transition occurs. The $M_s^2\mu^2$ term is the leading effect of the strange quark mass, and is common to all quark matter phases. The pressure of a color superconducting phase with less pairing than in the CFL phase would have a smaller coefficient of the $\Delta^2\mu^2$ term, and would also differ at order M_s^4, here lumped into a change in B_{eff}. Because pairing is a Fermi surface phenomenon, it only modifies the μ^2 term, leaving the larger μ^4 term untouched. However, it can nevertheless be important because at accessible densities the μ^4 term is largely canceled by B_{eff}, enhancing the importance of the μ^2 term (Alford and Reddy, 2003; Lugones and Horvath, 2002). Remarkably, and perhaps coincidentally, if we make the (reasonable) parameter choices $c = 0.3$, $M_s = 275$ MeV and $\Delta = 100$ MeV and choose B_{eff} such that nuclear matter gives way to CFL quark matter at the relatively low density $1.5\,n_0$, then over the entire range of higher densities relevant to neutron stars the quark matter equation of state (10.115) is almost indistinguishable from the nuclear equation of state due to Akmal, Pandharipande, and Ravenhall (APR) (Akmal et al., 1998) that is one of the stiffest nuclear equations of state in the compendium found in (Lattimer and Prakash, 2004, 2007). Neutron stars made entirely of nuclear matter with the APR equation of state and neutron stars with a quark matter core with the equation of state (10.115) with the parameters just described fall along almost indistinguishable curves on a mass versus radius plot, with the most significant difference being that the APR equation of state admits neutron stars with maximum mass $2.3M_\odot$, whereas the introduction of a quark matter core reduces the maximum mass slightly, to $2.0M_\odot$ (Alford et al., 2005a).

The recent observation that neutron star PSR J1614-2230 has mass $1.97 \pm 0.04\,\mathrm{M}_\odot$ (Demorest et al., 2010) puts strong constraints on the quark matter equation of state. An analysis using the ansatz (10.115) for quark matter and the APR equation of state for nuclear matter (Ozel et al., 2010) found that such a heavy star could have a quark matter core, if the transition occurs at a low density (1.5 n_0), and the quarks are strongly interacting ($c \gtrsim 0.33$) and the quadratic term is small (near cancelation between M_s^2 and $4\Delta^2$). It should be noted that mass as large as $2.0M_\odot$ is a challenge to theory, whether or not such a star contains a quark matter core. Nuclear equations of state which accomodate neutron stars as massive as this are known, but a star this heavy has a large enough central density (for example, about five times nuclear saturation density for the APR equation of state (Akmal et al., 1998)) that it is doubtful that the nuclear EoS can be trusted quantitatively in the core of the star.

10.8.2 Signatures of the compactness of neutron stars

If we could detect gravity waves from neutron stars spiraling into black holes in binary systems, the gravitational wave form during the last few orbits, when the neutron star is being tidally disrupted, will encode information about the density profile of the neutron star. For example, upon assuming a conventional density profile, the gravity wave form encodes information about the ratio M/R (Faber et al., 2002), essentially via encoding the value of the orbital frequency at which tidal deformation becomes significant. This suggests a scenario in which the presence of an interface separating a denser quark core from a less dense nuclear mantle could manifest itself via the existence of two orbital frequency scales in the wave form, the first being that at which the outer layers are deformed while the denser quark core remains spherical and the second being the time at which even the quark core is disrupted (Alford et al., 2001b). Although this idea needs further development, it seems clear that if LIGO sees events in which the tidal disruption of a neutron star occurs within the LIGO bandwidth, the gravity wave data will constrain the "compactness" of the neutron star, providing information about the density profile that is complementary to that obtained from a mass–radius relation.

If there is a "step" in the density profile at an interface, LIGO gravity waves may provide evidence for its presence. But, should a density step be expected if color superconducting quark matter is found in the core of a neutron star? There are two qualitatively distinct possibilities for the density profile, depending on the surface tension of the quark matter/nuclear matter interface σ. If σ is large enough, there will be a stable, sharp, interface between two phases having different densities (but the same chemical potential). If σ is small enough, it becomes favorable instead to form a macroscopic volume filled with a net-neutral mixture of droplets of negatively charged quark matter and positively charged nuclear matter, see Section 10.3.8, which allows a continuous density profile. The distinction between these two scenarios has been analyzed quantitatively for the case of a first-order phase transition from nuclear matter to

CFL quark matter (Alford et al., 2001b). This is the simplest possible phase diagram of QCD, with a single transition between the phases known to exist at nuclear density and at asymptotically high density. We have seen earlier in this review that this simple QCD phase diagram is obtained if Δ_{CFL} is large enough, allowing CFL pairing to fend off stresses that seek to split Fermi surfaces, all the way down in density until the nuclear matter takes over from quark matter. A sharp interface between the (electrically insulating) CFL phase and (electrically conducting) nuclear matter features charged boundary layers on either side of the interface, which play an important role in determining the σ above which this step in the density profile is stable (Alford et al., 2001b). The critical σ is about $40\,\mathrm{MeV/fm^3}$, lower than dimensional analysis would indicate should be expected, meaning that the sharp interface with a density step is more likely than a mixture of charged components. The increase in the density at the interface can easily be by a factor of two.

10.8.3 Cooling

The avenues of investigation that we have described so far may constrain the possible existence of quark matter within neutron star cores, but they are not sensitive to the differences among different color superconducting phases of quark matter. We turn now to the first of three observational signatures that have the potential to differentiate between CFL quark matter and other color superconducting phases.

Within less than a minute of its birth in a supernova, a neutron star cools below about 1 MeV and becomes transparent to neutrinos. For the next million years or so it cools mainly via neutrino emission from its interior. Photon emission from the surface becomes dominant only later than that. This means that information about properties of the interior, in particular its neutrino emissivity and heat capacity, can be inferred from measurements of the temperature and age of neutron stars. Because all forms of dense matter are good heat conductors (Shovkovy and Ellis, 2002), neutron star interiors are isothermal and the rate at which they cool is determined by the volume integrals over the entire interior of the local emissivity and the local specific heat. This means that the cooling tends to be dominated by the properties of whichever phase has the highest neutrino emissivity and whichever phase has the highest specific heat.

Different forms of dense matter fall into three categories, ordered by decreasing neutrino emissivity. The first category includes any phase of matter that can emit neutrinos via direct Urca processes, yielding an emissivity $\epsilon_\nu \propto T^6$. Examples include unpaired quark matter, phases of quark matter with some unpaired quarks including the crystalline phases, and the phases with single flavor pairing in Table 10.3, baryonic matter containing hyperons, nucleonic matter augmented by either a pion or a kaon condensate, and even ordinary nuclear matter at sufficiently high densities that the proton fraction exceeds about 0.1.

Ordinary nuclear matter at densities not too far above n_0, where the proton fraction is less than 0.1, falls into a second category in which there is no phase space for direct Urca processes and neutrino emission occurs only via modified

Urca processes like $n + X \to p + X + e + \bar{\nu}$ with X some spectator nucleon, giving an emissivity $\propto T^9$, with a much smaller value at typical neutron star temperatures $\sim 10^9$ K. Neutron stars whose interiors emit neutrinos at this rate, perhaps modified by effects of nucleon superfluidity, cool following a family of standard cooling curves (see (Page et al., 2004; Yakovlev et al., 2001, 2005) and references therein), taking 10^5 to 10^6 years to cool below 10^8 K.

CFL quark matter constitutes a third category. As we have seen in Section 10.7, it is unique among all phases of dense matter in having an emissivity $\propto T^{15}$ that is many orders of magnitude smaller than the rate for the Urca process in nuclear matter. Furthermore, whereas all other phases of dense matter have a specific heat $\propto T$, in the CFL phase the specific heat is controlled by bosonic excitations making it $\propto T^3$. This means that if a neutron star has a CFL core, the total neutrino emissivity and the total heat capacity of the star are both utterly dominated by the contributions of the outer layers, whether these are made of nuclear matter or of some phase that admits direct Urca reactions. The CFL core holds little heat, and emits few neutrinos, but is a good conductor and so stays at the same temperature as the rest of the star. The rest of the star controls how the star cools.

Finally, the single-flavor color superconducting phases are interesting because they represent a potential transition from the first to the third categories (Aguilera et al., 2005; Grigorian et al., 2005): their critical temperatures are so low that if some quarks can only pair in spin-one channels, they will not pair until after the star has cooled through an initial epoch of direct Urca emission; and, in certain cases (Aguilera et al., 2005; Marhauser et al., 2007; Schmitt et al., 2006) all quarks can be gapped below the critical temperature for color–spin locked pairing, meaning that these phases ultimately become like CFL quark matter, playing no role in the cooling of the star which at late times will be controlled by the modified Urca processes in the nuclear matter mantle.

There is now good evidence that, although the cooling of many neutron stars is broadly consistent with the standard cooling curves, some fraction of neutron stars cool much more quickly. (This evidence includes observations of neutron stars that are cold for their age (Page and Reddy, 2006), unsuccessful searches in many supernova remnants for the X-ray emission that should be seen if these supernovae had produced neutron stars cooling according to the standard cooling curve (Kaplan et al., 2006), and the low thermal luminosity of stars that are being heated by transient bouts of accretion during the quiescent periods between accretion episodes, indicating that heating by accretion must be balanced by rapid cooling by neutrino emission (Heinke et al., 2007; Jonker et al., 2007).) It is then reasonable to speculate that lighter neutron stars cool following the standard cooling curve and are composed of nuclear matter throughout, while heavier neutron stars cool faster because they contain some form of dense matter that can radiate neutrinos via the direct Urca process. This could be quark matter in one of the non-CFL color superconducting phases, but there are other, baryonic, possibilities. If this speculation is correct, then if neutron stars contain CFL cores they must be "inner cores", within an outer core made of whatever is responsible for the rapid neutrino emission.

10.8.4 *r*-modes limiting pulsar spins

A rapidly spinning neutron star will quickly slow down if it is unstable with respect to bulk flows known as Rossby modes, or *r*-modes, whose restoring force is due to the Coriolis effect and which transfer the star's angular momentum into gravitational radiation (Andersson, 1998; Andersson and Comer, 2001; Andersson et al., 1999a,b; Friedman and Morsink, 1998). For any given interior composition and temperature, above some critical spin frequency there is an instability which leads to an exponentially growing *r*-mode. This means that as a neutron star is spun up by accretion, its spin will be limited by a value very slightly above this critical frequency, at which the accretion torque is balanced by gravitational radiation from the *r*-mode flows (Andersson et al., 1999a,b; Bildsten, 1998; Lindblom et al., 1998; Owen et al., 1998). From a microphysical point of view, the *r*-mode instability is limited by viscous damping: the greater the damping, the higher the critical spin above which *r*-modes become unstable. The critical frequency is controlled by the shear viscosity in some regimes of temperature (typically lower) and by the bulk viscosity in others (typically higher). This means that the existence of pulsars with a given spin, as well as any observational evidence for an upper limit on pulsar spins, can yield constraints on the viscosities of neutron star interiors.

Current calculations of the bulk and shear viscosity of nuclear matter indicate that neutron stars could be made of purely nuclear matter: its shear viscosity of nuclear matter is just barely high enough to allow for the observed spin rates and temperatures of low-mass X-ray binaries (Jaikumar et al., 2008). For quark matter that is unpaired (or in the 2SC phase) the viscosities are higher, so existing data are also consistent with stars made of such matter, or with hybrid stars with an unpaired quark matter core (Jaikumar et al., 2008). For CFL quark matter the picture is less clear: the superfluid phonon contributes a high shear viscosity at low temperatures, but its mean free path is larger than the star at $T \lesssim 1$ MeV (Alford et al., 2010; Manuel et al., 2005), so a conventional hydrodynamic approach is inapplicable in the temperature range relevant for neutron stars.

It is a very interesting question, at present unresolved, whether the presence of a CFL quark matter core within an ordinary neutron star introduces unstable *r*-modes at low spin frequencies. If there is a density step at the nuclear/CFL interface, there may be oscillation modes localized near that interface. The question is whether there are *r*-modes that are sufficiently well localized on the CFL side of the interface that they are undamped, or whether the tails of the mode wave functions that extend into the nuclear matter side of the interface result in enough damping to prevent the modes from becoming unstable. If it were to turn out that a star with a CFL core is even close to as unstable with respect to *r*-modes as a star that is made entirely of CFL matter, the existence of pulsars spinning with hundreds of Hz frequencies would immediately rule out the possibility that these neutron stars have CFL cores.

10.8.5 Rigid quark matter and pulsar glitches

The existence of a rigid crystalline color superconducting core within neutron stars may have a variety of observable consequences. For example, if some agency (like magnetic fields not aligned with the rotation axis) could maintain the rigid core in a shape that has a non-zero quadrupole moment, gravity waves would be emitted. The LIGO non-detection of such gravity waves from nearby neutron stars (Abbott et al., 2007) already limits the possibility that they have rigid cores that are deformed to the maximum extent allowed by the shear modulus (10.116), upon assuming a range of possible breaking strains, and this constraint will tighten as LIGO continues to run (Haskell et al., 2007; Lin, 2007). (The analogous constraint on strange stars that are rigid throughout was obtained in (Owen, 2005).) Perhaps the most exciting implication of a rigid core, however, is the possibility that (some) pulsar "glitches" could originate deep within a neutron star, in its quark matter core.

A spinning neutron star observed as a pulsar gradually spins down as it loses rotational energy to electromagnetic radiation. But, every once in a while the angular velocity at the crust of the star is observed to increase suddenly in a dramatic event called a glitch. The standard explanation (Alpar, 1977; Alpar et al., 1984a, 1996, 1984b; Anderson and Itoh, 1975; Epstein and Baym, 1992; Jones, 1997; Link and Epstein, 1996; Link et al., 1993; Pines and Alpar, 1985) requires the presence of a superfluid in some region of the star which also features a rigid array of spatial inhomogeneities which can pin the vortices in the rotating superfluid. In the standard explanation of pulsar glitches, these conditions are met in the inner crust of a neutron star which features a neutron superfluid coexisting with a rigid array of positively charged nuclei that may serve as vortex pinning sites. We shall see below that a rigid core made of crystalline color superconducting quark matter also meets the basic requirements.

The viability of the standard scenario for the origin of pulsar glitches in neutron star crusts has recently been questioned (Link, 2007a). Explanation of the issue requires on understanding of how the basic requirements come into play in the generation of a glitch. As a spinning pulsar slowly loses angular momentum over years, since the angular momentum of any superfluid component of the star is proportional to the density of vortices, the vortices "want" to move apart. However, if within the inner crust the vortices are pinned to a rigid structure, these vortices do not move and after a time this superfluid component of the star is spinning faster than the rest of the star. When the "tension" built up in the array of pinned vortices reaches a critical value, there is a sudden "avalanche" in which vortices unpin, move outwards reducing the angular momentum of the superfluid, *and then re-pin*. As this superfluid suddenly loses angular momentum, the rest of the star, including in particular the surface whose angular velocity is observed, speeds up—a glitch. We see that this scenario requires superfluidity coexisting with a rigid structure to which vortices can pin that does not easily deform when vortices pinned to it are under tension. In very recent work, Link has questioned whether this scenario is viable, because once neutron vortices are moving through the inner crust, as must happen during a glitch, they are so resistant to bending that they may never re-pin (Link, 2007a). Link concludes that we do not have an understanding of any dynamics that could lead to the re-pinning of moving vortices, and hence that we do not currently understand glitches as a crustal phenomenon.

We have seen in Section 10.6.2 that if neutron star cores are made of quark matter but $\Delta_{\rm CFL}$ is not large enough for this quark matter to be in the CFL phase, then all of the quark matter core—and hence a significant fraction of the moment of inertia of the star—may be in one of the crystalline phases described in Section 10.6.2. By virtue of being simultaneously superfluids and rigid solids, the crystalline phases of quark matter provide all the necessary conditions to be the locus in which (some) pulsar glitches originate. Their shear moduli (10.100), namely,

$$\nu = 3.96 \times 10^{33}\,{\rm erg/cm^3} \left(\frac{\Delta}{10\ {\rm MeV}}\right)^2 \left(\frac{\mu}{400\ {\rm MeV}}\right)^2, \qquad (10.116)$$

with Δ the gap parameter in the crystalline phase as in Fig. 10.8, make this form of quark matter 20 to 1000 times more rigid than the crust of a neutron star (Mannarelli et al., 2007; Strohmayer et al., 1991), and hence more than rigid enough for glitches to originate within them. The crystalline phases are at the same time superfluid, and it is reasonable to expect that the superfluid vortices that will result when a neutron star with such a core rotates will have lower free energy if they are centered along the intersections of the nodal planes of the underlying crystal structure, i.e. along lines along which the condensate already vanishes, even in the absence of a rotational vortex. A crude estimate of the pinning force on vortices within crystalline color superconducting quark matter indicates that it is sufficient (Mannarelli et al., 2007). So, the basic requirements for superfluid vortices pinning to a rigid structure are all present. The central questions that remain to be addressed are the explicit construction of vortices in the crystalline phase and calculation of their pinning force, as well as calculation of the time scale over which sudden changes in the angular momentum of the core are communicated to the (observed) surface, presumably either via the common electron fluid or via magnetic stresses.

Much theoretical work remains before the hypothesis that pulsar glitches originate within a crystalline color superconducting neutron star core is developed efficiently to allow it to confront data on the magnitudes, relaxation time scales, and repeat rates that characterize glitches. Nevertheless, this hypothesis offers one immediate advantage over the conventional scenario that relied on vortex pinning in the neutron star crust. It is impossible for a neutron star anywhere within which rotational vortices are pinned to precess on $\sim$ year time scales (Link, 2006, 2007b; Sedrakian et al., 1999), and yet there is now evidence that several pulsars are precessing (Chukwude et al., 2003; Shabanova et al., 2001; Stairs et al., 2000). Since *all* neutron stars have crusts, the precession of any pulsar is inconsistent with the pinning of vortices within the crust, a requirement in the standard explanation of glitches. On the other hand, perhaps not all neutron stars have crystalline quark matter cores—for example, perhaps the lightest neutron stars have nuclear matter cores. Then, if vortices are never pinned in the crust but are pinned within a crystalline quark matter core, those neutron stars that do have a crystalline quark matter core can glitch but cannot precess, while those that don't can precess but cannot glitch.

Acknowledgments

We acknowledge helpful conversations with N. Andersson, M. Braby, D. Chakrabarty, T. Hatsuda, S. Hughes, D.L. Kaplan, B. Link, M. Mannarelli, C. Manuel, D. Nice, D. Page, A. Rebhan, S. Reddy, R. Sharma, I. Stairs, Q. Wang, and F. Wilczek. This research was supported in part by the Offices of Nuclear Physics and High Energy Physics of the Office of Science of the U.S. Department of Energy under contracts #DE-FG02-91ER40628, #DE-FG02-05ER41375 #DE-FG02-94ER40818, #DE-FG02-03ER41260.

References

Abbott, B., et al. (LIGO Scientific), 2007, *Phys. Rev.* **D76**, 042001.

Abuki, H., M. Kitazawa, and T. Kunihiro, 2005, *Phys. Lett.* **B615**, 102.

Abuki, H., and T. Kunihiro, 2006, *Nucl. Phys.* **A768**, 118.

Aguilera, D. N., 2007, *Astrophys. Space Sci.* **308**, 443.

Aguilera, D. N., D. Blaschke, M. Buballa, and V. L. Yudichev, 2005, *Phys. Rev.* **D72**, 034008.

Akmal, A., V. R. Pandharipande, and D. G. Ravenhall, 1998, *Phys. Rev.* **C58**, 1804.

Alford, M. G., J. Berges, and K. Rajagopal, 1999a, *Nucl. Phys.* **B558**, 219.

Alford, M. G., J. Berges, and K. Rajagopal, 2000, *Nucl. Phys.* **B571**, 269.

Alford, M. G., J. A. Bowers, J. M. Cheyne, and G. A. Cowan, 2003, *Phys. Rev.* **D67**, 054018.

Alford, M. G., J. A. Bowers, and K. Rajagopal, 2001a, *Phys. Rev.* **D63**, 074016.

Alford, M. G., M. Braby, and S. Mahmoodifar, 2010, *Phys. Rev.* **C81**, 025202.

Alford, M. G., M. Braby, M. W. Paris, and S. Reddy, 2005a, *Astrophys. J.* **629**, 969.

Alford, M. G., M. Braby, S. Reddy, and T. Schäfer, 2007, *Phys. Rev.* **C75**, 055209.

Alford, M. G., M. Braby, and A. Schmitt, 2008a, *J. Phys.* **G35**, 115007.

Alford, M. G., and G. A. Cowan, 2006, *J. Phys.* **G32**, 511.

Alford, M. G., and G. Good, 2004, *Phys. Rev.* **D70**, 036008.

Alford, M. G., P. Jotwani, C. Kouvaris, J. Kundu, and K. Rajagopal, 2005b, *Phys. Rev.* **D71**, 114011.

Alford, M. G., C. Kouvaris, and K. Rajagopal, 2004a, eprint hep-ph/0407257.

Alford, M. G., C. Kouvaris, and K. Rajagopal, 2004b, *Phys. Rev. Lett.* **92**, 222001.

Alford, M. G., C. Kouvaris, and K. Rajagopal, 2005c, *Phys. Rev.* **D71**, 054009.

Alford, M. G., and K. Rajagopal, 2002, *JHEP* **06**, 031.

Alford, M. G., and K. Rajagopal, 2006, in *Pairing in Fermionic Systems: Basic Concepts and Modern Applications* (World Scientific), pp. 1–36, eprint hep-ph/0606157.

Alford, M. G., K. Rajagopal, S. Reddy, and A. W. Steiner, 2006, *Phys. Rev.* **D73**, 114016.

Alford, M. G., K. Rajagopal, S. Reddy, and F. Wilczek, 2001b, *Phys. Rev.* **D64**, 074017.

Alford, M. G., K. Rajagopal, and F. Wilczek, 1998, *Phys. Lett.* **B422**, 247.

Alford, M. G., K. Rajagopal, and F. Wilczek, 1999b, *Nucl. Phys.* **B537**, 443.

Alford, M. G., and S. Reddy, 2003, *Phys. Rev.* **D67**, 074024.

Alford, M. G., and A. Schmitt, 2007, *J. Phys.* **G34**, 67.

Alford, M. G., A. Schmitt, K. Rajagopal, and T. Schafer, 2008b, *Rev. Mod. Phys.* **80**, 1455.

Alford, M. G., and Q.-h. Wang, 2005, *J. Phys.* **G31**, 719.

Alford, M. G., and Q.-h. Wang, 2006, *J. Phys.* **G32**, 63.

Alpar, M. A., 1977, *Astrophys. J.* **213**, 527.

Alpar, M. A., P. W. Anderson, D. Pines, and J. Shaham, 1984a, *Astrophys. J.* **278**, 791.

Alpar, M. A., H. F. Chau, K. S. Cheng, and D. Pines, 1996, *Astrophys. J.* **459**, 706.

Alpar, M. A., S. A. Langer, and J. A. Sauls, 1984b, *Astrophys. J.* **282**, 533.

Amore, P., M. C. Birse, J. A. McGovern, and N. R. Walet, 2002, *Phys. Rev.* **D65**, 074005.

Anand, J. D., N. Chandrika Devi, V. K. Gupta, and S. Singh, 2000, *Pramana* **54**, 737.

Anderson, P. W., 1963, *Phys. Rev.* **130**, 439.

Anderson, P. W., and N. Itoh, 1975, *Nature* **256**, 25.

Andersson, N., 1998, *Astrophys. J.* **502**, 708.

Andersson, N., and G. L. Comer, 2001, *Mon. Not. Roy. Astron. Soc.* **328**, 1129.

Andersson, N., K. D. Kokkotas, and B. F. Schutz, 1999a, *Astrophys. J.* **510**, 846.

Andersson, N., K. D. Kokkotas, and N. Stergioulas, 1999b, *Astrophys. J.* **516**, 307.

Andreev, A. F., and I. M. Lifshitz, 1969, *Sov. Phys. JETP* **29**, 2057.

Anglani, R., R. Gatto, N. D. Ippolito, G. Nardulli, and M. Ruggieri, 2007, *Phys. Rev.* **D76**, 054007.

Bahcall, J. N., and R. A. Wolf, 1965, *Phys. Rev.* **140**(5B), B1452.

Bailin, D., and A. Love, 1979, *J. Phys.* **A12**, L283.

Bailin, D., and A. Love, 1984, *Phys. Rept.* **107**, 325.

Balachandran, A. P., S. Digal, and T. Matsuura, 2006, *Phys. Rev.* **D73**, 074009.

Baldo, M., et al., 2003, *Phys. Lett.* **B562**, 153.

Bardeen, J., L. N. Cooper, and J. R. Schrieffer, 1957, *Phys. Rev.* **106**(1), 162.

Barrois, B. C., 1979, *Non-perturbative effects in dense quark matter*, Ph.D. thesis, California Institute of Technology, Pasadena, California, UMI 79-04847.

Baym, G., and D. K. Campbell, 1978, in *Mesons in Nuclei*, ed. M. Rho, D. Wilkinson, Vol III, 1031. Amsterdam: North Holland.

Baym, G., and S. A. Chin, 1976, *Phys. Lett.* **B62**, 241.

Beane, S. R., P. F. Bedaque, and M. J. Savage, 2000, *Phys. Lett.* **B483**, 131.

Bedaque, P. F., H. Caldas, and G. Rupak, 2003, *Phys. Rev. Lett.* **91**, 247002.

Bedaque, P. F., and T. Schäfer, 2002, *Nucl. Phys.* **A697**, 802.

Berges, J., and K. Rajagopal, 1999, *Nucl. Phys.* **B538**, 215.

Bildsten, L., 1998, *Astrophys. J.* **501**, L89.

Blaschke, D., S. Fredriksson, H. Grigorian, A. M. Oztas, and F. Sandin, 2005, *Phys. Rev.* **D72**, 065020.

Blaschke, D. B., D. Gomez Dumm, A. G. Grunfeld, T. Klahn, and N. N. Scoccola, 2007, *Phys. Rev.* **C75**, 065804.

Bowers, J. A., J. Kundu, K. Rajagopal, and E. Shuster, 2001, *Phys. Rev.* **D64**, 014024.

Bowers, J. A., and K. Rajagopal, 2002, *Phys. Rev.* **D66**, 065002.

Braaten, E., and R. D. Pisarski, 1992, *Phys. Rev.* **D45**, 1827.

Brown, G. E., 1995, in *Bose-Einstein Condensation*, ed. A. Griffin, et al. 438. Cambridge: Cambridge University Press.

Brown, W. E., J. T. Liu, and H.-c. Ren, 2000a, *Phys. Rev.* **D62**, 054013.

Brown, W. E., J. T. Liu, and H.-c. Ren, 2000b, *Phys. Rev.* **D61**, 114012.

Brown, W. E., J. T. Liu, and H.-c. Ren, 2000c, *Phys. Rev.* **D62**, 054016.

Buballa, M., 2005a, *Phys. Rept.* **407**, 205.

Buballa, M., 2005b, *Phys. Lett.* **B609**, 57.

Buballa, M., J. Hosek, and M. Oertel, 2003, *Phys. Rev. Lett.* **90**, 182002.

Buballa, M., and I. A. Shovkovy, 2005, *Phys. Rev.* **D72**, 097501.

Carlson, J., and S. Reddy, 2005, *Phys. Rev. Lett.* **95**, 060401.

Carter, G. W., and D. Diakonov, 1999, *Phys. Rev.* **D60**, 016004.

Casalbuoni, R., M. Ciminale, R. Gatto, G. Nardulli, and M. Ruggieri, 2006, *Phys. Lett.* **B642**, 350.

Casalbuoni, R., E. Fabiano, R. Gatto, M. Mannarelli, and G. Nardulli, 2002a, *Phys. Rev.* **D66**, 094006.

Casalbuoni, R., and R. Gatto, 1999, *Phys. Lett.* **B464**, 111.

Casalbuoni, R., R. Gatto, N. Ippolito, G. Nardulli, and M. Ruggieri, 2005a, *Phys. Lett.* **B627**, 89.

Casalbuoni, R., R. Gatto, M. Mannarelli, and G. Nardulli, 2001, *Phys. Lett.* **B511**, 218.

Casalbuoni, R., R. Gatto, M. Mannarelli, and G. Nardulli, 2002b, *Phys. Lett.* **B524**, 144.

Casalbuoni, R., R. Gatto, M. Mannarelli, G. Nardulli, and M. Ruggieri, 2005b, *Phys. Lett.* **B605**, 362.

Casalbuoni, R., R. Gatto, and G. Nardulli, 2002c, *Phys. Lett.* **B543**, 139.

Casalbuoni, R., and G. Nardulli, 2004, *Rev. Mod. Phys.* **76**, 263.

Casalbuoni, R., et al., 2003, *Phys. Lett.* **B575**, 181.

Casalbuoni, R., et al., 2004, *Phys. Rev.* **D70**, 054004.

Chandrasekhar, B. S., 1962, *Appl. Phys. Lett.* **1**, 7.

Chester, G. V., 1970, *Phys. Rev.* **A2**, 256.

Chukwude, A. E., A. A. Ubachukwu, and P. N. Okeke, 2003, *Astron. Astrophys.* **399**, 231.

Ciminale, M., G. Nardulli, M. Ruggieri, and R. Gatto, 2006, *Phys. Lett.* **B636**, 317.

Clogston, A. M., 1962, *Phys. Rev. Lett.* **9**, 266.

Demorest, P., T. Pennucci, S. Ransom, M. Roberts, and J. Hessels, 2010, *Nature* **467**, 1081.

Deryagin, D. V., D. Y. Grigoriev, and V. A. Rubakov, 1992, *Int. J. Mod. Phys.* **A7**, 659.

Dietrich, D. D., and D. H. Rischke, 2004, *Prog. Part. Nucl. Phys.* **53**, 305.

Dong, H., N. Su, and Q. Wang, 2007, *Phys. Rev.* **D75**, 074016.

Elze, H. T., W. Greiner, and J. Rafelski, 1983, *Phys. Lett.* **B124**, 515.

Elze, H.-T., W. Greiner, and J. Rafelski, 1984, *Z. Phys.* **C24**, 361.

Epstein, R. I., and G. Baym, 1992, *Astrophys. J.* **387**, 276.

Evans, N. J., J. Hormuzdiar, S. D. H. Hsu, and M. Schwetz, 2000, *Nucl. Phys.* **B581**, 391.

Faber, J. A., P. Grandclement, F. A. Rasio, and K. Taniguchi, 2002, *Phys. Rev. Lett.* **89**, 231102.

Ferrer, E. J., and V. de la Incera, 2007a, *Phys. Rev.* **D76**, 114012.

Ferrer, E. J., and V. de la Incera, 2007b, *Phys. Rev.* **D76**, 045011.

Ferrer, E. J., V. de la Incera, and C. Manuel, 2005, *Phys. Rev. Lett.* **95**, 152002.

Ferrer, E. J., V. de la Incera, and C. Manuel, 2006, *Nucl. Phys.* **B747**, 88.

Forbes, M. M., 2005, *Phys. Rev.* **D72**, 094032.

Forbes, M. M., and A. R. Zhitnitsky, 2002, *Phys. Rev.* **D65**, 085009.

Fraga, E. S., R. D. Pisarski, and J. Schaffner-Bielich, 2001, *Phys. Rev.* **D63**, 121702.

Freedman, B. A., and L. D. McLerran, 1977, *Phys. Rev.* **D16**, 1169.

Friedman, J. L., and S. M. Morsink, 1998, *Astrophys. J.* **502**, 714.

Fukushima, K., 2005, *Phys. Rev.* **D72**, 074002.

Fukushima, K., 2006, *Phys. Rev.* **D73**, 094016.

Fukushima, K., and K. Iida, 2005, *Phys. Rev.* **D71**, 074011.

Fukushima, K., and K. Iida, 2007, *Phys. Rev.* **D76**, 054004.

Fukushima, K., C. Kouvaris, and K. Rajagopal, 2005, *Phys. Rev.* **D71**, 034002.

Fukushima, K., and H. J. Warringa, 2008, *Phys. Rev. Lett.* **100**, 032007.

Fulde, P., and R. A. Ferrell, 1964, *Phys. Rev.* **135**, A550.

Gatto, R., and M. Ruggieri, 2007, *Phys. Rev.* **D75**, 114004.

Gerhold, A., and A. Rebhan, 2003, *Phys. Rev.* **D68**, 011502.

Gerhold, A., and A. Rebhan, 2005, *Phys. Rev.* **D71**, 085010.

Gerhold, A., and T. Schäfer, 2006, *Phys. Rev.* **D73**, 125022.

Gerhold, A., T. Schäfer, and A. Kryjevski, 2007, *Phys. Rev.* **D75**, 054012.

Giannakis, I., D.-f. Hou, and H.-C. Ren, 2005, *Phys. Lett.* **B631**, 16.

Giannakis, I., D.-f. Hou, H.-c. Ren, and D. H. Rischke, 2004, *Phys. Rev. Lett.* **93**, 232301.

Giannakis, I., J. T. Liu, and H.-c. Ren, 2002, *Phys. Rev.* **D66**, 031501.

Giannakis, I., and H.-c. Ren, 2003, *Nucl. Phys.* **B669**, 462.

Giannakis, I., and H.-C. Ren, 2005a, *Phys. Lett.* **B611**, 137.

Giannakis, I., and H.-C. Ren, 2005b, *Nucl. Phys.* **B723**, 255.

Glendenning, N. K., 1985, *Astrophys. J.* **293**, 470.

Glendenning, N. K., 1992, *Phys. Rev.* **D46**(4), 1274.

Gorbar, E. V., M. Hashimoto, and V. A. Miransky, 2006a, *Phys. Lett.* **B632**, 305.

Gorbar, E. V., M. Hashimoto, and V. A. Miransky, 2007, *Phys. Rev.* **D75**, 085012.

Gorbar, E. V., M. Hashimoto, V. A. Miransky, and I. A. Shovkovy, 2006b, *Phys. Rev.* **D73**, 111502.

Greiter, M., F. Wilczek, and E. Witten, 1989, *Mod. Phys. Lett.* **B3**, 903.

Grigorian, H., D. Blaschke, and D. Voskresensky, 2005, *Phys. Rev.* **C71**, 045801.

Gubankova, E., W. V. Liu, and F. Wilczek, 2003, *Phys. Rev. Lett.* **91**, 032001.

Halperin, B. i., T. C. Lubensky, and S.-k. Ma, 1974, *Phys. Rev. Lett.* **32**, 292.

Hands, S., and D. N. Walters, 2004, *Phys. Rev.* **D69**, 076011.

Hashimoto, M., 2006, *Phys. Lett.* **B642**, 93.

Hashimoto, M., and V. A. Miransky, 2007, *Prog. Theor. Phys.* **118**, 303.

Haskell, B., N. Andersson, D. I. Jones, and L. Samuelsson, 2007, *Phys. Rev. Lett.* **99**(23), 231101.

Hatsuda, T., M. Tachibana, N. Yamamoto, and G. Baym, 2006, *Phys. Rev. Lett.* **97**, 122001.

Heinke, C. O., P. G. Jonker, R. Wijnands, and R. E. Taam, 2007, *Astrophys. J.* **660**, 1424.

Heiselberg, H., and C. J. Pethick, 1993, *Phys. Rev.* **D48**, 2916.

Higgs, P. W., 1964, *Phys. Lett.* **12**, 132.

Hong, D. K., 2000a, *Nucl. Phys.* **B582**, 451.

Hong, D. K., 2000b, *Phys. Lett.* **B473**, 118.

Hong, D. K., 2005, eprint hep-ph/0506097.

Huang, M., 2005, *Int. J. Mod. Phys.* **E14**, 675.

Huang, M., 2006, *Phys. Rev.* **D73**, 045007.

Huang, M., and I. Shovkovy, 2003, *Nucl. Phys.* **A729**, 835.

Huang, M., and I. A. Shovkovy, 2004a, *Phys. Rev.* **D70**, 051501.

Huang, M., and I. A. Shovkovy, 2004b, *Phys. Rev.* **D70**, 094030.

Iida, K., and G. Baym, 2001, *Phys. Rev.* **D63**, 074018.

Iida, K., and G. Baym, 2002, *Phys. Rev.* **D66**, 014015.

Iida, K., and K. Fukushima, 2006, *Phys. Rev.* **D74**, 074020.

Ipp, A., A. Gerhold, and A. Rebhan, 2004, *Phys. Rev.* **D69**, 011901.

Ippolito, N. D., G. Nardulli, and M. Ruggieri, 2007, *JHEP* **04**, 036.

Iwamoto, N., 1980, *Phys. Rev. Lett.* **44**, 1637.

Iwasaki, M., and T. Iwado, 1995, *Phys. Lett.* **B350**, 163.

Jaikumar, P., M. Prakash, and T. Schäfer, 2002, *Phys. Rev.* **D66**, 063003.

Jaikumar, P., S. Reddy, and A. W. Steiner, 2006, *Phys. Rev. Lett* **96**, 041101.

Jaikumar, P., G. Rupak, and A. W. Steiner, 2008, *Phys. Rev.* **D78**, 123007.

Jones, P. B., 1997, *Phys. Rev. Lett.* **79**, 792.

Jonker, P. G., D. Steeghs, D. Chakrabarty, and A. M. Juett, 2007, *Astrophys. J.* **665**, L147.

Kaplan, D. B., and A. E. Nelson, 1986, *Phys. Lett.* **B175**, 57.

Kaplan, D. L., B. M. Gaensler, S. R. Kulkarni, and P. O. Slane, 2006, *Astrophys. J. Suppl.* **163**, 344.

Kim, E., and M. H. W. Chan, 2004a, *Nature* **427**, 225.

Kim, E., and M. H. W. Chan, 2004b, *Science* **305**, 1941.

Kitazawa, M., D. H. Rischke, and I. A. Shovkovy, 2008, *Phys. Lett.* **B663**, 228.

Kleinhaus, V., M. Buballa, D. Nickel, and M. Oertel, 2007, *Phys. Rev.* **D76**, 074024.

Kryjevski, A., 2003, *Phys. Rev.* **D68**, 074008.

Kryjevski, A., 2008, *Phys. Rev.* **D77**, 014018.

Kryjevski, A., D. B. Kaplan, and T. Schäfer, 2005, *Phys. Rev.* **D71**, 034004.

Kryjevski, A., and T. Schäfer, 2005, *Phys. Lett.* **B606**, 52.

Kryjevski, A., and D. Yamada, 2005, *Phys. Rev.* **D71**, 014011.

Kundu, J., and K. Rajagopal, 2002, *Phys. Rev.* **D65**, 094022.

Kurkela, A., P. Romatschke, and A. Vuorinen, 2010, *Phys. Rev.* **D81**, 105021.

Kurkela, A., P. Romatschke, A. Vuorinen, and B. Wu, 2010, eprint 1006.4062.

Landau, L. D., and E. M. Lifshitz, 1981, *Theory of Elasticity* (Pergamon, Oxford), 3rd edition.

Landau, L. D., and E. M. Lifshitz, 1987, *Fluid Mechanics* (Pergamon, Oxford), 2nd edition.

Larkin, A. I., and Y. N. Ovchinnikov, 1965, *Sov. Phys. JETP* **20**, 762.

Lattimer, J. M., and M. Prakash, 2001, *Astrophys. J.* **550**, 426.

Lattimer, J. M., and M. Prakash, 2004, *Science* **304**, 536.

Lattimer, J. M., and M. Prakash, 2007, *Phys. Rept.* **442**, 109.

Leggett, A. J., 1970, *Phys. Rev. Lett.* **25**, 1543.

Leibovich, A. K., K. Rajagopal, and E. Shuster, 2001, *Phys. Rev.* **D64**, 094005.

Lin, L.-M., 2007, *Phys. Rev.* **D76**, 081502.

Lindblom, L., B. J. Owen, and S. M. Morsink, 1998, *Phys. Rev. Lett.* **80**, 4843.

Link, B., 2006, *Astron. Astrophys.* **458**, 881.

Link, B., 2007a, talk given at INT workshop on The Neutron Star Crust and Surface, Seattle, and private communication.

Link, B., 2007b, *Astrophys. and Space Sci.* **308**, 435.

Link, B., and R. I. Epstein, 1996, *Astrophys. J.* **457**, 854.

Link, B., R. I. Epstein, and G. Baym, 1993, *Astrophys. J.* **403**, 285.

Litim, D. F., and C. Manuel, 2001, *Phys. Rev.* **D64**, 094013.

Lugones, G., and J. E. Horvath, 2002, *Phys. Rev.* **D66**, 074017.

Madsen, J., 1992, *Phys. Rev.* **D46**, 3290.

Malekzadeh, H., 2006, *Phys. Rev.* **D74**, 065011.

Mannarelli, M., K. Rajagopal, and R. Sharma, 2006, *Phys. Rev.* **D73**, 114012.

Mannarelli, M., K. Rajagopal, and R. Sharma, 2007, *Phys. Rev.* **D76**, 074026.

Manuel, C., A. Dobado, and F. J. Llanes-Estrada, 2005, *JHEP* **09**, 076.

Manuel, C., and F. J. Llanes-Estrada, 2007, *JCAP* **0708**, 001.

Manuel, C., and K. Rajagopal, 2002, *Phys. Rev. Lett.* **88**, 042003.

Marhauser, F., D. Nickel, M. Buballa, and J. Wambach, 2007, *Phys. Rev.* **D75**, 054022.

Matsuura, T., K. Iida, T. Hatsuda, and G. Baym, 2004, *Phys. Rev.* **D69**, 074012.

McLerran, L., and R. D. Pisarski, 2007, *Nucl. Phys.* **A796**, 83.

Migdal, A. B., 1971, *Zh. Eksp. Teor. Fiz.* **61**, 2209.

Miransky, V. A., and I. A. Shovkovy, 2002, *Phys. Rev. Lett.* **88**, 111601.

Nakano, E., M. Nitta, and T. Matsuura, 2008, *Phys. Rev.* **D78**, 045002.

Negele, J. W., and D. Vautherin, 1973, *Nucl. Phys.* **A207**, 298.

Neumann, F., M. Buballa, and M. Oertel, 2003, *Nucl. Phys.* **A714**, 481.

Nickel, D., J. Wambach, and R. Alkofer, 2006, *Phys. Rev.* **D73**, 114028.

Noronha, J. L., H.-c. Ren, I. Giannakis, D. Hou, and D. H. Rischke, 2006, *Phys. Rev.* **D73**, 094009.

Noronha, J. L., and I. A. Shovkovy, 2007, *Phys. Rev.* **D76**, 105030.

Ouyed, R., and F. Sannino, 2001, *Phys. Lett.* **B511**, 66.

Owen, B. J., 2005, *Phys. Rev. Lett.* **95**, 211101.

Owen, B. J., et al., 1998, *Phys. Rev.* **D58**, 084020.

Ozel, F., D. Psaltis, S. Ransom, P. Demorest, and M. Alford, 2010, *Astrophys. J.* **724**, L199.

Page, D., J. M. Lattimer, M. Prakash, and A. W. Steiner, 2004, *Astrophys. J. Suppl.* **155**, 623.

Page, D., and S. Reddy, 2006, *Ann. Rev. Nucl. Part. Sci.* **56**, 327.

Pines, D., and M. A. Alpar, 1985, *Nature* **316**, 27.

Pisarski, R. D., and D. H. Rischke, 1999a, *Phys. Rev. Lett.* **83**, 37.

Pisarski, R. D., and D. H. Rischke, 1999b, eprint nucl-th/9907094.

Pisarski, R. D., and D. H. Rischke, 2000a, *Phys. Rev.* **D61**, 074017.

Pisarski, R. D., and D. H. Rischke, 2000b, *Phys. Rev.* **D61**, 051501.

Prakash, M., J. M. Lattimer, J. A. Pons, A. W. Steiner, and S. Reddy, 2001, *Lect. Notes Phys.* **578**, 364.

Rajagopal, K., and A. Schmitt, 2006, *Phys. Rev.* **D73**, 045003.

Rajagopal, K., and R. Sharma, 2006a, *J. Phys.* **G32**, S483.

Rajagopal, K., and R. Sharma, 2006b, *Phys. Rev.* **D74**, 094019.

Rajagopal, K., and F. Wilczek, 2000, eprint hep-ph/0011333.

Rajagopal, K., and F. Wilczek, 2001, *Phys. Rev. Lett.* **86**, 3492.

Rapp, R., T. Schäfer, E. V. Shuryak, and M. Velkovsky, 1998, *Phys. Rev. Lett.* **81**, 53.

Rapp, R., T. Schäfer, E. V. Shuryak, and M. Velkovsky, 2000, *Annals Phys.* **280**, 35.

Ravenhall, D. G., C. J. Pethick, and J. R. Wilson, 1983, *Phys. Rev. Lett.* **50**(26), 2066.

Reddy, S., and G. Rupak, 2005, *Phys. Rev.* **C71**, 025201.

Reddy, S., M. Sadzikowski, and M. Tachibana, 2003a, *Phys. Rev.* **D68**, 053010.

Reddy, S., M. Sadzikowski, and M. Tachibana, 2003b, *Nucl. Phys.* **A714**, 337.

Ren, H.-c., 2004, eprint hep-ph/0404074.

Rischke, D. H., 2000a, *Phys. Rev.* **D62**, 054017.

Rischke, D. H., 2000b, *Phys. Rev.* **D62**, 034007.

Rischke, D. H., 2004, *Prog. Part. Nucl. Phys.* **52**, 197.

Rischke, D. H., D. T. Son, and M. A. Stephanov, 2001, *Phys. Rev. Lett.* **87**, 062001.

Rüster, S. B., and D. H. Rischke, 2004, *Phys. Rev.* **D69**, 045011.

Rüster, S. B., I. A. Shovkovy, and D. H. Rischke, 2004, *Nucl. Phys.* **A743**, 127.

Rüster, S. B., V. Werth, M. Buballa, I. A. Shovkovy, and D. H. Rischke, 2005, *Phys. Rev.* **D72**, 034004.

Sa'd, B. A., I. A. Shovkovy, and D. H. Rischke, 2007, *Phys. Rev.* **D75**, 125004.

Sawyer, R. F., 1972, *Phys. Rev. Lett.* **29**, 382.

Scalapino, D. J., 1972, *Phys. Rev. Lett.* **29**, 386.

Schäfer, T., 2000a, *Nucl. Phys.* **B575**, 269.

Schäfer, T., 2000b, *Phys. Rev.* **D62**, 094007.

Schäfer, T., 2002, *Phys. Rev.* **D65**, 094033.

Schäfer, T., 2003a, *Nucl. Phys.* **A728**, 251.

Schäfer, T., 2003b, eprint hep-ph/0304281.

Schäfer, T., 2006, *Phys. Rev. Lett.* **96**, 012305.

Schäfer, T., and K. Schwenzer, 2004a, *Phys. Rev.* **D70**, 114037.

Schäfer, T., and K. Schwenzer, 2004b, *Phys. Rev.* **D70**, 054007.

Schäfer, T., and K. Schwenzer, 2006, *Phys. Rev. Lett.* **97**, 092301.

Schäfer, T., D. T. Son, M. A. Stephanov, D. Toublan, and J. J. M. Verbaarschot, 2001, *Phys. Lett.* **B522**, 67.

Schäfer, T., and F. Wilczek, 1999a, *Phys. Rev. Lett.* **82**, 3956.

Schäfer, T., and F. Wilczek, 1999b, *Phys. Rev.* **D60**, 074014.

Schmidt, C., 2006, *PoS* **LAT2006**, 021.

Schmitt, A., 2005, *Phys. Rev.* **D71**, 054016.

Schmitt, A., 2010, *Lect. Notes Phys.* **811**, 1.

Schmitt, A., I. A. Shovkovy, and Q. Wang, 2006, *Phys. Rev.* **D73**, 034012.

Schmitt, A., Q. Wang, and D. H. Rischke, 2002, *Phys. Rev.* **D66**, 114010.

Schmitt, A., Q. Wang, and D. H. Rischke, 2003, *Phys. Rev. Lett.* **91**, 242301.

Schmitt, A., Q. Wang, and D. H. Rischke, 2004, *Phys. Rev.* **D69**, 094017.

Sedrakian, A., I. Wasserman, and J. M. Cordes, 1999, *Astrophys. J.* **524**, 341S.

Shabanova, T. V., A. G. Lyne, and J. O. Urama, 2001, *Astrophys. J.* **552**, 321.

Shankar, R., 1994, *Rev. Mod. Phys.* **66**, 129.

Shovkovy, I., and M. Huang, 2003, *Phys. Lett.* **B564**, 205.

Shovkovy, I. A., 2005, *Found. Phys.* **35**, 1309.

Shovkovy, I. A., and P. J. Ellis, 2002, *Phys. Rev.* **C66**, 015802.

Shovkovy, I. A., and P. J. Ellis, 2003, *Phys. Rev.* **C67**, 048801.

Shovkovy, I. A., and L. C. R. Wijewardhana, 1999, *Phys. Lett.* **B470**, 189.

Shuster, E., and D. T. Son, 2000, *Nucl. Phys.* **B573**, 434.

Son, D. T., 1999, *Phys. Rev.* **D59**, 094019.

Son, D. T., 2002, eprint hep-ph/0204199.

Son, D. T., 2005, *Phys. Rev. Lett.* **94**, 175301.

Son, D. T., and M. A. Stephanov, 2000a, *Phys. Rev.* **D61**, 074012.

Son, D. T., and M. A. Stephanov, 2000b, *Phys. Rev.* **D62**, 059902.

Son, D. T., and M. A. Stephanov, 2006, *Phys. Rev. A* **74**, 013614.

Son, D. T., M. A. Stephanov, and A. R. Zhitnitsky, 2001, *Phys. Rev. Lett.* **86**, 3955.

Stairs, I. H., A. G. Lyne, and S. L. Shemar, 2000, *Nature* **406**, 484.

Steiner, A. W., S. Reddy, and M. Prakash, 2002, *Phys. Rev.* **D66**, 094007.

Strohmayer, T., H. M. van Horn, S. Ogata, H. Iyetomi, and S. Ichimaru, 1991, *Astrophys. J.* **375**, 679.

Wang, Q., and D. H. Rischke, 2002, *Phys. Rev.* **D65**, 054005.

Warringa, H. J., 2006, eprint hep-ph/0606063.

Weber, F., 1999, *Pulsars as Astrophysical Laboratories for Nuclear and Particle Physics* (IOP Publishing Ltd., Bristol).

Weber, F., A. Torres i Cuadrat, A. Ho, and P. Rosenfield, 2006, *PoS* **JHW2005**, 018.

Yakovlev, D. G., A. D. Kaminker, O. Y. Gnedin, and P. Haensel, 2001, *Phys. Rept.* **354**, 1.

Yakovlev, D. G., and C. J. Pethick, 2004, *Ann. Rev. Astron. Astrophys.* **42**, 169.

Yakovlev, D. G., et al., 2005, *Nucl. Phys.* **A752**, 590.

Zarembo, K., 2000, *Phys. Rev.* **D62**, 054003.

11

The superfluid universe

G. E. Volovik

We discuss the phenomenology of quantum vacuum. Phenomenology of macroscopic systems has three sources: thermodynamics, topology, and symmetry. Thermodynamics of a self-sustained vacuum allows us to treat the problems related to the vacuum energy: cosmological constant problems. The natural value of the energy density of an equilibrium self-sustained vacuum is zero. Cosmology is discussed as the process of relaxation of a vacuum towards the equilibrium state. The present value of the cosmological constant is very small compared with the Planck scale, because the present Universe is very old and thus is close to equilibrium.

Momentum space topology determines the universality classes of fermionic vacua. The Standard Model vacuum both in its massless and massive states is topological medium. The vacuum in its massless state shares the properties of superfluid ^{3}He-A, which is topological superfluid. It belongs to the Fermi-point universality class, which has topologically protected gapless fermionic quasiparticles. At low energy they behave as relativistic massless Weyl fermions. Gauge fields and gravity emerge together with Weyl fermions at low energy. This allows us to treat the hierarchy problem in the Standard Model: the masses of elementary particles are very small compared with the Planck scale because the natural value of the quark and lepton masses is zero. The small non-zero masses appear in the infrared region, where the quantum vacuum acquires the properties of topological superfluid ^{3}He-B and 3+1 topological insulators. The other topological media in dimensions 2+1 and 3+1 are also discussed. In most cases, topology is supported by the discrete symmetry of the underlying microscopic system, which indicates the important role of discrete symmetry in the Standard Model.

11.1 Introduction. Phenomenology of quantum vacuum

11.1.1 Vacuum as a macroscopic many-body system

The aether of the 21st century is the quantum vacuum. The quantum aether is a new form of matter. This substance has very peculiar properties, strikingly different from the other forms of matter (solids, liquids, gases, plasmas, Bose condensates, radiation, etc.) and from all the old aethers. The new aether has equation of state, $p = -\epsilon$; it is Lorentz invariant; and as follows from recent cosmological observations its energy density is about 10^{-29} g/cm^3 (i.e. the quantum aether is by 29 orders of magnitude lighter than water) and it is actually anti-gravitating.

Quantum vacuum can be viewed as a macroscopic many-body system. The characteristic energy scale in our vacuum (analog of atomic scale in quantum liquids) is Planck energy, $E_{\rm P} = (\hbar c^5/G)^{1/2} \sim 10^{19}\,{\rm GeV} \sim 10^{32}\,{\rm K}$. Our present Universe has extremely low energies and temperatures compared with the Planck scale: even the highest energy in the present-day accelerators is extremely small compared with Planck energy: $E_{\rm max} \sim 10\,{\rm TeV} \sim 10^{17}\,{\rm K} \sim 10^{-15}\,E_{\rm P}$. The temperature of cosmic background radiation is much smaller, $T_{CMBR} \sim 1\,{\rm K} \sim 10^{-32}\,E_{\rm P}$.

Cosmology belongs to ultra-low frequency physics. Expansion of the universe is extremely slow: the Hubble parameter compared with the characteristic Planck frequency $\omega_P = (c^5/G\hbar)^{1/2}$ is $H \sim 10^{-60}\,\omega_P$. This also means that at the moment our Universe is extremely close to equilibrium. This is natural for any many-body system: if there is no energy flux from the environment, the energy will be radiated away and the system will be approaching an equilibrium state with vanishing temperature and motion.

According to Landau, although the macroscopic many-body system can be very complicated, at low energy and temperatures its description is highly simplified. Its behavior can be described in a fully phenomenological way, using symmetry and thermodynamic considerations. Later, it became clear that another factor also governs the low energy properties of a macroscopic system—topology. The quantum vacuum is probably a very complicated system. However, using these three sources—thermodynamics, symmetry, and topology—we can try to construct the phenomenological theory of the quantum vacuum near its equilibrium state.

11.1.2 Three sources of phenomenology: thermodynamics, symmetry, and topology

Following Landau, at low energy, $E \ll E_{\rm P}$, the macroscopic quantum system—superfluid liquid or our universe—contains two main components: vacuum (the ground state) and matter (fermionic and bosonic quasiparticles above the ground state). The physical laws which govern the matter component are more or less clear to us, because we are able to do experiments in the low-energy region and construct the theory. The quantum vacuum occupies the Planckian and trans-Planckian energy scales and it is governed by microscopic (trans-Planckian) physics, which is still unknown. However, using our experience with a similar two-component quantum liquid we can expect the quantum vacuum component also to obey the thermodynamic laws, which emerge in any macroscopically large system, relativistic or non-relativistic. This approach allows us to treat the cosmological constant problem. The cosmological constant was introduced by Einstein [1], and was interpreted as the energy density of the quantum vacuum [2, 3]. Astronomical observations (see, e.g. Refs. [4–6]) confirmed the existence of the cosmological constant whose value corresponds to the energy density of order $\Lambda_{\rm obs} \sim E^4_{\rm obs}$, with the characteristic energy scale $E_{\mathrm{obs}} \sim 10^{-3}$ eV. However, naive and intuitive theoretical estimation of the vacuum energy density as the zero-point energy of quantum fields suggests that vacuum energy must have the Planck energy scale: $\sim E_{\mathrm{P}}^4 \sim 10^{120} E_{\mathrm{obs}}$. The huge disagreement between the naive expectations and observation is resolved naturally using thermodynamics of the quantum vacuum, discussed in Sections 11.2–11.5 of this review. We shall see that intuitive estimation of the vacuum energy density as $\sim E_{\mathrm{P}}^4$ is correct, but the relevant vacuum energy which enters the Einstein equations as the cosmological constant is somewhat different, and its value in the fully equilibrium vacuum is zero.

The second element of the Landau phenomenological approach to macroscopic systems is symmetry. This is in the basis of the modern theory of particle physics—the Standard Model, and its extension to higher energy—the Grand Unification (GUT). The vacuum of the Standard Model and GUT obeys the fundamental symmetries, which become spontaneously broken at low energy, but which are restored when the Planck energy scale is approached from below.

This approach contains another huge disagreement between the naive expectations and observation. This concerns masses of elementary particles. The naive and intuitive estimation tells us that these masses should be in the order of the characteristic energy scale in our vacuum, which is the Planck energy scale, $M_{\mathrm{theor}} \sim E_{\mathrm{P}}$. However, the masses of observed particles are many orders of magnitude smaller, being below the electroweak energy scale $M_{\mathrm{obs}} < E_{\mathrm{ew}} \sim 1\,\mathrm{TeV} \sim 10^{-16} E_{\mathrm{P}}$. This is called the hierarchy problem. There should be a general principle, which would resolve this paradox. This is the principle of emergent physics based on the topology in momentum space. This approach supports our intuitive estimation of fermion masses of order E_{P}, however, this estimation is not valid for those vacua where the massless fermions are topologically protected.

Let us consider a fermionic quantum liquid, ^{3}He. In the laboratory we have four different states of this liquid. These are normal liquid ^{3}He, and three superfluid phases: ^{3}He-A, ^{3}He-B, and ^{3}He-A$_1$. Only one of these, ^{3}He-B, has fully gapped spectrum of fermionic excitations, and for this liquid the intuitive estimation of the gap (analog of mass) in terms of the characteristic energy scale is correct. However, the other liquids are gapless. The gaplessness in these fermionic systems is protected by the momentum space topology and thus is fundamental: it does not depend, therefore, on microscopic physics being robust to perturbative modification of the interaction between the atoms of the liquid.

11.1.3 Vacuum as a topological medium

Topology operates in particular with integer numbers—topological charges—which do not change under small deformations of the system. The conservation of these topological charges protects the Fermi surface and another object in momentum space—the Fermi point—from destruction. They survive when interaction between the fermions is introduced and modified. When the momentum of a particle approaches the Fermi surface or the Fermi point its energy necessarily vanishes. Thus the topology is the main reason why there are gapless quasiparticles in quantum liquids and (nearly) massless elementary particles in our Universe.

Topology provides the complementary anti-GUT approach in which the "fundamental" symmetry and "fundamental" fields of GUT gradually emerge, together with "fundamental" physical laws, when the Planck energy scale is approached from above [7, 8]. The emergence of the "fundamental" laws of physics is provided by that general property of topology—robustness to the detail of microscopic trans-Planckian physics. As a result, the physical laws which emerge at low energy are generic. They do not much depend on the details of the trans-Planckian subsystem, but are determined by the universality class, to which the whole system belongs.

In this scheme, fermions are primary objects. Approaching the Planck energy scale from above, they are transformed to Standard Model chiral fermions and give rise to secondary objects: gauge fields and gravity. Below the Planck scale, the GUT scenario intervenes, giving rise to symmetry-breaking at low energy. This is accompanied by formation of composite objects, Higgs bosons, and tiny Dirac masses of quarks and leptons.

In the GUT scheme, general relativity is assumed to be as fundamental as quantum mechanics, while in the second scheme, general relativity is a secondary phenomenon. In the anti-GUT scheme, general relativity is the effective theory, describing the dynamics of the effective metric, experienced by the effective low-energy fields. It is a side product of quantum field theory or of quantum mechanics in the vacuum with Fermi point.

Vacua with topologically protected gapless (massless) fermions are representative of the broader class of topological media. In condensed matter this includes topological insulators (see reviews [9, 10]), topological semimetals (see [11–18]), topological superconductors and superfluids, states which experience the quantum Hall effect, and other topologically non-trivial gapless and gapped phases of matter. Topological media have many peculiar properties: topological stability of gap nodes; topologically protected edge states including Majorana fermions; topological quantum phase transitions occurring at $T = 0$; topological quantization of physical parameters including Hall and spin-Hall conductivity; chiral anomaly; topological Chern–Simons and Wess–Zumino actions; etc.

It appears that the quantum vacuum of the Standard Model is topologically non-trivial, both in its massless and massive states. In the massless state, the quantum vacuum is topologically similar to the superfluid ^{3}He-A and a gapless semimetal. In the massive state, the quantum vacuum is topologically similar to the superfluid ^{3}He-B and a 3+1-dimensional topological insulator. This is discussed in Sections 11.6–11.8.

11.2 Quantum vacuum as a self-sustained medium

11.2.1 Vacuum energy and cosmological constant

There is a huge contribution to the vacuum energy density, which comes from the ultraviolet (Planckian) degrees of freedom, and is of order $E_{\rm P}^4 \approx \left(10^{28}\,{\rm eV}\right)^4$. The observed cosmological constant is smaller by many orders of magnitude and corresponds to the energy density of the vacuum $\rho_{\mathrm{vac}} \sim \left(10^{-3}\,\mathrm{eV}\right)^4$. In general relativity, the cosmological constant is an arbitrary constant, and thus its smallness requires fine tuning. If gravitation were a truly fundamental interaction, it would be hard to understand why the energies stored in the quantum vacuum would not gravitate at all [19]. If, however, gravitation were only a low-energy effective interaction, it could be that the corresponding gravitons as quasiparticles do not feel *all* microscopic degrees of freedom (gravitons would be analogous to small-amplitude waves at the surface of the ocean) and that the gravitating effect of the vacuum energy density would be effectively *tuned away*, and the cosmological constant would be naturally small or zero [8, 20].

11.2.2 Variables for a Lorentz invariant vacuum

A particular mechanism for nullification of the relevant vacuum energy works for those vacua which have the properties of a *self-sustained medium* [21–26]. A self-sustained vacuum is a medium with a definite macroscopic volume, even in the absence of an environment. A condensed matter example is a droplet of quantum liquid at zero temperature in empty space. The observed near-zero value of the cosmological constant compared with Planck-scale values suggests that the quantum vacuum of our Universe belongs to this class of systems. As with any medium of this kind, the equilibrium vacuum would be homogeneous and extensive. The homogeneity assumption is indeed supported by the observed flatness and smoothness of our Universe [27–29]. The implication from this is that the energy of the equilibrium quantum vacuum would be proportional to the volume considered.

Usually, a self-sustained medium is characterized by an *extensive conserved quantity*, whose total value determines the actual volume of the system [30, 31]. The quantum liquid at $T = 0$ is a self-sustained system because of the conservation law for the number of particles N, and its state is characterized by the particle density n which acquires a non-zero value, $n = n_0$ in the equilibrium ground state. As distinct from condensed matter systems, the quantum vacuum of our Universe is a relativistic invariant system. The Lorentz invariance of the vacuum imposes strong constraints on the possible form of the vacuum analog of the particle density n. An example of a possible vacuum variable is a symmetric tensor $q^{\mu\nu}$, which in a homogeneous vacuum is proportional to the metric tensor,

$$q^{\mu\nu} = q\, g^{\mu\nu}. \tag{11.1}$$

This variable satisfies the Lorentz invariance of the vacuum. Another example is the 4-tensor $q^{\mu\nu\alpha\beta}$, which in a homogeneous vacuum is proportional either to the fully antisymmetric Levi-Civita tensor,

$$q^{\mu\nu\alpha\beta} = q\, e^{\mu\nu\alpha\beta}, \tag{11.2}$$

or to the product of metric tensors such as,

$$q^{\mu\nu\alpha\beta} = q\left(g_{\alpha\mu}g_{\beta\nu} - g_{\alpha\nu}g_{\beta\mu}\right). \tag{11.3}$$

A scalar field is also a Lorentz invariant variable, but it does not satisfy another necessary condition of the self-sustained system: the vacuum variable q must obey some kind of conservation law. Below we consider some examples which satisfy the two conditions: Lorentz invariance of the perfect vacuum state and the conservation law.

11.2.3 Yang–Mills chiral condensate as an example

Let us first consider as an example the chiral condensate of gauge fields. This can be the gluonic condensate in QCD [32, 33], or any other condensate of Yang–Mills fields, if it is Lorentz invariant. We assume that the Savvidy vacuum [34] is absent, i.e. the vacuum expectation value of the color magnetic field is zero (we shall omit color indices):

$$\langle F_{\alpha\beta} \rangle = 0, \tag{11.4}$$

while the vacuum expectation value of the quadratic form is non-zero:

$$\langle F_{\alpha\beta} F_{\mu\nu} \rangle = \frac{q}{24} \sqrt{-g} e_{\alpha\beta\mu\nu}. \tag{11.5}$$

Here, q is the anomaly-driven topological condensate (see e.g. [35]):

$$q = \langle \tilde{F}^{\mu\nu} F_{\mu\nu} \rangle = \frac{1}{\sqrt{-g}} e^{\alpha\beta\mu\nu} \langle F_{\alpha\beta} F_{\mu\nu} \rangle. \tag{11.6}$$

In the homogeneous static vacuum state, the q-condensate violates the P and T symmetries of the vacuum, but it conserves the combined PT symmetry.

11.2.3.1 *Cosmological term*

Let us choose the vacuum action in the form,

$$S_q = \int d^4x \sqrt{-g} \epsilon(q), \tag{11.7}$$

with q given by (11.6). The energy–momentum tensor of the vacuum field q is obtained by variation of the action over $g^{\mu\nu}$:

$$T^q_{\mu\nu} = -\frac{2}{\sqrt{-g}} \frac{\delta S_q}{\delta g^{\mu\nu}} = \epsilon(q)\, g_{\mu\nu} - 2 \frac{\partial \epsilon}{\partial q} \frac{\partial q}{\partial g^{\mu\nu}}. \tag{11.8}$$

Using (11.5) and (11.6) one obtains,

$$\frac{\partial q}{\partial g^{\mu\nu}} = \frac{1}{2} q g_{\mu\nu}. \tag{11.9}$$

and thus,

$$T^q_{\mu\nu} = g_{\mu\nu} \rho_{\rm vac}(q), \quad \rho_{\rm vac}(q) = \epsilon(q) - q \frac{\partial \epsilon}{\partial q}. \tag{11.10}$$

In the Einstein equations this energy–momentum tensor plays the role of the cosmological term:

$$T^q_{\mu\nu} = \Lambda g_{\mu\nu}, \quad \Lambda = \rho_{\rm vac}(q) = \epsilon(q) - q \frac{\partial \epsilon}{\partial q}. \tag{11.11}$$

It is important that the cosmological constant is given, not by the vacuum energy as is usually assumed, but by the thermodynamic potential, $\rho_{\rm vac} = \epsilon(q) - \mu q$, where μ is thermodynamically conjugate to q variable, $\mu = d\epsilon/dq$. Below, when we consider dynamics we shall see that this fact reflects conservation of the variable q.

The crucial difference between the vacuum energy, $\epsilon(q)$ and the thermodynamic potential, $\rho_{\rm vac} = \epsilon(q) - \mu q$, is revealed when we consider the corresponding quantities in the ground state of quantum liquids, the energy density of the liquid $\epsilon(n)$ and the density of the grand canonical energy, $\epsilon(n) - \mu n$, which enters macroscopic thermodynamics due to conservation of particle number. The first, $\epsilon(n)$, has the value dictated by atomic physics, which is equivalent to $E_{\rm P}^4$ in the quantum vacuum. Contrary to this, the second one equals minus pressure, $\epsilon(n) - \mu n = -P$, according to the Gibbs–Duhem thermodynamic relation at $T = 0$. Thus its value is dictated not by microscopic physics, but by external conditions. In the absence of environment, the external pressure is zero, and the value of $\epsilon(n) - \mu n$ in a fully equilibrium ground state of the liquid is zero. This is valid for any self-sustained macroscopic system, including the self-sustained quantum vacuum, which suggests a natural solution of the main cosmological constant problem.

11.2.3.2 *Conservation law for q*

The equation for q in flat space can be obtained from the Maxwell equation, which in turn is obtained by variation of the action over the gauge field A_μ:

$$\nabla_\mu \left(\frac{\partial \epsilon}{\partial q} \tilde{F}^{\mu\nu} \right) = 0, \tag{11.12}$$

where ∇_μ is the covariant derivative. Since $\nabla_\mu \tilde{F}^{\mu\nu} = 0$, equation (11.36) is reduced to

$$\nabla_\mu \left(\frac{\partial \epsilon}{\partial q} \right) = 0. \tag{11.13}$$

The solution of this equation is

$$\frac{\partial \epsilon}{\partial q} = \mu, \tag{11.14}$$

where μ is an integration constant. In thermodynamics, this μ will play the role of the chemical potential, which is thermodynamically conjugate to q. This demonstrates that q obeys the conservation law and thus can be the proper variable for describing the self-sustained vacuum.

11.2.4 Four-form field as an example

Another example of a vacuum variable appropriate for a self-sustained vacuum is given by the four-form field strength [36–44], which is expressed in terms of q in the following way:

$$F_{\alpha\beta\gamma\delta} \equiv q\, e_{\alpha\beta\gamma\delta} \sqrt{-\det g} = \nabla_{[\alpha} A_{\beta\gamma\delta]}, \tag{11.15a}$$

$$q^2 = -\frac{1}{24} F_{\alpha\beta\gamma\delta} F^{\alpha\beta\gamma\delta}, \tag{11.15b}$$

where $e_{\alpha\beta\gamma\delta}$ is the Levi-Civita tensor density; and the square bracket around space–time indices complete anti-symmetrization.

Originally the quadratic action was used for this field [36, 37], which corresponds with the special case of (11.7), with $\epsilon(q) = \frac{1}{2} q^2$. For general $\epsilon(q)$ one obtains the Maxwell equation,

$$\nabla_\alpha \left(\sqrt{-\det g} \, \frac{F^{\alpha\beta\gamma\delta}}{q} \frac{\partial \epsilon(q)}{\partial q} \right) = 0. \qquad (11.16)$$

Using (11.15a) the Maxwell equation is reduced to

$$\nabla_\alpha \left(\frac{\partial \epsilon(q)}{\partial q} \right) = 0. \qquad (11.17)$$

The first integral of (11.17) with integration constant μ gives Eq. (11.14) again, which reflects the conservation law for q.

Variation of the action over $g^{\mu\nu}$ again gives the cosmological constant (11.11) with $\Lambda = \rho_{\text{vac}} = \epsilon(q) - \mu q$. This demonstrates the universality of the macroscopic description of the self-sustained vacuum: description of the quantum vacuum in terms of q does not depend on microscopic details of the vacuum or on the nature of the vacuum variable.

11.2.5 Aether field as an example

Another example of vacuum variable q may be through a four-vector field $u^\mu(x)$. This vector field could be the four-dimensional analog of the concept of shift in the deformation theory of crystals. (Deformation theory can be described in terms of a metric field, with the roles of torsion and curvature fields played by dislocations and disclinations, respectively; see, e.g. Ref. [45] for a review.) A realization of u^μ could also be a 4-velocity field entering the description of the structure of space–time. It is the 4-velocity of "aether" [46–49].

The non-zero value of the 4-vector in the vacuum violates the Lorentz invariance of the vacuum. To restore this invariance one may assume that $u^\mu(x)$ is not an observable variable, instead the observables are its covariant derivatives, $\nabla_\nu u^\mu \equiv u^\mu_\nu$. This means that the action does not depend on u^μ explicitly but only depends on u^μ_ν:

$$S = \int_{\mathbb{R}^4} d^4x \, \epsilon(u^\mu_\nu), \qquad (11.18)$$

with an energy density containing even powers of u^μ_ν:

$$\epsilon(u^\mu_\nu) = K + K^{\alpha\beta}_{\mu\nu} \, u^\mu_\alpha u^\nu_\beta + K^{\alpha\beta\gamma\delta}_{\mu\nu\rho\sigma} \, u^\mu_\alpha u^\nu_\beta u^\rho_\gamma u^\sigma_\delta + \cdots \qquad (11.19)$$

According to the imposed conditions, the tensors $K^{\alpha\beta}_{\mu\nu}$ and $K^{\alpha\beta\gamma\delta}_{\mu\nu\rho\sigma}$ depend only on $g_{\mu\nu}$ or $g^{\mu\nu}$ and the same holds for the other K-like tensors in the ellipsis of (11.19). In particular, the tensor $K^{\alpha\beta}_{\mu\nu}$ of the quadratic term in (11.19) has the following form in the notation of Ref. [46]:

$$K^{\alpha\beta}_{\mu\nu} = c_1 \, g^{\alpha\beta} g_{\mu\nu} + c_2 \, \delta^\alpha_\mu \delta^\beta_\nu + c_3 \, \delta^\alpha_\nu \delta^\beta_\mu, \qquad (11.20)$$

for real constants c_n. Distinct from the original aether theory in Ref. [46], the tensor (11.20) does not contain a term $c_4 \, u^\alpha u^\beta g_{\mu\nu}$, since such a term would depend explicitly on u^μ and would contradict the Lorentz invariance of the quantum vacuum.

The equation of motion for u^μ in flat space,

$$\nabla_\nu \frac{\partial \epsilon}{\partial u^\mu_\nu} = 0, \tag{11.21}$$

has the Lorentz invariant solution expected for a vacuum-variable q-type field:

$$u^q_{\mu\nu} = q\, g_{\mu\nu}, \quad q = \text{constant}. \tag{11.22}$$

With this solution, the energy density in action (11.18) is simply $\epsilon(q)$ in terms of contracted coefficients K, $K^{\mu\nu}_{\mu\nu}$, and $K^{\mu\nu\rho\sigma}_{\mu\nu\rho\sigma}$ from (11.19). However, just as for previous examples, the energy–momentum tensor of the vacuum field obtained by variation over $g^{\mu\nu}$ and evaluated for solution (11.22), is expressed again in terms of the thermodynamic potential:

$$T^q_{\mu\nu} = \frac{2}{\sqrt{-g}} \frac{\delta S}{\delta g^{\mu\nu}} = g_{\mu\nu}\left(\epsilon(q) - q\frac{d\epsilon(q)}{dq}\right) = \rho_{\rm vac}(q) g_{\mu\nu}, \tag{11.23}$$

which corresponds to the cosmological constant in Einstein's gravitational field equations.

11.3 Thermodynamics of quantum vacuum

11.3.1 Liquid-like quantum vacuum

The zeroth order term K in (11.19) corresponds to a "bare" cosmological constant, which can be considered as a cosmological constant in the "empty" vacuum—vacuum with $q = 0$:

$$\Lambda_{\rm bare} = \epsilon(q = 0). \tag{11.24}$$

The non-zero value $q = q_0$ in the self-sustained vacuum does *not* violate Lorentz symmetry, but leads to compensation of the bare cosmological constant $\Lambda_{\rm bare}$ in the equilibrium vacuum. This illustrates the important difference between the two states of vacua. The quantum vacuum with $q = 0$ can only exist with external pressure $P = -\Lambda_{\rm bare}$. By analogy with condensed matter physics, this kind of quantum vacuum may be called "gas-like" (Fig. 11.1). The quantum vacuum with non-zero q is self-sustained: it can be stable at $P = 0$, provided that a stable non-zero solution of equation $\epsilon(q) - q\, d\epsilon/dq = 0$ exists. This kind of quantum vacuum may then be called "liquid-like".

The universal behavior of the self-sustained vacuum in equilibrium suggests that it obeys the same thermodynamic laws as any other self-sustained macroscopic system described by the conserved quantity q, such as a quantum

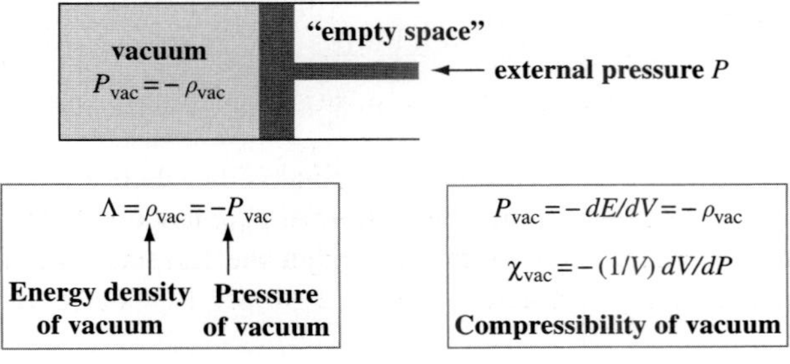

Fig. 11.1 Vacuum as a medium obeying macroscopic thermodynamic laws. Relativistic vacuum possesses energy density, pressure, and compressibility but has no momentum. In equilibrium, the vacuum pressure $P_{\rm vac}$ equals the external pressure P acting from the environment. The "gas-like" vacuum may exist only under external pressure. The "liquid-like" vacuum is self-sustained: it can be stable in the absence of external pressure. The thermodynamic energy density of the vacuum $\rho_{\rm vac}$ which enters the vacuum equation of state $\rho_{\rm vac} = -P_{\rm vac}$ does not coincide with the microscopic vacuum energy ϵ. While the natural value of ϵ is determined by the Planck scale, $\epsilon \sim E_{\rm P}^4$, the natural value of the macroscopic quantity, $\rho_{\rm vac}$, is zero for the self-sustained vacuum which may exist in the absence of environment, i.e. at $P = 0$. This may explain why the present cosmological constant, $\Lambda = \rho_{\rm vac}$, is small.

liquid. In other words, vacuum can be considered as a special quantum liquid which is Lorentz invariant in its ground state. This liquid is characterized by the Lorentz invariant "charge" density q—an analog of particle density n in non-relativistic quantum liquids.

Let us consider a large portion of such a vacuum liquid under external pressure P [21]. The volume V of quantum vacuum is variable, but its total "charge" $Q(t) \equiv \int d^3r\, q(\mathbf{r}, t)$ must be conserved, $\mathrm{d}Q/\mathrm{d}t = 0$. The energy of this portion of quantum vacuum at fixed total "charge" $Q = qV$ is then given by the thermodynamic potential,

$$W = E + PV = \int d^3r\, \epsilon\,(Q/V) + PV, \tag{11.25}$$

where $\epsilon\,(q)$ is the energy density in terms of charge density q. As the volume of the system is a free parameter, the equilibrium state of the system is obtained by variation over the volume V:

$$\frac{dW}{dV} = 0. \tag{11.26}$$

This gives an integrated form of the Gibbs–Duhem equation for the vacuum pressure:

$$P_{\mathrm{vac}} = -\epsilon(q) + q\,\frac{d\epsilon(q)}{dq} = -\rho_{\mathrm{vac}}(q), \tag{11.27}$$

whose solution determines the equilibrium value $q = q(P)$ and the corresponding volume $V(P, Q) = Q/q(P)$.

11.3.2 Macroscopic energy of quantum vacuum

Since the vacuum energy density is the vacuum pressure with a minus sign, Eq. (11.27) suggests that the relevant vacuum energy, which is revealed in the thermodynamics and dynamics of the low-energy Universe, is

$$\rho_{\mathrm{vac}}(q) = \epsilon(q) - q\,\frac{d\epsilon(q)}{dq}. \tag{11.28}$$

This is confirmed by Eqs. (11.11) and (11.23) for the energy–momentum tensor of a self-sustained vacuum, which demonstrates that it is $\rho_{\mathrm{vac}}\,(q)$ rather than $\epsilon\,(q)$, which enters the equation of state for the vacuum, and thus corresponds to the cosmological constant:

$$\Lambda = \rho_{\mathrm{vac}} = -P_{\mathrm{vac}}. \tag{11.29}$$

While the energy of microscopic quantity q is determined by the Planck scale, $\epsilon(q_0) \sim E_{\mathrm{P}}^4$, the relevant vacuum energy which sources the effective gravity is determined by a macroscopic quantity—the external pressure. In the absence of an environment, i.e. at zero external pressure, $P = 0$, one obtains that the pressure of pure and equilibrium vacuum is exactly zero:

$$\Lambda = -P_{\mathrm{vac}} = -P = 0. \tag{11.30}$$

Equation $\rho_{\mathrm{vac}}(q) = 0$ determines the equilibrium value q_0 of the equilibrium self-sustained vacuum. Thus, from thermodynamic arguments it follows that for any effective theory of gravity, the natural value of Λ is zero in equilibrium vacuum.

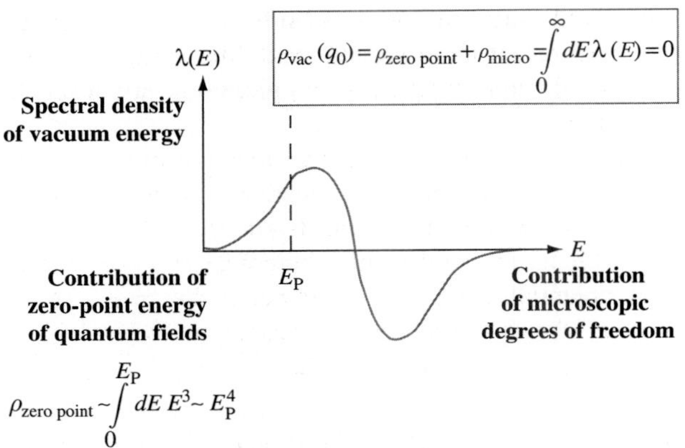

Fig. 11.2 Contribution of different energy scales into the macroscopic energy of the self-sustained system at $T = 0$. Zero-point energy of the effective bosonic and fermionic quantum fields gives rise to the diverging contribution to the energy of the system. In a quantum vacuum it is of order E_P^4. In equilibrium, this contribution is compensated without fine-tuning by microscopic degrees of freedom of the system (by trans-Planckian degrees of quantum vacuum correspondingly).

This result does not depend on the microscopic structure of the vacuum from which gravity emerges, and is actually the final result of the renormalization dictated by macroscopic physics. In a self-sustained quantum liquid the large contribution of the zero-point energy of the phonon field is naturally compensated by the microscopic (atomic) degrees of freedom of the quantum liquid. In the same manner, the huge contribution of the zero-point energy of macroscopic fields to the vacuum energy ρ_{vac} is naturally compensated by microscopic degrees of the self-sustained quantum vacuum: the vacuum variable q is adjusted automatically to nullify the macroscopic vacuum energy, $\rho_{\text{vac}}(q_0) = \rho_{\text{zero point}} + \rho_{\text{microscopic}} = 0$. The actual spectrum of the vacuum energy density (meaning the different contributions to ϵ from different energy scales) is not important for the cancelation mechanism, because it is dictated by thermodynamics. The particular example of the spectrum of the vacuum energy density is shown in Fig. 11.2, where the positive energy of the quantum vacuum, which comes from the zero-point energy of bosonic fields, is compensated by the negative contribution from trans-Planckian degrees of freedom [50].

Using the quantum-liquid counterpart of the self-sustained quantum vacuum as an example, one may predict the behavior of the vacuum after cosmological phase transition, when Λ is kicked from its zero value. The vacuum will readjust to a new equilibrium state with new q_0 so that Λ again approaches its equilibrium zero value [21]. The process of relaxation of the system to the equilibrium state depends on the details of the dynamics of the vacuum variable q and its interaction with matter fields, and we shall consider some examples of dynamical relaxation of Λ later.

11.3.3 Compressibility of the vacuum

Using the standard definition of the inverse of the isothermal compressibility, $\chi^{-1} \equiv -V\,dP/dV$ (Fig. 11.1), one obtains the compressibility of the vacuum by varying Eq. (11.27) at fixed $Q = qV$ [21]:

$$\chi_{\text{vac}}^{-1} \equiv -V\frac{dP_{\text{vac}}}{dV} = \left[q^2\,\frac{d^2\epsilon(q)}{dq^2}\right]_{q=q_0} > 0. \tag{11.31}$$

A positive value of the vacuum compressibility is a necessary condition for the stability of the vacuum. It is, in fact, the stability of the vacuum which is at the origin of the nullification of the cosmological constant in the absence of an external environment.

From the low-energy point of view, the compressibility of the vacuum $\chi_{\rm vac}$ is as fundamental a physical constant as the Newton constant $G_N = G(q = q_0)$. It enters equations describing the response of the quantum vacuum to different perturbations. While the natural value of the macroscopic quantity $P_{\rm vac}$ (and $\rho_{\rm vac}$) is zero, the natural values of the parameters $G(q = q_0)$ and $\chi_{\rm vac}(q = q_0)$ are determined by Planck physics and, correspondingly, are expected to be of order $1/E_{\rm P}^2$ and $1/E_{\rm P}^4$.

11.3.4 Thermal fluctuations of Λ and the volume of Universe

The compressibility of the vacuum $\chi_{\rm vac}$, although not measurable at the moment, can be used for estimation of the lower limit of the volume V of the Universe. This estimation follows from the upper limit for thermal fluctuations of the cosmological constant [51]. The mean square of thermal fluctuations of Λ equals the mean square of thermal fluctuations of the vacuum pressure, which in turn is determined by the thermodynamic equation [30]

$$\left\langle (\Delta\Lambda)^2 \right\rangle = \left\langle (\Delta P)^2 \right\rangle = \frac{T}{V\chi_{\rm vac}}. \tag{11.32}$$

Typical fluctuations of the cosmological constant Λ should not exceed the observed value: $\left\langle (\Delta\Lambda)^2 \right\rangle < \Lambda_{\rm obs}^2$. Let us assume, for example, that the temperature of the Universe is determined by the temperature $T_{\rm CMB}$ of the cosmic microwave background radiation. Then, using our estimate for vacuum compressibility $\chi_{\rm vac}^{-1} \sim E_{\rm P}^4$, one obtains that the volume V of our Universe highly exceeds the Hubble volume $V_H = R_H^3$—the volume of visible Universe inside the present cosmological horizon:

$$V > \frac{T_{\rm CMB}}{\chi_{\rm vac}\Lambda_{\rm obs}^2} \sim 10^{28} V_H. \tag{11.33}$$

This demonstrates that the real volume of the Universe is certainly not limited by the present cosmological horizon.

11.4 Dynamics of quantum vacuum

11.4.1 Action

In Section 11.2 a special quantity, the vacuum "charge" q, was introduced to describe the statics and thermodynamics of the self-sustained quantum vacuum. Now we can extend this approach to the dynamics of the vacuum charge. We expect to find some universal features of the vacuum dynamics, using several realizations of this vacuum variable. We start with the 4-form field strength [36–44] expressed in terms of q. The low-energy effective action takes the following general form:

$$S = -\int_{\mathbb{R}^4} d^4x \sqrt{|g|} \left(\frac{R}{16\pi G(q)} + \epsilon(q) + \mathcal{L}^{\rm M}(q, \psi) \right), \tag{11.34a}$$

$$q^2 \equiv -\frac{1}{24} F_{\kappa\lambda\mu\nu} F^{\kappa\lambda\mu\nu}, \quad F_{\kappa\lambda\mu\nu} \equiv \nabla_{[\kappa} A_{\lambda\mu\nu]}, \tag{11.34b}$$

$$F_{\kappa\lambda\mu\nu} = q\sqrt{|g|}\, e_{\kappa\lambda\mu\nu}, \quad F^{\kappa\lambda\mu\nu} = q\, e^{\kappa\lambda\mu\nu}/\sqrt{|g|}, \tag{11.34c}$$

where R denotes the Ricci curvature scalar; and $\mathcal{L}^{\mathrm{M}}$ is matter action. Throughout, we use the same conventions as in Ref. [52], in particular, those for the Riemann curvature tensor and the metric signature $(-+++)$.

The vacuum energy density ϵ in (11.34a) depends on the vacuum variable q which in turn is expressed via the 3-form field $A_{\lambda\mu\nu}$ and metric field $g_{\mu\nu}$ in (11.34b). The field ψ combines all the matter fields of the Standard Model. All possible constant terms in matter action (which include the zero-point energies from the Standard Model fields) are absorbed in the vacuum energy $\epsilon(q)$.

Since q describes the state of the vacuum, the parameters of the effective action—the Newton constant G and parameters which enter the matter action—must depend on q. This dependence results in particular in interaction between the matter fields and the vacuum. There are different sources of this interaction. For example, in the gauge field sector of the Standard Model, the running coupling contains the ultraviolet cutoff and thus depends on q:

$$\mathcal{L}^{\mathrm{G,q}} = \gamma(q) F^{\mu\nu} F_{\mu\nu}, \tag{11.35}$$

where $F_{\mu\nu}$ is the field strength of the particular gauge field (we have omitted the color indices). In the fermionic sector, q should enter parameters of the Yukawa interaction and fermion masses.

11.4.2 Vacuum dynamics

The variation of action (11.34a) over the three-form gauge field A gives the generalized Maxwell equations for the F-field,

$$\nabla_\nu \left(\sqrt{|g|}\, \frac{F^{\kappa\lambda\mu\nu}}{q} \left(\frac{d\epsilon(q)}{dq} + \frac{R}{16\pi} \frac{dG^{-1}(q)}{dq} + \frac{d\mathcal{L}^{\mathrm{M}}(q)}{dq} \right) \right) = 0. \tag{11.36}$$

Using (11.34c) for $F^{\kappa\lambda\mu\nu}$, we find that the solutions of the Maxwell equations (11.36) are still determined by the integration constant μ,

$$\frac{d\epsilon(q)}{dq} + \frac{R}{16\pi} \frac{dG^{-1}(q)}{dq} + \frac{d\mathcal{L}^{\mathrm{M}}(q)}{dq} = \mu. \tag{11.37}$$

11.4.3 Generalized Einstein equations

Variation over the metric $g^{\mu\nu}$ gives the generalized Einstein equations,

$$\frac{1}{8\pi G(q)} \left(R_{\mu\nu} - \frac{1}{2} R\, g_{\mu\nu} \right) + \frac{1}{16\pi}\, q\, \frac{dG^{-1}(q)}{dq}\, R\, g_{\mu\nu}$$
$$+ \frac{1}{8\pi} \left(\nabla_\mu \nabla_\nu G^{-1}(q) - g_{\mu\nu} \Box G^{-1}(q) \right) - \left(\epsilon(q) - q\, \frac{d\epsilon(q)}{dq} \right) g_{\mu\nu}$$
$$+ q \frac{\partial \mathcal{L}^{\mathrm{M}}}{\partial q} g_{\mu\nu} + T^{\mathrm{M}}_{\mu\nu} = 0, \tag{11.38}$$

where $\Box$ is the invariant d'Alembertian; and $T^{\rm M}_{\mu\nu}$ is the energy–momentum tensor of the matter fields, obtained by variation over $g^{\mu\nu}$ at constant q, i.e. without variation over $g^{\mu\nu}$, which enters q.

Eliminating dG^{-1}/dq and $\partial\mathcal{L}^{\rm M}/\partial q$ from (11.38) by use of (11.37), the generalized Einstein equations become,

$$\frac{1}{8\pi G(q)}\left(R_{\mu\nu}-\tfrac{1}{2}R\,g_{\mu\nu}\right)+\frac{1}{8\pi}\left(\nabla_\mu\nabla_\nu\,G^{-1}(q)-g_{\mu\nu}\,\Box\,G^{-1}(q)\right)$$
$$-\rho_{\rm vac}g_{\mu\nu}+T^{\rm M}_{\mu\nu}=0, \tag{11.39}$$

where

$$\rho_{\rm vac}=\epsilon(q)-\mu\,q. \tag{11.40}$$

For the special case when the dependence of the Newton constant and matter action on q is ignored, (11.39) reduces to the standard Einstein equation of general relativity with constant cosmological constant $\Lambda=\rho_{\rm vac}$.

11.4.4 Minkowski-type solution and Weinberg problem

Among the different solutions of equations (11.36) and (11.38) there is a solution corresponding to the perfect equilibrium Minkowski vacuum without matter. It is characterized by the constants in space and time values $q=q_0$ and $\mu=\mu_0$ which obey the following two conditions:

$$\left[\frac{{\rm d}\epsilon(q)}{{\rm d}q}-\mu\right]_{\mu=\mu_0\,,\,q=q_0}=0, \tag{11.41a}$$

$$\left[\epsilon(q)-\mu\,q\right]_{\mu=\mu_0\,,\,q=q_0}=0. \tag{11.41b}$$

The two conditions (11.41a)–(11.41b) can be combined into a *single* equilibrium condition for q_0:

$$\Lambda_0\equiv\left[\epsilon(q)-q\,\frac{{\rm d}\epsilon(q)}{{\rm d}q}\right]_{q=q_0}=0, \tag{11.42}$$

with the *derived* quantity

$$\mu_0=\left[\frac{{\rm d}\epsilon(q)}{{\rm d}q}\right]_{q=q_0}. \tag{11.43}$$

In order for the Minkowski vacuum to be stable, there is a further condition: $\chi(q_0)>0$, where χ corresponds to the isothermal vacuum compressibility (11.31) [21]. In this equilibrium vacuum the gravitational constant $G(q_0)$ can be identified with Newton's constant G_N.

Let us compare the conditions for the equilibrium self-sustained vacuum, (11.42) and (11.43), with the two conditions suggested by Weinberg, who used the fundamental scalar field ϕ for description of the vacuum. In this description there are two constant-field equilibrium conditions for a Minkowski vacuum, $\partial\mathcal{L}/\partial g_{\alpha\beta}=0$ and $\partial\mathcal{L}/\partial\phi=0$, see Eqs. (6.2) and (6.3) in [52]. These two conditions turn out to be inconsistent, unless the potential term in $\mathcal{L}(\phi)$ is fine-tuned (see also Sec. 2 of Ref. [53]). In other words, the Minkowski vacuum solution may exist only for a fine-tuned action. This is the Weinberg formulation of the cosmological constant problem.

The self-sustained vacuum naturally bypasses this problem [26]. Equation $\partial\mathcal{L}/\partial g_{\alpha\beta} = 0$ corresponds to the equation (11.42). However, the equation $\partial\mathcal{L}/\partial\phi = 0$ is *relaxed* in the q-theory of a self-sustained vacuum. Instead of the condition $\partial\mathcal{L}/\partial q = 0$, the conditions are $\nabla_\alpha(\partial\mathcal{L}/\partial q) = 0$, which allow for having $\partial\mathcal{L}/\partial q = \mu$ with an *arbitrary* constant μ. This is the crucial difference between a fundamental scalar field ϕ and the variable q describing a self-sustained vacuum. As a result, the equilibrium conditions for $g_{\alpha\beta}$ and q can be consistent without fine-tuning of the original action. For a Minkowski vacuum to exist, only one condition (11.42) must be satisfied. In other words, the Minkowski vacuum solution exists for arbitrary action provided that a solution of equation (11.42) exists.

11.4.5 Multiple Minkowski vacua

It is instructive to illustrate this using a concrete example. The particular choice for vacuum energy density function is considered in [26]:

$$\epsilon(q) = \Lambda_{\rm bare} + (1/2)\,(E_{\rm P})^4\, \sin\left[\, q^2/(E_{\rm P})^4 \,\right]. \qquad (11.44)$$

This contains the higher-order terms in addition to the standard quadratic term $\frac{1}{2}\,q^2$. With (11.44), the expressions for the equilibrium condition (11.42) and the stability condition (11.31) become

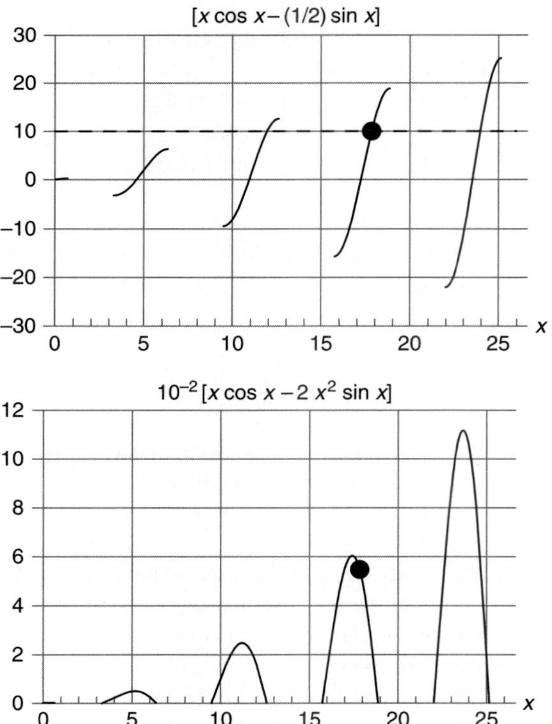

Fig. 11.3 A set of Minkowski equilibrium vacua emerging for a particular choice of the vacuum energy density function in (11.44). In each vacuum the huge bare cosmological constant $\Lambda_{\rm bare} \sim E_{\rm P}^4$ is compensated by the q-field. The curves of the top panel show the left-hand side of (11.45a) for those values of $x \equiv q^2/E_{\rm P}^4$ that obey the stability condition (11.45). The curves of the bottom panel show the corresponding positive segments of the inverse of the dimensionless vacuum compressibility $\chi E_{\rm P}^4$. Minkowski-type vacua are obtained at the intersection points of the curve in the top panel with a horizontal line at the value $\lambda_{\rm bare} \equiv \Lambda_{\rm bare}/E_{\rm P}^4$ (for example, the dashed line at $\lambda_{\rm bare} = 10$ gives the value $x_0 \approx 17.8$, corresponding to the heavy dot in the top panel. Each such vacuum is characterized, in part, by the corresponding value of the inverse vacuum compressibility from the bottom panel, shown by the heavy dot for the case chosen in the top panel). Minkowski vacua with positive compressibility are stable and become attractors in a dynamical context (cf. next section).

$$x \cos x - (1/2) \sin x = \lambda_{\text{bare}}, \tag{11.45a}$$

$$\chi^{-1} E_{\text{P}}^{-4} = x \cos x - 2x^2 \sin x > 0, \tag{11.45b}$$

where dimensionless quantities $x \equiv q^2/E_{\text{P}}^4$ and $\lambda_{\text{bare}} \equiv \Lambda_{\text{bare}}/E_{\text{P}}^4$ are introduced. A straightforward graphical analysis (Fig. 11.3) shows that, for any $\lambda_{\text{bare}} \in \mathbb{R}$, there are infinitely many equilibrium states of quantum vacuum, i.e. infinitely many values $q_0 \in \mathbb{R}$ which obey both (11.45a) and (11.45b). Each of these vacua has its own values of the Newton constant $G(q_0)$ and Standard Model parameters. But all these vacua have zero cosmological constant: the Planck-scale bare cosmological constant Λ_{bare} is compensated by the q field in any equilibrium vacuum. The top panel of Fig. 11.3 shows that the q values on the one segment singled-out by the heavy dot already allow for a complete cancelation of *any* Λ_{bare} value between $-15\, E_{\text{P}}^4$ and $+18\, E_{\text{P}}^4$.

11.5 Cosmology as an approach to equilibrium

11.5.1 Energy exchange between vacuum and gravity+matter

In the curved Universe and/or in the presence of matter, q becomes space–time dependent due to interaction with gravity and matter (see (11.37)). As a result, the vacuum energy can be transferred to the energy of the gravitational field and/or to the energy of the matter fields. This also means that the energy of matter is not conserved. The energy–momentum tensor of matter $T^{\text{M}}_{\mu\nu}$, which enters the generalized Einstein equations (11.39), is determined by variation over $g^{\mu\nu}$ at constant q. That is why it is not conserved:

$$\nabla_\nu T^{\text{M}\mu\nu} = -\frac{\partial \mathcal{L}^{\text{M}}}{\partial q} \nabla_\mu q. \tag{11.46}$$

The matter energy can be transferred to the vacuum energy due to interaction with the q-field. Using (11.37) and equation (11.40) for the cosmological constant, one obtains that the vacuum energy is transferred to both gravity and matter with a rate,

$$\nabla_\mu \Lambda \equiv \nabla_\mu \rho_{\text{vac}} = \left(\frac{d\epsilon(q)}{dq} - \mu \right) \nabla_\mu q = -\frac{R}{16\pi} \frac{dG^{-1}(q)}{dq} \nabla_\mu q + \nabla_\nu T^{\text{M}\mu\nu}. \tag{11.47}$$

The energy exchange between the vacuum and gravity+matter allows for relaxation of the vacuum energy and the cosmological "constant".

11.5.2 Dynamic relaxation of vacuum energy

Let us assume that we can apply a sharp kick to the system from its equilibrium state. For quantum liquids (or any other quantum condensed matter) we know the result of the kick: the liquid or superconductor starts to move back to the equilibrium state, and with or without oscillations it finally approaches equilibrium [54–58]. The same should happen with the quantum vacuum. Let us consider this behavior using the realization of the vacuum q field in terms of the 4-form field, when μ serves as the overall integration constant. We start with the fully equilibrium vacuum state, which is characterized by the values $q = q_0$ and $\mu = \mu_0$ in (11.41). The kick moves the variable q away from its equilibrium value, while μ remains the same, being the overall integration constant, $\mu = \mu_0$. In the non-equilibrium state, which arises immediately after the kick, the vacuum energy is non-zero and big. If the kick is very sharp, with a time scale of order $t_P = 1/E_P = \sqrt{G_N}$, the energy density of the vacuum can reach the Planck-scale value, $\rho_{vac} \sim E_P^4$.

For simplicity, we ignore the interaction between the vacuum and matter. Then from solution of the dynamic equations (11.37) and (11.39) with $\mu = \mu_0$, one finds that after the kick, q does return to its equilibrium value q_0 in the Minkowski vacuum. At late time the relaxation has the following asymptotic behavior [22]:

$$q(t) - q_0 \sim q_0 \frac{\sin \omega t}{\omega t}, \quad \omega t \gg 1, \tag{11.48}$$

where oscillation frequency ω is of the order of the Planck-energy scale E_P. The gravitational constant G approaches its Newton value G_N, also with power-law modulation:

$$G(t) - G_N \sim G_N \frac{\sin \omega t}{\omega t}, \quad \omega t \gg 1. \tag{11.49}$$

The vacuum energy relaxes to zero in the following way:

$$\rho_{vac}(t) \propto \frac{\omega^2}{t^2} \sin^2 \omega t, \quad \omega t \gg 1. \tag{11.50a}$$

For the Planck-scale kick, the vacuum energy density after the kick, i.e. at $t \sim 1/E_P$, has a Planck-scale value, $\rho_{vac} \sim E_P^4$. According to (11.50a), at the present time it must reach the value

$$\overline{\rho}_{vac}(t_{present}) \propto \frac{E_P^2}{t_{present}^2} \sim E_P^2 H^2, \tag{11.50b}$$

where H is the Hubble parameter. This value corresponds approximately to the measured value of the cosmological constant.

This, however, can be considered as an illustration of the dynamical reduction of the large value of the cosmological constant, rather than the real scenario of the evolution of the Universe. We have not taken into account quantum dissipative effects and the energy exchange between vacuum and matter. Indeed, matter field radiation (matter quanta emission) by the oscillations of the vacuum can be expected to lead to faster relaxation of the initial vacuum energy [59],

$$\rho_{vac}(t) \propto \Gamma^4 \exp(-\Gamma t), \tag{11.50c}$$

with a decay rate $\Gamma \sim \omega \sim E_P$.

Nevertheless, the cancelation mechanism and this example of relaxation provide the following lesson. The Minkowski-type solution appears without fine-tuning of the parameters of the action, precisely because the vacuum is characterized by a constant derivative of the vacuum field, rather than by a constant vacuum field itself. As a result, the parameter μ_0 emerges in (11.41a) as an *integration constant*, i.e. as a parameter of the solution rather than a parameter of the action. Since after the kick the integration constant remains intact, the Universe will return to its equilibrium Minkowski state with $\rho_{\rm vac} = 0$, even if in the non-equilibrium state after the kick the vacuum energy could reach $\rho_{\rm vac} \sim E_{\rm P}^4$. The idea that the constant derivative of a field may be important for the cosmological constant problem was earlier suggested by Dolgov [60, 61] and Polyakov [62, 63], where the latter explored the analogy with the Larkin–Pikin effect [64] in solid-state physics.

11.5.3 Minkowski vacuum as attractor

The example of relaxation of the vacuum energy in Section 11.5.2 has a principal drawback. Instead of fine-tuning of the action, which is bypassed in the self-sustained vacuum, we have fine-tuning of the integration constant. We have assumed that originally, before the kick, the Universe was in its Minkowski ground state, and thus the specific value of the integration constant $\mu = \mu_0$ was chosen to fix the value $q = q_0$ of the original Minkowski equilibrium vacuum. In the 4-form realization of the vacuum field, any other choice of integration constant ($\mu \neq \mu_0$) leads asymptotically to a de-Sitter-type solution [22].

To avoid this fine-tuning and to obtain natural relaxation of μ to μ_0, which we know occurs in quantum liquids, we must relax the condition on μ. It should not serve as an overall integration constant, while remaining the conjugate variable in thermodynamics. Then, using the condensed matter experience one may expect that the Minkowski equilibrium vacuum becomes an attractor and the de-Sitter solution with $\mu \neq \mu_0$ will inevitably relax to Minkowski vacuum with $\mu = \mu_0$. This expectation is confirmed in the aether-type realization of the vacuum variable in terms of a vector field, as discussed in Section 11.2.5.

There, the constant vacuum field q appears as the derivative of a vector field in the specific solution u^q_β corresponding to the equilibrium vacuum, $q\, g_{\alpha\beta} \equiv \nabla_\alpha\, u^q_\beta = u^q_{\alpha\beta}$. In this realization, $\mu \equiv d\epsilon(q)/dq$ plays a role of the effective chemical potential only for the equilibrium states (i.e. for their thermodynamical properties), but μ does not appear as an integration constant for the dynamics. Hence, the fine-tuning problem of the integration constant is overcome, simply because there is no integration constant.

The instability of the de-Sitter solution towards the Minkowski solution has already been demonstrated by Dolgov [61], who considered the simplest quadratic choices of the Lagrange density of $u_\beta(x)$. But his result also holds for the generalized Lagrangian with a generic function $\epsilon(u_{\alpha\beta})$ in Section 11.2.5 [26].

The Dolgov scenario does not require the variable gravitational coupling parameter, so we use $G(q) = \text{const}$. In this scenario, for a spatially flat Robertson–Walker metric with cosmic time t and scale factor $a(t)$, the initial de-Sitter-type expansion evolves towards the Minkowski attractor by the following $t \to \infty$ asymptotic solution for the aether-type field $u_\beta = (u_0(t), 0)$:

$$u_0(t) \to q_0\, t, \quad H(t) \to 1/t, \qquad (11.51)$$

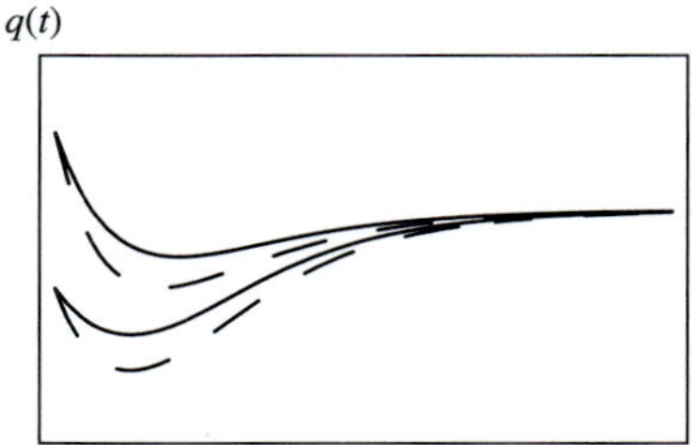

Fig. 11.4 Aether-field q evolution and Minkowski attractor in a spatially flat Friedmann–Robertson–Walker universe in the Dolgov model [61] (see [26] for details). The bare cosmological constant is $\Lambda_{\text{bare}} \sim E_{\text{P}}^4$. Four numerical solutions correspond to different boundary conditions, but all approach the Minkowski space time solution (11.51). The Minkowski vacuum is an attractor because the vacuum compressibility (11.31) is positive, $\chi(q_0) > 0$.

where the Hubble parameter $H \equiv [da/dt]/a$. At large cosmic times t, the curvature terms decay as $R \sim H^2 \sim 1/t^2$ and the Einstein equations lead to nullification of the energy–momentum tensor of the u_β field: $T_{\alpha\beta}[u] = 0$. Since (11.51) with $du_0/dt = H\,u_0$ satisfies the q-theory *Ansatz* $u_{\alpha\beta} = q\,g_{\alpha\beta}$, the energy–momentum tensor is completely expressed by the single constant q: $T_{\alpha\beta}(q) = [\epsilon(q) - q\,d\epsilon(q)/dq]\,g_{\alpha\beta}$. As a result, the equation $T_{\alpha\beta}(q) = 0$ leads to the equilibrium condition (11.42) for the Minkowski vacuum and to the equilibrium value $q = q_0$ in (11.51).

Figure 11.4 shows explicitly the attractor behavior for the simplest case of the Dolgov action, with the numerical value of q_0 in (11.51) appearing *dynamically*. This simple version of the Dolgov scenario does not appear to give a realistic description of the present Universe [65] and requires appropriate modification [66]. It nevertheless demonstrates that the compensation of a large initial vacuum energy density can occur dynamically and that Minkowski spacetime can emerge spontaneously, without setting a chemical potential. In other words, an "existence proof" has been given for the conjecture that the appropriate Minkowski value q_0 can result from an attractor-type solution of the field equations. The only condition for the Minkowski vacuum to be an attractor is a positive vacuum compressibility (11.31).

11.5.4 Remnant cosmological constant

Figure 11.5 demonstrates a possibly more realistic scenario with a step-wise relaxation of the vacuum energy density [67]. The vacuum energy density moves from plateau to plateau responding to the possible phase transitions or crossovers in the Standard Model vacuum and follows, on average, a steadily decreasing matter energy density. The origin of the current plateau with a small positive value of the vacuum energy density $\Lambda_{\text{present}} = \rho_{\text{vac}} \sim \left(10^{-3}\,\text{eV}\right)^4$ is still not clear. It may result from phenomena which occur in the infrared. It may come, for example, from anomalies in the neutrino sector of the quantum vacuum, such as the non-equilibrium contribution of light massive neutrinos to

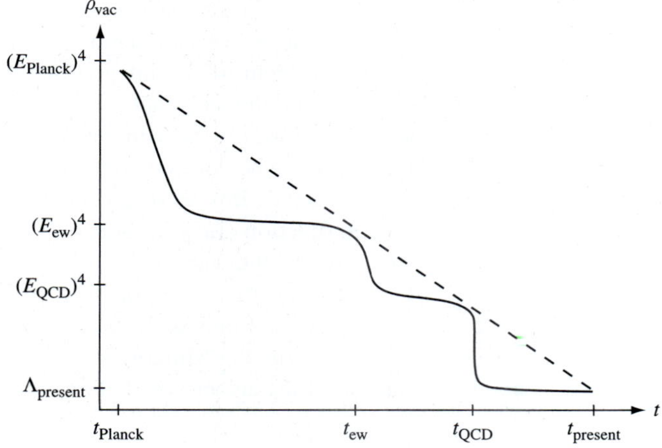

Fig. 11.5 *Dashed curve*: relaxation according to the relation $< \rho_{\text{vac}}(t) > \sim (E_{\text{Planck}})^2/t^2$ in Eq. (11.50a). *Full curve*: sketch of the relaxation of the vacuum energy density during the evolution of the Universe according to Ref. [67]. The origin of the current plateau in the vacuum energy Λ_{present} is discussed in Section 11.5.4.

the quantum vacuum [67]; re-entrant violation of Lorentz invariance [68] and Fermi point splitting in the neutrino sector [69, 70] (see Section 11.7.7). Other possible sources include the QCD anomaly [71, 25, 72–74]; torsion [75]; relaxation effects during the electroweak crossover [23]; etc. Most of these scenarios are determined by the momentum space topology of the quantum vacuum.

11.5.5 Summary and outlook

In studying the problems related to quantum vacuum one must search for a proper extension of the current theory of elementary particle physics—the Standard Model. However, many properties of the quantum vacuum can be understood by extending our experience with self-sustained macroscopic systems to the quantum vacuum. A simple picture of quantum vacuum is based on three assumptions: (i) the quantum vacuum is a self-sustained medium—a system which is stable at zero external pressure, like quantum liquids; (ii) the quantum vacuum is characterized by a conserved charge q, which is the analog of the particle density n in quantum liquids and which is non-zero in the ground state of the system, $q = q_0 \neq 0$; (iii) the quantum vacuum with $q = q_0$ is a Lorentz-invariant state. This is the only property which distinguishes the quantum vacuum from condensed matter quantum liquids.

These assumptions naturally solve the main cosmological constant problem without fine-tuning. In any self-sustained system, relativistic or non-relativistic, in thermodynamic equilibrium at $T = 0$ the zero-point energy of quantum fields, is fully compensated by the microscopic degrees of freedom, so that the relevant energy density is zero in the ground state. This consequence of thermodynamics is automatically fulfilled in any system, which may exist without an external environment. This leads to the trivial result for gravity: the cosmological constant in any equilibrium vacuum state is zero. The zero-point energy of the Standard Model fields is automatically compensated by the q–field which describes the degrees of freedom of the deep quantum vacuum.

These assumptions allow us to suggest that cosmology is a process of equilibration. From condensed matter experience we know that the ground state of the system serves as an attractor: starting far away from equilibrium, the quantum liquid finally reaches its ground state. The same should occur for the particular case of our Universe: starting far away from equilibrium in a very early phase of universe, the vacuum is moving towards the Minkowski attractor. We are now close to this attractor, simply because our Universe is old. This is a possible reason for the small remnant cosmological constant measured at the present time.

The q–theory transforms the standard cosmological constant problem into a search for the proper decay mechanism of the vacuum energy density and for the proper mechanism of formation of the small remnant cosmological constant. For that we need a theory of the dynamics of quantum vacuum. The latter is a new topic in physics waiting for input from theory and observational cosmology. Using several possible realizations of the vacuum variable q we are able to model some features of the vacuum dynamics in the hope that this will allow us to find the generic features and construct the phenomenology of equilibration.

11.6 Vacuum as a topological medium

11.6.1 Topological media

There are two schemes for the classification of states in condensed matter physics and relativistic quantum fields: classification by symmetry and classification by topology.

According to the first scheme, a given state of the system is characterized by a symmetry group H which is a subgroup of the symmetry group G of the relevant physical laws. The thermodynamic phase transition between equilibrium states is usually marked by a change of the symmetry group H. This classification reflects the phenomenon of spontaneously broken symmetry. In relativistic quantum fields the chain of successive phase transitions, in which the large symmetry group existing at high energy is reduced at low energy, is in the basis of the Grand Unification models (GUT) [76, 77]. In condensed matter the spontaneous symmetry breaking is a typical phenomenon, and the thermodynamic states are also classified in terms of the subgroup H of the relevant group G (see e.g. the classification of superfluid and superconducting states in Refs. [78, 79]). The groups G and H are also responsible for classification of topological defects, which are determined by the non-trivial elements of the homotopy groups $\pi_n(G/H)$ [80].

The second classification method deals with the ground states of the system at zero temperature ($T = 0$). In particle physics it is the classification of quantum vacua. Topological media are systems whose properties are protected by topology and thus are robust to deformations of the action. The universality classes of topological media are determined by momentum-space topology. The latter is also responsible for the type of the effective theory which emerges at low energy. In this sense, topological classification reflects the tendency which is opposite to GUT, called the anti Grand Unification (anti-GUT). In the GUT scheme, the fundamental symmetry of the vacuum state is primary and the phenomenon of spontaneous symmetry breaking gives rise to topological defects. In the anti-GUT scheme the topology is primary, while effective symmetry gradually emerges at low energy [7, 8].

Different aspects of the physics of topological matter have been discussed during the last decades, including topological stability of gap nodes; classification of fully gapped vacua; edge states; Majorana fermions; influence of disorder and interaction; topological quantum phase transitions; intrinsic quantum Hall and spin-Hall effects; quantization of physical parameters; experimental realization; connections with relativistic quantum fields; chiral anomaly; topological Chern–Simons and Wess–Zumino actions; etc.

11.6.2 Gapless topological media

There are two big groups of topological media: with fully gapped fermionic excitations and with gapless fermions.

In $3 + 1$ spacetime, there are four basic universality classes of gapless fermionic vacua protected by topology in momentum space [8, 81]:

(i) Vacua with fermionic excitations characterized by Fermi points (Dirac points, Weyl points, Majorana points, etc.)—points in 3D momentum space at which the energy of the fermionic quasiparticle vanishes. Examples are provided by the spin triplet p-wave superfluid ^{3}He-A, Weyl semimetals, and also by the quantum vacuum of the Standard Model above the electroweak transition, where all elementary particles are Weyl fermions with Fermi points in the spectrum. This universality class manifests the phenomenon of emergent relativistic quantum fields at low energy: close to the Fermi points the fermionic quasiparticles behave as massless Weyl fermions, while the collective modes of the vacuum interact with these fermions as gauge and gravitational fields.

(ii) Vacua with fermionic excitations characterized either by lines of zero energy in 3D momentum space or by point zeroes in 2D momentum space. We shall characterize zeros by their co-dimension—the dimension of **p**-space minus the dimension of the manifold of zeros. Lines in 3D momentum space and points in 2D momentum space have co-dimension 2: since $3 - 1 = 2 - 0 = 2$; compare this with zeros of class (i) which have co-dimension $3 - 0 = 3$. Zeros of co-dimension 2 are topologically stable only if some special symmetry is obeyed. Examples are provided by the vacuum of the high T_c cuprate superconductors where Cooper pairing into a d-wave state occurs [82] and graphene [83, 16–18]. Nodes in the spectrum are stabilized there by the combined effect of momentum-space topology and discrete symmetry.

(iii) Vacua with fermionic excitations characterized by Fermi surfaces. The representatives of this universality class are normal metals and normal liquid ^{3}He. This universality class also manifests the phenomenon of emergent physics, although non-relativistic: at low temperature all the metals behave in a similar way, and this behavior is determined by the Landau theory of a Fermi liquid—the effective theory based on the existence of a Fermi surface. A Fermi surface has co-dimension 1: in a 3D system it is the surface (co-dimension $= 3 - 2 = 1$), in a 2D system it is the line (co-dimension $= 2 - 1 = 1$), and in a 1D system it is the point (co-dimension $= 1 - 0 = 1$; in a one-dimensional system the Landau Fermi liquid theory does not work, but the Fermi surface survives).

(iv) The Fermi band class, where the energy vanishes in the finite region of the 3D momentum space, and thus zeros have co-dimension 0. Possible states of this kind has been discussed in [84–86]. In particle physics, the Fermi band or the Fermi ball appears in a $2 + 1$ dimensional non-relativistic quantum field theory which is dual to a gravitational theory in the anti-de Sitter background with a charged black hole [87]. A topologically stable flat band exists on the surface of the materials with lines of zeros in the bulk [88–90] and in the spectrum of fermion zero modes localized in the core of some vortices [91–93].

11.6.3 Fully gapped topological media

The gapless and gapped vacuum states are interrelated. For example, the quantum phase transition between fully gapped states with different topology occurs via an intermediate gapless state. The related phenomenon is that the interface between the fully gapped states with different values of topological invariant contains gapless fermions.

The most popular examples of fully gapped topological matter are topological insulators [94, 9, 10]. The first discussion of the possibility of $3+1$ topological insulators can be found in Refs. [95, 96]. The main feature of such materials is that they are insulators in bulk, where the electron spectrum has a gap, but there are $2+1$ gapless edge states of electrons on the surface or at the interface between topologically different bulk states, as discussed in Ref. [96]. The spin triplet p-wave superfluid ^{3}He-B is another example of the fully gapped $3+1$ matter with non-trivial topology in momentum space. It has $2+1$ gapless quasiparticles living at interfaces between vacua with different values of the topological invariant describing the bulk states of ^{3}He-B [97, 98]. The only difference from the topological insulators is that the gapless fermions living on the surface of the topological superfluid and superconductor or at the interface are Majorana fermions. The quantum vacuum of the Standard Model below the electroweak transition, i.e. in its massive phase, is the relativistic counterpart of the topological insulators and gapped topological superfluids [99].

Examples of $2+1$ topological fully gapped systems are provided by films of superfluid ^{3}He-A with broken time reversal symmetry [100, 101] and by a planar phase which is time reversal invariant [100, 101]. The topological invariants for $2+1$ vacua give rise to quantization of the Hall and spin-Hall conductivity in these films in the absence of an external magnetic field (the so-called intrinsic quantum and spin-quantum Hall effects) [100, 102], see Section 11.8.1.3.

11.6.4 Green's function as an object

For the study of the topological properties of condensed matter systems, ideal non-interacting systems are frequently used. Sometimes this is justified, if one can find the effective single-particle Hamiltonian, which emerges at low energy and which reflects the topological properties of the real interacting many-body system. However, in general the primary object for the topological classification of real systems is the one-electron propagator—Green's function $G(\omega, \mathbf{p})$. In principle one can construct the effective Hamiltonian by proper simplification of the Green's function at zero frequency, $H = G^{-1}(\omega = 0, \mathbf{p})$. Although in the interacting case, the propagator $G(\mathbf{p}, \omega = 0)$ only determines correctly the zero energy states, see e.g. [103], in some cases it can be used for construction of the topological invariants alongside the full Green's function $G(\omega, \mathbf{p})$. On the other hand, there are situations when the Green's function does not have poles (see [83, 104, 105]). In these cases, no well-defined energy spectrum exists, and the effective low energy Hamiltonian cannot be introduced. In particle physics, interaction may also lead to the anomalous infrared behavior of propagators. For example, the pole in the Green's function is absent for the so-called unparticles [106, 107]; the phenomenon of quark confinement in QCD can lead to the anomalous infrared behavior of the quark and gluon propagators [108–110]; a marginal Green's function of fermions may occur at the black hole horizon [111]; etc. Thus in interacting systems, all of the information on the topology is encoded in the topology of the Green's function matrix, and also in its symmetry. The latter is important, because symmetry supports additional topological invariants, which are absent in the absence of symmetry; see below.

Green's function topology has been used in particular for classification of topologically protected nodes in the quasiparticle energy spectrum of systems of different dimensions, including the vacuum of the Standard Model in its gapless state [7, 8, 81, 83]; for the classification of the topological ground states in the fully gapped $2+1$ systems, which experience intrinsic quantum Hall and spin-Hall effects [100, 112–114, 8, 83]; in relativistic quantum field theory of $2+1$ massive Dirac fermions [115–119] and $3+1$ massive Dirac fermions [120]; etc. (see also recent papers [121, 122]).

For topological classification of the gapless vacua, the Green's function is considered at imaginary frequency $\omega = ip_0$. This allows us to consider only the relevant singularities in the Green's function and to avoid the singularities on the mass shell, which exist in any vacuum, gapless or fully gapped.

11.6.5 Fermi surface as a topological object

Let us start with gapless vacua. The Green's function is generally a matrix with spin indices. In addition, it may have band indices (in the case of electrons in the periodic potential of crystals). The general analysis [81] demonstrates that topologically stable nodes of co-dimension 1 (Fermi surface in $3+1$ metal, Fermi line in $2+1$ system, or Fermi point in $1+1$ system) are described by the group Z of integers. The winding number N_1, which is responsible for the topological stability of these nodes, is expressed analytically in terms of the Green's function [8]:

$$N_1 = \mathbf{tr} \oint_C \frac{dl}{2\pi i} G(p_0, \mathbf{p}) \partial_l G^{-1}(p_0, \mathbf{p}). \tag{11.52}$$

Here the integral is taken over an arbitrary contour C around the Green's function singularity in the $D+1$ momentum–frequency space. See Fig. 11.6 for D $= 2$. An example of the Green's function in any dimension D is the scalar function $G^{-1}(\omega, \mathbf{p}) = ip_0 - v_F(|\mathbf{p}| - p_F)$. For $D = 2$, the singularity with winding number $N_1 = 1$ is on the line $p_0 = 0$, $p_x^2 + p_y^2 = p_F^2$, which represents the one-dimensional Fermi surface.

Because of a non-trivial topological invariant, the Fermi surface survives the perturbative interaction and exists even in marginal and Luttinger liquids without poles in the Green's function, where quasiparticles are not well defined.

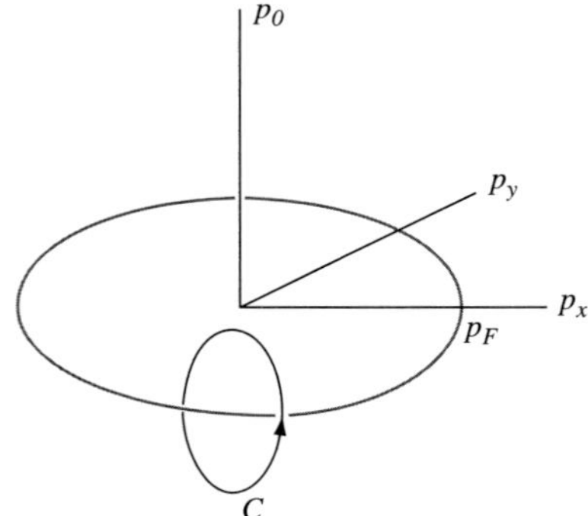

Fig. 11.6 Fermi surface in $2+1$ systems representing the nodes of co-dimension 1. In this case, the Green's function has singularities on line $p_0 = 0, p_x^2 + p_y^2 = p_F^2$ in the three-dimensional space (p_0, p_x, p_y). Stability of the Fermi surface is protected by the invariant (11.52), which is represented by the integral over an arbitrary contour C around the Green's function singularity. This is applicable to nodes of co-dimension 1 in any $D+1$ dimension. For $D = 3$ the nodes form a conventional Fermi surface in metals and in normal ^{3}He.

11.7 Vacuum in a semi-metal state

For our Universe, which obeys the Lorentz invariance, only those vacua are important that are either Lorentz invariant, or acquire the Lorentz invariance as an effective symmetry emerging at low energy. This excludes the vacua with Fermi surface and Fermi lines and leaves the class of vacua with Fermi point of chiral type, in which fermionic excitations behave as left-handed or right-handed Weyl fermions [7, 8], and the class of vacua with the nodal point obeying Z_2 topology, where fermionic excitations behave as massless Majorana neutrinos [81, 83].

11.7.1 Fermi points in 3 + 1 vacua

For the relativistic quantum vacuum of our 3 + 1 Universe the Green's function singularity of co-dimension 3 is relevant. They are described by the following topological invariant expressed via an integer-valued integral over the surface σ around the singular point in the 4-momentum space $p_\mu = (p_0, \mathbf{p})$[8]:

$$N_3 = \frac{e_{\alpha\beta\mu\nu}}{24\pi^2} \, \mathbf{tr} \int_\sigma dS^\alpha \; G\partial_{p_\beta} G^{-1} G\partial_{p_\mu} G^{-1} G\partial_{p_\nu} G^{-1}. \tag{11.53}$$

If the invariant is non-zero, the Green's function has point singularity inside the surface σ—the Fermi point. If the topological charge is $N_3 = +1$ or $N_3 = -1$, the Fermi point represents the so-called conical Dirac point, but actually describes the chiral Weyl fermions. This is a consequence of the so-called Atiyah–Bott–Shapiro construction [81], which leads to the following general form of expansion of the inverse fermionic propagator near the Fermi point with $N_3 = +1$ or $N_3 = -1$:

$$G^{-1}(p_\mu) = e^\beta_\alpha \Gamma^\alpha (p_\beta - p^{(0)}_\beta) + \cdots \tag{11.54}$$

Here $\Gamma^\mu = (1, \sigma_x, \sigma_y, \sigma_z)$ are Pauli matrices (or Dirac matrices in the more general case); the expansion parameters are the vector $p^{(0)}_\beta$ indicating the position of the Fermi point in momentum space where the Green's function has a singularity, and the matrix e^β_α; the ellipsis denote higher-order terms in expansion.

11.7.2 Emergent fermionic matter

Equation (11.54) can be continuously deformed to the simple form, which describes the relativistic Weyl fermions,

$$G^{-1}(p_\mu) = ip_0 + N_3 \boldsymbol{\sigma} \cdot \mathbf{p} + \cdots, \tag{11.55}$$

where the position of the Fermi point is shifted to $p^{(0)}_\beta = 0$ and the ellipsis denotes higher-order terms in p_0 and $\mathbf{p}$; the matrix e^β_α is deformed to unit matrix. This means that close to the Fermi point with $N_3 = +1$, the low energy fermions behave as right-handed relativistic particles, while the Fermi point with $N_3 = -1$ gives rise to the left-handed particles.

Equation (11.55) suggests the effective Weyl Hamiltonian,

$$H_{\text{eff}} = N_3 \boldsymbol{\sigma} \cdot \mathbf{p}\,. \tag{11.56}$$

However, the infrared divergences may violate the simple pole structure of the propagator in Eq. (11.55). In this case in the vicinity of the Fermi point one has,

$$G(p_\mu) \propto \frac{-ip_0 + N_3 \boldsymbol{\sigma} \cdot \mathbf{p}}{\left(p^2 + p_0^2\right)^\gamma}, \tag{11.57}$$

with $\gamma \neq 1$. This modification does not change the topology of the propagator: the topological charge of the singularity is N_3 for arbitrary parameter γ [83]. For fermionic unparticles, one has $\gamma = 5/2 - d_U$, where d_U is the scale dimension of the quantum field [106, 107].

For $N_3 = \pm 2$, the spectrum of (quasi)particles in the vicinity of singularity depends on symmetry. One may obtain either two Weyl fermions or exotic massless fermions with nonlinear dispersion at low energy: semi-Dirac fermions with linear dispersion in one direction and quadratic dispersion in the other two [68, 8, 83],

$$E(\mathbf{p}) \approx \pm\sqrt{c^2 p_z^2 + \left(\frac{p_\perp^2}{2m}\right)^2}\,. \tag{11.58}$$

Similar consideration for the $2+1$ systems may lead to semi-Dirac fermions and to fermions with quadratic dispersion at low energy [83, 123, 124],

$$E(\mathbf{p}) \approx \pm\frac{p^2}{2m}. \tag{11.59}$$

For the higher values of topological charge, the spectrum becomes even more interesting (see e.g. Refs. [125, 126] for $2+1$ systems). But if the relativistic invariance is obeyed, or under the special discrete symmetry, the non-zero invariant N_3 corresponds to N_3 species of Weyl fermions near the Fermi point.

The main property of the vacua with Dirac points is that, according to (11.55), close to the Fermi points the massless relativistic fermions emerge. This is consistent with the fermionic content of our Universe, where all the elementary particles—left-handed and right-handed quarks and leptons—are Weyl fermions. Such a coincidence demonstrates that the vacuum of the Standard Model in its massless phase is the topological medium of the Fermi point universality class. This solves the hierarchy problem, since the value of the masses of elementary particles in the vacua of this universality class is strictly zero.

Let us suppose, for a moment, that there is no topological invariant which protects massless fermions. Then the Universe is fully gapped and the natural masses of fermions must be in the order of the Planck energy scale: $M \sim E_{\mathrm{P}} \sim 10^{19}$ GeV. In such a natural Universe, where all masses are of order E_{P}, all fermionic degrees of freedom are completely frozen out because of the Boltzmann factor $e^{-M/T}$, which is about $e^{-10^{16}}$ at the temperature corresponding to the highest energy reached in accelerators. There is no fermionic matter in such a Universe at low energy. That we survive in our Universe is not the result of the anthropic principle (the latter chooses the Universes which are fine-tuned for life but have an extremely low probability). Our Universe is also natural and its vacuum is generic, but it belongs to a different universality class of vacua—the vacua with Fermi points. In such vacua the masslessness of fermions is protected by topology (combined with symmetry, see below).

11.7.3 Emergent gauge fields

Vacua with the Fermi point suggest a particular mechanism for the emergent symmetry. The Lorentz symmetry is simply the result of a linear expansion: this symmetry becomes better and better when the Fermi point is approached and the non-relativistic higher-order terms in Eq. (11.55) may be neglected. This expansion demonstrates the emergence of the relativistic spin, which is described by the Pauli matrices. It also demonstrates how gauge fields and gravity emerge together with chiral fermions. The expansion parameters $p^{(0)}_\beta$ and e^β_α may depend on the space and time coordinates and they actually represent collective dynamic bosonic fields in the vacuum with Fermi point. The vector field $p^{(0)}_\beta$ in the expansion plays the role of the effective $U(1)$ gauge field A_β acting on fermions.

For the Fermi points with topological charge $N_3 > 1$, the situation depends on the symmetry of the system. In the case, when the spectrum corresponds to several species of relativistic Weyl fermions, the shift $p^{(0)}_\beta$ becomes the matrix field; it gives rise to effective non-Abelian (Yang–Mills) $SU(N_3)$ gauge fields emerging in the vicinity of the Fermi point, i.e. at low energy [8]. For example, the Fermi point with $N_3 = 2$ may give rise to an effective $SU(2)$ gauge field in addition to the effective $U(1)$ gauge field,

$$G^{-1}(p_\mu) = e^\beta_\alpha \Gamma^\alpha \left(p_\beta - g_1 A_\beta - g_2 \mathbf{A}_\beta \cdot \boldsymbol{\tau}\right) + \text{ higher-order terms}, \qquad (11.60)$$

where $\boldsymbol{\tau}$ are Pauli matrices corresponding to the emergent isotopic spin. This is what happens in superfluid ^{3}He-A. In the case where symmetry leads to exotic fermions with the nonlinear spectrum $E \sim \pm p^{N_3}$, quantum electrodynamics with anisotropic scaling emerges [127, 128], which is similar to the quantum gravity with anisotropic scaling suggested by Hořava [129–131].

11.7.4 Emergent gravity

The matrix field e^β_α in (11.60) acts on the (quasi)particles as the field of vierbein, and thus describes the emergent dynamical gravity field. As a result, close to the Fermi point, matter fields (all ingredients of the Standard Model: chiral fermions and Abelian and non-Abelian gauge fields) emerge, together with geometry, relativistic spin, Dirac matrices, and physical laws: Lorentz and gauge invariance, equivalence principle, etc. In such vacua, gravity emerges together with matter. If this Fermi point mechanism of emergence of physical laws works for our Universe, then so-called "quantum gravity" does not exist. The gravitational degrees of freedom can be separated from all other degrees of freedom of quantum vacuum only at low energy.

In this scenario, classical gravity is a natural macroscopic phenomenon emerging in the low-energy corner of the microscopic quantum vacuum, i.e.

it is a typical and actually inevitable consequence of the coarse-graining procedure. It is possible to quantize gravitational waves to obtain their quanta—gravitons, since in the low energy corner the results of microscopic and effective theories coincide. It is also possible to obtain some (but not all) quantum corrections to the Einstein equations and to extend classical gravity to the semiclassical level. But one cannot obtain "quantum gravity" by quantization of the Einstein equations, since all other degrees of freedom of the quantum vacuum would be missed in this procedure.

11.7.5 Topological invariant for specific Fermi surface

If the symmetry which fixes the position of the conical (Dirac) point at zero energy level is violated, the conical point moves from the zero energy position upward or downward from the chemical potential and the Fermi surface is formed. This is shown in Fig. 11.7. This Fermi surface has a specific property: in addition to the local charge N_1 in (11.52), which characterizes singularities at the Fermi surface, it is described by the global charge N_3 in (11.53). The integral in (11.53) is now over the surface σ which embraces the whole Fermi sphere. The Fermi surface with global topological charge appears in superfluid ^{3}He-A in the presence of mass flow [8]; it is also discussed for the $2+1$ systems in relation to the gapless states on the surface of $3+1$ insulators [132].

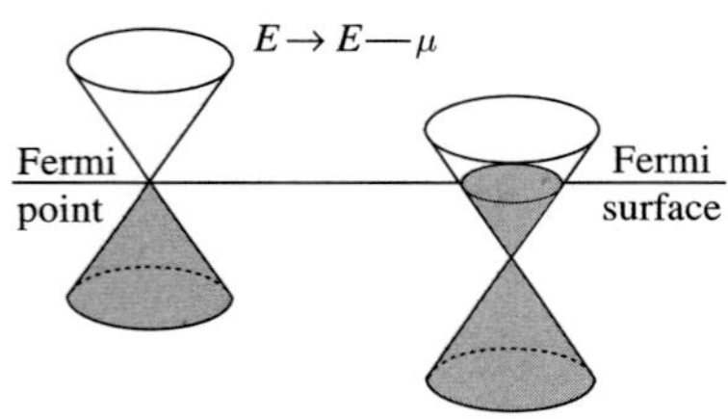

Fig. 11.7 A Fermi surface is formed from the Fermi point at a finite chemical potential of chiral fermions, when the Fermi point is moved away from the zero energy level.

The "collision" of the Fermi surfaces in momentum space leads to redistribution of the global topological charges N_3 between the Fermi surfaces when they touch each other, $(+1)+(-1) \to 0+0$ [69, 83]. Such collisions, at which the Fermi surface loses its global charge N_3, represent the topological quantum phase transition (see Fig. 11.8), one of numerous types of transition induced by topology in momentum space [83].

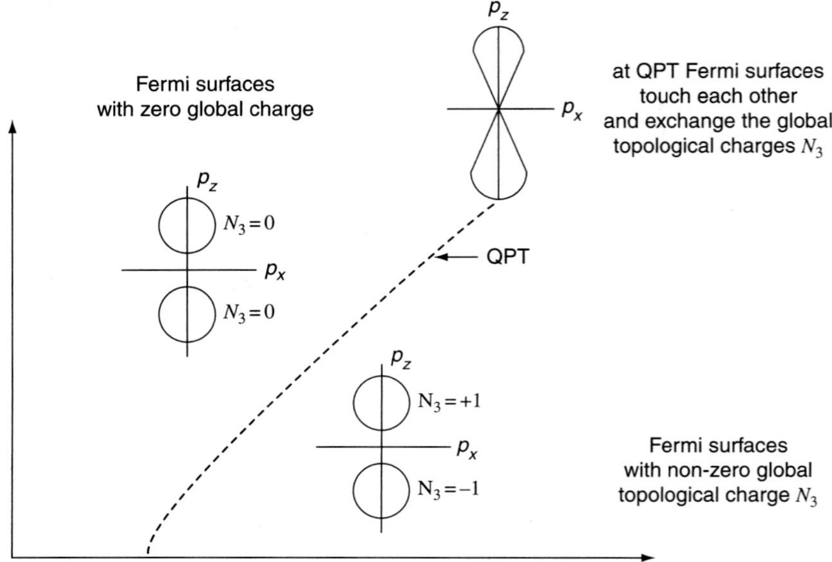

Fig. 11.8 Dashed line represents the topological quantum phase transition in the model [69]. The vacua on both sides of the transition have Fermi surfaces. On the right side of the transition, Fermi surfaces have non-zero global topological charges $N_3=+1$ and $N_3=-1$. At the transition, the Fermi surfaces "collide", and their topological global charges N_3 annihilate each other. On the left side of the transition, Fermi surfaces become globally trivial, $N_3=0$, but retain their local topological charge N_1 in (11.52).

11.7.6 Topological invariant protected by symmetry in semimetal state

We assume that the Standard Model contains equal numbers of right and left Weyl fermions, $n_R = n_L = 8n_g$, where n_g is the number of generations (we do not consider the Standard Model with Majorana fermions, and assume that in the insulating state of the Standard Model neutrinos are Dirac fermions). For such a Standard Model the topological charge in (11.53) vanishes, $N_3 = 8n_g - 8n_g = 0$. Thus the masslessness of the Weyl fermions is not protected by the invariant (11.53), and arbitrary weak interaction may result in massive particles.

However, there is another topological invariant, which takes into account the symmetry of the vacuum. The gapless state of the vacuum with $N_3 = 0$ can be protected by the following integral [8]:

$$N_3^K = \frac{e_{\alpha\beta\mu\nu}}{24\pi^2} \, \mathbf{tr} \left[K \int_\sigma dS^\alpha \; G\partial_{p_\beta} G^{-1} G\partial_{p_\mu} G^{-1} G\partial_{p_\nu} G^{-1} \right], \tag{11.61}$$

where K_{ij} is the matrix of some symmetry transformation, which either commutes or anticommutes with the Green's function matrix. In the Standard Model there are two relevant symmetries, both are the Z_2 groups, $K^2 = 1$. One of them is the center subgroup of the $SU(2)_L$ gauge group of weak rotations of left fermions, where the element K is the gauge rotation by angle 2π, $K = e^{i\pi\check{\tau}_{3L}}$. The other is the group of the hypercharge rotation by angle 6π, $K = e^{i6\pi Y}$. In the $G(224)$ Pati-Salam extension of the $G(213)$ group of the Standard Model, this symmetry comes as a combination of the Z_2 center group of the $SU(2)_R$ gauge group for right fermions, $e^{i\pi\check{\tau}_{3R}}$, and the element $e^{3\pi i(B-L)}$ of the Z_4 center group of the $SU(4)$ color group—the P_M parity (for the importance of discrete groups in particle physics see [133, 134] and references therein). Each of these two Z_2 symmetry operations changes the sign of the left spinor, but does not influence the right particles. Thus these matrices are diagonal, $K_{ij} = \mathrm{diag}(1, 1, \ldots, -1, -1, \ldots)$, with eigenvalues 1 for right fermions and -1 for left fermions.

In the symmetric phase of the Standard Model, both matrices commute with the Green's function matrix G_{ij}; as a result N_3^K is topological invariant: it is robust to deformations of the Green's function which preserve the symmetry K. The value of this invariant $N_3^K = 16n_g$, which means that all $16n_g$ fermions are massless if the symmetry K is obeyed.

11.7.7 Higgs mechanism versus splitting of Fermi points

The gapless vacuum of the Standard Model is supported by the combined action of topology and symmetry K, and also by the CPT and Lorentz invariance, which keep all the Fermi points at $\mathbf{p} = 0$.

Explicit violation or spontaneous breaking of one of these symmetries transforms the vacuum of the Standard Model into one of the two possible vacua. If, for example, the K symmetry is broken, the invariant (11.61) supported by this symmetry ceases to exist, and the Fermi point disappears. All $16\,n_g$ fermions become massive (Fig. 11.9 *bottom left*). This is assumed to happen below the

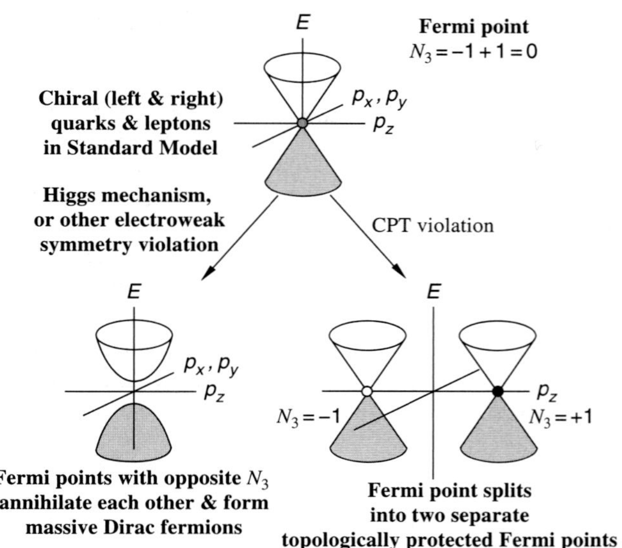

Fig. 11.9 (*top*): In the Standard Model the Fermi points with positive $N_3 = +1$ and negative $N_3 = -1$ topological charges are at the same point $\mathbf{p} = 0$, forming the marginal Fermi point with $N_3 = 0$. Symmetry K between the Fermi points prevents their mutual annihilation giving rise to the topological invariant (11.61) with $N_3^K = 16 n_g$. The figure illustrates the simple case with two Weyl fermions, one with $N_3 = +1$ and another with $N_s = -1$, when the invariant $N_3^K = 2$ protects the marginal Fermi point with $N_3 = +1 - 1 = 0$. (*bottom left*): If symmetry K is violated or spontaneously broken, Weyl points annihilate each other and a Dirac mass is formed. (*bottom right*): If Lorentz invariance is violated or spontaneously broken, the marginal Fermi point splits [135]. The topological quantum phase transition between the state with Dirac mass and the state with split Dirac points has been observed in cold Fermi gas [136].

symmetry breaking electroweak transition caused by the Higgs mechanism, where quarks and charged leptons acquire the Dirac masses.

If, on the other hand, the CPT symmetry is violated, the marginal Fermi point splits into topologically stable Fermi points with non-zero invariant N_3, which protects massless chiral fermions (Fig. 11.9 *bottom right*). Since the invariant N_3 does not depend on symmetry, further symmetry breaking cannot destroy the nodes. One can speculate that in the Standard Model the latter may happen with the electrically neutral leptons, the neutrinos [69]. Fermi point splitting in the neutrino sector may serve as an example of spontaneous breaking of the Lorentz symmetry [70, 137]. It may also provide a new source of T and CP violation in the leptonic sector, which may be relevant for creation of the observed cosmic matter–antimatter asymmetry [138]. Examples of splitting of Fermi and Majorana points in condensed matter are discussed in the review paper [83].

11.7.8 Fermi points in condensed matter

11.7.8.1 *Chiral superfluid ^{3}He-A*

The discovery of superfluid ^{3}He in 1972 [139, 79] marked the first condensed matter realization of a topological medium. Both phases of superfluid ^{3}He (gapless ^{3}He-A and fully gapped ^{3}He-B) are topological superfluids. The chiral superfluid ^{3}He-A with broken time reversal symmetry has the following simplified Green's function for each spin projection:

$$G^{-1}(\omega, \mathbf{p}) = ip_0 + \tau_3 \left(\frac{p^2}{2m} - \mu \right) + c(\tau_1 p_x + \tau_2 p_y), \tag{11.62}$$

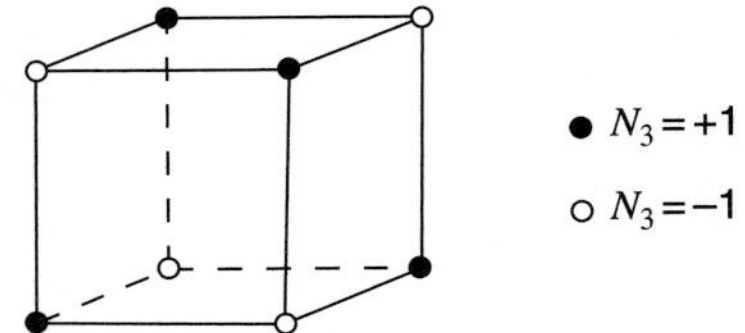

Fig. 11.10 Cube of Fermi points (Weyl points) in momentum space in the α-phase of a spin-triplet p-wave superfluid [79] and spin-singlet d-wave superconductor [78]. Filled circles denote Weyl points with positive topological charge $N_3 = +1$ and thus with right-handed fermions living in the vicinity of these Weyl points. Open circles denote the nodes with negative topological charge $N_3 = -1$ and the left-handed fermions. This is analogous to the four-dimensional graphene-like vacuum in a model of relativistic quantum field theory, which is characterized by Dirac points on the vertices of a 4D hypercube [140].

where τ_i are Pauli matrices of Bogolyubov–Nambu spin. For $\mu > 0$ there are two Weyl points at $p_x = p_y = 0$ and $p_z = \pm\sqrt{2m\mu}$ with $N_3 = \pm 1$. For $\mu < 0$ the vacuum is fully gapped. Thus at $\mu = 0$ there is a topological quantum phase transition from the gapless to gapped vacuum [101]. At the transition, i.e. at $\mu = 0$, there is a marginal (topologically trivial) Fermi point with $N_3 = 0$, situated at $\mathbf{p} = 0$, just as in Fig. 11.9 *top*. At $\mu > 0$, this marginal Fermi point splits into two topologically protected Weyl points with $N_3 = \pm 1$, Fig. 11.9 *bottom right*.

11.7.8.2 *Cube of Fermi points*

The vacuum of the Standard Model contains 16 Weyl points in each generation. An example of the condensed matter system with Weyl Dirac points is provided by the α-phase of a spin-triplet p-wave superfluid [79] and a spin-singlet d-wave superconductor [78]. The latter has the following simplified Green's function:

$$G^{-1}(\omega,\mathbf{p}) = ip_0 + \tau_3\left(\frac{p^2}{2m} - \mu\right) + \tau_1(2p_z^2 - p_y^2 - p_x^2) + \tau_2\sqrt{3}(p_y^2 - p_x^2). \qquad (11.63)$$

The spectrum has eight point nodes—Weyl points with $N_3 = +1$ and $N_3 = -1$ situated on the vertices of a cube in momentum space in Fig. 11.10 [78]. At $\mu = 0$ all eight Weyl points collapse to the marginal Fermi point with $N_3 = 0$, situated at $\mathbf{p} = 0$.

The α-phase is the analog of the $3 + 1$ "graphene" in relativistic quantum fields [140].

11.7.8.3 *Time-reversal invariant planar phase*

The planar phase of a spin-triplet superfluid/superconductor is characterized by the following simplified 4×4 Green's function matrix:

$$G^{-1}(\omega,\mathbf{p}) = ip_0 + \tau_3\left(\frac{p^2}{2m} - \mu\right) + \tau_1(\sigma_x p_x + \sigma_y p_y). \qquad (11.64)$$

As distinct from the chiral ^{3}He-A and α-phase, the planar phase obeys time-reversal invariance. Two nodes in the spectrum, at $p_x = p_y = 0$ and $p_z = \pm\sqrt{2m\mu}$, have zero topological charges (11.53), $N_3 = 0$. But these nodes are protected by the topological charge $N_3^K = \pm 2$ in (11.61), where the corresponding symmetry of the planar phase is $K = \tau_3\sigma_z$. This matrix K commutes with the Green's function matrix.

11.7.8.4 *Gapless 2+1 vacua*

In addition to the Fermi surface class of singularities of co-dimension 1, in $2 + 1$ systems there is a class of vacua with singularities of co-dimension 2: points in 2D momentum space. They corresponds to lines in $3 + 1$ vacua, which also have co-dimension 2. According to [81]: if no symmetry is imposed there is no singularity in Green's function, which is topologically stable. For real fermions, the Z_2 singularities of co-dimension 2 are possible [81]. This means that two such singularities may collapse, forming the fully gapped state, $1 + 1 = 0$. In $2 + 1$ dimension, the fermions near such a singularity behave as Majorana fermions. Some symmetries allow Fermi points of co-dimension 2 with group Z. The corresponding invariant protected by symmetry K [141, 83, 142] is,

$$N_2^K = \frac{1}{4\pi i} \mathbf{tr} \oint_C dl\, KG(\omega = 0, \mathbf{p})\partial_l G^{-1}(\omega = 0, \mathbf{p}), \tag{11.65}$$

where C is a contour around the Fermi point in 2D momentum space (p_x, p_y), or around the Fermi surface if the Fermi point expands to the Fermi surface. Examples are graphene and a d-wave cuprate superconductor. For the latter, the simplified Green's function has the form,

$$G^{-1}(\omega, p_x.p_y) = ip_0 + \tau_3 \left(\frac{p_x^2 + p_y^2}{2m} - \mu \right) + \tau_1 (p_y^2 - p_x^2), \tag{11.66}$$

with $K = \tau_2$, which anti-commutes with Green's function at zero frequency. The d-wave superconductor has four point nodes at $|p_x| = |p_y| = (m\mu)^{1/2}$ with $N_2^K = \pm 1$. The nodes do not disappear under deformation which preserves symmetry K. For example, the deformation which violates the 4-fold symmetry of (11.66)

$$G^{-1}(\omega, p_x.p_y) = ip_0 + \tau_3 \left(\frac{p_x^2 + p_y^2}{2m} - \mu \right) + \tau_1 (p_y^2 - ap_x^2), \tag{11.67}$$

does not destroy nodes, but shifts positions of nodes. The nodes disappear only at large deformation, when the deformation parameter a in (11.67) crosses zero, and the topological quantum phase transition occurs. At $a = 0$ nodes collapse forming two marginal nodes with $N_2^K = 0$ at $p_y = 0$, $p_x = \pm(2m\mu)^{1/2}$, and at $a < 0$ the fully gapped state is formed [83].

Note that the inverse propagator at $p_0 = 0$ has all the properties of a free-fermion Hamiltonian, whose topology was discussed in Ref. [143]. But this is actually the effective Hamiltonian, which emerges in the original interacting system (see [103, 144]).

Another class of $2 + 1$ Fermi points arises at the boundary between the $3 + 1$ gapped systems with different topological charges. Such points are described by the difference of bulk invariants [144]. This is analogous to the index theorem for fermion zero modes on strings [145] and vortices [8].

11.8 Vacuum in state of topological insulator

A special role in classification of topological systems is played by dimensional reduction. Dimensional reduction allows us to use for classification of gapped systems the scheme, which was suggested by Hořava for classification of the topologically non-trivial nodes in a spectrum [81]. The fully gapped vacua in $D + 1$ spacetime are described by the same invariants as nodes of co-dimension $D + 1$ [8]. For example, the winding number N_1 in (11.52) which describes zeros of co-dimension 1 (conventional Fermi surface in $D = 3$ momentum space), also describes $D = 0$ gapped systems. The integral (11.52) is now over an imaginary frequency,

$$\tilde{N}_1 = \mathbf{tr} \int \frac{dp_0}{2\pi i} G(p_0) \partial_{p_0} G^{-1}(p_0). \tag{11.68}$$

This integer-valued index now shows the difference between the numbers of positive and negative energy levels of a zero-dimensional system.

The classification must be supplemented by a symmetry consideration, which leads to additional invariants of the type,

$$\tilde{N}_1^K = \mathbf{tr} \int \frac{dp_0}{2\pi i} K G(p_0) \partial_{p_0} G^{-1}(p_0), \tag{11.69}$$

where K is the symmetry operator, which commutes or anti-commutes with the Green's function. The emergent symmetries might also appear within some topological classes.

11.8.1 2 + 1 fully gapped vacua

11.8.1.1 *^{3}He-A film: 2 + 1 chiral superfluid*

Let us start with $D = 2$. The fully gapped ground states (vacua) in $2 + 1$ or quasi $2 + 1$ thin films of ^{3}He-A are characterized by the invariant obtained by dimensional reduction from the topological invariant describing the nodes of co-dimension 3. This is the invariant N_3 for the Fermi point in (11.53), where the integration is now over the whole $(2 + 1)$-dimensional momentum–frequency space $(p_x.p_y, p_0)$:

$$\tilde{N}_3 = \frac{e_{ijk}}{24\pi^2} \mathbf{tr} \left[\int d^2p dp_0 \, G \partial_{p_i} G^{-1} G \partial_{p_j} G^{-1} G \partial_{p_k} G^{-1} \right]. \tag{11.70}$$

This equation (11.70) was first introduced in relativistic $2 + 1$ theories [115–117] and then independently for a film of ^{3}He-A in condensed matter [150, 100], where it was inspired by the dimensional reduction from the Fermi point, see [101]. The topological invariant for the general case of insulating relativistic vacua in even space dimension $D = 2n$ has been considered in [146–148]. In the simple case of a 2×2 matrix, the Green's function can be expressed in terms of the three-dimensional vector $\mathbf{d}(p_x, p_y)$,

$$G^{-1}(\omega, p_x.p_y) = ip_0 + \boldsymbol{\tau} \cdot \mathbf{d}(p_x, p_y). \tag{11.71}$$

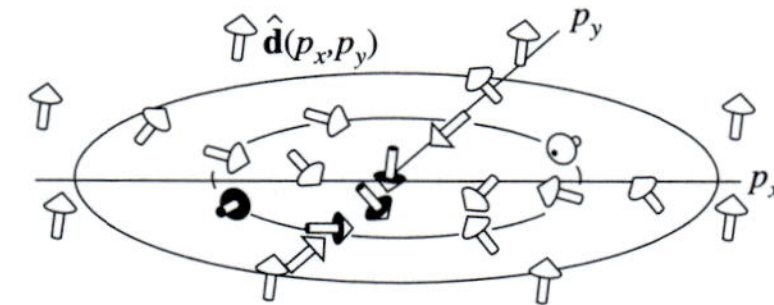

Fig. 11.11 Skyrmion in **p**-space with momentum space topological charge $\tilde{N}_3 = -1$ in (11.72). It describes topologically non-trivial fully gapped vacua in $2 + 1$ systems, which have a non-singular Green's function. Vacua with non-zero $\tilde{N}_3$ have topologically protected gapless edge states. The non-zero topological charge also leads to quantization of Hall and spin-Hall conductance.

An example of the **d**-vector configuration, which corresponds to the topologically non-trivial vacuum is presented in Fig. 11.11. This is the momentum-space analog of the topological object in real space—skyrmion (skyrmions in real space are described by the relative homotopy groups [149]). The winding number of the momentum-space skyrmion is [150]

$$\tilde{N}_3 = \frac{1}{4\pi} \int d^2p \, \hat{\mathbf{d}} \cdot \left(\frac{\partial \hat{\mathbf{d}}}{\partial p_x} \times \frac{\partial \hat{\mathbf{d}}}{\partial p_y} \right), \tag{11.72}$$

where $\hat{\mathbf{d}} = \mathbf{d}/|\mathbf{d}|$ is a unit vector. For a single layer of the ^{3}He-A film and for one spin projection, the simplified Green's function has the form,

$$G^{-1}(\omega, \mathbf{p}) = i\,p_0 + \boldsymbol{\tau} \cdot \mathbf{d}(\mathbf{p}) = ip_0 + \tau_3 \left(\frac{p_x^2 + p_y^2}{2m} - \mu \right) + \tau_1 p_x + \tau_2 p_y. \tag{11.73}$$

For $\mu > 0$ the topological charge $\tilde{N}_3 = 1$ and for $\mu < 0$ the topological charge is $\tilde{N}_3 = 0$. That is why at $\mu = 0$ there is a topological quantum phase transition between the topological superfluid at $\mu > 0$ and non-topological superfluid at $\mu < 0$ [101].

In the general case of multilayered ^{3}He-A, topological charge $\tilde{N}_3$ may take any integer value of group Z. This charge determines the quantization of Hall and spin-Hall conductance (see Section 11.8.1.3), and the quantum statistics of the topological objects—real-space skyrmions [100–102, 150]. For $\tilde{N}_3 = 4k + 1$ and $\tilde{N}_3 = 4k + 3$, the skyrmion is an anyon; for $\tilde{N}_3 = 4k + 2$ it is a fermion; and for $\tilde{N}_3 = 4k$ it is a boson [101]. This demonstrates the importance of the Z_2 and Z_4 subgroups of group Z in classification of topological matter; and also provides an example of the interplay of momentum-space and real-space topologies. For applications of topology in combined $(\mathbf{p}, \mathbf{r})$-space see in [8, 97, 121, 151–155], in particular it is responsible for the topologically protected spectrum of fermions living on topological objects such as walls, strings, and monopoles.

11.8.1.2 *Planar phase: time-reversal invariant gapped vacuum*

In the case when some symmetry is present, additional invariants appear, which correspond to dimensional reduction of invariant N_3^K in (11.61):

$$\tilde{N}_3^K = \frac{e_{ijk}}{24\pi^2} \, \mathbf{tr} \left[\int d^2p dp_0 \; K G \partial_{p_i} G^{-1} G \partial_{p_j} G^{-1} G \partial_{p_k} G^{-1} \right], \tag{11.74}$$

where as before, the matrix K commutes or anticommutes with the Green's function matrix. An example of the symmetric $2 + 1$ gapped state with $\tilde{N}_3^K$ is the film of the planar phase of superfluid ^{3}He [100, 101]. In the single layer case, the simplest expression for the Green's function is

$$G^{-1}(\omega, p_x, p_y) = ip_0 + \tau_3 \left(\frac{p_x^2 + p_y^2}{2m} - \mu \right) + \tau_1 (\sigma_x p_x + \sigma_y p_y), \tag{11.75}$$

with $K = \tau_3 \sigma_z$ commuting with the Green's function. This state is time-reversal invariant. For $\mu > 0$ it has $\tilde{N}_3 = 0$ and $\tilde{N}_3^K = 2$. For the general case of a quasi 2D film with multiple layers of the planar phase, the invariant $\tilde{N}_3^K$ belongs to the group Z, i.e. $\tilde{N}_3^K = 2k$.

11.8.1.3 *Quantum spin-Hall effect*

The topological invariants $\tilde{N}_3$ and $\tilde{N}_3^K$ give rise to quantization of Hall and spin-Hall conductance in $2 + 1$ gapped systems. There are several types of responses of spin current and electric current to transverse forces which are quantized in $2 + 1$ systems under appropriate conditions. The most familiar is the conventional quantum Hall effect (QHE) [156]. This is the quantized response of the particle current to a transverse force, say to a transverse gradient of chemical potential, $\mathbf{J} = \sigma_{xy}\hat{\mathbf{z}} \times \nabla\mu$. In electrically charged systems this is the quantized response of the electric current $\mathbf{J}^e$ to a transverse electric field $\mathbf{J}^e = e^2\sigma_{xy}\hat{\mathbf{z}} \times \mathbf{E}$.

The other effects involve spin degrees of freedom. An example is the mixed spin quantum Hall effect: the quantized response of the particle current $\mathbf{J}$ (or electric current $\mathbf{J}^e$) to a transverse gradient of magnetic field $\mathbf{H}$ interacting with Pauli spins (Pauli field in short) [100, 101]:

$$\mathbf{J} = \sigma_{xy}^{\text{mixed}}\hat{\mathbf{z}} \times \nabla(\gamma H^z), \quad \mathbf{J}^e = e\mathbf{J}. \tag{11.76}$$

Here γ is the gyromagnetic ratio. The related effect, which is determined by the same quantized parameter $\sigma_{xy}^{\text{mixed}}$, is the quantized response of the spin current, say current $\mathbf{J}^z$ of the z component of spin, to the gradient of the chemical potential [102]. In electrically charged systems this corresponds to the quantized response of the spin current to a transverse electric field:

$$\mathbf{J}^z = \sigma_{xy}^{\text{mixed}}\hat{\mathbf{z}} \times \nabla\mu = e\sigma_{xy}^{\text{mixed}}\hat{\mathbf{z}} \times \mathbf{E}. \tag{11.77}$$

This kind of mixed Hall effect is now used in spintronics [157].

Finally, there is a pure spin-Hall effect—the quantized response of the spin current to the transverse gradient of a magnetic field [100, 101, 158, 159]:

$$\mathbf{J}^z = \sigma_{xy}^{\text{spin/spin}}\hat{\mathbf{z}} \times \nabla(\gamma H^z). \tag{11.78}$$

All of these parameters σ_{xy} are quantized being expressed via topological charges $\tilde{N}_3$ and $\tilde{N}_3^K$.

11.8.2 3 + 1 fully gapped states: ^{3}He-B and quantum vacuum

In the asymmetric phase of the Standard Model, there is no mass protection by topology and all the fermions become massive, i.e. the Standard Model vacuum becomes a fully gapped insulator. In quantum liquids, the fully gapped three-dimensional system with time reversal symmetry and non-trivial topology is represented by another phase of superfluid ^{3}He—the ^{3}He-B. Its topology is also supported by symmetry and gives rise to 2D gapless quasiparticles living at interfaces between vacua with different values of the topological invariant or on the surface of ^{3}He-B [97, 98, 144, 120]. It is important that ^{3}He-B belongs to the same topological class as the vacuum of the Standard Model in its present insulating phase [99]. The topological classes of the ^{3}He-B states can be represented by the following simplified Green's function:

$$G^{-1}(\omega, \mathbf{p}) = ip_0 + \tau_3\left(\frac{p^2}{2m} - \mu\right) + \tau_1\left(\sigma_x c_x p_x + \sigma_y c_y p_y + \sigma_z c_z p_z\right). \tag{11.79}$$

In the isotropic ^{3}He-B all "speeds of light" are equal, $|c_x| = |c_y| = |c_z| = c$. The vacuum of free Dirac particles is obtained in the limit $1/m = 0$.

In fully gapped systems, the Green's function has no singularities in the whole 4-dimensional space $(p_0, \mathbf{p})$. That is why we are able to use the Green's function at $p_0 = 0$. The topological invariant relevant for ^{3}He-B and for quantum vacuum with massive Dirac fermions is

$$N^K = \frac{e_{ijk}}{24\pi^2} \, \mathbf{tr} \left[\int_{\omega=0} d^3p \; K \; G\partial_{p_i} G^{-1} G\partial_{p_j} G^{-1} G\partial_{p_k} G^{-1} \right]. \qquad (11.80)$$

with matrix $K = \tau_2$ which anti-commutes with the Green's function at $p_0 = 0$. In ^{3}He-B, the τ_2 symmetry is a combination of time-reversal and particle–hole symmetries; for the Standard Model the corresponding matrix $K = \gamma_5\gamma^0$. Note that at $p_0 = 0$ the symmetry of the Green's function is enhanced, and thus there are more matrices K, which commute or anti-commute with the Green's function, than at $p_0 \neq 0$.

Figure 11.12 shows the phase diagram of topological states of ^{3}He-B in the plane $(\mu, 1/m)$. On the line $1/m = 0$ one obtains free Dirac fermions ($n_L = n_R = 1$) with mass parameter $M = -\mu$. The conventional Dirac vacuum of free fermions has topological charge

$$N^K = \text{sign}(M). \qquad (11.81)$$

The real superfluid ^{3}He-B lives in the weak-coupling corner of the phase diagram: $\mu > 0, m > 0, \mu \gg mc^2$. However, in ultracold Fermi gases with triplet pairing the strong coupling limit is possible near the Feshbach resonance [160]. When μ crosses zero a topological quantum phase transition occurs, at which the topological charge N^K changes from $N^K = 2$ to $N^K = 0$.

There is an important difference between ^{3}He-B and Dirac vacuum. The space of the Green's function of free Dirac fermions is non-compact: G has different asymptotes at $|\mathbf{p}| \to \infty$ for different directions of momentum $\mathbf{p}$. As a result, the topological charge of the interacting Dirac fermions depends on the regularization at large momentum. ^{3}He-B can serve as regularization of the Dirac vacuum, which can be made in the Lorentz invariant way [99]. One can see from Fig. 11.12, that the topological charge of the free Dirac vacuum has intermediate value, between the charges of the ^{3}He-B vacua with compact Green's function. On the marginal behavior of free Dirac fermions see Refs. [161, 162, 8, 144].

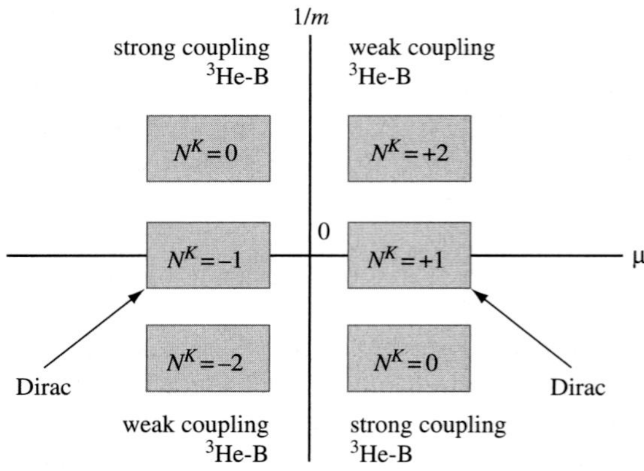

Fig. 11.12 Phase diagram of topological states of ^{3}He-B in equation (11.79) in the plane $(\mu, 1/m)$ for the speeds of light $c_x > 0$, $c_y > 0$, and $c_z > 0$. States on the line $1/m = 0$ correspond to the Dirac vacua, which Hamiltonian is non-compact. The topological charge of the Dirac fermions is intermediate between the charges of compact ^{3}He-B states. The line $1/m = 0$ separates the states with different asymptotic behavior of the Green's function at infinity: $G^{-1}(\omega = 0, \mathbf{p}) \to \pm\tau_3 p^2/2m$. The line $\mu = 0$ marks topological quantum phase transition, which occurs between the weak coupling ^{3}He-B (with $\mu > 0$, $m > 0$ and topological charge $N^K = 2$) and the strong coupling ^{3}He-B (with $\mu < 0$, $m > 0$ and $N^K = 0$). This transition is topologically equivalent to the quantum phase transition between Dirac vacua with opposite mass parameter $M = \pm|\mu|$, which occurs when μ crosses zero along the line $1/m = 0$. The interface which separates two states contains a single Majorana fermion in the case of ^{3}He-B, and a single chiral fermion in the case of relativistic quantum fields. The difference in the nature of the fermions is that in the Bogoliubov–de Gennes system the components of the spinor are related by complex conjugation. This reduces the number of degrees of freedom compared with the Dirac case.

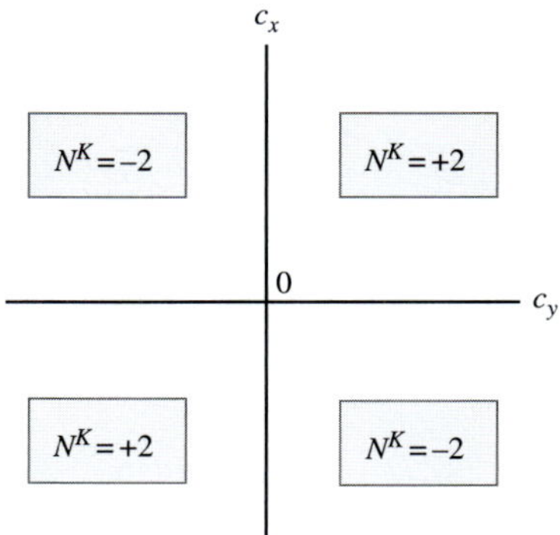

Fig. 11.13 Phase diagram of ^{3}He-B states at fixed $c_z > 0$, $\mu > 0$ and $m > 0$. At the phase boundaries the vacuum is gapless and corresponds to the $3+1$ planar phase. The interface between the gapped states with different winding number N^K contains Majorana fermions. In cosmology, such an interface would correspond to the vierbein wall, where the metric is degenerate [8].

The vertical axis in Fig. 11.12 separates the states with the same asymptote of the Green's function at infinity. The abrupt change of the topological charge across the line, $\Delta N^K = 2$, with fixed asymptote, shows that one cannot cross the transition line adiabatically. This means that all of the intermediate states on the line of this QPT are necessarily gapless. For the intermediate state between the free Dirac vacua with opposite mass parameter M this is well known. But this is applicable to the general case, with or without relativistic invariance: the gaplessness is protected by the difference between the topological invariants on the two sides of the transition.

Figure 11.13 shows the phase diagram of topological states of ^{3}He-B in the plane (c_x, c_y) at fixed $c_z > 0$, $\mu > 0$ and $m > 0$. When one of the components of the speed of light is nullified, the state becomes gapless. If, say, c_x approaches zero, two pairs of point nodes appear in this intermediate state at points $\mathbf{p} = \pm(0, 0, p_F)$, where $p_F = (2\mu m)^{1/2}$. The number of point nodes is related to the difference in topological invariant N^K of the states on two sides of the transition:

$$N_{\text{point nodes}} = |N^K(c_x > 0) - N^K(c_x < 0)| = 4. \tag{11.82}$$

The intermediate state is the $3+1$ planar phase in (11.64), and the corresponding gapless fermions in this state are characterized by the topological invariant N_3^K protected by discrete symmetry.

The gaplessness of the intermediate state on the phase diagram leads to the other related phenomena. The two-dimensional interface (brane), which separates two domains with different N^K, contains fermion zero modes, $2+1$ massless fermions. The number of zero modes is determined by the difference ΔN^K between the topological charges of the vacua on two sides of the interface [98, 144]. This is similar to the index theorem for the number of fermion zero modes on cosmic strings, which is determined by the topological winding number of the string [145].

Superfluid ^{3}He-B demonstrates that the interface between bulk states with the same topological charge may also contain fermion zero modes [97, 98]. For example, the state with positive c_x, c_y and the state with negative c_x, c_y in Fig. 11.13 have the same topological charge $N^K = +2$. But the interface between these bulk states is gapless. Moreover, the density of states of these irregular fermion zero modes is larger than the density of states of regular fermion zero modes living on the interface between the bulk states with different N^K. This is not very surprising, since the irregular gapless states of co-dimension 1 may appear in many systems. Nodes of co-dimension 1 are the most stable topological objects, and they may emerge even in bulk superconductors [163, 164]. They may also exist on the surface of non-topological insulators and superconductors, and at the interface with $\Delta N^K = 0$, as happens in ^{3}He-B.

The ^{3}He-B topology can be also extended to lattice models explored in QCD, where the four-dimensional Brillouin zone is used. The lattice model with fully gapped Wilson fermions is described both by $\tilde{N}_3$ and $\tilde{N}_5$ topological invariants, which give rise to two index theorems for gapless fermions in intermediate states [122, 165]. The topological invariant $\tilde{N}_5$ has also been used for description of fully gapped vacua in $4+1$ systems [8, 146–148]. They give rise to

$3+1$ gapless fermion zero modes living at the $3+1$ interfaces between the massive vacua. This provides another topological scenario for an emergent chiral relativistic fermion, accompanied by relativistic quantum field theory and gravity.

11.9 Discussion

There is a fundamental interplay between symmetry and topology in physics, both in condensed matter and relativistic quantum fields. Traditionally, the first role was played by symmetry (symmetry classification of crystals, liquid crystals, magnets, superconductors, superfluids, etc.). The phenomenon of spontaneously broken symmetry remains one of the major tools in physics. In particle physics, a chain of successive phase transitions is suggested in which the large symmetry group existing at high energy is reduced at low energy. This symmetry principle is in the basis of the Grand Unification Theory (GUT). It also gives rise to the classification of topological defects which arise due to spontaneous symmetry breaking. In this approach, symmetry is primary, while topology is secondary, being fully determined by the broken symmetry.

Past decades have demonstrated the opposite tendency in which topology is primary (reviews can be found in Refs. [101, 8, 83]). The unconventional properties of superfluid ^{3}He-A, which were found in the early 1970s, demonstrated that the topology in momentum space is the main characteristic of the ground states of the system at zero temperature ($T = 0$), in other words it is the characteristic of quantum vacua. It demonstrated that the gaplessness of fermions is protected by topology, and thus is not sensitive to the details of microscopic (trans-Planckian) physics. Irrespective of the deformation of the parameters of microscopic theory, the value of the gap (mass) in the energy spectrum of these fermions remains strictly zero. This solves the main hierarchy problem: for these classes of fermionic vacua the masses of elementary particles are naturally small.

The vacua, which have non-trivial topology in momentum space, are now called the topological matter (topological superfluids, superconductors, insulators, semimetals, etc.). Momentum-space topological invariants determine the universality classes of the topological matter and the type of the effective theory which emerges at low energy and low temperature. In many cases they also give rise to emergent symmetry. Examples are provided by the nodes of the energy spectrum in momentum space: if they are protected by topology they give rise to emergent symmetries such as Lorentz invariance, and to emergent phenomena such as gauge and gravitational fields. Contrary to the GUT scheme, in the anti-GUT scheme the symmetry is secondary, being emergent in the low-energy corner due to topology. If this is true, then it is the topology of the quantum vacuum, which gives rise to the fermionic matter in our Universe.

However, this is not the whole story. It appears that in many systems (including condensed matter and relativistic quantum vacua), the topological invariants are trivial, i.e. they have zero values. Nevertheless these systems remain as topological: the underlying discrete symmetry of the vacuum may transform the trivial topological invariant into the non-trivial one. This is the case when both topology and symmetry are equally important. This implies that if the Standard Model and gravity are effective theories, the underlying physics must contain discrete symmetries. Their role is extremely important. The main role is to prohibit the cancelation of the Fermi points with opposite topological charges in the Standard Model vacuum. As a side effect, in the low-energy corner, discrete symmetries are transformed into gauge symmetries and give rise to effective non-Abelian gauge fields. In particular, underlying Z_2 symmetry produces the effective $SU(2)$ gauge field [8]. Discrete symmetries also reduce the number of massless gauge bosons and the number of effective metric fields. To justify the Fermi point scenario of emergent physical laws, one should find such a discrete symmetry of the microscopic vacuum which leads in the low energy corner to one of the GUT or Pati-Salam models.

Discovery of the quantum Hall effect (QHE) in 1980 [156] triggered further consideration of condensed matter systems using topological methods [161, 166–169]. The topological invariant for the QHE effect in terms of the Green's function was introduced in relativistic $2+1$ quantum electrodynamics [115–117]. This invariant is responsible not only for ordinary QHE but also for intrinsic QHE, when the Hall conductance is quantized even in the absence of an external magnetic field [118].

Such intrinsic QHE can be realized for example in thin films of superfluid ^{3}He. These quasi-two-dimensional systems serve as an example of fully gapped $2+1$ systems, whose integer valued topological invariant is in the origin of quantization of physical parameters [100, 112, 150]. There are two phases which may exist in a thin film of liquid ^{3}He, both phases are fully gapped in film. These are chiral superfluid ^{3}He-A and the planar phase with time reversal symmetry. This symmetry supports the topological invariant which is responsible for the intrinsic quantum Hall and spin-Hall effects [100, 101], thus providing us with an important example of the interplay of topology and symmetry in a topological medium.

Recent observation of topological insulators [170] has given new impetus to the study of topological phases in $3+1$ systems, such as superfluid ^{3}He-B which is the condensed matter counterpart of Standard Model vacuum in its massive phase. Superfluid ^{3}He-B has Majorana bound states on the surface of the liquid. Andreev bound states on the surface or interface of ^{3}He-B have been discussed theoretically or probed experimentally [97, 171–177]. However, the Majorana signature of these states has not yet been reported experimentally. One of the possible tools is NMR, which requires an external magnetic field. The effect of magnetic field on Majorana fermions has recently been discussed in [120, 177]. Majorana fermions become massive, with mass being proportional to magnetic field, which violates the time reversal symmetry.

Vacuum of the Standard Model is a topological $3+1$ medium. Both known states of the quantum vacuum of the Standard Model have non-trivial topology. The insulating state is described by a non-zero value of topological invariant $\tilde{N}_3^K$ and the topology of this state is similar to that of superfluid ^{3}He-B. The semimetal state is described by a non-zero value of topological invariant N_3^K and the topology of this state is similar to that of a superfluid planar phase in 3 + 1 dimensions. Both invariants are supported by symmetry. Superfluid phases of ^{3}He (^{3}He-B, ^{3}He-A, and planar phase) thus provide a close connection with relativistic quantum fields.

These phases also give generic examples of different classes of topological media in dimensions 2 + 1 and 3 + 1 which experience topological stability of gap nodes; topological edge states on the surface of fully gapped insulators; Majorana fermions; topological quantum phase transitions; intrinsic quantum Hall and spin-Hall effects; quantization of physical parameters; chiral anomaly; topological Chern–Simons and Wess–Zumino actions; etc.

These examples of topological media demonstrate the important role of both topology and symmetry. That is why we need the general classification of topological matter in terms of its symmetry and topology. An attempt at such a classification was made in [143, 178]. However, only non-interacting systems were considered, and also not all symmetries were exploited, including the approximate symmetries. Examples of the latter are again provided by the superfluid phases of ^{3}He, where the symmetry is enhanced due to the relative smallness of spin–orbit interactions. One task should be to consider symmetry classes, including crystal symmetry classes, magnetic classes, superconductivity classes, etc., and to find out which topological classes of the Green's function are allowed within a given symmetry class. Then one should find out what happens when the symmetry is smoothly violated, etc. The Green's function matrices with spin and band indices must be used for this classification, since they take into account interaction. In this way we may finally obtain a full classification of topological matter, including insulators, superconductors, magnets, liquids, and vacua of relativistic quantum fields.

Finally, this experience must be used for the investigation of topologically non-trivial quantum vacua in relativistic quantum field theories, where the topology in momentum space is becoming the important tool (see e.g. [8, 81, 115, 116, 122, 140, 146–148, 165, 179]).

On the other hand, the gravitational properties of quantum vacuum do not depend much on the topology of the vacuum. In any self-sustained vacuum, gapless or gapped, topological or trivial, the gravitating energy of the vacuum is strictly zero if the vacuum is perfect and is isolated from the environment. This is a consequence of thermodynamics, which is not sensitive to the structure of the quantum vacuum. Nullification of $\rho_{\rm vac}$ occurs in any equilibrium system at $T = 0$, relativistic or non-relativistic. This solves the main cosmological constant problem: if the quantum vacuum belongs to self-sustained media, Λ is naturally small. According to this view, cosmology is the process of relaxation of the Universe to its equilibrium vacuum state, and this process does depend on the momentum space topology of the quantum vacuum.

Acknowledgments

This work is supported in part by the Academy of Finland, Centers of excellence program 2006–2011 and the Khalatnikov–Starobinsky leading scientific school (Grant No. 4899.2008.2).

References

[1] Einstein, A. Kosmologische Betrachtungen zur allgemeinen Relativitätstheorie, Sitzungsber. *Preuss. Akad. Wiss.* **1**, 142–152 (1917).

[2] Bronstein, M.P. On the expanding universe, *Phys. Z. Sowjetunion* **3**, 73–82 (1933).

[3] Zel'dovich, Y.B. Cosmological constant and elementary particles, *JETP Lett.* **6**, 316–317 (1967).

[4] Riess, A.G. et al., [Supernova Search Team Collaboration], Observational evidence from supernovae for an accelerating universe and a cosmological constant, *Astron. J.* **116**, 1009 (1998), arXiv:astro-ph/9805201.

[5] Perlmutter, S. et al., [Supernova Cosmology Project Collaboration], Measurements of Ω and Λ from 42 high-redshift supernovae, *Astrophys. J.* **517**, 565 (1999), arXiv:astro-ph/9812133.

[6] Komatsu, E. et al., Five-year Wilkinson Microwave Anisotropy Probe (WMAP) observations: Cosmological interpretation, *Astrophys. J. Suppl.* **180**, 330 (2009), arXiv:0803.0547.

[7] Froggatt, C.D. and Nielsen, H.B. *Origin of Symmetry*, World Scientific, Singapore, 1991.

[8] Volovik, G.E. *The Universe in a Helium Droplet*, Clarendon Press, Oxford (2003).

[9] Hasan, M.Z. and Kane, C.L. Topological Insulators, *Rev. Mod. Phys.* **82**, 3045–3067 (2010).

[10] Qi, Xiao-Liang and Zhang, Shou-Cheng. Topological insulators and superconductors, *Rev. Mod. Phys.* **83**, 1057–1110 (2011).

[11] Abrikosov, A.A. and Beneslavskii, S.D. Possible existence of substances intermediate between metals and dielectrics, *Sov. Phys. JETP* **32**, 699 (1971).

[12] Abrikosov, A.A. Quantum magnetoresistance, *Phys. Rev.* B **58**, 2788 (1998).

[13] Burkov, A.A. and Balents, L. Weyl semimetal in a topological insulator multilayer, *Phys. Rev. Lett.* **107**, 127205 (2011); Burkov, A.A. Hook, M.D. Balents, L. Topological nodal semimetals, *Phys. Rev.* B **84**, 235126 (2011), arXiv:1110.1089.

[14] Wan, Xiangang, Turner, A.M., Vishwanath, A., and Savrasov, S.Y. Topological semimetal and Fermi-arc surface states in the electronic structure of pyrochlore iridates, *Phys. Rev.* B **83**, 205101 (2011).

[15] Ryu, S. and Hatsugai, Y. Topological origin of zero-energy edge states in particle–hole symmetric systems, *Phys. Rev. Lett.* **89**, 077002 (2002).

[16] Manes, J.L., Guinea, F. and Vozmediano, M.A.H. Existence and topological stability of Fermi points in multilayered graphene, *Phys. Rev.* B **75**, 155424 (2007).

[17] Vozmediano, M. A. H. Katsnelson, M. I. and Guinea, F. Gauge fields in graphene, *Phys. Rept.* **496**, 109 (2010).

[18] Cortijo, A. Guinea, F., and Vozmediano, M.A.H. Geometrical and topological aspects of graphene and related materials, *J. Phys.* A **45**, 383001 (2012), arXiv:1112.2054.

[19] Nobbenhuis, S. Categorizing different approaches to the cosmological constant problem, *Found. Phys.* **36**, 613–680 (2006).

[20] Dreyer, O. Emergent general relativity. in: *Approaches to Quantum Gravity – toward a new understanding of space, time and matter*, edited by Oriti, D. Cambridge University Press (2007); arXiv:gr-qc/0604075.

[21] Klinkhamer, F.R. and Volovik, G.E. Self-tuning vacuum variable and cosmological constant, *Phys. Rev.* D **77**, 085015 (2008), arXiv:0711.3170.

[22] Klinkhamer, F.R. and Volovik, G.E. Dynamic vacuum variable and equilibrium approach in cosmology, *Phys. Rev.* D **78**, 063528 (2008), arXiv:0806.2805.

[23] Klinkhamer, F.R. and Volovik, G.E. Vacuum energy density kicked by the electroweak crossover, *Phys. Rev.* D **80**, 083001 (2009), arXiv:0905.1919.

[24] Klinkhamer, F.R. and Volovik, G.E. $f(R)$ cosmology from q-theory, *JETP Lett.* **88**, 289 (2008), arXiv:0807.3896.

[25] Klinkhamer, F.R. and Volovik, G.E. Gluonic vacuum, q-theory, and the cosmological constant, *Phys. Rev.* D **79**, 063527 (2009), arXiv:0811.4347.

[26] Klinkhamer, F.R. and Volovik, G.E. Towards a solution of the cosmological constant problem, *Pis'ma ZhETF* **91**, 279–285 (2010); arXiv:0907.4887.

[27] de Bernardis, P. et al., 2000 [Boomerang Collaboration], A flat universe from high-resolution maps of the cosmic microwave background radiation. *Nature* **404**, 955–959.

[28] Hinshaw, G. et al., [WMAP Collaboration] 2007 Three-year Wilkinson Microwave Anisotropy Probe (WMAP) observations: temperature analysis. *Astrophys. J. Suppl.* **170**, 288–334.

[29] Riess, A. G. et al., 2007 New Hubble Space Telescope discoveries of type Ia supernovae at $z > 1$: Narrowing constraints on the early behavior of dark energy, *Astrophys. J.* **659**, 98–121.

[30] Landau, L.D. and Lifshitz, E.M. *Statistical Physics, Part 1*, Oxford: Pergamon Press (1980).

[31] Perrot, P. *A to Z of Thermodynamics*, Oxford: University Press (1998).

[32] Shifman, M.A. *Vacuum Structure and QCD Sum Rules*, Elsevier, Amsterdam, 1992; Shifman, M.A., Vainstein, A.I., and Zakharov, V.I. *Nucl. Phys.* B **147**, 385 (1979); 448 (1979).

[33] Shifman, M.A. Anomalies in gauge theories, *Phys. Rep.* **209**, 341–378 (1991).

[34] Matinyan, S.G. and Savvidy, G.K. Vacuum polarization induced by the intense gauge field, *Nucl. Phys.* B **134**, 539 (1978); Savvidy, G.K. Infrared instability of the vacuum state of gauge theories and asymptotic freedom, *Phys. Lett.* B **71**, 133 (1977).

[35] Halperin, I. and Zhitnitsky, A. Can $\theta//N$ dependence for gluodynamics be compatible with 2π periodicity in θ? *Phys. Rev.* D **58**, 054016 (1998).

[36] Duff, M.J. and van Nieuwenhuizen, P. Quantum inequivalence of different field representations, *Phys. Lett.* B **94**, 179 (1980).

[37] Aurilia, A., Nicolai, H., and Townsend, P.K. Hidden constants: The theta parameter of QCD and the cosmological constant of $N = 8$ supergravity, *Nucl. Phys.* B **176**, 509 (1980).

[38] Hawking, S.W. The cosmological constant is probably zero, *Phys. Lett.* B **134**, 403 (1984).

[39] Henneaux, M. and Teitelboim, C. The cosmological constant as a canonical variable, *Phys. Lett.* B **143**, 415 (1984).

[40] Duff, M.J. The cosmological constant is possibly zero, but the proof is probably wrong, *Phys. Lett.* B **226**, 36 (1989).

[41] Duncan, M.J. and Jensen, L.G. Four-forms and the vanishing of the cosmological constant, *Nucl. Phys.* B **336**, 100 (1990).

[42] Bousso, R. and Polchinski, J. Quantization of four-form fluxes and dynamical neutralization of the cosmological constant, JHEP **0006**, 006 (2000), arXiv:hep-th/0004134.

[43] Aurilia, A. and Spallucci, E. Quantum fluctuations of a 'constant' gauge field, *Phys. Rev.* D **69**, 105004 (2004), arXiv:hep-th/0402096.

[44] Wu, Z.C. The cosmological constant is probably zero, and a proof is possibly right, *Phys. Lett.* B **659**, 891 (2008), arXiv:0709.3314.

[45] Dzyaloshinskii, I.E. and Volovik, G.E. Poisson brackets in condensed matter, *Ann. Phys.* **125**, 67 (1980).

[46] Jacobson, T. Einstein–aether gravity: Theory and observational constraints, arXiv:0711.3822 [gr-qc].

[47] (a) Will, C.M. and Nordvedt, K. Conservation laws and preferred frames in relativistic gravity: I. Preferred-frame, *Astrophys. J.* **177**, 757 (1972); (b) Nordvedt, K. and Will, C.M. Conservation laws and preferred frames in relativistic gravity: II. Experimental evidence, *Astrophys. J.* **177**, 775 (1972); (c) Hellings, R.W. and Nordvedt, K. Vector-metric theory of gravity, *Phys. Rev.* D **7**, 3593 (1973).

[48] (a) Gasperini, M. Singularity prevention and broken Lorentz symmetry, Class. *Quantum Grav.* **4**, 485 (1987); (b) Gasperini, M. Repulsive gravity in the very early Universe, *Gen. Rel. Grav.* **30**, 1703 (1998), arXiv:gr-qc/9805060.

[49] Jacobson, T. and Mattingly, D. Gravity with a dynamical preferred frame, *Phys. Rev.* D **64**, 024028 (2001), arXiv:gr-qc/0007031.

[50] Volovik, G.E. On spectrum of vacuum energy, *J. Phys.*: Conference Series **174**, 012007 (2009); arXiv:0801.2714.

[51] Volovik, G.E. On the thermodynamic and quantum fluctuations of the cosmological constant, *JETP Lett.* **80**, 531–534 (2004); gr-qc/0406005.

[52] Weinberg, S. The cosmological constant problem, *Rev. Mod. Phys.* **61**, 1 (1989).

[53] Weinberg, S. Theories of the cosmological constant, in: Turok, N. *Critical Dialogues in Cosmology* (World Scientific, Singapore, 1997), p. 195, arXiv:astro-ph/9610044.

[54] Volkov, A.F. and Kogan, S.M. Collisionless relaxation of the energy gap in superconductors, *JETP* **38**, 1018–1021 (1974).

[55] Barankov, R. A., Levitov, L. S., and Spivak, B. Z. Collective Rabi oscillations and solitons in a time-dependent BCS pairing problem, *Phys. Rev. Lett.* **93**, 160401 (2004).

[56] Yuzbashyan, A. Altshuler, B. L. Kuznetsov, V. B. and Enolskii, V. Z. Nonequilibrium cooper pairing in the nonadiabatic regime, *Phys. Rev.* B **72**, 220503 (2005).

[57] Yuzbashyan, E. A. Normal and anomalous solitons in the theory of dynamical Cooper pairing, *Phys. Rev.* B **78**, 184507 (2008).

[58] Gurarie, V. Nonequilibrium dynamics of weakly and strongly paired superconductors, *Phys. Rev. Lett.* **103**, 075301 (2009).

[59] Starobinsky, A.A. A new type of isotropic cosmological models without singularity, *Phys. Lett.* B **91**, 99 (1980).

[60] Dolgov, A.D. Field model with a dynamic cancellation of the cosmological constant, *JETP Lett.* **41**, 345 (1985),

[61] Dolgov, A.D. Higher spin fields and the problem of cosmological constant, *Phys. Rev.* D **55**, 5881 (1997); arXiv:astro-ph/9608175.

[62] (a) Polyakov, A.M. Selftuning fields and resonant correlations in 2–D gravity, *Mod. Phys. Lett.* A **6**, 635 (1991); (b) Klebanov, I. and Polyakov, A.M. Interaction of discrete states in two-dimensional string theory, *Mod. Phys. Lett.* A **6**, 3273 (1991); arXiv:hep-th/9109032.

[63] Polyakov, A.M. private communication.

[64] (a) Larkin, A.I. and Pikin, S.A. Phase transitions of first order close to the second order, *Sov. Phys. JETP* **29**, 891 (1969); (b) Sak, J. Critical behavior and compressible magnets, *Phys. Rev.* B **10**, 3957 (1974).

[65] Rubakov, V.A. and Tinyakov, P.G. Ruling out a higher spin field solution to the cosmological constant problem, *Phys. Rev.* D **61**, 087503 (2000); arXiv:hep-ph/9906239.

[66] Emelyanov, V. and Klinkhamer, F.R. Vector-field model with compensated cosmological constant and radiation-dominated FRW phase, *Int. J. Mod. Phys.* D; arXiv:1108.1995.

[67] Klinkhamer, F.R. and Volovik, G.E. Dynamics of the quantum vacuum: Cosmology as relaxation to the equilibrium state, *J. Phys.*: Conference Series **314**, 012004 (2011); arXiv:1102.3152.

[68] Volovik, G.E. Reentrant violation of special relativity in the low-energy corner, *JETP Lett.* **73**, 162–165 (2001); hep-ph/0101286.

[69] Klinkhamer, F.R. and Volovik, G.E. Emergent CPT violation from the splitting of Fermi points, *Int. J. Mod. Phys.* A **20**, 2795–2812 (2005); hep-th/0403037.

[70] Klinkhamer, F.R. and Volovik, G.E. Superluminal neutrino and spontaneous breaking of Lorentz invariance, *Pis'ma ZhETF* **94**, 731–733 (2011); *JETP Lett.* **94**, 673 (2011); arXiv:1109.6624.

[71] Schützhold, R. Small cosmological constant from the QCD trace anomaly? *Phys. Rev. Lett.* **89** 081302 (2002).

[72] Urban, F.R. and Zhitnitsky, A.R. The QCD nature of dark energy *Nucl. Phys. B* **835** 135 (2010).

[73] Ohta, N. Dark energy and QCD ghost, *Phys. Lett.* B **695**, 41–44 (2011); arXiv:1010.1339.

[74] Holdom, B. From confinement to dark energy, *Phys. Lett.* B **697**, 351–356 (2011); arXiv:1012.0551.

[75] Poplawski, N.J. Cosmological constant from quarks and torsion, *Annalen Phys.* **523**, 291 (2011); arXiv:1005.0893.

[76] Georgi, H. and Glashow, S.L. Unity of all elementary particle forces, *Phys. Rev. Lett.* **32**, 438 (1974).

[77] Georgi, H. Quinn, H.R. and Weinberg, S. Hierarchy of interactions in unified gauge theories, *Phys. Rev. Lett.* **33**, 451 (1974).

[78] Volovik, G.E. and Gorkov, L.P. Superconductivity classes in the heavy fermion systems, *JETP* **61**, 843–854 (1985).

[79] Vollhardt, D. and Wölfle, P. *The Superfluid Phases of Helium 3* (Taylor and Francis, London, 1990).

[80] Mermin, N.D. The topological theory of defects in ordered media, *Rev. Mod. Phys.* **51**, 591 (1979).

[81] Hořava, P. Stability of Fermi surfaces and K-theory, *Phys. Rev. Lett.* **95**, 016405 (2005).

[82] Campuzano, J. C. Norman, M. and Randeria M. in: *Superconductivity: Physics of Conventional and Unconventional Superconductors*, Vol. 1, edited by Bennemann, K. H. and Ketterson, J. B. (Springer, Berlin, 2008).

[83] Volovik, G.E. Quantum phase transitions from topology in momentum space, in: "Quantum Analogues: From Phase Transitions to Black Holes and Cosmology", eds. Unruh, W.G. and Schützhold, R. Springer Lecture Notes in *Physics* **718** (2007), pp. 31–73; cond-mat/0601372.

[84] Khodel, V.A. and Shaginyan, V.R. Superfluidity in system with fermion condensate, *JETP Lett.* **51**, 553 (1990).

[85] Volovik, G.E. A new class of normal Fermi liquids, *JETP Lett.* **53**, 222 (1991).

[86] Shaginyan, V.R. Amusia, M.Ya. Msezane, A.Z. and Popov, K.G. Scaling behavior of heavy fermion metals, *Phys. Rep.*, DOI: 10.1016/j.physrep.2010.03.001.

[87] Lee, Sung-Sik Non-Fermi liquid from a charged black hole: A critical Fermi ball, *Phys. Rev.* D **79**, 086006 (2009).

[88] Schnyder, A.P. and Ryu, S. Topological phases and flat surface bands in superconductors without inversion symmetry, arXiv:1011.1438; *Phys. Rev.* B **84**, 060504(R) (2011).

[89] Heikkilä, T.T. Kopnin, N.B. and Volovik, G.E. Flat bands in topological media, Pis'ma *ZhETF* **94**, 252–258 (2011); *JETP Lett.* **94**, 233–239(2011); arXiv:1012.0905.

[90] Schnyder, A. P. Brydon, P. M. R. and Timm, C. Types of topological surface states in nodal noncentrosymmetric superconductors, *Phys. Rev.* B **85**, 024522 (2012); arXiv:1111.1207.

[91] Kopnin, N. B. and Salomaa, M. M. Mutual friction in superfluid ^{3}He: Effects of bound states in the vortex core, *Phys. Rev.* B **44**, 9667–9677 (1991).

[92] Volovik, G.E. On Fermi condensate: near the saddle point and within the vortex core, *JETP Lett.* **59**, 830 (1994).

[93] Volovik, G.E. Flat band in the core of topological defects: bulk-vortex correspondence in topological superfluids with Fermi points, *Pis'ma ZhETF* **93**, 69–72 (2011); *JETP Lett.* **93**, 66–69 (2011); arXiv:1011.4665.

[94] Kane, C.L. and Mele, E. Z_2 topological order and the quantum spin Hall effect, *Phys. Rev. Lett.* **95**, 146802 (2005).

[95] Volkov, B.A. Gorbatsevich, A.A. Kopaev, Yu. V. and Tugushev, V.V. Macroscopic current states in crystals, *JETP* **54**, 391–397 (1981).

[96] Volkov B.A. and Pankratov, O.A. Two-dimensional massless electrons in an inverted contact, *JETP Lett.* **42**, 178–181 (1985).

[97] Salomaa, M.M. and Volovik, G.E. Cosmiclike domain walls in superfluid ^{3}He-B: Instantons and diabolical points in $(\mathbf{k},\mathbf{r})$ space, *Phys. Rev.* B **37**, 9298–9311 (1988).

[98] Volovik, G.E. Fermion zero modes at the boundary of superfluid ^{3}He-B, Pis'ma *ZhETF* **90**, 440–442 (2009); *JETP Lett.* **90**, 398–401 (2009); arXiv:0907.5389.

[99] Volovik, G.E. Topological invariants for Standard Model: from semi-metal to topological insulator, *JETP Lett.* **91**, 55–61 (2010); arXiv:0912.0502.

[100] Volovik, G.E. and Yakovenko, V.M. Fractional charge, spin and statistics of solitons in superfluid ^{3}He film, *J. Phys.: Condens. Matter* **1**, 5263–5274 (1989).

[101] Volovik, G.E. *Exotic Properties of Superfluid* ^{3}*He*, World Scientific, Singapore, 1992.

[102] Volovik, G.E. Fractional statistics and analogs of quantum Hall effect in superfluid ^{3}He films. In: *Quantum Fluids and Solids-1989* ed. by G.G. Ihas, Y. Takano (AIP Conference Proceedings, 1989) **194**, pp. 136–146.

[103] Haldane, F.D.M. Berry curvature on the Fermi surface: anomalous Hall effect as a topological Fermi-liquid property, *Phys. Rev. Lett.* **93**, 206602 (2004).

[104] Farid, B. and Tsvelik, A.M. Comment on "Breakdown of the Luttinger sum rule within the Mott-Hubbard insulator", arXiv:0909.2886.

[105] Giamarchi, T. *Quantum Physics in One Dimension*, Oxford: University Press (2004).

[106] Georgi, H. Another odd thing about unparticle physics, *Phys. Lett.* B **650**, 275–278 (2007); arXiv:0704.2457.

[107] Luo, M. and Zhu, G. Some phenomenologies of unparticle physics. *Phys. Lett.* B **659**, 341 (2008).

[108] Gribov, V.N. Quantization of non-Abelian gauge theories, *Nucl. Phys.* B **139**, 1–19 (1978).

[109] Chernodub, M.N. and Zakharov, V.I. Combining infrared and low-temperature asymptotes in Yang–Mills theories; *Phys. Rev. Lett.* **100**, 222001 (2008).

[110] Burgio, G. Quandt, M. and Reinhardt, H. Coulomb-gauge gluon propagator and the Gribov formula, *Phys. Rev. Lett.* **102**, 032002 (2009).

[111] Faulkner, T. Iqbal, N. Liu, Hong, McGreevy, J. and Vegh, D. From black holes to strange metals, arXiv:1003.1728.

[112] Yakovenko, V.M. Spin, statistics and charge of solitons in (2+1)-dimensional theories, *Fizika* (Zagreb) **21**, suppl. 3, 231 (1989); arXiv:cond-mat/9703195.

[113] Sengupta, K. and Yakovenko, V.M. Hopf invariant for long-wavelength skyrmions in quantum Hall systems for integer and fractional fillings, *Phys. Rev.* B **62**, 4586–4604 (2000).

[114] Read, N. and Green, D. Paired states of fermions in two dimensions with breaking of parity and time-reversal symmetries and the fractional quantum Hall effect, *Phys. Rev.* B **61**, 10267–10297 (2000).

[115] So, H. Induced topological invariants by lattice fermions in odd dimensions, *Prog. Theor. Phys.* **74**, 585–593 (1985).

[116] Ishikawa, K. and Matsuyama, T. Magnetic field induced multi component QED in three-dimensions and quantum Hall effect, *Z. Phys.* C **33**, 41–45 (1986).

[117] Ishikawa, K. and Matsuyama, T. A microscopic theory of the quantum Hall effect, *Nucl. Phys.* B **280**, 523–548 (1987).

[118] Matsuyama, T. Quantization of conductivity induced by topological structure of energy-momentum space in generalized QED_3, *Progr. Theor. Phys.* **77**, 711–730 (1987).

[119] Jansen, K. Domain wall fermions and chiral gauge theories, *Phys. Rept.* **273**, 1–54 (1996).

[120] Volovik, G.E. Topological superfluid ^{3}He-B in magnetic field and Ising variable, *Pis'ma ZhETF* **91**, 215–219 (2010); *JETP Lett.* **91**, 201–205 (2010); arXiv:1001.1514.

[121] Essin, A.M. and Gurarie, V. Bulk-boundary correspondence of topological insulators from their Green's functions, *Phys. Rev.* B **84**, 125132 (2011).

[122] Zubkov, M.A. and Volovik, G.E. Topological invariants for the 4*D* systems with mass gap, *Nuclear Physics* B **860**, Issue 2, 295–309 (2012); arXiv:1201.4185.

[123] Dietl, P. Piechon, F. and Montambaux, G. New magnetic field dependence of Landau levels in a graphenelike structure, *Phys. Rev. Lett.* 100, 236405 (2008).

[124] Banerjee, S. Singh, R. R. Pardo, V. and Pickett, W. E. Tight-binding modeling and low-energy behavior of the semi-Dirac point, *Phys. Rev. Lett.* **103**, 016402 (2009).

[125] Heikkilä, T.T. and Volovik, G.E. Fermions with cubic and quartic spectrum, *Pis'ma ZhETF* **92**, 751–756 (2010); *JETP Lett.* **92**, 681–686 (2010); arXiv:1010.0393.

[126] Heikkilä, T.T., and Volovik, G.E. Dimensional crossover in topological matter: Evolution of the multiple Dirac point in the layered system to the flat band on the surface, *Pis'ma ZhETF* **93**, 63–68 (2011); *JETP Lett.* **93**, 59–65 (2011); arXiv:1011.4185.

[127] Katsnelson, M.I. and Volovik, G.E. Quantum electrodynamics with anisotropic scaling: Heisenberg-Euler action and Schwinger pair production in the bilayer graphene, *Pis'ma ZhETF* **95**, 457–461 (2012); arXiv:1203.1578.

[128] Zubkov, M.A. Schwinger pair creation in multilayer graphene, *Pis'ma ZhETF* **95**, 540–543 (2012), arXiv:1204.0138.

[129] Hořava, P. Spectral dimension of the Universe in quantum gravity at a Lifshitz point, *Phys. Rev. Lett.* **102**, 161301 (2009).

[130] Hořava, P. Quantum gravity at a Lifshitz point, *Phys. Rev.* D **79**, 084008 (2009).

[131] Xu, Cenke, and Hořava, P. Emergent gravity at a Lifshitz point from a Bose liquid on the lattice, *Phys. Rev.* D **81**, 104033 (2010).

[132] Fu, L. Kane, C.L. and Mele, E.J. Topological insulators in three dimensions, *Phys. Rev. Lett.* **98**, 106803 (2007).

[133] Pepea, M. and Wieseb, U.J. Exceptional deconfinement in *G*(2) gauge theory, *Nucl. Phys.* B **768**, 21–37 (2007).

[134] Kadastik, M. Kannike, K. and Raidal, M. Dark matter as the signal of grand unification, *Phys. Rev.* D **80**, 085020 (2009).

[135] Klinkhamer, F.R. and Volovik, G.E. Emergent CPT violation from the splitting of Fermi points, *Int. J. Mod. Phys.* A **20**, 2795–2812 (2005); hep-th/0403037.

[136] Tarruell, L. Greif, D. Uehlinger, T. Jotzu, G. and Esslinger, T. Creating, moving and merging Dirac points with a Fermi gas in a tunable honeycomb lattice, *Nature* **483**, 302–306 (2012), arXiv:1111.5020.

[137] Klinkhamer, F.R. Superluminal neutrino, flavor, and relativity, *Phys. Rev.* D **85**, 016011 (2012), arXiv:1110.2146.

[138] Klinkhamer, F. R. Possible new source of T and CP violation in neutrino oscillations. *Phys. Rev.* D **73**, 057301 (2006).

[139] Osheroff, D.D. Richardson, R.C. and Lee, D.M. Evidence for a new phase of solid He3, *Phys. Rev. Lett.* **28**, 885–888 (1972).

[140] Creutz, M. Four-dimensional graphene and chiral fermions, *JHEP* 04 (2008) 017; arXiv:0712.1201.

[141] Wen, X.G. and Zee, A. Gapless fermions and quantum order, *Phys. Rev.* B **66**, 235110 (2002).

[142] Beri, B. Topologically stable gapless phases of time-reversal invariant superconductors, *Phys. Rev.* B **81**, 134515 (2010), arXiv:0909.5680.

[143] Kitaev, A. Periodic table for topological insulators and superconductors, AIP Conference Proceedings, Volume **1134**, pp. 22–30 (2009); arXiv:0901.2686.

[144] Volovik, G.E. Topological invariant for superfluid ^{3}He-B and quantum phase transitions, *JETP Lett.* **90**, 587–591 (2009); arXiv:0909.3084.

[145] Jackiw, R. and Rossi, P. Zero modes of the vortex-fermion system, *Nucl. Phys.* B **190**, 681–691 (1981).

[146] Kaplan, D.B. Method for simulating chiral fermions on the lattice, *Phys. Lett.* B **288**, 342–347 (1992); arXiv:hep-lat/9206013.

[147] Golterman, M.F.L. Jansen K. and Kaplan, D.B. Chern-Simons currents and chiral fermions on the lattice, *Phys. Lett.* B **301**, 219–223 (1993): arXiv: hep-lat/9209003.

[148] Kaplan, D.B. and Sun, Sichun Spacetime as a topological insulator, *Phys. Rev. Lett.* **108**, 181807 (2012), arXiv:1112.0302.

[149] Mineev, V.P. and Volovik, G.E. Planar and linear solitons in superfluid ^{3}He, *Phys. Rev.* B **18**, 3197–3203 (1978).

[150] Volovik, G.E. Analog of quantum Hall effect in superfluid ^{3}He film, *JETP* **67**, 1804–1811 (1988).

[151] Volovik, G.E. Gapless fermionic excitations on the quantized vortices in superfluids and superconductors, *JETP Lett.* **49**, 391–395 (1989).

[152] Grinevich, P.G. and Volovik, G.E. Topology of gap nodes in superfluid ^{3}He: π_4 homotopy group for $^3He - B$ disclination, *J. Low Temp. Phys.* **72**, 371–380 (1988).

[153] Teo, J.C.Y. and Kane, C.L. Majorana fermions and non-Abelian statistics in three dimensions, *Phys. Rev. Lett.* **104**, 046401 (2010).

[154] Teo, J.C.Y. and Kane, C.L. Topological defects and gapless modes in insulators and superconductors, *Phys. Rev.* B **82**, 115120 (2010).

[155] Silaev, M.A. and Volovik, G.E. Topological superfluid ^{3}He-B: fermion zero modes on interfaces and in the vortex core, *J. Low Temp. Phys.* **161**, 460–473 (2010); arXiv:1005.4672.

[156] Klitzing, K. v. Dorda, G., and Pepper, M. New method for high-accuracy determination of the fine-structure constant based on quantized Hall resistance, *Phys. Rev. Lett.* **45**, 494–497 (1980),

[157] Awschalom, D. and Samarth, N. Spintronics without magnetism, *Physics* **2**, 50 (2009).

[158] Haldane, F.D.M. and Arovas, D.P. Quantized spin currents in 2-dimensional chiral magnets, *Phys. Rev.* **B 52**, 4223 (1995).

[159] Senthil, T. Marston, J.B. and Fisher, M.P.A. The spin quantum Hall effect in unconventional superconductors, *Phys. Rev.* B **60**, 4245–4254 (1999).

[160] Gurarie, V. and Radzihovsky, L. Resonantly-paired fermionic superfluids, *Ann. Phys.* **322**, 2–119 (2007).

[161] Haldane, F.D.M. Model for a quantum Hall effect without Landau levels: Condensed-matter realization of the "Parity Anomaly", *Phys. Rev. Lett.* **61**, 2015–2018 (1988).

[162] Schnyder, A.P. Ryu, S. Furusaki A. and Ludwig, A.W.W. Classification of topological insulators and superconductors in three spatial dimensions, *Phys. Rev.* B **78**, 195125 (2008).

[163] Volovik, G.E. Zeroes in energy gap in superconductors with high transition temperature, *Phys. Lett.* A **142** 282–284 (1989).

[164] Gubankova, E. Schmitt, A. and Wilczek, F. Stability conditions and Fermi surface topologies in a superconductor, *Phys. Rev.* B **74**, 064505 (2006).

[165] Zubkov, M. A. Generalized unparticles, zeros of the Green function, and momentum space topology of the lattice model with overlap fermions, *Phys. Rev.* D **86**, 034505 (2012), arXiv:1202.2524.

[166] Nielsen, H.B. and Ninomiya, M. The Adler–Bell–Jackiw anomaly and Weyl fermions in a crystal, *Phys. Lett.* **130 B**, 389–396 (1983).

[167] Avron, J.E., Seiler, R., and Simon, B. Homotopy and quantization in condensed matter physics, *Phys. Rev. Lett.* **51**, 51–53 (1983).

[168] Semenoff, G.W. Condensed-matter simulation of a three-dimensional anomaly, *Phys. Rev. Lett.* **53**, 2449–2452 (1984).

[169] Niu, Qian, Thouless, D. J., and Wu, Yong-Shi. Quantized Hall conductance as a topological invariant, *Phys. Rev.* B **31**, 3372–3377 (1985).

[170] Hsieh, D. *The Experimental Discovery of Topological Insulators*, Barnes & Noble, 2009.

[171] Kopnin, N.B. Soininen, P.I., and Salomaa, M.M. Parameter-free quasiclassical boundary conditions for superfluid ^{3}He at rough walls. B-phase order parameter and density of states, *J. Low Temp. Phys.* **85**, 267–282 (1991).

[172] Castelijns, C.A.M., Coates, K.F., Guénault, A.M., Mussett, S.G., and Pickett, G.R. Landau critical velocity for a macroscopic object moving in superfluid ^{3}He-B: evidence for gap suppression at a moving surface, *Phys. Rev. Lett.* **56**, 69–72 (1986).

[173] Davis, J.P., Pollanen, J., Choi, H., Sauls, J.A., Halperin, W.P., and Vorontsov, A.B. Anomalous attenuation of transverse sound in ^{3}He, *Phys. Rev. Lett.* **101**, 085301 (2008).

[174] Nagai, K., Nagato, Y., Yamamoto M., and Higashitani, S. Surface bound states in superfluid ^{3}He, *J. Phys. Soc. Jap.* **77**, 111003 (2008).

[175] Murakawa, S., Tamura, Y., Wada, Y., Wasai, M., Saitoh, M., Aoki, Y., Nomura, R., Okuda, Y., Nagato, Y., Yamamoto, M., Higashitani, S., and Nagai, K. New anomaly in transverse acoustic impedance of superfluid ^{3}He-B with a wall coated by several layers of ^{4}He, *Phys. Rev. Lett.* **103**, 155301 (2009).

[176] Chung, Suk Bum and Zhang, Shou-Cheng. Detecting the Majorana fermion surface state of ^{3}He-B through spin relaxation, *Phys. Rev. Lett.* **103**, 235301 (2009); arXiv:0907.4394.

[177] Nagato, Y., Higashitani S., and Nagai, K. Strong anisotropy in spin susceptibility of superfluid He-3-B film caused by surface bound states, *J. Phys. Soc. Jpn.* **78**, 123603 (2009).

[178] Schnyder, A.P., Ryu, S., Furusaki, A., and Ludwig, A.W.W. Classification of topological insulators and superconductors, *AIP Conf. Proc.* **1134**, 10–21 (2009); arXiv:0905.2029.

[179] Nielsen, H.B. and Ninomiya, M. Absence of neutrinos on a lattice. I – Proof by homotopy theory, *Nucl. Phys.* B **185**, 20 (1981); Absence of neutrinos on a lattice. II – Intuitive homotopy proof, *Nucl. Phys.* B **193**, 173 (1981).

Author Index

Subject Index